GEOCHIMICA ET COSMOCHIMICA ACTA

SUPPLEMENT 4

PROCEEDINGS

OF THE

FOURTH LUNAR SCIENCE CONFERENCE

Houston, Texas, March 5–8, 1973

The Fourth Lunar Science Conference was held under the joint auspices of the National Aeronautics and Space Administration and The Lunar Science Institute in Houston, Texas, March 1973. Its purpose was to present the research of some 180 Principal Investigators and their co-workers on samples from eight lunar landing sites — Apollo 11, 12, 14, 15, 16, 17, and Luna 16 and 20. The Principal Investigators form an international group representing 14 countries. The reports of the Principal Investigators are supplemented by four papers from Russian scientists on samples gathered at the Luna 16 and 20 landing sites.

The results of experiments conducted or set up on the Moon's surface by the astronauts are also described in these volumes. Each Apollo mission has installed an ALSEP geophysics package on the lunar surface for long term observations. In addition, results obtained from the Command Module and the subsatellites are described.

The National Aeronautics and Space Administration, through The Lunar Science Institute, sponsored the publication of these volumes which contain a total of 228 papers. It is estimated that less than 5 per cent of the research results present in them have been reported in various journals; the rest of the material is making its first appearance in printed form.

The photographs on the front and back of the covers show the six Apollo landing sites (photos courtesy of NASA). On the front of Volume 1 is the Apollo 14 landing site (3.7°S, 17.5°W) and on the back, the Apollo 11 landing site (0.7°N, 23.5°E). On the front of Volume 2 is the landing site for Apollo 15 (26.1°N, 3.7°E) and on the back, Apollo 12 (3.2°S, 23.4°W). Apollo 16 landing site (8.9°S, 15.5°E) is on the front of Volume 3 and the Apollo 17 landing site (20°N, 30°E), on the back side.

GEOCHIMICA ET COSMOCHIMICA ACTA

Journal of The Geochemical Society and The Meteoritical Society

SUPPLEMENT 4

PROCEEDINGS

OF THE

FOURTH LUNAR SCIENCE CONFERENCE

Houston, Texas, March 5–8, 1973

Sponsored by

The NASA Johnson Space Center

and

The Lunar Science Institute

VOLUME 3

PHYSICAL PROPERTIES

PERGAMON PRESS

NEW YORK • OXFORD • TORONTO • SYDNEY • BRAUNSCHWEIG

Pergamon Press, Inc., Maxwell House, Fairview Park
New York, N.Y. 10523 USA Teletype 137328

Pergamon Press, Ltd. Headington Hill Hall
Oxford OX3 OBW England

Pergamon of Canada Ltd., 207 Queen's Quay West
Toronto 1 Canada

Vieweg & Sohn Gmbh Burgplatz 1, 33 Braunschweig, Germany

Pergamon House, 19a Boundary Street, Rushcutters Bay NSW2011, Australia

Type set by The European Printing Corporation, Ltd.
Dublin, Ireland

Printed by Publishers Production International
and bound by Arnold's Bindery
in the United States of America

Library of Congress Cataloging in Publication Data

Lunar Science Conference, 4th, Houston, Tex., 1973. Proceedings.

(Geochimica et cosmochimica acta, Supplement 4)
"Sponsored by the NASA Johnson Space Center and the Lunar
Science Institute."
CONTENTS: v. 1. Mineralogy and petrology.—v. 2. Chemical and
isotope analyses, organic chemistry.—v. 3. Physical properties.
1. Lunar petrology—Congresses. 2. Cosmochemistry—Con-
gresses. 3. Lunar geology—Congresses.
I. Lyndon B. Johnson Space Center. II. Lunar Science Institute.
III. Series.
QB592.L85   1973   559.9'1   73-15974

ISBN 0-08-017909-6
ISBN 0-08-017910-X (pbk.)

Supplement 4
GEOCHIMICA ET COSMOCHIMICA ACTA
Journal of The Geochemical Society and The Meteoritical Society

W. I. RIDLEY, The Lunar Science Institute, Houston, Texas 77058.

R. A. SCHMITT, Department of Chemistry, Oregon State University, Corvallis, Oregon 97331.

W. R. SILL, University of Utah, Salt Lake City, Utah 84112.

G. W. WETHERILL, Department of Planetary and Space Sciences, University of California, Los Angeles, California 90024.

EDITORIAL ASSISTANTS

C. A. BENDER, The Lunar Science Institute, Houston, Texas 77058.

J. H. HEIKEN, The Lunar Science Institute, Houston, Texas 77058.

J. M. SHACK, The Lunar Science Institute, Houston, Texas 77058.

Contents

Resonance Studies

Orbital Mapping

GEOCHIMICA ET COSMOCHIMICA ACTA

Supplement 4

PROCEEDINGS

OF THE

FOURTH LUNAR SCIENCE CONFERENCE

Houston, Texas, March 5–8, 1973

Proceedings of the Fourth Lunar Science Conference
(Supplement 4, *Geochimica et Cosmochimica Acta*)
Vol. 3, pp. 2275–2290

Surface irradiation and evolution of the lunar regolith

Narendra Bhandari, Jitendra Goswami, and Devendra Lal

Tata Institute of Fundamental Research, Homi Bhabha Road, Bombay

Abstract—Fossil tracks have been analyzed in Apollo 14, 15, and 16 samples. Principal results of the analyses are summarized below:

(i) Surface exposure ages of lunar rocks, designated as *suntan* ages are generally smaller than 3 m.y. indicating that rocks fragment and fresh surfaces are exposed at rates of the order of 0.3 mm/m.y. Differences in exposure age distribution for Apollo sites 11, 12, 14, 15, and 16 appear significant and probably reflect differences in the survival characteristics of different rocks.

(ii) Most surface rocks (80%) have had a complex fossil track irradiation history so that the exposure geometry and shielding varied in the past 10–100 m.y. Not more than twenty percent of the rocks exhibit a simple one stage exposure history corresponding to accumulation of most of the fossil tracks in a single geometry.

(iii) Fossil track data on Apollo 15, 16 surface samples and the 2.5 m drill stem of Apollo 15 core support our earlier model that generally the lunar regolith is made up of discrete layers of soil ejected in different cratering events. Typical surface irradiation ages of these layers range from 1–50 m.y. The throw-away layers keep their identity except to the extent of mixing up to depths of a few cms/(10–50) m.y. due to micrometeorite impacts. This short range mixing occurs till a layer is buried by blanketting from a fresh layer.

(iv) The formation ages of some major craters at Apollo 15 and 16 sites have been estimated on the basis of the irradiation ages of the ejecta.

INTRODUCTION

IN LUNAR samples one can clearly see distinct records of irradiation due to solar and galactic cosmic rays. Extensive fossil track data in lunar samples have been reported earlier (Arrhenius *et al.*, 1971; Bhandari *et al.*, 1971; Berdot *et al.*, 1972; Crozaz *et al.*, 1971; Comstock *et al.*, 1971; Price and O'Sullivan, 1970) and the work has been extended in recent years (cf. Lal, 1972).

Track densities observed in near surface region (<0.1 cm) of most lunar samples show a steep depth gradient with a negative slope in power law relation of almost unity, characteristic of a solar flare irradiation in equilibrium with erosional processes (Bhandari *et al.*, 1972a; Crozaz *et al.*, 1972; Barber *et al.*, 1971; Fleischer *et al.*, 1971). At depths exceeding 0.1 cm the track densities continue to decrease roughly as $(1/X)$, except for the rocks which had a complex exposure history. A comparison of the observed depth profile with the production profile of tracks (Bhandari *et al.*, 1971; Barber *et al.*, 1971) allows an understanding of the near surface irradiation and fragmentation processes. Furthermore the accumulation of tracks at $X = 0.1$ cm is generally linear with the surface irradiation time (for typical rates of erosion and exposure ages), allowing an exposure age to be calculated. Thus fossil track data have been useful in understanding the time scales of the evolution of the lunar regolith.

2275

In this paper we report analyses of fossil tracks in Apollo 14, 15, and 16 samples. These and the previous data lead to a fairly unified picture of cosmic ray irradiation and exposure history of the lunar regolith.

Experimental

Fossil track analyses in the major track forming silicate minerals viz. feldspar, pyroxene, and olivine have been made. In case of Apollo 16 samples only feldspars, which were most abundant, have been studied. We have developed suitable etchants for all the different silicate mineral types (Lal *et al.*, 1968; Krishnaswami *et al.*, 1971).

The experimental techniques used in sample preparation for fossil track work are the same as described earlier (Arrhenius *et al.*, 1971; Bhandari *et al.*, 1972b). Grain mount and thick-section techniques have been used in the case of lunar fines and rock samples respectively. In case of lunar fines, grain mounts were made from grains of size greater than 50 microns. The tracks were counted under optical microscope where densities up to 9×10^7 could be measured precisely ($\pm 10\%$). Higher densities were estimated ($\pm 50\%$) from amount of surface loss in etching which was calibrated with electron microscopy for a few typical samples.

Results and Discussions

The results of fossil track analyses are shown in Fig. 1 for Apollo 15 and 16 rocks and in Figs. 2, 3, and 4 for soil samples. In the case of soil, results are pooled together for all the minerals following the normalization factors determined by Bhandari *et al.* (1972b). The observed track densities in olivines have been multiplied by a constant factor of two to take into account the relatively higher threshold characteristics of this mineral as compared to pyroxene and feldspar group of minerals where track densities are similar; feldspars being only about 30% higher than pyroxenes (Bhandari *et al.*, 1972b).

Exposure history of the lunar rocks

The best fit track density profiles for four Apollo 15 (15058, 15118, 15535, and 15555) and four Apollo 16 rocks (60015, 62295, 65015, and 67016) are shown in Fig. 1 (a) and (b). Shock effects resulting in deformed shapes of etched tracks (Fleischer *et al.*, 1972) were visible in all Apollo 16 rocks but no evidence for significant track erasure was found.

In the case of rocks 15058, 62295, and 67016 the gradients in track densities at $X < 0.1$ cm are steep and similar to some of the rocks from Apollo 11, 12, and 14 missions. The rock 62295 following earlier discussions of Bhandari *et al.* (1971), is inferred to have had a simple surface exposure history with probably no shallow buried exposure in the regolith. The anorthositic rock 60015 has a glass splash on its surfaces. Our track data in this rock are restricted to feldspar grains and hence do not extend to the surface of this rock. However, the markedly low track density gradients, in contrast to rock 62295 at $(10^{-1}-1)$ cm depth interval indicates that most of the tracks in this rock accumulated when the rock was shielded by several cm of lunar material. It can be inferred from the steep track gradient on opposite

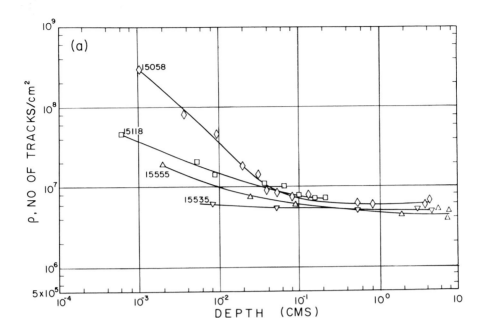

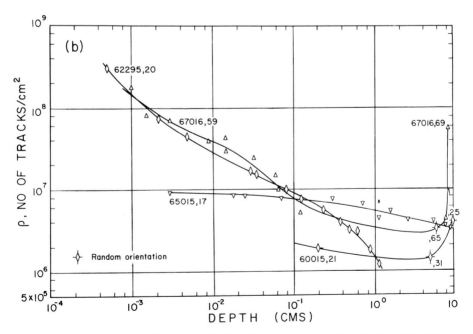

Fig. 1. Track density profiles for Apollo 15 rocks (a) and Apollo 16 rocks (b) shown as a function of depth measured from the exposed surface. Measurements were made in thick sections of known orientation except in a few cases indicated as random.

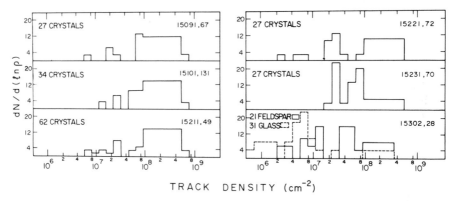

Fig. 2. Histograms showing track density frequency distribution in grains from Apollo 15 samples. The number of grains analyzed is indicated in each case but the histogram has been normalized to 100 grains to allow a comparison. Grains having track densities exceeding $9 \times 10^7 \, \mathrm{cm}^{-2}$ have been uniformly distributed between $\rho = 9 \times 10^7$ and $5 \times 10^8 \, \mathrm{cm}^{-2}$.

faces of the rock 67016 and its orientation on the lunar surface, as known from photodocumentation, that this rock has been exposed at least in two geometries on the moon.

We have earlier proposed (Bhandari et al., 1971, 1972a) that a convenient way to characterize the exposure history of a lunar rock is by describing two ages:

(i) the *suntan*, i.e., the surface exposure age, and

(ii) the *subdecimeter* age i.e., the approximate integrated dwell time of a rock within the upper 10 cm of regolith material.

The procedures for obtaining these ages and the track production parameters used have been described in detail earlier (Bhandari et al., 1972a). The *suntan* age is calculated from the track density at 0.1 cm depth from the exterior surface of the rock showing a track density gradient, typical of a surface exposure. When such a characteristic track density gradient is absent, the *suntan* ages are deduced

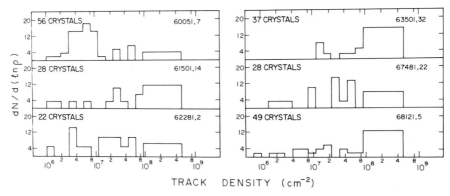

Fig. 3. Histograms showing track density frequency distribution in grains from Apollo 16 surface samples. Other details are as in Fig. 2.

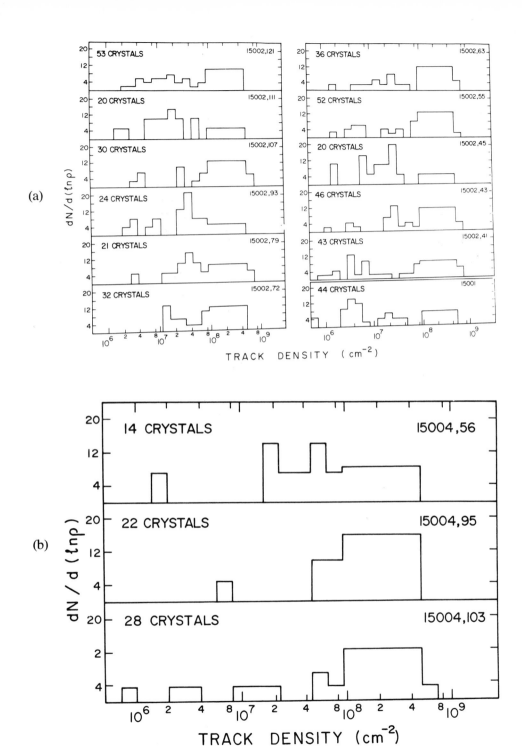

Fig. 4. Histograms showing track density frequency distribution in grains from Apollo 15 core stem 15002 and 15001 (a) and stem 15004 (b). Other details are as in Fig. 2.

2279

to be small (< 1 m.y.). If the track densities at $X > 0.1$ cm cannot be accounted for by the deduced exposure at the surface (suntan age), a *subdecimeter* age for the rock is calculated on the assumption that the rock has spent equal intervals of time within the first decimeter of the regolith. It may be mentioned here that the track production rates fall so steeply with depth that no information about the residence times of rocks in regions of $X > 10$ cms can be obtained from cosmic ray track studies unless the depth of exposure is accurately known.

Following these procedures, we have given in Table 1 the calculated suntan and subdecimeter ages of the Apollo 15 and 16 rocks. It should be mentioned here that the suntan age refers to the exposure of a particular face of the rock under consideration. The suntan ages are generally low for most of the samples and this indicates frequent tumbling and chipping off in these cases. The frequency distribution of the suntan ages for the lunar rocks from various missions is shown in Fig. 5. An interesting observation is that in a total of 25 rocks studied, only 4 rocks (12018, 12038, 14303, and 62295) have had a single surface exposure history indicating that about 80% of the lunar rocks undergo multiple exposures in the upper 10 cm of the regolith.

Table 1. Exposure ages of Apollo 15 and 16 rocks.

Sample	Mass (gm)	Ellipsoid semi-axes used (cms)	Sample detail	Range of track density* (10^6 cm^{-2})	Fossil track ages (m.y.)	
					Suntan age	Subdecimeter age
Apollo 15 Samples						
15058,49	2,672	$5\times6\times8$	Slice	6–500	2	10
15085,27	471	$4.25\times4\times2.5$	S.C.	6	<1	—
29			S.C.	12	<1	—
15265,14	314	$7\times3.5\times2.25$	S.C.	10	<1	—
15			S.C.	6	<1	—
15426,75	223	$3.5\times2\times2.5$	I.C.	7	?	15
15535,25	404	$5.5\times3.5\times1.75$	Slice	5	<1	10
15555,199	9,613	$13.5\times7.5\times7.5$	Slice	5–14	1	26
201						
203						
15557,19	2,518	$9\times5\times3.5$	S.C.	9	<1	—
37			I.C.	14		—
15118,3	27	$2\times1.5\times1$	S.C.	8–50	1.3	—
15388,5	Peanut	—	S.C.	5	<1	—
Apollo 16 Samples						
60015,21	5574	$14\times7.5\times5$	S.C.	2	<1	10
31			I.C.	1.5	—	
25			S.C.	4	1	
62295,20	251	$4.5\times3\times2$	S.C.	2–300	2.7	—
65015,17	1802	$9\times4.5\times5$	Slice	10–5	1.2	50
67016,59	4262	$12\times75\times5$	S.C.	200–5	1.2	
65			I.C.	4	—	15
69			S.C.	80–4	1	

S.C. = Surface chip. I.C. = Interior chip.
*Track densities in near surface grains refer to a depth of ~ 5 microns.

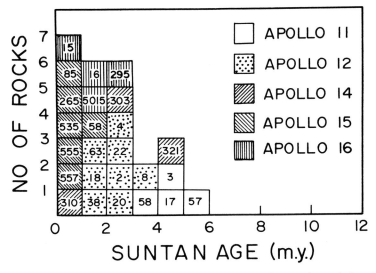

Fig. 5. Frequency distribution of Suntan ages for lunar rocks from various missions. The rocks are identified by the last digits in the histogram.

In spite of the differences in the irradiation histories, the suntan ages fall in a narrow group. About 75% of the rocks, for which data are available, have suntan ages of ≤ 3 m.y. as shown in Fig. 5. If the results of suntan ages are considered statistically significant, then it must be noted that Apollo 15, 12, and 11 rocks have successively higher suntan ages viz. (0–1), (1–3), and (3–6) m.y. respectively; three Apollo 14 rocks lie between 0–5 m.y. This behavior may reflect the survival properties of rocks against fragmentation by micrometeorite impacts. Although these characteristics have not been determined for all the rocks studied here, Hörz et al. (1971) and Gault et al. (1972) have calculated the survival times for several rocks before catastrophic rupture by micrometeorite impacts. It should be pointed out here that the Apollo 11 and 12 igneous rocks were highly cohesive whereas the Apollo 15 rocks, studied by us, were heavily fractured (Apollo 15 Lunar Sample Information Catalogue, 1971).

The energy spectrum of cosmic ray track forming nuclei can be obtained from the track density profiles observed in the lunar rocks provided the irradiation geometry and history of the rocks are precisely known. Earlier, based on Apollo 12 rocks, we have deduced the energy spectra of solar flare and galactic VH nuclei (Bhandari et al., 1971), on the assumption that the rocks showing steepest track density gradient represent a single stage exposure on the lunar surface. Similar analysis of several Apollo 14, 15, and 16 rocks, having the steepest track density gradients discussed above, yield an energy spectrum, in agreement with the one deduced earlier (Bhandari et al., 1971). The long term energy spectrum of combined solar and galactic VH and VVH nuclei based on fossil track analysis of lunar and meteoritic samples is discussed in detail elsewhere (Bhandari et al., 1973).

Surface irradiation history of lunar fines

Arrhenius *et al.* (1971) characterized the irradiation of lunar fines by two parameters; (i) N_H/N, the fraction of grains exposed on the very surface as determined by a steep track density gradient or a high track density ($> 10^8 \mathrm{cm}^{-2}$) and (ii) ρ_q, the quartile track density i.e., a value such that a quarter of the grains in a sample have track densities lower than ρ_q. These two parameters suffice to calculate:

(1) Surface exposure ages of fines, and
(2) Effective mixing depths during irradiation due to micrometeorite impacts.

This model assumes that most of the tracks are accumulated during the surface exposure of a particular layer, after the layer was deposited. The above model proposed by Arrhenius *et al.* (1971) was borne out by further observations on Apollo 14 core. Also the model is supported by a linear correlation between the track parameter (N_H/N) and the content of glassy fragments (in particular, agglutinates), as was pointed out by McKay *et al.* (1971). Since the formation of tracks and agglutinates both are surface phenomenon and a result of soil maturity, the correlation indicates a general validity of this model for the growth of regolith.

Exposure ages of surface fines. We have given in Table 2 the value of N_H/N, ρ_q and the calculated exposure ages of surface scoop samples from Apollo 15 and 16 sites. The exposure ages are based on an assumed scoop depth of 3 cm and calculations follow the earlier procedure of Arrhenius *et al.* (1971).

The N_H/N values in different scoop samples varies from 0.3 to 1.0 indicating extent of mixing of these samples during their exposure at the surface. The exposure ages vary from few million years to about hundred million years. The two soil samples from the rim and ejecta of North Ray Crater (Table 2) give ages of 25 and > 100 m.y. respectively. Husain and Schaeffer (1973) and Marti *et al.* (1973) have given similar age of 30 ± 5 and 50 m.y. respectively based on rare gas analysis in rock samples from the North Ray rim. The Buster Crater has exposure age of only 6 m.y. which supports the description (Interagency Report: Astrogeology 54) of this crater as fresh with sharp rim and angular blocky debris on the floor and parts of the wall. It is interesting to note that as mentioned earlier the rock 62295 collected within 40 meters of the Buster Crater also shows a similar suntan age of ≈ 3 m.y. with probably no buried exposure at shallow depth. The age of South Ray Crater, based on ejecta sample 68121,5 is estimated to be 17 m.y. (Table 2). However, the rare gas analysis in various rock fragments from the same area show a wide range of exposure ages, from 2 to 130 m.y. (Behrmann *et al.*, 1973; Husain and Schaeffer, 1973), probably because some of the rock fragments may not be associated with this event.

We have analyzed tracks in glassy fragments in soil sample 15302,28 and in the glass rich breccia 15426,75 collected from the same location (Station 7) at the Apollo 15 site. The track density distribution in the feldspar as well as in glass fragments from these samples are shown in Fig. 6. Our data in the breccia are in agreement with those obtained by Fleischer and Hart (1973a) but the mean soil track densities are two orders of magnitude higher than the soil sample 15401,

Table 2. Exposure ages of Apollo 15 and 16 surface samples.*

Sample No.	Grain Size	Location	ρ average ($\times 10^6$ cm^{-2}) for $\rho < 10^8$ cm^{-2}	N_H/N (No. of grains)	Quartile track density ($\times 10^6$ cm^{-2})	Surface irradiation age (m.y.)
Apollo 15 samples						
15091,67	<1 mm	Station 2	47.0	0.8(27)	38	34
15101,131	<1 mm	Station 2	50.4	0.79(34)	75	69
15211,49†	<1 mm	Base	33.7	0.79(62)	19	17
15221,72†	<1 mm	N.W. rim	27.7	0.74(27)	27	23
15231,70†	<1 mm	S. rim	40.6	0.58(27)	40	35
15302,28	<1 mm	Spur Crater	21.0	0.3(21)	11	10
			6.0	0.17(31 glass)	3.3	—
Apollo 16 samples						
60051,7	<1 mm	ALSEP	11.5	0.3(56)	4.7	4
61501,14	<1 mm	Plumb Crater rim?	27.5	0.71(28)	27	24
62281,2	<1 mm	Edge of Buster Crater	21.0	0.55(22)	6.7	6
63501,32	<1 mm	North Ray ejecta?	38.0	0.9(37)	>100	>100
64801,21	<1 mm	Station 4 crater rim	15.0	0.58(35)	6.7	6
67481,22	<1 mm	North Ray rim	26.6	0.7(27)	27	24
68121,5	<1 mm	South Ray ejecta?	19.2	0.76(49)	19	17

*Based on the model of Arrhenius *et al.* (1971). Assumed scoop depth = 3 cm.
†The position of these samples are given with respect to the big boulder found at Station 2.

from a nearby location, they have analyzed. This observation shows that the soil is mixed to different extents at different places. The track data in feldspar grains indicate a very short exposure age ($\sim$ 10 m.y.) for the sample 15302. If glass formation is a surface phenomenon resulting from impact vaporization and quenching, this low age is surprising since glass fragments and, more frequently, spherules constitute >80% of the grains. This observation therefore, must be related to other mechanisms of production of glass which are not well understood at the present time or to an enrichment of glassy fragments, by transport from a nearby glass splattered region.

The evolution of the lunar regolith at the Apollo 15 site. Our investigations refer to the principal layers of the core stem 15002 (Heiken and Duke, 1972) and a few other samples from various depths of the 2.5 meter long drill core taken at the Apollo 15 site (Bhandari *et al.*, 1972c). The results of track analyses along with the calculated exposure ages for individual layers, based on the model of Arrhenius *et al.* (1971) are given in Table 3. The observed variation in N_H/N for samples from different layers in the core is shown in Fig. 7. Some aspects of these results have been discussed earlier (Bhandari *et al.*, 1972c).

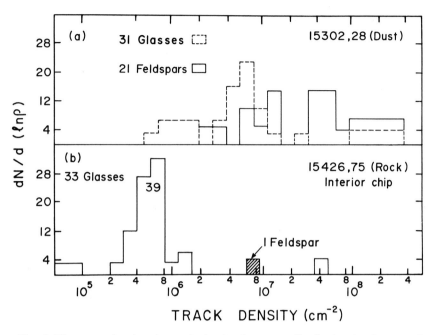

Fig. 6. Histogram showing the track density frequency distribution in glasses and feldspars for scoop sample 15302,28 and clod 15426, both collected at Station 7 at Apollo 15 site.

Table 3. Exposure ages of Apollo 15 core strata.*

Sample No.	Layer, thickness or location	Depth in core (meters)	N_H/N (No. of grains)	ρ average ($\times 10^6$ cm^{-2}) for $\rho < 10^8$ cm^{-2}	Quartile track density (10^6 cm^{-2})	Surface irradiation age (m.y.)
15004,56	IV, 12.8 cm	0.858	0.86(14)	36.1	27	—
15004,95	II, 6.9 cm	1.044	0.98(22)	53.3	$\sim 10^8$	—
15004,103	I, 14 cm	1.077	0.82(28)	25.8	53	—
15002,121	XI, 1.8 cm	1.63	0.6(53)	22.0	18	6
15002,111	X, 4.6 cm	1.66	0.4(20)	18.8	18	23
15002,107	IX, 4.3 cm	1.684	0.7(30)	33.32	64	35
15002,93	VIII B, 2 cm	1.74	0.41(24)	31.30	11	~ 0.5
15002,79	VII, 1.7 cm	1.804	0.9(21)	35.2	32	17
15002,72	VI, 4.8 cm	1.84	0.75(32)	32.0	64	40
15002,63	VC, 2.0 cm	1.88	0.75(36)	20.45	30	14
15002,55	IV E, 3.2 cm	1.912	0.69(52)	26.44	45	28
15002,45	III, 1.3 cm	1.966	0.4(20)	11.55	7	<1
15002,43	II, 0.8 cm	1.98	0.75(46)	26.35	45	18
15002,41	I, 1.5 cm	1.986	0.5(43)	12.0	7	<1
15001	?	2.33	0.4(44)	9.5	5	—

*Based on the model of Arrhenius *et al.* (1971).

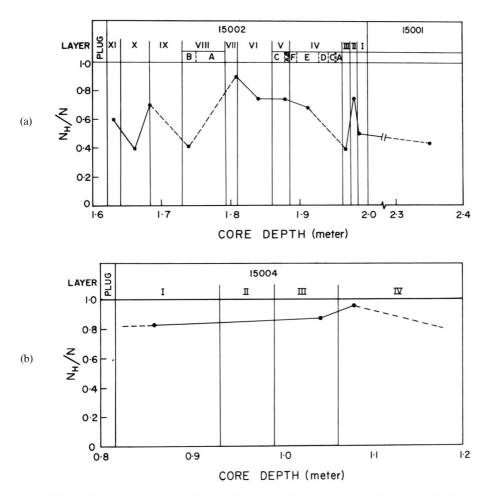

Fig. 7. Fraction of surface irradiated grains (N_H/N) for various strata of core stem 15002 and 15001 (a) and 15004 (b). The variation of N_H/N from layer to layer indicates that each layer had a different irradiation history.

The N_H/N value lies between 0.3 and 0.9 and the exposure ages range from a few million years to a few tens of million years. Results shown in Figs. 4 and 7 indicate that each layer in the core has a distinct individual irradiation pattern as was observed in case of Apollo 12 double core. The observed pattern can be explained in terms of surface exposure and impact gardening of individual layers for a period of a few million years before they are blanketted by a fresh layer. Similar considerations also apply for the samples from Apollo 14 trench and core (Fig. 8) and scoop samples from the Apollo 15 and 16 sites. The main feature of track data in Apollo 15 core can be summarized as follows:

(1) The surface irradiation ages for the 15002 layers, as also for the scoop samples, lie in the range 1–70 m.y. The duration of time involved in the deposition

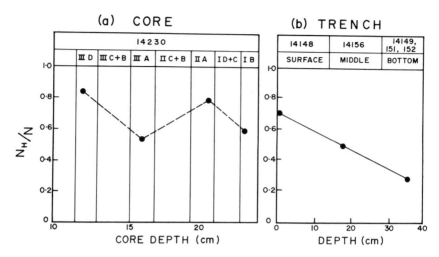

Fig. 8. Fraction of surface irradiated grains (N_H/N) for various samples taken from
Apollo 14 core 14230 and trench samples.

of drill stem 15002 is ~ 200 m.y. If each section of the core required similar deposition time one may place the date when the bottom of the core was laid as about 1
b.y. ago.

(2) Three of the layers of 15002 have surface exposure ages of ≤ 1 m.y., indicating their rapid episodic burial.

(3) The average rate of deposition of Apollo 15 core stem 15002 based on track
exposure ages in Table 3, is estimated to be 0.5 g cm^{-2} m.y.$^{-1}$. This does not
contradict the gadolinium data: the concentration of neutron produced Gd^{158} having a peak at about 200 g cm^{-2} (Russ et al., 1972, 1973) indicates that the rate of
deposition/vertical mixing is less than 2 g cm^{-2} m.y.$^{-1}$. Fleischer and Hart (1973b)
also deduced lower limits for deposition rates of 0.6–0.8 g cm^{-2} m.y.$^{-1}$ for other
drill stems of this core.

Impact gardening and micrometeorite influx rate

Finally, we wish to discuss the implications of fossil track data in lunar fines
with a view to study the correlation of irradiation dose with the small scale mixing
due to impact of micrometeorites.

Obviously a continuous micrometeoritic bombardment will garden the surface
samples and the core layers (during their surface exposure) to an extent which can
be determined from the value of the parameter, N_H/N i.e., the fraction of grains in
a particular layer that had a surface exposure. From simple considerations, one
would expect

(a) $N_H/N \propto T$ (for constant d)

(b) $N_H/N \propto 1/d$ (for constant T)

Therefore $N_H/N \propto T/d$ (both T and d varying) where T is the exposure age and d
the layer thickness.

The experimental data for the various scoop and core samples are shown in Figs. 9 and 10. It is interesting to note that in the case of scoop samples (where we have assumed a layer thickness of 3 cm), N_H/N is almost a linear function of T saturating for $N_H/N \approx 0.9$ at $T \geq 20$ m.y. In the case of the Apollo 15 core sample, the correlation of N_H/N and T/d is as expected (Fig. 10). The departure in case of Apollo 12 core may reflect that individual layers are not homogeneously mixed or that the core strata have not been well preserved. The possibility that the Apollo 12 core strata have been disturbed is indicated by large differences in 'actual' depths in the undisturbed soil on lunar surface and depths in the recovered core tube of various layers (Carrier *et al.*, 1971), whereas in case of Apollo 15 core samples, the differences are small (Carrier *et al.*, 1972). The scoop data suggest that the surface soil gets well mixed up to 3 cm in about 20 m.y. such that more than 80% of grains have individual surface exposure of 10^4–10^5 yr. The data on core stem 15002 also show a similar trend.

We have made model calculations based on the expected rates of mixing for different micrometeorite fluxes (Gault *et al.*, 1972; Neukum, 1973; Morrison *et al.*, 1972), and following crater dynamics determined by laboratory simulation data (Vedder, 1972). We deduce that the flux of micrometeorites of size <0.01 cm (Gault *et al.*, 1972), is sufficient to produce an efficient gardening for the top 1 mm layer before deeper mixing occurs due to larger size meteorites. Further, these calculations yield the value for the integrated flux of meteorites of 0.1–0.3 cm to

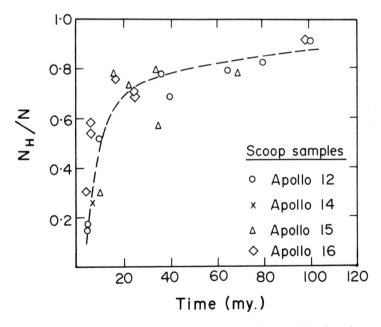

Fig. 9. Observed increase of surface irradiated grains (N_H/N) as a function of exposure time (T) for Apollo 12, 14, 15, and 16 scoop samples. N_H/N saturates at exposure times $\approx$ 20 m.y. for the 3 cm deep scoops.

N. Bhandari *et al.*

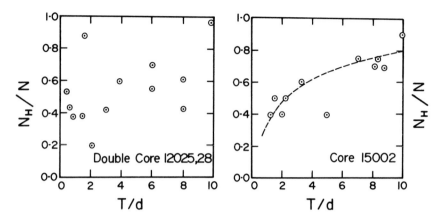

Fig. 10. Observed increase of surface irradiated grains (N_H/N) in each layer of Apollo 12
and 15 cores as a function of irradiation parameter T/d. To correct for the differences expected due to different thickness d of various layers, the exposure age T has been
divided by d to obtain the irradiation parameter.

be of the order of $(2\text{--}4) \times 10^{-4} \, \text{g cm}^{-2} \, \text{m.y.}^{-1}$. This flux is necessary to produce
N_H/N values in the range of 0.5–0.8 for layers of 2–3 cm thickness.

CONCLUSIONS

Based on extensive fossil track analyses in samples of rocks and fines from
Apollo 15 and 16 missions, we now confirm the earlier observations (Arrhenius *et
al.*, 1971; Bhandari *et al.*, 1971) that distinct solar and galactic irradiation patterns
are observed in most of the lunar samples and that this information allows one to
estimate the flux of track forming nuclei of solar and galactic origin (Bhandari *et
al.*, 1971, 1973). It also permits delineation of the principal processes occurring on
the lunar surface (erosion and fragmentation) and allows construction of a
plausible evolutionary model for the lunar regolith.

We have shown that the majority of lunar rocks have a complicated exposure
history. Moreover the continuous micrometeoritic impact results in erosion and
frequent chipping off of surfaces exceeding 0.1 cm in thickness in time periods of
the order of 3 m.y.

Track data in the long Apollo 15 core, observed to have a well preserved
stratification, indicates a discrete depositionary model for top meter thickness of
the regolith. The evolution of the lunar regolith can be thought of as a two stage
process: A cratering event leading to the deposition of a layer of soil excavated
from a depth of a few meters, followed by impact mixing due to micrometeorite
bombardment of this layer before it is buried by fresh layers due to another cratering event. The validity of this model on far away sites like Apollo 12 and 15 shows
that it is not only restricted to limited areas of the moon but may be applicable to
the lunar regolith, in general. The situation is, no doubt, more complicated than
envisioned here but nonetheless this model describes an important mechanism for
build up of the lunar regolith.

Fossil track data in scoop and core fines have been analyzed to deduce (i) exposure ages of lunar fines exposed on the surface at present, (ii) exposure ages of individual layers up to depths of few meters, and (iii) the micrometeorite influx rates for particles of size 0.1–0.3 cm.

Acknowledgments—We are grateful to NASA for making the precious lunar samples available to us. Our thanks are due to Mr. J. T. Padia for technical assistance.

REFERENCES

Arrhenius G., Liang S., MacDougall D., Wilkening L., Bhandari N., Bhat S., Lal D., Rajagopalan G., Tamhane A. S., and Venkatavaradan V. S. (1971) The exposure history of Apollo 12 regolith. *Proc. Second Lunar Sci. Conf., Geochim. Cosmochim. Acta*, Suppl. 2, Vol. 3, pp. 2583–2598. MIT Press.

Barber D. J., Cowsik R., Hutcheon I. D., Price P. B., and Rajan R. S. (1971) Solar flares, the lunar surface, and gas rich meteorites. *Proc. Second Lunar Sci. Conf., Geochim. Cosmochim. Acta*, Suppl. 2, Vol. 3, pp. 2705–2714. MIT Press.

Behrmann C., Crozaz G., Drozd R., Hohenberg C., Ralston C., Walker R., and Yuhas D. (1973) Radiation history of Apollo 16 site. In *Lunar Science—IV*, pp. 54–56. The Lunar Science Institute, Houston.

Berdot J. L., Chetrit G. C., Lorin J. C., Pellas P., and Poupeau G. (1972) Track studies of Apollo 14 rocks, and Apollo 14, Apollo 15 and Luna 16 soils. *Proc. Third Lunar Sci. Conf., Geochim. Cosmochim. Acta*, Suppl. 3, Vol. 3, pp. 2867–2881. MIT Press.

Bhandari N., Bhat S., Lal D., Rajagopalan G., Tamhane A. S., and Venkatavaradan V. S. (1971) High resolution time averaged (millions of years) energy spectrum and chemical composition of iron-group cosmic ray nuclei at 1 A.U. based on fossil tracks in Apollo samples. *Proc. Second Lunar Sci. Conf., Geochim. Cosmochim. Acta*, Suppl. 2, Vol. 3, pp. 2611–2619. MIT Press.

Bhandari N., Goswami J. N., Gupta S. K., Lal D., Tamhane A. S., and Venkatavaradan V. S. (1972a) Collision controlled radiation history of the lunar regolith. *Proc. Third Lunar Sci. Conf., Geochim. Cosmochim. Acta*, Suppl. 3, Vol. 3, pp. 2811–2829. MIT Press.

Bhandari N., Goswami J. N., Lal D., MacDougall D., and Tamhane A. S. (1972b) A study of the vestigial records of cosmic rays using thick section technique. *Proc. Ind. Acad. Sci.* **LXXVI** A, 27–52.

Bhandari N., Goswami J. N., and Lal D. (1972c) Apollo 15 regolith: A predominantly accretion or mixing model? In *The Apollo 15 Lunar Samples*, pp. 336–341. The Lunar Science Institute, Houston.

Bhandari N., Goswami J. N., and Lal D. (1973) Kinetic energy spectra of 5–1500 MeV/n VH nuclei long term averaged flux at 1–3 A.U. In *Proceedings of International Cosmic Ray Conference*, Paper 716, Denver.

Carrier W. D., III, Johnson S. W., Werner R. A., and Schmidt R. (1971) Disturbance in samples recovered with the Apollo core tubes. *Proc. Second Lunar Sci. Conf., Geochim. Cosmochim. Acta*, Suppl. 2, Vol. 3, pp. 1959–1972. MIT Press.

Carrier W. D., III, Johnson S. W., Carrasco L. H., and Schmidt R. (1972) Core sample depth relationship: Apollo 14 and 15. *Proc. Third Lunar Sci. Conf., Geochim. Cosmochim. Acta*, Suppl. 3, Vol. 3, pp. 3213–3221. MIT Press.

Comstock G. M., Evwaraye A. O., Fleischer R. L., and Hart H. R., Jr. (1971) The particle track record of lunar soil. *Proc. Second Lunar Sci. Conf., Geochim. Cosmochim. Acta*, Suppl. 2, Vol. 3, pp. 2569–2582. MIT Press.

Crozaz G., Walker R., and Woolum D. (1971) Nuclear track studies of dynamic surface processes on the moon and the constancy of solar activity. *Proc. Second Lunar Sci. Conf., Geochim. Cosmochim. Acta*, Suppl. 2, Vol. 3, pp. 2543–2558. MIT Press.

Crozaz G., Drozd R., Hohenberg C. M., Hoyt H. P., Jr., Ragan D., Walker R. M., and Yuhas D. (1972) Solar flare and galactic cosmic ray studies of Apollo 14 and 15 samples. *Proc. Third Lunar Sci. Conf., Geochim. Cosmochim. Acta*, Suppl. 3, Vol. 3, pp. 2917–2931. MIT Press.

Fleischer R. L., Hart H. R., Jr., and Comstock G. M. (1971) Very heavy solar cosmic rays: Energy spectrum and implications for lunar erosion. *Science* **171**, 1240–1242.

Fleischer R. L., Comstock G. M., and Hart H. R., Jr. (1972) Dating of mechanical events by deformation induced erasure of particle tracks. *J. Geophys. Res.* **77**, 5050–5052.

Fleischer R. L. and Hart H. R., Jr. (1973a) Particle track record of Apollo 15 green soil and rock. *Earth Planet. Sci. Lett.* **18**, 357–364.

Fleischer R. L. and Hart H. R., Jr. (1973b) Particle track record in Apollo 15 deep core from 54 to 80 cm depths. *Earth Planet. Sci. Lett.* **18**, 420–426.

Gault D. E., Hörz F., and Hartung J. B. (1972) Effects of microcratering on the lunar surface. *Proc. Third Lunar Sci. Conf., Geochim. Cosmochim. Acta*, Suppl. 3, Vol. 3, pp. 2713–2734. MIT Press.

Heiken G. and Duke M. (1972) Description and sampling of Apollo 15 core. NASA Preprint.

Hörz F., Hartung J. B., and Gault D. E. (1971) Micrometeorite craters and lunar rock surfaces. *J. Geophys. Res.* **76**, 5770–5798.

Husain L. and Schaeffer O. A. (1973) ^{40}Ar–^{39}Ar crystallization ages and ^{38}Ar–^{37}Ar cosmic ray exposure ages of samples from the vicinity of the Apollo 16 landing site. In *Lunar Science—IV*, pp. 406–408. The Lunar Science Institute, Houston.

Interagency Report: *Astrogeology* **54**. Apollo Field Geology investigation team U.S. Geological survey (1972) Published by U.S. Department of the interior geological survey.

Krishnaswami S., Lal D., Prabhu N., and Tamhane A. S. (1971) Olivines; revelation of tracks of charged particles. *Science*, **174**, 287–291.

Lal D. (1972) Hard rock cosmic ray archaeology, *Space Sci. Rev.* **14**, 3–102.

Lal D., Murali A. V., Rajan R. S., Tamhane A. S., Lorin J. C., and Pellas P. (1968) Techniques for proper revelation and viewing of etch tracks in meteoritic and terrestrial minerals. *Earth Planet. Sci. Lett.* **5**, 111–119.

Lunar Sample Information Catalogue—Apollo 15 (1971) *NASA* MSC 03209.

Marti K., Lightner B. D., Lugmair G. W., Osborn T. W., and Scheinin N. (1973) I. On the early history: Evidence from Pu^{244} and Nd^{143}. II. The age of North Ray Crater. In *Lunar Science—IV*, pp. 502–504. The Lunar Science Institute, Houston.

McKay D. S., Morrison D. A., Clanton U.S., Ladle G. H., and Lindsay J. F. (1971) Apollo 12 soil and breccia. *Proc. Second Lunar Sci. Conf., Geochim. Cosmochim. Acta*, Suppl. 2, Vol. 3, pp. 755–773. MIT Press.

Morrison D. A., McKay D. S., Heiken G. H., and Moore H. J. (1972) Microcraters on lunar rocks. *Proc. Third Lunar Sci. Conf., Geochim. Cosmochim. Acta*, Suppl. 3, Vol. 3, pp. 2767–2791. MIT Press.

Neukum G. (1973) Micrometeoroid flux, microcrater population development and erosion rates on lunar rocks, and exposure ages of Apollo 16 rocks derived from crater statistics. In *Lunar Science—IV*, pp. 558–560. The Lunar Science Institute, Houston.

Price P. B. and O'Sullivan D. (1970) Lunar erosion rate and solar flare paleontology. *Proc. Apollo 11 Lunar Sci. Conf., Geochim. Cosmochim. Acta*, Suppl. 1, Vol. 3, pp. 2351–2359. Pergamon.

Russ G. P., III, Burnett D. S., and Wasserburg G. J. (1972) Lunar neutron stratigraphy. *Earth Planet. Sci. Lett.* **15**, 172–182.

Russ G. P., III, Burnett D. S., and Wasserburg G. J. (1973) Regolith stratigraphy and neutron capture. In *Lunar Science—IV*, pp. 642–644. The Lunar Science Institute, Houston.

Vedder J. F. (1972) Craters formed in mineral dust by hypervelocity micrometersize projectiles. *J. Geophys. Res.* **77**, 4304–4308.

Proceedings of the Fourth Lunar Science Conference
(Supplement 4, *Geochimica et Cosmochimica Acta*)
Vol. 3, pp. 2291–2305

Charged-particle tracks in Apollo 16 lunar glasses and analogous materials

S. A. Durrani, H. A. Khan,* S. R. Malik, A. Aframian,
and J. H. Fremlin

Department of Physics, University of Birmingham,
Birmingham B15 2TT, England

J. Tarney

Department of Geology (with Geophysics), University of Birmingham,
Birmingham B15 2TT, England

Abstract—The main thrust of the paper is to study the effects of temperature and radiation damage on the registration and retention of heavily-charged particle tracks in lunar glasses and analogous materials in an effort to interpret the temperature and radiation history of lunar samples. Fossil-fission and external-particle tracks have been differentiated by a series of annealing experiments. The etch-pit diameter is the primary parameter in these experiments as well as in those involving the registration of artificially accelerated heavy ions of varying residual energies. It is also shown that glass detectors intensely irradiated with $\sim 3\,\text{MeV}$ protons (to fluences of up to $\sim 1.4 \times 10^{17}\,\text{cm}^{-2}$), either before or especially after exposure to heavy ions or fission fragments, are significantly less efficient in the registration of the latter highly-charged particles than are unirradiated detectors. The heavy-ion tracks yield much smaller etch pits if the detectors are subsequently irradiated with protons. These observations have an important bearing on the quantitative analysis of lunar track data. Also reported in the paper are useful track-etching parameters for lunar glasses; uranium-content values; and some age estimates.

Introduction

Charged-particle tracks in lunar rocks and soil have been used extensively in the recent past to glean a variety of information on the radiation history of the moon's surface (*see*, e.g., Bhandari *et al.*, 1972; Crozaz *et al.*, 1972; Hart *et al.*, 1972; Phakey *et al.*, 1972). We have concentrated on glass spherules and fragments (mostly from Apollo 16, but some also from Apollo 15 samples) to study these phenomena. Particular attention has been paid to the effects of lunar temperature and primary cosmic-ray proton bombardment on the registration and retention of heavily-charged particle tracks. These effects have been simulated by laboratory annealing experiments and by exposure of lunar glasses and analogous materials (terrestrial glasses, mica, olivine crystals, etc.) to energetic heavy ions (Fe beams, recoiling Au ions, fission fragments, etc.). The variation in the etch-pit diameters (under identical etching conditions) observed as a function of particle energy and charge, or of annealing temperature, has been used as the main parameter in these investigations.

---

*On leave of absence from the Pakistan Atomic Energy Commission

As a matter of general interest for workers in the field, track-etching parameters have been measured for lunar glasses. Uranium content, and total age (since last melting or severe heating) based on fossil-fission tracks in lunar glasses are amongst other measurements reported in the paper.

EXPERIMENTAL METHODS AND FACILITIES

A majority of the work reported here has been done with optical microscopy, though scanning-electron microscopy (SEM) has been used where appropriate.

Irradiations were carried out with the following machines: the University of Manchester Linear Accelerator (LINAC), which produces heavy-ion beams up to $\sim$ 10 MeV/nucleon; the 12 MeV Van de Graaff (Tandem) Generator at AERE, Harwell (heavy-ion acceleration to a total energy of $\sim$ 40 MeV); the University of Birmingham Nuffield Cyclotron (10 MeV protons); Rutherford High Energy Laboratory Synchrotron (7 GeV protons); and HERALD reactor at AWRE, Aldermaston (thermal neutrons). In addition, a Cf-252 spontaneous-fission source was extensively used.

For polishing and etching purposes, the lunar spherules were mounted, individually or in clusters, in epoxy resin on a transparent plastic backing, which allowed transmission microscopy. Glasses and mica were etched in 48.0 vol.% HF at 21°C; plastic in 6.25 N NaOH at 60°C; and olivine in "WN" etchant (Krishnaswami *et al.*, 1971) at 105°C. The etching was carried out in a constant-temperature bath, controlled to within ± 0.1°C. Polishing was done with different grades of diamond dust, ending with a hand-polishing step with 0–1 μm "Diadust" on micro-cloth. Other experimental details are given in appropriate sections below.

RESULTS AND DISCUSSION

Annealing behavior of tracks

The main annealing experiments were carried out on three medium-sized light-brown glass spherules from the Apollo 16 fines sample 66081,9. These were given the code numbers 1608 (diameter $\sim$ 350 μm), 1609 ($\sim$ 320 μm) and 1610 ($\sim$ 280 μm). The spherules were mounted individually and then ground to yield flat surfaces $\sim$ 20–30 μm from the top. After careful polishing and controlled etching in HF for 50 sec, the resulting etch-pit diameters were measured with an optical microscope at a magnification of 450.

Figure 1 shows the distribution of etch-pit diameters for spherule 1608. The ordinate represents the total numbers observed in two consecutive layers (with an intermediate polishing and re-etching step). A distinctly double-humped distribution of diameters is observed (*see* the fitted smooth curve). It was surmised that the high-diameter group labeled F represents fossil-fission events and the broad low-diameter group labeled E corresponds to the tracks produced by heavily-ionizing external particles. To test this hypothesis, two procedures were adopted. First, the successive etching and polishing method of Apollo 15 glasses (Durrani and Khan, 1972) was applied, in which the track density at and below an etched surface is measured at $\sim$ 2 μm depth intervals by removing successive layers by polishing. When all etched tracks have been polished away, a new etching step is performed and the procedure repeated. Had all the tracks encountered been fission events, distributed randomly with depth, the track density would have remained constant with depth down to $R \sin \theta_c$ (where θ_c is the minimum angle of

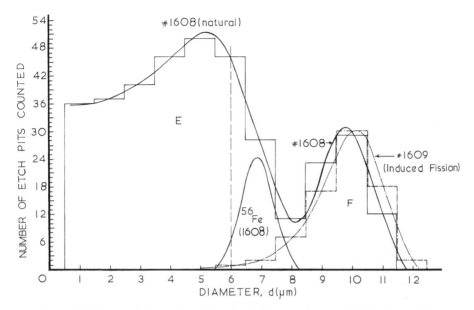

Fig. 1. Distribution of charged-particle etch-pit diameters in a natural light-brown glass spherule (1608) from sample 66081,9, and of induced fission tracks in a similar spherule (1609) which had been pre-annealed (at 600°C for 1 hr). The hump F in 1608 is attributed on this basis to natural fission, and the broad hump labeled E is believed to be due to external heavily-charged particles. Further calibration is provided by the ^{56}Fe peak (from the ~ 9.6 MeV/amu ion-beam exposure of annealed 1608) and by the vertical line deduced from ~40 MeV Fe irradiation of soda-lime glass. Etching for Figs. 1, 2, and 4 was always in 48.0 vol.% HF at 21.0±0.5°C for 50 sec; and magnification, 450×. Total area scanned in 1608 was 8.36×10^{-2} mm².

etching), and then fallen off steadily, reaching 0 at the full fission-fragment range R (~ 15 μm) below an etched surface. It was found, instead, that the track density declined sharply with depth below a given etch-surface (to a depth of ~ 5 μm), after which a plateau ensued, until at a depth of ~ 15 μm the etch pits suddenly ceased (*see* Fig. 2 of the above reference). It was therefore concluded that a short-track-length group (presumably due to external charged particles) was superimposed on the long-track-length group. The latter showed the same depth-profile behavior as did Cf-252 fission-fragment and induced-fission tracks in the spherule.

The second test for the identity of the putative fission group in Fig. 1 (labeled F) was to induce fresh fission in a similar brown-glass lunar spherule, No. 1609, after annealing it at 600°C for 1 hr to remove all pre-existing tracks in it. It is clear from Fig. 1 that the diameter distribution in 1609 (normalized for total number of tracks) is almost identical with that of group F in 1608. Furthermore, the diameter distribution of normally incident ^{56}Fe ions (~ 9.6 MeV/nucleon) in spherule 1608 is seen to peak at ~ 7 μm, distinctly below the fission peak ($\bar{d}$ ~ 10 μm).

From the above tests, based both on track lengths and on etch-pit diameters, it is concluded that the two humps in Fig. 1 do indeed represent the external charged

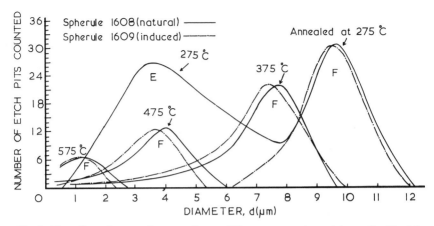

Fig. 2. The effect of successive annealing at different temperatures (always for 10 min) on the etch-pit diameters of natural tracks in 1608 and of induced-fission tracks in 1609. The external charged-particle hump E (the "soft component") is annealed out much more readily than the fission track hump F (the "hard component"), as is seen from the diminution both in numbers and in diameters. The histograms have been omitted and only the fitted smooth curves shown for clarity.

particles (group E) and the fossil fission events (group F), respectively. It was decided next to study the effects of annealing on the two groups.

Figure 2 displays the results of successive annealing steps (at different temperatures, but always for 10 min) on the natural tracks in spherule 1608 and the induced-fission tracks in spherule 1609. It is seen that annealing 1608 at 275°C for 10 min hardly affects the fission hump F (termed by us the "hard component"), while group E (the "soft component") is substantially reduced in numbers, the very low-diameter region, in fact, disappearing almost entirely. A further step of annealing at 375°C moves the fission hump to noticeably lower values (both in numbers and in mean diameter) while eliminating group E almost entirely. The fission track component is finally (and progressively) annealed out by two steps of 475 and 575°C.

The phenomenon of simultaneous fading and diametral reduction of tracks in various materials as a result of annealing is well known (Storzer and Wagner, 1969; Durrani and Khan, 1970; Khan and Durrani, 1973). In Fig. 3, the annealing behavior of fission tracks (both natural, i.e., the F component in Figs. 1 and 2, and induced) in lunar glasses is compared with that in an Australian tektite Au-12 (unpublished data of HAK and SAD). The correlations are fairly close, though the reduction in diameters in lunar glasses is somewhat greater than that in the tektite for a given degree of track annihilation. Using such curves, one can deduce the initial track density from a given reduction in diameters (as seen from a comparison with freshly induced fission tracks), and thus either infer the thermal history of a sample or make a correction for the thermal lowering of its age based on surviving tracks in the F component.

Spherule 1610 was found to have too great a track density for individual tracks

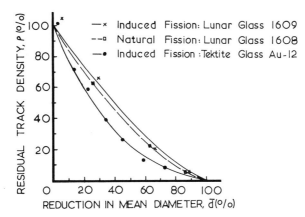

Fig. 3. Relationship between the residual track density ρ and the reduction in the mean etch-pit diameter $\bar{d}$ resulting from annealing lunar glasses and a tektite at various temperatures. The lunar glass spherule data are based on Figs. 1 and 2, and those for the australite Au-12 on the unpublished data of HAK and SAD.

to be studied with an optical microscope (at a magnification of 450). This high track density persisted down to a depth of $\sim 100\ \mu$m below the initial external surface when successive layers were revealed by polishing. It was therefore decided to anneal the sample to reduce the track density to manageable levels. On annealing the sample at 180°C for ~ 70 hr, individual tracks became just distinguishable (Fig. 4). The progressive diminution of both the track density and the mean diameter as further annealing steps of up to 375°C (for 10 min) were applied is also seen in Fig. 4.

The high-diameter (natural fission) hump is conspicuous by its absence in the diameter-distribution histogram of natural tracks in spherule 1610. To test whether this might have resulted from a change in the track registration properties of the spherule caused, say, by the high external-particle bombardment found in it, the spherule was exposed to a spontaneous-fission Cf-252 source. A hump F at the usual position (centered at $\sim 10.5\ \mu$m) resulted from this exposure (*see* Fig. 3 in Khan *et al.*, 1973). Subsequently, the spherule, after thermal annihilation of pre-existing tracks, was irradiated in a nuclear reactor. The induced fission hump is again seen (marked F in Fig. 4) at the expected position. Both these observations rule out radiation damage as the likely explanation for the missing natural-fission component. The prediction put forward in Khan *et al.* (1973) is, in fact, borne out in that the uranium content of this spherule is unusually low (found to be 0.07 ppm from a subsequent reactor irradiation experiment). The second point made in the paper just cited remains valid, too, namely, that the high external-particle contribution (observed prior to the laboratory annealing at 180°C) indicates that spherule 1610 had spent a greater proportion of its comparatively young age (estimated to be between ~ 3 and 8 times smaller than that of spherule 1608, even when the lower U-content of 1610 is taken into account, cf. Table 4) close to the top lunar surface. These tracks are likely to have been caused by low-damage

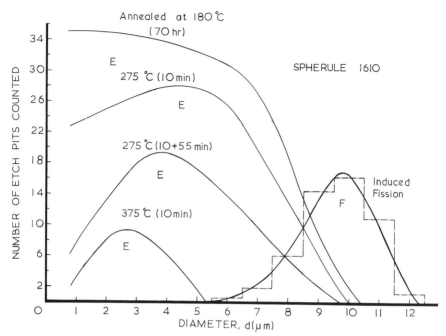

Fig. 4. The effect of successive annealing steps at different temperatures and for varying durations on a light-brown glass spherule (1610) from sample 66081,9. Both the number of tracks and the subsequent etch-pit diameters diminish progressively with temperature. No fossil-fission group of tracks is found in the natural sample, though thermal–neutron irradiation produces a hump labeled F at the expected position in the annealed sample. See text for discussion.

particles in view of their annealability at such low temperatures (180°C for a mere 70 hr).

Incidentally, assuming that spherules 1608 and 1610 have both come from comparable depths, so that the ambient temperature experienced by them on the moon was sensibly the same, the much longer age of 1608 (inferred from its substantial F component, as will be described later) may account for the depressed low-diameter edge of the E component in it (compare the track densities at ~ 1–4 μm diameter etch-pits in Figs. 1 and 4). It may also be mentioned, in passing, that a fourth lunar glass specimen, a fragment code-numbered 1611, showed a wider separation between humps F and E than was seen in spherule 1608.

Heavy-ion phenomena

In an effort to understand the heavily-charged external-particle phenomena in lunar materials, heavy-ion bombardments were carried out on a number of solid-state track detectors (SSTD). These included, in addition to lunar glasses themselves, terrestrial samples of soda-lime glass, muscovite mica, Makrofol*

---

*Manufactured by: Farbenfabriken Bayer A. G., Leverkusen, West Germany.

plastic, and olivine crystals. The beams used were those of ^{56}Fe (total energy ~ 40 MeV) and ^{16}O (total E ~ 35 MeV) accelerated in the Tandem generator at Harwell, and ^{56}Fe ($\cong 9.6$ MeV per nucleon: total E ~ 537 MeV) from the Manchester LINAC. In the Tandem generator, the ion beam, scattered by a thin Au foil (~ 100 μg/cm^2) inside an evacuated chamber, was made to fall perpendicularly at a series of SST detectors arranged in parallel strips (~ 1.2 cm wide). The LINAC beam, appropriately defocussed, fell directly on the SSTD. Track densities of $\sim 10^4$ to 10^6 cm^{-2} were generally obtained with both machines.

To investigate the effects of lunar erosion and churning (which may remove or interpose further layers of lunar material), as well as those of pre-existing outer layers, on the etchable damage produced in deeper layers by external charged particles, the detector strips were covered, during irradiation, with a step-wise (i.e., wedge-like) arrangement of different thicknesses of Al foil to degrade the ion energies by different known amounts. The mean etch-pit diameters in SSTD materials, corresponding to the respective residual energies of the ions, were then measured by etching each strip (without its Al covering) under standard conditions in a controlled-temperature bath. The etchants and the respective temperatures have already been given in the section on *Experimental Methods*; e.g., the soda-lime glass was etched for 5.5 sec (so that the thickness of the layer removed by etching was ~ 3.5 μm, cf. Table 1).

Figure 5 shows the variation in the mean etch-pit diameter, $\bar{d}$, as a function of the residual energy E of the ~ 40 MeV Fe ions in all the SSTD materials irradiated at Harwell. The mean diameter $\bar{d}$ is seen to fall gradually as the ionic charge diminishes with decreasing residual energy. The damage remains etchable down to Fe energies as low as ~ 2 MeV in all the SST detectors except olivine (which ceases to register tracks below $E \sim 10$ MeV). The curves for glass and mica in Fig. 5 also show one point each for recoiling Au ions. The ratios of the etch-pit diameters at equal-velocity values of $^{197}_{79}$Au and $^{56}_{26}$Fe (e.g., at 7 MeV and 2 MeV, respectively) are not far from those expected ($\sim 1.7:1$, cf. Northcliffe and Schilling, 1970) from the ($- \mathrm{d}E/\mathrm{d}x$) values when allowance has been made for the two (effective) z^2 values.

Figure 6 shows the behavior of etchable damage in soda-lime glass as a function of the residual energy of Fe ions from LINAC, degraded by varying thicknesses of Al foil. As the $^{56}_{26}$Fe ions slow down (initial total $E \sim 537$ MeV), the etchable damage increases at first, as expected from the Bragg effect. After a broad maximum, resulting from the opposing effects of slowing down and the steady diminution of the effective ionic charge, the etchable damage begins to fall off rapidly towards the end of the ions' range as their charge is progressively neutralized by electron pick-up. Similar curves for mean etch-pit diameter vs. residual energy have been obtained for Makrofol plastic (type KG), a stack of eight sheets, each ~ 22 μm thick, being used to record the progress of Fe ions of initial energy ~ 9.6 MeV/amu (*see* Malik *et al.*, 1973).

It may be emphasized that the etchable track-length of Fe ions of initial energy ~ 537 MeV (i.e., ~ 9.6 MeV/amu) in all three media studied by us, namely glass, mica and plastic, corresponded closely to the total range expected (McGowan *et al.*, 1969) in these media, namely ~ 27 mg/cm^2 (~ 108 μm and 100 μm, respec-

S. A. DURRANI *et al.*

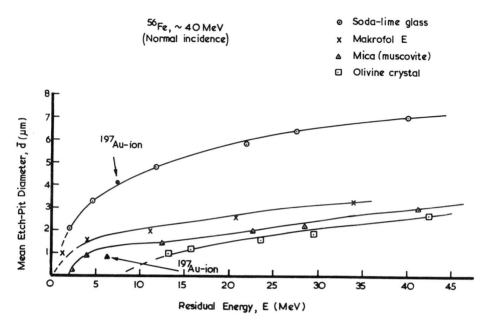

Fig. 5. The variation in the mean etch-pit diameter $\bar{d}$ in various SSTD materials as a function of the residual energy of ~ 40 MeV ^{56}Fe ions (degraded by different amounts with Al foils). The etching conditions were: glass, 48.0 vol.% HF at 21.0 ± 0.5°C for 5.5 sec; mica, as for glass, but etched for 25 min; Makrofol E, 6.25 N NaOH at 60°C for 60 min; olivine, the etchant "WN" (EDTA, NaOH, oxalic and orthophosphoric acids, pH6) at 105°C for 5.2 hr. The ^{197}Au ions originated from the scattering foil for the Fe beam. See text for discussion.

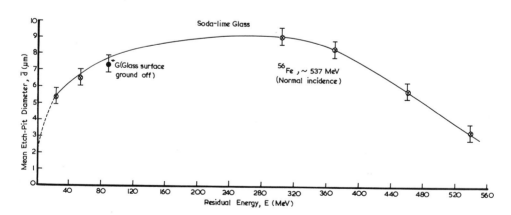

Fig. 6. The variation in the mean etch-pit diameter in soda-lime glass as a function of the residual energy of ^{56}Fe ions (initial total energy ~ 537 MeV), degraded by varying thickness of Al foil. Etching conditions as for Fig. 5. An extra point G on the curve is obtained by grinding off ~ 93 μm from the top surface of the bare glass detector. See text for discussion of changes in the ionization energy-loss rate as the Fe ions slow down.

tively) in glass and mica, and ~ 20 mg/cm^2 (~ 170 μm) in plastic. These values are very much greater than the etchable range of galactic cosmic-ray iron group nuclei assumed and extensively reported in the past (viz., ~ 10–16 μm in lunar feldspars and pyroxenes; *see*, e.g., Bhandari *et al.*, 1972; Plieninger *et al.*, 1972), but are not inconsistent with the results of Fleischer and Hart (1973) on Apollo 15 green glass, where cosmic ray tracks were found to be $\geqslant 22$ μm in length (the lower limit set by the etching conditions). Our values of the etchable range of Fe ions support the recent reports of Storzer *et al.* (1973) and Borg *et al.* (1973) concerning the much longer recordable ranges of "machine-made" heavy ions. It is possible that the fossil tracks of heavy ions have become very much shortened owing to lunar-temperature annealing or to radiation damage.

Effect of radiation damage on heavy-ion registration

In the light of the above-mentioned observations, a question of great current interest in charged-particle track studies in lunar crystals and glasses is to inquire into the effects of radiation damage on track registration. Thus the flux of solar-flare protons of energy > 3 MeV is $\sim 10^{12}$ cm^{-2} yr^{-1} (Haffner, 1967). A residence time of $\sim 10^5$–10^6 yr within the top 80 μm ($\sim$ the range of 3 MeV protons) of lunar surface corresponds to a total dose (or fluence) of 10^{17}–10^{18} protons/cm^2 for such particles. To investigate these effects, we carried out a series of experiments in which soda-lime glass slides were subjected to varying doses of ~ 3 MeV protons from a cyclotron, followed by exposure to ~ 4.14 MeV/amu beams of Fe ions, and vice versa. The glass slides, mounted on an Al target-holder, were cooled by air-blast during proton bombardment. A thermocouple at the back of the holder monitored the temperature of the target (which went up to $\sim 70°$C).

It is found (Fig. 7) that increasing doses of protons produce progressive reduction in the mean etch-pit diameter (under identical etching conditions) of the Fe tracks (total $E \sim 232$ MeV), whether the proton irradiation precedes or follows the Fe bombardment. The reduction in Fe diameters is, however, greater for a given proton fluence if the protons come last. These track-shrinkage effects are seen vividly in the photomicrograph (Fig. 8), which shows etch pits produced by the Fe ions on the same glass detector part of which had been subsequently irradiated with protons (fluence $\sim 3 \times 10^{16}$ cm^{-2}) while the rest had been masked. A reduction in Cf-252 fission-track diameters is also seen in Fig. 7 when proton-irradiated tektite or soda-lime glass is used for their detection.

Not only is the diameter of heavy ions and fission fragments reduced as a result of proton bombardment, but so is the track density registered. Table 1 records the relevant data for fission-fragment registration, when the same fluence of Cf-252 fission fragments ($\sim 2.56 \times 10^4$ cm^{-2}) was superimposed on identical soda-lime glass detectors which had previously received different doses of ~ 3 MeV protons. The observed fission-fragment track density progressively declines as the proton dose rises from 1.52×10^{15} to 1.37×10^{17} cm^{-2}. The efficiency of registration can be expressed in terms of the "critical angle of etching," θ_c. In the

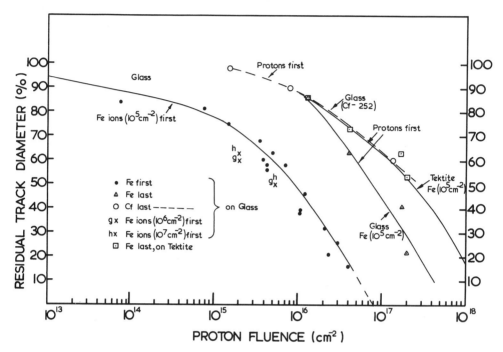

Fig. 7. The effect of $\sim 3\,\text{MeV}$ proton irradiation on the mean etch-pit diameter $\bar{d}$ produced by $\sim 232\,\text{MeV}$ Fe-56 ions and Cf-252 fission fragments in soda-lime glass and an Indochina tektite. The various proton irradiations either followed or preceded the exposure to the heavily-charged particles; the diminution in the etch-pit diameters is greater in the former case. The reduction in diameters was accompanied by a fall in the registration efficiency of the heavy particles (*see* Table 1). Etching in 48.0 vol.% HF at 21°C for 10 sec. The thickness of the layer removed by etching was $\sim 6\text{--}12\,\mu\text{m}$ in soda-lime glass, and $\sim 1\text{--}2\,\mu\text{m}$ in tektite glass (cf. Table 1, and Khan and Durrani, 1972b).

case of a thin, planar external source (e.g., the Cf-252 source used by us), the efficiency f (expressed as the "sensitive" fraction of the total solid angle) is given (Khan and Durrani, 1972a) by

$$f = 1 - \sin \theta_c \tag{1}$$

so that the track densities ρ_r and ρ_u in a reference (r) and an unknown (u) material for equal exposure are in the ratio

$$\rho_r/\rho_u = (1 - \sin \theta_{c,r})/(1 - \sin \theta_{c,u}) \tag{2}$$

By exposing parts of two identical glass slides to different proton fluences, and using the unirradiated portion as control or reference, the modified critical angle $\theta_{c,u}$ for subsequent Cf-252 fission-track registration was calculated from Eq. (2). In this calculation, $\theta_{c,r}$ for soda-lime glass was taken to be 35°30′ (Khan and Durrani, 1972a), whose etching conditions were somewhat different, but θ_c is

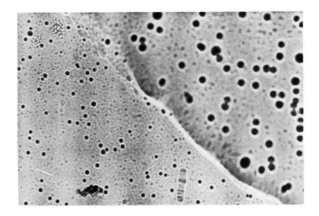

Fig. 8. Photomicrograph showing two regions of a soda-lime glass detector first exposed to a beam of $\sim$ 232 MeV Fe ions ($\sim$ 10^5 cm^{-2}) and then irradiated with $\sim$ 3 MeV protons ($\sim$ 3×10^{16} cm^{-2}). The region showing the larger etch-pits (somewhat out of focus in the photograph) had been masked by a thick Al foil during the irradiation with protons (to stop their passage). The diminution of Fe track diameters owing to proton radiation damage is obvious. Etching was as for Fig. 7.

Table 1. Changes in the registration efficiency of Cf-252 fission fragments in soda-lime glass as a result of prior irradiation with ($\sim$ 3 MeV) protons. The modified values of θ_c, V_g, and V_t are also measured or deduced.*

1† Detector (glass)	2 Proton fluence (cm^{-2})	3 Track density ρ	4 Registration efficiency(%)	5 Critical angle θ_c	6 V_g (measured) (μm/sec)	7 V_t (calculated) (μm/sec)
Control	0	1132	42.0	35°30′	0.61 ± 0.05	1.05
(a)	1.52×10^{15}	1110	41.3	35°57′	0.77	1.32
(b)	6.84×10^{15}	942	35.4	40°33′	0.90	1.38
(c)	1.37×10^{17}	625	23.6	49°49′	1.26	1.66

*The critical angle of etching θ_c for Cf-252 fission fragments was determined by the "comparison method" (Khan and Durrani, 1972a; see also Eq. (2) of text).

†Notes on column headings; 3. ρ is the fission track density observed (always for 10 sec of etching), normalized to 100 fields of view (at 450×, i.e., to $\sim 4.4 \times 10^{-2}$ cm^{-2}), although up to $\sim$ 200 fields of view were observed. The incident fluence of Cf-252 fragments was always constant ($\sim 2.56 \times 10^4$ cm^{-2}). 4. Registration efficiency has been normalized against that of the control, whose efficiency is based on the value obtained by Khan and Durrani (1972a). 5. θ_c for control is taken from the reference just quoted; that for detectors (a), (b), and (c) is based on values in column 4. 6. Etching of the irradiated area was done for 50 sec to obtain the general velocity of etching V_g. The error quoted for the control is based on three repeated measurements; that in the other detectors is assumed to be similar. 7. Track etching velocity V_t in the irradiated area is calculated from columns 5 and 6, using the relation $\sin \theta_c = V_g/V_t$.

essentially independent of them. As expected, θ_c goes up with proton fluence (cf. Table 1), making fission-fragment registration less efficient.

Since $\sin \theta_c = V_g/V_t$, where V_g and V_t are, respectively, the general and along-the-track velocities of etching in the medium under any given etching conditions, the observed variation in θ_c with proton fluence could have been due to a change in either V_g or V_t, or in both. To clarify this point, V_g was measured at each of the four positions (blank control plus three exposed areas) by prolonged etching (50 sec) and very careful initial and final thickness measurements with a micrometer (remembering that only one face of the detectors had been irradiated). V_g is found to increase (from 0.61 to 1.26 μm sec^{-1}) with increasing proton radiation damage (Table 1. Note that the glass thickness removed in 50 sec of etching was always within the range of ~ 3 MeV protons, viz. ~ 80 μm.) Krätschmer (1971) has also noted an increase in the value of V_g in SiO$_2$ glass as a result of electron irradiation. From the above values of V_g, and the θ_c values already calculated, V_t can be deduced. V_t is also seen (Table 1) to increase with proton fluence, but not as drastically as V_g (the former goes up by $\sim 60\%$ when the latter more than doubles at a fluence of 1.37×10^{17} protons cm^{-2}. Note that in these calculations we have assumed V_t to remain constant over the whole length of the fission track.) The effect on θ_c when the proton bombardment follows, rather than precedes, the Cf fission-fragment registration has subsequently been investigated and found to be much more severe (e.g., at a proton fluence of $\approx 2 \times 10^{16}$ cm^{-2}, $\theta_c \approx 61°40'$, i.e., $f \approx 12\%$, for soda-lime glass). Annealing of heavily irradiated glass (1.65×10^{17} protons cm^{-2}) at 450°C for 4.5 hr is found to restore V_g and f to their pre-irradiation values.

From the above experiments it is obvious that the track registration parameters are severely affected by radiation damage. The effect is likely to vary in degree from material to material and from one kind of radiation to another. It is therefore important to investigate these phenomena for each type of lunar crystal before accurate conclusions can be drawn. Bibring et al. (1973) have reported the conspicuous amorphous coating of a high proportion of fine lunar dust grains, ascribed by them to radiation damage from low-energy solar-wind particles. The track registration parameters close to external surfaces of lunar minerals are thus likely to be quite different from those well inside the samples. In our own glass spherules the top ~ 20 μm were usually ground off, which would remove regions damaged by low-energy particles (during the residence of a spherule at the very top lunar surface); but primary cosmic-ray proton damage is likely to have affected even the interior surfaces. These effects are being studied further in our laboratory. Lunar glasses and analogous materials have also been irradiated with 7 GeV protons by us to simulate galactic cosmic rays, and spallation tracks and their registration efficiency in various materials has been studied (SAD and HAK, to be published).

Track parameters of Apollo 16 lunar glasses

A number of parameters of general interest in charged-particle track analysis of lunar glasses have been measured, using in all about twenty light-brown and colorless glass spherules and fragments selected microscopically from the fines samples 66081,9 and 61281,2. The procedures followed have been described in previous reports (Durrani and Khan, 1972, 1973, for Apollo 15 and Apollo 16 investigations; *see* also Khan and Durrani, 1972a, b for general principles), and will not be repeated here. The results are recorded in Table 2 (values of V_g), Table 3 (θ_c), and Table 4 (U-content). While fairly constant values have been observed for the general velocity of etching V_g (0.144 ± 0.008 μm sec^{-1} for light-brown glasses; but note that the value for Apollo 15 light-green glasses was ~ 6 times as large, viz. 0.776 ± 0.014 μm sec^{-1}, as that for light-brown glasses, cf. Durrani and Khan, 1972), and for the critical angle θ_c for Cf-252 fission fragments ($37°22' \pm 0°22'$), the uranium content C_u is very variable from spherule to spherule. Thus C_u

Table 2. General velocity of etching, V_g, in glasses from sample 66081,9.*

Spherule No.	1604	1605	1606	1607	1608	1609
Time of etching (sec)	150	150	150	150	50	50
V_g (μm/sec)	0.147	0.143	0.150	0.140	0.130	0.144

Weighted mean $V_g = 0.144 \pm 0.008$ μm/sec.

*The etching was carried out under standard conditions (*see* text) for up to 150 sec. The spherules, mounted individually, were slightly ground and polished to yield plain initial surfaces. The thickness removed was measured with an optical microscope (at 450×).

Table 3. Critical angle of etching, θ_c,* for glasses from samples 66081,9 (marked (a)) and 61281,2 (marked (b)).

No.	θ_c	No.	θ_c	No.	θ_c
1604(a)†	36°42'	1614(a)	38°32'	1618(b)†	36°42'
1607(a)†	35°36'	1615(b)†	36°09'	1619(b)	39°08'
1612(a)	37°44'	1616(b)†	35°52'	1620(b)	38°32'
1613(a)	39°08'	1617(b)†	36°42'	1621(b)	37°22'

Mean for 12 particles: $37°22' \pm 0°22'$.

*The critical angle θ_c for fission fragments (from a Cf-252 source), was measured by the "comparison method," using Eq. (2) of text, and taking $\theta_{c,r}$ for the U-2 reference glass to be $31°45'$. For details see Durrani and Khan (1973).

†Those marked thus (†) are light brown glasses. The rest are colorless. θ_c for the colorless glasses (mean, $38°24' \pm 0°18'$) is somewhat larger than that for the light-brown glasses (mean, $36°17' \pm 0°12'$).

Table 4. Uranium content C_u (ppm)* in light-brown glasses from sample 66081,9.

Spherule No.	1604	1605	1606	1607	1608	1609	1610
U-content (ppm)	1.13	0.43	3.16	1.40	0.41	0.50	0.07

*The uranium content C_u (in ppm by weight) was determined by comparing the track densities in a Makrofol plastic detector resulting from the simultaneous reactor irradiation (fluence $\sim 10^{16}$ thermal neutrons/cm^2) of the spherules as well as a piece of reference glass (U-2, with 43 ± 1 ppm of U-content).

ranges from 0.07 ppm in spherule 1610 to 3.16 ppm in spherule 1606 (both from the same sample 66081,9), i.e., stretching over a factor of ~ 45. Similar variations have been reported by other authors (e.g., Walker and Zimmerman, 1972; and Fleischer and Hart, 1973, who also observe a variation by a factor of ~ 50 in C_u values).

The prolonged etching factor $f(t)$ (Khan and Durrani 1972b), defined as $f(t) = \rho(t)/\rho(0)$, where ρ is the density of internally recorded tracks as observed at times of etching t and 0 (the latter being found by backward extrapolation of the growth curve), has been shown graphically in Fig. 1 of Durrani and Khan (1973). The slope of the linear rise in track density with the time of etching (in 48.0 vol.% HF at $21.0 \pm 0.5°C$) is found to be $m = 0.47 \times 10^{-2}$ and 0.50×10^{-2} sec^{-1}, respectively, for spherules 1604 and 1607 (both from sample 66081,9), where $f(t) = 1 + mt$, so that, for example, $f(100\ \text{sec}) = 1.47$ and 1.50, respectively.

The apparent age of spherule 1608 is found to be $\sim 8.5 \times 10^8$ yr by using the track density in the natural-fission hump F (in Fig. 1) and the known U-content of the spherule (from its induced-fission track density). This is very similar to the age previously reported for an Apollo 15 glass spherule 11 from sample 15261,70, namely $\sim 7.5 \times 10^8$ yr (Durrani and Khan, 1972). No allowance has been made in this determination for track fading, radiation damage, extinct Pu-244 tracks, or induced fossil fission. The requisite correction for the last-named factor may be substantial if one accepts the estimate of Fields *et al.* (1973), who state that "there are 11.5 thermal neutron induced fissions for each lunar uranium spontaneous fission." Since we have based our age estimate on all natural fissions being spontaneous fissions of ^{238}U nuclei, our value (for age since the last severe heating or melting of the spherule) may have to be reduced by a factor of ~ 10 if the above correction is applied. If any substantial contribution has been made to the observed track density by ^{224}Pu fission, the above age estimate should also be lowered. On the other hand, corrections for track annealing by temperature and radiation effects on the moon would push the age up. These questions are being currently evaluated.

Acknowledgments—We wish to thank Mr. E. D. Lacy and Dr. G. L. Hendry for helpful discussions on geological aspects of the work. We are also grateful to the staff of the University of Manchester Linear Accelerator (in particular Dr. I. S. Grant and Mr. A. G. Smith) and of the AERE, Harwell, Tandem Generator (Dr. J. M. Freeman) for help with heavy-ion irradiations. It is a pleasure to thank Dr. R. L. Fleischer for his most helpful comments on the manuscript. Financial support by the Science Research Council and the Royal Society is also gratefully acknowledged. H. A. Khan wishes to thank the Pakistan Atomic Energy Commission for leave of absence to enable him to undertake this work.

REFERENCES

Bhandari N., Goswami J. N., Gupta S. K., Lal D., Tamhane A. S., and Vankatavaradan V. S. (1972) Collision controlled radiation history of the lunar regolith. *Proc. Third Lunar Sci. Conf., Geochim. Cosmochim. Acta*, Suppl. 3, Vol. 3, pp. 2811–2829. MIT Press.
Bibring J. P., Chaumont J., Comstock G., Maurette M., Meunir R., and Hernandez R. (1973) Solar wind and lunar wind microscopic effects in the lunar regolith (abstract). In *Lunar Science—IV*, pp. 72–74. The Lunar Science Institute, Houston.

Borg J., Dran J. C., Comstock G., Maurette M., Vassent B., and Duraud C. (1973) Nuclear particle track studies in lunar regolith: Some new trends and speculations (abstract). In *Lunar Science—IV*, pp. 82–84. The Lunar Science Institute, Houston.

Crozaz G., Drozd R., Hohenberg C. M., Hoyt H. P. Jr., Ragan D., Walker R. M., and Yuhas D. (1972) Solar flare and galactic cosmic ray studies of Apollo 14 and 15 samples. *Proc. Third Lunar Sci. Conf.*, *Geochim. Cosmochim. Acta*, Suppl. 3, Vol. 3, pp. 2917–2931. MIT Press.

Durrani S. A. and Khan H. A. (1970) Annealing of fission tracks in tektites: Corrected ages of bediasites. *Earth Planet. Sci. Lett.* 9, 431–445.

Durrani S. A. and Khan H. A. (1972) Charged-particle track parameters of Apollo 15 lunar glasses. In *The Apollo 15 Lunar Samples*, pp. 352–356. The Lunar Science Institute, Houston.

Durrani S. A. and Khan H. A. (1973) Track parameters of Apollo 16 lunar glasses (abstract). In *Lunar Science—IV*, pp. 202–204. The Lunar Science Institute, Houston.

Fields P. R., Diamond H., Metta D. N., and Rokop D. J. (1973) The reaction products of lunar uranium and cosmic rays (abstract). In *Lunar Science—IV*, pp. 239–241. The Lunar Science Institute, Houston.

Fleischer R. L. and Hart H. R. Jr. (1973) Particle track record of Apollo 15 green soil and rock. *Earth Planet. Sci. Lett.* 18, 357–364.

Haffner J. W. (1967) *Radiation and Shielding in Space*, p. 20. Academic Press.

Hart H. R. Jr., Comstock G. M., and Fleischer R. L. (1972) The particle track record of Fra Mauro. *Proc. Third Lunar Sci. Conf.*, *Geochim. Cosmochim. Acta*, Suppl. 3, Vol. 3, pp. 2831–2844. MIT Press.

Khan H. A. and Durrani S. A. (1972a) Efficiency calibration of solid state nuclear track detectors. *Nucl. Instr. and Meth.* 98, 229–236.

Khan H. A. and Durrani S. A. (1972b) Prolonged etching factor in solid state track detection and its applications. *Rad. Effects* 13, 257–266.

Khan H. A. and Durrani S. A. (1973) The annealing of latent damage trails in solid state nuclear track detectors. *Nucl. Instr. and Meth.* In press.

Khan H. A., Durrani S. A., and Fremlin J. H. (1973) Radiation history of some Apollo 16 lunar glasses by track annealing (abstract). In *Lunar Science—IV*, pp. 435–437. The Lunar Science Institute, Houston.

Krätschmer W. (1971) Die anätzbaren Spuren künstlich beschleunigter schwerer Ionen in Quarzglas. Doctoral thesis, Heidelberg.

Krishnaswami S., Lal D., Prabhu N., and Tamhane A. S. (1971) Olivine: Revelation of tracks of charged particles. *Science* 74, 287–291.

Malik S. R., Durrani S. A., Aframian A., and Fremlin J. H. (1973) Heavy-ion track spectra in lunar and analogous materials (abstract). In *Lunar Science—IV*, pp. 493–495. The Lunar Science Institute, Houston.

McGowan F. K., Milner W. T., Kim H. J., and Hyatt W. (1969) Reaction list for charged-particle-induced nuclear reactions. *Nuclear Data Tables* A7, 1–232.

Northcliffe L. C. and Schilling R. F. (1970) Range and stopping-power tables for heavy ions. *Nuclear Data Tables* A7, 233–463.

Phakey P. P., Hutcheon I. D., Rajan R. S., and Price P. B. (1972) Radiation effects in solids from five lunar missions. *Proc. Third Lunar Sci. Conf.*, *Geochim. Cosmochim. Acta*, Suppl. 3, Vol. 3, pp. 2905–2915. MIT Press.

Plieninger T., Krätschmer W., and Gentner W. (1972) Charge assignment to cosmic ray heavy ion tracks in lunar pyroxenes. *Proc. Third Lunar Sci. Conf.*, *Geochim. Cosmochim. Acta*, Suppl. 3, Vol. 3, pp. 2933–2939. MIT Press.

Storzer D., Poupeau G., Krätschmer W., and Plieninger T. (1973) The track record of Apollo 15, 16, and 20 samples and the charge assignment of cosmic ray tracks (abstract). In *Lunar Science—IV*, pp. 694–696. The Lunar Science Institute, Houston.

Storzer D. and Wagner G. A. (1969) Correction of thermally lowered fission track ages of tektites. *Earth Planet. Sci. Lett.* 5, 463–468.

Walker R. and Zimmerman D. (1972) Fossil track and thermoluminescence studies of Luna 16 material. *Earth Planet. Sci. Lett.* 13, 419–422.

Proceedings of the Fourth Lunar Science Conference
(Supplement 4, *Geochimica et Cosmochimica Acta*)
Vol. 3, pp. 2307–2317

Particle track record of Apollo 15 shocked crystalline rocks

ROBERT L. FLEISCHER, HOWARD R. HART, JR.,
and WALLACE R. GIARD

General Electric Research and Development Center, Schenectady, New York 12301

Abstract—Cosmic ray track densities in two mare basalts 15058 and 15555 are multivalued at each depth from the surface, and numerous indications of shock are present, suggesting that shock has lowered track densities in some crystals but not in others. From the minimum track densities, the maximum time since the last shock event is derived for each rock. In separate observations 15017, a black glass, has a surface age of ~ 14,000 years derived from impact pits, but because of low track retentivity a solar flare record of only 1 year.

INTRODUCTION

THE RECORD left by heavy cosmic ray nuclei in Apollo 15 rocks reveals complicated but interesting surface histories. We report here observations on crystalline igneous rocks, breccias, and a black glass shell displaying prominent impact pits. Our earlier observations of igneous rocks showed smooth variations of the track density from solar and galactic cosmic rays that decreased monotonically inward with very little scatter (Fleischer *et al.*, 1970, 1971a) as was predicted for simple objects that accumulate and retain tracks over some time period (Fleischer *et al.*, 1967a). On the other hand breccias often give widely varied track densities at each depth, indicating that many of the grains making up these rocks received cosmic ray bombardment prior to their being compacted into rock and that many of the tracks from the pre-irradiation survived the impacts which made the breccias (Hart *et al.*, 1972). In the present work we find a puzzle—a similar type of scatter can be seen in igneous rocks, which by their nature could not have had a pre-irradiation of the grains separately.

EXPERIMENTAL PROCEDURES

Except where specifically noted, groups of grains with typically >300 μ diameters were removed from specified locations in the rock samples reported here, mounted in epoxy, and polished prior to etching to reveal natural tracks. The procedures for mounting and polishing have been described in detail in a series of papers (Fleischer *et al.*, 1970, 1971a; Fleischer and Hart, 1973b). Reported track densities are increased relative to the raw data by dividing by 1.0, 0.7, 0.5, and 0.2 for feldspar, pyroxene, olivine, and glass respectively to allow for the differing etching and registration efficiencies of these detectors. Track densities reported are generally for tracks of length greater than 1.5 μ. Shorter tracks were placed in a separate category that will be discussed whenever considered. The "95% confidence minimum track density" is obtained by considering the feldspar or pyroxene with the lowest track density at a particular location in a rock and raising its track density by two standard deviations determined from the number of tracks counted in that crystal. Unless specific annealing data are available, lunar olivine and glass are normally excluded because of potential track fading problems.

Analysis Procedures

Most of the tracks of interest here are heavy cosmic rays of charge ~ 26, the so-called iron-group nuclei. If a sample has been buried at a known depth in a material, a knowledge of the flux of these nuclei plus the track density accumulated at that position allows the near surface exposure time to be computed.

We will make use of our previous observation that shock events commonly fragment and erase tracks (Fleischer *et al.*, 1972; Fleischer and Hart, 1973a) so that if a rock is shocked, some of the crystals will start recording tracks with a newly cleaned slate. The minimum track density among a group of grains identifies the one with the most recently and/or most completely erased tracks and allows the shock event to be dated. Similarly, finding the minimum track density at a known depth in a sample allows the last track-erasing shock to be dated. Most generally the age so found will be an upper limit, since the search for the most recently shocked material in which all tracks were erased may not have been successful. One may have found a crystal that was shocked in an earlier impact or in which only a fraction of the pre-existing tracks was erased. In order to insure that dislocations were not counted as tracks, we have considered only tracks $\leq 15\mu$ in length which were random in location and showed a wide variation in orientation. We note also that the normal uranium contents of the minerals considered here (Fleischer *et al.*, 1970) make fission tracks an unrealistic explanation of the high track densities observed.

Inferring Surface Ages

In using cosmic ray track densities to estimate rock residence times close to the lunar surface we will utilize as appropriate either the track production relation given in Fig. 1, which was computed for a grain of average orientation buried in a semi-infinite rock of overall specific gravity 3.4, or the calculations of Fleischer *et al.* (1967a) for spherical bodies of various sizes. The track production rate of Fig. 1 is based on the calculation of Comstock (1972). His production rate has been scaled up by a factor of two so as to utilize the same cosmic ray flux as was assumed by Fleischer *et al.* (1967a, b) for the interior of meteorites; this doubling is, as noted by Fleischer and Hart (1973a), consistent with recent calibration results for heavy ions. The steeply descending portion of the curve at depths of less than ~ 0.3 cm corresponds to solar flare particles. Since the Surveyor 3 observations used (Crozaz and Walker, 1971; Fleischer *et al.*, 1971b; Price *et al.*, 1971) were made over less than a full solar cycle, the long term average flux is not known accurately. Therefore ages inferred from grains buried by less than 0.3 cm will be uncertain to roughly a factor of two.

Results

Igneous rock 15058

This rock has been classified as a type I mare basalt (Brown *et al.*, 1972). Figure 2 indicates the location of our column and chip, Fig. 3 gives the densities of cosmic ray tracks in various minerals as a function of position, and Fig. 4 shows

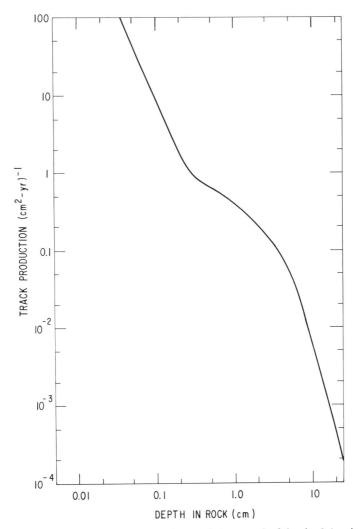

Fig. 1. Calculated average track production rate in lunar rock of density 3.4 gm/cc. The rate given here is twice that of Comstock (1972) as discussed in the text.

the density of short ($\leq 1\frac{1}{2}\,\mu$) tracks in pyroxenes. Fig. 3 has two interesting and obvious features. The first is the wide range of track densities at each depth. Even if only the most track retentive materials, the feldspars and pyroxenes, are considered the track densities vary by a factor of ten at most depths. The second striking feature is that the olivines are consistently lower than the other minerals in spite of the fact that they are corrected for their typical lesser sensitivity. Bhandari *et al.* (1972) report a smoothed fit to their data which is consistent with our data at 4.2 cm, where our samples join, but they do not indicate minimum or maximum track densities.

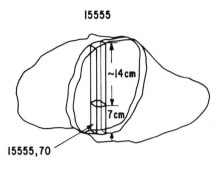

15555

~14 cm

7 cm

15555,70

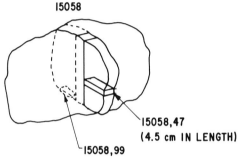

15058

15058,47
(4.5 cm IN LENGTH)

15058,99

Fig. 2. Positions of the samples studied from rocks 15058 and 15555.

Unless the classification of this rock is wrong, the tracks we observe are not inherited tracks from previous existence of the crystals as separate soil grains, and hence the tracks must have been altered by thermal (Fleischer *et al.*, 1965) or mechanical effects (Fleischer *et al.*, 1972). We believe that both have probably occurred.

Clear evidence of mechanical effects has been obtained, as shown in Fig. 5 by the electron micrographs of two etched pyroxenes from 15058, one with a mixture of coarse and fine deformation markings and one with duplex slip (deformation on two distinct sets of crystal planes). Twenty-one of eighty pyroxenes examined optically showed effects characteristic of deformation (Fleischer and Hart, 1973a)—oriented tracks, deformation markings, or both; and twenty of twenty-one replicas of pyroxenes that were examined with the transmission electron microscope displayed deformation markings.

When deformation has erased tracks by dividing them into short unresolvable segments, the long tracks that have accumulated after the deformation should allow the time of deformation to be measured (Fleischer *et al.*, 1972). If as in this case some of the grains have apparently not had their tracks totally erased, the minimum track densities become the best approximation to the post deformation track accumulation. Accordingly in Fig. 6 we have plotted the minimum track densities for each mineral type at each depth, and curves defining the envelope of

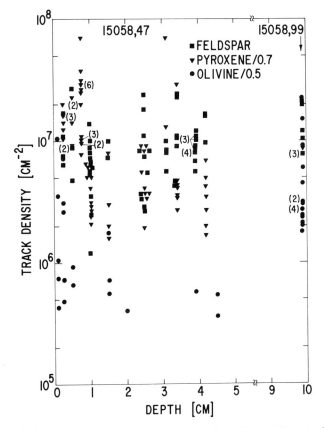

Fig. 3. Track densities observed in individual crystals of three different minerals of 15058,47 and 15058,99. The pyroxenes show evidence of shock.

the lower boundary for olivine and for feldspars plus pyroxenes. The fact that the olivines are low relative to the other minerals is consistent with their higher sensitivity to thermal effects. The near surface lowering also suggests that some brief but unidentified surface heating might have played a role in reducing the track densities in the olivines.

Using the minimum track density of $1.7 \times 10^6/cm^2$ at 1 cm depth we infer an age of not more than 7 million years since this rock was last shocked.

The short tracks could either be proton-induced spallation recoil tracks or fragmented cosmic ray tracks. The results unfortunately are compatible with there being no variation with position or with a factor of 3 or 4 increase with depth from 0 to 4 cm, so that it is not possible to decide on the basis of the depth variation whether these are spallation tracks or not. If they are from spallation, the median value of $\sim 2 \times 10^7/cm^2$ would imply a proton exposure 1.1×10^{16} protons/cm² or a near surface age of a few 100 m.y. (Fleischer et al., 1971a). Regardless of spallation tracks the maximum cosmic ray track densities of $\sim 7 \times$

R. L. FLEISCHER *et al.*

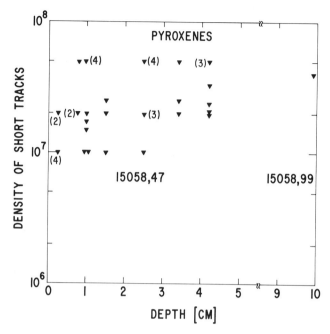

Fig. 4. Density of short tracks (≤1.5 μ in length) in pyroxenes from 15058.

$10^7/cm^2$ at 3 cm depth in the sample would imply ~2000 m.y. buried beneath 10 cm of soil, or a few 100 m.y. at a lesser depth. In short a long near surface pre-history is implied prior to the track erasing event ≤7 m.y. ago.

Rock 15555

This largest Apollo 15 rock is another mare basalt with somewhat similar properties to 15058, as Fig. 7 indicates for the sample sketched in Fig. 2. Both median and minimum track densities are plotted. Again, deformation markings are observed in pyroxenes. As for 15058 the median values scatter rather widely, but are consistent with a slight increase in track density from the bottom surface toward the center of the rock. The track density of 2×10^6 at 1 cm from the bottom surface indicates that surface has been exposed to space less than 5 m.y. since the last track erasure. This age is compatible with and much less than the 80 to 90 m.y. of total near surface cosmic ray exposure of this rock (Burnett *et al.*, 1972; Kenyon and Doyle, 1972; Lugmair and Marti, 1972).

15505: Exposure time to solar cosmic rays

This sample, a breccia taken from Station 9 near a crater, contained a minimum track density in pyroxene of $2 \times 10^6/cm^2$ at 1.5 mm below the surface. From a track production rate of $3.5/cm^2$–yr we compute a surface age of ≤600,000 years for this surface with 95% confidence.

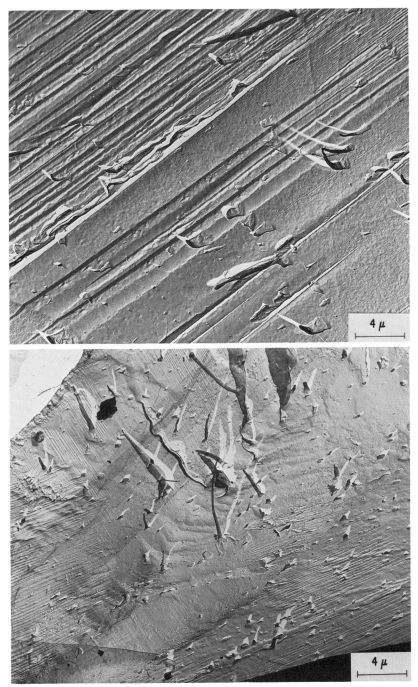

Fig. 5. Deformation markings in etched pyroxenes from 15058. The transmission electron micrographs are made of shadowed replicas of the etched crystals. Top: mixed slip plane spacings in a grain from a depth of 7 mm. Bottom: duplex slip in a crystal from within 2.5 mm of the surface.

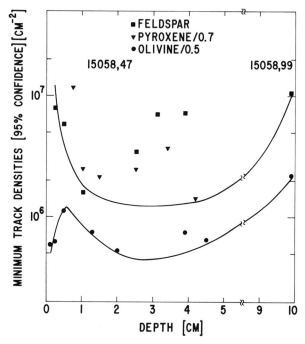

Fig. 6. Minimum track densities (95% confidence) at each position sampled in 15058. Curves show the smoothed lower limits for feldspars plus pyroxenes and for olivines.

15017 and impact pit counting

Sample 15017,6 is a vesicular, black, glass shell of varying thickness (1–5 mm) and irregular composition from point to point (Fabel *et al.*, 1972). Its exterior displays a low density of classic impact pits with interior craters surrounded by spall zones. The area of approximately 1 cm^2 has one pit of crater diameter $>200\,\mu$, 3 of $>100\,\mu$, and 6 of $>50\,\mu$, as viewed at 30 $\times$ in a stereo microscope. From the flux of hypervelocity micrometeorites inferred by Hartung *et al.* (1973) a surface exposure of $\sim 14{,}000$ years is implied.

In an attempt to obtain a track age to compare with this impact age we examined a 1 mm thick exterior fragment and performed a limited number of heating experiments to assess track fading. Track counts were made on the exterior face after two different etching times to observe tracks of cone angle $55°(\pm 5°)$ ending near $8.5\,\mu$ and $13\,\mu$ depths in the glass, and an in-profile view revealed tracks at $14\,\mu$ and $31\,\mu$ beneath the original surface. In Fig. 8 these limited data are plotted along with a curve that represents the density vs. depth relation expected if the sample were irradiated by cosmic rays for $2\frac{1}{2}$ months at the flux observed between the landing of Surveyor 3 and that of the Apollo 12 mission (Fleischer *et al.*, 1971b).

The discrepancy of a factor of 10^5 between the impact age and this apparent surface age is explained as thermal track fading. Using Cf-252 fission tracks we

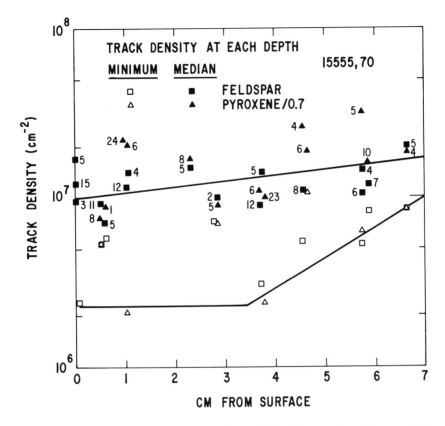

Fig. 7. Minimum (95% confidence) and median track densities at each position sampled in mare basalt 15555. Numbers adjacent to points indicate the number of separate measurements from which the medians were determined.

find that a 50 minute anneal at 230°C removes all of the tracks from most of the glass, with tracks being retained in some less rapidly etching portions (of therefore different composition from the majority). This fading is more rapid than that in the black glaze on rock 12017 where we inferred (Fleischer et al., 1971a) that fission tracks would be stable for ~500 yr on the lunar surface. Since cosmic ray tracks would be less tenaciously retained (Maurette, 1970; Price et al., 1973), a retention time ~ a year or less is consistent with the data.

If we consider the cosmic ray fluxes up to the time of the Apollo 15 mission we find that the fluence of protons over the $2\frac{1}{2}$ months just prior to the mission is much too low to provide the density of heavy particles we observed, assuming the iron/hydrogen ratio of $\sim 10^{-4}$ that we observed at ~ 1 MeV/amu (Fleischer and Hart, 1973c). Because the sun was more active prior to that period, we can however explain the observed dose if we assume that the glass retained the heavy particle tracks formed over a ~ 1 year time prior to its collection on Apollo 15.

R. L. FLEISCHER *et al.*

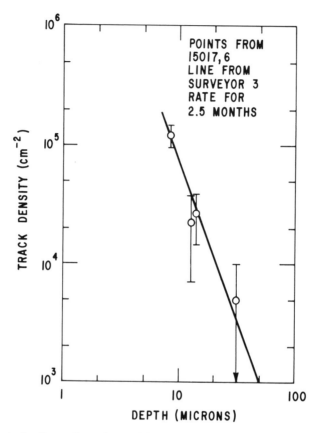

Fig. 8. Track density gradient observed in glass shell 15017, which was exposed to micrometeorites for >10,000 years. The slope is consistent with the Surveyor 3 solar flare energy spectrum determined from Surveyor 3 (Fleischer *et al.*, 1971b) using the fluence for a one year exposure, a discrepancy that is explained by track fading.

CONCLUSIONS

We have shown that in two mare basalts track densities are lowered by shock, by different amounts in different crystals. From the minimum track densities upper limits can be derived for the cosmic ray exposure times since the last shock event.

Acknowledgments—We are pleased to give thanks to E. Stella for experimental assistance. This work was supported in part by NASA under contract NAS9-11583.

REFERENCES

Bhandari N., Goswami J. N., and Lal D. (1972) Apollo 15 regolith: A predominantly accretion or mixing model? In *The Apollo 15 Lunar Samples*, pp. 336–341. The Lunar Science Institute, Houston.

Brown G. M., Emeleus C. H., Holland J. G., Peckett A., and Phillips R. (1972) Petrology, mineralogy, and classification of Apollo 15 mare basalts. In *The Apollo 15 Lunar Samples*, pp. 40–44. The Lunar Science Institute, Houston.

Burnett D. S., Huneke J. C., Podosek F. A., Russ G. P. III, Turner G., and Wasserburg G. J. (1972) The irradiation history of lunar samples (abstract). In *Lunar Science—III*, pp. 105–109. The Lunar Science Institute, Houston.

Comstock G. M. (1972) The particle track record of the lunar surface. *Proceedings of Conference on Lunar Geophysics, The Moon* (in press). This paper is available in preprint form as GE Report No. 71-C-190.

Crozaz G. and Walker R. (1971) Solar particle tracks in glass from the Surveyor III spacecraft. *Science* **171**, 1237–1239.

Fabel G. W., White W. B., White E. W., and Roy R. (1972) Structure of lunar glass by Raman and soft X-ray spectroscopy. *Proc. 3rd Lunar Sci. Conf., Geochim. Cosmochim. Acta*, Suppl. 3, Vol. 1, pp. 939–951. MIT Press.

Fleischer R. L., Price P. B., and Walker R. M. (1965) Effects of temperature, pressure, and ionization of the formation and stability of fission tracks in minerals and glasses. *J. Geophys. Res.* **70**, 1497–1502.

Fleischer R. L., Price P. B., Walker R. M., and Maurette M. (1967a) Origins of fossil charged particle tracks in meteorites. *J. Geophys. Res.* **72**, 331–353.

Fleischer R. L., Price P. B., Walker R. M., Maurette M., and Morgan G. (1967b) Tracks of heavy primary cosmic rays in meteorites. *J. Geophys. Res.* **72**, 355–366.

Fleischer R. L., Haines E. L., Hart H. R., Jr., Woods R. T., and Comstock G. M. (1970) The particle track record of the sea of tranquillity. *Proc. Apollo 11 Lunar Sci. Conf., Geochim. Cosmochim. Acta*, Suppl. 1, Vol. 3, pp. 2103–2120. Pergamon.

Fleischer R. L., Hart H. R., Jr., Comstock G. M., and Evwaraye A. O. (1971a) Particle track record of the ocean of storms. *Proc. Second Lunar Sci. Conf., Geochim. Cosmochim. Acta*, Suppl. 2, Vol. 3, pp. 2559–2568. MIT Press.

Fleischer R. L., Hart H. R., Jr., and Comstock G. M. (1971b) Very heavy solar cosmic rays: Energy spectrum and implications for lunar erosion. *Science* **171**, 1240–1242.

Fleischer R. L., Comstock G. M., and Hart H. R., Jr. (1972) Dating of mechanical events by deformation-induced erasure of particle tracks. *J. Geophys. Res.* **77**, 5050–5053.

Fleischer R. L. and Hart H. R., Jr. (1973a) Mechanical erasure of particle tracks: A tool for lunar microstratigraphic chronology. *J. Geophys. Res.* In Press.

Fleischer R. L. and Hart H. R., Jr. (1973b) Particle track record in Apollo 15 deep core from 54 to 80 cm depths. *Earth Planet. Sci. Letts.* **18**, 420–426. Summarized in *The Apollo 15 Lunar Samples*, 1972. pp. 371–373. The Lunar Science Institute, Houston.

Fleischer R. L. and Hart H. R., Jr. (1973c) Enrichment of heavy nuclei in the 17 April, 1972 solar flare. *Phys. Rev. Letters* **30**, 31–34.

Hart H. R., Jr., Comstock G. M., and Fleischer R. L. (1972) The particle track record of Fra Mauro. *Proc. 3rd Lunar Sci. Conf. Geochim. Cosmochim. Acta*, Suppl. 3, Vol. 3, pp. 2831–2844. MIT Press.

Hartung J. B., Aitken F. K., Blackmon J. W., and Hörz F. (1973) Microcrater population development on lunar rocks (abstract). In *Lunar Science—IV*, pp. 339–339b. The Lunar Science Institute, Houston.

Kenyon W. J. and Doyle A. J. (1972) ^{40}Ar–^{39}Ar ages of Apollo 14 and 15 samples (abstract). In *Lunar Science—III*, pp. 822–824. The Lunar Science Institute, Houston.

Lugmair G. W. and Marti K. (1972) Neutron and spallation effects in Fra Mauro regolith (abstract). In *Lunar Science—III*, pp. 495–497. The Lunar Science Institute, Houston.

Maurette M. (1970) Some annealing characteristics of heavy ion tracks in silicate minerals. *Rad. Effects* **5**, 15–19.

Price P. B., Hutcheon I., Cowsik R., and Barber D. J. (1971) Enhanced emission of iron nuclei in solar flares. *Phys. Rev. Lett.* **26**, 916–919.

Price P. B., Lal D., Tamhane A. S., and Perelygin V. P. (1973) Characteristics of tracks of ions with $14 \le Z \le 36$ in common rock silicates. *Earth Planet. Sci. Letts.* In press.

Proceedings of the Fourth Lunar Science Conference
(Supplement 4, *Geochimica et Cosmochimica Acta*)
Vol. 3, pp. 2319–2336

Irradiation history and accretionary processes in lunar and meteoritic breccias

D. Macdougall, R. S. Rajan, I. D. Hutcheon, and P. B. Price

Department of Physics, University of California, Berkeley, California 94720

Abstract—Particle track studies reveal an abundant record of fossil solar flare tracks in breccia components. Metamorphic events govern the degree to which this record is preserved, and studies of phases with different track retentivities allow limits to be placed on temperatures reached during a rock's history. The least affected breccias have never experienced temperatures as great as $\sim 300°C$. Two breccias which contain xenon from spontaneous fission of Pu^{244} (14301 and 14318) and Xe^{129} (14301 only) have never been heated above $\sim 700°C$. This and evidence for a surface irradiation of some of the breccia components support the surface implantation model (Behrmann *et al.*, 1973) for the origin of the xenon in these breccias. Green glass spheres in 15086 have not been heated above $\sim 300°C$ during or after breccia formation yet have retained fission tracks for less than 0.7 G.y. Argon ages of ~ 3.5 G.y. for similar green glass are at least five times as great and may indicate that the glass was not completely outgassed at its formation. Similarities between solar flare track retention in lunar breccias and gas-rich meteorites indicate that the latter may have been compacted in the regoliths of small bodies.

Introduction

To the surprise of many, most of the rocks brought back from the moon by the Apollo astronauts are breccias. This is particularly true of rocks from the highland sites. With the aid of hindsight, this does not seem such a startling observation since many of the oldest rocks in the solar system (meteorites) are breccias, very similar in some respects to those found on the moon. For this reason features which reflect the formation environment of these rocks can provide insights into the history of meteoritic as well as lunar breccias.

Particle track studies are unique in that they allow the radiation history of individual grains or coherent fragments to be investigated. Low energy heavy nuclei from solar flares are stopped in the outer microns of grains and fragments, so that high track densities (typically greater than $\sim 10^8/cm^2$) or track density gradients over distances of tens of microns are sensitive indicators of unshielded surface exposure. The presence of these features in grains now in the interior of breccias provides a fossil record of the surface exposure and the characteristics of the irradiation (the latter are discussed by Price *et al.*, 1973a).

In this paper we show that the extent of preservation of the fossil solar flare tracks depends upon the metamorphic events—thermal and mechanical—which have affected the breccia constituents during or after accretion. At the present time, few detailed quantitative data exist for the effects of pressure on track retention and we therefore have interpreted our observations in terms of thermal effects. This approach is justified by the fact that none of the rocks for which we report preserved solar flare tracks shows evidence of severe shock.

We have used the fact that different materials exhibit different track annealing characteristics to place some limits on the thermal history of the breccia components. For short time intervals—we consider hours, days or years—tracks fade from the major breccia constituents in the following order of increasing temperature: glasses, olivine, feldspar.

Parallels can be drawn between the presence of "track rich" (exhibiting solar flare tracks as described above) grains in the lunar breccias and those in gas-rich meteorites. Although such comparison is not definitive in characterizing the formation environment of the meteoritic breccias, it does narrow the field of speculation.

EXPERIMENTAL TECHNIQUES

In studying the lunar breccias we have concentrated our efforts in the high resolution methods of high and low voltage (replica) transmission electron microscopy (TEM) and scanning electron microscopy (SEM). A few data obtained by optical microscopy are also reported. The major improvement in technique employed in obtaining the results presented here is the use of extremely short etching times both for samples to be studied by SEM and for those to be replicated (the replica method is described by Macdougall *et al.*, 1971). To reveal tracks in feldspars we etch one to two minutes in boiling 1:2 NaOH and water solution which etches tracks in lunar feldspars to diameters of 300–500 Å, suitable for TEM replica study (Fig. 1a). Olivines etched for 10 minutes in the WN solution described by Krishnaswami *et al.* (1971) develop tracks of approximately the same size. Such light etching not only allows easy counting of more than 10^{10} tracks/cm² but also leaves fine grained areas of polished breccia sections intact and smooth so that plastic replicas do not tear or pull material from the section. After replication, an additional etching of four to five minutes in the sodium hydroxide solution for feldspars, or 30 minutes in WN for olivines, develops tracks to a size easily observed in the scanning electron microscope (Fig. 1b).

Breccia sections were also ion-thinned for examination with the Berkeley 650 keV electron microscope as described by Hutcheon *et al.* (1972). In the very fine-grained ($\leq 1\ \mu$m) regions this is the only method which can be used to observe solar flare tracks. Samples for optical microscopy were prepared and etched in the conventional manner (e.g., Lal *et al.*, 1968).

RESULTS AND DISCUSSION

Evidence for pre-accretion irradiation of breccia components

High track densities ($> 10^8$/cm²) and/or track density gradients within individual grains or fragments from the interior of a breccia are conclusive evidence that these components were exposed without shielding to solar flare radiation at some time before breccia formation. We have observed numerous grains containing well preserved solar flare tracks in the interiors of six of fourteen Apollo 14, 15, and 16 breccias examined (Table 1, Column 1). In an additional two of these breccias no solar flare tracks were seen in etched sections using the TEM replica method, or by SEM. However, very high densities (10^9–10^{10} tracks/cm²) of faint tracks were seen in ion-thinned sections using the high voltage TEM (Table 1, Column 2). Hart *et al.* (1972) report variable densities of etchable tracks in both of these rocks but their highest reported values (1.3×10^7 cm^{-2}) are well below the criteria we use for solar flare tracks ($> 10^8$/cm² or gradient). No fossil solar flare tracks were observed in the remaining five breccias (Table 1, Column 3).

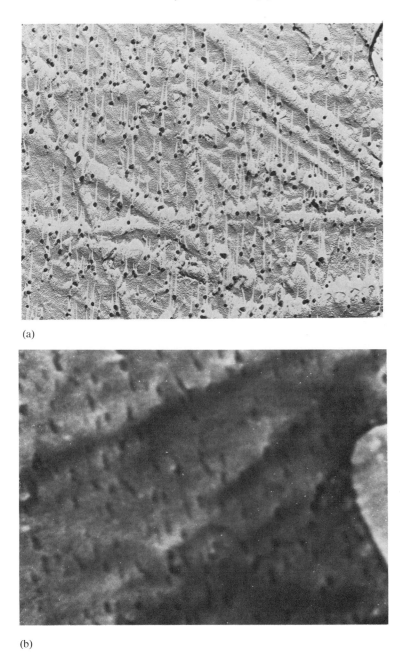

(a)

(b)

Fig. 1. (a) Transmission electron micrograph of a shadowed carbon replica of a feldspar grain from soil 15301. The grain was etched 2 minutes in boiling $1:2$ NaOH:H_2O. The measured track density is $\sim 1.5 \times 10^9$. The width of field in this photograph is 5.2 μm. (b) Same area as (a) seen directly by SEM after an additional etching of 5 minutes. The magnification is slightly greater (width of field is 4.4 μm), and the image is inverted with respect to (a). The track density measured on the TEM photo is somewhat higher because of the better resolution of the TEM.

Table 1. Solar flare tracks in Apollo 14, 15, and 16 breccias.

Preserved Etchable Solar Flare Tracks	Preserved Tracks Seen Only by HVEM	No Preserved Tracks Seen
14301	14311	15418
14315	14321	15445
14318	60255	61016
15086		66055
15426		67015
15459		

Others (Crozaz *et al.*, 1970; Dran *et al.*, 1972; Hart *et al.*, 1972; Hutcheon *et al.*, 1972; Berdot *et al.*, 1973) have also observed pre-compaction irradiation effects in Apollo 11 and Apollo 14 breccias.

In Figs. 2 and 3 we show examples of track-rich grains *in situ* in breccia sections as seen by the TEM replica method and SEM. Note the abundance of such grains and the track density gradients near their edges.

All breccias in which we have observed etchable solar flare tracks (Table 1, Column 1) have essentially detrital textures and contain undevitrified glass spheres or fragments. Some evidence of moderate shock exists for three of these rocks. Glass spheres in breccia 15459 are commonly shattered or heavily fractured, an apparent record of *in situ* shock or stress since the spheres have for the most part retained their original shapes. As discussed in a later section, crystalline components of the matrix of this rock show no evidence of extensive shock after breccia formation. Interstitial shock-induced glass and other evidence of moderate shock is present in breccias 14315 and 14318 (Hutcheon *et al.*, 1972). Samples listed in Table 1 exhibit progressively more strongly annealed or recrystallized textures going from Columns 1 to 3.

It is evident from these observations that (i) some fraction of the material making up at least the less thermally affected breccias was exposed to solar flare irradiation on the moon's surface before breccia compaction, and (ii) the sample groupings of Table 1, based on track retention characteristics, correlate in a general way with metamorphism as exhibited by textural features. Other features, such as solar rare gas in the low grade breccias (*see* Williams, 1972, for compilation) corroborate the track evidence for surface exposure. Finer details of metamorphic effects can be revealed by examining individual breccia phases as discussed in the following section.

Track annealing and thermal history

The fact that different breccia components exhibit different thermal track retention characteristics provides a rough thermometry which can be used to characterize heating events that took place during or after breccia formation (the two cases are not distinguishable solely from track retention data).

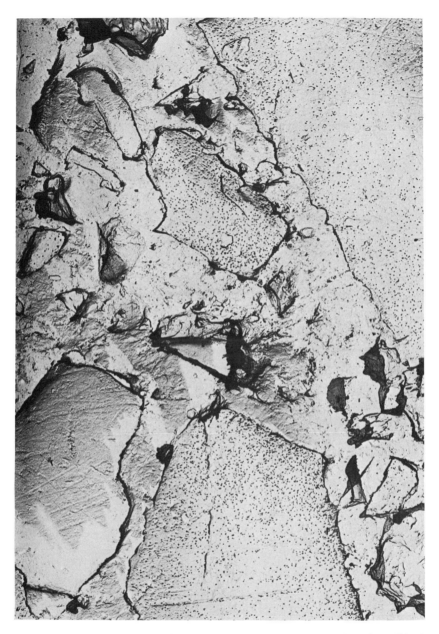

Fig. 2. Transmission electron micrograph of a shadowed carbon replica of a polished section of breccia 15086. Etching time was 2 minutes in boiling 1:2NaOH:H₂O. Note the numerous track-rich feldspars with gradients in this region. Tracks in this picture are small black dots. The width of field in this photograph is ~ 14 μm.

(a)

(b)

Fig. 3. (a) Scanning electron microscope photograph of an etched section of breccia 15086. The euhedral feldspar immediately upper left of center is track-rich, as are several other grains in the field of view (e.g., large grain upper right; smaller grain lower right partly etched away). Width of field is 210 μm. (b) A higher magnification SEM photograph of the euhedral feldspar shown in (a). Width of field is 21 μm. Note solar flare track density gradient.

Temperatures required to completely remove solar flare tracks from breccia constituents can be estimated from the numerous studies of track annealing which have been published (e.g., Fleischer et al., 1965; Price et al., 1973b). In this study, we consider only maximum temperatures, i.e., we assume short heating events. If some of the breccias were maintained at elevated temperatures over extended periods of time, temperatures required to remove tracks would be considerably lower than those given here. However, particularly for the rocks showing no or minor recrystallization, heating events must have been of short duration and followed by rapid cooling.

In Table 2 we list maximum temperatures or temperature intervals that we estimate have been experienced by the breccias. It should be emphasized that these are estimates based on the fact that solar flare tracks fade in a few hours at $\sim 300°C$ for glass, $\sim 400°C$ for olivine and $\sim 700°C$ for calcic plagioclase. The faint, unetchable tracks visible by the high voltage TEM disappear when samples are heated above $\sim 800°C$ for one hour. Wherever possible we have used matrix material rather than clasts for track observations. However, it is always possible that breccia heterogeneity has affected our findings in some cases. The result would always be that a rock is placed in a higher temperature group than that to which it really belongs.

Table 2. Temperatures experienced by Apollo 14, 15, and 16 breccias.

Sample Number	Solar Flare Tracks			Maximum Temperature or Temperature Range
	Glass	Olivine	Feldspar	
15086	yes	yes	yes	$\leqslant 300°C$
15426	yes	nd	yes	$\leqslant 300°C$
15459	nd	yes	yes	$\leqslant 400°C$
14301	no	no	yes	$400 \leqslant T \leqslant 700$
14315, 14318	nd	nd	yes	$\leqslant 700$
14311, 14321, 60255	nd	nd	HVEM only (unetched)	$700 \leqslant T \leqslant 800$
15418, 60017, 61016, 66055, 67015	nd	no	no	$\leqslant 800$

nd = Indicates no data.

Where direct comparisons can be made, the temperatures given in Table 2 are somewhat lower than those estimated by Williams (1972), although the sequence of increasing temperature which we obtain is in agreement with the metamorphic sequence of Williams. Samples in the last group listed in the table show considerable textural diversity; however, their track retention characteristics are similar.

Track density distributions in breccias 14301 and 15086

Rock 14301 is particularly interesting because it contains chondrules and chondrule-like bodies (King et al., 1972) and also because it contains both fission xenon from extinct Pu^{244} and radiogenic xenon from decay of I^{129} (Drozd et al.,

1972; Behrmann *et al.*, 1973). Sample 15086 was collected 60 m east of Elbow Crater as one of a set of radial samples selected to represent crater ejecta. The
measured Al^{26}/Na^{22} ratio indicates, however, that it has been exposed on the
surface only 0.2–0.5 m.y. (LSPET, 1972; Eldridge *et al.*, 1973).

Track densities measured by optical microscopy in hand-picked feldspar,
pyroxene and olivine crystals (50–300 μm) from the two rocks are shown in Fig. 4.
For ease of comparison, the histograms are drawn so that fractions of the total
number of crystals examined are plotted in each track density interval. All grains
having track density gradients and/or with density $> 10^8$ tracks/cm^2 are shown in
the highest interval. The fraction of grains in this interval is denoted N_H/N (Arrhenius *et al.*, 1971).

The track density distributions for the two rocks are quite different. The most
striking feature of the 14301 data is that while both feldspars and pyroxenes
exhibit a distinctly bimodal distribution, all olivines examined fall into a single
group having track densities of 7×10^5 to 2×10^6 tracks/cm^2. This is the "background" galactic cosmic ray track density, produced while the crystals resided
within the rock. Similarly an examination of approximately twenty glass spheres
and fragments (not shown in Fig. 4) revealed no evidence for solar flare tracks. In
contrast, approximately 25% of the feldspar and pyroxene crystals have track
densities $> 10^8$/cm^2 or gradients, products of solar flare irradiation of the grains as

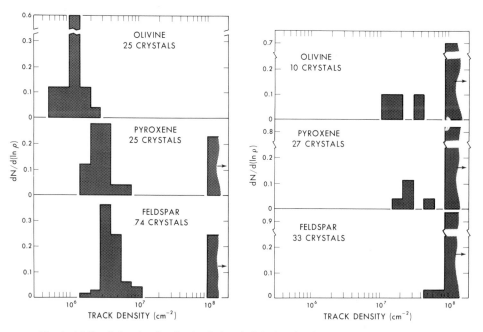

Fig. 4. (a) Track density distribution in hand picked grains from breccia 14301. Density
measurements were made by optical microscope. (b) Track density distribution in hand
picked grains from breccia 15086. As for the grains in (a), these range in size from
$\sim 50\ \mu$m to $\sim 300\ \mu$m.

individual fragments before incorporation into the rock. It is possible that the feldspar and pyroxene crystals in 14301 have had a different irradiation history than the glass and olivine, but we prefer the explanation that a mild thermal event has erased solar flare tracks from the latter without affecting the former. As indicated in Table 2, this would require temperatures in the range 400°C–700°C. This temperature constraint and the presence of solar flare tracks indicating surface exposure of some breccia components strongly support the surface implantation model (Behrmann *et al.*, 1973) for the observed $Xe^{129, 136}$ in 14301, as opposed to models requiring extensive heating.

The puzzle with this (and some other) breccias is not that solar flare tracks have been annealed from some phases, but that they are preserved in *any* phases since the presence of glass welding the breccia components (e.g., Agrell *et al.*, 1972; Lally *et al.*, 1972) would seem to indicate temperatures high enough to remove tracks from even feldspar and pyroxene. Dran *et al.* (1972) have suggested that some of the energy required to sinter breccias may be solar flare energy stored in amorphous grain coatings.

The data of Fig. 4 show that the value of N_H/N for the 14301 feldspar and pyroxene crystals is 0.25, considerably lower than that for 15086. However, the quartile track density (Arrhenius *et al.*, 1971) is high if the grains containing only galactic cosmic ray tracks are ignored. This situation suggests that heavily irradiated surface material was diluted approximately 3:1 with fresh, unirradiated material during the event which formed 14301.

Referring again to Fig. 4, the value of N_H/N in rock 15086 ranges from 0.7 for olivine crystals to 0.94 for feldspar, but within the statistical errors is the same for all three breccia components. Green glass fragments and spheres from the rock exhibit approximately the same value for N_H/N. Such values are typical of regolith soils which have surface exposure ages of 80–100 m.y. (Arrhenius *et al.*, 1971) and are consistent with the suggestion that 15086 is a soil breccia of local regolith origin (Brown *et al.*, 1972). Our unpublished studies of irradiation geometry as a function of crystal size suggest that the surface exposure age of 15086 components could be somewhat less than 80 m.y.

Large and small etch pits in 15086 green glass

An interesting feature of the track density distribution in the green glass from 15086 is the presence of two distinct track-size populations. Figure 5 is an SEM photograph of an etched green glass sphere from 15086 showing several large etch pits in a sea of smaller tracks with larger cone angles. Only two of thirty-two glass fragments examined did not reveal a high density of the small pits, which we attribute to solar flare iron nuclei. Visible gradients of the small pits in some spheres rule out spallation recoil nuclei tracks. Initially we thought that the large pits might be fission tracks because of the age (3.38 G.y., Huneke *et al.*, 1973) and the track-retentive nature (Fleischer and Hart, 1973) of similar green glass from another breccia, 15426. (A comparison of major element composition of green glass from the two rocks by SEM X-ray analysis revealed no chemical differ-

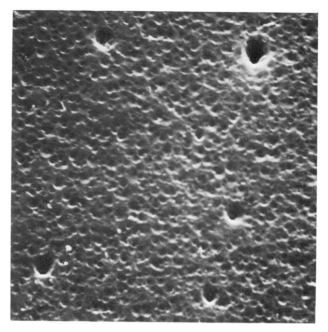

Fig. 5. Scanning electron microscope photograph of an etched green glass sphere from breccia 15086. The high density of small pits is due to solar flare iron nuclei. The large pits are attributed to nuclei of $Z > 26$. The width of field in this photograph is $\sim 26 \, \mu$m.

ences.) However, the large-pit density, which ranges approximately from 10^4 to 10^5 tracks/cm^2, does not correlate with the uranium content of the spheres, which ranges from 4 to 115 ppb. Counting statistics for the large pits are poor because of their low density and the small size of the spheres; nevertheless the non-correlation falls well outside of statistical viatiations. In addition, the large pits etch at a much less rapid rate than do pits from Cf^{252} fission fragments.

In a general way the large pit density correlates with the solar flare track density. Spheres having a high density of small tracks invariably have a high density of large pits. The ratio of large to small pits in Fig. 5 is 7×10^{-4}, similar to the ratio determined in meteorites for $[Z > 30]/[\text{Fe}]$ (Price *et al.*, 1968). Corrections for the etching efficiency of Fe vs. heavier ions would bring the ratio in the glass even closer to the value reported by Price *et al.* (1968). Thus, the majority of the large pits are almost certainly due to solar flare nuclei with $Z > 26$.

It is next necessary to account for the absence of fission tracks. The sphere with the highest uranium concentration would have a track age of ~ 0.7 G.y. only if all the large pits were assumed to be due to uranium fission, which, as we have just pointed out, does not seem to be the case. Thus, at a time more recently than ~ 0.7 G.y. ago, fission tracks (and any existing Fe and trans-Fe tracks) were erased in some thermal event. Subsequently the observed gradients and high densities of solar flare Fe tracks were produced during a surface exposure of the spheres, after which they were incorporated into the breccia.

The argon age of the green glass from 15426 is 3.4 G.y. (Huneke *et al.*, 1973). Assuming the same origin for the 15086 spheres, we emphasize that this age may not signify the time of glass formation, but may pertain to an earlier event. It seems entirely possible that the glass might actually have formed very recently without outgassing its original argon. This would explain the low density of fission tracks. It does not seem to have been emphasized that argon comes out of the spheres very slowly even at temperatures (e.g., 1660°C) well above the melting point. In fact, it is puzzling that the rate of release of argon is roughly independent of temperature over the entire range of temperatures in stepwise heating experiments where each heating stage lasts ~ 50 minutes (Podosek and Huneke, 1973). It seems to us unlikely that all argon would be outgassed in the fraction of a second during which these tiny spheres were molten, when some argon is retained for 50 minutes or more in the laboratory at temperatures above their melting point. One cannot rule out the possibility that both the green glass spheres and the orange glass spheres found at the Apollo 17 site were formed recently.

We would also like to point out that the large pits observed in green glass by Storzer *et al.* (1973) may have a solar flare origin as do those reported here, and therefore may not represent evidence for Pu^{244} spontaneous fission.

High voltage electron microscopy of 15445, 15459, and 66055

HVEM studies of a number of the breccias treated in this paper were described earlier (Hutcheon *et al.*, 1972). Here we report new observations for 15445, 15459, and 66055. In addition to the search for solar flare tracks in micron-sized grains (the results of which are incorporated into Tables 1 and 2) the HVEM studies are extremely useful in categorizing the shock history of the breccias.

Breccia 15445. We have examined two breccia clasts from this small (~ 300 g) rock picked up on an inner slope of Spur Crater. No solar flare tracks were seen in micron-sized grains from either clast. Figure 6a shows a typical unshocked feldspar from a chalky-white, very friable breccia clast (Clast A). No recrystallized zones, deformed crystals, multiple twinning, micro-fracturing or other metamorphic features were observed in grains from this clast. Crystals from the second clast exhibit dislocations and twinning, indicators of mild and presumably pre-brecciation shock.

Unfortunately, we have not been able to examine matrix material from this rock. From our experience with other breccias, the features just described from the two breccia clasts do not indicate thermal or shock events of sufficient magnitude to erase pre-existing solar flare tracks. Therefore it is possible that in the case of this breccia the small grains were never exposed on the surface. If this is true, the small grains in these breccia clasts are unique since we always find a high fraction of solar flare irradiated grains in this size fraction from regolith materials, including samples from the bottom of deep cores (Phakey *et al.*, 1972).

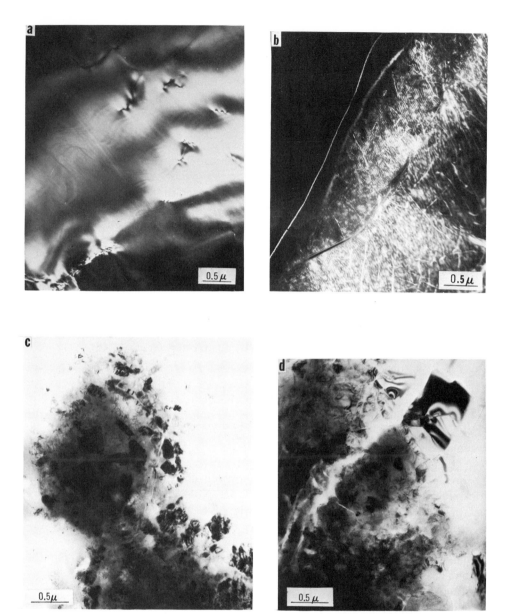

Fig. 6. (a) Interior of a micron sized plagioclase grain from a friable breccia clast of 15445 showing strain-free structure and clearly visible antiphase domain boundaries. DF micrograph. (b) Exsolution lamellae and dislocations typical of pyroxene crystals from a recrystallized clast of 15459. DF micrograph. (c) Heavily deformed region from the matrix of 15459. Individual recrystallized grains show evidence of moderate shocking after recrystallization. BF micrograph. (d) The complex structure of 66055 is illustrated by the juxtaposition of strain-free crystalline fragments with a deformed polycrystalline aggregate. BF micrograph.

Breccia 15459. This is a large (5.9 kg) breccia characterized by a diversity of clast types. We have examined two clasts, one igneous and one recrystallized breccia, as well as matrix fragments. Figure 6b shows a typical region of the recrystallized clast, with small undeformed grains and larger ($> 10 \mu$m) plagioclase crystals containing dislocations and other fine-scale features. The low track density in the clast ($\sim 2 \times 10^7$ tracks/cm^2) is the result of galactic cosmic ray irradiation after formation of the breccia.

Pyroxene and plagioclase crystals from the igneous clast exhibit moderate twinning and fracturing, suggestive of a low level of shock. Track densities of $\sim 10^8$ cm^2 imply solar flare irradiation of this clast before incorporation into the breccia.

Matrix fragments show a wide range of metamorphic effects. Features typical of strong shock (e.g., Fig. 6c) are memories of pre-breccia-formation events. Extremely high track densities ($\sim 3 \times 10^{10}$ tracks/cm^2) are common in matrix plagioclase crystals examined by HVEM. As shown in Table 2, track density gradients are seen in etched olivine crystals, indicating that this rock has not experienced temperatures greater than $\sim 400°$C during or after formation.

Breccia 66055. Our samples of this very complex 1.3 kg light gray matrix breccia are plagioclase-rich and contain abundant blebs or clasts of darker material which sometimes have basaltic textures but more frequently appear to be brown glass. Some are globular in shape and suggest melting. HVEM observations of these optically isotropic regions show, however, that they are not glass but rather tightly packed agglomerations of sub-micron sized crystallites. As described in the following section, uranium is concentrated in these fine grained areas (*see* Fig. 7(a) and (b)). Electron diffraction studies of the largest of these sub-micron grains indicate severe deformation, evidence of shocking after recrystallization. Deformation accompanying recrystallization has produced shock welding, the mechanism responsible for cohesion of much of the breccia studied. Figure 6d shows strain-free micron-sized crystalline fragments shock welded to a deformed polycrystalline aggregate. Larger ($\geqslant 10 \mu$m) plagioclase and pyroxene grains are remarkably unaffected by shock, showing only moderate twinning. Shock-produced interstitial glass is prevalent only in certain areas.

No solar flare tracks were observed in this rock and we conclude that it is among the most highly metamorphosed rocks that we have examined. Other studies of 66055 (McKay *et al.*, 1973 and private communication) also indicate substantial metamorphic effects and heating to greater than 500°C for a time period of days to weeks.

Uranium distribution in breccias

We have made Lexan "maps" (Kleeman and Lovering, 1967) of uranium distribution in sections of many of the breccias discussed in this paper, primarily for the purpose of locating uranium-rich phases suitable for fission track dating. Using an SEM X-ray analyzer we have found it relatively simple to locate large

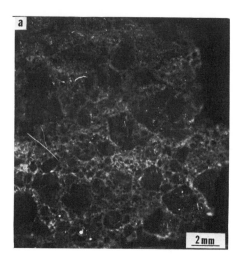
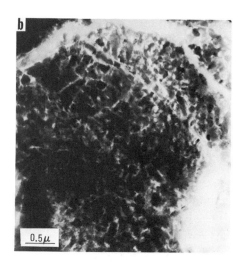

Fig. 7. (a) Lexan fission track map of the distribution of uranium in 66055 emphasizing the concentration of uranium in an optically isotropic interstitial mineral phase. Light areas are uranium-rich. Uranium is largely excluded from large plagioclase and pyroxene crystals. (b) High voltage electron micrograph of the uranium-rich phase. The optical isotropy is due to the composite structure of submicron crystals and miniature rock fragments with only a minor amount of glass. Electron diffraction shows the crystals to be heavily deformed. BF micrograph.

phosphates and zircons in breccia sections once the U-rich areas have been identified from the Lexan map. This work is still in progress and will be reported elsewhere.

The overall features of the uranium distribution in individual breccias seem to correlate with metamorphism. In general the higher the rock on the metamorphic scale, the more homogeneous the uranium distribution is. This feature is also evident in Table 1 of Graf *et al.* (1973) which shows that the percentage of uranium in "stars" decreases with increasing metamorphic grade.

The features of uranium distribution in a number of Apollo 14 breccias were discussed in an earlier paper (Hutcheon *et al.*, 1972). Apollo 16 breccias which we have recently studied show more variation in uranium concentrations, but generally the whole-rock value is lower than that for the Apollo 14 breccias by a factor of 5–20. Uranium in a partially recrystallized breccia, 60017, is uniformly distributed at about 0.2 ppm, although some moderate inhomogeneities have concentrations up to 0.4 ppm. In breccia 66055, a highly metamorphosed rock, uranium is concentrated and quite homogeneously distributed in glassy and very fine grained areas (Fig. 7 (a) and (b)). Plagioclase clasts and fragments are essentially devoid of uranium. In breccia 60255 uranium is also concentrated in fine grained areas, but is more heterogeneously distributed, reflecting the "unequilibrated" nature of this rock. Matrix regions contain uranium rich phosphates; a millimeter-sized feldspar-rich vein contains $\leqslant 0.1$ ppm uranium. In contrast to Apollo 14 breccias, it is a general feature of Apollo 16 breccias that plagioclase crystals do not exhibit uranium enrichments along cleavage planes.

Lunar breccias and gas-rich meteorites

A number of authors have already realized the importance of comparing track studies of lunar soils and breccias with those of gas-rich meteorites (e.g., Arrhenius *et al.*, 1971; Barber *et al.*, 1971; Berdot *et al.*, 1973; Dran *et al.*, 1972; Lal, 1972). For the most part, differences rather than similarities have been emphasized. In fact, it was not until 1972 that significant numbers of track-rich grains were reported from lunar breccias. This is at least partly due to the fact that earlier studies were made by optical microscopy using etching conditions sufficient to almost completely dissolve grains with very high track densities.

The extensive studies of lunar breccias reported in this paper and by Hutcheon *et al.* (1972) point also to several similarities between gas-rich meteorites and some lunar breccias. (i) The most obvious is the widespread preservation of low-energy irradiation records in grains of lunar breccias (now documented from Apollo 11, 14, and 15 sites) and in gas-rich meteorites (Lal and Rajan, 1969; Pellas *et al.*, 1969). Micron sized grains from both gas-rich meteorites (Barber *et al.*, 1971) and some lunar breccias (Hutcheon *et al.*, 1972) frequently contain $\geqslant 10^{10}$ tracks/cm^2. (ii) There appears to be nothing unusual about the fact that many of the irradiated grains in the gas-rich meteorites are euhedral (Pellas *et al.*, 1969); Fig. 3 shows a track-rich nearly euhedral feldspar from lunar breccia 15086, and we have observed many others. The gentleness of accretion necessary to preserve angular track-rich fragments can occur even on massive bodies such as the moon. (iii) On the other hand, chondrules, which occur in some gas-rich meteorites, and chondrule-like objects which in several of the Apollo 14 breccias discussed in this paper, presumably require energetic collisions and impact melting. Track-rich chondrules, which have been observed in Weston (Lal and Rajan, 1969; Pellas *et al.*, 1969) and Fayetteville (Macdougall *et al.*, 1973; Rajan, 1973) require on the order of 10^4 years irradiation time between formation and incorporation into the meteorite, assuming the same average solar flare intensity at that time as today. (iv) Mason and Melson (1970) have already pointed out textural similarities between lunar breccias and some meteorite types, and they use Kapoeta (a gas-rich meteorite containing track-rich grains) to illustrate their comparison.

The primary differences between track features in gas-rich meteorites and lunar breccias are in total radiation dose, frequency of track-rich grains, and details of irradiation geometry. All three features are related and do not constitute three independent differences. Most lunar breccias which we have examined contain a larger fraction of track-rich grains, often with higher track densities, than do gas-rich meteorites which have been studied. This simply requires longer exposure times for the components of lunar breccias. Lunar track-rich grains of $>50\ \mu$m size from both soils and breccias are typically more strongly irradiated on one side than the other. However, recent studies using high resolution SEM and TEM methods (Rajan *et al.*, 1973; Rajan, 1973; Macdougall *et al.*, 1973) indicate that the irradiation effects in the gas-rich meteorites are also not symmetrical, but that the anisotropy is much less marked. This feature suggests a much lower gravitational field and more frequent stirring in the irradiation environment of the track-rich meteorite grains. Evidence exists that even small asteroids have regoliths (Dunlap *et al.*, 1973); such an environment would seem suitable.

At the present time, there exist for the gas-rich meteorites no observations parallel to our discovery of partial track fading in the lunar breccias (i.e., tracks erased from glass and olivine in 14301; tracks visible only by HVEM in 14311 and 14321). The van Schmus and Wood (1967) classification purports to define degrees of metamorphism for the chondritic meteorites, yet preserved solar flare tracks are seen even in chondrites belonging to the highest metamorphic grade of this scheme (i.e., St. Mesmin, LL6, Pellas *et al.*, 1969). Thus, even type 6 must correspond in terms of temperature effects only to group 2 or perhaps group 3 in the metamorphic sequence of Warner (1972) for lunar breccias.

CONCLUSIONS

The studies described here indicate that low grade lunar breccias commonly preserve pre-compaction irradiation features. These features are similar to those observed in grains in the regolith to depths of several meters, and suggest that the low grade breccias are relatively local surface-ejecta products, perhaps consolidated in minor base surge processes as first postulated by McKay *et al.* (1970).

The degree of solar flare track retention for a given breccia is a function of its metamorphic history. While tracks are retained in all phases of soil breccias such as 15086, others such as 14301 show evidence of thermal events which have removed tracks from some but not all components. Highly metamorphosed breccias retain no solar flare tracks.

The argon retention age of green glass spheres from breccia 15426 is 3.4 G.y. (Podosek and Huneke, 1973), much greater than the fission track retention age of ≤ 0.7 G.y. for very similar green glass from breccia 15086. We raise the point that glass spherules, which freeze very quickly, may not completely outgas at their formation and may be considerably younger than the argon age indicates.

The presence of track rich grains in breccias 14301 and 14318 rules out very high temperature events ($> 700°C$) as a mechanism for homogenizing rare gas isotopes and supports the surface implantation model of Behrmann *et al.* (1973). A third chondrule-rich breccia which we have investigated (14315) also contains track-rich grains and should be studied for evidence of xenon from extinct nuclides.

Apollo 16 breccias which we have examined are more highly metamorphosed than our samples from the Apollo 14 and 15 sites, and we have not observed pre-compaction solar flare tracks in them.

Comparison of track features in lunar breccias with those of gas-rich meteorites suggests that the latter may be surface ejecta products formed in the regolith of small (asteroidal size) bodies.

Acknowledgments—We thank N. Peery for greatly appreciated help with sample preparation and M. Klash for skillful efforts in manuscript preparation. This research was supported by NASA grant NGL 05-003-410.

REFERENCES

Agrell S. O., Scoon J. H., Long J. V. P., and Coles J. N. (1972) The occurrence of goethite in a micro breccia from the Fra Mauro formation (abstract). In *Lunar Science—III*, pp. 7–9. The Lunar Science Institute, Houston.

Arrhenius G., Liang S., Macdougall D., Wilkening L., Bhandari N., Bhat S., Lal D., Rajagopalan G., Tamhane A. S., and Venkatavaradan V. S. (1971) The exposure history of the Apollo 12 regolith. *Proc. Second Lunar Sci. Conf., Geochim. Cosmochim. Acta*, Suppl. 2, Vol. 3, pp. 2583–2598. MIT Press.

Barber D. J., Cowsik R., Hutcheon I. D., Price P. B., and Rajan R. S. (1971) Solar flares, the lunar surface and gas-rich meteorites. *Proc. Second Lunar Sci. Conf., Geochim. Cosmochim. Acta*, Suppl. 2, Vol. 3, 2705–2714. MIT Press.

Behrmann C. J., Drozd R. J., and Hohenberg C. M. (1973) Extinct lunar radio-activities: Xenon from Pu^{244} and I^{129} in Apollo 14 breccias. *Earth Planet. Sci. Lett.* **17**, 446–455.

Berdot J. L., Chetrit G. C., Lorin J. C., Pellas P., and Poupeau G. (1973) Irradiation records in a compacted soil: Breccia 14307 (abstract). In *Lunar Science—IV*, pp. 63–65. The Lunar Science Institute, Houston.

Brown G. M., Emeleus C. H., Holland J. G., Peckett A., and Phillips R. (1972) Petrology, mineralogy and classification of Apollo 15 mare basalts. In *The Apollo 15 Lunar Samples*, pp. 40–44. The Lunar Science Institute, Houston.

Crozaz G., Haack U., Hair M., Maurette M., Walker R., and Woolum D. (1970) Nuclear track studies of ancient solar radiations and dynamic surface processes on the moon and constancy of solar activity. *Proc. Second Lunar Sci. Conf., Geochim. Cosmochim. Acta*, Suppl. 2, Vol. 3, 2543–2558. MIT Press.

Dran J. C., Duraud J. P., Maurette M., Durrieu L., Jouret C., and Legressus C. (1972) Track metamorphism in extraterrestrial breccias. *Proc. Third Lunar Sci. Conf., Geochim. Cosmochim. Acta*, Suppl. 3, Vol. 3, 2883–2903. MIT Press.

Drozd R., Hohenberg C. M., and Ragan D. (1972) Fission xenon from extinct ^{244}Pu in 14301. *Earth Planet. Sci. Lett.* **15**, 383–346.

Dunlap J. L., Gehrels T., and Howes M. L. (1973) Minor planets and related objects, IX. Photometry and polarimetry of (1685) Toro. *Astron. J.* **78**, In press.

Eldridge J. S., O'Kelley G. D., and Northcutt K. J. (1973) Concentrations of cosmogenic radionuclides in Apollo 15 rocks and soil. In *The Apollo 15 Lunar Samples*, pp. 357–359. The Lunar Science Institute, Houston.

Fleischer R. L. and Hart H. R. Jr. (1973) Particle track record of Apollo 15 green soil and rock. *Earth Planet. Sci. Lett.*, **18**, 420–426.

Fleischer R. L., Price P. B., and Walker R. M. (1965) Effects of temperature, pressure and ionization on the formation and stability of fission tracks in minerals and glasses. *J. Geophys. Res.*, **70**, 1497–1502.

Graf H., Hohenberg C., Shirck J., Sun S., and Walker R. (1973) Astrology of Apollo 14 extinct isotope breccias (abstract). In *Lunar Science—IV*, pp. 312–314. The Lunar Science Institute, Houston.

Hart H. R. Jr., Comstock G. M., and Fleischer R. L. (1972) The particle track record of Fra Mauro. *Proc. Third Lunar Sci. Conf., Geochim. Cosmochim. Acta*, Suppl. 3, Vol. 3, 2831–2844. MIT Press.

Huneke J. C., Podosek F. A., and Wasserburg G. J. (1973) An argon bouillabaisse including ages from the Luna 20 site (abstract). In *Lunar Science—IV*, pp. 403–405. The Lunar Science Institute, Houston.

Hutcheon I. D., Phakey P. P., and Price P. B. (1972) Studies bearing on the history of lunar breccias. *Proc. Third Lunar Sci. Conf., Geochim. Cosmochim. Acta*, Suppl. 3, Vol. 3, 2845–2865. MIT Press.

King E. A. Jr., Butler J. C., and Carman M. F. (1972) Chondrules in Apollo 14 samples and size analyses of Apollo 14 and 15 fines. *Proc. Third Lunar Sci. Conf., Geochim. Cosmochim. Acta*, Suppl. 3, Vol. 1, 673–686. MIT Press.

Kleeman J. D. and Lovering J. F. (1967) Uranium distribution studies by fission track registration in Lexan plastic prints. *Atomic Energy in Australia*, **10**, 3–8.

Krishnaswami S., Lal D., Prabhu N., and Tamhane A. S. (1971) Olivines: Revelation of tracks of charged particles. *Science*, **174**, 287–291.

Lal D. (1972) Accretion processes leading to formation of meteorite parent bodies. In *From Plasma to Planet* (Proc. Nobel Symp. 21, editor A. Elvius), pp. 49–64. Wiley.

Lal D. and Rajan R. S. (1969). Observations relating to space irradiation of individual crystals of gas-rich meteorites. *Nature*, **223**, 269–271.

Lal D., Murali A. V., Rajan R. S., Tamhane A. S., Lorin J. C., and Pellas P. (1968) Techniques for proper revelation and viewing of etch tracks in meteoritic and terrestrial minerals. *Earth. Planet. Sci. Lett.* **5**, 111–119.

Lally J. S., Fisher R. M., Christie J. M., Griggs D. T., Heuer A. H., Nord G. L. Jr., and Radcliffe S. V. (1972) Electron petrography of Apollo 14 and 15 rocks. *Proc. Third Lunar Sci. Conf., Geochim. Cosmochim. Acta*, Suppl. 3, Vol. 1, 401–422. MIT Press.

LSPET (Lunar Science Preliminary Examination Team) (1972) The Apollo 15 lunar samples: A preliminary description. *Science* **175**, 363–375.

Macdougall D., Lal D., Wilkening L., Liang S., Arrhenius G., and Tamhane A. S. (1971) Techniques for the study of fossil tracks in extraterrestrial and terrestrial samples. I. Methods of high contrast and high resolution study. *Geochemical Journal* (Japan) **5**, 95–112.

Macdougall D., Rajan R. S., and Price P. B. (1973) Gas-rich meteorites: Possible evidence for origin on a regolith. To be published.

Mason B. and Melson W. G. (1970) Comparison of lunar rocks with basalts and stony meteorites. *Proc. Apollo 11 Lunar Sci. Conf., Geochim. Cosmochim. Acta*, Suppl. 1, Vol. 1, 661–671. Pergamon.

McKay D. S., Greenwood W. R., and Morrison D. A. (1970) Origin of small lunar particles and breccia from the Apollo 11 site. *Proc. Apollo 11 Lunar Sci. Conf., Geochim. Cosmochim. Acta*, Suppl. 1, Vol. 1, 673–694. Pergamon.

McKay G., Kridelburgh S., and Weill D. (1973) A preliminary report on the petrology of microbreccia 66055 (abstract). In *Lunar Science—IV*, pp. 487–489. The Lunar Science Institute, Houston.

Pellas P., Poupeau G., Lorin J. C., Reeves H., and Adouze J. (1969) Primitive low-energy particle irradiation of meteoritic crystals. *Nature* **223**, 273–274.

Phakey P. P., Hutcheon I. D., Rajan R. S., and Price P. B. (1972) Radiation effects in soils from five lunar missions. *Proc. Third Lunar Sci. Conf., Geochim. Cosmochim. Acta*, Suppl. 3, Vol. 3, 2905–2915. MIT Press.

Podosek F. A. and Huneke J. C. (1973) Argon in Apollo 15 green glass spherules (15426): $^{40}Ar–^{39}Ar$ age and trapped argon. To be published.

Price P. B., Rajan R. S., and Tamhane A. S. (1968) The abundance of nuclei heavier than iron in the cosmic radiation in the geological past. *Astrophys. J* **151**, L109–L116.

Price P. B., Rajan R. S., Hutcheon I. D., Macdougall D., and Shirk E. K. (1973a) Solar flares, past and present (abstract). In *Lunar Science—IV*, pp. 600–602. The Lunar Science Institute, Houston.

Price P. B., Lal D., Tamhane A. S., and Perelygin V. P. (1973b) Characteristics of tracks of ions with $14 \leqslant Z \leqslant 36$ in common rock silicates. *Earth Planet. Sci. Lett.* In press.

Rajan R. S. (1973) On the irradiation history and origin of gas rich meteorites. Submitted to *Geochim. Cosmochim. Acta.*

Rajan R. S., Macdougall D., and Phakey P. P. (1973) Energy spectra of ancient solar flare particles and the origin of gas-rich meteorites (abstract). *Meteoritics* **8**, 64–65.

Storzer D., Poupeau G., Kratschmer W., and Plieninger T. (1973) The track record of Apollo 15, 16 and Luna 16, 20 samples and the charge assignment of cosmic ray tracks (abstract). In *Lunar Science—IV*, pp. 694–696. The Lunar Science Institute, Houston.

Van Schmus W. R. and Wood J. A. (1967) A chemical-petrological classification for the chondritic meteorites. *Geochim. Cosmochim. Acta* **31**, 747–765.

Warner J. L. (1972) Metamorphism of Apollo 14 breccias. *Proc. Third Lunar Sci. Conf., Geochim. Cosmochim. Acta*, Suppl. 3, Vol. 1, 623–643. MIT Press.

Williams R. J. (1972) The lithification and metamorphism of lunar breccias. *Earth Planet. Sci. Lett.* **16**, 250–256.

Proceedings of the Fourth Lunar Science Conference
(Supplement 4, *Geochimica et Cosmochimica Acta*)
Vol. 3, pp. 2337–2346

Indications for time variations in the galactic cosmic ray composition derived from track studies on lunar samples

T. PLIENINGER, W. KRÄTSCHMER, and W. GENTNER

Max-Planck-Institut für Kernphysik, Heidelberg, W.-Germany

Abstract—In order to improve the charge assignments to etchable VH cosmic ray ion tracks, pyroxene crystals from Apollo 11 lunar rock sample 10047,13 were irradiated with 9.6 MeV/n Fe ions at the Linac of the University of Manchester (England). Using the track-in-track technique, the lengths of artificial Fe and pre-existing galactic cosmic ray tracks within the same crystals were compared. By this, a direct comparison was possible without referring to etchable ranges, which were found to be difficult to develop completely by etching. Under the chosen etching conditions, the length distributions of artificial Fe and cosmic ray tracks coincide at a peak centered at about 16–18 μm. A more detailed evaluation of track lengths shows that annealing effects have shortened the mean length of galactic Fe tracks by about 0.5 μm. Additional peaks at 6 μm and 12 μm were ascribed to Ca and Cr cosmic ray ions. A primary ratio $(V + Cr + Mn)/Fe = 0.7 \pm 0.2$ for the energy range 300–600 MeV/n can be derived under the assumption of a maximum shielding depth of 5 cm. Length distributions measured in pyroxene crystals from the lunar regolith (sample 10084) show ratios ranging between 0.4 and 0.7. A comparison with recent data about the present day galactic cosmic ray composition indicates that the ratios obtained in rock and soil samples cannot be explained by the variation of the composition with energy. Therefore a variation with time of the VH cosmic ray composition is proposed.

INTRODUCTION

IN ORDER to study the history of the composition of the cosmic radiation much work has been invested to determine the VH ($20 \leq Z \leq 30$) element abundances of the ancient cosmic radiation from track length measurements. These attempts used the fact that the etchable portion of the ion range is a measure of its atomic number (*see*, for example Price and Fleischer, 1971). The first charge assignments were given by Lal, investigating galactic cosmic ray tracks by the track-in-track method (Lal, 1969; Lal *et al.*, 1970). They found in pyroxenes of the Patwar meteorite that most tracks have a length of about 12 μm. From abundance arguments, they ascribed these tracks to galactic Fe ions (Lal, 1969; Bhandari *et al.*, 1972). Other peaks in the length distributions at 5 μm, 9 μm, and 15 μm were assumed to be caused by Cr, Mn, and Co, respectively. The investigation of lunar rock samples yielded length distributions similar to those observed in meteorites (Lal *et al.*, 1970; Bhandari *et al.*, 1971a; Bhandari *et al.*, 1972). In pyroxenes from the lunar soil it was found however, that on the average the tracks in the 15 μm length region were significantly more abundant than 12 μm tracks (Bhandari *et al.*, 1971a; Plieninger *et al.*, 1972). This initially was interpreted as a fission track contribution to the cosmic 15 μm tracks (Bhandari *et al.*, 1971a).

The occurrence of a 15 μm peak in most length distributions, even at high cosmic ray track densities, and the low concentration of fissionable elements in pyroxenes, however, favors a cosmic ray origin of these tracks (Plieninger *et al.*,

1972; Fleischer and Hart, 1973). From length comparisons of cosmic ray and 2 MeV/n Ca tracks in lunar pyroxenes, we deduced that the 15 μm peak should be assigned to cosmic Fe ions and the other peaks to elements of even atomic number, i.e., Ca, Ti, and Cr (Plieninger et al., 1972).

The problem of charge assignment became more confused when additional data about track lengths of artificially accelerated heavier ions were available (Price et al., 1972; Plieninger, 1972; Plieninger and Krätschmer, 1972; Storzer et al., 1973). It turned out that the etchable range, or recordable length, of Fe ions must be greater than 20 μm for lunar pyroxenes (Price et al., 1972; Plieninger and Krätschmer, 1972). Since the observed lengths of most cosmic ray tracks are shorter, annealing processes have been proposed to resolve this apparent discrepancy. As thermal annealing alone seems to be ineffective (Plieninger et al., 1972; Plieninger, 1972) a so far unknown relaxation of the etchable radiation damage (Price et al., 1972) and annealing by shock effects were envisaged (Fleischer and Hart, 1973).

This discrepancy however, mainly arises from the assumption that a track length corresponding to the whole recordable length of the cosmic ray ions is really developed by etching. In the following it is shown that "complete etching" is not attainable for most tracks. Therefore only mean lengths, i.e., lengths averaged over different degrees of etching, are relevant. We have simultaneously compared lengths of cosmic ray and artificially produced 9.6 MeV/n Fe tracks in the same pyroxene crystals of rock sample 10047,13. The etching conditions for both kinds of tracks thus were identical. Despite the fact that the mean track lengths do not equal etchable ranges, it turns out that a charge assignment to cosmic tracks is possible. Based on this assignment an evaluation of element abundance ratios in the ancient cosmic radiation is made and the results are compared with the recent present day composition measurements by Webber et al. (1972).

EXPERIMENTAL PROCEDURES

Polished surfaces of pyroxene crystals (mare type clinopyroxenes) from rock sample 10047,13 were exposed perpendicularly to the 9.6 MeV/n Fe beam of the Linac at the University of Manchester (England). After irradiation, the crystals were polished again, parallel to the beam direction. All pyroxenes were etched simultaneously in a boiling NaOH solution (1 g NaOH in 1 g N_2O) for 4 hours. The track densities in the pyroxenes were found to be sufficient for applying the track-in-track technique (Lal, 1969). The lengths of tracks, revealed by this technique (TINTs) were measured by means of an optical microscope. Cosmic ray tracks (longer than 2 μm) and artificially produced tracks were measured under the same conditions. Furthermore, the distribution of incident angles was determined on tracks of cosmic ray ions where the direction of the incoming ion could be deduced from the shape of the track. Additional details about the techniques of measurement are described elsewhere (Plieninger et al., 1972).

RESULTS

Tracks of 9.6 MeV/n Fe ions in a pyroxene crystal from lunar rock sample 10047,13 are shown in Fig. 1. The etchable portion of the ion range lies completely within the crystal. Therefore, despite the different initial energies, these tracks are

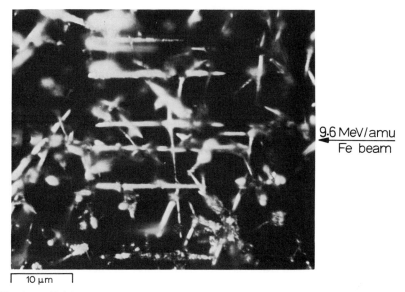

9.6 MeV/amu
Fe beam

10 μm

Fig. 1. Artificially produced Fe tracks in the interior of a pyroxene crystal from rock sample 10047,13. The Fe ions had an initial energy of 9.6 MeV/n and were stopped to rest in the pyroxene. Only near the end of range tracks can be developed by etching. The tracks were etched by the track-in-track method, decorated with silver and are photographed in reflected light.

directly comparable with cosmic ray ion tracks. The measured total range of the Fe ions is 65 ± 1 μm which agrees with the range of 66 μm, calculated after Henke and Benton (1967).

The track lengths show relatively large fluctuations since a certain amount of statistics is introduced by the track-in-track technique: (a) the effective etching time for TINTs depends on the span of time necessary for the formation of an etch-channel, and (b), because the etching rate varies with residual range, the track length depends on the position of the etch-channel along the etchable track. Obviously, the Fe tracks are not etched completely to their recordable length under the etching conditions employed. Therefore a relatively broad peak of the length distribution centered at 16–18 μm results (see Fig. 3 (a) and (c)).

Studying the length of artificial and fossil cosmic ray tracks as a function of etching time, it was found that it is impossible to obtain "completely etched" tracks under feasible etching conditions because of the low etching rates at the high energy parts of the tracks. Therefore, one generally cannot equate the measured track lengths with etchable ranges, especially in the case of tracks produced by Fe and heavier ions. For lighter ions, the differences between etchable ranges and mean track lengths are decreasing; Ca, for example, has a recordable range in pyroxenes of about 7–8 μm, while under the etching conditions used, a mean track length of about 6 μm results (see Fig. 3 (b) and (d)). (By a modified etching technique, the recordable length of Fe ions in terrestrial pyroxene (diopsite) was estimated to be 25–30 μm.)

The pyroxene crystals investigated were selected from the chipping residue of rock 10047 (El Goresy, 1973). Therefore, the locations of the crystals within the rock are not known. Since only very few track data for this rock exist, we have tried to get information about the cosmic ray irradiation conditions by measuring the track densities and the distribution of cosmic ray incident angles in the studied crystals themselves. Most of them show track densities ranging between 5×10^6 and 2×10^7 cm^{-2} without measurable density gradients in the scale of the crystal sizes ($\sim 200 \, \mu$m). The general shape of the angular distributions was found to be quite similar for all studied crystals, independent of the track density (shielding depth). As an example, Fig. 2 shows the distribution of cosmic ray incident angles in a crystal having a track density of 5×10^6 cm^{-2}. The distribution indicates that (a) the irradiation took place in a 2π geometry and (b) that the crystals were irradiated from two main directions. This we interpret as being due to shielding effects. The similarity of the angular distributions in all studied crystals indicates that the shielding probably is not caused by mm scale irregularities of the rock surface but by larger features of the rock itself.

There is evidence from track data that rock 10047 has been exposed to cosmic rays on the lunar surface (Crozaz et al., 1970). Since the track production rate is steeply decreasing with shielding depth, the crystals studied probably have received most of their tracks when the rock was situated on the surface of the regolith. Together with the dimensions of the rock ($3 \times 4 \times 7$ cm, after Schmitt et al. 1970), we regard a shielding depth of 5 cm as a reasonable upper limit. Therefore a shielding depth of 5 cm was assumed for the crystals having a track density of 5×10^6 cm^{-2}, while for the higher track densities, the shielding depths were calculated from galactic cosmic ray track production rates (Fleischer et al., 1967).

To obtain information about the influence of fragmentation of heavy primary nuclei in the rock material, track length distributions were investigated as function of track density, i.e., shielding depth. To improve statistics, the measurements

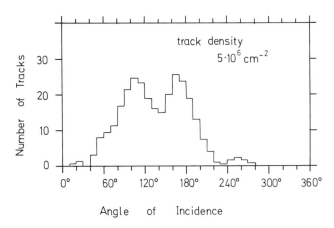

Fig. 2. Distribution of VH cosmic ray ion incident angles measured in a pyroxene from rock sample 10047,13.

were performed on two groups of crystals having track densities in the range between 1.5–$2 \times 10^7 \, cm^{-2}$ and 0.5–$0.8 \times 10^7 \, cm^{-2}$. These densities correspond to shielding depths between about 1–2 cm and 3–5 cm, respectively. The resulting track length distributions for fossil and artificial tracks are shown in Fig. 3. From the shape and position of the distribution of the artificially produced Fe tracks, the amount of fossil Fe tracks in the cosmic ray length spectrum was evaluated. It was found that a better fit with the Fe distributions could be obtained under the assumption that the fossil Fe tracks are 0.5 μm shorter than the artificial tracks. This indicates that the degree of annealing of fossil and artificial tracks is different. After subtraction of the fossil Fe component from the cosmic ray length distribution, the residual spectrum showed a second pronounced peak at 12 μm and smaller ones at 25 μm, 9 μm and 6 μm. In previous calibration measurements with 2 MeV/n Ca and 9.6 MeV/n Cu ions (Plieninger *et al.*, 1972; Plieninger and Krätschmer, 1972), it was shown that Ca and Cu under similar etching conditions have mean track lengths of 5–6 μm and 30–32 μm, respectively. This leads to the charge assignments indicated in Fig. 3 (b) and (d).

The 12 μm peak is probably mostly due to Cr ions with smaller contributions of V and Mn. Hence the abundance ratio of tracks in the 12 μm peak to those of the Fe peak represents the track abundance ratio (V + Cr + Mn)/Fe. Since the statistical errors are relatively high, our results only indicate an increase of this ratio with shielding depth, as shown in Fig. 4. From this, a primary ratio (V + Cr + Mn)/Fe = 0.7 ± 0.2 can be derived and the fragmentation parameter Fe→ (V + Cr + Mn) can be estimated. Taking the interaction mean free path of Fe in stony material as 6 cm (Fleischer *et al.*, 1967), a fragmentation parameter of about 0.3 results.

For comparison, pyroxene crystals from the lunar regolith showing about the same track densities and no gradients were chosen. It was found that the (V + Cr + Mn)/Fe ratio varies from crystal to crystal. Ratios ranging between 0.7 and 0.4 have been measured. A length distribution, from which a ratio of about 0.4 can be derived, is shown in Fig. 5.

DISCUSSION

In a first investigation of artificially produced Fe tracks we had assumed that the etchable range of fossil tracks was completely developed (Plieninger and Krätschmer, 1972). Therefore, we had selected from the artificial tracks only those which appeared to be etched to their full length. It turned out, however, that our comparison was based on the wrong assumption that the recordable ranges can be developed completely by the track-in-track method, when we found that even most fossil tracks were not completely etched. Therefore, as outlined in this work, we have performed a comparison of statistical mean lengths of fossil and Fe-calibration tracks in the same crystals.

Under the etching conditions used, one has to distinguish between track lengths and etchable ranges because a preferential, but low, track etching rate still exists over the part of the range which was assumed to not be etchable in previous

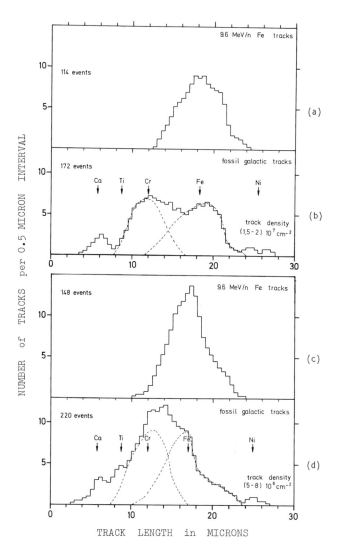

Fig. 3. Length distributions for artificially generated Fe and galactic VH ion tracks measured in the same pyroxene crystals under identical conditions, for different track densities, i.e., shielding depths. From the shape and position of the distribution of the calibration ions, the amount of Fe and of elements belonging to the Cr group (V + Cr + Mn) was evaluated for the galactic VH ions (dashed lines) in (b) and (d). The given charge assignments are based on calibrations with Ca and Cu ions. As one can see from (b) and (d), the resolution is poorer for lower track densities. Since all crystals were etched under the same conditions, the crystals with higher track densities are more "transparent" for the etching solution and therefore, especially, the Fe TINTs have a longer mean length.

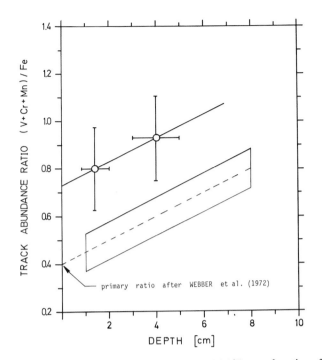

Fig. 4. The measured track abundance ratio (V + Cr + Mn)/Fe as a function of shielding depth (solid line). The shielding depths were estimated from track data and probably represent upper limits. From this, the primary ratio and the corresponding fragmentation parameter is evaluated. The dashed line shows the abundance ratio one would expect for the present galactic cosmic ray composition. These data are based on the primary ratio of 0.40 ± 0.08 measured by Webber *et al.* (1972) for the energy range 250–850 MeV/n.

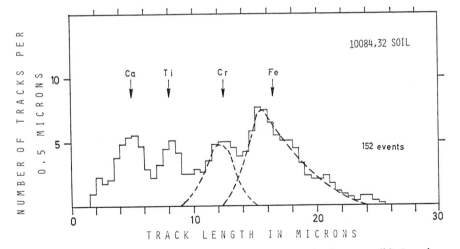

Fig. 5. Length distribution measured in a crystal from the lunar regolith (sample 10084,32). The dashed lines indicate the amount of Fe and (V + Cr + Mn). The abundance ratio derived is about 0.4. This ratio agrees with the results of Webber *et al.* (1972) about the present day galactic cosmic ray composition.

investigations (Bhandari *et al.*, 1971a; Plieninger *et al.*, 1972). This effect may at least partly account for the longer tracks found along cleavages in pyroxene crystals (Bhandari *et al.*, 1971b), since the effective etching times for these tracks are longer.

The amount of track shortening we observed in the investigated pyroxenes is comparable to the effect one would expect from thermal annealing by lunar surface temperatures alone (Plieninger, 1972), thus indicating that no severe annealing by other effects (e.g., by shock) has taken place during the storage time of tracks.

To convert the measured $(V + Cr + Mn)/Fe$ track abundance ratios into the corresponding element abundance ratios, the development and evaluation probabilities for the two groups of tracks must be taken into account. The development probability—i.e., the probability that a latent track intersects with an etch channel—increases with etchable range, and the evaluation probability—i.e., the probability that the length of a track can be measured within the crystal—decreases with track length. Thus the resulting probability for counting the track only weakly depends on the length of the track and especially the measured $(V + Cr + Mn)/Fe$ track abundance ratios are approximatively equal to the element abundance ratios.

The effects of lunar erosion have not yet been taken into account. A calculation based on the present day galactic VH energy spectrum shows that the primary $(V + Cr + Mn)/Fe$ ratio of 0.7 is lowered to 0.6 by the assumption of an erosion rate of 1 mm/m.y. (Crozaz *et al.*, 1972) effective for 90 m.y. (i.e., the spallation age of rock 10047 after Marti *et al.*, 1970). This small influence of erosion is due to the fact that the galactic track production rate is decreasing with shielding depth. Although the mean shielding depth is larger in this case, the $(V + Cr + Mn)/Fe$ ratio is predominantly determined by those tracks which were accumulated at low depths.

The shielding depths, assumed for the investigated pyroxenes, represent upper limits. The actual depth of the studied crystals during VH cosmic ray irradiation may have been smaller. If one assumes that they had a shielding depth of less than 1 cm, a primary $(V + Cr + Mn)/Fe$ ratio of about 0.8 would result, since in that case the effects of fragmentation are small.

In the region of depth of 1 cm and less, solar flare VH ions are contributing to an increasing extent to the production of tracks (Crozaz *et al.*, 1972). The available data about solar flare VH ion composition indicate that the abundance ratio $(V + Cr + Mn)/Fe$ is less than 0.1 (Crawford *et al.*, 1972). Provided that this value can be regarded as representative for all flares, a solar flare origin of the studied tracks can be ruled out. We therefore conclude, that the irradiation of the studied pyroxenes took place at least in the sub-cm and cm range.

Bhandari *et al.* (1972) have investigated galactic cosmic ray track length distributions as a function of shielding depth (2–6 cm) in pyroxenes from the Patwar meteorite. Although it is difficult to compare length distributions obtained under different etching conditions in samples of quite different origin, a reinterpretation of their charge assignements leads to a primary $(V + Cr + Mn)/Fe$ ratio close to our value of 0.7 for rock 10047.

Results based on track length distributions measured in lunar soil samples may be less conclusive because one has to consider their possible rather complex formation and irradiation histories. However, the measurements indicate that a variation in the (V + Cr + Mn)/Fe ratio between 0.4 and 0.7 exists. If one takes into account erosion and VH ion fragmentation, the value of 0.4 would represent an upper limit for the primary ratio.

Webber *et al.* (1972) have determined the present day galactic cosmic ray composition in the energy ranges 250–850 MeV/n and > 850 MeV/n. These energies correspond to ion ranges in lunar rocks of 1–8 cm and > 8 cm respectively. They obtained a ratio (V + Cr + Mn)/Fe = 0.40 ± 0.08 for 250–850 MeV/n and 0.23 ± 0.05 respectively for energies > 850 MeV/n. This indicates that the galactic (V + Cr + Mn)/Fe ratio gradually decreases with increasing energy. However, it seems questionable whether the present day energy dependence of this ratio can account for the observed cosmic ray compositional variations in lunar samples.

While the minimum value observed in lunar soils agrees with the present day ratio of 0.4, the higher ratios, especially the ratio measured in rock 10047, are difficult to explain on the basis of the present day cosmic ray composition.

If the pyroxenes from rock 10047 have received most of their tracks at depths in the cm range, which we regard as most probable, the measured (V + Cr + Mn)/Fe primary ratio of about 0.7 is higher than the present day value. Since fragmentation effects enlarge the (V + Cr + Mn)/Fe ratio, the assumption of larger shielding depths (8–10 cm) could account for this difference. However, these larger shielding depths seem improbable since the track densities observed in the studied crystals only can be accumulated at shielding depths of less than 5 cm. This estimate is based on a VH irradiation time of 90 m.y. (i.e., the spallation age) and an erosion rate of 1 mm/m.y. Therefore, we regard the difference between the measured "ancient" (V + Cr + Mn)/Fe ratio and the present day value as an indication for a time variation in the composition of the galactic cosmic radiation.

Although time variations are indicated, but not proven by our results, the consequences of time variations need to be taken into consideration when fossil track data are interpreted.

Acknowledgments—We are grateful to NASA for providing the lunar samples. We want to thank the LINAC staff of the Schuster Laboratory of the University of Manchester, especially Drs. A. G. Smith, I. S. Grant, R. B. Clark, R. King, V. Lowinger, and R. W. R. Hoisington, for their generous help during the irradiations. We appreciate the helpful discussions with Drs. A. El Goresy, T. Kirsten, D. Heymann, O. Müller, and D. Storzer.

REFERENCES

Bhandari N., Bhat S. G., Lal D., Rajagopalan G., Tamhane A. S., and Venkatavaradan V. S. (1971a) Spontaneous fission record of uranium and extinct transuranic elements in Apollo samples. *Proc. Second Lunar Sci. Conf., Geochim. Cosmochim. Acta,* Suppl. 2, Vol. 3, pp. 2599–2609. MIT Press.

Bhandari N., Bhat S. G., Lal D., Rajagopalan G., Tamhane A. S., and Venkatavaradan V. S. (1971b) Super-heavy elements in extraterrestrial samples. *Nature* **230**, 219–224.

Bhandari N., Bhat S. G., Goswami I. N., Lal D., Tamhane A. S., and Venkatavaradan V. S. (1972) Study of heavy cosmic rays in lunar silicates (abstract). In *Lunar Science—III,* pp. 65–67. The Lunar Science Institute, Huston.

Crawford H. J., Price P. B., and Sullivan J. D. (1972) Composition on and energy spectra of heavy nuclei with $0.5 < E < 40$ MeV/n in the 1971 January 24 and September 1 solar flares. *Ap. J.* **175**, L 149-L 153.

Crozaz G., Haack U., Hair M., Maurette M., Walker R., and Woolum D. (1970) Nuclear track studies of ancient solar radiations and dynamic lunar surface processes. *Proc. Apollo 11 Lunar Sci. Conf.*, *Geochim. Cosmochim. Acta*, Suppl. 1 Vol. 3, pp. 2051–2080. Pergamon.

Crozaz G., Drozd R., Hohenberg C. M., Hoyt H. P., Ragan D., Walker R. M., and Yuhas D. (1972) Solar flare and galactic cosmic ray studies of Apollo 14 and 15 samples. *Proc. Third Lunar Sci. Conf.*, *Geochim. Cosmochim. Acta*, Suppl. 3, Vol. 3, pp. 2917–2931. MIT Press.

El Goresy A. (1973) Private Communication.

Fleischer R. L., Price P. B., Walker R. M., and Maurette M. (1967) Origin of fossil charged particle tracks in meteorites. *J. Geophys. Res.* **72**, 331–353.

Fleischer R. L. and Hart H. R. (1973) Tracks from extinct radioactivity, ancient cosmic rays and calibration ions. *Nature* **242**, pp. 104–105.

Henke R. P. and Benton E. V. (1967) A computer code for the computation of heavy-ion range energy relationship in any stopping material. *Report USNRDL-TR-67-122.*

Lal D. (1969) Recent advances in the study of fossil tracks in meteorites due to heavy nuclei of the cosmic radiation. *Space Sci. Rev.* **9**, 623–650.

Lal D., MacDougall D., Wilkening L., and Arrhenius G. (1970) Mixing of the lunar regolith and cosmic ray spectra: Evidence from particle track studies. *Proc. Apollo 11 Lunar Sci. Conf.*, *Geochim. Cosmochim. Acta*, Suppl. 1, Vol. 3, pp. 2295–2303. Pergamon.

Marti K., Lugmair G. W., and Urey H. C. (1970) Solar wind gasses, cosmic ray spallation products and the irradiation history of Apollo 11 samples. *Proc. Apollo 11 Lunar Sci. Conf.*, *Geochim. Cosmochim. Acta*, Suppl. 1, Vol. 2, pp. 1357–1367. Pergamon.

Plieninger T. (1972) Minerale als Spurdetektor zur Untersuchung der schweren Komponente der kosmischen Primärstrahlung. Thesis, Max-Planck-Institut für Kernphysik, Heidelberg. Unpublished.

Plieninger T. and Krätschmer W. (1972) Registration properties of pyroxenes for various heavy ions and consequences to the determination of the composition of the cosmic radiation. In *Proc. Eighth Int. Conf. on Nucl. Photog. and Solid State Detectors*, Bucharest, 1972.

Plieninger T., Krätschmer W., and Gentner W. (1972) Charge assignment to cosmic ray heavy ion tracks in lunar pyroxenes. *Proc. Third Lunar Sci. Conf.*, *Geochim. Cosmochim. Acta*, Suppl. 3, Vol. 3, pp. 2933–2939. MIT Press.

Price P. B. and Fleischer R. L. (1971) Identification of energetic heavy nuclei with solid dielectric track detectors: Application to astrophysical and planetary studies. *Ann. Rev. Nucl. Sci.* **21**, 295–333.

Price P. B., Hutcheon I. D., Lal D., Perelygin V. P. (1972) Lunar crystals as detectors of very rare nuclear particles (abstract). In *Lunar Science—III*, pp. 619–621, The Lunar Science Institute, Houston.

Schmitt H. H., Lofgren G., Swann G. A., and Simmons G. (1970) The Apollo 11 samples: Introduction *Proc. Apollo 11 Lunar Sci. Conf.*, *Geochim. Cosmochim. Acta*, Suppl. 1, Vol. 1, pp. 1–54. Pergamon.

Storzer D., Poupeau G., Krätschmer W., and Plieninger T. (1973) The track record of Apollo 15, 16 and Luna 16, 20 samples and the charge assignment of cosmic ray tracks (abstract). In *Lunar Science—IV*, pp. 694–695. The Lunar Science Institute, Houston.

Webber W. R., Damle S. V., and Kish J. (1972) Studies of the chemical composition of cosmic rays with $Z = 3$–30 at high and low energies. *Astrophysics and Space Science* **15**, 245–271.

Proceedings of the Fourth Lunar Science Conference
(Supplement 4, *Geochimica et Cosmochimica Acta*)
Vol. 3, pp. 2347–2361

Low-energy heavy ions in the solar system

P. B. PRICE,* J. H. CHAN, I. D. HUTCHEON, D. MACDOUGALL,
R. S. RAJAN, E. K. SHIRK, and J. D. SULLIVAN

Department of Physics, University of California, Berkeley, California 94720

Abstract—In the previously inaccessible energy interval between $\sim 10\,\text{keV/nucleon}$ and $\sim 20\,\text{MeV/nucleon}$, we report measurements of the energy spectra of various ions from He to the elements heavier than Fe during various solar conditions ranging from nearly quiet times to periods of intense flares. Though the fluxes vary in absolute magnitude by many powers of ten, there is at all times an enrichment of heavy ions that monotonically increases with atomic number and decreases with energy. At sufficiently high energy ($\sim 20\,\text{MeV/nucleon}$ for intense flares, lower for weak flares) the composition approaches that in the photosphere. The enrichment of heavy ions results in very high track densities observed in the outer few microns of small grains within lunar soils and breccias and within gas-rich meteorites. It also should give rise to anomalously high ratios of heavy to light rare gases in such grains and these effects should be detectable by mass spectrometry if the outer 1 to 2 microns containing solar wind gas could be removed. At various epochs extending back some 4 G.y., the steepest profiles of solar flare track densities in lunar and meteoritic grains are similar to the Surveyor glass profile, suggesting that the distribution of energy in solar flares has not changed with time. The intensity of radiation damage in micron-size grains studied by transmission electron microscopy varies widely from site to site but not with depth at a given site. In the Apollo 15, 16, and 17 cores that sampled depths down to $\sim 250\,\text{cm}$, the radiation damage level does not vary significantly with depth.

INTRODUCTION

THE APOLLO PROGRAM has provided opportunities to study interplanetary charged particles in the previously inaccessible energy interval between $\sim 10\,\text{keV/nucleon}$ and $\sim 20\,\text{MeV/nucleon}$. It is well-known that at lower energy, $\sim 1\,\text{keV/nucleon}$, ions in the solar wind constantly stream outward from the sun. Hydrogen and helium in the solar wind are sufficiently abundant that they can be studied directly with detectors on satellites. The heavier rare gases can be detected by mass spectrometric analysis of metal foils placed on the lunar surface and of the outer 1000 Å or so of lunar soil grains. Electronic detectors capable of identifying interplanetary charged particles with energies above $\sim 20\,\text{MeV/nucleon}$ have been operating continuously on satellites for some years, and nuclear emulsions have been exposed in rockets during solar flares to study transient fluxes of such particles. One of the important results of emulsion studies is that at energies above $\sim 20\,\text{MeV/nucleon}$ the composition of solar flare particles seems to be similar to that measured spectroscopically in the sun's atmosphere.

In this paper, which is dedicated to Harold C. Urey on the occasion of his eightieth birthday, we survey the recent observations and implications of inter-

---

*Miller Institute Professor, 1972–1973.

planetary ions with energies ranging over the three orders of magnitude between ~10 keV/nucleon and ~20 MeV/nucleon.

Energy Dependence of the Solar Particle Composition

The first indication that the composition of solar flare particles is not always the same as that of the sun came from a comparison of Fe tracks in the Surveyor III camera glass with He counts in a satellite detector (Price *et al.*, 1971). During the period 1967–1970 when the Surveyor camera was on the moon, the ratio of (Fe flux)/(He flux) at energies of a few MeV/nucleon was inferred to be at least 20 times higher than the solar ratio, which is now estimated to be ~1/3000 (Aller, 1972). Price *et al.* speculated that the high Fe/He ratio resulted from the smaller fractional charge state and greater magnetic rigidity of heavy ions than of light ions at low energies, which might lead to a preferential escape of the heavy ions from an accelerating region.

This and several subsequent observations suffer from the drawbacks that the fluxes of heavy and light ions were measured in different detectors (Lanzerotti *et al.*, 1972) or in different energy intervals (Mogro-Campero and Simpson, 1972) or that individual species were not identified (Fleischer and Hart, 1973).

The use of Lexan plastic detectors on the Apollo 16 and 17 missions and in rockets fired from Fort Churchill during several major solar flares has provided a detailed picture of the composition of interplanetary particles during both quiet times and solar flares. The data and results are displayed in Figs. 1 to 6. To make it easy to visualize the trends of composition with energy, we have scaled the flux of each ion by an amount chosen so that its spectrum coincides with that of He at high energies.

It is clear from the figures why earlier detectors that could only study particles with energies above ~20 MeV/nucleon found that flare particles had normal solar composition. Only at lower energies does the composition become enriched in heavy elements.

In Fig. 7 we summarize the results for the ratios Fe/He and O/He in three flares and during a small interplanetary enhancement. Table 1 gives the details of the time intervals studied and the detectors used.

Several trends appear to characterize the composition of interplanetary ions at energies between ~10 keV/nucleon and ~20 MeV/nucleon. We can express those trends in terms of the enhancement factor $Q(Z, E) \equiv [Z/\mathrm{He}]_E/[Z/\mathrm{He}]_\odot$, where the subscript E refers to the relative abundance of interplanetary particles of atomic number Z at energy E and $\odot$ refers to the relative abundance in the solar atmosphere.

(1) Q is large at low energies and monotonically decreases with increasing energy. The asymptotic value at high energy is usually within a factor three of unity (Teegarden *et al.*, 1973; Bertsch *et al.*, 1973; Braddy *et al.*, 1973; Price *et al.*, 1973).

(2) At a given energy, Q monotonically increases with atomic number, Z. This appears to be true for all the elements we have studied: He, C, O, Si, Fe, and $Z \gtrsim 32$. We do not yet know whether it is true for all elements, independent of

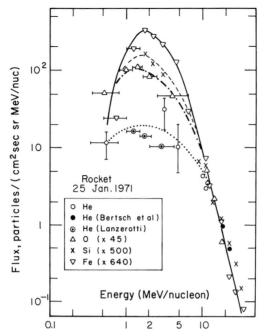

Fig. 1. Energy-dependent composition of heavy ions measured with Lexan carried on a rocket launched from Fort Churchill during an intense solar flare on 25 January 1971. In this and Figs. 2 to 6 the fluxes are scaled by amounts chosen so that the spectra coincide at high energies.

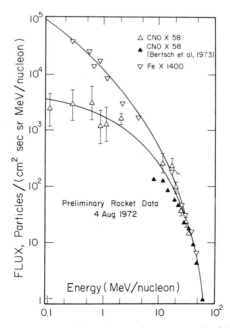

Fig. 2. Energy-dependent composition of heavy ions measured with Lexan carried on a rocket launched from Fort Churchill during the spectacular flare on 4 August 1972.

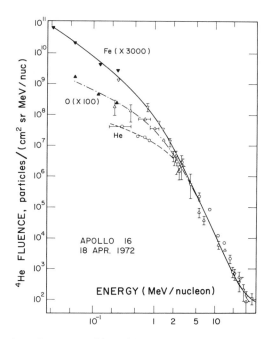

Fig. 3. Energy-dependent composition of heavy ions measured in Lexan stack on the
Apollo 16 Lunar Module and resulting from a weak flare on 16 April 1972.

such factors as ionization potential. There is no indication that it saturates at some
high value of Z. From Fig. 4 we see that Q is about a factor ten higher for the
elements $Z \gtrsim 32$ than for Fe. The value of Q for the elements $Z \gtrsim 50$ is unknown
because we do not yet have a direct calibration of the response of the glass to such
elements.

(3) The critical energy at which Q drops to unity appears to be an increasing
function of the strength of the flare, as we see in Fig. 7. A detailed look at the
spectra in Figs. 1, 2, 3, and 6 suggests that this critical energy is comparable to the
mean energy of the flare particles, which increases from about 1 MeV/nucleon for
the quiet time to ~10 or 20 MeV/nucleon for the 4 August 1972 flare.

(4) Heavy particle enhancements appear to be a permanent feature of solar
activity over at least the last few million years. The depth profiles of big and little
pits in glass 14148,5,6,6, shown in Fig. 10 of Hart *et al.* (1972) suggest that the
abundance ratio $[Z > 30]$/Fe decreases with energy. Results presented by Bhandari *et al.* (1973) at the Fourth Lunar Science Conference provide strong evidence
that the same ratio decreases from ~ five times the solar ratio at ~ 5 MeV/nucleon
to ~ the solar ratio at energies above ~ 20 MeV/nucleon during several million
years in which tracks were recorded in lunar fines.

None of the proposed mechanisms for preferential acceleration of heavy
nuclei (Korchak and Syrovatskii, 1958; Gurevich, 1960; Price *et al.*, 1971;
Cartwright and Mogro-Campero, 1972; Ramadurai, 1973) can account for the four
trends just enumerated. Somehow we must understand how the solar particles

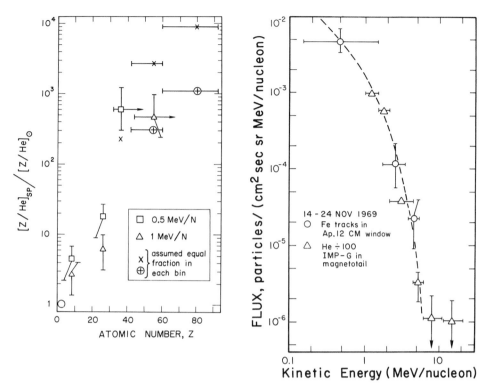

Fig. 4. Enhancement factor as a function of Z for heavy ions observed during the 16 April 1972 flare. The subscripts SP and $\odot$ refer to solar particles and the sun respectively. The data for $Z = 8$ and 26 are from Fig. 3. The data for heavier particles were obtained by measuring tracks in a window of the Apollo 16 command module. Beyond $Z \sim 40$ the identification is uncertain. The symbols $\times$ and $\oplus$ indicate how the enhancement would rise if the events with $Z \gtrsim 32$ were not bunched near the lower end of the charge interval but were distributed equally in the bins $[32 \leqslant Z \leqslant 40]$, $[41 \leqslant Z \leqslant 60]$, and $[Z > 60]$.

Fig. 5. Comparison of Fe flux measured from tracks in Apollo 12 command module window with He flux measured during the same time interval (14–24 November 1969) by the detector of L. J. Lanzerotti on the IMP-G satellite that was within the earth's magnetotail. There was a small interplanetary enhancement during the mission.

that reach the highest energy have approximately normal solar abundances whereas the particles of lower energy are enriched in heavy nuclei by an amount that increases with Z. This situation is not true of the solar wind composition, which, though it fluctuates wildly at times, is not normally enriched in heavy elements. The solar wind is simply an extension of the non-static solar corona, whereas the particles with energies 10 keV/nucleon to $\sim$20 MeV/nucleon that we are discussing probably originate in localized regions of the chromosphere where magnetic energy is suddenly converted to kinetic energy. Even during quiet times there are sunspots and active regions of the sun where energetic particles may originate.

To account qualitatively for the four trends it would seem that we are forced to

2352 P. B. PRICE *et al.*

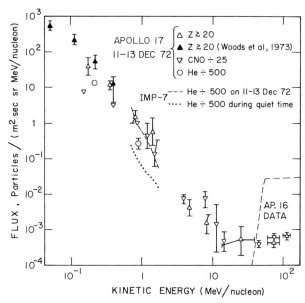

Fig. 6. Energy-dependent composition of heavy ions measured in Lexan stack on the Apollo 17 Lunar Module on 11 to 13 December 1972 when the sun was quiet except for an increase by ~5x in counting rates of protons and alpha particles at ~1 to 2 MeV/nucleon on 12 December.

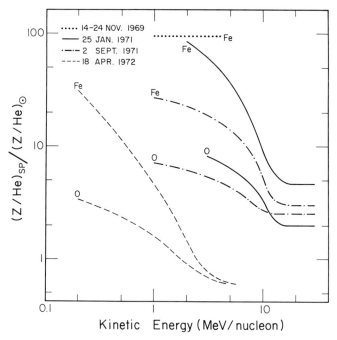

Fig. 7. Enhancement factors for Fe/He and O/He as a function of energy for several solar events. Data for the 2 September 1971 flare are from the paper by Price *et al.* (1973).

Table 1. Measurements of energy-dependent composition of interplanetary heavy ions.

Event	Time Interval Studied	Detector	Reference
Strong flare	1510 to 1514 UT, 25 Jan 1971	Lexan stack on rocket.	Crawford et al. (1972); Price et al. (1973).
Strong flare	0756 to 0760 UT, 2 Sept 1971	Lexan stack on rocket.	Price et al. (1973).
Strong flare	1914 to 1918 UT, 4 Aug 1972	Lexan stack on rocket.	Price et al. (1973).
Weak flare	17–19 April 1972	Lexan + SiO$_2$ glass on Apollo 16 LM; also CM window.	Braddy et al. (1973); Shirk and Price (1973).
Quiet sun + small interplanetary enhancement.	14–24 Nov 1969	SiO$_2$ glass in Apollo 12 CM window.	Chan et al. (1973).
Nearly quiet sun	11–13 Dec 1972	Lexan on Apollo 17 LM.	Price and Chan (1973).

find an acceleration or transport mechanism that does not depend strictly on Z/A. Several possibilities must be examined:

(1) The ratio Z^*/Z of ionic charge to atomic number may be a function of velocity and Z, as has already been suggested, but previous models have not utilized this property correctly. One is certainly not at liberty to use the empirical expression for the effective charge $Z^*(Z, \beta)$ of an ion passing through neutral matter, because of the presence of a plasma containing energetic free electrons that tend to increase the ionization state of ions in a flare region.

(2) Certain processes such as gravitational settling, thermal diffusion, ionic scattering, energy equipartition, ionization loss, and radiation pressure affect ions with the same Z/A but different mass differently. None of these processes has been incorporated into a model of solar particle emission. The first two processes form the ingredients of several calculations of the composition of the transition region, where the temperature is rapidly increasing from $\lesssim 10^{4\circ}$K in the chromosphere to $> 10^{6\circ}$K in the corona (Jokipii, 1966; Delache, 1967; Nakada, 1969; Alloucherie, 1970). All four authors reach the same qualitative conclusion that heavy ions preferentially diffuse upward in the strong temperature gradient in the transition region. Under certain assumptions the Fe/He abundance ratio may increase to a maximum value $\sim 10^2$ times the photospheric ratio. It is not yet possible observationally to rule out a strong heavy element enhancement in a region of the solar atmosphere localized either radially, laterally or temporally. It seems conceivable, then, that the energy-dependent composition of solar flare particles may reflect the actual source composition—the high-energy particles coming from regions of normal solar abundance and the low-energy particles coming from regions enriched in heavy elements.

IMPLICATIONS OF LONG-TERM HEAVY ELEMENT ENHANCEMENTS

The enhancement in the Fe/He ratio by as much as 50 to 100 times at energies below ~ 1 MeV/nucleon (Fig. 7) alleviates the problem (Barber et al., 1971) of accounting for the high track densities found in micron-sized grains at all depths throughout the top several meters of lunar soil. There is thus no longer any necessity to invoke an extra-lunar origin of the small size fraction of the soil as was at one time suggested by Barber et al. (1971). If the trend for heavy particle enhancements to extend to higher energies in stronger flares (Fig. 7) is true over a geologic time scale, then we ought to observe enhancements out to ~ 20 MeV/nucleon in lunar samples, since the occasional strong flares contribute far more tracks than the frequent weak ones, as shown in Fig. 8. There the Fe fluxes during several specific time intervals are compared with the Surveyor average over the years 1967–1970, which includes seven strong flares and many quiet periods. Notice that even on 18 April 1972, during the flare on the Apollo 16 mission, the Fe flux was well below the long-term average.

For heavy ions in silicate minerals an energy of ~ 20 MeV/nucleon corresponds to a range of ~ 200 microns. We might expect to find peculiarities of chemical composition of implanted solar flare ions extending down to this depth,

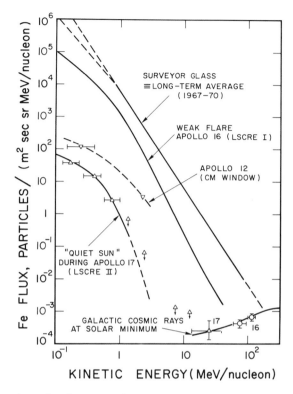

Fig. 8. Comparison of Fe fluxes at various times with the 2.6 year average obtained from Surveyor glass data. The labels "16" and "17" on the galactic cosmic ray data refer to measurements made during Apollo 16 and 17 missions.

provided we could avoid the intense contribution of solar wind ions implanted at very shallow depths (~0.1 micron). The signature of solar flare ions would be a heavy element enhancement that approaches solar composition at depths of ~200 microns. We have already cited the track observations of Hart *et al.* (1972) and Bhandari *et al.* (1973) as evidence for an enhancement in the abundance ratio of $[Z > 30]$/Fe.

We believe that enhanced abundances of heavy rare gases should be detectable by mass spectrometry, provided the outer portions of lunar grains containing solar wind gas can be chemically removed beforehand. In the most careful such experiment to date, Eberhardt *et al.* (1970) removed an amount ranging from 0.14 to 0.35 micron of material from the surfaces of ilmenite grains and found that the total gas concentration was reduced by about an order of magnitude. In examining their data we do not see any evidence for an increase in the abundance ratios $^{20}Ne/^{4}He$ or $^{36}Ar/^{20}Ne$ with amount of material removed, but we think it would be important to continue that approach, increasing the amount removed to ~1 or 2 microns and searching for a definite change in abundance ratios.

Our optimism that solar flare gases can be isolated and studied is based on a simple calculation of the arrival rates of heavy ions in the solar wind and in solar flares: Assuming a He/H ratio of 0.04 in the solar wind and an Fe/He ratio equal to that in the sun ($\sim 3 \times 10^{-4}$), we expect the flux of solar wind iron ions to be $\sim 0.5 \times (2 \times 10^8) \times 0.04 \times (3 \times 10^{-4}) \approx 10^3/\text{cm}^2$ sec., where the factor 0.5 accounts for the fact that the soil faces away from the sun half the time. These ions stop in the outer ~ 0.1 micron of a grain. To estimate the flux of solar flare iron ions we use the Surveyor average flux shown in Fig. 8, integrating from 0.1 to 20 MeV/nucleon. (Our estimate will be conservative since we neglect the contribution at energies less than 0.1 MeV/nucleon.) At energies less than 0.1 MeV/nucleon the flux is not known at all, and between 0.1 and ~ 1 MeV/nucleon it is rather uncertain. Using the upper dashed line we estimate an average flux of $300/\text{cm}^2$ sec. Using the lower dashed line we obtain a flux of at least $60/\text{cm}^2$ sec. Thus, for Fe the solar flares may contribute from 6% to 30% as much as the solar wind. Because of the preference for heavy elements this percentage will be lower for lighter ions and higher for heavier ions. Provided there has been no diffusion of the solar wind gas, leaching of surface material should increase the fractional contribution of solar flare gas and bring the heavy element components within range of detectability.

ENERGY SPECTRA OF SOLAR FLARE FE NUCLEI OVER THE LIFETIME OF THE SOLAR SYSTEM

We are studying solar flare track gradients in a variety of lunar and meteoritic samples that were irradiated at different epochs, in order to look for variations in the distribution of energy in solar flares over the lifetime of the sun. At present, we know of no way of detecting variations in the absolute intensity of solar flare radiation in the distant past. Radioactivities can only monitor exposure times dating back from the present. We can, however, look for systematic changes in track gradients in samples known to have been exposed to flares at different epochs.

Figure 9 summarizes our measurements of Fe track gradients in several kinds of samples. The curve labeled "average rock" is a composite of data compiled by Crozaz et al. (1972), including our own electron microscope observations on 12022. An attempt was made to reject data with appreciably flatter slopes than that shown, on the grounds that trivial processes such as attrition during handling can lead to artificially shallow slopes. There is now a consensus among track groups that track gradients in rocks are flatter than the Surveyor glass profile (also shown in Fig. 9) because of erosional processes. Crozaz et al. (1972) have emphasized that, in a steady state, uniform erosion converts a power law track profile to a new power law profile one power shallower in the exponent. Estimates of average erosion rates based on a comparison of rock profiles with the Surveyor profile range from ~ 1 to ~ 10 Å/year.

We find that lunar fines, grains within gas-rich meteorites and lunar breccias, and crystals at the bottom of vugs at the surfaces of rocks, often have track profiles similar to that in the Surveyor glass and much steeper than the average rock profile. We have not yet attempted the arduous task of finding the steepest

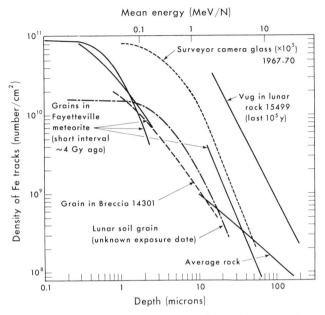

Fig. 9. Track density profiles in grains from meteoritic and lunar samples exposed at various epochs. These profiles are similar to the Surveyor glass profile and considerably steeper than the profiles in lunar rocks.

gradient in each grain selected at random, but have prepared sections through breccias and hand-picked soil grains and have measured only the steepest gradients. Some of these are shown in Fig. 9. We therefore cannot assess whether the shallower gradients we often see were due to such artifacts as sectioning through an off-center portion of a grain or were due to a genuine variation of the energy distribution in solar flares. We can draw two conclusions:

(1) The steepest gradients observed in grains measured *in situ* in the Fayetteville gas-rich chondrite and in lunar breccia 14301, in soil grains from sample 15601, and in a vug at the surface of rock 15499 (Hutcheon *et al.*, 1972) are similar to the gradient measured in the Surveyor glass. At various epochs in the past, extending from the last $\sim 10^5$ years to the time at least 4 G.y. ago when, it is thought by many cosmochemists that gas-rich meteorites such as Fayetteville were assembled, the *maximum* slope in solar flare spectra was similar to that determined from the Surveyor glass to be $\sim E^{-3}$. This does not rule out the existence of time periods in which the flare spectra were sometimes shallower. It does seem to rule out the possibility that the shallow gradients in rocks were caused by shallow solar flare energy spectra.

(2) The kind of erosional process that shapes the rock profiles is missing or insignificant or small in the grains found in soils and breccias and inside the vug in 15499. A tentative explanation is that micrometeorites are mainly responsible for rock erosion. Those soil grains that survive to be studied were the ones that escaped destruction by a direct impact by a micrometeorite. The vug was pointed

out of a plane of the ecliptic and thus was shielded from collisions by micrometeorites.

RADIATION DAMAGE OF SOLAR FLARE ORIGIN IN LUNAR SOILS

Table 2 updates our continuing survey of radiation damage in micron-size grains viewed by high-voltage electron microscopy. Such observations have been of interest to us even since Maurette and co-workers (Borg *et al.*, 1970) and we

Table 2. Radiation damage in micron-size soil grains.

Diffraction pattern	Strong spots	Weak spots; diffuse scattering	No spots; diffuse scattering
Contrast features in micrograph Track density (cm^{-2})	Tracks or blobs up to $\sim 5 \times 10^{11}$	Low contrast 5×10^{11} to few $\times 10^{12}$	Amorphous $>$few$\times 10^{12}$
Luna 16			
7 cm	7%	33%	60%
30 cm	9%	23%	68%
14259 (Apollo 14)	29%	45%	26%
12028 (Apollo 12)	43%	39%	18%
10084 (Apollo 11)	55%	29%	16%
15501 (Apollo 15)	65%	24%	11%
Deep core (Apollo 15)			
15006 (41 cm)	60%	27%	12%
15005 (83 cm)	65%	15%	20%
15004 (124 cm)	64%	12%	24%
15003 (165 cm)	67%	21%	12%
15002 (207 cm)	75%	12%	12%
15001 (246 cm)	75%	16%	9%
Deep core (Apollo 17)			
70006,9 (92 cm)	75%	13%	12%
70005,9 (133 cm)	60%	25%	15%
70004,9 (173 cm)	88%	6%	6%
70003,9 (213 cm)	70%	21%	9%
70002,9 (256 cm)	75%	15%	10%
70001,9 (289 cm)	85%	9%	6%
Deep core (Apollo 16)*			
60001 (bottom) (242±20 cm)	79%	15%	6%
60003 (158±20 cm)	82%	9%	9%
60004 (118±20 cm)	85%	9%	6%
60006 (58 cm)	79%	12%	9%
60007 (top) (22 cm)	76%	15%	9%
Luna 20 ($\sim$10 cm)	97%	3%	0%

*The estimated depths are approximate since a large amount of material in both 60005 and 60007 stems is known to be missing.

(Barber *et al.*, 1971) found that extremely high track densities are a ubiquitous feature of micron-sized soils. We have arranged the observations in order of decreasing radiation damage. Luna 16 and Luna 20 samples form end-members of the sequence.

Irradiation damage in micron-size grains is a useful parameter in characterizing the history of a soil. Before we can relate observations in such tiny grains to the stratum in which they are found, we must establish whether there can be large-scale transport of these grains either vertically or laterally. Such motion would tend to destroy their usefulness. From the data in Table 2 we can say that horizontal transport has not wiped out the striking differences in radiation damage level from one region of the moon to another. For example, the Luna 16 site has been far more heavily irradiated than any other site visited.

The question of vertical transport cannot be conclusively answered on the basis of our data. Consider the fraction of grains with weak diffraction spots (entries in column 3) as a measure of radiation damage. In the Apollo 15 core, the topmost layer 15006 appears to be more heavily irradiated than the rest of the layers studied. In the Apollo 16 core the extent of radiation damage is homogeneous along the core depth, which is most likely due to the fact that the core is known to have been violently disturbed (more than half the material is missing in both 60005 and 60007 stems; Hörz *et al.*, 1972). On the contrary, in the Apollo 17 core the extent of radiation damage in micron sized grains seems to be more stratified. Considering again the fraction of grains whose electron diffraction patterns show weak spots, we find that the layers 70005 and 70003 are moderately irradiated (21–25%) while the layers 70004 and 70001 are lightly irradiated (6–9%). Other observers (Heiken, 1972; Hörz *et al.*, 1972; Nagle, 1973) have reported numerous visual strata. The possibility of electrostatic transport of small grains on top of the lunar surface has been discussed by Gold (1971) and by Criswell (1972). When interpreting the data from Table 2 we must consider carefully whether some mechanism mixes the micron-size grains in the interstices between the bigger grains. In this connection it will be useful to have sections of a deep core impregnated with fixative so that grain-to-grain relationships can be studied. Other workers (Behrmann *et al.*, 1973; Crozaz *et al.*, 1972; Fleischer and Hart, 1973) including ourselves (Phakey *et al.*, 1972) have studied track densities in larger grains ($>100 \mu$) and have found similar trends.

Acknowledgments—We are grateful to Joan Steel and Nancy Peery for their technical support in all phases of this work.

REFERENCES

Aller L. H. (1972) The abundances of the elements. *Invited and Rapporteur Papers, 12th International Conference on Cosmic Rays*, Hobart, Australia, pp. 39–52.
Alloucherie Y. (1970) Diffusion of heavy ions in the solar corona. *J. Geophys. Res.* **75**, 6899–6914.
Barber D. J., Hutcheon I., and Price P. B. (1971) Extralunar dust in Apollo cores? *Science* **171**, 372–374.

Behrmann C., Crozaz G., Drozd R., Hohenberg C., Ralston C., Walker R., and Yuhas D. (1973) Radiation history of the Apollo 16 site (abstract). In *Lunar Science—IV*, pp. 54–56. The Lunar Science Institute, Houston.

Bhandari N., Goswami J., Lal D., and Tamhane A. (1973) Time averaged flux of very very heavy nuclei in solar and galactic cosmic rays. In *Lunar Science—IV*, pp. 69–71. The Lunar Science Institute, Houston.

Borg J., Dran J. C., Durrieu L., Jouret C., and Maurette M. (1970) High voltage electron microscope studies of fossil nuclear particle tracks in extraterrestrial matter. *Earth Planet. Sci. Lett.* **8**, 379–386.

Braddy D., Chan J., and Price P. B. (1973) Charge states and energy-dependent composition of solar-flare particles. *Phys. Rev. Letters* **30**, 669–671.

Cartwright B. G. and Mogro-Campero A. (1972) The preferential acceleration of heavy nuclei in solar flares. *Astrophys. J.* **177**, L43–L47.

Chan J. H., Price P. B., and Shirk E. K. (1973) Charge composition and energy spectrum of suprathermal solar particles. *Proc. 13th International Cosmic Ray Conference*, Denver, paper 364.

Crawford H. J., Price P. B., and Sullivan J. D. (1972) Composition and energy spectra of heavy nuclei with $0.5 < E < 40$ MeV per nucleon in the 1971 January 24 and September 1 solar flares. *Astrophys. J.* **175**, L149–L153.

Criswell D. R. (1972) Lunar dust motion. In *Proc. Third Lunar Sci. Conf., Geochim. Cosmochim. Acta*, Suppl. 3., Vol. 3, pp. 2671–2680. MIT Press.

Crozaz G., Drozd R., Hohenberg C. M., Hoty H. P., Ragan D., Walker R. M., and Yuhas D. (1972) Solar flare and galactic cosmic ray studies of Apollo 14 and 15 samples. In *Proc. Third Lunar Sci. Conf., Geochim. Cosmochim. Acta*, Suppl. 3, Vol. 3, pp. 2917–2931. MIT Press.

Delache P. (1967) Contribution a l'étude de la zone de transition chromosphere-couronne. *Ann. Astrophys.* **30**, 827–860.

Eberhardt P., Geiss J., Graf H., Grögler N., Krähenbühl V., Schwaller H., Schwarzüller J., and Stettler A. (1970) Trapped solar wind noble gases, exposure age and K/Ar-age in Apollo 11 lunar fine material. In *Proc. Apollo 11 Lunar Sci. Conf., Geochim. Cosmochim. Acta*, Suppl. 1, Vol. 2, pp. 1037–1070. Pergamon.

Fleischer R. L. and Hart H. R. Jr. (1973) Enrichment of heavy nuclei in the 17 April 1972 solar flare. *Phys. Rev. Letters* **30**, 31–34.

Fleischer R. L. and Hart H. R. (1973) Surface history of lunar soil and soil columns (abstract). In *Lunar Science—IV*, pp. 251–253. The Lunar Science Institute, Houston.

Gold T. (1971) Evolution of the mare surface. In *Proc. Second Lunar Sci. Conf., Geochim. Cosmochim. Acta*, Suppl. 2, Vol. 3, pp. 2675–2680. MIT Press.

Gurevich A. V. (1960) On the amount of accelerated particles in an ionized gas under various accelerating mechanisms. *Sov. Phys. JETP* **11**, 1150–1157.

Hart H. R., Comstock G. M., and Fleischer R. L. (1972) The particle track record of Fra Mauro. In *Proc. Third Lunar Sci. Conf., Geochim. Cosmochim. Acta*, Suppl. 3, Vol. 3, pp. 2831–2844. MIT Press.

Heiken G. (1972) Unpublished results.

Hörz F., Carrier W. D. III, Young J. W., Duke C. M., Nagle J. S., and Fryxel R. (1972) Apollo 16 special samples. *Apollo 16 Preliminary Science Report*. NASA SP-315, pp. 7-24 to 7-54.

Hutcheon I. D., Braddy D., Phakey P. P., and Price P. B. (1972) Study of solar flares, cosmic dust and lunar erosion with vesicular basalts (abstract). *The Apollo 15 Lunar Samples*, pp. 412–414. The Lunar Science Institute, Houston.

Jokipii J. R. (1966) Effects of diffusion on the composition of the solar corona and the solar wind. In *The Solar Wind*, Chap. 14, Pergamon.

Korchak A. A. and Syrovatskii S. I. (1958) On the possibility of a preferential acceleration of heavy elements in cosmic-ray sources. *Sov. Phys. Doklady* **3**, 983–985.

Lanzerotti L. J., Maclennan C. G., and Graedel T. E. (1972) Enhanced abundances of low-energy heavy elements in solar cosmic rays. *Astrophys. J.* **173**, L39–L43.

Mogro-Campero A. and Simpson J. A. (1972) The abundances of solar accelerated nuclei from carbon to iron. *Astrophys. J.* **177**, L37–L41.

Nagle J. S. (1973) Unpublished results.

Nakada M. P. (1969) A study of the composition of the lower solar corona. *Solar Physics* **7**, 302–320.

Phakey P. P., Hutcheon I. D., Rajan R. S., and Price P. B. (1972) Radiation effects in soils from five lunar missions. In *Proc. Third Lunar Sci. Conf., Geochim. Cosmochim. Acta*, Suppl. 3, Vol. 3, pp. 2905–2915. MIT Press.

Price P. B., Hutcheon I. D., Cowsik R., and Barber D. J. (1971) Enhanced emission of Fe nuclei in solar flares. *Phys. Rev. Letters* **26**, 916–919.

Price P. B. and Chan J. H. (1973) The nature of interplanetary heavy ions with $0.1 < E < 40$ MeV/nucleon. *Apollo 17 Preliminary Science Report*, NASA Special Publication. In press.

Price P. B., Chan J. H., Crawford H. J., and Sullivan J. D. (1973) Systematics of heavy ion enhancements in solar flares. *Proc. 13th International Cosmic Ray Conference*, Denver, paper 365.

Ramadurai S. (1973) The effect of ionisation loss on the energy spectra of cosmic ray nuclei undergoing Fermi acceleration. *Astrophys. Letters*. In press.

Shirk E. K. and Price P. B. (1973) Observation of trans-iron solar flare nuclei in and Apollo 16 command module window. *Proc. 13th International Cosmic Ray Conference*, Denver, paper 366.

Proceedings of the Fourth Lunar Science Conference
(Supplement 4, *Geochimica et Cosmochimica Acta*)
Vol. 3, pp. 2363–2377

Track-exposure and formation ages of some lunar samples

D. STORZER,* G. POUPEAU,* and W. KRÄTSCHMER

Max-Planck-Institut für Kernphysik, Heidelberg, W.-Germany

Abstract—The etchable ranges of 9.6 MeV/nucleon Fe-ion tracks in various detector materials are longer than previously assumed. They depend strongly on the etching conditions. This implies that the available cosmic ray track production rates have a high uncertainty and are correct only by order of magnitude. Using conventional track production rates, apparent track-exposure ages were determined for Apollo 15, 16, and 17 and Luna 16 samples in the age range between 10^1y and 10^7y. Fission track analyses in lunar glass spherules among green ones of Apollo 15, brown ones of Apollo 15, Luna 16, 20 and orange ones of Apollo 17 indicate that these spherules were formed in the early history of the moon and that they are probably related to the excavation of the lunar mare basins.

INTRODUCTION

FOSSIL TRACKS recorded in lunar samples are mainly due to very heavy nuclei of the cosmic radiation, e.g., iron-group ions of the solar flare and the galactic components (*see*, for example Fleischer *et al.*, 1970). Depending on the age of a given sample, nuclear fission of uranium-238 and plutonium-244 may also contribute to the track record (Hutcheon and Price, 1972). Cosmic ray tracks and, in special cases, fission tracks have been studied in several lunar samples from various sites. In order to interpret the fossil track record in terms of exposure and formation ages, a discrimination between cosmic ray tracks and fission tracks is necessary. Furthermore, cosmic ray track production rates must be known. The aim of this work is to reinvestigate these two prerequisites for age determinations on the basis of calibration experiments with artificially produced tracks.

EXPERIMENTAL PROCEDURES

Specimens of the following lunar samples were used for this investigation:

Apollo 15: 15101, 15301, 15421, 15601 soils, 15015, 15205, 15927 glass coated rocks, 15076 crystalline rock, 15426 green clod

Apollo 16: 63321, 63500, 64421, 67601 soils, 67455 breccia

Apollo 17: 74220 orange soil, 75055 basalt, 76055 breccia

Luna 16: A 3.6 (2–4 cm), C 19.118 (20–22 cm) core

Luna 20: (16–27 cm) core

Glasses and minerals were embedded individually in epoxy, polished and etched. Glasses and pyroxenes at 23°C in a mixture of (2 vol. 48% HF + 1 vol. 96% H_2SO_4 + 1 vol. 65% HNO_3 + 6 vol. H_2O) between 5 and 40 sec. and 1 and 5 min., respectively. Feldspars were etched at 135°C in (6 g NaOH + 8 g H_2O) between 1 and 12 min. The tracks were either directly observed with an optical microscope or on replicas with a scanning electron microscope. For uranium analyses the polished sections were covered with Lexan foils and irradiated with thermal neutrons at the Karlsruhe reactor together with

---

*Laboratoire de Minéralogie, Museum National d'Histoire Naturelle, Paris 5ème, France.

reference glasses of known uranium contents from the National Bureau of Standards (NBS—No. 617, 615). Thereafter the samples were repolished and etched under the same etching conditions as before. In the Lexan foils the induced fission tracks were etched at 70°C in 6 N NaOH for 9 min. Depending on the uranium content the samples were irradiated with integrated thermal neutron fluxes between 10^{16} and 10^{17} n/cm^2. Special procedures are described elsewhere (Storzer 1970). The uranium content can be measured within 5%.

For the calibration of cosmic ray tracks in lunar samples, polished sections of glasses and minerals were irradiated with 9.6 MeV/nucleon Fe-ions at the Linac of the University of Manchester. The irradiated sections were then repolished under an inclination angle of about 6° in order to cut progressively all depths of the latent Fe-tracks and to reveal simultaneously tracks along the total etchable range of the Fe-ions.

In order to determine solar flare track density gradients, the samples were embedded in epoxy together with Teflon spherules. The track densities were counted at increasing depths from the surface. After each step of polishing and etching, the respective depth was determined from the spherule calottes with a precision between 2 and 5 μm. The errors of the measured track densities range between 3% and 10%. For the calculation of the uranium and the plutonium-uranium fission track ages the following constants were used:

$$\lambda_{f\,U-238} = 8.42 \times 10^{-17}\,y^{-1}; \qquad \lambda_{d\,U-238} = 1.529 \times 10^{-10}\,y^{-1}$$
$$\lambda_{f\,Pu-244} = 1.058 \times 10^{-11}\,y^{-1}; \qquad \lambda_{d\,Pu-244} = 8.47 \times 10^{-9}\,y^{-1}$$

(Spadavecchia and Hahn, 1967; Fields et al. 1966).

RESULTS AND DISCUSSION

(I) Calibration experiments

Exposure ages of minerals and glasses on the lunar surface can be derived from their cosmic ray track densities if track production rates are known. However, the track production rates are still uncertain as the total etchable ranges of very heavy cosmic ray ions are not yet known. Therefore, the total etchable ranges of 9.6 MeV/nucleon Fe-ions were studied in various detectors (Table 1). Only for pyroxenes (and feldspars etched in acid solution) is this energy sufficient to produce the total etchable ion range within the crystal's interior. For glasses (and feldspars etched in NaOH) this energy is still too low, giving only minimum values for the total etchable ion ranges. In all cases however, the etchable ion ranges are longer than previously assumed (see, for example, Bhandari et al., 1971a, b; Fleischer et al., 1971a).

Under given etching conditions, the length of an etched track depends strongly on the residual range of the Fe-ion. The track lengths have a maximum near the end of the ion trail. This is illustrated in Figs. 1 and 2. The track etching rate and hence the radiation damage are not constant along the latent ion track.

In terrestrial bytownite, the variation of the apparent etchable range of Fe-ions was studied in detail by varying the etching conditions. It turned out that for a given etching solution the apparent range increases linearly with the logarithm of the etching time, probably reaching saturation at the total etchable range. In addition the apparent etchable range depends strongly on the etchant used. When etched with NaOH the apparent etchable ranges of Fe-ions increase with increasing etching times from 56 to >72 μm, but when etched with an acid solution only

Table 1. Etchable ranges of 9.6 MeV/n Fe-ions in different detectors.

Sample	Etchable range (μm)	Etching solvent
15426 "green clod"		
glass spherules	>90	acid
soda-lime glass	>85	acid
Glass with identical composition as the Surveyor-3 camera lens	>75	acid
Bytownite	>72	basic
	26	acid
Enstatite	~26	basic
Orthopyroxene	~25	basic
Augite	~25	acid
	~25	basic
Diopside	~30	basic

Fe-beam

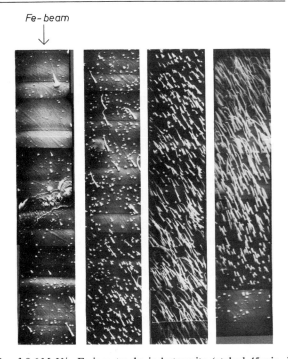

Fig. 1. Profile of 9.6 MeV/n Fe-ions tracks in bytownite (etched 45 min. in NaOH) at increasing residual ranges of the Fe-ions. The length of the individual tracks increase continuously from 0.1 μm at the beginning to about 17 μm near the end of the ion trails. The apparent etchable ion range is about 72 μm. The background of long tracks among the short ones is due to lower ion energies.

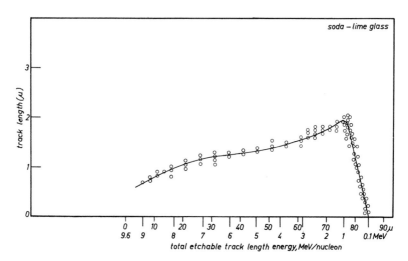

Fig. 2. Distribution of individual track lengths along the apparent etchable range of Fe-ions (9.6 MeV/n initial energy) in soda-lime glass, etched 40 sec. in acid solution.

an increase from 11 to 26 μm was found (Figs. 3 and 4). On the other hand, also the maximum track length (etched with NaOH) increase with the logarithm of the etching time (Figs. 3 and 4). An extrapolation indicates that for etching the maximum track length to about 72 μm, an etching time of about 80 hours is required. Those etching conditions would dissolve the feldspar crystal completely. Consequently, for reasonable etching conditions, the total etchable ion range can never be revealed completely; not even with the track in track method (Bhandari *et al.*, 1971a, b).

The total etchable ion range, however, determines the track density on a polished surface. As a surface cuts statistically all residual ranges of latent fossil tracks, the track density is a linear function of the total etchable ion range. This was proved by means of the galactic cosmic ray track record in feldspars and pyroxenes from subfloor basalt 75055. When counting also tracks <2 μm, the feldspars etched for 12 min. in NaOH respectively 4 sec. in acid solution have mean track densities of 1.5×10^7 respectively 5.5×10^6 tracks/cm^2. The pyroxenes etched for 5 min. in acid solution have a mean track density of 5.6×10^6 tracks/cm^2. This result is in good agreement with the calibrated etchable Fe-ion ranges in these minerals, ~ 25 μm for pyroxene and >72 μm respectively 26 μm for feldspars.

For an application of the fission track dating method to lunar samples it is necessary to distinguish between very heavy cosmic ray tracks and fission tracks. In order to compare the etching behaviour of both kinds of tracks, 9.6 MeV/nucleon Fe-ion tracks and thermal neutron induced fission tracks were etched under identical conditions in 15426 "green clod" glass spherules. Fission fragments produce stronger radiation damages than do the lighter Fe-ions. This results in a higher track etching rate and in a lower sensitivity against environmental parameters (e.g., thermal erasure). This fact provides three possibilities to

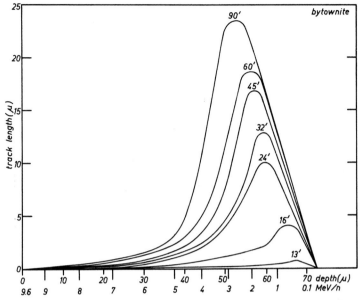

Fig. 3. Distribution of individual track lengths along the apparent etchable range of Fe-ions (9.6 MeV/n initial energy) in bytownite, etched in NaOH for increasing times up to 90 min.

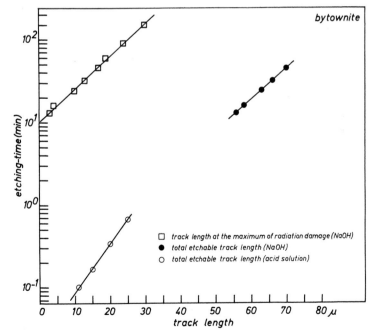

Fig. 4. Increase of the apparent etchable range of 9.6 MeV/n Fe-ions in bytownite as a function of the etching time and etching solution. For NaOH-etchant also the maximum track length is plotted vs. the etching time.

distinguish fission from cosmic ray tracks: (1) underetching, (2) size difference, (3) differential annealing.

(1) When samples containing both, fission and cosmic ray tracks, are etched stepwise, the fission tracks are revealed earlier. For instance, in "green clod" glass spherules fission tracks are visible after 5 sec. whereas the Fe-tracks appear progressively between 5 and 10 sec.

(2) After prolonged etching the diameters of fission tracks remain larger than those of Fe-tracks. In "green clod" glasses, for example, after an etching time of 20 sec. the etch pit diameters of fission tracks are about 30% larger than the largest Fe-track etch pits (Fig. 5).

(3) Fission tracks are markedly more stable against thermal erasure than Fe-tracks. In "green clod" glasses etched for 20 sec. all Fe-tracks are faded completely after 1 hour at 300°C, whereas about 43% of the fission tracks remain (Fig. 6).

From the above results we arrive at the following conclusions:

(a) For a given ion passing through a given detector material, neither a total etchable range nor an absolute registration threshold exists. Both are only apparent quantities and depend strongly on the experimental conditions (e.g., etching, microscopical observation).

(b) Short $<2\ \mu$m tracks in lunar pyroxenes have been attributed to cosmic ray spallation events (Fleischer *et al.*, 1970). Our experiments show that these short tracks may be partly due to tracks developed in the high energy portion of the total etchable range of very heavy cosmic ray ions. Therefore, a contribution of these tracks has to be taken into account when spallation track ages are calculated (Fleischer *et al.*, 1971b).

(c) Bhandari *et al.* (1971a, b) attributed $\sim 25\ \mu$m long tracks in cleavages of lunar and meteoritic pyroxenes to nuclear fission of extinct superheavy elements. Our results show that these $\sim 25\ \mu$m tracks can easily be explained as cosmic iron tracks.

(d) In Surveyor-3 glass the apparent etchable range of unfaded Fe-tracks is $>75\ \mu$m* and is probably not compatible with the $28\ \mu$m-$33\ \mu$m used for the

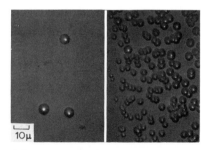

Fig. 5. Induced fission (left) and 9.6 MeV/n Fe-tracks (right) in green glass spherules of clod 15426. Fission track etch pits are about 30% larger than Fe-track etch pits.

---

*Cautious extrapolations indicate that the etchable range might be as long as $100\ \mu$m–$110\ \mu$m. The critical registration angle of 9.6 MeV/nucleon Fe-track etch pits decrease continuously from 49° at the surface to 35° at the maximum of radiation damage at $65\ \mu$m depth and increases then rapidly to 85° at the end of the etchable ion range at $77\ \mu$m depth.

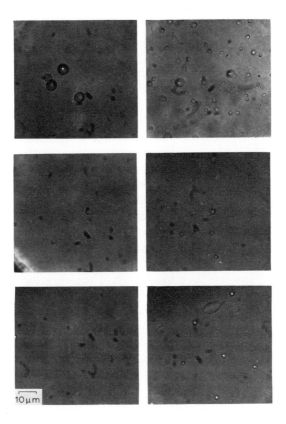

Fig. 6. Induced fission tracks (left column) and fossil fission and cosmic ray tracks (right column) in green glass spherules from soil 15301. Top series unannealed, middle series annealed at 300°C for 10 min, bottom series annealed at 300°C for 1 hour. Under these conditions all cosmic ray tracks are faded and only the fossil fission tracks remain (bottom series right).

calibration of the solar flare flux (Crozaz and Walker, 1971; Fleischer *et al.*, 1971a; Barber *et al.*, 1971). This would imply that the solar flare track production rate has to be lowered up to three times. The amount of this correction cannot be given precisely, as the Fe-track stability in this glass has not been studied so far. Furthermore, the track production rates are strongly dependent on the etching conditions used. Consequently, for the calculation of the solar flare exposure ages we will continue to use the conventional track production rates as given by Barber *et al.* (1971); but these ages are actually minimum ages.

(e) Also galactic cosmic ray track exposure ages should be considered cautiously. Both, the fossil track record in a given sample as well as the calibration of track production rates (by means of known exposure ages) depend on the fading degree of latent tracks, the etching conditions used and the microscopical observation. Moreover, there are indications that the galactic very heavy ion flux has changed during the last 100 m.y. (Plieninger *et al.*, 1973). This would imply that the galactic track production rates are not time-constant. Therefore, the galactic track exposure ages are only apparent and have probably to be recalibrated (e.g., by means of spallation gas ages, if the galactic proton flux was time-constant). Although available track production rates may be correct only within an order of magnitude, for the comparison with the track data of other authors, we have used conventional track production rates (Fleischer *et al.*, 1967) for the calculation of our galactic track exposure ages.

(II) *Solar flare tracks*

The cosmic ray track record of lunar soils and top surface (~ 1 mm) rocks is dominated by low energy solar flare ions. The percentage of solar flare irradiated soil particles is qualitatively correlated with the age of the soil and the local mixing rate. Solar flare tracks are identified either by their strong depth dependance or by their high densities ($\geq 10^8$ tracks/cm^2).

Among green and brown glass spherules, feldspars and pyroxenes, most particles from soils 15101 (Apennine Front), 15301 (Spur Crater), and 15601 (Hadley Rille) were exposed to solar flare irradiation at the very top of the regolith for times ranging typically between 10^2 and 10^6 y. The respective percentages of solar flare track bearing particles in these soils vary from 73 to 100%, 89 to 97%, 87 to 94%, and 91 to 98% (Table 2). Soil 15421 (Spur Crater) shows much lower percentages of solar flare track bearing particles. Only 16% of the green, 43% of the brown glass spherules, and 61% of the feldspars have solar flare tracks, whereas green spherules from a 1 mm agglomerate show no solar flare tracks. This might also be true for green spherules from the "clod" 15426. The about 2% of solar flare track bearing spherules may be due to a contamination from the soil. The larger percentages of solar flare track bearing particles among brown glass spherules and feldspars, however, about 20% respectively, means that these represent a con-

Table 2. Cosmic ray track record and apparent galactic cosmic ray track exposure ages of lunar samples.

Sample	Percentage of solar flare irradiated particles	Density of galactic cosmic ray tracks (10^6 tracks/cm^2)	Apparent exposure ages (m.y.)
Soil 15101,59			
Green glasses	100	—	—
Brown glasses	97	10	—
Feldspars	94	6–10	—
Pyroxenes	98	2	—
Soil 15301,79			
Green glasses	73	0.2–40	~2
Brown glasses	85	10–30	—
Feldspars	90	1–10	—
Pyroxenes	97	3	—

Table 2. (continued).

Sample	Percentage of solar flare irradiated particles	Density of galactic cosmic ray tracks (10^6tracks/cm^2)	Apparent exposure ages (m.y.)
Soil 15421,21			
Green glass from agglomerate	0	0.1–0.7	~2
Green glasses	16	0.1–20	~2
Brown glasses	43	1–40	—
Feldspars	61	1–30	—
Soil 15601,63			
Green glasses	86	1–2	—
Brown glasses	95	5	—
Feldspars	87	1–60	—
Pyroxenes	91	3–7	—
Clod 15426			
Green glasses	1.6	0.1–70	~2
Brown glasses	20	0.3–10	—
Feldspars	20	2–8	—
Soil 63321,12			
Feldspars	70	0.3–65	—
Soil 63500,5			
Feldspars	94	5–80	—
Soil 64421			
Feldspars	100	—	—
Soil 67601,16			
Feldspars	96	15–26	—
Breccia 67455			
Feldspars	~1	0.6–2.1	~30
Soil 74220,68			
Orange glass	~8	1–4	~9–11
Basalt 75055,11,2 Feldspars		10–24	—
Pyroxenes		3–9	
Breccia 76055,4,1 Feldspars		5–9	—
Luna 16 glasses			
A 3.6 (2–14 cm)	100	—	—
C 19.118(20–22 cm)	100	—	—
Luna 20 glasses			
(16–27 cm)	87	4–20	—

tamination with particles that received their solar flare irradiation before the "clod" was compacted. Apollo 15 green spherules show a rough correlation between trapped rare gas contents and solar flare irradiated particles (Fig. 7).

Among Apollo 16 soils studied, soils 63500 (North Ray Crater), 64421 (floor of a ~15 m crater) and 67601 (North Ray Crater) are heavily solar flare irradiated. The percentages of solar flare track bearing feldspars range from 94 to 100% (Table 2). The shadowed soil 63321 (North Ray Crater), on the other hand, contains 70% of solar flare irradiated feldspars and was in fact not perfectly shielded. Both single feldspars and microbreccias from breccia 67455 (North Ray Crater) contain traces of an ancient solar flare irradiation before breccia formation, indicating that this breccia was formed out of regolith. Among Apollo 17 "orange soil" glass spherules 74220 (Shorty Crater) only some 8% were exposed to a solar flare irradiation for times ranging between 10^3 and 10^4 y. Accordingly, the trapped rare gas content is also very low in this soil (Kirsten et al., 1973).

Glass spherules from two respectively one level of the Luna 16 and 20 cores are highly solar flare irradiated and are comparable to the irradiation history of the feldspars from the same levels (Poupeau et al., 1973; Berdot et al., 1972). Whereas all Luna 16 glasses have solar flare tracks, the Luna 20 glasses show a somewhat lower irradiation (Table 2).

In some crystals and in most of the solar flare irradiated glasses studied, the track densities showed a strong depth dependance, in that the densities P decrease as a function of the distance R from the sample's surface with $P = \text{const} \times R^{-\alpha}$. The α-values in lunar soils, except for one brown glass spherule of soil 15301 and a 1 mm glass lined pit of Luna 16 C 19.118, range between 0.8 and 1.4, which is low compared to the α-value of 2.5 found in the Surveyor-3 glass lens (Crozaz and Walker, 1971; Fleischer et al., 1971a; Barber et al., 1971). This discrepancy may be explained in terms of erosion and shielding effects or irradiation geometry of these samples. Therefore, we studied the depth dependence of solar flare tracks in lunar samples where erosion and shielding effects seemed very small on the basis of microcrater statistics (Schneider et al., 1972, 1973). The results are presented in

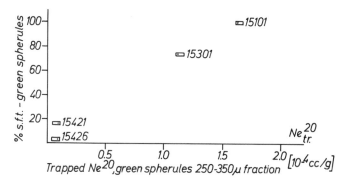

Fig. 7. Correlation between the percentage of solar flare irradiated particles and the amount of trapped Ne^{20} among Apollo 15 green glass spherules. (The Ne^{20} data are taken from Lakatos et al., 1973.)

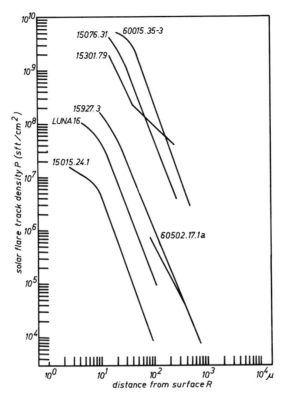

Fig. 8. Depth dependence of solar flare track densities in some lunar samples with minor erosion effects (deduced from crater statistics). The slopes range between -2.5 and -2.8 and are concordant with the track gradient in glass from the Surveyor-3 camera.

Fig. 8. The α-values of these samples range between 2.5 and 2.8, which agrees well with the Surveyor-3 value. From these results a mean erosion rate of $\leqslant 1$ Å/y can be derived for the last m.y. The solar flare track exposure ages range between 10 y and 5×10^5 y (Schneider *et al.*, 1973). As the solar flare track density gradients in various samples exposed at various times are similar to the Surveyor-3 track record it follows that the energy spectrum of solar flare iron group nuclei did not change significantly since 1 m.y. This time could be as long as several 100 m.y., because the Luna 16 glass lined pit was recovered from a depth of 20–22 cm.

(III) Galactic cosmic ray tracks

Contrary to solar flare ions the galactic very heavy cosmic ray component is recorded essentially down to some 20 cm below the lunar surface. Due to mixing in most lunar soils the galactic track density distributions are rather broad (Table 2). Only in special soil samples and near surface rocks residence times can be determined, provided their shielding conditions can be evaluated. Due to their relatively low track retentivity lunar glasses provide a tool for a depth evaluation in soils, as with increasing degree of fading the track size is reduced characteristically (Storzer and Wagner, 1969, 1972). Thermally unaffected galactic tracks are found only at depths below 5 cm which corresponds to a temperature of about 0°C (Creemers *et al.*, 1971).

Galactic tracks were studied in glasses excavated by the Spur Crater event. In green spherules of the agglomerate from 15421, the "clod" 15426, and some spherules of soil 15301, thermally unaffected galactic tracks were found (range of mean track densities: $4.5 - 5.5 \times 10^5$ tracks/cm^2). Assuming a mean burial depth of 5 cm for these spherules a surface residence time of about 2 m.y. is derived for the "green clod." This age is interpreted as the age of the Spur Crater. For "green clod" spherules a somewhat lower radiation age of 0.5 m.y. was found by Fleischer and Hart (1973).

Breccia 67455 was collected from a large boulder on the rim of North Ray Crater. Our sample comes from the deep interior of the recovered 6×10 cm sized rock. Feldspars of this breccia have a mean galactic track density of 1.5×10^6 tracks/cm^2 (only tracks $>2 \ \mu$m were considered). We do not know exactly the original location of our sample within the recovered rock but we can assume that our sample comes from 5 cm depth and was irradiated in a $\sim 1 \ \pi$ geometry (ALGIT, 1972). The corresponding surface residence time is about 30 m.y. (Table 2). Possibly this age is related to the North Ray Crater event.

Among glass spherules from "orange soil" 74220 collected at the rim of Shorty Crater, most spherules contain thermally unaffected galactic tracks. The mean density is 2.3×10^6 tracks/cm^2. At burial depths of 5 cm respectively 7.5 cm (ALGIT, 1973) this track density corresponds to exposure ages of 9 m.y. respectively 11 m.y. This time-span is interpreted as the formation time of Shorty Crater.

Subfloor basalt 75055 was sampled on the valley floor of Taurus Littrow. The mean galactic track density in pyroxenes of this basalt is 5.6×10^6 tracks/cm^2. As we do not yet know the original location of our sample within the recovered rock, a surface residence time cannot be evaluated for this basalt.

Also in case of highland breccia 76055 we do not yet know the original location of our sample to calculate the surface residence time. Feldspars from this breccia have a mean galactic track density of 6.7×10^6 tracks/cm^2 (also tracks $<2 \ \mu$m were counted).

(IV) *Fission tracks*

The fossil fission track record was studied in detail in lunar glasses where fission tracks are easily distinguishable from cosmic ray tracks due to their different etch pit size and their different annealing behavior. The results will be described in more detail elsewhere (Storzer *et al.*, 1973) and only a brief summary will be given here. The mean fission track densities and the uranium contents as well as the corresponding ages are compiled in Table 3. The ages were determined from the ratios of the mean fossil to induced track densities. In case of Luna 16 and 20 spherules also individual ages were determined. They are identical to the respective mean ages within $\sim 10\%$. All glass spherules studied have, however, apparent

Table 3. Fission track ages of lunar glass spherules.

Sample	Mean density of fossil fission tracks/cm^2	Mean Uranium content (ppm) (range)	Uranium fission track age (b.y.)	Plutonium Uranium fiss.tr. age(b.y.)	Fission track excess (%)
200 green spherules 15101, 15421, 15601	5.5×10^4 (a)	0.043 (0.024–0.060)	4.45	3.9	16.4
100 green spherules 15301	6.4×10^4 (a)	0.037 (0.016–0.066)	5.50	4.0	36.1
100 green spherules 15426	8.5×10^4 (a)	0.049 (0.026–0.071)	5.51	4.0	36.3
60 brown spherules 15101, 15301, 15601	5.8×10^5 (a)	0.37 (0.22–0.65)	5.14	3.95	30
100 orange spherules 74220	4.0×10^5 (b)	0.2 (c)	6.06	4.0	44
8 brown spherules Luna 16	8.9×10^5 (a)	0.5 (0.27–0.83)	5.61	4.0	38
7 brown spherules Luna 20	1.0×10^6 (b)	0.7 (0.48–1.10)	4.82	3.95	24

(a) Corrected for an integrated neutron flux of 4×10^{16} n/cm^2, Russ et al., 1972.
(b) Corrected for 10^{16} n/cm^2 (assumed).
(c) U-content from Morrison et al., 1973, and the same track etching efficiency as in green spherules assumed.

fission track ages unrealistically high, indicating a substantial fission track excess. Neither a recent ($<5 \times 10^8$ y) neutron induced fission of uranium-235 nor high energy proton fission can account for the track excess. Also a contribution of tracks from very very heavy cosmic ray ions which are as heavily ionizing as fission fragments can be neglected ($Z \geq 40$/Fe $\sim 3.7 - 7 \times 10^{-5}$) (Price et al., 1971). If we exclude a thermal neutron irradiation of these spherules soon after their formation in the early history of the moon, a contribution of another spontaneously fissioning element must be called upon. Assuming that the track excess is due to plutonium-244 fission, and that the ratio ^{244}Pu/^{238}U was 0.013, 4.58 b.y. ago (Podosek, 1970; Turner, 1969), the corrected fission track ages range between 3.9 and 4.0 b.y.

The observed fission track excess in the lunar glass spherules studied can only be understood if these spherules were formed in the early history of the moon. Whatever the source for this substantial excess might have been, spontaneous fission of plutonium-244 and/or thermal neutron induced fission of uranium-235, in any case the spherules must be older than 3 b.y. As this track excess occurs in spherules from various lunar sites (Apollo 15, 17; Luna 16, 20; preliminary results on Apollo 14, 16) their formation may be related to major impacts probably the excavation of the lunar mare basins.

Acknowledgments—We are grateful to NASA for providing us with lunar samples and to Drs. A. El Goresy, D. Heymann, and P. Pellas for their collaboration in lending us further samples. We wish to acknowledge helpful suggestions from Drs. W. Gentner, D. Heymann, P. Horn, T. Kirsten, P. Pellas, and G. A. Wagner. The assistance of A. Haidmann is appreciated. Also we wish to thank the Linac crew of the University of Manchester for the irradiations. Financial support from the Deutsche Forschungsgemeinschaft and the Deutscher Akademischer Austauschdienst is gratefully acknowledged.

REFERENCES

Apollo Lunar Geology Investigation Team (ALGIT), U.S. Geological Survey (1972) Documentation and environmennt of the Apollo 16 samples. Interagency report. *Astrogeology* **51**.

Apollo Lunar Geology Investigation Team (ALGIT), U.S. Geological Survey (1973) Documentation and environment of the Apollo 17 samples: Interagency report. *Astrogeology* **71**.

Barber D. J., Cowsik R., Hutcheon I. D., Price P. B., and Rajan R. S. (1971) Solar flares, the lunar surface and gas-rich meteorites. *Proc. Second Lunar Sci. Conf., Geochim. Cosmochim. Acta,* Suppl. 2, Vol. 3, pp. 2705–2714. MIT Press.

Berdot J. L., Chetrit G. C., Lorin J. C., Pellas P., and Poupeau G. (1972) Irradiation studies of lunar soils: 15100, Luna 20 and compacted soil from breccia 14307 (abstract). In *The Apollo 15 Lunar Samples,* pp. 333–335. The Lunar Science Institute, Houston.

Bhandari N., Bhat S., Rajagopalan G., Tamhane A. S., and Venkatavaradan V. S. (1971a) Superheavy elements in extraterrestrial samples. *Nature* **230**, 219–224.

Bhandari N., Bhat S., Lal D., Rajagopalan G., Tamhane A. S., and Venkatavaradan V. S. (1971b) Spontaneous fission record of uranium and extinct transuranic elements in Apollo samples. *Proc. Second Lunar Sci. Conf., Geochim. Cosmochim. Acta,* Suppl. 2, Vol. 3, pp. 2599–2609. MIT Press.

Creemers C. J., Birkebak R. C., and White J. E. (1971) Lunar surface temperatures from Apollo 12. *The Moon* **3**, 346–351.

Crozaz G. and Walker R. M. (1971) Solar particle tracks in glass from Surveyor 3 spacecraft. *Science* **171**, 1237–1239.

Fields P. R., Friedmann A. M., Milsted J., Lerner J., Stevens C. M., Metta D., and Sabine W. K. (1966) Decay properties of plutonium-244, and comments on its existence in nature. *Nature* **212**, 131–134.

Fleischer R. L., Price P. B., Walker R. M., and Maurette M. (1967) Origins of fossil charged-particle tracks in meteorites. *J. Geophys. Res.* **72**, 331–353.

Fleischer R. L., Haines E. L., Hart H. R., Woods R. T., and Comstock G. M. (1970) The particle track record of the Sea of Tranquillity. *Proc. Apollo 11 Lunar Sci. Conf., Geochim. Cosmochim. Acta,* Suppl. 1, Vol. 3, pp. 2103–2120. Pergamon.

Fleischer R. L., Hart H. R., and Comstock G. M. (1971a) Very heavy solar cosmic rays: Energy spectrum and implications for lunar erosion. *Science* **171**, 1240–1242.

Fleischer R. L., Hart H. R., Comstock G. M., and Evwaraye A. O. (1971b) The particle track record of the Ocean of Storms. *Proc. Second Lunar Sci. Conf., Geochim. Cosmochim. Acta,* Suppl. 2, Vol. 3, pp. 2559–2568. MIT Press.

Fleischer R. L. and Hart H. R. (1973) Particle track record of Apollo 15 green soil and rock. *Earth Planet. Sci. Lett.* **18**, 357–364.

Hutcheon I. D. and Price P. B. (1972) Plutonium-244 fission tracks: Evidence in a lunar rock 3.95 billion years old. *Science* **176**, 909–911.

Kirsten T., Horn P., Heymann D., Hübner W., and Storzer D. (1973) Apollo 17 crystalline rocks and soils: rare gases, ion tracks, and ages. *EOS, Trans. Amer. Geophys. Union* **54**, 595–596.

Lakatos S., Heymann D., and Yaniv A. (1973) Green spherules from Apollo 15: Inferences about their origin from inert gas measurements. *The Moon* **7**, 132–148.

Morrison G. H., Nadkarni R. A., Jaworski J., Botto R. B., Roth J. R., and Turekian K. K. (1973) Report at the Apollo 17 session of the Fourth Lunar Sci. Conf.

Plieninger T., Krätschmer W., and Gentner W. (1973) Indications for long-time variations in the galactic cosmic ray composition derived from track studies on lunar samples. *Proc. Fourth Lunar Sci. Conf. Geochim. Cosmochim. Acta.*

Podosek F. A. (1970) The abundance of ^{244}Pu in the early solar system. *Earth Planet. Sci. Lett.* **8**, 183–187.

Poupeau G., Berdot J. L., Chetrit G. C., and Pellas P. (1973) Etude par la méthode des traces nucleaires du sol de la mer de fécondité (Luna 16). *Geochim. Cosmochim. Acta.* In press.

Price P. B., Rajan R. S., and Shirk E. K. (1971) Ultra-heavy cosmic rays in the moon. *Proc. Second Lunar Sci. Conf., Geochim. Cosmochim. Acta,* Suppl 2, Vol. 3, pp. 2621–2627. MIT Press.

Russ III G. P., Burnett D. S., and Wasserburg G. J. (1972) Lunar neutron stratigraphy. *Earth Planet. Sci. Lett.* **15**, 172–186

Schneider E., Storzer D., and Fechtig H. (1972) Exposure ages of Apollo 15 samples by means of microcrater statistics and solar flare particle tracks (abstract). In *The Apollo 15 Lunar Samples,* pp. 415–419. The Lunar Science Institute, Houston.

Schneider E., Storzer D., Mehl A., Hartung J. B., Fechtig H., and Gentner W. (1973) Microcraters on Apollo 15 and 16 samples and corresponding cosmic dust fluxes. *Proc. Fourth Lunar Sci. Conf., Geochim. Cosmochim. Acta.*

Spadavecchia A. and Hahn B. (1967) Die Rotationskammer und einige Anwendungen. *Helv. Phys. Acta* **40**, 1063–1079.

Storzer D. (1970) Spaltspuren des 238-Urans und ihre Bedeutung für die geologische Geschichte natürlicher Gläser. Doctor's Thesis, Heidelberg.

Storzer D. and Wagner G. A. (1969) Correction of thermally lowered fission track ages of tektites. *Earth Planet. Sci. Lett.* **5**, 463–468.

Storzer D. and Wagner G. A., (1972) Track analyses and uranium contents of Apollo 14 glasses (extended abstract). *Trans. Amer. Nucl. Soc.* **15**, 119–120.

Storzer D., Müller H. W. and Poupeau G. (1973) The fission track record of lunar glass spherules. In preparation.

Turner G. (1969) Thermal histories of meteorites by the ^{39}Ar-^{40}Ar method. In *Meteorite Research* (editor P. M. Millman), pp. 407–417. Reidel Publ. Co.

Proceedings of the Fourth Lunar Science Conference
(Supplement 4, *Geochimica et Cosmochimica Acta*)
Vol. 3, pp. 2379–2389

Cosmic ray track production rates in lunar materials

R. WALKER and D. YUHAS

Washington University Laboratory for Space Physics, St. Louis, Mo. 63130

Abstract—A particularly favorable rock from the Apollo 16 mission (68815) has been used to derive an "empirical track production energy spectrum" suitable for calculating track exposure ages from galactic cosmic rays in lunar materials. The spectrum is a long term average for the last ~2m.y. and is independent of assumptions about the detailed nature of track registration in lunar minerals. The spectrum is close to that previously estimated from measurements of contemporary cosmic rays and does not support the recent suggestion that track exposure ages should be revised upwards.

INTRODUCTION

FOSSIL TRACK data have been used to measure the exposure ages of various types of lunar samples. Conversion of the track data to ages requires that the rates of track production by VH nuclei ($20 < Z \leqslant 28$) be known as a function of depth. At shallow depths ($< 1.5 \text{ g/cm}^2$) the track production is dominated by tracks produced in solar flares; in this region the analyses of a glass filter from Surveyor 3 can be used to give empirical track production rates (Crozaz and Walker, 1971; Fleischer *et al.*, 1971a; Price *et al.*, 1971). At deeper depths, galactic cosmic rays dominate and it has been the customary procedure to assume a long-term energy spectrum for cosmic rays based on balloon and satellite measurements of the contemporary VH flux. The general agreement with data from the St. Séverin meteorite has been taken as the basic justification of this approach (Cantelaube *et al.*, 1967; Lal *et al.*, 1968; Maurette *et al.*, 1968). However, this agreement was obtained by using the unknown ablation distance as an adjustable parameter. St. Séverin also undoubtedly spent much of its exposure life much further from the sun than 1 A.U.

Based on an analysis of the track length spectrum, Comstock (1971) has suggested that previously calculated track production rates should be lowered by a factor of ~2, with a consequent doubling of previously quoted particle track exposure ages.

The spectrum heretofore used predicts a large gradient of the galactic cosmic ray track density in rocks whose dimensions are large compared to 1 cm. However, when careful measurements have been made, most lunar rocks show track profiles that are either flat or that contain sufficient scatter to accommodate a flat profile.* These flat profiles can be interpreted either as reflecting a complex irradi-

---

*Some rocks have been found with finite track gradients but for one reason or another are not suitable for deriving empirical track production rates as is done in this paper. Rock 10057 (Crozaz *et al.*, 1970), for example, has a long exposure age and is probably affected by erosion. Its orientation on the lunar surface is also not known.

ation history or as arising from an inadequacy in the assumed cosmic ray spectrum (Crozaz *et al.*, 1972).

In the present paper we describe our work on a particularly favorable sample obtained from the Apollo 16 mission. This sample is a well-documented chip from a large, angular boulder found at Station 8. It has a very young exposure age of 2.0 ± 0.2 m.y. as measured by rare gas techniques (Behrmann *et al.*, 1973). A complex irradiation history is unlikely for this sample and we use it to derive an empirical track production energy spectrum in lunar materials.

This spectrum is derived from a long term average and is thus independent of assumptions about the ratio of the present solar cycle data to long term values. It is further independent of assumptions about the relative abundances of the nuclear species that produce the tracks. Exposure ages calculated from this curve are valid to the extent that cosmic rays are constant in time, and to the extent that different lunar feldspars record tracks in the same way as those in rock 68815.

The derived track production rates are close to those previously used by us and others and do not support the revision suggested by Comstock (1971).

DESCRIPTION OF SAMPLE 68815

Rock 68815 is a chip (1.8 kg) from a large boulder at Station 8. The surface documentation of the chip was excellent and shows that the flat, fracture surface was oriented at $45° \pm 10°$ to the vertical on the lunar surface. Measurements on a plaster cast and photographs provided by the curator allow us to specify the location and orientation of our samples with respect to the rock and determine the overall geometry of irradiation. Measurements of the radius of curvature around the surface of the chip showed that the rock could be well approximated by two conjoining spheres with radii of 5.7 ± 1 cm and 13.7 ± 1 cm. The local surface relief on the rock surfaces as measured on the plaster cast was approximately ± 1 mm.

Three samples were received, 68815,74; 68815,109; and 68815,113. The first was a surface chip containing visible impact pits. This chip extended to a total depth of 5 mm. The second was a chip from a depth of 2.8 cm ± 0.3 mm, and the third was at a depth of 5.5 cm ± 0.3 mm.

The boulder from which 68815 was removed is highly angular and does not possess a well developed fillet. Rare gas measurements described in an accompanying paper (Behrmann *et al.*, 1973) give a spallation age of 2.0 ± 0.2 m.y. At this age we would expect the boulder to have suffered relatively few large scale motions such as extensive chipping or rolling, although a constant fine scale erosion may be present. Thus, a priori, we expect that, contrary to most lunar rocks thus far studied, 68815 should have had a relatively simple radiation history.

Preliminary measurement of the track profiles showed a rather steep gradient. Erosion, chipping, or rolling can flatten a track profile but not steepen it. The observed gradient thus gives qualitative confirmation of a rather simple one-stage exposure history.

The lack of a solar gas component and the monotonic decrease of track density with increasing depth indicate that, although 68815 is a breccia, the individual crystals retain no pre-irradiation history such as is commonly seen in soil breccias.

MEASUREMENT OF TRACK PROFILE IN 68815

Slices from the individual chips were mounted in epoxy, polished, and then etched in our standard etching solution (3 g NaOH, 4 g H_2O) for 10–12 min. to reveal tracks in the feldspar component of the rock. The orientation of the polished sections was at $90°$ to the surface of the rock. Track densities were measured in an optical microscope in transmitted light. Only tracks $\geq 2 \mu m$ were counted.

The crystals were not the best that we have worked with, but enough good crystals were found at any one location to give reliable counting results. The data for the rock are shown in Fig. 1.

Track length measurements gave a maximum track length of 9 μm and an average track length of all tracks counted of 4.9 ± 0.7 microns.

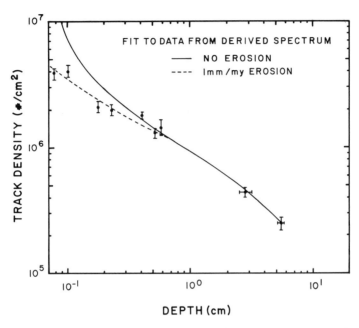

Fig. 1. Fit to the track data on rock 68815. The solid curve assumes no erosion. The dashed curve assumes a 1 mm/m.y. erosion rate. Both fit equally well below 5 mm.

DERIVATION OF THE EMPIRICAL SPECTRUM

The production rate of tracks as a function of depth D below the surface is given by the following expression:*

$$\frac{d\rho(D)}{dT} = \int \frac{dN}{dE}\left(\frac{dE}{dR}\right)\bigg|_{R_0} (1 + F\psi R_0)\overline{\Delta R_C} \exp(-\psi R_0)\left|\hat{R}_0 \cdot \hat{a}\right| d\Omega$$

$\dfrac{d\sigma(D)}{dT}$ is the number of tracks per unit area per unit time

$\dfrac{dN}{dE}$ is the differential energy spectrum

$\dfrac{dE}{dR}$ is the slope of the energy–range curve

ψ is one over the mean interaction path

---

*For a more detailed discussion of this equation, see Fleischer *et al.* (1967).

F is the weighted fragmentation parameter which gives the fraction of total interactions that yield a product nucleus still capable of registering tracks

$\overline{\Delta R_C} = \Sigma \ A_i \Delta R_C$ where the sum is over the VH group, A_i being the abundance of the ith ion relative to the VH group and ΔR_{C_i} the maximum etchable track length of the ith ion

$R_0 = R_0(D, \theta, \phi)$ is the distance from the point of observation to the rock surface. The explicit form of R_0 is of course determined by rock geometry

$|\hat{R}_0 \cdot d\hat{a}|$ gives the component of area normal to the flux where $d\hat{a}$ is a unit vector normal to the plane of observation

For a given depth D the integration is carried out over 2π solid angle for rocks sitting on the lunar surface.

There are several fundamental problems that make it difficult to convert the track data in 68815 directly to a true time averaged energy spectrum for a specified nuclear species.

The most important of these is the value to be assigned to $\overline{\Delta R_C}$. As written, the equation assumes that a given species has a fixed, maximum etchable track length. However, recent work (Price et al., 1973) has shown that this concept is not strictly correct. Increased etching can lead to increased track lengths. Also, particles of a given energy give a spread of track lengths when incident isotropically on a crystalline material. The equation is strictly valid only for a specified set of etching conditions and a suitably defined average $\overline{\Delta R_C}$.

Much more important, it has recently become clear (Plieninger and Krätschmer, 1972; Burnett et al., 1972; Price et al., 1973) that fresh tracks produced by slowing down iron nuclei (these are expected to be the dominant species capable of producing tracks in lunar materials) are longer than those found in lunar samples; the simplest explanation of this effect is that the temperature of lunar rocks is sufficiently high to partially anneal the tracks.

However, the problem of the indentification of the nuclear species and the correct assignment of $\overline{\Delta R_C}$ values can be avoided if $\overline{\Delta R_C}$ is treated as an experimental parameter. It is then possible to derive an "empirical track production energy spectrum." This will be valid for all lunar materials provided the tracks are basically produced by the same mechanism in different samples. We return to this point below.

The samples of 68815 that were given to us did not cover the depth range as extensively as we would have preferred. In particular, there is a gap in the data between 5 mm and 2 cm. In order to derive a spectrum starting at 5 mm it is necessary to make certain assumptions about the contribution of solar flare particles. Erosion of the rock will also play a role if it attains the level of $\sim 1 \text{ mm}/10^6$ yr previously inferred by us (Crozaz et al., 1970; Crozaz et al., 1971; Crozaz et al., 1972).

To estimate the solar contribution we have taken the average of the spectra determined by Fleischer et al. (1971a) and Crozaz and Walker (1971). (The individual spectra differ by 30%.) We have further assumed that the Surveyor period was 1.5 times as active as the long term average. This is based on the measured ratios of Al^{26} [$T_{1/2} = 7 \times 10^5$ yr] and Na^{22} [$T_{1/2} = 2.6$ yr] by Finkel *et al.* (1971) on Apollo 12 samples. The Na^{22} has been corrected for a contribution from a 1956 flare. No account has been taken of the possible effects of erosion on the Al^{26} value; this could raise the estimate of the long term average flux by perhaps as much as 20%. The final value adopted for the long term average solar flare contribution is $3.1 \times 10^6 E^{-3}$ particles cm^{-2} yr^{-1} $ster^{-1}$ $(MeV/nuc)^{-1}$. An assumed erosion rate of 1 mm/m.y. was taken; this has only a small effect on the calculated values at depths ≥ 5 mm.

The interaction probability ψ was taken as 0.054 $(gm/cm^2)^{-1}$. The fragmentation parameter F was chosen as 0.25.

The resulting effective track production spectrum is shown in Fig. 2. We give the results in this form to enable others to calculate the track production rates for different geometry samples. The fit between theory and experiment for rock 68815 using this spectrum, is shown in Fig. 1. Using the above spectrum and interaction parameters, we calculate the track production rate vs. depth for an infinite slab with no erosion. This is shown in Fig. 3. The depth is given in gm/cm^2 and the chemical composition of the shielding material is that of rock 12063 (Wänke *et al.*, 1971). As opposed to nuclear reaction rates, track production rates do not depend critically on the assumed chemistry.

DISCUSSION

We now consider the degree to which the curves in Figs. 2 and 3 can be used to reliably estimate exposure ages. Above ~ 200 MeV/nuc, corresponding to depths

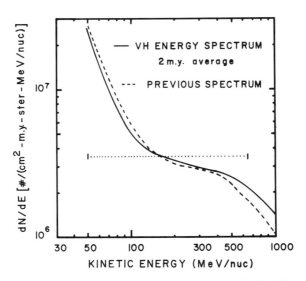

Fig. 2. Solid curve is the effective track production spectrum derived from rock 68815 using the 2 m.y. Kr–Kr age. The dashed curve shows the spectrum previously used. The dotted bar represents the energy interval defined by measured track densities.

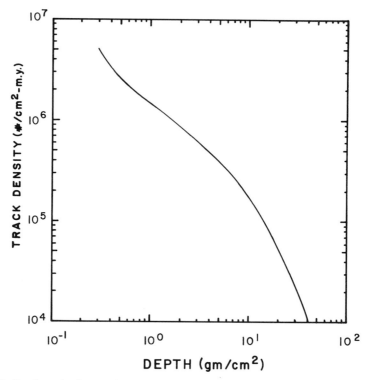

Fig. 3. Track production as a function of depth for a semi-infinite plane using the derived
energy spectrum and a $\overline{\Delta R_c}$ of 10 microns. This calculation was done for plane of
observation perpendicular to the rock surface.

$\geqslant 3\,\mathrm{gm/cm^2}$, the solar contribution to the 68815 data becomes negligible ($< 10\%$).
The effects of our assumed erosion rate of $1\,\mathrm{mm/10^6}$ yr also becomes negligible
($< 3\%$) at this depth. At deep depths therefore the track production values are
essentially determined by our track counts on rock 68815 and the measured Kr–Kr
age of this rock.

To the extent that all lunar materials register and store tracks to the same
degree as the feldspar crystals in 68815, the basic accuracy of the method at
depths $> 3\,\mathrm{gm/cm^2}$ should be the combined error in the Kr–Kr age, the track
counting statistics, the measurement of ΔR_c and the uncertainties in the geometry
of 68815. These contribute a combined error with a one sigma deviation of $\pm 25\%$.

The empirical track production energy spectrum is strictly valid only over the
energies corresponding to the depths measured in 68815—$\sim 50\,\mathrm{MeV/nuc}$ to $\sim$
$650\,\mathrm{MeV/nuc}$. This is the energy range of interest for most lunar samples. To
estimate rates at deeper depths shown in Fig. 3 we have assumed an energy spectrum of the form

$$\frac{\mathrm{d}N}{\mathrm{d}E} = \frac{6.43 \times 10^6}{(1+E)^{2.2}} \sim 10 \qquad E > 0.65\,\mathrm{GeV/nuc}.$$

This agrees within 30% to the best fit to the St. Séverin meteorite data which extend to 34 gm/cm^2. At larger depths, calculated track production rates are dominated by the nuclear interaction terms and the major uncertainty arises from the fragmentation parameter, which is poorly known.

How realistic is it to assume that 68815 is representative of all lunar materials? We have already mentioned that tracks in lunar material are shortened either by annealing or mechanical effects, as demonstrated by Fleischer and Hart (1973a). Samples that have had different thermal and shock histories might therefore be expected to show different track length distributions. This partial annealing can be taken into account, in a first approximation, by actually measuring the total track length distribution in any sample and substituting the experimental value $(\overline{\Delta R_C})_{exp}$ for $\overline{\Delta R_C}$ in Eq. 1.

In Fig. 3 we have given the track production rate curve normalized to a $\overline{\Delta R_C} = 10 \, \mu$m. To get the appropriate value for a particular sample this curve should be multiplied by $(\overline{\Delta R_C})_{exp}/10 \, \mu$m. The estimated track production rates should also be corrected for scanning efficiencies. Thus in the case of a measurement on a polished surface, where tracks $<2 \, \mu$m are excluded, the track production rates should be multiplied by $[(\overline{\Delta R_C})_{exp} - 2 \, \mu\text{m}]/[\overline{\Delta R_C}]_{exp}$.

Measurements of total track length distributions by the TINT method (Price *et al.*, 1973) in different lunar and meteoritic minerals give an average value of 11.0 μm with values for different samples varying from 9.3 μm to 12.6 μm. The average value is thus close to our measured value of 9 μm for rock 68815. The observed constancy of $(\overline{\Delta R_C})_{exp}$ also means that Fig. 3 can be used directly (up to 15 g/cm^2) without introducing more than an additional 20% error.

The fact that the lengths of fossil tracks in lunar and meteoritic samples are shorter than those of fresh tracks is potentially disturbing. Uniform $(\overline{\Delta R_C})_{exp}$ values would be obtained if the tracks were in thermal equilibrium, with as many fading away as being introduced. If this were the case, track exposure ages would be lower than the true values. However, since 68815 has a typical average track length and a short exposure age, this explanation would predict that apparent track ages much beyond 2.0 m.y. could not be observed.

This is not the case. Several lunar rocks have track exposure ages of greater than 25 m.y. (Crozaz *et al.*, 1970; Fleischer *et al.*, 1970; Behrmann *et al.*, 1972). Furthermore, in a companion paper (Behrmann *et al.*, 1973) we show that a track age of 29 m.y. is found on a rock from North Ray Crater. The difference between the Kr–Kr age of 50 m.y. is best explained by assuming rock erosion; this explanation gives a much better fit to the depth profile than assuming that the difference is due to track fading. There is thus no evidence of any serious loss of tracks up to 50 m.y.

We attribute the fact that samples with different ages, and hence different thermal histories, have approximately the same track lengths to the special nature of track annealing in crystals. Numerous studies beginning with Maurette *et al.* (1964) have shown that the early stages of track annealing are characterized by low activation energies. As the annealing progresses, however, the activation energies increase considerably; the last vestiges of tracks are very difficult to remove. We might thus expect a fairly rapid initial shortening that would result in a stabilized track that would be difficult to remove.

Both Borg *et al.* (1973) and Price *et al.* (1973) have noted a further effect—fossil tracks of the same length are more difficult to remove than partially annealed fresh tracks of the same length. Apparently some long term track stabilization process occurs in nature.

The region below 3 gm/cm^2 is increasingly influenced by solar flare particles. This is also a region where erosional effects and the details of the surface topography play increasing roles. Using experimental values from 10 μm to 1 mm in depth, "reasonable" values of erosion rates ranging from 2 Å/yr to 7 Å/yr have been obtained by several groups (Barber *et al.*, 1971; Crozaz *et al.*, 1971; Fleischer *et al.*, 1971a; Fleischer *et al.*, 1971b). These values have been renormalized with the current assumption that the activity of the sun during the Surveyor period was 1.5 times the long term average activity. It should be remembered that this estimate is based on measurements of short-lived and long-lived radioactive isotopes produced by protons. We cannot currently prove that it is valid for the VH nuclei responsible for the tracks. Further track work on relatively fresh rock surfaces with unsaturated solar-flare-produced Al^{26} will be needed to resolve this point. In any event, the present uncertainties in the long term average solar flare values probably limit the absolute accuracy of solar flare derived exposure ages, even in cases where erosion is not important, to no better than a factor of two.

Track production rates at depths less than 10 μm are not known satisfactorily at the present time. Although the Surveyor glass data showed very little increase from 10 μm to the surface, more recent results on a solar flare observed during the Apollo 16 mission show a large increase in track density in the range from 0 to 10μm (Burnett *et al.*, 1972; Price *et al.*, 1972; Fleischer and Hart, 1972; Fleischer and Hart, 1973b; Braddy *et al.*, 1973).

Shown for comparison in Fig. 2 is the cosmic ray spectrum that was previously used by us as well as the General Electric and Berkeley groups to estimate exposure ages. It can be seen that the deviation from the present effective track production rate spectrum is rather small. The recent suggestion by Comstock (1971) that track ages calculated from the previous spectrum should be raised by a factor of two is thus not substantiated. It should be realized however that many exposure ages are model dependent; the same track production rate curves can give different ages depending on the assumptions concerning the conditions of irradiation.

The empirical spectrum does not agree with that proposed by Bhandari *et al.* (1971). Our spectrum lies below theirs by about a factor of 1.6 over the energy range 100 MeV/nuc to 500 MeV/nuc; below 100 MeV/nuc it rises rapidly while theirs remains fairly flat down to 10 MeV/nuc. The region from 0.8 to 3 mm covers the approximate energy range of 50 MeV/nuc to 100 MeV/nuc. Because the spectra cross in this range, the calculated track production rates will probably not differ much. The "sun-tan" ages quoted by Bhandari *et al.* (1971), which are based on track measurements at shallow depths, are thus probably not very different from what we would obtain from the same data. However, their track production rates for depths $\geqslant 5$ mm will be higher by ~ 1.6 than our values. We intend to review the subject of track exposure ages in a separate paper.

It must also be kept in mind that track counting involves subjective factors. The track consortium report on rock 14310 (Yuhas *et al.*, 1972), in which five groups participated, showed that track densities measured in different laboratories generally agreed to $\pm 20\%$. This is probably the best measure of what can be expected in comparing data from different laboratories. Any individual laboratory however should be able to make reproducible measurements on most samples at the $\pm 10\%$ level.

The spectrum that has been used heretofore was based on measurements of contemporary cosmic rays. The basic agreement with the present empirical spectrum carries with it the implication that the iron to proton ratio has not changed appreciably in the last 2×10^6 yr. However, the degree to which limits can be set on this variation is a somewhat complicated issue involving such factors as the proper averaging over the solar cycle modulation and the question of the track etching characteristics of lunar minerals for different very heavy ions. We will discuss these questions in detail in a succeeding publication. The question of constancy is a timely and important one, since recent cosmic ray work has shown differences in the high energy spectra of iron nuclei and protons. This has led to the suggestion (Ramaty *et al.*, 1972) that iron may have a different origin than the protons.

The curves that we have derived for track production rates are applicable only to lunar feldspar crystals etched and measured in the manner described. To ensure that the curves are applicable in other cases, it is necessary to verify that the measured track length distributions are essentially the same as these measured here. Lunar pyroxenes give consistently lower values than feldspars by $\sim 30\%$. The reason for this difference is not known but could either be due to a difference in etching efficiency (Fleischer *et al.*, 1970) or to different annealing characteristics (Crozaz *et al.*, 1970). For minerals other than feldspars, the curves given here should be considered as upper limits on the track production rates and the exposure ages as corresponding lower limits.

Acknowledgment—This work was supported by NASA Grant NGL 26-008-065.

REFERENCES

Barber D. J., Cowsik R., Hutcheon I. D., Price P. B., and Rajan R. S. (1971) Solar flares, the lunar surface, and gas-rich meteorites. *Proc. Second Lunar Sci. Conf., Geochim. Cosmochim Acta*, Suppl. 2, Vol. 3, pp. 2705–2714. MIT Press.

Behrmann C., Crozaz G., Drozd R., Hohenberg C. M., Ralston C., Walker R. M., and Yuhas D. (1972) Rare gas and particle track studies of Apollo 15 samples: Hadley Rille and special soils. In *The Apollo 15 Lunar Samples*, pp. 329–332. The Lunar Science Institute, Houston.

Behrmann C., Crozaz G., Drozd R., Hohenberg C. M., Walker R. M., and Yuhas D. (1973) Cosmic ray exposure history of North Ray and South Ray materials. *Proc. Fourth Lunar Sci. Conf., Geochim. Cosmochim. Acta.*, Suppl. 4, Vol. 3.

Bhandari N., Bhat S., Lal D., Rajagopalan G., Tamhane A. S., and Venkatavaradan V. S. (1971) High resolution time averaged (millions of years) energy spectrum and chemical composition of iron-group cosmic ray nuclei at 1 A.U. based on fossil tracks in Apollo samples. *Proc. Second Lunar Sci. Conf., Geochim. Cosmochim. Acta,* Suppl. 2, Vol. 3, pp. 2611–2619. MIT Press.

Borg J., Dran J. C., Comstock G., Maurette M., Vassent B., Duraud J. P., and Legressus C. (1973) Nuclear particle track studies in the lunar regolith—some new trends and speculations (abstract). In *Lunar Science—IV,* pp. 82–84. The Lunar Science Institute, Houston.

Braddy D., Chan J., and Price P. B. (1973) Charge states and energy-dependent composition of solar-flare particles. *Phys. Rev. Lett.* **30,** 669–671.

Burnett D., Hohenberg C. M., Maurette M., Monnin M., Walker R., and Woolum D. (1972) Cosmic ray, solar wind, solar flare, and neutron albedo measurements. *Apollo 16 Preliminary Science Report,* NASA SP-315, pp. 15-19 to 15-32.

Cantelaube Y., Maurette M., and Pellas P. (1967) Etude des traces d'ions lourds dans les mineraux de la chondrite de Saint Séverin. *Symposium of Radioactive Dating and Methods of Low-Level Counting,* Monaco, I.A.E.C., pp. 215–229.

Comstock G. (1971) The particle track record of the lunar surface. *Proc. IAU Symp.* **47,** *Newcastle.*

Crozaz G., Haack U., Hair M., Maurette M., Walker R., and Woolum D. (1970) Nuclear track studies of ancient solar radiations and dynamic lunar surface processes. *Proc. Apollo 11 Lunar Sci. Conf. Geochim. Cosmochim. Acta,* Suppl. 1, Vol. 3, pp. 2051–2080. Pergamon.

Crozaz G. and Walker R. M. (1971) Solar particle tracks in glass from the Surveyor 3 spacecraft. *Science* **171,** 1237–1239.

Crozaz G., Walker R., and Woolum D. (1971) Nuclear track studies of dynamic surface processes on the moon and the constancy of solar activity. *Proc. Second Lunar Sci. Conf., Geochim. Cosmochim. Acta,* Suppl. 2, Vol. 3, pp. 2543–2558. MIT Press.

Crozaz G., Drozd R., Hohenberg C. M., Hoyt H. P. Jr., Ragan D., Walker R. M., and Yuhas D. (1972) Solar flare and galactic cosmic ray studies of Apollo 14 and 15 samples. *Proc. Third Lunar Sci. Conf., Geochim. Cosmochim. Acta,* Suppl. 3, Vol. 3, pp. 2917–2931. MIT Press.

Finkel R. C., Arnold J. R., Imamura M., Reedy R. C., Fruchter J. S., Loosli H. H., Evans J. C., and Delany A. C. (1971) Depth variation of cosmogenic nuclides in a lunar surface rock and lunar soil. *Proc. Second Lunar Sci. Conf., Geochim. Cosmochim. Acta,* Suppl. 2, Vol. 2, pp. 1773–1789. MIT Press.

Fleischer R. L., Price P. B., Walker R. M., and Maurette M. (1967) Origins of fossil charged particle tracks in meteorites. *J. Geophys. Res.* **72,** 331–353.

Fleischer R. L., Haines E. L., Hart H. R. Jr., Woods R. T., and Comstock G. M. (1970) The particle track record of the Sea of Tranquillity. *Proc. Apollo 11 Lunar Sci. Conf., Geochim. Cosmochim. Acta,* Suppl. 1, Vol. 3, pp. 2103–2120. Pergamon.

Fleischer R. L., Hart H. R. Jr., and Comstock G. M. (1971a) Very heavy solar cosmic rays: Energy spectrum and implications for lunar erosion. *Science* **171,** 1240–1242.

Fleischer R. L., Hart H. R. Jr., Comstock G. M., and Evwaraye A. O. (1971b) Particle track record of the ocean of storms. *Proc. Second Lunar Sci. Conf., Geochim. Cosmochim. Acta,* Suppl. 2, Vol. 3, pp. 2559–2568. MIT Press.

Fleischer R. L. and Hart H. R. Jr. (1972) Composition and energy spectra of solar cosmic ray nuclei. *Apollo 16 Preliminary Science Report,* NASA SP-315, pp. 15-2 to 15-11.

Fleischer R. L. and Hart H. R. Jr. (1973a) Possible explanation of tracks in lunar materials. *Nature* **242,** 104–105.

Fleischer R. L. and Hart H. R. Jr. (1973b) Enrichment of heavy nuclei in the 17 April 1972 solar flare. *Phys. Rev. Lett.* **30,** 31–33.

Lal D., Lorin J. C., Pellas P., Rajan R. S., and Tamhane A. S. (1968) On the energy spectrum of iron-group nuclei as deduced from fossil-track studies in meteoritic minerals. In *Meteorite Research* (editor P. Millman). D. Reidel.

Maurette M., Pellas P., and Walker R. M. (1964) Etude des traces de fission fossiles dans le mica. *Bull. Soc. Franc. Miner. Crist.* **87,** 6–17.

Maurette M., Thro P., Walker R., and Webbink R. (1968) Fossil tracks in meteorites and the chemical abundance and energy spectrum of extremely heavy cosmic rays. In *Meteorite Research* (editor P. Millman), pp. 286–315. D. Reidel.

Plieninger T. and Krätschmer W. (1972) Registration properties of pyroxenes for various heavy ions and consequences to the determination of the composition of the cosmic radiation. *8th Int. Conf. on Nucl. Photog. and Solid State Detectors*, Bucharest.

Price P. B., Hutcheon I., Cowsik R., and Barber D. J. (1971) Enhanced emission of iron nuclei in solar flares. *Phys. Rev. Lett.* **26**, 916.

Price P. B., Braddy D., O'Sullivan D., and Sullivan J. D. (1972) Composition of interplanetary particles at energies from 0.1 to 150 MeV/nucleon. *Apollo 16 Preliminary Science Report*, NASA SP-315, pp. 15-11 to 15-19.

Price P. B., Lal D., Tamhane A. S., and Perelygin V. P. (1973) Characteristics of tracks of ions with $14 \leq Z \leq 36$ in common rock silicates. *Earth Planet. Sci. Lett.* Submitted for publication.

Ramaty R., Balasubrahmanyan V. K., and Ormes J. F. (1972) Cosmic ray sources: Evidence for two acceleration mechanisms. Goddard Space Flight Center, Preprint X-661-72-479.

Wänke H., Wlotzka F., Baddenhausen H., Balacescu A., Spettel B., Teschke F., JaGoutz E., Kruse H., Quijano-Rico M., and Rieder R. (1971) Apollo 12 samples: Chemical composition and its relation to sample locations and exposure ages, the two component origin of the various soil samples and studies on lunar metallic particles. *Proc. Second Lunar Sci. Conf., Geochim. Cosmochim. Acta*, Suppl. 2, Vol. 3, pp. 1187–1208. MIT Press.

Yuhas D. E., Walker R. M., Reeves H., Poupeau G., Pellas P., Lorin J. C., Chetrit G. C., Berdor J. L., Price P. B., Hutcheon I. D., Hart H. R. Jr., Fleischer R. L., Comstock G. M., Lal D., Goswami J. N., and Bhandari N. (1972) Track consortium report on rock 14310. *Proc. Third Lunar Sci. Conf., Geochim. Cosmochim. Acta*, Suppl. 3, Vol. 3, pp. 2941–2947. MIT Press.

Proceedings of the Fourth Lunar Science Conference
(Supplement 4, *Geochimica et Cosmochimica Acta*)
Vol. 3, pp. 2391–2401

Solar wind and terrestrial atmosphere effects on lunar sample surface composition

D. A. CADENHEAD, B. R. JONES*, W. G. BUERGEL, and J. R. STETTER

Department of Chemistry, State University of New York at Buffalo, Buffalo, N.Y. 14214

Abstract—Samples returned from the Apollo missions have been shown to have undergone a partial surface oxidation with the degree of oxidation being dependent on the intensity and duration of exposure to a terrestrial or other oxidizing atmosphere. Exposure to atomic hydrogen at room temperature, or molecular hydrogen above 100°C results in a surface reduction. The adsorption of water vapor on a test sample was found to be only slightly dependent on the state of surface oxidation, a situation consistent with the formation of hydroxyl groups on the surface when a sample is exposed to hydrogen. That hydroxyl groups are indeed formed is substantiated by the release of water vapor (and by release of heavy water following exposure to deuterium) indicating that water vapor can be synthesized from solar wind hydrogen and sample oxygen. Observation of trace amounts of methane indicate that the reduction process is by no means restricted to the formation of water vapor.

INTRODUCTION

WHEN LUNAR samples are returned to a terrestrial atmosphere, there is a constant danger of surface contamination and of possible changes in surface composition. During the Apollo missions such contamination almost certainly occurred for the first time during the approach and landing of the lunar vehicle to be subsequently followed by exposure to vapors emitted by the astronaut and the only partially outgassed plastic and metal sample containers (60 hours at 120°C and 10^{-7} torr). While samples are traditionally shipped in dry nitrogen, this will inevitably involve exposure to trace impurities of more actively adsorbed species. Finally, deliberate lengthy exposure of samples to terrestrial atmospheres occurs in most investigations, being frequently necessitated by the very nature of these studies. When one considers that adsorption of a gas at one atmosphere pressure and having a sticking coefficient of unity, requires an exposure of only one microsecond in order to achieve monolayer coverage (de Boer, 1953) the inevitability of *some* exposure and contamination becomes evident.

It would seem a reasonable postulate that lunar rocks and soil particles, exposed on the lunar surface may be regarded as having relatively clean surfaces. These samples undergo a prolonged high energy bombardment and outgassing at approximately 10^{-12} torr and up to about 150°C for prolonged periods of time, with one important proviso: that among the high energy particles are a large number of protons (Hundhausen *et al.*, 1967; Geiss *et al.*, 1972). These protons which do not undergo transmutation will, after electron capture to form hydrogen atoms, either combine with each other to form molecular hydrogen or combine chemically with

---

*Present Address: Bell Research Corp. Allentown, Pa.

any other available material. Most of this molecular hydrogen presumably is ionized and escapes from the lunar atmosphere as protons to complete the cycle, however, a certain amount suffers entrapment and under more restricted conditions, may also reduce the lunar material. In any reduction process which occurs, however, the hydrogen atom is the primary reducing agent.

One further point should be emphasized in connection with this process and that is that reduction need not necessarily be confined to the immediate outer surface of lunar materials. Thus, Bibring *et al.* (1972) have drawn attention to the fact that solar wind bombardment creates an amorphous region some 500–1000 Å thick. The penetrating power of the constituent high energy particles coupled with the well known ability of hydrogen (atoms) to diffuse in defect structure (Fast, 1972) suggests that reduction will occur in depth. Recent reports confirm such a suggestion (Leich *et al.*, 1973). Moreover, while the precise length of time required for complete reduction will depend on the physical state of an exposed soil or rock particle and its particular situation, penetration times will be short (days, weeks, or months) compared to typical exposure times.

The question then arises as to what the effects will be of exposure to a terrestrial atmosphere. We may reasonably assume that a typical surface will present a wide range energy spectrum of chemisorption sites for an oxidizing or "active" adsorbate (Ross, 1971). However, penetration of oxygen into defect structure may well be a slow process (compared to hydrogen) and we may expect that when brief exposures take place, oxidation may well be confined to the more active external sites. Extensive or observable oxidation (if it occurs at all) might only be expected with samples that have undergone considerable exposure. On this basis, we chose to initiate our investigation with a sample which had undergone considerable exposure including initial exposure to the high oxygen content atmosphere of the command module: 14163,111 (NASA Report, 1971).

EXPERIMENTAL AND RESULTS

Two high vacuum systems were used in this study. A volumetric system previously described (Cadenhead *et al.*, 1972) was used to titrate the surface of several samples with atomic hydrogen. In order to do this, the system was modified with a special sample chamber incorporating a tungsten filament to dissociate the molecular hydrogen. Further details of this system have been provided elsewhere (Cadenhead and Jones, 1973). A gravimetric microbalance system, also previously described (Cadenhead and Wagner, 1968, 1971), was similarly adapted by incorporation of a tungsten filament, care being taken to avoid excessive thermomolecular effects (Fig. 1).

In addition, use was made of an ultra high vacuum system capable of attaining a vacuum of approximately 10^{-10} torr. Essentially, the system consisted of an all stainless steel volumetric adsorption system with an incorporated quadrupole mass spectrometer (E.A.I. Model 1110 A), pressures being measured with both ion and millitorr gauges. Details of the system are illustrated in Fig. 2 and are described in the accompanying legend.

Typical procedures for atomic hydrogen adsorption (Cadenhead and Jones, 1972, 1973) involve sample outgassing at 150°C (selected for removal of physisorbed material and consistency with maximum lunar noon temperature conditions). Current-voltage conditions sufficient to establish a suitable rate of hydrogen atom formation had been previously established. A thermocouple imbedded in the sample monitored the sample temperature to ensure that excessive deviations from room temperature

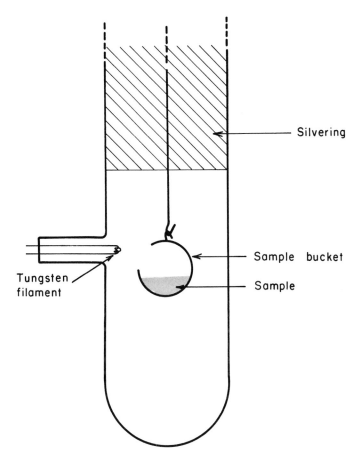

Fig. 1. Arrangement for formation of atomic hydrogen in the microbalance gravimetric adsorption system. System used to obtain data illustrated in Fig. 3.

did not occur when the tungsten filament was switched on and to indicate the precise temperature when adsorption measurements were made. Preliminary reports have been made on two samples (one slightly and one extensively exposed or contaminated sample). The results indicated (Cadenhead and Jones, 1972, 1973) that brief exposure to air (minutes or hours) would result in some oxidation (approximately 150 Å^2/adsorbed hydrogen atom). Extensive exposure (weeks or months) could significantly increase the degree of oxidation, (50 Å^2/adsorbed hydrogen atom). Dissociative chemisorption of molecular hydrogen was found to occur above 100°C, on typical lunar samples, the rate of adsorption increasing rapidly with increasing temperature. This method of reduction was used in experiments involving the ultra-high vacuum system already described.

An evaluation of the effect of surface reduction on the subsequent adsorption of water vapor was made by obtaining water vapor isotherms under otherwise identical conditions on lunar fines 15101,68. The results are shown in Fig. 3. The two isotherms show small differences in the sub-monolayer region but were identical from monolayer coverage to saturation vapor pressure conditions. The surface area and helium density of 15101,68 were determined previously in the

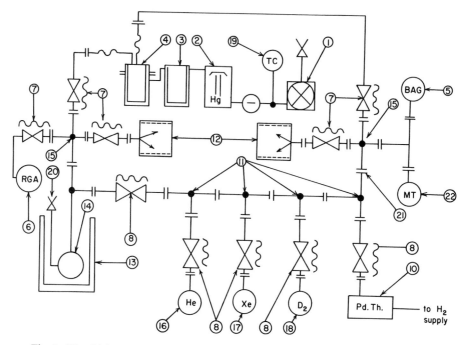

Fig. 2. Ultra-high vacuum system. Numbers 1 through 22 on the figure are described below:

1. Rotary two-stage mechanical pump.
2. Mercury three-stage diffusion pump (General Electric).
3. Foreline trap (glass).
4. Foreline trap (stainless steel).
5. Varian calibrated U.H.V. ion gauge.
6. E.A.I. Quadrupole 1110 A residual gas analyzer.
7. Granville Phillips gold seal valve (1 inch bore).
8. Varian $\frac{3}{4}$ inch mini-valve.
9. Accessory.
10. Hydrogen purifier (Pd–Ag alloy thimble).
11. Varian $\frac{3}{4}$ inch mini-hardware.
12. Varian 8 liter-Sec super vacion pumps.
13. Dewar vessel or Sample furnace.
14. Sample chamber (glass).
15. Varian $1\frac{1}{2}$ inch stainless steel cross.
16. Helium supply (glass bulb).
17. Xenon supply (glass bulb).
18. Deuterium supply (glass bulb).
19. Thermocouple gauge.
20. Varian mini flange/blank flange.
21. Varian conflat flange connections.
22. Varian calibrated millitorr gauge.

All other symbols are those approved by the American Vacuum Society. This system was used to obtain the data illustrated in Figs. 4 and 5.

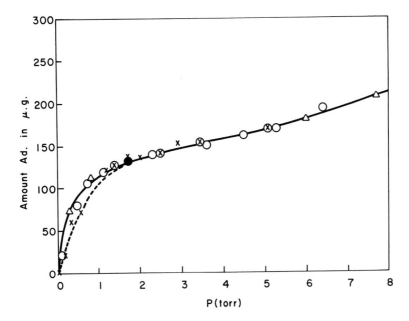

Fig. 3. Adsorption of water vapor on lunar fines 15101,68 ($\leqslant 0.1$ mm) at 15°C.
Before reduction ○ Adsorption △ Desorption
After reduction × Adsorption
A sample weight of 0.4933 g was used in the study. See Fig. 1 for details of the system.

volumetric adsorption system and were 0.65 m²/g and 3.1 g/cc respectively. A sample weight of these fines ($\leqslant 0.1$ mm) of 0.4933 g was introduced into an unsilvered sample bucket in the gravimetric system and counterweighed with a combination of platinum and aluminum weights having the same density and weight. The sample was outgassed at 150°C for 24 hours after which the initial (non-reduced) isotherm in Fig. 3 was determined. The system was then re-evacuated and the sample again outgassed for 24 hours at 150°C to remove all traces of physically adsorbed water. After thoroughly outgassing the tungsten filament used for hydrogen dissociation, the sample was exposed to a succession of pulses of hydrogen (0.024, 0.040, 0.058, and 0.80 torr) during three minute tungsten filament activation periods at approximately 1500°K. Because a small zero shift occurred each time the filament was activated it was difficult to decide when the adsorption process was complete. Previous studies in a volumetric adsorption system, however, indicated that conditions chosen were more than adequate to ensure complete reduction (Cadenhead and Jones, 1972, 1973). Following this procedure, the second water adsorption isotherm was obtained with the results seen in Fig. 3.

Studies carried out in the ultra-high vacuum (U.H.V.) system involved exceptional sample (and system) preparative procedures. The first outgassing of each sample (14163,111 and 15565,3G) was carried out at 150°C for 20 hours. Outgassing of the sample chamber at a higher temperature would have resulted in a

D. A. CADENHEAD *et al.*

loss of information concerning the initial gas content of the sample. Regretably, it was not possible to outgas the bulk of the U.H.V. system (excluding the sample chamber) at a higher temperature because of the possibility of distilling impurities into the sample chamber. However, the U.H.V. system made little contribution to the resultant composition of the gas evolved when the temperature (system plus sample) was raised from 150° to 400°C as indicated on taking the system alone through the same procedure. The gas composition observed for gases evolved from 15565,3G is indicated in Fig. 4.

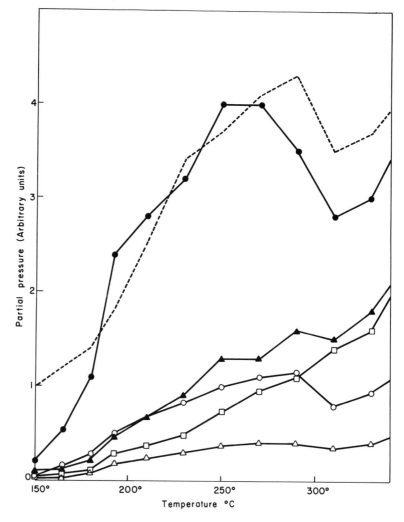

Fig. 4. Initial gas evolution as a function of temperature for 15565-3(G) (After 20 hours outgassing at 150°C) ● H_2O; ▲ N_2, CO; □ CO_2; ○ H_2; △ CH_4; ----- indicates the total pressure (with continuous pumping) ranging from 2×10^{-8} torr (left hand ordinate) to 8×10^{-8} torr (right hand ordinate). The system illustrated in Fig. 2, was used in this study.

Subsequently, the sample and system were subjected to a second and third outgassing (25–400°C). On this final out-gassing only trace water was seen at the highest temperature reached (400°C). The sample was then equilibrated with hydrogen at 150°C and after lowering the temperature to 25°C the sample was outgassed (25–400°C). The primary gases evolved are shown in Fig. 5. After equilibration with deuterium at 150°C, the sample was again outgassed (25–400°C) and the primary composition of the gas evolved noted (Fig. 5). Finally, a Xenon

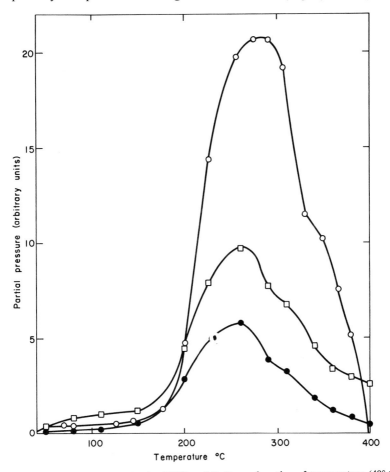

Fig. 5. Partial pressure of H_2O, also HDO and D_2O as a function of temperature (40° to 400°C) after exposure of 15,563-3G to H_2 (initially) or D_2 (subsequently): ○ H_2O; □ HDO; and ● D_2O. The absolute pressure rose from approximately 10^{-8} torr at 100°C to 8×10^{-8} at 300°C. An approximate calibration of the arbitrary scale of the ordinate may be made based on an estimated effective 5 l/sec pumping capacity and a 2.5 l U.H.V. system volume. Prior to H_2 or D_2 exposure, levels for all three vapors were zero over the entire temperature range, except for H_2O where traces were seen at (and above) 400°C. The H_2O curve has been corrected for these amounts so that it shows only the difference before and after H_2 exposure. The temperature was increased approximately 20° at 1 hour intervals.

surface area was measured and was evaluated as 0.07 m^2/g. There are indications, however, that a somewhat higher value would be obtained with nitrogen since slow equilibration with this gas are indicative of molecular sieve effects. A preliminary report on part of this work has already been published (Cadenhead and Buergel, 1973).

DISCUSSION

(a) *Exposure to air*

Without further investigation, it is clear that any sample that has undergone surface reduction (either external or internal surface) will undergo partial reoxidation on exposure to either oxygen or air. Nitrogen exposure had no noticeable effect. Moreover, it is clear that brief exposures of minutes or hours produce smaller effects than prolonged exposures (weeks or months). Energetically uniform, or near uniform, surfaces are possessed by very few substances such as freshly cleaved mica or graphitized carbon blacks (Ross, 1971). Most adsorbents possess highly heterogeneous surfaces and there is no reason to believe that lunar samples will be any exception to the rule, quite the contrary, where pore and defect structure are likely to play a role (Bibring *et al.*, 1972), a wide range of energetic adsorption sites may be predicted.

Another factor is that while "internal" surface sites may well be more desirable energetically, they may be available only to adsorbates capable of activated diffusion. Atomic hydrogen may be able to reach such sites within minutes or hours, but diffusion of oxygen would be a much slower process and, indeed, hydrogen and oxygen may well see significantly different surface areas.

All of the above is consistent with a fairly rapid adsorption and penetration of atomic hydrogen on both external and internal surfaces. Brief exposure to oxygen or air will effect a partial oxidation (primarily on the external surface). With prolonged exposure, oxidation of internal surface sites will also occur, though presumable, the final surface "seen" will be less than that for hydrogen. Typically, after exposure of a newly created sample surface (e.g., by meteorite impact) in a lunar environment any reduction that can take place, takes place quite rapidly. Exposure to a terrestrial atmosphere should lead to a rapid partial oxidation with subsequent oxidation taking place quite slowly. On this basis, there would seem to be little hope of maintaining a completely pristine sample, however, if a small degree of surface contamination should be acceptable, further brief exposure would not seem to be excessively deleterious.

(b) *Adsorption on exposed surfaces*

Changes in the state of oxidation of adsorbent surface should have little effect where adsorbate–adsorbent interactions involve only dispersive forces (noble gases) or possibly multipole interaction (nitrogen). On this basis, water was selected as a suitable adsorbate for study with the oxidized and reduced forms of the same adsorbent sample; since surface oxidation might conceivably increase the number of hydrophillic sites. While only one sample (15015,68) has been so examined, there would appear to be relatively little effect, though it should be realized that this particular sample did not undergo prolonged exposure to air. It is also of interest that neither low pressure hysteresis nor water retention were observed under these conditions for this particular sample (Holmes *et al.* 1973).

That adsorption beyond monolayer coverage should be relatively unaffected is only to be expected. That adsorption below monolayer coverage should be so little affected is somewhat surprising. One reasonable explanation, though not the only one possible, is that all available surface oxygen has been converted to surface hydroxyl groups. Such a situation would explain the relatively small change in the hydrophillic nature of the sample surface.

(c) *Formation of water vapor*

The formation of hydroxyl groups from surface oxygen by reduction with atomic hydrogen postulated above, suggests that on desorption or outgassing, water vapor rather than hydrogen would be desorbed. Before carrying out such experimentation, however, it was felt important to consider carefully the elimination of possible contaminant water vapor and the initial procedure adopted was to outgas a sample for a prolonged period at high temperatures. This outgassing was carried out until essentially no water was evolved over the temperature range to be studied.

Since, as was pointed out, some contamination had already occurred, much of the water vapor evolved during the initial outgassing could be assumed to be non-lunar in origin. A typical heterogeneous surface should consist (Ross, 1971) of a small number of high energy sites and a large number of low energy sites. This would lead, on exposure to contaminant water vapor, to a small amount of chemisorbed plus a large amount of physisorbed water. The initial 20 hour outgassing at 150°C was undertaken in the hope that most of the physisorbed contaminating water would be removed below 150° leaving only a small amount of chemisorbed contaminant water to be evolved at higher temperatures (along with any water formed from hydroxyl groups created by lunar surface reduction). It should, of course, be realized that lowering this outgassing temperature would have meant retention of significant amounts of contaminant physisorbed water, while raising it above 150° would have resulted in the loss of some of the "lunar" water. The water vapor evolved in the temperature range 150–400° (Fig. 4, sample 15565-3G) we consider to have been primarily "lunar" with only a small amount of contaminant material. Similar results have been obtained by us on 14163,111 and by others on a number of samples (Gibson and Moore, 1972). The two successive outgassings, first with the sample only, then with the sample plus system, (25–400°C) ensured that, within this temperature range, sample gas evolution was essentially zero (background pressure 10^{-9} torr).

Having ensured that all water vapor capable of removal by outgassing below 400°C had been removed, the sample was exposed to molecular hydrogen at 150°C, outgassed, exposed to molecular deuterium and once again, outgassed. The data illustrated in Fig. 5 clearly demonstrates our contention that water vapor is being formed from chemisorbed hydrogen and sample oxygen. What is not clear is the precise origin of the oxygen: whether it was part of the original complex silicate material, or of solar wind origin, or oxygen chemisorbed after an original lunar reduction. Further study is presently being carried out to clearly establish this point. Other studies concern themselves with a quantitative evaluation of the amount of water formed and of the possible existence of a mineral effect. Preliminary results show that most complex silicates will produce some water in this way and gas-solid interactions clearly must be considered when considering the evolution of an atmosphere on any planet possessing a reducing atmosphere and oxygen rich materials. Under lunar conditions, some water vapor must be produced in this way.

The high efficiency of atomic hydrogen to act as a reducing agent suggests that other elements besides oxygen should be reduced. Specifically, the suggestion of Zeller *et al.* (1970), concerning the possible formation of hydrocarbons from sample carbon appears to be realistic under typical lunar conditions. We have observed small amounts of methane in the initial outgassing (Fig. 4) and trace amounts following hydrogen exposure. Regretably, it was not possible to demonstrate the synthesis of deuterated methane because of the overwhelming water and heavy water peaks (Fig. 5). Nevertheless, we can state that the methane observed did not originate from the stainless steel U.H.V. system since blank tests with hydrogen showed no hydrocarbon formation whatsoever. Unless the hydrogen chemisorption, in some way, releases methane that was previously trapped and that resisted rigorous high temperature outgassing, we must consider that the observed methane has been synthesized.

Zeller and Ronca (1967) have previously predicted that solar wind protons should interact with exposed lunar material to produce water vapor and hydrocarbons. Later, it was demonstrated that proton irradiation of terrestrial silicate materials produced significant amounts of water vapor (Zeller *et al.*, 1970; Gibson and Moore, 1972). Using deuterated acids on lunar samples Cadogen *et al.* (1972) showed that CH_4 was obtained and postulated that this gas, unlike accompanying deuterated hydrocarbons, must have been released but not synthesized by the action of the acid. As has been indicated, our expectations of water vapor formation were based on our previously obtained hydrogen atom and water vapor adsorption studies. Qualitatively, we have now established the following points:

1. The water vapor formed was synthesized from chemisorbed hydrogen and sample oxygen under conditions designed to exclude the possibility of contamination by terrestrial water vapor.

2. The active reducing agent, prior to water formation, was the hydrogen atom. Molecular hydrogen proved ineffective at temperatures too low to dissociate the molecule.

3. The synthesis of hydrocarbons primarily methane under the same conditions appears a likely event. The amount found, however, is small and further work is required for complete confirmation.

Acknowledgment—We wish to express our appreciation for financial support provided through NASA grants NGR 33-183-004 and NGR 33-183-013. We also acknowledge the financial support of NSF through grant GP 24058 for the purchase of the quadrupole mass spectrometer.

REFERENCES

Bibring J. P., Duraud J. P., Durrieu L., Jouret C., Maurette M., and Meunier R. (1972) Ultrathin amorphous coatings on lunar dust grains. *Science* **175**, 753–755.

Cadenhead D. A. and Buergd W. G. (1973) Water vapor from a lunar breccia: Implications for evolving planetary atmospheres. *Science* **180**, 1166–1168.

Cadenhead D. A. and Jones B. R. (1972) The adsorption of atomic hydrogen on 15101,68. In *The Apollo 15 Samples*, pp. 272–274. The Lunar Science Institute, Houston.

Cadenhead D. A. and Jones B. R. (1973) The surface reduction of lunar fines 14163,111. *J. Colloid Interface Sci.* **42**, 650–653.

Cadenhead D. A. and Wagner N. J. (1968) *J. Phys. Chem.* **72**, 2775–2781.

Cadenhead D. A. and Wagner N. J. (1971) In *Vac. Microbalance Tech.*, Vol. 8 (editor A. W. Czanderna), pp. 97–109. Plenum.

Cadogen P. H., Eglington G., Firth J. N. M., Maxwell J. R., Mays B. J., and Pillinger C. T. (1972) In *Proc. Third Lunar Sci. Conf., Geochim. Cosmochim. Acta*, Suppl. 3, Vol. 2, pp. 2069–2090. MIT Press.

de Boer J. H. (1953) *The dynamical character of adsorption*, Oxford University Press.

Geiss J., Buehler F., Cerutti H., Eberhardt P., and Filleux, Ch. (1972). In *Apollo 16 Preliminary Science Report*, N.A.S.A. SP-315, Chapter 14.

Gibson E. K. Jr. and Moore G. W. (1972) In *Proc. Third Lunar Sci. Conf., Geochim. Cosmochim. Acta*, Suppl. 3, Vol. 2, pp. 2029–2040. MIT Press.

Holmes H. F., Fuller E. L. Jr., and Gammage R. B. (1973) Interaction of gases with lunar materials. Apollo 12, 14, and 16 samples (abstract). In *Lunar Science—IV*, pp. 378–380. The Lunar Science Institute, Houston.

Hundhausen A. J., Asbridge J. R., Bame S. J., Gilbert H. F., and Strong I. B. (1967) *J. Geophys. Res.* **72**, 87–100.

Leich D. A., Tombrello T. A., and Burnett D. S. (1973) The depth distribution of hydrogen distribution of hydrogen in lunar materials (abstract). In *Lunar Science—IV*, pp. 463–465. The Lunar Science Institute, Houston.

N.A.S.A. Report (1971) Fines samples collected by Apollo 14.

Ross S. (1971) Monolayer adsorption on crystalline surfaces. *Progress in Surface and Membrane Science* Vol. 4, (editors J. F. Danielli, M. D. Rosenberg, and D. A. Cadenhead), pp. 377–415. Academic Press.

Zeller E. J. and Ronca L. B. (1967) *Icarus* **7**, 372–379.

Zeller E. J., Dreschhoff G., and Kevan L. (1970) *Modern Geology* **1**, 141–148.

Proceedings of the Fourth Lunar Science Conference
(Supplement 4, *Geochimica et Cosmochimica Acta*)
Vol. 3, pp. 2403–2411

The relative density of lunar soil

W. David Carrier, III,* James K. Mitchell,† and Arshud Mahmood†

Abstract—The specific gravity and minimum and maximum bulk densities were determined for three one-gram lunar soil samples containing particles less than 1 mm in diameter: 14163,148, 14259,3 and 15601,82. The specific gravity varied from 2.90 to 2.93 for the Apollo 14 samples to 3.24 for the Apollo 15 sample. The difference is attributed to the higher proportion of agglutinates and breccias and fewer mineral fragments and basalts in the Apollo 14 soils. The minimum bulk densities varied from 0.87 to 1.10 g/cm^3; and the maximum densities varied from 1.51 to 1.89 g/cm^3. The ranges in values were due to differences in the specific gravity, re-entrant intra-granular voids, particle shape, surface texture and grain arrangements. The *in situ* lunar soil in the plains areas of the moon can have a low to medium relative density at the surface, increasing rapidly to a very high relative density at depths greater than 10 to 20 cm. The sub-surface soil may be overconsolidated and the variation of relative density with depth may be a strong indicator of the deposition age.

Introduction

The relative density, D_R, of a soil deposit is defined as:

$$D_R = \frac{e_{max} - e}{e_{max} - e_{min}} \times 100\%$$

where $e = $ *in situ* void ratio of soil deposit

$e_{max} = $ maximum void ratio (corresponding to minimum bulk density) at which the soil can be placed

$e_{min} = $ minimum void ratio (corresponding to maximum bulk density) at which the soil can be placed

Void ratio and bulk density of a soil, ρ, are related according to:

$$\rho = \frac{G\rho_w}{1 + e}$$

where $G = $ specific gravity of soil particles
    $\rho_w = $ density of water at 4°C $= 1$ g/cm^3

Consequently, relative density can also be defined in terms of bulk density according to:

$$D_R = \frac{\rho_{max}}{\rho} \frac{\rho - \rho_{min}}{\rho_{max} - \rho_{min}} \times 100\%$$

If $\rho = \rho_{min}$ (or $e = e_{max}$), then $D_R = 0\%$ and the soil deposit would be exceptionally loose; if $\rho = \rho_{max}$ (or $e = e_{min}$), then $D_R = 100\%$ and the soil would be extremely

---

*Johnson Space Center, NASA, Houston, Texas 77058
†University of California, Berkeley, California 94720

compact. Physical properties such as thermal conductivity, sonic velocity, penetration resistance, shear strength, compressibility, and dielectric constant are dependent on the *in situ* relative density as well as the absolute bulk density. Some soil properties may vary several orders of magnitude between a relative density of 0% and 100%. The factors which affect ρ_{max} and ρ_{min} are primarily: grain size distribution, specific gravity, grain shape distribution, particle surface texture, and grain arrangement. Two soils can have the same absolute densities and yet quite different relative densities, and as a result, dissimilar behavior. Conversely, two soils can have the same relative density and different absolute densities but exhibit similar behavior. Additional discussion on the significance of relative density may be found in Mitchell *et al.*, 1972a.

The minimum and maximum densities of three lunar soil samples (two from Apollo 14 and one from Apollo 15) have been measured. The purpose of this paper is to discuss these results and to draw conclusions concerning the relative density of lunar soil *in situ*.

EXPERIMENTAL METHODS

In terrestrial geotechnical investigations, samples of several hundred to a few thousand grams of soil are normally used for determination of minimum and maximum densities according to carefully prescribed test procedures (ASTM, 1972). Since lunar samples of this size are not readily available, the writers developed a procedure for relative density determination that requires only one-gram samples (Carrier *et al.*, 1973). Small graduated cylinders of 1.0 and 1.5 cm^3 capacity were used to measure sample volumes after placement in loose and dense states. The loosest condition was obtained by pouring the sample from a small height in a single, continuous operation. To obtain the maximum density, the cylinders were filled with soil and tapped 90 times by dropping 4 to 5 cm in nearly free fall onto a table. It was found that the maximum densification that could be obtained was reached after 90 taps of the sample. Minimum and maximum densities determined by this procedure on a ground basalt lunar soil simulant agreed within 9% of the values obtained with the standard ASTM tests. The method developed by the writers, while arbitrary, provides a well-defined repeatable approach.

MINIMUM AND MAXIMUM DENSITY

Minimum and maximum density tests were performed on three lunar samples, each of one gram and containing less than millimeter size particles: two from Apollo 14 (14163,148 and 14259,3) and one from Apollo 15 (15601,82). The specific gravity of each of these samples was also determined by means of conventional water immersion micropycnometry techniques. The results of these tests are presented in Table 1; values obtained by other investigators are also listed. In those cases where the specific gravity is known, the maximum and minimum void ratios have also been calculated.

The Apollo 11 densities reported by Costes *et al.* (1970) were determined as part of a study of penetration resistance. Cremers *et al.* (1970), Cremers and Birkebak (1971), Cremers (1972), and Cremers and Hsia (1973) found minimum densities for Apollo 11, 12, 14, and 15 samples as part of their investigation of thermal conductivity and noticed that they were not able to place the Apollo 11 and 12 samples at as low an absolute bulk density as the Apollo 14 sample. The

Table 1. Minimum and maximum density of lunar soils.

Mission	Sample Number	Sample* Weight (g)	Density ρ_{min} (g/cm³)	Density ρ_{max} (g/cm³)	Specific Gravity G	Void Ratio e_{max}	Void Ratio e_{min}	Reference
Apollo 11	10084	565	1.36	1.80	3.01†	1.21	0.67	Costes et al. (1970)
	10084,68	5	1.26		3.01†	1.39		Cremers et al. (1970)
Apollo 12	12001,19	6	1.30	1.93				Cremers and Birkebak (1971)
	12029,3	1.3	1.15					Jaffe (1972)
Apollo 14	14163,133	5	1.10		2.9±0.1‡	1.64		Cremers (1972)
	14163,148	0.97	0.89±0.03	1.55±0.03	2.90±0.05	2.26	0.87	This paper
	14259,3	1.26	0.87±0.03	1.51±0.03	2.93±0.05	2.37	0.94	This paper
Apollo 15	15031,38	5	≤1.30	1.89±0.03	3.24±0.05	1.94	0.71	Cremers and Hsia (1973)
	15601,82	0.96	1.10±0.03					This paper
Luna 16	—	~10	1.12	1.79				Gromov et al. (1971)

*All tests performed on −1 mm size fraction.
†Duke et al. (1970b): suspension in a density gradient.
‡Cadenhead et al. (1972): helium pycnometry.

densities determined by Jaffe (1972) were for a sample returned inside the scoop of the Surveyor III spacecraft and were part of a study on penetration resistance. The densities of the Luna 16 sample were determined by Gromov *et al.* (1971) in connection with penetrometer, oedometer, and direct shear tests. Their sample represented approximately 10% of the entire Luna 16 returned sample. The minimum and maximum density measurements to date have been by a variety of methods. As it is well known that the measured density is dependent on the placement method, the values reported by these investigators are difficult to compare directly. Consequently the writers will limit this discussion to their own data.

The significantly lower specific gravities of the two Apollo 14 samples studied by the writers are undoubtedly due to the higher proportion of agglutinates and breccias and fewer mineral fragments and basalts than the Apollo 15 soil (cf. McKay *et al.*, 1972; Clanton *et al.*, 1972). The lower minimum and maximum bulk densities of the Apollo 14 soils is obviously partly due to these lower specific gravities. This cannot be the entire explanation, however, otherwise the maximum and minimum void ratios would be comparable, and the Apollo 14 soils have greater void ratios than the Apollo 15 soil. The submillimeter grain size distributions for the three samples are all quite similar: well-graded silty-sands to sandy-silts with average particle sizes by weight ranging from 0.045 to 0.085 mm; consequently the explanation for the difference in void ratios must lie elsewhere. One likely possibility is that the higher proportion of agglutinates and breccias in the Apollo 14 soils contributes more re-entrant, intra-granular voids than the Apollo 15 soil. If it is assumed that the Apollo 15 soil has no re-entrant voids, which is probably not true, then based on the minimum void ratios in Table 1, the Apollo 14 soils would have a re-entrant component of void ratio of about $0.91 - 0.71 = 0.2$, which is a significant amount. Even so, this cannot be the entire explanation, since the difference in maximum void ratios is even higher: $2.32 - 1.94 = 0.38$.

It appears, therefore, that the other factors which affect relative density (e.g., particle shape, surface texture, and grain arrangement) must vary significantly from lunar soil to lunar soil. It should also be noted that the maximum and minimum void ratios of a ground basalt simulant with the same grain size distribution as the lunar soils are 1.32 and 0.48, respectively. These void ratios are significantly less than even the Apollo 15 soil, which means that the results of some of the simulation studies reported in the literature should be corrected and re-stated in terms of relative densities. Obviously, it is imperative to make minimum and maximum density tests on a variety of lunar soils, because with these values and the *in situ* density it is possible to calculate the relative density on the lunar surface. As will be discussed in the following section, relative density is important for both engineering and geological applications.

RELATIVE DENSITY

The core tube samplers used on Apollos 11, 12, and 14 produced disturbed to badly disturbed samples (Carrier *et al.*, 1971, 1972). The core tubes used on Apollo

15 and subsequent missions had far thinner walls which resulted in nearly undisturbed samples of the lunar soil and 90–95% core recovery. Besides being extremely important for studies of geology, sedimentology, fabric, cosmic ray exposure, etc., these core samples have also provided the first direct, accurate estimates of *in situ* bulk density of the lunar soil. If the minimum and maximum density were known for each of these core tube samples, the relative density could be determined for each of the core locations on Apollos 15, 16, and 17.

The one Apollo 15 sample for which ρ_{max} and ρ_{min} are known, 15601,82 (*see* Table 1), was taken at Station 9A at the rim of Hadley Rille, less than 10 meters from a double core tube sample: 15011/15010. Although no index properties are available for the core sample itself, since it has not yet been opened, nor is it known how these properties might vary with depth, the 15601,82 data can be used to estimate relative density vs. depth at this one location on the lunar surface.

Only the average density for each of the two sections of the double core tube are known: 1.69 ± 0.03 g/cm^3 for the top 29.2 ± 0.5 cm and 1.85 ± 0.06 g/cm^3 for the next 34.9 cm (Carrier *et al.*, 1972), shown graphically in Fig. 1. The corresponding

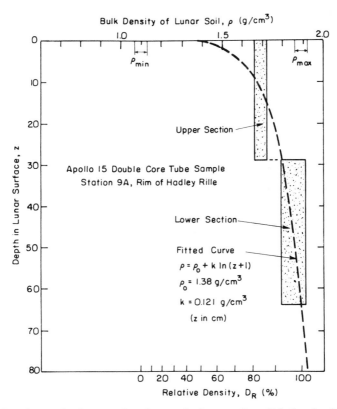

Fig. 1. Density vs. depth at one location on the lunar surface. Relative density scale and ρ_{min} and ρ_{max} are based on data for sample 15601,82 (Table 1) which was taken at the same station.

average relative densities are 87% and 94%, respectively. While these values are somewhat arbitrary, they do indicate a high relative density at this location. This had previously been predicted by Mitchell *et al.* (1972a) on the basis of the high number of hammer blows required to drive the core tube and the fact that the soil surface surrounding the tube heaved slightly during driving.

An idealized density profile is also shown in Fig. 1 which was calculated by assuming that the density increases logarithmically with depth, z, from a finite value, ρ_0, at the surface. The form of the expression is:

$$\rho = \rho_0 + k \ln (z + 1) \qquad [z \text{ in cm}]$$

ρ_0 and k, a constant multiplying factor, can be determined explicitly from the given data and are 1.38 g/cm^3 and 0.121 g/cm^3, respectively. The calculated density increases rapidly for the first 10 to 20 cm and then slowly thereafter. The relative density is 48% at the surface, 82% at 10 cm, 93% at 30 cm, and 99% at 60 cm.

In addition, the average relative density for the top 5 to 10 cm of the lunar surface at the Apollo 15 site can be estimated on the basis of the astronaut boot-print studies reported by Houston *et al.* (1972) and Mitchell *et al.* (1973b). Their results for the Apollo 15 site indicate an average porosity, n, of 43.6% with a standard deviation of 2.8%; the maximum and minimum porosities are 58.3% and 31%, respectively. Porosity can be converted to void ratio by the relationship $e = n/(1-n)$; the relative density, D_R, can then be calculated from the equation given in the Introduction. The result is $D_R = 66\%$ with a standard deviation of about 10%. This value agrees extremely well with the idealized curve in Fig. 1.

It is important to note that if a density-depth relationship is arbitrarily chosen such that $\rho_0 = \rho_{\min}$, that is, the surface is at 0% relative density, the effect is to have even higher relative densities at shallower depths than given by the idealized profile. Consequently, one is led to the inescapable conclusion that while the surface at this location may be at a low to medium relative density, the soil just 10 to 20 cm down is at a very high relative density, much higher than would be required to support the very small overburden stress in the low lunar gravity.

There is additional evidence to indicate that a high relative density at a shallow depth is a general condition for all of the lunar plains areas:

(1) The average absolute density at the Apollo 14 core tube sites was previously predicted to lie in the range of 1.4^5 to 1.6 g/cm^3 to a depth of 30 to 60 cm (Carrier *et al.*, 1972). Referring to Table 1, it can be seen that this would imply a high average relative density of 92% to 106%. Thus, even though the absolute densities at the Apollo 14 site are less than at Apollo 15, the *in situ* relative densities are very similar.

(2) The average relative densities of the top 5 to 10 cm of soil at the Apollo 11, 12, 14, and 17 sites, based on the bootprint studies by Houston *et al.* (1972) and Mitchell *et al.* (1973b), are remarkably similar to that at Apollo 15.

(3) Penetrometer studies of the lunar surface (Mitchell *et al.*, 1972a, 1972b, 1973a) have shown that the penetration resistance, or relative density, in the plains areas is relatively high and falls in a fairly narrow band (although the penetration

resistance is lower and extremely variable on slopes: downslope movements probably de-densify the soil).

(4) The absolute densities in all of the Apollo 15, 16, and 17 core tubes increase rapidly with depth (Mitchell et al., 1972b, 1973b), in a manner similar to the core sample shown in Fig. 1.

The one possible exception is the plains area at the Apollo 16 site. Mitchell et al. (1972b) found that the average surface relative density based on bootprints is slightly less at the Apollo 16 site than at the other sites. In addition, the higher drill penetration rate observed at the Apollo 16 site indicates a lower relative density at a depth of a few meters than at either the Apollo 15 or 17 drill sites (Mitchell et al., 1973b). However, both of these observations may be due to the recent addition of loose material from the South Ray event 3 to 5 m.y. ago, which has not yet been densified by prolonged exposure to meteorite impact. This interpretation is consistent with the larger proportion of +10 mm particles found in the Apollo 16 soils. It has been presumed that these particles have not yet been worked into the soil matrix by meteorite impact (Mitchell et al., 1972b); approximately 100 m.y. of exposure are required to produce the "steady-state" grain size distribution found in the majority of the returned lunar soil samples (Carrier, 1973).

Obviously, minimum and maximum density tests are needed on samples from the other core tube sites before definitive statements can be made. However, the apparent mechanism controlling the relative density of lunar soil in the plains areas seems to be that the constant meteorite bombardment maintains a loose, stirred-up surface; but directly beneath the surface, the vibrations due to innumerable shock waves shake and densify the soil to a very high relative density. The sub-surface soil may even be overconsolidated; i.e., the soil may have been densified under a greater confining stress at some time in the past than is presently applied to it by the overlying soil.

One geologic implication of these results is that the variation of relative density of a lunar soil with depth may be a strong indicator of how long ago the soil was deposited. The most important engineering implication is obviously that the lunar soil can be quite hard and incompressible at shallow depths, which had not been previously appreciated and consequently led to problems with the heat flow bore stem drilling on Apollo 15.

CONCLUSIONS

The specific gravity and minimum and maximum densities were determined for three one-gram, submillimeter lunar soil samples: 14163,148, 14259,3, and 15601,82. The specific gravity was determined by conventional water immersion micropycnometry techniques and varied from 2.90 to 2.93 for the Apollo 14 samples to 3.24 for the Apollo 15 sample. The difference in specific gravities is attributed to the higher proportion of agglutinates and breccias and fewer mineral fragments and basalts in the Apollo 14 soils.

The minimum and maximum bulk densities were determined by a new procedure specifically developed for use with small lunar samples. The minimum densities varied from 0.87 g/cm³ (14259) to 1.10 g/cm³ (15601); and the maximum densities varied from 1.51 g/cm³ (14259) to 1.89 g/cm³ (15601). The ranges in values are due to differences in the specific gravity, re-entrant intra-granular voids, particle shape, surface texture, and grain arrangements. The differences in the grain size distributions of the three samples are minimal and thus do not contribute to the ranges in the minimum and maximum densities.

Sample 15601 was taken at Station 9A, the same station as double core tube sample 15011 and 15010, and consequently relative density vs. depth can be calculated for this one location. The results suggest that the *in situ* lunar soil in the plains areas can have a low to medium relative density at the surface, increasing rapidly to a very high relative density at depths greater than 10 to 20 cm. This is evidently due to the meteorite impacts which maintain a loose, stirred-up surface but shake and densify the soil directly beneath the surface; the sub-surface soil may even be overconsolidated. The variation of relative density of a lunar soil with depth may be a strong indicator of the deposition age.

Acknowledgments—Portions of this study were supported by the National Aeronautics and Space Administration under Contract NAS 9-11266. David W. Strangway, Johnson Space Center, Houston, permitted the first two writers to become Co-Investigators on his experiment, Electrical Properties of Returned Lunar Samples, for the purpose of testing the three one-gram lunar samples. The writers appreciate comments made by Charles H. Simonds of the Lunar Science Institute.

References

ASTM (1972) Standard method of test for relative density of cohesionless soils. *American Society for Testing and Materials*, Method D2049, Standards, Part II.

Cadenhead D. A., Wagner N. J., Jones B. R., and Stetter J. R. (1972) Some surface characteristics and gas interactions of Apollo 14 fines and rock fragments. *Proc. Third Lunar Sci. Conf., Geochim. Cosmochim Acta*, Suppl. 3, Vol. 3, pp. 2243–2257. MIT Press.

Carrier W. D., III (1973) Lunar soil grain size distribution. *The Moon.* In press.

Carrier W. D., III, Johnson S. W., Werner R. A., and Schmidt R. (1971) Disturbance in samples recovered with the Apollo core tubes. *Proc. Second Lunar Sci. Conf., Geochim. Cosmochim. Acta*, Suppl. 2, Vol. 3, pp. 1959–1972. MIT Press.

Carrier W. D., III, Johnson S. W., Carrasco L. H., and Schmidt R. (1972) Core sample depth relationships: Apollo 14 and 15. *Proc. Third Lunar Sci. Conf., Geochim. Cosmochim. Acta*, Suppl. 3, Vol. 3, pp. 3213–3221. MIT Press.

Carrier W. D., III, Mitchell J. K., and Mahmood A. (1973) The nature of lunar soil. *ASCE J. Soil Mech. Found. Div.* In press.

Clanton U. S., McKay D. S., Taylor R. M., and Heiken G. H. (1972) Relationship of exposure age to size distribution and particle types in the Apollo 15 drill core. In *The Apollo 15 Lunar Samples*, pp. 54–56. The Lunar Science Institute, Houston.

Costes N. C., Carrier W. D., III, Mitchell J. K., and Scott R. F. (1970) Apollo 11: Soil mechanics results. *ASCE J. Soil Mech. Found. Div.* **96**, 2045–2080.

Cremers C. J. (1972) Thermal conductivity of Apollo 14 fines. *Proc. Third Lunar Sci. Conf., Geochim. Cosmochim. Acta*, Suppl. 3, Vol. 3, pp. 2611–2617. MIT Press.

Cremers C. J., Birkebak R. C., and Dawson J. P. (1970) Thermal conductivity of fines from Apollo 11. *Proc. Apollo 11 Lunar Sci. Conf., Geochim. Cosmochim. Acta*, Suppl. 1, Vol. 3, pp. 2045–2050. Pergamon.

Cremers C. J. and Birkebak R. C. (1971) Thermal conductivity of fines from Apollo 12. *Proc. Second Lunar Sci. Conf., Geochim. Cosmochim. Acta*, Suppl. 2, Vol. 3, pp. 2311–2315. MIT Press.

Cremers C. J. and Hsia H. S. (1973) Thermal conductivity of Apollo 15 fines at low density (abstract). In *Lunar Science—IV*, pp. 164–166. The Lunar Science Institute, Houston.

Duke M. B., Woo C. C., Bird M. L., Sellers G. A., and Finkelman R. B. (1970) Lunar soil: Size distribution and mineralogical constituents. *Science* **167**, 648–650.

Gromov V. V., Leonovich A. K., Lozhkin V. A., Rybakov A. V., Pavlov P. S., Dmitriev A. D., and Shvarev V. V. (1971) Results of investigations of the physical and mechanical properties of the lunar sample from Luna 16. Paper K13 at the *14th COSPAR Session*, Seattle, Wash., USA.

Houston W. N., Hovland H. J., and Mitchell J. K. (1972) Lunar soil porosity and its variation as estimated from footprints and boulder tracks. *Proc. Third Lunar Sci. Conf., Geochim. Cosmochim. Acta*, Suppl. 3, Vol. 3, pp. 3255–3263. MIT Press.

Jaffe L. D. (1972) Bearing strength of lunar soil. *The Moon* **3**, 337–345.

McKay D. S., Heiken G. H., Taylor R. M., Clanton U. S., and Morrison D. A. (1972) Apollo 14 soils: Size distribution and particle types. *Proc. Third Lunar Sci. Conf., Geochim Cosmochim. Acta*, Suppl. 3, Vol. 1, pp. 983–994. MIT Press.

Mitchell J. K., Bromwell L. G., Carrier W. D., III, Costes N. C., Houston W. N., and Scott R. F. (1972a) Preliminary analysis of soil behavior. *Apollo 15 Preliminary Science Report*, NASA SP-289, pp. 7-1 to 7-28.

Mitchell J. K., Carrier W. D., III, Houston W. N., Scott R. F., Bromwell L. G., Durgunoglu H. T., Hovland H. J., Treadwell D. D., and Costes N. C. (1972b) Soil mechanics. *Apollo 16 Preliminary Science Report*, NASA SP-315, pp. 8-1 to 8-29.

Mitchell J. K., Carrier W. D., III, Costes N. C., Houston W. N., and Scott R. F. (1973a) Surface soil variability and stratigraphy at the Apollo 16 site (abstract). In *Lunar Science—IV*, pp. 525–527. The Lunar Science Institute, Houston.

Mitchell J. K., Carrier W. D., III, Costes N. C., Houston W. N., Scott R. F., and Hovland H. J. (1973b) Soil mechanics. *Apollo 17 Preliminary Science Report*. In press.

Proceedings of the Fourth Lunar Science Conference
(Supplement 4, *Geochimica et Cosmochimica Acta*)
Vol. 3, pp. 2413–2423

Interaction of gases with lunar materials: Apollo 12, 14, and 16 samples*

H. F. HOLMES and E. L. FULLER, JR.

Reactor Chemistry Division
Oak Ridge National Laboratory
Oak Ridge, Tennessee 37830

R. B. GAMMAGE

Health Physics Division
Oak Ridge National Laboratory
Oak Ridge, Tennessee 37830

Abstract—Surface properties of lunar fines samples from the Apollo 12, 14, and 16 missions have been investigated by studying the adsorption of nitrogen, argon, oxygen, carbon monoxide (all at 77°K), and water vapor (at 20 or 22°C) on the samples. Initially the samples were all nonporous and had a uniformly low specific surface area (0.3 to 0.6 m²/g). Water interacts strongly with the surface of lunar fines, chemisorbing at low pressures followed by a massive adsorption at high pressures. Nitrogen adsorption measurements after the interaction with water showed the surface properties had undergone a severe alteration as a result of the attack by water vapor. This alteration consisted of a marked increase in the specific surface area and the creation of a pore system. The results are interpreted on the basis of a penetration of water into the damage tracks.

INTRODUCTION

THE INTERACTION of gases with lunar particles provides information on surface characteristics at the molecular level of dimensions. The BET treatment (Brunauer *et al.*, 1938) of the adsorption data provides a quantitative measure of the specific surface area. In addition, if pores are present in sizes ranging up to a few hundred angstroms their size and distribution can be ascertained from the adsorption isotherms. No other reliable methods are available for characterizing pore structures in this size range (Gregg and Sing, 1967).

In addition to providing basic information concerning the state of subdivision of lunar soils and their porosity, the specific surface area is a quantitative measure of their capacity to adsorb reactive molecules (e.g., water and carbon dioxide) from an environment. Such information would, for example, be vital for the establishment of life support stations on the lunar surface. A more immediate problem is the possible weathering and deterioration of lunar samples by the terrestrial atmosphere. Lunar fines from the Apollo 11 and 12 missions have been shown to interact extensively with water vapor (Fuller *et al.*, 1971; Holmes *et al.*, 1973). It is important to determine if this strong interaction with water vapor is a general characteristic of lunar fines and to correlate the interaction with, for

---

*Research sponsored by NASA under Union Carbide contract with the U.S. Atomic Energy Commission.

example, the radiation damaged nature of lunar materials. In general, the surface properties of lunar fines will have to be considered in theories concerning the formation and history of such samples.

EXPERIMENTAL

The adsorption-desorption measurements were made with existing vacuum microbalance systems which have been described in detail (Fuller *et al.*, 1965, 1972). Both systems were equipped with a device for maintaining a constant pressure of water vapor during equilibration at each chosen pressure (Fuller *et al.*, 1972). Background blank corrections have been applied to all of the data. These corrections, and their critical importance for accurate acquisition of adsorption data, have been discussed recently (Holmes *et al.*, 1973).

Experimental procedures were generally the same as we have used for our studies of adsorption on thorium oxide (Holmes *et al.*, 1968; Gammage *et al.*, 1972). Prior to an adsorption experiment each sample was always outgassed for a minimum of 16 hours (overnight). Outgassing pressures were in the range of 10^{-5} to 10^{-6} torr (measured on a 25 mm O.D. manifold leading directly to the balance chamber) and the temperatures ranged from 20 to 1000°C. Measurement of the isotherms required 15 to 20 minutes for equilibration at each pressure except for water vapor at high relative pressures which required an overnight waiting period, or longer, for equilibration.

Results reported in this paper were obtained with 0.3 or 0.4 g aliquots of lunar fines samples 12001,151, 12070,218, 14003,60, and 63341,8. These samples were the fine sieve fraction (<1 mm) of lunar soil and were used without further classification. When measuring adsorption at 77°K (liquid nitrogen bath) by gravimetric techniques it is necessary to apply a buoyancy correction to the data. In order to calculate this correction we have used the mean value (2.95 ± 0.09 g/cm^3) of the reported densities (Greene *et al.*, 1971) of selected particles from Apollo 12 lunar fines. Weight determinations are estimated to be reliable to ± 2 micrograms over the extended time interval involved in the experiments.

RESULTS

Isotherms for the adsorption of carbon monoxide, at 77°K, on lunar samples 12001,151, 14003,60, and 63341,8 are shown in Fig. 1. The amount of adsorption, in mg of adsorbate per g of sample, is shown as a function of the relative pressure, P/P_0, where P_0 is the saturation vapor pressure of carbon monoxide at 77°K. The shape of these isotherms is quite common and fits the general classification of Type II isotherms found for adsorption on nonporous solids (Brunauer, 1943). There was exact reversibility on desorption with no indication of hysteresis or other complicating factors. Isotherms for the adsorption of nitrogen, argon, and oxygen (all at 77°K) on these three lunar samples were near duplicates of the carbon monoxide isotherms.

It should be emphasized that all of the isotherms were completely reversible over the entire pressure range from high vacuum to saturation. All of the data for the adsorption of the four gases on these three lunar samples were subjected to a standard BET treatment (Brunauer *et al.*, 1938) to obtain monolayer capacities and "C" constants. Monolayer capacities were converted to specific surface areas by means of recommended cross-sectional areas (McClellan and Harnsberger, 1967) for the adsorbates (the values we have used are given in Table 1). Results from the BET treatment of the data are summarized in Table 1. Agreement of the specific surface areas should be tempered by the fact that cross-sectional areas for

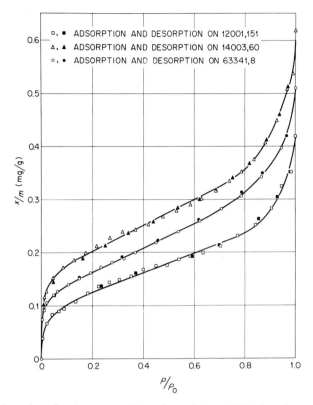

Fig. 1. Adsorption of carbon monoxide on lunar fines at 77°K. Samples outgassed at 25°C.

Table 1. Specific surface areas and BET "C" constants for lunar samples outgassed at 25°C.

Adsorbate	Sample 12001,151		Sample 14003,60		Sample 63341,8	
	Σ	C	Σ	C	Σ	C
N_2^a	0.33	65	0.51	84	0.42	57
Ar^b	0.34	12	0.54	37	0.38	31
CO^c	0.35	82	0.60	100	0.49	113
O_2^d	—	—	0.54	12	0.42	21

Σ = Specific surface area in m^2/g.
[a] Co-area = 16.2 $Å^2$.
[b] Co-area = 16.6 $Å^2$ and P_0 is that of supercooled liquid.
[c] Co-area = 16.8 $Å^2$.
[d] Co-area = 17.5 $Å^2$.

adsorbed molecules are usually selected to give agreement with nitrogen adsorption data. For example, the range of values commonly used for oxygen is 14 to 18 $\mathring{A}^2$ (Gregg and Sing, 1967). If these co-areas are applied to the oxygen data for sample 14003,60 the resulting range of specific surface area is 0.43 to 0.56 m^2/g which is still reasonable agreement. The BET "C" constant is a measure of the average net heat of adsorption for about the last half of monolayer completion (Brunauer, 1961), that is, for adsorption on the least energetic sites on the surfaces of the lunar particles.

Without outgassing the samples at an elevated temperature water vapor isotherms were measured at 20 or 22°C on these same lunar samples. The complete isotherm for 63341,8 and the adsorption data for 12001,151 are shown in Fig. 2. A characteristic feature of all water isotherms which we have measured on lunar fines is a general hysteresis over the entire pressure range including vacuum retention (irreversibly adsorbed water which could not be removed by prolonged evacuation at the conclusion of the isotherm).

An additional common feature is the massive adsorption of water vapor at high relative pressures (above 0.9). Specific surface areas calculated from the water

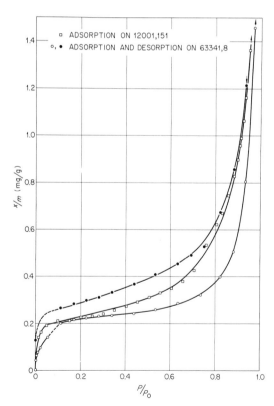

Fig. 2. Adsorption of water vapor on lunar fines. Samples outgassed at 20°C (63341,8) and 22°C (12001,151). Isotherms measured at these respective temperatures.

data are uncertain because of the general irreversibility of the isotherms coupled with the specific nature of water adsorption (this latter factor gives rise to a large uncertainty in the cross-sectional area of an adsorbed water molecule). However, a visual comparison of the disparity between the adsorption and desorption branches of the water isotherms (e.g., sample 63341,8 in Fig. 2 and Holmes *et al.*, 1973) indicates that the capacity for reversible physical adsorption of water has increased as a result of the adsorption-desorption cycles.

After several additional adsorption-desorption cycles in water vapor the adsorption of nitrogen (at 77°K) on these samples was remeasured. Figure 3 shows nitrogen adsorption on sample 12001,151 before and after the water adsorption experiments. Supporting experiments with samples not treated with water vapor have shown that the changes were brought about by the action of water vapor and not by raising the outgassing temperature from 25 to 300°C. Corresponding nitrogen adsorption isotherms for lunar sample 63341,8 are shown in Fig. 4. As a result of the water treatment the amount of nitrogen adsorption is greater at all pressures and the isotherms have well defined hysteresis loops at relative pressures above 0.5. The specific surface area of sample 12001,151 increased from 0.33 to 0.97 m²/g because of exposure to water vapor during the adsorption experiments. The corresponding increase for sample 63341,8 is from 0.42 to 0.74 m²/g. Results for sample 14003,60 are in essential agreement with the data presented in Figs. 3 and 4.

Nitrogen adsorption (at 77°K) on samples 12001,151 and 12070,218 was meas-

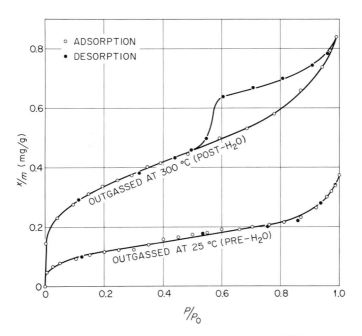

Fig. 3. Adsorption of nitrogen on 12001,151 at 77°K.

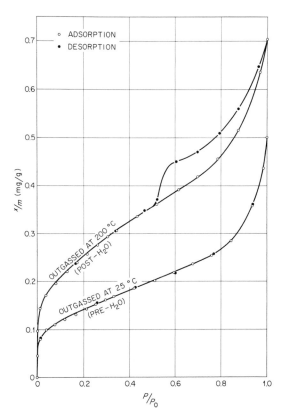

Fig. 4. Adsorption of nitrogen on 63341,8 at 77°K.

ured after outgassing the samples at increasing temperatures from 500 to 1000°C. The resulting isotherms are shown in Fig. 5 (for sample 12001,151) and Fig. 6 (for sample 12070,218). (For the sake of clarity no data points are shown for 12001,151 but the precision of the data is well within the estimated error bar shown on the isotherm which was measured after outgassing the sample at 800°C.) There is a general decrease in the amount of adsorption (at all pressures) as the outgassing temperature is increased. This decrease is reflected in the specific surface areas which are given in Table 2. The high pressure hysteresis loop induced by the reaction with water vapor was absent for samples 12070,218 and 12001,151 after outgassing at 700° and 800°C, respectively.

DISCUSSION

Data tabulated in Table 1 are in essential agreement with the uniformly low specific surface areas which have been reported for lunar fines samples (Fuller *et al.*, 1971; Holmes *et al.*, 1973; Cadenhead *et al.*, 1972; Grossman *et al.*, 1972). By assuming spherical or cubical particles one can calculate an effective particle size

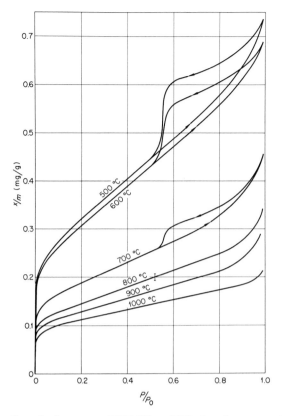

Fig. 5. Adsorption of nitrogen on 12001,151 at 77°K. Sample outgassed at indicated temperatures (Post-H₂O).

Table 2. Specific surface area as a function of outgassing temperature.

Outgassing Temperature (°C)	Sample 12001,151 Σ(m²/g)	Sample 12070,218 Σ(m²/g)
500	0.90	1.15
600	0.87	—
700	0.52	0.59
800	0.42	—
900	0.37	0.49
1000	0.31	0.37

from the specific surface areas given in Table 1. Mean particle sizes calculated in this manner range from 3 to 6 microns. This is in reasonable agreement with particle sizes deduced from sedimentation studies (e.g., Gold *et al.*, 1972). Since sedimentation studies give the Stokes diameter for an equivalent sphere this agreement indicates the absence of gross surface roughness and/or internal area.

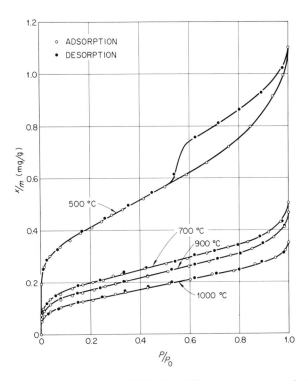

Fig. 6. Adsorption of nitrogen on 12070,218 at 77°K. Sample outgassed at indicated temperatures (Post-H$_2$O).

Average net heats of adsorption calculated from the "C" constants given in Table 1 range from 380 (C = 12) to 730 cal/mole (C = 113). These values are in qualitative agreement with the expected trend based on polarizability, quadrupole, and dipole contributions to adsorption energies for these four adsorbates. This trend in adsorption energies has been observed for terrestrial materials such as titanium oxide (Smith and Ford, 1965).

The reversibility of the initial isotherms is a clear indication that these samples do not have the type of porosity which gives rise to capillary condensation hysteresis loops. It should be emphasized that we are referring to the porosity of individual soil particles and not, for example, to the type of porosity which results from the random packing of particles. In order to cause hysteresis effects in adsorption isotherms the pore size must be a few hundred angstroms or less (Gregg and Sing, 1967). Adsorption in larger pores is essentially like adsorption on open surfaces. These results, coupled with previous work (Fuller *et al.*, 1971; Holmes *et al.*, 1973), indicate that the approximate surface characteristics of lunar fines (prior to water treatment) are independent of their location on the lunar surface. Most probably there are processes occurring on the lunar surface to produce a lunar soil whose dynamic equilibrium properties include a relatively low specific surface area coupled with a lack of porosity. These processes certainly include micrometeorite impact, solar wind sputtering, and radiation damage (e.g., Dran *et al.*, 1972; Phakey *et al.*, 1972).

The contrasting nature of water adsorption on these samples, as compared to adsorption of the "inert" gases, is evident from a comparison of Figs. 1 and 2. Retention of water in high vacuum and the low pressure hysteresis can be explained on the basis of chemisorption. However, the massive uptake at high relative pressures (above 0.9) cannot be attributed to simple multilayer physical adsorption of water vapor on open surfaces. Desorption branches of the isotherms clearly indicate the presence of an internal pore system which gives rise to a capillary condensation hysteresis loop. Quite obviously there has been a severe alteration of the surface properties. This agrees with our previous result for sample 12070,218 (Holmes *et al.*, 1973) which had been outgassed at 300°C prior to the adsorption of water. Present results show that the reaction of water vapor with these samples does not require prior "activation" at an elevated temperature. It is now apparent that exposure of lunar samples to the normal laboratory atmosphere will result in chemical and physical adsorption of water. Attributing experimental difficulties to adsorbed water is clearly justified (Tittmann *et al.*, 1972). Nitrogen adsorption isotherms on the same sample before and after the reaction with water vapor (Figs. 3 and 4) give a clearer indication of the changes caused by adsorbed water. Obviously the adsorptive capacity has increased by as much as a factor of three. According to de Boer (1958), hysteresis loops such as those in Figs. 3 and 4 are due to either slit-shaped pores with parallel walls or wide pores with narrow necks. The blocking effect of irreversibly adsorbed water (Holmes *et al.*, 1973) favors pores with narrow necks. The isotherms in Figs. 3 and 4 are essentially the same as those reported for 12070,218 (Holmes *et al.*, 1973) and unreported data for 14003,60.

The temperature stability of the water induced porosity has been measured (Figs. 5 and 6). Heating at temperatures of 700 to 800°C eliminates the porosity and markedly reduces the specific surface area (Table 2). This type of sintering behavior is similar to that observed for materials such as some silica-alumina catalysts (Ries, 1952). Significant surface area remained after the 1000°C outgassing (the temperature limit imposed by the microbalance system). One might have expected more drastic sintering on the basis of initial melting temperatures of about 1150°C which have been reported for lunar fines (Gibson and Moore, 1972).

It presently appears that lunar fines, independent of their original location on the lunar surface, will suffer the same general type of alteration by interaction with water vapor at high relative pressures. A common feature to all fines samples is the extensive radiation damage they have suffered. This, the radiation damage, is the basis for our postulated mechanism for the attack by water vapor. According to the "ion explosion spike" model of Fleischer *et al.* (1965), track damage in nonconductors consists of regions of heavy damage (~ 100 Å) separated by trails of relatively little damage (a few atoms displaced). The damaged material is considerably more soluble than the surrounding material, a fact that has found application in radiation dosimetry (e.g., Becker, 1972) as well as in numerous studies of radiation damage in lunar samples. We postulate that when sufficient water has been adsorbed (at a relative pressure of about 0.9) the damaged material starts to dissolve. The solution process lowers the vapor pressure of the adsorbed water which leads to increased sorption of water and the entire process is enhanced. Because of the concentration gradient between dissolved material in the damage track and the outer layers of adsorbed water the dissolved material migrates from the damaged region and thereby creates a pore. According to the "ion explosion spike" model the geometry of the pore would be a wide void with a narrow opening. This is one of the two shapes which, according to de Boer (1958), give capillary condensation hysteresis loops like those we observe in nitrogen isotherms on water treated lunar fines. Work in progress should provide further support for this postulated mechanism for the alteration of lunar fines by adsorbed water.

Acknowledgment—This research was sponsored by NASA under Union Carbide contract with the U.S. Atomic Energy Commission.

REFERENCES

Becker K. (1972) Dosimetric applications of track etching. *Topics in Radiation Dosimetry*, Suppl. 1, pp. 79–142. Academic Press.

Brunauer S. (1943) The adsorption of gases and vapors. *Physical Adsorption*, Vol. I. Princeton University Press.

Brunauer S. (1961) Solid surfaces and the solid-gas interface. *Advances in Chemistry Series*, No. 33, pp. 5–17. American Chemical Society.

Brunauer S., Emmett P. H., and Teller E. (1938) Adsorption of gases in multimolecular layers. *J. Amer. Chem. Soc.* **60**, 309–314.

Cadenhead D. A., Wagner N. J., Jones B. R., and Stetter J. R. (1972) Some surface characteristics and gas interactions of Apollo 14 fines and rock fragments. *Proc. Third Lunar Sci. Conf., Geochim. Cosmochim. Acta*, Suppl. 3, Vol. 3, pp. 2243–2257. MIT Press.

de Boer J. H. (1958) The shapes of capillaries. *The Structure and Properties of Porous Materials* (editors D. H. Everett and F. S. Stone), pp. 68–94. Academic Press.

Dran J. C., Duraud J. P., Maurette M., Durrieu L., Jouret C., and Legressus C. (1972) Track metamorphism in extraterrestrial breccias. *Proc. Third Lunar Sci. Conf., Geochim. Cosmochim. Acta*, Suppl. 3, Vol. 3, pp. 2883–2903. MIT Press.

Fleischer R. L., Price P. B., and Walker R. M. (1965) Ion explosion spike mechanism for formation of charged-particle tracks in solids. *J. Appl. Physics* **36**, 3645–3652.

Fuller E. L., Jr., Holmes H. F., and Secoy C. H. (1965) Gravimetric adsorption studies of thorium dioxide surfaces. *Vacuum Microbalance Tech.* **4**, pp. 109–125. Plenum Press.

Fuller E. L., Jr., Holmes H. F., Gammage R. B., and Becker K. (1971) Interaction of gases with lunar materials: Preliminary results. *Proc. Second Lunar Sci. Conf., Geochim. Cosmochim. Acta*, Suppl. 2, Vol. 3, pp. 2009–2019. MIT Press.

Fuller E. L., Jr., Holmes H. F., Gammage R. B., and Secoy C. H. (1972) Gravimetric adsorption studies of thorium oxide, IV. System evaluation for high temperature studies. *Progress in Vacuum Microbalance Techniques*, Vol. 1, pp. 265–274. Heyden and Son.

Gammage R. B., Fuller E. L., Jr., and Holmes H. F. (1972) Adsorption on porous thorium oxide modified by water. *J. Colloid Interfac. Sci.* **38**, 91–96.

Gibson E. K., Jr. and Moore G. W. (1972) Thermal analysis-inorganic gas release studies on Apollo 14, 15, and 16 lunar samples. *The Apollo 15 Lunar Samples*, pp. 307–310. The Lunar Science Institute, Houston.

Gold T., Bilson E., and Yerbury M. (1972) Grain size analysis, optical reflectivity measurements, and determination of high-frequency electrical properties for Apollo 14 lunar samples. *Proc. Third Lunar Sci. Conf., Geochim. Cosmochim. Acta*, Suppl. 3, Vol. 3, pp. 3187–3193. MIT Press.

Greene C. H., Pye L. D., Stevens H. J., Rase D. E., and Kay H. F. (1971) Compositions, homogeneity, densities, and thermal history of lunar glass particles. *Proc. Second Lunar Sci. Conf., Geochim. Cosmochim. Acta*, Suppl. 2, Vol. 3, pp. 2049–2055. MIT Press.

Gregg S. J. and Sing K. S. W. (1967) *Adsorption, Surface Area, and Porosity*. Academic Press.

Grossman J. J., Mukherjee N. R., and Ryan J. A. (1972) Microphysical, microchemical, and adhesive properties of lunar material, III. Gas interactions with lunar material. *Proc. Third Lunar Sci. Conf., Geochim. Cosmochim. Acta*, Suppl. 3, Vol. 3, pp. 2259–2269. MIT Press.

Holmes H. F., Fuller E. L., Jr., and Secoy C. H. (1968) Gravimetric adsorption studies of thorium oxide, III. Adsorption of water on porous and nonporous samples. *J. Phys. Chem.* **72**, 2293–2300.

Holmes H. F., Fuller E. L., Jr., and Gammage R. B. (1973) Alteration of an Apollo 12 sample by adsorption of water vapor. *Earth Planet. Sci. Lett.* **19**, 90–96.

McClellan A. L. and Harnsberger H. F. (1967) Cross-sectional areas of adsorbed molecules. *J. Colloid Interfac. Sci.* **23**, 577–599.

Phakey P. P., Hutcheon I. D., Rajan R. S., and Price P. B. (1972) Radiation effects in soils from five lunar missions. *Proc. Third Lunar Sci. Conf., Geochim. Cosmochim. Acta*, Suppl. 3, Vol. 3, pp. 2905–2915. MIT Press.

Ries H. E., Jr. (1952) Structure and sintering properties of cracking catalysts and related materials. *Advances in Catalysis*, Vol. IV, pp. 87–149. Academic Press.

Smith W. R. and Ford D. G. (1965) Adsorption studies on heterogeneous titania and homogeneous carbon surfaces. *J. Phys. Chem.* **69**, 3587–3592.

Tittmann B. R., Abdel-Gamad M., and Housley R. M. (1972) Elastic velocity and Q factor measurements on Apollo 12, 14, and 15 rocks. *Proc. Third Lunar Sci. Conf., Geochim. Cosmochim. Acta*, Suppl. 3, Vol. 3, pp. 2565–2575. MIT Press.

Proceedings of the Fourth Lunar Science Conference
(Supplement 4, *Geochimica et Cosmochimica Acta*)
Vol. 3, pp. 2425–2435

Downslope movement of lunar soil and rock caused by meteoroid impact

W. N. Houston, Y. Moriwaki, and C.-S. Chang

Department of Civil Engineering, University of California, Berkeley, California 94720

Abstract—The relative importance of downslope movement of lunar soil and rock caused by meteoroid-impact-induced vibrations as a mode of lunar soil "erosion" has been assessed. Magnitudes of downslope movements were estimated by superimposing meteoroid-impact-induced dynamic stresses on existing static stresses in slopes of various inclinations. Accelerations in excess of the yield accelerations were double-integrated to obtain an estimate of the movements. It was found that only the very steep lunar slopes have experienced significant downslope movements due to shaking from meteoroid impacts alone and that lunar slope degradation must arise primarily by other mechanisms.

Introduction

The purpose of this study was to assess the relative importance of downslope movement of lunar soil and rock caused by meteoroid-impact-induced vibrations as a mode of lunar soil "erosion." Other, possibly concurrent, modes of erosion were excluded from this study. Gravity loads induce shear stresses along shallow surfaces which are parallel to steep lunar slopes. A seismic disturbance such as a meteoroid impact induces additional body forces on the soil and rock on lunar slopes. These body forces are typically cyclic in nature and may act in any direction in general. In particular when they act downslope the combined gravity (static) stresses and seismic (dynamic) stresses may exceed the shearing resistance of the lunar soil. When this occurs a mass of material is accelerated downslope. However, movement is normally arrested when the dynamic stresses reverse direction and act upslope. Therefore, downslope movement accumulates in small increments—often several increments per second. The magnitude of cumulative downslope movement has been estimated for a wide range of lunar slope angles by computing initial gravity (static) stresses, (Lambe and Whitman, 1969) adding meteoroid-impact-induced dynamic stresses, and comparing with computed soil strength. The general procedure used in the analysis is as follows:

1. Compute initial static stresses due to gravity.
2. Estimate magnitude, duration, and frequency content of near-surface disturbances due to meteoroid impacts and associated dynamic stress history.
3. Superimpose dynamic stresses on static stresses to determine yield acceleration. Double integrate accelerations which are in excess of yield acceleration to get movement for one impact.
4. Sum up the movements for the significant range in meteoroid size, range in distance, and age of lunar surface.

Static Stresses

Average lunar soil profile development

Computation of average static stress vs. lunar slope angle required the development of an average lunar soil profile. This development was based on data from the following sources:

1. returned core tube samples (Mitchell *et al.*, 1972),
2. self-recording penetrometer and trench failures (Mitchell *et al.*, 1972),
3. estimates of near-surface porosity from correlations with footprint depth (Houston *et al.*, 1972),
4. compression and shear parameters obtained for lunar soil simulant (Houston and Namiq, 1971).

The profile obtained is presented in Fig. 1 and a summary of physical properties of lunar soil as used herein is as follows.

$$G_s = 3.1$$

$$\phi = \tan^{-1} \frac{0.7}{e} \tag{1}$$

$$c = 0.0982 \times 10^{(1.14 - e)/0.45} \tag{2}$$

where G_s = specific gravity of solid

e = void ratio $= \dfrac{n}{1 - n}$

n = porosity
ϕ = angle of friction
c = cohesion in kN/m²

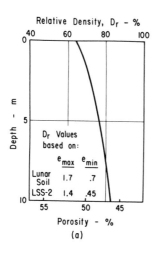

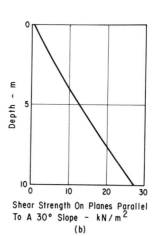

Fig. 1. Variation of relative density and shear strength with depth for "average" lunar soil profile.

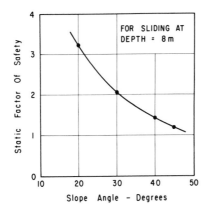

Fig. 2. Variation of static factor of safety with slope angle for "average" lunar profile.

When a soil with moderately low cohesion exhibits a marked increase in shear strength with depth as does the lunar soil, sliding on long slopes tends to occur on relatively shallow planes parallel to the ground surface. The static slope stability analyses were made according to generally accepted procedure for long slopes (Lambe and Whitman, 1969).

After substitution for c and ϕ from Eqs. (1) and (2), Eq. (3) was obtained for factor of safety.

$$\text{F.S.}_{\text{sliding}} = \frac{(1+e)(1.935)10^{(1.14-e)/0.45}}{d \sin \alpha} + \frac{0.7}{e \tan \alpha} \tag{3}$$

where d = thickness of sliding layer measured perpendicular to the surface
α = inclination of long slope
e = void ratio

Slope stability for sliding along a plane parallel to the ground surface was studied using Eq. (3) and the profile given in Fig. 1. The most critical depth for sliding was found by trial and error to be about 8 m and the variation of minimum factor of safety with slope angle is given in Fig. 2.

SEISMIC DISTURBANCE CHARACTERISTICS

Maximum accelerations

The primary source of data for maximum horizontal ground accelerations consists of measured results on terrestrial alluvium (Air Force Systems Command, 1967) for both buried and surface explosions. Data presented by Latham et al. (1970), showing that impact accelerations were 2 or 3 times greater than for surface explosions, but only 30 or 40 percent as great as for buried explosions were used as a basis for interpolating between results for surface explosions and buried explosions to get results for impacts—shown in Fig. 3.

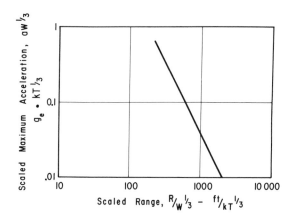

Fig. 3. Maximum radial acceleration vs. distance, scaled to 1 kT of TNT.

Acceleration values in Fig. 3 were further adjusted to account for differences in P-wave velocity, V_p, for the lunar soil profile and the terrestrial alluvium profile. In the upper several hundred meters, estimated values of V_p for lunar soil (Kovach *et al.*, 1973; Latham *et al.*, 1972) were significantly less than for terrestrial alluvium. Measured accelerations (Air Force Systems Command, 1967) on terrestrial alluvium, "soft" rock, and "hard" rock indicate that maximum acceleration increases roughly in proportion to V_p. Adjustments to the values of acceleration in Fig. 3 were made by assuming that maximum accelerations are directly proportional to V_p. Other assumptions were tried in developing upper and lower bounds on the final results, which are presented in the summary section.

Adjusted values from Fig. 3 were taken as the maximum accelerations adjacent to the rim of the crater and it was assumed that attenuation with distance from the crater was in inverse proportion to the distance from the crater, as suggested by Latham *et al.* (1973). The resulting family of attenuation curves is shown in Fig. 4.

Crater diameter, D, in meters, was related to, W, equivalent yield of TNT in kT by

$$D = 123.5 \, W^{1/3} \qquad (4)$$

This relationship is based on data presented by Vinson and Mitchell (1970) and Sauer *et al.* (1964) using a scaled depth of burial $= 14 \, \text{m/kT}^{1/3}$ as suggested by Nordyke (1961) and Shoemaker (1959).

Duration and frequency content

Estimates of duration and frequency contents were based primarily on data presented in Preliminary Science Reports for the Active and Passive Seismic Experiments. In addition, a seismic record for a small meteoroid impact was recorded as a part of the Active Seismic Experiment and was made available by the Principal Investigator (Kovach, 1972). This record, shown in Fig. 5, was used for scaling to obtain the desired acceleration, frequency, and duration characteristics.

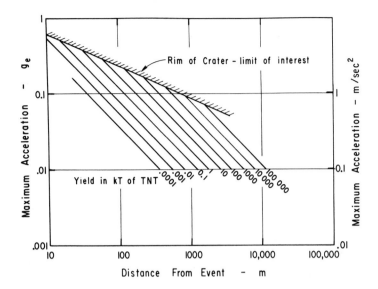

Fig. 4. Variation of maximum acceleration with distance and equivalent yield of TNT.

All ordinate values in the record were amplified as required to produce the maximum acceleration given by Fig. 4 for each impact event being considered.

Available data suggest that the predominant frequency of seismic disturbances due to meteoroid impacts, in the meteoroid size range of interest, may vary from about 20 or 30 Hz as recorded. Thus the time scale had to be expanded to obtain the desired frequency characteristics.

Parts of the "strong shaking" segment of the record were repeated to obtain the desired duration of "strong shaking." For the purpose of this analysis "strong shaking" was defined as that segment of the record where accelerations of about 30 percent of the maximum acceleration occasionally occur. Using this definition, it appears that the duration of strong shaking may vary from only a few seconds adjacent to the crater to more than 10 or 20 minutes at distances of 100 km or more. A parabolic distribution was assumed between these limits; i.e., the logarithm of the duration was assumed proportional to the logarithm of the distance from the crater. However, the form of this relationship was not critically

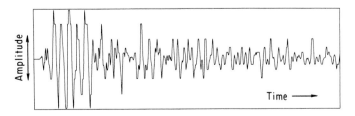

Fig. 5. Seismic signal for a small meteor impact.

important because the range of primary interest turned out to be less than 1 or 2 km, thus the duration of strong shaking was typically 10 sec or less.

Alternatively, duration of "strong shaking" may be defined as that period during which yield accelerations for a given slope are being exceeded. By this definition duration of strong shaking increases with meteroid size (and thus crater size) as would be expected.

SLOPE MOVEMENT COMPUTATIONS

Slope movement for single meteoroid impact

The downslope movement of a layer of soil is analogous to the incremental sliding of a block down an inclined plane, as illustrated in Figs. 6 and 7. Without dynamic loading the frictional resistance, *F*, in Fig. 6 may be sufficient to prevent sliding. This means that the static factor of safety against sliding is more than one. However, when the block is accelerated and inertial forces are added to static forces, the frictional resistance, *F*, may be exceeded and the block slides down the slope. If the accelerations are cyclic, movement will be arrested when the inertial force acts upslope. Thus downslope movement accumulates in increments.

The surface acceleration records discussed in the preceding section were used in conjunction with a shear wave propagation solution (Schnabel *et al.*, 1971) to compute both acceleration and shear stress histories at various depths within the profile. It was found that the accelerations within the top 8 m differed only slightly from the input accelerations at the surface. A depth of 8 m was found to be the critical depth for sliding. The dynamic stresses at this depth were superimposed on the static stresses and compared with the shear strength to obtain a yield stress, or, alternatively, a yield acceleration to be used in Fig. 8. The accelerations in excess of the yield accelerations were double-integrated to obtain increments of movement until accelerations decayed to less than the yield values.

Using this procedure the family of curves shown in Fig. 9 was developed for an average predominant frequency of about 15 Hz and a very short duration of strong shaking. Values taken from Fig. 9 were then adjusted as appropriate to account for variation in duration of strong shaking with distance from crater.

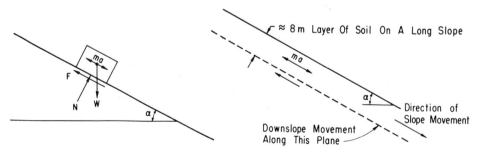

Fig. 6. Block on inclined plane analogy.   Fig. 7. Slope profile.

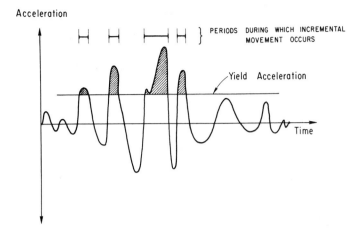

Fig. 8. Double integration of accelerograms for incremental movement determinations.

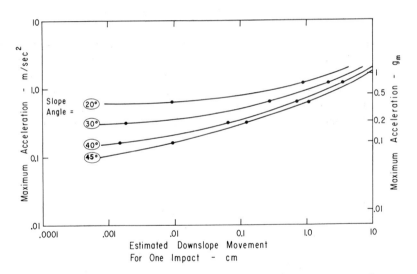

Fig. 9. Variation of movement with maximum acceleration and slope angle.

Summation of movements

Movement values from Fig. 9 account for only one impact. The next step in the analysis required integration of all impact events—with respect to distance from the point of consideration, meteoroid size, and time.

Meteoroid flux rates were used to relate number of events, meteoroid size, and time, for a unit area. Before Apollo landings, Hawkins (1963) estimated meteoroid flux rates, for the range of meteoroid sizes of interest. Gault (1970) translated these rates into cumulative crater-size frequency distributions for the lunar surface. When the ages of Apollo 11 and 12 rocks became available, Gault concluded

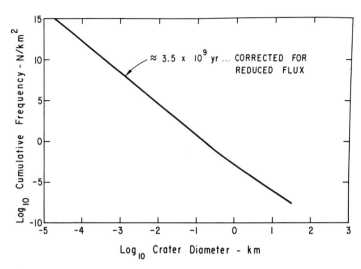

Fig. 10. Calculated size frequency distributions for craters on lunar surface.

that the Hawkins flux rates should be reduced by a factor of about 30 for consistency. Latham *et al.* (1973) concluded that the reduction factor should be about 35. A factor of 30 was used to produce the distribution shown in Fig. 10.

Figure 11 illustrates how integration with respect to area was accomplished. Each "ring" was assigned an average range from the point of interest (point on the slope being considered). The area of each ring times the meteoroid flux, Fig. 10,

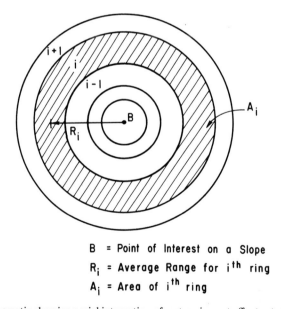

B = Point of Interest on a Slope

R_i = Average Range for i^{th} ring

A_i = Area of ith ring

Fig. 11. Schematic showing aerial integration of meteor impact effects at all distances.

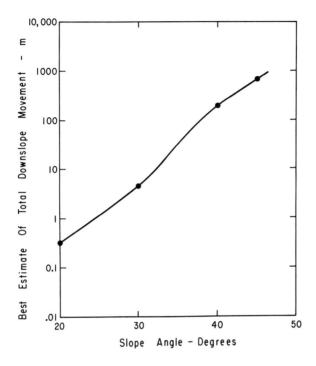

Fig. 12. Variation of downslope movement with lunar slope angle.

gave number of impacts. This computation was repeated for many small increments in meteoroid size. For each increment in meteoroid size, the average
meteoroid size and average range, used in conjunction with Figs. 4 and 9, yielded
an increment of movement. This procedure was repeated for all meteoroid sizes
of interest, all ranges of interest, and for slopes of 20 to 45 degrees. For ranges
smaller than a critical size and meteoroids larger than a critical size the frequency
of occurrence was too low to produce significant movements. Likewise, for sufficiently large ranges and small meteoroid size, movements become insignificant.
The limits of integration were established in each case so as to omit only insignificant movements in comparison with the total. The final results are shown in Fig.
12.

Summary and Conclusions

Figure 12 indicates that the flattest slope on which significant downslope
movement due to meteoroid impact may have occurred is about 25 degrees, as
listed in Table 1. The flattest slope likely to have undergone significant slope reduction by the mechanism being studied herein is about 48 degrees.

In view of the many uncertainties affecting these results—in particular the
meteoroid flux rates—it is necessary to place rather wide bounds on the best
estimates of movements. In terms of the slopes affected, upper and lower bounds
are suggested in Table 1.

Table 1. Summary of results.

	Flattest Slope on Which Significant (1 m) Downslope Movement Has Occurred	Flattest Slope Which Has Undergone Significant (1° or more) Slope Reduction
Upper Bound	35	55
Best Estimate	25	48
Lower Bound	20	45

The results of this study point to the conclusion that only the very steep lunar slopes have experienced significant downslope movements due to shaking from meteoroid impacts alone. Lunar slope degradation arises primarily by other mechanisms.

REFERENCES

Air Force Systems Command (1967) Effects of air burst, ground shock, and cratering on hardened structures. AFSCM 500-8, March 1967, USAF.

Gault D. (1970) Saturation and equilibrium conditions for impact cratering on the lunar surface: criteria and implications. *Radio Science* **5**, 273–291.

Hawkins G. S. (1963) Impacts on the earth and moon. *Nature* **197**, 781.

Houston W. N. and Namiq L. I. (1971) Penetration resistance of lunar soils. *Journal of Terramechanics* **8**, 59–69.

Houston W. N., Hovland H. J., Mitchell J. K., and Namiq L. I. (1972) Lunar soil porosity and its variation as estimated from footprints and boulder tracks. *Proc. Third Lunar Sci. Conf., Geochim. Cosmochim. Acta*, Suppl. 3, Vol. 3, pp. 3255–3263. MIT Press.

Kovach R. L. (1972) Personal communication.

Kovach R. L., Watkins J. S., and Talwani P. (1973) Active Seismic Experiment. *Apollo 16 Preliminary Science Report*, NASA SP-315, pp.10-1–10-14.

Lambe T. W. and Whitman R. V. (1969) *Soil Mechanics*, pp. 192–193, 354–356. John Wiley.

Latham G. V., MacDonald W. G., and Moore H. J. (1970) Missile impacts as sources of seismic energy on the moon. *Science* **168**, 242–245.

Latham G. V., Ewing M., Press F., Sutton G., Dorman J., Nakamura Y., Toksöz N., Lammlein D., and Duennebier F. (1972) Passive seismic experiment. *Apollo 15 Preliminary Science Report*, NASA SP-289, pp. 8-1–8-25.

Latham G. V., Ewing M., Press F., Sutton G., Dorman J., Nakamura Y., Toksoz N., Lammlein D., and Duennebier F. (1973) Passive seismic experiment. *Apollo 16 Preliminary Science Report*, NASA SP-315, pp. 9-1–9-29.

Mitchell J. K., Houston W. N., Scott R. F., Costes N. C., Carrier W. D. III, and Bromwell L. G. (1972) Mechanical properties of lunar soil: Density, porosity, cohesion and angle of internal function. *Proc. Third Lunar Sci. Conf., Geochim. Cosmochim. Acta*, Suppl. 3, Vol. 3, pp. 3235–3253. MIT Press.

Nordyke M. (1961) On cratering: A brief history, analysis and theory of cratering. UCRL-6578, Lawrence Radiation Laboratory, Livermore, Calif.

Sauer F. M., Clark G. B., and Anderson D. C. (1964) Cratering by nuclear explosives, DASA-1285(IV), Nuclear Geoplosics, Part Four, Empirical analysis of ground motion and cratering. The Defense Atomic Support Agency, Washington, D.C.

Schnabel P., Seed H. B., and Lysmer J. (1971) Modifications of seismograph records for effects of local soil conditions. *Earthquake Engineering Research Center Report*, No. EERC 71-8, University of California, Berkeley, December.

Shoemaker E. M. (1959) Impact mechanics at meteor crater, Arizona. Open File Report, July 1959, *U.S. Geological Survey.*

Vinson T. S. and Mitchell J. K. (1970) Deduction of lunar surface material strength parameters from lunar slope failures caused by impact events—feasibility study. *Lunar Surface Engineering Properties Experiment Definition, Vol. II, Final Report*, Contract NAS 8-21432, University of California, Berkeley, pp. 2-1–2-32.

Proceedings of the Fourth Lunar Science Conference
(Supplement 4, *Geochimica et Cosmochimica Acta*)
Vol. 3, pp. 2437–2445

Surface soil variability and stratigraphy at the Apollo 16 site

J. K. Mitchell

Department of Civil Engineering, University of California, Berkeley, California 94720

W. D. Carrier III

NASA Johnson Space Center, Houston, Texas 77058

N. C. Costes

NASA Marshall Space Flight Center, Huntsville, Alabama 35812

W. N. Houston

Department of Civil Engineering, University of California, Berkeley, California 94720

R. F. Scott

California Institute of Technology, Pasadena, California 91109

Abstract—The results of penetration tests, analyses of footprint and Lunar Roving Vehicle track depths, and core tube sample data have been used to deduce details of near-surface stratigraphy (to depths of several tens of cm) and lateral variability in soil conditions. Local variations (meter scale) in penetration resistance and porosity may be large, and soil stratigraphy may be complex. Since average properties are about the same at all sites, these variations probably reflect individual cratering and depositional events. These local variations cannot be anticipated on the basis of surface appearance or behavior.

Introduction

Measurements of penetration resistance to depths of several tens of centimeters using a Self-Recording Penetrometer (SRP), analyses of footprint depths, study of Lunar Roving Vehicle (LRV)-soil interaction, and core tube samples have enabled determination of some aspects of the local soil variability and stratigraphy at the Apollo 16 landing site. It has been established that local variability in soil physical properties may be great and that the vertical stratigraphy of the near-surface soil may be complex even though relative uniformity may be suggested by the surface appearance. The purpose of this paper is to indicate the nature of some of these variations and to indicate how stratigraphic variability may be inferred from the results of rather simple tests and observations.

Methods

The SRP was the main quantitative data source for *in situ* soil physical properties measurements during the Apollo 16 mission. This device, which has been described by Mitchell *et al.* (1973), indicates

2437

penetration resistance vs. depth to a maximum depth of 76 cm and a maximum penetration force of 215 N. Four cone penetration tests were made at Station 4 in the Descartes Material on Stone Mountain, and five cone penetration tests were done in the Cayley Plains in the Station 10-ALSEP area.

The core drive tubes used for Apollo 16 were of the same type as used for Apollo 15 and consisted of thin-walled tubes 37.5 cm long with an inside diameter of 4.13 cm and outside diameter of 4.38 cm. The tubes were used both singly and in combination (double-core tubes). X-radiographs of the unopened core tube samples have provided direct indication of stratigraphy. Interpretation of Apollo 16 radiographs are presented by Hörz *et al.* (1973).

A total of 309 footprints in the Apollo 16 photographs was studied, and porosity variations were analyzed statistically using the method of Houston *et al.* (1972). Values determined in this way reflect average porosities for the top 15 cm, but stress distribution considerations indicate the values are influenced most strongly by soil conditions in the upper 5 to 10 cm. Porosity estimates were also made based on analysis of LRV track depths using the methods of Costes (1973).

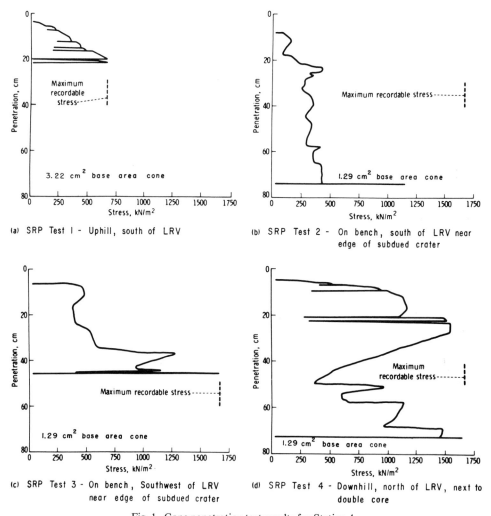

(a) SRP Test 1 – Uphill, south of LRV

(b) SRP Test 2 – On bench, south of LRV near edge of subdued crater

(c) SRP Test 3 – On bench, Southwest of LRV near edge of subdued crater

(d) SRP Test 4 – Downhill, north of LRV, next to double core

Fig. 1. Cone penetration test results for Station 4.

PENETROMETER TEST RESULTS

Curves showing penetration resistance as a function of depth are shown in Fig. 1 for cone-penetration tests at Station 4 and Fig. 2 for cone-penetration tests at Station 10 (no data were recorded for one of the five tests at Station 10). The spikes shown on several of the penetration curves reflect a sudden unloading and reloading and have been shown (Mitchell *et al.*, 1973) to be a consequence of test procedure and not of the soil conditions. As may be seen in Figs. 1 and 2, an intercept for zero stress on the penetration axis is present in each test. This resulted from apparatus and test procedure effects. The penetration curves do not reflect soil conditions above the intercept point, but they are correct below it.

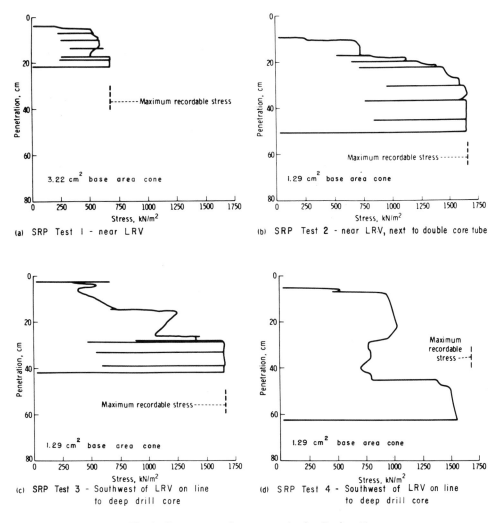

Fig. 2. Cone-penetration test results for Station 10.

DISCUSSION

Stratigraphy

Figures 1 and 2 may be used to infer details of the lunar soil to depths of a few decimeters. The difference in penetration curve shapes are a direct indication of local variability. Some characteristics at the different test locations may be noted.

Station 4 (*Stone Mountain*). Figure 1(a) for penetration at a point uphill south of the LRV shows relatively homogeneous soil to a depth of about 20 cm. Much softer soil was encountered a few meters south of the LRV near the edge of a subdued crater, Fig. 1(b). A layer of higher resistance is indicated at a depth of approximately 24 cm. On the bench southwest of the LRV a highly resistant layer was encountered at a depth at 45 cm, Fig. 1(c).

Downhill to the north of the LRV the penetration curve, Fig. 1(d) indicates a very dense and resistant layer to a depth of 27 cm. Softer soil was located below this layer to a depth of approximately 50 cm, where firm material was again encountered. A simulation test was done on a model soil composed of a soft, weak layer sandwiched between two firm layers to test this interpretation. The curve for penetration resistance as a function of depth for this condition is shown in Fig. 3. It is comparable to the curve in Fig. 1(d). Figures 1(b) and 1(c) also suggest the presence of this soft layer at depths of 25 and 55 cm and 15 to 35 cm, respectively.

A double-core drive tube sample was taken at a point about 1 m from the test shown in Fig. 1(d). The X-radiograph of this core sample reveals layers that

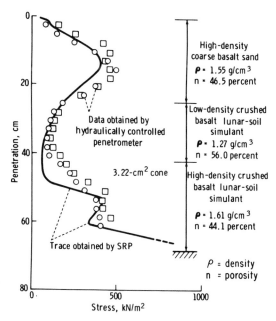

Fig. 3. Penetration resistance as a function of depth for a soft soil sandwiched between two firm layers.

correlate well with the penetration curve. Interpretations of the soil stratigraphy in the core tube by J. S. Nagle of the LSPET are shown in Fig. 4. It has been suggested by Nagle that the coarse layer with abundant rock fragments decreasing with depth is South Ray Crater material and that the Descartes deposit underlies the South Ray layer at a depth of about 50 cm.

Station 10-*ALSEP area.* The ground surface along the line from the LRV at Station 10 southwest to the ALSEP area was generally level and free of large rock fragments. From (1) the shapes of the penetration curves, Fig. 2; (2) X-radiographs of the drill core stem at the ALSEP site; and (3) the X-radiographs of the Station 10 core sample, a preliminary stratigraphic section has been prepared as shown in Fig. 5. Five layers have been detected with varying thicknesses: surficial, hard, soft, harder, and rocky. This complex stratigraphy may reflect a series of depositional events associated with several meteor impacts.

Variability in lateral direction

Whereas vertical stratigraphy can be inferred from the penetration curves presented above, lateral variations in soil conditions to depths of several decimeters can be deduced through comparison of penetration curves and drive tube and drill

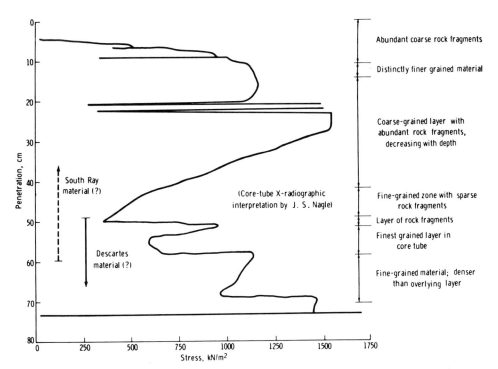

Fig. 4. Correlation of X-radiograph interpretation of Station 4 double core-tube stratigraphy with adjacent SRP test downhill, north of LRV.

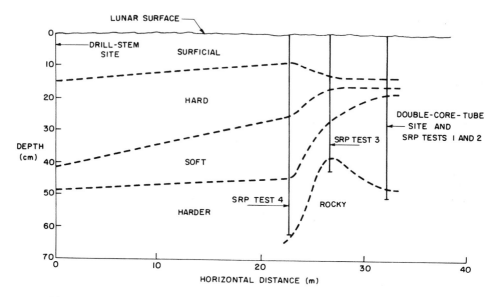

Fig. 5. Approximate soil profile between Station 10 double-core-tube site and deep-drill-stem site in ALSEP area.

stem samples from different locations. Near-surface variability can also be deduced from variations in footprint and LRV track depth.

Figure 6 shows envelopes within which all Apollo 16 penetration curves fall. Two zones have been identified; tests at Station 4 on Stone Mountain and tests near Station 10 and the ALSEP area on the Cayley Plains. These same data are included in Fig. 7 along with penetration data from Apollo 14 (Fra Mauro), Apollo 15 (Hadley-Apennine) and Lunokhod 1. While several different penetrometer configurations were used to obtain the data in Fig. 7, it is apparent that greater soil variability was encountered on the slopes of Stone Mountain than in the Plains area, and that the range at Station 4 is about as great as has yet been encountered at any site on the moon. It may be seen also that the average penetration resistance on the Plains is greater than on Stone Mountain. Because the variations in penetration resistance at Station 4 on Stone Mountain appear to bear little relationship to local slope or surface appearance, generalizations concerning the strength of soils on sloping terrain are not possible.

Drive tube, drill stem, and penetrometer data all indicate a general increase in density and decrease in porosity with depth, with the greatest rate of change occurring at shallow depths. Measured densities are in the range of about 1.4 to 2.0 g/cm^3, with most values in the range of 1.5 to 1.8 g/cm^3. The densities at depths greater than a few centimeters are greater on the Cayley Plains than on Stone Mountain.

Statistical analysis of 309 footprints on the lunar surface at the Apollo 16 site has given indication of the variation porosity in the upper 5 to 10 cm of the lunar soil. Figure 8 shows the variability observed. Separate analyses were made for

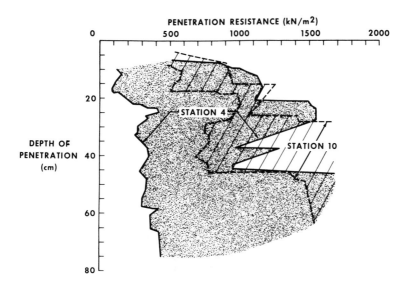

Fig. 6. Envelopes for the Apollo 16 penetration test results.

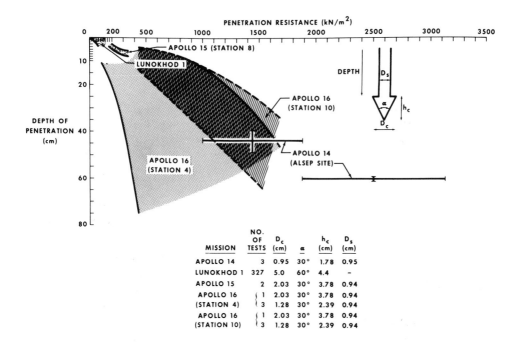

MISSION	NO. OF TESTS	D_c (cm)	α	h_c (cm)	D_s (cm)
APOLLO 14	3	0.95	30°	1.78	0.95
LUNOKHOD 1	327	5.0	60°	4.4	–
APOLLO 15	2	2.03	30°	3.78	0.94
APOLLO 16 (STATION 4)	1	2.03	30°	3.78	0.94
	3	1.28	30°	2.39	0.94
APOLLO 16 (STATION 10)	1	2.03	30°	3.78	0.94
	3	1.28	30°	2.39	0.94

Fig. 7. Penetration resistance of the lunar surface at different locations.

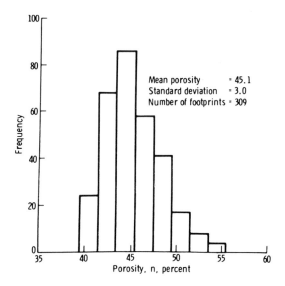

Fig. 8. Histogram of porosity variations at the Apollo 16 site as deduced from footprint depths.

different stations (Mitchell *et al.*, 1973) as well as the composite analysis shown in Fig. 8. From these analyses, as well as from analyses of LRV track depths, it appears that:

(1) The average porosity of the surface soil varies little among different locations.

(2) The average porosity (45% by footprint analysis and 41% by LRV track analysis) of intercrater areas at the Apollo 16 site is slightly greater than that at four previous Apollo sites (43.3% by footprint analysis).

(3) The average porosity on crater rims at the Apollo 16 site (46.1% by footprint analysis) is almost the same as that at the previous Apollo sites (46.7%).

(4) The standard deviation of porosities for the intercrater areas is the same as the average standard deviation (2.8%) for all previous Apollo missions.

(5) The standard deviation of porosities of crater rims is 4.6% at the Apollo 16 site as compared with the average of 4.0% for all previous sites.

Preliminary analysis of Apollo 17 photographs has also yielded results consistent with the above findings. It should be noted that while lunar soil variability may be described in terms of density and porosity, as done herein, the physical behavior, e.g., strength, compressibility, penetration resistance, will also depend strongly on the relative density; i.e., the looseness or denseness of the soil relative to maximum and minimum porosity values. Relative density considerations are considered in more detail by Carrier *et al.* (1973).

Conclusions

Soil mechanics data from all the Apollo missions support the general conclusion that global lunar processes, such as meteorite impact and the solar wind, control average properties such as grain size distribution and porosity, which are nearly the same at all sites regardless of composition or topography. Significant local (meter scale) variations may exist which probably reflect individual cratering and depositional events. Near surface soil stratigraphy may be complex and generally cannot be anticipated on the basis of surface appearances or behavior, but is readily detectable from the results of penetration tests and study of core samples.

References

Carrier W. D. III, Mitchell J. K., and Mahmood A. (1973) The relative density of lunar soil (abstract). In *Lunar Science—IV*, pp. 118–120. The Lunar Science Institute, Houston.

Costes N. C. (1973) Regional variations in physical and mechanical properties of lunar surface regolith (abstract). In *Lunar Science—IV*, pp. 159–161. The Lunar Science Institute, Houston.

Hörz F., Carrier W. D. III, Young J. W., Duke C. M., Nagle J. S., and Fryxell R. (1973) Apollo 16 special samples. *Part B of Sec. 7 Apollo 16 Preliminary Science Report*, NASA SP-315.

Houston W. N., Hovland H. J., Mitchell J. K., and Namiq L. I. (1972) Lunar soil porosity and its variation as estimated from boulder tracks. In *Proc. Second Lunar Sci. Conf., Geochim. Cosmochim. Acta*, Suppl. 3, Vol. 3, pp. 3255–3263. MIT Press.

Mitchell J. K., Carrier W. D. III, Houston W. N., Scott R. F., Bromwell L. G., Durgunoglu H. T., Hovland H. J., Treadwell D. D., and Costes N. C. (1973) Soil mechanics. *Sec. 8. Apollo 16 Preliminary Science Report*, NASA SP-315.

Proceedings of the Fourth Lunar Science Conference
(Supplement 4, *Geochimica et Cosmochimica Acta*)
Vol. 3, pp. 2447–2452

Solar albedo and spectral reflectance for Apollo 15 and 16 lunar fines

R. C. BIRKEBAK

Department of Mechanical Engineering, University of Kentucky, Lexington, Kentucky 40506

J. P. DAWSON

Oklahoma Foundation for Research and Development Utilization,
Oklahoma City, Oklahoma

Abstract—The spectral directional reflectance of Apollo 15 and 16 fines were obtained for bulk densities of approximately 1000 and 1600 kg/m³. The solar albedo as a function of angle of illumination was calculated from these results. Comparison of solar albedos show that Apollo 11, 12, and 15 soils fall into one group and the Apollo 14 and 16 soil results into a second group corresponding to mare or lunar highland materials.

INTRODUCTION

THE DIRECTIONAL solar albedo is a required thermophysical property if heat transfer calculations are to be made on the lunar surface. It is defined as

$$\rho_s(\theta) = \int_{\lambda=0.3}^{\lambda=6} \rho(\lambda, \theta) S_\lambda \, d\lambda \Big/ \int_{\lambda=0.3}^{\lambda=6} S_\lambda \, d\lambda \tag{1}$$

where $\rho(\lambda, \theta)$ is the directional spectral reflectance, S_λ the spectral solar energy distribution, and θ the angle of illumination.

Cremers *et al.* (1971a, b; 1972) have used the directional solar albedo along with the emittance, thermal conductivity, and specific heats of lunar fines to calculate the variation of both lunar surface temperature and sub-surface temperatures as a function of the lunation period.

Results are presented for Apollo 15 sample 15041,38 and Apollo 16 sample 68501,18. The Apollo 15 sample was collected near the Alsep, Station 8, and is from the top of the trench dug at this location. The Apollo 16 sample was obtained at Station 8 and is situated on a blanket of light-colored ejector from South Ray Crater. The spectral directional reflectances were obtained for soil bulk densities of approximately 1000 and 1600 kg/m³. The larger bulk density used falls near the average value of that of the core-tube samples, LSPET (1972) and Carrier *et al.* (1973). Further increase in bulk density cause only minor changes in the reflectance values. The soils were illuminated at angles from 10 to 60 degrees.

The Experiment

The spectral directional reflectances were measured using an integrating sphere reflectometer with the sample mounted centrally in the interior of the sphere. A spectrophotometer was used as the energy source and the reflected energy was measured with a lead sulfide detector. For details see Birkebak *et al.* (1970). The system was coupled to a minicomputer system which logged the data from the spectrophotometer scans and performed the necessary calculations to obtain the reflectance.

The sample holder consisted of a Teflon cup with a sample area 25 mm in diameter by 9 mm in thickness. Each sample was weighed to the appropriate number of grams corresponding to the desired density and then carefully loaded into the cell. The sample surface was smoothed with a stainless steel spatula until it was level with the top of the Teflon cup. Some settling occurred during the measurements using a sample with a bulk density of 1000 kg/m³ and this necessitated several sample loadings until the desired results were obtained.

Results and Discussion

The spectral directional reflectance for the Apollo 15 sample at a bulk density of 1615 kg/m³ is shown in Fig. 1 and for the Apollo 16 sample at a bulk density of 1630 kg/m³ in Fig. 2. The reflectance of both samples increase with wavelength and angle of illumination. The estimated error over the entire wavelength regions is less than ±1 percent. The increased reflectance with angle of illumination has a

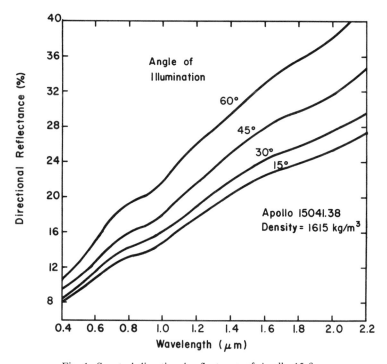

Fig. 1. Spectral directional reflectance of Apollo 15 fines.

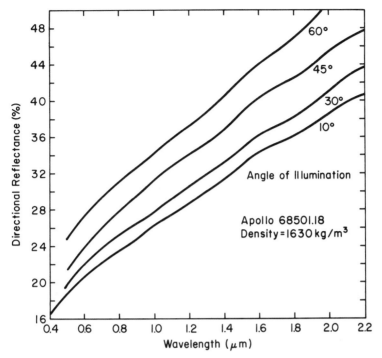

Fig. 2. Spectral directional reflectance of Apollo 16 fines.

pronounced effect on the lunar surface temperature. This effect is discussed in greater detail in reference to the solar albedo.

The Apollo 15 results indicate an absorption band centered approximately at 0.96 μm. The Apollo 16 sample results show no indication of any absorption band between 0.9 and 1 μm. The presence or absence of absorption bands as discussed by Adams and McCord (1972, 1973) is a function of the dark glass content and the crystal to glass ratio.

The variation in spectral reflectance between the Apollo samples is associated with their chemical composition and glass content. As the lunar fines become lighter in color, we find fewer opaque materials in the fines. Adams and McCord (1972) have discussed the lighter appearance of Apollo 14 fines and relate it to the fines having lower overall iron oxide and titanium dioxide content. In general the color of the Apollo 15 and 16 samples follows this general trend with the Apollo 15 sample having a greater percentage of TiO_2 and FeO and a lower percentage of Al_2O_3 than the Apollo 16 sample.

The present spectral reflectance results are compared to Apollo 11, 12, and 14 fines for similar bulk densities in Fig. 3. From these results the Apollo 11, 12, and 15 samples can be classified as mare material and the Apollo 14 and 16 as lunar highland material.

The solar albedo was calculated from the spectral results for a wavelength range of 0.3 to 6.0 μm. Approximately 93 percent of the solar energy falls between

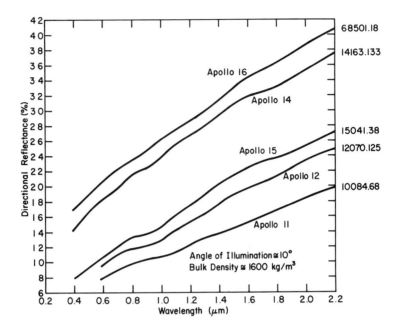

Fig. 3. Spectral directional reflectance of fines from the Apollo missions.

the wavelengths 0.3 to 2.2 μm. Data from 2.0 to 6.0 μm were obtained from spectral emittance results to be presented elsewhere and the inclusion of this has little effect on the calculated albedo. The results of our calculations are presented in Table 1 as a function of density and angle of illumination. Over the range of bulk densities from 1000 to 1600 kg/m^3, the Apollo 16 sample albedo is approximately twice that of the Apollo 15 sample and is due mainly to the differences of TiO_2, FeO, and Al_2O_3 contents of the samples. The increase in albedo with increasing density is due to the decrease in porosity of the sample (*see* Birkebak, 1973).

Comparisons of the solar albedo for samples from Apollo 11, 12, 14, 15, and 16 missions are presented in Table 2. As would be expected the results are divided into two groups, the samples from mare regions and those from the highlands.

Table 1. Solar albedo for Apollo 15 and 16 lunar fines.

Angle of illumination (degrees)	Apollo 15 Density (kg/m^3)			Apollo 16 Density (kg/m^3)		
	1000	1615		1000	1310	1630
15	0.105	0.137	10	0.204	0.218	0.238
30	0.113	0.146	30	0.220	0.225	0.255
45	0.13	0.162	45	0.242		0.278
60	0.148	0.193	60	0.281	0.280	0.312

To facilitate the use of the directional solar albedo in lunar heat transfer calculations, the results were fitted to the following equation

$$\rho(\theta) = A + B\theta + C\theta^2 + D\theta^3 + E\theta^4 \tag{2}$$

and when $\theta = 90°$, the reflectance is one. The results of this curve fitting procedure are given in Table 3. The change in albedo as a function of angle of illumination is characteristic of dielectric materials. The albedo changes rapidly for angles greater than approximately 70 degrees. Until measurements are obtainable for angle of illumination greater than 60 degrees, extrapolation of the 4th order curve can be used to estimate the albedo.

Cremers *et al.* (1972) have used the directional albedo as described by Eq. (2) in their heat transfer calculations to predict the lunar surface temperature variation with time for Apollo 12 fines. Over the lunation period where experimental directional albedo data are available, the lunar surface temperatures are lower. For an angle of illumination of 60 degrees the temperatures are approximately 8°K lower than for the constant albedo case. The maximum difference occurs for a lunation fraction of 0.24 and 0.76 where, as present in Table 4, the variable property calculated temperatures are 43.7°K and 64.1°K lower than the constant

Table 2. Comparison of solar albedo for lunar fines.
(Density = 1600 kg/m³)

Angle of illumination (degrees)	Apollo Mission				
	11	12	14	15	16
10	0.099	0.119	0.213	0.137	0.238
30	0.107	0.115	0.221	0.146	0.255
45	0.113	0.136	0.262	0.162	0.278
60	0.133	0.181	0.297	0.193	0.312

Table 3. Coefficients for directional reflectance equations $\rho(\theta) = A + B\theta + C\theta^2 + D\theta^3 + E\theta^4$.

	Apollo 15 15041,38 Density (kg/m³)		Apollo 16 68501,18 Density (kg/m³)		
	1000	1615	1000	1310	1630
A	0.283	0.2606	0.267	0.1469	0.3064
$B \times 10$	-0.2842	-0.1768	-0.1122	0.1063	-0.1219
$C \times 10^2$	0.1151	0.0846	0.0609	-0.0395	0.0655
$D \times 10^4$	-0.2126	-0.1609	-0.1215	0.0425	-0.1286
$E \times 10^6$	0.1391	0.10996	0.0864		0.0894

Table 4. (From Cremers *et al.* 1972). Comparison of
temperature for case of variable surface properties with
that of constant surface properties Apollo 12 lunar fines.

| Fraction of Lunation | Temperature (°K) | |
	variable $\rho(\theta)$, $\epsilon(T)$	constant ρ, ϵ Cremers *et al.* (1971)
0 (noon)	389.3	389.4
0.167*	315.0	324.5
0.24	161.2	204.9
0.25 (sunset)	134.4	147.5
0.50 (midnight)	94.7	96.8
0.75 (sunrise)	86.1	87.8
0.76	125.4	189.5
0.833*	317.5	325.5

ϵ = total emittance of the surface.
*Equivalent to 60° angle of illumination.

property solution temperatures. However, these temperatures are based on extrapolated directional albedo values and should be considered only as approximate
values. The importance of the directional albedo, however is clearly evident in the
results of the thermal calculations.

Acknowledgment—This research was supported by NASA Grant NGR18-001-062.

References

Adams J. B. and McCord T. B. (1972) Optical evidence for regional cross-contamination of highland
and mare soils (abstract). In *Lunar Science—III*, pp. 1–3. The Lunar Science Institute, Houston.
Adams J. B. and McCord T. B. (1973) Further evidence for vitrification darkening of lunar soils
(abstract). In *Lunar Science—IV*, pp. 7–8. The Lunar Science Institute, Houston.
Birkebak R. C. (1973) Thermal radiation properties of lunar materials from the Apollo mission. In
Advance in Heat Transfer (editor J. P. Hartnett and T. F. Irvine). Academic Press. In press.
Birkebak R. C., Cremers J. C., and Dawson J. P. (1970) Directional spectral and total reflectance of
lunar material. *Proc. Apollo 11 Lunar Sci. Conf., Geochim. Casmochim. Acta*, Suppl. 1, Vol. 3, pp.
1993–2000. Pergamon.
Carrier W. D., Mitchell J. K., and Mohmood A., (1973) The relative density of lunar soil (abstract). In
Lunar Science—IV, pp. 118–120. The Lunar Science Institute, Houston.
Cremers C. J., Birkebak R. C., and White J. E. (1971a) Lunar surface temperatures at tranquillity base.
AIAA J. 9, 1899–1903.
Cremers C. J., Birkebak R. C., and White J. E. (1971b) Lunar surface temperatures from Apollo 12.
The Moon 3, 346–351.
Cremers C. J., Birkebak R. C., and White J. E. (1972) Thermal characteristics of the lunar surface
layer. *Int. J. Heat Mass Transfer* 15, 1045–1055.
LSPET (Lunar Sample Preliminary Examination Team) (1972) *Apollo 15 Preliminary Science Report*,
NASA SP-289.

Proceedings of the Fourth Lunar Science Conference
(Supplement 4, *Geochimica et Cosmochimica Acta*)
Vol. 3, pp. 2453–2458

Thermoluminescence of lunar fines (Apollo 12, 14 and 15) and lunar rock (Apollo 15)

U. Brito, C. Lalou, and G. Valladas

Centre des Faibles Radioactivités (CNRS/CEA) 91190 Gif-sur-Yvette, France

T. Ceva and R. Visocekas

Laboratoire de Luminescence II—Tour 13 Université de Paris VI—4, place Jussieu 75230 Paris Cedex
05—France
Equipe de recherche associée au CNRS

Abstract—A study is made of the spectrum of the thermoluminescent emission of fines samples of Apollo 12, 14, and 15 missions and rock sample of Apollo 15 mission. Differences in intensity of the whole TL from one site to another have been found that may be explained by differences in mineralogical composition. In the Apollo 15 drill core sample, the percentage of TL carriers seems different from one level to another. To make precise studies of TL gradients, it is necessary to work with normalized TL on mineralogical separates.

Introduction

In a previous paper (Lalou *et al.*, 1972) we have shown on sample 14163,147 that its TL, natural as well as artificial, is composed of light emitting in two wavelengths: predominantly green and yellow light below 400°C and blue and violet light above 400°C.

This sample was very poorly thermoluminescent and did not allow a detailed study of the whole spectrum using interferential filters. Nevertheless, this first study was encouraging for us as Hoyt *et al.* (1972) have shown, using a single grain technique, that the low temperature peak is probably due to cosmic ray radiation alone, whereas the high temperature peak is due to cosmic ray radiation and to internal natural radioactivity.

We have had the opportunity, due to the kindness of Dr. Zimmerman to have a much more thermoluminescent sample from Apollo 12 (12033,49) which shows a TL about a hundred times greater, as may be seen on Fig. 1.

The measurements have been made with the apparatus described in a previous paper. The only modification is the detector of light output which is here a EMI 9635 QB photomultiplier. Thus, the signal/noise ratio is improved.

The use of either a colored filter (3800 Å, $\Delta\lambda = 800$ Å) or an interferential filter (4390 Å, $\Delta\lambda = 230$ Å) differentiates three peaks, two of which cannot be resolved without a filter. Moreover, the use of an interferential filter at 5380 Å ($\Delta\lambda = 200$ Å) resolves completely the first peak (Fig. 2). This was a good confirmation of our previous study.

It was then interesting, on samples from cores, fines or rocks, to see if the variation with depth of the different peaks was parallel or different.

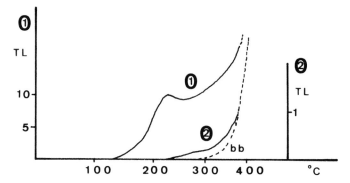

Fig. 1. Natural glow curves for (1) sample 12033,49 (2) sample 14163,147. TL is given in arbitrary units, left scale for curve 1 and right scale for curve 2. Sample weight: 1 mg.

CORES OF APOLLO 15 FINES

For this study, we received 5 levels:

15002,248
15003,256
15004,184
15005,151
15006,181

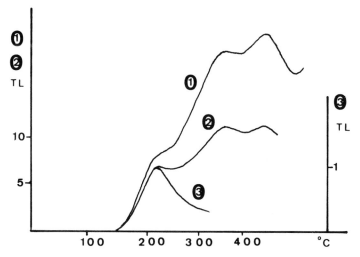

Fig. 2. Natural glow curves for sample 12033,49 (1 mg); (1) with a coloured filter centered around 3800 Å; (2) with an interferential filter centered around 4390 Å; (3) with an interferential filter centered around 5380 Å; TL is given in arbitrary units, left scale for curves 1 and 2, right scale for curve 3. No correction has been done for filter transmission.

Unfortunately, we didn't receive any sample in the top part of the core where the greatest variations may be expected.

Figure 3 gives the spectral analysis of the 5 levels for the peak at 220°C and Fig. 4, the same study for the peak at 380°C. To draw these curves, we used a set of 15 interferential filters whose principal wavelengths are: 3565, 3680, 3715, 3835, 3990, 4240, 4390, 4620, 4790, 4980, 5190, 5380, 5550, 5790, and 6000 Å. Δλ being about 70 Å until 3835 Å, and 200 Å from 3990 to 6000 Å. The transmission varies from 18 to 60 percent; the measurements have been corrected for each filter.

Owing to the small amount of available material, the number of samples was strictly limited. The procedure has been: first, a measurement of the total spectrum, with all of the filters but with only one measurement with each of them. And then, the most characteristic wavelengths having been chosen, six measurements were done for each of them and each level. The error bars appearing on Figs. 3, 4, and 5 show the mean square error on the six measurements made.

These two figures show that the two peaks do not have the same wavelength spectrum, the 380°C peak is due to light emission about 4000 Å whereas the 220°C peak shows a broader spectrum. But, as TL is about 15 times lower than that of sample 12033,49, it was not possible to resolve completely this peak as was done for 12033,49 sample (Fig. 2).

The variation in TL intensity may be due to two main reasons:

(1) Difference in ionization rates, due either to the attenuation of cosmic rays with depth or to differences in radionuclides content.

(2) Difference in the quantity of TL carriers.

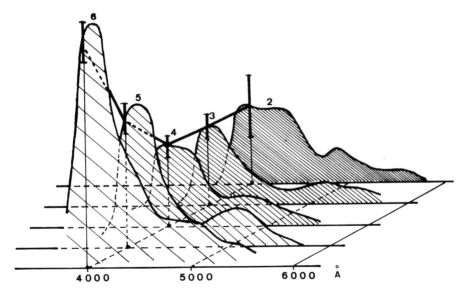

Fig. 3. TL emission spectrum for the five levels of Apollo 15 fines core at 220°C, level 2 is the deepest and level 6 the highest. Corrections for transmission of filters have been made. Sample weight: 1 mg.

U. Brito *et al.*

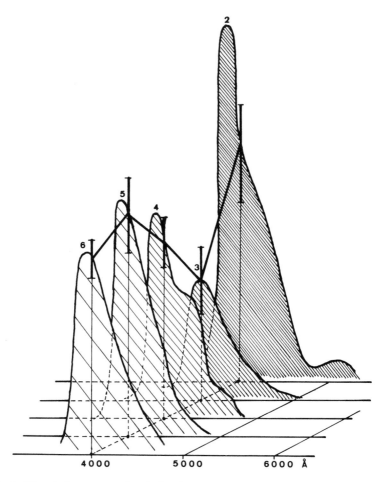

Fig. 4. TL emission spectrum for the five levels of Apollo 15 fines core at 380°C, level 2 is
the deepest and level 6 the highest. Corrections for transmission of filters have been
made. Sample weight: 1 mg.

The second reason may be investigated by using normalized TL, that is the
ratio between natural TL and artificial TL obtained by irradiation.

As we saw, specially for the 380°C peak, an anomalous increase of TL in level
2, which is the deepest, we irradiated the 5 levels with 500 krads of ^{60}Co. The
corresponding responses obtained with a blue filter ($\lambda = 3900$ Å, $\Delta\lambda = 220$ Å) were
found to be practically the same within 20 percent. The curves showing the
variation with depth of normalized TL are given on Fig. 5.

A few remarks can be made:

(1) The discrepancy of the level number 2, visible on Fig. 4, disappears.

(2) The normalized TL for level 6—highest level—has a low value.

(3) We are led to consider the variations of the TL of various levels in Fig. 4 as
related with the percentage of TL carriers.

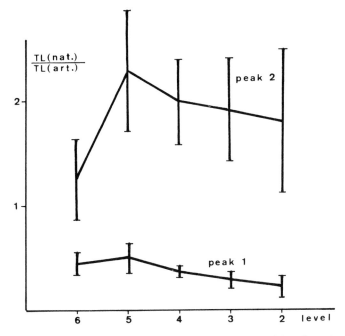

Fig. 5. TL output (normalized TL) variation with depth in core of dust from Apollo 15.
Sample weight: 1 mg.

(4) For level 2, the various TL output measurements on different samples show a great dispersion, probably indicating a great heterogeneity of the sample. Besides, they indicate that very few minerals would be responsible for the TL.

Rock 15555

In this rock, we received 6 samples, taken from the center to the outside along a radius whose orientation is badly defined.

The TL output is about 150 times inferior to the one of sample 12033,49, that is inferior to the one of sample 14163,147, and presents only one badly defined peak at 400°C.

Nevertheless, we were able to measure the TL emission spectrum at 400°C given in Fig. 6.

We tried to study the gradient in this rock. A great heterogeneity of the material was found, and the error bars, based on five different measurements are too large to allow any conclusion.

Conclusions

This study does not bring definite conclusions, but some results may be pointed out and may be used for further studies:

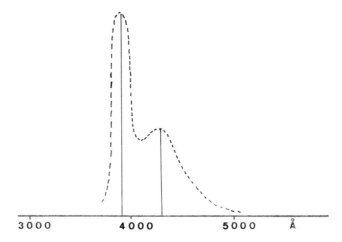

Fig. 6. TL emission spectrum of rock 15555 at 400°C. Correction for transmission of filters have been made. Sample weight: 1 mg.

(1) There is a great difference in intensity and in the shape of TL curves from one point to another on the lunar surface. The intensity is certainly dependent on the mineralogical composition; that is, on the percent of TL carriers.

Referred to TL of sample 12033,49:

TL of rock 15555 is about 150 times lower; TL of fines 14163,147 is about 100 times lower; TL of Apollo 15 core is about 15 times lower.

(2) The use of filters allows a separation of the various TL peaks. Differences in the shape of TL curves must be due to differences in TL carriers.

(3) For an accurate study of TL variations, it is necessary to work with normalized TL or, better, with normalized TL on mineralogical separates.

REFERENCES

Hoyt H. P. Jr., Walker R. M., Zimmerman D. W., and Zimmerman J. (1972) Thermoluminescence of individual grains and bulk samples of lunar fines. *Proc. Third Lunar Sci. Conf., Geochim. Cosmochim. Acta*, Suppl. 3, Vol. 3, pp. 2997–3007. MIT Press.
Lalou C., Valladas G., Brito U., Henni A., Ceva T., and Visocekas R. (1972) Spectral emission of natural and artificially induced thermoluminescence in Apollo 14 lunar sample 14163,147. *Proc. Third Lunar Sci. Conf., Geochim. Cosmochim. Acta*, Suppl. 3, Vol. 3, pp. 3009–3020. MIT Press.

Proceedings of the Fourth Lunar Science Conference
(Supplement 4, *Geochimica et Cosmochimica Acta*)
Vol. 3, pp. 2459–2464

Thermal conductivity and diffusivity of Apollo 15 fines at low density

C. J. CREMERS and H. S. HSIA

Department of Mechanical Engineering, University of Kentucky, Lexington, Kentucky 40506

Abstract—The thermal conductivity of the Apollo 15 fines, sample 15031,38, was measured under vacuum conditions as a function of temperature. Measurements were made for a sample density of 1300 kgm^{-3}. The conductivity was found to vary from about 0.57×10^{-3} Wm^{-1}K^{-1} at 95°K to about 1.36×10^{-3} Wm^{-1}K^{-1} at 406°K. The data are compared with the correlation using a cubic temperature dependence and also with data from samples gathered during prior Apollo missions. The thermal diffusivity is obtained for the sample by calculation using the given density and measured thermal conductivity along with specific heats from the literature.

INTRODUCTION

THE THERMAL CONDUCTIVITY and diffusivity are material properties which appear in thermal energy conservation equations. Consequently, knowledge of these properties is required if one is to calculate the effect of given thermal processes on the temperatures or heat fluxes extant in a particular material. In the case of the moon, these properties must be known in order to permit the calculation of lunar surface temperatures and surface layer heat fluxes. Knowledge of these properties is also essential for proper thermal design of systems to be used on the moon.

The thermal conductivity is defined by the Fourier–Biot law of heat conduction. That is

$$\mathbf{q} = -k\nabla T. \tag{1}$$

Here $\mathbf{q}$ is the heat flux arising from the temperature gradient ∇T. The thermal conductivity k is simply the proportionality factor arising from the phenomenological relation given by Equation (1). Over limited temperature ranges, k may be taken as a constant. However, in most cases of interest it must be treated as a function of temperature.

The thermal diffusivity appears in time-dependent heat flow problems and is related to the conductivity by the expression:

$$\alpha = k/\rho c \tag{2}$$

where ρ is the density and c the specific heat of the material. The dimensions of α are length^2 time^{-1} and so there is no reference to the property being diffused.

Heat transfer calculations involving the lunar surface layer for the determination of either temperatures or heat fluxes concern primarily the lunar fines. This is true at least for the regions so far visited by the Apollo manned missions as well as the Surveyor, Luna, and Lunakhod unmanned missions. Photographic, seismic, and sampling studies suggest a lunar surface layer composed of a generally deep deposit of compact but finely divided material. Rocks, boulders, and escarpments are present; however, they are relatively infrequent and scattered. Consequently, from a general thermal point of view, heat transfer in the lunar surface layer should be dominated by the properties of the fines and so the present property measurements have been concentrated on these.

The porous nature of the lunar soil has a significant effect on the mechanism of internal heat transfer. Any porous dielectric material with an internal temperature gradient will transfer heat through a complicated interaction between phonon conduction through the particles and their contact surfaces and thermal radiation which will be scattered in the voids and absorbed and emitted by the solid material. As there is no gaseous conduction or convection at the low atmospheric pressures on the lunar surface (about 10^{-12} torr), the internal heat transfer is a combination of conduction and radiation.

The heat flux in the evacuated lunar fines will be taken as an entity, rather than considering separate radiative and conductive components. Then the Fourier–Biot law may be used provided that the thermal conductivity so defined is recognized to be only an effective one rather than a basic property of the material. Elementary theory, e.g., Watson (1964) and Clegg *et al.* (1966), shows that such an effective conductivity of a particulate medium can be represented by the sum of a constant term, representing phonon conduction, plus a term proportional to temperature to the third power, representing radiation. That is

$$k = A + BT^3 \tag{3}$$

where A and B are constants.

As in the previous studies by Cremers *et al.* (1970), Cremers and Birkebak (1971), and Cremers (1972), the present investigation assumes that the heat flux can be represented in total by Equation (1) and an experiment to measure k is chosen accordingly. The data which are obtained are presented in this paper as a function of temperature and a correlation curve of the form of Equation (3) is fitted to the data. The curves of this form for the earlier Apollo samples are presented for comparison along with such a curve for terrestrial basalt. In all cases the density is 1300 kgm^{-3}.

The thermal diffusivity was calculated on a point by point basis for the given density using conductivities calculated from the correlating equation. The specific heats necessary for this exercise were taken from the literature.

The sample used for the experiments reported in this paper was lunar fines sample 15031,38 as catalogued by the Curator of the Lunar Sample Analysis Program at the Johnson Space Center, NASA, Houston.

The Experiment

The method chosen for the thermal conductivity measurements was the line heat-source technique which lends itself well to application to small samples of powdered materials. Briefly, the application of this method requires that a long thin heat-source (length to diameter ratio greater than 30) of constant strength be imbedded in a medium of extent large compared with the dimensions of the heat source. For such a source of strength per unit length q, it can be shown that the temperature change at a given point in the medium during the time interval from t_1 to t_2 (where t is measured from the time the heat source is "switched on") is given approximately by

$$T_2 - T_1 = (q/4\pi k) \ln (t_2/t_1). \tag{4}$$

The approximation implicit in Equation (4) becomes better as time t_1 increases and the duration of this initial period will increase with distance from the source. A practical upper limit on t_1 will be imposed by the finite extend of the sample. The details of the line heat-source method as used in the present investigation are described by Cremers (1971).

The ambient pressure for the experiments was on the order of 10^{-6} torr. Cremers (1973) demonstrated that gaseous effects on internal heat transfer were eliminated at pressures below 10^{-2} torr. The experiments reported were conducted with the Apollo 11 and 12 fines samples, however, the results should be sufficiently general to be applicable here.

RESULTS AND DISCUSSION

The thermal conductivity data for a sample density of 1300 kgm^{-3} are shown by the circles in Fig. 1. The temperature range covered is 95–406°K. This is roughly the range of diurnal temperatures on the lunar equator. It is apparent that the conductivity, which is just an "effective" value because of the inclusion of radiative effects, is a function of temperature. In fact it doubles in value over the temperature range in question.

An attempt was made to correlate the data with Equation (3). This is shown by the solid line in Fig. 1. The coefficients A and B were obtained by a least-square analysis of the data shown. The resultant thermal conductivity expression in units

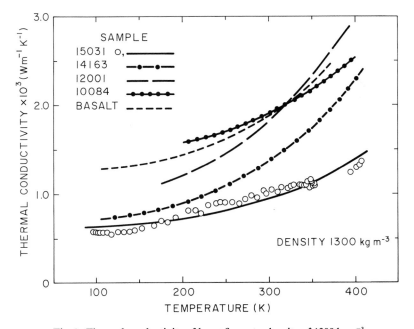

Fig. 1. Thermal conductivity of lunar fines at a density of 1300 kgm^{-3}.

of $Wm^{-1}K^{-1}$ is

$$k = 0.6246 \times 10^{-3} + 0.1192 \times 10^{-10}T^3. \tag{5}$$

The fit is not outstanding nor is it bad. The standard error of the estimate is $0.6618^{-4}Wm^{-1}K^{-1}$ which is on the order of 10 percent.

There is a great temptation to fit a straight line through the data. The fit through all but the highest and lowest groupings would be quite good. However, there is no real physical reason for doing this. Previous investigations of Apollo 12 and 14 samples by Cremers and Birkebak (1971) and Cremers (1972) as well as studies on powdered terrestrial basalt by Fountain and West (1970) showed that temperature cubed correlation fit the respective data sets quite well. The Apollo 11 data of Cremers *et al.* (1970) for the density of 1300 kgm^{-3} were too limited in number and temperature range for any such inference.

Also shown on Fig. 1 are the least-squares fits of Equation (3) to the Apollo 11, Apollo 12, and Apollo 14 data as well as to the data for powdered terrestrial basalt. All of these samples were at or very near a density of 1300 kgm^{-3}. In all cases the thermal conductivity is higher than for the Apollo 15 sample. Of particular interest is the magnitude of the radiative component of the conductivity. Note that as the temperature increases the conductivity of the Apollo 15 sample does not increase nearly so fast as it does for the other samples. In this behavior it is most similar to the Apollo 11 sample which had a thermal conductivity twice as large as the Apollo 15 sample but which did not have a large radiative component.

One might infer from these comparisons that the Apollo 11 and 15 fines samples were similar to one another in particle size distribution, probably having a larger fraction of small particles which fill the larger voids and so tend to interfere with radiative transport. At the same time, the lower absolute values of the Apollo 15 data would indicate a greater surface irregularity and edge sharpness for the particles causing smaller intersurface contact areas and therefore a greater heat flow resistance.

The thermal diffusivity was calculated using Equation (2) and employing thermal conductivity values calculated with Equation (3). Again the density of 1300 kgm^{-3} was used. Specific heat values presented by Robie *et al.* (1970) for the Apollo 11 sample were employed in the calculations as this is the most extensive data set published to date. Later results of a more limited nature published for other Apollo samples have shown excellent agreement with these earlier data as they should, considering the chemical similarity of the various samples. The resultant thermal diffusivity curve is shown in Fig. 2. No comparison is made with diffusivities calculated for the earlier Apollo samples as the resulting curves would differ from one another only in the same way that the conductivities of Fig. 1 do as the same densities and specific heats were used.

The curve in Fig. 2 represents a fifth degree polynomial whose coefficients were determined by a least-squares analysis of the calculated diffusivities. The fifth degree was chosen because it represented the data better than lesser or greater degree polynomials. The calculations were made by using the thermal conductivity calculated from Equation (5) along with a density of 1300 kgm^{-3} and

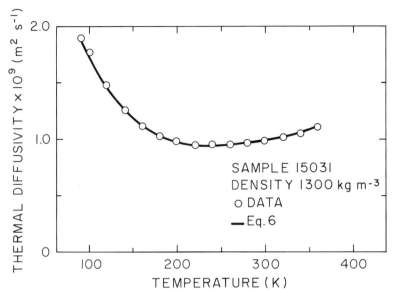

Fig. 2. Thermal diffusivity of lunar fines sample 15031 at a density of 1300 kgm^{-3}.

smoothed values of the specific heat data. Temperature intervals of 20°K were chosen starting at 90°K. The resulting empirical expression, in units of m^{2}s^{-1} is

$$\alpha = 0.4061 \times 10^{-8} - 0.2922 \times 10^{-10} T + 0.2467 \times 10^{-13} T^2$$
$$+ 0.5382 \times 10^{-15} T^3 - 0.2019 \times 10^{-17} T^4 + 0.2177 \times 10^{-20} T^5. \qquad (6)$$

Attempts to measure α directly have as yet met with little success.

Acknowledgment—This research was supported by the National Aeronautics and Space Administration under Grant NGR18-001-060.

REFERENCES

Clegg P. E., Bastin J. A., and Gear A. E. (1966) Heat transfer in lunar rock. *Mon. Not. Roy. Astr. Soc.,* **133**, 63–6.
Cremers C. J. (1971) Thermal conductivity cell for small powdered samples. *Rev. Sci. Inst.* **42**, 1694–1696.
Cremers C. J. (1972) Thermal conductivity of Apollo 14 fines. *Proc. Third Lunar Sci. Conf., Geochim. Cosmochim. Acta,* Suppl. 3, Vol. 3, pp. 2611–2617. MIT Press.
Cremers C. J. (1973) Thermophysical properties of Apollo 12 fines. *Icarus.* In press.
Cremers C. J., Birkebak R. C., and Dawson J. P. (1970) Thermal conductivity of fines from Apollo 11. *Proc. Apollo 11 Lunar Sci. Conf., Geochim. Cosmochim. Acta,* Suppl. 1, Vol. 3, pp. 2045–2050. Pergamon.
Cremers C. J. and Birkebak R. C. (1971) Thermal conductivity of fines from Apollo 12. *Proc. Second Lunar Sci. Conf., Geochim. Cosmochim. Acta,* Suppl. 2, Vol. 3, pp. 2311–2315. MIT Press.
Fountain J. A. and West E. A. (1970) Thermal conductivity of particulate basalt as a function of density in simulated lunar and martian environments. *J. Geophys. Res.* **75**, 4063–4069.

Robie R. A., Hemingway B. S., and Wilson W. H. (1970) Specific heats of lunar surface materials from 90° to 350°K. *Proc. Apollo 11 Lunar Sci. Conf., Geochim. Cosmochim. Acta*, Suppl. 1, Vol. 3, pp. 2361–2367. Pergamon.

Watson K. (1964) The thermal conductivity measurements of selected silicate powders in vacuum from 150° to 300°K. Part 1, Ph.D. Dissertation, California Institute of Technology, Pasadena.

Proceedings of the Fourth Lunar Science Conference
(Supplement 4, *Geochimica et Cosmochimica Acta*)
Vol. 3, pp. 2465–2479

Thermoluminescence of some Apollo 14 and 16 fines and rock samples*

S. A. Durrani, W. Prachyabrued,† and F. S. W. Hwang

Department of Physics, University of Birmingham, Birmingham B15 2TT, England

J. A. Edgington

Department of Physics, Queen Mary College, London, England

I. M. Blair

Nuclear Physics Division, Atomic Energy Research Establishment, Harwell, England

Abstract—Natural as well as β, γ, and UV induced thermoluminescence (TL) has been studied in some Apollo 14 and 16 samples in the temperature interval 20–550°C. TL parameters of the trapping states have been measured by the "initial rise" and "isothermal annealing" techniques. From these and related data, the effective lunar storage temperature and subsurface depths of samples have been calculated. Effects of ultraviolet radiation on the emptying and filling of peak I ($\sim$ 160°C) and peak II ($\sim$ 450°C) traps in natural, γ-irradiated, and annealed samples; spectral composition of the light emitted as a function of temperature; the TL efficiency of lunar samples relative to meteorites and tektites; and the question of non-thermal leakage of trapped carriers are also discussed.

Introduction

Thermoluminescence (TL), both natural and that induced by γ and UV irradiation, has been studied in Apollo 14 fines (comprehensive fines 14259,78, designated Sample A; and bulk fines 14163,113, Sample B) and powdered rock chip 14321,147 (subdivided into a dark fraction termed Sample C, and a light-colored fraction, Sample D).‡ The basic aim has been the same as that in the case of Apollo 12 samples (Durrani *et al.* 1972a, hereinafter referred to as paper I), namely, to draw inferences regarding the thermal and radiation history of the lunar surface. This has necessitated the determination of the TL parameters of the high-temperature peaks by the "initial rise" method, the natural or "equilibrium" dose in the samples, and the rate of growth of TL by artificial irradiation. Some new results are also reported on the natural as well as β and γ induced TL in the Apollo 16 fines sample 62281,11 (for previous results on this and the fines samples 66081,9 and 68501,21 *see* Durrani and Prachyabrued, 1973).

---

*Paper dedicated to Professor J. H. Fremlin on the occasion of his 60th birthday.

†Now at the Department of Physics, Mahidol University, Bangkok 4, Thailand.

‡Comprehensive fines come from the top 1 cm of lunar surface. Bulk fines are those contained in a scoopful of lunar material. The "rock chip" was a coarse-grained breccia with dark and light fragments, and a crystalline matrix of feldspar-rich basalt.

Other aspects examined include temperature sensitization of samples, non-thermal leakage of electron traps, UV effects, and the production efficiency and spectral composition of TL.

Experimental Procedure

The apparatus used (set A of paper I) consists essentially of a heating chamber in which sieved samples of ≈ 2 mg each are heated on a tantalum strip at an accurately controlled rate (normally 5°C sec^{-1}) in a nitrogen atmosphere. The light output, normally transmitted through an Ilford "Bright Spectrum Blue" filter No. 622 (375–530 nm transmission band), is seen by a quartz photomultiplier (EMI: 6256 SQ) in a cooled housing. The PM output, measured by a sensitive electrometer, is registered on a chart recorder. The γ and β irradiations are carried out with a ~ 200 Ci Co-60 and a ~ 0.5 Ci Sr-90 source, respectively. The error in dose measurement is $\sim \pm 10\%$.

Results

Natural and induced glow curves

Figure 1 shows some typical glow curves for two of the Apollo 14 samples studied (Samples B and C), both in their natural state and after irradiating the natural samples with increasing doses of γ-rays. Figure 2 shows the results for the natural and β-irradiated Apollo 16 fines sample 62281,11.

The first peak is found to have been almost completely drained in all Apollo 14 and 16 natural samples, as has been the case for samples from all other Apollo

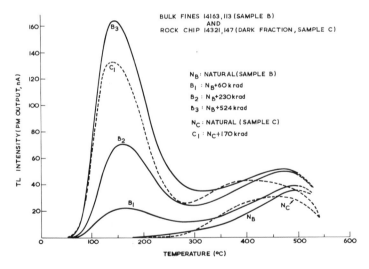

Fig. 1. Typical glow curves from Apollo 14 natural and γ-irradiated (Co-60) Samples B (grain diameters d, 63–106 μm) and C($d < 63\ \mu$m). Notice the drained first peak in the natural samples. In this and subsequent figures (unless otherwise stated): the rate of heating $\beta = 5$°C sec^{-1}; sample weight ≈ 2 mg ($\pm 10\%$); a new sample used for each irradiation; glow seen through Ilford broad-band (375–530 nm) filter No. 622; black-body radiation has been subtracted (though with increasing uncertainty beyond ~ 500°C).

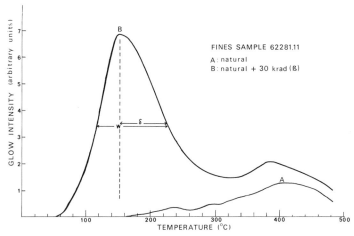

Fig. 2. Typical glow curves from Apollo 16 natural and β-irradiated (Sr-90) sample 62281,11 (grain diameters d, 63-106 μm). The rate of heating, 4°C sec^{-1}; sample weight, 1.2 mg. Parameters δ and W are explained in Discussion (*see* Equation 5) and define the Halperin–Braner factor.

missions studied by us except for a fines sample from Apollo 15 (15261, 70) which had come from the bottom of a trench ~ 20 cm deep (Durrani *et al.* 1972b). Peak I in all samples is gradually built up (at ~ 160°C) as increasing amounts of γ and β doses are imparted to them. Peak II is a broad peak, which occurs between ~ 375 and 475°C in different samples.

The general similarity of the glow-curve structure in all samples suggests an identity of TL phosphors in them. These curves are also similar to those previously reported for Apollo 12, 15, and 16 samples by us (Durrani *et al.*, 1972a, b; Durrani and Prachyabrued, 1973).

Response to γ irradiation and natural dose

In Fig. 3 is shown the growth of the TL "glow area" with γ-ray dose for Apollo 14 Samples A–D. The TL output has been integrated over the temperature interval 300–550°C to cover peak II (growth curves for the interval 20–300°C, to cover peak I, are not shown). Sample A was actually given a γ-ray dose of up to 4.5 Mrad, and Samples B and D up to 3.7 Mrad. The Apollo 16 fines sample 62281,11 was given a γ-ray dose of up to 3 Mrad (figure not shown).

In all samples peak II proceeds towards saturation in an exponential manner:

$$n = N(1 - e^{-0.693R/R_{1/2}}) \tag{1}$$

where n is the number of filled traps for a total dose R (including the natural dose), N is the total number of available traps, and $R_{1/2}$ is the dose required to fill half the remaining traps at any instant (and termed "half-dose" in paper I). The value of $R_{1/2}$ for each of the Apollo 14 samples, as obtained from Fig. 3, is shown in Table 1.

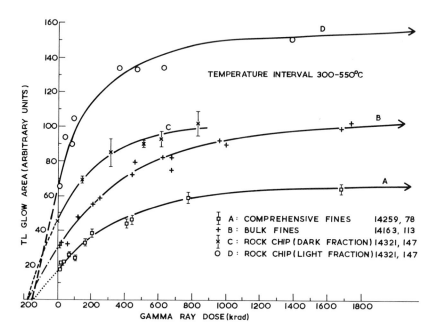

Fig. 3. Growth of peak II (300–500°C) in Apollo 14 Samples A–D with γ-ray dose. Grain diameters: Samples A and B, 38–63 μm; C and D, <63 μm. The arrows at the ends of curves signify higher doses not shown. The error bars result from repeated measurements, intrinsic inhomogeneities of samples, etc. By backward extrapolation of the initial trend of the TL growth curves, the natural (or "equilibrium") dose in each sample may be estimated (see text for discussion).

The growth of TL (over both peaks) with γ-ray dose proceeds at different rates in the four Apollo 14 samples: the growth rate increases steadily from Samples A to D. Taking the response of Sample A as 1, the relative TL response for Samples B, C, and D is ~ 1.2, 3.0, and 5.5 (for peak I, 20–300°C), and ~ 1.8, 2.6, and 3.8 (for peak II, 300–500°C), respectively, at points sufficiently far removed from saturation. Peak II saturation is practically attained in all Apollo 14 samples at ~ 800 krad (and in the Apollo 16 sample, at ~ 1 Mrad).

As in the case of other Apollo samples (and of meteorites, Christodoulides *et al.*, 1970, and tektites, Durrani *et al.*, 1970), it was observed that heating the Apollo 14 lunar samples during a TL readout (to 550°C) makes them more sensitive in their response to subsequent gamma irradiation. Thus, "heated" Sample B (14163,113) exhibited a TL area under peak II (300–550°C) which was ~40% higher at an imparted dose of 800 krad than that for the unheated sample containing the same total dose; the increase was ~80% for peak I (20–300°C). On heating the sample twice, no *further* increase was observed (i.e., the enhancement in TL output relative to the unheated sample remained constant). Similarly, a part of the Apollo 16 fines sample 62281,11, given a γ-ray dose of 1.4 Mrad and then annealed at 450°C for 1 hr, showed a 50% enhancement of TL response to a subsequent standard dose. The fines sample 68501,21 has already been reported (Durrani and

Table 1. Trap parameters for peak II and the calculated "effective" lunar storage temperature T_{eff} (assuming a dose rate of 10 rad/yr).†

1 Sample	2‡ $T^*(II)$ (°C)	3 E (eV)	4 s sec^{-1}	5 $R_{1/2}$ (krad)	6 R_{eq} (krad)	7 N/n_{eq}	8 T_{eff} (°K)	9 $\tau_{1/2}(T_{eff})$ (yr)	10 $\tau_{1/2}(20°C)$ (yr)	11 d (cm)
14259,78 (A; comp. fines)	475	1.60 ± 0.05	$\sim 1 \times 10^{10}$	300	140	4.1	372 ± 5	1.2×10^4	6.7×10^9	0–0.1
14163,113 (B; bulk fines)	475	1.60 ± 0.05	$\sim 1 \times 10^{10}$	450	170	3.7	369 ± 5	1.3×10^4	6.7×10^9	0–0.2
14321,147 (C; rock, dark fr.)	425	1.60 ± 0.10	$\sim 7 \times 10^{10}$	200	180	2.2	355 ± 10	1.8×10^4	9.7×10^8	~ 1.4
14321,147 (D; light fr.)	425	1.60 ± 0.10	$\sim 7 \times 10^{10}$	200	150	2.4	356 ± 10	1.5×10^4	9.7×10^8	~ 1.3

†See text for explanation of symbols, and paper I for method of calculation.

‡The column headings denote: 2. $T^*(II)$, Peak II temperature (at a heating rate of 5°C sec^{-1}). 3. E, trap depth. 4. s, frequency factor. 5. $R_{1/2}$, half-dose. 6. R_{eq}, equilibrium (natural) dose. 7. N/n_{eq}, ratio of traps filled at saturation and at equilibrium. 9 and 10. $\tau_{1/2}(T_{eff})$ and $\tau_{1/2}(20°C)$, half-life of peak II at T_{eff} and at 20°C, respectively. 11. d, mean depth of sample below lunar surface.

Prachyabrued, 1973) to show an enhancement of $\sim 100\%$. Such temperature sensitization and predose effects, while not fully understood, seem to be a general property of TL phosphors (Zimmerman et al., 1965) and may be related to phase-changes in the phosphors. Our observations place an upper limit ($\sim 450°C$) on the lunar surface temperature attained during the TL storage.

By extrapolating backwards the initial trend of the growth curves in Fig. 3, the natural dose can be determined for each sample. This is termed the "equilibrium dose" R_{eq} as reached in the prevailing radiothermal environment on the moon. The natural (or equilibrium) dose corresponding to peak II is found for all Apollo 14 samples (A–D) to be ~ 150–200 krad (see Table 1), which is similar in magnitude to that found by us in Apollo 15 (~ 150 krad) and Apollo 16 (~ 100 krad) fines samples, while being considerably smaller than that determined for the Apollo 12 samples (~ 900 krad for fines and ~ 2800 krad for rock chip). Parameters such as R_{eq} and $R_{1/2}$ are used in estimating the lunar storage temperature of the samples studied (see Discussion below).

TL emission spectra

The TL emission spectra from natural as well as γ- and β-irradiated samples were studied using sets of interference filters (Filtraflex B40, made by Balzers, typical transmission band-width at half-height ~ 10 nm) and colored-glass absorption filters (Ilford Spectrum Filters, typical fwhh ~ 50 nm). The method has been

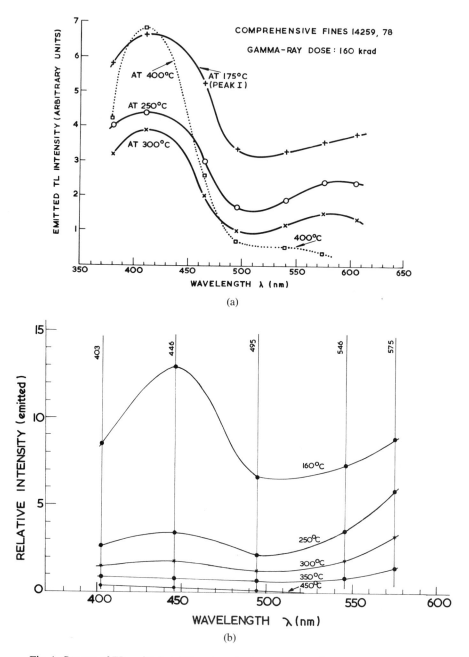

Fig. 4. Spectra of TL emitted at different readout temperatures from (a) the γ-irradiated Apollo 14 Sample A (grain diameters d, 38–63 μm) and (b) β-irradiated (0.92 Mrad) Apollo 16 sample 62281,11 (grain diameters, $d < 63$ μm). The emitted intensities have been normalized to 2 mg sample weights. In (a), a set of six Ilford Spectrum Filters (with transmission peaks at the experimental points marked on the curves) and in (b) a set of interference filters (Filtraflex B40 by Balzers) have been used. See text for the method.

described in paper I, which explains how account is taken of the quantum efficiency of the (quartz-window) photomultiplier and the transmission characteristics of the filters in reducing the data.

Figures 4(a) and (b) show typical sets of TL emission curves for irradiated
samples. Figure 4(a) is for the Apollo 14 fines Sample A (14259,78) irradiated with
160 krad of γ-rays, and Fig. 4(b) for the Apollo 16 fines sample 62281,11 irradiated
with 0.92 Mrad of β particles. There are discernible differences between the two
sets of spectra, despite broad similarities. While in both samples the blue peak
(~ 400–450 nm) predominates at low readout temperatures (175 and 160°C) and the
long wavelength component (>500 nm) is nearly missing at high temperatures
(400 and 450°C), the proportion of the long wavelength component is a good deal
higher (and indeed exceeds all other wavelengths) at intermediate temperatures
(250–350°C) in the case of the Apollo 16 sample as compared with the Apollo 14
sample. The implications are considered in the section on Discussion.

The relative intensity of *total* TL emission at various readout temperatures depends, of course, on the amount of radiation dose in the sample, which enhances
the TL output under peak I considerably more than it does the high-temperature
TL (hence the relatively large height of the emission curve at 160°C in Fig. 4(b)).
Thus the TL emission spectrum from the Apollo 14 and 16 natural samples (figure
not shown), which were depleted in low-temperature TL owing to natural drainage, yielded a much higher emission curve at 450°C (peaking at ~ 400 nm) than at
the lower readout temperatures.

Spectral composition of UV-irradiated Apollo 14 and 16 samples was not
studied, but Apollo 15 samples showed (Durrani *et al.*, 1972b) general similarity to
γ-irradiated spectra.

Effect of UV irradiation

The study of ultraviolet irradiation of lunar samples is of interest both intrinsically and also as a possible clue for transient lunar phenomena (TLP). Effects of
exposure of UV radiation (centered at 254 nm; power input 10^{-2} W cm^{-2} at the
position of the samples, which were spread in single-grain layers at a distance of
10 cm from the source) were studied in natural, γ-irradiated, and annealed Apollo
14 fines as well as powdered rock samples. The results are shown in Fig. 5. It is
seen from curves A and B in Fig. 5(a) that UV radiation induces TL in the region
of peak I (previously empty), while draining some of the filled traps in peak II. If,
however, both peaks are initially populated, UV radiation drains some of the traps
in each (cf. curves A' and B' in Fig. 5(b); but note that part of the emptying of peak
I in A' may be due to thermal drainage over the 44 hr spent in UV irradiation).
When both peaks are initially empty (annealed samples), UV radiation induces TL
throughout (curves C and C'), but fills peak I preferentially (as is the case with
γ-rays). It will be noticed, however, that the shape of UV-induced first peak is
quite distinct (the rise being much less steep) from its γ-irradiated counterpart.

To test whether UV radiation was helping to transfer some of the trapped electrons from peak II to peak I (*see* curve B in Fig. 5(a)), the following experiment

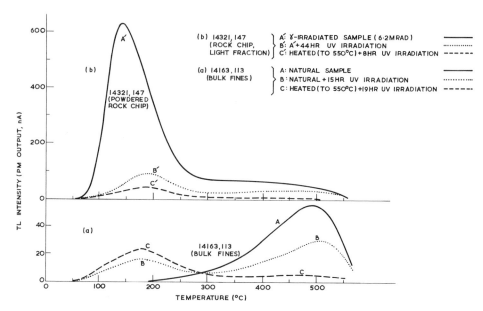

Fig. 5. Effects of UV irradiation on Apollo 14 natural, γ-irradiated, and annealed samples. Grain diameters: A, B, 106–180 μm; C, 38–250 μm; A', B', C', <63 μm; sample weights: A, B, C, 2.0 mg; A', B', 1.7 mg; C', 2.0 mg. UV radiation is seen to fill empty traps while emptying filled traps. Notice the difference in the shapes of glow peaks (first peaks in (b)) induced by UV and γ radiations, respectively.

was carried out. The fines sample 14163,113 was irradiated with UV (Fig. 6) in two distinct conditions: (i) the natural sample was irradiated to saturation with γ-rays and then annealed at 200°C for $\frac{1}{2}$ hr to empty peak I but leaving peak II saturated (curves a, b, and c in Fig. 6), and (ii) the sample had previously been completely drained in a readout i.e., heated up to 550°C (curve d). On irradiating each of the above specimens (i) and (ii) with UV for $6\frac{1}{2}$ hr, the height of UV-induced peak I was found to be practically the same in the two cases (cf. peak I in curves c and d). Thus (ignoring any temperature sensitization for UV irradiation in curve d) it seems to make no difference to the height of peak I whether or not a large number of trapped electrons is available for possible transfer from peak II.

The "half-dose" of UV radiation for filling TL traps (in Sample C, grain size <63 μm) in the readout interval 20–300°C is again found, as in the case of Apollo 15 fines samples (Durrani *et al.*, 1972b), to be that imparted in ~1 hr of irradiation (calculated to be ~400 Mrad of UV); this produces a TL glow equivalent to that resulting from ~30 krad of γ rays (the effective relative efficiency for producing TL being thus ~10^{-4} for UV vs. γ rays). Saturation of TL output in peak I is reached after only a few hours of UV irradiation. The value of trap depth E corresponding to peak I resulting from UV irradiation was also determined by the "initial rise method" (*see* below) and was found to be ~1.1 eV. This lies just below the E value for γ-induced peak I, so that the electron traps filled by UV and γ radiations seem to form a continuum extending from ~1.1 to 1.3 eV.

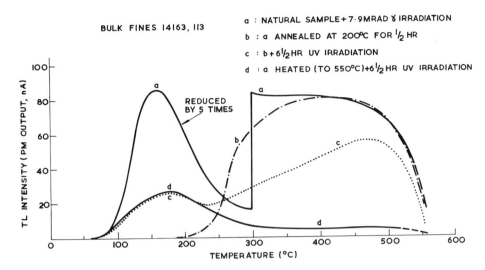

BULK FINES 14163, 113

a : NATURAL SAMPLE + 7·9 MRAD ϒ IRRADIATION

b : a ANNEALED AT 200°C FOR ½ HR

c : b + 6½ HR UV IRRADIATION

d : a HEATED (TO 550°C) + 6½ HR UV IRRADIATION

Fig. 6. Test of transfer of trapped electrons from peak II to peak I by UV irradiation. Grain diameters, < 63 μm. The scale of peak I in curve a (heavily irradiated sample) has been reduced by a factor of 5. It is seen that whether or not peak II is initially occupied, the height of the UV-induced peak I is about the same (curves c and d).

Non-thermal leakage

The question of non-thermal leakage (Garlick and Robinson, 1972) was examined in the case of the Apollo 14 sample 14163,113 (bulk fines) and the Apollo 16 fines sample 62281,11. According to the "first order kinetics" model of Randall and Wilkins (1945), the trapped carriers obey the formula

$$n = n_0 e^{-\lambda t}, \tag{2}$$

where

$$\lambda = s e^{-E/kT}, \tag{3}$$

n being the number of filled traps at time t in a body kept at an absolute temperature T, n_0 the number at time $t = 0$, s and E the "frequency factor" and the "trap depth" concerned, and k the Boltzmann constant.

The half-life of the trapped carriers at temperature T can be determined (cf. Eq. (2)) by observing the time needed to empty half the initially filled traps ($\tau_{1/2} = \tau \times \ln 2$, where $\tau(T) = \lambda^{-1}$). Figure 7 shows the results of "isothermal annealing" (or storage) at temperatures varying from -196 to 90°C for the sample 62281,11, which had been irradiated with 1 Mrad of β particles. Temperature dependence of TL decay is obvious from Fig. 7.

The fines sample 14163,113 was irradiated with 550 krad of γ-rays and held at 5°C for varying lengths of time up to 275 hours, after which the residual TL was observed through normal readout. While the TL output in the readout temperature interval 20–300°C was found to decay with storage time with a half-life of

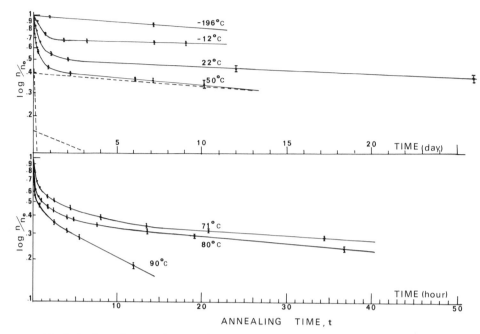

Fig. 7. Isothermal decay of TL (total area under the glow curve) at various temperatures in the Apollo 16 sample 62281,11 irradiated with ~ 1 Mrad of β's. The decay curves can, in most cases, be decomposed into at least three linear components; one such decomposition is shown for the 50°C curve. Grain diameters, 63–106 μm; glow intensity normalized to 1 mg of sample weight and 1 rad of β dose. See text for discussion.

~ 260 hours, there was no detectable fall off in the 300–550°C TL output (figure not shown). The former decay was compatible with the thermal drainage of traps of depth $E \sim 1$ eV and the frequency factor $s \sim 10^{12} - 10^{13}$ sec^{-1}, that would be expected at a storage temperature of $T = 278°$K from Eq. (2) above. Thus no evidence for non-thermal leakage of trapped electrons was found. The point is considered further in Discussion below.

TL parameters

To derive quantitative values for lunar storage temperature and related phenomena, it is essential to determine trap parameters E and s (*see* Eq. (3) above) for lunar samples. This was done by using the "initial rise method" (Garlick and Gibson, 1948; cf. paper I) in the case of the Apollo 14 samples. The values of these parameters for peak II (the only one used in lunar storage-temperature calculations) are recorded in Table 1 for Samples A–D.

In the case of the Apollo 16 sample, the values of E and s were derived mainly by the "isothermal annealing" technique (Garlick and Gibson, 1948; Durrani *et al.*, 1970). It is clear from Eq. (3) that

$$\ln \lambda = \ln s - E/kT. \qquad (4)$$

Thus, by first determining the values of the half-lives for a given trap at different storage (or annealing) temperatures T, and then plotting the corresponding values of $\ln \lambda$ against $1/T$, one can easily obtain the value of E from the slope $(-E/k)$ and of s from the intercept made by the straight line on the ordinate ($\ln \lambda$), provided that discrete, independent traps of constant E and s values are involved in each "isothermal annealing" experiment.

It is seen from Fig. 7 that the plots of $\ln (n/n_0)$ against t are not straight lines. The decay curves can be decomposed (though not very reliably) into several linear components, each representing a discrete decay constant λ (the slope of the straight line). One such decomposition is shown in Fig. 7 for the 50°C decay curve. By decomposing each of the decay curves of Fig. 7 into three linear components (and thereby splitting the first or major peak of sample 62281,11 into three constituent peaks at 140, 165, and 200°C), the values of E and s ranging from ~ 1.06 to $\sim 1.34 \pm 0.10$ eV, and $\sim 10^{12}$ to $\sim 10^{13}$ sec^{-1}, respectively, were obtained.

DISCUSSION

Lunar storage temperature and sample depth

Using the procedures and concepts explained in paper I, the "effective" lunar storage temperature T_{eff} has been calculated from peak II in the Apollo 14 Samples A–D. The results, together with the relevant data, are given in Table 1. A dose rate of 10 rad/yr (chiefly from galactic cosmic rays) has been assumed (Haffner, 1967).

Since the temperature at or near the lunar surface is not constant but undergoes cyclic variation during a lunation, the calculated value of T_{eff} is a "Boltzmann-average" (according to the factor $e^{-E/kT}$) over the temperature cycle. With the values of parameters under discussion, T_{eff} is found to be $\sim 16°$K below the maximum temperature T_{max} reached at the storage position of the samples. T_{max}, in turn, can be used to estimate the subsurface depth of the sample concerned, if the amplitude of the diurnal temperature wave (say 150°C) superimposed on the "natural" temperature of the lunar surface (say 240°K) and the thermal conductivity of the lunar fines (or of surface rocks) are known. Assuming a value of "diffusivity" $\epsilon = K/\rho c$ (where K is the thermal conductivity, ρ the density and c the specific heat) $\approx 2 \times 10^{-5}$ cm^2 sec^{-1} for lunar fines (corresponding to a wavelength $\lambda \approx 24$ cm for the lunar cyclic heat wave), the mean subsurface depth d for the fines Samples A and B is calculated to be ≤ 1 mm and ≤ 2 mm, respectively. (Table 1; an uncertainty of 5°K in T_{eff} corresponds to an uncertainty of ~ 1 mm in d for $\lambda \sim 24$ cm.) On using a value of $\epsilon \sim 1.7 \times 10^{-4}$ cm^2 sec^{-1} (based on the data of Langseth et al., 1973, for the top ~ 2 m of lunar regolith, i.e., $K \sim 1.7 \times 10^{-4}$ W cm^{-1} K^{-1}, yielding $\lambda \sim 70$ cm) for the rock chip Samples C and D (14321,147), the mean subsurface depth is found to be ~ 1.4 cm. But since a very thin layer of fine dust can attenuate the temperature wave drastically, the last quoted value for the rock chip is effectively an upper limit. Conversely, assuming a higher value of ϵ would increase the estimated depth (e.g., $\epsilon = 6 \times 10^{-4}$ cm^2 sec^{-1} gives $d = 2.6$ cm). An uncertainty of $\pm 10°$K in T_{eff} produces an uncertainty of ± 1 cm (when $\lambda \sim 70$ cm).

Non-thermal leakage

In the above calculations of T_{eff}, we have ignored the possibility of non-thermal leakage of electrons from peak II traps. Such leakage in the case of lunar samples was first pointed out by Garlick and Robinson (1972) and may result either from quantum mechanical "tunneling" or from "distant pair" theory (Thomas *et al.*, 1964). As pointed out in an earlier section, we have failed to observe any significant amount of non-thermal leakage of traps in γ- and β-irradiated Apollo 14 and 16 samples.

It is worth noting that a composite peak may display an apparently faster-than-expected decay at a given temperature. Thus, the major ("first") peak in the sample 62281,11 has been shown above by the isothermal annealing method to consist of at least three components, each with a different set of E and s values. The fact that this main peak is in reality a multiple peak is also indicated by the analysis proposed by Halperin and Braner (1960) who have derived a numerical factor to test such a possibility. The Halperin–Braner factor, μ_g, is defined as

$$\mu_g = \frac{T_2 - T_m}{T_2 - T_1} = \frac{\delta}{W}, \tag{5}$$

where T_1 and T_2 are, respectively, the lower and higher temperatures at half-height of a peak which has a maximum at a temperature T_m. Values of μ_g around 0.43 specify first-order (Randall–Wilkins type) kinetics and around 0.53, second-order kinetics (which involve re-trapping of ejected carriers, and hence a non-exponential decay); but in both cases such magnitudes for the value of the Halperin–Braner factor indicate a single trap depth. From Fig. 2, the value of μ_g for the prominent peak induced at room temperature is found to be ≈ 0.68, which suggests a multiple trapping level or peak to be present.

The question of non-thermal leakage of traps can be considered in the light of Fig. 7. It is observed that the rate of decay of TL (most of it in the main peak at 140–200°C) is temperature-dependent; the loss increases as the storage temperature is raised from -196 to 22°C (and eventually to 90°C). No decay was detectable after 7 hr of storage at -196°C. These results are at variance with those reported by Blair *et al.* (1972), who carried out their irradiations of lunar samples at -196°C (with 160 MeV protons) and observed the decay of the ~ 100°C peak. The disagreement may result partly from the fact that our irradiations were at room temperature (and with β's or γ's rather than fast protons).

The phenomenon of "non-thermal leakage," if it occurs, has profound implications for TL studies and must be investigated further. Such considerations, however, do not diminish the importance of keeping lunar samples in deep freeze conditions pending their distribution to TL groups, a procedure adopted by NASA for some Apollo 17 samples at the recommendation of one of us (Durrani, 1972).

TL efficiency and spectral composition

The TL efficiency of lunar material seems to vary greatly from sample to sample. The "energy (conversion) efficiency" is defined as the ratio of the energy emitted as TL to the radiation energy deposited in the material. Table 2 shows the γ-irradiation efficiency values computed for the Apollo 12 to 16 samples so far studied by us (for comparison, we have included the values for some meteorites and tektites). The absolute values shown are probably accurate only to within a factor of 10 in view of difficulties in computing the light collection efficiency and amplification of the detection system, uncertainties of spectral composition of TL emission, etc. The relative values of the TL energy efficiency are, however, probably accurate to within a factor of 2. It is seen from the table that both fines and rock-chip samples from Apollo 14 are the most thermoluminescent of the lunar samples listed. Thus the light fraction of rock 14321,147 (a feldspar-rich basalt) has a TL efficiency ~ 75 times that of rock 12051,15 (an olivine basalt).

The lunar samples are seen from Table 2 to be much more thermoluminescent than tektites, and much less so than meteorites. Our lunar TL efficiency values may be compared with that of the plagioclase fraction of Apollo 11 fines 10084,3 and 4, irradiated with 160 MeV protons, namely $\sim 4.3 \times 10^{-6}$, reported by Edgington and Blair (1970). (The TL efficiency values for Apollo 12 and 14 fines irradiated with 160 MeV protons have more recently been found to be higher by a factor of ~ 2 and ~ 4, respectively, than the Apollo 11 values.) Our TL efficiency values are also much higher than those reported by Nash and Greer (1970) and Nash and Conel (1972) for irradiation of Apollo 11 and 12 surface and core samples with 40 keV X-rays, namely 10^{-12}–10^{-10}. These differences probably result

Table 2. Gamma-ray induced TL energy efficiency for the first peak* in various materials of cosmological interest.

Material	Absolute efficiency†	Material	Absolute efficiency†
Lunar samples		*Lunar samples*	
12070,112 (fines)	2×10^{-5}	62281,11 (fines)	5×10^{-5}
12051,15 (rock chip)	4×10^{-6}	66081,9 (fines)	3×10^{-5}
14259,78 (comprehensive fines)	1×10^{-4}	68501,21 (fines)	5×10^{-5}
14163,113 (bulk fines)	1×10^{-4}	*Meteorites*	
14321,147 (rock chip, dark fraction)	2×10^{-4}	Dimmitt (olivine-bronzite chondrite)	1×10^{-3}
14321,147 (rock chip, light fraction)	3×10^{-4}	Lost City (olivine-bronzite chondrite)	5×10^{-3}
15261,70 (fines) 15271,90 (fines) 15301,84 (fines)	6×10^{-5}	*Tektites* Bediasite B55 (Sommerville, Texas)	1.5×10^{-6}
		Indochinite S169 (N.E. Thailand)	2×10^{-6}
15601,78 (fines)	1×10^{-5}	Moldavite (Moravia)	3×10^{-6}

*The temperature interval between 100 and 300°C, and the TL output in the pass band 375–530 nm (Ilford Bright Spectrum Filter No. 622) only are taken into account.

†The absolute values of the energy-conversion efficiency are accurate to a factor of ~ 10 only (if the whole glow spectrum as well as the total temperature interval 20–550°C were taken into account, the TL efficiency would go up); the relative values are probably accurate to within a factor of 2.

from variations in the plagioclase content (the most dominant source of lunar TL, cf. Geake *et al.*, 1970) of the various samples.

Some light on the variation of the TL efficiency of lunar samples is thrown by the observed differences in the spectral composition of the emitted TL. Thus we have noted above (*see* Results) the larger long-wavelength component (greenish yellow) observed in Apollo 16 fines at intermediate readout temperatures than in the Apollo 14 samples. According to Geake *et al.* (1972), the lunar plagioclase content shows three main emission peaks in direct luminescence, namely in the blue, green, and near infra-red regions (at ~ 450, 560, and 780 nm). Our spectral composition (Fig. 4) is certainly compatible with these observations (though we did not investigate the IR component).

The proposition that ultraviolet radiation effects may conceivably be responsible for some of the transient lunar phenomena (TLP) was examined by us. The hypothesis was that some of the lunar surface materials might accumulate ~ 100 krad of γ-equivalent dose over a few hours towards a lunar sunset (the total radiant energy of the sun in wavelengths below 400 nm received at ~ 1 A.U. is $\sim 1.4 \times 10^{-2} \text{W cm}^{-2}$ (Tousey, 1966), which is quite comparable to the value used in our UV experiment described above). The lunar surface material might then either stay saturated during the lunar night and emit the stored energy as TL at the start of the next sunrise, or release it by meteoritic impact during the night. Calculation, however, showed that the order of magnitude of the (thermo-) luminescence would be about a million times below that of the background light. It should also be borne in mind that whereas the TLP glows are generally reddish or pinkish in color (*see* for example, Mills, 1970), both the γ- and β-induced and the UV-induced (Durrani *et al.*, 1972b; Lalou *et al.*, 1972) TL is predominantly in the blue or green region. In view of these considerations, the hypothesis that UV effects were responsible for TLP was rejected. The same remarks would apply to meteoritic impact releasing the cosmic-ray induced natural TL in freshly exposed core material

Acknowledgment—We wish to thank Professor J. H. Fremlin (to whom this paper is warmly dedicated on the occasion of his 60th birthday) for many stimulating discussions. The financial support of these investigations by the Science Research Council and the Royal Society is also gratefully acknowledged.

REFERENCES

Blair I. M., Edgington J. A., Chen R., and Jahn R. A. (1972) Thermoluminescence of Apollo 14 lunar samples following irradiation at $-196°C$. *Proc. Third Lunar Sci. Conf., Geochim. Cosmochim. Acta,* Suppl. 3, Vol. 3, pp. 2949–2954 MIT Press.

Christodoulides C., Durrani S. A., and Ettinger K. V. (1970) Study of thermoluminescence in some stony meteorites. *Modern Geology* 1, 247–259.

Durrani S. A. (1972) Refrigeration of lunar samples destined for thermoluminescence studies. *Nature* **240**, 96–97.

Durrani S. A. and Prachyabrued W. (1973) Thermoluminescence of Apollo 16 fines (abstract). In *Lunar Science—IV*, pp. 199–201. The Lunar Science Institute, Houston.

Durrani S. A., Christodoulides C., and Ettinger K. V. (1970) Thermoluminescence in tektites. *J. Geophys. Res.* **75**, 983–995.

Durrani S. A., Prachyabrued W., Christodoulides C., Fremlin J. H., Edgington J. A., Chen R., and Blair I. M. (1972a) (Paper I) Thermoluminescence of Apollo 12 samples: Implications for lunar temperature and radiation histories. *Proc. Third Lunar Sci. Conf., Geochim. Cosmochim. Acta*, Suppl. 3, Vol. 3, pp. 2955–2970. MIT Press.

Durrani S. A., Prachyabrued W., Edgington J. A., and Blair I. M. (1972b) Thermoluminescence of Apollo 15 lunar samples: (II) 20 to 550°C. In *The Apollo 15 Lunar Samples*, pp. 457–461. The Lunar Science Institute, Houston.

Edgington J. A. and Blair I. M. (1970) Luminescence and thermoluminescence induced by bombardment with protons of 159 million electron volts. *Science* **167**, 715–717.

Garlick G. F. J. and Gibson A. F. (1948) The electron trap mechanism of luminescence in sulphide and silicate phosphors. *Proc. Phys. Soc. Lond.* **60**, 574–590.

Garlick G. F. J. and Robinson I. (1972) The thermoluminescence of lunar samples. In *The Moon* (editors H. C. Urey and S. K. Runcorn), pp. 324–329. International Astronomical Union.

Geake J. E., Dollfus A., Garlick G. F. J., Lamb W., Walker G., Steigman G. A., and Titulaer C. (1970) Luminescence, electron paramagnetic resonance and optical properties of lunar material from Apollo 11. *Proc. Apollo 11 Lunar Sci. Conf., Geochim. Cosmochim. Acta*, Suppl. 1, Vol. 3, pp. 2127–2147. Pergamon.

Geake J. E., Walker G., Mills A. A., and Garlick G. F. J. (1972) Luminescence of lunar material excited by electrons. *Proc. Third Lunar Sci. Conf., Geochim. Cosmochim. Acta*, Suppl. 3, Vol. 3, pp. 2971–2979. MIT Press.

Haffner J. W. (1967) *Radiation and Shielding in Space*, p. 279. Academic Press.

Halperin A. and Braner A. A. (1960) Evaluation of thermal activation energies from glow curves. *Phys. Rev.* **117**, 408–415.

Lalou C., Valladas G., Brito V., Henni A., Ceva T., and Visocekas R. (1972) Spectral emission of natural and artificially induced thermoluminescence in Apollo 14 lunar sample 14163,147. *Proc. Third Lunar Sci. Conf., Geochim. Cosmochim. Acta*, Suppl. 3, Vol. 3, pp. 3009–3020. MIT Press.

Langseth M. G., Chute J. L., and Keihm S. (1973) Direct measurement of heat flow from the moon (abstract). In *Lunar Science—IV*, pp. 455–456. The Lunar Science Institute, Houston.

Mills A. A. (1970) Transient lunar phenomena and electrostatic glow discharges. *Nature* **225**, 929–930.

Nash D. B. and Greer R. T. (1970) Luminescence properties of Apollo 11 lunar samples and implications for solar-excited lunar luminescence. *Proc. Apollo 11 Lunar Sci. Conf., Geochim. Cosmochim. Acta*, Suppl. 1, Vol. 3, pp. 2341–2350. Pergamon.

Nash D. B. and Conel J. E. (1972) Luminescence and reflectance of Apollo 12 samples. *Proc. Second Lunar Sci. Conf., Geochim. Cosmochim. Acta*, Suppl. 2, Vol. 3, pp. 2235–2244. MIT Press.

Randall J. T. and Wilkins M. H. F. (1945) Phosphorescence and electron traps. *Proc. Roy. Soc. London* A **184**, 366–407.

Thomas D. G., Hopfield J. J., and Colbow K. (1964) *Radiative Recombination in Semiconductors*, p. 67. Dunod, Paris.

Tousey R. (1966) The radiation from the sun. In *The Middle Ultraviolet: Its Science and Technology* (editor A. E. S. Green), p. 30. Wiley.

Zimmerman D. W., Rhyner C. R., and Cameron J. R. (1965) Thermal annealing effects on the thermoluminescence of LiF. *Health Phys.* **12**, 525.

Proceedings of the Fourth Lunar Science Conference
(Supplement 4, *Geochimica et Cosmochimica Acta*)
Vol. 3, pp. 2481–2487

Specific heats of lunar soils, basalt, and breccias from the Apollo 14, 15, and 16 landing sites, between 90 and 350°K*

B. S. HEMINGWAY, R. A. ROBIE, and W. H. WILSON

U.S. Geological Survey
Silver Spring, Md. 20910

Abstract—The specific heats of lunar samples 14163 (soil), 14321 (breccia), 15301 (soil), 15555 (basalt), and 60601 (soil) were measured over the temperature range 90 to 350°K. These materials behave similarly, having specific heats which rise monotonically from about 0.21 $Jg^{-1}K^{-1}$ at 95°K to approximately 0.88 $Jg^{-1}K^{-1}$ at 350°K.

Based on our specific heat data and upon a number of studies of the thermal conductivity of lunar soil materials (in vacuum) the thermal parameter, γ, varies from 0.063 to 0.023 $m^2sec^{1/2}KJ^{-1}$ between 100 and 350°K.

THE SPECIFIC HEATS of lunar soils 14163, 15301, and 60601, basalt 15555, and breccia 14321 from the Apollo 14, 15, and 16 landing sites have been measured between 90 and 360°K as a part of a continuing study of the thermal properties of lunar surface materials. The magnitudes of the specific heats of these samples are similar to the results reported by Robie *et al.* (1970), Robie and Hemingway (1971), and Hemingway and Robie (1972) for materials from Tranquillitatus and The Ocean of Storms.

The equipment and procedures used in these experiments have been described elsewhere (Robie and Hemingway, 1972). The experimental data for samples 14163, 14321, 15301, 15555, and 60601 are listed in Tables 1 through 5 respectively, and are shown graphically in Figs. 1 through 4. Smoothed values at integral temperatures are listed for each sample in Table 6. Table 7 is a list of sample weight and percentage of the observed heat capacity represented by the sample container at 100°K and 340°K for each experiment. The heat capacities of the empty sample containers were determined in a separate set of measurements at ten degree intervals between 90 and 370°K.

The experimental specific heat data for samples 14163, 15301, and 60601 were combined with our previous measurements, Robie *et al.* (1970), on 10084 and all of the data were fitted by least squares in order to generate a simple analytical expression which would represent the specific heat of an *average* lunar soil over the temperature range 90 to 350°K.

The equation

$$Cp = -2.3173 \times 10^{-2} + 2.1270 \times 10^{-3}T + 1.5009 \times 10^{-5}T^2 - 7.3699 \times 10^{-8}T^3 + 9.6552 \times 10^{-11}T^4 \tag{1}$$

represents the experimental data to better than 10 percent between 90 and 350°K.

---

*Publication authorized by the Director, U.S. Geological Survey.

Table 1. Experimental specific heat measurements for lunar sample 14163,186 (> 1 mm fines) from Fra Mauro.

Temperature °K	Specific heat J/(gram · K)	Temperature °K	Specific heat J/(gram · K)	Temperature °K	Specific heat J/(gram · K)
93.34	0.2510	187.22	0.5485	259.68	0.6996
101.77	0.2820	196.75	0.5778	268.92	0.7171
104.64	0.2912	205.69	0.5916	278.72	0.7347
112.07	0.3180	211.38	0.6058	288.37	0.7531
125.80	0.3707	219.94	0.6251	297.86	0.7673
133.79	0.3950	228.32	0.6406	304.91	0.7782
142.28	0.4222	236.92	0.6586	314.89	0.7920
152.82	0.4481	245.66	0.6740	324.70	0.8050
169.10	0.5010	254.37	0.6891	334.30	0.8184
177.70	0.5217				

Table 2. Experimental specific heat measures for lunar sample 14321,153 (breccia) from Fra Mauro.

Temperature °K	Specific heat J/(gram · K)	Temperature °K	Specific heat J/(gram · K)	Temperature °K	Specific heat J/(gram · K)
96.45	0.2427	185.96	0.5255	269.55	0.6971
104.15	0.2682	194.47	0.5456	278.32	0.7121
112.15	0.2937	203.13	0.5648	287.10	0.7276
120.49	0.3226	211.49	0.5845	295.90	0.7414
129.14	0.3552	220.06	0.6025	304.55	0.7548
133.46	0.3761	228.81	0.6205	308.97	0.7640
143.74	0.4084	237.36	0.6385	317.81	0.7795
154.73	0.4389	245.74	0.6544	327.13	0.7958
165.55	0.4740	252.74	0.6673	336.96	0.8117
168.91	0.4874	260.94	0.6828	346.40	0.8284
177.58	0.5067				

Table 3. Experimental specific heat measurements for lunar sample 15301,20 (soil) from Hadley-Apennine Base.

Temperature °K	Specific heat J/(gram · K)	Temperature °K	Specific heat J/(gram · K)	Temperature °K	Specific heat J/(gram · K)
83.98	0.2218	198.34	0.5540	299.50	0.7669
95.46	0.2569	208.28	0.5774	303.96	0.7778
104.47	0.2870	217.89	0.6004	310.91	0.7847
114.25	0.3197	227.13	0.6205	306.01	0.7782
125.20	0.3498	237.00	0.6427	315.65	0.7958
135.82	0.3824	247.16	0.6636	325.09	0.8125
146.20	0.4171	257.76	0.6862	334.25	0.8280
156.58	0.4443	268.32	0.7063	343.42	0.8452
167.16	0.4724	278.68	0.7276	352.89	0.8619
177.64	0.5004	289.07	0.7481	363.10	0.8765
188.03	0.5280				

Table 4. Experimental specific heat measurements for lunar sample 15555,159 (basalt) from Hadley-Apennine Base.

Temperature °K	Specific heat J/(gram · K)	Temperature °K	Specific heat J/(gram · K)	Temperature °K	Specific heat J/(gram · K)
83.56	0.2088	187.06	0.5180	293.09	0.7556
94.05	0.2414	197.83	0.5456	303.77	0.7703
103.71	0.2715	208.68	0.5736	313.96	0.7895
113.58	0.3038	218.95	0.5983	323.97	0.8071
124.82	0.3389	230.05	0.6230	333.68	0.8222
136.51	0.3749	240.98	0.6481	343.36	0.8397
143.07	0.3946	249.20	0.6648	353.28	0.8565
154.35	0.4276	260.22	0.6874	363.53	0.8728
165.24	0.4586	271.21	0.7084		
176.18	0.4879	281.80	0.7330		

Table 5. Experimental specific heat measurements for lunar sample 60601,31 (soil) from Lunar Highlands.

Temperature °K	Specific heat J/(gram · K)	Temperature °K	Specific heat J/(gram · K)	Temperature °K	Specific heat J/(gram · K)
83.57	0.2218	189.82	0.5356	285.53	0.7519
94.97	0.2577	201.17	0.5644	296.10	0.7715
105.29	0.2904	213.39	0.5954	304.06	0.7895
115.24	0.3230	224.23	0.6209	313.63	0.8063
125.39	0.3569	234.21	0.6439	323.47	0.8230
135.83	0.3883	244.40	0.6653	333.23	0.8406
146.65	0.4201	254.90	0.6878	343.17	0.8577
157.86	0.4519	265.60	0.7100	353.09	0.8761
168.95	0.4833	274.91	0.7289	363.04	0.8916
179.61	0.5121				

In two earlier papers, Robie *et al.* (1970), Robie and Hemingway (1971), we have estimated values for the thermal parameter γ (that is, the inverse of the thermal inertia) for lunar soils based upon our specific heat measurements and *estimates* of the thermal conductivity of the lunar soil in vacuum.

Recently measurements have become available for the thermal conductivity (in a vacuum of 10^{-6} torr) of basaltic powders, (Fountain and West, 1970), and of several lunar soils, (Cremers *et al.*, 1970; Cremers and Birkebak, 1971; Cremers, 1971; Cremers and Hsia, 1973), at temperatures between 100 and 350°K. The most extensive results are for powders at a density of 1300 kg m^{-3}.

Examination of the conductivity measurements indicates that a linear fit represents the data at least as well as the theoretical cubic equation of Watson (1964).

B. S. Hemingway *et al.*

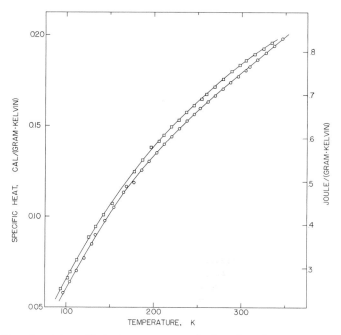

Fig. 1. Experimental specific heat measurements for lunar breccia 14321, hexagons, and soil 14163, squares, from Fra Mauro. Solid curves are the least squares fit to the data.

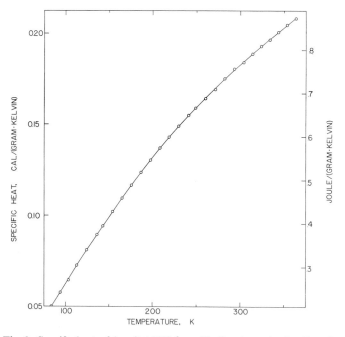

Fig. 2. Specific heat of basalt 15555 from Hadley-Apennine landing site.

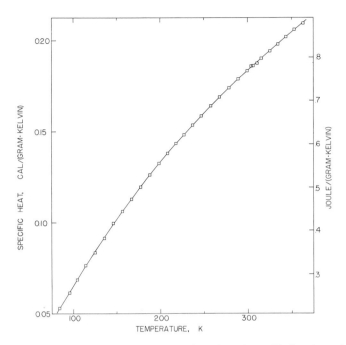

Fig. 3. Specific heat of fine soil (< 1 mm) 15301 from Spur Crater Hadley-Apennine area.

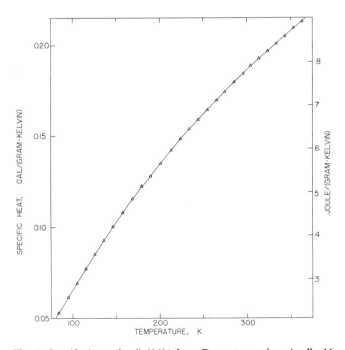

Fig. 4. Specific heat of soil 60601 from Descartes region, Apollo 16.

Table 6. Smoothed values for the specific heats of Apollo 14, 15, and 16 soils, breccia and basalt.

Temperature °K	Specific heat J/(gram · K)				
	14321,153 breccia	14163,186 soil	15555,159 basalt	15301,20 soil	60601,31 soil
90	(0.2234)	(0.2343)	0.2289	0.2406	0.2418
100	0.2536	0.2736	0.2920	0.2724	0.2741
120	0.3226	0.3494	0.3548	0.3356	0.3385
140	0.3950	0.4138	0.4151	0.3966	0.4008
160	0.4590	0.4724	0.4720	0.4385	0.4586
180	0.5121	0.5297	0.5255	0.5071	0.5121
200	0.5586	0.5812	0.5757	0.5573	0.5623
220	0.6021	0.6251	0.6230	0.6046	0.6100
240	0.6435	0.6627	0.6669	0.6489	0.6556
260	0.6812	0.7004	0.7084	0.6908	0.6991
280	0.7150	0.7376	0.7468	0.7301	0.7406
300	0.7485	0.7711	0.7828	0.7673	0.7799
320	0.7832	0.7987	0.8171	0.8037	0.8171
340	0.8171	0.8255	0.8502	0.8389	0.8531
360	(0.8506)	(0.8590)	0.8862	0.8724	0.8870

Table 7. Contributions of sample container to total observed heat capacity.

Sample number	weight grams	Percentage of observed heat capacity represented by the empty calorimeter at:	
		100°K	340°K
14163	40.00	48	36
14321	40.98	52	41
15301	39.91	43	36
15555	39.66	37	36
60601	30.04	51	43

Table 8. Temperature dependence of the thermal parameter γ for lunar soil at a density of 1300 kg m^{-3}.

Temperature °K	Thermal conductivity k $W\ m^{-1}K^{-1}$	Specific heat Cp $J\ kg^{-1}\ K^{-1}$	Thermal parameter γ $m^2\ sec^{1/2}\ KJ^{-1}$
100	0.0007	275.7	0.06313
150	0.0008	433.9	0.04707
250	0.0011	672.4	0.03225
300	0.0014	758.1	0.02692
350	0.0017	848.9	0.02309

In Table 8 we have calculated the thermal parameter for a typical lunar soil, for temperatures between 100 and 350°K, based upon our *average* specific heat, Equation (1), and using the thermal conductivities of sample 14163 measured by Cremers (1972) at a density of 1300 kg m^{-3}. This calculation supercedes the estimates of γ made by Robie *et al.* (1970), (0.037 to 0.021 m^2 sec$^{1/2}$ KJ^{-1}).

The values for the thermal parameter listed in Table 8 should be of considerable importance in quantifying the results of the Apollo 15 and 17 lunar heat flow experiments, (Langseth *et al.*, 1972, 1973), and in the interpretation of the Infrared Scanning Radiometer (ISR) experiment carried on board the Apollo 17 command module, (Low and Mendell, 1973).

REFERENCES

Cremers C. J. (1972) Thermal conductivity of Apollo 14 fines. *Proceedings Third Lunar Sci. Conf.*, *Geochim. Cosmochim. Acta*, Suppl. 3, pp. 2611–2617. MIT Press.

Cremers C. J. and Birkebak R. C. (1971) Thermal conductivity of fines from Apollo 12. *Proceedings Second Lunar Science Conf.*, *Geochim. Cosmochim. Acta*, Suppl. 2, Vol. 3, pp. 2311–2315. MIT Press.

Cremers C. J. and Hsia H. S. (1973) Thermal conductivity of Apollo 15 fines at low density (abstract). In *Lunar Science—IV*, pp. 164–166. The Lunar Science Institute, Houston.

Cremers C. J., Birkebak R. C., and Dawson J. P. (1970) Thermal conductivity of fines from Apollo 11. *Proceedings Apollo 11 Lunar Sci. Conf.*, *Geochim. Cosmochim. Acta*, Suppl. 1, Vol. 3, pp. 2045–2050. Pergamon.

Fountain J. A. and West E. A. (1970) Thermal conductivity of particulate basalt as a function of density in simulated lunar and martian environments. *J. Geophys. Res.* **75**, 4063–4069.

Hemingway B. S. and Robie R. A. (1972) The specific heats of Apollo 14 soils (14163) and breccia (14321) between 90 and 350°K (abstract). In *Lunar Science—III*, p. 369. The Lunar Science Institute, Houston.

Langseth M. G., Jr., Clark S. P., Jr., Chute J. L., Jr., Keihm S. J., and Wechsler A. E. (1972) Heat-flow experiment. *Apollo 15 Preliminary Science Report*, NASA SP-289, pp. 11-1 to 11-23.

Langseth M. G., Jr., Chute J. L., Jr., and Keihm S. J. (1973) Direct measurement of heat flow from the moon (abstract). In *Lunar Science—IV*, pp. 455–457. The Lunar Science Institute, Houston.

Low F. J. and Mendell W. W. (1973) An overview of early data from the Apollo 17 infrared scanning radiometer (abstract). In *Lunar Science—IV*, pp. 481–482. The Lunar Science Institute, Houston.

Robie R. A. and Hemingway B. S. (1971) Specific heats of the lunar breccia (10021) and olivine dolerite (12018) between 90 and 350° Kelvin. *Proceedings Second Lunar Sci. Conf.*, *Geochim. Cosmochim. Acta*, Suppl. 2, Vol. 3, pp. 2361–2365. MIT Press.

Robie R. A. and Hemingway B. S. (1972) Calorimeters for heat of solution and low temperature heat capacity measurements. U.S. Geol. Survey Prof. Paper 755, 32 pp.

Robie R. A., Hemingway B. S., and Wilson W. H. (1970) Specific heats of lunar surface materials from 90 to 350°K. *Proceedings Apollo 11 Lunar Sci. Conf.*, *Geochim. Cosmochim. Acta*, Suppl. 1, Vol. 3, pp. 2361–2367. Pergamon.

Watson K. I. (1964) Thermal conductivity measurements of selected silicate powders in vacuum from 150–350°K, II. An interpretation of the Moon's eclipse and lunation cooling curve as observed through the earth's atmosphere from 8–14 microns. Thesis, California Institute of Technology.

Proceedings of the Fourth Lunar Science Conference
(Supplement 4, *Geochimica et Cosmochimica Acta*)
Vol. 3, pp. 2489–2502

Solar flare proton spectrum averaged over the last $5×10^3$ years

H. P. HOYT, JR., R. M. WALKER, and D. W. ZIMMERMAN

Washington University
Laboratory for Space Physics
St. Louis, Missouri 63130

Abstract—Thermoluminescence (TL) measurements in the top centimeter of lunar rocks show a strong depth dependence that we have previously shown to be semi-quantitatively consistent with that expected from solar flare irradiation. In this paper we derive from the TL data the solar flare differential energy spectrum and integral proton flux above 10 MeV averaged over the last several thousand years. The dose-rate depth profile is obtained using a new TL equilibrium technique which is independent of the TL decay kinetics. The dose-rate depth profile produced by solar flare protons with a differential energy spectrum of the form $dJ/dE = KE^{-\gamma}$ is calculated for arbitrary γ. The best fit to the TL data in rock 14310 is obtained for $\gamma = 2.3 \pm 0.2$ and an omnidirectional (4π) integral flux above 10 MeV of 40–80 prot/cm² sec. The TL half-life is determined to be 2×10^3 yr. These results are compared to those for ^{22}Na ($T_{1/2} = 2.6$ yr) and ^{26}Al ($T_{1/2} = 7.4 \times 10^5$ yr) obtained by Wahlen *et al.* (1972) and Rancitelli *et al.* (1972) and we conclude that the spectral shape and flux of protons in the interval 25–100 MeV is the same within experimental errors when averaged over these three very different time periods. Our integral flux values do not agree with those obtained from measurements of ^{14}C ($T_{1/2} = 5.7 \times 10^3$ yr) by Begemann *et al.* (1972) and Boeckl (1972). One possible explanation is that the assumed cross section for production of ^{14}C is inaccurate but a more interesting alternative suggested by Begemann *et al.* (1972) is that the ^{14}C is surface correlated and represents direct implantation of ^{14}C from the sun.

INTRODUCTION

THIS PAPER reports the differential energy spectrum of solar protons averaged over the last $\sim 5 \times 10^3$ yr ($T_{1/2} = 2 \times 10^3$ yr) as measured by thermoluminescence (TL). In previous studies of solar flare effects in lunar rocks, fossil nuclear particle track gradients have been used to determine the average spectral shape of solar flare Fe group nuclei (Barber *et al.*, 1971; Crozaz *et al.*, 1971; Fleischer *et al.*, 1971). The depth profiles of bombardment produced radionuclides have been used to obtain solar proton spectra averaged over time periods varying with the half-life of the particular nuclide (Rancitelli *et al.*, 1972; Wahlen *et al.*, 1972). However, the time interval bracketed by ^{22}Na ($T_{1/2} = 2.6$ yr) and ^{26}Al ($T_{1/2} = 7.4 \times 10^5$ yr) is not covered by any routinely measured radionuclide. Measurements of ^{39}Ar ($T_{1/2} = 270$ yr) do cover this time span but only the integral flux of high energy solar protons ($E > 200$ MeV) is obtained (Fireman, 1972). Two measurements of ^{14}C ($T_{1/2} = 5.7 \times 10^3$ yr) have been made (Begemann *et al.*, 1972; Boeckl, 1972) but there is some question as to the absolute particle flux over this time span because of uncertainties in the cross section for ^{14}C production. The present work thus covers a time interval for which little was known about the average behavior of solar flares.

The nature of TL in crystalline lunar rocks is illustrated by the glow curves of Fig. 1 where we show the natural TL emitted by an internal sample of rock 14311. Glow curves for equal aliquots that have been artificially irradiated before measurement are also shown. Clearly it is possible to store much more TL in a sample than is represented by the natural signal. Such considerations, and in particular the depth variation of TL in lunar cores lead to the conclusion that the low temperature TL is in equilibrium (Dalrymple and Doell, 1970; Doell and Dalrymple, 1971; Hoyt *et al.*, 1970; Hoyt *et al.*, 1971). The rate of acquisition of trapped charge is balanced by the rate of loss by thermal draining; differences between equilibrium levels for various samples reflect differences in average ionization dose-rate and temperature.

Small lunar rocks ($\leqslant 10$ cm) are at any given time nearly isothermal, as we will discuss below. Therefore the variation of TL with depth is particularly simple, reflecting only the variation of dose rate. In the top centimeter of surface rocks the ionization dose-rate from solar flare particles dominates the other sources (galactic cosmic rays and internal radioactivity) and is rapidly attenuated with depth. The stored TL accordingly decreases with depth. We have reported this effect in several rocks which had been diced and the TL measured on a millimeter scale (Hoyt *et al.*, 1971; Crozaz *et al.*, 1972). Figure 2 shows the normalized TL as a function of depth in rock 14310. (This data was previously published (Crozaz *et*

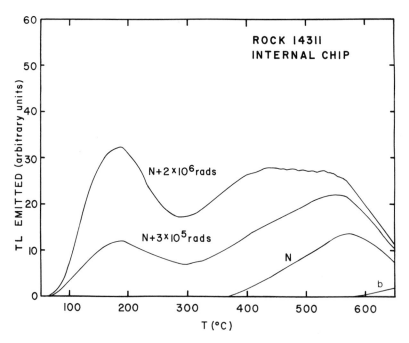

Fig. 1. Natural TL glow curve, *N*, for an internal sample of rock 14311. Glow curves for equal aliquots that have been artificially irradiated before measurement are also shown. Curve *b* is the thermal background.

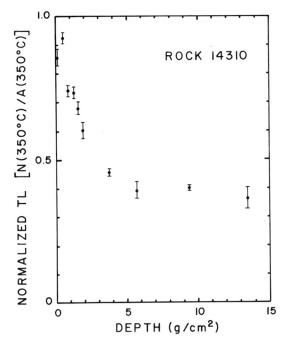

Fig. 2. Normalized TL as a function of depth in rock 14310. The points were obtained by dividing the natural TL intensity at 350°C by the intensity found in a second heating of the sample after an irradiation of 38 krad. The error bars are one σ based on repeated measurements. The rock density is taken to be 2.88 g/cm³.

al., 1972); it is presented here using the correct rock density, 2.88 g/cm³.) In this paper we will derive from this change in TL with depth, the solar flare proton spectrum that produced it. Optical bleaching of TL above ~250°C in the glow curve has been reported (Hoyt *et al.*, 1972). There is also the possibility that micro-meteorite impacts may influence the near surface TL. For example, Finkel *et al.* (1971) have reported a surface friability that decreases with depth for the first few millimeters in rock 12002. This effect was found only on exposed surfaces and is presumably caused by micro-meteorite impact. To avoid problems from such sources, we will restrict ourselves to depths greater than 0.75 g/cm² (proton energies >25 MeV).

A quantitative treatment of temperature effects in lunar rocks is obtained if we let t_c be the characteristic time for relaxation of thermal gradients within a rock. This time depends on the size of the rock and its thermal parameters. If t_a is the time scale for changes in the surface temperature produced by changes in the incident radiation flux, then the rock will remain isothermal if $t_c \ll t_a$. It has been shown by Roelof (1968) that this inequality holds, except for about one hour at sunrise, for rocks ≤ 10 cm. We calculate that this would give a thermal gradient averaged over a lunation of <0.1 deg/cm which can be ignored for our purposes.

Deriving the average solar flare spectrum involves two steps. We first calculate a general relationship between a given solar flare spectrum and the dose rate as a function of depth that it produces. Secondly the dose rate as a function of depth is obtained from the TL data. These steps are treated in separate sections.

EXPERIMENTAL TECHNIQUE

The experimental procedures and apparatus for glow curve measurements are the same as described previously with the exception that samples are mounted on 5 mm silver disks for ease of handling (Hoyt *et al.*, 1970; Hoyt *et al.*, 1971).

For equilibrium experiments, a small oven-radiation source combination is used. The temperature in this oven can be adjusted and is regulated to better than $\pm 1.5°C$. The radiation is from a 50 mCi $^{90}Sr - ^{90}Y$ beta source. By adjusting the position of this source, the dose rate to the sample is continuously variable over two orders of magnitude with a maximum dose rate of ~ 2.5 krad/min. The dose rates were measured by inter-comparison with a calibrated ^{60}Co source using $CaSO_4$ and LiF TL dosimeters. The ^{60}Co dose rate was measured with an NBS calibrated ionization chamber. Our absolute dose rates are determined within $\sim 5\%$, the relative values within $\sim 3\%$. Dose rates quoted in this paper are calculated for the chemistry of rock 14310 and are 4% less than the dose rate measured in LiF.

DOSE RATES FROM SOLAR FLARE PARTICLES

Solar cosmic rays are produced in short bursts or flares with most of the particles being contributed by a few large events around solar maximum. These particles are predominately protons with typical energies of tens of MeV and their average flux dominates that of galactic cosmic rays below several hundred MeV. Solar particle energy distributions are sometimes conveniently described as either a rigidity distribution or a power law distribution. The rigidity distribution is

$$\frac{dJ}{dR} \sim \exp\left(-R/R_0\right)$$

where R is the rigidity of the particle (momentum per unit charge in units of mega volts, MV) and R_0 is a shape parameter. For flares measured during the last two solar cycles, typical values of R_0 have been 50–150 MV for the energy range 30–100 MeV (Webber *et al.*, 1963; King, 1972). The power law distribution is

$$\frac{dJ}{dE} \sim E^{-\gamma}$$

Typical values for the shape parameter γ have been 1.5–3.5 for the same energy interval. Spectra for some flares are best fit with a rigidity distribution and others with a power law in energy, but the difference in shape is not significant over the energy interval to which our measurements are restricted.

In this section we calculate the dose rate as a function of depth produced by solar flare irradiation for arbitrary γ. Ionization energy-loss is the dominant mechanism by which solar particles are slowed down. We first obtain the energy flux E_x transferred by solar particles at a depth x below the surface of a semi-infinite slab. The rate at which energy is dissipated as a function of depth (dose rate vs. depth) is then just the negative derivative of this quantity.

We consider first only protons incident vertically. For a differential kinetic energy spectrum $dJ/dE = KE^{-\gamma}$ (prot/cm^2 sec str MeV) and range energy relation $R = kE^{1/\alpha}$ (g/cm^2), the flux of protons with range in dR at R is

$$dJ = \frac{dJ}{dE}\frac{dE}{dR}\,dR = \alpha K k^{\alpha(\gamma-1)}R^{\alpha(1-\gamma)-1}\,dR$$

Then the flux of protons at depth x with residual range in dr at r is

$$dJ = \alpha K k^{\alpha(\gamma-1)}(r+x)^{\alpha(1-\gamma)-1}\,dr$$

To obtain the energy flux, E_x (MeV/cm^2 sec), at depth x (g/cm^2) transferred by protons incident vertically, we multiply by the residual energy $(r/k)^\alpha$ and integrate

$$E_x = \alpha K k^{\alpha(\gamma-2)}\int_0^\infty r^\alpha(r+x)^{\alpha(1-\gamma)-1}\,dr$$

The energy dissipated per unit mass (MeV/g sec) is then given by $-dE_x/dx$

$$\frac{-dE_x}{dx} = [\alpha^2(\gamma-1)+\alpha]K k^{\alpha(\gamma-2)}\int_0^\infty r^\alpha(r+x)^{\alpha(1-\gamma)-2}\,dr$$

This can be rearranged:

$$\frac{-dE_x}{dx} = [\alpha^2(\gamma-1)+\alpha]K k^{\alpha(\gamma-2)}x^{\alpha(2-\gamma)-1}\int_0^\infty \frac{Z^\alpha\,dZ}{(1+Z)^{2+\alpha(\gamma-1)}}$$

where $Z = r/x$ and integration yields

$$\frac{-dE_x}{dx} = f(\alpha,\gamma)x^{\alpha(2-\gamma)-1}\frac{\text{MeV}}{\text{g sec str}}$$

Here $f(\alpha,\gamma)$ is a constant given by

$$f(\alpha,\gamma) = [\alpha^2(\gamma-1)+\alpha]K k^{\alpha(\gamma-2)}B(\alpha+1,\alpha(\gamma-2)+1)$$

where B is the beta function and can be defined in terms of gamma functions

$$B(p,q) = \Gamma(p)\Gamma(q)/\Gamma(p+q)$$

The calculation for a 2π isotropic irradiation is particularly simple. If $-dE_x^T/dx$ is the total rate at which ionization energy is dissipated at depth x for 2π irradiation, then

$$\frac{-dE_x^T}{dx} = 2\pi f(\alpha,\gamma)\int_0^{\pi/2}\left(\frac{x}{\cos\theta}\right)^{\alpha(2-\gamma)-1}\sin\theta\,d\theta$$

and our final result is

$$\frac{-dE_x^T}{dx} = \frac{\pi}{\alpha(\gamma-2)+2}f(\alpha,\gamma)x^{\alpha(2-\gamma)-1}\frac{\text{rad}}{y} \tag{1}$$

It should be noted that $\alpha \simeq 1/2$, and thus the depth dependence of dose rate, D, is proportional to $x^{-\gamma/2}$.

Solar flare particles consist mainly of protons with α particles the second most abundant species. For the contemporary flux, measured values of the p/α ratio for particles of equal energy/nucleon vary from flare to flare and even within a single flare. At ~ 30 MeV/nuc the ratio is usually greater than 10 and an estimate of the average is $p/\alpha \simeq 25$ (Lanzerotti, 1972). Alpha particles and protons have approximately the same range energy/nuc relation but the rate of energy loss for α particles is four times greater than for protons. We therefore expect the energy dissipated by α particles to be $\sim 16\%$ of that by protons. This should decrease at higher energy/nucleon because the p/α ratio increases. At 100 MeV/nuc a good average value is probably $p/\alpha \simeq 150$ (Lanzerotti, 1972). Including the alpha particle contributions to the dose rate and using these ratios produces a change in our derived value of γ of less than 5%.

Little is known about the long-term average α flux at these energies. There is evidence that the long-term flux below ~ 10 MeV/nuc is not very different from the present-day flux through the August 1972 flare (Lanzerotti et al., 1973). In this paper we will ignore contributions from α particles. The error in our derived value of γ introduced by this assumption is probably $< 5\%$.

Equilibrium Experiments

Basic considerations

We turn now to the question of how the TL variations with depth can be used to obtain the dose rate profile. We cannot simply take the dose rate to be directly proportional to the natural TL equilibrium level. This is equivalent (as will be shown below) to assuming the kinetics of TL decay are first order, which they might not be. Furthermore, although experiments could be performed to try to determine the order of kinetics, the results in cases like this where there is not an isolated glow peak to work with are often inaccurate and ambiguous.

In our work this problem is eliminated with an empirical method based on a series of equilibrium experiments. The basic idea is to measure the natural TL at a given depth and then in a subsequent experiment determine the dose rate necessary to re-establish this equilibrium level while the sample is held at a constant temperature. A series of such experiments on samples from different depths gives us the dose-rate profile directly, independent of the kinetics. These experiments are performed at elevated temperatures to shorten the time necessary to reach equilibrium. All TL values are measured at 350°C in the glow curve. The temperature used is somewhat arbitrary; high enough to give an easily measurable signal, low enough for the natural TL to be well below saturation.

The rationale for the equilibrium experiments is demonstrated by the following simple model for the TL process. The rate of filling of traps by ionizing radiation is taken to be proportional to the number of unfilled traps $(N - n)$ where N is the total number of existing traps and n is the number of filled traps. If the thermal decay process at 350°C in the glow curve can be characterized by a single equivalent activation energy, the rate of decay of filled traps can be written as $-\beta(T)f(n)$. Here $\beta(T)$ is a temperature dependent rate constant given by

$$\beta(T) = S \exp(-Q/kT)$$

where k is the Boltzmann constant, Q is the thermal activation energy of filled traps, and S is a constant frequency factor. The exact functional dependence $f(n)$ will depend on the order of kinetics which we do not want to specify. For equilibrium we can write

$$dn/dt = 0 = \alpha D(N-n) - \beta(T)f(n)$$

where D is the radiation dose rate and α is the probability per unit dose of filling an empty trap. At low glow curve temperatures, well below saturation, $N \gg n$ and

$$D = \beta(T)\frac{f(n)}{\alpha N} \tag{2}$$

Thus the dose rate D is an unknown function of the TL equilibrium value n, dependent on the kinetics.

In the equilibrium experiment the dose rate is adjusted in the laboratory to reproduce the natural equilibrium level for a particular depth x_1. We can write the ratio of the two dose rates as follows:

$$\frac{D_{lab}(x_1)}{D_{moon}(x_1)} = \frac{\beta(T_{lab})}{\beta(T_{moon})} = \frac{D_{lab}(x_2)}{D_{moon}(x_2)}$$

or

$$\frac{D_{moon}(x_1)}{D_{moon}(x_2)} = \frac{D_{lab}(x_1)}{D_{lab}(x_2)}$$

Measurements of $D_{lab}(x)$ in this way give a direct measure of the shape of the dose rate vs. depth curve on the lunar surface. This procedure is independent of the kinetics of recovery and the spectral shape deduced should be independent of T_{lab}. This is checked by repeating the experiment at several values of T_{lab}. If the model is not strictly valid, then the deduced value of γ may vary with T_{lab}. If this is the case then the method can still be considered an empirical one, γ determined as a function of T_{lab} and then extrapolated to the effective lunar temperature. As will be seen later, the initial results indicate that γ does not vary measurably with temperature.

Natural equilibrium dose-rate determination

Figure 3 shows the build-up of TL with irradiation time for two different dose rates when the sample is held at 239°C. This data is obtained by first measuring the natural TL to ~ 450°C in the normal way. The natural TL intensity at a glow curve temperature of 350°C is indicated by the dashed line. The sample is then held at 239°C and irradiated for the indicated time; the sample is removed and a second glow curve is recorded. The intensity at a glow curve temperature of 350°C is shown for varying irradiation times.

At 9 rad/sec the equilibrium level is much higher than the natural level but at 2 rad/sec the level reached is within 10% of the natural. We can correct for this small difference by linear extrapolation and obtain the natural equilibrium dose

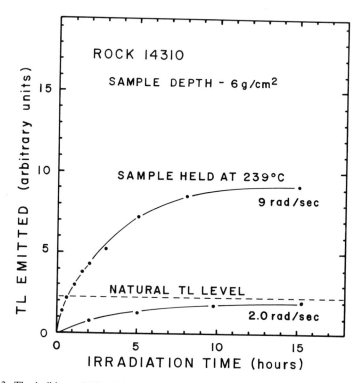

Fig. 3. The build-up of TL with irradiation time for two different dose rates when the sample is held at 239°C. The points plotted are the TL intensity at 350°C for glow curves obtained after irradiation for the time and dose rate indicated. The natural TL intensity at a glow curve temperature of 350°C is indicated by the dashed line.

rate for samples irradiated at 239°C. Samples from other depths are measured similarly, the dose rate for each being adjusted to match the natural levels. Thus the variation of natural equilibrium dose rate with depth is measured directly for a given temperature.

Solar flare spectral shape

Figure 4 shows the natural equilibrium dose rate vs. depth obtained at 266°C for two to three samples at each measured depth of rock 14310. This curve reflects ionization from solar flare particles, galactic cosmic rays, and internal radioactivity. The galactic contribution is expected to change by less than 20% in the top 15 g/cm^2 (Haffner, 1967; Ryan, 1964) and the internal radioactivity should be constant with depth. Therefore the background is assumed constant with depth and a computer routine is used which gives the best fit allowing both the background and solar flare spectrum to vary. The best fit is obtained for $\gamma_{slab} = 2.0 \pm 0.1$ (semi-infinite slab geometry assumed; best fit slope of 1.0) and a background of 6.7 ± 0.4 rad/sec. The assumption of slab geometry underestimates γ for this rock.

We have approximated the effect of rock geometry by averaging the dose-rate depth variation in hemispheres with radii of 5 and 10 cm. This amounts to a 15% correction. Thus the spectral parameter that gives the best fit for 14310 is $\gamma_{rock} = 2.3 \pm 0.2$ and the dose-rate profile is shown in Fig. 4.

As discussed above, we expect the same depth variation in dose rate at different temperatures. At the present time we have data obtained at a second temperature of 239°C for three depths. These values are normalized and plotted in Fig. 4 for comparison with the 266°C data. The results for these depths are the same within experimental error which suggests that the deduced depth profile and thus γ are indeed independent of temperature used in the equilibrium experiment. It should also be noted that if the simple assumption of first order kinetics is made, then the data of Fig. 2 give an incorrect value of γ 30% lower than that derived above.

Solar flare integral flux

We obtain the absolute solar particle flux by estimating the background ionization rate and comparing this number to the background obtained in the fit to the data of Fig. 4. This defines the normalization at lunar conditions and we can calculate the constant terms in Equation (1). The dose rate from galactic cosmic rays is taken to be 4–13 rad/yr which covers the range of values measured on Apollo flights (T. White, NASA JSC, private communication) as well as most calculated values. We estimate the effective dose rate from internal radioactivity to be ~ 1 rad/yr. These values give the omnidirection (4π) integral flux above 10 MeV, $J_{10} = 40$–80 prot/cm² sec.

TL equilibrium time

We now turn to the question of the time required for the TL to reach equilibrium. The mean lifetime for a trapped electron is $\tau = 1/\beta$ and from Equation (2) we may write

$$\tau_{moon} = \frac{D_{lab}\tau_{lab}}{D_{moon}}$$

From Fig. 3 we have $\tau_{lab} = 6$ hr from $D_{lab} = 2$ rad/sec. These values give $\tau_{moon} \simeq 3 \times 10^3$ yr for $D_{moon} = 16$ rad/yr or $T_{1/2} = (\ln 2)\tau_{moon} = 2 \times 10^3$ yr. A second way to approach this problem is to calculate the time for the normalized TL to reach equilibrium in the absence of thermal draining. Taking $N/A \simeq 0.4$ for the deep samples in Fig. 2 obtained for an artificial dose of 38 krad we estimate that this natural signal could be built up in $\sim 2 \times 10^3$ yr. A third method of estimating τ_{moon} is to measure τ as a function of T_{lab} and extrapolate to lunar conditions. This method however would require more data at other temperatures than are presently available.

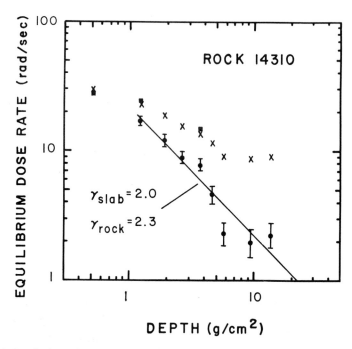

Fig. 4. Depth dependence of equilibrium dose rate (x) for samples of 14310 held at 266°C. The solar contribution (●) obtained by subtracting a constant background of 6.7 rad/sec is also shown along with the best fit for an assumed solar flare power law spectrum (*see* text). Also shown for comparison are data obtained at 239°C (■), multiplied by 4.6 for normalization. Error bars, shown only on the solar points, are one σ based on repeated measurements.

DISCUSSION

While solar protons are mainly attenuated by ionization loss, they do undergo some nuclear interactions. The depth profile of bombardment produced radio-nuclides has been used to derive average solar particle spectra for time periods comparable to the mean life of the particular nuclide. Table 1 summarizes solar proton parameters determined from measurements of radionuclides covering periods relevant to this work. Where results are quoted for exponential rigidity spectra, we show an approximate γ for the interval 25–100 MeV. While this is not rigorously correct, it will serve to compare the two representations. The spectra are shown in Fig. 5. Solar-proton-produced activity is negligible at depths greater than ~ 10 g/cm^2 and reaction threshold energies range from ~ 15 to 40 MeV/nuc. We have therefore chosen to compare the spectra in the interval 25–100 MeV/nuc which also corresponds to the energy interval we are restricted to. That this is the important energy interval is also apparent if we consider the flux of particles as a function of depth in the moon (Reedy and Arnold, 1972).

Our spectral shape as shown in Fig. 5 is about the same as that obtained by

Table 1

		$T_{1/2}$ (years)	$\dfrac{dJ}{dR} \sim \exp(-R/R_0)$	$\dfrac{dJ}{dE} \sim E^{-\gamma}$	Integral Flux (4π) $E > 10$ MeV (prot/cm^2 sec)
			Spectral Parameter		
This work		2×10^3		$\gamma = 2.3 \pm 0.2$	40–80
Wahlen	^{26}Al	7.4×10^5	$R_0 = 100$	$(\gamma \sim 2.2)^*$	80
et al.	^{55}Fe	2.6	$R_0 = 100$	$(\gamma \sim 2.2)^*$	100
(1972)	^{22}Na	2.6	$R_0 = 85$	$(\gamma \sim 2.4)^*$	110
Rancitelli	^{26}Al	7.4×10^5		$\gamma = 3.1$	120
et al.	^{22}Na	2.6		$\gamma = 3.1$	120
(1972)					
Boeckl	^{14}C	5.7×10^3	$R_0 = 100$	$(\gamma \sim 2.2)^*$	200
(1972)					
Begemann	^{14}C	5.7×10^3	$R_0 = 80$–100	$(\gamma \sim 2.5$–$2.2)^*$	300–500
et al.					
(1972)					

*Approximate fit for the interval 25–100 MeV.

Wahlen et al. (1972) for both short and long half-lives. It is also clear that our data are systematically lower by $\sim 50\%$. However, their flux values are not determined more precisely than this and it is difficult to conclude that the differences are real. Also shown in Fig. 5 is the spectrum given by Rancitelli et al. (1972) using depth data for radionuclide production that are somewhat less detailed than that of Wahlen et al. (1972). Although the spectral index is quite different, because of the limited energy range the differential flux in the 25–100 MeV range is within a factor of ± 2 of that given here. It should be noted that while each of these groups determines a different spectrum from their radiochemical measurements, the spectrum they determine is essentially the same for both short and long half-lives.

Begemann et al. (1972) have measured ^{14}C activity in three broad depth regions in rock 12053 and Boeckl (1972) has determined somewhat more detailed depth information for ^{14}C activity in rock 12002. Both find that their data can be fit by a production profile for a proton spectral shape similar to that found for ^{22}Na and ^{26}Al by Wahlen et al. (1972) but that a much higher integral flux is required. Our results do not support the latter conclusion and one possibility is that the discrepancy arises from inaccuracies in the assumed cross section for production of ^{14}C. A more interesting possibility suggested by Begemann et al. (1972) is that the ^{14}C is surface correlated and represents direct implantation of ^{14}C from the sun.

The radionuclide ^{39}Ar ($T_{1/2} = 270$ yr) is produced from Fe, Ti, Ca, and K by high energy protons. By looking at the depth profile of ^{39}Ar measured in rocks with different abundances of these elements Fireman (1972) obtains the ^{39}Ar produced

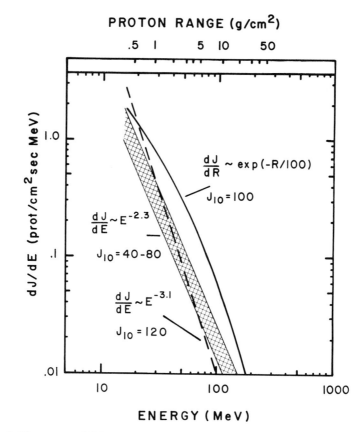

Fig. 5. Time averaged differential flux of solar protons at the lunar surface as determined by different laboratories (*see* text—hatched area in this work). The parameter J_{10} is the omnidirectional (4π) integral flux above 10 MeV in free space (prot/cm² sec).

in Fe. He finds an amount of Fe→^{39}Ar which is constant with depth and in excess of that expected from galactic cosmic rays. The excess is attributed to production by solar flare protons with energy >200 MeV. Thus he concludes that solar cycle 19 which included a number of solar flares with very hard spectra has been more typical of the last 10^3 yr than has the present solar cycle (cycle 20). While this result is for energies above those of this work, it is relevant in that it also covers the time span bracketed by ^{22}Na and ^{26}Al

Spectral parameters for the contemporary solar flux have been measured by balloon and satellite experiments. For cycle 19 (1954–1964) average values obtained from compilations by Webber *et al.* (1963) are $R_0 = 100$ MV and $J_{10} = 110$ prot/cm² sec. Average values for the present solar cycle are $R_0 = 80$ MV and $J_{10} = 130$ prot/cm² sec (8 yr) (from compilations by King, 1972). It is clear that the long-term solar proton spectrum determined now by different techniques and averaging over three very different time spans extending over the last 10^6 yr is about the same as this contemporary flux. This is quite surprising given that the flux is produced predominantly by a few events in each cycle and the characteristics of these events are quite variable.

Acknowledgments—We thank A. Feldman, Washington University School of Medicine, for assistance with dose-rate calibrations. We also wish to thank our colleagues E. Zinner and L. Benofy for useful discussion. H. Ketterer has been indispensible in the preparation of this manuscript. This work was supported by NASA Grant NGL 26-008-065.

REFERENCES

Barber D. J., Cowsik R., Hutcheon I. D., Price P. B., and Rajan R. S. (1971) Solar flares, the lunar surface, and gas-rich meteorites. *Proc. Second Lunar Sci. Conf., Geochim. Cosmochim. Acta,* Suppl. 2, Vol. 3, pp. 2705–2714. MIT Press.

Begemann F., Born W., Palme H., Vilcsek E., and Wänke H. (1972) Cosmic-ray produced radioisotopes in Apollo 12 and Apollo 14 samples. *Proc. Third Lunar Sci. Conf., Geochim. Cosmochim. Acta,* Suppl. 3, Vol. 2, pp. 1693–1702. MIT Press.

Boeckl R. S. (1972) A depth profile of ^{14}C in the lunar rock 12002. *Earth Planet. Sci. Lett.* 16, 269–272.

Crozaz G., Walker R., and Woolum D. (1971) Nuclear track studies of dynamic surface processes on the moon and the constancy of solar activity. *Proc. Second Lunar Sci. Conf., Geochim. Cosmochim. Acta,* Suppl. 2, Vol. 3, pp. 2543–2558. MIT Press.

Crozaz G., Drozd R., Hohenberg C. M., Hoyt H. P. Jr., Ragan D., Walker R. M., and Yuhas D. (1972) Solar flare and galactic cosmic ray studies of Apollo 14 and 15 samples. *Proc. Third Lunar Sci. Conf., Geochim. Cosmochim. Acta,* Suppl. 3, Vol. 3, pp. 2917–2931. MIT Press.

Dalrymple G. B. and Doell R. R. (1970) Thermoluminescence of lunar samples from Apollo 11. *Proc. Apollo 11 Lunar Sci. Conf., Geochim. Cosmochim. Acta,* Suppl. 1, Vol. 3, pp. 2081–2092. Pergamon.

Doell R. R. and Dalrymple G. B. (1971) Thermoluminescence of Apollo 12 lunar samples. *Earth Planet. Sci. Lett.* 10, 357–360.

Finkel R. C., Arnold J. R., Imamura M., Reedy R. C., Fruchter J. S., Loosli H. H., Evans J. C., and Delany A. C. (1971) Depth variation of cosmogenic nuclides in a lunar surface rock and lunar soil. *Proc. Second Lunar Sci. Conf., Geochim. Cosmochim. Acta,* Suppl. 2, Vol. 2, pp. 1773–1789. MIT Press.

Fireman E. L. (1972) Depth variation of Ar^{37} and Ar^{39} in lunar material. In *The Apollo 15 Lunar Samples,* pp. 364–367. The Lunar Science Institute, Houston.

Fleischer R. L., Hart H. R. Jr., and Comstock G. M. (1971) Very heavy solar cosmic rays: Energy spectrum and implications for lunar erosion. *Science* 171, 1240–1242.

Haffner J. W. (1967) *Radiation and Shielding in Space,* pp. 275–280. Academic Press.

Hoyt H. P. Jr., Kardos J. L., Miyajima M., Seitz M. G., Sun S. S., Walker R. M., and Wittels M. C. (1970) Thermoluminescence, X-ray and stored energy measurements of Apollo 11 samples. *Proc. Apollo 11 Lunar Sci. Conf., Geochim. Cosmochim. Acta,* Suppl. 1, Vol. 3, pp. 2269–2287. Pergamon.

Hoyt H. P. Jr., Miyajima M., Walker R. M., Zimmerman D. W., Zimmerman J., Britton D., and Kardos J. L. (1971) Radiation dose rates and thermal gradients in the lunar regolith: Thermoluminescence and DTA of Apollo 12 samples. *Proc. Second Lunar Sci. Conf., Geochim. Cosmochim. Acta,* Suppl. 2, Vol. 3, pp. 2245–2263. MIT Press.

Hoyt H. P. Jr., Walker R. M., Zimmerman D. W., and Zimmerman J. (1972) Thermoluminescence of individual grains and bulk samples of lunar fines. *Proc. Third Lunar Sci. Conf., Geochim. Cosmochim. Acta,* Suppl. 3, Vol. 3, pp. 2997–3007. MIT Press.

King J. H. (1972) Study of mutual consistancy of IMP 4 solar proton data. National Space Science Data Center NSSDC 72-14.

Lanzerotti L. J. (1972) Solar flare particle radiation. In *Proc. National Sym. on Natural and Manmade Radiation in Space* (editor E. A. Warman), pp. 193–208, NASA TM X-2440.

Lanzerotti L. J., Reedy R. C., and Arnold J. R. (1973) Alpha particles from solar cosmic rays over the last 80,000 years. *Science* **179**, 1232–1234.

Rancitelli L. A., Perkins R. W., Felix W. D., and Wogman N. A. (1972) Lunar surface processes and cosmic ray characterization from Apollo 12–15 lunar sample analyses. *Proc. Third Lunar Sci. Conf.*, *Geochim. Cosmochim. Acta*, Suppl. 3, Vol. 2, pp. 1681–1691. MIT Press.

Reedy R. C. and Arnold J. R. (1972) Interaction of solar and galactic cosmic-ray particles with the moon. *J. Geophys. Res.* **77**, 537–555.

Roelof E. C. (1968) Thermal behavior of rocks on the lunar surface. *Icarus* **8**, 138–159.

Ryan J. A. (1964) Corpuscular radiation produced crystalline damage at the lunar surface. In *The Lunar Surface Layer* (editors J. W. Salisbury and P. E. Glaser), pp. 265–312. Academic Press.

Wahlen M., Honda M., Imamura M., Fruchter J. S., Finkel R. C., Kohl C. P., Arnold J. R., and Reedy R. C. (1972) Cosmogenic nuclides in football-sized rocks. *Proc. Third Lunar Sci. Conf., Geochim. Cosmochim. Acta*, Suppl. 3, Vol. 2, pp. 1719–1732. MIT Press.

Webber W. R., Benbrook J. R., Thomas J. R., Hunting H., and Duncan R. (1963) An evaluation of the radiation hazard due to solar-particle events. Boeing Report D2-90469.

Proceedings of the Fourth Lunar Science Conference
(Supplement 4, *Geochimica et Cosmochimica Acta*)
Vol. 3, pp. 2503–2513

Surface brightness temperatures at the Apollo 17 heat flow site: Thermal conductivity of the upper 15 cm of regolith

S. J. KEIHM and M. G. LANGSETH, JR.

Lamont-Doherty Geological Observatory of Columbia University, Palisades, New York 10964

Abstract—Lunar surface brightness temperatures derived as part of the Apollo 17 heat flow experiment are reported. Nighttime surface temperatures, calculated from the data provided by two thermocouples suspended about 15 cm above the lunar surface, are used to determine the conductivity profile of the upper 15 cm of regolith at the ALSEP site. The surface reaches a maximum temperature of 384 (± 6)°K at lunar noon and cools to a minimum temperature of 102 ($\pm 1\frac{1}{2}$)°K at the end of the lunar night. The nighttime cool-down curve is best fit by an essentially two-layer conductivity model of the lunar regolith. Conductivities of the order of 1.5×10^{-5} W/cm-°K are postulated for a 2 cm porous surface layer overlying more compact regolith material with conductivities in the range of 1.0–1.5×10^{-4} W/cm-°K between 2 and 15 cm.

A mean surface temperature of 216 (± 5)°K is deduced from the thermocouple data. The 256°K temperature measured by the probe sensors at 130 cm thus indicates that a large mean temperature gradient exists at the Apollo 17 site. A strongly temperature-dependent thermal conductivity, at least in the upper few centimeters, accounts for the large mean temperature gradient. A ratio of radiative to conductive heat transfer = 2.0 at 350°K is required for at least the upper few centimeters to produce the observed mean temperature gradient.

INTRODUCTION

UNTIL THE Surveyor missions remote observations of the moon's infrared radiation constituted the primary data used to determine lunar surface brightness temperatures and the thermal properties of the near-surface layer. Infrared observations within the last ten years have included the work of Murray and Wildey (1964), Shorthill and Saari (1965), Low (1965), and Mendell and Low (1970). Besides the inherent limitations on the accuracy of such measurements, especially during the lunar night, the earth-based observations also suffered the effects of a finite antenna beam-width which measured the integrated signal over a finite area of the lunar surface. Near-terminator observations were especially difficult and, as will be pointed out later, subsequent determinations of near-surface thermal properties from the nighttime cool-down curve could not be obtained unambiguously without accurate data for the first 40 hours following sunset.

As part of the Apollo 17 heat flow experiment, two thermocouples (one at each probe site) were embedded in a section of the electronics cable located about 15 cm from the top of the drill stem and suspended above the lunar surface (*see* Fig. 1). These thermocouples are in radiative balance with the lunar surface, the solar flux and space and hence provide a measurement of the surface brightness temperature throughout the lunation.

L-DGO Contribution #1980

Fig. 1. Photograph of probe # 2 borestem protruding from the lunar surface. The heat flow experiment housing is in the background. The thermocouple is in the black portion of cable about 10 cm from the top of the stem.

Surface Temperatures Deduced from Thermocouple Measurements

The flux balance equation governing the thermocouple temperature is:

$$2\pi a_c d\ell \epsilon_c \sigma T_c^4 = 2\pi a_c d\ell F_{c-m}\epsilon_m \alpha_{cir}\sigma T_m^4 + 2a_c d\ell S\alpha_{cs}\sqrt{1-p^2}$$
$$+ 2\pi a_c d\ell SA\alpha_{cs}F_{c-m}\cos\lambda\,\sin\phi \qquad (1)$$

where T_c = thermocouple temperature

$\quad T_m$ = lunar surface brightness temperature

$\quad a_c$ = radius of thermocouple cable

$\quad d\ell$ = elemental length of cable

$\quad \epsilon_c$ = infrared emissivity of cable

$\quad \alpha_{cir}$ = infrared absorptivity of cable

$\quad \alpha_{cs}$ = absorptivity of cable to solar flux

F_{c-m} = view factor of cable to the lunar surface, including the surrounding mountains

$\quad p$ = cosine of the angle between the sun line and the cable axis, a function of cable orientation and lunar phase angle

$\quad \lambda$ = selenographic latitude at Taurus Littrow

ϕ = lunar phase angle measured from local sunrise
ϵ_m = infrared emissivity of the lunar surface = 1.0
S = the mean solar constant = 0.1352 W/cm^2
A = the lunar albedo = 0.08
σ = the Stefan–Boltzmann constant

The first term on the right-hand side of Equation (1) represents flux into the cable element from the lunar surface; the second term represents direct flux from the sun; the third term represents solar energy reflected diffusely from the lunar surface and impinging on the cable.

The radiative properties of the cable, ϵ_c, α_{cir}, and α_{cs}, were determined by laboratory measurement prior to the Apollo 17 mission. The cable orientations for both probe locations were determined from ALSEP photographs.

The choice of ϵ_m = 1.0 implies that the surface temperatures deduced correspond to blackbody brightness temperatures. Measurements of the infrared emissivity of the lunar surface material indicate that the emissivity is actually in the range 0.90–0.95 (Logan *et al.*, 1972). If the value 0.93 is used, an upward correction of $(1.0/0.93)^{1/4}$, approximately 1.7%, should be applied to obtain actual surface temperatures. In any case, thermal properties determined from the nighttime cool-down temperature curve are very nearly independent of emissivity as long as the same value of ϵ_m is used in thermal model calculations as is used in the thermocouple reduction equation.

Solving Equation (1) for the surface brightness temperature yields:

$$T_m = \left[\frac{\epsilon_c \sigma T_c^4 - (S/\pi)\alpha_{cs}\sqrt{1-p^2} - SAF_{c-m} \cos \lambda \sin \phi}{F_{c-m}\epsilon_m \alpha_{cir}\sigma} \right]^{1/4} \qquad (2)$$

During the lunar night, using $\epsilon_c = \alpha_{cir}$, Equation (2) reduces to

$$T_m = \left[\frac{T_c^4}{F_{c-m}} \right]^{1/4} \qquad (3)$$

Equations (2) and (3) assume that the surrounding mountains at Taurus Littrow are at the same temperature as the surface throughout the lunation. The deviation from this assumption, especially during the lunar day, may be quite large. However, both of the thermocouples have view factors to the mountains about 1/12 of their view to the surface. Thus, even large anomalous temperatures on the slopes of surrounding mountains will produce only negligible errors in the surface temperature determinations.

Shown in Fig. 2 is a full lunation plot of deduced lunar surface brightness temperatures at Taurus Littrow. Vertical bars represent estimated error bounds. The daytime temperatures were determined solely from the temperature data of the exposed probe # 2 thermocouple for two reasons. First, the orientation of the thermocouple at probe # 1 was much more difficult to obtain from the photographs; second, the probe # 1 thermocouple appeared to have a more substantial

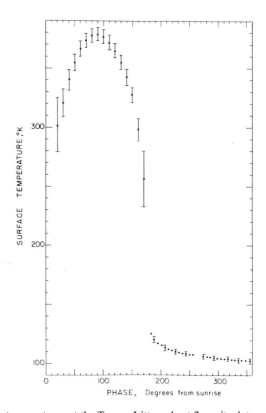

Fig. 2. Surface temperatures at the Taurus Littrow heat flow site determined from thermocouple measurements. Vertical bars are estimates of the error limits.

view of the radiation shield atop the drill stem; the radiation shield is a highly
reflective square of aluminized mylar which could add a substantial unknown factor to Equation (2). The relatively large error bounds for the daytime temperatures
are primarily due to uncertainties in the direct solar contribution to the thermocouple temperatures and the effects of local surface roughness.

The nighttime surface temperatures are not subject to the errors due to
uncertainties of the cable orientations and the data shown are an average of the
two thermocouple reductions. Nighttime surface temperatures deduced from each
of the thermocouples differed by no more than 2°K throughout the lunar night.
The mid-lunar night temperature of 106 (±2)°K agrees with Saari's (1964) estimate
of 104°K based on infrared observations. The measured pre-sunrise temperature
of 102 (±2)°K lies within the range of cold terminator temperatures measured by
Mendell and Low (1970). Comparison with a similar surface temperature measurement at the Apollo 15 heat flow site (Keihm et al., 1973) indicates that the
Taurus Littrow surface temperatures are about 10°K higher than the Apollo 15
surface temperatures through most of the lunar night.

The data gaps shown near sunset and immediately following sunrise in Fig. 2
correspond to times of rapid temperature changes. During these periods, Equation

(2) loses its validity as the finite time constant of the cable must be taken into account.

MEAN-TEMPERATURE GRADIENT AT TAURUS LITTROW

From the data shown in Fig. 2, a mean surface temperature of 216 (±5)°K was calculated for the Taurus Littrow site, indicating a mean temperature rise of ~40°K between the surface and the 256°K temperature at 130 cm measured by the heat flow probes (Langseth *et al.*, 1973). Only a small part of this mean temperature rise (no more than 5°K) can be accounted for by the temperature gradient produced by the measured heat flow. The mean temperature rise is due mainly to the non-linear effects of radiative heat transfer within the highly porous dust layer about 2–3 cm thick at the surface. During the warm lunar day heat is transferred more effectively into the lunar surface than it can be transferred out during the cold lunar night. To conserve net flux over a lunation, a mean-temperature gradient is established mainly confined to the porous surface layer. A similar phenomenon was observed at the Apollo 15 heat flow site where a mean temperature rise of 45°K was measured between the surface and the top probe sensor.

The radiative contribution to the diurnal heat transfer in the near-surface layer can be examined quantitatively by postulating an effective thermal conductivity which is a function of temperature:

$$K(T) = K_c + K_r T^3 \qquad (4)$$

where K_c represents conductive transfer across the grain boundaries and $K_r T^3$ represents radiative transfer across the evacuated space between particles. The functional form of Equation (4) has been verified experimentally by Watson (1964) for silicate powders in vacuum. Cremers and Birkebak (1971) and Cremers and Hsia (1973) have also found the conductivity of returned lunar fines to fit a functional relationship of the form of Equation (4). The parameter R_{350} which equals $K_r \cdot 350^3 / K_c$, first used by Linsky (1966) in examining this phenomenon, represents a measure of the radiative contribution to the heat transfer in the porous surface layer. By using one-dimensional models of the lunar regolith, one finds that R_{350} must be within the range 1.7–2.2 for the Apollo 17 site and between 2.5 and 3.0 for the Apollo 15 site to produce the observed mean temperature gradients. The range of values for the Apollo 17 site is in reasonable agreement with the value of 1.48 obtained by Cremers and Birkebak (1971) for Apollo 12 sample 12001,19. Similar measurements on Apollo 11 sample 10084,68 (Cremers and Birkebak, 1971) and Apollo 15 sample 15031,38 (Cremers and Hsia, 1973) yield R_{350} values of 0.50 and 0.83 respectively. It is important to note, however, that even slight disturbances to the *in situ* configuration of the porous lunar fines may have relatively large effects on their bulk thermal properties.

THERMAL CONDUCTIVITY OF THE UPPER 15 CM OF REGOLITH AT TAURUS LITTROW

It is fortunate that our more accurate surface temperature determinations are made during the lunar night for it is the post-sunset cool-down data which are most strongly constrained by the thermal properties within the top 15 cm of the surface.

To compare the measured surface temperatures with theoretical models of thermal property profiles, one must solve the heat conduction equation in one dimension with a flux boundary condition at the surface. To do this, a standard forward difference technique was used incorporating a depth dependent density, a temperature dependent specific heat and a temperature and depth dependent thermal conductivity of the form of Equation (4). Solar flux input to the surface is assumed to follow the cosine law for a full half lunation. Effects of macroscopic shadowing near sunrise and sunset have not been considered since the primary surface relief lies to the north and south of the ALSEP site. The density profile for the Apollo 17 site was determined from preliminary examination of returned core tube samples (D. Carrier, personal communication). The specific heat as a function of temperature was modeled by a least square fit to the data of Robie et al. (1970) on Apollo 11 fine samples. Similar measurements on Apollo 15 soil sample 15301,20 and Apollo 16 soil sample 60601,31 by Hemingway and Robie (1973) yielded nearly identical values of specific heat vs. temperature as the Apollo 11 sample. Since the ratio of the radiative to conductive component of the thermal conductivity was constrained by the observed mean temperature gradient discussed earlier, only the conductive component, K_c, remained to be varied as a function of depth to fit the cool-down surface data. Figure 3 shows on expanded scale the reduced surface nighttime temperatures at the Apollo 17 site together with the best fitting theoretical curve produced by the mean conductivity and density profiles shown in the inset. (Mean conductivity refers to the effective conductivity at a given depth at the mean temperature of that depth.) The Apollo 15 density profile is based on inferences drawn from drill core penetration rates and surface disturbance due to astronaut activity by J. Mitchell of the Apollo 15 soil mechanics team (personal communication). In both the 15 and 17 models, a low conductivity layer of about 2 cm thickness, with mean conductivity values between 0.9 and 1.5×10^{-5} W/cm-°K, is required to fit the steep drop in surface temperature for the first 40–60 hours following sunset. The low conductivity values near the surface for the Apollo 15 model are also supported by surface temperature measurements made during the total lunar eclipse of August 6, 1971. The Apollo 15 model then requires a steep, but not discontinuous, rise in conductivity with depth down to 15 cm to produce the increased flattening of the cool-down curve through the lunar night. The Apollo 17 model, however, requires a very sharp jump in conductivity at a depth of about 2 cm to produce the abrupt flattening of the cool-down curve at about 190° phase angle. The subsequent increase in conductivity with depth is slight enough so that the Apollo 17 model may be considered essentially a two-layer model over the top 15 cm. The large jump in conductivity at 2 cm is also supported by the preliminary density profile which indicates a fairly high density quite close to the surface. The Apollo 15 density profile, on the other hand, supports the possibility that a substantial conductivity gradient exists over the upper 10–15 cm of the regolith. The mean conductivity values determined for the near-surface layer are in good agreement with measure-

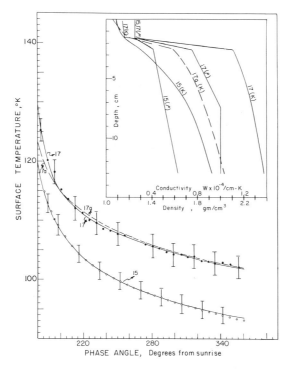

Fig. 3. Surface temperatures during the lunar night at the Apollo 17 (solid dots) and Apollo 15 (open circles) heat flow sites. Vertical bars are estimated errors. The continuous curves are theoretical curves derived from the thermal property models shown in the inset.

ments on returned lunar fine samples. Cremers and Birkebak (1971) measured conductivities in the range $1.2–1.6 \times 10^{-5}$ W/cm-°K at 220°K on lunar fine samples 10084,68 and 12001,19. Cremers and Hsia (1973) measured a conductivity of $\sim 0.8 \times 10^{-5}$ W/cm-°K at 220°K on lunar fine sample 15031,38. The large rise in conductivity below the surface layer is supported by the *in situ* conductivity measurements made as part of the heat flow experiments at the Apollo 15 and 17 sites (Langseth *et al.*, 1973).

It is important to emphasize that the most critical surface temperature data required for the purpose of determining regolith thermal profiles is that during the 10–40 hours immediately following sunset. Surface temperature data during this period have been the most difficult to obtain from remote infrared brightness scans. The level and steepness of the cool-down data immediately following sunset is controlled almost entirely by the thermal properties of the upper 2 cm. If the very early nighttime data are not sufficiently accurate to constrain the thermal properties of the upper 2–3 cm of dust layer within ±30%, then subsequent attempts to determine deeper conductivity values unambiguously from the flattened part of the cool-down curve will not be possible. For example, the broken-line curve (17a) of Fig. 3 fits the nighttime data after 192° phase well within the error bounds of the data. However, discrepancies in the early post-sunset fit produced by different conductivities within the upper 2 cm lead to discrepancies up to a factor of 2 in conductivity determinations for depths below 2 cm. (*See* curve 17a in inset of Fig. 3.)

Using the deduced thermal property profiles at the Apollo 15 and 17 sites, the information shown in Figs. 4 and 5 results from numerical solution of the heat conduction equation. (Thermal properties below 15 cm are assumed to follow the trends deduced for the 5–15 cm depths.) It is seen that 80% of the mean-temperature rise and 50% attenuation of the surface variation is realized within the first 3 cm at both the Apollo 15 and 17 sites.

The thermal properties model we have deduced is sufficient to explain the observed surface brightness temperatures at the Taurus Littrow site. However, it is not unique and other effects such as centimeter scale surface roughness (Bastin and Gough, 1969) have not been considered and could change the parameters of

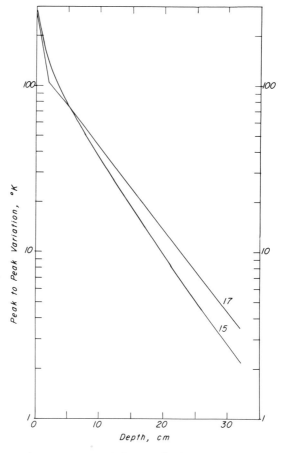

Fig. 4. Peak-to-peak temperature variations vs. depth produced by the Apollo 15 and 17 thermal property models.

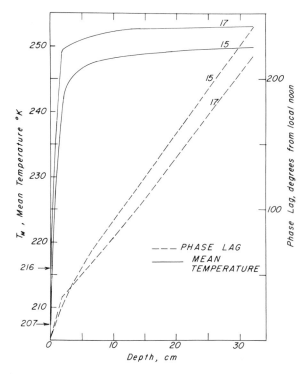

Fig. 5. Mean temperature and phase lag vs. depth produced by the Apollo 15 and 17 thermal property models.

our model to a small extent if applied. If our model is essentially correct, it is not necessarily wholly representative of undisturbed regolith in the ALSEP area. The upper few centimeters of a portion of the surface have been compacted and disrupted by the astronaut and could have properties significantly different from the upper few centimeters of undisturbed lunar soil. We feel, however, that there is strong evidence indicating that the effects of surface disturbance made by the astronaut are not large. First, the nighttime temperature curves and the basic thermal property model deduced from them are in good agreement with the results of earth-based infrared and microwave observations. Second, nighttime temperatures measured by the thermocouples which viewed different areas of the lunar surface were in good agreement at each of the heat flow sites. The two above-surface thermocouples, separated by about 11 m at the Apollo 17 site, agreed within 2°K throughout the lunar night. At the Apollo 15 site, five thermocouples lay on or just above different portions of the lunar surface and measured nighttime temperatures which differed by no more than 5°K throughout the lunar night. A large part of the 5°K variation could be explained by the effects of variable amounts of dust covering the cable at the thermocouple locations. The agreement of the various thermocouples which viewed different portions of the surface possessing varying degrees of disruption leads us to believe that the overall effects of

surface disturbance do not significantly alter the general applicability of our results to the mare regions typified by the Apollo 15 and 17 ALSEP sites.

Conclusions

From our surface temperature measurements at the Apollo 15 and 17 sites, we
draw the following conclusions regarding the thermal properties of the lunar surface material in these mare regions. The upper 2–3 cm consists of a porous layer
of low conductivity material in which radiative heat transfer plays an important
role. The steepness of the surface cool-down curve immediately following sunset
indicates that the near-surface material possesses low conductivities in the range
$0.9–1.5 \times 10^{-5}$ W/cm-°K. The large mean-temperature gradients at both sites indicate ratios of radiative to conductive heat transfer in the ranges 2.50–3.0 and
1.7–2.2 at the highest surface temperatures for the Apollo 15 and 17 sites respectively. The flattening of the cool-down curves through the lunar night indicates
significant increases in conductivity below the surface. Conductivity values on the
order of $1.0–1.4 \times 10^{-4}$ W/cm-°K at depths below 10 cm are supported by *in situ*
conductivity measurements made by the Apollo 15 and 17 heat flow experiments.
Planned surface temperature measurements at the Apollo 15 and 17 sites during
times of lunar eclipse will provide a valuable addition to our knowledge of the
near-surface thermal properties.

Acknowledgments—The authors thank Professor John L. Chute, Jr. of Lehman college for his help in
the reduction of thermocouple temperatures. The work of Bendix Corporation in the determination of
the cable radiative properties is appreciated. Computer facilities were made available through the
courtesy of Professor Robert Jastrow, Director of the Goddard Institute for Space Studies. This work
was supported by NASA under contract NAS9-6037.

References

Bastin J. A. and Gough D. O. (1969) Intermediate scale lunar roughness. *Icarus* **11**, 289–319.
Cremers C. J. and Birkebak R. C. (1971) Thermal conductivity of fines from Apollo 12. *Proc. Second
Lunar Sci. Conf., Geochim. Cosmochim. Acta*, Suppl. 2, Vol. 3, pp. 2311–2315. MIT Press.
Cremers C. J. and Hsia H. S. (1973) Thermal conductivity of Apollo 15 fines at low density (abstract).
In *Lunar Science—IV*, pp. 164–166. The Lunar Science Institute, Houston.
Hemingway B. S. and Robie R. A. (1973) Specific heats of lunar basalt, 15555, and soils 15301 and
60601, from 90 to 350°K (abstract). In *Lunar Science—IV*, pp. 355–356. The Lunar Science Institute,
Houston.
Keihm S. J., Peters K., Langseth M. G., Jr., and Chute J. C., Jr. (1973) Apollo 15 measurement of lunar
surface brightness temperatures: Thermal conductivity of the upper $1\frac{1}{2}$ meters of regolith. *Earth
Planet. Sci. Lett.* **19**, pp. 337–351.
Langseth M. G., Chute J. L., and Keihm S. J. (1973) Direct measurements of heat flow from the moon
(abstract). In *Lunar Science—IV*, pp. 455–456. The Lunar Science Institute, Houston.
Linsky J. L. (1966) Models of the lunar surface including temperature dependent thermal properties.
Icarus **5**, 606–634.
Logan L. M., Hunt G. R., Balsamo S. R., and Salisbury J. W. (1972) Midinfrared emission spectra of
Apollo 14 and 15 soils and remote compositional mapping of the moon. *Proc. Third Lunar Sci. Conf.,
Geochim. Cosmochim. Acta*, Suppl. 3, pp. 3069–3076. MIT Press.
Low F. J. (1965) Lunar nighttime temperatures measured at 20 microns. *Astrophys. J.* **142**, 806–808.

Mendell W. W. and Low F. J. (1970) Low-resolution differential drift scans of the moon at 22 microns. *J. Geophys. Res.* **75**, 3319–3324.

Murray B. C. and Wildey R. L. (1964) Surface temperature variations during the lunar nighttime. *Astrophys. J.* **139**, 734–750.

Robie R. A., Hemingway B. S., and Wilson W. H. (1970) Specific heats of lunar surface materials from 90 to 350 degrees Kelvin. *Science* **167**, 749–750.

Saari J. M. (1964) The surface temperature of the antisolar point of the moon. *Icarus* **3**, 161–163.

Shorthill R. W. and Saari J. M. (1965) Radiometric and photometric mapping of the moon through a lunation. *Ann. N. Y. Acad. Sci.* **123**, 722–739.

Watson K. (1964) I. Thermal conductivity measurements of selected silicate powders in vacuum from 150–350°K. II. An interpretation of the moon's eclipse and lunation cooling curve as observed through the earth's atmosphere from 8–14 microns. Ph.D. Thesis, California Institute of Technology.

Proceedings of the Fourth Lunar Science Conference
(Supplement 4, *Geochimica et Cosmochimica Acta*)
Vol. 3, pp. 2515–2527

Moonquakes, meteoroids, and the state of the lunar interior*

G. Latham, J. Dorman, F. Duennebier, M. Ewing, D. Lammlein,
and Y. Nakamura

Earth and Planetary Sciences Division, Marine Biomedical Institute,
University of Texas Medical Branch, Galveston, Texas 77550

Abstract—Analysis of data returned from the four stations of the Apollo Seismic Network has revealed that the lunar interior can be divided into two major zones: a rigid, dynamically inactive outer shell, about 1000 km thick (the lunar lithosphere); and a relatively weak central zone (the lunar asthenosphere) in which partial melting is probable. The transition between these two zones is gradual. Seismic activity within the moon is far below that of the earth. The small moonquakes that do occur originate near the base of the lithosphere, and appear to fall within two major belts. Tidal energy appears to be an important, if not the dominant, source of energy released as moonquakes. A secular component of moonquake energy release may result from slight thermal expansion or contraction of the moon, weak convection in the asthenosphere, or secular recession of the moon from the earth. Lack of shallow moonquake activity implies that the moon is neither expanding nor contracting at an appreciable rate at present. The mass flux of meteoroids that collide with the moon to produce detectable seismic signals, appears to be between 1 and 3 orders of magnitude lower than that predicted from earth-based measurements. The largest meteoroid impacts have occurred each year during the months of April through July. This clustering suggests the presence of a distinct meteoroid population of as yet unexplained origin.

Introduction

The Apollo seismic network includes stations installed at the landing sites of missions 12, 14, 15, and 16. It spans the near-face of the moon in an approximate equilateral triangle with 1100 km spacing between stations (Stations 12 and 14 are 181 km apart at one corner of the triangle). The oldest of these stations, Apollo 12, has now operated for over three years, and the entire network has been in operation for one year, as of April, 1973. Four seismometers are included at each station (Latham *et al.*, 1972a). Three long-period components form a triaxial set (one sensitive to vertical motion and two sensitive to horizontal motion), with sensitivity to ground motion sharply peaked at 0.45 Hz (2.2 sec) and a band-width of 1 to 3 sec. A fourth seismometer (short-period component) is sensitive to vertical motion with peak sensitivity at 8 Hz and a band-width of 1 to 16 Hz. These instruments can detect vibrations of the lunar surface as small as one-half angstrom at maximum sensitivity. All but two of the sixteen separate seismometers are presently operating properly. The short-period component at Station 12 has failed to operate since initial activation, and the long-period vertical seismometer at Station 14 became unstable after one year of operation.

Moonquakes have been detected by the long-period seismometers of each station at average rates of between 600 and 3000 per year, depending upon the

---

*Earth and Planetary Sciences Division, Marine Biomedical Institute Contribution No. 24.

station. All of the moonquakes are quite small by terrestrial standards (Richter body-wave magnitude of 2 or less). Thousands of even smaller moonquakes are detected by the short-period seismometers. Meteoroid impacts are detected by the long-period seismometers at average rates of between 70 and 150 per year. Although less numerous than moonquakes, meteoroid impacts generate the largest signals detected.

Analysis of the seismic signals from the man-made impacts and from natural sources has led to a model for the moon that is quite different from that of the earth (Latham *et al.*, 1972a; Nakamura *et al.*, 1973; Lammlein, 1973). Refinements in the present model can be expected as data accumulate from natural events. In this report, the major findings to date from the Passive Seismic Experiment are summarized.

STRUCTURE AND STATE OF THE LUNAR INTERIOR

The surface of the moon is covered by a highly heterogeneous zone in which, probably owing to the nearly complete absence of volatiles, seismic waves propagate with very little attenuation. Scattering and low attenuation of seismic waves in this zone account for the prolongation of lunar seismic signals and the complexity of the recorded ground motion relative to typical terrestrial signals (Latham *et al.*, 1970; 1972a). Most of the scattering occurs in the outer several hundred meters, but significant scattering may occur to depths as great as 10 to 20 km. The "granularity" within the scattering zone ranges from micron-size fragments to heterogeneity on a scale of at least several kilometers. The roughness of the lunar surface primarily resulting from meteoroid impacts undoubtedly contributes to the scattering of seismic waves.

Knowledge of lunar structure below the scattering zone, derives mainly from the analysis of seismic signals generated by nine impacts of two types of Apollo space vehicles: the LM (Lunar Module) ascent stage and the third, or S-IVB, stage of the Saturn booster. The LM's were guided to impact following the return of the surface crew to the CSM (Command Service Module) in lunar orbit. The S-IVB stages were directed by remote control from earth to planned impact points following separation from the Apollo spacecraft. Seismic signals from these impacts were recorded at ranges of from 67 to 1750 km. The Apollo 17 LM impact was also recorded by the geophones of the Active Seismic Experiment, located at the Apollo 17 landing site, at a range of 9 km. These data, combined with data from laboratory measurements at high-pressure on returned lunar samples, provide information on lunar structure to a depth of about 150 km. Below this depth, information on lunar structure derives principally from analysis of signals from deep moonquakes and distant meteoroid impacts. Analysis of the signals generated by the man-made impacts has revealed a major discontinuity at a depth of between 55 km and 65 km in the eastern part of the Oceanus Procellarum (*see* Toksöz *et al.*, this volume, for a detailed discussion of these results). By analogy with the earth, we refer to the zone above the discontinuity as the crust, and to the zone below as the mantle. Whether the crust is regional or global cannot be determined from the present seismic network. However, the early formation of a crust by igneous processes on a global scale would appear to explain such observations as the unexpectedly high heat flow (Langseth *et al.*, 1972; 1973) and the presence of large scale petrological provinces inferred from orbital X-ray fluorescence data (Adler *et al.*, 1973) and lunar sample analysis.

Below the crust, extending to a depth of about 1000 km, the velocity of seismic waves is nearly constant, suggesting that this is a relatively homogeneous zone (Nakamura *et al.*, 1973; Lammlein, 1973). The average velocity of compressional waves in the homogeneous zone is about 8 km/sec. A slight decrease in velocity with depth probably occurs in the lower half of this zone. Attenuation of both compressional waves and shear waves is very low in this zone precluding the presence of any appreciable melting. Poisson's ratio is about 0.25 at the top of the mantle.

A striking contrast has been found between signals that originate on the near-side of the moon and those from far-side sources. Direct shear waves, normally prominent in signals from near-side moonquakes, and weakly defined in near-side meteoroid impact signals, cannot be identified in the seismograms recorded at several of the seismic stations from far-side events. For example, as shown in Fig. 1, short-period shear waves from a large meteoroid impact that struck the far-side near the Crater Moscoviense, arrive at Station 15 at the expected time, but they are missing at Stations 14 and 16. Much later phases at these stations arrive at times predicted for surface reflected shear waves (SS) that travel through the upper mantle of the moon. Similarly, long-period shear waves can be identified at Stations 15 and 16 from far-side moonquakes, but are missing at Station 14 as shown in Fig. 2.

Although available data are not sufficient to derive a detailed seismic velocity model for the deep interior, these observations can be explained by introducing a "core" with a radius of between 600 and 800 km, in which shear waves do not propagate or are highly attenuated (Q of about 100 or less) (Nakamura *et al.*, 1973). The compressional wave velocity within this zone may be slightly lower than that in the mantle. The maximum velocity decrease for compressional waves is about 0.3 km/sec.

Seismic wave attenuation is strongly temperature dependent, showing rapid increase with increasing temperatures, and increasing sharply with the onset of melting (Jackson and Anderson, 1970). Thus, temperatures approaching the solidus in the lunar interior may account for the lack of shear wave transmission indicated by the seismic data. Assuming an interior of silicate composition, this would require temperatures of between 1450 and 1650°C at a depth of about 1000 km. This model is in substantial agreement with several thermal models recently proposed by Toksöz *et al.* (1972). Partial melting of silicate material is considered to be a possible cause of the low-Q, low-velocity zone of the upper mantle of the earth (Jackson and Anderson, 1970; Solomon, 1972). A completely molten core of the size indicated, however, is not likely because a decrease of the compressional wave velocity exceeding the value of 0.3 km/sec obtained by our preliminary analysis would be expected. Other possibilities, such as increased volatiles in the

G. LATHAM *et al.*

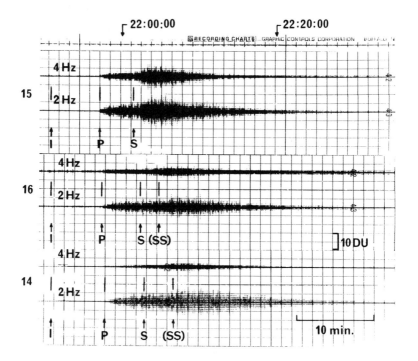

Fig. 1. Filtered short-period seismograms of seismic events detected on July 17, 1972. The signal is believed to be from a meteoroid impact on the far-side of the moon. Numbers at far left indicate Apollo stations. Two traces for two different filter settings are shown for each station. The frequencies given, 4 and 2 Hz, are the center frequencies of the narrow band-pass filters used in the data playback. At these frequencies, the rise and decay times of the seismic wave-trains are sufficiently short so that the compressional- and shear-wave arrivals appear as characteristic bulges in the seismograms. *I* indicates the estimated time of impact, *P* the observed arrival time of the direct compressional wave, and *S* the expected arrival time of direct shear wave (estimated to arrive at about 1.7 minutes before the peak of the wavetrain in the 4 Hz seismogram). Note that the characteristic shear-wave bulge, clearly visible at Station 15, is missing at Stations 16 and 14 when expected.

deep interior of the moon, however, cannot be ruled out at present.

The core radius of about 700 km, inferred from the seismic data, is inconsistent with a high-density molten metallic core material both by moment-of-inertia and seismic wave velocity considerations.

From the foregoing, it appears that the transition at about 1000 km from an outer lithosphere to an inner asthenosphere may be similar to, though much deeper than, its terrestrial analog. The lunar lithosphere is a relatively rigid outer shell, while the less rigid asthenosphere of the moon has properties similar to those of the asthenosphere (low-velocity zone) of the earth. Present observations suggest that the transition from the lithosphere to the asthenosphere is gradual (Nakamura *et al.*, 1973; Lammlein, 1973).

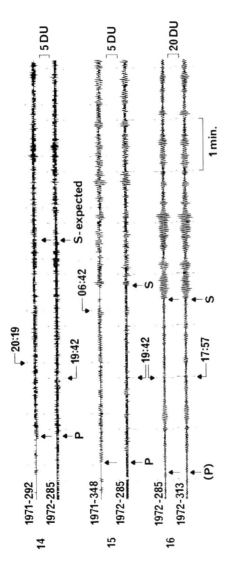

Fig. 2. Signals from A_{33} moonquakes detected at Stations 14, 15, and 16. The signals were too weak to be detected at Station 12. Only the horizontal-component seismogram is shown for each event at each station. These moonquakes originated at the same point within the far half of the moon. Note that the shear wave is not observable at Station 14 when expected (marked *S-expected*).

MOONQUAKES AND LUNAR TECTONISM

Thermal moonquakes (Latham et al., 1971b; Duennebier, 1972)

Thousands of small seismic signals have been detected by the short-period seismometers of Stations 14, 15, and 16 (the short-period seismometer of Station 12 failed to operate). These signals are not recorded by the long-period seismometers owing to the restricted bandwidth of these instruments. The short-period signals are generated by: (1) thermoelastic stresses within the LM descent stage and other equipment left on the lunar surface at each site; (2) small meteoroid impacts at near ranges (a few tens of km and less); and (3) small moonquakes (micromoonquakes) that originate within a few km of each station. Micromoonquake activity begins abruptly about 2 days after lunar sunrise and decreases rapidly after sunset. As in the case of the distant moonquakes (next section) micromoonquakes are also recognized by the repetition of nearly identical signals, implying a highly localized source for each set. Forty-eight sets of matching micromoonquake signals have been identified at the Station 14 site, and 245 sets at Station 15. Micromoonquake activity at Station 16 is very low in comparison with Stations 14 and 15. Signals of each set occur at monthly intervals; usually one event per month, and at the same time relative to sunrise to within a few hours. The strong correlation with sunrise and sunset indicates that micromoonquakes are of thermal origin. Two possible mechanisms for thermal moonquakes are suggested: (1) cracking or movement of rocks along zones of weakness; and (2) slumping of soil on lunar slopes triggered by thermal stresses. The unipolarity of the signals and slight changes in waveforms with time imply that motion is always in the same direction and that some thermal moonquake sources change position slightly from one lunar day to the next, but the source mechanism is not yet understood.

Distant moonquakes (Latham et al., 1971b; Lammlein, 1973)

Comparison has revealed that many of the long-period lunar seismic signals match each other in nearly every detail throughout the entire wavetrain. Forty-one sets of matching events have been identified thus far. Matching signals of each set are generated by repetitive moonquakes which occur at monthly intervals at one of forty-one moonquake hypocenters. As in the case of thermal moonquakes, the repetition of seismic signals of identical waveforms indicates that each hypocenter remains fixed, probably to within a few kilometers, and that the source mechanism does not change with time. Matching seismic signals from these 41 hypocenters account for ten percent of the total number of signals believed to be of moonquake origin. The similarity between the character and occurrence of the weaker signals and the larger amplitude matching signals suggests that the smaller ones originate at many additional hypocenters.

Each hypocenter is active for only a few days per month at a characteristic phase of the tidal cycle. The number of moonquakes observed in an active period ranges from none to four. Approximately equal numbers of hypocenters are

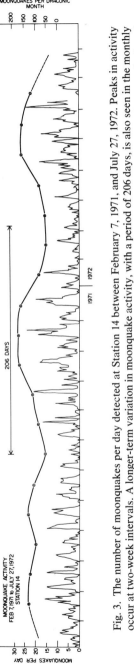

Fig. 3. The number of moonquakes per day detected at Station 14 between February 7, 1971, and July 27, 1972. Peaks in activity occur at two-week intervals. A longer-term variation in moonquake activity, with a period of 206 days, is also seen in the monthly activity plot shown above the daily count. This corresponds to the period of the tidal variation introduced by solar perturbation of the lunar orbit.

active at nearly opposite phases of the monthly tidal cycle, thus accounting for the observed semi-monthly peaks in moonquake activity (Fig. 3). A $7\frac{1}{2}$ anomalistic month variation in moonquake activity, corresponding to a solar perturbation of the lunar orbit, is also apparent in this plot. The activity of the A_1 moonquake hypocenter, Fig. 4, shows not only the monthly and the $7\frac{1}{2}$ month periodicities, but also a longer-term variation. This long-term decrease in activity appears to correspond to the 6-year variation in tidal stress which results from variations in the relative phase relationships among several of the lunar orbital parameters. The strong correlation between moonquake occurrence times and energy release and lunar tidal amplitudes and periodicities suggests that tidal energy is an important, if not the dominant source of the energy released as moonquakes. The uniform polarity of matching moonquake signals from individual hypocenters implies that a secular effect is also present. This secular component may be due to either slight thermal expansion or contraction of the moon, or relaxation of tidal deformation due to the recession of the moon from the earth.

Moonquakes can be located on the basis of body wave (compressional (P) and shear (S) waves) arrival times at three or four of the Apollo seismic stations and a knowledge of body wave travel times within the moon. To calculate travel times within the deep lunar interior, the seismic velocity model for the upper 150 km of the moon (Latham *et al.*, 1972a; Toksöz *et al.*, this volume) has been extrapolated to great depth on the basis of observed travel times of body waves from deep moonquakes and distant meteoroid impacts.

Signals from 18 of the 41 moonquake hypocenters have been large enough to provide data necessary to compute the source location. Epicentral locations can be determined in 9 additional cases if focal depths are assumed. The moonquake epicenters located thus far, shown in Fig. 5, fall in two long belts of activity lying approximately along arcs of great circles: a western belt trending approximately north-south and extending about 2200 km, and an eastern belt trending east-northeast to west-southwest and extending about 1800 km. The 18 hypocenters for

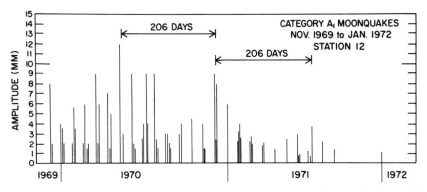

Fig. 4. History of occurrence of A_1 moonquakes as recorded at Station 12. The length of each bar is proportional to the maximum amplitude of the recorded signal. Monthly and 206-day cycles are evident in this plot along with longer-term variation that may correspond to the 6-year variation in tidal stress.

Fig. 5. Map showing the locations of the stations of the Apollo seismic network and 26 moonquake epicenters. Station numbers indicate the Apollo missions in which the stations were installed. The epicenters are the points on the surface immediately above the active zones (foci) in which moonquakes originate. Solid circles indicate foci for which the depth of the focus can be determined. Open circles correspond to cases in which data are not sufficient for determination of depth. A depth of 800 km has been assumed in these cases. The number at each epicenter is an arbitrary identification code used by the experiment team. Note that in two cases (epicenters 1 and 6, and 18 and 32), the epicenters are so closely spaced that their separation cannot be distinguished at the scale plotted. The A_{33} epicenter, located on the far-side of the moon (6°N, 104°E), is not shown.

which depths can be determined are concentrated at depths of 800 to 1000 km, as shown in Fig. 6.

According to our present model, moonquakes occur within the base of a thick, rigid shell (the lithosphere) immediately above a relatively weak central zone (the asthenosphere). If this model is correct, then the depth of the moonquake zone may simply be a consequence of the differing elastic properties within the two zones. The strength of material will decrease with depth owing to the increase of temperature with depth. Tidal deformation within the two zones will differ resulting in stress concentrations at the boundary. Strain energy accumulated at the base of the lithosphere is released by moonquakes when the shear strength of the

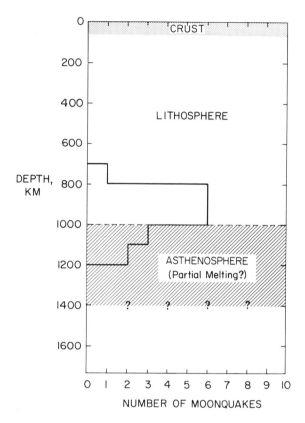

Fig. 6. The depth distribution of moonquakes for which present data are sufficient for determination of focal depths (18 cases). Question marks in the zone labeled "asthenosphere" indicate that present data are not sufficient to define the inward extent of this zone. The asthenosphere, in which partial melting is believed to occur, may extend to the center of the moon, or it may be restricted to a thin shell at a depth of about 1000 km.

material is exceeded. This situation may be analogous to the high stress concentration that occurs along the perimeter of a hole within a stressed plate. Fatigue cracks occur in the overstressed portions during stress cycling (Timoshenko, 1951). Dissipation of tidal energy at this depth would account for the close correlation between moonquake energy release and lunar tides. Kaula (1963, 1964, 1966) found that the dissipation of tidal energy due to solid friction within a homogeneous moon increases with depth. Partial melting would serve to further increase the amount of tidal energy dissipated at great depth. The injection by tidal forces of fluids or partially molten material into the moonquake zone from below may influence dislocation. The importance of fluids on the occurrence of earthquakes has recently been noted (Scholz *et al.*, 1973). Tidal deformation within the asthenosphere may be largely accommodated by inelastic creep as a result of the lower yield strength of the material of that zone.

The alignment of moonquake epicenters in long belts is difficult to understand.

These may be a consequence of large-scale variations in the mechanical properties of the lithosphere or the asthenosphere. For example, they may be zones of increased temperature or they may reflect compositional heterogeneity. Alternately, the moonquake belts may be manifestations of weak, large-scale convective motions; or residual stress dating from the time of formation of the moon. Earlier suggestions of spatial correlations between moonquakes and mare rims are no longer supported by the increased data now at hand. More precise delineation of the zones of moonquake activity may provide additional clues as to their significance.

Surface features that appear to be tensional in nature have been taken as evidence of slight past expansion as the lunar interior grew warmer. However, lack of moonquakes at shallow and intermediate depths suggests that, presently, the moon is neither expanding nor contracting at an appreciable rate. Hence, the moon must be close to thermal equilibrium at the present time, i.e., the rate of heat flow out of the moon must be about equal to the rate of internal heat production.

METEOROID FLUX

Seismic signals that display characteristics similar to those of the LM and S-IVB signals are believed to be generated by meteorids impacting on the surface of the Moon (Latham *et al.*, 1971a, b, 1972a, b). Meteoroid impact signals are detected by the long-period seismometers of the Apollo seismic network at average rates of between 70 and 150 per year (Latham *et al.*, 1972a). They appear to be generated by objects in the mass range 100 gm to 1000 kg. Results obtained to date have been derived by a method using only the statistical distribution of maximum amplitudes of seismic signals recorded from meteoroid impacts. The average flux estimated from data of more than one year is

$$\log N = -1.62 - 1.16 \log m$$

where N is the cumulative number of meteoroids of mass m (in gm) and greater, which strike the moon per year per km^2. This flux estimate is 1 to 3 orders of magnitude lower than that derived from earlier earth-based measurements. Our estimate is lower than the average flux estimated from the distribution of crater sizes on the youngest lunar maria. This is consistent with a hypothesis that the population of small fragments in the solar system decreases with time as they are gathered up by collisions with the planets. The seismic data predict that a meteoroid of mass 7 to 10 kg can be detected by the least sensitive station (Station 12) from any point on the moon. About 50% of the impacts detected by a station occur more than 1000 km from it. The total data appear to contain at least two distinct meteoroid populations: The normal distribution of fragments that varies little throughout the year, and a population of relatively large objects that intersect the lunar orbit during the months of April through July each year. As the latter are detectable seismically from anywhere on the moon, the Apollo network affords greater exposure to these rare events than any other method of measurement.

SUMMARY AND CONCLUSIONS

Presently, the moon is nearly inert compared to the earth. It has a rigid, dynamically inactive, outer shell, about 1000 km thick, surrounding a "core" of markedly different elastic properties. It is likely that core temperatures are at, or near, the melting point, and that the core is much weaker than the outer shell. If so, we may regard the outer shell as the lunar lithosphere, and the weak central zone as the asthenosphere. Present moonquake activity is concentrated at great depth near the boundary between these two zones. Tidal energy appears to be an important, if not the dominant, source of the energy released as moonquakes. A secular component of moonquake energy may result from slight thermal expansion or contraction of the moon, weak convection in the asthenosphere, or secular recession of the moon from the earth. In contrast, earthquake activity is concentrated at shallow depth, caused by movements of a relatively thin, mobile lithosphere, presumably driven by internal heat energy. Thus, the great thickness of the lunar lithosphere relative to that of the earth, and the much lower thermal energy available within the lunar interior, accounts for the widely differing tectonism of these two bodies.

The presence of a thick lunar crust suggests early, intense heating of the outer shell of the moon. Lack of shallow seismic activity indicates that the moon is neither expanding nor contracting appreciably at the present time. Thus, the rate of heat flow out of the moon must be approximately equal to the rate of internal heat production.

The presence of a relatively hot lunar interior suggests that we may have discovered something very basic about the physics of the interiors of the terrestrial planets. If we consider the earth and the moon as end members in the size distribution of the terrestrial planets, and both have molten or partially molten interiors 4.6 billion years after their formation, then we may reasonably expect that Mars, Venus, and Mercury also have hot interiors.

Acknowledgments—This research was supported by NASA under contract NAS 9-13143. The authors wish to thank Y. P. Aggarwal for providing a copy of his manuscript (Scholz *et al.*, 1973) prior to publication.

REFERENCES

Adler I., Trombka J., Lowman P., Schmadebeck R., Blodget H., Eller E., Yin L., Gerard J., Gorenstein P., Bjorkholm P., Gursky H., Harris B., and Osswald G. (1973) Results of the Apollo 15 and 16 X-ray fluorescence experiment (abstract). In *Lunar Science—IV*, pp. 9–10. The Lunar Science Institute, Houston.
Duennebier F. (1972) Moonquakes and meteoroids: Results from the Apollo seismic experiment short period data. Ph.D. Dissertation, University of Hawaii.
Jackson D. and Anderson D. (1970) Physical mechanisms of seismic wave attenuation. *Rev. Geophysics*, **8**, pp. 1–64.
Kaula W. M. (1963) Tidal dissipation in the Moon. *J. Geophys. Res.* **68**, 4959–4965.

Kaula W. M. (1964) Tidal dissipation by solid friction and the resulting orbital evolution. *Rev. Geophys.* **2**, 661–685.

Kaula W. M. (1966) Thermal effects of tidal friction. *The Earth-Moon System*, pp. 46–51. Plenum Press.

Lammlein D. (1973) Lunar seismicity, structure, and tectonics. Ph.D. Dissertation, Columbia University.

Langseth M., Clark S., Chute J., Keihm S., and Wechsler A. (1972) Heat flow experiment. *Apollo 15 Preliminary Science Report*, NASA SP-289, Sec. 11.

Langseth M., Chute J., and Keihm S. (1973) Direct measurements of heat flow from the Moon (abstract). In *Lunar Science—IV*, pp. 455–456. The Lunar Science Institute, Houston.

Latham G., Ewing M., Press F., Sutton G., Dorman J., Nakamura Y., Toksöz N., Wiggins R., Derr J., and Duennebier F. (1970) Apollo 11 passive seismic experiment. In *Proc. Apollo 11 Lunar Sci. Geochim. Cosmochim. Acta*, Suppl. 1, Vol. 3, pp. 2309–2320. Pergamon.

Latham G., Ewing M., Press F., Sutton G., Dorman J., Nakamura Y., Toksöz N., Duennebier F., and Lammlein D. (1971a) Passive seismic experiment. *Apollo 14 Preliminary Science Report*, NASA SP-272, 133.

Latham G., Ewing M., Dorman J., Lammlein D., Press F., Toksöz N., Sutton G., Duennebier F., and Nakamura Y. (1971b) Moonquakes. *Science* **174**, 687–692.

Latham G., Ewing M., Press F., Sutton G., Dorman J., Nakamura Y., Toksöz N., Lammlein D., and Duennebier F. (1972a) Passive seismic experiment. *Apollo 16 Preliminary Science Report*, NASA SP-315, Sec. 9.

Latham G., Ewing M., Press F., Sutton G., Dorman J., Nakamura Y., Toksöz N., Lammlein D., and Duennebier F. (1972b) Passive seismic experiment, *Apollo 15 Preliminary Science Report*, NASA SP-289, Sec. 8.

Nakamura Y., Lammlein D., Latham G., Ewing M., Dorman J., Press F., and Toksöz N. (1973) New seismic data on the state of the lunar interior. *Science*. In press.

Scholz C. H., Sykes L. R., and Aggarwal Y. P. (1973) Earthquake prediction: a physical basis. *Science* **181**, 803–810.

Solomon S. (1972) Seismic wave attenuation and partial melting in the upper mantle of North America. *J. Geophys. Res.*, **77**, pp. 1483–1502.

Timoshenko S. (1951) *Theory of Elasticity*, 506 pp.

Toksöz N., Solomon S., Minear J., and Johnson J. (1972) Thermal evolution of the Moon. *The Moon*, **4**, 190–213.

Proceedings of the Fourth Lunar Science Conference
(Supplement 4, *Geochimica et Cosmochimica Acta*)
Vol. 3, pp. 2529–2547

Velocity structure and evolution of the moon

M. Nafi Toksöz, Anton M. Dainty, Sean C. Solomon, and Kenneth R. Anderson

Department of Earth and Planetary Sciences, Massachusetts Institute of Technology, Cambridge, Massachusetts 02139

Abstract—Seismic data from the Apollo Passive Seismic Network stations are analyzed to determine the velocity structure and to infer the composition and physical properties of the lunar interior. Data from artificial impacts (SIVB booster and LM-ascent stage) cover a distance range of 9–1750 km. Travel times and amplitudes, as well as theoretical seismograms, are used to derive a velocity model for the outer 150 km of the moon. The *P*-wave velocity model confirms our earlier report of a lunar crust in the eastern part of Oceanus Procellarum.

The crust is about 60 km thick and may consist of two layers in the mare regions. Possible values for the *P*-wave velocity in the uppermost mantle are between 7.6 km/sec and 9.0 km/sec. The 9 km/sec velocity represents either a localized heterogeneous unit, or a thin layer less than about 40 km in thickness. The elastic properties of the deep interior as inferred from the seismograms of natural events (meteoroid impacts and moonquakes) occurring at great distances indicate that there is an increase in attenuation and a possible decrease of velocity at depths below about 1000 km.

Models of the thermal evolution of the moon that fit the chronology of igneous activity on the lunar surface, the velocity structure, the stress history of the lunar lithosphere implied by the presence of mascons, and the surface concentrations of radioactive elements involve extensive differentiation early in lunar history. This differentiation may be the result of rapid accretion and large-scale melting at the time of formation of the moon. If the Apollo 15 and 17 heat flow measurements of about 30 ergs/cm² sec are representative of the lunar surface, the average uranium concentration in the moon is 50–70 ppb. This is consistent with an achondritic bulk composition (between howardites and eucrites) for the moon.

Introduction

Seismic signals from Saturn SIVB booster and LM-ascent stage impacts have been recorded by the Apollo Lunar Seismic Network. These impacts have provided 22 seismograms, covering a source-to-station distance range of 9–1750 km. Travel times, amplitudes, and seismogram shapes of all events with identifiable *P*-waves have been used to derive a compressional wave velocity structure for the outer 150 km of the moon. In this paper these results are combined with those derived from distant meteorite impacts and moonquakes (Latham *et al.*, 1973a,b; Nakamura *et al.*, 1973) to infer the structure and physical conditions in the lunar interior.

In the second part of this paper, the thermal evolution of the moon is discussed using constraints imposed by a wide spectrum of lunar data including the new seismic results. In the thermal history calculations, procedures used in our earlier studies (Toksöz *et al.*, 1972d; Solomon and Toksöz, 1973; Toksöz and Solomon, 1973) are followed.

VELOCITY MODELS AND LUNAR STRUCTURE

The general characteristics of the impact seismograms recorded by the Apollo Passive Seismic Experiment network have been discussed extensively in previous papers (Latham et al., 1971, 1972a,b,c, 1973a,b; Toksöz et al., 1972a,b,c, 1973a,b). Here we will present the data briefly without details, except in the case of the newest impact seismograms.

The station network and the distribution of impact sites shown in Fig. 1 cover a fairly wide region on the front side of the moon. The distances between impact sites and stations are listed in Table 1. In the study of seismograms we confined our analysis to the first few minutes of each record, where the direct body wave phases are not dominated by scattered arrivals. Examples of some records are shown in Fig. 2. The general shape of the wave train envelope and the role of seismic scattering has been discussed in detail by Dainty et al. (1973).

The travel times of compressional waves were determined from the unfiltered vertical components of the seismograms. Up to a distance of 350 km, the first arrivals from the SIVB impacts can be identified clearly and the travel times read to an accuracy of ±0.5 sec. At greater distances the signal to noise ratio is generally very low, and it is difficult to determine whether a given phase is a first arrival or a later phase with greater amplitude. This introduces uncertainties in the velocity models for the upper mantle. The extent of this uncertainty is discussed later in this section.

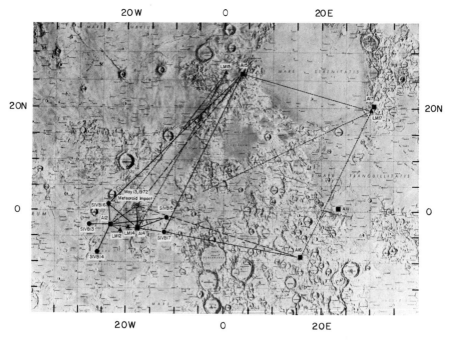

Fig. 1. Map showing the Apollo Passive Seismic Network stations (squares), impact sites (open and closed circles and triangles) and paths for seismic rays.

Table 1. Coordinates, distances, and azimuths of stations and impacts (from Latham *et al.*, 1973b).

Location	Coordinates[a]	Distance and azimuth from Apollo seismic stations			
		12	14	15	16
Apollo 12 station	3.04°S 23.42°W		181 km 276°	1188 km 226°	1187 km 276°
Apollo 14 station	3.65°S 17.48°W	181 km 96°	—	1095 km 218°	1007 km 277°
Apollo 15 station	26.08°N 3.66°E	1188 km 40°	1095 km 33°	—	1119 km 342°
Apollo 16 station	8.97°S 15.51°E	1187 km 100°	1007 km 101°	1119 km 160°	—
Apollo 12 LM impact point	3.94°S 21.20°W	73 km 112°	—	—	—
Apollo 13 SIVB impact point	2.75°S 27.86°W	135 km 274°	—	—	—
Apollo 14 SIVB impact point	8.09°S 26.02°W	172 km 207°	—	—	—
Apollo 14 LM impact point	3.42°S 19.67°W	114 km 96°	67 km 276°	—	—
Apollo 15 SIVB impact point	1.51°S 11.81°W	355 km 83°	184 km 69°	—	—
Apollo 15 LM impact point	26.36°N 0.25°E	1130 km 36°	1048 km 29°	93 km 276°	—
Apollo 16 SIVB impact point	1.3 ± 0.7°N 23.8 ± 0.2°W	132 km 355°	243 km 308°	1099 km 231°	—
Apollo 17 SIVB impact point	4.21°S 12.31°W	338 km 96°	157 km 96°	1032 km 209°	850 km 278°
Apollo 17 LM impact point	19.96°N 30.50°E	1750 km 64°	1598 km 61°	770 km 98°	985 km 27°

[a]Listed coordinates are derived from the Manned Space Flight Network Apollo tracking data. Locations based on these data are referenced to a mean spherical surface and may differ by several kilometers from coordinates referenced to surface features. Premature loss of tracking data reduced the accuracy of the estimate of the Apollo 16 SIVB impact point. The listed coordinates for this impact are estimated from seismic data.

The travel time-distance plots for the first arrivals and distinct later arrivals are shown in Figs. 3 and 4. Also shown in Fig. 4 are the theoretical travel times for models with different upper mantle velocities. To a distance of 360 km (Fig. 3) the travel time curve is well-constrained by the first as well as later arrival phases. The overall characteristics of the curve show rapidly increasing velocities near the surface, an intermediate zone with a nearly constant velocity, and a significant velocity discontinuity below this depth. The amplitudes of the refracted (direct) and reflected arrivals also support this model (Fig. 3). The triplication of the travel time curve between 170 and 650 km (Fig. 4) is due to the velocity jump at the base of the lunar crust. The effect of the rapid velocity increase at the base of the lunar crust on the seismic rays is shown by the ray path plots in Fig. 3.

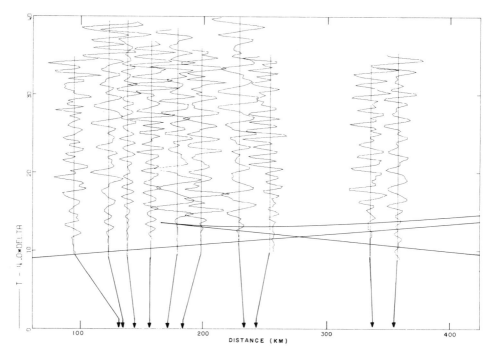

Fig. 2. Record section illustrating Long Period records of some impacts for ranges less
than 360 km. From the top, the records are SIVB-16 @12Z; SIVB-13 @12Z; Day 134, 1972
meteor impact @14R; SIVB-17 @14R; SIVB-14 @12Z; SIVB-15 @14Z; Day 134, 1972
meteor impact @12Z; SIVB-16 @14R; SIVB-17 @12Z; SIVB-15 @12Z. In the above, the
number after "SIVB" indicates the Apollo mission on which the impacts occurred, the
number after "@" indicates the Alsep station at which the signal was received, Z =
vertical component, R = horizontal component. The records are presented in a reduced
time plot, the reduction being T–4△ where △ is in degrees. The travel time lines shown
are those of Fig. 3. The seismograms have been separated for clarity; arrows indicate the
true position.

At far distances the amplitudes of the first recognizable phases are very small.
Suitable seismograms at distances greater than 400 km were produced only by the
Apollo 17 SIVB impact. Because of tracking difficulties, the coordinates and time
of the SIVB-16 impact were not determined independently; thus the travel times
could not be incorporated in this study. Other events (LM impacts at far distances) did not produce seismograms of sufficient amplitude to be usable. The two
arrivals from SIVB-17 at far distances (850 and 1040 km) correspond to an average
velocity of 7.7 km/sec. Whether these are first arrivals or surface reflected later
arrivals has not been determined. The velocity models constructed for the lunar
mantle incorporate this uncertainty.

The amplitudes and character of first and later arrivals strongly constrain the
velocity gradients and the position and size of any discontinuities in velocity. For
this purpose theoretical seismograms were computed using generalized ray theory
(Helmberger, 1968; Helmberger and Wiggins, 1971) and compared with the

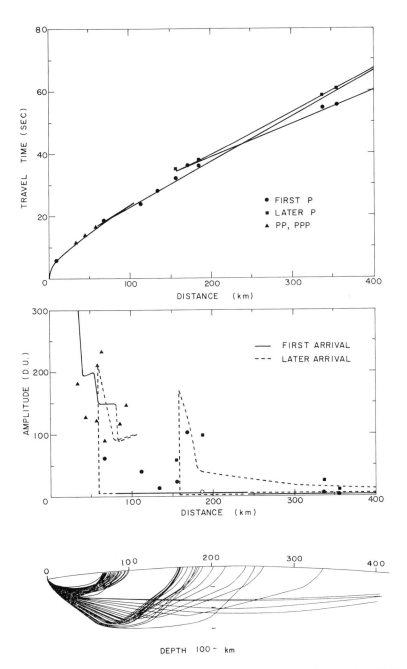

Fig. 3. Top part, travel time-distance plot for compressional waves from LM and SIVB impacts as recorded at Apollo Passive Seismic Network stations for ranges 0–400 km. The points are observations, the lines are theoretical curves using the model of Fig. 6. PP and PPP refer to surface reflected phases. The middle part shows observed and ray theory amplitudes. The bottom part is a ray path diagram for the model of Fig. 6.

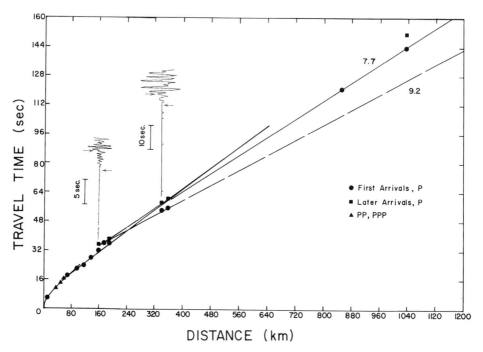

Fig. 4. Travel time-distance plot for ranges 0–1200 km. The two theoretical curves "7.7" and "9.2" indicate the range of possible upper mantle velocities; the numbers are the corresponding compressional wave velocities. Two seismograms showing refracted (first) and reflected arrivals from the base of the lunar crust are shown (SIVB-17 @ 14, Short period Z at 160 km, SIVB-15 @ 12, Long period Z at 360 km).

observed seismograms. Velocity models were adjusted until a good fit was obtained between the theoretical and observed seismograms (Fig. 5).

The final velocity model is shown in Fig. 6. This model is similar to those described in earlier publications (Toksöz *et al.*, 1972a,b,c, 1973a,b). It includes a two-layered lunar crust with a total thickness of about 60 km. In the upper few kilometers the velocity increases rapidly because of the transition from regolith, to fractured rocks, to competent crustal materials. The shallow layering for the Apollo 17 site, determined by Kovach *et al.* (1973), is also shown on the figure.

Below the crust there are two possible velocity models depending on how the SIVB-17 data at far stations are interpreted. These two models have velocities of 7.7 and 9.0 km/sec, respectively, immediately below the crust. As inferred from the amplitudes of *P*-waves, both models have negative velocity gradients with depth; velocity decreases by about 0.2 km/sec between 65 and 150 km depth.

The nature of the 9.0 km/sec velocity requires some discussion. We see evidence for such a high velocity from two independent seismograms from SIVB-17 and SIVB-15 impacts recorded at Station 12. What we observed, however, are "refraction" arrivals along an unreversed profile. If the crust-mantle interface is dipping, the 9 km/sec apparent velocity could be higher than the actual velocity in

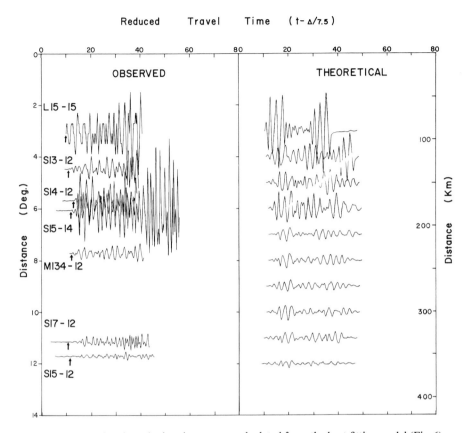

Fig. 5. Observed and synthetic seismograms calculated from the best-fitting model (Fig. 6).

the mantle. If the 9 km/sec layer indeed exists, the absence of observable refracted arrivals through it at Stations 16 and 15 (distances 850–1040 km) may be due to two possible causes: (a) such a layer is localized and not a moon-wide phenomenon, or (b) it is of relatively small thickness (less than about 40 km) so that refracted arrival amplitudes are small and decay rapidly with distance (Levin and Ingram, 1962). At this stage none of these three possibilities can be ruled out.

COMPOSITIONAL IMPLICATIONS

The compositional implications of the lunar velocity models can be explored with the aid of high-pressure laboratory measurements on lunar and terrestrial rocks. Velocity measurements have been made on lunar soils, breccias, and igneous rocks from four Apollo missions (Anderson *et al.*, 1970; Kanamori *et al.*, 1970, 1971; Wang *et al.*, 1971, 1973; Chung, 1972, 1973; Warren *et al.*, 1971, 1972; Mizutani *et al.*, 1972a; Mizutani and Newbigging, 1973; Tittman *et al.*, 1972; Todd *et al.*, 1972, 1973). Regardless of composition, these rocks are characterized by very

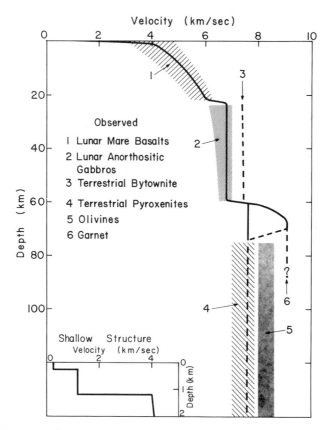

Fig. 6. Comparison between the compressional velocity profile for the lunar crust and upper mantle derived from seismic measurements, and the velocities of several types of lunar and terrestrial rocks. The velocity model is shown by a heavy line (or dashed heavy line where the model is uncertain). The lunar basalts for which velocities have been measured include samples from different sites. Region 2 ("anorthositic gabbro") is bounded by Apollo 16 aluminous rocks 68415 at the high velocity bound and 62295 and 65015 at the low velocity bound, from Wang *et al.* (1973). Terrestrial bytownite velocities are from Wang *et al.* (1973). Pyroxenites cover a wide range of composition. The garnet field is an average value for high-pressure phases of anorthite-rich rocks. Olivines cover the range 75–100% forsterite (Chung, 1971).

low velocities at low pressures relative to terrestrial rocks. This can be attributed to the absence of water in the lunar rocks combined with the effects of porosity and micro-cracks. Laboratory measurements on terrestrial igneous rocks have demonstrated this effect (Nur and Simmons, 1969; Chung, 1972). In dry-state conditions appropriate to the moon, the pressure gradient of compressional velocity is very high at low pressures, a behavior very similar to that of the lunar *P* velocity profile.

The measured velocities of appropriate lunar samples are shown in a generalized form along with the observed compressional velocity profile in Fig. 6.

Compressional velocities of terrestrial pyroxenites and olivines shown in Fig. 6 specify general bounds between which most values fall (*see* Anderson and Liebermann, 1966; Press, 1966 for tabulations; the olivines are taken from Chung, 1971).

From the comparison of the laboratory data and the lunar velocity profile, the following units can be identified:

(i) Near the surface the extremely low seismic velocities (about 100 m/sec near the surface) are very similar to those of lunar fines (soils). The velocity of about 250 m/sec indicates that broken or granulated material extends to a depth of about 300 m at the Apollo 17 site (Kovach *et al.*, 1973). This indicates a layered but fairly thick regolith. Material below this may be fractured and broken, as indicated by a *P*-wave velocity of 1.2 km/sec.

(ii) Below a depth of about a kilometer the measured velocities of lunar basaltic rocks fit the seismic velocity profile to a depth of about 20 km. In Fig. 6, the upper and lower boundaries of laboratory data on lunar basalts are defined by velocities of samples 14053 (Mizutani *et al.*, 1972a) and 15555 (Chung, 1973). The rapid increase of velocity with depth can be explained by the pressure effect on dry rocks having micro- and macro-cracks (Todd *et al.*, 1973). Whether this layer consists of a series of flows or fairly thick intrusive basalt cannot be resolved with our data. Nor is it possible to rule out other compositions or rock types which may have similar velocities under a lunar environment. Again it is important to keep in mind that this structure is defined primarily in the eastern part of Oceanus Procellarum as shown in Fig. 1.

There are other data indicating similar values for the thickness of mare basalts. The elevation difference between Oceanus Procellarum (particularly Mare Cognitum) and the adjacent highlands is about 4 km, as deduced from laser altimetry (Kaula *et al.*, 1972). A model based on isostatic compensation between basaltic mare and feldspathic highland requires approximately 25 km of basalt in Oceanus Procellarum to explain the elevation difference. The gravity data from Apollo 15 subsatellite observations fits a model of about 20 km of basalt filling in mascon basins (Sjogren *et al.*, 1972).

(iii) The second layer of the lunar crust (20–60 km) appears to be made of competent rock. The increase in pressure affects velocities very little (velocity at the bottom is about 7.0 km/sec). The petrological interpretation of the velocity curve is not very simple. It is clear from Fig. 6 that velocities of lunar "anorthositic gabbros" defined by the laboratory data from samples 15415 (Mizutani and Newbigging, 1973; Chung, 1973), 62295 and 68415 (Wang *et al.*, 1973), when corrected for microfracturing effects, are close to the observed velocity profile. There are aluminous basalts such as 65015 that also approach our velocity model (Mizutani and Newbigging, 1973). Terrestrial bytownites have somewhat higher velocities (Wang *et al.*, 1973). While the observed velocity profile is consistent with anorthositic gabbros, other compositions, especially some basalts, cannot be ruled out. The lower bound for terrestrial pyroxenite is slightly higher than the observed *P* velocity curve. Dunite velocities are definitely higher than the observations, and these rocks cannot be considered as serious contenders.

Lunar samples, Surveyor data and orbital results favor a feldspathic ("anorthositic gabbro") composition for the lunar highlands (Wood *et al.*, 1971; Gast, 1972; Reid *et al.*, 1972; Turkevich, 1971; Adler *et al.*, 1972). It is not clear, however, whether the same composition extends to the base of the lunar crust. The seismic velocity profile is compatible with a crust of anorthositic gabbro but it does not limit the composition uniquely. The simplest model is a moon-wide feldspathic initial crust cratered by impacts and flooded by basalts in the maria. Thus, the discontinuity at 20 km depth would exist only in mare regions and not in highland areas. Unfortunately there are no independent seismic data in highland regions suitable for regional crustal studies.

(iv) The discontinuity at about 60 km depth is required on the basis of all seismic data. The velocities below this depth however are not uniquely fixed. As shown in Fig. 6, several compositions must be considered as possible models. These include:

(a) A pyroxene-rich lunar upper mantle as inferred from a 7.7 km/sec P-wave velocity. A pyroxene-olivine composition could also be compatible with this velocity. The velocity density systematics of this composition imply density $\rho = 3.4 \, \text{g/cm}^3$ as required by moment of inertia considerations (Solomon and Toksöz, 1973; Solomon, 1973). Such a composition is favored by Ringwood and Essene (1970) and Green *et al.* (1972) as the source of lunar basalts.

(b) If the high apparent velocity (9.0 km/sec) is not due to a steeply dipping interface or local heterogeneity under the site investigated, then models with intrinsically high velocities must be found. Within the general petrological constraints, two possible models can be advanced. One requires an abundance of spinel (at least 25%) in addition to olivine rich rocks. This would imply an olivine-pyroxene-spinel upper mantle composition with spinel as a prominent phase. This suggestion has been advanced by Warren *et al.* (1972).

The other possible model would be garnet rich and would correspond to a high pressure phase of anorthosite (i.e., anorthite → grossularite + kyanite + quartz). The phase equilibria of such models are discussed by Boettcher (1971) and the velocity characteristics by Anderson and Kovach (1972). Experimental data on the above phase transformation are available only at higher pressures and temperatures than those pertinent to the lunar crust (Hays, 1967; Hariya and Kennedy, 1968; Raheim *et al.*, 1973). However, if it is assumed that the phase boundary can be extrapolated toward lower pressures and temperatures, then a high density and high velocity layer becomes a feasible model. This was illustrated by Toksöz *et al.* (1973b). It is possible to have a layer of high pressure phases between 60 and 100 km depth. Below this depth, if the composition remains the same, the material will revert back to the low pressure form (i.e., anorthosite or plagioclase) because of the increased temperature. Details of such models are discussed by Anderson (1973) and Solomon and Toksöz (1973).

The only qualifying constraint to the high velocity (i.e., 9 km/sec) upper mantle that the seismic observations impose is that if such high velocities indeed exist, the thickness of such a layer could not be greater than about 40 km. Otherwise, theoretical amplitude calculations show that the first arrivals from such a high velocity zone would be observed at distances of 850 and 1032 km from the SIVB–17 impact (Alseps 16 and 15 respectively). Such arrivals have not been observed. Thus the velocity must decrease as a result of temperature effects, compositional change or both.

Extension of the structural model to the deep lunar interior will have to be done with seismic data from moonquakes and meteorite impacts. The occurrence of moonquakes at depths of 700–1200 km (Latham *et al.*, 1972a,b,c, 1973a,b) and the recording of well-defined *S*-waves from these moonquakes imply that the lunar mantle must be sufficiently rigid to 1000 km depth to prevent appreciable *S*-wave attenuation.

The attenuation of *S*-waves that have penetrated deeper than about 1000 km into the moon (Nakamura *et al.*, 1973) indicates a "softening" of the material in the central region below 1000 km. (These *S*-waves were generated by a farside impact and by moonquakes.) Any softening which would reduce *Q* to less than about 500 (*Q* in the outer regions of the moon is greater than 3000) would explain the observations. This could be achieved by temperatures approaching the solidus, by a very small amount of partial melt or by other mechanisms (such as perhaps an increase in the amount of volatiles in the deep lunar interior). Total melting is not a preferred model for 1000 km depth, but cannot be ruled out at the very center of the moon.

By analogy to the earth, the deep interior of the moon may resemble the earth's asthenosphere in its rheological properties. The 1000 km thick outer shell would be the "lunar lithosphere."

The major units of lunar structure are shown schematically in Fig. 7. From the displacement of the center of mass from the center of figure and differences

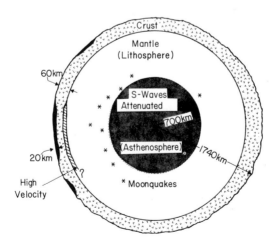

Fig. 7. Schematic diagram of lunar structure. The near side is to the left of the figure—maria are shown by solid shading. The possible limited extent of the high-velocity zone is indicated.

between the principal moments of inertia, a case can be made for possible thickening of the crust to about 150 km on the backside of the moon (Wood, 1973). Due to an absence of mare basalts, the crust on the backside would probably be a single layer of "anorthositic gabbro" composition. The lunar "lithosphere" and "asthenosphere," shown in Fig. 7, should be considered as preliminary results until the properties of these regions are verified by additional data.

THERMAL MODELS AND THE EVOLUTION OF THE MOON

The structure and the elastic properties of the lunar interior described in the preceding section indicate a differentiated moon and impose constraints on the thermal state of the lunar interior. Magmatization, differentiation, and evolution of the moon are controlled by its thermal history. In this section we briefly describe some thermal evolution models that satisfy both the present day temperature and heat flow requirements, as well as the constraints related to the early history of the moon.

Thermal evolution models for the moon have been calculated by a number of investigators, among them—Urey and MacDonald, 1971; Papanastassiou and Wasserburg, 1971; Hays, 1972; Wood, 1972; Reynolds *et al.*, 1972; McConnell and Gast, 1972; Toksöz *et al.*, 1972d; Tozer, 1972; Hanks and Anderson, 1972; Solomon and Toksöz, 1973; Toksöz and Solomon, 1973. In this paper we will give some results without going into detail. The computational techniques were described in our earlier papers (Toksöz *et al.*, 1972d; Toksöz and Solomon, 1973). Basically we use a finite difference scheme to compute temperatures for a radially heterogeneous moon taking into account temperature-dependent conductivity, differentiation, partial melting and crystallization, and, indirectly, convection occurring when melting or partial melting takes place.

The constraints that must be satisfied by the thermal models include: (1) the differentiation history and chronology of lunar igneous activity; (2) development early in lunar history of a lithosphere strong enough to support mascons; (3) lunar heat flow values at the Apollo 15 and 17 sites; (4) the electrical conductivity distribution inside the moon; (5) very low tectonic and moonquake activity at the present time; and (6) the relatively greater attenuation of S-waves below 1000 km as discussed in the previous section. The initial conditions, heat sources, and mechanisms of heat transfer must be specified in the calculations. These must be adjusted as required by the constraints.

The chronology of lunar igneous activity starts from the formation of the original crust about 4.6 b.y. ago (Wasserburg and Papanastassiou, 1971; Tatsumoto, 1970; Tatsumoto *et al.*, 1972; Silver, 1970). The crust probably differentiated within a relatively short time (100 m.y. or so). Although the record of pre-mare volcanism has been obscured by subsequent events, there is evidence of magmatic activity in the time between the appearance of the original crust and the filling of the mare basins (Papanastassiou and Wasserburg, 1972; Schonfeld and Meyer, 1972). The filling of mare basins has been well documented by age-dating of lunar samples and spans a time interval between 3.16 and 3.80 b.y. (Tera *et al.*,

1973). The above chronology indicates some activity in the moon from the time of formation of the original crust to the time of the emplacement of mare basalts, although there may have been a cataclysmic jump in activity at about 3.95 b.y. The span of igneous activity is shown in Fig. 9.

The failure to discover lunar igneous rocks younger than about 3.16 b.y. is an important constraint on the moon's thermal history (Papanastassiou and Wasserburg, 1971). If melting occurred since that time, it must have been very localized in extent or confined to depths in excess of several hundred kilometers.

The Apollo 15 and 17 heat flow values are both about 30 ergs/cm² sec (Langseth *et al.*, 1972, 1973). These place narrow limits on the total radioactivity, and to a lesser extent, on the temperatures in the moon. We will utilize the measured heat flow values as a strong constraint in our thermal models. Other constraints were discussed in detail by Toksöz and Solomon (1973) and will not be repeated here.

To calculate the thermal evolution it is necessary to specify the initial conditions. The initial temperature distribution in the moon is a function of the process of lunar formation and of the immediate environment of the moon during and shortly after its origin. The evidence favoring an early episode of extensive near-surface melting in the moon has been outlined above. Probably the most important energy source which was available to heat the moon to solidus or near-solidus temperatures was the gravitational energy liberated during accretion. Tidal dissipation (Hallam, 1973), solar wind flux, short-lived radioactivity and adiabatic compression are lesser sources.

Taking a time-dependent accretion rate (Hanks and Anderson, 1969; Mizutani *et al.*, 1972b), the accretion time may be adjusted so as to give an initial temperature profile in excess of the solidus in the outer portions of the moon. To differentiate a 60 km crust composed largely of feldspar requires melting to a depth perhaps as great as 500 km; i.e., involving at least half of the moon's volume. This can be achieved if the total accretion time is about 100 years. The duration of accretion may be made arbitrarily long, however, by appropriately lowering the assumed emissivity of the lunar surface.

A number of thermal history models were calculated with different initial conditions and heat sources. These were discussed in detail by Toksöz and Solomon (1973). Here, we will show the temperature curves for a typical model (Fig. 8) which satisfies most of the constraints. The initial temperature profile is from the accretion model with a starting temperature of 800°C and a total accretion time of 100 years (Mizutani *et al.*, 1972b). The last differentiation of radioactive heat sources was taken to occur 4.0 b.y. ago. The present day uranium concentration averages 60 ppb.

It may be seen that the model satisfies the major constraints upon lunar evolution. The zone of melting deepens with time. The lithosphere during the first 2 b.y. thickens at the rate of about 120 km/b.y. In particular, the shallowest melting progresses from 120 to 170 km depth during the period of mare filling, in agreement with the depth of origin of mare basalts (Ringwood and Essene, 1970) and with the need for a reasonably thick lithosphere to sustain the stresses associated with mascon gravity anomalies. At present the temperatures below a depth of

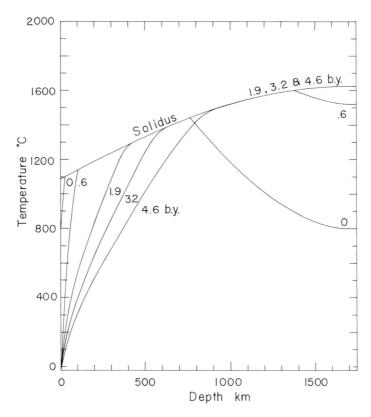

Fig. 8. Thermal evolution in a moon accreted in 100 years at a base temperature of 800°C with average present day uranium concentration equal to 60 ppb. The numbers on the curves indicate time after lunar origin.

about 1000 km reach that of the basalt solidus, but the whole moon is cooling. The heat flow value for this model is about 30 ergs/cm² sec.

The thermal evolution of the lunar interior as a function of time is shown in Fig. 9. Some of the major constraints are also included in the figure. During the first two billion years of lunar history, the lunar upper mantle underwent sufficient melting to account for the differentiation of the crust and the subsequent lunar volcanism and mare filling. The disappearance of this melting coincides roughly with the termination of magmatic events. The deep interior is hot enough at the present time to be partially molten. This could account for the attenuation of S-waves in this zone. If indeed partially molten, the rheologic properties of this zone may be similar to the earth's asthenosphere and the zone may have convective motions. Such convection could not induce sufficient stresses in the 1000 km thick lunar "lithosphere" to cause large moonquakes.

From the calculations of a large number of models the following can be

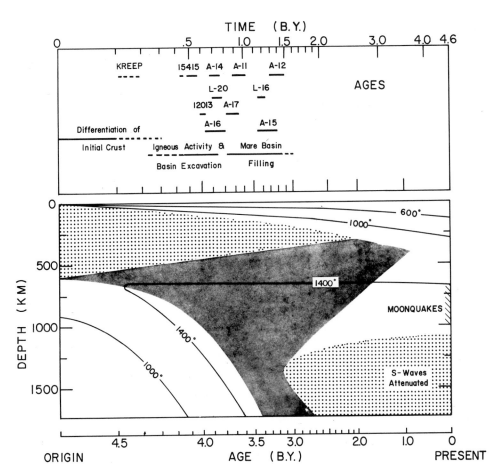

Fig. 9. (Above). A summary of igneous activity at the lunar surface from Rb–Sr and Ar^{39}–Ar^{40} ages of returned lunar samples. Samples from different missions are designated by "A" for Apollo (A-11 for Apollo 11) and "L" for Luna (L-16 for Luna 16). Special rocks are designated by numbers. Sources for this summary are given partly in the text, and partly in Toksöz and Solomon (1973). (Below). Evolution of the lunar interior as a function of time for a thermal history model similar to Fig. 8. Isotherms are in °C. Light and heavy stippling denote regions of partial and complete melting. The abscissa is linear in the square root of time. Present zones of moonquakes and low S-wave Q are also shown.

summarized regarding the thermal evolution of the moon:

(1) Models of the thermal evolution of the moon that fit the chronology of surface igneous activity, the need for a cool, rigid lithosphere since the formation of mascons, and the surface concentrations of radioactive elements all require extensive differentiation and upwards concentration of radioactive heat sources early in lunar history.

(2) A model of the moon which is initially hot, and extensively or totally molten can satisfy most of the constraints described in this section. A totally molten moon cools faster than other models due to complete and early differentiation.

(3) The early history of the moon is characterized by strong magmatic and tectonic activity. Convection must have played an important role in lunar differentiation and dynamics.

(4) If the Apollo heat flow values of about 30 erg/cm^2 sec (Langseth *et al.*, 1973) are representative of the moon, then the average concentration of uranium in the moon is currently 50–70 ppb, assuming the Th/U and K/U ratios of lunar surface rocks hold throughout the interior. This uranium concentration is intermediate between the average values for howardites and eucrites, although closer to howardites.

Acknowledgments—We thank David Johnston and Seth Stein for their contributions to the computations, and to this study. This research was supported by NASA Grant NGL 22-009-187 and by NASA Grant NAS9-12334.

References

Adler I., Gerard J., Trombka J., Schmadebeck P., Lowman P., Blodget H., Yin L., Eller E., Lamothe R., Gorenstein P., Bjorkholm P., Harris B., and Gursky H. (1972) The Apollo 15 X-ray fluorescence experiment. *Proc. Third Lunar Sci. Conf., Geochim. Cosmochim. Acta*, Suppl. 3, Vol. 3, pp. 2157–2178. MIT Press.

Anderson D. L. (1973) The composition and origin of the moon. *Earth Planet. Sci. Lett.* **18**, 301–316.

Anderson D. L. and Kovach R. L. (1972) The lunar interior. *Phys. Earth Planet. Interiors* **6**, 116–122.

Anderson O. L. and Liebermann R. C. (1966) Sound velocities in rocks and minerals. *Vesiac State-of-the-Art Report*, # 7887-4-X, Willow Run Laboratories, University of Michigan, 189 pp.

Anderson O. L., Scholz C., Soga N., Warren N., and Schreiber E. (1970) Elastic properties of a micro-breccia, igneous rock and lunar fines from Apollo 11 mission. *Proc. Apollo 11 Lunar Sci. Conf., Geochim. Cosmochim. Acta*, Suppl. 1, Vol. 3, pp. 1959–1973. Pergamon.

Boettcher A. L. (1971) The nature of the crust of the earth, with special emphasis on the role of plagioclase. In *The Structure and Properties of the Earth's Crust* (editor J. G. Heacock), *Amer. Geophys. Un. Mono.* **14**, 264–278.

Chung D. H. (1971) Elasticity and equations of state of olivines in the Mg_2SiO_4-Fe_2SiO_4 system. *Geophys. J. Roy. Astr. Soc.*, **25**, 511–538.

Chung D. H. (1972) Laboratory studies on seismic and electrical properties of the moon. *The Moon* **4**, 356–372.

Chung D. H. (1973) Elastic wave velocities in anorthosite and anorthositic gabbros from Apollo 15 and 16 landing sites (abstract). In *Lunar Science—IV*, pp. 141–142. The Lunar Science Institute, Houston.

Dainty A. M., Toksöz M. N., Anderson K. R., Pines P. J., Nakamura Y., and Latham G. (1973) Seismic scattering and shallow structure of the moon in Oceanus Procellarum. *The Moon.* In press.

Gast P. W. (1972) The chemical composition and structure of the moon. *The Moon* **5**, 121–148.

Green D. H., Ringwood A. E., Ware N. G., and Hibberson W. O. (1972) Experimental petrology and petrogenesis of Apollo 14 basalts. *Proc. Third Lunar Sci. Conf., Geochim. Cosmochim. Acta*, Suppl. 3, Vol. 1, pp. 197–206. MIT Press.

Hallam M. (1973) Heat sources for early differentiation of the lunar interior (abstract). *Trans. Amer. Geophys. Un.* **54**, 344.

Hanks T. C. and Anderson D. L. (1969) The early thermal history of the earth. *Phys. Earth Planet. Interiors* **2**, 19–29.

Hanks T. C. and Anderson D. L. (1972) Origin, evolution and present thermal state of the moon. *Phys. Earth Planet. Interiors*, **5**, 409–425.

Hariya Y. and Kennedy G. C. (1968) Equilibrium study of anorthite under high pressure and high temperature. *Amer. Jour. Sci.* **266**, 193–203.

Hays J. F. (1967) Lime-alumina-silica. *Carnegie Inst. of Washington Yearbook* **65**, 234–239.

Hays J. F. (1972) Radioactive heat sources in the lunar interior. *Phys. Earth Planet. Interiors* **5**, 77–84.

Helmberger D. V. (1968) The crust-mantle transition in the Bering Sea. *Bull. Seism. Soc. Amer.* **58**, 179–214.

Helmberger D. V. and Wiggins R. A. (1971) Upper mantle structure of midwestern United States. *J. Geophys. Res.* **76**, 3229–3245.

Kanamori H., Nur A., Chung D. H., Wones D., and Simmons G. (1970) Elastic wave velocities of lunar samples at high pressures and their geophysical implications. *Science* **167**, 726–728.

Kanamori H., Mizutani H., and Hamano Y. (1971) Elastic wave velocities of Apollo 12 rocks at high pressures. *Proc. Second Lunar Sci. Conf., Geochim. Cosmochim. Acta*, Suppl. 2, Vol. 3, pp. 2323–2326. MIT Press.

Kaula W. M., Schubert G., Lingenfelter R. E., Sjogren W. L., and Wollenhaupt W. R. (1972) Analysis and interpretation of lunar laser altimetry. *Proc. Third Lunar Sci. Conf., Geochim. Cosmochim. Acta*, Suppl. 3, Vol. 3, pp. 2189–2204. MIT Press.

Kovach R. L., Watkins J. S., Nur A., and Talwani P. (1973) The properties of the shallow lunar crust: an overview from Apollo 14, 16, and 17 (abstract). In *Lunar Science—IV*, pp. 444–445. The Lunar Science Institute, Houston.

Langseth M. G., Jr., Clark S. P., Jr., Chute J. L., Jr., Keihm S. J., and Wechsler A. E. (1972) The Apollo 15 lunar heat-flow measurement. *The Moon* **4**, 390–410.

Langseth M. G., Chute J. L., and Keihm S. (1973) Direct measurements of heat flow from the moon (abstract). In *Lunar Science—IV*, pp. 455–456. The Lunar Science Institute, Houston.

Latham G., Ewing M., Dorman J., Lammlein D., Press F., Toksöz N., Sutton G., Duennebier F., and Nakamura Y. (1971) Moonquakes. *Science* **174**, 687–692.

Latham G., Ewing M., Dorman J., Lammlein D., Press F., Toksöz N., Sutton G., Duennebier F., and Nakamura Y. (1972a) Moonquakes and lunar tectonism. *The Moon* **4**, 373–382.

Latham G., Ewing M., Dorman J., Lammlein D., Press F., Toksöz N., Sutton G., Duennebier F., and Nakamura Y. (1972b) Moonquakes and lunar tectonism results from the Apollo passive seismic experiment. *Proc. Third Lunar Sci. Conf., Geochim. Cosmochim. Acta*, Suppl. 3, Vol. 3, pp. 2519–2526. MIT Press.

Latham G. V., Ewing M., Press F., Sutton G., Dorman J., Nakamura Y., Toksöz N., Lammlein D., and Duennebier F. (1972c) Passive seismic experiment. In *Apollo 16 Preliminary Science Report*, NASA SP-315, pp. 9-1 – 9-29.

Latham G., Dorman J., Duennebier F., Ewing M., Lammlein D., and Nakamura Y. (1973a) Moonquakes, meteoroids, and the state of the lunar interior (abstract). In *Lunar Science—IV*, pp. 457–459. The Lunar Science Institute, Houston.

Latham G. V., Ewing M., Press F., Dorman J., Nakamura Y., Toksöz N., Lammlein D., Duennebier F., and Dainty A. (1973b) Results from the Apollo Passive Seismic Experiment. In *Apollo 17 Preliminary Science Report*, NASA Special Publication.

Levin F. K. and Ingram J. D. (1962) Heat waves from a bed of finite thickness. *Geophysics* **27**, 753–765.

McConnell R. K., Jr. and Gast P. W. (1972) Lunar thermal history revisited. *The Moon* **5**, 41–51.

Mizutani H., Fujii N., Hamano Y., and Osako M. (1972a) Elastic wave velocities and thermal diffusivities of Apollo 14 rocks. *Proc. Third Lunar Sci. Conf., Geochim. Cosmochim. Acta*, Suppl. 3, Vol. 3, pp. 2557–2564. MIT Press.

Mizutani H., Matsui T., and Takeuchi H. (1972b) Accretion process of the moon. *The Moon* **4**, 476–489.

Mizutani H. and Newbigging D. F. (1973) Elastic wave velocities of Apollo 14, 15 and 16 rocks and thermal conductivity profile of the lunar crust (abstract). In *Lunar Science—IV*, pp. 528–530. The Lunar Science Institute, Houston.

Nakamura Y., Lammlein D., Latham G., Ewing M., Dorman J., Press F., and Toksöz N. (1973) New seismic data on the state of the deep lunar interior. *Science.* **181**, 49–51.

Nur A. and Simmons G. (1969) The effect of saturation on velocity in low porosity rocks. *Earth Planet. Sci. Lett.* **7**, 183–193.

Papanastassiou D. A. and Wasserburg G. J. (1971) Lunar chronology and evolution from Rb–Sr studies of Apollo 11 and 12 samples. *Earth Planet. Sci. Lett.* **11**, 37–62.

Papanastassiou D. A. and Wasserburg G. J. (1972) The Rb–Sr age of crystalline rock from Apollo 16. *Earth Planet. Sci. Lett.* **16**, 289–298.

Press F. (1966) Seismic velocities. In *Handbook of Physical Constants* (editor S. P. Clark), *Geol. Soc. Amer. Mem.* **97**, 195–218.

Raheim A., Green D. H., Ringwood A. E., and Ware N. (1973) Experimental petrology of lunar high-alumina compositions (post-deadline abstract). *Fourth Lunar Science Conference*, Houston, March 5–8.

Reid A. M., Warner J., Ridley W. I., Johnston D. A., Harmon R. S., Jakes P., and Brown R. W. (1972) The major element compositions of lunar rocks as inferred from glass compositions in the lunar soils. *Proc. Third Lunar Sci. Conf., Geochim. Cosmochim. Acta*, Suppl. 3, Vol. 1, pp. 363–378. MIT Press.

Reynolds R. T., Fricker P. E., and Summers A. L. (1972) Thermal history of the moon. In *Thermal Characteristics of the Moon* (editor J. W. Lucas), pp. 303–337. MIT Press.

Ringwood A. E. and Essene E. (1970) Petrogenesis of Apollo 11 basalts, internal constitution and origin of the moon. *Proc. Apollo 11 Lunar Sci. Conf., Geochim. Cosmochim. Acta*, Suppl. 1, Vol. 1, pp. 769–799. Pergamon.

Schonfeld E. and Meyer C., Jr. (1972) The abundances of components of the lunar soils by a least-squares mixing model and the formation age of KREEP. *Proc. Third Lunar Sci. Conf., Geochim. Cosmochim. Acta*, Suppl. 3, Vol. 2, pp. 1397–1420. MIT Press.

Silver L. T. (1970) Uranium-thorium-lead isotopes in some Tranquillity Base samples and their implications for lunar history. *Proc. Apollo 11 Lunar Sci. Conf., Geochim. Cosmochim. Acta*, Suppl. 1, Vol. 2, pp. 1533–1574. Pergamon.

Sjogren W. L., Muller P. M., and Wollenhaupt W. R. (1972) Apollo 15 gravity analysis from the S-band transponder experiment. *The Moon* **4**, 411–418.

Solomon S. C. (1973) Density within the moon and implications for lunar composition. *The Moon.* In press.

Solomon S. C. and Toksöz M. N. (1973) Internal constitution and evolution of the moon. *Phys. Earth Planet. Interiors* **7**, 15–38.

Sonett C. P., Colburn D. S., Dyal P., Parkin C. W., Smith B. F., Schubert G., and Schwartz K. (1971) Lunar electrical conductivity profile. *Nature* **230**, 359–362.

Tatsumoto M. (1970) Age of the moon: An isotopic study of U-Th-Pb systematics of Apollo 11 lunar samples—II. *Proc. Apollo 11 Lunar Sci. Conf., Geochim. Cosmochim. Acta*, Suppl. 1, Vol. 2, pp. 1595–1612. Pergamon.

Tatsumoto M., Hedge C. E., Doe B. R., and Unruh D. M. (1972) U-Th-Pb and Rb–Sr measurements on some Apollo 14 lunar samples. *Proc. Third Lunar Sci. Conf., Geochim. Cosmochim. Acta*, Suppl. 3, Vol. 2, pp. 1531–1555. MIT Press.

Tera F., Papanastassiou D. A., and Wasserburg G. J. (1973) A lunar cataclysm at ~3.95 AE and the structure of the lunar crust (abstract). In *Lunar Science—IV*, pp. 723–725. The Lunar Science Institute, Houston.

Tittman B. R., Abdel-Gawad M., and Housley R. M. (1972) Elastic velocity and Q factor measurements on Apollo 12, 14, and 15 rocks. *Proc. Third Lunar Sci. Conf., Geochim. Cosmochim. Acta*, Suppl. 3, Vol. 3, pp. 2565–2575. MIT Press.

Todd T., Wang H., Baldridge W. S., and Simmons G. (1972) Elastic properties of Apollo 14 and 15 rocks. *Proc. Third Lunar Sci. Conf., Geochim. Cosmochim. Acta*, Suppl. 3, Vol. 3, pp. 2577–2586. MIT Press.

Todd T., Wang H., Richter D., and Simmons G. (1973) Unique characterization of lunar samples by physical properties (abstract). In *Lunar Science—IV*, pp. 731–733. The Lunar Science Institute, Houston.

Toksöz M. N., Press F., Anderson K., Dainty A., Latham G., Ewing M., Dorman J., Lammlein D., Nakamura Y., Sutton G., and Duennebier F. (1972a) Velocity structure and properties of the lunar crust. *The Moon* **4**, 490–504.

Toksöz M. N., Press F., Anderson K., Dainty A., Latham G., Ewing M., Dorman J., Lammlein D., Sutton G., Duennebier F., and Nakamura Y. (1972b) Lunar crust: structure and composition. *Science* **176**, 1012–1016.

Toksöz M. N., Press F., Dainty A., Anderson K., Latham G., Ewing M., Dorman J., Lammlein D., Sutton G., and Duennebier F. (1972c) Structure, composition, and properties of lunar crust. *Proc. Third Lunar Sci. Conf., Geochim. Cosmochim. Acta*, Suppl. 3. Vol. 3, pp. 2527–2544. MIT Press.

Toksöz M. N., Solomon S. C., Minear J. W., and Johnston D. H. (1972d) Thermal evolution of the moon. *The Moon* **4**, 190–213; **5**, 249–250.

Toksöz M. N. and Solomon S. C. (1973) Thermal history and evolution of the moon. *The Moon* **7**, 251–278.

Toksöz M. N., Press F., Dainty A. M., and Anderson K. R. (1973a) Lunar velocity structure and compositional and thermal inferences. *The Moon*. In press.

Toksöz M. N., Press F., Dainty A. M., Solomon S. C., and Anderson K. R. (1973b) Lunar structure, compositional inferences and thermal history (abstract). In *Lunar Science—IV*, pp. 734–736. The Lunar Science Institute, Houston.

Tozer D. C. (1972) The moon's thermal state and an interpretation of the lunar electrical conductivity distribution. *The Moon* **5**, 90–105.

Turkevich A. L. (1971) Comparison of the analytical results from the Surveyor, Apollo and Luna missions. *Proc. Second Lunar Sci. Conf., Geochim. Cosmochim. Acta*, Suppl. 2, Vol. 2, pp. 1209–1215. MIT Press.

Urey H. C. and MacDonald G. J. F. (1971) Origin and history of the moon. In *Physics and Astronomy of the Moon*, 2nd ed. (editor Z. Kopal), pp. 213–289. Academic Press.

Wang H., Todd T., Weidner D., and Simmons G. (1971) Elastic properties of Apollo 12 rocks. *Proc. Second Lunar Sci. Conf., Geochim. Cosmochim. Acta*, Suppl. 2, Vol. 3, pp. 2327–2336. MIT Press.

Wang H., Todd T., Richter D., and Simmons G. (1973) Elastic properties of plagioclase aggregates and seismic velocities in the moon (abstract). In *Lunar Science—IV*, pp. 758–760. The Lunar Science Institute, Houston.

Warren N., Schrieber E., Scholz C., Morrison J. A., Norton P. R., Kumazawa M., and Anderson O. L. (1971) Elastic and thermal properties of Apollo 11 and Apollo 12 rocks. *Proc. Second Lunar Sci. Conf., Geochim. Cosmochim. Acta*, Suppl. 2, Vol. 3, pp. 2345–2360. MIT Press.

Warren N., Anderson O. L., and Soga N. (1972) Applications to lunar geophysical models of the velocity-density properties of lunar rocks, glasses and artificial lunar glasses. *Proc. Third Lunar Sci. Conf., Geochim. Cosmochim. Acta*, Suppl. 3, Vol. 3, pp. 2587–2598. MIT Press.

Wasserburg G. J. and Papanastassiou D. A. (1971) Age of an Apollo 15 mare basalt; lunar crust and mantle evolution. *Earth Planet. Sci. Lett.* **13**, 97–104.

Wood J. A., Marvin U. B., Reid J. B., Jr., Taylor G. J., Bower J. F., Powell B. N., and Dickey J. S., Jr. (1971) Mineralogy and petrology of the Apollo 12 lunar sample. *Smithsonian Astrophys. Obs. Spec. Rep.* **333**, 272 pp.

Wood J. A. (1972) Thermal history and early magmatism in the moon. *Icarus* **16**, 229–240.

Wood J. A. (1973) Asymmetry of the moon (abstract). In *Lunar Science—IV*, pp. 790–792. The Lunar Science Institute, Houston.

Proceedings of the Fourth Lunar Science Conference
(Supplement 4, *Geochimica et Cosmochimica Acta*)
Vol. 3, pp. 2549–2560

The structure of the lunar crust at the Apollo 17 site

ROBERT L. KOVACH

Department of Geophysics
Stanford University
Stanford, California 94305

JOEL S. WATKINS

Earth and Planetary Sciences Division
The Marine Biomedical Institute
The University of Texas Medical Branch
Galveston, Texas 77550

Abstract—A seismic profiling experiment was successfully executed on the lunar surface during the Apollo 17 mission allowing a determination of the structure of the lunar crust to a depth of several kilometers. The most outstanding feature of the seismic velocity variation, in the Taurus–Littrow region, is the stepwise increase with depth. A total thickness of about 1200 meters for the infilling mare basalts at the 17 landing site is also indicated from the seismic results. The apparent velocity is high (about 4 km/sec for *P* waves) in the material below the basalts.

INTRODUCTION

WITH THE successful installation of a geophysical station at the Taurus–Littrow landing site of the Apollo 17 mission an exciting period of manned lunar exploration was brought to an end. Prior to the Apollo 17 mission there was a surprising gap in our knowledge concerning the nature of the properties of the upper 10 km of the lunar crust, because of the absence of pertinent seismic travel time data at distances (Δ) closer than 30 km. Travel times of seismic waves are inverted in a classical fashion to determine seismic velocity structure and as a result provide the direct means of probing the lunar interior.

On the Apollo 17 mission a Lunar Seismic Profiling Experiment (LSPE) was successfully executed. Eight explosive charges were armed and placed on the lunar surface by the Apollo 17 astronauts at various points along the geological traverses. In addition, the impact of the Lunar Module (LM) ascent stage, at a distance of 8.7 km from the 17 landing site, was recorded on an array of miniature seismometers. In this paper we describe the seismic data, the velocity structure in the lunar crust beneath the Taurus–Littrow area and the implications for lunar history. The interpretation is more expansive than in previous reports (Kovach *et al.*, 1973; Kovach and Watkins, 1973a). Post-mission analyses of the Apollo 17 lunar surface photography has necessitated some adjustments in the absolute positions of the deployed explosive charges.

SEISMIC DATA

The LSPE recorded the LM ascent stage impact on December 15, 1972. The location of the impact was at 19.91 degrees north latitude and 30.51 degrees east longitude, at a distance 8.7 km southwest of the Apollo 17 landing site. Pertinent parameters for the LM impact are shown in Table 1.

A portion of the seismic signal from the Apollo 17 LM impact is shown in Fig. 1. The impact signal is similar in character to previous impact signals, possessing

Table 1. Parameters of Apollo 17 LM impact.

Impact parameters	LM
Day, G.m.t.	December 15, 1972
Range time,* G.m.t.	
hr:min:sec	06:50:20.84
Real time, G.m.t.	
hr:min:sec	06:50:19.60
Velocity, km/sec	1.67
Mass, kg	2260
Kinetic energy, ergs	3.15×10^{16}
Heading, deg	283

*Range time is the time that the signal of the event was observed on Earth.

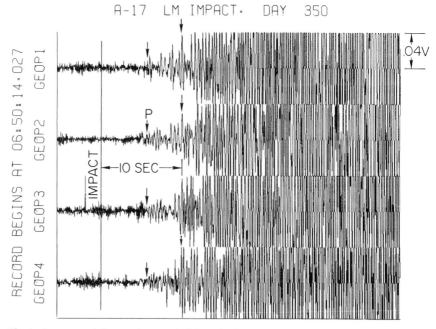

Fig. 1. Compressed time scale record of the seismic signal received from the LM impact of the Apollo 17 mission. Arrows point to the onset of first and second arrivals. The first arrow indicates a travel time of 5.75 seconds.

an emergent beginning and a long duration. The travel-time of the first P wave is 5.75 sec. A prominent second arrival with a travel time of 10 sec can also be identified on the seismograms. Whether this arrival represents a shear wave arrival or a later P wave arrival cannot be resolved from only a vertical component seismogram.

The observed amplitude of the impact signal is of interest when compared with the P wave amplitudes for previous impact signals. Comparison of previous LM impact signal amplitudes of SIVB impacts (Latham et al., 1972) demonstrated that the LM impact amplitude data had to be adjusted upwards by a factor of 17.4 to be directly compared to the amplitudes observed for SIVB impacts. Extrapolating the observed amplitudes of the earlier LM impacts to a distance of 8.7 km leads to a predicted peak to peak amplitude of 26 nm. The Apollo 17 LM impact signal is centered at 4 Hz and has a measured peak to peak amplitude of 400 nm. It is interesting that, if we multiply the predicted amplitude of 26 nm by the scale factor 17.4 we obtain 452 nm which is in good agreement with the observed amplitude of 400 nm.

Analyses of previous lunar seismic impact signals (Latham et al., 1973) have demonstrated that many of their characteristics (signal rise time, duration of signal, and lack of coherence between horizontal and vertical components of motion) can be explained by scattering. Seismic energy is considered to spread with a diffusivity ξ. The larger the value of diffusivity the smaller the amount of scattering. For a surface impact one can show that the signal rise time (the time from signal onset to its maximum value) is given by R^2/ξ where R is the range.

The Apollo 17 LM impact seismic signal has a rise time of 56 seconds leading to a diffusivity of 1.35 km^2/sec, a larger value than that inferred at the Apollo 15 and 16 sites from analysis of the seismic signals generated by the lunar roving vehicle out to distances of about 4 km (Latham et al., 1973). The implication is that the Apollo 17 site is more homogeneous (for dimensions on the order of the wavelengths of seismic waves considered, i.e., approximately 25 m) than either the Apollo 15 or 16 landing areas. A difference in the near-surface properties of these landing sites may be due to differing ages for the areas and the effects of varying amounts of comminution and gardening by meteoroid impacts.

Figure 2 shows the seismogram recorded on the LSPE geophone array from the detonation of one of the explosive charges placed on the lunar surface at the 17 landing site. The arrows point to the onset of the first seismic arrival.

The locations of the explosive packages with respect to the geophone array are shown on the map in Fig. 3. Explosive charge weights ranged from 1/8 to 6 lbs. The one-pound explosive package (EP-6) was deployed on the outgoing traverse to the south and the one-half pound charge (EP-7) was positioned on the return leg to the lunar module. EP-4 (1/8 lb) and EP-8 (1/4 lb) were placed west of the geophone array. EP-2 (1/4 lb) and EP-3 (1/8 lb) were positioned east of the geophone array. EP-1 (6 lbs) was placed at a distance of 3 km NW from the LM and EP-5 (3 lbs) at a distance of 2.1 km NE of the LM.

The observed travel times from EP's 4 and 8 to the west of the geophone array and EP's 2 and 3 to the east are shown in Fig. 1 of Watkins and Kovach (1973).

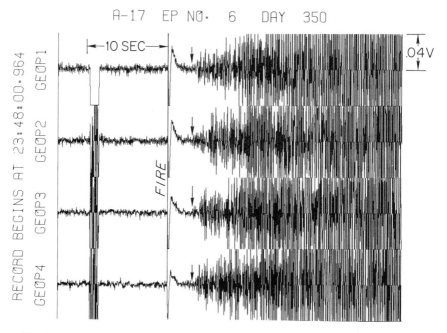

Fig. 2. Seismic signals produced by the detonation of explosive package 6 on the lunar surface as recorded on Apollo 17 geophone array. Arrows point to the onset of the seismic arrival.

These data yield a partially reversed profile and are discussed in detail in Watkins and Kovach (1973).

It is useful at this point to restate the assumptions (Ewing *et al.*, 1939) underlying the method of interpreting a seismic refraction profile: (1) the top and bottom of any layer is planar and propagates seismic waves with constant velocity; (2) at a layer interface the path of a seismic wave is governed by Snell's law; (3) a seismic wave propagating in a layer with a velocity V, which is incident on the surface of the layer at an angle α with the normal, travels with an apparent velocity $V/\sin \alpha$ along the surface; (4) travel time is the same if the shot point and recording position are interchanged. Departures from these stated assumptions are revealed by the departure of the observed travel time data from straight lines drawn on the travel time graph.

Figure 4 shows the travel time data obtained from the detonation of the eight explosive packages. The digital sampling interval for the seismic data is 0.0085 sec. Using expanded scale time plots the onset of the seismic first arrivals could be picked to an accuracy of ± 0.05 sec. Subsequent array processing techniques on the digital data may contribute some additional information on apparent velocities, but it is not believed that subsequent processing will have any major effect on the travel time data discussed here. Inasmuch as complete reversal information was not obtained we shall interpret the travel time data on the assumption of zero dip of all interfaces. When treated *in toto* the close in shot

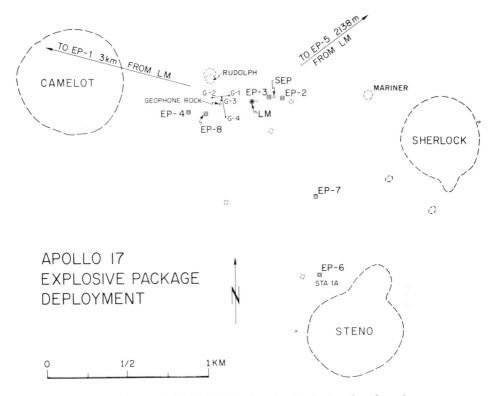

Fig. 3. Planimetric map of the Apollo 17 landing site showing location of geophone array and positions of explosive packages.

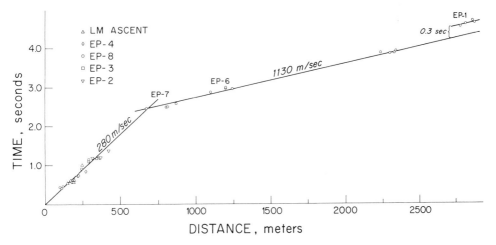

Fig. 4. Travel time distance graph for *P* wave pulses recorded from detonation of explosive packages and thrust of LM ascent engine.

points (EP's 2, 3, 4, and 8) define an *average* velocity arrival of 280 m/sec. Travel times from EP's 7, 6, and 5 define an apparent velocity of 1130 m/sec. The calculated thickness of the 280 m/sec layer is 264 m.

One can note that the travel time data for EP 1 are offset by about 0.3 sec with respect to the 1130 m/sec line. Examination of the path between EP 1 and the LSPE geophone array reveals that the seismic path is affected by the presence of the 600 m diameter crater Camelot. We can explain the observed time delay by postulating that the low-velocity 280 m/sec material extends to a greater depth beneath the crater Camelot than along the remainder of the travel time path. Figure 5 shows diagrammatically a simple model approximation for Camelot crater which can be constructed to explain the observed travel time delay. Subsequent digital velocity filtering (beam steering) of the seismic signals from EP 1 should allow more detailed models to be considered.

The travel time data from the explosive charges can be combined with the observed travel time for the LM impact to provide information about the seismic velocity to a depth of several km. Travel time data from the seismic signals produced by the LM impact and the explosive charges are shown in Fig. 6. A line with an apparent velocity of 4 km/sec can be fit through the LM impact data point to intersect very close to the corrected travel time data points for EP 1.

Inasmuch as the LM impacted at an elevation of 1.2 km above the valley floor at the 17 landing site the LM impact travel time should be adjusted to the same reference elevation as the LSPE geophone array. The effect of a difference in elevations is to contribute an additional delay time equal to the ratio of the elevation difference to the seismic velocity of the material traversed times the cosine of the angle of incidence that the particular seismic arrival under consideration departed the source (impact point). We shall assume that the underlying 4000 m/sec material in the Taurus–Littrow valley extends to the surface beneath the LM impact point in the South Massif. If we further assume that at a distance of 8.7 km (LM impact point) on the travel time graph the seismic velocity increases to 4500 m/sec (say) the angle of incidence for a refraction at this depth and the related angle of incidence at the source can be straightforwardly calculated.

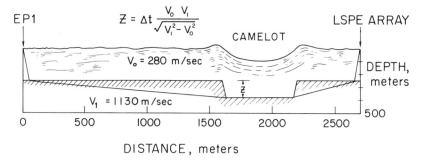

Fig. 5. Model approximation for seismic ray path from explosive package 1 to geophone array. Ray path crosses Camelot crater and observed time delay is produced by presence of low-velocity material beneath crater.

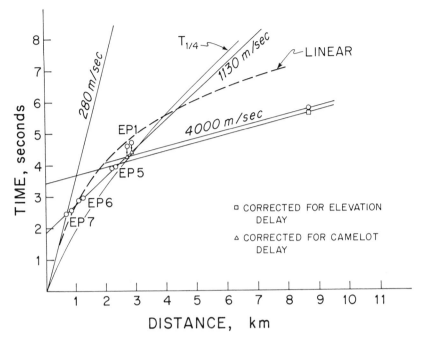

Fig. 6. Seismic travel times from LM impact and explosive charges. Travel time for EP 1 has been corrected for Camelot crater delay and LM impact travel time for 1.2 km elevation difference between impact point and LSPE geophone array. These corrections shift position of 4 km/sec apparent velocity line slightly downward as shown.

Inserting appropriate values for our case leads to a delay time correction of 0.14 sec. This correction will shift the position of the 4 km/sec apparent velocity line downward as shown in Fig. 6 such that its zero distance time intercept is decreased. There are obvious uncertainties in correcting for the elevation delay but the *important first order* consideration is that high velocity material (~4 km/sec) must lie beneath the 1130 m/sec material. The end effect of the elevation correction is to decrease the derived thickness of the 1130 m/sec material from 1006 m to 923 m.

Various power laws for the variation of seismic velocity with depth in the moon have been proposed (Gold and Soter, 1970; Gangi, 1972). Among these are that the seismic velocity increases as either the fourth or sixth power of the depth. It is of interest to compare the Apollo 17 *observed* travel times with some of these proposed velocity-depth variations.

The fourth power model with velocity variation of the form $v = v_0(z/z_0)^{1/4}$ gives a travel time relationship

$$t = \frac{2\pi z_0}{v_0}[2x/3\pi z_0]^{3/4},$$

where v_0 is the assigned velocity at a depth of z_0. Normal practice is to choose v_0

such that the observed travel time agrees with that observed at a distance x. We shall select $z_0 = 1$ km and the observed travel time of 4.2 sec at a distance x of 2.75 km. In this case $v_0 = 0.998$ km/sec. Using these parameters a travel time curve has been calculated and is shown as the curved marked $T_{1/4}$ in Fig. 6. It is seen that the fourth root velocity variation predicts travel times that are too slow at distances greater than 3 km, implying that the velocities are too low at depth.

We can also examine a uniform increase of velocity with depth. In this case $v = v_0 + Kz$, where v_0 = velocity at surface, K = rate of increase with depth and

$$t = \frac{2}{K} \sinh^{-1}\left(\frac{Kx}{2v_0}\right).$$

Let us assume a surface velocity of 280 m/sec and a value for K of 0.92/sec. This linear gradient model gives a travel time graph marked linear in Fig. 6. The agreement is quite good out to a distance of about 1 km but at greater distances the predicted velocity is again too slow. Nevertheless, it is to be emphasized that both the fourth-power and linear gradient models cannot satisfy the well defined linear segments of the observed travel times.

Discussion

Prior to the Apollo 17 mission our best estimates of the seismic velocity variation in the upper 20 km of the moon were as depicted by lunar models 1 or 2 (Latham et al., 1973; Toksoz et al., 1972) in Fig. 7. The seismic velocity was known to increase very rapidly from values of 100–300 m/sec in the upper 100 m or so to a value of ~ 4 km/sec at a depth of 5 km. Even though the seismic velocity variation was taken to be a smooth increase with depth it was surmised (Kovach and Watkins, 1973b) that such a rapid increase of velocity (~ 2 km/sec/km) could not be solely explained by the pressure effect on dry rocks with macrocracks and microcracks nor by the self-compression of any rock powder.

Laboratory velocity measurements on returned lunar soils (Kanamori et al., 1970, 1971; Anderson et al., 1970; Mizutani et al., 1972; Warren et al., 1971) and recent measurements under hydrostatic pressure conditions on terrestrial sands and basaltic ash (Talwani et al., 1973) have indicated velocity-depth gradients of 0.4 to 0.8 km/sec/km but such gradients occur only to pressures of 50 bars (lunar depth of ~ 1 km). An examination of these experimental data led to the inference that compositional or textural changes must be important in the upper 5 km of the moon (Kovach and Watkins, 1973b).

The lunar seismic profiling results have shown that, at least beneath the Taurus–Littrow site, the seismic velocity increases with depth in a stepwise manner in the upper several km. It is of interest to examine our in situ velocity information in the light of the surface geological investigations at the 17 site, laboratory velocity measurements from returned lunar samples and seismic velocity measurements on terrestrial lunar analogs.

Underlying the regolith at the 17 site the dominant rock type observed is a

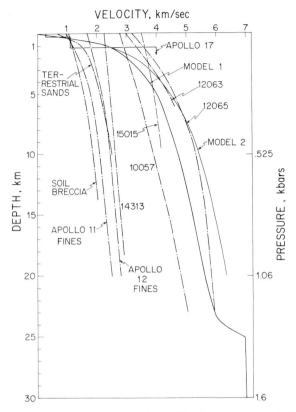

Fig. 7. Inferred compressional wave velocity profiles for moon and velocities of lunar and terrestrial rocks measured in the laboratory as a function of pressure. Lunar rocks are identified by a sample number. Lunar models 1 and 2 are based on results available through Apollo 16 (Toksoz *et al.*, 1972). Apollo 17 results reveal marked stepwise increase in seismic velocity in upper 2 km.

medium grained vesicular basalt believed to be primarily a mare type basalt. Observations by the crew in crater walls revealed textural variations suggesting that individual flow units are involved. Our seismic observations have indicated 264 m of material with an average velocity of 280 m/sec overlying 923 m of 1130 m/sec material. We suspect the 280 m/sec layer contains several higher velocity lava flow units (Watkins and Kovach, 1973).

The abrupt change in seismic velocity and, by inference in other physical properties, from 280 m/sec to 1130 m/sec is suggestive of a major change in the nature of the evolution or deposition of the Apollo 17 subfloor basalts. However, a similar range of seismic velocities is observed with refraction surveys on lava flows on earth. Some insight can be gained when we consider specific lava flows which have been examined in some detail as possible lunar analogs: the Southern Coulee, the SP flow and the Kana-a flow (Watkins, 1966; Watkins *et al.*, 1972).

The Southern Coulee is a recent lava flow near Mono Craters in eastern California. Seismic velocities range at the surface, from 160 m/sec to 2000 m/sec at depth. The higher velocities are found in more competent denser lava which underlies higher porosity, lower density surface material. The SP flow is a blocky basalt flow located in the northern part of the San Francisco volcanic field near Flagstaff, Arizona. Vesicularity ranges from 5 to 50% and *in situ* seismic velocities range from 700 to 1100 m/sec. The Kana-a flow, also located near Flagstaff, is an olivine basalt flow intermingled with ash; velocities range from 700 to 1200 m/sec.

We see that observed velocities on terrestrial lava flows bracket the velocities measured at the 17 site and therefore are consistent with the presence of lava flows in the Taurus–Littrow valley. The 280 m/sec velocity is probably representative of several flows separated by lower velocity layers of ash or ejecta. Individual flows may be fractured or brecciated which would further decrease their seismic velocities. Surface layers of fractured loose blocky material merging into more welded flows are common occurrences on earth. We therefore believe that the sum of the 280 m/sec and the 1130 m/sec materials (1187 m) represent the full thickness of the subfloor basalts at the Apollo 17 site.

The nature of the material underlying the basalts with a seismic velocity of ~ 4 km/sec is difficult to assign unambiguously to any particular rock type. It seems likely, based on the geological evidence, that the highland massif material which rings the narrow graben-like valley at the 17 site underlies the basalt flow(s). Several rock types were recognized in the North and South Massifs but the dominant rock type is apparently a coherent breccia believed to be similar to breccias sampled at the Apennine Front (Apollo 15) and Descartes (Apollo 16).

Laboratory velocity measurements have been reported for two Apollo 15 breccias, 15418 and 15015 (Todd *et al.*, 1972). Sample 15418 is described as a dark gray crystalline breccia of chemical composition similar to anorthite rich gabbro. Sample 15015 is a more friable breccia of unknown composition. The *in situ* value of ~ 4 km/sec is close to the values measured in the laboratory for sample 15015.

CONCLUSIONS

Prior to the Apollo 17 mission the question of how the *P*-wave velocity increased from 100 to 300 m/sec near the surface to 6 km/sec at a depth of 15 to 20 km was most uncertain. The main reason for this uncertainty was the gap in travel time data between the ranges of a few hundred meters (previous Active Seismic Experiments) and 67 km (Apollo 14 LM impact as recorded by the Apollo 14 PSE). The Apollo 17 Lunar Seismic Profiling results have demonstrated that the seismic velocity increases in a sharp stepwise manner in the upper 2.5 km. A surface layer with a seismic velocity of 280 m/sec overlies a layer with a velocity of 1130 m/sec. Beneath the 1130 m/sec layer the seismic velocity increases sharply to a velocity of 4000 m/sec. The velocities of 280 m/sec and 1130 m/sec are in agreement with those observed for basaltic lava flows on earth and therefore suggest a total thickness of about 1200 m for the infilling mare basalts at Taurus–Littrow. When the Apollo 17 results are combined with earlier travel time data for direct and surface reflected arrivals from LM and SIVB impacts it will be possible to construct a velocity model for the upper lunar crust believed to be representative for a mare basin.

Acknowledgments—This work was supported by the National Aeronautics and Space Administration under contract NAS9-5632 and grant NGL05-020-232.

REFERENCES

Anderson O. L., Scholz C., Soga N., Warren N., and Schreiber E. (1970) Elastic properties of a micro-breccia, igneous rock and lunar fines from Apollo 11 mission. *Proc. Apollo 11 Lunar Sci. Conf., Geochim. Cosmochim. Acta*, Suppl. 1, Vol. 3, pp. 1959–1973. Pergamon.

Ewing M., Woollard G., and Vine A. (1939) Geophysical investigations in the emerged and submerged Atlantic coastal plain: Part III: Barnegat Bay, New Jersey, section. *Bull. Geol. Soc. Amer.* **50**, 257–296.

Kanamori H., Nur A., Chung D. H., Wones D., and Simmons G. (1970) Elastic wave velocities of lunar samples at high pressures and their geophysical implications. *Science* **167**, 726–728.

Kanamori H., Mizutani H., and Hamano Y. (1971) Elastic wave velocity of Apollo 12 rocks at high pressures. *Proc. Second Lunar Sci. Conf., Geochim Cosmochim. Acta*, Suppl. 2, Vol. 3, pp. 2323–2326. MIT Press.

Kovach R. L. and Watkins J. S. (1973a) Apollo 17 seismic profiling—probing the lunar crust. *Science* **180**, 1063–1064.

Kovach R. L. and Watkins J. S. (1973b) The velocity structure of the lunar crust. *The Moon* **7**, 63–75.

Kovach R. L., Watkins J. S., and Talwani P. (1973) Seismic profiling experiment. *Apollo 17 Preliminary Science Report*. In press.

Latham G., Ewing M., Press F., Sutton G., Dorman J., Nakamura Y., Toksoz N., Lammlein D., and Duennebier F. (1972) Passive seismic experiment. *Apollo 15 Preliminary Science Report*, NASA SP-289, section 8.

Latham G., Ewing M., Press F., Sutton G., Dorman J., Nakamura Y., Toksoz N., Lammlein D., and Duennebier F. (1973) Passive seismic experiment. *Apollo 16 Preliminary Science Report*, NASA SP-315, section 9.

Mizutani H., Fujii N., Hamano Y., Osako M., and Kanamori H. (1972) Elastic wave velocities and thermal diffusivities of Apollo 14 rocks (abstract). In *Lunar Science—III*, p. 547. The Lunar Science Institute, Houston.

Gangi A. F. (1972) The lunar seismogram. *The Moon* **4**, 40–48.

Gold T. and Soter S. (1970) Apollo 12 seismic signal: Indications of a deep layer of powder. *Science* **169**, 1071–1075.

Talwani P., Nur A., and Kovach R. L. (1973) Compressional and shear wave velocities in granular materials to 2.5 kilobars. *J. Geophys. Research*. In press.

Todd T., Wang H., Baldridge W., and Simmons G. (1972) Elastic properties of Apollo 14 and 15 rocks. *Proc. Third Lunar Sci. Conf., Geochim. Cosmochim. Acta*, Suppl. 3, Vol. 3, pp. 2577–2586. MIT Press.

Toksoz M. N., Press F., Dainty A., Anderson K., Latham G., Ewing M., Dorman J., Lammlein D., Sutton G., and Duennebier F. (1972) Structure, composition and properties of lunar crust. *Proc. Third Lunar Sci. Conf., Geochim. Cosmochim. Acta*, Suppl. 3, Vol. 3, pp. 2527–2544. MIT Press.

Warren H., Schreiber E., Scholz C., Morrison J. A., Norton P. R., Kumazala M., and Anderson O. L. (1971) Elastic and thermal properties of Apollo 12 and Apollo 11 rocks. *Proc. Second Lunar Sci. Conf., Geochim. Cosmochim. Acta*, Suppl. 2, Vol. 3, pp. 2345–2360. MIT Press.

Watkins J. S. (1966) Annual Report. Investigation of *in situ* physical properties of surface and subsurface site materials by engineering geophysical techniques. NASA Contract T-25091 (G).

Watkins J. S. and Kovach R. L. (1973) Seismic investigation of the lunar regolith. *Proc. Fourth Lunar Sci. Conf., Geochim. Cosmochim. Acta.* In press.

Watkins J. S., Walters L. A., and Godson R. H. (1972) Dependence of *in situ* compressional-wave velocity on porosity in unsaturated rocks. *Geophysics* **37**, 29–35.

Proceedings of the Fourth Lunar Science Conference
(Supplement 4, *Geochimica et Cosmochimica Acta*)
Vol. 3, pp. 2561–2574

Seismic investigation of the lunar regolith*

Joel S. Watkins

Earth and Planetary Sciences Division, University of Texas, Galveston, Texas 77550

Robert L. Kovach

Department of Geophysics, Stanford University, Stanford, California 94305

Abstract—Seismic investigation of the lunar regolith shows that the regolith consists of a layer with an average velocity of 105 m/sec. Seismically determined regolith thicknesses in the highlands average 10 m, about twice the thickness deduced from astronaut observations and photographs from the maria. The difference in thickness is thought to be due to a depositional rate about one order of magnitude greater during the interval between formation of the circular maria and the extrusion of mare basalts than during the post-extrusion interval.

Seismic velocities observed in the regolith indicate that the regolith consists primarily of impact derived ejecta. The regolith at the Apollo 17 site is anomalously thick, probably because the site is in the midst of a crater field and because of irregularities in the surface of the underlying subfloor basalt. The Apollo 17 seismic data do not support the compacted powder model of the lunar near surface but show that the lunar crust to a depth of over 1 km consists of horizontal or subhorizontal strata comprised of breccia, lava flows, and rock.

Introduction

Seismic data pertaining to the lunar regolith were obtained at all but the Apollo 11 site. On missions 12, 14, and 15, the Passive Seismic Experiment (PSE) provided seismic data from the regolith; on missions 14 and 16 the Active Seismic Experiment (ASE) provided the seismic data, and on mission 17, the Lunar Seismic Profiling Experiment (LSPE) provided the data.

The ASE included a low energy source of seismic energy designed primarily to obtain seismic data from the regolith. The other experiments gleaned data incidentally in the course of their primary missions of collecting seismic data from rocks beneath the regolith.

Data analyzed to date consist entirely of *P*-wave travel times. From these data it has been possible to deduce velocities of *P*-waves traversing the regolith and/or regolith thicknesses at each site except the Apollo 17 site. At this site it was possible only to obtain a maximum thickness.

The seismic data have been interpreted in terms of rock properties and rock structure. The interpretation is based on properties of similar terrestrial rocks (e.g., basaltic lavas, fragmental rocks and breccias at Meteor Crater, Arizona), theoretical models, and geological evidence. In this paper, we review seismic data obtained during the Apollo missions, discuss the model of the regolith which we feel best explains the data, and discuss some of the implications of the data.

---

*Earth and Planetary Sciences Division, Marine Biomedical Institute. Contribution No. 18.

Experimental Results

Astronauts deployed the PSE on Apollo missions 12, 14, 15, and 16. On missions 12, 14, and 15, the PSE short period vertical (SPZ) seismometer recorded the LM ascent. In each instance, a regolith arrival was observed (Latham *et al.*, 1970, 1971, 1972). Regolith P-wave velocities were 108 m/sec, 104 m/sec, and 92 m/sec. Since each ascent was recorded by only one instrument, it was not possible to deduce the thickness of the regolith.

On Apollo 14 and 16, astronauts deployed the ASE, an experiment which depends on explosives for seismic energy. The ASE consists of a thumper, a mortar package assembly (MPA), a linear array of 3 geophones, and associated electronics (Kovach, 1967; McAllister *et al.*, 1969; Watkins *et al.*, 1969). The thumper, which was designed to investigate the regolith, is an astronaut-activated staff containing 21 small explosives in its base. Firing of an explosive produces seismic energy by driving the base plate into the regolith or "thumping" the regolith surface. A pressure transducer in the base detects the instant of detonation. The geophones detect seismic waves radiating away from the thump via the regolith and subjacent strata. ASE electronics compress the geophone data and transmit it to the earth in real time.

Mortar launches provide additional data on the seismic properties of the regolith. Although the MPA was designed to provide seismic data from rocks below the regolith, mortar launches generate a detectible signal which travels through the regolith and immediately subjacent rocks. These signals were used in analysis

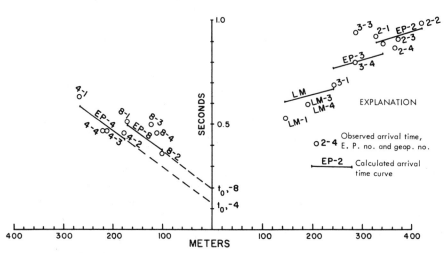

Fig. 1. Arrival time observed from headwaves refracted from EP-2, -3, -4, -8, and LM ascent events; and arrival time ranges calculated for headwaves refracted by a layer 17.4 m beneath geophone 3 and dipping 3° East.

of Apollo 16 data (Kovach *et al.*, 1973a). LM ascent data were recorded by the ASE at the Apollo 16 site (Kovach *et al.*, 1973a) and used in the interpretation. ASE data from Apollo 14 and 16 sites yielded regolith velocities of 104 and 114 m/sec, respectively, and regolith thicknesses of 8.5 and 12.2 m, respectively.

The LSPE consisted of 4 geophones deployed in a triangular array, eight explosive packages (EP), and associated electronics. All EPs were deployed too far from the array to obtain first arrivals of the regolith *P*-wave. It is possible that examination of second arrivals will yield the velocity of the regolith in the vicinity of the LSPE array but examination of second arrivals to date has been cursory and inconclusive.

The first arrivals from explosives detonated near the Apollo 17 LSPE array at Taurus-Littrow are shown in Fig. 1. The pattern of arrivals differs in two important respects from the pattern of arrivals recorded by the ASE at the Apollo 14 and 16 sites.

The first important difference is that velocities calculated from waves traversing the array are significantly higher than the velocities calculated from arrivals at more distant EPs (850 m/sec vs. 280 m/sec). This phenomenon we attribute to the "thin-layer" effect. When a layer has a velocity higher than that of layers above and below, arrivals from the layer will mask arrivals from lower layers. The extent of masking depends on velocities and thicknesses of the layers. If the high velocity is thin (less than a wave length thick) energy will leak from the bottom of the layer at a relatively high rate. As a result, the amplitude of the refracted wave will attenuate more rapidly with distance than it normally would. The refracted wave will quickly die out and disappear into the background noise (Press and Dobrin, 1956). At this point, the refracted arrival from a previously masked lower layer becomes the first arrival. Terrestrial examples of this phenomenon have been reported by Press and Dobrin (1956), Domzalski (1956), and Watkins and Spieker (1972) among others.

The thin-layer phenomenon is not unexpected in mare basalts. Stratigraphy in some terrestrial basalts indicates that thin-layer attenuation should occur. For example, cores from the Kana-a basalt flow near Flagstaff, Arizona (Watkins *et al.*, 1964), revealed a stratigraphy consisting about 2 m of cinders (velocity 300 m/sec) overlying 8 m of basalt and more cinders. The upper 5 m of basalt had a velocity of 900 m/sec while the lower 3 m had a velocity of 2700 m/sec. The difference in velocity is probably due to increased fracturing and vesicularity in the upper 5 m. The cinders had a velocity of 300 m/sec. Another core hole in the same flow successively penetrated 1 m of cinders (350 m/sec), 5 m of basalt (1–350 m/sec; 3–900 m/sec; 1–730 m/sec) and 6 m of cinders (730 m/sec). Only high frequency energy could propagate in these flows because of their small thicknesses.

Stratigraphy observed in Hadley Rille during Apollo 15 would lead to thin layer attenuation. Photographs of Hadley Rille showed 3 sequences of outcropping rocks in the upper 50 m of the rille wall (Swann *et al.*, 1972). These outcrops consisted of a sequence of massive units sandwiched between discontinuous exposures of less massive units. The massive sequence is about 25 m thick. The velocity of mare basalt at the Apollo 17 site is in excess of 800 m/sec and the dominant frequency less than 20 Hz. These values would require 40 m or more in thickness in order to efficiently transmit the seismic energy, or almost twice the thickness of the Hadley Rille sequence. Actually, the effective thickness of the massive interval in Hadley Rille may be significantly less than the total thickness if fracturing and vesicularity decrease the seismic velocity in the upper few meters. We suggest that a similar stratigraphy at Taurus-Littrow causes the anomalously high velocities observed in data from EP -2, -3, -4, and -8.

In a companion paper devoted to a discussion of deeper structure (Kovach and Watkins, 1973c) it was necessary to estimate an average velocity for the zone containing the thin, high-velocity layers because seismic refraction theory does not permit unique inversion of data where high-velocity layers are intercalated with low-velocity layers. Average shot point-array travel times for EPs -2, -3, -4, -8, the LM ascent, and the travel time to the geophone nearest EP-7 indicate that the average velocity in this zone is about 280 m/sec.

The second important difference derives from the fact that the EPs were detonated over a dipping subsurface layer. The dipping refractor causes arrivals from EPs detonated on the updip (western) side to be earlier than arrivals from EPs detonated at comparable distances on the downdip side. The earlier arrivals result from a smaller thickness of low-velocity regolith traversed by waves generated on the updip side. A dipping refractor coupled with the array-EP configuration causes the "shingle" effect (an echelon pattern) of arrival times shown in Fig. 1. The shingle effect does not occur unless the refracting layer dips. The analysis of these data differs slightly from analysis of conventional data. Analysis procedures are discussed below.

Analysis of the data in Fig. 1 is complicated by noise originating in the ALSEP electronics (probably in the LSPE firing-pulse transmitter) and picked up by LSPE detectors and/or cables. The noise makes it difficult to precisely pick the first arrivals from all EPs. Picks were especially difficult in EP-2 and -3 data. Therefore, EP-2 and -3 data are treated with less confidence than data from EP-4 and -8 where first arrivals are more distinct. Analysis of the noise suggests that some of it can be removed. Therefore, it may be possible to improve on the following interpretation at a later date.

In spite of the scatter in the first arrivals it is clear that for given distances arrivals from EPs detonated east of the array are later than arrivals from EPs detonated west of the array. As previously mentioned, this difference in arrival times suggests that the refractor is tilted to the east causing refracted waves from EP-2, -3 and the LM ascent to traverse a thicker low velocity layer (probably the regolith) than waves from EP-4 and -8. The dipping refractor suggestion is further supported by the obvious shingling of arrivals from EP-4 and -8. The shingling of EP-2, -3 and LM arrivals is less obvious but still apparent.

The shingling effect is explained in terms of the travel time equations for a two layer case with dipping refractor:

$$t = t_0 + mx$$

t = travel time; t_0 = shot-intercept time as shown in Fig. 1; m = slope; and x = distance from shot point to detector, where,

$$t_o = \frac{2z \cos \theta \cos \phi}{v_1} \quad \text{and} \quad m = \frac{\sin (\theta \pm \phi)}{v_2}$$

z = depth of refractor beneath the EP; v_1, v_2 = velocity in surface layer and underlying refractor, respectively; $\theta = \sin^{-1}(v_1/v_2)$; and ϕ = dip of the refractor; the sign in the expression $(\theta \pm \phi)$ is negative when shooting updip and positive when shooting downdip. All updip data have the same slope and all downdip data have the same slope. The downdip slope is always greater than the updip slope. The shingling effect derives from the fact that changes in z alone affect the t_0 terms since ϕ, v_1, and θ are all constant for a given model.

If arrivals from EP-2, -3 and the LM were better, the travel time data could be uniquely inverted to give the depth of the refractor and its dip. However, values from EPs east of the array were not considered sufficiently well established for inversion, so depths and dip were calculated by the approximate method discussed below.

If $v_2 \geqslant 600$ m/sec, $v_1 = 105$ m/sec, and the $\phi < 5°$, then $\cos \theta \cos \phi = 0.99 \pm 0.01$. It has previously been shown that the regolith velocity v_1 is relatively constant 105 m/sec. Analysis indicates that $v_0 \geqslant 600$ m/sec and $\phi < 5°$. Thus, z can be accurately estimated for EPs-4 and -8 from the t_0 term alone. Estimates of z were not made for EP -2, -3 and LM ascent arrivals.

A least squares fit of the EP-4 data (which had the least scatter of all data) yielded an apparent velocity v_2' of 560 m/sec and an intercept time (t_0) of 0.115 sec or a depth (z) of 6.2 m. Projecting a line with a slope of 1/560 through the mean position of EP-8 data points results in $t_0 = 0.220$ sec and $z = 11.8$ m. The distance from EPs -4 and -8 to the center of the array (geophone 3) were 215 and 122 m, respectively. Hence the dip is about 3.4°. Use of 3.4° in the "m" term yields a true velocity for the refractor of 910 m/sec which is well within the range of velocities observed in terrestrial basalts and close to the velocity observed in the next deeper refractor (1130 m/sec) at Taurus-Littrow (Kovach and Watkins, 1973c). The agreement of velocities support our interpretation of the layer as a basalt flow. Use of the above values to calculate arrival time from EP-2, -3, and LM ascent yielded approximately correct values. Better agreement (that shown in Fig. 1) was obtained by reducing the average dip to 3° which changed the average velocity of the refractor to 850 m/sec. The recalculated depth of the refractor beneath the center of the array (geophone 3) is 17.4 m.

Electrical properties of lunar near-surface rocks at the Apollo 17 site are consistent with at least two different model interpretations (Simmons et al., 1973). One model bears no resemblance to the seismic model of the regolith and subjacent layering, but the other model consists of an eastward dipping layer about 20 m below the ALSEP, a second interface about 300 m below the surface and no other interfaces up to depths of 1.5 km or more. This model closely resembles the Apollo 17 seismic model.

No craters penetrate the regolith within the area encompassing the geophone array and EPs -2, -3, -4, and -8, therefore we have no direct geological evidence of regolith thickness from the area of where seismic data were obtained. Outside the area, however, astronaut observations suggest that the regolith varies in depth from 5–20 m (ALGIT, 1973). For example, rims of craters less than 30 m in diameter or near LRV-1 (west of the array) were block free except for one crater 10–15 m in diameter whose rim included blocks with diameters up to 1 m (ALGIT, 1973). Data reported by Baldwin (1949) and Short (1970) indicate that small craters excavate to depths equal to 0.25–0.35 of their diameters. Therefore, regolith is probably 3–10 m thick in the LRV-1 area, a range of depths consistent with LSPE data. East of the array, 15–20 m of surficial fragmental material were exposed in the walls of Van Serg Crater. These data agree qualitatively with our interpretation of the seismic data to the extent that the regolith is shallower west of the array than it is east of the array. The dip of the refractor indicated by the seismic data is somewhat greater than the dip of the subfloor basalt indicated from crater data, however.

We believe that the answer to these apparently conflicting dip data derive from the constricted, sloping nature of the Taurus-Littrow valley floor coupled with an anomalously high density of medium-sized craters surrounding the array. The valley of Taurus-Littrow slopes from an elevation of about 4800 m in the northwest to an elevation of about 4220 m in the southeastern corner. Lee scarp, a prominent feature in the western part of the valley, slopes eastward at angles up to 17°, which is much steeper than the 3° indicated by the LSPE data. The undulating, dipping surface suggests that a local increase in slope may be responsible for the change in depth of the refractor beneath the LSPE.

The anomalous average depth of the refractor is probably due to an anomalous regolith thickness. The LSPE array and the four closest explosives are within an area with an unusually high density of both young craters and old, subdued craters. The array is underlain by ejecta from Camelot, Horatio, Trident, Sherlock, and others. The ejecta from these craters and others near the array is probably responsible for the anomalously large (17.4 m) regolith thickness observed by the astronauts and deduced from LSPE data.

We conclude that the substrate is probably subfloor basalt and that it has an undulating surface. The LSPE array is located over a portion of the subfloor basalt where the dip is slightly oversteepened to the east. Ejecta from an anomalously large number of medium-sized craters in the vicinity of the array smoothed the topography and created an unusually thick regolith.

Discussion

The regolith is a moonwide phenomenon. Photography from orbiting satellites reveals everywhere a smoothed, pockmarked lunar surface now associated with the mantling effect of the regolith. Bodies of rock larger than a few meters are exposed only in small areas or inside craters. Apollo seismic data suggest that the regolith observed in different lunar geological habitats is also remarkably uniform in its properties moonwide.

Low, uniform P-wave velocity

The P-wave velocity of the regolith is very low and uniform. The average velocity determined from Apollo data is 105 m/sec and the standard deviation is 8 m/sec (Table 1). Velocities in terrestrial soils have much greater variability. For example, Watkins and Spieker (1972) found that unconsolidated surface materials in the Miami River Valley of southwestern Ohio had velocities ranging from 460 m/sec to over 1000 m/sec. Standard deviations in specific areas of more or less homogeneous surface material ranged from 9 to 34% of the mean velocity. *In situ* velocity measurements of terrestrial rock outcrops show similar variations. Watkins *et al.* (1972) reported standard deviations equal to 14% of the mean velocity in outcrops of the welded, rhyolitic Bishop Tuff in eastern California, 27% of the mean velocity in Navajo Sandstone outcrops in northern Arizona, and 22% of the mean velocity in Amboy basalt outcrops in the Mojave Desert, California. Figure 2 shows scatter in lunar regolith and selected terrestrial rocks.

Brecciation and high porosity are probably the major causes of the extremely low velocities observed in the lunar regolith. Watkins and Kovach (1972) compared velocity data from brecciated rock (Apollo 14 regolith, Apollo 11 returned samples and ejecta from Meteor Crater, Arizona) with velocity data from terrestrial rocks of comparable porosity (Fig. 3). They found that the lunar regolith and Meteor Crater ejecta had significantly lower velocities than common terrestrial rocks with comparable porosities, a relationship which they attributed to brecciation of the Apollo 14 regolith and Meteor Crater ejecta.

Watkins *et al.* (1972) reported that only very fine, extremely porous (60%) terrestrial materials had velocities as low as 100 m/sec. The velocity in basaltic cinders was 300 m/sec, a value significantly greater than that of the lunar regolith. The low regolith velocities thus indicate that the regolith probably contains no significant amounts of volcanic ash at Apollo 12, 14, 15, and 16 sites. The regolith velocity was not established at the Apollo 17 site but absence of ash in returned samples from the Apollo 17 site (ALGIT, 1973) seems to preclude its existence there, too.

Table 1. P-wave velocity and thickness of the regolith at the Apollo sites. Although mare basalt underlies the Apollo 17 site, the site is in a small valley surrounded by highlands and is located within a local crater field which probably accounts for the unusual range of regolith thickness values.

Mission	Velocity (m/sec)		Thickness (meters)		Geologic Setting
	PSE	ASE	Craters	Seismic	
11	—	—	3–6	—	Mare
12	108	—	2–4	—	Mare
14	104	104	—	8.5	Highlands
15	92	—	5	—	Mare
16	—	114	—	12.2	Highlands
17	—	—	3–20	6.2–36.9	Restricted Mare

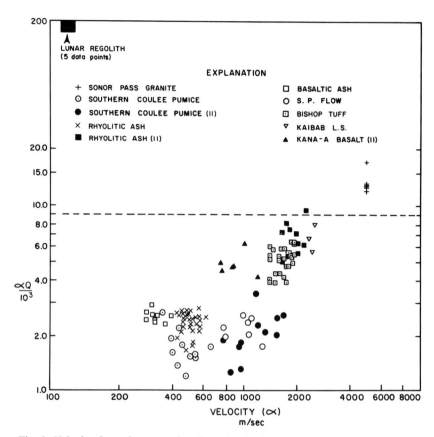

Fig. 2. Velocity-Q product as a function of velocity (α) for the lunar regolith and selected unsaturated *in situ* terrestrial rocks. Note the small scatter in the lunar regolith velocities relative to the scatter observed in the terrestrial rocks. A high Q is characteristic of lunar crustal rocks. (Q of the lunar regolith was provided by Y. Nakamura, personal communication, 1973.)

Regolith thickness

Variation in the thickness of the regolith is small. At the Apollo 11 site the regolith is estimated to be 3–6 m thick (LSPET, 1970); at the Apollo 12 site it is estimated to be 2–4 m thick (LSPET, 1970); at the Apollo 15 site measurements from photographs indicate that it is 5 m thick; ASE data indicate it is 8.5 m thick at the Apollo 14 site and 12.2 m thick at the Apollo 16 site; and anomalous Apollo 17 seismic data indicate an average depth of 10–15 m (*see* Table 1). The range of thickness is thus less than one order of magnitude. When maria and highland thicknesses are considered separately, the ranges in thickness are even less; the maria determinations differing by no more than a factor of 3 and the two highland determinations differing by a few tens of percent.

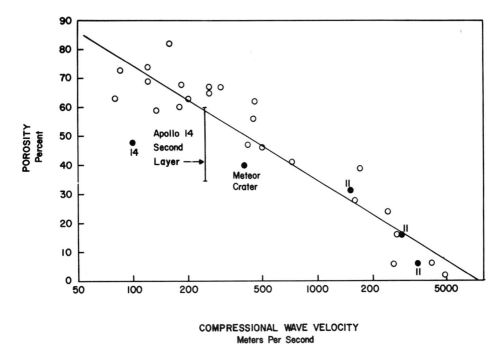

Fig. 3. Terrestrial velocity-porosity data (open circles) and lunar velocity-porosity data (solid circles). Numbers refer to Apollo missions that provided data. Note the progressively lower seismic velocities observed in shocked rocks (Apollo and Meteor Crater data) relative to the least-squares fit calculated for all terrestrial data. (Apollo 11 data are from Kanamori *et al.*, 1970).

Impact ejecta models and the origin of the regolith

The thickness of the regolith is an important parameter in attempts to deduce models of regolith formation, cratering mechanics, and meteorite flux. The many investigators who have studied the problem of crater formation and ejecta thicknesses have reached diverse conclusions. Part of the diversity derives from the use of different assumed crater and/or meteorite populations. Other differences derive from differing assumptions regarding the impact mechanism and the regolith formation process.

Among the investigators whose estimates agree reasonably well with observations of the regolith thickness are Gault (1970), Short and Forman (1972), and Hu (1971). Gault, using the Naumann–Hawkins meteorite flux rate to estimate the equilibrium ejecta accumulation due from craters up to 2.5 km in diameter, obtained an average ejecta thickness of 6 m at Sinus Medii. Short and Forman (1972), using volumes excavated from all lunar craters, estimated that the ejecta thickness would be 1–20 m on the maria and about 2 km in the highlands. Their estimate of ejecta thickness on the maria agree with observed regolith thicknesses, and their highland thickness estimate is in approximate agreement with the depth to "hard rock" (4 km/sec) determined from LSPE data. Hu (1971) developed an empirical model of the variation of ejecta thickness as a function of crater size and distance from the center of the crater. His estimates of total ejecta thicknesses from post-Imbrian impacts near Apollo 11, 12, and 14 sites agreed closely with observed thicknesses and support a post-Imbrian age for the regolith at these locations. Hu's calculated thicknesses and stratigraphic sequences agreed well with observed thicknesses and stratigraphic sequence in the Apollo 12 drive sample. This close agreement indicates that mixing of ejecta is not as effective as some investigators have suggested.

Age of the regolith

Since the oldest portion of the regolith overlies mare basalts ranging in ages from 3.1 to 3.7 billion years (Tera *et al.*, 1973), mare regolith ranges in age from 0–3.7 billion years. The regolith in the highlands appears to include slightly older ejecta.

Radiometric ages of soil from the Apollo 16 site fall mainly in a narrow range of ages between 3.9 and 4.2 billion years (Tera *et al.*, 1973; Hussain and Schaeffer, 1973). The soil from which these ages were determined probably consists mainly of ejecta from the 250 m/sec layer immediately subjacent to the regolith. Although we cannot calculate the exact thickness of the 250 m/sec layer, the fact that it transmits seismic energy peaking at 8 Hz indicates that it is at least 30 m thick. Since most nearby craters bottom at lesser depths, we can conclude that the 250 m/sec layer is the source of most of the surface fragments.

At the Apollo 14 site, the Fra Mauro Formation is thought to be 3.9 billion years (Alexander *et al.*, 1973). Thus the maximum age of overlying regolith ejecta is very nearly the same as that at the Apollo 16 site.

The greater thickness of the lunar regolith on the highlands relative to the younger regolith in the maria regions suggests that the rate of deposition of regolith was almost in order of magnitude greater during the period between the great cataclysm about 4 billion years ago (Tera *et al.*, 1973) and the middle of the period of maria formation (about 3.4 billion years ago) than during the subsequent time interval.

Compacted powder model

Unusual seismic properties of the lunar regolith and underlying crustal breccias led Gold and Soter (1970) and Gangi (1972) to suggest that the lunar near-surface consists of a deep layer of powder. Their powder model is gravitationally self-compressed in such a way as to provide a linear velocity increase with depth (Gold and Soter, 1970) or a velocity which increases as the sixth root of the depth (Gangi, 1972).

ASE and LSPE data do not support this model. Figure 4 shows velocity-depth models deduced from Apollo 14, 16, and 17 data. Depth is plotted on a logarithmic scale so that a power law velocity-depth function appears as straight line. From

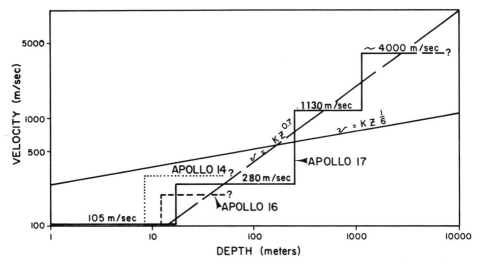

Fig. 4. Velocity-depth models deduced from Apollo 14, 16, and 17 data. Depth is plotted on a logarithmic scale in order to show power law velocity-depth functions as straight lines.

this figure it is clear that if one attempts to fit a sixth root velocity-depth function to the regolith and immediately subjacent rocks and breccia, the curve is badly in error at greater depths. Conversely, a sixth root function fitted in the 1130 and 4000 m/sec region is badly in error in the 100–300 m/sec region. It is possible to fit a power law velocity-depth function ($v = kz^{0.7}$) over much of the range of the lunar crust but this is probably coincidental. A linear increase of velocity with depth fits the data worse than the sixth root curve.

A major reason for the recurrence of the compacted dust hypothesis is the extremely low seismic velocity of near-surface rocks. Although seismic velocities in lunar near-surface rocks are low by terrestrial standards, comparable velocities have been observed in terrestrial rocks with extremely high porosities (Fig. 3).

Brecciation tends to enhance the decrease in velocity accompanying an increase in porosity. Todd et al. (1973) have observed anomalously low seismic velocities in shocked rock samples from the Ries impact crater and from lunar samples, and Watkins et al. (1972) observed similar results during in situ measurements of velocities in shocked rocks near Meteor Crater.

In Fig. 3, velocities reported by Kanamori et al. (1970) proportionately average only slightly less than velocities observed in unshocked terrestrial rocks of comparable porosity; Meteor Crater rock velocities average about 50% of comparable unshocked rock; and the lunar regolith velocities average about 20% of comparable unshocked rock. These data indicate that the relative decrease in velocity due to brecciation increases in magnitude with increasing porosity.

Interpolation of Meteor Crater and Apollo regolith data suggests that unshocked rock comparable to the 299 m/sec and 250 m/sec layers observed at Apollo 14 and 16 sites would have velocities between 500 and 800 m/sec. These velocities are comparable to velocities observed in unsaturated terrestrial soils. A layer of very porous soil overlying a layer of normal soil would thus constitute a good velocity analog of unshocked lunar regolith and unshocked subjacent breccia at the Apollo 14 and 16 sites. Without the reduction in velocity due to brecciation, the velocities of the lunar regolith and substrate at the Apollo 14 and 16 sites are within the normal range of velocities observed in terrestrial soils.

We conclude that the anomalously low *in situ* velocities observed in the lunar regolith and in lunar breccias result primarily from shock induced microcracks and high porosity. Without microcracks, velocities in these rocks would be comparable to velocities in terrestrial soils.

SUMMARY AND CONCLUSIONS

Synthesis of seismic, geological, and crater ejecta calculations suggest the following conclusions.

1. The regolith consists primarily of impact ejecta having an average compressional wave velocity of 105 m/sec. Pyroclastics do not appear to comprise a significant portion of the ejecta in the areas investigated.

2. In the maria areas the regolith averages between four and five meters in thickness and represents the sum of the post-maria ejecta blankets. The regolith was deposited during the interval of 0 to 3.7 billion years ago.

3. In the highlands the regolith thickness averages about 10 m. The highland regolith represents the sum of post-cataclysm ejecta blankets and was formed during the interval between 0 and 4 billion years ago. Differences in thicknesses of highland and maria suggest that the regolith deposition rate was approximately an order of magnitude greater in the period between 3.5 and 4 billion years ago than during the period 3.5 billion years ago to present.

4. The regolith in the valley of Taurus-Littrow (Apollo 17) is anomalously thick, ranging from 6.2 to 36.9 m in the area of investigation. The anomalous thickness is attributed to the combined effects of an unusually high concentration of medium-sized craters and irregularities in the surface of underlying mare basalts.

5. Apollo seismic data do not support the compacted power model of the lunar crust but indicate that lunar near-surface regolith, breccia, and crust consist of discrete layers. Anomalously low seismic velocities observed in the regolith and subjacent breccia are due to impact fracturing.

REFERENCES

Alexander E. C., Jr., Davis P. K., Reynolds J. H., and Srinauason B. (1973) Age exposure history, and trace element composition of some Apollo 14 and 15 rocks as determined from rare gas analysis (abstract). In *Lunar Science—IV*, pp. 27–28. The Lunar Science Institute, Houston.

Apollo Lunar Geology Investigation Team (1973) Documentation and environment of the Apollo 17 samples: A preliminary report. U.S. Geol. Survey Interagency Rpt., *Astrogeology* 71, 322 pp.

Domzalski W. (1956) Some problems of shallow refraction investigation. *Geophys. Prosp.* 4, 140–166.

Gangi A. F. (1972) The lunar seismogram. *The Moon* 4, 40–48.

Gold T. and Soter S. (1970) Apollo 12 seismic signal: Indication of a deep layer of powder. *Science* **169**, 1071–1075.

Hu T. (1971) The crater ejecta component in the lunar regolith. Univ. North Carolina Master's Thesis, 41 pp.

Hu T. and Watkins J. S. (1971) The crater ejecta component in the lunar regolith. *Trans. Am. Geophys. Union* **52**, 273.

Hussain L. and Schaeffer O. A. (1973) ^{40}Ar–^{39}Ar crystallization ages and ^{38}Ar–^{37}Ar cosmic ray exposure ages of samples from the vicinity of the Apollo 16 landing site (abstract). In *Lunar Science—IV*, pp. 406–407. The Lunar Science Institute, Houston.

Kanamori H., Nur A., Chung D., Wones D., and Simmons G. (1970) Elastic wave velocities of lunar samples at high pressures and their geophysical implications. *Science* **167**, 726–728.

Kovach R. L. (1967) Lunar seismic exploration. *Physics of the Moon* (editor S. F. Singer). *Am. Astronautical Soc.* **13**, 189–198.

Kovach R. L., Watkins J. S., and Landers T. (1971) Active seismic experiment. In *Apollo 14 Preliminary Science Report*, NASA SP-272, pp. 163–174.

Kovach R. L., Watkins J. S., and Talwani P. (1973a) Active seismic experiment. In *Apollo 16 Preliminary Sciences Report*, NASA SP-315, Chap. 10, 14 pp.

Kovach R. L., Watkins J. S., Nur A., and Talwani P. (1973b) The properties of the shallow lunar crust: An overview from Apollo 14, 16 and 17 (abstract). In *Lunar Science—IV*, pp. 444–445. The Lunar Science Institute, Houston.

Kovach R. L. and Watkins J. S. (1973c) The structure of the lunar crust at the Apollo 17 site. *Proc. Fourth Lunar Sci. Conf., Geochim. Cosmochim. Acta.* This volume.

Latham G. V., Ewing M., Press F., Sutton G., Dorman J., Nakamura Y., Toksöz N., Wiggins R., and Kovach R. (1970) Passive seismic experiment. In *Apollo 12 Preliminary Science Report*, NASA SP-235, pp. 39–54.

Latham G. V., Ewing M., Press F., Sutton G., Dorman J., Nakamura Y., Toksöz N., Duennebier F., and Lammlein D. (1971) Passive seismic experiment. In *Apollo 14 Preliminary Science Report*, NASA SP-272, pp. 133–162.

Latham G. V., Ewing M., Press F., Sutton G., Dorman J., Nakamura Y., Toksöz N., Lammlein D., and Duennebier F. (1972) Passive seismic experiment. In *Apollo 15 Preliminary Science Report*, NASA SP-289, Chap. 8, 25 pp.

Lunar Sample Preliminary Examination Team (1970) Preliminary examination of lunar samples from Apollo 12. *Science* **167**, 1325–1339.

McAllister D. B., Kerr J., Zimmer J., Kovach R. L., and Watkins J. S. (1969) A seismic refraction system for lunar use. *Trans. IEEE,* **GE-7**, pp. 164–171.

Press F. and Dobrin J. (1956) Seismic wave studies over a high-speed surface layer. *Geophysics* **21**, 285–298.

Short N. M. (1970) Anatomy of a meteorite impact crater: West Hawk Lake, Manitoba, Canada. *Bull. Geol. Soc. Am.* **81**, 609–648.

Short N. M. and Forman M. L. (1972) Thickness of impact crater ejecta on the lunar surface. *Mod. Geol.* **3**, 69–91.

Simmons G., Baker R. H., Bannister L. A., Brown R., Kong J. A., LaTorraca G. A., Tsang L., Strangway D. W., Cubley H. D., Annan A. P., Redman J. D., Rossiter J. R., Watts R. D., and England A. W. (1973) Electrical structure of Taurus-Littrow (abstract). In *Lunar Science—IV*, p. 675. The Lunar Science Institute, Houston.

Swann G. A., Bailey N. G., Batson R. M., Freeman V. L., Haiit M. H., Head J. W., Holt H. E., Howard K. A., Irwin J. B., Larson K. B., Muehlberger W. M., Reed V. S., Rennilson J. J., Schaber G. G., Scott D. R., Silver L. T., Sutton R. L., Ulrich G. E., Wilshire H. G., and Wolfe E. W. (1972) Preliminary Geologic Investigation of the Apollo 15 landing site. In *Apollo 15 Preliminary Science Report*, NASA SP-289, Chap. 5, 112 pp.

Tera F., Papanastassiou D. A., and Wasserburg G. J. (1973) A lunar cataclysm at 3.95 Ae and the structure of the lunar crust (abstract). In *Lunar Science—IV*, pp. 723–725. The Lunar Science Institute, Houston.

Todd T., Wang H., Richter D., and Simmons G. (1973) Unique characterizations of lunar samples by physical properties (abstract). In *Lunar Science—IV*, pp. 731–733. The Lunar Science Institute, Houston.

Watkins J. S. and Kovach R. L. (1972) Apollo 14 active seismic experiment. *Science* 175, 1244–1245.

Watkins J. S. and Spieker A. M. (1972) Seismic refraction survey of pleistocene drainage channels in the lower Great Miami River Valley, Ohio. *U.S. Geol. Survey Prof. Paper* 605-B, 17 pp.

Watkins J. S., Loney R. A., and Godson R. H. (1964) Investigation of *in situ* physical properties. *Proj. Ann. Rept.*, 1964. *U.S. Geol. Survey* open-file rpt., 78 pp.

Watkins J. S., Walters L. A., and Godson R. H. (1972) Dependence of *in situ* compressional-wave velocities on porosity in unsaturated rocks. *Geophysics* 37, 29–35.

Watkins J. S., Whitcomb J. H., Weeks E. L., Graves T. J., Kovach R. L., and Godson R. H. (1969) A lunar engineering seismic system. *Bull. Assoc. Eng. Geol.* 6, 119–130.

Proceedings of the Fourth Lunar Science Conference
(Supplement 4, *Geochimica et Cosmochimica Acta*)
Vol. 3, pp. 2575–2590

Shock compression of a recrystallized anorthositic rock from Apollo 15*

THOMAS J. AHRENS,[1] JOHN D. O'KEEFE,[2] and REX V. GIBBONS[1]

[1]Seismological Laboratory, California Institute of Technology, Pasadena, California 91109
[2]Department of Space Sciences, University of California, Los Angeles, California 90000

Abstract—Hugoniot measurements on 15,418, a recrystallized and brecciated gabbroic anorthosite, yield a value of the Hugoniot elastic limit (HEL) varying from ~ 45 to 70 kbar as the final shock pressure is varied from 70 to 280 kbar. Above the HEL and to 150 kbar the pressure-density Hugoniot is closely described by a hydrostatic equation of state constructed from ultrasonic data for single-crystal plagioclase and pyroxene. Above ~ 150 kbar, the Hugoniot states indicate that a series of one or more shock-induced phase changes are occurring in the plagioclase and pyroxene. From Hugoniot data for both the single-crystal minerals and the Frederick diabase, we infer that the shock-induced high-pressure phases in 15,418 probably consist of a 3.71 g/cm^3 density, high-pressure structure for plagioclase (An$_{93}$) and a 4.70 g/cm^3 perovskite-type structure (En$_{64}$) for pyroxene. Using the Kelly–Truesdell mixture theory we separately calculated the entropy production in each phase, and predict incipient and complete melting in the plagioclase occurs upon release from ~ 500 and ~ 600 kbar. For the pyroxene component, incipient and complete melting occurs upon release from 700 and 850 kbar. The onset of shock-induced vaporization will occur upon release from ~ 1300 kbar and would require the impact of an iron meteoroid traveling at a velocity of ~ 8 km/sec.

INTRODUCTION

THIS IS THE first report of shock wave data taken on a rock sample returned from the moon. The major motivation for determining the shock wave equation of state of such rocks stems from the widespread occurrence of impact effects on both large and small scales on the lunar surface. Although the large scale cratering features on the surface of the moon were recognized well before the U.S. and Soviet programs of sample collection, the wide variety of effects resulting from the passage of an intense pressure pulse associated with impact have only been recognized upon analysis of returned rock and soil samples over the last four years. Among the shock effects observed in these samples are melting (e.g., Chao *et al.*, 1972; von Engelhardt *et al.*, 1971; King *et al.*, 1972; Chao *et al.*, 1970), cataclastic crushing (von Engelhardt *et al.*, 1972; LSPET, 1973), intense crushing, twinning and solid-state reactions (von Engelhardt, 1970, 1971, 1972). Also, geochemical evidence points to shock vaporization as a likely explanation for the apparent mobility of Pb, K, Rb, Th, and U in the regolith (Silver, 1972; Doe and Tatsumoto, 1972). Undoubtedly, the most important shock effect on the moon is melting.

---

*Contribution No. 2142, Division of Geological and Planetary Sciences, California Institute of Technology.

The shock measurements on lunar materials which are important in relating impact parameters (shock impedance and impact velocity of the meteoroid) to their resulting effects on the moon include: (a) the Hugoniot elastic limit, or, the maximum stress achievable in the rock under one-dimensional compression without internal rearrangement taking place at the shock front. Although some twinning and other planar deformation features result from shock stresses lower than this critical stress, large scale shock-produced deformation and flow takes place only above the HEL, and (b) the high pressure equation of state along the deformational Hugoniot. Knowledge of the equation of state of the rock in terms of its component minerals permits correlation of the shock pressure with the onset and completion of melting and vaporization of the constituent minerals. In the case of many of the lunar surface materials, the unique mineralogic and physical characteristics, as compared to terrestrial materials, e.g., high ferrosilite content of pyroxene combined with appreciable porosity, have made the high pressure Hugoniot difficult to predict theoretically.

In this paper we summarize our experimental results for lunar gabbroic anorthosite and describe its Hugoniot data and that of a somewhat similar terrestrial gabbro in terms of equations of state of the major constituents, plagioclase and pyroxene.

Samples and Experimental Method

Rock 15,418, collected from site 7 (LSPET, 1972) on the Hadley delta, appears to be a previously shocked and recrystallized rock of gabbroic anorthositic composition. It is quite possible that this rock was in fact ejected via one or more impacts from an initial position in the lunar highlands. Petrographic analysis yields the following mineralogic composition: plagioclase $74 \pm 5\%$, pyroxenes plus a small fraction olivine, $25 \pm 5\%$, opaque minerals (iron-nickel, troilite) $0.1 \pm 0.05\%$ and vesicles $1.0 \pm 0.5\%$. Microprobe analysis of the plagioclase (An_{93}), the ortho- and clino-pyroxenes, and olivine are given in Table 1.

Twelve individual samples, approximately 1 cm square and 3.5 mm thick, were machined out of 15,418. Upon carefully measuring the weighing, we determined their average density as 2.825 g/cm³. The density of a single sample did not differ from the mean by more than 0.015 g/cm³. The above density includes the volume of vesicles.

After internal inspection using X-ray radiography, the samples were mounted with flat mirrors and an inclined mirror to measure shock and free surface velocities on 1.5 mm thick polycrystalline tungsten or aluminum alloy (2024) plates (Table 2). These assemblies were impacted with 2.5 mm thick flyer plates of the same material at speeds ranging from 0.8 to 2.2 km/sec. A typical record resulting from the streak photography is shown in Fig. 1. In all cases an elastic precursor, whose amplitude is taken to represent the Hugoniot elastic limit (HEL), was detected. The final shock state was determined using the impedance match method (Rice et al., 1958). The detailed analysis of the resulting streak camera film was similar to that used by Ahrens et al. (1968).

Experimental Results

The velocity of the elastic precursor varies from 5.9 to 6.3 km/sec for the different samples. The variation undoubtedly reflects slight variations in crack porosity and crack geometry along the 3.5 mm shock propagation path. These val-

Table 1. Analysis of constituent minerals in 15,418.*

Plagioclase		Clinopyroxene		Orthopyroxene		Olivine	
(oxide)	(wt.%)	(oxide)	(wt.%)	(oxide)	(wt.%)	(oxide)	(wt.%)
SiO_2	44.7 ± 0.1	SiO_2	52.3 ± 0.9	SiO_2	51.7	SiO_2	36.4 ± 0.6
Al_2O_3	35.5 ± 0.8	FeO	22.3 ± 0.2	CaO	21.1	FeO	39.3 ± 0.8
CaO	19.90 ± 0.08	MgO	21.4 ± 0.7	MgO	14.2	MgO	24.9 ± 1.0
Na_2O	0.39 ± 0.02	CaO	1.6 ± 0.1	FeO	10.2	CaO	0.7 ± 0.6
FeO	0.25 ± 0.08	Al_2O_3	1.5 ± 0.6	Al_2O_3	2.0	MnO	0.48 ± 0.01
MgO	0.06 ± 0.04	TiO_2	0.54 ± 0.05	TiO_2	1.1	Al_2O_3	0.4 ± 0.4
K_2O	0.019 ± 0.01	MnO	0.39 ± 0.01	Cr_2O_3	0.85	Cr_2O_3	0.06 ± 0.01
BaO	0.01 ± 0.01	Cr_2O_3	0.24 ± 0.01	MnO	0.2	Na_2O	0.003 ± 0.003
		Na_2O	0.08 ± 0.03	Na_2O	0.04		

*A. Chodos, California Institute of Technology, Analyst.

ues are significantly higher than the elastic precursor velocity of similar porosity terrestrial basalts reported by Ahrens and Gregson (1964) and Bass et al. (1963). The amplitude of the elastic precursor (HEL) (varying from 42 to 71 kbar) is also higher than somewhat similar terrestrial materials, where values between 40 and 50 kbar are reported. Although we have not carried out detailed analysis, we attribute these differences to the lack of hydrous mineral alteration products and the higher CaO content of the lunar material. In addition, there appears to be a moderate strain-rate effect on the value of the HEL (Fig. 2). For the 3 shots in which a final amplitude below 90 kbar was achieved, the average HEL value is 48 kbar, while the seven values of the HEL measured for experiments in which final shock states were above 120 kbar had an average value of 66 kbar. Similar effects have been noted in studies of the shock compression of quartz (Wackerle, 1962; Ahrens and Duvall, 1966) and feldspar (Ahrens and Liu, 1973).

Above the HEL to 150 kbar, the achieved deformational shock states agree closely with those reported earlier for Vacaville basalt (Ahrens and Gregson, 1964) and Centreville and Frederick diabase (McQueen et al., 1967). (See Fig. 3.) In the next section we show that these states are also in good agreement with the theoretical Hugoniot predicted from the mineralogic constituent equations of state. Above ~ 150 kbar the data suggests that one or more phase changes take place in the lunar (and terrestrial) rocks. Since above a pressure of ~ 135 and ~ 140 kbar the data for pyroxene (Ahrens and Gaffney, 1971; McQueen et al., 1967) and plagioclase feldspars (Ahrens et al., 1969a) demonstrate the occurrence of phase changes, this result is expected. The possible nature of these phase changes is discussed below.

A THEORETICAL DESCRIPTION OF THE EQUATION OF STATE

In order to theoretically describe the equation of state of a rock such as 15,418, which is composed of essentially two minerals, plagioclase and pyroxene, a

Table 2. Hugoniot data, lunar sample 15,418.

Shot No.	Initial Density (g/cm³)	Flyer Plate Velocity (km/sec)	Elastic Shock Velocity (km/sec)	Free-Surface Velocity (km/sec)	Hugoniot Elastic Limit (kb)	Final Shock Pressure (kb)	Final Shock Density (g/cm³)
270	2.821	1.618ᵃ ±0.005	5.88 ±0.09	0.84	70±4	204±4	3.82±0.07
276	2.834	1.318ᵃ ±0.001	6.02 ±0.12	*	*	155±4	3.69±0.08
268	2.813	2.166ᵃ ±0.005	6.30 ±0.01	0.60	65±10	282±6	4.25±0.04
269	2.822	1.992ᵃ ±0.005	6.10 ±0.05	*	*	261±5	4.07±0.04
279	2.846	1.139ᵇ ±0.005	5.94 ±0.10	0.49	42±5	88±3	3.22±0.06
280	2.823	0.850ᵇ ±0.0015	6.04 ±0.05	0.56	48±5	65±2	3.08±0.05
281	2.812	0.803ᵇ ±0.002	6.18 ±0.02	0.61	53±2	63±1	3.03±0.01
277	2.821	1.020ᵃ ±0.005	5.99 ±0.04	0.84	71±10	129±7	3.37±0.06
287	2.823	1.17ᵃ ±0.006	6.24 ±0.11	0.65	57±6	148±8	3.48±0.08
288	2.806	1.108ᵃ ±0.005	6.14 ±0.02	0.80	69±11	145±7	3.38±0.04

ᵃPolycrystalline W, 19.3 g/cm³.
ᵇAluminum alloy, 2024.
*Not measured.

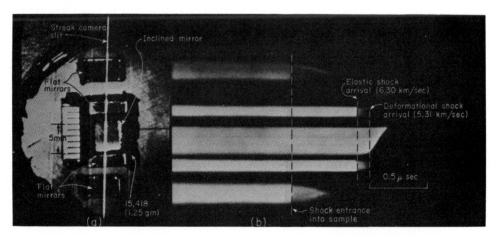

Fig. 1. Streak camera photograph, shot 268, Apollo sample 15,418. (a) Still photograph of specimen viewed through streak camera. Slit image is swept across film. Mirror reflectivity is destroyed by shock wave. (b) Resulting streak photograph showing streak cutoffs of mirror reflectivities. Shock pressures of 65±10 and 282±6 kbar were determined for the elastic and deformational shock wave from this record.

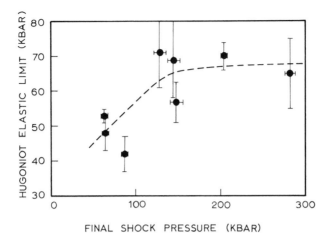

Fig. 2. Hugoniot Elastic Limit (HEL) as a function of final shock pressure in 15,418.

theoretical framework was chosen which uses an appropriate mixture of the equations of state of the constituents. The extension of the mixture theory of Truesdell (1957) by Kelly (1964) has been used here to synthesize the Hugoniots of Frederick diabase and lunar gabbroic anorthosite and elsewhere to synthesize the Hugoniots of a number of other rocks (O'Keefe and Ahrens, 1973). Similar analyses have been carried out on water-saturated tuff (Garg and Kirsh, 1971) and composite materials (Tsou and Chou, 1969, 1970; Torvik, 1969; Davis and Wu,

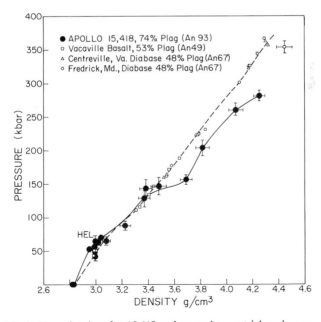

Fig. 3. Hugoniot data for 15,418 and several terrestrial analogues.

1972). In using this theory, each of the mineralogical constituents of a rock is modeled as a separate continuum and the interactions between continua accounted for. The pertinent theory and accompanying assumptions are outlined below. Application to a terrestrial analog of the lunar sample and the lunar sample is also presented.

We assume that the total mass density of the rock is expressed as the sum of the mass densities of each mineral per unit volume of total mixture,

$$\rho = \sum_{\alpha=1}^{N} \rho_{(\alpha)} \tag{1}$$

where N is the number of minerals. Upon being encompassed by a shock, the mean velocity of the rock is given by the mass weighted average

$$V^i \equiv C_{(\alpha)} V^i_{(\alpha)} \tag{2}$$

where $V^i_{(\alpha)}$ is the velocity component of the S_α mineral and $C_{(\alpha)}$ is the mass fraction defined by

$$C_{(\alpha)} \equiv \frac{\rho_{(\alpha)}}{\rho} \tag{3}$$

The vector component of diffusion velocity U^i_α of a single mineral relative to the mean rock particle velocity is

$$U^i_{(\alpha)} \equiv V^i_{(\alpha)} - V^i \tag{4}$$

The component of the total stress tensor is

$$t^{ij} \equiv \sum_{\alpha=1}^{N} (t^{ij}_{(\alpha)} - \rho_{(\alpha)} U^i_{(\alpha)} U^i_{(\alpha)}) \tag{5}$$

where $t^{ij}_{(\alpha)}$ is called the partial stress tensor for the S_α constituent. The internal energy, ϵ, is defined as

$$\epsilon \equiv \sum_{\alpha=1}^{N} C_{(\alpha)} (e_{(\alpha)} + \tfrac{1}{2} U_{(\alpha)i} U^i_{(\alpha)}) \tag{6}$$

where $e_{(\alpha)}$ is the internal energy per unit mass within a given mineral.

In the case of a shock front in a rock, the balance of mass for each mineral is

$$[\rho_{(\alpha)} U_{(\alpha)}] = \beta_{(\alpha)} \tag{7}$$

where β is the surface supply of mass, and the notation, $[\Psi_{(\alpha)}] \equiv \Psi^+_{(\alpha)} - \Psi^-_{(\alpha)}$ implies that Ψ^+ and Ψ^- are the limiting values of the function in brackets at two points on either side of the shock propagating at velocity U_s, and

$$U_{(\alpha)} \equiv U_s - V_{(\alpha)N} \tag{8}$$

where $V_{(\alpha)N}$ is the particle velocity of the mineral, S_α, normal to the shock surface.

The balance of momentum for each constituent is

$$[\rho_{(\alpha)} U_{(\alpha)} V^i_{(\alpha)}] + [t^{ij}_{(\alpha)}] N_j = \gamma^i_{(\alpha)} \tag{9}$$

where $\gamma^i_{(\alpha)}$ is the surface supply of momentum and N_j is the component of a unit vector normal to the shock surface.

The balance of energy is given by

$$[\rho_{(\alpha)}(e_{(\alpha)} + \tfrac{1}{2} U_{(\alpha)i} U^j_{(\alpha)}) U_{(\alpha)}] + [t^{ij}_{(\alpha)} V_{(\alpha)i}] N_j = \hat{\epsilon}_{(\alpha)} \tag{10}$$

where $\hat{\epsilon}_{(\alpha)}$ is the surface supply energy for mineral S_α and the assumption has been made that no heat transfer occurs within a given mineral.

The above equations are in terms of mixture constituent mass densities and partial stresses. We will now relate these quantities to the mineral properties.

The mass density in terms of the crystal density $\hat{\rho}_{(\alpha)}$, is defined as

$$\hat{\rho}_{(\alpha)} \equiv n_{(\alpha)} \rho_{(\alpha)} \tag{11}$$

where $n_{(\alpha)}$ is the ratio of the volume of a constituent to the volume of the mixture. The volume fraction has the following relationship

$$\sum_{\alpha=1}^{N} n_{(\alpha)} = 1 \tag{12}$$

The partial stress tensor in terms of the crystal stress tensor $t^{ij}_{(\alpha)}$, is given by

$$t^{ij}_{(\alpha)} = n_{(\alpha)} \hat{t}^{ij}_{(\alpha)} \tag{13}$$

The crystal internal energy $\hat{\epsilon}_{(\alpha)}$ is the same as the mixture internal energy since they are both defined per unit mass, i.e.,

$$\epsilon_{(\alpha)} = \hat{\epsilon}_{(\alpha)} \tag{14}$$

We assume that the stress state behind the shock is hydrodynamic

$$\hat{t}^{ij}_{(\alpha)} = -P_{(\alpha)} I^{ij} = -n_{(\alpha)} P I^{ij} \tag{15}$$

where I^{ij} is the unit tensor, $P_{(\alpha)}$ is the partial pressure and P is the total pressure which is the same for all minerals. In addition to the above equations, the equations of state of each constituent are required. The Mie–Grüneisen equation of state was used for each constituent

$$P_{(\alpha)} = P_{(\alpha)}(T = 300°K) + \rho_{0(\alpha)} \gamma_{(\alpha)} \{ e_{(\alpha)}(\rho_{(\alpha)}, T_{(\alpha)}) - e_{TR(\alpha)} - e_{0(\alpha)} \} \tag{16}$$

where $P_{(\alpha)}$ $(T = 300°K)$ is the 300°K isotherm, $e_{TR(\alpha)}$, the transition energy in the case of complete phase changes, and $\gamma_{(\alpha)}$, the Grüneisen parameter (Ahrens et al., 1969b). We assume that the 300°K isotherm is described by the Birch-Murnaghan equation (Birch, 1952) in which K_0 is the zero pressure isotherm bulk modulus and ξ is

$$3 - \frac{3}{4} \frac{\partial K_0}{\partial P}$$

The internal energy as a function of density and temperature was calculated from Debye theory

$$e_{(\alpha)}(\rho_{(\alpha)}, T_{(\alpha)}) = \int_{T=0°K}^{T(\alpha)} C_V\left(\frac{\theta_D}{T}\right) dT \tag{17}$$

T. J. Ahrens *et al.*

where C_V is the Debye form of specific heat at constant volume. The dependence of the Debye temperature, θ_D, and the Grüneisen parameter on volume is given by Davies' (1973) extension of Eulerian filnite strain theory as

$$\theta_{D(\alpha)} = \theta_{DO(\alpha)}(1 + g_{(\alpha)}f_{(\alpha)} + \tfrac{1}{2}i_{(\alpha)}f_{(\alpha)}^2)^{1/2} \tag{18}$$

and

$$\gamma_{(\alpha)} = \frac{(1 + f_{(\alpha)})(g_{(\alpha)} + i_{(\alpha)}f_{(\alpha)})}{6(1 + g_{(\alpha)}f_{(\alpha)} + \tfrac{1}{2}i_{(\alpha)}f_{(\alpha)}^2)} \tag{19}$$

where we assumed for the high pressure phases that

$$\theta_{DO(\alpha)} = 164 \cdot \rho_{O(\alpha)} \qquad \text{(Anderson, 1965)} \tag{20}$$

and $f_{(\alpha)}$ is defined as a strain parameter linear in atomic displacements

$$f_{(\alpha)} \equiv \left(\frac{\rho_{O(\alpha)}}{\rho_{(\alpha)}}\right)^{1/3} - 1 \tag{21}$$

and

$$g_{(\alpha)} = -6\gamma_{O(\alpha)} \tag{22}$$

$$i_{(\alpha)} = g_{(\alpha)}\left[3\left(\frac{d \ln}{d \ln V_{(\alpha)}}\right)_0 + g_{(\alpha)} + 2\right] \tag{23}$$

We also assume that (Davies and Gaffney, 1973)

$$\left(\frac{d \ln \gamma}{d \ln V_{(\alpha)}}\right) = 1 \tag{24}$$

The individual constituent balance equations when summed over all constituents are required to give the well known Rankine-Hugoniot equations for a single continuum. This requires the additional set of constraints

$$\sum_{\alpha=1}^{N} \beta_{(\alpha)} = 0$$

$$\sum_{\alpha=1}^{N} \gamma_{(\alpha)}^i = 0 \tag{25}$$

$$\sum_{\alpha=1}^{N} \hat{\epsilon}_{(\alpha)} = 0$$

Because individual minerals have different equations of state, the shock temperatures are, in general, different. If the shock wave has a duration much greater than the relaxation time for thermal transport between minerals then the constituents will be in thermal equilibrium, that is, they would all have the same temperature. However, meteoroid impacts involving centimeter-sized objects and laboratory experiments have shock wave durations of the order of a microsecond, which precludes significant thermal transport between minerals. With the goal in mind of examining the present data, we will assume that there is no thermal transport between constituents,

$$\hat{\epsilon}_{(\alpha)} = 0 \tag{26}$$

In addition, we have assumed that the rock is in dynamic equilibrium,

$$U^i_{(\alpha)} = 0 \tag{27}$$

that is, all of the particle velocities behind the shock wave are the same, and the diffusion velocities of Eq. 4 are zero. The assumption that the rock is in dynamic equilibrium is very reasonable for rocks of low porosity. A violation of this assumption would imply that in a shocked rock dissimilar minerals separate relative to their unshocked positions. This condition has not been observed in thin section analysis of shocked specimens. The balance equations (Eqs. 7–10) along with the equations of state (Eq. 18) and the assumptions (Eqs. 26, 27) can be shown to result in $2N + 2$ equations in $2N + 3$ unknowns. In calculating a mixture Hugoniot, we have chosen to fix the pressure, P, and calculate the $2N + 2$ unknowns, $\rho_{(\alpha)}, T_{(\alpha)}, U_s,$ and V. The use of the above synthesis technique requires a knowledge of the equation of state parameters ($K_0, K', \gamma_0,$ and ξ_0) of each mineral and, implicitly, the crystal structure. The crystal structures and equation of state parameters and the uncertainties in their values are discussed by Ahrens *et al.* (1969b) and Davies and Gaffney (1973). In the following analysis, we make calculations using several possible hpp crystal structures and compare these to the experimental data. However, non-uniqueness with regard to resolving the hpp crystal structures of multi-mineralic rocks exists with the present treatment.

THEORETICAL EQUATIONS OF STATE AND PREDICTED SHOCK CONDITIONS FOR MELTING AND VAPORIZATION

We have used the mixture theory described in the previous section to synthesize the Hugoniots of Frederick diabase and the lunar gabbroic anorthosite. Previous studies of terrestrial basalts, diabases, and other plagioclase-bearing rocks (Ahrens and Gregson, 1964; McQueen *et al.*, 1967; Ahrens *et al.*, 1969b) have demonstrated that the Hugoniot can be considered as describing states in essentially three regimes, a low pressure (untransformed) regime, a mixed phase regime, and a fully transformed or high pressure phase regime (Ahrens *et al.*, 1969b).

Because our experimental Hugoniot data for 15,418 extend only over the low pressure and mixed phase regimes, a mixture theory calculation was carried out for the high pressure phase regime. In order to have confidence in this calculation, we computed the Hugoniot of a similar but less plagioclase-rich rock, Frederick diabase. For this rock, which is well-characterized mineralogically, Hugoniot data are available in the mixed and high pressure phase regime (McQueen *et al.*, 1967).

The crystal density of 15,418 was calculated by using both (a) the results of our microprobe analysis (Table 1) of plagioclase, clino and ortho-pyroxenes and olivine and the normative compositions from the LSPET (1972) analysis and (b) the LSPET normative compositions and the whole rock analysis upon which these are based. Using the molar volumes in Clark (1967), zero-pressure crystal densities of 2.92 g/cm^3 and 2.93 g/cm^3 are obtained for cases (a) and (b), respectively. The porosity was determined to be 3.25% ($m = 1.0325$). The 2.92 g/cm^3 value for the density was adopted in the construction of the equation of state model

T. J. Ahrens *et al.*

Table 3. Volume fractions, observed and assumed, in theoretical models.

Mineral	Frederick Diabase[a]		Apollo 15,418	
	Observed[a]	Model	Observed	Model
Anorthite	0.45	0.50 (an$_{93}$)	0.74	0.74 (an$_{93}$)
Pyroxene	0.49	0.50 (en$_{93}$)	0.25	0.26 (en$_{64}$)
Mica	0.03	—	—	—
Olivine	0.01	—	—	—
Iron–Nickel troilite	—	—	0.01	—

[a]McQueen *et al.*, 1967.

described below. The observed and modeled volume fractions are listed in Table 3. Frederick diabase is approximately 50% plagioclase and 50% pyroxene (by volume). For simplicity, we adopted these values and a non-porous zero-pressure density of 3.01 g/cm^3. The plagioclase components of the rocks were modeled as the anorthosite shocked by McQueen *et al.* (1967). The pyroxene components were modeled as a solid solution of enstatite and ferrosilite and the fraction of these chosen so that the zero-pressure value of the density of rock was matched. The observed volume fraction of plagioclase of both rocks was exactly matched and the Fe to Mg ratio of the model pyroxene was adjusted to exactly match the zero-pressure density, i.e., pyroxene was substituted for the minor quantities of other minerals (mica and olivine) present. The equation of state parameters for the plagioclase and pyroxene components of both rocks are listed in Table 4. (The equations of state parameters for the pyroxene were calculated by taking a mass fraction average of the properties of enstatite and ferrosilite.)

The calculated Hugoniots for Frederick diabase along with the experimental data are shown in Fig. 4. In contrast to the case for 15,418, there are no

Table 4. Equation of state constants used in theoretical calculations.

Rock	Mineral	Phase	m	ρ_0 (g/cm^3)	K_0 (Mbar)	ξ	γ_0
Frederick	Plagioclase	lpp	1.0	2.73	0.54	−0.3	0.45
Diabase	Pyroxene	lpp	1.0	3.29	1.2	−0.6	1.0
	Plagioclase	hpp	1.0	3.71	2.08	1.2	0.7
(ρ_0 = 3.01 g/cm^3)	Pyroxene	Majorite	1.0	3.68	1.9	−0.27	1.5
		Ilmenite	1.0	3.89	2.3	−0.26	1.5
		Perovskite	1.0	4.49	3.78	−1.3	1.5
Apollo 15,418	Plagioclase	lpp	1.0325	2.73	0.54	−0.3	0.45
	Pyroxene	lpp	1.0325	3.45	1.2	−0.6	1.0
	Plagioclase	hpp	1.0325	3.71	2.08	1.2	0.7
(ρ_0 = 2.92 g/cm^3)	Pyroxene	Majorite	1.0325	3.83	1.9	−0.33	1.5
		Ilmenite	1.0325	4.08	2.38	−0.19	1.5
		Perovskite	1.0325	4.7	3.75	−1.9	1.5

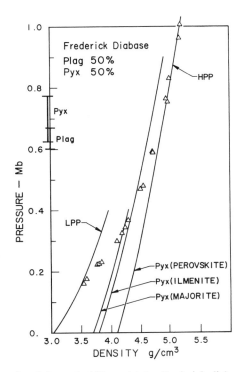

Fig. 4. Experimental and theoretical Hugoniot for Frederick diabase. The bars on the abscissa indicate the melting ranges of the minerals. The hpp curves are calculated assuming a 3.71 g/cm³ plagioclase component *and* the hpp components indicated for the pyroxene.

experimental shock data for the low pressure phase (lpp). The intersection of the trend of the data in the mixed phase regime with the theoretical Hugoniot for the lpp indicates that there is an onset of a phase change(s) at pressures of ~ 140 kbar.

A series of high-pressure phase Hugoniots (hpp) were calculated assuming that the phase changes were complete. The plagioclase in all cases was assumed to go to an unknown crystal structure having the zero-pressure density of 3.71 g/cm² inferred by Davies and Anderson (1971). However, we feel that this assumption is difficult to defend on the basis of evidence, other than the good fit to the data of McQueen *et al.* (1967), since no hpp corresponding to this density has yet been recovered from either static or dynamic experiments. Also this assumption was made only after we were unsuccessful in fitting Hugoniot data above 600 kbar for the gabbroic rocks with the hollandite-type structure zero-pressure density of 3.84 g/cm³. The response of pyroxene is quite complex as it is thought to undergo a series of transformations with increasing shock pressure (Ahrens *et al.*, 1969b; Davies and Gaffney, 1973). The hpp crystal structures assumed in the calculations were the majorite, ilmenite, and perovskite structures. Referring to Fig. 4, the high-pressure Hugoniot calculated by assuming the perovskite structure for the high-

pressure phase agrees with the experimental data quite well. The Hugoniots calculated assuming the hpp's had majorite and ilmenite structures are not directly related to the shape of the Hugoniot in the mixed phase regime, but are indicative of
when phase transformations occur and go to completion. The description of the
Hugoniot in the mixed phase regime is complicated by the fact that the phases are
probably not in thermodynamic equilibrium, and the degree of reaction is probably controlled by the kinetics of the phase transformation.

The shock pressures required to melt the individual minerals were determined
by calculating the entropy increase along the shock and comparing these values to
the experimental values at one atmosphere (Ahrens and O'Keefe, 1972). This procedure has the implied assumption that the release path from the shock state is
isentropic. The shock temperatures and entropies of each of the minerals are
listed in Table 5 and the melting regimes plotted in Fig. 5. Referring to Fig. 4 for
Frederick diabase, the shock pressure for incipient and complete melting of
plagioclase is 0.6 and 0.67 Mb respectively, and the pressure for incipient and
complete melting of pyroxene is 0.625 and 0.775 Mb respectively, and overlaps the
melting range of plagioclase.

The calculated Hugoniot of 15,418 is shown in Fig. 5. The theoretical Hugoniot
for the hpp correlates well with the experimental data shown, and passes through

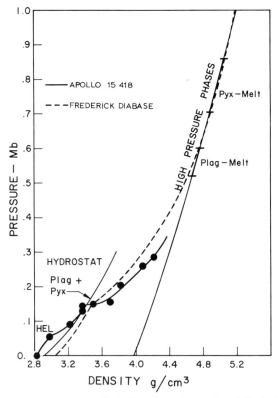

Fig. 5. Experimental and theoretical Hugoniots of 15.418 and Frederick diabase.

Table 5. Calculated temperatures and entropies of minerals in shocked Frederick diabase and Apollo 15,418.

| | Frederick diabase | | | | Apollo 15,418 | | | |
| | PLAG | | PYX | | PLAG | | PYX | |
Pressure (Mbar)	Temperature ($°K \times 10^{-3}$)	Entropy (ergs/g-°K $\times 10^{-7}$)	Temperature ($°K \times 10^{-3}$)	Entropy (ergs/g-°K $\times 10^{-7}$)	Temperature ($°K \times 10^{-3}$)	Entropy (ergs/g-°K $\times 10^{-7}$)	Temperature ($°K \times 10^{-3}$)	Entropy (ergs/g-°K $\times 10^{-7}$)
0.5	2.16	2.44	1.61	2.20	2.48	2.6	1.54	2.03
0.6	2.64	2.67	2.02	2.45	3.0	2.83	1.97	2.29
0.7	3.12	2.86	2.44	2.66	3.57	3.02	2.4	2.51
0.8	3.59	3.03	2.89	2.83	4.14	3.18	2.85	2.69
0.9	4.09	3.15	3.33	3.02	4.73	3.33	3.31	2.86
1.0	4.58	3.27	3.76	3.16	5.32	3.45	3.78	3.0

the cusp in the experimental data which indicates the presence of a phase change at ~ 150 kbar. The hpp Hugoniot is similar to the Hugoniot of Frederick diabase. However, the temperatures and entropies of the minerals are different (Table 5). At a pressure of 1.0 Mbar, the temperature of the plagioclase in 15,418 was 700°K greater than in Frederick diabase. This difference in temperature is attributed to the porosity of 15,418. The temperature of the pyroxenes in the two rocks is very similar. This is a result of the greater crystal density of pyroxene in 15,418, which would tend to lower the shock temperature relative to Frederick diabase, being compensated by the porosity of the rock, which would increase the temperature. The shock pressures for incipient and complete melting of plagioclase are ~ 0.5 and 0.6 Mbar respectively, and the pressures for incipient and complete melting of 15,418 pyroxene are ~ 0.7 and 0.85 Mbar respectively. Note, as opposed to Frederick diabase, the minerals in 15,418 have non-overlapping melting regimes. The shock pressure required for incipient vaporization is ~ 1.3 Mbar.

The meteoroid impact velocities required for incipient melting and vaporization were estimated using the methods of Ahrens and O'Keefe (1972). The impact velocities of an iron meteoroid required to induce incipient melting in the plagioclase component is ~ 6.0 km/sec and in the pyroxene component, ~ 7.0 km/sec, and to induce vaporization for plagioclase is ~ 8.0 km/sec.

CONCLUSIONS AND DISCUSSIONS

The dynamic yield stress under shock compression (the HEL) is observed to vary from ~ 45 to ~ 70 kbar, apparently depending on final shock pressure. These values, which probably give a lower bound to the shock pressure range above which major shock-induced deformations occur, are somewhat higher than the ~ 45–50 kbar HEL values previously reported for terrestrial basalts. The reason for this may be the lack of a small component of usually hydrous alteration minerals often present in terrestrial rocks. The higher than terrestrial value of the HEL implies that a welded rock such as 15,418 may require more energy to achieve comminution than a terrestrial counterpart. Above the HEL and to ~ 150 kbar our data agree closely with the theoretical pressure-density hydrostat inferred from ultrasonic data on pyroxene and plagioclase single crystals. The Hugoniot points are also in approximate agreement with earlier data for a pyroxene and amphibole-rich basalt (Vacaville basalt). Since the data for both single-crystal pyroxene (Ahrens and Gaffney, 1971; McQueen *et al.*, 1967) and various plagioclases (McQueen *et al.*, 1967; Ahrens *et al.*, 1969) demonstrate that shock-induced phase changes take place in these materials above ~ 135 and ~ 140 kbar, respectively, it is not surprising that a mixed phase regime occurs above ~ 140 kbar in the 15,418. However, above 150 kbar and extending to 282 kbar the present pressure-density data lie at significantly greater densities (by about $0.12 \, g/cm^3$) than the earlier results for Vacaville basalt and Frederick and Centreville diabases (McQueen *et al.*, 1967).

We suggest that the higher apparent compression of 15,418 in the mixed phase region arises from the higher iron content of its pyroxene in contrast to the terrestrial basalts. Although we have yet to describe the transformation kinetics in detail, we suspect that the higher Fe^{++} content allows the extent of reaction of the shock-induced phase change to be greater for a Fe^{++}-rich pyroxene, for a given shock pressure than for a more Mg^{++}-rich pyroxene. (For Mg–Fe solid-solution silicates transformation to the high-pressure denser polymorph always occur at a significantly lower pressure for the Fe-member.)

We have examined the Hugoniot data for Frederick diabase with the Kelly–Truesdell theory of mixtures using mineral equations of states obtained from the mineral properties' systematics of Davies and Gaffney (1973), with the result that the high pressure phase data could be closely reproduced in the range of 0.6–1.0 Mbar. The plagioclase component was assumed to be equivalent to the anorthosite shocked by McQueen *et al.* (1967), which is assumed to transform into an unknown structure at high pressures (inferred zero-pressure density of $\sim 3.71 \, g/cm^3$). Although the equilibrium phase assemblages resulting from the application of high static pressures and high temperatures to the Ca- and Na-rich plagioclase are quite different, similar behavior of these minerals under shock condition is observed, i.e., the Hugoniots of albite and anorthite are observed to nearly coincide (Ahrens *et al.*, 1969). The pyroxene component of the diabase was assumed to be in the perovskite structure (zero-pressure density of $4.7 \, g/cm^3$).

We did not carry out calculations in the mixed phase regime (150 to 600 kbar); however, the calculations we performed do suggest that the "intermediate" high-pressure phases having a garnet structure (majorite) and ilmenite structure might form at pressures less than 600 kbar. A theoretical hydrostat was calculated for the low-pressure phases and from the intersection of the trend in the data in the mixed phase regime a phase change at 140 kbar is inferred. As far as we know, all the phase changes are reversible. They are, however, of considerable importance in calculating the shock conditions necessary for impact-induced melting or vaporization.

A calculation similar to that for the Frederick diabase was carried out for the low- and high-pressure regimes of 15,418. The results predict the shock-induced phase transformation implied by the data should go to completion by ~ 400 kbar.

We note that although the pyroxene in 15,418 is more iron-rich than that in terrestrial rocks, because of the higher plagioclase content of this rock, the zero-pressure density was slightly less than that of previously studied terrestrial diabases.

Acknowledgments—We appreciate the help of A. Chodos in characterizing our sample and the excellent experimental assistance of H. Richeson and D. Johnson. Supported under NGL-05-005-105 and NGL-05-007-002 at the California Institute of Technology and the University of California, respectively.

REFERENCES

Ahrens T. J. and Gregson V. G. (1964) Shock compression of crustal rocks: Data for quartz, calcite and plagioclase rocks. *J. Geophys. Res.* **69**, 4389–4874.

Ahrens T. J. and Duvall G. E. (1966) Stress relaxation behind elastic shock waves in rocks. *J. Geophys. Res.* **71**, 4349–4360.

Ahrens T. J. and Gaffney E. S. (1971) Dynamic compression of enstatite. *J. Geophys. Res.* **76**, 5504–5513.

Ahrens T. J. and Liu H. P. (1973) A shock-induced phase change in orthoclase. *J. Geophys. Res.* **78**, 1274–1278.

Ahrens T. J., Gust W. H., and Royce E. B. (1968) Material strength effect in the shock compression of alumina. *J. Appl. Phys.* **39**, 4610–4616.

Ahrens T. J., Petersen C. F., and Rosenberg J. T. (1969) Shock compression of feldspars. *J. Geophys. Res.* **74**, 2727–2746.

Ahrens T. J., Anderson D. L., and Ringwood A. E. (1969) Equations of state and crystal structures of high-pressure phases of shocked silicates and oxides. *Rev. Geophysics* **7**, 667–707.

Ahrens T. J. and O'Keefe J. D. (1972) Shock melting and vaporization of lunar rocks and minerals. *The Moon* **4**, 214–249.

Anderson O. (1965) Determination and some uses of isotropic elastic constants of polycrystalline aggregates using single crystal data. *Physical Acoustics* (editor W. P. Mason), pp. 43–95. Academic Press.

Bass R. C., Chabai R., and Arms D. (1963) Hugoniot data for some geological materials. *Sandia Corp. Report*, SC 4903.

Birch F. (1952) Elasticity and constitution of the earth's interior. *J. Geophys. Res.* **57**, 227–286.

Chao E. C. T., Minkin J. A., and Best J. B. (1972) Apollo 14 breccias: General characteristics and classification. *Proc. Third Lunar Sci. Conf., Geochim. Cosmochim. Acta*, Suppl. 3, Vol. 1, pp. 645–659. MIT Press.

Chao E. C. T., Boreman J. A., Minkin J. A., James O. B., and Desborough G. A. (1970) Lunar glasses of impact origin: Physical and chemical characteristics and geologic implications. *J. Geophys. Res.* **75**, 7445–7479.

Clark S. P. Jr. (1967) *Handbook of Physical Constants*. Geol. Soc. Am. Memoir **97**.

Davies G. F. (1973) Quasi-harmonic finite strain equations of state of solids. *J. Phys. Chem. Solids*. In press.

Davies G. F. and Anderson D. L. (1971) Revised shock-wave equations of state for high-pressure phases of rocks and minerals. *J. Geophys. Res.* **76**, 2617–2627.

Davies G. F. and Gaffney E. S. (1973) Identification of high pressure phases of rocks and minerals from Hugoniot data. *Geophys. J. Roy. Astr. Soc.*

Davis R. O. Jr. and Wu J. H. (1972) Prediction of shock response for several composite materials. *J. Composite Materials* **6**, 126–135.

Doe. B. R. and Tatsumoto M. (1972) Volitized lead from Apollo 12 and 14 soils (abstract). In *Lunar Science—III*, p. 178. The Lunar Science Institute, Houston.

Garg S. K. and Kirsh J. W. (1971) Hugoniot analysis of composite materials. *J. Composite Materials* **5**, 428–445.

Kelly P. D. (1964) A reacting continuum. *Int. J. Engrg. Sci.* **2**, 129–153.

King E. A. Jr., Butler J. C., and Carman M. F. (1972) Chondrules in Apollo 14 samples and size analyses of Apollo 14 and 15 fines. *Proc. Third Lunar Sci. Conf., Geochim. Cosmochim. Acta,* Suppl. 3, Vol. 1, pp. 673–686. MIT Press.

Lunar Sample Preliminary Examination Team (LSPET) (1972) The Apollo 15 Lunar Samples: A preliminary description. *Science* **175**, 363–374.

Lunar Sample Preliminary Examination Team (LSPET) (1973) The Apollo 16 Lunar Samples: A petrographic and chemical description. *Science* **179**, 23–34.

McQueen R. A., Marsh S. P., and Fritz J. N. (1967) Hugoniot equation of state of twelve rocks. *J. Geophys. Res.* **72**, 4999–5036.

O'Keefe J. D. and Ahrens T. J. (1973) Equations of state of multimineralic rocks. In preparation.

Rice M. H., McQueen R. A., and Walsh J. M. (1958) Compression of solids by strong shock waves. *Solid State Phys.* **6**, 1–63.

Silver L. T. (1972) Uranium-thorium-lead isotopes and the nature of the mare surface debris at Hadley-Apennine. In *The Apollo 15 Lunar Samples*, pp. 388–390. The Lunar Science Institute, Houston.

Torvik P. J. (1969) Shock propagation in a composite material. *J. Composite Matls.* **4**, 296–309.

Truesdell C. (1957) Sulle basi della termorneccanica, *Rend. Lincei* **22**, 33–38, 158–166.

Tsou F. K. and Chou P. C. (1969) Analytical study of Hugoniot in unidirectional Fiber Reinforced Composites. *J. Composite Matls.* **3**, 500–514.

Tsou F. K. and Chou P. C. (1970) The control-volume approach to Hugoniot of macroscopically homogeneous composites. *J. Composite Matls.* **4**, 526–537.

von Engelhardt W., Arndt J., Muller W. F., and Stöffler D. (1970) Shock metamorphism of lunar rocks and the origin of the regolith at the Apollo 11 landing site. In *Proc. Apollo 11 Lunar Sci. Conf., Geochim. Cosmochim. Acta,* Suppl. 1, Vol. 1, pp. 363–384. Pergamon.

von Engelhardt W., Arndt J., and Muller W. F. (1971) Shock metamorphism and origin of the regolith at the Apollo 11 and Apollo 12 landing sites. *Proc. Second Lunar Sci. Conf., Geochim. Cosmochim. Acta,* Suppl. 2, Vol. 1, pp. 833–854. MIT Press.

von Engelhardt W., Arndt W. J., Stöffler D., and Schneider H. (1972) Apollo 14 regolith and fragmental rocks, their compositions and origin by impacts. *Proc. Third Lunar Sci. Conf., Geochim. Cosmochim. Acta,* Suppl. 3, Vol. 1, pp. 753–770. MIT Press.

Wackerle J. (1962) Shock-wave compression of quartz. *J. Appl. Phys.* **33**, 922–930.

Proceedings of the Fourth Lunar Science Conference
(Supplement 4, *Geochimica et Cosmochimica Acta*)
Vol. 3, pp. 2591–2600

Elastic wave velocities in anorthosite and anorthositic gabbros from Apollo 15 and 16 landing sites

DAE H. CHUNG

Weston Geophysical Observatory, Weston, Massachusetts 02193
Massachusetts Institute of Technology, Cambridge, Massachusetts 02139

Abstract—Laboratory measurements of ultrasonic velocities in lunar samples 15065, 15555, 15415, 60015, and 61016 as well as in synthetic materials corresponding to compositions of anorthositic gabbros are presented as a function of hydrostatic pressure to about 7 kb. Combining these laboratory data with the observed variations of the seismic velocities with depth in the eastern part of Oceanus Procellarum as reported by Toksöz *et al.* (1972a, b, 1973a, b), the author examined the seismic velocity distributions in the moon with reference to the variations to be expected in a homogeneous medium. The lunar mantle begins about 60 km and the velocity of *P* waves in this area is about 7.7 km/sec, however, one velocity model suggests a thin layer of $V_P \sim 9$ km/sec immediately below the crust, which is attributed to a laterally placed garnet layer. Variation of the seismic parameter with depth in the upper crust (about 20 km thick) is much too rapid to be explained by compression of a uniform material and the departure from expectation is so great that no reasonable adjustment of the material parameters can bring agreement; therefore, this author concludes that this result in this region of the moon is not due to self-compression but to textural gradients. In the lower crust (about 40 km thick), the region is shown to be relatively homogeneous, consisting probably of anorthositic rocks, consistent with interpretation by Toksöz *et al.* (1972a). Below the 60 km depth, lunar mantle models consisting of predominantly pyroxene with some olivine in an anorthositic matrix are favored with our laboratory measurements. The composition of such combinations would meet the known geophysical constraints imposed by the mean density and moment of inertia as well as the *in situ* seismic velocity distributions. And further, such a composition of lunar mantle is essentially in agreement with geochemical and petrological analyses of Ringwood *et al.* (1970) and Green *et al.* (1972) as the source of lunar basalts.

INTRODUCTION

THE INTERPRETATION of the variations of seismic velocities with depth in the moon is important in any discussion of the internal constitution and the history of the moon. The essential basis for such interpretations is laboratory results of high-pressure/high-temperature velocity measurements on the returned lunar samples and their earth analogs in a simulated lunar condition and their extrapolation by means of equations of state. High-pressure acoustic studies used with various equations of state, as done by most investigators thus far, give some indication of the expected variation of elastic properties in a self-compressed layer. Kanamori *et al.* (1970), reported the first measurements of the elastic wave velocities in the returned lunar samples from the Apollo 11 mission. Schreiber *et al.* (1970) also reported similar data in the meantime. Kanamori *et al.* (1971), Wang *et al.* (1971), and Warren *et al.* (1971) reported velocity measurements made on returned Apollo 12 lunar samples. For Apollo 14 lunar samples, such measurements have been made by Todd *et al.* (1972), Mizutani *et al.* (1972), and Warren *et al.* (1972).

As part of a continuing study of the geophysical properties of returned lunar samples, we report in this paper our experimental results on seismic *P* and *S* wave

velocities as a function of hydrostatic pressure to about 7 kb at ambient temperature in selected lunar samples of anorthosite and anorthositic gabbros. The following lunar samples 15065,27, 15555,88, 15415,57, 60015,29, and 61016,34 are studied
under the present program.

Lunar Samples and Experimental

The chemical and mineralogical description of samples 15065, 15459, 15555, and 15415 may be
found in various reports, e.g., LSPET (1972) and Chamberlain and Watkins (1973) and of samples
60015, 60315, 61016, and 62235 in LSPET (1973) and also various reports found in Chamberlain and
Watkins (1973); readers are referred to these reports for detailed information about the lunar samples.
A brief description of their classification and rock type along with apparent density is made in Table 1.

Methods of transporting, handling, heating, and drying lunar samples, along with other experimental details utilized in the present work were exactly the same as those described earlier in Kanamori *et
al.* (1970) and also by Chung (1972) and, therefore, will not be repeated here. We measured both
compressional (P) and shear (S) wave velocities in a given sample by a Birch-type pulse-transmission
(time-of-flight) method in which the transit time of a high-frequency pulse through a dimension of the
sample is measured as a function of pressure. The range of measurements where pressure is variable
was from 1 bar to about 7 kbar under a pressure-controlled environment. The usual procedure to
encapsulate the lunar sample with a Sylgard material (obtainable from the Dow-Corning Corporation),
as outlined in Kanamori *et al.* (1970), was followed in the present work.

Velocity Data

All velocity data as a function of confining pressure are tabulated in Table 2.
These data are plotted also in Fig. 1 for both P and S wave velocities as a function
of hydrostatic pressure. The estimated accuracy is 2% for P waves and 5% for S
waves. These accuracies are considerably lower than those normally achieved because of the size of our lunar samples and the high attenuation. The pressure variation of both P and S wave velocities in these samples is very similar to the
behavior observed for dry lunar crystalline rocks like 10020, 12002, 12022, and
14310, e.g., *see* Kanamori *et al.* (1970) and Wang *et al.* (1971), Warren *et al.* (1971),
Todd *et al.* (1972), and Mizutani *et al.* (1972). As discussed earlier in Chung (1972),
the change in the P and S velocities at low pressures is common to most terrestrial
rocks containing microcracks; even for lunar samples, this is now well-recognized

Table 1. Lunar samples and their material classification.

Sample Number	ρ (gm/cm^3)	Classification of Rock Type
15065,27	2.86	gabbroic basalt
15555,88	3.10	gabbroic basalt
15415,57	2.70	anorthosite
60015,29	2.76	feldspar-rich basalt
61016,34	2.79	feldspar-rich basalt

Table 2. Elastic wave velocities as a function of hydrostatic pressure.

Sample Number	Mode	Confining Pressure (in kb)									
		0.5	1.0	1.5	2	3	4	5	6	7	(10)*
15415	P	5.0	5.6	6.02	6.40	6.65	6.70	6.78	6.83	6.85	6.87
	S	2.0	2.5	2.90	3.26	3.42	3.54	3.56	3.58	3.61	3.69
15065	P	3.9	4.7	5.25	5.62	6.20	6.52	6.76	6.84	6.98	7.10
	S	2.5	2.8	3.06	3.29	3.50	3.68	3.75	3.86	3.90	3.97
15555	P	5.6	6.1	6.45	6.66	6.90	7.02	7.14	7.25	7.30	7.42
	S	2.6	3.0	3.24	3.45	3.66	3.76	3.87	3.94	4.01	3.12
60015	P	5.5	6.0	6.27	6.52	6.75	6.86	6.90	6.94	6.97	7.02
	S	2.6	2.9	3.21	3.40	3.58	3.68	3.74	3.86	3.88	3.91
61016	P	5.6	6.2	6.30	6.60	6.77	6.87	6.91	6.96	6.99	7.02
	S	2.4	3.1	3.22	3.36	3.58	3.69	3.74	3.86	3.88	3.90

*Estimated by a linear extrapolation of high-pressure velocity data.

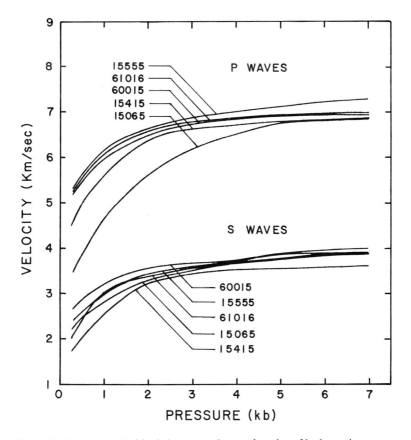

Fig. 1. P and S wave velocities in lunar samples as a function of hydrostatic pressure.

observation at low pressures and needs no further elaboration. The more abundant concentration of microcracks in most lunar samples than in terrestrial rocks is also well-recognized, but the origin of these microcracks in lunar samples has not been adequately explained. This author speculated in Chung (1972) that these abundant microcracks may have originated from effects due either to thermal fluctuations or to shock metamorphism, or to a combination of both. Stöffler (1971), in his study of shock metamorphism, provided a means of classifying shocked rocks collected at impact craters and clearly showed that intense microfracturing without a phase change is very common in shocked rocks at impact craters. In their work with lunar rocks, Ahrens and O'Keefe (1972) reported various phase changes and glass formation from shocking lunar samples and noted that the shocked samples may have severely cracked prior to the observed phase changes and glass formation. Thus, the origin of the abundant microcracks in lunar crystalline rocks being attributed to the effects of shock metamorphism seems to be justifiable. The question then is what effects on these shocked lunar crystalline rocks are expected from thermal cycling between the temperatures of the lunar day and lunar night? In an attempt to answer this question, the following experiment was conducted.

Lunar sample 61016,34 was heated in vacuum to 400°C for 10 hours and cooled down to ambient temperature. Velocity measurements were conducted in the usual way as described above and the results are tabulated in Table 3 and then illustrated in Fig. 2. Note that a very drastic change at low pressures ($P \leq 5$ kb) of the behavior of both the P and S wave velocities with varying confining pressures. The velocities at ambient conditions were severely affected by this heat-treatment of the sample, although at higher pressures ($P \geq 6$ kb) no difference was observed. The author attributes this observation to effects of thermally induced microcracks

Table 3. Elastic wave velocities in sample 61016 as a function of hydrostatic pressure, comparison of velocity data with "wet" and "dry" states.

Condition of Sample	Mode	0.5	1.0	1.5	2	3	4	5	6	7	(10)*
As Received	P	5.6	6.2	6.30	6.60	6.77	6.87	6.91	6.96	6.99	7.02
(from Table 2)	S	2.4	3.1	3.22	3.36	3.58	3.69	3.74	3.86	3.88	3.90
	V_P/V_S	2.3	2.0	1.96	1.96	1.89	1.86	1.84	1.80	1.80	1.80
"dry" (1)	P	3.75	5.23	5.32	5.72	6.23	6.41	6.70	6.97	7.00	7.02
	S	2.23	3.04	3.09	3.25	3.52	3.64	3.72	3.86	3.88	3.90
	V_P/V_S	1.68	1.72	1.72	1.76	1.77	1.76	1.80	1.80	1.80	1.80
"wet" (2)	P	6.5	6.70	6.73	6.75	6.78	6.88	6.91	6.97	6.99	7.03
	S	2.4	3.10	3.23	3.36	3.58	3.69	3.74	3.86	3.88	3.90
	V_P/V_S	2.7	2.16	2.09	2.01	1.89	1.86	1.84	1.80	1.80	1.80

*Estimated by a linear extrapolation of high-pressure velocity data.
(1) The term "dry" refers to the state of sample 61016 as it was heated in vacuum at 400°C for 10 hours and slowly cooled down to ambient temperature.
(2) The term "wet" refers to a water-saturated sample 61016.

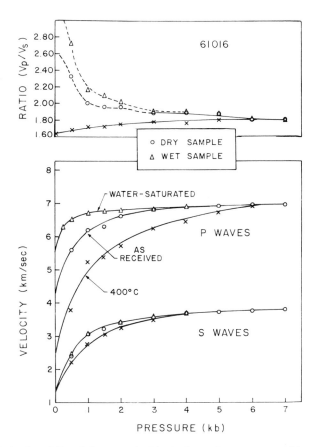

Fig. 2. Behavior of *P* and *S* wave velocities in dry and water-saturated lunar sample
61016 as a function of pressure.

that are already present in the sample and that originate from shock metamorphism.

The effects of the "dryness" of lunar samples on elastic wave velocities are
also studied with the sample 61016,34. These experimental results are also tabulated in Table 3. As shown in Fig. 2, effects on the *P* wave velocity of a
water-saturated sample 61016 is pronouncedly different for the velocity data
measured on the dry sample. Note in the lower pressure region how the (V_P/V_S)
ratio varies with the change in the physical state of this sample. It is already seen
that water-saturation on the *S* wave velocity has no observable effect, essentially
in agreement with an observation made by Nur and Simmons (1969) on terrestrial
rocks.

The last column of Table 2 lists velocity data to 10 kb estimated by a linear
extrapolation in the manner described by Birch (1961). These velocity values represent probably the crack-free values of elastic wave velocities, free from the
microcrack effects. To model elasticity and composition of the interior of the

moon, these values with an appropriate correction for vugs and pores as outlined earlier (*see* Chung, 1971c) should find a wide application.

Application to the Moon

Based on the observed variation of the seismic velocities with depth to about 120 km, in the Fra Mauro region of the moon Toksöz *et al.* (1972a, b) concluded that the moon has a crust about 60–65 km thick and divided into an upper crust (about 20–25 km thick) and a lower crust (about 40 km thick). The lunar mantle begins about 60 km deep as determined from the velocities of P waves in this region which are about 7–8 km/sec and possible with a thin high-velocity layer ($V_P \sim 9$ km/sec) at about the 60 km depth (Toksöz *et al.*, 1972a, b). Combining our laboratory data with the observed variations of the seismic velocities with depth in this region, it is possible to examine at least two aspects: one related to the constitution of the lunar crust and mantle and the other to relative homogeneity of materials present in these various layers within the moon.

In regard to lunar composition, our laboratory data are plotted in Fig. 3.

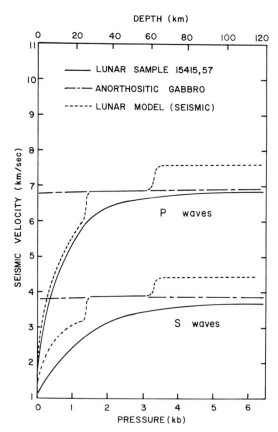

Fig. 3. Superposition of laboratory seismic velocities with *in situ* seismic velocity distributions in the moon. The lunar seismic model is due to Toksöz *et al.* (1972b).

Superimposed on this figure is the observed seismic velocity distribution within the eastern Oceanus Procellarum of the moon. In the lunar interior between about 25 and 60 km, the seismic velocity distribution as observed by Toksöz et al. (1972a, b) is consistent with our laboratory measurements of elastic wave velocities on anorthosite and anorthositic gabbros and this is shown in Fig. 3. The intrinsic seismic velocities measured in the laboratory on the synthetic anorthositic gabbros (resembling the chemical composition of lunar samples 15418 and 68415), e.g., $V_P = 6.95$ and $V_S = 3.79$ (in km/sec) with the density of 3.05 gm/cm³ are very consistent with the compositional interpretation of the lower crust by Toksöz et al. (1972a, b).

Below 60 km deep, it was noted earlier (Chung, 1973a) that except for a thin 9 km/sec layer, the lunar mantle models consisting of some olivine and pyroxene in a gabbroic anorthosite matrix are favored because of laboratory measurements of elastic wave velocities in olivine (Chung, 1971a) and pyroxene (Chung, 1971b). Such lunar models favoring a pyroxene-rich mantle were independently postulated by Solomon (1973) and Toksöz et al. (1973a) during the LSI conference on Geophysical and Geochemical Exploration of the moon and other planets, 10–12 January 1973.

Next, based on the observed variations of the seismic velocities with depth in the Fra Mauro region of the moon as reported by Toksöz et al. (1972a, b; 1973a, b), the author examines the seismic velocity distributions in the moon with reference to the variations to be expected in a homogeneous medium. Consider a uniform gravitational layer in which there is an arbitrary gradient of temperature. An expression (*see*, for example Birch, 1952) for the rate of change of the seismic parameter ϕ with radius r is

$$1 - g^{-1}\frac{d\phi}{dr} = \left(\frac{\partial K}{\partial P}\right)_T - 5T\alpha\gamma - \frac{2\tau\alpha\phi}{g} \qquad (1)$$

where $\phi = (V_P^2 - 4V_S^2/3)$, derived from the variations of the P and S wave velocities with depth in the moon observed by Toksöz et al. (1972b). α is the coefficient of volume expansion, usually in the order of 10^{-5} per degree for most rocks, γ is Grüneisen's ratio, about 1–2 for most rocks, τ is the gradient of temperature denoting the difference between the actual gradient of temperature and the adiabatic gradient, and g is the acceleration of gravity of the moon. The term $(\partial K/\partial P)_T$ is the isothermal pressure derivative of the isothermal incompressibility of material. Rewriting Eq. (1) into a more manageable form, we have

$$1 - g^{-1}\frac{d\phi}{dr} = \left(\frac{\partial K_S}{\partial P}\right)_S + 1 - \delta_S\left(\frac{\tau\alpha\phi}{g}\right) \qquad (2)$$

where

$$\delta_S = -\frac{1}{\alpha}\left(\frac{\partial \ln K_S}{\partial T}\right)_P$$

is the well-known Grüneisen's anharmonic parameter arising from temperature effects (Grüneisen, 1912, p. 278; Birch, 1952, p. 250). The term given by $1 - g^{-1}(d\phi/dr)$ is the quantity derived from the observed seismic velocity distributions due to Toksöz et al. (1972b). For most minerals and rocks at ambient conditions, $(\partial K_s/\partial P)$ has a value on the order of 4–7 and δ, on the order of 3–6. ϕ ranges from 29 for anorthosite, 31 for pyroxene (90 En), 33 for aluminous basalts, 39 for olivine (90 Fo) and 40–45 for garnets. Because of g, then, the last term on the right-hand side of Eq. (2) is about *six* times larger on the moon than that on the earth.

Evaluating term by term of Eq. (2), we find the results shown in Fig. 4. As noted in an earlier report (*see* Chung, 1973a,b), variation of the seismic parameter ϕ with depth in the upper crust (about 20–25 km thick) is much too rapid to be explained by compression of a uniform material and the departure from expectation as shown in Fig. 4 is so great that no reasonable adjustment of the material parameters can bring to an agreement; therefore, this author concludes that this region is not due to self-compression but to a result of textural gradients as was stated in Chung (1973a). In the lower crust, the region is seen to be relatively homogeneous, consisting probably of anorthositic rocks.

Recent seismic evidence as reported by Toksöz *et al.* (1972a, b; 1973a, b) also

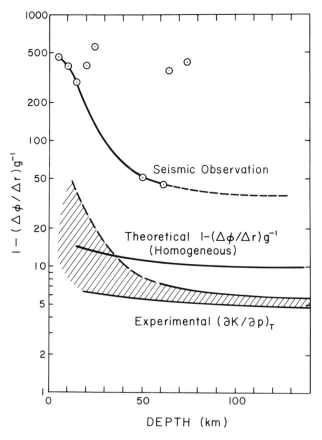

Fig. 4. Test on the homogeneity of the lunar interior.

suggests a slight decrease in density and velocity due possibly to temperature effects including partial melting at about a 1000 km depth. With the combined knowledge of the probable ranges in both laboratory and *in situ* seismic velocity distribution and density of the lunar mantle, some useful statements can be made about the composition of the lunar mantle. Whereas a predominantly "olivine mantle" can be ruled out, a predominantly "pyroxene mantle" seems to fit the presently available seismic velocity models. The author, however, favors lunar models consisting of an olivine and pyroxene composition in a gabbroic anorthosite matrix in view of our laboratory experience. Such compositional models will have a density in the neighborhood of 3.4 g/cm³ as required by moment of inertia considerations (*see* Solomon, 1973; Solomon and Toksöz, 1973). And further, in accordance with Ringwood and Essence (1970) and Green *et al.* (1972), geochemical and petrological analyses do indicate a pyroxene–olivine mantle as the source of lunar basalts. Further refinement of this idea concerning these compositional models would require laboratory data on temperature and pressure dependences of elastic wave velocities in the returned lunar samples and their laboratory analogs. Such experiments on lunar samples are presently in progress.

Acknowledgments—Support of NASA over the last four years are gratefully acknowledged. The author has enjoyed many discussion sessions with M. N. Toksöz, S. C. Solomon, A. Dainty, and K. Anderson of the MIT team on lunar seismology.

REFERENCES

Ahrens T. J. and O'Keefe J. D. (1972) Shock melting and vaporization of lunar rocks and minerals. *The Moon* **4**, 214–249.

Anderson D. L. and Kovach R. L. (1972) The lunar interior. *Phys. Earth Planet. Interiors* **6**, 116–122.

Birch F. (1952) Elasticity and constitution of the earth's interior. *J. Geophys. Res.* **57**, 227–286.

Birch F. (1961) The velocity of compressional waves in rocks to 10 kb, Part 2. *J. Geophys. Res.* **66**, 2199–2224.

Chamberlain J. W. and Watkins C. (editors) (1973) In *Lunar Science—IV*. The Lunar Science Institute, Houston.

Chung D. H. (1971a) Elasticity and equations of state of olivines in the Mg₂SiO₄–Fe₂SiO₄ system. *Geophys. J. Roy. Astr. Soc.* **25**, 511–528.

Chung D. H. (1971b) Equations of state of pyroxene in the MgSiO₃–FeSiO₃ system. *Eos, Trans. Am. Geophys. Un.* **52**, 919.

Chung D. H. (1971c) Pressure coefficients of elastic constants for porous materials: Correction for porosity and discussion on literature data. *Earth Planet. Sci. Lett.* **10**, 316–324.

Chung D. H. (1972) Laboratory studies on seismic and electrical properties of the moon. *The Moon* **4**, 356–372.

Chung D. H. (1973a) Elasticity and constitution of the lunar crust and mantle: Laboratory studies. *The Moon.* In press.

Chung D. H. (1973b) Elastic wave velocities in anorthosite and anorthositic gabbros from Apollo 15 and 16 landing sites. In *Lunar Science—IV*, pp. 141–142. The Lunar Science Institute, Houston.

Green D. H., Ringwood A. E., Ware N. G., and Hibberson W. O. (1972) Experimental petrology and petrogenesis of Apollo 14 basalts. *Proc. Third Lunar Sci. Conf., Geochim. Cosmochim. Acta*, Suppl. 3, Vol. 1, pp. 197–206. MIT Press.

Grüneisen E. (1912) Theory of the state of solids (in German). *Ann. Phys.* **39**, 257–295.

Kanamori H., Mizutani H., and Hamano Y. (1971) Elastic wave velocities of Apollo 12 rocks at high pressures. *Proc. Second Lunar Sci. Conf., Geochim. Cosmochim. Acta*, Suppl. 2, Vol. 3, pp. 2323–2326. MIT Press.

Kanamori H., Nur A., Chung D. H., Wones D., and Simmons G., (1970) Elastic wave velocities of lunar samples at high pressures and their geophysical implications. *Science* **167**, 726–728.

LSPET (Lunar Sample Preliminary Examination Team) (1973) Preliminary examination of lunar samples from Apollo 16. *Science* **179**, 23–24.

Mizutani H., Fuji N., Hamano Y., and Osako M. (1972) Elastic wave velocities and thermal diffusivities of Apollo 14 rocks. *Proc. Third Lunar Sci. Conf., Geochim. Cosmochim. Acta*, Suppl. 3, Vol. 3, pp. 2557–2564. MIT Press.

Nur A. and Simmons G. (1969) The effect of saturation on velocity in low porosity rocks. *Earth Planet. Sci. Lett.* **7**, 183–193.

Ringwood A. E. and Essence E. (1970) Petrogenesis of Apollo 11 basalts, internal constitution and origin of the moon. *Proc. Apollo 11 Lunar Sci. Conf., Geochim. Cosmochim. Acta*, Suppl. 1, Vol. 1, pp. 769–799. Pergamon.

Schreiber E., Anderson O. L., Soga N., Warren N., and Scholz C. (1970) Sound velocity and compressibility for lunar rocks 17 and 46 and for glass spheres from the lunar soil. *Science* **167**, 732–734.

Solomon S. C. (1973) Density within the moon and implications for lunar composition. *The Moon*. In press.

Solomon S. C. and Toksöz M. N. (1973) Thermal history and evaluation of the moon. *Phys. Earth Planet Interiors*. In press.

Stöffler D. (1971) Progressive metamorphism and classification of shocked and brecciated crystalline rocks at impact craters. *J. Geophys. Res.* **76**, 5541–5551.

Todd T., Wang H., Baldridge W. S., and Simmons G. (1972) Elastic properties of Apollo 14 and 15 rocks. *Proc. Third Lunar Sci. Conf., Geochim. Cosmochim. Acta*, Suppl. 3, Vol. 3, pp. 2577–2586. MIT Press.

Toksöz M. N., Press F., Anderson K., Dainty A., Latham G., Ewing M., Dorman J., Lammlein D., Sutton G., Duennebier F., and Nakamura Y. (1972a) Lunar crust: Structure and composition. *Science* **176**, 1102–1116.

Toksöz M. N., Press F., Dainty A., Anderson K., Latham G., Ewing M., Dorman J., Lammlein D., Sutton G., and Duennebier F. (1972b) Structure, composition, and properties of lunar crust. *Proc. Third Lunar Sci. Conf., Geochim. Cosmochim. Acta*, Suppl. 3, Vol. 3, pp. 2527–2544. MIT Press.

Toksöz M. N., Press F., Dainty A. M., and Anderson K. R. (1973a) Lunar velocity structure and compositional and thermal inferences. *The Moon*. In press.

Toksöz M. N., Dainty A. M., Solomon S. C., and Anderson K. R. (1973b) Velocity structure and evolution of the moon. *Proc. Fourth Lunar Sci. Conf., Geochim. Cosmochim. Acta*. In press.

Wang H., Todd T., and Simmons G. (1971) Elastic properties of Apollo 12 rocks. *Proc. Second Lunar Sci. Conf., Geochim. Cosmochim. Acta*, Suppl. 2, Vol. 3, pp. 2327–2336. MIT Press.

Warren N., Schreiber E., Scholz C., Morrison J. A., Norton P. R., Kumazawa M., and Anderson O. L. (1971) Elastic and thermal properties of Apollo 11 and 12 rocks. *Proc. Second Lunar Sci. Conf., Geochim. Cosmochim. Acta*, Suppl. 2, Vol. 3, pp. 2345–2360. MIT Press.

Warren N., Anderson O. L., and Soga N. (1972) Applications to lunar geophysical models of the velocity-density properties of lunar rocks, glasses, and artificial lunar glasses. *Proc. Third Lunar Sci. Conf., Geochim. Cosmochim. Acta*, Suppl. 3, Vol. 3, pp. 2587–2598. MIT Press.

Proceedings of the Fourth Lunar Science Conference
(Supplement 4, *Geochimica et Cosmochimica Acta*)
Vol. 3, pp. 2601–2609

Elastic wave velocities of Apollo 14, 15, and 16 rocks*

Hitoshi Mizutani† and David F. Newbigging

Seismological Laboratory, California Institute of Technology, Pasadena, California 91109

Abstract—Elastic wave velocities of two Apollo 14 rocks, 14053 and 14321, three Apollo 15 rocks, 15058, 15415, and 15545, and one Apollo 16 rock 60315 have been determined at pressures up to 10 kb. For sample 14321, the variation of the compressional wave velocities with temperature has been measured over the temperature range from 27 to 200°C. Overall elastic properties of these samples except sample 15415 are very similar to those of Apollo 11, 12, and 14 rocks and are concordant with Toksöz *et al.*'s (1972) interpretation that lunar upper crust is of basaltic composition. Temperature derivative of the *P* wave velocity for sample 14321 is a half to one order of magnitude larger than that for single crystalline minerals. This suggests that the seismic velocity in the lunar crust may be affected significantly by the temperature distribution.

Introduction

THE LABORATORY EXPERIMENT of measuring seismic-wave velocities in rocks can now simultaneously cover pressures up to 10 kb and temperatures up to ~ 500°C with little difficulty. Therefore little or no extrapolation of either pressure or temperature is required to reach the conditions expected in the upper two hundred kilometers in the moon.

In this paper we report new experimental results of ultrasonic wave velocities in Apollo 14, 15, and 16 rocks at pressures up to 10 kb. Temperature variation of the compressional wave velocities of sample 14321 is also measured in the temperature range from 27 to 200°C.

Experimental Method

Elastic wave velocities were determined with the pulse transmission method (Birch, 1960). The pressure system used in the present experiment is the same one as used by Spetzler (1970). The pressure medium is conventional argon gas which is pumped into the thick-walled vessel in two stages using two pumps and an intensifier.

In the case of a liquid pressure system, a copper jacket with soft-soldered caps at the ends or a rubber tubing has been successfully used in preventing the pressure medium from penetrating into the sample. However, these jacketing methods were found to be unsuccessful in the gaseous high-pressure system (some difficulties in sealing in the gaseous high pressure system are described by Birch, 1943). Eventually we devised a successful sample assemblage as shown in Fig. 1. Three grooves are cut in the steel buffer rod. In one groove, a neoprene or a silicone-rubber O-ring is fitted to seal low pressure. The other two grooves are designed to seal high pressure by intimate contact with the copper tubing and the walls of the grooves; the intimate contact is assured by the unsupported areas at the bottoms of the

---

*Contribution no. 2223, Division of Geological and Planetary Sciences, California Institute of Technology, Pasadena.
†Now at Geophysical Institute, University of Tokyo, Hongo, Tokyo, Japan.

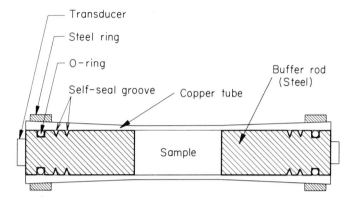

Fig. 1. Arrangements of specimen, buffer rods, copper jacket, and transducers. Outside diameter of the annealed copper tubing is slightly tapered toward the central part where thickness of the tubing is 0.015 inch. Three grooves, one for O-ring and the other two for self-sealing, are cut in the buffer rod. The steel rings are forced with interference fit over the copper to get effective initial sealing.

grooves. This jacketing method is found to work at high pressures, at least, up to 10 kb and temperatures up to 400°C.

A sample is usually moulded in high-temperature epoxy (Tra-Bond 2212, Tra-Con, Inc.) and then shaped into a cylindrical shape of 1.40 cm in diameter. To get efficient acoustic coupling, Nonaq stopcock grease was used at the boundaries between the sample and the buffer rods. The PZT transducers of 2 MHz were also bonded to the buffer rods by the Nonaq. The pressure is read by a bridge consisting of manganin coils and by a Heise pressure gage. The temperatures at the samples are measured with two chromel-alumel thermocouples. The temperature gradient across the sample is in general less than 2°C.

The estimated accuracy of the velocity determination reported here is ±1% in room temperature measurements, and is about ±2% in the measurements at higher temperatures.

SAMPLE DESCRIPTION

Sample 14053 is a fine-grained holocrystalline basalt, which differs from basaltic rocks returned from the Apollo 11 and 12 sites in being much richer in plagioclase and poorer in ilmenite (LSPET, 1971). The sample 14321 is a coarse grained breccia which contains abundant fragmental lithic clasts. Samples 15058 and 15545 are a porphyritic clinopyroxene basalt and a porphyritic olivine basalt respectively (LSPET, 1971). The chemical compositions of these rocks are generally similar to those of the Apollo 12 igneous rocks and classified as mare basalts. Sample 15415 is an anorthosite consisting of over 99% plagioclase (An 95~98%) grains. Since the anorthosite is considered to be a major component of highland material and is also suggested to be an important material in the lunar interior (Toksöz *et al.*, 1972; Anderson and Kovach, 1973), it is important to determine the ultrasonic velocity of this rock. Sample 60315 has the bulk composition similar to those of the KREEP basalts found at the Apollo 12, 14, 15 sites, and is classified as the KREEP basalt.

In Table 1 are listed the chemical compositions (Hubbard *et al.*, 1972; Maxwell

Table 1. Chemical compositions, normative mineral compositions, mean atomic weights, and ideal densities.

	14053[a]	14321[b]	15058[c]	15415[a]	15545[d]	60315[e]
	Chemical composition (weight %)					
SiO_2	46.4	47.78	48.47	44.1	45.72	45.61
Al_2O_3	13.7	15.20	8.90	35.5	7.54	17.18
FeO	16.8	12.25	19.75	0.23	24.24	10.53
MgO	8.48	10.73	9.56	0.09	11.11	13.15
CaO	11.2	9.94	10.23	19.7	9.18	10.41
MnO	0.26	0.17	0.27	—	0.29	0.12
Na_2O	—	0.72	0.28	0.34	0.29	0.56
K_2O	0.10	0.62	0.04	—	0.04	0.35
TiO_2	2.64	2.06	1.60	0.02	2.36	1.27
P_2O_5	0.09	0.41	0.05	0.01	0.06	0.45
Cr_2O_3	—	0.26	0.66	—	0.73	—
S	0.14	0.07	0.06	—	0.07	0.14
H_2O^+	—	0.05		—		—
Total	99.7	100.26	99.84	100.0	100.13	99.77
Mean atomic weight	22.80	22.20	22.99	21.40	23.04	22.01
	Normative mineral composition (mole %)					
Or	0.32	2.12	0.12	—	1.09	1.31
Ab	—	3.76	1.24	2.98	23.44	3.18
An	19.97	21.15	11.30	92.74	5.79	27.41
En	31.50	36.41	32.53	0.60	2.00	27.42
Fs	30.07	19.57	34.37	0.80	2.42	10.69
Wo	9.95	7.49	13.73	2.33	23.10	5.26
Fo	—	3.29	—	—	23.31	15.78
Fa	—	1.77	—	—	28.27	6.15
Il	4.95	4.17	2.75	0.07	0.58	2.80
Ch	—	0.27	0.60	—	—	—
Qz	3.25	—	3.37	0.49	—	—
Theoretical density (g/cm³)	3.192	3.093	3.270	2.763	3.252	3.063

[a]Hubbard et al. (1972).
[b]Scoon (1972).
[c]Willis et al. (1972).
[d]Maxwell et al. (1972).
[e]LSPET (1973).

et al., 1972; Scoon, 1972; Willis *et al.*, 1971; LSPET, 1973), calculated normative mineral compositions, mean atomic weights, and ideal densities for the lunar samples studied in the present paper.

Besides the lunar samples we have measured compressional wave velocities of a terrestrial anorthosite collected by B. F. Windley in Fiskenaesset complex, West Greenland. Plagioclase composition in this sample is An 85.6% (Windley, personal communication, 1972). Proportion of mafic minerals in the sample was estimated from the calculation of densities with the same procedure as that used by Birch (1961). Assuming the density of a mafic component to be 3.2 to 3.4 g/cm^3, the proportion of the mafic component is found to be 10 to 15% by volume. The result is in good agreement with CIPW norm of this sample by Windley and Smith (1973).

Results and Discussion

All the experimental results are summarized in Table 2. The compressional wave velocities of samples 14053,32; 15058,57; and 15545,24 as functions of pressure are given in Fig. 2. These three rocks give compressional velocities fairly close to each other at high pressures. The high-pressure value of ~ 7.0 km/sec at 10 kb is consistent with the mineral compositions of these rocks and close to the velocities obtained for Apollo 12 and 14 crystalline rocks (Kanamori *et al.*, 1971; Todd *et al.*, 1972). The sample 15545,24 was compressed up to 10 kb several times before we get the result in Fig. 2. The bulk density of this sample listed in Table 2 and Fig. 2 is the original one before compression. Actual density of the sample 15545,24 at the time of the measurement was probably about 3.0 g/cm^3.

In Fig. 3 are shown both P and S wave velocities of sample 60315,33. Like other lunar samples, the velocities of this rock also increase very rapidly with

Table 2. Smoothed values of ultrasonic velocities of lunar rocks.

						Pressure (kb)				
			0.0	0.5	1.0	2.0	3.0	5.0	7.0	9.0
14053,32 ($\rho = 3.18$)	(27°C)	V_p	3.02	4.54	5.32	6.02	6.35	6.71	6.86	6.93
14321,93 ($\rho = 2.40$)	(27°C)	V_p	4.41	4.68	4.86	5.10	5.27	5.53	5.69	5.84
	(100°C)	V_p				4.82	5.06	5.23	5.59	5.79
	(200°C)	V_p				4.46	4.71	5.09	5.37	5.68
15058,57 ($\rho = 2.99$)	(27°C)	V_p	4.03	5.33	5.54	5.85	6.12	6.49	6.65	6.73
15415,96 ($\rho = 2.76$*)	(27°C)	V_p	5.98	6.20	6.28	6.43	6.56	6.75	6.89	7.01
15545,24 ($\rho = 2.56$†)	(27°C)	V_p	5.60	6.10	6.37	6.63	6.76	6.90	6.98	7.02
60315,33 ($\rho = 3.05$)	(27°C)	V_p	4.15	4.81	5.40	6.13	6.38	6.88	7.03	7.11
	(27°C)	V_s	2.0	2.48	2.79	3.10	3.26	3.50	3.68	3.83
132025‡ ($\rho = 2.80$)	(27°C)	V_p	5.42	6.64	6.84	7.03	7.10	7.14	7.16	

*Density for pure anorthite.
†Density for the uncompressed sample (*see* text).
‡Terrestrial anorthosite from Fiskenaesset complex, West Greenland.

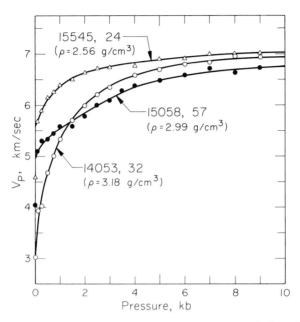

Fig. 2. Compressional wave velocities of samples 14053,32; 15058,57; and 15545,24 as functions of pressure.

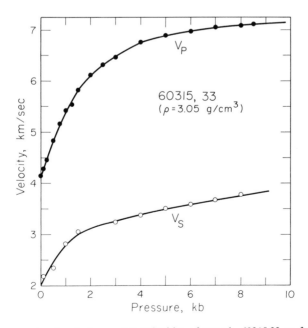

Fig. 3. Compressional and shear wave velocities of sample 60315,33 as functions of pressure.

pressure. The velocity-pressure (depth) curves of this rock fit very well both P and S wave seismic velocity profiles from 5 to 25 km depth (Toksöz *et al.*, 1972). The coincidence of the velocity of the sample 60315,33 with the seismic structure could be fortuitous but nevertheless it is interesting that the coincidence is obtained for the KREEP-basalt "whose major element abundances correspond to those of liquidus known to have been produced by partial melting of the lunar interior or terrestrial interior" (LSPET, 1972).

In Fig. 4 is shown the variation of the compressional wave velocity of sample 14321,93 with pressure and temperature. Temperature dependence $(\partial V_P/\partial T)_P$ observed for the sample is $(2\sim3)\times10^{-3}$ km/sec/°K in the temperature range 27 to 200°C and in the pressure range from 3 to 10 kb. The temperature derivative of the P wave velocity of the sample 14321,93 is a half to one order of magnitude larger than those of single crystalline silicates or oxides (Anderson *et al.*, 1968) and two to ten times larger than those of compact terrestrial rocks (Hughes and Maurette, 1957; Birch 1958). The discrepancy of the temperature dependence of the P wave velocity between a single crystal or a compact terrestrial rock and the porous lunar rock may be due to the effect of the large porosity of the lunar sample or due to the newly formed cracks in the sample with increase of temperature. It should be noted that the seismic wave velocity in the upper part of the moon may be affected significantly by the temperature distribution in the moon.

The compressional wave velocities of the lunar anorthosite 15415,96 and the terrestrial anorthosite from Fiskenaesset complex, West Greenland, are shown in Fig. 5. In order to obtain the compressional wave velocities for the plagioclase alone, a correction for the content of mafic minerals was applied to the data of the Greenland anorthosite using the same procedure as that used by Birch (1961) (V_P and density of the mafic minerals are assumed to be 7.9 km/sec and $3.2\sim3.4$ g/cm^3 respectively). The velocity so found for the plagioclase of An 85% is

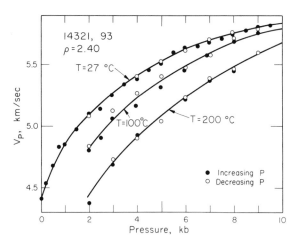

Fig. 4. Compressional wave velocity of sample 14321,93 as a function of pressure and temperature.

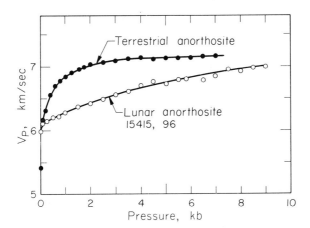

Fig. 5. Compressional wave velocities of lunar anorthosite 15415,96 and a terrestrial anorthosite as functions of pressure.

7.08–7.12 km/sec at 10 kb. This result is in very good agreement with that obtained by Birch (1961) for plagioclase feldspars and is fairly close to the value of the lunar anorthosite at 10 kb. But even after correction for the content of mafic minerals, the terrestrial anorthosite has higher P wave velocities than the lunar anorthosite at pressures below 10 kb. The sample shape of the sample 15415,96 as we received it was very irregular and the potting material (epoxy) may have entered into interstices of the sample. Therefore, the data of the sample 15415,96 may be affected by the potting material used. However, the very good agreement of the present data with those by Chung (1973) for the same sample as the present one and those by Wang *et al.* (1973) for Apollo 16 lunar anorthosites indicates that the effect of the potting material on the velocity of the sample 15415,96 in Fig. 5 might be negligible. If this is the case, the lower velocities of the lunar anorthosites than terrestrial anorthosites at 0 to 8 kb should be attributed to the difference of textures of the samples.

In fact, the lunar anorthosite 15415,96 contains numerous fracture surfaces which indicate cataclastic events subsequent to its crystallization (e.g., *see* Fig. 5 in LSPET, 1972). Therefore the anorthosites without micro- and macro-cracks would give compressional wave velocities of about 7.0 km/sec which is consistent with the seismic velocity observed from 25 to 65 km depth (Toksöz *et al.*, 1972).

Compressional wave velocities of all the lunar rocks which we have measured ultrasonically so far are unanimously smaller than 8~9 km/sec which corresponds to the seismic velocity of the lunar mantle below 70 km depth. It is apparent that we have not obtained the rocks corresponding to the lunar mantle. Considering the extensive excavation of the lunar crust by intense bombardment of the planetesimals it seems a little strange that we have not sampled the mantle rocks yet. Dunites from a large clast in one of the boulders at Apollo 17 site may have the origin in the lunar mantle.

Acknowledgments—We acknowledge Professor D. L. Anderson for his support and constant encouragement. Dr. H. Spetzler offered valuable suggestions on the sealing problems of the sample. We are also grateful to Dr. Brian Windley for his courtesy in sending the anorthositic samples from Greenland and in sending the chemical analysis data of the samples before publication and to the director of the Geological Survey of Greenland for permission to publish the results on the survey material. The comments of Drs. H. Kanamori, M. N. Toksöz, and H. Wang are greatly acknowledged. This work was partially supported by National Aeronautics and Space Administration Contract NGL-05-002-069.

References

Anderson D. L. and Kovach R. L. (1973) The lunar interior. *Phys. Earth Planet. Interiors* **6**, 116–122.

Anderson O. L., Schreiber E., Liebermann R. C., and Soga N. (1968) Some elastic constant data on minerals relevant to geophysics. *Rev. Geophys.* **6**, 491–524.

Birch F. (1943) Elasticity of igneous rocks at high temperatures and pressures. *Bull. Geol. Soc. Am.* **54**, 263–286.

Birch F. (1958) Interpretation of seismic structure of the crust in the light of experimental studies of wave velocities in rocks. In *Contribution in Geophysics* (editors H. Benioff, M. Ewing, B. F. Howell, Jr., and F. Press), pp. 158–170 Pergamon.

Birch F. (1960) The velocity of compressional waves in rocks to 10 kilobars, Part 1. *J. Geophys. Res.* **65**, 1083–1102.

Birch F. (1961) The velocity of compressional waves in rocks to 10 kilobars, Part 2. *J. Geophys. Res.* **66**, 2199–2224.

Chung D. (1973) Elastic wave velocities in anorthosite and anorthositic gabbros from Apollo 15 and Apollo 16 landing sites (abstract). In *Lunar Science—IV*, pp. 141–142. The Lunar Science Institute, Houston.

Hubbard N. J., Gast P. W., Rhodes J. M., Bansal B. M., and Wiesmann H. (1972) Nonmare basalts: Part II. *Proc. Third Lunar Sci. Conf., Geochim. Cosmochim. Acta*, Suppl. 3, Vol. 2, pp. 1161–1179. MIT Press.

Hughes D. S. and Maurette C. (1957) Variation of elastic wave velocities in basic igneous rocks with pressure. *Geophysics* **22**, 22–31.

Kanamori H., Mizutani H., and Hamano Y. (1971) Elastic wave velocities of Apollo 12 rocks at high pressures. *Proc. Second Lunar Sci. Conf., Geochim. Cosmochim. Acta*, Suppl. 2, Vol. 3, pp. 2323–2326. MIT Press.

LSPET (Lunar Sample Preliminary Examination Team) (1971) Preliminary examination of lunar samples from Apollo 14. *Science* **173**, 681–693.

LSPET (Lunar Sample Preliminary Examination Team) (1972) The Apollo 15 lunar samples: A preliminary description. *Science* **175**, 363–375.

LSPET (Lunar Sample Preliminary Examination Team) (1973) The Apollo 16 lunar samples: A petrographic and chemical description of samples from the lunar highlands. *Science* **179**, 23–34.

Maxwell J. A., Bouvier J. L., and Wiik H. B. (1972) Chemical composition of some Apollo 15 lunar samples. In *The Apollo 15 Lunar Samples*, pp. 233–238. The Lunar Science Institute, Houston.

Scoon J. H. (1972) Chemical analyses of lunar samples 14003, 14311, and 14321. *Proc. Third Lunar Sci. Conf., Geochim. Cosmochim. Acta*, Suppl. 3, Vol. 2, pp. 1335–1336. MIT Press.

Spetzler H. (1970) Equation of state of polycrystalline and single-crystal MgO to 8 kilobars and 800°K. *J. Geophys. Res.* **75**, 2073–2087.

Todd T., Wang H., Baldridge W. S., and Simmons G. (1972) Elastic properties of Apollo 14 and 15 rocks. *Proc. Third Lunar Sci. Conf., Geochim. Cosmochim. Acta*, Suppl. 3, Vol. 3, pp. 2577–2586. MIT Press.

Toksöz M. N., Press F., Anderson K., Danity A., Latham G., Ewing M., Dorman J., Lammlein D., and Nakamura Y. (1972) Velocity structure and properties of the lunar crust. *The Moon* **4**, 490–504.

Wang H., Todd T., Richter D., and Simmons G. (1973) Elastic properties of plagioclase aggregates and seismic velocities in the moon (abstract). In *Lunar Science—IV*, pp. 758–760. The Lunar Science Institute, Houston.

Willis J. P., Erlank A. J., Gurney J. J., and Ahrens L. H. (1971) Geochemical features of Apollo 15 materials. In *The Apollo 15 Lunar Samples*, pp. 268–271. The Lunar Science Institute, Houston.

Windley B. F. and Smith J. V. (1973) The Fiskenaesset complex, West Greenland. Part 2. General mineral chemistry from Qeqertarssuatsiaq island. *Gronlands Geol. Unders. Rep.* (also Meddr. Gronland). In press.

Proceedings of the Fourth Lunar Science Conference
(Supplement 4, *Geochimica et Cosmochimica Acta*)
Vol. 3, pp. 2611–2629

Rock physics properties
of some lunar samples*

N. WARREN, R. TRICE, N. SOGA,† and O. L. ANDERSON

Institute of Geophysics and Planetary Physics University of California,
Los Angeles, California 90024

Abstract—Linear strains and acoustic velocity data for lunar samples under uniaxial and hydrostatic loading are presented. Elastic properties are presented for 60335,20; 15555,68; 15498,23; and 12063,97. Internal friction data are summarized for a number of artificial lunar glasses with compositions similar to lunar rocks 12009, 12012, 14305, 15021, and 15555. Zero porosity model-rock moduli are calculated for a number of lunar model-rocks, with mineralogies similar to Apollo 12, 14, and 16 rocks.

Model-rock calculations indicate that rock types in the troctolitic composition range may provide reasonable modeling of the lunar upper mantle. Model calculations involving pore crack effects are compatible with a strong dependence of rock moduli on pore strain, and therefore of rock velocities on nonhydrostatic loading. The high velocity of rocks under uniaxial loading appears to be compatible with, and may aid in, interpretation of near-surface velocity profiles observed in the active seismic experiment. High values of lunar Q are compatible with results for glass and with recent measurements on rocks.

INTRODUCTION

THE NEW lunar seismic profile detail which is becoming available (Kovach *et al.*, 1973; Toksöz *et al.*, 1973) provides an increasingly detailed framework against which lunar rock data and model calculations can be compared. It is becoming increasingly reasonable to consider in some detail such effects as the relations between velocity and loading conditions (hydrostatic and nonhydrostatic), pore and crack structure, and mineralogy.

In this paper, the effects of mineralogy and crack and pore structure and loading conditions are considered using model calculations and observed rock data. The techniques used are preliminary but appear to be promising.

Data are presented from linear strain and velocity measurements under both uniaxial and hydrostatic loading. Model rock calculations are made and compared to returned lunar rock data and lunar *in situ* velocity profiles. Results of attenuation measurements on artificial lunar glasses are presented and compared to recent preliminary rock internal friction measurements.

---

*Contribution No. 1195, Institute of Geophysics and Planetary Physics, University of California, Los Angles, CA 90024.

†Department of Industrial Chemistry, Faculty of Engineering, Kyoto University, Sakyo-ku, Kyoto, Japan.

ELASTIC PROPERTIES

Samples

Table 1 summarizes descriptions of lunar rock samples for which new elastic properties data are presented in this paper. The mean mineralogies were obtained from various references as indicated.

The terrestrial basalt, to which lunar samples are compared, is aphanitic and nonvesicular with nodules of olivine, from Vulcan's Throne in the Grand Canyon.

Samples were parallelopipeds of either approximately 2 cm × 1.5 cm × 1.5 cm or 2 × 1 cm × 1 cm. In this paper, sampled directions are denoted 3–3', 2–2', and 1–1'. Direction 3–3' is always the longest. Sample faces were ground using dry Silicon Carbide powder (120 through 600 grit), and to a parallelness of about $\approx \pm 0.001''$ over opposing faces. The sample surfaces were cleaned of grinding residue with petroleum ether or isopentane.

Experimental

Acoustic velocities (compressional V_p, and shear V_s) and linear strain were measured for samples under uniaxial and hydrostatic pressure.

Linear strains were measured in the 3–3' direction. Gauges were mounted directly to the samples using a two-part epoxy (Epoxi-Patch, Dexter Corporation). This epoxy is dissolvable with exposure to hydrogen peroxide, allowing removal of the strain gauges.

Strain gauges were BLH SR-4 epoxy backed: 350 Ω and 120 Ω resistances. Strains were measured using either a Wheatstone bridge circuit (Anderson et al., 1970; Brace, 1965), or a simple active resistance series circuit in the case of some of the linear loading experiments.

Gauge lengths were chosen to cover as much of the mounting surface of the samples as feasible. Gauges of 0.94 cm and 0.46 cm lengths were used.

For velocity (V_p and V_s) measurements, $\frac{1}{4}''$, $\frac{3}{8}''$, and $\frac{1}{2}''$ diameter PZT transducers were used (either coaxially plated or sandwich plated).

Generally, 3 MHz transducers were used throughout all experiments.

Acoustic velocities were determined using pulse transmission method. For some experiments, signals were recorded on an x-y plotter, using a boxcar integrator (Keithley, 881/882). Time-of-flight was also measured using the scaling wave technique (Anderson et al., 1970; Mattaboni and Schreiber, 1967). In the case of long rise time signals, first arrival was determined by simultaneously displaying both the arrival signal and its inverse. The signal would be picked at the tip of the "v" where the two signals started to emerge from the baseline noise (Warren and Anderson, 1973).

The hydrostatic press is that described by Anderson et al. (1970) which uses a pressure medium of 50-50 mixture of kerosene and petroleum ether.

Hydrostatic pressures were measured using a Heise gauge correlated against a manganin coil. The uniaxial press was a simple single-ram, two-piston press.

Samples were unjacketed for the velocity and strain measurements under uniaxial loading. The two pistons of the press were similar to those in the press described by Warren and Anderson (1973). The acoustic transducers were mounted in stress-free cavities above aluminum piston heads. Acoustic pulse transmission time through the pistons as a function of pressure was calibrated.

Linear loads were determined from the output of a load gauge constructed on a simple Wheatstone bridge configuration of strain gauges mounted on a 2'' diameter steel cylinder. The load cell was calibrated by using a Heise pressure gauge connected to the loading ram. Uniaxial pressures on the sample were estimated from the load divided by the sample cross-section area.

Most of the experiment's sample preparation was done under normal lab temperature and humidity conditions. This included preparation for velocity and strain measurements under uniaxial loading. However, for hydrostatic measurements, the samples were jacketed either with 0.005'' thick copper shim, or with sylgard epoxy. Those jacketed in sylgard epoxy were encapsulated in a low vacuum used to outgas the sylgard.

Table 1. Sample descriptions.

Sample	Mineralogy	%	Ref.*
60335			
holocrystalline/metaigneous	Host Rock	75	13, 16
completely recrystallized	Plagioclase (An 90)	70	18, 21
breccia	Olivine	25	
two penetrative fracture	(Fo 80–Fo 82)		
systems, perpendicular and	Pyroxene (Cpx)	1	
parallel to 3–3′ direction	Mesostasis	3–4	
Vuggy	Xenocrysts	5	
	Plagioclase	100	
	Xenoliths	20	
	Troctolite	1	
	Breccia	95	
15555			
Vuggy basalt (mare)	Pyroxene	58–65	3, 8
Porphyritic	(augite and		15, 16
set of fractures generally	pigeonite)		17
E & W (lab orientation)	Plagioclase (An 90)	25	
dipping 45 N	Olivine (Fo 58)	5–8	
15555,68: 2–2′ direction is	Ilmenite ⎧Opaques⎫	1	
E–W; 3–3′ direction is	Cr-Spinel ⎨ 3–4 ⎬	0.5	
N45B–S45T	Ulvöspinel ⎩ ⎭	2	
	Cristobalite⎫		
	Tridymite ⎬	0.3–2	
	⎭		
15498			
Breccia	Matrix	84	7
high degree of	(glassy to fine		
recrystallization	granular)		
shock deformation in clasts	Clasts		
few non-penetrative fractures	Basalt	10	
some clasts show melting on	(pyroxene rich)		
rims	Plagioclase	3	
	(anorthosite)		
	Breccia	2	
	Mafic silicate	1	
	(pyroxenite)		
12063			
Type III porphyritic basalt	Pyroxene	35–60	20
	Olivine	2–20	
	Plagioclase	10–25	
	Opaques	5–15	
	Cristobalite	0–2	

*References listed in Table 2.

Data

Hydrostatic. Hydrostatic pressure velocity data are summarized in Fig. 1. Some strain measurement data are shown in Fig. 2.

In the velocity experiments compressional and shear velocity were measured in the 3–3′ direction of the sample. The precision of measuring time duration between pulse and first arrival is estimated as better than a few percent to better than 0.5% depending on the pressure range and the particular sample. The same precision estimates hold for the uniaxial velocity measurements. The data shown in Fig. 1 are not corrected for linear strain.

Linear strain corrections for rocks similar to 60335 can be estimated from the data in Fig. 2(a) and (b). Figure 2b shows two pressure runs made on the terrestrial basalt sample which served as a standard and as a precision check for strain-pressure measurements. In two separate runs, two different strain gauges (120 and 350 Ω) were mounted to two faces of the basalt, using two different types of epoxy (Epoxi-Patch and Epox-E (Duro)). The plotted data are from both the increasing and decreasing portions of the two separate pressure runs. The terrestrial sample was unjacketed.

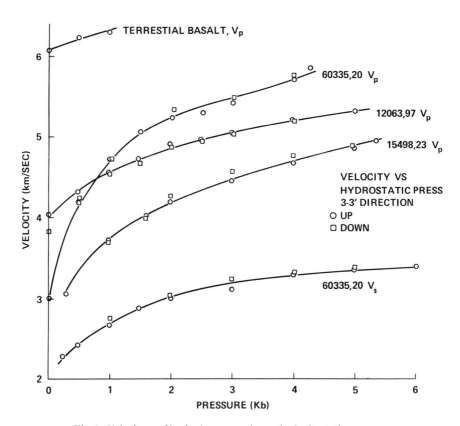

Fig. 1. Velocity profiles for lunar samples under hydrostatic pressure.

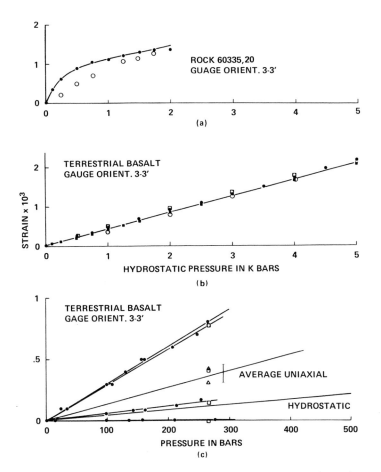

Fig. 2. Linear strain measurements under uniaxial and hydrostatic loading. Figures (a)
and (b) show results for hydrostatic pressure runs on terrestrial basalt (unjacketed), and
60335,20 (jacketed). Figure (c) shows results for terrestrial basalt under uniaxial loading
(2 runs) compared to hydrostatic loading.

Results for linear strains for 60335,20 (jacketed) under hydrostatic pressure
are shown in Fig. 2a. Black dots are from increasing pressure, open circles from
decreasing pressure. The sample was badly fractured on removal from the jacket.
Apparently some epoxy had entered fractures in the rock. From the data, it is felt
that the sample started to fail near 2 Kb.

Nevertheless, estimates of linear strain correction to velocity can be made by
extrapolating the strain curve in Fig. 2a based on the up-pressure data run to about
1.5 Kb. At 4 Kb, the velocity-strain correction is probably about or less than half a
percent.

Uniaxial. Measurements made on lunar samples under uniaxial loading provide some estimate of the magnitude of the effect of nonhydrostatic loading on

lunar seismic properties (Warren *et al.*, 1971). Linear strain data for samples under uniaxial loading are summarized in Figs. 2c and 3, and uniaxial velocity results are shown in Fig. 4.

Uniaxial strain measurements were made on the terrestrial basalt sample and compared to hydrostatic strain results as a check on the calibration of uniaxial pressure (P_L) against hydrostatic pressure (P_H). The relation between P_L and P_H follows from simple elasticity theory.

For equal 3–3′ strain in the two stress fields, P_L and P_H should be related by

$$P_H = \frac{1}{1-2\sigma} P_L \qquad (1)$$

where σ is the Poisson's ratio.

Figure 2c shows results for two separate runs on terrestial basalt. In the two runs shown, different gauges were attached using different epoxies, bonding to different faces of the sample.

Data points are for linear strains measured in the 3–3′ direction with uniaxial loading on the 3–3′ faces, for different orientations of rotation of the sample about the 3–3′ axis. Stress loading and strain of the sample are obviously nonuniform, due to, in part, the nonparallelness of the piston assembly. However, the average

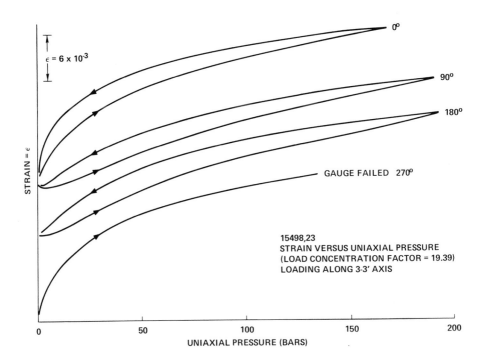

Fig. 3. Strain results under uniaxial loading for 15498,23. Sample was rotated about the 3–3′ axis. Degree marks refer to rotation orientation. The gauge failed due to a small crack which opened up under it.

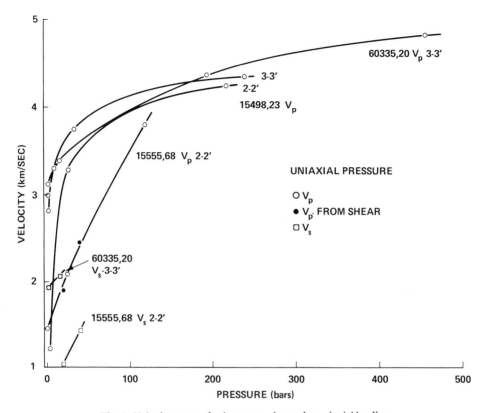

Fig. 4. Velocity curves for lunar samples under uniaxial loading.

strain-uniaxial pressure curve is in reasonable agreement with that estimated from the hydrostatic strain results.

Better results are shown for the case of lunar rock 15498,23 (Fig. 3). The curves in Fig. 3 are traced from analogue data recorded on an $x - y$ plotter. The degree marks are for different orientations of the sample, rotated about the 3–3′ axis.

Over the first few bars, nonuniform strain is recorded, similar to that observed for the terrestrial basalt. At higher pressures, however, the strains become quite consistent, independent of orientation. This observation is interpreted in terms of the large value of the compliance of the sample.

As well noted in the literature, the compliances of lunar samples under hydrostatic pressure are grossly larger than those of good terrestrial rocks (Anderson *et al.*, 1970; Stephens and Lilley, 1970). Under uniaxial loading, strains are even larger. Equation (1) probably gives a lower estimate of P_H/P_L for very heterogeneous samples.

The large strains observed for uniaxial loading are consistent with observed large changes in velocity for samples under uniaxial loading, since for a highly fractured or porous material pore deformation dominates the moduli-pressure relation of the sample.

Uniaxial velocity data are summarized in Fig. 4. Pressure corresponds to the ram pressure multiplied by the ratio of ram to sample effective area. The ratios are 21.56 (15555,68), 19.38 (15498,23: 2–2′), 27.04 (14598,23 3–3′), and 56.06 (60335,20). Velocities plotted in Fig. 4 are not corrected for strain. For 15498,23, strain correction to the velocity at 150 bars is a few percent.

Model rocks

Velocities and Poisson's ratios were obtained for model solid rocks which could be compared to the values expected from experimental measurements. The moduli and velocities were calculated using a computer program involving the self-consistent approximation method of Budiansky (1965). The program, based on an independent derivation by Korringa (1973), was compiled and donated by Thompson and Runga, Chevron Oil Co., La Habra, California.

Korringa's method provides the Hashin and Shtrikman bounds (Hashin and Shtrikman, 1963) for the moduli of a polycrystalline aggregate as well as average moduli values for the case of spherical grains, with or without spherical pores. The program yields the result of Kröner (Kröner, 1967; Thomsen, 1972) in its first reiterative step.

Model mineralogies and average moduli for the minerals are given in Table 2. Calculated values of velocities, density, and Poisson's ratios are presented in Table 3, and compared to observed values where possible.

Besides obvious difficulties due to inhomogeneities in the mineralogy of different portions of single rock samples, a strong limitation on lunar rock modeling, at this stage, is that published modal analyses of lunar rocks are often not detailed enough to allow confident parameterization of the rock model program. For example, both mineralogy and corresponding VRH elastic moduli may need to be arbitrarily assigned to minerals simply identified as "opaques."

The models presented are calculated directly from the model mineralogies and crystal elastic properties, and have not been reiterated to any "best fit" to the measured rock elastic properties.

Discussion

Rock structure: modeling. Measured velocity, strain, and model rock parameters begin to allow better quantitative understanding and modeling of lunar rocks and lunar *in situ* geophysical structure.

An approach to quantitatively modeling lunar rocks and *in situ* geophysical parameters is demonstrated using the results for 60335,20. A crude crack aspect ratio distribution for 60335,20 can be obtained using the strain data presented in Fig. 2a, and checked in terms of its expected effect on elastic compressibility.

Crack porosity and crack aspect ratios may be derived from the strain data on 60335,20 (up-pressure run), using the results of Walsh (1965) as done by Brace (1965). The estimate is very approximate, since the rock started to fail, apparently near the maximum pressure achieved during the experiment.

Table 2. Rock model compositions and mineral properties.

(ρ in gm/cm^3; bulk modulus K; and shear modulus μ in megabars)

Sample Number		Plagioclase	Pyroxene	Olivine	"Opaques"	Silica	Ref.
12065		(An 90)	(Augite-Pigeonite)	(Fo 72)	(Spinel)	(Cristobalite)	11, 20
	Vol. %	17.44	67.78	2.76	11.24	0.78	
	ρ	2.73	3.32	3.53	3.6	2.3	
	K	0.806	0.957	1.266	2.0	0.309	
	μ	0.376	0.58	0.729	1.1	0.266	
10017		(An 80)	(Augite-Pigeonite)		(Ilmenite)		4, 20
	Vol. %	25.1	59.4		14.5		
	ρ	2.72	3.32		4.9		
	K	0.775	0.957		1.75		
	μ	0.366	0.58		0.74		
12038		(An 90)	(Augite)		(Ilmenite)	(Cristobalite)	6, 11,
	Vol. %	33.0	57.0		10.0	5.0	20
	ρ	2.73	3.32		5.0	2.32	
	K	0.806	0.957		1.75	0.309	
	μ	0.370	0.58		0.742	0.266	
12018		(An 90)	(Augite-Pigeonite)	(Fo 75)	(Spinel)	(Cristobalite)	2, 11,
	Vol. %	17.7	55.7	18.7	6.95	0.61	19, 20
	ρ	2.73	3.32	3.5	3.8	2.32	
	K	0.806	0.957	1.267	2.0	0.309	
	μ	0.376	0.58	0.735	1.0	0.266	
62295		(An 90)	(Augite)	(Fo 85)	(Ilmenite)		1, 9, 14
	Vol. %	73.0	5.0	11.0	3.0		16, 18
	ρ	2.73	3.32	3.386	4.0		
	K	0.806	0.957	1.272	2.0		
	μ	0.376	0.58	0.76	1.1		
					(Spinel)		
					(Mg 90)		
					8.0		
					3.62		
					2.02		
					1.17		
60335		(An 90)	(Bronzite)	(Fo 80)			13, 16
	Vol. % $\begin{cases}\end{cases}$	(a) 71.8	(a) 8.7	(a) 19.5			
		(b) 79.5	(b) 0.8	(b) 19.7			18, 21
	ρ	2.73	3.38	3.44			
	K	0.806	1.022	1.264			
	μ	0.376	0.607	0.748			
14310 Model A		(An 90)	(Bronzite)		⎛ Ulvöspinel ⎞		3, 5, 10
	Vol. %	54.0	44.0		⎜ Ilmenite ⎟		12
	ρ	2.73	3.38		⎜ Troilite ⎟		
	K	0.806	1.022		⎝ Apatite ⎠		
	μ	0.376	0.607				
14310 Model B		(An 70)					
	Vol. %	54.0			2.0 ⎤		
	ρ	2.71			4.0 ⎟		
	K	0.749	Same as Model A		2.0 ⎟		
	μ	0.356			1.1 ⎦		

Table 2. (continued).

(ρ in gm/cm^3; bulk modulus K; and shear modulus μ in megabars)						
Sample Number	Plagioclase	Pyroxene	Olivine	"Opaques"	Silica	Ref.
End Member Troctolite (Spinel enriched)			(Fo 90)	(Spinel)		
Vol. %			85.0	15.0		
ρ			3.3	3.62		
K			1.274	2.02		
μ			0.772	1.17		

Mineralogy Reference List for Tables 1 and 2

1. Agrell et al., 1973.
2. Brown et al., 1971.
3. Brown et al., 1972.
4. Brown et al., 1970.
5. Brown and Peckett, 1971.
6. Christie et al., 1971.
7. Christie et al., 1973.
8. Czank et al., 1973.
9. Hodges et al., 1973.
10. Kushiro et al., 1972.
11. Kushiro et al., 1971.
12. Longhi et al., 1972.
13. LSPET, 1972, pp. 114–116.
14. LSPET, 1972, pp. 181–182.
15. Mason et al., 1972.
16. Nord et al., 1973.
17. Rhodes et al., 1973.
18. Walker et al., 1973.
19. Walter et al., 1971.
20. Warner, 1971.
21. Wilshire et al., 1973.

Elastic Moduli Reference List for Table 2

Akimoto, 1971.
Anderson and Liebermann, 1966.
Anderson et al., 1968.
Birch, 1961.
Chung, 1971.
Manghnani, 1969.
Simmons and Wang, 1971.

The crack aspect ratio α (major to minor axis) can be determined from the results of Walsh, using the observation from Fig. 2a that crack closure occurs at a pressure (P) of about 0.8 to 1 Kb, and using the model calculation of a Young's modulus (E) of about 1.15 Mb

$$\alpha = E/P$$

giving $\alpha = 1.0 \times 10^3$ to 1.5×10^3. Crack porosity, ω_c, may be estimated from Fig. 2a as described by Brace (1965). In this case, since linear strain was measured only in the 3–3′ direction ω_c is estimated simply as $3 \times$ the strain to the point of intersection of the flat portion of the strain curve through the ordinate: $\omega_c = 2.0$ to 2.4×10^{-3}. From the ratio of measured to model density (Table 3) the total porosity, ω_t, can be obtained. The total porosity is about 0.1103 and can be considered as the sum of a crack porosity, ω_c, and a spherical pore porosity, ω_s; ($\omega_s = \omega_t - \omega_c$).

These porosities and aspect ratios can be used to estimate the relative compressibility β/β_0 (Warren, 1973a). Using a computer program, the estimated relative bulk modulus is

$$K/K_0 = (\beta/\beta_0)^{-1} = 0.18$$

Table 3. Model-rock lunar samples properties.

Model	Sample Number	ρ	V_p	V_s	σ	Ref.
		(gm/cm³)	(km/sec)			
Mare Basalt	12065	3.246	7.39	4.22	0.258	
	(Experimental) $P = 10$ Kb	3.26	6.96	3.86	0.272	(a)
	10017	3.399	7.08	3.99	0.268	
	12038	3.247	6.969	3.937	0.2656	
	12018	3.284	7.42	4.22	0.261	
High Alumina to						
Anorthositic	14310A	3.041	7.118	3.949	0.2776	
	14310B	3.031	7.004	3.899	0.2755	
	(Experimental) $P = 5$ Kb	2.86	6.68	3.66	0.285	(b)
	62295	2.94	7.278	3.988	0.2854	
	(Experimental) $P = 5$ Kb	2.86	6.9	3.6	0.313	(c)
	60335A	2.925	7.144	3.914	0.2856	
	60335B	2.909	7.127	3.898	0.2867	
	(Experimental) $P = 4$ Kb	2.589	5.84	3.26	0.273	(d)
Troctolite End						
Member		3.348	8.57	4.95	0.249	

(a) Kanamori et al., 1971.
(b) Todd et al., 1972.
(c) Wang et al., 1973.
(d) This paper.

Estimated "experimental" value of K/K_0 can be obtained using model values for K_0 (Table 3), and calculated K from zero pressure velocity results at $P = 0$. The dynamic relative modulus obtained is

$$K/K_0 = 0.1$$

The estimated decrease in K of 82% vs. the observed decrease of about 90% is fairly reasonable, considering the uncertainties involved in the calculations.

This result suggests that proper choice of model crack, porosity and mineralogy parameters allows reasonable predictive modeling of bulk elastic properties.

The values of the parameters obtained for 60335,20 are in reasonable agreement with values proposed for lunar geophysical models (e.g., Warren et al., 1972).

The calculations here, however, do not take into account details of the crack aspect ratio distribution, which controls the shape of the velocity-pressure profile. Baldridge et al. (1972) have compared velocity-pressure gradients of thermally cycled terrestrial diabase to those of lunar rock 14310,72. Todd et al. (1973) have shown strong similarity of velocity and strain profiles between returned lunar rocks and meteorite-shocked terrestrial basalt.

Nonhydrostatic loading

Comparison of the calculated and observed results for 60335,20 also show the extreme control of rock elastic properties by pore and crack effects. Strong dependence on crack aspect ratios, and hence on crack strains, implies that for the lunar materials, velocity profiles may be very different under nonhydrostatic loading conditions vs. hydrostatic loading.

Although the uniaxial data presented in this paper are very possibly affected by nonuniformity in the applied stress, they are indicative of the extent to which velocity-depth or velocity-stress (pressure) profiles may be steepened if stresses are not hydrostatic.

This strong effect of nonhydrostatic stress on lunar rocks has possible consequences for lunar seismic interpretation.

Because of the apparent strengths and extensive fracture and dryness of the lunar crustal material, stress distribution within the first few kilometers of depth may not be hydrostatic, and if so, velocity profiles may not be simply modeled on results from hydrostatic pressure experiments.

In Fig. 5, the recent near-surface velocity profile by Kovach and Watkins (1973) is plotted along with early (Apollo 11) suggested velocity profiles (Schreiber *et al.*, 1970) and results from the Passive Seismic Experiments (Toksöz *et al.*, 1972a).

Below about 0.8 km, Kovach and Watkins observe a velocity of 4 km/sec. Comparison of their result to the model curves and to the results of the uniaxial pressure experiments suggest that velocity models, for, say, highly fractured rock and breccia (model II) can be increased significantly assuming nonhydrostatic

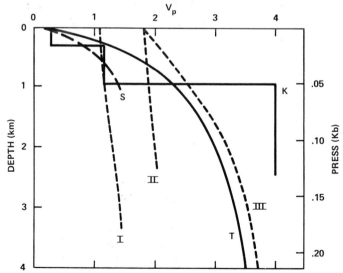

Fig. 5. Comparison of velocity (km/sec)-depth lunar profiles and model profiles. T = Toksöz *et al.*, 1972; K = Kovach and Watkins, 1973; I, II, III are soil, breccia, and rock velocity models (Schreiber *et al.*, 1970) redrawn from Warren *et al.* (1972).

stress. For uniaxial pressures in the range of 0.05 Kb, sample velocities are on the order of 3.5–4.0 km/sec. The observed upper crust velocity profile may be consistent with nonhydrostatic stresses acting on dry lunar breccia and fractured basalts.

Mineralogy and moduli of the lower lithosphere

In situ lunar velocity profiles (depths $\geq$ 25 km), and extrapolated experimental results on returned lunar samples can be compared to results for zero porosity model rocks.

There are expected uncertainties in model results, due to assumptions about mineralogy; mineral-moduli values; isotropic, homogeneous, spherical grains; and isotropic, homogeneous aggregation. Nevertheless, both the models and observed data indicate that for plagioclase (An) rich rocks, Poisson's ratios for the solid rocks tend toward 0.28–0.29, while the basaltic rocks are closer to 0.25–0.26.

A high value of σ for An-rich rocks may be a constraint on the interpretation of rock types from seismic results. It is not clear at present that interpretation of the 25–65 km zone as rich in anorthosite is consistent with the inferred *in situ* Poisson's ratio of 0.25–0.26 (Toksöz *et al.*, 1972b).

For depths greater than 60 km, model calculations support possible interpretations based on troctolite compositions. Troctolite samples have been returned in the Apollo 16 and 17 samples. Sample 62295 (Tables 2 and 3) is a plagioclase end troctolite. Prinz *et al.* (1973) have identified a troctolite lithic fragment (olivine 69, spinel 5, plagioclase 26) in sample 67435 and interpret the fragment as being a possible primative residual from primative anorthositic crustal material. In Table 3 the spinel-rich troctolite end-member rock model is the same as that mentioned in Warren *et al.* (1972). Although it is an extreme model, it is suggestive of values of the elastic properties of possible refractive residual material from anorthositic rock types.

The point to be made about possible velocity models for depths over 60 km, based on troctolitic composition, is that they have reasonable mineralogical and chemical agreement with observed crustal material; have densities which do not violate lunar requirements; and also have low Poisson's ratios, with values close to those observed *in situ*. Indeed, mineralogy trends toward An-rich troctolites may also provide values for velocities observed in the 25–65 km range, if σ must be lower than observed for anorthosites.

ATTENUATION

Experimental

The internal friction for a number of artificial lunar glasses, whose compositions were similar to those of lunar rock samples from Apollo 12, 14, and 15, was obtained. They were 12009, 12012, 14065, 14305, 15021, and 15555.

Figure 6 shows their chemical compositions in comparison with the ranges of the chemical compositions of Apollo 12, 14, and 15 rocks, soils, and breccias reported by LSPET. In addition to these six components, a small amount of K_2O and/or Na_2O were added as in the case for the original rocks.

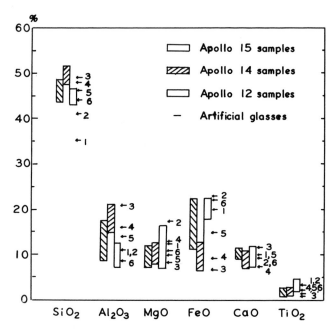

Fig. 6. The compositions of artificial lunar glasses in comparison with those of Apollo 12, 14, and 15 rocks. Their compositions are similar to: 1,12009; 2,12012; 3,14065; 4,14305; 5,15021; 6,15555.

Artificial lunar glasses were used since we have not yet been able to determine with reasonable accuracy the internal friction of lunar glass spheroids.

The glasses were melted in the same manner as described in Warren *et al.* (1972). The glass samples obtained were shaped into spheres about 3 mm in diameter for the internal friction measurements by using the sphere resonant technique (Fraser and LeCraw, 1964) employed previously (Soga and Anderson, 1967). The temperature range over which measurements were made was 77 to 420°K. This temperature range covers the temperature variation on the lunar surface at day and at night.

The arrangement of an electronic system for measuring Q is shown in Fig. 7. The specimen is placed between two shear transducers (PZT-7). When the frequency sent from the generator to the transmitting transducer through a solid-state switch matches a resonant frequency of the sphere (f_n), the sphere continues to vibrate even if the sending signal is cut off, so that a decay pattern appears on the scope. The logarithmic decrement of this decay is obtained by superimposing it with a variable RC-decay curve from the RC-decay generator on the other channel of the scope. The values of Q for the sphere can be calculated from this logarithmic decrement and f_n. These values can be checked by determining the half amplitude frequency range Δf and calculating Q from the well-known relation $1/Q = 4f/f_n \cdot \sqrt{3}$. Among the numerous modes of free oscillation of a glass sphere, the $_2S_1$ mode whose resonant frequency is about 1 MHz was used in the present study.

Data

Results are shown in Fig. 8. It seems clear that the values of Q for all the samples are almost the same in the temperature range from 77 to 300°K, above which they tend to decrease slightly. When a glass contains some alkali oxides, the internal friction peaks appear usually above 250°K. Those positions depend upon

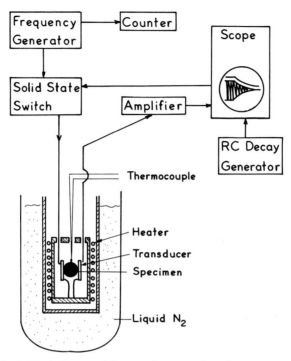

Fig. 7. Block diagram of the setup for measuring sphere resonance.

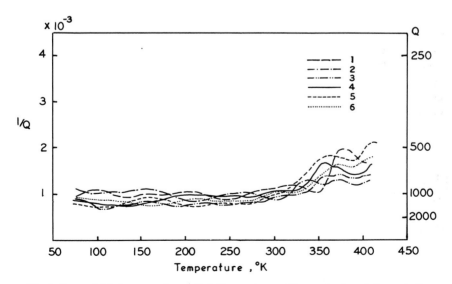

Fig. 8. Internal friction curve for artificial lunar glasses. The numbering of glasses is the same as in Fig. 6.

the types of alkali oxides (e.g., Day *et al.*, 1969). Since the alkali contents in these lunar glasses are very small (less than 1%), these alkali peaks are not distinct, but they may contribute partly to the increase in $1/Q$ above 300°K. A similar increase has been observed by Ryder and Rindone (1960) in alkali-free alkaline earth silicate glasses. Although the frequencies used in these works were on the order of 1 Hz compared with about 1 MHz in the present study, the baselines of the internal friction for these glasses were similar to the values obtained here.

DISCUSSION

The values of Q obtained for these glass spheres are fairly typical of glass. Recent work by Warren (1973b) and Tittmann and Housley (1973) have indicated that values of Q of rock under hard vacuum conditions are also in the order of thousands, even though the rock may have pore and fracture damage.

For example, a thermally pulsed terrestrial basalt, in hard vacuum, showed a change in Q from about 100 to 1600 while Young's modulus decreased about 24%. After exposure to moisture, Q returned to 100, while the modulus stayed depressed by 24%.

Such experimental results indicate that rock structure may strongly affect elastic modulus properties, while not being, *per se*, a dominant attenuation source in the rock.

One implication is the lunar observed Q values of 3000–5000 may represent reasonable values for intrinsic Q in clean, anhydrous igneous rock. This would imply that, at least for some classes of igneous rock, "internal" and "structure" related attenuation mechanisms (such as thermal elastic dissipation near crack tips, Savage, 1966; grain boundary effects, Knopoff, 1964; Peselnick and Zietz, 1959; and dislocation densities, Mason, 1971) may not be the dominant loss mechanisms lowering Q to typical rock values within a frequency range of a few Hertz through kilohertz.

Values of Q typical of glass may be indicative of values of intrinsic Q in rock under lunar conditions. This would imply that while extensive glass is not necessary to explain high Q, glass and rock models cannot be resolved on the basis of Q alone.

It is interesting to speculate that planetary bodies, such as asteroids, may also be very high Q material.

Acknowledgment—The authors wish to thank Drs. R. J. Runga and D. D. Thompson for the copy of the aggregate moduli averaging program, Dr. H. Mizutani for help when we were first rebuilding the 10 Kb pressure vessel, and Mr. R. Nashner for computer programming. This work was supported by NASA Grant NGL-05-007-330.

REFERENCES

Agrell S. O., Agrell J. E., Arnold A. R., and Long J. V. P. (1973) Some observations on rock 62295 (abstract). In *Lunar Science—IV*, pp. 15–17. The Lunar Science Institute, Houston.
Akimoto S. I. (1971) The system MgO-FeO-SiO₂ at high pressures and temperatures—phase equilibria and elastic properties. *Tectonophysics* 13, 161–187.

Anderson O. L. and Liebermann R. C. (1966) Sound velocities in rocks and minerals, VESIAC State-of-the-art report 7885-4-X. Willow Run Laboratory, Univ. of Michigan.

Anderson O. L., Schreiber E., Liebermann R. C., and Soga N. (1968) Some elastic constant data on minerals relevant to geophysics. *Rev. Geophys.* **6**, 491–524.

Anderson O. L., Scholz C., Soga N., Warren N., and Schreiber E. (1970) Elastic properties of a micro-breccia, igneous rock and lunar fines from Apollo 11 mission. *Proc. Apollo 11 Lunar Sci. Conf., Geochim. Cosmochim. Acta*, Suppl. 1, Vol. 3, pp. 1959–1973. Pergamon.

Baldridge W. S., Miller F., Wang H., and Simmons G. (1972) Thermal expansion of Apollo lunar samples and Fairfax diabase. *Proc. Third Lunar Sci. Conf., Geochim. Cosmochim. Acta*, Suppl. 3, Vol. 3, pp. 2599–2609. MIT Press.

Birch F. (1961) The velocity of compressional waves in rocks to 10 kilobars, Part 2. *J. Geophys. Res.* **66**, 2199–2224.

Brace W. F. (1965) Some new measurements of linear compressibility of rocks. *J. Geophys. Res.* **70**, 391–398.

Brown G. M., Emeleus C. H., Holland J. G., and Phillips R. (1970) Mineralogical, chemical, and petrological features of Apollo 11 rocks and their relationship to igneous processes. *Proc. Apollo 11 Lunar Sci. Conf., Geochim. Cosmochim. Acta*, Suppl. 1, Vol. 1, pp. 195–219. Pergamon.

Brown G. M. and Peckett A. (1971) Selective volatilization on the lunar surface: Evidence from Apollo 14 feldspar-phyric basalts. *Nature* **234**, 262–266.

Brown G. M., Emeleus C. H., Holland J. G., Peckett A., and Phillips R. (1971) Picrite basalts, fer-robasalts, feldspathic norites and rhyolites in a strongly fractionated lunar crust. *Proc. Second Lunar Sci. Conf., Geochim. Cosmochim. Acta*, Suppl. 2, Vol. 1, pp. 583–600. MIT Press.

Brown G. M., Emeleus C. H., Holland J. G., Peckett A., and Phillips R. (1972) Mineral-chemical variations in Apollo 14 and Apollo 15 basalts and granitic fractions. *Proc. Third Lunar Sci. Conf., Geochim. Cosmochim. Acta*, Suppl. 3, Vol. 1, pp. 141–157. MIT Press.

Budiansky B. (1965) On the elastic moduli of some heterogeneous materials. *J. Mech. Phys. Solids* **13**, 223–227.

Christie J. M., Lally J. S., Heuer A. H., Fisher R. M., Griggs D. T., and Radcliffe S. V. (1971) Comparative electron petrography of Apollo 11, Apollo 12 and terrestrial rocks. *Proc. Second Lunar Sci. Conf., Geochim. Cosmochim. Acta*, Suppl. 2, Vol. 1, pp. 69–89. MIT Press.

Christie J. M., Heuer A. H., Radcliffe S. V., Lally J. S., Nord G. L., Fisher R. M., and Griggs D. T. (1973) Lunar Breccias: An electron petrographic study (abstract). In *Lunar Science—IV*, pp. 133–135. The Lunar Science Institute, Houston.

Chung D. H. (1971) Elasticity and equations of state of olivines in the Mg_2SiO_4–Fe_2SiO_4 system. *Geophys. J. R. Astr. Soc.* **25**, 511–538.

Czank M., Girgis K., Gubser R. A., Harnik A. B., Laves F., Schmid R., Schulz H., and Weber L. (1973) Temperature dependence of the diffuseness of c-reflections in Apollo 15 plagioclases (abstract). In *Lunar Science—IV*, pp. 169–171. The Lunar Science Institute, Houston.

Day D. E. and Steinkamp W. E. (1969) Mechanical damping spectrum for mixed-alkali R_2O Al_2O_3/$6SiO_2$ glasses. *J. Am. Ceram. Soc.* **52**, 571–574.

Fraser D. B. and LeCraw R. C. (1964) Novel method of measuring elastic and anelastic properties of solids. *Rev. Sci. Instr.* **35**, 1113–1115.

Hashin Z. and Shtrikman S. (1963) A variational approach to the theory of the elastic behavior of multi-phase materials. *J. Mech. Phys. Solids* **11**, 127.

Hodges F. N., Kushiro I., and Seitz M. G. (1973) Petrology of lunar highland rocks of Apollo 16 (abstract). In *Lunar Science—IV*, pp. 371–375. The Lunar Science Institute, Houston.

Kanamori H., Mizutani H., and Hamano Y. (1971) Elastic wave velocities of Apollo 12 rocks at high pressures. *Proc. Second Lunar Sci. Conf., Geochim. Cosmochim. Acta*, Suppl. 2, Vol. 3, pp. 2323–2326. MIT Press.

Knopoff L. (1964) Q. *Rev. Geophys* **2**, 625–660.

Korringa J. (1973) Theory of elastic constants of heterogeneous media. *J. Math. Phys.* **14**, 509–513.

Kovach R. L., Watkins J. S., Nur A., and Talwani P. (1973) The properties of the shallow lunar crust: An overview from Apollo 14, 16, and 17 (abstract). In *Lunar Science—IV*, pp. 144–145. The Lunar Science Institute, Houston.

Kovach R. L. and Watkins J. S. (1973) Apollo 17 seismic profiling, probing the lunar crust. *Science* **80**, 1063–1064.

Kröner E. (1967) Elastic moduli of perfectly disordered composite materials. *J. Mech. Phys. Solids* **15**, 319.

Kushiro I., Ikeda Y., and Nakamura Y. (1972) Petrology of Apollo 14 high-alumina basalt. *Proc. Third Lunar Sci. Conf., Geochim. Cosmochim. Acta*, Suppl. 3, Vol. 1, pp. 115–129. MIT Press.

Kushiro I., Nakamura Y., Kitayama K., and Akimoto S. (1971) Petrology of some Apollo 12 crystalline rocks. *Proc. Second Lunar Sci. Conf., Geochim. Cosmochim. Acta*, Suppl. 2, Vol. 1, pp. 481–495. MIT Press.

Longhi J., Walker D., and Hays J. F. (1972) Petrography and crystallization history of basalts 14310 and 14072. *Proc. Third Lunar Sci. Conf., Geochim. Cosmochim. Acta*, Suppl. 3, Vol. 1, pp. 131–139. MIT Press.

LSPET (1972) Lunar Sample Information Catalog: Apollo 16, MSC 03210, 114–116, 181–182.

Manghnani M. H. (1969) Elastic constants of single-crystal rutile under pressure to 7.5 kilobars. *J. Geophys. Res.* **74**, 4317–4328.

Mason W. P. (1971) Internal friction in moon and earth rocks. *Nature* **234**, 461–462.

Mason B., Jarosewich E., Melson W. G., and Thompson G. (1972) Mineralogy, petrology, and chemical composition of lunar samples 15085, 15256, 15271, 15471, 15475, 15476, 15535, 15555, and 15556. *Proc. Third Lunar Science Conf., Geochim. Cosmochim. Acta*, Suppl. 2, Vol. 1, pp. 785–796. MIT Press.

Mattaboni P. and Schreiber E. (1967) Method of pulse transmission measurements for determining sound velocities. *J. Geophys. Res.*, **72**, 5160–5163.

Nord G. L., Lally J. S., Christie J. M., Heuer A. H., Fisher R. M., Griggs D. T., and Radcliffe S. V. (1973) High voltage electron microscopy of igneous rocks from Apollo 15 and 16 (abstract). In *Lunar Science—IV*, pp. 564–566. The Lunar Science Institute, Houston.

Peselnick L., and Zietz I. (1959) Internal friction of fine grained limestones at ultrasonic frequencies. *Geophysics* **24**, 285–296.

Prinz M., Dowty E., Keil K., and Bunch T. E. (1973) Spinel troctolite and anorthosite in Apollo 16 samples. *Science* **179**, 74–76.

Rhodes J. M., Ridley W. I., and Brett R. (1973) Classification and petrogenesis of Apollo 15 mare basalts (abstract). In *Lunar Science—IV*, pp. 622–624. The Lunar Science Institute, Houston.

Ryder R. J. and Rindone G. E. (1960) Internal friction of simple alkali silicate glasses containing alkaline-earth oxides: I. *J. Am. Ceram. Soc.* **43**, 662–669.

Savage J. C. (1966) Thermoelastic attenuation of elastic waves by cracks. *J. Geophys. Res.* **71**, 3929–3938.

Schreiber E., Anderson O. L., Soga N., Warren N., and Scholz C. (1970) Sound velocity and compressibility for lunar rocks 17 and 46 and for glass spheres from the lunar soil. *Science* **167**, 732–734.

Simmons G. and Wang H. (1971) Single crystal elastic constants and calculated aggregate properties; a handbook. MIT Press.

Soga N. and Anderson O. L. (1967) Elastic properties of tektites measured by resonant sphere technique. *J. Geophys. Res.* **72**, 1733–1739.

Stephens D. R. and Lilley E. M. (1970) Loading-unloading pressure-volume curves to 40 kilobars for lunar crystalline rock, microbreccia and fines. *Proc. of Apollo 11 Lunar Sci. Conf., Geochim. Cosmochim. Acta*, Suppl. 1, Vol. 3, pp. 2427–2434. Pergamon.

Thomsen L. (1972) Elasticity of polycrystals and rocks. *J. Geophys. Res.* **77**, 315–327.

Tittmann B. R. and Housley R. H. (1973) High Q (low internal friction) observed in a strongly outgassed terrestrial analog of lunar basalt. *Phys. Stat. Solidi*, (b) **56**, K109–K111.

Todd T., Wang H., Baldridge W. S., and Simmons G. (1972) Elastic properties of Apollo 14 and 15 rocks. *Proc. Third Lunar Sci. Conf., Geochim. Cosmochim. Acta*, Suppl. 3, Vol. 3, pp. 2577–2586. MIT Press.

Todd T., Wang H., Richter D., and Simmons G. (1973) Unique characterization of lunar samples by physical properties (abstract). In *Lunar Science—IV*, pp. 731–733. The Lunar Science Institute, Houston.

Toksöz M. N., Press F., Dainty A. M., Solomon S. C., and Anderson K. R. (1973) Lunar structure, compositional inferences and thermal history (abstract). In *Lunar Science—IV*, pp. 734–736. The Lunar Science Institute, Houston.

Toksöz M. N., Press F., Anderson K., Dainty A., Latham G., Ewing M., Dorman J., Lammlein D., Nakamura Y., Sutton G., and Duennebier F. (1972a) Velocity structure and properties of the lunar crust. *The Moon* **4**, 490–504.

Toksöz M. N., Press F., Dainty A., Anderson K., Latham G., Ewing M., Dorman J., Lammlein D., Sutton G., Duennebier F. (1972b) Structure, composition, and properties of lunar crust. *Proc. Third Lunar Sci. Conf.*, *Geochim. Cosmochim. Acta*, Suppl. 2, Vol. 3, pp. 2527–2544. MIT Press.

Walker D., Longhi J., and Hays J. F. (1973) Petrology of Apollo 16 metaigneous rocks (abstract). In *Lunar Science—IV*, pp. 752–755. The Lunar Science Institute, Houston.

Walsh J. B. (1965) The effect of cracks on the uniaxial compression of rocks. *J. Geophys. Res.* **70**, 399–411.

Walter L. S., French B. M., Heinrich K. J. F., Lowman P. D. Jr., Doan A. S., and Adler I. (1971) Mineralogical studies of Apollo 12 samples. *Proc. Second Lunar Sci. Conf.*, *Geochim. Cosmochim. Acta*, Suppl. 2, Vol. 1, pp. 343–358. MIT Press.

Wang H., Todd T., Richter D., and Simmons G. (1973) Elastic properties of plagioclase aggregates and seismic velocities in the moon (abstract). In *Lunar Science—IV*, pp. 758–760. The Lunar Science Institute, Houston.

Warner J. L. (1971) Lunar crystalline rocks: Petrology and geology. *Proc. Second Lunar Sci. Conf.*, *Geochim. Cosmochim. Acta*, Suppl. 2, Vol. 1, pp. 469–480. MIT Press.

Warren N. (1973a) Theoretical calculations of the compressibility of porous media. *J. Geophys. Res.* **78**, 352–362.

Warren N. (1973b) Brief note on effects of thermal pulses on elastic moduli and Q of rock. *Earth Planet. Sci. Lett.* In press.

Warren N. and Anderson O. L. (1973) Elastic properties of granular materials under uniaxial compaction cycles. *J. Geophys. Res.* **78** (29).

Warren N., Anderson O. L., and Soga N. (1972) Applications to lunar geophysical models of the velocity-density properties of lunar rocks, glasses, and artificial lunar glasses. *Proc. Third Lunar Sci. Conf.*, *Geochim. Cosmochim. Acta*, Suppl. 3, Vol. 3, pp. 2587–2598. MIT Press.

Warren N., Schreiber E., Scholz C., Morrison J. A., Norton P. R., Kumazawa M., and Anderson O. L. (1971) Elastic and thermal properties of Apollo 11 and 12 rocks. *Proc. Second Lunar Sci. Conf.*, *Geochim. Cosmochim. Acta*, Suppl. 2, Vol. 3, pp. 2345–2360. MIT Press.

Wilshire H. G., Stuart-Alexander D. E., and Jackson E. D., (1973) Petrology and classification of the Apollo 16 samples (abstract). In *Lunar Science—IV*, pp. 784–786. The Lunar Science Institute, Houston.

Proceedings of the Fourth Lunar Science Conference
(Supplement 4, *Geochimica et Cosmochimica Acta*)
Vol. 3, pp. 2631–2637

Internal friction of rocks and volatiles on the moon

B. R. TITTMANN, R. M. HOUSLEY, and E. H. CIRLIN

Science Center, Rockwell International, Thousand Oaks, California 91360

Abstract—Internal friction quality factors Q up to 2200 have been observed in a strongly outgassed terrestrial analog of lunar basalt. This was accomplished by successively cycling a bar shaped sample vibrating in its fundamental longitudinal mode at 15 kHz to higher and higher temperatures in a vacuum between 10^{-7} and 10^{-8} torr. After each cycle, Q measured at room temperature in the vacuum was observed to decrease with time suggesting that gas re-absorption was taking place even at these low pressures.

A study of the effect of exposing a sample to a variety of gases and vapors showed that of the volatiles most likely to be present in the lunar environment H_2O was by far the most effective in lowering Q.

The fact that only a very small amount of volatile material is required to lower the Q of rocks suggests that with further study of the loss mechanism, the lunar seismic Q values may be used to place strong constraints on the outgassing history of the moon and on any possible prehistoric lunar atmospheres.

INTRODUCTION

WE HAVE recently reported (Tittmann and Housley, 1973) dramatic changes in the internal friction of a terrestrial basalt when it was outgassed under conditions of high vacuum and thermal cycling. The quality factor Q at room temperature increased from below 75 in normal laboratory air to above 1500 after outgassing at elevated temperature in a vacuum better than 10^{-7} torr, then again fell to about 100 on re-exposing the sample to laboratory air.

This dramatic increase in Q was not unexpected. Since Latham *et al.* (1970) had indicated that the *in situ* seismic Q of lunar rocks is in the range 3000–5000, it has seemed inescapable to many people that the Q of rocks really free of volatiles, measured in the laboratory, should be similar. However, for reasons which are discussed in the experimental section, this conclusion had not previously been confirmed.

The demonstration that the major part of the internal friction of rocks is caused by absorbed atmospheric volatiles allows us to conclude that all previously suggested internal friction mechanisms in rocks, including the popular sliding friction across microfractures (Knopoff and MacDonald, 1958; Walsh, 1966), plastic flow at contact points (Mason, 1971), and thermoelastic relaxation (Savage, 1966) are relatively unimportant, at least for this type of rock under ordinary laboratory conditions. It then forces us to conclude that previous laboratory studies of internal friction in rocks are not suitable as guides to thinking about the seismic behavior of lunar rocks. Further studies of the frequency, temperature, composition, and texture dependence of Q in thoroughly outgassed rocks will be required.

It seems likely that the high *in situ* Q values observed on the moon can be used to place strong constraints on the absorbed volatile content of lunar rocks and hence on the integrated outgassing history of the lunar interior and on any past lunar atmospheres. In order to do this, however, extensive measurements on thoroughly outgassed rocks will have to be made to see if the effect of volatiles is still strong at seismic frequencies and if all major lunar rock types have high Q values when volatile free. We will also have to determine which gases present on the moon, or likely to have been outgassed from the interior, are effective in lowering the Q and to establish the quantitative relationship between amount absorbed and reduction in Q at temperatures appropriate for the lunar subsurface.

MEASUREMENT TECHNIQUE

Sample preparation

All our absolute Q measurements so far have been done on either lunar rock 14310,86 or on olivine basalt designated W-8 from Shasta County, California, which was chosen to be similar to some lunar basalts in mineralogy and which behaved similarly to 14310,86 in early outgassing experiments.

These rocks were attached to metal blocks with low melting point wax and sawed into bars with an oil-cooled diamond saw. Preliminary attempts to saw rocks into bars with a gas-cooled diamond saw always resulted in breaking of the bars.

After sawing, the bulk of the wax and oil was removed from the samples by washing in acetone at room temperature. This procedure typically left the Q at about 20. It was generally followed with a cleaning procedure in which the sample was washed first in boiling trichloethylene for about an hour, then in boiling ethanol, and finally in boiling distilled water. Lately we have supplemented or partially bypassed the above cleaning procedure by boiling the samples in 30% H_2O_2 for up to several hours, until excess bubbling stopped, indicating that oxidizable organics had all been consumed. Although no systematic comparison has yet been made on the outgassing behavior of similar samples cleaned by different procedures our general impression is that the H_2O_2 treatment is the most effective. With these procedures we have generally been able to raise the Q to about 75–100.

Both our longitudinal and our transverse bar resonance apparatuses employ a magnetic drive and detection system utilizing small iron buttons cemented to the ends of the bars (Wegel and Walther, 1935). Initially we used an epoxy to cement the buttons to the bars but eventually learned that, at least for small samples, losses in the epoxy dominated the other loss mechanisms above room temperature. We now use either a fast drying general purpose household cement, which is satisfactory at room temperature, or a high temperature silicate cement. The latter is cured by gradually warming the sample to about 150°C in air and holding it there for about a day. This drying procedure raises Q at room temperature to as high as ~ 400.

Longitudinal resonance apparatus

The basic longitudinal resonance apparatus has recently been described (Tittmann, 1972). Since then we have slightly modified the support system to accommodate larger samples and have rounded the support points somewhat to minimize the risk of sample breakage.

Transverse resonance apparatus

The basic apparatus has been thoroughly described by Schlein and Shen (1969). The procedure of clamping the sample with sharp pointed set screws which was satisfactory for metals and plastics proved to be unsatisfactory for brittle materials such as rocks, so various alternatives including strings, wires, and knife edges were tried. In a fairly satisfactory holder which we currently use the sample sits at its nodal points on small ceramic roller bearings, which allow it complete freedom to expand or contract with temperature. This apparatus is now being modified to permit measurements in pressure down to 10^{-7} torr and in controlled atmospheres.

RESULTS

Strongly outgassed rocks

New results obtained with the longitudinal mode apparatus on a W-8 bar $12.4 \times 0.317 \times 0.317$ cm in dimensions are presented in Figs. 1 and 2. Figure 1 summarizes Q data during one series of heat treatments and plots the measured Q

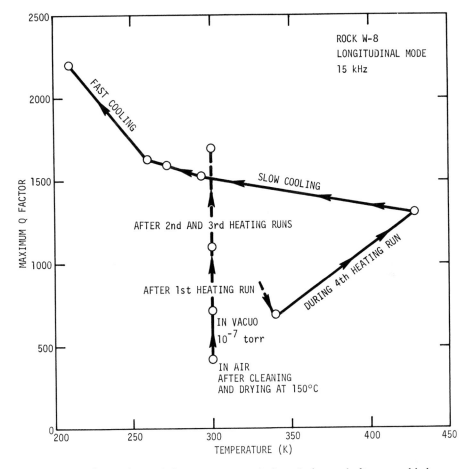

Fig. 1. Maximum observed Q vs. temperature before, during, and after several bake-out cycles in a vacuum of between 10^{-8} and 10^{-7} torr. The data are for a terrestrial analog of lunar basalt in the shape of bar driven in longitudinal resonance. The highest value achieved was $Q \approx 2200$.

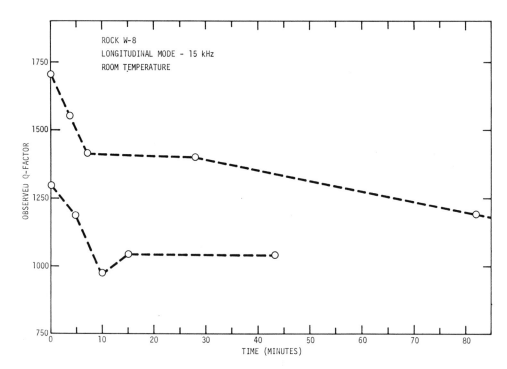

Fig. 2. Room temperature Q as a function of time immediately after temperature stability was achieved at the end of each of two heating cycles. The data indicate first a rapid,
then a more gradual decrease in Q with time suggesting that the rock re-absorbs volatile
impurities which were removed during the heating cycle even at pressure less than 10^{-7}
torr.

vs. the absolute temperature. Since after each heating cycle the Q decreased
rapidly with time, as will be discussed below, the Q values plotted in Fig. 1 are the
initial maximum values obtained immediately after temperature equilibrium was
achieved and in turn represent lower limits on the intrinsic Q at each temperature.
The data begin with $Q \approx 400$ for the bar after thorough cleaning and drying as
discussed. After a several hours exposure to a vacuum of 1×10^{-7} torr resulting in
a Q increase to about 700, the heat treatment of the rock was initiated.

The heaters each consisted of a Nichrome wire coil inside a 9 mm diameter
glass tube surrounded by a thick-walled Cu cylinder which smoothed out thermal
gradients and reflected the radiant energy back to the sample. Two of these small
tubular furnaces surrounded the sample except in the immediate vicinity of the
ends and at the center support point. Temperatures were varied with a variable
auto-transformer and measured with a thermocouple attached to one of the Cu
tubes.

Because of the high speed of the vacuum pump (15 liter/minute) the vacuum
pressure during the heat treatment typically stayed below 5×10^{-7} torr. In order to
minimize re-absorption of vapors by the bar during cooling, the walls of the glass

and stainless steel vacuum chamber were outgassed during the peak of the sample heating cycle by means of heating tapes wrapped around the structure. During the initial part of the heating cycle, the Q typically dropped to about 500 whereas cooling in the last part of the cycle always produced a Q increase. Each successive heating cycle produced an increase in room temperature Q depending strongly on the cooling speed and somewhat on the highest temperature achieved during the heating run. The first heating run with peak temperatures near 100°C produced a rise in room temperature Q to about 1100, the second and third heating runs with peak temperatures near 140 and 170°C respectively, brought the Q successively higher to 1300 and to 1700. In addition to raising the Q, each successive heating cycle widened the temperature range of high Q values, so that in the fourth heating run even the high temperature Q was 1300.

After each cooling cycle, the Q was observed to decrease with time as demonstrated in Fig. 2, which plots the room temperature Q as a function of the time in minutes after temperature equilibrium was achieved. The initial decrease in Q is rapid, about $\Delta Q \approx 300$ in the first 10 minutes which is fortunately still slow enough to obtain meaningful measurements (each measurement takes less than $\frac{1}{2}$ minute). This rapid time dependence of the Q followed by a continued more gradual decrease strongly suggests that re-absorption of volatiles takes place even in a well-baked out vacuum chamber at 10^{-8} torr. This behavior probably also explains the necessity for rapid cooling at the end of a heating cycle. This is exemplified in the data of Fig. 1 for the fourth heating run. Here a slow cooling gave a Q well below the previous high of $Q \approx 1700$, but when the cooling rate was suddenly increased, the Q also increased so that a maximum value of ~ 2200 was obtained immediately upon temperature equilibration. This run and previous experiments with slow and fast cooling suggest that re-absorption takes place even during the cooling cycle and severely limits the maximum Q value observed. Upon re-exposure to air and water vapor the Q decreased again to ~ 100.

In the light of the above results it is appropriate to briefly discuss previously reported (Tittmann, 1972) data on lunar rock 14310,86. The measurements above and below room temperature were done in separate series of runs. For those below room temperature the sample had not been outgassed at elevated temperatures. Those above room temperature were taken during warming only. It now seems likely that these measurements were all dominated by remaining absorbed volatiles and do not represent intrinsic properties of the rock.

Effect of various volatiles

In order to obtain some idea of which volatiles are effective in contributing to the internal friction in rock, a W-8 sample open to laboratory air on a low humidity day in the transverse mode apparatus was exposed to a variety of gases and vapors. Blowing one's breath on the sample resulted in an immediate substantial reduction in Q which we attribute to water vapor. The sample returned to its initial Q value within a few minutes. The gases except for HCl and NH_3 were gently blown on the sample through a rubber hose. Exposure to HCl and NH_3 was

Table 1. Effect of gases and vapors
on Q.

Big Effect	No Effect
Ethanol	N_2
Methanol	CO_2
Trichloroethane	O_2
Acetone	H_2
Carbon tetrachloride	NH_3
Water	HCl
	Ar
	He

achieved by holding beakers of saturated solutions of these gases in water below the sample. Exposure to vapors was obtained by painting a film of the corresponding liquid over the sample holder with a cotton swab. Following complete evaporation Q always returned to its initial value within minutes. The results of these simple qualitative tests are given in Table 1. All vapors tested proved to be of comparable effectiveness in reducing Q while all gases showed no effect and hence were at least an order of magnitude less effective.

DISCUSSION

The temperature cycling results shown in Fig. 1 indicate that the Q of a rock at any particular temperature and vacuum depends to a great extent on its previous outgassing history including temperature, time, and pressure, so that the highest observed $Q \approx 2200$ can best be viewed as a lower bound on the Q of W-8 rock in the complete absence of any absorbed volatiles. The mechanism by which the absorbed volatiles interact with the elastic wave to cause absorption is not understood at this time.

Table 1 and previous studies (Tittmann and Housley, 1973) strongly suggest that of the likely lunar volatiles examined thus far, water is by far the most effective in reducing Q. Although no quantitative data on reduction in Q vs. amount absorbed are yet available, the difficulty of removing enough to achieve a high Q state implies that very small amounts are effective in lowering Q.

The high lunar seismic Q values (Latham *et al.*, 1970) together with the above argument suggest that there are no extensive regions near any of the seismic stations where the lunar rocks are even a little bit damp and that, on the average, water loss from the surface rocks into space has exceeded water injection due to outgassing of the lunar interior. This last conclusion, however, would be stronger if more information were available on the possible frequency dependence of the loss.

Acknowledgment—We thank Dr. M. Abdel-Gawad for providing and characterizing our terrestrial basalt sample. This work was partially supported by NASA Contract NAS9-11539.

REFERENCES

Knopoff L. and MacDonald G. J. F. (1958) Attenuation of small amplitude stress waves in solids. *Rev. of Mod. Phys.* **30**, 1178–1192.

Latham G. V., Ewing M., Dorman J., Press F., Toksoz N., Sutton G., Meissner R., Duennebier F., Nakamura Y., Kovach R., and Yates M. (1970) Seismic data from man-made impacts on the moon. *Science* **170**, 620–626.

Mason W. P. (1971) Internal friction in moon and earth rocks. *Nature* **234**, 461–463.

Savage J. C. (1966) Thermoelastic attenuation of elastic waves by cracks. *J. Geophys. Res.* **71**, 3929–3938.

Schlein H. and Shen M. (1969) Computerized acoustic spectrometer for polymeric solids at low temperatures. *Rev. Sci. Instr.* **40**, 587–591.

Tittmann B. R. (1972) Rayleigh wave studies in lunar rocks. *IEEE Ultrasonics Symposium Proceedings,* Catalog No. 72 CH0708–8SU, 130–134.

Tittmann B. R. and Housley R. M. (1973) High *Q* (low internal friction) observed in a strongly outgassed terrestrial analog of lunar basalt. *Phys. Stat. Solidi* **56**, K109–K111.

Walsh J. B. (1966) Seismic attenuation in rock due to friction. *J. Geophys. Res.* **71**, 2591–2599.

Wegel R. and Walther H. (1935) Internal dissipation in solids for small cyclic strains. *Physics* **6**, 141–157.

Proceedings of the Fourth Lunar Science Conference
(Supplement 4, *Geochimica et Cosmochimica Acta*)
Vol. 3, pp. 2639–2662

Unique characterization of lunar samples by physical properties

Terry Todd, Dorothy A. Richter, and Gene Simmons

Department of Earth and Planetary Sciences
Massachusetts Institute of Technology
Cambridge, Massachusetts 02139

Herbert Wang

Department of Geology and Geophysics
University of Wisconsin
Madison, Wisconsin 53706

Abstract—The measurement of compressional velocity, shear velocity, static compressibility, and thermal expansion of (1) a suite of shocked rocks from the Ries impact crater in Germany, (2) a suite of samples cracked by thermal cycling to high temperatures, (3) many terrestrial igneous rocks, and (4) lunar basalts, gabbroic anorthosites, and breccias, indicate that shock metamorphism is the primary cause for values of physical properties of lunar rocks being different from their intrinsic values. Large scale thermal metamorphism, thermal cycling between temperatures of lunar day and night, large thermal gradients, or thermal fatigue could possibly cause minor cracking in the top few centimeters of the lunar regolith, but are probably not important mechanisms for extensively changing values of physical properties of lunar rocks.

Values of physical properties as functions of depth or radial distance from the center of large impact craters reflect the intensity of the shock event. The profile for seismic velocity vs. depth over the first 25 km of the moon indicates decreasing shock metamorphism with depth. A depth of 25 km probably represents the maximum depth of shock metamorphism or the depth at which shock-produced microfractures are closed or annealed.

Introduction

BASIC DIFFERENCES exist between the values of physical properties of lunar rocks and those of terrestrial rocks of similar composition. These differences have been attributed to intense microfracturing in lunar rocks caused by large temperature fluctuations and gradients on the lunar surface and to shock metamorphism.

On the surface of the moon the presence of large thermal gradients and temperature fluctuations between lunar day and night can introduce crack porosity in lunar rocks (Thirumalai and Demou, 1970). Thermally induced microseismic activity on the moon (Duennebier and Sutton, 1972) indicates that thermal energy may play an important role in the metamorphism of lunar rocks. Thermal metamorphism can alter values of physical properties of lunar rocks.

Shock metamorphism differs substantially from other metamorphic processes in the earth and moon. Regional thermal metamorphism on earth generally occurs over millions of years, throughout large regions of the earth, at pressures less than

2639

10 kb and temperatures less than $10^{3\circ}C$. The rates of pressurization and heating are slow and chemical equilibrium is generally approximated.

Shock metamorphism however occurs on a time scale of microseconds, at pressures up to megabars, and at temperatures up to $10^{4\circ}C$. Shock metamorphism is usually a local surface process. For very large impacts, the effects of shock metamorphism are not likely to reach more than a few tens of kilometers depth. Because the rates of pressurization and heating are extremely fast and of short duration, chemical equilibrium cannot be attained. Shock metamorphism therefore introduces crack patterns and mineralogy (e.g., glasses) distinctly different from those produced by regional metamorphic or igneous processes on earth (von Engelhardt and Stöffler, 1968; Short, 1966).

Stöffler (1971) classifies shocked rocks on the basis of six zones of increasing shock metamorphism. Rocks in zone 0 are shocked to peak shock pressures sufficient to cause microfracturing but not phase changes. Rocks in zone 1 are shocked to pressures within the two phase region of the Hugoniot. Such rocks generally contain diaplectic (stress disordered) areas within crystals of the constituent minerals, as well as other shock deformational features (e.g., slip bands, mosaicism, etc.). In zone 2, peak shock pressures are sufficient to reach the high pressure phase region of the Hugoniot so that nearly amorphous disorder (diaplectic glass) is produced. Zone 3 is characterized by selective melting of some component minerals, and zone 4 by melting of all component minerals. Rocks in zone 5 include all materials condensed from rocks vaporized by the shock event. This classification scheme is similar to that compiled by von Engelhardt and Stöffler (1968) for the Ries impact crater in Germany. Other classification schemes are those of Chao (1968), Dence (1968), James (1969), and Short (1969).

In an effort to better understand how thermal cycling and shock processes influence the values of physical properties of lunar rocks, we have measured values for compressional velocity, shear velocity, static compressibility, and thermal expansion coefficient for several thermally cycled terrestrial rocks, for a number of shocked rocks recovered from the Ries impact crater in Germany, and for lunar breccias, basalts, and gabbroic anorthosites.

EXPERIMENTAL METHOD

Velocities were measured using the standard pulse transmission technique of Birch (1960, 1961). Sample preparation and jacketing techniques are similar to those previously described by Kanamori *et al.* (1970), Wang *et al.* (1971), and Todd *et al.* (1972). Velocity measurements were made in a simple piston-cylinder high-pressure vessel. The estimated accuracy of the measurements for lunar and intensely shocked terrestrial rocks is 2–3% for P waves and 5% for shear waves. The accuracy is lower than those normally obtained for terrestrial rocks because of small sample size (1–2 cm^3) and high attenuation at low confining pressures.

Linear static compressibilities were measured (when possible) on the same samples on which velocity measurements were made. A method similar to that of Brace (1965) was used. Strain gauges (BLH Co. constantan foil-type FAE–50–12S6) were epoxied directly to the sample. The sample and gauge were then jacketed in Sylgard, and the gauges seated to the sample under 5 kb pressure, prior to a compressibility run. A correction of $+ 0.54 \times 10^{-7}$ bar^{-1} (Brace, 1964) was added to each compressibility value to compensate for the pressure effect on the strain gauges. Strains were measured point by point on a BLH type 120 C strain indicator (BLH Corp., Waltham, Mass.). At low pressure where large strains were recorded, readings were taken every 25 to 50 bars; at high pressures where strain changes were less and are approximately linear with pressure, readings were taken every few hundred bars. Compressibility was measured in one direction only, i.e., anisotropy was not investigated. Reported volume compressibility values are simply three times the linear compressibility. The estimated accuracy in the linear compressibility (slope of the strain–pressure curve) ranges from 10% at pressures less than 1 kb to about 1–2% at pressures greater than 1 kb.

Thermal expansions were measured at atmospheric pressure for the same rocks used in velocity and compressibility measurements (when possible) using a Brinkmann model TDIX dilatometer modified according to Baldridge *et al.* (1972). The linear expansion was measured only along the longest dimension of the samples, and anisotropy was not investigated. The data for each rock were expressed as relative expansion ($\Delta L/L_0$), fitted to a third degree polynomial, adjusted to a reference temperature of 25°C and corrected for the expansion of the fused quartz measuring system of the dilatometer. The dilatometer was calibrated with a core of monocrystalline quartz, cut parallel to the c-axis, using the data of Buffington and Latimer (1926) and Kozu and Takane (1929) for the expansions of quartz between 25°C and 600°C. The uncertainty in each ΔL measurement is about 2%.

Crack porosities were measured using the method of Walsh (1965). The total crack porosity (η_0) is obtained by extrapolating linearly the high pressure segment of the strain ($\Delta V/V$)-pressure (P) curve to zero pressure.

$$\eta_0 = \frac{\Delta V}{V_0} - \beta P \tag{1}$$

β is the intrinsic compressibility at high pressure. All cracks in unshocked terrestrial rocks are generally closed by 2 kb confining pressure beyond which the strain–pressure curve is linear. Extrapolating the linear portion of this curve to zero pressure gives crack porosities accurate to ± 0.05. We infer that cracks in shocked terrestrial rocks and lunar rocks are not all closed by 5 kb confining pressure because at pressures less than 5 kb, the strain–pressure curve is nowhere linear with pressure. A linear extrapolation of the strain–pressure curve from 5 kb to zero pressure provides a lower bound for crack porosity. The true values could be as much as 10% higher. The data from Stephens and Lilley (1971) can be used to provide better estimates of crack porosities because their pressure range extended above 5 kb to pressures as great as 40 kb.

The volume distribution of cracks with aspect ratio α [$h(\alpha)$] was calculated from Morlier's (1971) expression

$$h(\alpha) = -\frac{\pi E}{4(1-v^2)} P \frac{d^2\eta(P)}{dP^2} \tag{2}$$

where E and v are Young's modulus and Poisson's ratio at pressure P, respectively, and $\eta(P)$ is the crack porosity as a function of pressure. The value of $\eta(P)$ is obtained by subtracting the actual measured strain at pressure P from that predicted by a linear extrapolation of the high pressure segment of the strain–pressure curve. Values of Young's modulus and Poisson's ratio as functions of pressure were estimated from shear and compressional velocity data using the relations of infinitesimal elasticity. Because dynamically calculated elastic properties differ somewhat from statically calculated elastic properties (Simmons and Brace, 1965) and because the calculation of $(d^2\eta(P))/dP^2$ possibly involves large errors, it should be emphasized that values for the distribution function are probably only accurate to within 20%.

EFFECT OF SHOCK METAMORPHISM ON PHYSICAL PROPERTIES

There have been relatively few measurements of physical properties of shocked rocks despite the return to earth of lunar samples, the recognition of an increasing number of ancient impact craters on the earth's surface, and the interest in the effects of large nuclear explosions on the deformation of rocks. One set of data (Short, 1966) includes atmospheric pressure values for compressional velocity, fracture strength, crack densities, permeability, etc. for a suite of granodiorite samples shocked to different peak pressures. Short's samples were selected from drill core taken at different depths in a borehole drilled towards the center of the cavity produced by the Hardhat nuclear explosion. The peak shock pressure for which he reports data is about 45 kb. Rocks experiencing these peak shock pressures are equivalent to rocks from zone 0 in Stöffler's (1971) classification for shocked rocks. The effect of increased microfracturing at higher shock pressures is clearly evident in Short's data; compressional velocity and fracture strength decrease and permeability increases as peak shock pressure increases. It is improbable that any glass is present in the samples for which he reports physical property measurements.

Velocities

The complicated structure of the Ries impact crater in Germany has been studied by Angenheister and Pohl (1969), using the techniques of both reflection and refraction seismology. Their velocity profiles show low velocities associated with layers of post impact sediments, suevites, and breccias within the 20 km wide impact crater. Velocities in the fractured crystalline rock beneath the brecciated layers (1.7 km/sec) increase steadily with depth until velocities equivalent to those of the surrounding country rocks (3.2 km/sec) are reached near 2 km depth. Velocity values in the cracked crystalline rock surrounding the crater increase with radial distance and reach values equivalent to those of surrounding unshocked rocks at distances of 2–3 crater diameters. We infer from these facts either (1) that 2 km may be the maximum depth at which peak shock pressures were capable of cracking rocks beneath the Ries crater (in which case shock metamorphism alters rock properties far more extensively with radial distance than with depth), (2) that new cracks were formed, but pressures at 2 km are sufficient to close them, or (3) that cracks are present at 2 km but saturation is complete and velocity values are high (Nur and Simmons, 1969), reaching values equal to those of the surrounding country rock. However, it is improbable that the effects of an impact, capable of producing a 20 km diameter crater, would only extend to 2 km depth, and it is equally improbable that pressure (less than 1 kb) at 2 km depth is capable of closing all the cracks produced by the shock event. [Birch's (1960) data indicate that pressures greater than 1 kb, and often approaching 2 kb, are required to close all cracks in most igneous rocks.] We therefore conclude that the velocity increase with depth in the Ries Crater is due to saturation. In *dry* lunar craters, the velocity increase would be much more gradual.

Our data for compressional velocity, shear velocity, static compressibility, thermal expansion, and density of the Ries samples are given in Tables 1 and 2. Compressional velocity–pressure profiles for four Ries samples are shown in Fig. 1. These four samples were chosen to illustrate the effect of increasing shock metamorphism on elastic properties.

Table 1. Elastic properties of samples.

Sample	Density (g/cc)	Crack Porosity	Elastic Property*	1	100	250	500	750	1000	1500	2000	3000	4000	5000
10065,24	2.34	3.45	P†											
			β	120.0	70.0	38.0	20.0	15.2	11.9	8.5		6.3	5.6	4.8
12002,58	3.30	1.35	P‡											
			β	45.0	20.1	13.5	8.8	5.9	4.0	2.8	2.0	1.8	1.6	1.4
14318,30	2.81	3.55	P	2.47	2.79	3.14	3.62	3.96	4.16	4.42	4.71	5.11	5.45	5.67
			β	90.0	49.0	33.0	22.7	16.0	11.7	7.5	6.3	5.4	4.6	3.8
62295,18	2.83	0.69	P	4.39	4.73	5.11	5.54	5.81	5.98	6.23	6.42	6.66	6.82	6.92
			S	2.00	2.40	2.61	2.83	2.98	3.11	3.28	3.41	3.56	3.68	3.75
			β	15.0	10.9	7.8	5.4	4.3	3.5	2.8	2.3	2.2	2.1	2.0
65015,9	2.97	0.68	P	4.77	4.99	5.25	5.58	5.80	5.98	6.17	6.33	6.53	6.72	6.90
			S	2.38	2.49	2.63	2.79	2.91	3.02	3.19	3.29	3.45	3.54	3.63
			β	25.0	9.5	6.5	4.5	3.5	2.8	2.2	1.8	1.6	1.5	1.4
68415,54 A direction	2.78	0.83	P	4.70	5.02	5.29	5.63	5.89	6.09	6.37	6.54	6.76	6.85	6.94
			S	2.59	2.69	2.80	2.94	3.05	3.13	3.26	3.35	3.43	3.47	3,54
			β	22.0	15.2	11.1	7.7	5.8	4.7	3.8	3.4	2.7	2.6	2.5
B direction			P	4.95	5.25	5.57	5.92	6.11	6.27	6.49	6.64	6.80	6.92	7.04
			S	2.48	2.60	2.73	2.88	3.00	3.09	3.23	3.31	3.41	3.46	3.54
Ries 929	2.07		P§	3.49	3.86	4.17	4.48	4.72	4.90	5.15	5.29	5.44	5.49	5.53
			S§	—	2.08	2.40	2.70	2.84	2.93	3.05	3.12	3.19	3.22	3.23
Ries 931	2.56	1.90	P	2.94	3.50	3.77	4.06	4.24	4.39	4.66	4.80	5.01	5.25	5.49
			β	45.0	29.0	15.2	9.6	8.1	7.1	6.3	5.7	5.3	4.8	4.3
Ries 934	2.58	0.22	P	4.73	5.05	5.33	5.58	5.72	5.80	5.90	5.99	6.07	6.12	6.16
			S	2.97	3.11	3.24	3.33	3.39	3.42	3.46	3.48	3.52	3.55	3.57
			β	7.8	4.81	3.63	3.30	2.92	2.74	2.40	2.31	2.26	2.20	2.15
Ries 936	2.57	0.66	P	3.14	4.14	4.64	5.01	5.23	5.38	5.60	5.78	6.00	6.12	6.18
			S	1.80	2.15	2.52	2.74	2.87	2.96	3.09	3.19	3.27	3.32	3.36
			β	41.0	14.0	8.4	4.9	3.5	2.8	2.6	2.3	2.2	2.2	2.2
Rock 613	2.42	0.63	P‖											
			S	2.45	2.56	2.65	2.74	2.77	2.79	2.81	2.83	2.87	2.91	2.94
			β	13.1	10.0	8.2	7.2	6.71	6.33	6.07	5.93	5.70	5.40	5.20
Rock 652	2.52	0.33	P‖											
			S	2.38	2.51	2.65	2.74	2.77	2.79	2.82	2.85	2.87	2.90	2.93
			β	11.2	6.1	4.8	4.48	4.21	3.99	3.61	3.50	3.27	3.17	3.00
Westerly Granite $T_{max} = 108°C$	2.60	0.13	P	4.41	4.99	5.39	5.63	5.73	5.81	5.87	5.90	5.98	6.05	6.10
			S	2.96	3.16	3.39	3.52	3.58	3.62	3.65	3.69	3.76	3.82	3.84
			β	4.2	3.65	3.30	3.08	2.78	2.50	2.43	2.38	2.33	2.25	2.20
$T_{max} = 319°C$	2.60	0.13	P‖											
			S	2.86	3.11	3.37	3.52	3.58	3.62	3.65	3.69	3.76	3.82	3.84
			β	4.20	3.65	3.30	3.08	2.78	2.50	2.43	2.38	2.33	2.25	2.20
$T_{max} = 514°C$	2.58	0.16	P	2.79	4.11	4.90	5.37	5.58	5.68	5.78	5.85	5.93	6.00	6.07
			S	2.07	2.90	3.16	3.42	3.51	3.56	3.62	3.67	3.76	3.82	3.84
			β	4.30	2.68	3.30	3.08	2.78	2.50	2.43	2.38	2.33	2.25	2.20
$T_{max} = 714°C$	2.55	0.32	P‖											
			S	1.3	2.12	2.7	3.13	3.29	3.38	3.49	3.56	3.59	3.62	3.65
			β	14.0	9.2	5.2	3.62	3.18	2.91	2.83	2.72	2.60	2.47	2.38
$T_{max} = 953°C$	2.47	1.90	P‖											
			S	0.53	1.18	1.71	2.16	2.46	2.69	3.00	3.19	3.40	3.48	3.56
			β	33.0	27.0	20.0	12.5	8.0	6.6	4.8	3.60	3.01	2.76	2.61

Table 1. (continued).

Sample	Density (g/cc)	Crack Porosity	Elastic Property*	Confining Pressure (Bars)										
				1	100	250	500	750	1000	1500	2000	3000	4000	5000
Fairfax diabase														
$T_{max} = 200°C$	3.00	0.06	$P^{\parallel}$											
			S	3.66	3.72	3.78	3.86	3.91	3.94	3.96	3.97	3.98	3.99	4.00
			β	2.20	2.10	1.97	1.78	1.61	1.49	1.44	1.43	1.40	1.39	1.38
$T_{max} = 310°C$	3.00	0.07	P	5.77	5.98	6.20	6.34	6.54	6.62	6.70	6.72	6.73	6.73	6.74
			$S^{\parallel}$											
			β	2.30	2.14	2.03	1.84	1.68	1.49	1.44	1.43	1.40	1.39	1.38
$T_{max} = 588°C$	2.98	0.13	$P^{\parallel}$											
			$S^{\parallel}$											
			β	7.0	4.4	3.4	2.58	2.06	1.75	1.59	1.49	1.41	1.39	1.38
$T_{max} = 998°C$	2.92	0.72	$P^{\parallel}$											
			$S^{\parallel}$											
			β	21.1	13.5	9.2	5.5	3.8	2.79	2.22	1.83	1.50	1.40	1.38

T_{max} is the maximum temperature to which a sample was heated. All measurements were made at room temperature.

*P is compressional velocity (km/sec), S is shear velocity (km/sec), and $β$ is static compressibility (Mb^{-1}).

[†]Data previously given by Kanamori *et al.* (1970).
[‡]Data previously given by Wang *et al.* (1971).
[§]Velocities are not corrected for length change under pressure.
[‖]Data previously given by Todd *et al.* (1972).

Ries sample 934 is a weakly shocked granite from zone 0. Minor microfracturing has possibly occurred, but in thin section there is little or no evidence of any shock metamorphism (Fig. 2). The velocity–pressure profile is very similar to that of most terrestrial granites (compare with Birch, 1960). All cracks are closed by 1–2 kb confining pressure and velocity reaches intrinsic values near 2 kb.

Ries sample 936 is of similar composition to, but more extensively cracked than sample 934. Again there is no evidence of glass in thin section. The intense cracking is manifested by lower velocities at low pressure. Cracks are closed and velocity values reach intrinsic values near 4 kb. The intrinsic velocity is high, indicating that no glass has been formed. The distribution of crack aspect ratios and the total numbers of microcracks in samples 934 and 936 are significantly different because pressures as high as 4 kb are required to completely close all cracks in sample 936, while only 2 kb were sufficient to close the cracks in sample 934. Cracks of larger aspect ratio have been formed in sample 936. Because the velocities of 936 are lower than the velocities of 934 at pressures below 2 kb, sample 936 contains more cracks of low aspect ratio.

Ries sample 931 is a diorite from zone 1. This sample contains abundant microfractures and shock deformational features (Fig. 2). The intense microfracturing and the presence of glass cause low velocities throughout the entire 5 kb range. Because the slope of the velocity–pressure profile is steep at 5 kb, intrinsic velocities are not reached at 5 kb. An even wider distribution of crack aspect ratios

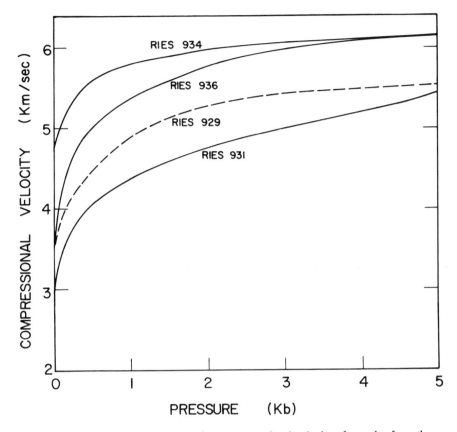

Fig. 1. Effect of shock metamorphism on compressional velocity of samples from the
Ries Crater, Germany.

exists in sample 931 than in samples 936 or 934 because cracks are still not closed
at this pressure. The very low velocities near room pressure indicate that cracks
of low aspect ratio have been formed. At some pressure greater than 5 kb where
all cracks are closed, intrinsic velocity values will be reached. Because the sample
contains some glass, these intrinsic velocities will be lower than those of crystalline rock of similar composition. (Birch (1960) shows that the velocity of a diabase
glass is about 10% less than the intrinsic velocity of a crystalline diabasic rock of
similar composition.)

Ries sample 929 is a zone 3 or 4 glassy breccia. The breccia is composed of
25% crystalline fragments in a matrix of glass. Vesicles and flow structure
(schlieren) are evident throughout the sample. Low velocity values at low pressures are due to the presence of the glass matrix, microcracks, and pores. Low
intrinsic velocity values are due to the presence of the glass matrix and pores.
Intrinsic values of velocity are probably not reached until pressures near 5 kb,
indicating again, a wide distribution of crack aspect ratios.

(a)

(b)

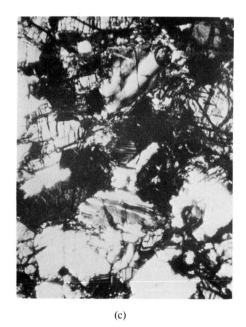

(c)

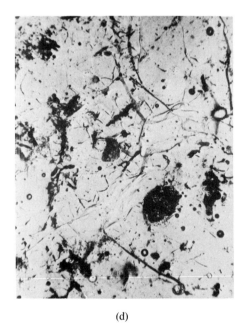

(d)

Table 2. Calculated and measured values of the mean volume thermal expansion coefficient over the range 25–200°C. Units are °C^{-1}.

Sample	$\alpha_{calc.}(\times 10^{6})$ $\left\{\alpha_v = \dfrac{\Sigma_i \alpha_i K_i V_i}{\Sigma_i K_i V_i}\right\}$	$\alpha_{meas.}(\times 10^{6})$ $\left\{\alpha_v = \dfrac{\Delta V}{V_0 \Delta t}\right\}$
Ries 934	23.5	25.6
Ries 936	23.3	19.3
Barre Granite		
$T_{max} = 426°C$	23.2	24.3
$T_{max} = 532°C$	23.2	20.8
$T_{max} = 570°C$	23.2	20.0
Frederick Diabase		
$T_{max} = 250°C$	15.8	19.0
$T_{max} = 350°C$	15.8	19.3
$T_{max} = 400°C$	15.8	18.2
$T_{max} = 450°C$	15.8	20.0
$T_{max} = 500°C$	15.8	19.5
$T_{max} = 550°C$	15.8	20.0
$T_{max} = 600°C$	15.8	19.2
62295	16.9	6.8
65015	18.3	10.9
68415	14.3	4.8
10020*	22.5	16.2
10057*	21.6	14.7
12022*	23.7	15.9
10046*	22.1	22.2

*Data from Baldridge *et al.* (1972).

Fig. 2a. Photomicrograph of Ries sample 934, a granite from zone 0, crossed nicols. Most of the grains are quartz and slightly altered orthoclase. (Width of field is approximately 2.5 mm.)

Fig. 2b. Photomicrograph of Ries sample 936, a granite from zone 0, crossed nicols. The minerals are quartz, orthoclase, and plagioclase. Note the abundance of irregular trans-granular cracks (some of which are now filled with iron oxides). (Width of field is approximately 2.5 mm.)

Fig. 2c. Photomicrograph of Ries sample 931, a hornblende diorite from zone 1. The minerals are plagioclase, quartz, and hornblende. Note the abundance of cracks. Some plagioclase grains are partially isotropic (diaplectic). (Width of field is approximately 2.5 mm.)

Fig. 2d. Photomicrograph of Ries sample 929, a glassy breccia from zone 3 or 4, plane polarized light. Scattered in the glass are a few fragments of crystals and diaplectic glass; and gas bubbles. Note the tiny perlitic cracks. (Width of field is approximately 2.5 mm.)

THERMAL EXPANSION

Thermal expansion curves for the range 25 to 350°C are shown in Fig. 3 for Ries samples 936 and 934. The two samples have quite different thermal expansion behavior which we attribute to the differences in microcracks. Note that compositions (Table 3) are similar but that sample 936 contains more microcracks, and also has a wide distribution of crack aspect ratios.

In Table 2 are listed the values of mean volume coefficient of thermal expansion for each sample,

$$\alpha_v = \frac{3\Delta L}{L_0 \Delta T} \tag{3}$$

calculated from the relative expansion data. Also listed in Table 2 are the intrinsic values $\alpha_v(\text{int})$ for the volume coefficient of thermal expansion expected for a crack free rock. The $\alpha_v(\text{int})$ are calculated from Turner's equation (Kingery, 1960) for the coefficient of thermal expansion for a multiphase aggregate.

$$\alpha_v(\text{int}) = \frac{\Sigma_i \alpha_i K_i V_i}{\Sigma_i K_i V_i} \tag{4}$$

where α_i, K_i, and V_i are the volume coefficient of expansion, the bulk modulus, and the volume fraction, respectively, of the ith phase. Values of the α_i are from Skinner (1966) and values for the K_i are from Simmons and Wang (1971). Baldridge *et al.* (1972) discuss the applicability of Turner's equation to low porosity rocks.

The measured expansion coefficient α_v for the relatively uncracked sample 934 is approximately equal to the intrinsic expansion coefficient. However, α_v for the intensely cracked sample 936 is about 17% less than $\alpha_v(\text{int})$. We attribute the

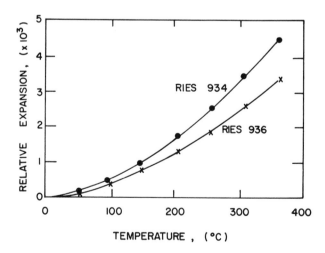

Fig. 3. Effect of shock metamorphism on thermal expansion of samples from the Ries Crater, Germany.

Table 3. Modal analyses of samples.

Sample	qtz	feld	plag	amph	px	biot	musc	chl	opq	ol	acc	other
Barre Granite	31.5	19.8	36.5			7.3	4.3					
Westerly Granite	28.3	35.0	29.9			3.1	2.5		0.7		0.5	
Ries Granite 934	25.9	35.0	31.2				5.6	2.3				
Ries Granite 936	23.2	28.2	30.9			10.2	7.1					
Frederick Diabase			45.0		48.1				4.2	1.0		
Fairfax Diabase	1.8	3.0	49.0		42.4	1.0		0.3	2.0			
Ries 931	3.7	1.5	28.5	55.7				4.8	2.1			3.7 dqtz
Ries 929	13.4	3.9		0.6					4.7			2.9 d gl
												74.5 gl
												tr coes
62295*			57		24						3	16 gl
65015*			70		29				1			
68415*			76.6		18					2.1		3.3 mes

*Lunar Sample Information Catalog—Apollo 16. Descriptions of samples 652 and 613 are in Todd *et al.* (1972).

qtz = quartz	opq = opaque
feld = K-feldspar	ol = olivine
plag = plagioclase	acc = accessories
amph = amphibole	dqtz = diaplectic quartz
px = pyroxene	d gl = diaplectic glass
biot = biotite	gl = glass
musc = muscovite	coes = coesite
chl = chlorite	mes = mesostasis

difference between intrinsic and measured thermal expansion coefficients in sample 936 to the presence of cracks produced by shock processes. Apparently, the presence of large concentrations of cracks along grain boundaries and within individual grains allows many grains to expand into the pore space without contributing to the overall expansion of the aggregate.

We have not as yet been able to obtain satisfactory thermal expansion data for glassy breccias from the Ries. However, Baldridge *et al.* (1972) have measured thermal expansion on glassy (20% glass) lunar breccia 10046 and found that the measured thermal expansion is approximately equal to the intrinsic value. They attribute the agreement between measured and intrinsic thermal expansion to the effect of glass. Because the glasses have comparatively high thermal expansions, or the glasses tend to cement grains together, thermal expansions of glassy breccias are close to intrinsic values.

CONCLUSIONS

Although we have illustrated the effects of increasing shock metamorphism with data for compressional velocity and thermal expansion, similar effects are seen in the data for shear velocity and compressibility. We base our conclusions on the data for all four parameters. We now summarize these conclusions.

For rocks exposed to low shock pressures (zone 0–1) the values of physical properties are influenced primarily by the presence of microcracks produced by the shock event. The new cracks formed by the shock contain both cracks with aspect ratios similar to those found in typical igneous rocks (10^{-4} to 10^{-3}) and cracks of larger aspect ratios (10^{-3} to 10^{-2}). When cracks are closed at high pressure, values of elastic properties reach intrinsic values and are equal to those of unshocked rocks at comparable pressures. For high shock pressures (zones 3–4), the presence of glass and spherical pores lowers velocity values below those expected for the crack free crystalline aggregate (intrinsic values). The high microcrack density results in low velocity values at low confining pressures.

Thermal Effects on Physical Properties

Thermal cycling is a process capable of cracking rocks. We have found that crack production is particularly striking for quartz-rich rocks. Intense cracking in quartz-rich rocks is caused by the unique thermal expansion properties of quartz near the α–β quartz transition at 573°C (Fig. 4). The expansion of most other minerals is much smaller than that of quartz. Consequently rocks not containing quartz (including lunar basalts and anorthosites) do not crack extensively until temperatures greater than 500–600°C are reached.

Our data for thermally cycled rocks are given in Tables 1 and 2. We previously reported data (Todd *et al.*, 1972) for compressional and shear velocity of thermally cycled Westerly granite and Fairfax diabase. Some of our data are illustrated in Figs. 5 and 6.

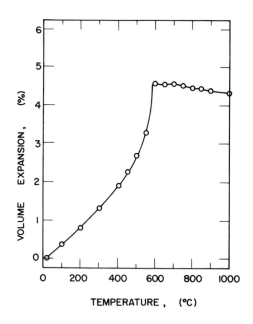

Fig. 4. Volume expansion of quartz (from Kozu and Takane, 1929).

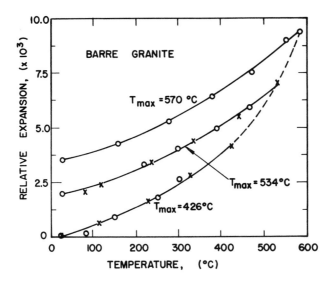

Fig. 5. Steady state relative expansion curves of the Barre granite. Steady state curves are those for which increasing expansion values (crosses) are equal to decreasing expansion values (open circles). The steady state curve is generally reached after 4–5 cycles to a maximum temperature (T_{max}). The dashed curve approximates the expansion curve for a single ascent to 570°C.

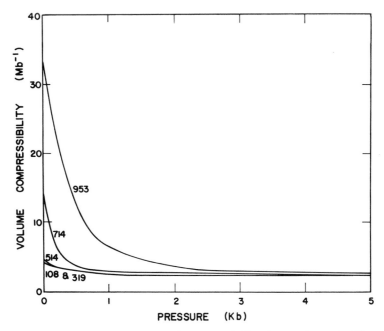

Fig. 6. Effect of thermal cycling on static compressibility of Westerly granite. The numbers on each curve are the maximum temperatures to which the sample was heated. All measurements were made at room temperature.

Thermal Expansion

When a rock is cycled to temperatures sufficiently high to cause cracking, the cracking is manifested by a permanent length change after the sample is returned to room temperature. Recycling the sample to the same temperature often introduces further permanent length changes. The steady state, usually reached after 4–5 cycles, represents the curve for which no further permanent offset occurs.

In Fig. 5, the expansion of the quartz-rich Barre granite is shown as a function of temperature for three cycles to different maximum temperatures. The cycles to maximum temperatures of 426 and 534°C represent steady state expansion–contraction curves; the cycle to maximum temperature of 570°C represents only the decreasing temperature curve after two cycles to 570°C.

In the cycle to 426°C there was no permanent length change $> 0.02\%$ when the Barre granite was returned to room temperature. This fact is in opposition to the data of Thirumalai and Demou (1970) who report permanent length changes of 0.08% in a quartz-rich granodiorite cycled to 325°C. The rate of heating in Thirumalai and Demou's experiments is about equal to that in our experiments (5°C/min). The granodiorite contained about 40% quartz, compared to 31% quartz in the Barre granite. The cracking in Thirumalai and Demou's sample could possibly be due to the presence of additional quartz. The resolution of our dilatometer is adequate for us to record permanent length changes of less than 0.02%.

After repeated cycling to 534°C, a permanent length change of about 0.2% is evident. We attribute this permanent length change to the introduction of cracks into the rock matrix. The cracks were probably caused by the *large volume expansion* of quartz just prior to the α–β transition (Fig. 4). Cycling the Barre granite to even higher temperatures near 570°C introduces further cracks, also probably due to the large volume expansion of quartz relative to the feldspar. Similar data for the Frederick diabase [a rock containing no quartz (Table 3)] indicate that no permanent volume change and therefore no cracking occurs until temperatures greater than 500°C are reached.

Compressibility

In Fig. 6 we have plotted values for the static compressibility of five samples of Westerly granite thermally cycled to maximum temperatures between 108 and 953°C. Because only a small change in the value of compressibility occurs for $T_{max} = 319°C$, we infer that only a small volume of cracks has been introduced by thermal stresses at this temperature. However, because large changes occur in compressional and shear velocity, many low aspect ratio (and therefore low volume) cracks must have been produced. At higher temperatures ($T_{max} = 514$–953°C), intense cracking induced by the large expansion of quartz near the α–β quartz and β quartz-tridymite transitions is manifested by large increases in the static compressibilities measured at low pressure.

The ten-fold increase in room pressure values for compressibility of samples cycled to high temperatures as compared with those of samples cycled to $T_{max} \leq 514°C$ indicates many new cracks are formed. Even in the most intensely cracked samples, however the value of compressibility reaches values close to those of the virgin rock near 2 kb confining pressure. Apparently, most cracks are closed at 2 kb confining pressure, and therefore have low aspect ratios. Consequently, thermally induced cracks differ from shock induced cracks, which have higher aspect ratios.

THERMAL GRADIENTS

On the lunar surface, large thermal gradients might also crack rocks. The temperatures between lunar day and night range from -200 to $+100°C$, and the maximum gradient in temperature with respect to time between lunar day and night at the lunar surface is about 10°C per minute (Stimson and Lucas, 1972). Gradients can be higher during lunar eclipses.

Using acoustic emission techniques, Warren and Latham (1970) listened to cracks forming in several materials subjected to large thermal gradients. Their data indicate that gradients of 10–100°C per minute crack rocks but that very few cracking events occur for gradients less than 10°C per minute. The maximum gradients during our experiments were about 5°C/min. Furthermore, Apollo 15 temperature profiles of Langseth *et al.* (1972) indicate that depth of only a few tens of centimeters are sufficient for all thermal fluctuations to disappear completely. We conclude that the diurnal temperature changes on the lunar surface could possibly crack rocks but that the cracking due to this process would extend only a few centimeters below the surface.

We do not preclude the fact that larger temperature fluctuations could have occurred in the past on the lunar surface (Todd *et al.*, 1972), and that cracking near contacts of hot breccia or lava flows could be extensive.

THERMAL FATIGUE

Data on thermal fatigue of rocks are rare. Warren and Latham (1970) found that after an initial cycle had induced cracking in firebrick, the amount of cracking decreased on each subsequent cycle. After 15–20 cycles all cracking ceased. Similar observations hold for the Barre granite. Our thermal expansion data indicate that no further cracking in the Barre granite occurs after 4–5 cycles to the same temperature.

Although no one has yet looked carefully at the effect of thermal fatigue, Hardy and Chugh (1970) and Haimson (1972) have produced comprehensive results for the effect of stress-induced fatigue in rocks. Using acoustic emission techniques, hysteresis in stress–strain curves, and other methods, they have found that no new cracks are formed in rocks until stresses equivalent to about 60% of the fracture strength are reached, even though the number of cycles is 10^6. Repeated cycling to stresses only slightly above the stress at which cracking first occurs eventually leads to fracture of the rock. Apparently a few new cracks are formed on each cycle, and these cracks eventually weaken the rock enough to cause failure. A low stress limit appears to exist, below which no cracks are formed. We infer that the stresses produced by tidal forces and other low stress phenomena, even though cyclic for periods of 10^6 years, do not crack rock. Assuming an analogous fatigue limit exists for thermally induced stress in rocks, we would expect no cracking to occur on subsequent cycles, if no cracking occurs on the first cycle. Interpreted in terms of our results for Barre granite and Fairfax diabase, thermal fatigue does not occur below a maximum temperature of a few hundred °C.

Conclusions

We conclude that: (1) at temperatures above a few hundred °C, thermal stresses induced by differential thermal expansion of constituent minerals are sufficient to crack rocks, (2) stresses produced by large thermal gradients could possibly crack rocks within the first few centimeters of the lunar regolith, (3) the cracks formed by thermal stresses have lower aspect ratios than shock induced cracks, and (4) thermal fatigue does not occur for cycles to temperatures below a few hundred °C. These conclusions are based on our new data for compressibility of Westerly granite and thermal expansion of Barre granite, additional compressional velocity, shear velocity, thermal expansion, and Q data given in Todd *et al.* (1972), Baldridge *et al.* (1972), Kissell (1972), and Tables 1 and 2.

Discussion

Velocity

In Fig. 7 we compare values of compressional velocity for a number of lunar samples with those of terrestrial shocked rocks and thermally cycled Fairfax diabase. Note the shape of the velocity profile for the three lunar rocks (basalt 12002, gabbroic anorthosite 62295, and breccia 14318) and the two Ries shocked samples (granite 936 and breccia 929). For all samples there is a continuous change in the slope of the velocity profile throughout the 5 kb pressure range. The change in slope indicates that cracks with a wide range of aspect ratios close continuously throughout the pressure range. In terms of crack aspect ratio distribution, the lunar samples appear similar to terrestrial shocked samples.

The elastic properties of thermally cycled samples with $T_{max} < 514°C$ are significantly different from those of lunar rocks. For example, the Fairfax diabase sample with $T_{max} = 200°C$ reaches a relatively constant slope at 1–2 kb. Much lower pressures are sufficient to close cracks in this sample, indicating that thermally-induced cracks have low aspect ratios. If $T_{max} > 500°C$, the velocity profiles for lunar samples can be approximated by those of thermally cycled samples. Therefore, there remains the possibility that temperature fluctuations to temperatures greater than 500°C previously generated the cracks that we presently see in lunar rocks (Todd *et al.*, 1972). However, because present lunar surface temperatures are substantially less than 500°C, we conclude that shock metamorphism is the primary cause for elastic properties of lunar rocks differing

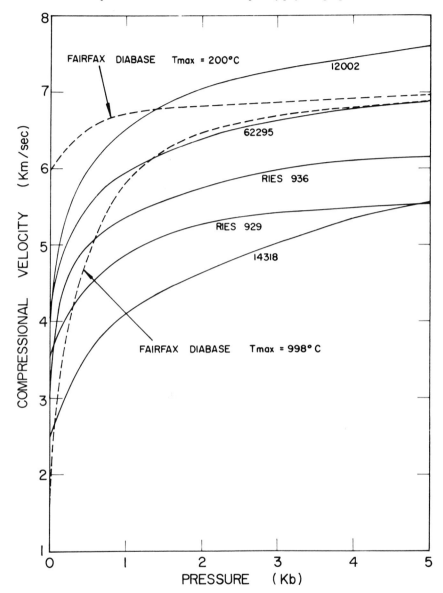

Fig. 7. Compressional velocity for lunar samples 12002 (basalt), 62295 (gabbroic anorthosite), and 14318 (breccia), Ries samples 929 and 936, and Fairfax diabase cycled to maximum temperatures of 200 and 998°C. All measurements were made at room temperature.

from intrinsic values. The data on shear velocity and compressibility given in Table 1 support this interpretation.

At high confining pressures cracks in rocks are closed. Once cracks are closed, velocity values depend on the elastic properties of the materials composing the matrix of the rock. Intrinsic velocities for crystalline silicate rocks are high compared to glassy aggregates of the same composition. The high pressure velocity values for basalt 12002, gabbroic anorthosite 62295, Fairfax diabase, and Ries sample 936 are near the intrinsic values expected for crack free aggregates of similar composition. The high pressure velocities of breccias 14318 and Ries 929, however, are low compared to the values for the crystalline rocks. Lower intrinsic velocity values are expected because of the presence of glass in the breccias.

Thermal expansion

Measured and calculated intrinsic thermal expansion coefficients for lunar basalts (from Baldridge *et al.*, 1972), lunar anorthosites, and shocked Ries samples are listed in Table 2. For *all* crystalline lunar samples the measured expansion coefficients are considerably less than the calculated intrinsic values. Apparently cracks in lunar samples provide void space into which some mineral grains can expand. These grains do not contribute to the overall expansion of the aggregate, thus accounting for the fact that measured expansions of crystalline samples are less than intrinsic values.

We have also measured thermal expansion coefficients for thermally cycled rocks with T_{max} between 200 and 600°C. We have found that thermal cycling of a quartz-rich granite (Barre, Vermont) to 570°C can reduce the expansion coefficient and that the reduction in the expansion coefficient is about equal to that produced by shock in the Ries granites. However, we have also found that thermal cycling of two diabases (Fairfax, Virginia, and Frederick, Maryland) to maximum temperatures of 220°C (Fairfax diabase) and 600°C (Frederick diabase) does not reduce α_v, but rather increases the expansion coefficient slightly. We do not yet understand the cause of high measured expansion coefficients in the diabases. However, because present lunar surface temperatures are less than 500°C and are insufficient to reduce the thermal expansion coefficient due to thermally induced cracking, we conclude that thermal cycling alone cannot explain the low expansion coefficients measured for lunar rocks.

We also list expansion coefficients for two shocked Ries samples in Table 2. We have found that shock metamorphism, and in particular shock produced cracks, reduced the expansion coefficient of a granitic rock by 17% below intrinsic values. In this sense, the thermal expansion of shocked terrestrial rocks appears similar to the thermal expansion of lunar rocks. We conclude that shock induced cracking is the more likely cause for the low expansions measured on lunar samples.

Microcracking

We used Morlier's (1971) technique to obtain the volume distribution of cracks as a function of aspect ratio, and show in Fig. 8 the results for lunar basalt 12002, Ries granites 931 and 936, and three samples of Fairfax diabase thermally cycled to 108, 700, and 998°C. The same scale has been used for each sample. It is apparent that lunar basalt 12002 and Ries samples 931 and 936 contain many more

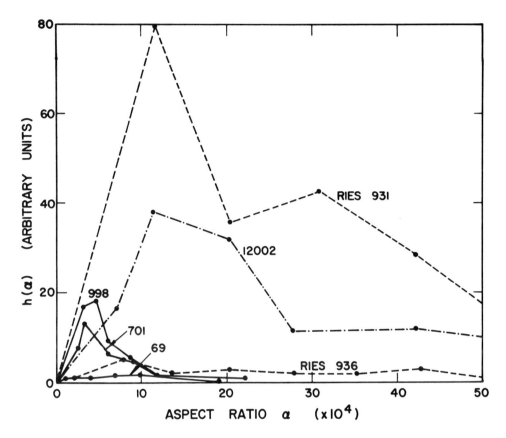

Fig. 8. The distribution function of aspect ratios $h(\alpha)$ for lunar basalt 12002, Ries samples 931 and 936, and Fairfax diabase cycled to maximum temperatures of 69, 701, and 998°C. All measurements were made at room temperature.

cracks of large aspect ratio than do thermally cycled samples. Because thermally cycled samples have many low aspect ratio cracks, the velocities of these samples are low at low pressure. Because shocked samples have many cracks of large aspect ratios in addition to low aspect ratio cracks, velocities of shocked rocks are significantly lower at both high and low pressures. For instance, even though the porosity of Ries sample 936 is about equal to that of the Fairfax diabase sample with $T_{max} = 998$°C, the diabase has a much lower velocity at low pressure (Fig. 7) because the sample has a greater *number* of cracks than the shocked sample. It is therefore not primarily crack porosity which determines velocities at low pressures, but rather the relative numbers of cracks with low and high aspect ratios. Shocked and thermally cycled samples differ in the relative number of cracks at low and high aspect ratios. And again it is apparent that lunar samples and Ries shocked samples are similar in terms of crack densities and crack aspect ratios. Lunar samples and thermally cycled samples are correspondingly dissimilar in terms of crack parameters.

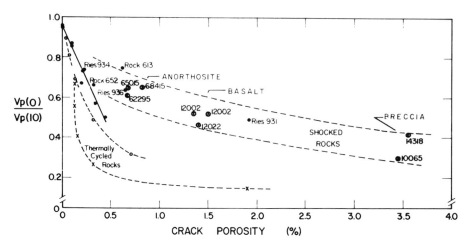

Fig. 9. Compressional velocity normalized to 10 kb values $[V_p(0)/V_p(10)]$ as a function of crack porosity for terrestrial igneous rocks, thermally cycled rocks, and shocked terrestrial and lunar rocks. The data for the terrestrial igneous rocks are from Nur and Simmons (1970). [OL = Oak Hall limestone, FD = Frederick diabase, RQ = Rutland quartzite, TG = Troy granite, WG = Westerly granite, WD = Webatuck dolomite, SG = Stone Mountain granodiorite, CG = Casco granite.] The data for the thermally cycled rocks (Fairfax diabase and Westerly granite) and the shocked rocks are included in Table 2. Samples Ries 931, 934, and 936 are shocked granitic rocks from the Ries Crater in Germany; samples 613 and 652 are terrestrial breccias, samples 12002 and 12022 are lunar basalts, samples 65015, 62295, and 68415 are lunar gabbroic anorthosites, and samples 14318 and 10065 are lunar breccias.

In an attempt to correlate compressional velocity with crack parameters, Nur and Simmons (1970) plotted compressional velocity, normalized to 10 kb or intrinsic values, as a function of crack porosity (η_c). Because of normalization to intrinsic values, the ratio of velocity at zero confining pressure to the velocity at 10 kb confining pressure, $V_p(0)/V_p(10)$, should contain only information about the crack dependent part of velocity. In Fig. 9, we have extended the plot of Nur and Simmons (1970) to include data for thermally cycled rocks (Westerly granite and Fairfax diabase), lunar and terrestrial shocked rocks, and lunar and terrestrial breccias.

Nur and Simmons previously found that normalized velocity is an approximate linear function of crack porosity. Their samples (Fig. 9) included a wide assortment of igneous rocks and contained only cracks produced by natural processes that occurred while the rocks were in the earth. Apparently the cracks produced by the mismatch of thermal and elastic properties across grain boundaries when igneous rocks are transported to the surface of the earth are similar from one rock type to the next, so that a common velocity-crack porosity relation exists for all samples.

When rocks are thermally cycled, under vacuum (or at constant pressure), only the relative thermal expansions of the mineral grains composing the rock influence crack production. We might expect the parameters of the cracks produced by the thermal cycling process to differ from the parameters of cracks produced by typical igneous processes under both pressure and temperature. Indeed, in Fig. 9, very different curves fit the data for virgin igneous rocks. The profiles for the thermally cycled samples are similar to each other, but fall substantially to the low velocity side of the curve defined by Nur and Simmons' data.

Shock metamorphism is a mechanism of crack production entirely different from the processes already discussed, and hence, we would expect different sorts of crack parameters. Figure 9 includes velocity and crack porosity data for an assortment of shocked lunar and terrestrial rocks and two unshocked terrestrial analogues of lunar breccias. The very weakly shocked Ries sample 934 falls near Nur and Simmons' curve for terrestrial igneous rocks. The more intensely shocked Ries samples 936 and 931 fall to the high velocity side of their curve; the intensely shocked lunar breccias 14318 and 10065 fall far from their curve. Moderately shocked lunar basalts and gabbroic anorthosites fall along a curve defined by the data for the Ries shocked rocks and the breccias. The curve defined by the shocked rocks, both lunar and terrestrial, departs greatly from the curves for typical igneous rocks and for thermally cycled rocks. The curve for shocked rocks appears to be unique to shocked rocks. These observations, perhaps more than any of our other data, indicate that physical properties of lunar rocks are determined primarily by shock processes.

Lunar crust

We can now speculate as to the effect of shock metamorphism on the physical properties of the lunar crust. Consider only compressional velocity. Interpreted in terms of the velocity profiles for the Ries Crater (Angenheister and Pohl, 1969) and our laboratory data, we conclude that on the lunar surface (1) velocities are low in craters where breccias from the cratering event itself or from other impact events are present, (2) velocities in the immediate vicinity of craters are low due to intense microfracturing and the presence of glass, and (3) velocities increase with radial distance from the crater, reaching values equal to those of the surrounding unshocked rock at 2–3 crater diameters horizontally and perhaps one crater radius with depth. Other physical properties will be affected by the intensity of shock metamorphism correspondingly.

The velocity profile around the Ries Crater (about 20 km in diameter) can perhaps be scaled upwards to describe larger diameter events of, say, the size of Ritter or Sabine on the moon (30–32 km in diameter). However, it is difficult to infer the effects of shock metamorphism on a crater the size of the ringed maria (diameter of Imbrium = 700 km) using the Ries data. For these extremely large events, large volumes of lava were either formed by the impact event itself (Urey and McDonald, 1971) or brought from the subsurface through the cracks formed by the event (Wise and Yates, 1970). The fact that (1) the lava can help fill or anneal cracks produced by the shock event, (2) the heat produced by the lava over long periods of time can help recrystallize glassy breccias and anneal cracks, and (3) the pool itself will crystallize upon cooling—all tend to give high velocities within and directly adjacent to the crater. The breccias ejected from the crater and the microfracturing around the crater should keep velocities low within a few crater radii. However, a very different velocity profile will probably be obtained across the flood basalt filled craters, when compared with the smaller craters.

The current model of seismic velocity with depth (Toksöz *et al.*, 1972) indicates decreasing shock metamorphism with depth over the first 25 km of the moon. A sharp velocity increase occurs near 25 km. The region between 25 and 60 km has a very small increase (if any) in velocity with depth, indicating that few open microfractures are present. We infer that the depth of 25 km probably represents the maximum depth for shock metamorphism or the depth at which pressure and temperature are sufficient to close or anneal cracks.

CONCLUSIONS

Our laboratory data for compressional velocity, shear velocity, static compressibility, and thermal expansion for lunar rocks, shocked terrestrial rocks, and thermally cycled rocks indicates that shock metamorphism is the primary cause of the differences between measured and intrinsic values of physical properties of lunar rocks. Large scale thermal metamorphism and thermal cycling between temperatures equivalent to lunar day and night probably are not important mechanisms in changing the values of physical properties of lunar rocks. Large thermal gradients or thermal fatigue could possibly cause minor cracking in the upper few centimeters of the lunar regolith.

Velocity profiles with depth or radial distance from large impact craters on the moon should show a graded scale of shock effects. Because the largest impact craters contain layers of basalts formed by essentially igneous processes, we infer that the physical properties of rocks associated with such craters differ substantially from those of rocks associated with smaller craters.

The seismic velocity vs. depth curve over the first 25 km of the moon indicates decreasing shock metamorphism with depth. The 25 km depth probably represents either the maximum depth of shock metamorphism or the depth at which shock produced microcracks are closed or annealed.

We have chosen our samples from the Ries Crater to illustrate the effects of increasing shock metamorphism, and have found that increasing shock metamorphism systematically alters physical properties. We suggest that one may be able to invert the problem, and use measurements of physical properties of shocked rocks to deduce the intensity of shock metamorphism. Then, if cratering mechanisms are understood, and the pressure and temperature distribution is known for a given event, the origin in space of shocked samples can be determined from values of physical properties.

Acknowledgments—We wish to thank Frank Guzikowski for helping with the thermal expansion measurements, and Amos Nur for collecting the Ries samples. NASA contract NGR-22-009-540 supported this research.

REFERENCES

Angenheister G. and Pohl J. (1969) Die seismischen Messungen im Ries von 1948–1969. *Geologica Bavarica* **61**, 304–326.

Baldridge W. S., Miller F., Wang H., and Simmons G. (1972) Thermal expansion of Apollo lunar samples and Fairfax diabase. *Proc. Third Lunar Sci. Conf., Geochim. Cosmochim. Acta*, Suppl. 3, Vol. 3, pp. 2599–2609. MIT Press.

Birch F. (1960) The velocity of compressional waves in rocks to 10 kilobars, Part 1. *J. Geophys. Res.* **65**, 1083–1102.

Birch F. (1961) The velocity of compressional waves in rocks to 10 kilobars, Part 2. *J. Geophys. Res.* **66**, 2199–2224.

Brace W. F. (1964) Effects of pressure on electric-resistance strain gauges. *Exp. Mech.* **4**, 212–216.

Brace W. F. (1965) Some new measurements of linear compressibility of rocks. *J. Geophys. Res.* **70**, 391–398.

Buffington R. M. and Latimer W. M. (1926) The measurement of coefficients of expansion at low temperatures, some thermodynamic applications of expansion data. *J. Amer. Chem. Soc.* **48**, 2305–2319.

Chao E. C. T. (1968) Pressure and temperature histories of impact-metamorphised rocks-based on petrographic observations. In *Shock Metamorphism of Natural Materials* (editors B. M. French and N. M. Short), pp. 135–158. Mono.

Dence M. R. (1968) Shock zoning in Canadian craters; petrology and structural implications. In *Shock Metamorphism of Natural Materials* (editors B. M. French and N. M. Short), pp. 169–184. Mono.

Duennebier F. K. and Sutton G. W. (1972) Thermal moonquakes. *EOS Trans. Am. Geophys. Union* **53**, p. 1041.

Haimson B. C. (1972) Mechanical behavior of rock under cyclic loading. *Annual Technical Progress Report*, Dept. Metallurgical and Mineral Engineering, University of Wisconsin, Madison.

Hardy H. R. Jr. and Chugh Y. P. (1970) Failure of geologic materials under low cycle fatigue. Paper presented at Sixth Canadian Symposium on Rock Mechanics, Montreal, Quebec.

James O. B. (1969) Shock and thermal metamorphism of basalt by nuclear explosion, Nevada Test Site. *Science* **166**, 1615–1620.

Kanamori H., Nur A., Chung D., and Simmons G. (1970) Elastic wave velocities of lunar samples at high pressures and their geophysical implications. *Proc. Apollo 11 Lunar Sci. Conf., Geochim. Cosmochim. Acta*, Suppl. 1, Vol. 3, pp. 2289–2293. Pergamon.

Kingery W. D. (1960) *Introduction to ceramics*, p. 478. Wiley.

Kissell F. N. (1972) Effect of temperature on internal friction in rocks. *J. Geophys. Res.* **77**, 1420–1423.

Kozu S. and Takane K. (1929) Influence of temperature on the axial ratio, the interfacial angle and the volume of quartz. *Sci. Rep. Tokoku Univ.*, **3** (3rd series), 239–246.

Langseth M. G., Clark S. P., Chute J. L., Keihm S. J., and Wechsler A. E. (1972) Heat-flow experiment. In *Apollo 15 Preliminary Science Report*. NASA.

Morlier P. (1971) Description de l'état de fissuration d'une roche à partir d'essais non-destructifs simples. *Rock Mechanics* **3**, 125–138.

Nur A. and Simmons G. (1969) The effect of saturation on velocity in low porosity rocks. *Earth Planet. Sci. Lett.* **7**, 99–108.

Nur A. and Simmons G. (1970) The origin of small cracks in igneous rocks. *Int. J. Rock Mech. Min. Sci.* **7**, 307–314.

Short N. M. (1966) Effects of shock pressures from a nuclear explosion on mechanical and optical properties of granodiorite. *J. Geophys. Res.* **71**, 1195–1215.

Short N. M. (1969) Shock metamorphism of basalt. *Modern Geol.* **1**, 81–95.

Simmons G. and Brace W. F. (1965) Comparison of static and dynamic measurements of compressibility of rocks. *J. Geophys. Res.* **70**, 5649–5656.

Simmons G. and Wang H. (1971) *Single Crystal Elastic Constants and Calculated Aggregate Properties; A Handbook*, 2nd edition, 370 pp. MIT Press.

Skinner B. J. (1966) Thermal expansion. In *Handbook of Physical Constants* (editor S. Clark), Section 6, pp. 75–96. *Geol. Soc. Amer. Mem.* **97**.

Stephens D. R. and Lilley E. M. (1971) Pressure-volume properties of two Apollo 12 basalts. *Proc. Second Lunar Sci. Conf., Geochim. Cosmochim. Acta*, Suppl. 2, Vol. 3, pp. 2165–2172. MIT Press.

Stimson L. D. and Lucas J. W. (1972) Lunar thermal aspects from Surveyor data. In *Thermal Characteristics of the Moon* (editor J. W. Lucas), pp. 121–150. MIT Press.

Stöffler D. (1971) Progressive metamorphism and classification of shocked and brecciated crystalline rocks in impact craters. *J. Geophys. Res.* **76**, 5541–5551.

Thirumalai K. and Demou S. G. (1970) Effect of reduced pressure on thermal-expansion behavior of rocks and its significance to thermal fragmentation. *J. Applied Physics* **41**, 5147–5151.

Todd T., Wang H., Baldridge W. S., and Simmons G. (1972) Elastic properties of Apollo 14 and 15 rocks. *Proc. Third Lunar Sci. Conf., Geochim. Cosmochim. Acta*, Suppl. 3, Vol. 3, pp. 2577–2586. MIT Press.

Toksöz M. N., Press F., Dainty A., Anderson K., Latham G., Ewing M., Dorman J., Lammlein D., Sutton G., and Duennebier F. (1972) Structure, composition, and properties of lunar crust. *Proc. Third Lunar Sci. Conf., Geochim. Cosmochim. Acta*, Suppl. 3, Vol. 3, pp. 2527–2544. MIT Press.

Urey H. C. and G. J. F. McDonald (1971) Origin and history of the moon. In *Physics and Astronomy of the Moon*, 2nd edition (editor Z. Kopal), pp. 481–523. Academic.

Von Englehart W. and Stöffler D. (1968) Stages of shock metamorphism in crystalline rocks of the Ries basin, Germany. In *Shock Metamorphism of Natural Materials* (editors B. M. French and N. M. Short), pp. 1159–168. Mono.

Walsh J. B. (1965) The effect of cracks on the compressibility of rocks. *J. Geophys. Res.* **70**, 381–389.

Wang H., Todd T., Weidner D., and Simmons G. (1971) Elastic properties of Apollo 12 rocks. *Proc. Second Lunar Sci. Conf., Geochim. Cosmochim. Acta*, Suppl. 2, Vol. 3, pp. 2327–2336. MIT Press.

Warren N. W. and Latham G. V. (1970) An experimental study of thermally induced microfracturing and its relation to volcanic seismicity. *J. Geophys. Res.* **75**, 4455–4464.

Wise D. V. and Yates M. T. (1970) Mascons as structural relief on a lunar "Moho." *J. Geophys. Res.* **75**, 261–268.

Proceedings of the Fourth Lunar Science Conference
(Supplement 4, *Geochimica et Cosmochimica Acta*)
Vol. 3, pp. 2663–2671

Elastic properties of plagioclase aggregates and seismic velocities in the moon

HERBERT WANG

Department of Geology and Geophysics, University of Wisconsin, Madison, Wisconsin 53706

TERRY TODD, DOROTHY RICHTER, and GENE SIMMONS

Department of Earth and Planetary Sciences, Massachusetts Institute of Technology, Cambridge, Massachusetts 02139

Abstract—The compressional velocities of Apollo 16 gabbroic anorthosites in which the cracks have been closed match the seismic velocity of 7 km/sec in the 25 to 65 km depth region of the moon beneath the Imbrium Basin. The intrinsic velocities of plagioclase aggregates indicate that a velocity of 7 km/sec in a highly calcic gabbroic anorthosite is consistent only with a very small pyroxene component. Because mare basalts and gabbroic-anorthosites both have intrinsic velocities of 7 km/sec, the laboratory velocity data do not require a compositional change from basalt to anorthosite at the 25 km discontinuity. The laboratory velocity data only imply that the 25 km seismic discontinuity is one of microcrack density. The physical rather than the chemical or mineralogical state is constrained.

INTRODUCTION

SEVERAL APOLLO 16 gabbroic anorthosites of sufficient size and coherence were available to us for making laboratory velocity measurements as a function of pressure. Laboratory measurements of lunar samples, when compared to lunar seismic velocity profiles, provide a first order test of hypotheses on the composition of the lunar interior.

In mare regions the seismic profile for the 1 to 25 km depth range matches the laboratory velocity vs. pressure curves of Apollo 11 and 12 basalts (Kanamori *et al.*, 1970; Toksöz *et al.*, 1972). At 25 km the compressional velocity changes from 6 to 7 km/sec. Then at 65 km another discontinuity from 7 to possibly 9 km/sec occurs. In this paper we investigate the laboratory velocity evidence for the 25 to 65 km region being anorthosite or gabbroic anorthosite. A companion paper (Todd *et al.*, 1973) discusses the effect of shock impacts upon velocities.

Compressional and shear-wave velocity measurements were made on Apollo 16 gabbroic anorthosites (62295,18; 65015,9; 68415,54) and on terrestrial plagioclase aggregates. We emphasize at the outset that agreement between a laboratory velocity measurement on a particular material and a seismic velocity does not prove in itself that the moon is made up of that material. The point is nicely illustrated by Schreiber and Anderson (1970). We believe that petrological and other data must also be convincing before a particular layer is assigned a mineralogical composition. Specifically, two problems are encountered in the interpretation of the laboratory velocity data. The first problem is that variable amounts of microcracking dominate velocity behavior in the first 2 kb of pressure (or 36 km in the moon). The second problem is that the velocities of lunar basalts and lunar gabbroic anorthosites are nearly identical at pressures sufficient to close all cracks.

EXPERIMENTAL

The travel times of 1-MHz compressional and shear-waves were measured by the pulse-transmission technique using a mercury delay line (Birch, 1960). The specific method of sample potting used by us on the lunar samples is described by Kanamori *et al.* (1970), Wang *et al.* (1971), and Todd *et al.* (1972).

For the short lunar samples (less than 2-cm path length) the largest source of error is in determining the mercury column length equivalent to zero sample length. An error of 0.020 cm of mercury in the zero implies a 3% error for a velocity of 7 km/sec and a sample length of 2 cm. The precision of measurement is about 1%.

APOLLO 16 GABBROIC ANORTHOSITES

The velocity vs. pressure data for the three Apollo 16 gabbroic anorthosites we measured are given in Table 1. Also included in the table are our previous results for samples 14310 and 15418, rocks which can be classified as gabbroic

Table 1. Velocities (km/sec) of lunar gabbroic anorthosites.

Sample	Density (g/cm³)	Mode	Confining Pressure (kb)						
			0.0	0.5	1.0	1.5	2.0	3.0	5.0
62295,18	2.83	P	4.39	5.54	5.98	6.23	6.42	6.66	6.92
		S	2.20	2.83	3.11	3.28	3.41	3.56	3.75
65015,9	2.97	P	4.77	5.58	5.98	6.17	6.33	6.53	6.90
		S	2.38	2.79	3.02	3.19	3.29	3.45	3.63
68415,54 (A-direction)	2.78	P	4.70	5.63	6.09	6.37	6.54	6.76	6.94
		S	2.59	2.94	3.13	3.26	3.35	3.43	3.54
68415,54 (B-direction)	2.78	P	4.95	5.92	6.27	6.49	6.64	6.80	7.04
		S	2.48	2.88	3.09	3.23	3.31	3.41	3.54
Average of* Apollo 16 Samples	2.86	P	4.70	5.70	6.10	6.30	6.50	6.70	6.90
		S	2.40	2.90	3.10	3.20	3.30	3.40	3.60
14310,72 (A-direction)	2.88	P	3.93	4.96	5.51	5.86	6.08	6.33	6.68
		S	2.08	2.63	2.95	3.15	3.30	3.46	3.66
14310,72 (B-direction)	2.88	P	3.84	4.91	5.55	5.92	6.14	6.39	6.79
		S	2.07	2.60	2.94	3.11	3.26	3.42	3.63
15418,43	2.80	P	4.85	5.50	6.02	6.33	6.50	6.64	6.75
		S	2.82	3.08	3.28	3.42	3.50	3.58	3.69

*Only two significant digits.

anorthosites. Sample 14310 is very similar petrologically to 68415 (Gancarz *et al.*, 1972). Rocks 62295 and 68415 are crystalline gabbroic anorthosites and rock 65015 is a brecciated gabbroic anorthosite (LRL, 1972). The average density of the three samples was 2.86 g/cm^3, the average plagioclase to pyroxene ratio was about 70 to 30, and the plagioclase composition was about An$_{90}$ (Agrell *et al.*, 1973; Albee *et al.*, 1973; Helz and Appleman, 1973).

Only sample 68415,54 was large enough and regular enough in shape to measure the velocities in two mutually perpendicular directions (termed A and B). At 5 kb confining pressure the velocities in the A and B directions differed by 0.1 km/sec. Since the velocity vs. pressure data are very similar for all three samples, average measurements are summarized in Table 1 and plotted in Fig. 1.

At 5 kb the average V_p value we report for Apollo 16 gabbroic anorthosites is 6.9 km/sec. Chung (1973) and Mizutani and Newbigging (1973) report V_p values of

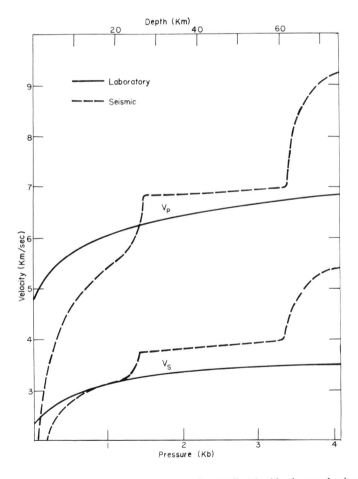

Fig. 1. Laboratory velocity vs. pressure curves for Apollo 16 gabbroic-anorthosites. The seismic profiles (Toksöz *et al.*, 1972) are superimposed.

H. WANG *et al.*

6.78 km/sec and 6.75 km/sec, respectively, for the lunar anorthosite 15415. The 2%-lower velocity is not significant, but it can be explained by the 30% pyroxene content of the gabbroic anorthosites since sample 15415 does not contain any appreciable pyroxene. We interpret all the measurements to mean that a crack-free rock with the composition of an average Apollo 16 gabbroic anorthosite has a compressional velocity of about 7 km/sec.

TERRESTRIAL PLAGIOCLASES

The elastic properties of plagioclases are of interest in interpreting the Apollo 16 rock measurements. We measured velocities on five different terrestrial plagioclase aggregates ranging from An_0 to An_{100}. Previously, Birch (1960, 1961) measured the compressional velocity of some plagioclase aggregates from An_0 to An_{100} and Ryzhova (1964) measured the elastic constants of single-crystal plagioclases from An_9 to An_{53}.

The seismic compressional velocity in the 25 to 65 km depth region is a nearly constant 7 km/sec. The laboratory velocities of lunar gabbroic-anorthosites are consistent with the seismic velocities only if they are crack free. Velocities at pressures less than 5 kb are dependent on the past microcracking history of the rock. Therefore our main interest is in the 5 kb velocities of the terrestrial plagioclases (Table 2). For anorthite content greater than 50% the compressional velocity ranges from 7.2 to 7.6 km/sec. The An_{50} sample with V_p equal to 7.47 km/sec and the average of three An_{75} samples with V_p equal to 7.50 km/sec are about 0.5 km/sec greater than values previously reported for similar anorthite contents (Birch 1960, 1961). Birch (1961) suggested velocities of 6.5 km/sec and 7.2 km/sec for pure albite and pure anorthite, respectively, with a linear velocity dependence on composition between these end members.

Table 2. Intrinsic velocities of terrestrial plagioclases.

Sample	Plag. content (Vol. %)	Density (g/cm^3)	V_p at 5 kb (km/sec)	V_s at 5 kb (km/sec)
An_0. Branchville, Conn.	99	2.57	6.60*	—
Parallel to foliation	99	2.57	7.68	—
Perpendicular to foliation	99	2.57	5.53	—
An_{50}. Mt. Davies, S. Austr.	94	2.66	7.47	3.87
An_{55}. Adirondack Mtn., N.Y.	92	2.66	7.18	4.09
An_{75}. Crystal Bay, Minn.	96	2.71	7.50†	—
Core 1	98	2.81	7.47	—
Core 3	98	2.66	7.72	—
Core 4	92	2.68	7.31	—
An_{100}. Pelham, Mass.	25	2.85	7.15	—

*Average of velocities parallel and perpendicular to foliation.
†Average of 3 cores from different hand specimens.

Velocities in plagioclase single crystals are highly anisotropic as shown in Table 3 (Ryzhova, 1964) and preferred orientation may exist in our terrestrial samples. However, the petrographic work we did was primarily to determine plagioclase content, plagioclase composition (by extinction angles), texture, and alteration. Crystallographic orientation of grains was not done. Brief petrographic descriptions follow. The An contents as determined by extinction angles are only approximate to ± 5 in An number.

An_0. *Albite. Branchville, Connecticut.* The albite aggregate was highly foliated. It consisted of sheaf-like aggregates and is the variety of albite known as cleavelandite because the crystals have pronounced elongations preferentially oriented along the [010]-direction as determined from twinning. The grains are virtually unaltered.

An_{50}. *Calcic Andesine. Mt. Davies, S. Australia.* Most grains are subhedral and of uniform size. Crystals are complexly twinned; lamellae are frequently bent. Cleavage is prominent and grain boundaries are altering to clay.

An_{55}. *Labradorite. Adirondack Mtns., New York.* The grains are equidimensional and anhedral with very minor alteration along cracks. The composition is 92% plagioclase, 3.5% microcline, 1.5% quartz, 2% clinopyroxene, and 1% epidote.

An_{75}. *Bytownite. Crystal Bay, Minnesota.* The rock is medium-grained with prominent cleavage and transgranular cracking. The 3% alteration occurs along grain boundaries and cracks.

An_{100}. *Anorthite. Pelham, Massachusetts.* The plagioclase is extremely altered in a very fine network of white mica and clay. The original texture is almost completely obscured.

Correlation between velocity anisotropy and preferred orientation of grains was possible only for the cleavelandite (An_0). At 5 kb velocities were 7.68 km/sec parallel to the foliation and 5.53 km/sec perpendicular to it. The thin section used for the petrographic description was cut perpendicular to the direction in which the 5.53 km/sec velocity was measured. Ryzhova's velocities are 7.26 km/sec in the [010]-direction, 5.31 km/sec in the [101]-direction, and 5.42 km/sec in the [100]-direction. The anisotropy is just about that in the foliated sample. We did not keep track of the rotational orientation of the thin section relative to the sample,

Table 3. Compressional-velocities in single-crystal plagioclase
(An_9) (Ryzhova, 1964).

Propagation direction	V_p (km/sec)
[001]	7.125
[110]	6.380
[010]	7.260
[101]	5.310
[100]	5.420
[011]	6.200

but the similar anisotropy in the aggregate and the single crystal suggests that the elongations were in the fast-measured direction of the aggregate.

The high degree of anisotropy in single-crystal plagioclase results in a large spread between the Voigt and Reuss averages. The Voigt average is 6.3 km/sec and the Reuss average is 5.8 km/sec. The arithmetical average of 6.05 km/sec is about 0.5 km/sec lower than Birch's value of 6.5 km/sec. The fastest velocity in Ryzhova's An_9 is 7.26 km/sec which is also about 0.5 km/sec lower than the fast velocity of 7.68 km/sec in the cleavelandite. Since Ryzhova does not mention that his specimens are of gem quality, we think that the lower single crystal values are due to minor cracks and imperfections which reduce the values from the intrinsic. But the single-crystal data are useful to us in interpreting in terms of anisotropy the present terrestrial plagioclase data.

The data of Table 3 are indicative of the magnitude of the velocity anisotropy in plagioclase single crystals. The difference between the fastest and slowest velocity is about 30%. Though we do not have crystallographic orientation data to explain the high velocity in the Mt. Davies (An_{50}) aggregate, preferred orientations could result in such a velocity. The three Adirondack anorthosites (An_{55}) all have the same petrographic description, but the 5 kb velocities range from 7.3 to 7.7 km/sec which can be due also to anisotropy. In each of these cored samples, we measured the velocity only in one direction. The An_{100} sample has V_p equal to about 7.2 km/sec in agreement with Birch's value for that end member. But the agreement must be fortuitous in light of the highly altered nature of the sample. In hand specimen the rock was almost pure white and gave every appearance of being a nice An_{100} aggregate.

VELOCITY-DENSITY RELATION

Calcium-rich plagioclase rocks have compressional velocities up to 1 km/sec greater than predicted by Birch's linear velocity-density-mean atomic weight relation (Birch, 1961; Simmons, 1964). The anorthosite data of Birch (1960, 1961) showed less than 2% anisotropy at 5 kb and he corrected the rock data for just the feldspar component. He suggested that V_p be 6.5 km/sec for the end-member albite (An_0) and that V_p be 7.2 km/sec for the end-member anorthite (An_{100}). Velocities are linear with anorthite content and are probably correct to about 1%.

Birch (1961) further combined these plagioclase values with that for pyroxenite ($\rho = 3.4$ g/cm^3; $V_p = 7.9$ km/sec) to calculate a velocity for gabbros. The calculated values agreed with the measured ones to within 2%. The gabbros had feldspar volume percentages of 50, 60, and 67% and anorthite mole percentages of 60, 60, and 65% for the feldspar component.

A contoured ternary diagram (Fig. 2) is a useful method for estimating velocities of lunar gabbroic-anorthosites. The velocity contours in Fig. 2 are volume averages over the three components albite (Ab), anorthite (An), and orthopyroxene (Opx). The V_p value of 7.8 km/sec we used for the orthopyroxene end component of Fig. 2 is the arithmetical average of the Voigt and Reuss values of single-crystal bronzite (Kumazawa, 1969). A volume average over the plagioclase

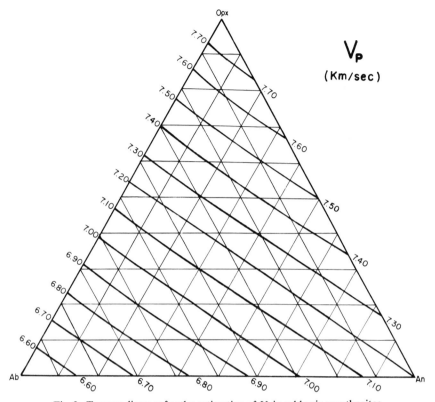

Fig. 2. Ternary diagram for the estimation of V_p in gabbroic anorthosites.

and pyroxene components predicts the intrinsic velocities in lunar gabbroic anorthosite to within 0.2 km/sec.

Similar ternary diagrams for density, ρ, and shear velocity, V_s, in principle, uniquely determine a composition. However, we caution against such a procedure. The Apollo 16 gabbroic anorthosites all have a complicated recrystallization history because of partial or total impact melting (Gancarz *et al.*, 1972; Helz and Appleman, 1973). Only careful and detailed petrological studies can place the composition of the hypothesized anorthositic layer within the ternary plot (Fig. 2). Then the velocity can be estimated from the figure and compared to seismic values. The inverse procedure of inferring composition in the triangle from seismic velocities has little significance because of the 0.2 km/sec uncertainties. The situation is analogous to similar calculational schemes for predicting lower mantle iron contents in the earth. We have commented in detail on this subject (Wang and Simmons, 1972).

LUNAR STRUCTURE

Compressional velocities of *crack-free* rocks with the composition of typical Apollo 16 gabbroic anorthosites are *consistent* with the seismic profile in the 25 to

65 km depth range. However, the importance of the words *crack-free* and *consistent* must be emphasized.

The 25 to 65 km depth region in the moon corresponds to pressures between 1.5 and 3.5 kb since the confining pressure is related to the lithostatic pressure by the ratio 18 km/kb. The velocity vs. pressure curve for lunar gabbroic anorthosites is shown in Fig. 1 with the lunar seismic profile superimposed. The gabbroic-anorthosite velocity data are not flat in the 1.5 to 3.5 kb region whereas the seismic profiles are. Thus, if the layer is one of gabbroic anorthosite, the rocks *in situ* cannot be cracked. Small compositional trade offs exist around the 7 km/sec contour within Fig. 2. A highly calcic gabbroic anorthosite must have a small pyroxene component in order to keep the velocity down to 7 km/sec. The velocity contours indicate that the area near the lower right hand corner is where consistency exists between the seismic velocities and Apollo 16 rock compositions. Everybody's petrologic model for the layer is incorporated within the right hand corner area because allowances of 0.2 km/sec velocity error mean that the pyroxene component may vary by 20% and the anorthite component by 30%.

A more major compositional variation is possible because at 5 kb many typical mare basalts have V_p equal to about 7 km/sec (Kanamori *et al.*, 1970; Wang *et al.*, 1971). Thus, if they are crack-free, both lunar basalts and lunar gabbroic anorthosites match the seismic profile. A portion of the 25 to 65 km region could thus be crack-free basalt. On the other hand anorthosite could be present at depths shallower than 25 km if its microcracking state lowered the velocity to match the seismic values.

We repeat the point that the basalt/anorthosite boundary need not coincide with the 25 km seismic discontinuity (Wang *et al.*, 1973). Neither petrologic nor gravity data constrain the boundary to be at exactly this depth (Anderson *et al.*, 1972; Taylor *et al.*, 1972). On the basis of laboratory velocity data on lunar basalts and anorthosites, we can only say that the seismic discontinuity represents a discontinuity in microcrack density. Microcracks are open above the discontinuity and are closed below it, regardless of whether the rock is basalt or anorthosite.

Acknowledgments—Several of the plagioclase aggregates were loaned to us by the U.S. National Museum. W. S. Baldridge helped with the petrography of some of the samples. The research was supported by NASA contract NGR-22-009-540.

REFERENCES

Agrell S. O., Agrell J. E., Arnold A. R., and Long J. V. P. (1973) Some observations on rock 62295 (abstract). In *Lunar Science—IV*, pp. 15–17. The Lunar Science Institute, Houston.

Albee A. L., Gancarz A. J., and Chodos A. A. (1973) Sanidinite facies metamorphism of Apollo 16 sample 65015 (abstract). In *Lunar Science—IV*, pp. 24–26, The Lunar Science Institute, Houston.

Anderson A. T., Braziunas T. F., Jacoby J., and Smith J. V. (1972) Thermal and mechanical history of breccias 14306, 14063, 14270, and 14321. *Proc. Third Lunar Sci. Conf., Geochim. Cosmochim. Acta*, Suppl. 3, Vol. 1, pp. 819–835. MIT Press.

Birch F. (1960) The velocity of compressional waves in rocks to 10 kilobars, Part 1. *J. Geophys. Res.* **65**, 1083–1102.

Birch F. (1961) The velocity of compressional waves in rocks to 10 kilobars, Part 2. *J. Geophys. Res.* **66**, 2199–2224.

Chung D. H. (1973) Elastic wave velocities in anorthosite and anorthositic gabbros from Apollo 15 and 16 landing sites (abstract). In *Lunar Science—IV*, pp. 141–142. The Lunar Science Institute, Houston.

Gancarz A. J., Albee A. L., and Chodos A. A. (1972) Comparative petrology of Apollo 16 sample 68415 and Apollo 14 samples 14276 and 14310. *Earth Planet. Sci. Lett.* **16**, 307–330.

Helz R. T. and Appleman D. E. (1973) Mineralogy, petrology and crystallization history of Apollo 16 sample 68415 (abstract). In *Lunar Science—IV*, pp. 352–354. The Lunar Science Institute, Houston.

Kanamori H., Nur A., Chung D., and Simmons G. (1970) Elastic wave velocities of lunar samples at high pressures and their geophysical implications. *Proc. Apollo 11 Lunar Sci. Conf., Geochim. Cosmochim. Acta*, Suppl. 1, Vol. 3, pp. 2289–2293. Pergamon.

Kumazawa M. (1969) The elastic constants of single-crystal orthopyroxene. *J. Geophys. Res.* **74**, 5973–5980.

Lunar Receiving Laboratory (1972) Lunar Sample Information Catalogue, Apollo 16. Manned Spacecraft Center, Houston.

Mizutani H. and Newbigging D. F. (1973) Elastic-wave velocities of Apollo 14, 15, and 16 rocks and thermal conductivity profile of the lunar crust (abstract). In *Lunar Science—IV*, pp. 528–530. The Lunar Science Institute, Houston.

Ryzhova T. V. (1964) Elastic properties of plagioclase. *Bull. Acad. Sci. USSR, Geophys. Ser.*, English Transl., no. 7, 633–635.

Schreiber E. and O. L. Anderson (1970) Properties and composition of lunar materials: Earth analogies. *Science* **168**, 1579–1580.

Simmons G. (1964) Velocity of compressional waves in various minerals at pressures to 10 kilobars. *J. Geophys. Res.* **69**, 1117–1121.

Taylor G. J., Marvin U. B., Reid J. B., and Wood J. A. (1972) Noritic fragments in the Apollo 14 and 12 soils and the origin of Oceanus Procellarum. *Proc. Third Lunar Sci. Conf., Geochim. Cosmochim. Acta*, Suppl. 3, Vol. 1, pp. 995–1014. MIT Press.

Todd T., Wang H., Baldridge W. S., and Simmons G. (1972) Elastic properties of Apollo 14 and 15 rocks. *Proc. Third Lunar Sci. Conf., Geochim. Cosmochim. Acta*, Suppl. 3, Vol. 3, pp. 2577–2586. MIT Press.

Todd T., Wang H., Richter D., and Simmons G. (1973) Unique characterization of lunar samples by physical properties (abstract). In *Lunar Science—IV*, pp. 731–733. The Lunar Science Institute, Houston.

Toksöz M. N., Press F., Dainty A., Anderson K., Latham G., Ewing M., Dorman J., Lammlein D., Sutton G., and Duennebier F. (1972) Structure, composition, and properties of lunar crust. *Proc. Third Lunar Sci. Conf., Geochim. Cosmochim. Acta*, Suppl. 3, Vol. 3, pp. 2527–2544. MIT Press.

Wang H., Simmons G., and Todd T. (1973) Microcracks and the 25-km lunar discontinuity. *Earth Planet. Sci. Lett.* Submitted for publication.

Wang H. and Simmons G. (1972) FeO and SiO_2 in the lower mantle. *Earth Planet. Sci. Lett.* **14**, 83–86.

Wang H., Todd T., Weidner D., and Simmons G. (1971) Elastic properties of Apollo 12 rocks. *Proc. Second Lunar Sci. Conf., Geochim. Cosmochim. Acta*, Suppl. 2, Vol. 3, pp. 2327–2336. MIT Press.

Proceedings of the Fourth Lunar Science Conference
(Supplement 4, *Geochimica et Cosmochimica Acta*)
Vol. 3, pp. 2673-2684

On the formation of the lunar mascons

JAFAR ARKANI-HAMED*

Lunar Science Institute, Houston, Texas USA

Abstract—A new mascon hypothesis is proposed which accounts for: (1) the existence of lunar mascons for more than 3 b.y., (2) the evidence for extensive volcanic activity from 3.7 to 3.2 b.y. ago, (3) the existence of negative gravity anomaly rings, (4) insufficient mare material in Mare Orientale, and (5) the lack of mascons associated with craters smaller than 200 km diameter. Moreover, it provides a simple mechanism for mass transfer into the basins. The hypothesis is based on the perturbations introduced into a spherically symmetric thermal evolution model of the moon by a giant impact.

INTRODUCTION

SINCE the discovery of the lunar mascons (Muller and Sjogren, 1968) several hypotheses have been proposed for their formation. These are mainly concerned with the creation of excess mass in a circular basin. However, recent analysis of the lunar gravity field and radioactive age determination of the lunar rocks have provided further criteria that must be fulfilled by a plausible mascon hypothesis. These criteria are:

I. Late volcanic activity: Radioactive age determination indicates extensive volcanism on the moon from 3.7 to 3.2 b.y. ago (Papanastassiou and Wasserburg, 1971; Fouad *et al.*, 1973). This volcanism was independent of the early differentiation process which probably took place within the first 0.1 b.y. of the lunar history (Fouad *et al.*, 1973). Also, it did not follow the giant impact events immediately, but rather the impact sites remained dry for about 100–500 m.y. (Shoemaker, 1964; Baldwin, 1970). After this elapsed time the volcanism started and filled the basins by successive superposed flows of mare material (Ronca, 1971). This evidence implies that in some places the lunar interior became hot around 3.7 b.y. ago and stayed hot for about 0.5 b.y. The latest flows in a mare are typically tongue-shaped which probably originated from the circumferential region and flowed into the basin (Ronca, 1973). This indicates that the hot places were likely beneath the highlands surrounding the basins. Notice that we use basin to denote the excavation zone of an impact site, i.e., the region inside the inner ring of a mare while highland denotes the surrounding area whether part of it is presently overlayed by mare material, as for example in Mare Imbrium, or whether it is not, as in Mare Orientale.

II. Negative gravity anomalies: The spherical harmonic analysis of lunar orbiter data (Michael *et al.*, 1969; Michael and Blackshear, 1971) showed that the positive gravity anomalies associated with circular maria are surrounded by nega-

---

*Permanent address: Department of Physics, Arya-Mehr University of Technology, Tehran, Iran.

tive ones (the negative rings). The existence of the negative rings was confirmed by the S-band transponder experiments of the Apollo missions and subsatellites of Apollos 15 and 16 (Sjogren *et al.*, 1972a, b, c; Sjogren *et al.*, 1973). Figure 1 illustrates a gravity profile deduced from Apollo 15 data (Sjogren *et al.*, 1972b). Notice that the magnitude and the width of the negative anomalies are comparable with those of the positive ones. Since the sources of the positive anomalies are most likely located at very shallow depths (Phillips *et al.*, 1972) the sources of the negative ones are also probably at shallow depths.

Besides the negative rings, it has been shown that, with the exception of Grimaldi, all craters studied with diameters less than 200 km have negative gravity anomalies (Sjogren *et al.*, 1973).

III. Prolonged existence of the mascons: The mascons have existed at least since the latest volcanic activity, 3.2 b.y. ago. This indicates that the upper 200–400 km of the lunar interior has been strong enough to support stress differences of more than 100 bars for this long time period (Arkani-Hamed, 1973a,b). From the Apollo 15 gravity and laser altimetry data it is concluded that the positive gravity anomaly associated with Mare Serenitatis has decayed with an average relaxation time of at least 5 b.y. (Arkani-Hamed, 1973c), in close agreement with a value given by Kaula (1969).

IV. Mass transfer mechanism: Besides the foregoing criteria a plausible mascon hypothesis should give an appropriate mechanism for mass transfer to the maria.

Notice that criteria I and II are not consistent unless two separate regions of the moon are related with the associated phenomena. It is hypothesized in this paper that the region directly beneath a circular basin is responsible for prolonged existence of the associated mascon while the region beneath the surrounding highlands provided the material for the volcanic flows.

Existing mascon hypotheses are classified as external and internal. The external hypotheses, involving deeply buried iron meteorites (Stripe, 1968), or remnants of basin-forming projectiles (Urey, 1968; Urey and McDonald, 1971) do not create negative gravity rings nor do they explain a lag of 100–500 m.y. between

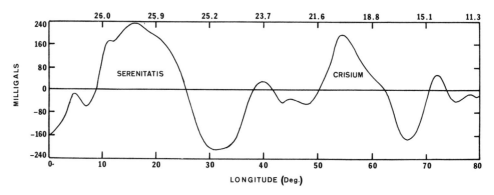

Fig. 1. Gravity profile deduced from Apollo 15 data (Sjogren *et al.*, 1972b).

formation of the basin and the volcanic activity. Thus they conflict with criteria I and II. Some of the internal hypotheses (Conel and Holstrom, 1968; Baldwin, 1968; O'Hara *et al.*, 1970) do not satisfy criterion II, nor do they provide a mechanism for mass transfer. The mascon hypotheses proposed by Wood (1970) and by Wise and Yates (1970) give somewhat similar mechanisms for mass transfer. They do not, however, satisfy criterion II, nor do they satisfy criteria I and III simultaneously. A transfer of mass by a global excess pressure produced by the thermal contraction of the outer part of the moon with respect to the inner part as suggested by Kaula (1969) cannot produce negative gravity anomalies surrounding the positive ones and thus cannot satisfy criterion II. Moreover, it does not explain a delay of 100–500 m.y. between the impact event and the volcanic activity.

In the present paper a new hypothesis is proposed for the formation of the lunar mascons which not only satisfies all the foregoing criteria but is consistent with the small amount of mare material in Mare Orientale and with the lack of mascons associated with craters smaller than 200 km diameter. According to this hypothesis mascons are produced by perturbations introduced into an otherwise spherically symmetric lunar model by the giant impacts.

IMPACT PERTURBATIONS

Here we briefly outline the perturbations produced by a giant impact. Pertinent perturbations are:

(1) The formation of a basin by ejection of material from the impact site. This will introduce the relatively low surface temperature to a deeper part of the moon at the basin. It also displaces substantial amounts of radioactive material from the basin onto the surrounding area and enhances the rate of heat generation in the highland region relative to that in the basin.

(2) The reduction of the effective thermal conductivity of the region beneath the highland by impact induce fractures. However, beneath the basin because of pre-impact high temperatures and because of the heat energy of the impact, the fractures would close.

(3) The coverage of the highland with a thick ejecta blanket of low thermal conductivity. Both the blanket and the impact-induced fractures, cause thermal insulation in the highland. Therefore, the region beneath the basin cools much faster than that beneath the highland.

MASCON HYPOTHESIS

In this section, the effects of the perturbations on the thermal evolution of the moon are determined and the formation of a mascon as a consequence of this evolution is described. Mare Imbrium is chosen as a numerical example. It is assumed that a spherically symmetric lunar model is subjected to the Imbrium impact at 4 b.y. ago. The estimated age of the formation of the Imbrium basin is 3.9 b.y. (Baldwin, 1971a). Toksöz and Solomon's (1973) thermal model is adopted for the symmetric lunar model because the detailed information of its parameters

Table 1. Physical properties and geometrical parameters of the lunar model. Differentiation depth denotes the depth above which the upward concentration of the radioactive material has taken place. Below this depth the radioactive material is distributed uniformly. This upward concentration reduces to $1/e$ of its surface value at the skin depth.

Radius	1740 km
Density	3.34 g/cm^3
Heat of fusion	400 joules/g
Specific heat	1.2 joules g/°C
U	60 ppb
K/U	2000
Th/U	4
Surface temperature	0
Thermal conductivity	0.03 watt/cm deg
Differentiation depth	800 km
Skin depth	100 km

are available. The model consists of a solid surface layer of 100 km, the lithosphere, underlain by a partially molten layer of 800 km, the mantle, which in turn overlies a uniform solid sphere. Table 1 shows the physical parameters, chosen by Toksöz and Solomon (1973) with the exception of thermal conductivity for which we have chosen a constant value. It is, however, demonstrated that using a temperature dependent thermal conductivity such as the one used by Toksöz and Solomon has little effect on the lateral variations of temperature in the lunar interior (Arkani-Hamed, 1973d). The Imbrium impact is inferred to have produced a basin 600 km in diameter and 50 km in depth, in close agreement with estimates by Baldwin (1970). Figure 2 illustrates the basin immediately after its formation. Notice that (1) the ejecta blanket is assumed to be a uniform layer of 5 km thick. The thermal insulation of this blanket is less than the plausible wedge-shaped actual blanket of Imbrium, (2) the fractured zone is confined to the upper 20 km of the highland. The hydrostatic pressure below 20 km depth is greater than 1 kb which is sufficient to close the fractures at the temperature conditions existing in this region (Walsh and Decker, 1966). The effect of fractures and ejecta blanket is incorporated into the upper 20 km of the highland by reducing its thermal conductivity by a factor of 2. (3) A mantle plug of 20 km thickness and a molten layer of 10 km thickness on the basin floor are introduced instantaneously for the simplicity of computation. They are meant to present the average effect of continuous isostatic adjustment and volcanic flow respectively. The detailed description of the model is presented elsewhere (Arkani-Hamed, 1973e).

Assuming that there is only one impact basin, the post-impact thermal state of the moon is azimuthally symmetric about the Z-axis (Fig. 2). Therefore, the relevant diffusion equation is

$$C\rho \frac{\partial}{\partial t} T = \frac{1}{r^2} \frac{\partial}{\partial r} \left(r^2 K \frac{\partial}{\partial r} T \right) + \frac{1}{r^2 \sin \theta} \frac{\partial}{\partial \theta} \left(K \sin \theta \frac{\partial}{\partial \theta} T \right) + A$$

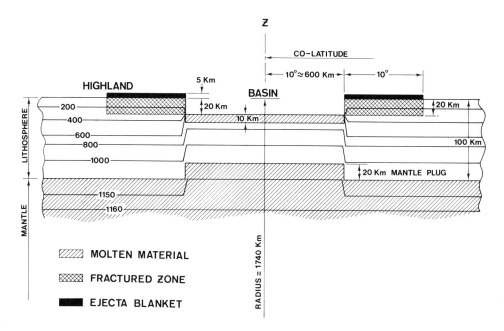

Fig. 2. A portion of the initial lunar model near the impact site. The isotherms are concentric spherical surfaces except in the lithosphere and the uppermost part of the mantle beneath the basin where they have been displaced upward by 20 km.

where C = heat capacity, ρ = density, A = heat generation rate, t = time, r = radial distance from the center of the moon, θ = co-latitude, K = thermal conductivity, and T = Temperature. Equation (1) is solved under the following boundary and initial conditions:

1. Regularity of the solution at the center of the moon,
2. Vanishing of the temperature at the lunar surface, and
3. At time $t = 0$ the temperature distribution inside the moon is the same as Toksöz and Solomon's model.

The numerical procedures employed are similar to those presented in a previous paper (Arkani-Hamed, 1973d). A liquid state convection process similar to that adopted by Reynolds *et al.* (1966) is also used in the present calculations.

Figure 3 illustrates the variations of temperature with depth beneath the center of the basin at different times after the impact. The molten layer on the basin becomes completely solidified within the first 1 m.y., while the mantle plug takes more than 100 m.y. This is because (1) the temperature at the surface of the molten layer is much lower than that at the top of the mantle plug, and (2) the upward heat flux by fluid convection into the mantle plug enhances the amount of heat available there, while such excess heat is not available in the molten layer. The thermally insulating ejecta blanket and fractured zone elevate the temperature beneath the highland relative to that beneath the basin. Figure 4 displays the

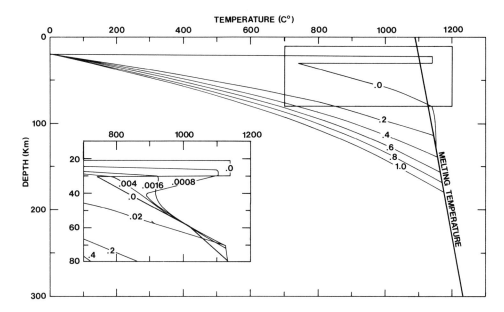

Fig. 3. Temperature variation with depth beneath the center of Imbrium basin. The numbers on the curves denote the time after the impact in b.y. The early stages of the temperature variations are illustrated in an expanded scale for clarity.

lateral variations of temperature at 0.5 b.y. after the impact which illustrates the influence of the impact on the thermal evolution of the moon. Included in the figure is the 1160°C isotherm which does not undulate, indicating that the impact-induced perturbations of temperature do not penetrate the partially molten region, i.e., deeper than 140 km within 0.5 b.y.

Because of the highland-basin temperature difference, the lithosphere thickens at different rates under these regions. Figure 5 illustrates the lateral variations of the base of lithosphere, at different times. The lithosphere beneath the basin thickens monotonically while that beneath the highland becomes molten to a depth of

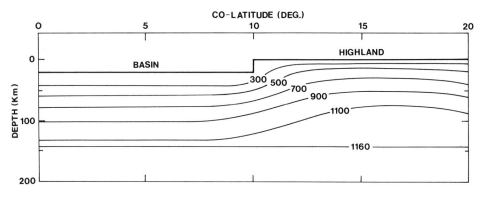

Fig. 4. Lateral variations of temperature in the upper part of the moon 0.5 b.y. after the impact. Units are in °C.

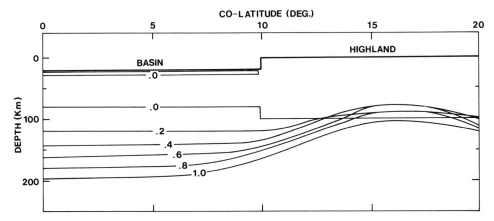

Fig. 5. Lateral variations of the base of lithosphere. The numbers on the curves denote
the time after the impact in b.y.

80 km within the first 0.5 b.y. and then starts to thicken. The stresses associated
with the volume increase due to the remelting process enhance the global excess
pressure locally and introduce new fractures and/or open the impact-induced fractures in the overlying lithosphere, causing magmatization and volcanic activity
which transfers the mare material from beneath the highland onto the basin.
Moreover, the prolonged existence of the molten phase beneath the highland,
probably enhances the differentiation of the underlying region and the upward
concentration of the lighter material which results in the present day observed
negative gravity rings. On the contrary, since the region beneath the basin sol-
idifies relatively faster, no significant differentiation takes place and the lithosphere has the same density as the solidified mantle material.

It remains to demonstrate the prolonged existence of the mascon. The initial
depth of the basin, 50 km, implies that the total mass of the mascon associated
with Mare Imbrium is about an order of magnitude less than the total mass of the
mantle plug plus the mare filling (Arkani-Hamed, 1973e). This indicates that the
mascon was probably formed at the latest stage of volcanism, i.e., when the
lithosphere beneath the basin was about 150 km thick and the underlying molten
mantle was about 700 km thick. Bearing in mind that, because of the molten mantle, the stress differences produced by the mascon were confined to the lithosphere (Arkani-Hamed, 1973a), the response of the lunar model to these stresses
could be approximated by that of a two layered spherical body; a viscous
lithosphere overlying an incompressible fluid interior. Figure 6 illustrates the
viscosity-thickness relationship required for the lithosphere in order that the
stress differences decay with a relaxation time of 5 b.y. The figure is constructed
from the tabulated results of Shimazu (1966) for a surface load specified by a
spherical harmonic of degree 8. This harmonic is very close to the principal term
of the spherical harmonic analysis of the Imbrium mascon.* It is clear from

---

*A feature of diameter L on a sphere of radius a has a spherical harmonic expression whose
principal term is of degree $n = \pi a/L$. In the case of Imbrium $L = 600$ km, $a = 1740$, and thus $n = 9$.

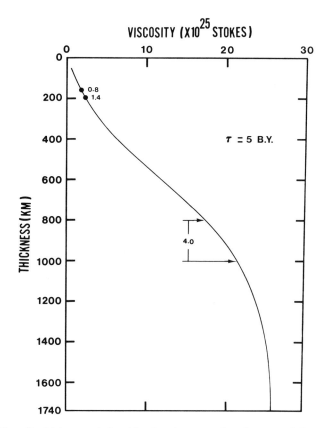

Fig. 6. Viscosity-thickness relationship of a viscous surface layer overlying an incompressible fluid interior, constructed from the tabulated results of Shimazu (1966).

the figure that a viscosity of at least 1.5×10^{-25} stokes ($\approx 5 \times 10^{25}$ poise) is required for the lithosphere in order that the stresses produced by the freshly formed mascon decay with the lower limit of the relaxation time, 5 b.y. This viscosity value is not far from the lower limit of 1.1×10^{25} poise proposed for the lunar crust by the analysis of the diameter-depth relationship of the lunar craters (Baldwin, 1971b). However, as the lithosphere thickens, a higher viscosity is required in order to maintain a relaxation time of 5 b.y. Numbers of the curve denote the times after the impact at which the lithosphere beneath the basin acquires the associated thicknesses. For example, a lithosphere 200 km thick would form about 1.4 b.y. after the impact or about 0.6 b.y. after the formation of the mascon. Or, according to the lunar passive seismic data (Latham et al., 1973) the thickness of the lithosphere at the present time is about 800–1000 km and a viscosity of about $5.6 - 7.3 \times 10^{26}$ poise is required if the stresses associated with the observed mascon decay with a relaxation time of 5 b.y. The increase of viscosity required by the thickening of the lithosphere is provided by the cooling of the lithosphere.

In this paper we have adopted a lunar model with an initially hot upper part, and it appears that the presence of mascons can be made compatable with such a model. However, even though the hypothesis has provided us with an efficient procedure for elevating temperatures beneath the highland, it remains for future study to demonstrate whether it is possible to produce a partially molten local region in an initially solid moon by this procedure. If it is possible, then this hypothesis will remove the objection made of the cold moon theory in the light of volcanic activity between 3.7 and 3.2 b.y. ago.

MARE ORIENTALE AND SMALL CRATERS

In this section we point out the consistency of the proposed mascon hypothesis with the small amount of mare material in Mare Orientale and with the lack of mascons associated with craters smaller than 200 km diameter.

The Orientale impact probably took place 3.6 b.y. ago (Hartmann and Wood, 1971; Baldwin, 1971a) and produced a basin of about 300 km diameter. Using the diameter/depth ratio of the Imbrium basin, the depth of the Orientale basin is about 25 km, however, we take it to be 30 km since Orientale basin was probably formed by a high velocity projectile (Urey and McDonald, 1971). For the same reason described for Imbrium basin, a 10 km mantle plug and a 10 km molten layer in the basin are introduced immediately after the impact event. Despite the small size of the Orientale impact, its effect on the thermal conductivity of the surrounding highland, both due to its ejecta blanket and impact induced fractures, are assumed to be the same as that of Imbrium impact. This enhances the effect of the Orientale impact in elevating temperature beneath the highland to some extent but not drastically. Figure 7 illustrates the lateral variations of the base of the lithosphere at different times after the Orientale impact. The calculation procedure is the same as that employed for Imbrium. Although the thermal conductivity of the highland is the same as adopted for Imbrium, there is only a very limited remelting at the base of the lithosphere beneath the highland. This is because of the reduction of radioactive material through their decay processes between 4 and 3.6 b.y. ago and also because of the thick pre-impact lithosphere, which is taken to

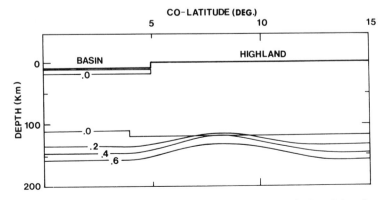

Fig. 7. Lateral variations of the base of the lithosphere beneath Mare Orientale.

be 120 km (This is the thickness of the lithosphere at a place far from Mare Imbrium, where the Imbrium impact had no effect on the temperature distribution, $\theta > 90°$, at 0.4 b.y. after the Imbrium impact). A similar reason was also suggested by Kaula (1971). The slow thickening of the lithosphere beneath the basin is, however, due to the small size of the basin. The limited remelting produces stress differences small enough that they cannot produce new fractures and/or open the impact-induced ones in the overlying lithosphere effectively. Therefore, only a limited amount of mare material could flow onto the basin. Wise and Yates (1970) attributed the lack of extensive volcanism on the back side of the moon to the discrepancy between the center of figure and center of mass of the moon. The present mascon hypothesis does not contradict their suggestion but rather it takes the advantage of the suggestion in explaining the small amount of filling of the Orientale basin as well as of other large impact basins of the lunar back side. It is clear from the figure that there was no prolonged molten state at about 100 km depth beneath the highland and therefore, there was possibly no extensive upward differentiation of light material in order to produce the observed broad negative gravity anomaly ring (Sjogren *et al.*, 1972c). However, the small amount of mare material that flowed into the basin, did not fill the interval between the first and the second rings and thus could not reduce the negative gravity anomaly of the surface topography. Therefore, the broad negative gravity anomaly associated with this mare is partially, if not totally, due to the surface topography.

To apply the present mascon hypothesis to small craters, it is assumed that a crater with 180 km diameter and 15 km depth is formed by an impact 4 b.y. ago. The other parameters of the crater are obtained by suitable scaling of those of Mare Imbrium, except the mantle plug and molten layer on the basin which are not introduced for small craters. Figure 8 illustrates the lateral variations of temperature produced by the crater at 0.5 b.y. after its impact. It is clear from the figure that the perturbations introduced to an otherwise spherically symmetric moon are negligible and they cannot produce any appreciable undulations of the base of the lithosphere and subsequent magmatization and volcanism. Therefore, according

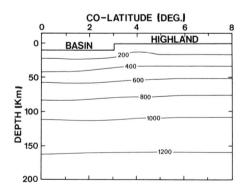

Fig. 8. Lateral variations of temperature beneath a crater of 150 km diameter 0.5 b.y. after impact. Units are in °C.

to the mascon hypothesis, such a crater is not expected to be filled by mare material and has no positive gravity anomaly.

Summary

In the present paper it is shown that a giant impact causes thermal insulation in the highland surrounding a basin by inducing fracture and also by covering the region with a thick and low-conductive ejecta blanket. Consequently, a laterally heterogeneous thermal regime develops in the lunar interior, which results in fast cooling of the region beneath the basin and remelting of the base of the lithosphere beneath the highland. The former results in a thick lithosphere strong enough to support the associated mascon and the latter produces high stresses which create fractures in the overlying lithosphere and cause volcanic activity and subsequent transfer of mass from beneath the highland into the basin.

It is also concluded that the small amount of filling of Mare Orientale is mainly due to its impact time, late in the thermal history of the moon. The lack of mascons associated with craters smaller than 200 km diameter is because of their small sizes.

Acknowledgments—I would like to thank Dr. D. W. Strangway for reading the manuscript and also for his helpful discussions. This work was supported by the Lunar Science Institute which is operated by the University Space Research Association under Contract NSR-09-051-001 with the National Aeronautics and Space Administration. This paper constitutes the Lunar Science Institute contribution number 149.

References

Arkani-Hamed J. (1973a) Stress differences in the Moon as an evidence for a cold Moon. *The Moon* **6**, 135–163.
Arkani-Hamed J. (1973b) Density and stress distribution in the Moon. *The Moon* **7**, 84–126.
Arkani-Hamed, J. (1973c) Viscosity of the Moon, I: After mare formation. *The Moon* **6**, 100–111.
Arkani-Hamed, J. (1973d) Effect of a giant impact on the thermal history of the Moon. *The Moon*. In press.
Arkani-Hamed, J. (1973e) Lunar mascons as consequences of giant impacts. *The Moon*. In press.
Baldwin R. B. (1968) Lunar mascons: Another interpretation. *Science* **162**, 1407–1408.
Baldwin R. B. (1970) Summary of arguments for a hot Moon. *Science* **170**, 1297–1300.
Baldwin R. B. (1971a) On the history of lunar impact cratering: The absolute time scale and the origin of planetesimals. *Icarus* **14**, 36–52.
Baldwin R. B. (1971b) The question of isostasy on the Moon. *Phys. Earth Planet. Interior* **4**, 167–179.
Conel J. E. and Holstrom G. B. (1968) Lunar mascons: A near-surface interpretation, *Science* **162**, 1403–1404.
Fouad T., Papanastassiou D. A., and Wasserburg G. J. (1973) A lunar cataclysm ~ 3.95 AE and the structure of the lunar crust (abstract) In *Lunar Science—IV*, pp. 723–725. The Lunar Science Institute, Houston.
Hartmann W. K. and Wood C. A. (1971) Moon: origin and evolution of multi-ring basins. *The Moon* **3**, 3–78.
Kaula W. M. (1969) Interpretation of the lunar mass concentrations. *Phys. Earth Planet. Interior* **2**, 123–137.
Kaula W. M. (1971) Interpretation of the lunar gravitational field. *Phys. Earth Planet. Interior* **4**, 185–192.

Latham G., Dorman J., Duennebier F., Ewing E., Lammlein D., and Nakamura, Y. (1973) Moon-quakes, meteoroids, and the state of the lunar interior. *Proc. 4th Lunar Sci. Conf., Geochim. Cosmo-chim. Acta*, Suppl. 4, Vol. 3, pp. 2515–2527. Pergamon.

Michael W. H. Jr. and Blackshear W. T. (1971) Recent results on the mass, gravitational field, and moments of inertia of the Moon. *The Moon* 3, 388–402.

Michael W. H. Jr., Blackshear W. T., and Gapcynski J. P. (1969) Results of the mass and the gravitational field of the moon as determined from dynamics of lunar satellites. Presented at the *12th Plenary Meeting of COSPAR*, Prague.

Muller P. M. and Sjogren W. L. (1968) Mascons: Lunar mass concentrations, *Science* 161, 680–684.

O'Hara M. J., Biggar G. M., and Richardson S. W. (1970) Experimental petrology of lunar material: The nature of mascons, seas, and the lunar interior. *Science* 167, 605–607.

Papanastassiou D. A. and Wasserburg G. J. (1971) Lunar chronology and evolution from Rb-Sr studied of Apollo 11 and 12 samples. *Earth Planet. Sci. Lett.* 11, 37–62.

Phillips R. J., Conel J. E., Abbott E. A., Sjogren W. L., and Morton J. B. (1972) *J. Geophys. Res.* 77, 7106–7114.

Reynolds R. T., Fricker P. E., and Summers A. L. (1966) Effects of melting upon thermal models of the earth. *J. Geophys. Res.* 71, 573–582.

Ronca L. B. (1971) Age of lunar mare surfaces. *Geol. Soc. Am. Bull.* 82, 1743–1748.

Ronca L. B. (1973) The filling of the lunar mare basins. *The Moon* 7, 239–248.

Shimazu Y. (1966) Survival time of lunar surface irregularities and viscosity distribution within the Moon. *Icarus* 5, 455–458.

Shoemaker E. M. (1964) The geology of the Moon. *Sci. Am.* 211, 38–47.

Sjogren W. L., Gottlieb P., and Muller P. M. (1972a) Lunar gravity via Apollo 14 doppler radio tracking. *Science* 175, 165–168.

Sjogren W. L., Muller P. M., and Wollenhaupt W. R. (1972b) Apollo 15 gravity analysis from the S-band transponder experiments. *The Moon* 4, 411–418.

Sjogren W. L., Muller P. M., and Wollenhaupt W. R. (1972c) S-band transponder experiment. In *Apollo 16 Prel. Sci. Rept.*, pp. 24-1, 24-7. NASA SP-315.

Sjogren W. L., Wollenhaupt W. R., and Wimberly R. N. (1973) Lunar gravity via the Apollo 15 and 16 subsatellite. *The Moon*. In press.

Stripe J. G. (1968) Mascons and the history of the moon. *Science* 161, 1402–1403.

Toksöz M. N. and Solomon S. C. (1973) Thermal history and evolution of the Moon. *The Moon* 7, 251–278.

Urey H. C. (1968) Mascons and the history of the Moon, *Science* 162, 1408–1410.

Urey H. C. and McDonald G. J. F. (1971) Origin and history of the Moon. In *Physics and Astronomy of the Moon, 2nd ed.* (editor Z. Kopal) pp. 213–289, Academic Press.

Walsh J. B. and Decker E. R. (1966) Effect of pressure and saturating fluid on the thermal conductivity of compact rock. *J. Geophys. Res.* 71, 3053–3061.

Wise D. U. and Yates M. T. (1970) Mascons as structural relief on a lunar moho. *J. Geophys. Res.* 75, 261–268.

Wood J. A. (1970) Petrology of the lunar soil and geophysical implications. *J. Geophys. Res.* 75, 6497–6513.

Proceedings of the Fourth Lunar Science Conference
(Supplement 4, *Geochimica et Cosmochimica Acta*)
Vol. 3, pp. 2685–2696

Viscous flow and crystallization behavior of selected lunar compositions

M. Cukierman, L. Klein, G. Scherer, R. W. Hopper, and D. R. Uhlmann

Department of Metallurgy and Materials Science, Center for Materials Science and Engineering
Massachusetts Institute of Technology, Cambridge, Mass.

Abstract—The flow characteristics of lunar compositions 15555 and 68502 have been determined over a wide range of viscosity. The temperature ranges covered by the measurements are 1201–1410°C and 622–695°C for the 15555 composition and 1261–1515°C and 725–840°C for 68502 composition. Reliable data could not be obtained over the intermediate ranges of temperature because of the occurrence of crystallization. The experimental data in the high temperature regions are found to be in close agreement with predictions of the semi-empirical model of Bottinga and Weill. The results on these compositions are compared with previous data on other lunar compositions and on anorthite; and the importance of the flow behavior in interpreting lunar flows and phase morphologies is emphasized.

Data are also reported on the kinetics of crystallization of the 15555 composition over the temperature interval from 700 to 1020°C. The growth rate data are combined with the viscosity data to construct a reduced growth rate vs. undercooling relation. The form of this relation, which exhibits positive curvature, is suggestive of growth by a surface nucleation mechanism; but a plot of logarithm (growth rate × viscosity) vs. $1/T\Delta T$ does not display the straight-line form expected by the standard models for such a growth mechanism. It does, however, have the form suggested by recent computer simulations of crystal growth. The combined data are also used to construct a time-temperature-transformation curve corresponding to a barely-detectable degree of crystallinity. From this curve, the cooling rate required to form a glass of composition 15555 and the maximum thickness obtainable as a glass are estimated. This composition is found to be more difficult than previously investigated lunar compositions to form as a glass; and it is unlikely that specimens of this or closely similar compositions can be obtained as amorphous solids for sizes larger than about 0.5 cm.

Introduction

The viscous flow behavior and crystallization kinetics of three lunar compositions have previously been determined over wide ranges of temperature (Cukierman *et al.*, 1972; Scherer *et al.*, 1972; Cukierman and Uhlmann, 1972). The results were originally used to understand the nature of the crystallization process in the materials and to evaluate the likelihood of their formation as glassy solids. The data are presently being employed to understand the rheology of lava flows on the moon as well as to infer information about thermal and mechanical histories of specimens from their observed phase assemblages.

In all cases, the viscous flow data in the high-temperature regions were well represented (within ±0.20 in $\log_{10} \eta$) by the semi-empirical model of Bottinga and Weill (1972). In all cases, appreciable curvature was noted in the overall $\log \eta$ vs. $1/T$ relations, with the apparent activation energies for flow at low temperatures being larger than those at high temperatures by factors of about 2.5–3.

The forms of the crystal growth rate vs. temperature relations for the lunar

2685

compositions are similar to those observed with other oxide glass-forming materials, passing through maxima in the range about 50–150°C below the liquidus temperatures. Combining growth rate with viscosity data, the fraction of preferred growth sites on the crystal-liquid interface was found to increase with increasing undercooling. The forms of these variations were suggestive of crystallization by a surface nucleation growth mechanism; but the log (growth rate × viscosity) vs. $1/T\Delta T$ plots exhibited pronounced curvature rather than the straight lines predicted by the standard models for such growth.

The minimum cooling rates required to form glasses of the lunar compositions were estimated by constructing time-temperature-transformation curves corresponding to a just-detectable fraction crystallized. From these curves, which were constructed using the measured growth rate and viscosity data, the critical cooling rates required to form glasses of the lunar compositions were estimated to lie in the range of 1–50°C sec^{-1}, and the maximum thicknesses obtainable as glasses in the range 0.2–2 cm.

The three lunar compositions investigated previously—14259, 14310, and 15418—are shown in Table 1. All are anorthite-rich, low in TiO_2, and range in FeO content from 5.4 to 10 wt.%. The last two were found as crystalline rocks on the moon, and the first as a soil with a large glass component in the smaller size fractions. To broaden the base of these kinetic investigations, it seemed advisable to determine the flow and crystallization behavior of a lunar composition (15555) which is much richer in FeO and poorer in Al_2O_3 than those studied previously, and a composition (68502) which was found as a soil fraction rich in crystalline material. The crystallization behavior of 68502 was similar to that of the compositions previously studied, and its description will be deferred to a subsequent review paper (Scherer *et al.*, 1973). The other results, on the crystallization and flow of 15555 and the flow of 68502, will be reported in the present paper.

Table 1. Compositions investigated (in wt.%).

	15555 olivine-phyric mare basalt	68502 highland soil	14259 Fra Mauro soil	14310 Fra Mauro basalt	15418 shock melted gabbroic anortho-site	74420 orange glass	Anor-thite
SiO_2	44.2	45.8	48.0	50.0	45.0	38.6	43.1
Cr_2O_3	0.7	0.1			0.1		
TiO_2	2.3	0.4	1.8	1.3	0.3	8.8	
Al_2O_3	8.5	27.3	18.0	20.0	26.7	6.3	36.65
FeO	22.5	5.0	10.0	7.7	5.4	22.0	
MgO	11.2	5.3	9.2	8.0	5.4	14.45	
CaO	9.45	15.3	11.0	11.0	16.1	7.7	20.16
Ba_2O	0.2	0.6	0.5	0.6	0.3	0.4	
K_2O		0.1	0.5	0.5		0.1	
MnO	0.3				0.3	0.3	

EXPERIMENTAL PROCEDURE

The materials studied in this investigation, lunar compositions 15555 and 68502, were prepared from the reagent grade raw material powders shown in Table 1. Weighed powders of the component materials were milled together for about 24 hours prior to melting. The melting was carried out under conditions of low oxygen activity to simulate the Fe^{2+}/Fe^{3+} ratios of the lunar material. The melting procedure as well as the sample preparation method for the viscous flow study have been described previously (Cukierman *et al.*, 1972). The only significant difference in the present study was occasioned by the rapid crystallization of the 15555 composition. The rapid quenching required to make glass samples for the low-temperature viscosity study was obtained by pouring the molten liquid onto copper blocks under a protective atmosphere of 5% H_2 forming gas (95% N_2–5% H_2). Even with this procedure, samples thicker than a few mm could not be obtained as glasses.

Viscosities in the molten range were determined using a rotating-cylinder viscosimeter, in which the outer cylinder containing the specimen was rotated at a uniform angular velocity and the torque on a bob suspended in the specimen was measured. Viscosities greater than about 10^9 poise were determined using a bending-beam viscosimeter, in which the rate of viscous deformation of a simple beam was measured. Both instruments have been described in detail elsewhere (Cukierman *et al.*, 1972); both were provided with an atmosphere of 5% H_2 forming gas; and both were calibrated using NBS reference materials.

The Fe^{2+} and total Fe contents of the 15555 composition were determined for specimens both prior to and following the viscous flow measurements. The analyses were carried out in Chemical Analytical Facility of the Center for Materials Science and Engineering at MIT. The results for both sets of specimens indicated that approximately 94% of the iron in the glass is present as Fe^{2+}. The remaining iron is largely if not entirely in the form of small metal particles, 50–200 Å in diameter, observed in electron microscope studies (Cukierman and Uhlmann, 1973a).

The measurements of crystal growth rates were carried out on small glass specimens (typically $3 \text{ mm} \times 3 \text{ mm} \times 3 \text{ mm}$) cut from the sample material used in the viscous flow study. The growth rates were determined from the measured thicknesses of the external crystal layers on specimens held at a given crystallization temperature for various periods of time. The detailed experimental procedure has been described previously (Scherer *et al.*, 1972). All the present crystallization runs were carried out in an atmosphere of dry nitrogen.

RESULTS AND DISCUSSION

Viscous flow

The viscosity vs. temperature relation for the 15555 liquid is shown in Fig. 1 and that for the 68502 liquid in Fig. 2. Reliable data for the 15555 composition could only be obtained at temperatures above 1201°C and below 695°C, and for the 68502 composition at temperatures above 1261°C and below 840°C. In both cases, crystallization prevented the obtaining of reliable information at intermediate temperatures, since holding times at temperature of 30–60 minutes were used to assure thermal equilibrium during the measurements. Data in the high temperature range below the liquidus in each case were obtained by super-heating the specimens well above the respective liquidus temperatures prior to cooling.

In the case of the 15555 compostion, the samples contained a small volume fraction (in the range of 1 pct. or less) of small metal particles. In such small concentrations, the presence of the particles would not be expected to have a significant effect on the measured viscosity. Further, any differences between

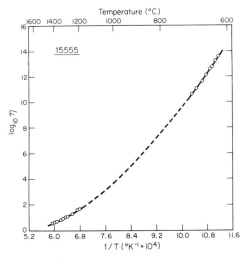

Fig. 1. Viscosity vs. temperature relation for lunar composition 15555.

the iron oxidation state of the present samples and that of the natural lunar material are also expected to have an insignificant effect on the measured viscosity. As an example of this, samples of the 15555 composition with only 76 pct. of the total iron in the Fe^{2+} state (prepared by melting under partially oxidizing conditions) exhibit a viscosity-temperature relation in the low-temperature region which is closely similar to that shown in Fig. 1 (*see* Cukierman and Uhlmann, 1973a).

The glass transition temperature ($\eta = 10^{13}$ poise) for the 68502 composition is about 740°C, in the same range as that of the lunar compositions investigated pre-

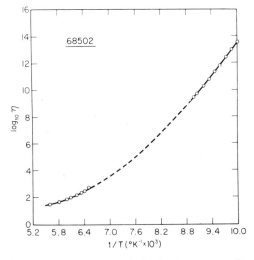

Fig. 2. Viscosity vs. temperature relation for lunar composition 68502.

viously. The glass transition of the 15555 composition occurs at a significantly lower temperature, about 635°C. The apparent activation energies for flow in the high-temperature and low-temperature regions for the 68502 composition are about 60 and 200 kcal (g at.)$^{-1}$, respectively, while those for the 15555 composition are about 55 and 150 kcal (g at.)$^{-1}$. These values are also in the range found for the other lunar compositions. The pronounced curvature in the overall log η vs. $1/T$ relation for both these liquids—and for all other lunar compositions as well—suggests, however, that flow of the lunar compositions should not be regarded as a simply-activated process, and hence that caution must be exercised in attaching physical significance to the derived values of the apparent activation energy.

The high temperature flow behavior predicted by the Bottinga and Weill model falls within the widths of the experimental points shown in Figs. 1 and 2. In detail, the model consistently overestimates the viscosity by a small factor, typically about $10^{0.1}$, and closely predicts the variation of viscosity with temperature. The quality of the agreement for the present compositions is similar to that found previously for other lunar liquids. The agreement in the case of the 15555 composition indicates that the values of the parameters used in the model are appropriate for liquids containing FeO concentrations as large as 22 wt.%, at least when these compositions are prepared and tested under conditions of low oxygen activity.

Semi-empirical models of the Bottinga and Weill type are not presently available to describe flow in the low-temperature region. The variability in form of the log η vs. $1/T$ relations for the lunar compositions in this region is similar to the results found on other glass-forming liquids, both organic and inorganic (*see* discussion and references in Laughlin and Uhlmann, 1972). The flow behavior in the low-temperature region cannot be satisfactorily described by any available theoretical model. It seems clear from the data, however, that flow in this region involves a significant degree of cooperative motion.

For nearly all glass forming liquids, pronounced curvature is observed in the overall log η vs. $1/T$ relations. In the high-temperature region (for viscosities smaller than about 10^4 poise) free volume models seem to provide a close representation of available experimental data (*see*, e.g., Cukierman and Uhlmann, 1973b). The data on all lunar compositions investigated to date seem consistent with this suggestion, although the occurrence of crystallization in the intermediate regions of temperature preclude a detailed evaluation.

The viscous flow behavior of liquid anorthite, which can be regarded as an end-member for lunar rocks, particularly highland rocks, has also been determined in this laboratory (Cukierman and Uhlmann, 1973c). In this case, the wider accessible range of experimental data combined with the extension of experimental results using the Bottinga and Weill model permitted detailed testing of the free volume model in the high-temperature region. The results indicate that the free volume picture provides a close description of experimental data for viscosities smaller than about 10^4 poise. The magnitudes of the constants in the free volume description of the data, even the pre-exponential constant, were in close agreement with those predicted theoretically and those found in studies of viscous flow of other materials. The model is also expected to provide a close description of the high-temperature flow of the plagioclase-rich lunar compositions.

According to the free volume model, the critical step in flow involves opening up a hole of sufficient size to permit molecular transport. The hole is viewed as being opened by a redistribution of the free volume in the system; and the temperature dependence of the viscosity can then be associated with the temperature dependence of the free volume. The model predicts a relation of the form:

$$\log \eta = A T^{1/2} \exp (\delta V_M / V_f) \tag{1}$$

Here A and δ are constants; V_M is the molecular volume; and V_f is the average free volume, given by $V_f = V_M - V_o$, where V_o is the Van der Waals volume of the molecule.

The viscous flow behavior of the two lunar compositions studied in the present investigations can usefully be compared with the flow behavior of other lunar compositions and of anorthite. Such a comparison is shown in Fig. 3 where data from the previously cited references are combined with the present data and with those on lunar composition 74220 which will be published elsewhere (Uhlmann *et*

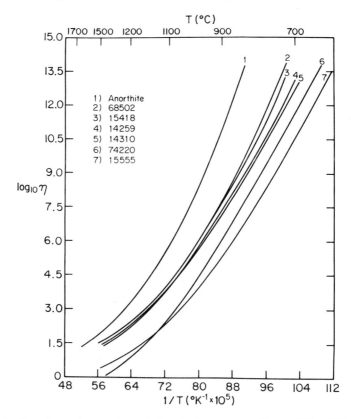

Fig. 3. Viscosity vs. temperature relation for lunar compositions 14259, 14310, 15418, 15555, 68502, 74220, and for anorthite.

al., 1973). Continuous curves are drawn through the full range of viscosity for each of the liquids. It should be recalled, however, that experimental data could generally not be obtained for viscosities between about 10^3–10^4 poise and about 10^9–10^{10} poise because of the occurrence of crystallization; and the curves drawn through these regions of viscosity represent the smooth interpolations shown in the original figures.

From the data shown in Fig. 3, it is apparent that the flow behavior of lunar compositions 14259, 14310, 15418, and 68502 are closely similar in the high-temperature fluid range (in this range the viscosity of the 68502 composition is within experimental error identical to that of the 15418 composition). More pronounced differences among the four compositions are noted, however, in the low-temperature region as the glass transition is approached. As shown in Table 1, these compositions have a similar range of $SiO_2 + Al_2O_3$ contents, 66–73 wt.% (62–68 g at.%), and a similar range of $MgO + CaO + FeO$ contents, 26–30 wt.% (31–36 g at.%).

At any given temperature, liquid anorthite has an appreciably higher viscosity than any of the lunar compositions, and the temperature dependence of its viscosity is more pronounced. The two compositions which are relatively rich in $CaO + MgO + FeO$ and poor in $SiO_2 + Al_2O_3$ (compositions 15555 and 74220) have the smallest viscosities at a given temperature and the lowest glass transition temperatures.

While a simple structural model based on concepts of network-forming (Si and Al) and network-modifying (Mg, Ca, and Fe) cations can usefully rationalize the three principal groupings shown in Fig. 3—i.e., anorthite; 68502, 15418, 14259, 14310, 74220, and 15555—it cannot describe the flow behavior within the groups (either the viscosity at a given temperature or the variation of viscosity with temperature). More generally, the relation between structural features of oxide liquids and their viscous flow characteristics remains to be elucidated in any satisfactory detail.

From these data, it is apparent that many of the compositions returned from the lunar surface are fluid only at quite elevated temperatures (in the range of 1500°C and higher). Even at these temperatures, their viscosities are more like that of glycerin than of water under ambient conditions (e.g., 25°C). Further, since liquid viscosity increases significantly with increasing pressure, the viscosities of the lunar liquids at any appreciable depth on the moon should be appreciably higher than the values shown in Fig. 3. In light of these considerations and the marked increase in viscosity with decreasing temperature below the fluid range, any model which attempts to describe the transport of such liquids from the lunar interior to its surface must take account of the fact that the flowing phase will be quite viscous over the range of temperature where transport takes place. The times required for flow at the assumed stress levels must then be compared with the times for cooling of the liquid phase and for crystallization to take place. These comparisons and their implications for the observed occurrence of selected lunar samples will be explored in a subsequent paper (Hopper and Uhlmann, 1973).

A lunar interior which is (or was) only partially molten, with the liquid-phase composition in the range of the lunar compositions studied to date, would be characterized by appreciably higher viscosities than those shown in Fig. 3. Further, the viscosities of such material should increase more strongly with falling temperature as additional crystallization takes place.

Qualitatively similar arguments would maintain for the flow generated under impact conditions on the lunar surface. Assuming that the plasticizing effects of dissolved or second-phase gases and water can be neglected for the lunar environment, these estimates on viscosity should be realistic, and the constraints which they impose on possible lunar flows should be incorporated in any discussion of the formation of structural features on the moon.

Crystallization behavior

At all temperatures, the crystallization of composition 15555 was observed to proceed uniformly from the external surfaces of the samples. In a number of cases, where the specimens cracked from thermal stresses during insertion into the furnace, crystallization was also observed to proceed from the crack surfaces. The thickness of the crystal layer was observed to increase linearly with time at all temperatures. The growth rates were determined from the slopes of least square lines drawn through the thickness vs. time-at-temperature plots. The temperature range covered by the present investigation (approximately 700–1000°C) lies below the estimated solidus temperature for the 15555 composition (J. F. Hays, private communication). This is consistent with the observed similarity of the crystallization products (olivine, spinel, clinopyroxene, and plagioclase) over the entire range of the crystallization study.

The growth rate was found to vary with temperature as shown in Fig. 4. The observed decrease in growth rate with decreasing temperature over the range covered by the present data reflects the corresponding decrease in the molecular mobility. The largest growth rate observed in the present study was about 10^{-1} cm min^{-1}. This is larger by about an order of magnitude than the maximum growth rates for lunar compositions investigated previously.

The variation of growth rate with undercooling below the liquidus can be combined with the viscosity data to obtain information about the nature of the interface kinetic process. In making such an evaluation, the observed fact of similar products of crystallization over the range of temperature investigated will be used together with the assumption that the overall driving force for crystallization increases linearly with increasing undercooling below the liquidus. The liquidus temperature will be taken as the value (1310°C) determined by Longhi *et al.* (1972) for the 15555 composition. Using the kinetic data, the reduced growth rate, U_R, can be evaluated over a wide range of temperature. The reduced growth rate is defined:

$$U_R \equiv \frac{u\eta}{\Delta T} \qquad (2)$$

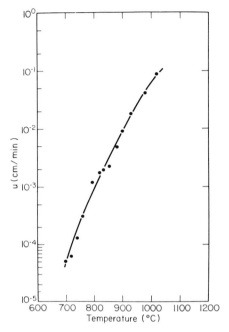

Fig. 4. Crystal growth rate vs. temperature relation for lunar composition 15555.

Here u is the growth rate, η is the viscosity, and ΔT is the undercooling ($\Delta T = T_{\mathrm{liquidus}} - T_{\mathrm{experiment}}$).

The variation of U_R with undercooling directly indicates the temperature dependence of the interface site factor (the fraction of sites on the interface where growth preferentially takes place). This relation for the 15555 composition, shown in Fig. 5, exhibits positive curvature of the type expected for surface nucleation growth. In evaluating the applicability of such a growth mechanism, the $\log(u\eta)$ vs. $1/T\Delta T$ relation has been constructed. The results, shown in Fig. 6, indicate appreciable curvature in the overall relation. According to the standard models of surface nucleation growth, this relation should be a straight line of negative slope provided the edge surface energy of the surface nucleus is independent of temperature (see Uhlmann, 1972a).

Reduced growth rate vs. undercooling relations exhibiting positive curvature, combined with curved $\log(u\eta)$ vs. $1/T\Delta T$ relations have been found in many studies of crystal growth for both organic and inorganic materials (e.g., Uhlmann, 1972a, Scherer and Uhlmann, 1972); and in particular, such behavior has been found for each of the lunar compositions investigated previously (Scherer et al., 1972). The data shown in Fig. 6 very likely reflect inadequacies of the standard crystal growth models; and recent computer simulations of crystal growth (Gilmer et al., 1973) have indicated that such curvature should be expected when the entropy of fusion is large ($\Delta S/R \geqslant 5$).

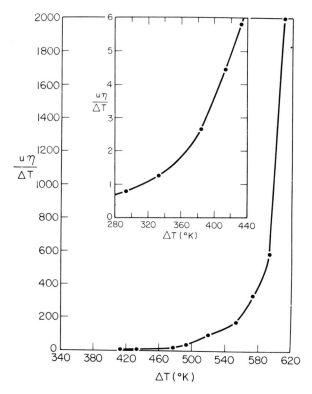

Fig. 5. Reduced growth rate vs. temperature relation for lunar composition 15555. U_R in units of cm poise min^{-1} °K^{-1}.

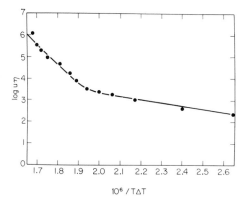

Fig. 6. Logarithm (growth rate × viscosity) vs. $1/T\Delta T$ relation for lunar composition 15555. $u\eta$ in units of cm poise min^{-1}.

Glass formation

The kinetic model for glass formation described in detail elsewhere (Uhlmann, 1972b) can be employed with the combined data on viscosity and crystallization kinetics to evaluate the critical conditions for forming glasses of the 15555 composition. In doing this, a time-temperature-transformation (TTT) curve corresponding to a barely-detectable volume fraction of crystals (10^{-6}) has been constructed using the present data on growth rate and viscosity. As seen in Fig. 7, the nose of the TTT curve for the 15555 composition occurs at an undercooling of about 375°C and corresponds to a time of about 10^{-1} sec. From these values, the critical cooling rate requiring to form a glass of this composition is estimated as about 3750°C sec^{-1}, and the corresponding maximum thickness obtainable as a glass is estimated as about 0.2 mm. In the latter calculation, a thermal diffusivity of 4×10^{-3} cm^2 sec^{-1} was assumed. These numbers indicate that the 15555 composition is a much poorer glass former than the lunar compositions investigated previously, whose critical cooling rates were estimated in the range of 1–50°C sec^{-1} and maximum glass thicknesses in the range of 2–20 mm. This finding is in accord with experience in preparing synthetic glasses, where much more severe cooling had to be employed to make glassy samples of the 15555 composition than of the other lunar materials.

In detail, the estimated critical cooling rates are somewhat higher and the estimated thicknesses obtainable as glasses are somewhat lower than those found in the laboratory. As noted previously, the estimates are not intended as valid to better than order-of-magnitude accuracy. Since actual growth rate data were used in the calculations leading to Fig. 7, the greater ease of glass formation for the 15555 composition relative to predictions based on the figure suggests that the rates of homogeneous nucleation of the crystalline phases in the liquid are smaller than those inferred from the viscosity data together with the standard theoretical model for such nucleation and experience with other materials.

The present results on glass formation of the 15555 composition indicate clearly that specimens of a size even approaching that of the lunar sample cannot

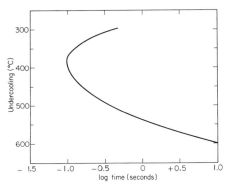

Fig. 7. Time-temperature-transformation curve for lunar composition 15555 corresponding to a volume fraction crystallized of 10^{-6}.

be obtained in the glassy state by *any* cooling conditions. More generally, for any of the lunar compositions investigated to date, the maximum size obtainable as glasses lies in the range of a few cm; and one should not expect to find larger amorphous samples of these or any similar compositions.

Acknowledgments—The authors are happy to acknowledge the experimental assistance of Mr. G. Wicks of MIT and Dr. D. Walker of Harvard University, as well as stimulating discussions with Professor J. H. Hays of Harvard and Dr. J. A. O'Keefe of the Goddard Space Flight Center. Appreciation is also due to the National Aeronautics and Space Administration who provided financial support for the present work under Grant NGR 22–009–646 and to Owens-Illinois Inc., who provided one of the authors (G. S.) with Owens-Illinois Fellowship in Materials Science.

REFERENCES

Bottinga Y. and Weill D. F. (1972) The viscosity of magmatic silicate liquids: A model for calculation. *Amer. J. Sci.* **272**, 438–475.

Cukierman M. and Uhlmann D. R. (1972) Viscous flow of lunar compositions. In *The Apollo 15 Lunar Samples*, pp. 57–59. The Lunar Science Institute, Houston.

Cukierman M. and Uhlmann D. R. (1973a) Effects of iron oxidation state on viscosity, lunar composition 15555. To be published.

Cukierman M. and Uhlmann D. R. (1973b) High temperature flow behavior of glass-forming liquids, a free volume interpretation. *J. Chem. Phys.* In press.

Cukierman M. and Uhlmann D. R. (1973c) Viscosity of liquid anorthite. *J. Geophys. Res.* Accepted for publication.

Cukierman M., Tutts P. M., and Uhlmann D. R. (1972) Viscous flow behavior of lunar compositions 14259 and 14310. *Proc. Third Lunar Sci. Conf., Geochim. Cosmochim. Acta*, Suppl. 3, Vol. 3, pp. 2619–2625. MIT Press.

Gilmer G., Leamy H. J., and Jackson K. A. (1973) Crystal growth kinetics, a computer simulation. To be published.

Hopper R. W. and Uhlmann D. R. (1973) The occurrence of lunar rocks and the flow of parent liquids. To be published.

Laughlin W. T. and Uhlmann D. R. (1972) Viscous flow in simple organic liquids. *J. Phys. Chem.* **76**, 2317–2325.

Longhi J., Walker D., Stolper E. N., Grove T. L., and Hays J. F. (1972) Petrology of mare/rille basalts 15555 and 15065. In *The Apollo 15 Lunar Samples*, pp. 131–134. The Lunar Science Institute, Houston.

Scherer G. and Uhlmann D. R. (1972) Crystallization behavior of α-phenyl o-cresol. *J. Crystal Growth* **15**, 1–10.

Scherer G., Hopper R. W., and Uhlmann D. R. (1972) Crystallization behavior and glass formation of selected lunar compositions. *Proc. Third Lunar Sci. Conf., Geochim. Cosmochim. Acta*, Suppl. 3, Vol. 3, pp. 2627–2637. MIT Press.

Scherer G., Klein L., Jorgensen P., Hopper R. W., and Uhlmann D. R. (1973) Crystallization of lunar compositions, a review. To be published.

Uhlmann D. R. (1972a) Crystal growth in glass-forming systems, a review. In *Advances in Nucleation and Crystallization in Glasses* (editors L. Hench and S. W. Freiman), pp. 91–115. American Ceramic Society.

Uhlmann D. R. (1972b) A kinetic treatment of glass formation. *J. Non-Cryst. Solids* **7**, 337–348.

Uhlmann D. R., Cukierman M., Scherer G., and Hopper R. W. (1973) Viscous flow, crystallization behavior and thermal history of orange soil material. *EOS Trans. Am. Geophys. Union* **54**, 617–618.

Proceedings of the Fourth Lunar Science Conference
(Supplement 4, *Geochimica et Cosmochimica Acta*)
Vol. 3, pp. 2697–2707

Mössbauer search for ferric oxide phases in lunar materials and simulated lunar materials

D. W. FORESTER

Naval Research Laboratory, Washington, D.C. 20375

Abstract—Mössbauer studies were carried out on lunar fines and on simulated lunar glasses containing "magnetite-like" precipitates with the primary objective of determining how much, if any, ferric oxide is present in the lunar soils. Although unambiguous evidence of lunar Fe^{3+} phases was not obtained, an upper limit was estimated from different portions of the Mössbauer spectra to be between 0.1 and 0.4 wt.% (as Fe_3O_4). A $<62\ \mu$m fraction of 15021,118 showed ~ 0.5 wt.% ferromagnetic iron at 300°K in as-returned condition. After heating to 650°C in an evacuated, sealed quartz tube for ~ 1400 hours, the same sample exhibited ~ 1 wt.% ferromagnetic iron at room temperature. An accompanying decrease in "excess" absorption area near zero velocity was noted. Thus, the result of the vacuum heat treatment was to convert fine grained iron to larger particles, apparently without the oxidation effects commonly reported. The Ni content of this "consolidated" iron was estimated to be ~ 2 wt.%. A conversion electron backscatter experiment was carried out on a $>147\ \mu$m fraction of 15021,118. A strongly enhanced "excess area" near zero velocity was observed in this mode, tending to corroborate the evidence of Housley *et al.* (1973) that the occurrence of "excess area" is surface associated.

INTRODUCTION

CONTROVERSY continues over whether or not ferric (Fe^{3+}) iron phases exist as ubiquitous components in the lunar regolith fines. A large part of this controversy arises from differing interpretations of the "characteristic" resonance observed in electron spin resonance (ESR) studies of lunar fines. Although this "characteristic" resonance is quite intense and is present in all the fines material studied thus far, it has been interpreted by some authors (Weeks *et al.*, 1970; Griscom and Marquardt, 1973 a,b; Griscom *et al.*, 1973) in terms of Fe^{3+} in "magnetite-like" phases and by others (Tsay *et al.*, 1971 a,b, 1973) in terms of spherical metallic iron particles. [The term "magnetite-like" is used to describe a ferrimagnetic Fe^{3+} phase, e.g., a titanomagnetite.]

It is well established (e.g., *see* Housley *et al.*, 1972) that on the order of 0.5 wt.% metallic iron is present in the lunar soils. Partly for this reason it has been tempting to explain the "characteristic" resonance *completely* in terms of iron metal. On the other hand, Griscom and Marquardt (1973) attribute as much as 20% of this resonance to Fe^{3+} in "magnetite-like" phases. Their interpretation of this resonance would require concentrations of ~ 0.01 to 0.4 wt.% "magnetite-like" phases (in rough proportion to soil maturity) in the lunar fines. Previous Mössbauer results (Housley *et al.*, 1972) have indicated an upper limit of 0.08 wt.% (as Fe_3O_4) for a highly magnetic separate of 10084,85 soil. It should be noted that there is overlap between this upper limit and the required "magnetite" phase concentration in the ESR studies. The only serious conflict occurs for the larger wt.% indicated by ESR data for the more mature fines. Further, Griscom *et al.* (1973) point out that the metallic iron hypothesis requires up to ~ 10 wt.% Ni alloyed in the fine-grained spherical iron to explain the observed ESR resonance shape. Such large Ni concentrations are in conflict with Mössbauer data (e.g., Gibb *et al.*, 1972; Housley *et al.*, 1972; Forester *et al.*, 1973) and with a direct-reduction mechanism postulated by Grant *et al.* (1973).

Because of the important selenological implications of a widespread distribution of Fe^{3+} in lunar soils, the present Mössbauer search for ferric phases in lunar fines material and simulated lunar material exhibiting strong "characteristic" type resonances was undertaken. Temperature and vacuum annealing studies were made on a < 62 μm fraction of lunar soil 15021,118 which has the largest "characteristic" resonance intensity yet observed (Griscom *et al.*, 1973). A series of powdered simulated lunar glasses (Griscom *et al.*, 1972, 1973a) which had been annealed at 650°C in various atmospheres and showed a strong resonance of the "characteristic" type were also investigated. A coarser fraction > 147 μm of 15021,118 was studied using a conversion-electron, backscattering Mössbauer detector to determine the surface contribution to the elusive "excess-iron" (Housley *et al.*, 1972, 1973) component near zero velocity observed in transmission spectra of lunar material.

APPARATUS

Mössbauer transmission experiments were obtained using a conventional, constant acceleration drive system. The samples studied were used as absorbers at varying temperatures with a Co^{57} in Cu source at room temperature. Calibrations were made using an Fe^{57} enriched iron foil and an Fe_3O_4 crystal, both of which were standardized with high precision nuclear magnetic resonance techniques using a ground state nuclear magnetic moment of 0.0903 nm. The conversion electron backscattering measurements were made with the lunar soil lightly packed into a spectroscopically pure graphite holder. This sample was placed inside a gas-flow proportional counter using helium $+ 10\%$ methane (quench gas) to detect the emitted conversion electrons. The data were analyzed using conventional least-squares computer analysis.

SIMULATED LUNAR GLASSES

It has been shown by Griscom and Marquardt (1972, 1973a) that the ESR spectra of powdered simulated lunar glasses, heated at 650°C in air pressures as low as ~ 0.5 torr for periods of hours bear a strong resemblance to the lunar "characteristic" resonance. Subsequent growth of this resonance with annealing time in vacuo at 650°C has been interpreted by Griscom and Marquardt (1972, 1973a) as evidence for possible crystallization of ferrite phases. Further, they propose that the "characteristic" resonance of lunar fines may be due to similar ferrite-like phases.

A Mössbauer spectrum of as-quenched simulated glass GS-64 containing ~ 10 wt.% FeO and described earlier (Griscom and Marquardt, 1972) displayed a strong asymmetric Fe^{2+} doublet. This spectrum is shown at the top of Fig. 1a along with the resultant computer fitted curve assuming two symmetric Fe^{2+} doublets with quadrupole splittings (Q.S.) of 2.20 mm/sec and 1.57 mm/sec and Isomer

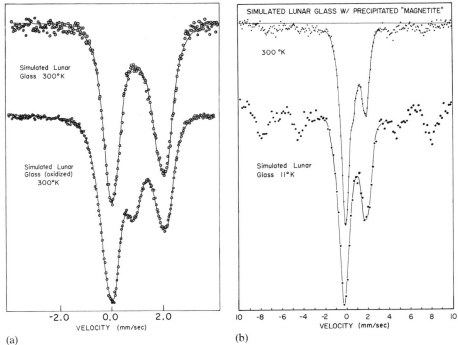

Fig. 1. Mössbauer transmission spectra of simulated lunar glass (Sample GS-64). (a) The top spectrum is of the as-quenched glass before oxidation. The lower spectrum is after oxidation in 0.5 torr air at 650°C for 67 hours. (b) These spectra are for the same *oxidized* sample in 1(a) except after a vacuum anneal at 650°C for ~2000 hours. Note the apparent decrease in quadrupole splitting of the Fe^{3+} component at 300°K and the reduction of superparamagnetic ferric component near zero velocity at 11°K. The magnetically split spectrum is obviously from Fe^{3+} and is probably a mixture of Fe_2O_3 and Fe_3O_4.

shifts (I.S.), with respect to iron metal, of 1.16 mm/sec and 1.11 mm/sec, respectively. In addition there is a weak component (~0.1 wt.% as Fe_2O_3) near zero velocity attributed to Fe^{3+}.

After heating to 650° in air at ~0.5 torr for 67 hours, in addition to the original Fe^{2+} doublet, a large Fe^{3+} doublet is observed (*see* lower curve in Fig. 1a). The Fe^{2+} hyperfine parameters remain the same and the Fe^{3+} parameters are Q.S. = 0.84 mm/sec and I.S. = 0.57 mm/sec. These parameters are consistent with Fe^{3+} ions localized at octahedral (network-modifying) sites in the glass (e.g., *see* Kurkjian and Sigety, 1968). From the area ratios, approximately 40% of the iron is in the Fe^{3+} state or ~4 wt.% as FeO. The possibility that some of this Fe^{3+} is contained in incipient superparamagnetic precipitates is supported by a reduction of the Fe^{3+} area near zero velocity upon cooling to 3°K. However, any magnetically-split spectral features were too weak or too broadened to be detected.

After a subsequent anneal at 650°C for ~2000 hours in an evacuated sealed silica tube, the Mössbauer spectrum showed evidence of Fe^{3+} precipitation in a "magnetite-like" phase (Fig. 1b at top). The "magnetite-like" spectrum has a somewhat reduced splitting and is considerably broadened in comparison with mineral Fe_3O_4 spectra. The area of this spectrum at 300°K is equivalent to ~ 2 wt.% Fe_3O_4, which is consistent with conclusions reached on the basis of ESR, X-ray diffraction and susceptibility measurements (Griscom et al., 1973). This annealed sample exhibited an ESR spectrum nearly identical in shape and intensity to the "characteristic" resonance of typical sub-mm mare fines samples but ~ 10 times as intense as the resonance obtained before the 2000 hour vacuum anneal.

Upon cooling the annealed simulated sample to 11°K, an increase in area and sharpness of the "magnetite" spectrum was observed. Because of a general Fe^{2+} component broadening, the apparent increase in "magnetite" spectrum area is exaggerated. It is estimated that an amount equivalent to ~ 3 wt.% Fe_3O_4 is associated with this spectrum at 11°K. The "magnetite-like" spectrum is much broader than mineral Fe_3O_4 at the same temperature. This most probably reflects a distribution of spinel or other oxide phase compositions within the sample. For example, Banerjee et al. (1967) have observed very broad spectra in terrestrial FeTi spinels with distributions in composition. This broadening must be taken into account when placing limits on the amount of "magnetite" phases in lunar soil samples (Forester et al., 1973).

LUNAR FINES—15021,118

Figure 2 shows Mössbauer absorption spectra at 300°K and 25°K for a < 62 μm fraction of 15021,118 which exhibited an ESR "characteristic" resonance approximately twice as intense as typical unsorted, sub-mm fines from Apollo 11 and 12. The data represent counting statistics of more than 2×10^7 counts per channel. Following the approach of Gibb et al. (1972) the spectrum at 300°K was analyzed using least-squares fitting assuming three symmetric doublets with Lorentzian line-shapes for Fe^{2+} components in (1) ilmenite (2) M2 pyroxene sites + glass and (3) olivine + M1 pyroxene sites + glass, and assuming 6 Lorentzian lines for the iron metal spectrum. An additional single-line component was used to account for the "excess" area (Housley et al., 1972) near zero velocity. The hyperfine parameters obtained from this analysis are in agreement with those generally obtained in lunar samples (e.g., see Gibb et al., 1972) with the exception of the relative intensities for the various spectra. Of particular interest is the amount of ferromagnetic iron metal which is computed to be ~ 0.5 wt.% Fe. The observed average iron hyperfine field is 330.9 kOe compared with 330 kOe for the calibration foil. There is also a small (-0.016 mm/sec) isomer shift of the lunar iron spectrum.

When the sample is cooled to 25°K the area due to ferromagnetic iron increases slightly and the area near zero velocity decreases. A five-fold expansion of the data at 25°K reveals some weak structure and the arrow near 8.5 mm/sec indicates a possible "magnetite" spectral feature. With counting statistics $\sim 2 \times 10^7$ counts/channel the noise spread in the background count may be dominated by considerations other than statistics (e.g., drive or electronic characteristics).

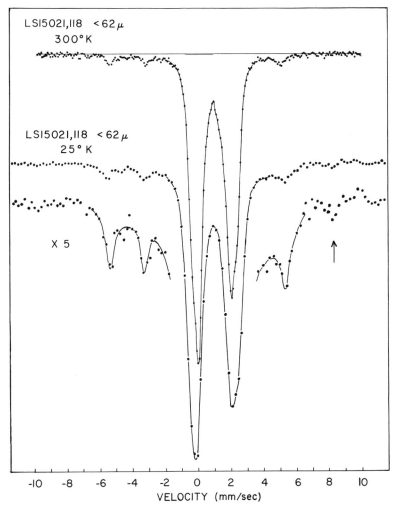

Fig. 2. Mössbauer transmission spectra from a <62 μm fraction of 15021,118 at 300°K and 25°K. The lower partial spectrum is a 5× magnification of the iron metal lines at 25°K. An arrow marks the location of a possible peak due to a "magnetite-like" phase.

However, structure similar to that near 8.5 mm/sec was not observed in other comparable experimental runs on different materials. This peak was used to place an upper limit on the amount of "magnetite-like" phase in this sample. Using the line positions and widths observed in Fig. 1b this upper limit is equivalent to ~0.4 wt.% Fe_3O_4. At negative velocities near −8 mm/sec, the data may be used to estimate an upper limit for the amount detectable by Mössbauer spectroscopy. This upper limit is ~0.1 wt.% Fe_3O_4 which is slightly greater than that deduced by Housley *et al.* (1972). The spectral feature near 8.5 mm/sec remains an intriguing piece of evidence that Fe^{3+} phases may be present in the lunar fines in concentration up to a few tenths wt.% Fe.

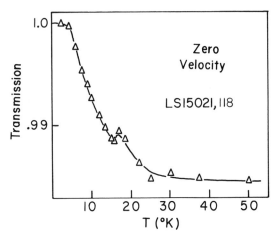

Fig. 3. Intensity of transmitted 14.4 keV gamma rays through 15021,118 at zero relative
velocity between source and absorber.

It is generally assumed that Fe^{2+} components play a passive role in the behavior of Fe metal spectra in the lunar soil. However, we find that broadening and
irregular features of the spectral components, including Fe metal, begin to
develop at low temperatures along with the development of Fe^{2+} magnetic ordering or relaxation rate slowing down. Fe^{2+} magnetic ordering or relaxation has been
observed in Mössbauer studies of other lunar samples (Schwerer et al., 1972;
Huffman et al., 1971). Spectra taken on 15021,118 $<62\,\mu m$ from 50°K to 2.5°K
show little change in structure until about 25°K. Below 25°K the Fe^{2+} spectra
broaden considerably and develop partially resolved peaks near 2.5°K. A measure
of this development is given in Fig. 3 where the intensity of gamma rays transmitted at zero relative velocity was monitored as a function of temperature. Increased transmission indicates broadening or removal of absorption lines from
zero velocity. The break at about 18°K appeared in a number of scans but its
origin is unknown. This break or the steep rise below 10°K of Fig. 3 may be
indicative of a low-temperature magnetic transition (e.g., see Nagata et al., 1972).
It is also interesting that the ESR linewidths (Griscom et al., 1973) go through a
maximum near 18°K. The broadening effects illustrated in Fig. 3 were the reason
for choosing temperatures no lower than 25°K to take the spectra in Fig. 2.

Heat treatment of lunar fines in vacuo

In the ESR studies (Griscom and Marquardt, 1973b) a number of lunar fines
samples were given vacuum heat treatments at 650°C. Special precautions were
taken to eliminate residual water vapor and air from the quartz tube holders before the samples were sealed in place. Even so, similar vacuum annealing treatment has at times produced undesired oxidation effects (e.g., Collinson et al.,
1972). To test this procedure, a sample of 15021,118 $<62\,\mu m$ was sealed and

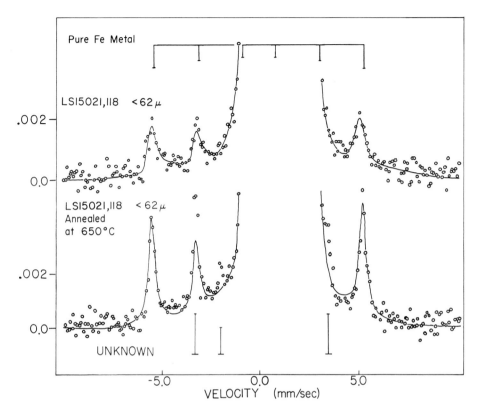

Fig. 4. Mössbauer absorption spectra of LS15021,118 ($<62\,\mu$m) before and after a vacuum anneal at 650°C for 1400 hours. Only the iron metal lines are shown on this amplified scale and a bar diagram of pure iron metal line positions is given at the top. In the upper spectrum note an inward tailing of the spectral lines indicative of super-paramagnetic relaxation. After annealing the lines have sharpened and increased intensity by a factor of 2. The three vertical bars at the bottom correspond to one or more unknown spectra.

annealed at 650°C for 1400 hours. Surprisingly, all components of the Mössbauer spectra are *sharpened*. This is shown quite dramatically in Fig. 4 where spectra taken at 300°K both with and without annealing are compared on an expanded scale to emphasize the Fe metal lines. The overall intensity of the Fe metal lines is *doubled*. Computer analysis shows that the increase from 0.5 wt.% Fe before the anneal to at least 1 wt.% Fe after anneal can be accounted for almost completely by removal of excess area near zero velocity. Previous studies (Housley *et al.*, 1972) of iron metal in lunar fines have shown no more than ~ 0.6 wt.% Fe with some possible paramagnetic Fe (<45 Å in size) remaining even at 11°K. The results of the present experiment strongly indicate that vacuum annealing at 650°C has produced growth of Fe metal particles (probably by diffusion to the surface) to sizes which appear magnetically ordered at 300°K. The average Fe hyperfine field is 332.3 kOe corresponding to ~ 2 wt.% Ni in the iron (Johnson *et al.*, 1961).

The linewidths are still somewhat broadened (~0.45 mm/sec) and a weak hyperfine pattern corresponding to about 0.2 of the total Fe with up to 5 wt.% Ni in Fe cannot be excluded. Three additional weak but sharp lines are observed and labeled as "unknown" at the bottom of Fig. 4. If these lines are part of a single pattern they have a hyperfine field $H_e \sim 207$ kOe. This spectrum does not correspond to FeS or other likely candidates but may be an FeS compound with appropriate vacancy distribution. If it is due to Fe^{3+} then a magnetic transition near and above 300°K is required.

CONVERSION ELECTRON BACKSCATTERING

The technique of conversion electron backscattering can be used to monitor the fluorescent yield of both atomic electrons and X-rays which are emitted during internal conversion processes of the resonant Mössbauer nuclei. For iron metal, the 7 keV K-shell conversion electrons penetrate only a few hundred angstroms (Krakowski and Miller, 1972) and therefore can be used to sample upper surface layers. In lunar soil this penetration depth is about 5 times greater due to density differences.

Backscattering spectra taken with a helium (methane) gas-flow proportional detector are shown in Fig. 5. A 147 μm separated sample of LS 15021,118 was

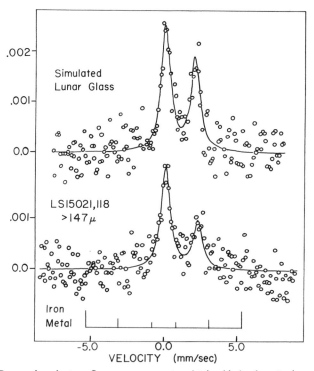

Fig. 5. Conversion electron fluorescence spectra obtained in backscattering geometry.

chosen so that the particle size was much greater than the penetration depth of 7 keV conversion electrons ($\sim 0.25\ \mu$m). Similar particle sizes of an unoxidized simulated lunar glass were used as a control sample. Although the signal to noise ratio is relatively small, it is clear that there is a larger absorption peak near zero velocity for the lunar fines sample than for the simulated lunar sample. This difference was further confirmed by transmission studies on the same 147 μm sample. It was not possible to determine whether this excess area is due to Fe metal or Fe^{3+}. The results do show, however, that the excess area near zero velocity in transmission studies is due in large part to Fe metal or Fe^{3+} within a few thousand angstroms of the surface of individual particles. Hence, one would expect a larger "excess" area near zero velocity for smaller particle sizes where the surface to volume ratio is larger. These effects are exactly those observed by Housley *et al.* (1973).

Conclusions

No conclusive evidence was found for "magnetite-like" phases in mature lunar soil 15021,118. However, a possible upper-limit range of from 0.1 to 0.4 wt.% (as Fe_3O_4) for these phases was determined from different portions of the Mössbauer spectra; these amounts are sufficient to account for the narrower parts of the "characteristic" ferromagnetic resonance spectrum of this sample (Griscom *et al.*, 1973).

A 62 μm fraction of a mature lunar soil, 15021,118 was found to contain ~ 0.5 wt.% ferromagnetic iron when the Mössbauer spectra of the as-returned sample were obtained at room temperature. An additional superparamagnetic component was present which could not be estimated by cooling to 2.5°K because of the obscuring effects of magnetic ordering in the Fe^{2+}-rich phases. A sample heated to 650°C in an evacuated, sealed quartz tube for ~ 1400 hours displayed a ferromagnetic iron concentration of ~ 1 wt.% with little further evidence of super-paramagnetism. Several conclusions are:

(1) Such heat treatment causes fine grain iron to coalesce into coarser grained particles showing bulk properties at room temperature.

(2) The amount of superparamagnetic iron in the as-returned sample was equal to the amount of ferromagnetic iron, provided none was oxidized in the heat treatment.

(3) Probably very little, if any, oxidation by terrestrial contaminants actually took place, since no additional ferric oxides were observed; (production of ≥ 0.1 wt.% Fe_3O_4 could have been detected).

(4) Since the Fe metal lines were sharpened and since demagnetizing corrections to the hyperfine field are very small for coarse grained iron, it was possible to determine the average Ni content of the metal to be ~ 2 wt.%.

Mössbauer conversion-electron backscatter experiments show that a large part of the "excess" area near zero velocity arises from Fe^{3+} or Fe metal within a few thousand angstroms of the surface of individual soil particles. This explains the increase of "excess" area with decreasing particle size as observed by Housley *et al.* (1973).

Acknowledgments —The author acknowledges many contributions from the Principal Investigator, D. L. Griscom and from Co-Investigator, C. L. Marquardt. These contributions include sample preparation and treatments and numerous discussions. L. J. Swartzendruber of the National Bureau of Standards (Gaithersburg, Md.) provided use of a conversion electron backscatter detector and gracious assistance with the backscatter experiments. I would also like to thank W. A. Ferrando for assistance during part of the data collection. This research was sponsored by NASA under Contract no. T-4735A.

REFERENCES

Banerjee S. K., O'Reilly W., and Johnson C. E. (1967) Mössbauer effect measurements in FeTi spinels with local disorder. *J. Appl. Phys.* **38**, 1289–1290.

Collinson D. W., Runcorn S. K., Stephenson A., and Manson A. J. (1972) Magnetic properties of Apollo 14 lunar samples. In *Proc. Third Lunar Sci. Conf., Geochim. Cosmochim. Acta*, Suppl. 3, Vol. 3, pp. 2343–2361. MIT Press.

Forester D. W., Marquardt C. L., and Griscom D. L. (1973) Mössbauer search for ferric oxide phases in lunar materials and simulated lunar materials (abstract). In *Lunar Science—IV*, pp. 257–259. The Lunar Science Institute, Houston.

Gibb T. C., Greatrex R., Greenwood N. N., and Battey M. H. (1972) Mössbauer studies of Apollo 14 lunar samples. In *Proc. Third Lunar Sci. Conf., Geochim. Cosmochim. Acta*, Suppl. 3, Vol. 3, pp. 2479–2493. MIT Press.

Grant R. W., Housley R. M., and Paton N. E. (1973) Origin and characteristics of excess Fe metal in lunar glass (abstract). In *Lunar Science—IV*, pp. 315–316. The Lunar Science Institute, Houston.

Griscom D. L. and Marquardt C. L. (1972) Evidence of lunar surface oxidation processes: Electron spin resonance spectra of lunar materials and simulated lunar materials. In *Proc. Third Lunar Sci. Conf., Geochim. Cosmochim Acta*, Suppl. 3, Vol. 3, pp. 2397–2415. MIT Press.

Griscom D. L. and Marquardt C. L. (1973a) Electron spin resonance studies of ferrimagnetic phases precipitated in simulated lunar glasses heat treated in the presence of oxygen. In *Amorphous Magnetism* (editors H. O. Hooper and A. M. deGraaf), pp. 95–102. Plenum.

Griscom D. L. and Marquardt C. L. (1973b) The origin and significance of the "characteristic" ferromagnetic resonance spectrum of lunar soils: Two views (abstract). In *Lunar Science—IV*, pp. 320–322. The Lunar Science Institute, Houston.

Griscom D. L., Friebele E. J., and Marquardt C. L. (1973) Evidence for a ubiquitous, sub-microscopic "Magnetite-like" constituent in the lunar soils. In *Proc. Fourth Lunar Sci. Conf., Geochim. Cosmochim. Acta*, In press.

Housley R. M., Grant R. W., and Abdel-Gawad M. (1972) Study of excess Fe metal in the lunar fines by magnetic separation, Mössbauer spectroscopy, and microscopic examination. In *Proc. Third Lunar Sci. Conf., Geochim. Cosmochim. Acta*, Suppl. 3, Vol. 1, pp. 1065–1076. MIT Press.

Housley R. M., Cirlin E. H., and Grant R. W. (1973) Characterization of fines from the Apollo 16 site (abstract). In *Lunar Science—IV*, pp. 381–383. The Lunar Science Institute, Houston.

Huffman G. P., Dunmyre G. R., Fisher R. M., Wasilewski P. J., and Nagata T. (1971) Mössbauer and supplementary studies of Apollo 11 lunar samples. *Mössbauer Effect Methodology*, Vol. 6, pp. 209–224. Plenum.

Johnson C. E., Ridout M. S., Cranshaw T. E., and Madsen P. E. (1961) Hyperfine field and atomic moment of iron in ferromagnetic alloys. *Phys. Rev. Letters* **6**, 450–451.

Krakowski R. A. and Miller R. B. (1972) An analysis of backscatter Mössbauer spectra obtained with internal conversion electrons. *Nucl. Instr. and Methods* **100**, 93–105.

Kurjian C. R. and Sigety E. A. (1968) Co-ordination of Fe^{3+} in glass. *Phys. Chem. Glasses* **9**, 73–83.

Nagata T., Fisher R. M., Schwerer F. C., Fuller M. D., and Dunn J. R. (1972) Rock magnetism of Apollo 14 and 15 materials. In *Proc. Third Lunar Sci. Conf., Geochim. Cosmochim. Acta*, Suppl. 3, Vol. 3, pp. 2423–2447. MIT Press.

Schwerer F. C., Huffman G. P., Fisher R. M., and Nagata T. (1972) Electrical conductivity and Mössbauer study of Apollo lunar samples. In *Proc. Third Lunar Sci. Conf., Geochim. Cosmochim. Acta*, Suppl. 3, Vol. 3, pp. 3173–3185. MIT Press.

Tsay F. D., Chan S. I., and Manatt S. L. (1971a) Magnetic resonance studies of Apollo 11 and Apollo 12 samples. In *Proc. Second Lunar Sci. Conf., Geochim. Cosmochim. Acta*, Suppl. 2, Vol. 3, pp. 2515–2528. MIT Press.

Tsay F. D., Chan S. I., and Manatt S. L. (1971b) Ferromagnetic resonance of lunar samples. *Geochim. Cosmochim. Acta* **35**, 865–875.

Tsay F. D., Manatt S. L., Live D. H., and Chan S. I. (1973) Electron spin resonance studies of Apollo 16 fines (abstract). In *Lunar Science—IV*, pp. 737–739. The Lunar Science Institute, Houston.

Weeks R. A., Kolopus J. L., Kline D., and Chatelain A. (1970) Apollo 11 lunar material: Nuclear magnetic resonance of ^{27}Al and electron resonance of Fe and Mn. In *Proc. Apollo 11 Lunar Sci. Conf., Geochim. Cosmochim Acta*, Suppl. 1, Vol. 3, pp. 2467–2490. Pergamon.

Proceedings of the Fourth Lunar Science Conference
(Supplement 4, *Geochimica et Cosmochimica Acta*)
Vol. 3, pp. 2709–2727

Evidence for a ubiquitous, sub-microscopic "magnetite-like" constituent in the lunar soils

D. L. Griscom, E. J. Friebele, and C. L. Marquardt

Naval Research Laboratory, Washington, D.C. 20375

Abstract—Electron spin resonance (ESR) has been employed in a study of the ferromagnetic constituents of a wide variety of soils from six sampled regions of the moon as well as glasses made to simulate lunar compositions. The technique is most sensitive to both highly spherical particles of Fe metal $\leqslant 320$ Å and equant particles of magnetite. A significant result has been that "magnetite-like" phases (magnetic iron spinel) precipitated in and on simulated lunar glasses as a result of sub-solidus oxidation yield room-temperature ESR spectra virtually identical with the line shape predicted for spherical, single domain particles of metallic Fe. It is shown that such "magnetite-like" phases can nevertheless be distinguished from metallic iron on the basis of the temperature dependence of the ESR intensity. Thus, the ESR spectrum of a 2-mg separate of green glass balls from 15426,78 is interpreted as arising *entirely* from "magnetite-like" phases of the same genre as those produced in the laboratory. The total "magnetite" concentration in this specimen is estimated to be ~ 0.01 wt.%.

Well established methods (e.g., computer simulation) and a variety of innovative techniques have been used to partially disentangle the effects of metallic iron and "magnetite" in the other samples. It is inferred that all sub-mm fines samples examined may contain "magnetite-like" components in amounts ranging up to ~ 0.4 wt.%, roughly in proportion to the soil maturity. The probable morphologic expressions of these phases are suggested by scanning electron micrographs of the (sub-microscopic) laboratory analogs. Preliminary static magnetic studies of these synthetic "magnetite-like" precipitates indicate that lunar "magnetites" of similar origin may be important carriers of natural remanent magnetism (NRM).

Introduction

THE LUNAR FINES are magnetically distinguished from the crystalline rocks by a high abundance of ferromagnetic phases in superparamagnetic and single domain particle sizes (e.g., Nagata *et al.*, 1970). Most investigators have assumed that these phases are wholly comprised of metallic iron, since the usual tests for magnetite proved negative. Large grain pure magnetite would have been detected by its Curie temperature at 850°K as well as by the Verwey transition at 120°K (e.g., Runcorn *et al.*, 1970). It is known, however, that superparamagnetic effects in fine grained particles can suppress observation of a well-defined Curie temperature (Kneller, 1962) and that variations in stoichiometry can also produce this effect (Nagata, 1961) while washing out the low temperature transition as well (Smit and Wijn, 1959). Thus, the possibility of magnetite existing in the lunar soil has not been eliminated and it is appropriate to seek and evaluate new sources of evidence.

One such source is the technique of electron spin resonance (ESR). Previous difficulties with obtaining specific information from ESR were due to an initial dichotomy in interpreting the data. The so-called "characteristic" resonance spectrum of the lunar fines has been attributed to metallic iron on the one hand (e.g., Tsay *et al.*, 1971a,b) and to ferric oxides on the other (e.g., Weeks *et al.*, 1970). Tsay and coworkers correctly observed that the resonances of Apollo 11 and Apollo 12 soils behaved much as expected for metallic iron over the temperature range 77–300°K. It was subsequently demonstrated, however, that "ferric oxides" precipitated in simulated lunar glasses behave in essentially the *same* manner in this temperature range (Griscom and Marquardt, 1972a,b). Moreover, samples returned by Apollo 14 and subsequent missions exhibited substantially narrower linewidths than were expected for pure metallic iron.

At the Fourth Lunar Science Conference, the linewidth data emerged as one of the important keys to identifying the source(s) of the resonance. Tsay *et al.* (1973) pointed to an inferred correlation between linewidth and nickel content and proposed the presence of sub-micron Fe–Ni containing up to ~ 10 wt.% Ni. However, the latter explanation appeared to be in conflict with Mössbauer data (e.g., Gibb *et al.*, 1972; Housley *et al.*, 1972; Forester, 1973) which show the average nickel content of lunar irons to be $\leq 3\%$, and also with the rapid reduction mechanism postulated by Grant *et al.* (1973) to explain the excess fine grained iron in the soils. On the other hand, an apparent correlation between linewidth and TiO_2 content suggested a titanomagnetite origin for the narrowest parts of the resonance (Griscom and Marquardt, 1973a,b). (As indicated by the central, shaded regions in Fig. 1, the "narrow parts" of the resonance comprise $\sim 20\%$ of the total absorption intensity.) The observation of discontinuities in the temperature dependence of the linewidth (Weeks, 1973) provided an independent basis for inferring the presence of ferric oxides.

The present paper details additional evidence for the existence of "magnetite-like" constituents in the lunar soils. A significant new result is a well-defined peak in the curve of ESR intensity vs. temperature which corresponds with the magnetic behavior observed for many magnetic iron spinels (Smit and Wijn, 1959). On this basis, the ESR contributions of lunar "magnetite-like" phases appear to be separable from those due to metallic iron. Thus, the previous questions regarding the interpretation of the ESR data seem to be resolved and the present results complement other evidence for magnetite in lunar materials (Runcorn *et al.*, 1971; Olhoeft *et al.*, 1973).

BASIC MAGNETIC PROPERTIES MEASURED

ESR spectra were obtained at X-band (9.12–9.52 GHz) and Ka-band (34.5–35.5 GHz) frequencies and temperatures between 5 and 573°K by means of a Varian E-9 spectrometer and associated accessories. The experimental data consist of x-y recorder traces of the first derivative of the microwave absorption as a function of the applied d.c. magnetic field (0 to 15 kOe). The microwave absorption is proportional to the imaginary part of the complex r.f. susceptibility χ'', which is itself proportional to the initial reversible magnetic susceptibility χ_0 multiplied by a normalized lineshape function (Pake, 1962). As will be discussed below, the line shape functions are amenable to analysis by computer simulation

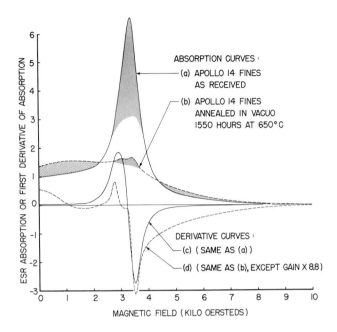

Fig. 1. ESR spectra obtained at 9 GHz and 300°K for a sample of 14230,92 before (a, c) and after (b, d) heat treatment at 650°C in an evacuated sealed quartz tube for ~3000 hours. The derivative curves (c, d) are x-y recorder traces; the absorption curves (a, b) were obtained by numerical integration (*see* footnote c to Table 1). Shaded areas centered near 3.5 kOe are the "narrow parts" of the absorption which wholly determine the linewidths measured between the positive and negative derivative extrema.

techniques leading to accurate measurements the anisotropy field, $2K_1/M_s$, for spherical ferromagnetic particles in single domain sizes and smaller. (Here, K_1 is the first order cubic magnetocrystalline anisotropy constant and M_s is the saturation magnetization.) For a randomly oriented ensemble of spherical, single domain particles, the anisotropy field relates to the coercive force, H_c, according to $H_c \approx 0.64\ K_1/M_s$ (Morrish, 1965). In terms of the present experiments, $H_c \approx 0.2\ W_{pp}$, where W_{pp} is the linewidth in Oersteds measured between the positive and negative peaks of the first-derivative ESR spectrum (*see* below). In turn, the double numerical integral of the first-derivative spectrum is approximately proportional to χ_0. (The reason χ_0 is measured, in spite of the large d.c. field values, is related to the fact that the microwave magnetic field, H_1, has a peak amplitude of $\leqslant 0.1$ Oe and is applied *perpendicular* to the d.c. field.) In Table 1, the specific intensity "near $g \approx 2$" corresponds roughly to χ_0 for the spherical particles in single domain and superparamagnetic sizes; the "zero-field" intensity arises from acicular and multidomain particles. Isolated paramagnetic ions or magnetic structures $\leqslant 30$ Å make negligible contribution to these intensities at temperatures $\geqslant 10$°K (Tsay *et al.*, 1971a,b; Housley *et al.*, 1972).

Table 1. Ferromagnetic resonance intensities and peak-to-peak first-derivative linewidths for sub-mm lunar fines, as obtained at spectrometer frequencies near 9.53 GHz and temperatures near 20°C.

Sample No.	Major Morphologic Unit Sampled	Sub-unit[a]	Size Fraction	Specific ESR Intensity[b]			Number of Sub-samples Averaged	Linewidth (Oe)
				Total[c]	Near $g \approx 2$[d]	Zero-Field[e]		
15021,118	Mare Imbrium	Mature inter-crater surface	<62 μ	740	512	228	1	720
			<1 mm	539	389	150	1	720
			>147 μ	438	303	135	1	720
Lunar 16, A-41	Mare Fecunditatus	Mature (inter-crater?) surface	<1 mm	404	322	82	1	795
10084,241	Mare Tranquillitatus	Mature inter-crater surface	<1 mm	339	258	81	1	790
12001,15	Oceanus Procellarum	Mature inter-crater surface	<1 mm	326	234	92	1	720
6961,31	Cayley Plains	Mature fines from underneath boulder. Probable South Ray ejecta	<62 μ	356	205	151	1	550
			<1 mm	321	208	113	1	562
			105–147 μ	350	187	163	1	572
69941,33	Cayley Plains	Surface fines, probably including South Ray ejecta	<1 mm	320	173	147	1	550
14230,88 92 96	Fra Mauro	Mature core tube fines from 15–20 cm depths	<1 mm	229^{+78}_{-49}	188^{+59}_{-38}	42^{+42}_{-21}	6	592^{+38}_{-32}
65901,11	Stone Mountain (Descartes)	15 cm below surface	<1 mm	270	180	90	1	550
62241,15	Cayley Plains	Rim of Buster	<1 mm	222	134	88	1	575
15302,27	Apennine Front	Rim of Spur	<1 mm	207	138	69	1	697
61141,11	Cayley Plains	Reworked (?) ejecta from Plum	<1 mm	163	97	66	1	555
63501,26	Cayley/North Ray	Surface of ejecta blanket,~5 m west of Shadow Rock	<1 mm	111^{+13}_{-24}	72^{+5}_{-8}	40^{+10}_{-17}	3	548^{+2}_{-3}
63341,6	Cayley/North Ray	Ejecta blanket: shadowed fines, below surface	<1 mm	106^{+10}_{-9}	73^{+10}_{-18}	28^{+8}_{-14}	3	531^{+10}_{-5}
63321,9	Cayley/North Ray	Ejecta blanket: shadowed fines	<1 mm	96^{+11}_{-17}	64^{+17}_{-23}	33^{+6}_{-6}	3	541^{+9}_{-4}
67601,28	Cayley/North Ray	Crater rim rake soil	<1 mm	95	50	45	1	540
15426,78	Apennine Front	Ejecta from Spur (green clods)	<1 mm	20	17	3	1	580

[a] The sampled sub-unit was considered "mature" if the less-than-1-man fines had a mean grain size ≤100 μ (McKay et al., 1972). The appropriate estimates were based on tabulations in McKay, et al. (1972) and Vinogradov (1971) and the authors' own sieve fractionation of 69961, 31.

[b] Intensity divided by sample mass. In all cases samples weighing 1-4 mg (usually~2 mg) were encapsulated in Varian precision bore, 1 mm I.D. fused silica tubes for observation, the microwave power level was set at 10 mW and the modulation amplitude was 16 Oe. Units are arbitrary but are roughly equivalent to the wt% paramagnetic Fe_2O_3 which would produce the same intensity under these experimental conditions (Griscom and Marquardt, 1972b).

[c] Intensity determined by double numerical integration (Ayscough, 1967), beginning at the highest field for which a non-zero first derivative amplitude could be discerned (generally 6-9 k Oe) and ending at zero field. Determination of the zero derivative level (or "baseline") is critical to this procedure.

[d] Total integrated specific intensity minus the area of a "wedge" whose height is the amplitude of the absorption at zero field and whose base is the range of field covered by the integration. This quantity is independent of errors in determining the derivative baseline.

[e] Area of a "wedge" whose height is the absorption amplitude at zero field and whose base is the length of the region of integration. This quantity is extremely sensitive to the choice of baseline.

The fact that the measured X-band linewidths, W_{pp}, are as narrow as those listed in Table 1 puts rigid constraints on the types of particles which contribute in the restricted spectral region "near $g \approx 2$". On the basis of the room temperature linewidths, Weeks (1972) has calculated that if the particles are metallic iron they must be spherical within an accuracy of $<10\%$. Considering the narrower widths measured at 570°K (Griscom and Marquardt, 1972b), this sphericity restriction tightens to $<5\%$, and a value $\leqslant 2\%$ would be in better agreement with the line shape analyses to be described below. In addition, when the spectrometer frequency is 9 GHz, the iron particles contributing to the narrow parts of the resonance are only those which are single domain ($\leqslant 320$ Å). Small, *multidomain* particles of iron give far broader resonances at this frequency (Griscom and Marquardt, 1972c, 1973c).

In contrast, all equant grains of magnetite, irrespective of size, will contribute to the narrow parts of the X-band spectrum. The sphericity restriction is relaxed in the ratio of the saturation magnetizations for iron and magnetite (~ 3.5), and the laboratory applied fields of $\geqslant 3$ kOe exceed the maximum coercive force $2\pi M_s$ for magnetite, assuring magnetic saturation.

SIMULATED LUNAR GLASSES

Modeling experiments were carried out on a number of simulated lunar glasses produced to rigid specifications by C. E. Schott at the Owens Illinois Technical Center. Sample GS-64, upon which the bulk of the present studies were performed, has been thoroughly described previously (Griscom and Marquardt, 1972b). GS-64 was melted in a platinum crucible under nominally reducing conditions and was found to contain ~ 0.1 wt.% Fe^{3+} as received (Griscom and Marquardt, 1972b). Recently delivered samples were melted in molybdenum crucibles under 99% CO–1% CO_2 and were noted by visual examination to contain some metallic iron. The ESR spectra of the latter materials were similar to the spectrum of a mm-sized lunar glass chip, 10084,170,6 (Griscom and Marquardt, 1972b) and were interpreted as indicating that most of the iron was present in multidomain sizes and/or non-spherical shapes. The Fe^{3+} content of a reduced glass (GS-78) similar in composition to an Apollo 12 soil was determined by standard ESR methods to be $\leqslant 10$ ppm (Sigel and Ginther, 1968; Griscom and Marquardt, 1972b).

A number of experiments carried out on the newer, low-ferric-iron samples has confirmed the previous conclusion (Griscom and Marquardt, 1972a,b, 1973a) that ferric oxide precipitates form in and on these glasses as a consequence of atmospheric oxidation near 650°C, irrespective of the presence or absence of Fe^{3+} or metallic Fe in the as-quenched glass. Conversely, *no* ferric oxide precipitates are formed by 650°C treatments lasting more than 1000 hours *in evacuated, sealed quartz tubes*, whether or not some Fe^{3+} or metallic Fe is in the starting material.

Scanning electron micrographs of sample GS-64 before and after oxidation at 650°C and subsequent annealing are shown in Fig. 2. Mössbauer studies (Forester et al., 1973; Forester, 1973) coupled with X-ray diffraction and magnetic susceptibility measurements (Griscom and Marquardt, 1973a) indicate that the sub-micron

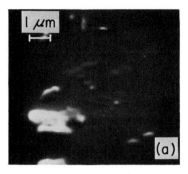

Fig. 2. Scanning electron micrographs of the surfaces of particles of simulated lunar glass GS-64. (a) Freshly exposed surface of as-quenched glass. (b) Surface of a glass particle heated to 650°C in air at ~0.5 torr for 67 hours, followed by ~2000 hours at the same temperature in an evacuated, sealed quartz tube. Sub-micron mounds in (b) have been identified as "magnetite-like" precipitates (*see* text).

structures in Fig. 2b are magnetic iron spinels ("magnetite-like" phases) which have precipitated out of the host glass. Corroborative evidence comes from electron microprobe scans which have shown the mounds of Fig. 2b to be enriched in iron relative to the host. A few mounds observed on the untreated sample (Fig. 2a) proved to be of the same composition as their surroundings. Room temperature static magnetic parameters for the sample of which Fig. 2b is representative are $M_s = 0.8$ e.m.u./g and $M_R/M_s = 0.09$, where M_s is the saturation magnetization and M_R is the saturation remanence. Figure 2b should be compared with SEM images of lunar fines and exposed rock surfaces (e.g., in Asunmaa *et al.*, 1970; Housley *et al.*, 1972) where similar structures are observed.

The ESR spectrum of the sample shown in Fig. 2b was found to be virtually identical with the "characteristic" resonance of an Apollo 12 fines sample when both spectra were obtained at room temperature (spectrum "C" in Griscom and Marquardt, 1973a). The temperature dependence of the two spectra were qualitatively similar (*see* below).

ESR EXPERIMENTS, RESULTS, AND DISCUSSION

Heat treatment of lunar fines in vacuo

A total of 16 fines samples (~2 mg each) have been dried overnight at 130°C under a vacuum of $\leq 10^{-5}$ torr, sealed off at this pressure in quartz tubes of small volume (~0.5 cm³) and annealed at 650°C for various times up to ~3000 hours. The object of this experiment has been to erase all memory of the samples' thermal experiences on the moon so that the dependences of resonance properties on chemical composition may be disentangled from the effects of thermal history. In terms of the specific magnetic phases under consideration, the effects of such anneals would be to homogenize the spinel phases (650°C is above the solvus dome for the series Fe_2TiO_4–Fe_3O_4 (Nagata, 1961)) and to cause the unmixing of

Fe–Ni which contains $\geqslant 3\%$ Ni into a homogeneous α phase and a nonmagnetic γ phase (Griscom and Marquardt, 1973b).

Experimentally, these heat treatments have caused the narrow parts of the resonance spectrum to become distinctively sharper in the first-derivative display (Fig. 1d) and to decrease in intensity (compare the shaded areas of Figs. 1a and b). Much of the intensity lost from the narrow parts of the resonance reappears in the high and low field sides of the spectrum (hatched regions in Fig. 1) so that the loss in *total* intensity after 3000 hours is $\leqslant 25\%$. Based on extensive ESR studies of simulated lunar glasses given the same treatments (*see* below) and lunar fines heated in a dynamic vacuum of $\leqslant 10^{-7}$ torr, and on Mössbauer studies of a vacuum-heated lunar fines sample (Forester, 1973), it has been concluded that none of the described effects are related to oxidation by terrestrial contaminants.

Line shape analysis of the narrow parts of the resonance

The timely introduction of computer simulation methods by Tsay and co-workers (Manatt *et al.*, 1970; Tsay *et al.*, 1971a,b) has been a major factor in understanding the ferromagnetic resonance spectra of lunar soils. In the application of these methods, the following limitations should be recognized, however: (i) It is only the "narrow parts" of the resonance that are being simulated (shaded regions of Fig. 1). (ii) In the cases of as-returned lunar fines samples, it is difficult to tell from the simulations whether the magnetocrystalline anisotropy has axial or cubic symmetry (Griscom and Marquardt, 1973b). (iii) While the algebraic sign of the anisotropy constant is unambiguously determined by this method, positive values are not sufficient proof that the narrow parts of the resonance must arise from metallic iron (Griscom and Marquardt, 1972b, 1973b).

The heavy curve at the top of Fig. 3 shows the narrow parts of the "characteristic" resonance of an Apollo 14 fines sample after sharpening by heat treatment *in vacuo*. The lighter unbroken curve is a computer simulation based on the resonance condition for first-order *cubic* anisotropy treated as a first-order perturbation on the Zeeman energy (Tsay *et al.*, 1971b). The computational algorithm used here was developed by Taylor and Bray (1968, 1970); it first computes the powder absorption spectrum (bottom of Fig. 3), then convolutes this with a "single-crystal" broadening function (either Gaussian or Lorentzian) and finally obtains the first derivative of the convoluted "powder pattern." Exact corrections described by Schlömann (1958) have been applied to the first-order-perturbation calculations with the results indicated by the light dashed curves of Fig. 3. The dotted curve, which is the difference between the dashed simulation and the experimental spectrum, shows the presence of a broad, slowly varying resonance which underlies the narrow parts (*see* below).

Accepting the influence of the underlying resonance, the computer fit of Fig. 3 must be considered excellent. Moreover, computer simulations of this spectrum using a resonance condition for *axial* anisotropy were clearly inferior (D. L. Griscom, unpublished). Since the dependence of the "sharpened" spectrum on temperature is very similar to that for the "characteristic" resonance of as-

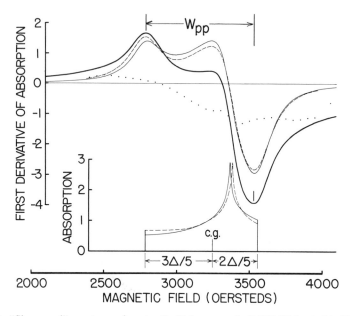

Fig. 3. "Sharpened" spectrum of an Apollo 14 fines sample (14230,92) heated to 650°C in evacuated sealed quartz tube for 880 hours (heavy curve). Lighter-weight curves are computer simulations. The dotted curve is the difference between the dashed simulation and the experimental spectrum. The unbroken absorption curve is the computer-generated "powder pattern" for first-order magnetocrystalline anisotropy treated as a first-order perturbation. The center of gravity of this curve is indicated by "c.g."; this is the point at which the g value is measured. Δ is defined as $5/3(2K_1/M_s)$. Dashed absorption curve incorporates some added sophistication prescribed by Schlömann (1958). Greater deviations from the "first-order" powder pattern are to be expected for larger values of Δ and the center of gravity will shift to lower fields with respect to the powder-pattern edges (Schlömann, 1958). Relevant parameters for this simulation are
$$2K_1/M_s = +465 \text{ Oe}, \ g = 2.083, \ \sigma_{pp} = 155 \text{ Oe}.$$

returned samples, it has been inferred that the same magnetic phases are responsible for both, and that they are *cubic* (Griscom and Marquardt, 1973b). This eliminates hexagonal ferrites such as hematite and pyrrhotite from further consideration as possible sources of the narrow parts of the resonance. Only metallic iron and spinel remain as strong contenders.

As pointed out by Tsay *et al.* (1971a,b), the ESR linewidth measured between positive and negative derivative peaks (W_{pp}) is approximately equal to $5/3$ $(2K_1/M_s)$. By means of a further computer simulation study, it has now been demonstrated that this relation is valid to $\pm 7\%$ accuracy for both as-returned (Fig. 4, region A) and vacuum annealed samples (Fig. 4, region B). The decrease in apparent "single-crystal" linewidth (σ_{pp}) for the heated samples is interpreted as a narrowing of the statistical distribution of anisotropy constants present in the as-returned samples. This would be expected for the homogenization of partially unmixed phases.

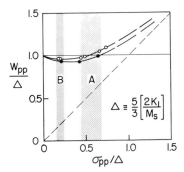

Fig. 4. Computer generated "data" for a polycrystalline ferrite with positive anisotropy field $2K_1/M_s$. W_{pp} is the peak-to-peak derivative linewidth of the computed spectrum. σ_{pp} is the peak-to-peak derivative linewidth of the Lorentzian convolution function, i.e., the "single-crystal" linewidth. Filled circles: computed with first-order powder pattern of Fig. 3. Open circles: computed with dashed powder pattern of Fig. 3. Regions "A" and "B" indicate the apparent σ_{pp}/Δ values which characterize as-returned and vacuum-heated lunar fines, respectively; these were determir•d by visual comparison of super-posed experimental and computed spectra.

Line shape analysis of the broad parts of the resonance

There is currently no existing computer program for simulating the parts of the lunar ferromagnetic resonances that are spectrally broad, since the mechanisms which lead to this broadening are too numerous to be anticipated and generally too complex for accurate computation on a limited budget. However, these broad resonances can be analyzed by comparison with a catalog of broad resonances from a wide variety of well characterized sources.

In Fig. 5, the broad, slowly varying, "daughter" resonance for a vacuum heated Apollo 12 fines sample is compared with several other broad, slowly varying resonances obtained in a simulated lunar glasses by various treatments, including a final episode of annealing *in vacuo*. As indicated on the drawing, resonances (a), (b), and (d) are attributed to ferric oxide, ferric oxide plus metallic iron, and metallic iron, respectively. Here, several points bear emphasizing: First, all of the spectra of Fig. 5 are carefully referenced to the first derivative zero level. Thus, for instance, curve (d) cannot be matched more closely with curve (c) merely by translating it upward; the only adjustable parameter here is "spectrometer gain" which varies the amplitude of the spectrum while preserving its original relationship to the zero level. Secondly, spectrum (a) was found to be typical of *dozens* of oxidized simulated lunar glass samples with regard to the spectral behavior above 5 kOe. Over the ranges 0–2 and 8–10 kOe, spectrum (d) is similarly typical of several reduced glasses containing metallic iron. Finally, the broad parts of spectrum (c) are typical of all 16 lunar fines samples which have been heated to 650°C *in vacuo* for >1000 hours.

It would appear that spectra (a) and (d) are in large degree diagnostic of the respective magnetic phases indicated in Fig. 5 and that spectrum (c) is diagnostic of the lunar fines. If this pattern continues to be repeatable, the results of Fig. 5

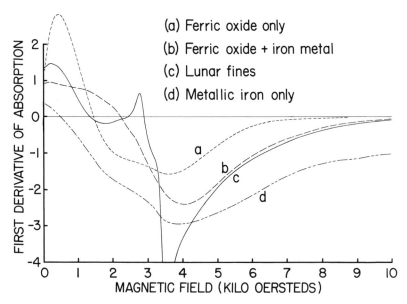

Fig. 5. ESR spectra obtained at 9 GHz and 300°K for simulated and actual lunar materials annealed at 650°C in evacuated, sealed quartz tubes (*in vacuo*) for > 1000 hours. (a)
Simulated lunar glass GS-64 oxidized 1 hour at 650°C in air at ~ 0.5 torr, followed by 1400
hours at 650°C *in vacuo*. (b) Simulated lunar glass S-17-3B reduced in the melt, followed
by reducing treatment at 800°C, followed by oxidation for 1/2 hour at 650°C in air at
~ 0.5 torr, followed by 1400 hours at 650°C *in vacuo*. (c) Lunar fines from 12001,15, baked
dry 16 hours at 130°C in dynamic vacuum of ~ 10^{-6} torr, followed by 1800 hours at 650°C
in vacuo. (d) Simulated lunar glass GS-78, reduced in the melt, ground and sieved to
< 62 μm, and subsequently given the same treatments as 12001,15 in (c).

alone will provide cogent evidence for the co-presence of both metallic iron and
ferric oxides in the lunar soil. Thus, spectrum (c) (exclusive of its narrow parts)
may be a weighted average of spectral types (a) and (d). Alternatively, curves (b)
and (c) may both be representative of a composite magnetic phase comprising ferric oxide in intimate contact with metallic iron.

Linewidth and intensity as functions of temperature

The linewidths and (numerically integrated) intensities of the ESR spectra of a
number of lunar fines samples were studied over the range 5–570°K, and comparisons were made with the corresponding quantities for the "magnetite-like" phases
of Fig. 2b. The results are summarized in Fig. 6. Several aspects of the linewidth
data are worthy of special comment. First, it can be noted that all curves of Fig. 6a
reach maxima in the range 10–25°K, with the exception of the annealed sample of
14230. This maximum is possibly associated with magnetic ordering of Fe^{2+} in the
bulk of the sample (Forester, 1973) and hence may not be a property intrinsic to
the ferromagnetic phases giving the resonance. If this is true, the vacuum heat
treatment of 14320 has magnetically decoupled these ferromagnetic phases from
the surrounding ferrous ions.

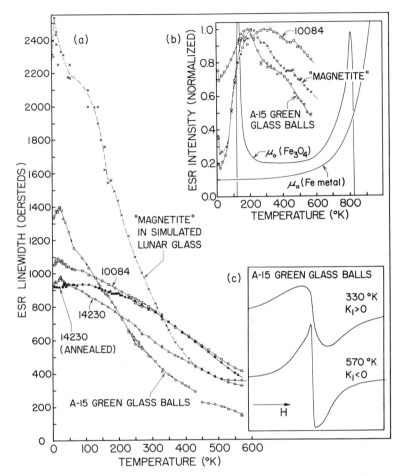

Fig. 6. Temperature dependences of ESR linewidth (a) and (numerically integrated) intensity (b) for several lunar soil samples and one simulated lunar glass containing "magnetite-like" precipitates. Data are shown for unsorted fines from 10084,241 and 14230,92, before (open symbols) and after (filled symbols) annealing in an evacuated sealed quartz tube at 650°C for ~3000 hours. The "A-15 green glass balls" (GGB) were separated from a green and gray clod, 15426,78. Inset (c) shows that the GGB spectrum changes its asymmetry in the high temperature range. Numerical integrations for data points in (b) spanned the range ~2–5 kOe, and were carried out in the sense of footnote d in Table 1.

Possibly the most striking feature of Fig. 6a is the qualitative similarity between the curve for the synthetic "magnetite" and the curve for a 2-mg separate of green glass balls (GGB) (hand picked from 15426,78 and ultrasonically cleaned in freon). This similarity suggests that the ferromagnetic phases in the GGB may also be "magnetite-like" in nature. The different behavior of the GGB vis-à-vis the other lunar samples could result from their having experienced a distinctly different thermal history on the moon. If the data for the annealed and unannealed samples of 14230 are any indication, it is inferred that the GGB may have cooled more rapidly than the as-returned sample of 14230.

The next aspect of the linewidth data to catch the eye is the large number of cusps and discontinuities. At least one of these, occurring near 150–175°K, is clear in the data for the GGB, the vacuum annealed sample of 14230 and the synthetic "magnetite." Similar features were first reported by Weeks (1973) and are best interpreted as manifestations of phase transitions within the ferromagnetic phases themselves. Cusps near 250–300°K in 10084 and the GGB were observed only on cooling. If real, these thermal hysteresis effects may result from cooling from room temperature to 90°K in the residual field of the laboratory magnet (~100 Oe) and rewarming. Arrowheads on the curves indicate the measurement sequences; data below room temperature were acquired first.

The discontinuity near 470°K in the linewidth for the GGB is real and results from the magnetocrystalline anisotropy constant of a fraction of the phases becoming zero or negative in that temperature regime. This is readily concluded from the reversal in spectral asymmetry shown in Fig. 6c (*see* Tsay *et al.*, 1971b). Such behavior is typical of many titanomagnetites (e.g., Syono and Ishikawa, 1964) but unknown for metallic iron.

In Fig. 6b it can be seen that the *intensities* for the GGB and the synthetic "magnetite" exhibit dramatic maxima near 200°K. Some of this behavior is also evident in the "characteristic" resonance of 10084. As pointed out by Tsay *et al.* (1973), superparamagnetic relaxation effects are not observed in ESR due to the high observation frequencies (9–35 GHz). Thus, the temperature dependence of the resonance intensity should always be that of the r.f. susceptibility of the bulk *ferromagnetic* material, which in turn will be similar to the temperature dependence of the initial permeability, $\mu_0 = 1 + 4\pi\chi_0$. Accordingly, the initial permeabilities of Fe_3O_4 (Smit and Wijn, 1959) and metallic Fe (Spooner, 1927) are included for purposes of comparison. It is readily inferred that the observed low-temperature intensity peak is an expected signature of "magnetite-like" phases and is inconsistent with the behavior predicted for metallic iron. Additional magnetic transitions near 450°K and Curie temperatures >570°K are inferred for all three samples of Fig. 6b.

Intensity as a function of collection site

The data of Table 1 are sufficiently extensive to permit tentative conclusions about the distribution of "characteristic" magnetic phases over the sampled regions of the moon. The fact that the four mare materials rank 1-2-3-4 in intensity is perhaps misleading. Probably a more significant observation is that the resonance intensity is greatest in the most mature soils and the least in those which are the least reworked. Thus, mature Cayley fines from the vicinity of South Ray show intensities quite comparable to most of the mare samples.

An inverse correlation of resonance intensity with particle size is noted for 15021,118, though not for 69961,31. This apparent inconsistency may be related to the observation of Housley *et al.* (1973) that coarser fractions of mature sub-mm

Apollo 16 fines appear to be mainly indurated finer material, in contrast with 15021 where 33% of the particles 0.5–1.0 mm were basaltic grains (LRL, 1971).

To secure an idea of how the resonance intensity might be related to such additional factors as thermal and radiation history, several sets of "shadowed fines" were investigated. Fines from underneath a boulder at Station 9 on the Cayley Plains were found to exhibit virtually the same intensities as nearby surface fines (compare 69941,33 with 69961,31 in Table 1). Similarly, two shadowed samples from Shadow Rock (63341,6 and 63321,9) showed intensities similar to those of exposed soils from other locations on the North Ray ejecta blanket (63501,26 and 67601,28). It would appear, then, that most of the "characteristic" resonance intensity was present in the North Ray soils at the time of the emplacement of Shadow Rock, since the shadowed fines subsequently have resided at lower temperatures and have been shielded from other forces which could have chemically or physically altered them. Carrying the inference one step further, much of the resonance may have been *created* by the North Ray event. The surface debris at Station 13 were undoubtedly excavated from considerable depth and it is known that crystalline (or recrystallized) lunar rocks generally do not display a narrow "characteristic" resonance (Kolopus *et al.*, 1971; Weeks, 1972).

In Table 1, it can be noted that the "zero-field" intensity (the broad part of the resonance) does not correlate with the intensity "near $g \approx 2$" (the narrow parts). It is believed to be significant that the exposed fines from North Ray Crater show larger average values of "zero-field" intensity than do their shadowed counterparts. Based on careful spectral comparisons it has been determined that crater rim fines 67601,28 differ from shadowed fines 63341,6 by a relatively broad ESR

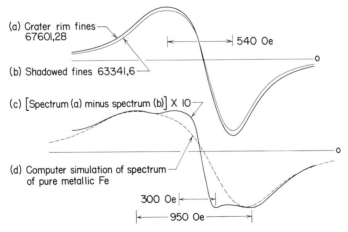

Fig. 7. ESR spectra of fines from (a) the rim and (b) the ejecta blanket of North Ray Crater. (Frequency = 9.538 GHz and temperature ≈ 300°K.) This figure shows that the shape of the spectrum for the rim material differs from that of the shadowed material by a curve having the shape (c), which in turn can be decomposed into a "metallic-iron-like" component (d) plus a sharper resonance with a 300 Oe width. Computer simulation of the spectrum of "pure" metallic iron involved the following parameters: $2K_1/M_s = +590$ Oe and $\sigma_{pp} = 460$ Oe.

component which is identical with the computer simulated spectrum of pure metallic iron (Fig. 7). Additional iron in the rim fines could result from micrometeoroid bombardment according to the mechanism proposed by Grant et al., (1973). It is the thesis of the present paper that the *narrowest* parts of the resonance are due to "magnetite-like" phases produced in the course of an oxidizing event, such as the excavation of North Ray Crater by a comet (Gibson and Moore, 1973).

Linewidth as a function of chemistry

Figure 8 exhibits the ESR linewidths of a wide range of sub-mm fines samples plotted so as to emphasize possible correlations with soil chemistry. The "predicted wt.% Ni in metallic Fe" has been indicated on the righthand ordinate to help evaluate the Ni–Fe hypothesis of Tsay et al. (1973). This calibration was derived from the anisotropy data of Tarasov (1939) according to the relation $W_{pp} = (0.93)(5/3)(2K_1/M_s)$ (*see* Fig. 4).

A remarkable feature of Fig. 8 is the fact that the data points for the GGB and the Apollo 17 orange soil (Weeks, private communication) form the anchor points on the curve of linewidth vs. titanium content for the as-returned samples, whereas the GGB point does not fit the Ni–Fe correlation scheme. Significantly, both the GGB and the orange soil appear to contain "magnetite" (*see* above, and Olhoeft et al., 1973).

The behaviors of the heated samples (filled circles and crosses) are phenomenologically consistent with expectation for the thermal homogenization of diverse "magnetites" which had become partially unmixed due to episodes of heating and slow cooling on the moon (Griscom and Marquardt, 1973d). However, they do not appear to be consistent with the predicted unmixing of Fe–Ni alloys (Griscom and Marquardt, 1973b). If the measured spectra arise from titanomagnetites, the apparent correlation with TiO_2 is expected to remain and perhaps improve with still longer anneals. However, a persistent hiatus between mare and highlands materials is a possibility, since Mg and Al are also likely to be incorporated into "magnetite-like" phases precipitated out of lunar dust grains.

The possibilities of any other potential linewidth correlations (e.g., with ESR intensity) are rejected by inspection of Table 1. Only the titanium correlation appears significant.

How much magnetite on the moon?

The simulated lunar glass sample of Fig. 2b was shown by low-temperature Mössbauer measurements to contain ~3 wt.% ferric oxide phases (Forester, 1973); a major fraction of these are expected to be identical with the spinel phases indicated by X-ray diffraction (Griscom and Marquardt, 1973a). The saturation magnetization measured at room temperature was consistent with ~1 wt.% Fe_3O_4 in single domain sizes and larger. A reasonable estimate, then, is that ~1–2 wt.% "magnetite-like" phases were present. At room temperature, the total ESR inten-

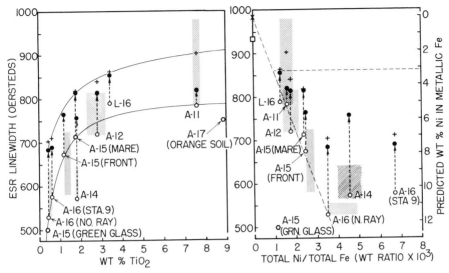

Fig. 8. Linewidths of the "characteristic" resonance of fines from seven sampled regions of the moon are plotted vs. the local abundance of TiO_2 (left) and the ratio of the local abundances of Ni and Fe (right). Shaded and hatched areas correspond to the extreme ranges of both ESR and chemical data for a large number of as-returned sub-mm fines. ESR data from the present investigation were supplemented by data from Weeks *et al.* (1970a), Kolopus *et al.* (1971), Weeks and Kolopus (1972), and Weeks (private communication—Apollo 17 orange soil). Chemical data are from LSPET (1972, 1973) and other sources. All ESR data were obtained at temperatures near 300°K and frequencies near 9 GHz (with the exception of certain Apollo 11 data which were obtained at both 9 and 35 GHz: Weeks *et al.*, 1970a). Open circles pertain to unsorted subsamples of 67601,28, 69961,31, 15302,27, 15021,118, 14230,88, 12001,15, 10084,241, and Luna 16, A-41. Filled circles pertain to the same subsamples, following heat treatment at 650°C for 3190 hours in evacuated, sealed quartz tubes. Crosses indicate *maximum* linewidth values attained at earlier stages of the same heating experiment. Open hexagons pertain to special samples which were not heated. Chemical data for the GGB were taken from Taylor *et al.* (1972) and Agrell *et al.* (1973). The TiO_2 and FeO contents of the orange soil were given by Phinney (1973). Curves are drawn to emphasize apparent and/or predicted trends (*see* text and Griscom and Marquardt, 1973b). The open square represents an experimental value for the linewidth of spherical, single domain particles of pure metallic iron (Griscom and Marquardt, 1973c).

sity of this sample was finally established to be only slightly greater than the *total* ESR intensity of 10084,241 (an incorrect estimate was given in Forester *et al.*, 1973). The data of Fig. 6b represent the temperature dependence of 55% of the total ESR intensity of 10084,241 (due to a method of integration that excluded the broader parts of the resonance comprising the other 45%). These data were then decomposed graphically into a component having the temperature dependence of the synthetic "magnetite-like" phases ($0.37 \times 55\%$) and a temperature independent, metallic iron component ($0.63 \times 55\%$). Thus, it is estimated that the "magnetite" concentration of 10084,241 is $0.37 \times 0.55 \times (1–2 \text{ wt.}\%) = 0.2–0.4 \text{ wt.}\%$. These values are higher than the upper limit estimated by Housley *et al.* (1972) for

a magnetic fraction of 10084 (0.08 wt.% Fe_3O_4) but are consistent with other estimates of the Mössbauer detectability threshold for "magnetite-like" phases in
lunar materials (Forester *et al.*, 1973; Forester, 1973).

The Apollo 15 GGB, on the other hand, exhibit a resonance spectrum that
appears to be completely dominated by "magnetite-like" phases (*see* Fig.
6). However, the total specific intensity of the sample investigated was only about
2 in the units of Table 1, leading to an estimate of ~0.01 wt.% "magnetite" in (or
on) the green glasses. It is suggested that the iron-rich platelets observed by
McKay and Clanton (1973) on the surfaces of the GGB may in fact be the
"magnetite-like" phases detected by ESR.

The "magnetite" contents of the other lunar soils investigated have not yet
been estimated by the precise means applied to 10084,241. However, they are
expected to vary in rough proportion to the specific intensities "near $g \approx 2$" given
in Table 1.

CONCLUSIONS

On the basis of the temperature dependence of the ESR intensity over the
range 5–570°K, it has been demonstrated that the principal ferromagnetic constituent in a separate of Apollo 15 green glass balls is probably a magnetic iron
spinel (i.e., a "magnetite-like" phase). This demonstration was based on a dramatic peak near 200°K in the intensity vs. temperature curves of both the GGB
and sub-microscopic "magnetite-like" precipitates produced by sub-solidus oxidation of simulated lunar glasses. This behavior is similar to prediction for Fe_3O_4
and counter to expectation for Fe metal. These results, shown in Fig. 6b, are
worthy of close scrutiny by specialists in rock magnetism, for they promise the
first workable method for disentangling the effects of co-mingled sub-microscopic
"magnetites" and metallic iron.

Should the method indeed prove to be diagnostic, then the present data would
be interpreted as indicating that a typical mare soil, 10084, contains up to 0.4 wt.%
"magnetite-like" phases. Similar intensity vs. temperature curves have been obtained for other lunar soils (e.g., Kolopus *et al.*, 1971; Griscom and Marquardt,
1973b), showing that 10084 would not be unusual in this regard. Moreover, the
ESR linewidths measured at room temperature for a wide range of returned lunar
soils fall in a smooth band when plotted vs. the TiO_2 contents of the samples. The
Apollo 15 green glasses and the Apollo 17 orange soil, both of which now seem to
contain magnetite, are apparent end members on this diagram (Fig. 8). Taken together, these results and others described above strongly suggest that sub-
microscopic titanomagnetite constituents may be ubiquitous in the lunar soils.
Studies of simulated lunar glasses reported here and elsewhere (Griscom and
Marquardt, 1972a,b, 1973a; Forester *et al.*, 1973; Forester, 1973) indicate the
presence of such phases could be a consequence of oxidation occurring during
ash-flow-type transport on the moon.

In view of the evidence presented here, the possible existence of sub-
microscopic "magnetite-like" phases in the lunar soils should be a consideration

in any detailed interpretation of the chemical or magnetic histories of the lunar surface.

Note added in proof: It has been stated above that the integrated ESR intensity should be proportional to the initial susceptibility χ_0 of the resonating species. While strictly true for *para* magnetic resonance (Pake, 1962), this assertion may not apply in general to the case of *ferro* magnetic resonance (FMR). Assuming a sample which is magnetically saturated, Morrish (1965) presents a simple theoretical treatment showing the FMR intensity to vary as the saturation magnetization M_s. However, below the Verwey temperature, magnetite is quite difficult to saturate (Domenicali, 1950) and is evidently far from saturation over most of the field range scanned in an X-band ESR experiment. Thus, for a spectrometer frequency near 9 GHz, the FMR intensity of magnetite might be expected to behave like M_s above $\sim 120°K$ but to drop off substantially when spectra are obtained at lower temperatures. In this light, the data of Fig. 6b are even more easily understood in terms of the presence of "magnetite-like" phases, and the basic conclusions drawn above are strengthened further.

Acknowledgments—The authors are deeply indebted to the following individuals for their generous and important contributions to this work: P. C. Taylor for adapting his computer program to the ferromagnetic resonance problem; E. J. Brooks and F. W. Fraser for electron microscopy and probe work on simulated glass samples; C. E. Schott for initial preparation of these glasses; J. H. Schelleng for magnetic susceptibility measurements; G. H. Sigel, Jr., for experiments involving heating lunar samples in very high vacua; D. W. Forester for numerous helpful discussions of his Mössbauer results; and F. L. Carter and W. C. Sadler for X-ray diffraction studies of simulated samples. R. A. Weeks is acknowledged for kindly pointing out the work of Syono and Ishikawa (1964), for communicating the orange soil linewidth datum, and for other invaluable exchanges of information. R. M. Housley brought the work of Tarasov (1939) to the authors' attention and also provided simulated lunar glass S-17-3B (cf., Fig. 5). This research was supported in part by NASA Order No. T-4735A.

REFERENCES

Agrell S. O., Agrell J. E., and Arnold A. R. (1973) Observations on Glasses from 15425, 15426, and 15427 (abstract). In *Lunar Science—IV*, pp. 12–14. The Lunar Science Institute, Houston.

Asunmaa S. K., Liang S. S., and Arrhenius G. (1970) Primordial accretion; inferences from the lunar surfaces. In *Proc. Apollo 11 Lunar Sci. Conf., Geochim. Cosmochim. Acta*, Suppl. 1, Vol. 3, pp. 1975–1985. Pergamon.

Ayscough P. B. (1967) *Electron Spin Resonance in Chemistry*, p. 442. Methuen.

Domenicali C. A. (1950) Magnetic and electric properties of natural and synthetic single crystals of magnetite. *Phys. Rev.* **78**, 458–467.

Forester D. W. (1973) Mössbauer search for ferric oxide phases in lunar materials and simulated lunar materials. In *Proc. Fourth Lunar Sci. Conf., Geochim. Cosmochim. Acta.* In press.

Forester D. W., Marquardt C. L., and Griscom D. L. (1973) Mössbauer search for ferric oxide phases in lunar materials and simulated lunar materials (abstract). In *Lunar Science—IV*, pp. 257–259. The Lunar Science Institute, Houston.

Gibb T. C., Greatrex R., Greenwood N. N., and Battey M. H. (1972) Mössbauer studies of Apollo 14 lunar samples. In *Proc. Third Lunar Sci. Conf., Geochim. Cosmochim. Acta*, Suppl. 3, Vol. 3, pp. 2479–2493. MIT Press.

Gibson E. K. Jr. and Moore G. W. (1973) Volatile-rich lunar soil: Evidence of possible cometary impact. *Science* **179**, 69–71.

Grant R. W., Housley R. M., and Paton N. E. (1973) Origin and characteristics of excess Fe metal in lunar glass (abstract). In *Lunar Science—IV*, pp. 315–316. The Lunar Science Institute, Houston.

Griscom D. L. and Marquardt C. L. (1972a) Electron spin resonance studies of iron phases in lunar glasses and simulated lunar glasses (abstract). In *Lunar Science—III*, pp. 341–343. The Lunar Science Institute, Houston.

Griscom D. L. and Marquardt C. L. (1972b) Evidence of lunar surface oxidation processes: Electron spin resonance spectra of lunar materials and simulated lunar materials. In *Proc. Third Lunar Sci. Conf., Geochim. Cosmochim. Acta*, Suppl. 3, Vol. 3, pp. 2397–2415. MIT Press.

Griscom D. L. and Marquardt C. L. (1972c) Ferromagnetic resonance of small, multidomain iron particles in an 0.5-cm fragment of lunar glass, 15434,62 (abstract). In *The Apollo 15 Lunar Samples*, pp. 435–437. The Lunar Science Institute, Houston.

Griscom D. L. and Marquardt C. L. (1973a) Electron spin resonance studies of ferrimagnetic phases precipitated in simulated lunar glasses heat treated in the presence of oxygen. In *Amorphous Magnetism* (editors H. O. Hooper and A. M. deGraaf), pp. 95–102. Plenum.

Griscom D. L. and Marquardt C. L. (1973b) The origin and significance of the "characteristic" ferromagnetic resonance spectrum of lunar soils: Two views (abstract). In *Lunar Science—IV*, pp. 320–322. The Lunar Science Institute, Houston.

Griscom D. L. and Marquardt C. L. (1973c) Ferromagnetic resonance of fine grained metallic iron precipitates in reduced silicate glasses. *Bull. Am. Phys. Soc.* **18**, 542.

Griscom D. L. and Marquardt C. L. (1973d) Thermal histories of lunar soils recorded in their ferromagnetic resonance spectra. *EOS, Trans. Am. Geophys. Union* **54**, 359.

Housley R. M., Grant R. W., and Abdel-Gawad M. (1972) Study of excess Fe metal in the lunar fines by magnetic separation. Mössbauer spectroscopy, and microscopic examination. In *Proc. Third Lunar Sci. Conf., Geochim. Cosmochim. Acta*, Suppl. 3, Vol. 1. pp. 1065–1076. MIT Press.

Kneller E. (1962) *Ferromagnetismus*, 791 pp. Springer Verlag.

Kolopus J. L., Kline D., Chatelain A., and Weeks R. A. (1971) Magnetic resonance properties of lunar samples: Mostly Apollo 12. In *Proc. Second Lunar Sci. Conf., Geochim. Cosmochim. Acta*, Suppl. 2, Vol. 3, pp. 2501–2514. MIT Press.

LRL (Lunar Receiving Laboratory) (1971) Lunar sample information catalog—Apollo 15 MSC 03209, Manned Spacecraft Center, Houston.

LSPET (Lunar Sample Preliminary Examination Team) (1972) The Apollo 15 lunar samples: A preliminary description. *Science* **175**, 363–375.

LSPET (Lunar Sample Preliminary Examination Team) (1973) Preliminary examination of lunar samples from Apollo 16. *Science* **179**, 23–34.

McKay D. S. and Clanton U. S. (1973) Surface morphology of Apollo 15 green glass spheres (abstract). In *Lunar Science—IV*, pp. 484–486. The Lunar Science Institute, Houston.

McKay D. S., Heiken G. H., Taylor R. M., Clanton U. S., Morrison D. A., and Ladle G. H. (1972) Apollo 14 soils: Size distribution and particle types. In *Proc. Third Lunar Sci. Conf., Geochim. Cosmochim. Acta*, Suppl. 3, Vol. 1, pp. 983–994. MIT Press.

Manatt S. L., Elleman D. D., Vaughan R. W., Chan S. I., Tsay F.-D., and Huntress W. T. Jr. (1970) Magnetic resonance studies of some lunar samples. *Science* **167**, 709–711; and *Proc. Apollo 11 Lunar Sci. Conf., Geochim. Cosmochim. Acta*, Suppl. 1., Vol. 3, pp. 2321–2323. Pergamon.

Morrish A. H. (1965) *The Physical Principles of Magnetism*. Wiley.

Nagata T. (1961) *Rock Magnetism*. Maruzen.

Nagata T., Ishikawa Y., Kinoshita H., Kono M., Syono Y., and Fisher R. M. (1970) Magnetic properties and natural remanent magnetization of lunar materials. In *Proc. Apollo 11 Lunar Sci. Conf., Geochim. Cosmochim. Acta*, Suppl. 1, Vol. 3, pp. 2325–2340. Pergamon.

Olhoeft G. R., Strangway D. W., Pearce G. W., Frisillo A. L., and Gose W. A. (1973) Electrical and magnetic properties of Apollo 17 soils. *EOS, Trans. Am. Geophys. Union* **54**, 601.

Pake G. E. (1962) *Paramagnetic Resonance*. W. A. Benjamin.

Phinney W. C. (1973) Summary of Apollo 17 preliminary examination team results. Preprint.

Runcorn S. K., Collinson D. W., O'Reilly W., Battey M. H., Stephenson A., Jones J. M., Manson A. J., and Readman P. W. (1970). Magnetic properties of Apollo 11 lunar samples. In *Proc. Apollo 11 Lunar Sci. Conf., Geochim. Cosmochim. Acta*, Suppl. 1, Vol. 3, pp. 2369–2387. Pergamon.

Runcorn S. K., Collinson D. W., O'Reilly W., Stephenson A., Battey M. H., Manson A. J., and Readman P. W. (1971) Magnetic properties of Apollo 12 lunar samples. *Proc. Roy. Soc. London* Ser. A. **325**, 157–174.

Schlömann E. (1958) Ferromagnetic resonance in polycrystalline ferrites with large anisotropy—I. *J. Phys. Chem. Solids* **6**, 257.

Sigel G. H. Jr. and Ginther R. J. (1968) The effect of iron on the ultraviolet absorption of high purity soda-silica glass. *Glass Technology* **9**, 66–69.

Smit J. and Wijn H. P. J. (1959) *Ferrites.* Wiley.

Spooner T. (1927) *Properties and Testing of Magnetic Materials,* p. 173. McGraw-Hill.

Syono Y. and Ishikawa Y. (1964) Magnetocrystalline anisotropy and magnetostriction of $x\mathrm{Fe_2TiO_4 \cdot (1-x)Fe_3O_4}$ $(x > 0.5)$. *J. Phys. Soc. Japan* **19**, 1752.

Tarasov L. (1939) Ferromagnetic anisotropy of low nickel alloys of iron. *Phys. Rev.* **56**, 1245–1246.

Taylor P. C. and Bray P. J. (1968) Lineshape program manual—Computer simulations of magnetic resonance spectra observed in powdered and glassy samples. Brown University, Providence, R. I. Available upon request.

Taylor P. C. and Bray P. J. (1970) Computer simulations of magnetic resonance spectra observed in polycrystalline and glassy samples. *J. Mag. Res.* **2**, 305.

Taylor S. R., Gorton M., Muir P., Nance W., Rudowski R., and Ware N. (1972) Composition of the lunar highlands II. The Apennine Front (abstract). In *The Apollo 15 Lunar Samples,* pp. 262–264. The Lunar Science Institute, Houston.

Tsay F.-D., Chan S. I., and Manatt S. L. (1971a) Magnetic resonance studies of Apollo 11 and Apollo 12 samples. In *Proc. Second Lunar Sci. Conf., Geochim. Cosmochim. Acta,* Suppl. 2, Vol. 3, pp. 2512–2528. MIT Press.

Tsay F.-D., Chan, S. I., and Manatt S. L. (1971b) Ferromagnetic resonance of lunar samples. *Geochim. Cosmochim. Acta* **35**, 865–875.

Tsay F.-D., Manatt S. L., Live D. H., and Chan S. I. (1973) Electron spin resonance studies of Apollo 16 fines (abstract). In *Lunar Science—IV,* pp. 737–739. The Lunar Science Institute, Houston.

Vinogradov A. P. (1971) Preliminary data on lunar ground brought to Earth by automatic probe "Luna-16." In *Proc. Second Lunar Sci. Conf., Geochim. Cosmochim. Acta,* Suppl. 2, Vol. 1, pp. 1–16. MIT Press.

Weeks R. A. (1972) Magnetic phases in lunar material and their electron magnetic resonance spectra: Apollo 14. In *Proc. Third Lunar Sci. Conf., Geochim. Cosmochim. Acta,* Suppl. 3, Vol. 3, pp. 2503–2517. MIT Press.

Weeks R. A. (1973) Ferromagnetic resonance properties of lunar fines: Apollo 16 (abstract). In *Lunar Science—IV,* pp. 772–774. The Lunar Science Institute, Houston.

Weeks R. A. and Kolopus J. L. (1972) Magnetic phases in lunar material and their electron magnetic resonance spectra: Mostly Apollo 14 (abstract). In *Lunar Science—III,* p. 791. The Lunar Science Institute, Houston.

Weeks R. A., Chatelain A., Kolopus J. L., Kline D., and Castle J. G. (1970a) Magnetic properties of some lunar material. *Science* **167**, 704–707.

Weeks R. A., Kolopus J. L., Kline D., and Chatelain A. (1970b) Apollo 11 lunar material: Nuclear magnetic resonance of $^{27}\mathrm{Al}$ and electron resonance of Fe and Mn. In *Proc. Apollo 11 Lunar Sci. Conf., Geochim. Cosmochim. Acta,* Suppl. 1, Vol. 3, pp. 2467–2490. Pergamon.

Proceedings of the Fourth Lunar Science Conference
(Supplement 4, *Geochimica et Cosmochimica Acta*)
Vol. 3, pp. 2729–2735

Characterization of fines from the Apollo 16 site

R. M. HOUSLEY, E. H. CIRLIN, and R. W. GRANT

Science Center, Rockwell International, Thousand Oaks, California 91360

Abstract—We describe the characteristics of Apollo 16 fines samples 61281,8; 65701,13; 66031,6; 67701,26; and 67712,16 observed microscopically during the course of size and magnetic separations. Sample 67712,16 is unique in that almost all grains are rounded and no glass welded aggregates are present. All samples except 67712,16 contained about 0.5 wt.% of metal fragments 45 μm or more in diameter including occasional spheres and large single crystals. Some of this metal showed rust spots.

Mössbauer spectra showed all the samples to be high in olivine compared to samples from other Apollo sites and to vary significantly in modal composition from each other. The fine grained Fe metal content and the excess absorption area near zero velocity in the Mössbauer spectra both vary with particle size and regolith maturity in the way one would predict from our model of Fe metal reduction and agglomeration during glass aggregate formation. The excess area is interpreted as being due to Fe^0 atoms which have been reduced by solar wind bombardment, but have not yet aggregated into small grains as a result of micrometeorite impacts.

INTRODUCTION

THE FINES at any site must contain a wealth of information about lunar rock types and about the processes which have been important in the evolution of the regolith. Because of their very complex nature, however, it appears that progress in deciphering this information can most effectively be made by applying a number of experimental techniques to the same samples in an integrated way.

The physical separations which we describe here and the binocular microscope descriptions of the separates were made in preparation for such a group of studies. Mössbauer spectra have concurrently been taken of most of the separates to provide further information on their Fe content and modal mineralogy.

A special objective of the Mössbauer spectroscopy has been to provide information which will be useful in evaluating a model of the origin of excess Fe metal in glass welded aggregates which we have recently discussed (Grant *et al.*, 1973).

PHYSICAL SEPARATIONS

We initially dry sieved our five Apollo 16 samples into five size fractions each. Fractions greater than 45 μm in size were in turn ultrasonically washed in ethanol, dried and resieved so that the final distribution reported in Table 1 is characteristic of clean coherent material.

Samples 61281,8; 65701,13; 66031,6 show a nearly identical size distribution which suggests that they represent mature regolith while 67701,26 shows substantial differences, particularly a deficiency of particles in the smaller sizes suggesting that it is not yet in equilibrium. Sample 67712,6, nominally a 1–2 mm size

Table 1. Mass fraction vs. size distribution Apollo 16 samples.

Size μm	61281,8 < 1 mm	65701,13 < 1 mm	66031,6 < 1 mm	67701,26 < 1 mm	67712,16 < 1–2 mm
>420	0.082	0.059	0.055	0.133	0.592
150–420	0.156	1.141	0.138	0.204	0.062
75–150	0.145	1.145	0.15	0.15	0.046
45–75	0.142	1.136	0.134	0.119	0.042
<45	0.476	0.507	0.522	0.394	0.258

fraction, contained a number of sugary appearing white clots consisting largely of plagioclase that broke up during our sieving procedure. This sugary material also strongly adhered to most of the firmer grains and not all of it came off during the ultrasonic washing.

Examination of the greater than 420 μm fraction of each sample with the binocular microscope revealed general differences from samples previously studied. Gray recrystallized breccia fragments are common in all samples whereas recognizable samples of mare basalt are quite rare. Glass welded aggregates are abundant in samples 61281,8; 65701,13; and 66031,6, but are minor in sample 67701,26 and entirely absent in 67712,16. Almost all the larger transparent grains in all the samples are maskelynite.

Sample 67712,16 is unusual in that almost all the grains, from the largest to the smallest observable in an immersion oil mount at 1000X, show a high degree of rounding. Sharp angles are hard to find. The absence of glass welded aggregates shows that surface erosional processes are not responsible for this rounding and suggests that all the necessary abrasion may have taken place during a major impact event, perhaps the one which produced North Ray Crater.

About half the size fractions have been magnetically separated as described by Housley et al. (1972) into 7 magnetic fractions each. The distributions obtained are shown in Tables 2 and 3. The 8 mm height setting produced the weakest magnetic force which we could achieve with the present separator design. Microscopic examination of the grains collected at that setting showed half or more of them to be nearly pure metal including occasional spheres and euhedral crystalline

Table 2. Mass distribution of larger grain size material among magnetic fractions.

Height (mm)	61281,8 420–1000 μm	65701,13 420–1000 μm	150–420 μm	66031,6 420–1000 μm	150–420 μm	67701,26 420–1000 μm
8.0	0.025	0.04	0.014	0.007	0.016	0.012
5.0	0.065	0.097	0.047	0.168	0.05	0.007
3.5	0.199	0.217	0.269	0.248	0.215	0.045
2.5	0.209	0.251	0.13	0.113	0.149	0.088
1.5	0.186	0.192	0.229	0.259	0.204	0.319
1.0	0.198	0.143	0.15	0.102	1.268	1.268
Residue	0.119	0.06	0.161	0.104	0.225	0.262

Table 3. Mass distribution of 45–75 μm material among magnetic fractions.

Height (mm)	61281,8	65701,13	66031,6	67701,26
8.0	0.01	0.009	0.008	0.005
5.0	0.027	0.041	0.055	0
4.0	0.071	0.146	0.124	0.011
3.0	0.121	0.142	0.116	0.048
2.0	0.129	0.204	0.17	0.091
1.0	0.296	0.256	0.263	0.237
Residue	0.348	0.203	0.265	0.608

forms. Rusty looking spots on the metal are not rare. It appears that in all the samples except 67712,16 grains of Fe metal larger than 45 μm in diameter constitute about 0.5% or more of the mass. A more precise estimate cannot be given on the basis of this data since the separation from silicates was not complete.

Hand sorting the magnetic separates from the 420–1000 μm size fraction of 61281,8 showed that glass welded aggregates were mainly limited to the 2.5–5 mm height fractions. The relative scarcity of material in 67701,26 which could be collected in this range reflects its low content of glass welded aggregates.

MÖSSBAUER SPECTROSCOPY

Mössbauer spectra of the 0–45, 45–75, and 75–150 μm size fractions from each sample were taken using low velocity ranges. As illustrated in Fig. 1 no clearly discernable change in modal composition with size for a given sample is apparent. Quite pronounced changes are seen between samples as illustrated in Fig. 2. The fraction of Fe in olivine is generally considerably higher than for samples from any of the previous Apollo missions.

Mössbauer spectra for all samples were also taken over a larger velocity range, as illustrated in Fig. 3, and nominal Fe metal content was determined from the lowest energy Fe metal line as we have previously described (Housley et al., 1971). This is presented in Table 4 along with the equivalent weight of Fe metal needed to account for the excess area (Housley et al., 1971) near zero velocity obtained from narrower velocity scan runs on the same samples. No attempt has been made to estimate the average Ni content in the metal because of complications which arise when a single domain particles are involved (Housley et al., 1972).

It is immediately apparent from Table 4 that Fe metal contents determined by the above analysis are generally considerably lower than those estimated directly from the magnetic separation.

This reflects a limitation on the interpretation of Mössbauer data which has previously been ignored in the discussion of results on lunar samples. If a Mössbauer absorber is inhomogeneous with respect to nonresonant absorption of the gamma ray beam then the regions with strong absorption will receive too low a weight in the average spectrum obtained experimentally. In the case of Fe metal

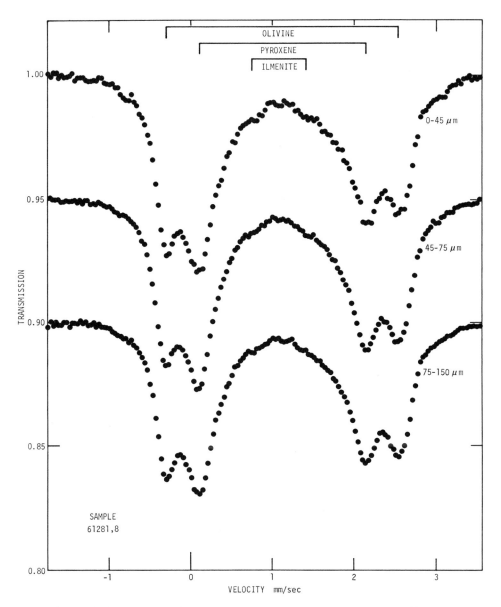

Fig. 1. Mössbauer spectra showing grain size independence of modal composition. Velocity scale is referenced to Fe metal at 23°C.

grains in a powdered rock matrix this effect begins to be important for grains as small as 5 μm in diameter and makes grains bigger than about 25–30 μm in diameter practically opaque and hence unobservable. All reported Mössbauer results on Fe metal in lunar samples including those in this paper should be interpreted with this in mind. They really indicate only the content of Fe metal in particles less than about 25 μm in diameter.

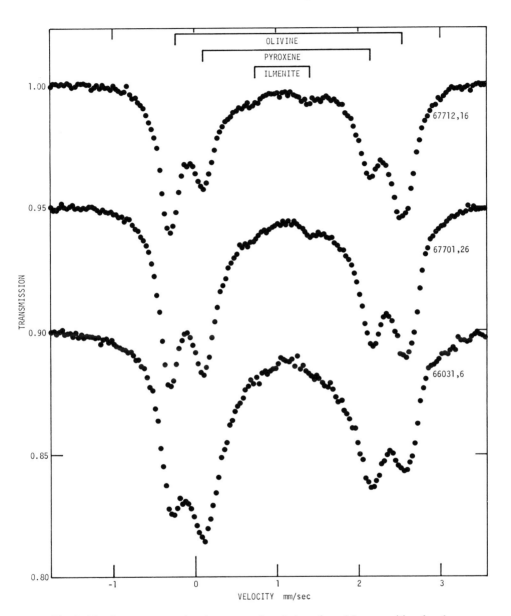

Fig. 2. Mössbauer spectra showing range of variation of modal composition for the samples studied. Velocity scale is referenced to Fe metal at 23°C.

DISCUSSION

The results in Table 4 generally support our model of Fe metal reduction during glass aggregate formation (Grant *et al.*, 1973). The mature regolith samples 65701,13 and 66031,6 which are highest in glass welded aggregates also have the highest concentrations of fine Fe metal.

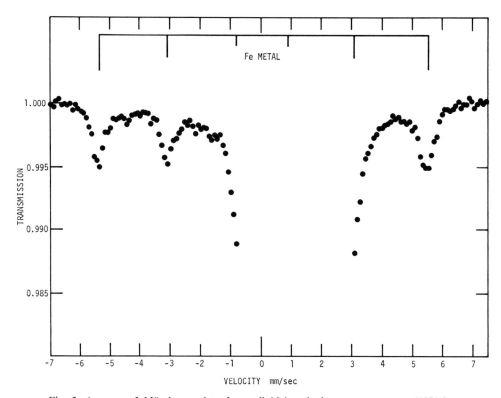

Fig. 3. Average of Mössbauer data from all high velocity range runs on 61281,8; 65701,13; and 66031,6. Velocity scale is referenced to Fe metal at 23°C.

Table 4. Fe metal contents in different size fractions.

Sample	Size μm	Fe Metal wt.%	Excess Area as Fe Metal
61281,8	<45	0.186 ± 0.026	0.196 ± 0.007
	45–75	0.170 ± 0.016	0.098 ± 0.003
	75–150	0.161 ± 0.015	0.071 ± 0.003
65701,3	<20	0.297 ± 0.022	0.365 ± 0.005
	<45	0.242 ± 0.027	0.254 ± 0.007
	45–75	0.231 ± 0.02	—
	75–150	0.177 ± 0.02	0.104 ± 0.004
66031,6	<45	0.265 ± 0.03	0.226 ± 0.006
	45–75	0.196 ± 0.014	0.123 ± 0.004
67701,26	<20	0.057 ± 0.027	0.123 ± 0.006
	<45	0.100 ± 0.026	0.114 ± 0.005
	45–75	0.100 ± 0.015	0.055 ± 0.004
	75–150	0.147 ± 0.017	0.033 ± 0.006
Crushed	75–150	—	0.056 ± 0.004
67712,16	<45	0.024 ± 0.013	0.029 ± 0.005
60017,84	<150	—	0.047 ± 0.003

The excess area near zero velocity shows a consistent inverse dependence on grain size, which is quite strong in the samples of mature regolith. We now suggest that this excess area is largely due to Fe^0 atoms in the surface layers of the particles which have been reduced by solar wind bombardment.

The now well established grain size dependence rules out disproportionation of Fe^{++} in glass and Fe^{+++} in minerals such as anorthite as possible explanations of the excess area.

Acknowledgment—This work was supported by NASA contract NAS9-11539.

REFERENCES

Grant R. W., Housley R. M., and Paton N. E. (1973) Origin and characteristics of excess Fe metal in lunar glass welded aggregates (abstract). In *Lunar Science—IV*, pp. 315–316. The Lunar Science Institute, Houston.

Housley R. M., Grant R. W., Muir A. H. Jr., Blander M., and Abdel-Gawad M. (1971) Mössbauer studies of Apollo 12 samples. *Proc. Second Lunar Sci. Conf., Geochim. Cosmochim. Acta*, Suppl. 2, Vol. 3, pp. 2125–2136. MIT Press.

Housley R. M., Grant R. W., and Abdel-Gawad M. (1972) Study of excess Fe metal in the lunar fines by magnetic separation, Mössbauer spectroscopy, and microscopic examination. *Proc. Third Lunar Sci. Conf., Geochim. Cosmochim. Acta*, Suppl. 3, Vol. 1, pp. 1065–1076. MIT Press.

Proceedings of the Fourth Lunar Science Conference
(Supplement 4, *Geochimica et Cosmochimica Acta*)
Vol. 3, pp. 2737-2749

Origin and characteristics of excess Fe metal in lunar glass welded aggregates

R. M. HOUSLEY, R. W. GRANT, and N. E. PATON

Science Center, Rockwell International, Thousand Oaks, California 91360

Abstract—We show that the characteristic features of lunar glass welded aggregates including their irregular shapes, vesicularity and content of submicron Fe metal in the welding glass can be explained by a model in which they are predominantly formed by micrometeorite impacts into the solar wind saturated topmost surface of the regolith. The Fe metal is reduced from silicates by the solar wind gases. Other possible mechanisms of Fe metal formation are discussed and shown to play at most minor roles.

We also show that surface tension forces control vesicularity and that the low gravity and high vacuum conditions prevailing on the moon are unimportant. Consequences of this surface tension control to possible lunar eruptive volcanism and to the $^{40}Ar-^{39}Ar$ dating of impact events are discussed.

We have studied the Fe metal particles in the welding glass by transmission electron microscopy and found them to be single crystal spheres mostly less than 250 μm in diameter. The implications of these results with respect to the magnetic and microwave resonance properties of fines and breccias are discussed.

INTRODUCTION AND MODEL

IN ADDITION TO fragments resulting from the comminution of local rocks the lunar fines at any site may contain material representing more distant rock types This may for example have been brought to the collection site by single or multiple impact events or down slope migration of loose material from nearby more elevated regions. These fines in general will have been exposed to electromagnetic and charged particle irradiation from the sun and bombardment by micrometeorites. They thus are expected to contain a great deal of information on the composition of lunar rocks, on the radiation and meteorite bombardment history of the moon, and on processes such as vaporization and condensation or oxidation and reduction, which might accompany irradiation and meteorite bombardment. The challenge is to learn to read this information from the fines in as clear and unambiguous a manner as possible. This discussion of the origin of excess Fe metal in the glass welded aggregates is intended to be a step toward that goal.

Glass welded aggregates are abundant in most of the lunar fines samples and almost all of them, at least almost all greater than 45 μm in diameter, in a variety of samples from all Apollo sites are vesicular and contain fine-grained, frequently submicroscopic, Fe metal. We believe that these facts and a number of other observations are most consistently explained by a model in which the majority of glass welded aggregates are assumed to have formed directly by the impact of micrometeorites into the regolith as schematically illustrated in Fig. 1. It is further assumed that the surface of the regolith which was impacted was saturated with solar wind gases which provided the volatiles necessary for the generation of

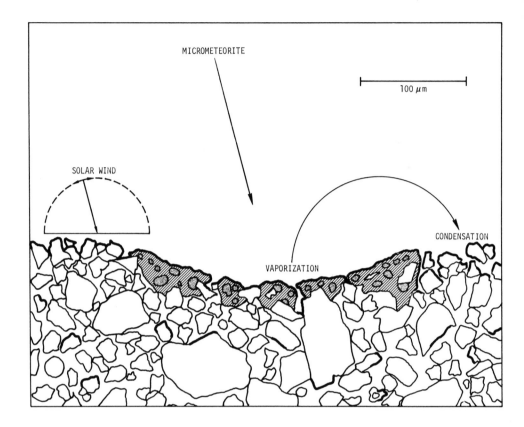

Fig. 1. Schematic cross section through the lunar regolith illustrating the formation of glass welded aggregates (shaded) by micrometeorite impact. Notice that because of lunar rotation and mixing processes, many surfaces of particles in the top several layers (dark lines) are saturated with solar wind gases which form the vesicles and reduce Fe from the silicates to metal during aggregate formation. These same surfaces also provide sites for recondensation of volatilized material. For clarity particles less than 10 μm in diameter which in reality are abundant have been omitted.

vesicles and that the evolution of solar wind H and C during or before the impact was partly as H_2O and CO which reduced Fe from the silicates accounting for the universal presence of Fe metal in these aggregates.

Data and arguments in support of this model from our own work and from the literature are discussed in the following section. In the concluding section a number of consequences of the model, and the data and arguments used in support of it, to other problems in lunar evolution are discussed. These problems include volcanism vs. impact as a source of certain lunar features and materials, the dating of impact events, and possible mechanisms by which lunar breccias might acquire remanent magnetization.

Characteristics of glass welded aggregates

Glass welded aggregates are abundant in the Apollo 11 fines and their physical characteristics including magnetism, vesicularity, and the frequent presence of microscopically visible metal spheres, were described by a number of workers. Particularly clear descriptions were given by Agrell *et al.* (1970), Duke *et al.* (1970), and Frondel *et al.* (1970). Using the techniques of magnetic separation (Housley *et al.*, 1972) and optical microscopy we have now shown that glass welded aggregates from the fines samples 10084,85; 12042,38; 14163,52; 14003,22; 15101,92; 15301,85; 61281,8; 65701,13; 66031,6; 67701,26; 71501,21; 72141,11; 72501,53; 74220,107; 75081,26; and 76501,43, representing all the Apollo sites are almost universally magnetic and vesicular.

Importance of vesicles

The importance of vesicles in lunar samples seems to have received little attention up till now, probably because it is easy to assume that in the low gravity and high vacuum conditions prevailing on the moon very little volatile material would be required to produce them. This is not true! Surface tension is by far the dominant force controlling nucleation and growth of small vesicles so that gas densities required to nucleate vesicles on the moon are almost identical to those which would be required on earth.

The pressure p inside a spherical vesicle of radius r in a fluid of surface tension σ due to the surface tension forces alone is

$$p = \frac{2\sigma}{r}.$$

A good estimate of σ is a silicate melt at 1200°C is 350 ± 50 dynes/cm (Elliott *et al.*, 1963). A melt temperature much higher than 1200°C does not seem reasonable because of the abundance of unmelted crystalline fragments in the aggregates and in any case σ is not a strong function of temperature.

A log–log plot of p vs. r using $\sigma = 350$ dyns/cm is shown in Fig. 2, where it can be seen for example that the pressure inside a 1 μm vesicle is 7.0 atm while that inside a 10 μm vesicle is 0.7 atm. The atom densities obtained from the pressures using the ideal gas law are also shown in Fig. 2 which suggests that densities approaching liquid density would be required to guarantee rapid-nucleation of vesicles (for a discussion of nucleation theory see for example Blander *et al.*, 1971).

The only readily apparent source of volatile materials on the moon in the high concentrations necessary to nucleate the observed high density of vesicles is in the surface layers of particles exposed at some time on the lunar surface to solar wind bombardment. McKay and Ladle (1971) have also commented on the universal occurrence of vesicles in glass welded aggregates and attributed it to the

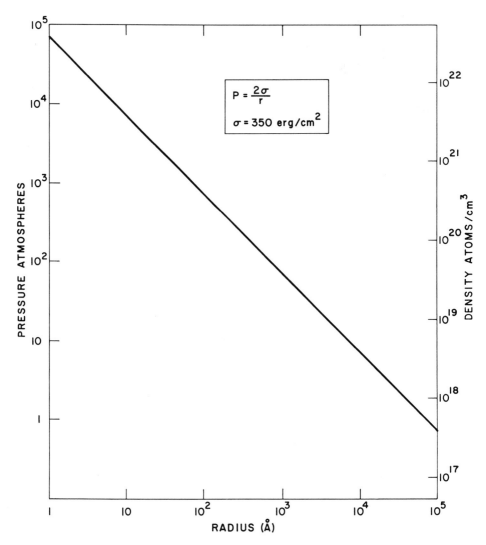

Fig. 2. Approximate pressure and density of gas necessary to support a spherical vesicle
in a silicate melt at 1200°C.

evolution of solar wind gases from regolith material. It seems plausible that high
densities of gas could be retained in such layers. Indeed Lord (1968) has shown
that in laboratory implantation experiments saturation does not occur until there
is about a one to one correspondence between hydrogen atoms and host atoms.
Using an estimate of $2 \times 10^8\ cm^{-2}\ sec^{-1}$ for the average solar wind proton flux
(Tilles, 1965) this concentration of hydrogen would build up in less than 100 years,
so it is reasonable to suppose that a good fraction of the grains in the lunar fines
near the surface do contain high surface densities of implanted gases. This supposition is supported by a number of studies of H_2, N_2, C and nobel gas distribu-

tion (*see* for example, Cadogan *et al.*, 1972; Eberhardt *et al.*, 1970; Hintenberger *et al.*, 1970; Moore *et al.*, 1970) and total abundance (*see* for example, Epstein and Taylor, 1971; Moore *et al.*, 1971).

Since high densities of gas are necessary to nucleate and support vesicles during the formation of glass welded aggregates and since such gas densities are known to be present in solar wind implanted grains, but are so far unknown otherwise on the moon, we conclude that the glass welded aggregates formed during impacts which did not penetrate the regolith and are hence generally relatively local in origin. Conversely, since there is no driving force to remove vesicles from a glass during free fall, we believe that the homogeneous glass spheres must have formed in large impacts which penetrated the regolith, and hence they could come to any site from large distances.

One of the most direct bits of evidence supporting our view that solar wind implanted gases are responsible for vesicles in lunar glasses is well illustrated in the photo documentation of impact features on rocks by Bloch *et al.* (1971). The glass lining impact pits on igneous rocks is largely vesicle free while that lining pits on gas rich breccias is highly vesicular.

Examination of the surfaces of lunar rocks (Hörz *et al.*, 1971) has revealed an abundance of micrometeorite craters ranging in diameter from a few mm downward, even on rocks with relatively short surface exposure age. These micrometeorites mostly in the 10^{-8} to 10^{-4} g size range (Gault *et al.*, 1972) impinging directly into the most gas rich surface layer of the regolith as shown in Fig. 1 must produce a large fraction of the glass welded aggregates.

This same idea has evolved from the work of many other people. Apparently the first comprehensive discussion of it was given by McKay *et al.* (1972) who also demonstrated a good correlation between the glass welded aggregate content in fines samples and other measures of their surface exposure age. This discussion of the significance of vesicles provides further strong support for it. In addition we have examined thousands of glass welded aggregates from all Apollo sites under the binocular microscope and find that the vast majority have shapes either suggesting, or at least compatible with a micrometeorite origin. A few have more regular shapes or other features suggesting an origin in larger events.

A micrometeorite impacting loosely consolidated fines like the lunar regolith at 20 km/sec is expected to throw out an amount of material about 1000 times its own mass (Vedder, 1972). This should provide a good mechanism for stirring the upper few mm of the regolith and exposing fresh grains to solar wind bombardment. Such an impact at 20 km/sec also is expected to melt an amount of material equal to 5 or 10 times the meteorite mass and to vaporize itself plus 1 or 2 equivalent masses of the impacted material (Gault *et al.*, 1972).

Characteristics of excess Fe *metal*

We have previously shown using Mössbauer spectroscopy that the Apollo 11 fines 10084,85 contain a factor of 5–10 more Fe metal than can be accounted for by comminution of the local rocks (Housley *et al.*, 1970) and with the aid of physical separations and optical examination have shown that almost all of this excess Fe metal is finely dispersed in the glass welded aggregates. Using magnetic separations (Housley *et al.*, 1972) and Mössbauer spectroscopy we have now shown that this is also true of the other mare fines in the group previously listed. The highland fines frequently contain appreciable metal as large grains and/or associated with thermally metamorphosed breccias.

We have shown (Housley *et al.*, 1971) in agreement with the interpretation of magnetic measurements (Nagata *et al.*, 1970; Runcorn *et al.*, 1970) that much of the Fe metal in the 10084,85 fines is present as single domain size grains less than about 300 Å in diameter and that most of it is essentially Ni free (Housley *et al.*, 1972). Optical examination of polished sections of 45–75 μm glass welded aggregates from the 10084,85 and 15101,92 fines showed that far less than half contain microscopically visible Fe metal. Magnetic measurements as recently summarized by Nagata *et al.* (1972) have also shown that the fines at mare sites contain 5–10 times as much Fe metal as the igneous rocks, most of which is very fine grained and has negligible Ni content.

To gain further insight into the mode of occurrence and characteristics of this excess Fe metal we have examined by transmission electron microscopy a number of fragments from the most magnetic fraction of 45–75 μm fines separated from samples 10084,85 and 15101,92. Samples were prepared by gently crushing, then suspending the fragments on thin carbon films supported on copper grids. Micrographs were taken through thin edges of fragments, less than about 1000 Å thick, with a Philips EM-300 electron microscope. Regions, similar to those first described by Agrell *et al.* (1970), containing high densities of very small Fe metal spheres were abundant in the welding glass from both fines samples. A typical micrograph of such a region is shown in Fig. 3. Stereoviewing of several

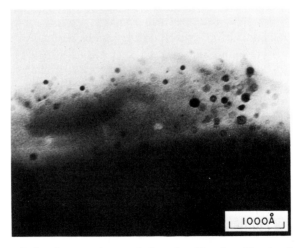

Fig. 3. Transmission electron micrograph through a thin edge of fractured welding glass in crushed 45–75 μm glass welded aggregates from the 15101,92 fines. All spheres are believed to be Fe metal single crystals. Differences in contrast depend on the relative orientation of the crystallographic axes with respect to the electron beam direction.

such regions showed that the Fe metal spheres are distributed throughout the bulk of the glass, not just on the surface. Preliminary data on the size distribution of Fe metal spheres obtained from one region in the 10084,85 sample are shown in Fig. 4.

Because of the high energy associated with an Fe metal-silicate interface a distribution of fine grained Fe metal in glass such as shown in Fig. 3 is very unstable thermodynamically and its preservation in the samples indicates that the metal production process was followed by an extremely rapid quench.

Origin of excess Fe metal

The relatively high abundance, small grain size distribution, low Ni content, and dissemination through the welding glass together suggest reduction of silicates as the most probable source for this Fe metal. The importance of solar wind gases in forming the vesicles in the glass welded aggregates supports the suggestion (Housley *et al.*, 1970) that they play a vital role in the reduction process. Carter and McKay (1972) have also suggested reduction involving solar wind gases to account for the Ni poor metal mounds frequently observed on glass welded aggregates.

Zeller *et al.* (1966) have shown by infrared absorption measurements that the implantation of 0.5 MeV protons into silicate glasses leads to the formation of OH^- radicals with an efficiency estimated to be between 5 and 100%. Lord (1968) has shown that when doses near saturation for 1.8 keV protons have been

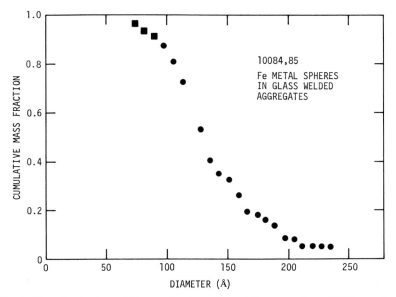

Fig. 4. Size distribution of Fe metal spheres in welding glass from 45–75 μm glass welded aggregates in fines 10084,85. Data were taken from transmission electron micrographs similar to Fig. 3. Square dots are for sizes less than 40 Å too small to be counted with certainty.

achieved, gas release in forsterite samples begins below 100°C on heating. He apparently did not look for evolved H_2O. More recently, Cadenhead *et al.* (1973) have shown that when atomic hydrogen is chemi-sorbed on clean lunar samples at 150°C, H_2O is observed in the gases evolved over the temperature range from 40 to 400°C.

The above observations suggest two slightly different mechanisms by which the solar wind could reduce Fe^{++} in the silicates to metal. (1) In grains saturated with gases, hydrogen could diffuse out as H_2O during the diurnal temperature cycle leaving behind Fe^0 as isolated atoms or small atomic clusters in the surface layers of the grains. These would behave paramagnetically or super-paramagnetically in magnetic measurements. (2) During micrometeorite impacts as illustrated in Fig. 1, H and C could be released as H_2O and CO reducing corresponding amounts of Fe^{++} to metal. These micrometeorite impacts would also allow the aggregation of Fe^0 atoms already present into grains.

In order to evaluate the effectiveness of the above processes quantitatively it is important to know how thick a surface region is affected by the solar wind gases. In etching experiments, Eberhardt *et al.* (1970) have shown that solar wind rare gases reside in the outer 2000 Å skin of grains. High voltage electron micro-scope observations by Dran *et al.* (1970) and Borg *et al.* (1971) have shown that many of the crystalline grains in the lunar fines have amorphous surface layers up to about 1000 Å thick. They have suggested that this amorphous coating is the result of solar wind damage. Direct measurements of the H vs. depth profile in lunar grains by Leich *et al.* (1973) do show high concentrations of H to depths greater than 1000 Å.

In the micrometeorite bombardment process we visualize to be responsible for the production of the glass welded aggregates roughly one third as much material is vaporized as the amount of glass produced. Much of this material must recon-dense on or become re-implanted in the grains exposed on the regolith surface. In fines containing 20% glass roughly 6% of the mass must have vaporized and recondensed. Assuming specific surfaces in the range 0.1–1.0 m²/g (Cadenhead *et al.*, 1972; Holmes *et al.*, 1973) this amount of material corresponds to an average surface coating 200–2000 Å in thickness. This material being deposited while the grains are exposed on the regolith surface will all have been exposed to solar wind gases. It is also possible that some oxygen loss could have occurred during the vaporization-recondensation process.

It sometimes appears that the regions of welding glass rich in metal grains are considerably larger than the thicknesses of the surface regions originally saturated with solar wind gases. It is easy to imagine a process by which Fe^0 generated by either of the above processes could become redistributed and partially equilized in concentration throughout the glass extremely rapidly, before precipitation of the metal grains. It is not required that the Fe^0 atoms actually diffuse through the glass. Electron migration compensated by a small amount of general cation–anion counter diffusion would allow Fe^{++} to be reduced to Fe^0 anywhere tending, to equalize the Fe^0/Fe^{++} ratio throughout the glass. Migration of Ni from the glass to the metal spheres would require diffusion and hence would be much slower.

When considered quantitatively the above processes appear reasonable. With a solar wind flux of 2×10^8 protons/cm^{2}sec the amount of Fe which could be reduced from Fe^{++} to metal in 3.6×10^9 years assuming sufficient fresh material is periodically brought to the surface, is about 10^3 g/cm^2. The Fe metal content in Apollo 11 fines is 0.58 wt.% (Nagata *et al.*, 1970). Conservatively estimating a regolith thickness and density of 10 m and 2 g/cm^3 respectively we find a total Fe metal content of 11.6 g/cm^2. Therefore the solar wind micrometeorite reduction process operating with an overall efficiency of about 1% could have produced all the observed Fe metal in the fines.

Finally we note that the H_2 (Epstein and Taylor, 1970), and C (Moore *et al.*, 1970) presently found in the 10084 fines is sufficient to reduce about half as much Fe^{++} to metal as the amount of metal actually observed. Considering that only a small fraction of the sample would be composed of fully saturated surface material and that some of this surface material must be recycled, this again seems to be in reasonable agreement with our model.

Other reduction mechanisms

It has been suggested (Pearce *et al.*, 1972) that the excess Fe metal in breccias and fines formed by subsolidus reduction in hot ejecta blankets. They demonstrated in the laboratory that such reduction actually takes place when the samples are immersed in a reducing atmosphere at *one atmosphere total pressure*. However, since the vapor pressure of Fe above a melt of lunar composition is much greater than that of O_2 (De Maria *et al.*, 1971; Sato and Hickling, 1973) the process could be significant on the moon only to the extent to which reducing gases are available in the ejecta blanket. Since a hot ejecta blanket could only be formed in a large impact event only a small fraction of the material could be regolith material saturated with solar wind gases. Therefore it appears that this mechanism can be important only to the extent that other sources of reducing gases in adequate amounts can be found.

The fragile nature of many of the glass welded aggregates, their location on the lunar surface, the distribution of Fe metal throughout the welding glass together with its absence on surfaces of adhering crystals, and the absence of recrystallization of the glass all seem inconsistent with an origin by subsolidus reduction in a hot ejecta blanket.

Disproportionation of Fe^{++} to Fe^0 and Fe^{+++} has been suggested by Rao and Cooper (1972) as a major source of Fe metal on the moon. If this disproportionation is assumed to occur on cooling then it cannot account for the excess Fe metal in the glass welded aggregates since the glass in them is quenched from high temperature whereas the starting material lower in Fe metal was more slowly cooled. If it is assumed to occur on heating it would be inconsistent with much data on igneous rocks showing continued formation of Fe metal grains over a range of temperatures during crystallization of the rocks. Direct Mössbauer measurements by Housley *et al.* (1970) and Herzenberg and Riley (1970) have shown that less than 1% of the total Fe in lunar rocks is Fe^{+++}. In previously unpublished work we have also shown that less than 1% of the Fe is Fe^{+++} in a number of synthetic lunar rocks and lunar glasses made in closed Fe metal crucibles and corresponding in composition to Apollo 11, 12, and 15 igneous rocks.

CONCLUSION AND CONSEQUENCES

We have provided strong evidence in support of a model of glass welded aggregate formation, by micrometeorite impacts into solar wind saturated regolith, which qualitatively accounts for their general characteristics of shape, appearance, and vesicularity and which can qualitatively and quantitatively account for the location, small grain size distribution, total amount, and low Ni content of the metal contained in them. We have also discussed reasons for concluding that other reduction processes are of minor importance in producing this excess Fe metal. It is hoped that this model will prove to be a useful guide in planning further studies of regolith evolution.

As a consequence of our discussion on the nucleation of vesicles, we conclude that eruptive volcanism such as the production of cinder cones and volcanic ash would require just as high a volatile content in lunar magmas as it does in terrestrial ones. Therefore since all lunar rocks so far examined have been characterized by an extremely low volatile content (*see* for example, Frondel *et al.*, 1970) we conclude that the probability that eruptive volcanism has occurred on the moon is very low.

As another consequence we conclude that low densities of radiogenic gas would be unlikely to nucleate bubbles and would have to escape from a melt by the much slower process of diffusion. Therefore we feel that the interpretation of $^{40}Ar-^{39}Ar$ measurements on impact generated glass beads (Husain, 1972) requires considerable caution and in fact might be more indicative of the age of the source rocks than of the impact event unless the beads formed by evaporation and recondensation.

The single crystal Fe metal spheres which we see in our electron micrographs, Fig. 3, were assumed by Tsay *et al.* (1971) in their computer simulation of the ferromagnetic resonance spectra seen in Apollo 11 fines. Our data therefore strengthen the proof that Fe metal is present in the form and amount needed to account for the qualitative features and intensity of the ferromagnetic resonance in the fines. It should be noted, however, that the grains are not all perfectly spherical, are close enough together so that interaction effects are expected to be important, and are not free but imbedded in a paramagnetic matrix. Therefore it cannot be expected that the first order theory used in the computer simulation will adequately account for all features of the resonance.

Dunn *et al.* (1973) have shown that shock metamorphism above a certain threshold increased the saturation magnetization of lunar fines and from measurements of the initial susceptibility have concluded that this increase was caused by the production of Fe metal particles in the 160–170 Å diameter size range. We would interpret this as being due either to the agglomeration of atomically isolated Fe^0 atoms produced by solar wind bombardment without subsequent micrometeorite melting, or perhaps due to coarsening of very small Fe metal grains in welding glass as shown in Fig. 3. This suggestion can easily be tested since if it is correct shock should then not produce such an effect in lunar or synthetic analog material not having a previous history of solar wind exposure or micrometeorite bombardment.

Acknowledgments—We thank Dr. M. Blander for enthusiastic participation in the early stages of this work. Work was supported by NASA contract NAS9-11539.

REFERENCES

Agrell S. O., Sconn J., Muir I. D., Long J. V. P., McCornell J. D., and Peckett A. (1970) Observations on the chemistry, minerology and petrology of some Apollo 11 samples. *Proc. Apollo 11 Lunar Sci. Conf., Geochim. Cosmochim. Acta*, Suppl. 1, Vol. 1, pp. 93–128. Pergamon.

Blander M., Hengstenberg D., and Katz J. L. (1971) Bubble nucleation in *n*-pentane, *n*-hexane + hexadecane mixtures, and water. *J. Phys. Chem.* **75**, p. 3613–3619.

Bloch R., Fechtig H., Gertner W., Neukum G., Schneider E., and Wirth H. (1971) Natural and simulated impact phenomena: A photo-documentation. MPI H-1971 V-, Max-Planck-Institute für Kernphsik, Heidelberg.

Borg J., Maurette M., Durrieu L., and Jouret C. (1971) Ultramicroscopic features in micron-sized lunar dust grains and cosmophysics. *Proc. Second Lunar Sci. Conf., Geochim. Cosmochim. Acta*, Suppl. 2, Vol. 3, pp. 2027–2040. MIT Press.

Cadenhead D. A. and Jones B. R. (1972) The adsorption of atomic hydrogen on 15101,68. In *The Apollo 15 Lunar Samples*, pp. 272–273. The Lunar Science Institute, Houston.

Cadenhead D. A., Jones B. R., Buergel W. G., and Stetter J. R. (1973) The effects of a terrestrial atmosphere on lunar sample surface compositions and the formation of lunar water vapor (abstract). In *Lunar Science—IV*, pp. 109–111. The Lunar Science Institute, Houston.

Cadogan P. H., Eglinton G., Firth J. N. M., Maxwell J. R., Mays B. J., and Pillinger C. T. (1972) Survey of carbon compounds II: The carbon chemistry of Apollo 11, 12, 14 and 15 samples. *Proc. Third Lunar Sci. Conf., Geochim. Cosmochim. Acta*, Suppl. 3, Vol. 2, pp. 2069–2090. MIT Press.

Carter J. L. and McKay D. S. (1972) Metallic mounds produced by reduction of material of simulated lunar composition and implications on the origin of metallic mounds on lunar glasses. *Proc. Third Lunar Sci. Conf., Geochim. Cosmochim. Acta*, Suppl. 3, Vol. 1, pp. 953–970. MIT Press.

De Maria G., Balducci G., Guido M., and Piacente V. (1971) Mass spectrometric investigation of the vaporization process of Apollo 12 lunar samples. *Proc. Second Lunar Sci. Conf. Geochim. Cosmochim. Acta*, Suppl. 2, Vol. 2, pp. 1367–1380. MIT Press.

Dran J. C., Durrieu L., Jouret C., and Maurette M. (1970) Habit and texture studies of lunar and meteoritic material with a 1 Mev electron microscope. *Earth Planet. Sci. Lett.* **9**, 391–400.

Duke M. B., Woo C. C., Sellars G. A., Bird M. L., and Finkelman R. B. (1970) Genesis of lunar soil at Tranquillity Base. *Proc. Apollo 11 Lunar Sci. Conf., Geochim. Cosmochim. Acta*, Suppl. 1, Vol. 1, pp. 347–361. Pergamon.

Dunn J., Fisher R., Fuller M., Lally S., Rose F., Schwerer F., and Wasilewski P. (1973) Shock remanent magnetization of lunar soil (abstract). In *Lunar Science—IV*, pp. 194–195. The Lunar Science Institute, Houston.

Eberhardt P., Geiss J., Graf H., Grögler N., Krähenbühl U., Schwaller H., Schwarzmüller J., and Stettler A. (1970) Trapped solar wind noble gases, exposure age and Kr/Ar-age in Apollo 11 lunar fine material. *Proc. Apollo 11 Lunar Sci. Conf., Geochim. Cosmochim. Acta*, Suppl. 1, Vol. 2, pp. 1037–1070. Pergamon.

Elliott J. F., Gleiser M., and Ramakrishna V. (1963) In *Thermochemistry for Steelmaking: Thermodynamic and Transport Properties*, Vol. II. Addison-Wesley.

Epstein S. and Taylor H. P. Jr. (1970) The concentration and isotopic composition of hydrogen carbon and silicon in Apollo 11 lunar rocks and minerals. *Proc. Apollo 11 Lunar Sci. Conf., Geochim. Cosmochim. Acta*, Suppl. 1, Vol. 2, pp. 1085–1096. Pergamon.

Epstein S. and Taylor J. P. Jr. (1971) O^{18}/O^{16}, Si^{30}/Si^{28}, D/H, and C^{13}/C^{12} ratios in lunar samples. *Proc. Second Lunar Sci. Conf., Geochim. Cosmochim. Acta*, Suppl. 2, Vol. 2, pp. 1421–1441. MIT Press.

Frondel F., Klein C. Jr., Ito J., and Drake J. C. (1970) Mineralogical and chemical studies of Apollo 11 lunar fines and selected rocks. *Proc. Apollo 11 Lunar Sci. Conf., Geochim. Cosmochim. Acta*, Suppl. 1, Vol. 1, pp. 445–474. Pergamon.

Gault D. E., Hörz F., and Hartung J. B. (1972). Effects of microcratering on the lunar surface. *Proc. Third Lunar Sci. Conf., Geochim. Cosmochim. Acta*, Suppl. 3, Vol. 3, pp. 2713–2734. MIT Press.

Herzenberg C. L. and Riley D. L. (1970) Analysis of the first returned lunar samples by Mössbauer spectrometry. *Proc. Apollo 11 Lunar Sci. Conf., Geochim. Cosmochim. Acta*, Suppl. 1, Vol. 3, pp. 2221–2241. Pergamon.

Hintenberger H., Weber H. W., and Takaoka N. (1970) Concentrations and isotopic abundances of the rare gases, hydrogen and nitrogen in Apollo 11 lunar matter. *Proc. Apollo 11 Lunar Sci. Conf., Geochim. Cosmochim. Acta*, Suppl. 1, Vol. 2, pp. 1269–1282. Pergamon.

Holmes H. F., Fuller E. L. Jr., and Gammage R. B. (1973) Interaction of gases with lunar materials. Apollo 12, 14, and 16 samples (abstract). In *Lunar Science—IV*, pp. 378–380. The Lunar Science Institute, Houston.

Hörz F., Hartung J. B., and Gault D. E. (1971) Micrometeorite craters on lunar rock surfaces. *J. Geophys. Res.* **76**, 5770–5798.

Housley R. M., Blander M., Abdel-Gawad M., Grant R. W., and Muir A. H. Jr. (1970) Mössbauer spectroscopy of Apollo 11 samples. *Proc. Apollo 11 Sci. Conf., Geochim. Cosmochim. Acta*, Suppl. 1, Vol. 3, pp. 2251–2268. Pergamon.

Housley R. M., Grant R. W., Muir A. H. Jr., Blander M., and Abdel-Gawad M. (1971) Mössbauer studies of Apollo 12 Samples. *Proc. Apollo 11 Lunar Sci. Conf., Geochim. Cosmochim Acta*, Suppl. 1, Vol. 3, pp. 2125–2136. Pergamon.

Housley R. M., Grant R. W., and Abdel-Gawad M. (1972) Study of excess Fe metal in the lunar fines by magnetic separation, Mössbauer spectroscopy, and microscopic examination. *Proc. Third Lunar Sci. Conf., Geochim. Cosmochim. Acta*, Suppl. 3, Vol. 1, pp. 1065–1076. MIT Press.

Husain L. (1972) The ^{40}Ar-^{39}Ar and cosmic ray exposure ages of Apollo 15 crystalline rocks, breccias and glasses. In *The Apollo 15 Lunar Samples*, pp. 374–377. The Lunar Science Institute, Houston.

Leich D. A., Tombrello T. A., and Burnett D. S. (1973) The depth distribution of hydrogen in lunar materials (abstract). In *Lunar Science—IV*, pp. 463–465. The Lunar Science Institute, Houston.

Lord H. C. (1968) Hydrogen and helium ion implantation into olivine and enstitite: Retention coefficients, saturation concentrations, and temperature release profiles. *J. Geophys. Res.* **73**, 5271–5280.

McKay D. S. and Ladle G. H. (1971) Scanning electron microscope study of particles in the lunar soil. In *Proc. Fourth Annual Scanning Electron Microscope Symposium*, IIT Research Institute, Chicago.

Moore C. B., Gibson E. K., Larimer J. W., Lewis C. F., and Nichiporuk W. (1970) Total carbon and nitrogen abundances in Apollo 11 lunar samples and selected achondrites and basalts. *Proc. Apollo 11 Lunar Sci. Conf., Geochim. Cosmochim. Acta*, Suppl. 1, Vol. 2, pp. 1375–1382. Pergamon.

Moore C. G., Lewis C. F., Larimer J. W., Delles F. M., Gooley R. C., and Nichiporuk W. (1971) Total carbon and nitrogen abundances in Apollo 12 lunar samples. *Proc. Second Lunar Sci. Conf., Geochim. Cosmochim. Acta*, Suppl. 2, Vol. 2, pp. 1343–1350. MIT Press.

Nagata T., Ishikawa Y., Kinoshita H., Kono M., Syono Y., and Fisher R. M. (1970). Magnetic properties and natural remanent magnetization of lunar materials. *Proc. Apollo 11 Lunar Sci. Conf., Geochim. Cosmochim. Acta*, Suppl. 1, Vol. 3, pp. 2325–2340. Pergamon.

Nagata T., Fisher R. M., and Schwerer F. C. (1972) Lunar rock magnetism. *The Moon* **4**, 160–186.

Pearce G. W., Williams R. J., and McKay D. S. (1972) The magnetic properties and morphology of metallic iron produced by subsolidus reduction of synthetic Apollo 11 composition glasses. *Earth Planet. Sci. Lett.* **17**, 95–104.

Rao K. J. and Cooper A. R. (1972) Optical properties of lunar glass spherules from Apollo 14 fines. *Proc. Third Lunar Sci. Conf., Geochim. Cosmochim. Acta*, Suppl. 3, Vol. 3, pp. 3143–3155. MIT Press.

Runcorn S. K., Collinson D. W., O'Reilly W., Battey M. H., Stephenson A., Jones J. M., Manson A. J., and Readman P. W. (1970) Magnetic properties of Apollo 11 lunar samples. *Proc. Apollo 11 Lunar Sci. Conf., Geochim. Cosmochim. Acta,* Suppl. 1, Vol. 3, pp. 2369–2387. Pergamon.

Sato M. and Hickling N. (1973) Oxygen fugacity values of some lunar rocks (abstract). In *Lunar Science—IV,* pp. 650–652. The Lunar Science Institute, Houston.

Tilles D. (1965) Atmospheric noble gases: Solar wind bombardment of extraterrestrial dust as a possible source mechanism. *Science* **148**, 1085–1087.

Tsay F. D., Chan S. I., and Manatt S. L. (1971) Ferromagnetic resonance of lunar samples. *Geochim. Cosmochim. Acta,* **35**, 865–875.

Vedder J. F. (1972) Craters formed in mineral dust by hypervelocity microparticles. *J. Geophys. Res.* **77**, 4304–4309.

Zeller E. J., Ronca L. B., and Levy P. W. (1966). Proton induced hydroxyl formation on the lunar surface. *J. Geophys. Res.* **71**, 4855–4860.

Proceedings of the Fourth Lunar Science Conference
(Supplement 4, *Geochimica et Cosmochimica Acta*)
Vol. 3, pp. 2751–2761

Metallic Fe phases in Apollo 16 fines: Their origin and characteristics as revealed by electron spin resonance studies

FUN-DOW TSAY* and STANLEY L. MANATT

Space Sciences Division, Jet Propulsion Laboratory, California Institute of
Technology, Pasadena, California 91103

and

DAVID H. LIVE and SUNNEY I. CHAN

Arthur Amos Noyes Laboratory of Chemical Physics,† California Institute of
Technology, Pasadena, California 91109

Abstract—The intense electron spin resonance (ESR) signals ($g = 2.08 \pm 0.03$) detected in the Apollo 16 fines from three sites (61141,4, Station 1; 64501,22, South Ray Crater; 67601,20, North Ray Crater) are found to be essentially similar in g-value, in lineshape asymmetry and in temperature dependence to those previously observed for the Apollo 11–15 fines. On the basis of these similarities, it is concluded that these ESR signals like those detected in the Apollo 11–15 fines are principally ferromagnetic in nature arising from metallic Fe phases having the body-centered cubic structure, and not from hematite, magnetite, or any other ferric oxides. It is shown that a quantitative correlation exists between the ESR linewidth observed for the Apollo 11–16 fines and their average Ni contents in the metallic Fe phases as determined by other means. For the three Apollo 16 fines investigated, the ESR linewidths are found to be essentially identical. This together with a high Ni content in the metallic Fe phases of these samples as determined from ESR linewidth correlation indicates a common source of meteoritic origin for the metallic Fe phases of these samples. Significant variations are observed in the metallic Fe content as well as in the total Ni content for these samples, in particular, between the fines from South Ray Crater and that from North Ray Crater. These variations appear to correlate with the surface exposure ages of the samples. It is also shown that the first-order crystalline anisotropy energy, $2K_1/M_s$, determined from ESR linewidth measurements is essentially equivalent to the remanence coercive force, H_{RC}, obtained in static magnetic susceptibility measurements for the lunar fines.

INTRODUCTION

THIS PAPER describes our continuing investigations with electron spin resonance (ESR) technique of the nature and origin of the metallic Fe phases in the returned lunar samples. Apollo 16 fines (< 1 mm) from three sites (61141,4, Station 1; 64501,22, South Ray Crater; 67601,20, North Ray Crater) were investigated. In our previous reports (Tsay *et al.*, 1971a, b; 1973a, b), ESR evidence has been presented to indicate that the intense ESR signals ($g = 2.08 \pm 0.03$) detected in all the Apollo 11–16 fines arise from ferromagnetic centers of metallic Fe and not from hematite, magnetite, or any other ferric oxides as suggested by others (Griscom and Marquardt, 1972; Weeks, 1972). In addition, on the basis of (1) an effective g-value of 2.08 as compared to the free electron g-value of 2.0023, (2) an

---

*NRC-NASA Resident Research Associate 1971–1973.
†Contribution No. 4686.

asymmetric lineshape with a narrower appearance on the high-field side, and (3) no eddy current effects together with an intrinsic linewidth of 300–400 G at room temperature, it is possible to ascertain with certainty that the metallic Fe particles in the Apollo 11–16 fines are essentially spherical in shape, having a body-centered cubic structure and a diameter in the range of 30Å–$1\,\mu\text{m}$. The size, shape, crystallinity, and abundance of the metallic Fe as determined from our ESR lineshape analyses and intensity measurements have been recently confirmed by Grant et al. (1973) in their transmission electron microscope results. We have noted that on the ESR time scale of observation (10^{-9}–10^{-11} sec), these metallic Fe particles remain essentially in the ferromagnetic state at room temperature, although on the Mössbauer and static magnetic susceptibility time scales of observation ($> 10^{-9}$ sec), any metallic Fe particle with a diameter smaller than $130\,\text{Å}$ is expected to exhibit superparamagnetic behavior at room temperature (Housley et al., 1972; Nagata et al., 1972).

However, the impression still exists, particularly after the return of orange soil by the Apollo 17 mission, that at least some portion if not all of the intense ESR signals detected in the Apollo 11–16 fines thus far investigated is due to ferric oxides (Griscom and Marquardt, 1973; Weeks, 1973). We show here that in spite of the fact that the ESR linewidth observed for the Apollo 16 fines is much narrower than that observed for the Apollo 16 fines, its temperature behavior is essentially similar to that observed for other Apollo samples and for metallic Fe as well, and in no way resembles that observed for magnetite, a ferrite phase (Weeks, 1972), or the "unknown ferric oxide" phase reported by Griscom and Marquardt (1972, 1973). This narrower linewidth is attributable to a higher Ni content in the metallic Fe phases of the Apollo 16 fines.

With the resolution of the metallic-Fe-or-ferric-oxide controversy which persisted between our group and others using ESR technique, it is now appropriate to demonstrate the uniqueness of ESR technique for studying the metallic Fe phases of the lunar samples. This includes the ease with which the ESR technique can detect and identify various demagnetization effects such as cubic crystalline anisotropy and uniaxial shape anisotropy. We are able to show that for the lunar fines, and for the low-grade fragmental rocks and breccias as well, the remanence coercive force H_{RC} determined by other investigators using static magnetic susceptibility measurements arises from first-order crystalline anisotropy energy $2K_1/M_s$, and as such, is expected to correlate with the Ni and Co content in the metallic Fe phases of these samples.

EXPERIMENTAL

The spectra were obtained on a Varian model 4500 Electron Spin Resonance Spectrometer, operating at X-band (9.1 GHz) and equipped with a dual rectangular cavity system. One cavity housed the sample and variable temperature dewar, while the second cavity contained either a DPPH or Mn^{2+} doped calcite sample (Tsay et al., 1972) for g-value, linewidth and intensity calibration. An Air Products LTD-3-110 Heli-Tran Liquid Transfer System was used for the variable temperature experiments. The temperature was measured by an Iron doped Gold-Chromel thermocouple located about 4 inches from the sample. Samples (8–10 mg) were run in evacuated, sealed quartz tubes.

RESULTS AND DISCUSSION

Ferromagnetic resonance of metallic Fe

Figure 1 shows the ESR spectra observed at room temperature and at X-band frequency (9.1 GHz) for the Apollo 16 fines (61141,4; 64501,22; 67601,20). The g-value and the lineshape asymmetry are essentially identical to those already observed for the Apollo 11–15 fines (Tsay *et al.*, 1971a, b; Griscom and Marquardt, 1972; Weeks, 1972). A comparison is given in Fig. 2 of the temperature behavior of the ESR linewidths observed for the lunar fines, metallic Fe, magnetite, Ni ferrite, and the "unknown ferric oxide" phase of Griscom and Marquardt (1972). On the basis of this comparison, it is evident that the intense ESR signals observed for the Apollo 11–16 fines cannot be interpreted as arising from a ferrite phase or from an "unknown ferric oxide" phase as suggested by Weeks (1972, 1973) and Griscom and Marquardt (1972, 1973), respectively. It is seen in Fig. 2 that the observed temperature dependence of the linewidth for the Apollo 11–16 fines is in agreement with that expected for metallic Fe over the temperature range of 73–298°K. As has been suggested (Tsay *et al.*, 1973a), the apparent disagreement below 73°K probably arises from the uncertainty involved in the calculated linewidth for metallic Fe where extrapolated values of K_1 (Bozorth, 1961) are used.

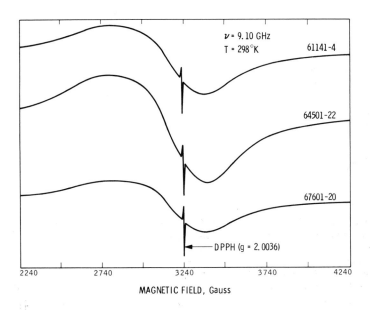

Fig. 1. Electron spin resonance spectra of metallic Fe in Apollo 16 fines. Spectra were recorded at 298°K and at X-band frequency (9.10 GHz). Reproduction of actual spectra demonstrating the great intensity of these signals. The sharp line ($g = 2.0036$) marks the resonance position of a standard DPPH sample.

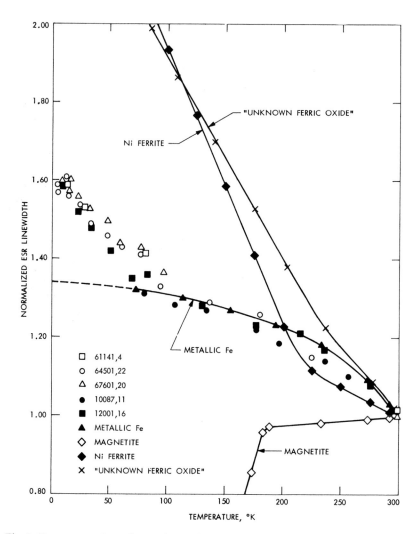

Fig. 2. Temperature dependence of ESR linewidth observed for lunar fines, metallic Fe
and some ferric oxides. Normalized ESR linewidth, the linewidth relative to that observed at room temperature, was used in plot. Room temperature linewidths for these
samples are: 61141, 581 G; 64501, 585 G; 67601, 572 G; 10087, 805 G; 12001, 770 G;
metallic Fe, 820 G; magnetite, 776 G; Ni ferrite, 1000 G; "unknown ferric oxide," 650 G.
Linewidth variations for metallic Fe and for magnetite were calculated using the relation
$\Delta H = \frac{5}{3}(2K_1/M_s)$ (Tsay *et al.* 1973a) and the data given by Bozorth (1961) and Bickford
(1950). Observed linewidth variations were taken from Griscom and Marquardt (1972)
("unknown ferric oxide" phase), Kolopus *et al.* (1971) (12001), Lin and Neaves (1962) (Ni
ferrite), and Tsay *et al.* (1971a, 1973b) (Apollo 11, 16) as well as present study (Apollo 16).

Correlation of ESR linewidth with Ni *and/or* Co *content*

The ESR linewidth observed for the Apollo 16 fines (580 ± 40 G) was found to be close to that for the Apollo 14 fines (606 ± 40 G) (Weeks *et al.*, 1972), both from highlands, but was narrower than those for the Apollo 11 (800 ± 40 G), 12 (770 ± 40 G), and 15 fines (720 ± 40 G) (Griscom and Marquardt, 1972). On the basis of the literature value of $2K_1/M_s$ for pure metallic Fe (+ 500 G), Ni (− 220 G), and Co (+ 7000 G) in their ferromagnetic states (Kittel, 1948, 1967), the ESR linewidth observed for the randomly oriented ferromagnetic centers of metallic Fe, which is found to be proportional to $2K_1/M_s$ (Tsay *et al.*, 1971a), is expected to decrease with increasing Ni content and to increase with increasing Co content in the metallic Fe phases. Because of the presence of Ni and/or Co in the metallic Fe phases analyzed from the lunar samples, we have suggested (Tsay *et al.*, 1973a) that the variation in linewidth observed for the lunar fines correlates with the Ni and/or Co content in the metallic Fe phases of these samples. This correlation is illustrated in Fig. 3. As seen in the left of this figure (Fig. 3a) where ESR linewidth vs. total Ni contents are plotted for some Apollo 11–16 fines for which ESR and chemical data are available, a linear correlation exists only for those fines having almost the same amount of metallic Fe. This prompted us to consider in Fig. 3b

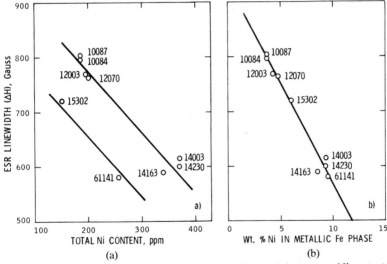

Fig. 3. Correlation between the observed ESR linewidths and the average Ni contents in the Apollo 11–16 fines. (a) Linewidth vs. total Ni content. (b) Linewidth vs. wt.% Ni in the metallic Fe phase. Data for total Ni content were taken from Rose *et al.* (1972) for the average Ni content of Apollo 11, 12 and 14 fines (10087, 12003, 14230); Schonfeld and Meyer (1972) (10084, 12070); Taylor *et al.* (1972) (14003, 14163). Morgan *et al.* (1972) for 15432 (Apennine Front) assumed similar to 15302; LSPET (1973) for 61501 (Station 1) assumed similar to 61141. Data for wt.% Ni were calculated assuming all the Ni was present in the metallic Fe phase, where the metallic Fe content was obtained by ESR intensity measurements. ESR data for linewidth and metallic Fe content were taken from Tsay *et al.* (1971a, b) (Apollo 11 and 12 fines) and Weeks (1972) and Griscom and Marquardt (1972) (Apollo 14 and 15 fines).

linewidth vs. wt.% Ni in the metallic Fe phases assuming that all the Ni content in the lunar sample is present in the metallic Fe phase. It is evident that a quantitative correlation does exist between the observed ESR linewidth and the Ni content in the metallic Fe phases of the lunar fines. It is to be noted that this quantitative correlation is in agreement with the experimental data for K_1 and M_s given by Hofmann (1967) for Fe–Ni alloys, where K_1 is found to decrease linearly with increasing Ni content up to 30 wt.% while M_s is independent of Ni content in the range of 0–12 wt.%. Thus, on the basis of these experimental data, it is expected that the ESR linewidth observed for the lunar fines, which is proportional to $2K_1/M_s$, should decrease linearly with increasing Ni content in the metallic Fe phase up to 12 wt.% Ni.

The linewidth obtained by extrapolation to 0 wt.% Ni using the correlation given in Fig. 3b is found to be about 950 G, which is somewhat greater than both the experimentally observed (700–800 G; Bagguley, 1955) and theoretically calculated linewidth (800–860 G; Bozorth, 1961, and Hofmann, 1967) known for pure metallic Fe. This excess in linewidth for the lunar fines is likely to be due to the presence of a small amount of Co (about 0.5 wt.%) in the metallic Fe phases of these samples.

The Ni contents for the Apollo 16 fines as determined from ESR linewidth correlation are summarized in Table 1, together with the metallic Fe contents obtained by our present ESR intensity measurements. The total Ni contents available for these samples or for the same type of samples as determined by other means are also given in Table 1 for comparison.

For the three Apollo 16 fines we have investigated, the Ni contents in the metallic Fe phases are found to be essentially identical (about 9.6 wt.%), suggesting a common origin for the metallic Fe phases of these samples. A high Ni content for all these metallic Fe phases further indicates that they are meteoritic in origin (Goldstein and Axon, 1973). In addition, we find a significant variation in the metallic Fe content as well as in the total Ni content in these samples. In particular, the fines 64501,22 from South Ray Crater are found to be about a factor of 3 higher in metallic Fe content and in total Ni content than the fines from North

Table 1. Metallic Fe and Ni contents for some Apollo 16 fines.

Fines (< 1 mm)	Locality	Observed ESR Linewidth (Gauss)	Metallic Fe Content (wt.%)[a]	Wt.%Ni in Metallic Fe[b]	Total Ni Content (ppm)	
					This Work[c]	Literature[d]
61141,4	Station 1	581 ± 40	0.27 ± 0.05	9.6 ± 1.5	260	256
64501,22	South Ray Crater	585 ± 40	0.42 ± 0.05	9.4 ± 1.5	400	350
67601,20	North Ray Crater	572 ± 40	0.15 ± 0.05	9.8 ± 1.5	150	120

[a]Determined by ESR intensity measurements.
[b]Determined from ESR linewidth correlation with other samples.
[c]Obtained assuming all Ni in metallic Fe phase.
[d]Taken from LSPET (1973) for 61501 (61141), Morrison et al. (1973) (64501), and Wänke et al. (1973) for 67461 (67601).

Ray Crater. As suggested (Tsay *et al.*, 1971b), an important factor determining the meteoritic Fe content obviously should be the surface exposure age. Thus, less metallic Fe as well as less total Ni content in the fines from North Ray Crater implies that this sample has been on the surface a shorter period of time. This interpretation is in accord with the exposure ages determined by Kirsten *et al.* (1973) for these samples. For example, the exposure age for the fines 67601 from North Ray Crater is found to be 55 m.y. as compared to 235 m.y. for the fines 64501 from South Ray Crater.

Crystalline anisotropy energy and remanence coercive force

Perhaps the greatest advantage of ESR over other methods for studying the metallic Fe phases in the lunar samples is its ability to detect and to easily distinguish various demagnetization effects such as cubic crystalline anisotropy and uniaxial shape anisotropy. We find that the first order crystalline anisotropy energy $2K_1/M_s$ determined by ESR linewidth measurements is essentially equivalent to the remanence coercive force, H_{RC}, determined by other investigators using static magnetic susceptibility measurements for the lunar fines as well as for the lowgrade fragmental rocks and breccias. Since H_{RC} is an important factor determining the relaxation time or the stability of the natural remanent magnetization (NRM) detected in the returned lunar samples, our finding may prove to be of importance in further understanding the nature and origin of this NMR.

Table 2 summarizes the first-order crystalline anisotropy energy, $2K_1/M_s$, and the remanence coercive force, H_{RC}, for the Apollo 11–16 fines. The data in this

Table 2. Comparison of first-order crystalline anisotropy energy $2K_1/M_s$ with remanence coercive force H_{RC} for Apollo 11–16 fines.

Fines (< 1 mm)	Average $2K_1/M_s$ at 300°K (Oe)[a]	H_{RC} at 300°K (Oe)
Apollo 11 fines	480 ± 40	460[b]
Apollo 12 fines	460 ± 40	450[b]
Apollo 14 fines	360 ± 40	350[c]
Apollo 15 fines	430 ± 40	—
Apollo 16 fines	350 ± 40	350[d]

[a]Obtained using the relation $\Delta H = \frac{5}{3}(2K_1/M_s)$ (Tsay *et al.*, 1973a) and the average ESR linewidth ΔH taken from Tsay *et al.* (1971a, b; 1973b) (Apollo 11 fines, 800 Oe; 12 fines, 770 Oe; 16 fines, 580 Oe), Weeks (1972) (Apollo 14 fines, 606 Oe), and Griscom and Marquardt (1972) (Apollo 15 fines, 720 Oe).
[b]Taken from Nagata *et al.* (1971) for 10084, 12070.
[c]Average of Nagata *et al.* (1972) (300 Oe) and Collinson *et al.* (1972) (400 Oe) for 14259.
[d]Average of Brecher *et al.* (1973) for 62295 (300 Oe) and Nagata *et al.* (1973) for 65016 (400 Oe).

table indicates that, in addition to the equivalence between $2K_1/M_s$ and H_{RC} for the same type of fines, both $2K_1/M_s$ and H_{RC} are found to be consistently lower for the highland material (Apollo 14 and 16) than those for the fines from mare regions (Apollo 11 and 12). It appears that H_{RC} like $2K_1/M_s$ correlates with the Ni and/or Co content in the metallic Fe phases of these samples. This correlation is further demonstrated in Fig. 4 with the available chemical data on samples in which H_{RC} has been determined by others (Gose *et al.*, 1972; Nagata *et al.*, 1972). As seen in Fig. 4, the samples 10048, 14301, and 14313, all of which are low metamorphic grade fragmental rock, have a higher H_{RC} than that for the fines 10084 and 14259, respectively, from the same Apollo site. This increase in H_{RC} can be attributed to the fact that the metallic Fe phases of these fragmental rocks have either a low Ni content as in the case of 10048, or a high Co content as in the case of 14301 and 14313. Since the relaxation time or the stability of remanent magnetization increases exponentially with increasing H_{RC} as shown by Néel (1949), it is suggested that the stable component of the NRM detected in the returned lunar samples is probably carried by those single domain metallic Fe particles which have a high Co and/or low Ni content and consequently have a relatively higher H_{RC} than the average H_{RC} (300–500 Oe) observed for the bulk of the metallic Fe particles in the returned lunar samples.

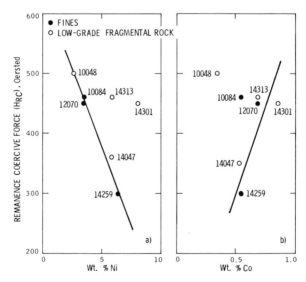

Fig. 4. Correlation between remanence coercive force H_{RC} and Ni and Co content in the metallic Fe phases of some lunar samples. (a) H_{RC} vs. wt.% Ni; (b) H_{RC} vs. wt.% Co. Data for wt.% Ni and Co were calculated assuming all the Ni and Co content were present in the metallic Fe phase. Data for total Ni and Co content were taken from Helmke *et al.* (1972) (14313); Rose *et al.* (1972) (10048, 14301, 14259); Schonfeld and Meyer (1972) (10084, 12070); Taylor *et al.* (1972) (14047). Data for H_{RC} and metallic Fe content determined by static magnetic susceptibility measurements were taken from Nagata *et al.* (1971, 1972) except 14313 from Gose *et al.* (1972).

Conclusions

We have shown that our ESR data provide a wide range of information about the metallic Fe phases in the returned lunar samples. This information includes the abundance, crystallinity, and grain size and shape of metallic Fe as well as its Ni and/or Co content. No ESR signals attributable to ferric oxide phases have been identified in the Apollo 11–16 fines. In addition to a correlation that exists between the ESR linewidth and the Ni and/or Co content in the metallic Fe phases of lunar fines, we have been able to show that the remanence coercive force, H_{RC}, obtained in static magnetic susceptibility measurements arises from first-order crystalline anisotropy energy, $2K_1/M_s$, and correlates with the Ni and/or Co content in the metallic Fe phases of the samples. Significant variations in the metallic Fe content as well as in the total Ni content have been noted for the three Apollo 16 fines we have investigated, in particular, between the fines from South Ray Crater and that from North Ray Crater.

Acknowledgments—This work was supported by NASA under Contract No. NAS 7-100 to the Jet Propulsion Laboratory, California Institute of Technology. We thank Dr. R. M. Housley of North American Rockwell Science Center for helpful discussions.

References

Bagguley D. M. S. (1955) Ferromagnetic resonance absorption in colloidal suspensions. *Proc. Roy. Soc. (London)* **228A**, 549–567.

Bickford L. R. Jr. (1950) Ferromagnetic resonance absorption in magnetite single crystals. *Phys. Rev.* **78**, 449–475.

Bozorth R. M. (1961) *Ferromagnetism*, pp. 54, 568. Van Nostrand Co.

Brecher A., Vaughan D. J., Burns R. G., Cohen D., and Morash K. R. (1973) Magnetic and Mössbauer studies of Apollo 16 rock chips (abstract). In *Lunar Science—IV*, pp. 88–90. The Lunar Science Institute, Houston.

Collinson D. W., Runcorn S. K., Stephenson A., and Manson A. J. (1972) Magnetic properties of Apollo 14 rocks and fines. *Proc. Third Lunar Sci. Conf., Geochim. Cosmochim. Acta*, Suppl. 3, Vol. 3, pp. 2342–2361. MIT Press.

Goldstein J. I. and Axon H. J. (1973) Metallic particles from 3 Apollo 16 soils (abstract). In *Lunar Science—IV*, pp. 299–301. The Lunar Science Institute, Houston.

Gose W. A., Pearce G. W., Strangway D. W., and Larson E. E. (1972) Magnetic properties of Apollo 14 breccias and their correlation with metamorphism. *Proc. Third Lunar Sci. Conf., Geochim. Cosmochim. Acta*, Suppl. 3, Vol. 3, pp. 2387–2395. MIT Press.

Grant R. W., Housley R. M., and Paton N. E. (1973) Origin and characteristics of excess Fe metal in lunar glass welded aggregates (abstract). In *Lunar Science—IV*, pp. 315–316. The Lunar Science Institute, Houston.

Griscom D. L. and Marquardt C. L. (1972) Evidence of lunar surface oxidation processes: Electron spin resonance spectra of lunar materials and simulated lunar materials. *Proc. Third. Lunar Sci. Conf., Geochim. Cosmochim. Acta*, Suppl. 3, Vol. 3, pp. 2397–2415. MIT Press.

Griscom D. L. and Marquardt C. L. (1973) The origin and significance of the "characteristic" ferromagnetic resonance of lunar soils: Two views (abstract). In *Lunar Science—IV*, pp. 320–322. The Lunar Science Institute, Houston.

Helmke P. A., Haskin L. A., Korotev R. L., and Ziege K. E. (1972) Rare earths and other trace elements in Apollo 14 samples. *Proc. Third Lunar Sci. Conf., Geochim. Cosmochim. Acta*, Suppl. 3, Vol. 2, pp. 1275–1292. MIT Press.

Hofmann U. (1967) Die magnetischen Kristallenergie Konstanten K_1, K_2, K_3 von Nickel-Eisen-Legierungen. *Z. Angew. Phys.* **22**, 106–111.

Housley R. M., Grant R. W., and Abdel-Gawad M. (1972) Study of excess Fe metal in the lunar fines by magnetic separation, Mössbauer spectroscopy, and microscopic examination. *Proc. Third Lunar Sci. Conf., Geochim. Cosmochim. Acta*, Suppl. 3, Vol. 1, pp. 1065–1076. MIT Press.

Kirsten T., Horn P., and Kiko J. (1973) Ar^{40}–Ar^{39} dating of Apollo 16 and 15 rocks and rare gas analysis of Apollo 16 fines (abstract). In *Lunar Science—IV*, pp. 438–440. The Lunar Science Institute, Houston.

Kittel C. (1948) On the theory of ferromagnetic resonance absorption. *Phys. Rev.* **73**, 155–161.

Kittel C. (1967) *Introduction to Solid State Physics*, 3rd ed., pp. 491, 525. Wiley.

Kirsten T., Horn P., and Kiko J. (1973) Ar^{40}–Ar^{39} dating of Apollo 16 and 15 rocks and rare gas analysis of Apollo 16 fines (abstract). In *Lunar Science—IV*, pp. 438–440. The Lunar Science Institute, Houston.

Kolopus J. L., Kline D., Chatelain A., and Weeks R. A. (1971) Magnetic resonance properties of lunar samples: Mostly Apollo 12. *Proc. Second Lunar Sci. Conf., Geochim. Cosmochim. Acta*, Suppl. 2, Vol. 3, pp. 2501–2514. MIT Press.

Lin C. J. and Neaves O. (1962) Ferromagnetic resonance in very fine particles ferrites. *Proc. Int. Conf. on Mag. Cryst., J. Phys. Soc. Japan*, Vol. 17, Supp. B-1, pp. 389–392.

LSPET (Lunar Sample Preliminary Examination Team) (1973) Preliminary examination of lunar samples from Apollo 16. *Science* **179**, 23–24.

Morgan J. W., Krähenbühl U., Granapathy R., and Anders E. (1972) Trace elements in Apollo 15 samples: Implications for meteorite influx and volatile depletion on the moon. *Proc. Third Lunar Sci. Conf., Geochim. Cosmochim. Acta*, Suppl. 3, Vol. 2, pp. 1361–1376. MIT Press.

Morrison G. H., Nadkarni R. A., Jaworski J., Botto R. B., and Roth J. R. (1973) Elemental abundances of Apollo 16 samples (abstract). In *Lunar Science—IV*, pp. 543–545. The Lunar Science Institute, Houston.

Nagata T., Fisher R. M., Schwerer F. C., Fuller M. D., and Dunn J. R. (1971) Magnetic properties and remanent magnetization of Apollo 12 lunar materials and Apollo 11 lunar microbreccia. *Proc. Second Lunar Sci. Conf., Geochim. Cosmochim. Acta*, Suppl. 2, Vol. 3, pp. 2461–2476. MIT Press.

Nagata T., Fisher R. M., Schwerer F. C., Fuller M. D., and Dunn J. R. (1972) Rock magnetism of Apollo 14 and 15 materials. *Proc. Third Lunar Sci. Conf., Geochim. Cosmochim. Acta*, Suppl. 3, Vol. 3, pp. 2423–2447. MIT Press.

Nagata T., Schwerer F. C., Fisher R. M., Fuller M. D., and Dunn J. R. (1973) Magnetic properties and natural remanent magnetization of Apollo 15 and 16 lunar rocks (abstract). In *Lunar Science—IV*, pp. 552–554. The Lunar Science Institute, Houston.

Néel L. (1949) Theorie des trainage magnétiques des ferromagnétiques au grains fin avec applications aux terrestres cuites. *Ann. Géophys.* **5**, 99–136.

Rose H. J. Jr., Cuttitta F., Annell C. S., Carron M. K., Christian R. P., Dwornik E. J., Greenland L. P., and Ligon D. T. Jr. (1972) Compositional data for twenty-one Fra Mauro lunar materials. *Proc. Third Lunar Sci. Conf., Geochim. Cosmochim. Acta*, Suppl. 3, Vol. 2, pp. 1215–1229. MIT Press.

Schonfeld E. and Meyer C. Jr. (1972) The abundances of components of the lunar soils by a least-squares mixing model and the formation age of KREEP. *Proc. Third Lunar Sci. Conf., Geochim. Cosmochim. Acta*, Suppl. 3, Vol. 3, pp. 1397–1420. MIT Press.

Taylor S. R., Kaye M., Muir P., Nance W., Rudowski R., and Ware N. (1972) Composition of the lunar uplands: Chemistry of Apollo 14 samples from Fra Mauro. *Proc. Third. Lunar Sci. Conf., Geochim. Cosmochim. Acta*, Suppl. 3, Vol. 3, pp. 1231–1249. MIT Press.

Tsay F. D., Chan S. I., and Manatt S. L. (1971a) Ferromagnetic resonance of lunar samples. *Geochim. Cosmochim. Acta* **35**, 865–875.

Tsay F. D., Chan S. I., and Manatt S. L. (1971b) Magnetic resonance studies of Apollo 11 and 12 samples. *Proc. Second Lunar Sci. Conf., Geochim. Cosmochim. Acta*, Suppl. 2, Vol. 3, pp. 2515–2528. MIT Press.

Tsay F. D., Manatt S. L., and Chan S. I. (1972) Electron spin resonance of manganous ions in frozen methanol solution. *J. Chem. Phys. Lett.* **17**, 223–226.

Tsay F. D., Manatt S. L., and Chan S. I. (1973a) Magnetic phases in lunar fines: Metallic Fe or ferric oxides? *Geochim. Cosmochim. Acta* **37**, 1201–1211.

Tsay F. D., Manatt S. L., Live D. H., and Chan S. I. (1973b) Electron spin resonance studies of Apollo 16 fines (abstract). In *Lunar Science—IV*, pp. 737–739. The Lunar Science Institute, Houston.

Wänke H., Baddenhausen H., Dreibus G., Quijano-Rico M., Palme H., Spettel B., and Teschke F. (1973) Multielement analysis of Apollo 16 samples and about the composition of the whole moon (abstract). In *Lunar Science—IV*, pp. 761–763. The Lunar Science Institute, Houston.

Weeks R. A. (1972) Magnetic phases in lunar material and their electron magnetic resonance spectra: Apollo 14. *Proc. Third Lunar Sci. Conf., Geochim. Cosmochim. Acta*, Suppl. 3, Vol. 3, pp. 2503–2417. MIT Press.

Weeks R. A. (1973) Ferromagnetic resonance properties of lunar fines: Apollo 16 (abstract). In *Lunar Science—IV*, pp. 772–774. The Lunar Science Institute, Houston.

Proceedings of the Fourth Lunar Science Conference
(Supplement 4, *Geochimica et Cosmochimica Acta*)
Vol. 3, pp. 2763–2781

Ferromagnetic phases of lunar fines and breccias: Electron magnetic resonance spectra of Apollo 16 samples*

R. A. WEEKS

Solid State Division, Oak Ridge National Laboratory, Oak Ridge, Tennessee

Abstract—Electron magnetic resonance measurements have been made at 9 GHz and at temperatures from 1.2 to 400°K and 35 GHz (300°K) on samples of fines and breccias from Apollo 11–16. Unsorted Apollo 16 fines (< 1 mm) have ΔH (average) = 580 G and specific intensities that have the same range as fines from the other Apollo collections. The temperature dependence of ΔH dΔH/d$T \sim 1.8$ from 1.2 to 400°K for most Apollo 11–16 samples, but for some dΔH/d$T > 1.8$ at the lowest temperatures. A magnetic transition or compensation point between 160 and 130°K is deduced from the temperature dependence of ΔH. The magnetic properties of the "characteristic" resonance are not in accord with those of iron particles. On the bases of the properties of the "characteristic" resonance as a function of temperature and Apollo site, laboratory heat treatments on synthetic materials and lunar crystalline rocks and a comparison with the "characteristic" resonance of the resonance spectra of breccia specimens for which iron particle sizes have been determined from other measurements, it is suggested that some fraction ($\sim 20\%$) of the "characteristic" resonance is due to sub-micron particles of ferric oxide phases. Possible sources for ferric oxide phases in fines are (1) oxidation of fines by cometary impacts, (2) oxides derived from ancient volcanic activity, and (3) remanents of carbonaceous chondrites which have impacted the moon.

INTRODUCTION

ELECTRON MAGNETIC resonance spectra of samples of fines from all of the Apollo collections are remarkably uniform with respect to one component which, because of its ubiquitous character, has been called "characteristic" (Weeks *et al.*, 1970; Kolopus *et al.*, 1971; Weeks, 1973; Griscom and Marquardt 1972, 1973). The maximum absorption of this component occurs at a field near which the resonance absorption of a free electron would occur; i.e., a field, H, for which the relation $g = h\nu/\beta H = 2.0023$, where h, ν, and β have their customary meanings. In most cases, $2.005 \leqslant g \leqslant 2.17$. The intensity of the absorption is almost independent of temperature, whereas the line width, ΔH (measured usually between the maximum and minimum of the first derivative of the absorption), increases with decreasing temperature (Kolopus *et al.*, 1971; Manatt *et al.*, 1970). The specific intensity ranges over less than two orders of magnitude for unsorted samples from any one collection, whereas for selected fractions and particles it may be of the same intensity as the unsorted sample or may range to below detectable limits (Tsay *et al.*, 1971; Weeks *et al.*, 1972; Weeks, 1972, 1973). The intensity does not appear to be correlated with any particular mineral mode (Weeks, 1972), although the data relating these two parameters are fragmentary. A correlation between breccia

---

*Research sponsored by the U. S. Atomic Energy Commission and supported by NASA Contract MSC-T-76458.

type and intensity has been established (Weeks, 1972) for Apollo 14 breccias with intensity decreasing with increasing consolidation and recrystallization. A correlation of intensity with wt.% metallic iron has not been found. One parameter of this resonance, the width ΔH, is a function of the Apollo site from which the samples were collected (Weeks, 1972; Griscom and Marquardt, 1972). A correlation with wt.% TiO_2 in the fines has been indicated by Griscom and Marquardt (1972) with ΔH decreasing with decreasing average wt.% TiO_2. Tsay *et al.* (1973) show that there is apparently a decrease in ΔH with increase in wt.% of Ni in FeNi particles. (These correlations will be discussed below.)

Measurements of the electron magnetic resonance spectra of several fines and breccia samples from the Apollo 16 collection (Table 1 summarizes some data on these samples), and of a sample of 12021-55 heat treated in a closed system have been made as a function of frequency (9 and 35 GHz) and temperature (1.2 to 400°K) in homodyne-type spectrometers (Weeks *et al.*, 1970). Additional measurements have been made on the temperature dependence of line width, ΔH, of the "characteristic" resonance in fines from Apollo 11, 14, and 15. Some of the data on Apollo 11, 12, 14, and 15 fines are summarized.

The temperature dependence of line width, ΔH, was measured in a Varian Associates N_2 gas flow system. Samples, weighing ~ 3 mg and sealed in quartz tubes, were placed in the stream of gas and centered in the microwave cavity. The temperature was measured by a thermocouple placed in the gas stream 1 cm above the sample position. For a 10 or 20° change in the setting of the temperature control of the gas flow system the sample temperature was at the new setting within 2 minutes. The EMR spectrum was recorded between 4 and 5 minutes after a change in temperature. The maximum temperature fluctuation during recording was ±0.2°C. The temperature measured by the thermocouple was 2° higher than the sample temperature. The temperature data have not been corrected for this systematic error.

EXPERIMENTAL RESULTS

Spectra of fines

The reason for labeling the dominant spectral component of the fines "characteristic" is evident in Fig. 1, in which are shown typical spectra of fines from Apollo 11 through Apollo 16 collections. The differences between the various curves are subtle variations in shape (e.g., the curve for 67701-29-1 has a less asymmetrical shape than the one for 10005-5 cm), variation in width ΔH measured between the minimum and maximum of the derivative curve, and variations in g, $2.005 \leq g \leq 2.17$, measured at the absorption maximum.

In addition to these variations in the "characteristic" resonance, there are other variations in the spectra observable when the amplifier gain is ten times greater. An initial rapid increase in absorption is observed in most samples between 0 and 500 gauss. The amplitude of this shoulder of the first derivative curve has not been found to have a correlation with the amplitude of the "characteristic" resonance (Kolopus *et al.*, 1971). A secondary maximum of absorption or in

Table 1. Some characteristics of fines and breccias on which EMR measurements have been made.

Specimen	Comments[a]	Station	FeO (wt.%)	TiO$_2$ (wt.%)	Ni (ppm)
60007	Medium grey; probable origin North Ray (NR) overlain by South Ray (SR); vertically heterogeneous.	ALSEP	0.42	0.39	
60051	Light grey; may represent reworked SR ejecta.	ALSEP	4.5[b]	0.44[b]	270[b]
60603	Medium grey; SR ejecta.	10[e] (LA)			
(60601)			(4.31)[f]		(400)[f]
61181	Medium grey; rim of Plume	1			340[c]
(61141)	Crater; origin?		(3.85)[f]		(400)[f]
62242	Grey; rim of Buster Crater.	2	4.04[d] (as Fe)	0.16[d] (as Ti)	397[d]
64421	Grey; SR ejecta.		4.7[e] 3.75[f]	0.32[f]	
65701	Grey; from below SR ejecta.		4.30[f]		420[g]
66043	Grey; edge of Cayley Plain.				
66081	Grey; edge of Cayley Plain; agglutinates and glass ~ 70%.		6.15[h]	0.66[h]	
66031	Grey.		4.25[d]	0.36[d]	417[d]
67701	Light grey; NR ejecta.				
61243					
66043					
66095	Dark matrix, light clast[d] B$_4$.		7.42[i]		477[i]

[a]Apollo Lunar Geology Investigation Team (1972) Documentation and environment of the Apollo 16 samples. Interagency Report: *Astrogeology* **51** (May 26, 1972).

[b]B. Mason *et al.* (1973) Abstract. In *Lunar Science—IV*, (*LS—IV*), pp. 505–507. The Lunar Science Institute, Houston.

[c]S. R. Taylor *et al.*, *LS—IV*, pp. 720–722.

[d]A. O. Brunfelt *et al.*, *LS—IV*, pp. 100–102.

[e]W. D. Ehmann *et al.*, *LS—IV*, pp. 212–214.

[f]H. Wänke *et al.*, *LS—IV*, pp. 761–763.

[g]H. C. J. Taylor and J. L. Carter, *LS—IV*, pp. 713–714.

[h]H. J. Rose *et al.*, *LS—IV*, pp. 631–633.

[i]A. R. Duncan *et al.*, *LS—IV*, pp. 190–192.

[j]H. G. Wilshire *et al.*, *LS—IV*, pp. 784–786.

some cases a shoulder on the characteristic resonance at $g = 4.3$ (9 and 35 GHz) is observed in samples listed in Table 1. The same shoulder has been observed in the spectra of fines samples from other missions (Weeks, 1972). Another feature of the microwave absorption of fines samples is absorption at zero field at a frequency of 9 GHz. A few absorption curves from Apollo 16 samples, along with curves for an Apollo 15, 14, and 11 sample and one for iron particles suspended in a diamagnetic insulating matrix, are shown in Fig. 2. The amplitude of this zero field absorption and the peak absorption are given in Table 2 measured relative to the absorption at 10 kG. The absorption curves for two samples, 60051-14-2 and

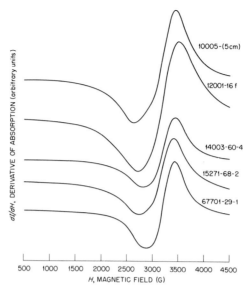

Fig. 1. Electron magnetic resonance (EMR) spectra typical of the spectra of fines from Apollo 11, 12, 14, 15, and 16 collections measured at 9 GHz and 300°K.

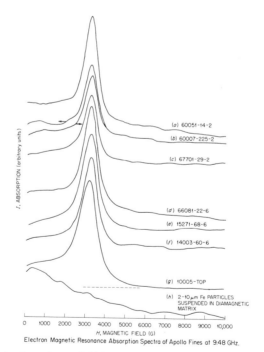

Fig. 2. Absorption intensity vs. field for representative samples from various Apollo collections measured at 9.48 GHz and 300°K. The dashed line in (g) shows the base line for the absorption of 10005-Top measured at 10 kG. The arrows in (b) indicate the direction in which the field was changing. A hysteresis at zero field is evident.

Table 2. Ratio of absorption amplitude at zero field to peak absorption of "characteristic" resonance.

Sample	$A(H=0)$*	A(max)*	$A(H=0)/A$(max)
60051-14-2 ($H=0 \rightarrow H=10\,\mathrm{kG}$)	45	170	0.265
60051-14-2 ($H=10\,\mathrm{kG} \rightarrow H=0$)	25	151	0.165
60007-225-2 ($H=0 \rightarrow H=10\,\mathrm{kG}$)	8	92	0.087
60007-225-2 ($H=10\,\mathrm{kG} \rightarrow H=0$)	23	108	0.21
67701-29-6 ($H=0 \rightarrow 10\,\mathrm{kG}$)	10	96	0.104
66081-22-6 ($H=0 \rightarrow 10\,\mathrm{kG}$)	35	160	0.218
15271-68-6 ($H=0 \rightarrow 10\,\mathrm{kG}$)	3	130	0.023
14003-60-6 ($H=0 \rightarrow 10\,\mathrm{kG}$)	8	128	0.063
10005-Top	30	154	0.195
Fe Particles ($H=0 \rightarrow 10\,\mathrm{kG}$)	70	Maximum absorption at $H=0$	

*$A=0$ set by signal at 10 kG. In some cases $dA/dH \neq 0$ at 10 kG, and hence these values must be considered as approximate, with the error undetermined.

60007-225-2, were obtained by increasing the field to 10 kG and then decreasing to 0. Curve b in Fig. 2 illustrates the hysteresis which is observed. It is evident from these two curves that zero field absorption is a function of the magnetic history of a sample, and hence the data in Table 2 must be considered in this light.

Table 3 summarizes some of the data on the "characteristic" resonance. Line width, ΔH, at 9 GHz is given, as is ΔH for some measurements at 35 GHz. The method of calculating the specific intensity, given in Table 3, reduces the influence of zero field absorption, which varies independently of the absorption due to the characteristic resonance (compare the ratio $A(0)/A$(peak) in Table 2 with the specific intensities given in Table 3).

Kolopus *et al.* (1971) measured ΔH of the "characteristic" resonance in a sample of 12001 as a function of temperature down to 10°K. Below 60°K they found that $d\Delta H/dT$ (this symbolism is used for convenience and means $[\Delta H(T_1) - \Delta H(T_2)]/[T_1 - T_2]$) was greater than for temperatures >60°K. Measurements of ΔH as a function of temperature have now been made on fines samples from Apollo 12, 14, 15, and 16 down to 4°K and down to 1.2°K for one Apollo 16 sample. These data are shown in Fig. 3. In the case of the Apollo 14 and 15 samples, $d\Delta H/dT$ (100 to 4°K) $> d\Delta H/dT$ (300 to 200°K), which is similar to that observed for the Apollo 12 sample, whereas for the two Apollo 16 samples $d\Delta H/dT$ (100 to 4°K) $\sim d\Delta H/dT$ (300 to 200°K). Another feature of the temperature dependence of ΔH is the decrease in $d\Delta H/dT$ in the temperature range 130 to 170°K. This

Table 3. Line width and specific intensity of Apollo 16 fines.

Sample (<0.1 mm)	Line Width,[a] ΔH (at 9 GHz, gauss)	Line Width,[a] ΔH (at 35 GHz, gauss)	Intensity,[b] I (mg^{-1}, arbitrary units)
60051-14-1	588	833	9
60007-225-1	575	700	7
60603-2-1	575		8
61181-9-1	560	859	9
62242-2-1	640		25
64421-30-1	560		15
65701-17-1	592		14
66043-3-1	565		10
66081-22-1	600		19
66081-22-2	600		19
66031-7-1	613	984	22
67701-29-1	540		4

[a]The line width, ΔH, is measured between the inflection points (maximum and minimum amplitude of dI/dH curve) of the absorption.

[b]Intensity is calculated from the relation $I = \Delta H^2 A/\omega$ where ΔH, A, and ω are line width, amplitude, measured in chart units, of dI/dH curve between inflection points.

decrease is most evident in the data points for the Apollo 11 sample, for which $d\Delta H/dT \rightarrow 0$ at T between 150 and 140°K and then increases at $T < 140$°K. In the case of one of the Apollo 16 samples, 66081-22-2, $d\Delta H/dT \rightarrow 0$ at ~ 150°K was <0 between 150 and 140°K and then increased again, whereas in the other Apollo 16 sample, 67701-29-1, $d\Delta H/dT \rightarrow 0$ but did not become <0 in this temperature range and then at a lower temperature returned approximately to its value between 200 and 150°K. A similar behavior of ΔH is found in the data for the Apollo 15 sample, but with $d\Delta H/dT = 0$ between 180 and 190°K. A considerable scatter in the data is evident in Fig. 3, and hence some question might be raised about the decrease in $d\Delta H/dT$ to zero or to negative values in the indicated temperature ranges. In support of the suggestion that these changes in $d\Delta H/dT$ are real effects, it is noted that a best-fit straight line drawn through the points between 200 and 300°K is neither collinear nor parallel with a line drawn through the points between 90 and 130°K. The offset between these two lines is greater than the error in the measurements. (The error in ΔH is ± 12 gauss and in T ± 2°K.) Only one sample, 66081-22-2, has been measured at 1.2°K. An increase in ΔH between 4 and 1.2°K was observed that was greater than the error in the measurement. As can be seen in Fig. 4, curves c, d, and e, there is little change in the shape of the "characteristic" resonance at 4°K. Comparison of these curves with those in Fig. 1 shows that the asymmetry in shape is greater and the peak of the absorption is at lower fields at 4°K.

One of the specimens, 66095, a dark-matrix, light-clast breccia (B₄) (Wilshire *et al.*, 1973) is a "rusty" rock containing goethite, hematite, and magnetite (El Goresy *et al.*, 1973; Taylor *et al.*, 1973). Hence it is of interest to compare the spectra of various fragments of a sample of 66095, 66095-58, with the spectra of fines and other breccias. The first derivatives of the absorption of three groups of frag-

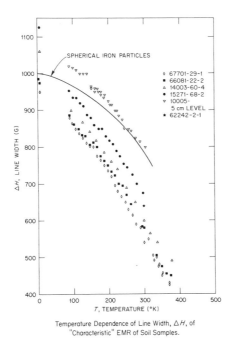

Temperature Dependence of Line Width, ΔH, of "Characteristic" EMR of Soil Samples.

Fig. 3. Temperature dependence of ΔH of the "characteristic" EMR resonance in the spectra of samples of fines. ΔH is measured between the inflection points of the absorption line, the maximum and minimum of the first derivative of the absorption peak.

ments, one with abundant red stains, one with no stains visible, and one composed of glass fragments extracted from glassy "veins" in the breccia, are shown in Fig. 5. The sample with "rust" stains has one component, labeled the A component in Weeks (1972), a shoulder at 1500 gauss, and an absorption peak, labeled the F peak, with a width $\Delta H \approx 500$ gauss and a peak at $H = 3133$ or $g = 2.055$ (Fig. 5, curve a). The spectrum of the sample without visible strains (Fig. 5, curve b) has an A component with the high-field inflection point of the absorption occurring at a lower field (~ 3800 gauss as compared to 4340 gauss), a shoulder at ~ 670 gauss, and an F component with an intensity approximately one-half the intensity in the "rusty" fragments. The black, glassy pieces have a spectrum, Fig. 5, curve c, which has an A component ($\Delta H \sim 5300$ G), shoulders at ~ 333 gauss, ~ 1333, and ~ 7500 gauss, and a component, labeled I, with a shape not heretofore observed in lunar materials. The separation of the low-field inflection point from the high-field inflection point of this component is $\Delta H = 700$ gauss at 300°K, and the peak absorption is at $g = 2.14$.

All of the components in these fragments have a temperature dependence expected of ferri- or ferromagnetic phases, as can be seen by comparing curve c with curve d and curve e with curve f in Fig. 5; i.e., no increase in intensity and increasing width with decreasing temperature.

Breccia fragments, selected from samples of coarse fines, 62242-2, 61243-3, 66043-3, have spectra similar to those shown in Fig. 4, a and b, in which the A

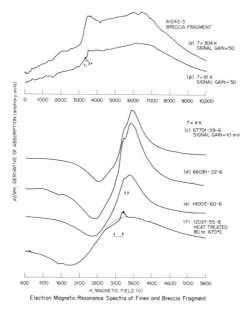

Fig. 4. The first derivative of the "characteristic" resonance is shown for various samples at 4°K and 9.29 GHz. The brackets with arrows below the curve in (c) and (d) indicate paramagnetic resonance components not observed at higher temperatures.

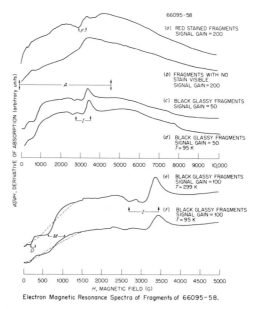

Fig. 5. First derivative of the electron magnetic resonance spectrum of various fragments of "rusty" rock 66095-58 at 9.01 GHz and 300 and 95°K. The dashed lines in (e) and (f) are estimates of that portion of the spectrum due to the "A" component indicated by a bracket with arrows in (b). Other components are indicated in a similar way.

component has a width $\Delta H \approx 6500\,G$. A resonance similar to the "characteristic" resonance is resolved at 300°K (Fig. 4, a). Those breccias with a high white-clast content, presumed to be primarily anorthite (LSPET, 1973), also had paramagnetic components due to Fe^{3+} and Ti^{3+} (Weeks, 1973). A Ti^{3+} component is indicated in Fig. 4, b by the arrows.

Heat treatment of a crystalline rock: 12021-55

A sample of 12021-55, a porphyritic basalt, was crushed and sieved, and particles with dimensions $<43\,\mu m$ were sealed in a quartz tube at a pressure of $\sim 10^{-2}\,mm\,Hg$. The number of oxygen molecules in the tube was $\sim 5 \times 10^{18}$. The sample, weighing 25 mg, contained ~ 0.05 wt.% Fe (Herzenberg et al., 1971, estimated on the basis of data for a similar rock, 12020) or $\sim 2 \times 10^{17}$ iron atoms as Fe^{0} plus 19.4 wt.% iron atoms as Fe^{2+} in various minerals (Engel et al., 1971). The initial EMR spectrum of the sample is shown in Fig. 6, a. This type of spectrum is due primarily to the particles of iron in the sample (Weeks, 1972). Even after a short time (1 hour) at 673°C, the spectrum was altered to that shown in Fig. 6,b. The spectrum is composed of two components, the A component with a width $\Delta H = 3500\,gauss$, and the second appearing as a shoulder between 2000 and 3000 G. This component continued to grow with increase in heating times until, after a total time of 90 hours, it is the only resolvable component, Fig. 6, c. The width $\Delta H = 1115\,gauss$ and $g = 2.10$. The first derivative curve, Fig. 6, e, is almost symmetrical; however, the slight asymmetry which is present is similar to that of the "characteristic" resonance. A distinct red color was apparent at the end of 90 hours.

The temperature dependence of ΔH is given in Fig. 7. The slope $d\Delta H/dT \approx 3$ between 300 and 150°K, $= 0$ between 150 and 115°K, and <3 and decreasing with decreasing temperature between 112 and 4°K. The first derivative of the absorption at 4°K is shown in Fig. 4, f. A new spectral component is detected, indicated by the arrows in Fig. 4, f. The dashed line is an estimate of the curve for the ferrimagnetic component. The new component is paramagnetic, since it is not detected at higher temperatures. A similar absorption is observed in the spectrum of plagioclase fraction of 67455-16 (Weeks, unpublished data). Hence this component is tentatively attributed to a paramagnetic state of the plagioclase fraction of 12021-55.

A second sample sealed in a quartz tube at a pressure 10^4 times less and given the same heat treatment showed only small alterations in the spectrum (Fig. 6, a), no new component was detected, and the color did not change.

DISCUSSION

Perhaps the most crucial question for which an answer is sought in the data on the "characteristic" resonance of lunar fines is whether all of the absorption intensity is due to only sub-micron particles of iron and of iron–nickel–cobalt alloys or whether some part of the resonance is due to sub-micron particles of ferric oxide

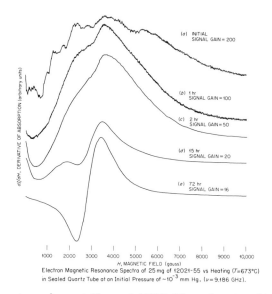

Electron Magnetic Resonance Spectra of 25 mg of 12021-55 vs Heating (T=673°C)
in Sealed Quartz Tube at an Initial Pressure of ~10⁻³ mm Hg. (ν = 9.186 GHz).

Fig. 6. EMR spectrum of a crystalline rock (12021-55) as a function of heating at 673°C for increasing periods of time at 9.186 GHz and 300°K. The 0.25 mg sample (<43 μm particle size) was sealed in a quartz tube containing $\sim 5 \times 10^{18}$ O_2 molecules.

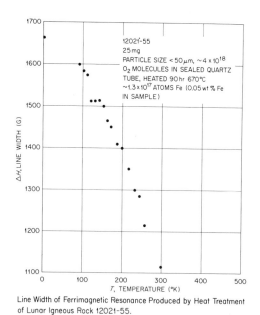

Line Width of Ferrimagnetic Resonance Produced by Heat Treatment
of Lunar Igneous Rock 12021-55.

Fig. 7. Temperature dependence of ΔH of the EMR resonance in the spectrum of 12021-55 particles after heating for 90 hours at 673°C. Measurements were made at 9 GHz.

phases. The reason for restricting the particle size in this latter case to sub-micron dimensions is that such ferric oxide particles with dimensions greater than a micron have a very small abundance in any lunar sample (Ramdohr and El Goresy, 1970; El Goresy *et al.*, 1973; Agrell *et al.*, 1972; Jedwab, 1973), with the exception of a few Apollo 16 samples (El Goresy *et al.*, 1973; Taylor *et al.*, 1973) and possibly the orange soil from the Apollo 17 collection (Strangway, 1973).

Since the weight fraction of iron in many samples of lunar fines is of the order of 0.5 wt.% (Nagata *et al.*, 1973; Goldstein and Axon, 1973), a major fraction of the microwave absorption must be due to this ferromagnetic phase. However, in order that this phase have EMR spectra consistent with the properties observed for the "characteristic" resonance at 9 GHz and 300°K, the particles must be almost perfectly spherical. Further, if the particles are almost pure Fe (i.e., the wt.% Ni is <2 wt.%), then the minimum line width measured between the inflection points of the absorption curve (the minimum and maximum points of the first derivative of the absorption) is $\Delta H = \frac{5}{3}(2K_1/M_s) \sim 1070 \, \text{G}$ (Griscom and Marquardt, 1972) where K_1 and M_s are the crystalline anisotropy constant and saturation magnetization, respectively and $(2K_1/M_s) = +640$ (Tarasov, 1939). If the particles deviate from sphericity, ΔH is larger (Weeks, 1972). With increasing amounts of Ni, K_1 and M_s decrease, with the fractional decrease in K_1 being larger than the decrease in M_s (Morrish, 1965; Tarasov, 1939; Hofmann, 1967); hence ΔH will decrease with increasing Ni. Near 30 wt.% Ni, K_1 is small (Hofmann, 1967) compared to its value for pure iron. In the case of iron and FeNi alloys, Bagguley (1953, 1955) found $800 < \Delta H < 2300$ gauss for particle sizes which he estimated to be $\sim 100 \, \text{Å}$. In Fig. 8, ΔH vs. wt.% TiO_2 and ppm Ni is plotted for several samples (Table 1). It is evident that there is no correlation of ΔH with ppm Ni. The correlation with wt.% TiO_2 is better.

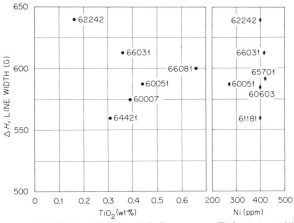

Line Width of Characteristic Resonance vs Ti Content and Ni Content of Apollo 16 Fines.

Fig. 8. The line width, ΔH, is shown as a function of wt.% TiO_2 and as a function of Ni in ppm by weight for several Apollo 16 samples.

Another parameter which is characteristic of a ferri- or ferromagnetic material is the temperature dependence of K_1 and M_s and hence of ΔH. The temperature dependence of ΔH for magnetically isolated iron particles is shown as a solid curve in Fig. 4. The critical point to be noted is that $d\Delta H/dT \approx 0$ between 10 and 1°K (Morrish, 1965). It is obvious that the data in Fig. 3 for the temperature dependencies of ΔH do not agree with those for iron or FeNi alloys in several respects. (1) For most of the samples $d\Delta H/dT$ between 300 and 150°K is approximately twice that expected for iron particles. (The Apollo 11 sample has a value of $d\Delta H/dT$ closest to that for iron.) (2) The value of $d\Delta H/dT$ does not go to zero between 150 and 130°K for iron or FeNi alloys as it does for most of the samples. (3) Between 100 and 1.2°K, $d\Delta H/dT$ decreases to zero for iron, but for most of the samples $d\Delta H/dT \sim 2$. This temperature dependence is quite unusual, since it is necessary that $d\Delta H/dT = 0$ at 0°K. Hence the decrease in $d\Delta H/dT$ for the samples must occur below 1.2°K for one of the samples (66081-22-2) and below 4°K for the others. The increase in ΔH may be due to an interaction between the magnetic phase whose resonance is observed and some other phase for which changes in magnetic properties continue at temperatures below 4°K. An exchange anisotropy (Meiklejohn, 1962) may have such an effect on the resonance line width. It is also possible that there is an interaction, as yet undiscovered in laboratory experiments on the ferromagnetism of iron, that produces the observed effects. Although the temperature dependence of ΔH might be explained by such interactions, the apparent magnetic transition at 150°K cannot be so explained.

Another difficulty in assigning all of the resonance intensity in the "characteristic" resonance to spherical iron or FeNi particles is the variation in ΔH (300°K) from site to site for unsorted fines and for particles selected from the fines, e.g., glassy spheroids. If iron were the only source of this component, then the minimum $\Delta H \approx 830$ gauss. The minimum ΔH which has been measured is 417 gauss (Table 2, glassy layer from 14318-36). Tsay *et al.* (1973) attribute this variation of $(2 K_1/M_s)$ to variable Ni content of the iron particles. The median Ni content in iron particles in all of the soil samples (Apollo 12, 14, 15, and 16) on which measurements have been made is ~ 5 wt.% (Gibb *et al.*, 1972; Wlotzka *et al.*, 1972, 1973; Goldstein and Axon, 1971; Goldstein *et al.*, 1972). All of the particles on which these measurements have been made were $>1 \mu$m, whereas the particles contributing to the resonance are $<1 \mu$m and most probably <1000 Å. The Ni content of these smaller particles is unknown. There are some data that indicate that Ni <2 wt.% in these particles (Gibb *et al.*, 1972).

The value of K_1 for Fe with 5 wt.% Ni is approximately $0.9 K_1$ for pure iron (Tarasov, 1939). Hence ΔH for spherical particles of this composition is $\sim 0.9 \Delta H$ for pure iron, assuming that the decrease in M_s is small ($<2\%$). Since the fraction of particles with Ni content >8 wt.% in some Apollo 16 and 14 soils is $<2\%$ (Goldstein and Axon, 1973), and if it is assumed that this is the case for the soil samples listed in Table 1, then it is not possible to explain the decrease in ΔH for soils from different sites by this mechanism. The lack of a correlation with Ni content is also evident in Fig. 8. If a mechanism proposed by Grant *et al.* (1973) for the formation of iron particles with dimensions ~ 100 Å is active, then most of the particles of this size will have Ni contents $< 5\,\text{wt.\%}$. Hence, if the "characteristic" resonance is only due to iron, then, as suggested above, novel new interactions will have to be sought to explain variations in ΔH.

The "characteristic" resonance has been observed in the spectrum of green glass spherules from soil specimen 15271-68 with $\Delta H = 467$ gauss (Table 4 and Weeks, 1972). The specific intensity is low compared to unsorted fines, but is of the same order as is observed in spheroids of other colors. These results, along with the several difficulties in assuming that the iron-particle hypothesis accounts for all of the resonance intensity of the "characteristic" resonance discussed above, point to other sources for at least some fraction of the intensity of the "characteristic" resonance. These sources must have (1) a positive K_1, (2) K_1 must have a strong temperature dependence, or M_s must decrease with decreasing temperature, or a combination of these two, (3) K_1 must be a function of some chemical property that is, to some degree, Apollo site dependent, and (4) the

Table 4. Intensities and line width of a ferromagnetic resonance characteristic of lunar soils and selected fragments from Apollo 11, 12, 14, and 15

Sample (comments)	Line Width, ΔH (gauss)[d]	Relative Intensity[d] $I(\text{mg}^{-1})$
A-11	950	
A-12 (average of 4 samples)	750-768-785[b]	1.5–26
A-12 (magnetic fraction[a])	791	18
A-14 (average of 5 samples)	600-613-634[b]	8–18
A-14 (magnetic fraction average of 4 samples[a])	692-710-724[b]	9–18
15271,68	670	5.0
15601,80	734	4.5
15271,68 (dark red spheroid[c])	600	0.7
15271,68 (30 green spheroids[c])	467	~0.3
14003,60 (6 spheroids[c])	568	~0.1
14003,60 (plagioclase-rich fragments[c])	510	~0.1
14003,60 (grey and yellow glasses[c])	750	3.4
14318,36 (glassy layer[c])	417	1.5

[a]The magnetic separation was made with a small alnico horseshoe magnet. The magnet was placed on the outside of a petri dish and the sample was distributed uniformly over the bottom of the dish. The particles were attracted by the magnet from a distance of $\sim 1\,\text{cm}$. Hence they were strongly ferromagnetic. Lustrous metallic particles were observed in the samples.

[b]The first value is the minimum, and the third the maximum observed, the second value is the average of all values.

[c]These samples were washed in ethanol to remove dust adhering to their surfaces. Some dust was observed on the surface of most particles after washing.

[d]The intensity of the absorption was found from the relation $I = \Delta H^2 A / W$, with $\Delta H = H_{\text{max}} - H_{\text{min}}$ and $A = $ amplitude of the dI/dH curve between the field, H_{max}, at which the maximum amplitude of the curve occurs and the field, H_{min}, at which the minimum amplitude occurs, and W is the sample weight in mg. The magnetic field modulation amplitude was constant for all these measurements. These values have all been normalized to a constant amplifier setting.

compound must have a magnetic phase change or compensation point between 160 and 130°K. These requirements are met in part by ferrite compounds such as $x\mathrm{Fe_3O_4}:(1-x)\mathrm{Fe_2TiO_4}$, for which K_1 is a function of x and for some values $K_1 = 0$ between 160 and 120°K (Syono and Ishikawa, 1963, 1964). Other ferric oxide compounds with the required properties have been produced in simulated lunar glasses (containing metal particles) by heating the glasses ($\sim 700°C$) in oxygen at pressures below atmospheric pressure (Griscom and Marquardt, 1972). Although tektites are probably not of lunar origin, their composition is similar to some lunar soils, and iron spheroids have been detected in some of them (Chao, 1963). Heating crushed samples of these under the conditions used in the heat treatment of 12021-55 has produced a ferric oxide phase whose resonance has properties similar to the "characteristic" resonance (Weeks *et al.*, 1969). The heating experiments, carried out on a sample of a lunar crystalline rock, 12021-55, illustrate a possible mechanism for the production of ferric oxide phases in lunar soil.

The changes in the spectrum of the sample heated in a low-pressure oxygen atmosphere are attributed to oxidation of the metallic iron, since the spectrum due to these particles disappeared and this spectrum in the second sample heated at a pressure 10^{-4} times less did not exhibit any significant changes. Also, oxidation and precipitation of ferric oxide phases in some of the other minerals probably occurred (Weeks *et al.*, 1969). Between 150 and 130°K, $d\Delta H/dT \approx 0$, as is the case for most of the lunar samples. Between 130 and 4°K, $d\Delta H/dT$ decreases and hence does not exhibit the increase shown by the Apollo 11, 14, and 15 samples (Fig. 3). The decrease is greater than for the two Apollo 16 samples, for which the increase in ΔH at these temperatures is less than was the case for other samples. However, this component in the spectrum of the heat-treated and partially oxidized crystalline rock does have a positive K_1, a magnetic phase between 150 and 130°K, and a strong temperature dependence over the range 300 to 4°K, and hence meets some of the requirements for the "characteristic" resonance in the spectrum of lunar fines.

Griscom and Marquardt (1972) have plotted ΔH for fines from each site vs. average $\mathrm{TiO_2}$ content of the fines from each site and have found a correlation. The average Ti content of the fines, as $\mathrm{TiO_2}$, is a function of the Apollo site from which the specimens were collected. Hence this correlation and the weak one exhibited by Apollo 16 fines (Fig. 8) give added support to the suggestion made above that a compound such as $x\mathrm{Fe_3O_4}:(1-x)\mathrm{Fe_2TiO_4}$ might be present in lunar fines. Since the ferric oxide phases in particle sizes $> 1\ \mu\mathrm{m}$ which have been detected in Apollo 11 (Ramdohr and El Goresy, 1970), Apollo 14 specimens (Agrell *et al.*, 1972), and Apollo 16 specimens (El Goresy *et al.*, 1973; Taylor *et al.*, 1973) are rare, if they are present in fines, then they must be predominantly in particles $< 1\ \mu\mathrm{m}$ in size and may exist as independent particles and as a surface layer on the sub-micron-size iron particles or other iron-rich minerals, e.g., troilite. If the predominant anisotropy is crystalline, i.e., due to K_1, as is the case with some ferrites (Morrish, 1965), rather than shape, e.g., iron, then the shape of the resonance will be determined by K_1 despite variations in particle shapes.

Since shape anisotropy is crucial in determining the ΔH of the resonance of iron particles, deviations from sphericity of particles even by small amounts (10%) increase ΔH by $\sim 20\%$ even for sub-micron-size particles. This effect is illustrated in the spectra of some breccias from the Apollo 14 collection. Gose et al. (1972) have found that the major fraction of iron particles in samples from specimens 14321, 14311, and 14303 have dimensions < 1 μm. The most intense component in the resonance spectra of other samples from these specimens is the A component (Weeks, 1972) with $\Delta H \approx 5000$ gauss, and hence the A component is due to iron particles < 1 μm. Gose et al. (1972) also have data on 14301 from which they derive particle sizes < 500 Å. The EMR spectrum of a sample of this specimen at 9 GHz does have a component similar to the "characteristic" resonance, but with $\Delta H = 900$ gauss and $g = 2.17$; however, the zero-field absorption is high. At 35 GHz the most intense component is a line with $\Delta H \sim 5000$ G, and its intensity is at least two orders of magnitude more intense than the absorption attributable to the "characteristic" resonance (Weeks, 1972). Hence iron particles < 500 Å have a $\Delta H \sim 5000$ G. These data do indicate that line widths of the spectral component due to iron particles, < 500 Å in size, may be quite large.

Although there is evidence for spheroidal-shaped particles of iron (Grant et al., 1973; Masson et al., 1971), perfect spheres are required if a minimum ΔH is to be observed in the EMR spectra, provided some unusual and unknown magnetic interaction effect which produces a decrease in ΔH is not operative. ΔH has been observed to decrease with decreasing particle sizes (Morrish and Valstyn, 1962) for a ferrite in which crystalline anisotropy dominates shape anisotropy. Whether this condition applies to iron particles with a mean diameter of 140 Å (Grant et al., 1973) is uncertain. Bagguley's results (1955) indicate that it does not apply; however, additional experiments should be made on samples in which iron particles sizes have been determined.

The spectrum of a sample of 66095, the "rusty" rock, is dominated by the A component due to particles of iron; however, a weak ferromagnetic component, the F component, is observed which is more intense in the red-stained fragments than in the fragments with no visible stain. The shape of the component may be indicative of a negative K_1, and if so, then "reduced" hematite (Weeks et al., 1970) or magnetite may be the source. The spectrum of black glassy fragments from this sample is also dominated by the A component, but other components are also present and are indicated by the brackets in Fig. 5, d. The peak of the M component occurs at a field, $H = 900$ gauss, and has a width $\Delta H \sim 400$ gauss. The shape, width, and peak of this component are in accord with those observed for relatively large particles (< 43 μm) of magnetite (Weeks et al., 1972) at the same spectrometer frequency, and hence are tentatively attributed to such particles.

The I component has not been observed before in lunar samples. Although the maximum width, $\Delta H = 700$ gauss, at 299°K and the g-value are in approximate agreement with those of the "characteristic" resonance, its shape is distinctly different. The shape is in excellent agreement with a shape calculated by Griscom and Marquardt (1973) using cubic crystalline anisotropy and a first-order approximation with $2 K_1/M_s = \frac{3}{5}\Delta H = +420$ gauss and the width of a single crystal resonance σ_{pp} (Lorentzian) ~ 100 gauss. The shape corresponds to the shape which is obtained after vacuum annealing fines (Griscom and Marquardt, 1973). Based on this agreement of shape, it is tentatively concluded that the black glassy fragments may have experienced a similar annealing history, i.e., a temperature of $\sim 700°C$ for a time >800 hours and low oxygen pressure ($< 10^{-6}$ torr). The increase in ΔH with decrease in temperature is less than for the fines, increasing from 700 gauss at 300°K to 840 gauss at 95°K, giving $d\Delta H/dT \approx 0.7$, for the Apollo 16 fines $d\Delta H/dT \approx 1.86$. The curve for iron particles given in Fig. 3 has $d\Delta H/dT \approx 0.8$ over this temperature interval in reasonable agreement with the I component. The value of $2 K_1/M_s$ deduced from ΔH is too low for iron, for which $2 K_1/M_s = +640$ gauss. Perfect metal spheres of FeNi with between 5 and 10 wt.% Ni and $< 1 \mu m$ in size might be expected to exhibit these properties, but so would a ferric oxide phase of suitable composition.

The discussion has indicated the various reasons for ascribing a fraction of the "characteristic" resonance intensity to ferri- or ferromagnetic phases other than iron or FeNi alloys. Various laboratory experiments on a lunar crystalline rock, on synthetic lunar glasses, and on thermal treatments of lunar fines have indicated some oxidation mechanisms which produce ferric oxide phases with resonance properties similar to those of the fines. In some Apollo 16 specimens, relatively large amounts (compared to previous Apollo collections) of goethite, hematite, and magnetite have been observed (El Goresy et al., 1973; Taylor et al., 1973). EMR spectral components have been tentatively assigned to some of these ferric oxides. Particles of ferric oxides in a variety of oxidation states with sizes $< 1 \mu m$ would be expected, and hence there should be contributions to resonance absorption in the region of the "characteristic" resonance.

Various lunar oxidizing processes have been suggested (Weeks, 1972; Griscom and Marquardt, 1972) by which such oxide phases could be produced. In addition to these sources of oxides, there is some evidence for a component (~ 2 wt.%) in the soils derived from carbonaceous chondrites (Morgan et al., 1972). Although impact heating may reduce some of the ferric oxides in chondrites impacting the lunar surface, a remnant would remain. The initial and subsequent impacts would reduce the particle sizes by shock effects and reduction reactions producing submicron sizes. The release of H_2O from an Apollo 16 sample (Gibson and Moore, 1973) and the detection of molecular species attributable to H_2O by the SIDE experiments (Freeman et al., 1972) indicate that fumarolic activity (Krähenbühl et al., 1973) or cometary impact may also be a source of oxidizing events. Williams and Gibson (1972) have also pointed out that these phases should be present in medium and high-grade breccias. They should also be present in soil particles derived from such breccias and in soil particles which have not been subjected to breccia-forming processes.

Acknowledgments—Discussions with D. L. Griscom have been particularly useful. His careful reading of this paper and subsequent comments are appreciated. R. Housley is thanked for his comments on the mechanisms for formation of iron particles in lunar soils and for the reference to Hofmann's paper.

REFERENCES

Agrell S. O., Scoon J. H., Long J. V. P., and Coles J. N. (1972) The occurrence of goethite in a microbreccia from the Fra Mauro formation (abstract). In *Lunar Science—III*, pp. 7–9, The Lunar Science Institute, Houston.

Bagguley D. M. S. (1953) Ferromagnetic resonance in colloidal suspensions. *Proc. Phys. Soc.* **A66**, 765–767.

Bagguley D. M. S. (1955) Ferromagnetic resonance absorption in colloidal suspensions. *Proc. Roy. Soc.* **A228**, 549–567.

Chao E. C. T. (1963) The petrographic and chemical characteristics of tektites. In *Tektites* (editor J. A. O'Keefe), pp. 51–93. Univ. of Chicago Press.

Engel A. E. J., Engel C. G., Sutton A. L., and Myers A. T. (1971) Composition of five Apollo 11 and Apollo 12 rocks and one Apollo 11 soil and some petrogenic considerations. *Proc. Second Lunar Sci. Conf., Geochim. Cosmochim. Acta*, Suppl. 2, Vol. 1, pp. 439–488. MIT Press.

Freeman J. W. Jr., Hills H. K., and Vondrak P. R. (1972) Water vapor, whence comest thou? *Proc. Third Lunar Sci. Conf., Geochim. Cosmochim. Acta*, Suppl. 3, Vol. 3, pp. 2217–2230. MIT Press.

Gibb T. C., Greatrex R., and Greenwood N. N. (1972) Mössbauer studies of Apollo 14 lunar samples. *Proc. Third Lunar Sci. Conf., Geochim. Cosmochim. Acta*, Suppl. 3, Vol. 3, pp. 2479–2493. MIT Press.

Gibson E. K. and Moore G. W. (1973) Volatile-rich lunar soil; evidence of possible cometary impact. *Science* **179**, 69–71.

Goldstein J. I. and Yakowitz H. (1971) Metallic inclusions and metal particles in the Apollo 12 lunar soil. *Proc. Second Lunar Sci. Conf., Geochim. Cosmochim. Acta*, Suppl. 2, Vol. 1, pp. 177–191. MIT Press.

Goldstein J. I. and Axon H. J. (1973) Metallic particles from 3 Apollo 16 soils (abstract). In *Lunar Science—IV*, pp. 299–301. The Lunar Science Institute, Houston.

Goldstein J. I., Axon H. J., and Yen C. F. (1972) Metallic particles in the Apollo 14 soil. *Proc. Third Lunar Sci. Conf., Geochim. Cosmochim. Acta*, Suppl. 3, Vol. 1, pp. 1037–1064. MIT Press.

Goresy A. El., Ramdohr P., and Medenbach O. (1973) Lunar samples from the Descartes site: Mineralogy and geochemistry of the opaques (abstract). In *Lunar Science—IV*, pp. 222–224. The Lunar Science Institute, Houston.

Gose W. A., Pearce H. E., Strangway D. W., and Larson E. E. (1972) Magnetic properties of Apollo 14 breccias and their correlation with metamorphism. *Proc. Third Lunar Sci. Conf., Geochim. Cosmochim. Acta*, Suppl. 3, Vol. 3, pp. 2387–2395. MIT Press.

Gose W. A., Strangway D. W., Pearce G. W., and Carnes J. G. (1973) The time dependent magnetization of lunar breccias of low metamorphic grade (abstract). In *Lunar Science—IV*, pp. 309–311. The Lunar Science Institute, Houston.

Grant R. W., Housley R. M., and Paton N. E. (1973) Origin and characteristics of excess Fe metal in lunar glass welded aggregates (abstract). In *Lunar Science—IV*, pp. 315–317. The Lunar Science Institute, Houston.

Griscom D. L. and Marquardt C. L. (1972) A means of studying the thermal and weathering histories of the lunar soil: Ferromagnetic resonance spectra of returned samples. *Semiannual Tech. Prog. Rep. for Period Ending Sept. 30, 1972*, Solid State Div., N.R.L.

Griscom D. L. and Marquardt C. L. (1972) Evidence of lunar surface oxidation processes: Electron spin resonance spectra of lunar materials and simulated lunar materials. *Proc. Third Lunar Sci. Conf., Geochim. Cosmochim. Acta*, Suppl. 3, Vol. 3, pp. 2397–2415. MIT Press.

Griscom D. L. and Marquardt C. L. (1973) The origin and significance of the "characteristic" ferromagnetic resonance of lunar soils: Two views (abstract). In *Lunar Science—IV*, pp. 320–322. The Lunar Science Institute, Houston.

Herzenberg C. L., Moler R. B., and Riley D. L. (1971) Mössbauer instrumental analysis of Apollo 12 lunar rock and soil samples. *Proc. Second Lunar Sci. Conf., Geochim. Cosmochim. Acta*, Suppl. 2, Vol. 3, pp. 2103–2123. MIT Press. (The estimate is based on a measurement on a similar rock, 12020.)

Hofman U. (1967) Die magnetischen Kristallenergie Konstanten K_1, K_2, K_3, von Nickel–Eisen–Legierungen. *Z. Angew Phy.* **22**, 106–111.

Jedwab J. (1973) Some rare minerals in lunar soils (abstract). In *Lunar Science—IV*, pp. 412–414. The Lunar Science Institute, Houston.

Jedwab J., Herbosch A., Wollast R., Naessers G., and Van Gaen-Peers N. (1970) Search for magnetite in lunar rocks and fines. *Science* 167, 618–619.

Kolopus J. L., Kline D., Chatelain A., and Weeks R. A. (1971) Magnetic resonance properties of lunar samples: Mostly Apollo 12. *Proc. Second Lunar Sci. Conf., Geochim. Cosmochim. Acta*, Suppl. 2, Vol. 3, pp. 2501–2514. MIT Press.

Krähenbühl U., Ganapathy R., Morgan J. W., Kimura K., and Anders E. (1973) Volatile and siderophile metals on the moon: Reappraisal in the light of Apollo 16 and Luna 20 data (abstract). In *Lunar Science—IV*, pp. 446–448. The Lunar Science Institute, Houston.

Manatt S. L., Eleman D. D., Vaughan R. W., Chan S. I., Tsay F. D., and Huntress W. T. Jr. (1970) Magnetic resonance studies of lunar samples. *Proc. First Lunar Sci. Conf., Geochim. Cosmochim. Acta*, Suppl. 1, Vol. 3, pp. 2321–2323. Pergamon.

Masson C. R., Götz J., Jamieson W. D., McLachlan J. L., and Volborth A. (1971) Chromatographic and mineralogical study of lunar fines and glass. *Proc. Second Lunar Sci. Conf., Geochim. Cosmochim Acta*, Suppl. 2, Vol. 1, pp. 957–971. MIT Press.

Meiklejohn W. H. (1962) Exchange anisotropy—a review. *J. Appl. Phys.* 33, 1328.

Morgan J. W., Krähenbühl U., Ganapathy R., Anders E., and Marvin U. B. (1972) Trace elements in Apollo 15 material and the ancient meteoritic component (abstract). In *Lunar Science—IV*, pp. 537–539. The Lunar Science Institute, Houston.

Morrish A. H. (1965) *The Physical Principles of Magnetism*, Chapter 7, pp. 332–431. Wiley.

Morrish A. H. and Valstyn E. P. (1962) Ferrimagnetic resonance of iron-oxide micropowders. *Proc. Int. Conf. Mag. Cryst., J. Phys. Soc. Japan* 17, Suppl. B-1, pp. 392–397.

Nagata T., Schwerer F. C., Fisher R. M., Fuller M. D., and Dunn J. R. (1973) Magnetic properties and natural remanent magnetization of Apollo 15 and 16 lunar rocks (abstract). In *Lunar Science—IV*, pp. 552–554. The Lunar Science Institute, Houston.

Ramdohr P. and El Goresy A. (1970) Opaque minerals of the lunar rocks and dust from Mare Tranquillitatis. *Science* 167, 615–618.

Runcorn S. K., Collinson D. W., O'Reilly W., Stephenson A., Battey M. H., Manson M. H., and Readman P. W. (1971) Magnetic properties of Apollo 12 lunar samples. *Proc. Roy. Soc. London Ser. A* 325, 157–174.

Strangway D. W. (1973) Report at Fourth Lunar Science Conference, March 8–11, 1973, Houston.

Syono Y. and Ishikawa Y. (1963) Magnetocrystalline anisotropy of xFe$_2$TiO$_4$:$(1-x)$Fe$_3$O$_4$. *J. Phys. Soc. Japan* 18, 1230–1231.

Syono Y. and Ishikawa Y. (1964) Magnetocrystalline anisotropy and magnetostriction of xFe$_2$TiO$_4$:$(1-x)$Fe$_3$O$_4$ $(x>0.5)$. *J. Phys. Soc. Japan* 19, 1752–1753.

Tarasov L. P. (1939) Ferromagnetic anisotropy of low-nickel alloys of iron. *Phys. Rev.* 56, 1245–1246.

Taylor L. A., Mao H. K., and Bell P. M. (1973) Apollo 16 "rusty" rock 66095 (abstract). In *Lunar Science—IV*, pp. 715–717. The Lunar Science Institute, Houston.

Tsay F-D, Chan S. I., and Manatt S. L. (1971) Ferromagnetic resonance of lunar samples. *Geochim. Cosmochim. Acta* 35, 865–875.

Tsay F.-D., Manatt S. L., Live D. H., Chan S. I., and Noyes A. A. (1973) Electron spin resonance studies of Apollo 16 fines (abstract). In *Lunar Science—IV*, pp. 737–739. The Lunar Science Institute, Houston.

Warner J. L. (1972) Metamorphism of Apollo 14 breccias. *Proc. Third Lunar Sci. Conf., Geochim. Cosmochim. Acta*, Suppl. 3, Vol. 3, pp. 623–643. MIT Press.

Weeks R. A. (1972a) Ferromagnetic and paramagnetic resonance of magnetic phases and Fe^{3+} in Apollo 15 samples. In *The Apollo 15 Lunar Samples*, pp. 182–186. The Lunar Science Institute, Houston.

Weeks R. A. (1972b) Magnetic phases in lunar material and their electron magnetic resonance spectra: Apollo 14. *Proc. Third Lunar Sci. Conf., Geochim. Cosmochim. Acta*, Suppl. 3, Vol. 3, pp. 2503–2517. MIT Press.

Weeks, R. A. (1973) Ferromagnetic resonance properties of lunar fines: Apollo 16 (abstract). In *Lunar Science—IV*, pp. 772–774. The Lunar Science Institute, Houston.

Weeks R. A. (1973) Paramagnetic states of Apollo 16 plagioclases (abstract). In *Lunar Science—IV*, pp. 775–777. The Lunar Science Institute, Houston.

Weeks R. A. and Kolopus J. L. (1969) Paramagnetic resonance spectra of some silicate materials. ORNL-CF-69-3-5 (*Prog. Report for Period Ending December 31, 1968*, submitted to NASA, Lunar Sample Analysis Program, March 3, 1969.

Weeks R. A., Kolopus J. L., and Arafa S. (1972) Ferromagnetic and paramagnetic resonance spectra of lunar material: Apollo 12. *The Moon* **4**, 271–295.

Weeks R. A., Kolopus J. L., Kline D., and Chatelain A. (1970) Apollo 11 lunar material: Nuclear magnetic resonance of ^{27}Al and electron resonance of Fe and Mn. *Proc. First Lunar Sci. Conf., Geochim. Cosmochim. Acta*, Suppl. 1, Vol. 3, pp. 2467–2490. Pergamon.

Williams R. J. and Gibson E. K. (1972) The origin and stability of lunar goethite, hematite, and magnetite. *Earth and Plant. Sci. Let.* **17**, 84–88.

Wilshire H. G., Stuart-Alexander D. E., and Jackson E. D. (1973) Petrology and classification of the Apollo 16 samples (abstract). In *Lunar Science—IV*, pp. 784–786. The Lunar Science Institute, Houston.

Wlotzka F., Spettel B., and Wänke H. (1973) On the trace element content of metal particles from fines 60601 (abstract). In *Lunar Science—IV*, pp. 787–789. The Lunar Science Institute, Houston.

Wlotzka F., Jagoutz E., Spettel B., Baddenhausen H., Balacescu A., and Wänke H. (1972) On lunar metallic particles and their contribution to the trace element content of Apollo 14 and 15 soils. *Proc. Third Lunar Sci. Conf., Geochim. Cosmochim. Acta*, Suppl. 3, Vol. 1, pp. 1077–1084. MIT Press.

Proceedings of the Fourth Lunar Science Conference
(Supplement 4, *Geochimica et Cosmochimica Acta*)
Vol. 3, pp. 2783–2791

Results of the Apollo 15 and 16 X-ray experiment

I. Adler, J. I. Trombka, R. Schmadebeck, P. Lowman,
H. Blodget, L. Yin, and E. Eller

NASA Goddard Space Flight Center

M. Podwysocki, J. R. Weidner, A. L. Bickel, and R. K. L. Lum

University of Maryland

J. Gerard

Eastman Kodak Research Center

P. Gorenstein, P. Bjorkholm, and B. Harris

American Science and Engineering

Abstract—Except for some minor modifications the Apollo 16 X-ray fluorescence experiment was similar to that flown aboard Apollo 15 (Adler *et al.*, 1972). The Apollo 16 provided data for a number of features not previously covered such as Mare Cognitum, Mare Nubium, Ptolemaeus, Descartes, Mendeleev as well as other areas. Many data points were obtained by the X-ray experiments so that comparisons could be drawn between Apollo 15 and 16 flights. The agreement was generally within about 10 percent. Al/Si concentration ratios ranged from 0.38 percent in Mare Cognitum to 0.67 percent in the Descartes area highlands. A comparison of the Apollo 16 data Al/Si values with optical albedo values along the ground tracks showed the same positive correlation as in the Apollo 15 flight. A reexamination of the detector and collimator geometries showed that the spatial resolution was better by almost a factor of two than the initial estimates. The data are presently being examined using smaller integration intervals in an effort to prepare concentration contours and enhanced spatial resolution.

INTRODUCTION

THE APOLLO 16 Command and Service Module carried an integrated geochemical package, which except for minor modifications was identical to that flown aboard Apollo 15. This package included the X-ray, gamma-ray, and alpha particle spectrometers. These experiments were flown to extend (our) observations to larger areas of the moon and to allow us to extrapolate from the data obtained on the surface to the rest of the moon. The Apollo 16 mission provided data for a number of features not previously covered, for example: Mare Cognitum, Mare Nubium, Ptolemaeus, the Descartes area, and Mendeleev as well as other areas. Many data points were obtained by the X-ray experiment so that comparisons could be drawn between the orbital measurements and the analyzed returned materials.

Unlike the high inclination orbit of Apollo 15, the Apollo 16 flight path was nearly equatorial (9 deg. inclination) so that the projected areas covered were somewhat smaller than during the Apollo 15 flight.

Table 1. Overlap between the Apollo 15 and 16 ground tracks.

Feature (a)	Apollo 16 concentration ratio		Apollo 15 concentration ratio	
	Al/Si ± 1σ	Mg/Si ± 1σ	Al/Si ± 1σ	Mg/Si ± 1σ
Mare Fecunditatis	0.41 ± 0.05	0.26 ± 0.05	0.36 ± 0.06	0.25 ± 0.03
Mare Smythii	0.45 ± 0.08	0.25 ± 0.05	0.45 ± 0.06	0.27 ± 0.06
Langrenus area	0.48 ± 0.07	0.27 ± 0.06	0.48 ± 0.11	0.24 ± 0.06
Highlands west of Smythii	0.57 ± 0.07	0.21 ± 0.03	0.55 ± 0.06	0.22 ± 0.03
Western border of Smythii	0.58 ± 0.08	0.22 ± 0.04	0.52 ± 0.06	0.22 ± 0.06
Eastern border of Smythii	0.61 ± 0.09	0.20 ± 0.06	0.60 ± 0.10	0.21 ± 0.03

(a) The overlap between corresponding areas of the Apollo 16 and 15 ground tracks is not exact, so that differences for the same area may be real.

There was some overlap of orbital coverage between the two missions so that the reproducibility of our measurements could be determined by a comparison of the results. The region of overlap between the Apollo 15 and Apollo 16 tracks was mainly between 50 and 100°E longitude and covered such areas as Mare Fecunditatis, Mare Smythii, Langrenus, and the highlands west of Smythii. For these areas the Al/Si and Mg/Si concentration ratios agreed to better than 10 percent (*see* Table 1). This agreement made it possible to draw close comparisons between

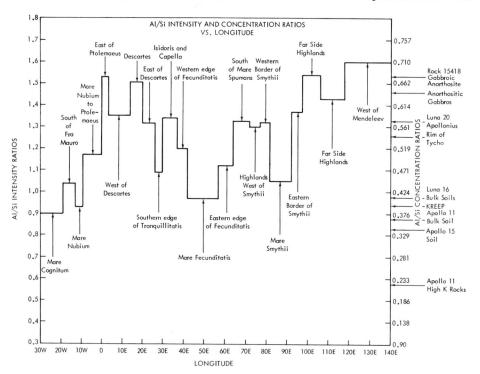

Fig. 1. Al/Si intensity and concentration ratios vs. longitude for Apollo 16 ground track.

the two flights. It also demonstrated that the sun's X-ray spectral distribution, which produces the lunar fluorescent X-rays was about the same on both missions.

RESULTS

The data obtained during the Apollo 16 flight were reduced to chemical concentration rates using the procedures described in reducing the Apollo 15 data (Adler *et al.*, 1972). Figure 1 shows a plot of Al/Si intensity ratios for the Apollo 16 ground track, from Mendeleev on the east to Mare Cognitum on the west. The values for various analyzed lunar materials are shown along the left hand axis. As in Apollo 15 we see the Al/Si values range from a high in the eastern highlands, corresponding to plagioclase rich gabbros to low values in the western Maria (Mare basalts). A summary of the results is shown in Table 2. Included for comparison are some selected lunar samples (*see* Table 3).

One of the results to emerge from the Apollo 15 mission was the excellent association between the Al/Si values and the optical albedo values, both following each other well. This observation was particularly significant in view of the long-standing discussion as to whether the albedo differences were solely representative of topographic differences or were also a reflection of compositional variations among surface materials. Early investigators recognized convincing evidence for compositional changes where sharp albedo changes occur. Chemical differences related to albedo were first confirmed by the alpha scattering experiments of Surveyor V, VI, and VII. The Surveyor results and analysis of returned lunar samples confirmed that albedo is indeed affected by compositional as well as topographic differences. The X-ray fluorescence experiment on Apollo 15 and 16 has now provided the means to correlate regional albedo with surface composition (for selected elements).

Data locations for selected Apollo 15 and 16 orbits covered by the X-ray fluorescence experiment were plotted on the map of the normal albedo of the moon by Pohn, Wildey, and Sutton (1970). Average albedo for each 3° area for which the X-ray data were available was plotted against the Al/Si intensity ratios. The positive correlation between the albedo and the Al/Si values is strong. In the Apollo 15 plots, the main anomalies were observed where an occasional small Copernican crater occurred, which produced an abnormally high albedo value. The brightness of these craters is generally considered to be due to highly reflective, finely divided ejecta rather than compositional changes. A similar anomaly is noted in the Apollo 16 data at approximately 27°E in a Tranquillitatis embayment north of Theophilus (Fig. 2). Four Apollo 16 revolutions were plotted and revolutions 58 and 60 show the expected decrease in Al/Si values with decreasing albedo. Revolutions 55 and 59, on the other hand, show an occasional increase in Al/Si values although the albedo decreases. This may record the existence of old "weathered" rays consisting of aluminum-rich highland derived material that has lost its high reflectivity.

The results of the Apollo 16 experiment generally support the main conclu-

Table 2. Concentration ratios of Al/Si and Mg/Si for various features.

Feature	N (a)	Concentration ratio	
		Al/Si ± 1σ	Mg/Si ± 1σ
Mare Cognitum	8	0.38 ± 0.11	0.40 ± 0.29
Upper part of Mare Nubium (9 to 13°W)	8	0.39 ± 0.12	0.20 ± 0.05
Mare Fecunditatis (42 to 57°E)	80	0.41 ± 0.05	0.26 ± 0.05
South of Fra Mauro (13 to 19°W)	9	0.45 ± 0.07	0.26 ± 0.04
Mare Smythii (82 to 92.5°E)	24	0.45 ± 0.08	0.25 ± 0.05
Southern edge of Mare Tranquillitatis, Torricelli area (26 to 30°E)	21	0.47 ± 0.09	0.23 ± 0.05
Eastern edge of Fecunditatis Langrenus area (57 to 64°E)	44	0.48 ± 0.07	0.27 ± 0.06
Ptolemaeus (4°W to 0.5°E)	17	0.51 ± 0.07	0.21 ± 0.04
Highlands west of Ptolemaeus to Mare Nubium (4 to 9°W)	16	0.51 ± 0.11	0.25 ± 0.12
Highlands west of Mare Fecunditatis (37.5 to 42°E)	29	0.52 ± 0.07	0.24 ± 0.05
Highlands west of Smythii (72 to 77°E)	35	0.57 ± 0.07	0.21 ± 0.03
Western border of Smythii (77 to 82°E)	33	0.58 ± 0.08	0.22 ± 0.04
Highlands east of Descartes (20.5 to 26°E)	23	0.58 ± 0.07	0.21 ± 0.04
South of Mare Spumans (64 to 72°E)	45	0.58 ± 0.07	0.25 ± 0.04
Isidorus and Capella (30 to 37.5°E)	38	0.59 ± 0.11	0.21 ± 0.05
Highlands west of Descartes (3 to 14°E)	44	0.59 ± 0.11	0.21 ± 0.05
Eastern border of Mare Smythii (92.5 to 97.5°E)	17	0.61 ± 0.09	0.20 ± 0.06
Far-side highlands (106 to 118°E)	29	0.63 ± 0.08	0.16 ± 0.05
Descartes area, highlands, Apollo 16 landing site (14 to 20.5°E)	30	0.67 ± 0.11	0.19 ± 0.05
East of Ptolemaeus (0.5 to 3°E)	12	0.68 ± 0.14	0.28 ± 0.09
Highlands (97.5 to 106°E)	31	0.68 ± 0.11	0.21 ± 0.05
Far-side highlands west of Mendeleev (118 to 141°E)	30	0.71 ± 0.11	0.16 ± 0.04

(a) N is the number of individual data points used to determine the average Al/Si and Mg/Si values ± 1 standard deviation and was obtained from the various passes over each feature.

sions of the Apollo 15 mission. The good correlation between the aluminum and magnesium contents inferred from the X-ray data and the returned samples implies that the X-ray measurements are a reliable guide, at least to this aspect of the surface composition of the moon. The strong correlation between optical albedo and Al/Si ratios suggests that if allowances are made for features, the brightness of which is caused primarily by physiographic youth (such as Copernican craters), the albedo is a reasonable guide to highland composition, particularly for its plagioclase content. Together these conclusions imply that the plagioclase-rich

Table 3. Concentration ratios of Al/Si and Mg/Si selected lunar samples.

Selected Lunar Samples	Concentration ratio	
	Al/Si	Mg/Si
Apollo 12, Oceanus Procellarum type AB rocks, average	0.22	0.22
Apollo 11, Mare Tranquillitatis high potassium rocks, average	0.23	0.24
Apollo 12, Oceanus Procellarum type A rocks, average	0.24	0.31
Apollo 15, Hadley–Apennine soil	0.34	0.30
Luna 16, Mare Fecunditatis rocks	0.35	0.21
Apollo 14, Fra Mauro rocks, average	0.38	0.26
Apollo 11 and 12, potassium, rare-earth elements, and phosphorus (KREEP), average	0.39	0.21
Apollo 14, Fra Mauro soil	0.41	0.26
Apollo 12, norite material, average	0.42	0.20
Luna 16, Mare Fecunditatis bulk soil	0.42	0.27
Surveyor VII, rim of Tycho, regolith	0.55	0.20
Luna 20, Apollonius Highlands	0.58	0.26
Apollo 11 and 12, anorthositic gabbro	0.64	0.21
Apollo 15, rock 15418, gabbroic anorthosite	0.67	0.15
Apollo 11 and 12, gabbroic anorthosite	0.82	0.074
Apollo 11 and 12, anorthosite	0.89	0.038
Apollo 15, rock 15415, anorthosite, Genesis Rock	0.91	0.003

highland and crust is global in extent. The data from the highland areas covered by the Apollo 16 X-ray spectrometer have maximum Al/Si ratios that lie between anorthositic gabbro and gabbroic anorthosites returned from previous Apollo missions.

SPATIAL RESOLUTION

In the results presented above no attempt has been made to achieve the minimum resolution element. The results shown are based on integral spectra for periods of anywhere from 24 to 64 seconds. In order to achieve the best spatial resolution one would need to work with the 8 second spectra despite the increased statistical uncertainties. Another factor to be considered is the resolution of the X-ray spectrometer. The latter parameter was recently examined more carefully in an effort to obtain a more realistic value of its angular response. The need for this was made obvious as a result of the X-ray astronomy measurements performed during the trans-earth coast. It was found that the variation of X-ray intensities with sighting angle proved to be greater than one would predict from a simple slat collimator. The effect was found to be explainable by the interaction of the collimator with the template covering the proportional counter windows, producing a modulation of the spectrometer transmission and a resulting reduction in the solid angle of the instrument for lunar fluorescence observations. It can be

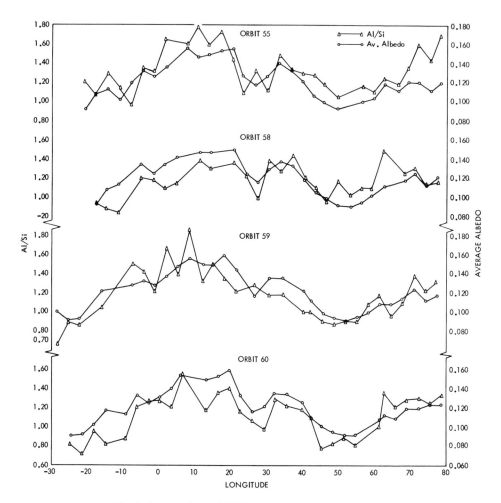

Fig. 2. A comparison of Al/Si variation with average albedo.

shown that about 68 percent of the total radiation would come from an area approximately 37 nautical miles on edge at orbital altitudes. While these figures are approximate, the implication is that the instrumental resolution was nearly a factor of two better than the original estimates.

DETAILED MAPPING

Having produced relatively coarse maps of the chemical distributions across the lunar surface it is now our objective to prepare more detailed maps relating the X-ray results to small lunar morphological features, and to various parameters independently determined such as photo-geologic observations, gravitational anomalies and the electromagnetic sounder results to mention a few.

Various methods of contouring are being considered. One preliminary approach has been to perform trend surface studies of the lunar data. Topological surfaces described by polynomial expressions are fitted to the observed data by a least squares technique. The significance of the surfaces is tested by statistical means. Those surfaces which are shown to be statistically valid are considered to be an adequate regional model. Deviations between the observed values and the model values are plotted as residuals which can then be interpreted as local anomalies with respect to the model. Figure 3 is the 2nd order trend surfaces for Al/Si in Tranquillity. As we had previously established the trend of the surfaces is toward high values of Al/Si approaching the highlands and low values in the mare areas. Figure 4 shows selected areas in which negative residuals predominated. This is in contrast to other areas where the residuals were of a random character. The negative residuals shown here represent several flight paths giving us more confidence that the variations were not instrumental. The residuals plot indicates distinctive areas in Tranquillity of lower Al/Si values. This is particularly true relative to the areas of random residuals. One can make a similar statement about

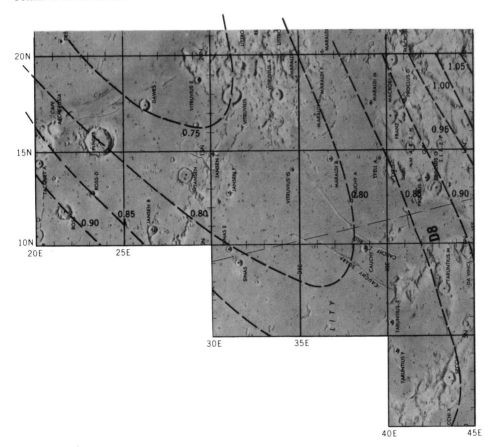

Fig. 3. Mare Tranquillitatis Al/Si—2nd degree surface.

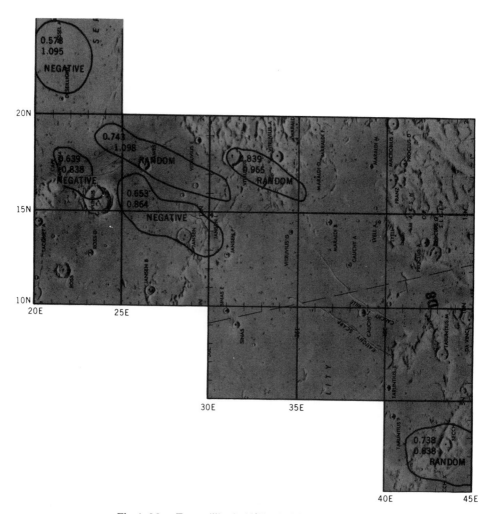

Fig. 4. Mare Tranquillitatis Al/Si—2nd degree residuals.

the Mg/Si although with less certainty. We emphasize at this point that these observations are based on 8 second data. It is obvious that to attempt to use this 8 second data on a point by point basis requires discretion.

The above results indicate that one can now begin to prepare more detailed compositional maps from the X-ray data. It is important to attempt to reconcile these data with observations. Such a program is now underway.

REFERENCES

Adler I., Gerard J., Trombka J., Schmadebeck R., Lowman P., Blodget H., Yin L., Eller E., Lamothe R., Gorenstein P., Bjorkholm P., Harris B., and Gursky H. (1972) The Apollo 15 X-ray fluorescence experiment. *Proc. Third Lunar Sci. Conf., Geochim. Cosmochim. Acta,* Suppl. 3, Vol. 3, pp. 2157–2178. MIT Press.

Pohn H. A. and Wildey R. L. (1970) A photoelectric photographic study of the normal albedo of the moon, accompanied by an albedo map of the moon by H. A. Pohn, R. L. Wildey, and G. E. Sutton (Contributions to Astro-geology) U.S. Geological Survey Professional Paper, 599-E.

Proceedings of the Fourth Lunar Science Conference
(Supplement 4, *Geochimica et Cosmochimica Acta*)
Vol. 3, pp. 2793–2802

Distribution of ^{222}Rn and ^{210}Po on the lunar surface as observed by the alpha particle spectrometer

PAUL J. BJORKHOLM, LEON GOLUB, and PAUL GORENSTEIN

American Science and Engineering, 955 Massachusetts Avenue, Cambridge, Mass. 02139

Abstract—The distribution of ^{222}Rn and her daughter products has been observed from orbit during the Apollo 15 and Apollo 16 missions. Decays of ^{222}Rn were observed locally over the crater Aristarchus and more generally over Oceanus Procellarum and Mare Imbrium. Decays of ^{210}Po (a delayed daughter of ^{222}Rn) were observed over most of the eastern hemisphere on both missions. All observations indicate a lack of radioactive equilibrium between ^{222}Rn and ^{210}Po implying a time dependent process for the production of radon at the lunar surface. In addition, the ^{210}Po shows a remarkable correlation with the Mare edges.

INTRODUCTION

RADON-222 is a radioactive noble gas which is produced during the decay of uranium. Because it is a noble gas, under certain conditions it may diffuse through the lunar regolith and reach the surface before decaying by alpha particle emission. The alpha particle spectrometers aboard the Apollo 15 and 16 spacecrafts were designed to detect and identify the alpha particles emitted by ^{222}Rn and her daughter products during decay. The theory and experimental techniques have been described previously by Gorenstein and Bjorkholm (1972). The observed signal strengths relate to the concentration of the parent uranium and the depth from which radon may reach the surface prior to decay.

In order to fully appreciate the observations one must have at least some understanding of the processes involved in transporting radon to the lunar surface. Therefore included below is a brief summary of these possible processes. The ultimate source of the observed radon is the natural radioactive decay chain of uranium. The radon is produced at the site of its parents, i.e., within mineral grains. In order for the radon to reach the surface two distinct processes must occur. First the radon must be released to the voids between the mineral grains. This process is commonly referred to as emanation. Emanation can occur by diffusion within the solid grains or by recoil out of the grain from the alpha decay of its parent. Neither process is expected to be efficient on the lunar surface. In particular, to have the recoil process be efficient one requires a stopping medium between the mineral grains. On earth the atmosphere provides some stopping power. The lack of a lunar atmosphere must significantly reduce the efficiency of this process. One can postulate other mechanisms to allow some residual emanation even without any intergrain stopping power available but these effects will be very weak. Once a neutral radon atom is in the lunar regolith it must migrate to the surface to be observable by the spectrometer. This can occur by a random walk through the regolith or by transport within some other medium passing through the regolith.

Any spatial and/or temporal variation in the observed signal at the lunar surface is a reflection of the variations in these processes of emanation and migration. Therefore any mechanisms postulated to describe these processes must fit the observed boundary conditions, i.e., the data to be presented below.

OBSERVATIONS

The observed data can be grouped in several different ways. They will be discussed here by considering first the general trends and then the specific correlation with lunar features. Observations of ^{210}Po and ^{222}Rn will be discussed simultaneously. The important point to bear in mind is that the decay of ^{222}Rn represents activity that was occurring at the time of observation. Decay of ^{210}Po represents activity that is caused by ^{222}Rn reaching the lunar surface prior to the observation (from about 10 days to 60 years). Therefore the observed ^{222}Rn to ^{210}Po ratio is an indication of the secular history of the radon emanation.

GLOBAL TRENDS

Figure 1 shows the observed count rate for ^{222}Rn (plus her two prompt daughters) as a function of longitude for the Apollo 15 ground track. The data are displayed in four panels, each containing 90° of lunar longitude, the top panel starting at 0° and proceeding to the East (towards Mare Crisium). There is a background level, and the middle two panels (the lunar farside) are consistent with this background level (the dashed line represents the whole moon average). The average count rate in each quadrant is indicated on the figure. While the ability to observe detailed structure in these data is limited by the counting statistics, there is an increase in the amount of ^{222}Rn observed over the western quadrant (Oceanus Procellarum and Mare Imbium) and to a lesser extent over the eastern quadrant (Maria Serenitatis, Tranquillitatis, Fecunditatis, and Crisium). The signal bin labeled "Arist." (for Aristarchus) in Fig. 1 has been excluded from the average and will be discussed below. The general trend of the count rate is in agreement with the observed uranium and thorium concentrations over the same region (Metzger *et al.*, 1972).

The ^{210}Po distributions show a considerably larger spatial variation and are shown in Figs. 2 (Apollo 15) and 3 (Apollo 16). The Apollo 15 ^{210}Po distribution (including a cosmic ray background) shows an increase in signal between 40°E and 180°E. The average count rate between 40°E and 180°E is 0.072 ± 0.002 counts/sec and 0.062 ± 0.001 counts/sec elsewhere. The excess count rate between 40°E and 180°E corresponds to a ^{210}Po decay rate of $(4.6 \pm 1.4) \times 10^{-3}$ disintegration/cm^2-sec (dis/cm^2-sec).

The Apollo 16 ^{210}Po distribution (Fig. 3) shows an even more striking spatial variation. In general there is an increase starting at 90°W (Oceanus Procellarum) continuing across the front side of the moon and reaching a maximum over Mare

SPATIAL VARIATION OF ^{222}Rn*

APOLLO 15 GROUND TRACK

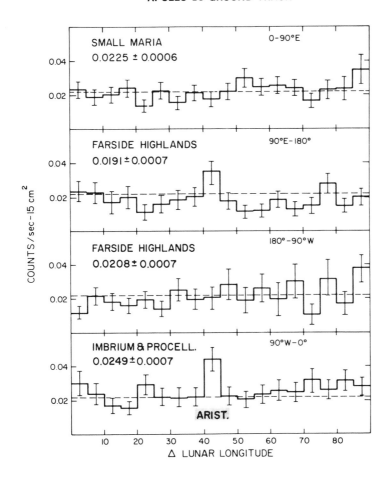

* plus two prompt daughters

Fig. 1. Distribution of the observed ^{222}Rn (plus two prompt daughters) count rate as a function of longitude for the Apollo 15 ground track. The dashed line represents the average decay rate for the total observed data (including background). The average count rate for each quadrant is indicated in each panel. The error bars are counting statistics only.

Fecunditatis and Mare Smythii. Beyond this point the rate drops dramatically and remains low over the lunar farside. Because the Apollo 16 mission had a fairly low inclination orbit, the statistics are sufficient to generate a two-dimensional display. This is shown in Fig. 4. Here the data are grouped into $10° \times 10°$ bins and displayed relative to the background counting rate obtained from data taken with the

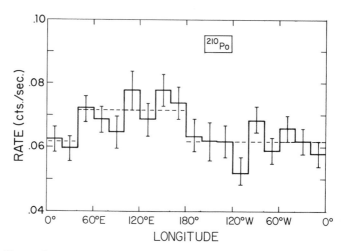

Fig. 2. Observed count rate at the energy of lunar ^{210}Po as a function of longitude for the Apollo 15 ground track. The dashed lines represent the average count rate in two regions (40–180°E, and 0–40°E plus 180°W–0°). The error bars are counting statistics only. The decay rate at the lunar surface in dis/cm²-sec is 0.43 times the experiment count rate.

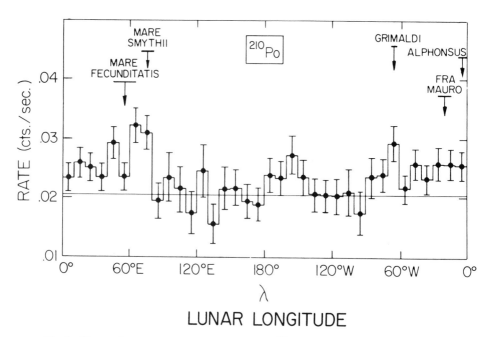

Fig. 3. Observed count rate at the energy of lunar ^{210}Po as a function of lunar longitude for the Apollo 16 ground track. The solid line is the observed background level in the detectors. Location information is included for reference only. The error bars are counting statistics only. The decay rate at the lunar surface in dis/cm²-sec is 0.36 times the experiment count rate.

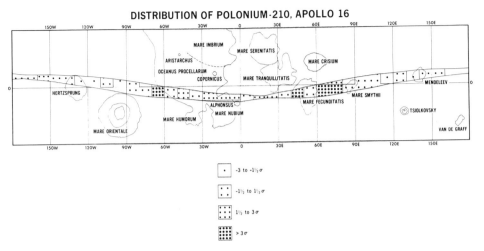

Fig. 4. Observed count rate of lunar ^{210}Po displayed on a schematic lunar map. The data are grouped in $10° \times 10°$ bins. The symbols are given on the figure and refer to standard deviations (counting statistics only) above (or below) the measured background level.

instrument in a non-lunar orientation. These data are discussed in more detail in Bjorkholm *et al.* (1973).

The observed ^{210}Po count rates, relative to background, from both missions are shown overlayed on a lunar map in the frontispiece. Because the amount of data gathering time (and therefore statistical accuracy) varies from region to region, the data are displayed in terms of the number of standard deviations (counting statistics only) above or below background. In the region of ground track overlap between the two missions, the data are combined. Within these regions the observed count rates above background from the two missions were in statistical agreement.

The observed ^{222}Rn and ^{210}Po decay rates are not in radioactive equilibrium. This implies that the processes causing the variation are time varying.

CORRELATIONS WITH TOPOGRAPHICAL FEATURES

The ^{222}Rn distribution as a function of longitude (Fig. 1) showed one 5° bin which was significantly in excess of the surrounding region. This corresponded to the region including the crater Aristarchus. Figure 5 shows a subset of these data, including only those times when Aristarchus was within the instrument's field of view, overlayed on a photograph of the moon. The dashed line represents the average ground track during this time interval and the solid line is the ^{222}Rn count rate referenced to the dashed line. There is an increase in count rate over the crater Aristarchus which is 4.3 standard deviations (σ) with respect to the mean counting rate for the moon. Based on Poisson counting statistics the probability that this increase is due to a statistical fluctuation coincident with Aristarchus is 10^{-4}. The excess over background is $(10.1 \pm 3.3) \times 10^{-3}$ dis/cm^2-sec. The increase is

Fig. 5. The observed count rate of ^{222}Rn (plus two prompt daughters) superimposed on a photograph of the moon. The dashed line represents the average ground track for that portion of the Apollo 15 mission when the crater Aristarchus was within the alpha particle spectrometer's field of view. The data are referenced to this line.

centered over the crater Aristarchus, but the statistics are insufficient to distinguish among Aristarchus, Schröter's Valley, and Cobrahead as the center of the effect. These data are discussed in more detail in Gorenstein and Bjorkholm (1973).

A similar feature was seen in the Apollo 16 ^{210}Po data. Here we see a very strong increase over the crater Grimaldi. This is shown in Fig. 6. The dashed line represents the average ground track and the data are plotted with respect to this line assuming it to be the background level. The ^{210}Po decay rate directly over the crater Grimaldi exceeds the background rate by $(3.3 \pm 1.1) \times 10^{-3}$ dis/cm^2-sec.

Another example of this type of spatially localized concentration of ^{210}Po is shown in Fig. 7. The count rate for ^{210}Po decays is shown as the ground track crosses Mare Fecunditatis. The highlands to either side of the Mare, the edges of the Mare, and the central region are indicated on the figure. The solid line is the background level. The observed rate falls to background levels on either side of

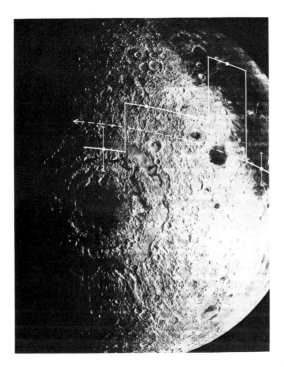

Fig. 6. The observed ^{210}Po count rate superimposed on a photograph of the moon. The dashed line represents the average ground track and is used to represent the background counting level. The crater directly beneath the local increase in count rate is the crater Grimaldi.

the Mare and in the center. The increase is associated strictly with the edges.

Figure 8 shows the energy spectrum observed over the edges (solid line) and over the center (dashed line) of the Mare. The lower panel shows the differential spectrum (edge—center). This clearly shows that the excess is due to ^{210}Po and shows that radioactive equilibrium does not exist between ^{222}Rn and ^{210}Po at the time of observation.

While Mare Fecunditatis shows this effect most dramatically, it is not unique. Table 1 gives the count rate at all observed Mare edges relative to background. The data are divided into five categories based on the distance from the Mare edge. The terms "In" and "Out" are used to refer to data taken over the Mare surface and over the surrounding highlands respectively. Certain entries are indicated as "no data" either because the ground track did not cover that region or because the Mare itself is less than 500 kilometers in diameter. The data from the two missions for the same Mare need not agree since different portions of the Mare were observed. All of the observed Maria, with the exception of Serenitatis, show an enhanced decay rate for ^{210}Po at their edges. In particular, Maria Fecunditatis and Crisium show this effect most dramatically.

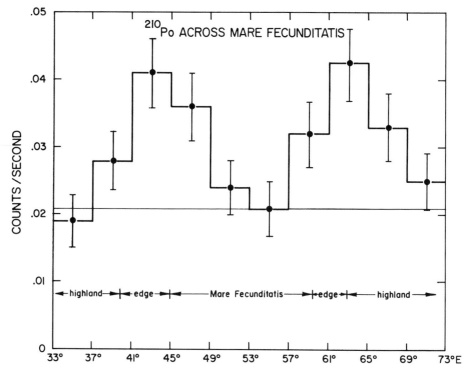

Fig. 7. The observed ^{210}Po count rate as a function of longitude as the ground track crossed Mare Fecunditatis (Apollo 16). The solid line is the observed background level. The highland, mare, and edge regions are indicated. The finite width to the "edge" is caused by the oblique crossing of the Mare-highland interface by the ground track.

Discussion

The observed ^{222}Rn and ^{210}Po distributions over the lunar surface are remarkably inhomogeneous. Variations in decay rate occur over distances of five degrees or less and most of the variations can be directly correlated with lunar surface features. Further, even though the data obtained over Aristarchus indicated that ^{222}Rn was in radioactive equilibrium with her prompt daughters, none of the observed data indicated radioactive equilibrium between ^{222}Rn and ^{210}Po, her delayed daughter. This implies that the processes responsible for radon transportation to the lunar surface are also time varying.

The spatial and implied temporal variations of the observed signals require a similar type of variation for the radon transport mechanisms in the lunar regolith. Any proposed mechanisms for radon transport must then be capable of producing these types of observations. On the other hand, the spatial correlations are so specific that a study of other lunar surface phenomena might yield some insight into possible radon transport mechanisms. This will be considered further in the following paper.

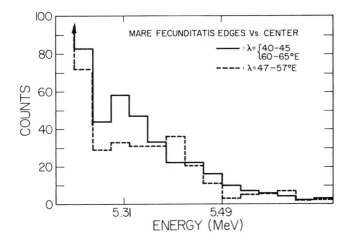

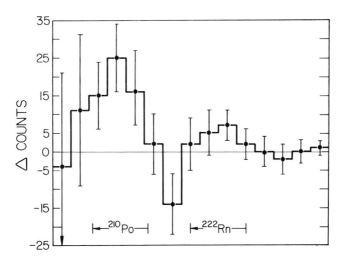

Fig. 8. Top panel: Energy spectra observed at Mare Fecunditatis. The solid line is data obtained over the edges and the dashed line is data obtained over the center of the Mare. Bottom panel: The (time normalized) differential energy spectrum (edge—center). Any energy region having a higher count rate over the edge region will appear as a positive excess in this spectrum. The excess due to ^{210}Po decays is clearly visible. The ^{222}Rn excess is of marginal statistical significance but the ratio between the ^{210}Po and ^{222}Rn excesses establishes secular variability.

P. J. BJORKHOLM *et al.*

Table 1. ^{210}Po decay rate at Maria edges[a].

Mare	In: 200–350 km	In: 75–200 km	Edge ±75 km	Out: 75–200	Out: 200–350 km
A. Apollo 15					
Fecunditatis	−0.1±4.3	11.1±4.4	12.5±3.5	8.1±5.8	no data
Crisium	no data	no data	20.1±5.2	4.1±5.6	−5.6±8.0
Smythii	no data	−5.6±10.1	14.1±4.2	9.0±6.3	6.2±7.5
Procellarum	−7.7±5.3	1.5±3.8	5.4±4.1	4.0±4.2	−6.7±6.2
Imbrium	3.4±4.5	2.3±4.2	4.6±3.1	no data	no data
Tranquillitatis	2.1±6.3	−7.7±2.8	8.0±2.6	no data	no data
Serenitatis	no data	5.1±4.4	4.0±2.9	4.5±3.4	no data
Total—Apollo 15	−0.1±2.5	1.0±1.6	8.4±1.3	5.3±2.1	−1.9±4.1
B. Apollo 16					
Fecunditatis	no data	6.8±2.1	19.5±3.1	7.1±2.8	8.3±7.1
Smythii	no data	3.4±4.7	9.5±3.1	10.9±3.0	7.5±5.3
Tranquillitatis	no data	no data	11.6±3.8	4.7±2.4	9.3±2.7
Nubium	no data	no data	11.5±3.6	8.1±2.8	3.9±3.4
Cognitum	no data	6.6±2.8	11.2±2.7	7.3±3.7	no data
Total—Apollo 16	no data	6.5±1.7	13.3±1.5	7.7±1.3	7.6±1.9
Total A15 & A16	−0.1±2.5	2.9±1.3	9.8±1.0	6.7±1.2	4.9±1.8

[a]Rates are $\times 10^{-3}$ cts/sec $= 0.36 \times 10^{-3}$ dis/cm^2-sec at the lunar surface.

REFERENCES

Bjorkholm P., Golub L., and Gorenstein P. (1973) Detection of a non-uniform distribution of ^{210}Po on the Moon with the Apollo 16 alpha particle spectrometer. *Science* **180**, 957–959.

Gorenstein P. and Bjorkholm P. (1972) Observation of lunar radon emanation with the Apollo 15 alpha particle spectrometer. *Proc. Third Lunar Sci. Conf., Geochim. Cosmochim. Acta*, Suppl. 3, Vol. 3, pp. 2179–2187. MIT Press.

Gorenstein P. and Bjorkholm P. (1973) Detection of radon emanation from the crater Aristarchus by the Apollo 15 alpha particle spectrometer. *Science* **179**, 792–794.

Metzger A. E., Trombka J. I., Peterson L. E., Reedy R. C., and Arnold J. R. (1972) A first look at the lunar orbital gamma-ray data. *Proc. Third Lunar Sci. Conf., Geochim. Cosmochim. Acta*, Suppl. 3, Vol. 3, frontispiece. MIT Press.

Proceedings of the Fourth Lunar Science Conference
(Supplement 4, *Geochimica et Cosmochimica Acta*)
Vol. 3, pp. 2803–2809

Spatial features and temporal variability in the emission of radon from the moon: An interpretation of results from the alpha particle spectrometer

PAUL GORENSTEIN, LEON GOLUB, and PAUL J. BJORKHOLM

American Science and Engineering, 955 Massachusetts Avenue, Cambridge, Mass. 02139

Abstract—Observations of ^{222}Rn and ^{210}Po on the lunar surface with the orbiting Apollo Alpha Particle Spectrometer reveal a number of features in their spatial distribution and indicate the existence of time variations in lunar radon emission. Localized ^{222}Rn or ^{210}Po around the craters Aristarchus and Grimaldi and the edges of virtually all maria indicating time varying radon emission suggest a correlation between alpha "hot spots" and sites of transient events as seen from the earth. In a gross sense, the slower variations of ^{222}Rn seem to correlate with the distribution of gamma activity.

The distribution of radon gas in the lunar atmosphere is characterized by an enhancement near the terminator. On the dark side of the moon radon apparently does not migrate freely; it is stored in the regolith or condensed out upon the surface.

INTRODUCTION

THE PRECEDING paper, Bjorkholm *et al.* (1973) described several observations pertaining to the distribution of ^{222}Rn and her daughter product ^{210}Po upon the lunar surface. Those results plus another concerning the distribution of ^{222}Rn near the terminator are summarized in Table 1. ^{222}Rn has been detected at a level of about 10^{-3} of terrestrial values. It is one of perhaps only two gaseous species in the lunar atmosphere that is native to the moon; the other being Argon-40 as reported by Hoffman *et al.* (1973) at this conference. It is reasonable to expect that the rather short half-life of ^{222}Rn gas (3.8d) could lead to the preservation of features which appear initially in the spatial distribution of ^{222}Rn and her daughter product ^{210}Po. The features include localized emission that is associated with certain well-known lunar craters and maria edges, in addition to a diffuse component of ^{222}Rn and ^{210}Po that exists more or less over the entire moon. Thus a general interpretation of

Table 1. Results from the observation of lunar radon emanation with the Apollo Alpha Particle Spectrometer.

Spatial Variation
  Finite level of ^{222}Rn, $\sim 10^{-3}$ terrestrial
Hot spots apparently associated with sites of "transient lunar phenomena"
  examples: Aristarchus, Grimaldi
^{222}Rn higher in maria than highlands
Correlation of higher Rn activity with sunrise terminator
Temporal Variation
  ^{210}Po levels exceed amount in equilibrium with ^{222}Rn
  ^{210}Po concentration highest in Sea of Fertility
  Maria "edge effect," enhancement of ^{210}Po at boundaries of maria

radon emission from the moon should address itself to factors which determine local conditions at particular "hot spots," as well as the factors which determine the global activity level. Although we have not yet arrived at a complete general theory that explains both, it is worthwhile to discuss interpretations.

The models must satisfy the following set of constraints:

(1) The process of radon emission takes place over a large portion of the moon as evidenced by the widespread distribution of ^{222}Rn and ^{210}Po. However, it need not act simultaneously in many places.

(2) There exist spatial variations of ^{222}Rn and ^{210}Po activity across specific regions and across maria edges which are more rapid than the variation of the concentration of uranium, the source of radon.

(3) The process is time dependent and varying on a time scale between ~ 10 days and ~ 60 years. The evidence for time dependence is the detection of ^{210}Po activity which exceeds that of its progenitor ^{222}Rn at various locations and on the moon as a whole.

(4) The process of nonuniform radon emission is operating at the present time and increases in the rate of emanation have occurred during the last 21 years (effective half-life of the ^{210}Po surface deposit).

(5) Any *external* process which relies on impulsive release of radon (e.g., meteor bombardment, thermal seismic activity, mass slumping) must satisfy the severe constraint that a thickness of several tenths g/cm^2 of material *per year* is involved in the release over those areas where we detect an enhancement in ^{222}Rn or ^{210}Po.

In view of the persistence of lunar surface features on a time scale of 10^9 years, the last constraint appears to rule out external processes as a source of surface concentrations of radon. This set of constraints, (2) and (3) in particular, seems to require that the mechanism responsible for the emission of radon be a time varying, internal process.

LOCALIZATION

Several of the radon or polonium features observed from Apollo seem to be localized to within 5° of lunar longitude, i.e., within 150 km. Examples are the craters Aristarchus and Grimaldi. It is possible that an instrument of larger area and better spatial resolution would show even finer localization. The ^{210}Po maria edge effect (e.g., Fig. 7 in Bjorkholm *et al.* (1973)) may also be a feature no greater than 5°. With successive Apollo orbits the longitude of the intersection of the ground track with the mare-highland interface changes. Thus any sharp effect at the edge appears to spread when plotted against longitude. The observed extent is not inconsistent with the assumption that the maria edge effect occurs in a narrow perimeter around the mare. The observed localization implies that free radon atoms stick on the lunar surface a large fraction of the time. If radon atoms did not stick we would expect the spatial extent of features to be much larger even on the dark side of the moon. On the dark side the temperature of the surface is about 100°K. The mean diameter of a feature should be approximately equal to two times the square root of the average number of trajectories a radon atom undergoes during its half-life multiplied by the average displacement per trajectory. At a temperature of 100°K initially localized radon atoms undergoing successive free trajectories without sticking would have spread to a diameter of about 10^3 km at the end of their 3.8 day half-life. This is several times the apparent size of the features. Thus at 100°K radon atoms emitted from the surface are in free flight only a fraction of the time. This result has implications for the migration of radon through the regolith. A sticking effect would be consistent with the fact that the emission of radon from the lunar regolith is so low compared to the earth in spite of similar uranium concentrations. It also has possible implications for the transport of other gases across the lunar surface, although other gases are much lighter than radon and presumably have a lower heat of absorption at the lunar surface.

CORRELATION WITH LUNAR FEATURES

Sites of transient lunar phenomena

Two of the brightest alpha particle "hot spots" occur in the vicinity of the well known craters Aristarchus and Grimaldi. The former is seen in the ^{222}Rn data while the latter is seen in the ^{210}Po data. The significance of these two craters is that over a long period they have been reported repeatedly by ground based observers to be sites of transient optical events. In the case of Aristarchus the reports go back several hundred years. Middlehurst (1967) and co-workers have summarized these reports. If the correlation between alpha hot spots and lunar transient events is not merely coincidental we can predict other specific correlations. For example, the crater Alphonsus is another frequently reported site of transient optical events. However, the Apollo 15/16 Alpha Particle Spectrometer did not pass over it directly so there is no data from Alphonsus. The second largest ^{222}Rn feature occurs in the Apollo 15 data starting at about a lunar longitude of 130°E (Fig. 1 in Bjorkholm *et al.*, 1973). The feature is not as statistically significant with respect to the lunar average as the Aristarchus feature (4.3 standard deviations) but is about 3 standard deviations with respect to the lower local average over the farside highlands. The feature is close to the crater Tsiolkovsky which could not have been seen as a site of transient optical events because it is on the farside of the moon. Thus it is not possible to make the next two most direct tests of the prediction. However, there is evidence that the correlation does indeed go much further. Along with the discrete repeating sites of transient optical events such as Aristarchus, Grimaldi, and Alphonsus there are a large number of reports of smaller events at non-specific sites. Figure 1 illustrates the distribution of the events which are judged to be most reliable. It is clear that the distribution correlates with the edges of the maria. In light of the fact that our preceding paper described a rather striking ^{210}Po effect at the edges of the maria we suggest that there is indeed a real spatial correlation between radon hot spots and sites of transient lunar phenomena.

The correlation between radon features and transient events suggests that there may be a mechanism which is common to both processes. We have already concluded that radon emission is a time varying, internal process. Now we con-

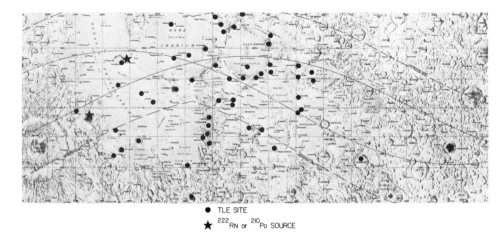

● TLE SITE
★ ^{222}Rn or ^{210}Po SOURCE

Fig. 1. Black dots indicate sites where the more reliable transient events have been reported. Stars indicate localized concentrations of ^{222}Rn or ^{210}Po detected in either Apollo 15 or Apollo 16. Reports of transient events have been summarized by Middlehurst (1967).

sider the hypothesis that radon is merely a trace component of more common gases when transient emission occurs. On the earth radon comprises only one part in 10^{14} of soil gas (Kraner *et al.*, 1964). Certainly the diffusion of other more abundant gases through lunar material will promote the diffusion of radon and it is difficult to imagine an increase in the rate of radon diffusion without the involvement of other gases. The fact that the process is spatially localized and variable in time may make it difficult to detect these putative other gases unless the detector happens to be at the appropriate place and at the appropriate time. Suggestions have been made for associating gas releases and visual events (Middlehurst, 1967).

Enhancement near lunar terminator

Since radon is a gas, the solar heating effects are expected to influence its distribution on the lunar surface. Heymann and Yaniv (1971) have described a model of radon transport which predicts that the peak activity will occur at the sunrise terminator. The model is based on the hypothesis that free radon atoms will diffuse rapidly on the sunlit (warm) side and condense as soon as they enter on the dark (cold) side. Figure 2 shows the observed radon activity for Apollo 15/16 as a function of angle from the sunrise terminator. There is indeed a peak in the vicinity of the terminator; however, it seems to be somewhat past the sunrise terminator where the lunar surface has been warmed slightly by the sun. Any attempt to correlate ^{222}Rn data with the terminator is complicated by the fact that there are other spatial features to contend with such as the localized hot spot at Aristarchus and a diffuse component which is slowly varying. Figure 2 also shows some very preliminary results from another model of radon outgassing that is

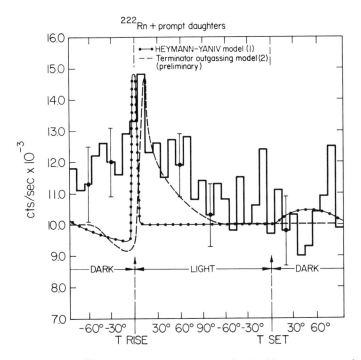

Fig. 2. Distribution of ^{222}Rn during Apollo 15 and Apollo 16 with respect to terminator. Two theoretical models are shown, that of Heymann and Yaniv (1971) and another being considered by us.

presently being considered by us. In this model the radon concentration in the regolith just below the surface builds up while the moon is cold. Little radon leaves the surface until the sun heats it to a certain critical temperature. At that temperature the layer is suddenly purged of all the radon accumulated during the entire time it has been cold. As it is heated to still higher temperatures there is no storage; radon escapes rapidly and the observed activity should be low once more. This model predicts that peak activity will be seen in the region where the temperature has reached the critical value. This should be somewhat past the sunrise terminator. Due to the complication mentioned above and the fact that our counting statistics are limited it will be difficult to discriminate between the two models on the basis of these data.

CORRELATION WITH OTHER MEASUREMENTS

Several of the observations of the Alpha Particle Spectrometer correlate with other data. We have already discussed the correlation between our most highly localized features and reports of lunar transient events. Another correlation may be seen between the slower spatial variation of ^{222}Rn and the spatial variation of gamma ray activity across the Apollo ground track as reported by Metzger *et al.*

(1973). Figure 1 in Bjorkholm et al. (1973) shows the average counting rate of ^{222}Rn during Apollo 15 (including a cosmic ray background) in four quadrants. The highest levels are seen in the quadrant containing Mare Imbrium and Oceanus Procellarum and the lowest over the farside highlands. The ordering of the gamma ray data is similar. Both the alpha and gamma signals depend on the variation of uranium so that it is not surprising that a correlation exists. However, the local rate of alpha emission also depends on the local radon conductance through the regolith. As we have noted this varies as a function of time. If there were sufficient counting statistics a detailed comparison of the alpha and gamma results would locate the regions of high conduction.

Our results pertaining to ^{222}Rn in the lunar atmosphere could be compared to those from instruments that detect other lunar gases. We have hypothesized that an increase in the release of radon locally will involve a concurrent release of other gases. Instruments stationed on the lunar surface such as the mass spectrometer and the ion gauges may be able to detect an increase in the other gases. There are probably at least several such events during a 21-year period. If the ratio of other gases relative to radon is similar to terrestrial values and if the event is near the station they may be detectable. Finally there may be a "sticking" or condensation effect with other heavy gases on the dark side such as we have noted for radon which reduces their concentration in the lunar atmosphere.

We have interpreted our localized ^{210}Po concentrations in excess of ^{222}Rn as the result of a time varying internal process. The only time varying internal process known at the present time are the seismic events (Latham et al., 1972). It is difficult to find a mechanism whereby rather weak seismic events originating at great depths could influence a surface effect such as the release of radon. However, there is a terrestrial correlation between seismic events and increases in the release of radon (e.g., Okabe, 1956). On this rather circumstantial basis, we would hope to find a lunar correlation. Should the edges of the maria turn out to be a key feature of the seismic events as they are in the ^{210}Po data it would greatly increase the probability that there is a connection between the two processes.

REFERENCES

Bjorkholm P., Golub L., and Gorenstein P. (1973) The distribution of ^{222}Rn and ^{210}Po on the lunar surface as observed with the alpha particle spectrometer. Proc. Fourth Lunar Science Conf., Geochim. Cosmochim. Acta. This volume.

Heymann D. and Yaniv A. (1971) Distribution of ^{222}Rn on the surface of the moon. Nature Physical Science 233, 37–39.

Hoffman J. H., Hodges R. R. Jr., and Evans D. E. (1973) Lunar atmospheric composition results from Apollo 17. Proc. Fourth Lunar Science Conf., Geochim. Cosmochim. Acta. This volume.

Kraner H. W., Schroeder G. L., and Evans R. D. (1964) Measurements of atmospheric variables on radon-222 flux and soil-gas concentrations. The Natural Radiation Environment, (editors Adams and Lowder). University of Chicago Press.

Latham G. V., Ewing M., Press F., Sutton G., Dorman J., Nakamura Y., Toksöz N., Lammlein D., and Duennebier F. (1972) Passive seismic experiment. Apollo 16 Preliminary Science Report NASA SP-315, Sec. 9.

Metzger A. E., Trombka J. I., Peterson L. E., Reedy R. C., and Arnold J. R. (1973) Lunar surface radioactivity: Preliminary results of the Apollo 15 and Apollo 16 gamma-ray spectrometer experiments. *Science* **179**, 800–883.

Middlehurst B. M. (1967) An analysis of lunar events. *Rev. Geophys.* **5**, 173–189.

Okabe S. (1956) Time variation of the atmospheric radon-content near the ground surface with relation to some geophysical phenomena. *Kyoto Univ. Coll. Sci. Mem., Ser. A,* **28**, 99–115.

Proceedings of the Fourth Lunar Science Conference
(Supplement 4, *Geochimica et Cosmochimica Acta*)
Vol. 3, pp. 2811–2819

Lunar topography from Apollo 15 and 16 laser altimetry

W. M. KAULA, G. SCHUBERT, and R. E. LINGENFELTER

University of California, Los Angeles

W. L. SJOGREN

Jet Propulsion Laboratory, Pasadena, California

W. R. WOLLENHAUPT

Lyndon B. Johnson Space Center, Houston, Texas

Abstract—In the orbital plane of Apollo 15 the mean lunar radius is 1737.3 km, the mean altitude of terrae above maria is about 3 km, and the center-of-figure is displaced from the center-of-mass by about 2 km away from longitude 25°E.

The Apollo 16 laser altimeter obtained a total of about $7\frac{1}{2}$ revolutions of partially overlapping data. The principal difference in results from Apollo 16 is the absence of any great farside basin similar to the 1400-km wide feature found by Apollo 15, 1200 km to the south. This absence of a farside depression in the Apollo 16 orbital plane largely accounts for a greater mean radius: 1738.1 km; a greater mean altitude of terrae above maria: about 4 km; and a greater offset of centers: about 3 km, also away from 25°E.

In the Apollo 16, as well as Apollo 15, data the farside terrae are much "rougher" than the nearside terrae. Mare surfaces are generally smooth to within ± 150 m, and have slopes of 1:500 to 1:2000 persisting over distances as great as 500 km.

INTRODUCTION

THIS PAPER is essentially a progress report intermediate between the discussion of results from the initial altimetry flight (Kaula *et al.*, 1972) and the final analysis and interpretation which will be made after all Apollo 17 data have been received and analyzed. Earlier reports of Apollo 16 altimetry have been made by Wollenhaupt and Sjogren (1972b) and Sjogren and Wollenhaupt (1973). Kaula *et al.* (1972) give references to earlier key papers on the lunar figure.

In the Apollo 16 mission, laser altimeter measurements were obtained during lunar revolutions 3–4, 17–18, 28–29, 37–39, 46–48, 54–60, and 63. Most of the time, the altimeter was operated simultaneously with the mapping camera. Approximately six complete revolutions of data were obtained, plus other segments for a total of seven and a half revolutions. Although some modifications were incorporated to improve the lifetime of the instrument over that experienced during the Apollo 15 mission, performance was somewhat degraded during revolutions 17–18 of Apollo 16. This degradation continued until the laser failed during revolution 63. A total of 2372 laser firings were made, of which 69 percent were valid.

Spacecraft positions necessary to calculate altitudes with respect to the center-of-mass of the moon were obtained by fitting orbits to the earth-based Doppler radio tracking, as described in Kaula *et al.* (1972). For orbits determined using a five parameter model of the lunar gravitational field, the radial position uncertainties are estimated to average 400 m. However, for the difference in radial position between successive points (about 33 km apart), the uncertainty is only about 10 m. Hence slopes of smooth terrain can be determined along-track quite accurately, but adjacent orbital ground tracks will normally show offsets suggesting a spurious cross-track slope. These offsets in turn should normally be appreciably less than 400 m since two adjacent orbits will be affected quite similarly by gravity model errors and inadequacies. An offset of this type of about 60 m shows quite clearly in Fig. 6 below.

The nearest thing to an independent check of the *a priori* accuracy estimate is the intersection of two orbital ground tracks of significantly different inclinations over a smooth area. Several Apollo 15 and 16 ground tracks intersect over Mare Smythii. Except for one Apollo 15 arc, the agreement appears to be within 100 m, good enough to attempt a contour map (Fig. 5). Since this agreement amounts to only a single comparison for gravity model effects, it does not have much statistical significance.

Tracks central to the bands covered by Apollo 15 and 16 are shown in Fig. 1.

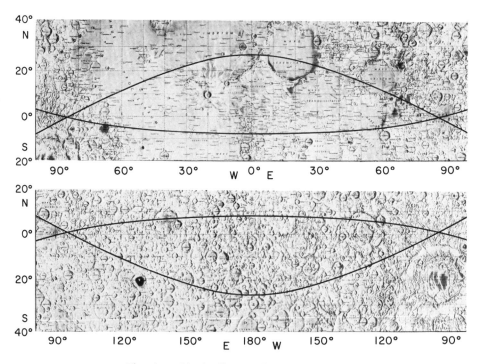

Figs. 1a and b. Apollo 15 and 16 ground tracks.

APOLLO 15   NEAR SIDE TOPOGRAPHY

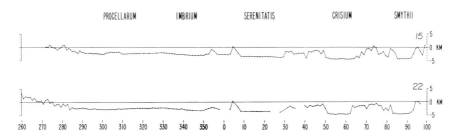

FAR SIDE TOPOGRAPHY

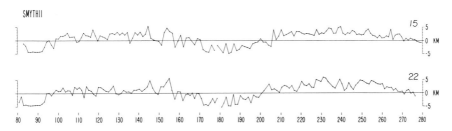

Fig. 2. Profiles of Apollo 15 elevations with respect to a 1738 km sphere about the center of mass. Based on revised orbit determinations.

BROAD SCALE FEATURES

The main value of the altimeter is to give the variations in the shape of the moon on a global scale, since the metric photography will give much greater detail on a local or regional scale. The main results from Apollo 15 are contained in Fig. 2, and from Apollo 16 in Figs. 3 and 4. Figure 2 differs up to 0.5 km from Fig. 2 of Wollenhaupt and Sjogren (1972a) and Fig. 2 of Kaula *et al.* (1972) because different orbit determinations have been used. These orbit determinations differ mainly in their gravity models. The largest discrepancies occur in the middle of the farside. The least-squares fit of a second degree curve to a revolution of Apollo 15 data obtains for elevation h in km with respect to a 1738 km sphere about the center-of-mass.

$$h = -0.7 - 2.1 \cos (\lambda - 25°) + 0.7 \cos 2(\lambda - 90°) \tag{1}$$

The constant term in Equation (1) constitutes a difference in mean radius; the first degree term, an offset of center-of-figure from center-of-mass; and the second degree term, the amplitude of the difference of the ellipse from a circle.

The same fitting to a revolution of Apollo 16 data obtains

$$h = +0.1 - 2.9 \cos (\lambda - 25°) + 1.2 \cos 2(\lambda - 50°) \tag{2}$$

W. M. Kaula *et al.*

APOLLO 16   NEAR SIDE TOPOGRAPHY

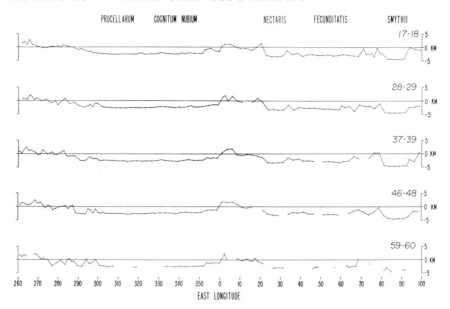

Fig. 3. Profiles of Apollo 16 nearside elevations with respect to a 1738 km sphere about the center-of-mass.

APOLLO 16   FAR SIDE TOPOGRAPHY

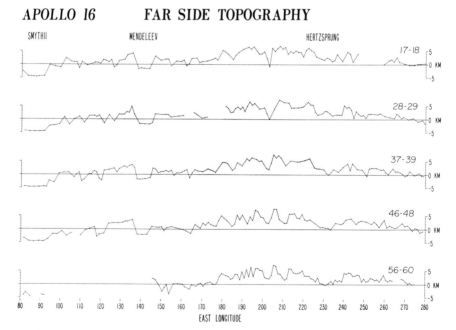

Fig. 4. Profiles of Apollo 16 farside elevations with respect to a 1738 km sphere about the center-of-mass.

The main reason for the larger mean radius and greater offset of center-of-figure from center-of-mass inferred from the Apollo 16 data is the absence of a large basin in the farside northern latitudes, such as was found in the southern hemisphere by Apollo 15. The farside northern hemisphere topography is appreciably higher in longitudes 180–260° than it is in longitudes 90–170°, which probably accounts for the larger mean ellipse oriented on the line 50–230° longitude.

The larger offset of the center-of-figure from center-of-mass in the Apollo 16 plane compared to the Apollo 15 plane suggests that the center-of-figure is displaced northward with respect to the center-of-mass. This displacement is contrary to that inferred from grazing occultation (Van Flandern, 1970). This discrepancy is probably not significant, because of the limited coverage of both types of data.

The large offset of centers suggests large scale convective motions associated with the crustal differentiation very early in the moon's history, as discussed by Lingenfelter and Schubert (1973).

If the 1.2 km best-fitting ellipse of the Apollo 16 ground track were the equator of an ellipsoidal mass distribution on a homogeneous sphere, the equatorial coefficient J_{22} of the gravity field would be 1.5×10^{-4}. The observed value is about 0.2×10^{-4} (Sjogren, 1971); the difference indicates of course that isostatic compensation prevails. Hence a more meaningful characterization of the elevations would be by the terrain types associated with the variations in lithology necessary for isostasy. A simple classification of areas traversed by Apollo ground tracks is: farside terrae; nearside terrae; ringed maria (Serenetatis, Crisium, Smythii, and Nectaris); and other maria (Procellarum, outer Imbrium, Cognitum, Nubium, Fecunditatis, and Tranquillitatis). The resulting values for the mean elevations of each of these terrain classes with respect to a 1738 km sphere are

	Apollo 15 Track (km)	Apollo 16 Track (km)
Farside Terrae	+1.9	+2.1
Nearside Terrae	−1.7	−1.2
Ringed Maria	−4.1	−4.1
Other Maria	−2.0	−2.5

The +1.9 km for the farside terrae of the Apollo 15 track is a combination of a −2.1 km mean for the great basin at longitudes 160–205° and a +3.1 km mean for other areas. Deviations from these means range from 0.7 km for ringed maria to about 4 km for farside terrae.

The mean elevation of terrae with respect to maria for the Apollo 16 data is about 4 km, considerably more than the 3 km for the Apollo 15 data. Now that it is known that the maria were filled considerably subsequent to the creation of a lunar crust, the explanation for the presence of maria on the nearside and their absence on the farside appears to be mainly a difference in the thickness of a pre-existing low density crust. We adapt the explanation for the height of volcanic accumulations on earth (Eaton and Murata, 1960). For a lava of density ρ_L to penetrate from a source depth D through mantle of density ρ_M and crust of thickness T and density ρ_c to height h requires

$$(h + D)\rho_L = T\rho_c + (D - T)\rho_M \qquad (3)$$

i.e., isostatic equilibrium exists at depth D, and volcanic accumulations are possible because melt density ρ_L is lower than mantle density ρ_M and possibly crust density ρ_c. With imperfect real materials, an efficiency factor $\epsilon < 1$ should be applied

$$(h + D)\rho_L = \epsilon[T\rho_c + (D - T)\rho_M] \qquad (4)$$

whence

$$h = [(\epsilon\rho_M - \rho_L)D - \epsilon(\rho_M - \rho_c)T]/\rho_L \qquad (5)$$

or, for differences Δh in the height to which magma has risen

$$\Delta h \approx \Delta(\epsilon D) - \epsilon(\rho_M - \rho_c)\Delta T/\rho_L, \qquad (6)$$

obtained by differentiating Equation (5) and neglecting the small terms ΔD and $T\Delta\epsilon$. For the nearside, the maximum height of extensive mare basalts in isostatic balance is observed to be about $h = -2$ km. For source depth of mare basalts, speculations range around $D = 300$ km (McConnell and Gast, 1972). For the thickness of the anorthositic gabbro layer below the mare basalt, Toksöz *et al.* (1972) obtained about $T = 40$ km from seismological data. Using these estimates with

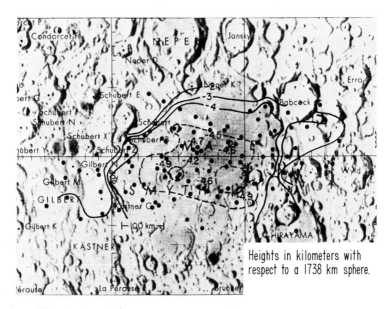

Heights in kilometers with respect to a 1738 km sphere.

Fig. 5. Elevations in Mare Smythii based on Apollo 15 and 16 altimetry.

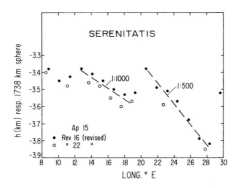

Fig. 6. Elevations in Mare Serenitatis, approximate latitude 22°N.

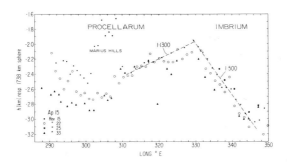

Fig. 7. Elevations in northern Oceanus Procellarum and southern Mare Imbrium, approximate latitude 20°N.

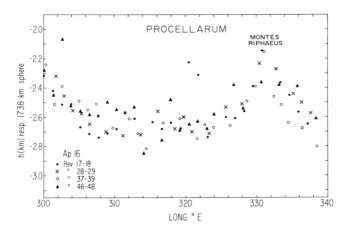

Fig. 8. Elevations in southern Oceanus Procellarum, approximate latitude 6°S.

$\rho_L = \rho_c = 3.00 \, \text{g cm}^{-3}$, $\rho_M = 3.35 \, \text{g cm}^{-3}$, we get a factor ϵ of 0.892. The offset of centers indicates that the mean crustal thickness T of the farside is about 60 km (Kaula *et al.*, 1972). Using a ΔT of 20 km with the same ϵ and the other numbers unchanged in Equation (6) gives a height difference $\Delta h = -2$ km for farside maria below nearside. The deepest parts of the farside basin are about 3 km lower than the surface of Oceanus Procellarum. Hence either a slightly lower product ϵD, efficiency times source depth, is appropriate for the farside, or the farside basin is maintained to some extent by the strength of the lithosphere. We predict the farside basin will have a negative gravity anomaly, but not as great in absolute magnitude as the nearside positive mascons. From the infinite plate formula

$$\Delta g = 2\pi G \rho \delta h \qquad (7)$$

Δg is about -125 milligals for a height difference δh of -1 km.

FINE SCALE FEATURES

In Kaula *et al.* (1972) the altimetry data were crudely analyzed for "roughness" measured by the change of elevation between successive points about 33 km apart, and for "levelness" measured by the variations about the mean within a specified segment. These analyses gave the results that the altimetry interval was too great to indicate much about the terrae other than that the predominant horizontal scale of variations was something less than the interval, and that the variations were somewhat larger in the farside terrae than in the nearside.

The analyses also indicated, however, that the characteristic horizontal scale of variations in the maria is often some hundreds of kilometers. While the altimetry cannot hope to match the metric photography in detail, it is still of some interest to examine it for prevailing slopes.

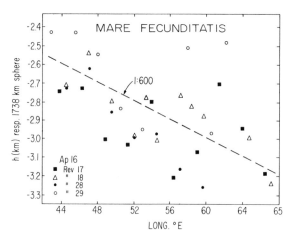

Fig. 9. Elevations in Mare Fecunditatis, approximate latitude 6°S. Note greater variability about mean slope than in younger maria.

The only area in which there are sufficient data to draw a contour map is Mare Smythii, where the Apollo 15 and 16 orbital paths intersected: *see* Fig. 5. Mare Smythii appears to be level within ± 150 m over a region 200 km in extent, with slopes less than 1:500. For other areas, only profiles can be drawn: *see* Figs. 6–9. These indicate that maria slopes of 1:500 or less may persist over several 100 km; that there may be fairly sharp changes in slope; that smooth maria surfaces may be abruptly interrupted by promontories such as the Marius Hills; and that the amplitude of variations about a mean slope is correlated with the age of the mare.

These slopes are not inconsistent with the hypothesis that the lower viscosity of lunar lavas compensates for the lower gravity as compared to the earth, as proposed by Murase and McBirney (1970), Danes (1972), and Schaber (1973).

Acknowledgment—A portion of this work was supported by NASA contract NAS 9-12757.

REFERENCES

Danes Z. F. (1972) Dynamics of lava flows. *J. Geophys. Res.* **77**, 1430–1432.

Eaton J. P. and Murata I. J. (1960) How volcanoes grow. *Science* **132**, 925–938.

Kaula W. M., Schubert G., Lingenfelter R. E., Sjogren W. L., and Wollenhaupt W. R. (1972) Analysis and interpretation of lunar laser altimetry. *Proc. Third Lunar Sci. Conf., Geochim. Cosmochim. Acta*, Suppl. 3, Vol. 3, pp. 2189–2204. MIT Press.

Lingenfelter R. E. and Schubert G. (1973) Evidence for convection in planetary interiors from first-order topography. *The Moon* **7**, 172–180.

McConnell R. K. Jr. and Gast P. W. (1972) Lunar thermal history revisited. *The Moon* **5**, 41–51.

Murase T. and McBirney A. R. (1970) Viscosity of lunar lavas. *Science* **167**, 1491–1493.

Schaber G. C. (1973) Lava flows in Mare Imbrium: Geologic evaluation from Apollo orbital photography (abstract). In *Lunar Science—IV*, pp. 653–655. The Lunar Science Institute, Houston.

Sjogren W. L. (1971) Lunar gravity estimate: Independent confirmation. *J. Geophys. Res.* **76**, 7021–7026.

Sjogren W. L. and Wollenhaupt W. R. (1973) Lunar shape via the Apollo laser altimeter. *Science* **179**, 275–278.

Toksöz M. N., Press F., Anderson K., Dainty A., Latham G., Ewing M., Dorman J., Lammlein D., Sutton G., Duennebier F., and Nakamura Y. (1972) Lunar crust: Structure and composition. *Science* **176**, 1012–1016.

Van Flandern T. C. (1970) Some notes on the use of the Watts limb-correction charts. *Astron. J.* **75**, 744–746.

Wollenhaupt W. R. and Sjogren W. L. (1972a) Comments on the figure of the moon based on preliminary results from laser altimetry. *The Moon* **4**, 337–347.

Wollenhaupt W. R. and Sjogren W. L. (1972b) Apollo 16 laser altimeter. *Apollo 16 Prelim. Sci. Rep.*, NASA SP-315, (30) 1–5.

Proceedings of the Fourth Lunar Science Conference
(Supplement 4, *Geochimica et Cosmochimica Acta*)
Vol. 3, pp. 2821–2831

The Apollo 17 Lunar Sounder

R. J. Phillips,[a][†][‡] G. F. Adams,[b] W. E. Brown, Jr.,[a][*]
R. E. Eggleton,[c][†] P. Jackson,[b] R. Jordan,[a] W. J. Peeples,[d]
L. J. Porcello,[b][†] J. Ryu,[d] G. Schaber,[c] W. R. Sill,[d]
T. W. Thompson,[a] S. H. Ward,[d][*] and J. S. Zelenka[b]

(a) Jet Propulsion Laboratory
(b) Environmental Research Institute of Michigan
(c) U.S. Geological Survey, Flagstaff
(d) University of Utah

Abstract—The Apollo Lunar Sounder Experiment, a coherent radar operated from lunar orbit during the Apollo 17 mission, has scientific objectives of mapping lunar subsurface structure, surface profiling, surface imaging, and galactic noise measurement. Representative results from each of the four disciplines are presented. Subsurface reflections have been interpreted in both optically and digitally processed data. Images and profiles yield detailed selenomorphological information. The preliminary galactic noise results are consistent with earlier measurements by other workers.

Introduction

The Apollo Lunar Sounder Experiment (ALSE) was a three-frequency wideband coherent radar system operated from lunar orbit during the Apollo 17 mission.

The scientific objectives of the experiment are: (1) mapping of subsurface structure via radar sounding, (2) surface profiling, (3) surface imaging, and (4) galactic noise measurement. The first three objectives are discussed below.

The concept of *radar sounding* is quite analogous to active seismic profiling. That is, a short pulse of (electromagnetic) energy is propagated into the lunar subsurface, and that energy reflected at geologic interfaces is detected. The moon is continually pulsed as the Command-Service Module (CSM) moves in lunar orbit and thus a profile of subsurface structure is assembled. The ability to map the lunar interior is dependent upon a low electrical loss subsurface; that is, the attenuation of the signal must not preclude a significant depth of exploration. Pre-mission estimates of path loss, based on measurements from returned lunar samples, indicated a maximum depth of exploration of about 1 km. Further, in order to reflect energy at a geologic boundary, an electrical property contrast must exist across the interface. The controlling electrical property is the dielectric constant, ϵ. Pre-mission estimate of the mean interface ratio was 1.25.

---

*Principal Investigator.
†Co-Investigator.
‡ALSE Team Leader.

Profiling is accomplished by maintaining an absolute timing reference between the firing of the radar transmitter and the reception of the energy returned from the lunar surface. The rate of transmitter firing is high enough to make the profiling essentially continuous. In addition to acquiring the global profile for selenodesy studies, detailed profiles were acquired and may be utilized to address local selenomorphological problems.

The quality of *imagery* is dependent, for a given frequency, on the diffuse backscattering properties of the lunar surface. One trace of the surface backscattered energy is built from each transmitted pulse. A sequence of pulses then creates the image. The image is typically bounded on one side by the spacecraft nadir topographic profile, and the combination of profile and image is quite useful for surface geological interpretation.

The ALSE system operated in three frequency ranges, 5 MHz (HF1), 15 MHz (HF2), and 150 MHz (VHF). HF1 provides the deepest exploration. HF2 operated simultaneously with HF1, providing partial overlap in depth of exploration, trading off depth of exploration for improved resolution. VHF was designed for shallow sounding and for surface imaging. All three frequencies are capable of surface profiling.

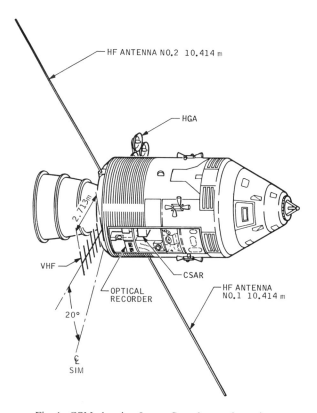

Fig. 1. CSM showing Lunar Sounder configuration.

The ALSE had four major hardware components (Fig. 1). At the heart of the system is the *Coherent Synthetic Aperture Radar* (CSAR) containing the transmitting and receiving elements. There were separate antenna systems for HF and VHF. The receiver video is fed to an *optical recorder* where the input voltage modulates the light intensity output of a cathode ray tube (CRT). The CRT output was written on continuously moving film. The film, representing the prime data storage medium, was recovered during the trans-Earth EVA.

The CSAR receiver also monitored the average reflected power and telemetered this data, along with other engineering data, via a 51.2 kb/s telemetry channel. The average reflected power data is used for calibration purposes, as well as an aid in interpretation.

The CSAR also had a receive-only (non-sounding) capability to measure the noise background in HF. This data was also telemetered via the 51.2 kb/s channel.

This paper provides only a summary of typical results of the experiment. Detailed discussions on instrumentation, data processing, and theory may be found in Phillips *et al.* (1973).

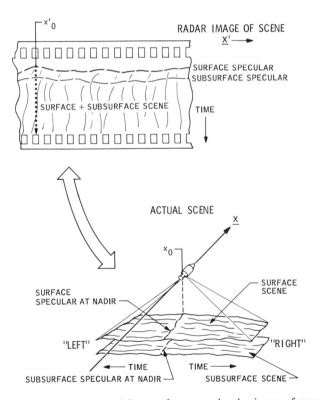

Fig. 2. Relationship between actual image of scene and radar image of scene. When spacecraft is at position x_0, the lunar return for one pulse is exposed *across* the film at position x_0'. As the spacecraft moves in the X direction, successive returns are exposed along the X' direction of the film.

DATA FORMAT

Figure 2 schematically illustrates the relationship of the actual geometry of the experiment and how that geometry appears in the data format. Basically, the signal is recorded as a function of time, and thus elements of the lunar surface and subsurface are recorded according to their two-way time delay to the CSM. The pulse repetition rate of the radar is high enough so that the CSM may be considered in a fixed position, x_0, during one transmit-receive cycle of the radar. The data processing techniques allow one to consider that the return from a single radar pulse is the result of backscattering from a plane, at x_0, orthogonal to the spacecraft trajectory direction, X. Except in regions of mountainous topography, the first return for a single pulse is the specular reflection from the nadir point, directly beneath the CSM. As time increases, the return is the lunar surface backscatter at an increasing distance out from the nadir point. If there is a subsurface interface, then the subsurface specular reflection arrives at some later time than the surface specular reflection, followed by subsurface backscatter at an increasing distance out from the subsurface nadir point. The return is written on film by the CRT at position x_0'.

As the spacecraft moves along the X axis, the surface and subsurface reflection voltages are recorded along the X' direction of the moving film, with the time delays of the reflections actually building up the nadir elevation profiles of the two surfaces. Further, the non-nadir backscatter of the two surfaces superimposes to give the radar "scene." The distance along the film is related to the distance along the lunar surface by a simple linear relationship involving spacecraft velocity and film velocity.

In general, the scene to the left of the nadir point will superimpose upon the scene to the right, the so-called left-right ambiguity problem. However, the axis of the VHF antenna was oriented 20° off nadir to the right, greatly reducing this effect.

Fig. 3. Apollo photograph of Mare Tranquillitatis in the vicinity of the Jansen craters. In this orientation, south is to the top of the photo, east to the left. The mare ridge examined with the VHF data is in the upper left quadrant, interrupted by the crater Jansen L.

EXAMPLES OF SUBSURFACE REFLECTIONS

VHF sounding

Figure 3 shows an Apollo 17 mapping camera photograph of a region in northern Mare Tranquillitatis. The small mare ridge, interrupted by the crater Jansen L, has been examined with the VHF data. Figure 4 shows the VHF profile and image of this ridge. A subsurface reflection appears to arise from within the ridge at a depth of about 24 m. A possible interpretation for this ridge is a volcanic flow and the reflection is from a layer of the flow.

HF2 sounding

Figure 5 shows a 150 km section of HF2 image film in western Mare Serenitatis. The broad line is the main lobe of the specular surface return. Faint lines parallel to and above the surface main lobe are negative time sidelobes of the surface pulse. (The processed signal is of the form $(\sin x/x)^2$). The positive time region is dominated by scattering from random reflectors; this is termed *clutter* or *clutter noise*. There may also be seen a faint event, occasionally broken up,

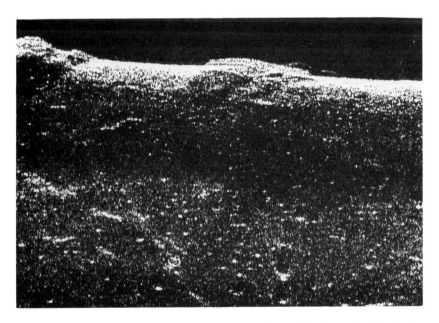

Fig. 4. VHF image of mare ridge (center) pointed out in previous figure. In this orientation, east is to the left of the image with the scene imaged to the north (bottom). A second mare ridge appears on the extreme left and may be identified in the Apollo photo. The rim of Jansen L is apparent behind the center mare ridge. The double bright line in this ridge is interpreted as reflection from the surface of the ridge and from an interface about 24 m deep into the ridge.

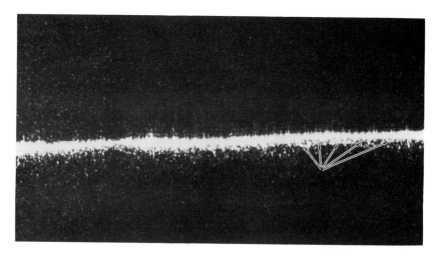

Fig. 5. Candidate subsurface feature at a depth of about 100 m, as seen in HF2 imagery.
Region is approximately 150 km along ground track in western Mare Serenitatis.

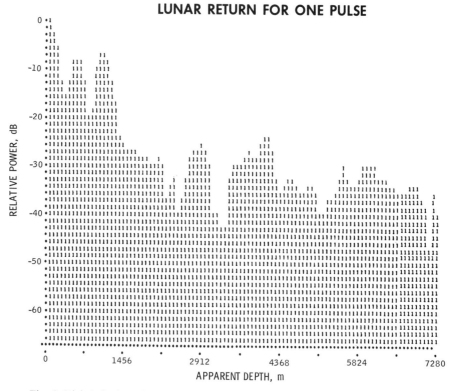

Fig. 6. Digital display of pulse return from lunar surface, showing high clutter level.
Depth (apparent) may be converted to surface time delay by dividing by the two-way
assumed velocity of 75 m μ sec^{-1}.

approximately parallel to the surface. This event is interpreted as a subsurface layer about 100 m deep.

HF1 sounding

To examine the full dynamic range inherent in the experiment, the data must be examined digitally. Once the data is digitized, the data may be integrated coherently pulse to pulse, rejecting the non-coherent clutter noise (Fig. 6).

The example result given is a 60 km segment of data collected on revolution 16, at the eastern boundary of Mare Serenitatis (Fig. 7). Figure 8 shows an isometric block diagram of the apparent subsurface cross section in this region. Each individual trace represents the coherent integration of 100 pulse returns, or about 5 km along track. From each trace, we have plotted only those events that were no more than 45 dB weaker than the surface return, the 45 dB figure being a conservative estimate of the dynamic range of this data set. Further, the events have been normalized to the same strength. The type of coherent integration used here preserves only those subsurface interfaces parallel to the lunar surface, the latter of which has been mapped onto a straight line in the diagram. As an *interpretation* we have emphasized those events that are parallel to the surface and aligned for at least 15 km. These features are either reflecting horizons in the subsurface or linear reflecting features that are fortuitously parallel to the space-

Fig. 7. Ground track (revolution 16) for digital data reduction test area, shown on Apollo photograph. Area is located at eastern boundary of Mare Serenitatis.

APPARENT CROSS SECTION
[VELOCITY ASSUMED = 150m μsec⁻¹]
EASTERN SERENITATIS 0 Km ≈ 20°N, 28.5°E
REV 16-APOLLO 17

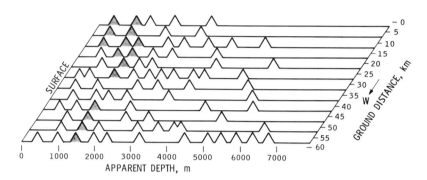

Fig. 8. Isometric block diagram of apparent cross section of HF1 digital data. The events shown are subsurface reflections or surface reflections from linear features parallel to the spacecraft ground track. It is also possible that a time varying systematic phase error in the ALSE instrumentation could cause a similar response. To date, indications primarily from pre-launch testing have indicated that this is probably not the case. This possibility will be further investigated by comparing results from multiple orbits over similar lunar regions. Events aligned for more than 15 km have been emphasized.

craft ground track. Examination of Apollo 15 and Apollo 17 photography reveals no such features and we conclude that the Lunar Sounder is mapping real subsurface stratigraphy (Fig. 9). The significance of our mapping will perhaps only become clear when we have examined all of the Serenitatis data (8 spacecraft passes) at the various ALSE frequencies.

APPARENT CROSS SECTION OF
LINEAR SURFACE FEATURES

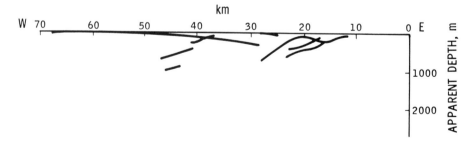

Fig. 9. Apparent cross section of (approximately) linear surface features. Features within one lunar degree of the ground track shown in Fig. 7 have been projected into an apparent cross section assuming a subsurface velocity of 150 m μsec⁻¹. The surface features show no correspondence to the linear features mapped in Fig. 8.

EXAMPLES OF IMAGERY AND PROFILES

HF imagery

Figure 10 shows an HF1 image-profile on the lunar farside, unadjusted for the ellipticity of the CSM orbit. As a scale of reference, the crater Aitken is about 80 km wide under the orbital track and 5–6 km from rim to floor. We observe that the profile from 114 to 147°E is relatively flat compared to the remainder of the profile farther east. This flat region is deeply notched by two large depressions. One is the 75 km diameter crater Marconi which was profiled roughly half way from the center toward the southern rim. Part of the floor shows in the profile and the southern rim appears in the imagery. The other depression has been recognized by El-Baz (1972) in Apollo 16 orbital photography; he recommended the name Necho for it. This depression is approximately 180 km in diameter and the profile in Fig. 10 is a nearly central section of it. The crater is very old; roughly 50% of its interior is overlapped by the interiors of superposed craters larger than 20 km in diameter. The moderately rough plain between Marconi and "Necho" appears to be very primitive in the sense of being a little-dissected remnant of an old plain situated between many 25 to 100 km craters.

High points along the profile in Fig. 10 near 160°E, 177.5°W, and 170°W may represent the high parts of very ancient multi-ringed basins tentatively recognized during the preparation of geologic maps for planning the Lunar Sounder observations. In this partial, preliminary mapping, the two basins in question can best be fitted by a 630 km diameter circle centered north of the track of revolution 16–17 near 9°S, 164°E and circles of 690, 930, and 1140 km centered south of the track near 32°S and 159°E.

VHF imagery

Figure 11 shows a VHF image and profile of a central pass over the crater Hevelius. The north rim of the crater is imaged along with the smaller crater

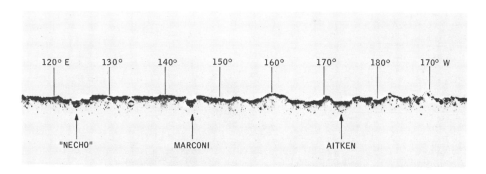

Fig. 10. A segment of HEl imagery in the western half and central part of the lunar farside obtained on revolutions 16–17. The longitudes shown are preliminary and probably correct to within about $\frac{1}{2}°$.

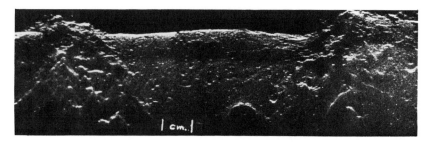

Fig. 11. VHF image of the crater Hevelius. The "cm" scale indicates an along track
distance of 11.1 km and a profile distance of 2.325 km.

Hevelius A. The profile shows a rim-to-rim distance for Hevelius of about 110 km.
The maximum vertical distance from rim to floor is approximately 2 km. The floor
is arched and shows tension features. The maximum amount of arching is about
450 m. A tentative interpretation of the arching is rebound following the
Hevelius impact event.

Galactic Noise Data

The receive-only data showed a very high terrestrial component on the lunar
nearside. The nighttime portion of the lunar farside provided, however, an opportunity to measure the galactic noise.

The noise brightness intensity measure on the nighttime lunar farside is
approximately 8×10^{-21} W m^{-2} Hz^{-1} sr^{-1} for both the HF1 and HF2 channels. This
result is in generally good agreement with measurements taken by other workers
(e.g., Alexander *et al.*, 1969). By contrast the nearside brightness intensities for
both HF frequencies are approximately 2.3×10^{-19} W m^{-2} Hz^{-1} sr^{-1}.

The ALSE results must be taken as preliminary as the CSM guidance and
navigation system is a source of interference in the HF range whose contribution
is being investigated.

Summary

In this paper we have attempted to briefly describe the Apollo Lunar Sounder
Experiment. We have given a few basic examples of the ability of the experiment
to sound the lunar subsurface, to image and profile the lunar surface, and to
measure the galactic noise background. Future papers shall deal with the synthesis of ALSE data with other types of data to help answer a variety of lunar
questions.

References

Alexander J. K., Brown L. W., Clark T. A., Stone R. G., and Weber R. R. (1969) The spectrum of the
cosmic radio background between 0.4 and 6.5 MHz. *Astrophys. J. (Letters)* **157**, L163.

El-Baz F. (1972) Discovery of two lunar features. In *Apollo 16 Preliminary Science Report*, NASA SP-315, 29-33 to 29-38.

Phillips R. J., Adams G. F., Brown W. E. Jr., Eggleton R. E., Jackson P., Jordan R., Linlor W. I., Peeples W. J., Porcello L. J., Ryu J., Schaber G., Sill W. R., Thompson T. W., Ward S. H., and Zelenka J. S. (1973) Apollo Lunar Sounder Experiment, to appear in *Apollo 17 Preliminary Science Report*.

Proceedings of the Fourth Lunar Science Conference
(Supplement 4, *Geochimica et Cosmochimica Acta*)
Vol. 3, pp. 2833–2845

Subsatellite measurements of the lunar magnetic field

C. T. Russell, P. J. Coleman, Jr.

Institute of Geophysics and Planetary Physics, University of California, Los Angeles 90024

B. R. Lichtenstein, G. Schubert, and L. R. Sharp

Department of Planetary and Space Science, University of California, Los Angeles 90024

Abstract—The Apollo 15 subsatellite magnetometer data have been used to map the lunar magnetic field over a narrow band of the lunar surface. Within this band the magnetic field is generally stronger and more variable over the farside highlands than the nearside maria. The correspondence between the strong variable lunar field regions and the source regions for limb compressions suggests that limb compressions arise as the result of the deflection of the solar wind just upstream of the terminator by the lunar magnetic field. Using this apparent relationship between field strength and limb compressions source regions, we deduce that the field strength in the northern farside highlands is not as strong as in the southern hemisphere at similar longitudes. This agrees with Apollo 16 subsatellite measurements in the northern hemisphere but closer to the equator.

Fourier analysis of the orbital plane components of the lunar magnetic field at 100 km altitude places an upper limit of 2.1×10^{18} Γ-cm^3 on the present day lunar magnetic dipole moment in the orbital plane. If the total magnetic dipole moment is parallel to the rotation axis of the moon, this measurement places an upper limit of 4.4×10^{18} Γ-cm^3 on the lunar moment. This is 5×10^{-8} of the earth's magnetic moment.

Simultaneous measurements of the interplanetary magnetic field obtained by Explorer 35 and the Apollo 15 subsatellite above the dayside hemisphere are essentially identical. Thus, both instruments are measuring the undisturbed interplanetary field. Exceptions to this rule occur near the lunar limbs where the occasional occurrence of limb compressions and the possible influence of upstream wave phenomena cause differences between the two sets of data. Thus, if the regions of the terminators and of the lunar cavity are avoided the subsatellite magnetometer measurements can be used as a measure of the driving function in lunar conductivity studies.

Introduction

Our present knowledge of the lunar magnetic field is derived from three sources: analysis of the remanent magnetization of the returned lunar samples; *in situ* measurements at the Apollo landing sites; and orbital measurements with the Explorer 35 satellite and the Apollo 15 and 16 subsatellites. The magnetic fields observed by the orbital and surface magnetometers are almost certainly due to the remanent magnetization of near surface material. The observed magnetic field varies markedly from site to site, from about 6γ at the Apollo 15 landing site (Dyal *et al.*, 1972a) to over 300γ at the Apollo 16 site (Dyal *et al.*, 1973). Significant field changes, even reversals have been observed during the Apollo 14 and 16 surface traverses with the lunar portable magnetometers (Dyal *et al.*, 1971; Dyal *et al.*, 1973). Furthermore, the lunar field does not vary with distance from the center of the moon as would be expected for a planetary field, but rather varies strongly with altitude measured from the surface, as would be expected for sources due to magnetization of surface material (Sharp *et al.*, 1973).

Studies of the remanent magnetization of lunar rocks can be used to deduce the strength of the ancient lunar field, while analysis of the Apollo subsatellite data can be used to infer the strength of the present field. This places important constraints on the theories of lunar formation and evolution. While only a polar lunar orbit would permit the mapping of the entire surface, a large enough region of the moon has been magnetically mapped to allow us to draw several important conclusions. Furthermore, although these magnetic maps can be constructed only from data obtained while the moon is in the geomagnetic tail, they can apparently be extended qualitatively by use of the occurrence of limb compressions at the terminators when the moon is in the solar wind.

In the following sections, we present a brief review of the mapping of the lunar field with the Apollo 15 subsatellite magnetometer, the details of which can be found elsewhere (Sharp *et al.*, 1973). We then discuss the magnitude of the present lunar dipole field and the implications of this measurement. Next, we examine the use of the occurrence rate of limb compressions, when the moon is in the solar wind, to extend our magnetic field maps qualitatively. Finally, we close with some remarks on the use of the subsatellite data in sounding studies of the electrical conductivity of the lunar interior.

INSTRUMENTATION

The Apollo 15 and 16 Particles and Fields subsatellites are essentially identical small satellites which were placed in lunar orbit from their respective command modules at an altitude of about 100 km. The resulting orbit periods are approximately two hours. Magnetometer observations continued on the Apollo 15 mission for seven months until a telemetry system malfunction terminated the magnetometer data words, though the satellite and other parts of the telemetry system continued to function. Apollo 16 subsatellite operations ceased after only 35 days when the subsatellite crashed into the moon. The orbits of the Apollo 15 and 16 subsatellites were inclined by 28° and 10°, respectively, to the lunar equator.

Each magnetometer has two sensors, one parallel to the spin axis and one in the spin plane. The spin plane sensor returns the magnitude and phase of the field in the spin plane. Computer processing later permits the construction of the three components of the vector field. The subsatellites and the magnetometers have been described in previous reports (Coleman *et al.*, 1972a, b; 1973).

MAGNETIC FIELD MAPS

The optimum location for mapping the lunar field is in the lobes of the geomagnetic tail. Here, there is very little plasma to distort the lunar field and temporal fluctuations of the ambient magnetic field are small in contrast to the conditions in the solar wind, in the magnetosheath and in the plasma sheet which lies at the center of the tail roughly parallel to the ecliptic plane. Such favorable conditions for mapping occur on only about 4 days per lunation. Thus, although the moon completely rotates under the satellite's orbit each month, only a small band of the lunar surface can be mapped. If the orbit plane of the subsatellite were fixed in inertial space, this surface band would change from month to month. Unfortunately for the mapping studies, the subsatellite orbit plane processed in such a way that similar ground tracks were covered each month.

In our early results, obtained before computer processing of the data was initiated, we presented field maps derived from the magnitude of the transverse field in the spin plane (Coleman *et al.*, 1972b). On these maps, the zero field contour was arbitrarily chosen to occur above the northern rim of the farside lunar crater Van de Graaff. Regions of overlap of different orbits were then used to reference all orbits to the Van de Graaff zero level. While this technique permitted a qualitative survey of the behavior of the lunar field in different regions of the moon, it was subject to systematic biases as the distance of the closest approach of an orbit to Van de Graaff increased.

With the initiation of computer processing of these data a different approach has been used in which each orbit provides an independent sample of the lunar field. First, the average field in solar ecliptic coordinates for each orbit is subtracted from each measurement during the orbit. This separates the geomagnetic and lunar fields, if the geomagnetic field is constant for the two hour orbital period and if there are no lunar sources of constant field around the orbit. These residual fields are then expressed in a local selenographic system of radial, east and north components, referenced to a common 100 km attitude, and used to construct contour maps of the lunar field.

Maps of the radial component of the magnetic field have been published by Sharp *et al.* (1973). At 100 km altitude, this field is small, typically less than 1 γ. In the region surveyed, the nearside field maps show features with generally larger scale sizes and smaller amplitudes than the farside maps. The nearside ground tracks generally cross mare basins and the farside tracks irregular highland terrain. To this extent, the magnetic field and the terrain are correlated in these regions. Finally, we note that the maps of Sharp *et al.* (1973) were constructed from only part of the data which will eventually become available and will be revised and updated in the future.

THE DIPOLE MAGNETIC FIELD

A magnetic dipole moment perpendicular to the orbit plane of the subsatellite would contribute a constant field around the orbit. In our procedure, we would remove any such steady component of the field. Thus, with our present techniques we cannot measure the magnetic moment perpendicular to the orbit plane. However, we can measure the dipole moment in the orbit plane. Figure 1 shows the longitudinal variation of the radial component of the magnetic field averaged over the latitude band mapped by the Apollo 15 subsatellite. The first harmonic of this variation is a measure of the dipole moment. Figure 2 shows the Fourier harmonics of the radial and of the horizontal components. The amplitude of the first harmonic is about 0.05 γ, and is quite similar in amplitude to neighboring harmonics.

Two sources of error in this determination are the power introduced by the process of digitizing the data and the power in natural fluctuations of the field of the earth's magnetotail at half the orbital period. Since the digital error of an

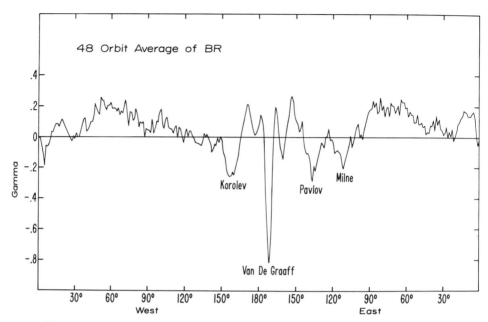

Fig. 1. Forty-eight orbit average of the radial component, B_R, of the lunar magnetic field. Positive values indicated fields directed radially outward, (after Sharp *et al.*, 1973).

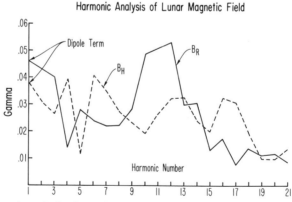

Fig. 2. Harmonic analysis of field in orbit plane. Fourier components of both the radial field, B_R, and the field in the orbit plane measured parallel to the surface, B_H, are shown.

individual vector component measurement is 0.4γ and there are approximately 300 such measurements per orbit, the error due to digitization in determining the amplitude of the first harmonic for an individual orbit would be about 0.01γ. Averaging 48 orbits together would reduce the error even further. Thus, the error introduced by the digitization process is insignificant.

The power in natural fluctuations in the tail field at periods of one hour has not been investigated to our knowledge. Therefore, it is difficult to estimate the contribution of this source of error. However, our 48 orbit average would reduce the contribution of this source by a factor of 7. Thus, a 0.05γ first harmonic could be produced by an average wave amplitude of tail fluctuations at an hour period of 0.35γ. The existence of such small fluctuations in the quiet data, which we have selected for the mapping and Fourier analysis, cannot be ruled out. Hence, this measurement of a 0.05γ first harmonic may, in fact, be a measure of our noise level and therefore it can only provide an upper limit to the present day dipole moment.

A first harmonic variation of 0.05γ, accounting for the attenuation caused by subtracting the mean field for each orbit, is equivalent to a dipole moment of 2.1×10^{18} Γ-cm^3. If we assume that the overall dipole moment is parallel to the axis of lunar rotation, then the moment along the lunar spin axis, which would give the observed projection in the subsatellite orbit inclined 28° to the lunar equator, is 4.4×10^{18} Γ-cm^3. This is 5×10^{-8} of the earth's magnetic moment.

It is interesting to compare this value with the range of estimates of the ancient lunar magnetic field, of from 1500γ to 5000γ, deduced from studies of the returned lunar samples (Helsley, 1970, 1971). If, for example, we use the recent determination of the paleointensity of 2100γ by Gose *et al.* (1973) using the low metamorphic grade breccia 15498, and assume this was the equatorial field of a lunar dipole, then the ancient lunar magnetic moment was 1.1×10^{23} Γ-cm^3. Thus, the ratio of the present to hypothesized ancient magnetic moment is less than 4×10^{-5}.

Runcorn and Urey (1973) have suggested that an early lunar dipole moment developed as cold magnetized material accreted during lunar formation, but that much of this magnetization has been removed by subsequent heating of the lunar interior. We can use our ratio to determine the thickness of the remaining shell of magnetization assuming that the original dipole moment was due to uniform magnetization. The remaining shell must be less than 23 m thick. This is, of course, equivalent to saying none of this hypothesized lunar dipole field now exists whether it was due to remanent magnetization or an internal dynamo. While "magcons" or magnetic features could be due either to islands of magnetization, or holes in an otherwise uniformly magnetized crust caused by meteorite impact or local heating (Coleman *et al.*, 1972b), our data are not consistent with holes in a layer of the Urey-Runcorn primordial material, since, if this material ever existed, its average thickness over the lunar surface is vanishingly small. However, the existence of islands of such uniformly magnetized material is not inconsistent with our data. A scheme for the formation of magnetized islands has been proposed by Strangway *et al.* (1973). In this model the outer crust of the moon always had a low magnetization with the exception of small regions of enhanced iron content.

Finally, we note the distinction between the lunar dipole discussed above and the magnetic dipole possibly induced in permeable lunar surface material by the geomagnetic field as discussed by Parkin *et al.* (1973). While both measurements are made in the geomagnetic tail and result in a similar amplitude surface field, the induced lunar dipole is fixed in solar magnetospheric coordinates rather than selenographic coordinates. In other words, this latter induced dipole is fixed with respect to the direction of the geomagnetic tail field while the planetary dipole that we have investigated is fixed in the moon. Since the direction of such an induced dipole moment reverses between the north and south lobes of the geomagnetic field, this induced field would average to zero in our analysis which includes data from both lobes.

LIMB COMPRESSIONS

When the moon is in the solar wind the subsatellite magnetometer detects the occasional presence of very large increases in the magnetic field over the lunar limb region, which we refer to as limb compressions (Coleman *et al.*, 1972b; Sharp *et al.*, 1973; Russell and Lichtenstein, 1973). The interaction of the solar wind with the moon, and the lunar cavity were first investigated with the Explorer 35 satellite (Colburn *et al.*, 1967; Lyon *et al.*, 1967; Ness *et al.*, 1967). The limb compression phenomenon was not treated in the early papers but has since been extensively investigated by both Explorer 35 magnetometer investigator groups (cf. Ness *et al.*, 1968; Colburn *et al.*, 1971). At the present time, there is very little agreement between these two groups as to the basic nature of this phenomenon (Sonett and Mihalov, 1972b; Ness and Whang, 1972).

Limb compressions observed at the subsatellite altitude appear to be much stronger than those observed at the higher altitudes of Explorer 35 ($\geq$850 km). Examples of these features have been shown by Sharp *et al.* (1973). Such features may be seen at both limbs, either limb, or neither limb. When they do occur at a particular limb crossing, they generally occur on successive crossings of the same lunar limb, even though there may be no limb compression at the other limb crossing on those same orbits. Furthermore, limb crossings on successive orbits often reveal similar fine structure while there is a large variety of limb compression signatures seen over the subsatellite lifetime.

Limb compressions measured by the subsatellite magnetometers vary in magnitude from less than 1 gamma (observable during periods of very quiet solar wind) to strengths approaching those observed at the earth's bow shock (approximately 20γ). This wide range of magnitude and structure may be due either to a wide range of source strengths or to the fact that there are several mechanisms responsible for the observed increases in magnetic field magnitude. In a previous study (Sharp *et al.*, 1973) we have examined the possibility that the lunar limb regions were a controlling factor in the appearance of large limb compressions. This was done by plotting, on a map of the lunar surface, the lunar limbs associated with increases in the magnitude of magnetic field in the ecliptic plane of greater than 2γ, using the first three months of data from the Apollo 15 subsatellite. The 2γ criterion was chosen because we were primarily concerned with investigating the source of large limb compressions, and to avoid the problem of identifying small ($<1\gamma$) features in a background of solar wind magnetic fields whose variability was often greater than 1γ. The frequency of limb compressions as a function of the selenographic position of the lunar limbs for each subsatellite orbit was also calculated.

We have made a similar study using the one month of data produced by the Apollo 16 subsatellite before it crashed into the lunar surface. The same criteria were used as in the earlier analysis using Apollo 15 subsatellite data. The results of this study together with the previous data are shown in Fig. 3. The Apollo 16 coverage of lunar limbs is in two latitude bands at 8°N and 8°S, while the Apollo 15 coverage is in two bands at 28°N and S.

These maps show that large limb compressions almost never occur over mare regions. Also, large limb compressions have a very high probability of occurring when certain highland regions are at a lunar limb position, but that there are other highland regions which are not associated with large limb compressions. This map of limb compression sources is similar to one derived from Explorer 35 magnetometer data by Sonett and Mihalov (1972a) and strongly supports their conclusion that some property of the lunar surface at the limb region is related to the source of an observed large limb compression. Care should be taken when comparing these maps because of the different altitudes of observation and the differ-

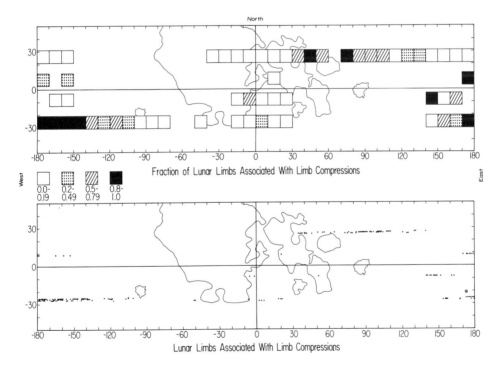

Fig. 3. Limb compression occurrence map derived from the first three months of Apollo 15 and all Apollo 16 subsatellite data. The selenographic positions of the intersection of the orbit plane with the aberrated solar wind terminator have been found for each limb crossing suitable for analysis. These positions are plotted in the lower panel for orbits with limb compression effects. The top panel normalizes these occurrences by the total number of observing periods in each 10° longitude box. At least four observing periods were required before a box was drawn in the upper panel. The outline of the mare regions is shown on both panels.

ent criteria used to define a large limb compression. We consider it significant that both studies show a small probability of observing a large compression when mare regions are at the limb positions and that both satellites identified a strong source region on the far side at southern equatorial latitudes.

In light of surface magnetic measurements revealing fields up to $300\,\gamma$ (Dyal *et al.*, 1973) the most likely candidate for the source of the limb compressions is deflection of the solar wind by magnetic fields associated with regions of strong magnetization. This hypothesis, first put forward by Mihalov *et al.* (1971) and later expanded upon by Barnes *et al.* (1971), remains entirely consistent with all observations. However, experimental results have not ruled out alternate hypotheses for the source of the interaction which also depend on the physical properties of the lunar regions at the limbs. Among these are deflection due to hot photoelectrons (Criswell, 1973) and deflection by a lunar atmosphere (Siscoe and Mukherjee, 1972). In addition, Pérez de Tejada (1973) has presented an MHD boundary layer calculation which predicts magnetic field increases of approximately $1\,\gamma$ over the lunar limb regions. It appears this interaction is distinct from the large $(>2\gamma)$ limb compressions discussed in this section, but may be responsible for the small increases which are occasionally observed.

While we cannot rule out any of the above mechanisms, we can show further evidence of the consistency of the observed large limb compressions with the magnetic deflection hypothesis. There is a small amount of overlap between the regions mapped in Fig. 3 and the band on the lunar surface for which we have contour maps (Sharp *et al.*, 1973). This occurs in two regions; on the farside at longitudes from 140° to 180°E at about 25°S latitudes and on the nearside at longitudes from 10° to 40°W at 25°N latitude. Figure 4 compares the limb compression occurrence rate and the radial component of the lunar magnetic field at 100 km at these two locations. In the northern region, which is over nearside mare terrain, there are no limb compressions in 36 observing periods, and the lunar field is weak. In the southern region, which is over farside highland terrain, the occurrence rate is quite variable ranging from 0 to 90%. The lunar field is much stronger here than in the north and correspondingly quite variable. This similarity suggests that it is magnetic deflection of the solar wind which causes limb compressions.

If this is so, then the limb occurrence compression map provides a means of extending at least qualitatively our knowledge of the lunar magnetic field outside the narrow band mapped directly. Using this hypothesis, the large occurrence rate from 170°E to 140°W longitude at 25°S indicates that the strong fields observed in the Van de Graaff-Aitken region extend much further to the east. Furthermore, it indicates that the magnetic field from 140°E to 140°W at 25°N is very weak despite the superficial similarity of this terrain with the terrain at the same longitude at 25°S latitude. This agrees with the observations of a much weaker field made on the Apollo 16 subsatellite at latitudes from 0 to 10°N and a similar longitude on the farside (Coleman *et al.*, 1973; Sharp *et al.*, 1973). Thus, it appears that, while maria regions tend to have low magnetic fields, highland regions can have either low or high field strengths.

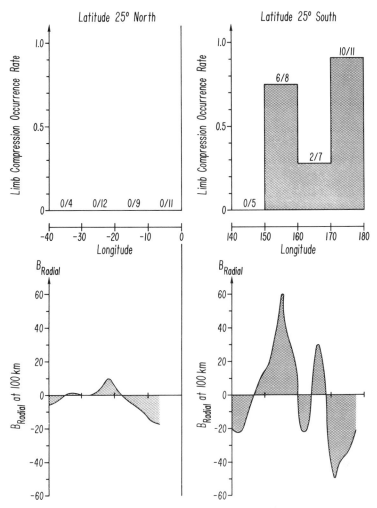

Fig. 4. The occurrence rate of limb compressions and the radial component of the lunar field at 100 km altitude in the two regions of overlap of the limb compression occurrence map (Fig. 3) and the lunar field maps of Sharp *et al.* (1973).

LUNAR ELECTRICAL CONDUCTIVITY STUDIES

The electrical conductivity of the moon has been probed by studying its response to interplanetary time-dependent magnetic fields with periods ten to hundreds of seconds and wavelengths of the order of a lunar radius or larger (cf. Dyal *et al.*, 1972b; Sonett *et al.*, 1972). Such studies require a measure of the driving function, or the undisturbed field external to the moon and the induced field plus driving field on the moon. In past studies, the driving function was assumed measured by the Ames Explorer 35 magnetometer data upstream from the moon during periods of wave trains or large discontinuities in the solar wind or magnetosheath. The induced field was measured together with the driving field and any local surface field by one of the lunar surface magnetometers. One of the objectives of the subsatellite magnetometer experiment was to pursue such sounding studies of the lunar conductivity (Coleman *et al.*, 1972a). However, it was unclear from early measurements in the solar wind whether some of the induced field was present at the altitude of the subsatellite or whether the interplanetary magnetic field was essentially unaffected by the presence of the moon. If the former were true, then the assumption of total dayside confinement of the induced lunar field by the dynamic pressure of the solar wind at the lunar surface, which was used in lunar electrical conductivity studies such as those of Sonett *et al.* (1972) would have to be modified. If induced field were clearly observable at the subsatellite altitude over the dayside hemisphere, this would be a significant modification. However, if the latter were true, then the subsatellite measurements could be used in place of the Explorer 35 measurements as a measure of the input to the moon.

Simultaneous Explorer 35 and Apollo 15 subsatellite data are now available for

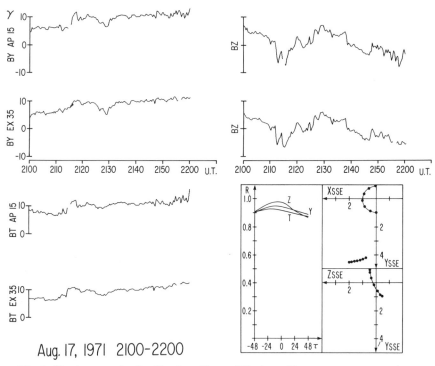

Fig. 5. Simultaneous Apollo 15 subsatellite and Explorer 35 magnetometer records on August 17, 1971 in solar ecliptic coordinates. The lower right-hand panel shows the cross correlation coefficients, *R*, between Explorer 35 and the subsatellite for the *Y* and *Z* components and the magnitude of the field (Schubert *et al.*, 1973). The lag τ is given in seconds. This panel also shows the relative positions of the two spacecraft every ten minutes. Distances are in lunar radii and the motion of both spacecraft is retrograde.

the first lunation after the Apollo 15 launch. They show that, when the moon is in the solar wind, over much of the subsatellite orbit the subsatellite measurements are essentially identical to those of Explorer 35, so that both instruments are measuring an undisturbed interplanetary field (Schubert *et al.*, 1973). Figure 5 shows the data from a subsatellite pass across the sunlit hemisphere of the moon on August 17, 1971 together with the simultaneous measurements of the interplanetary magnetic field by Explorer 35. Cross-correlations of the data and the positions of both satellites during the measurement are also shown. (*See* Schubert *et al.* (1973) for a discussion of this and other examples.) The excellent agreement between the two sets of data is typical of such dayside passes. These data are not identical, however. Most of this difference occurs near the dawn terminator. As discussed in the previous section, deflection of the solar wind can occur in the terminator region. Often, a well defined limb compression is observed. However, it is possible that weak interaction regions at the limb or strong interaction regions upstream from the limb would result in an increased variability of the field here. Alternatively, it is possible that some information about the solar wind-moon interaction in the form of high frequency waves is moving upstream against the solar wind flow near the terminator region. Over the rest of the orbit the differences are minor and may be simply due to differences in the data handling and in the sampling time relative to the arrival of the various discontinuities at the two satellites. Thus, it appears that, if the region of the terminators and, of course, the lunar cavity are avoided, the subsatellite magnetometer measurements can be used as a measure of the driving function in lunar conductivity studies.

SUMMARY AND CONCLUSIONS

The Apollo 15 subsatellite magnetometer data have been used to map the lunar magnetic field over a narrow band of the lunar surface. These maps cover less than 10% of the lunar surface. Within this band, the field is generally weaker and more slowly varying over the maria than in the highlands. Comparison of these field maps with the occurrence of phenomena over the lunar limbs known as limb compressions, indicates that the solar wind is deflected at the limbs when magnetized regions are at the terminators. If we take the occurrence rate of limb compressions to be indicative of the character of the lunar magnetic field at the source region of the limb compression we can draw some qualitative inferences about the magnetic field external to the narrow band actually mapped. In particular, the data suggest that the northern ($\sim 25°$N) farside highlands are only weakly magnetized.

Although the present data analysis techniques permit only a calculation of the dipole moment in the spin plane, they do provide an estimate of the total moment if it lies along the lunar spin axis. We estimate that the present dipole moment is less than $4.4 \times 10^{18} \, \Gamma\text{-cm}^3$ or 5×10^{-8} of the earth's magnetic moment. We note that this is less than 200 times the size of the magnetic moment of the Van de Graaff-Aitken region, and 4×10^{-5} of the ancient lunar dipole moment whose amplitude has been inferred from the magnetization of the returned lunar samples. Thus, if a large lunar dipole moment existed in the past, whether due to dynamo action or cold accretion of magnetized material, essentially none of this field exists today. Our present measurements are also consistent with the hypothesis that the primordial magnetizing field responsible for the magnetization of the returned lunar samples was of external origin, e.g., the early geomagnetic field.

Finally, although the presence of the moon affects the measurements of the interplanetary magnetic field in the limb regions and in the lunar wake region, comparison of simultaneous Explorer 35 and Apollo 15 subsatellite data shows very little if any upstream influence of the moon over the dayside hemisphere in the subsatellite magnetic field data. Thus, with care, the subsatellite data may be used as the driving function in studies of the lunar electrical conductivity.

Acknowledgments—The authors wish to acknowledge many useful discussions of these data with Drs. S. K. Runcorn, G. Siscoe, and C. P. Sonett. This work was supported by the National Aeronautics and Space Administration under contract NAS 9-12236 and by NASA grant NGR 05-007-351. One of us (GS) would like to thank the Guggenheim Foundation for support of his work.

References

Barnes A. P., Cassen S. D., Mihalov J. D., and Eviator A. (1971) Permanent lunar surface magnetism and its deflection of the solar wind. *Science* **172**, 716–718.

Criswell D. R. (1973) Photoelectrons and solar wind/lunar limb interaction. *The Moon* **7**, 202.

Colburn D. S., Mihalov J. D., and Sonett C. P. (1971) Magnetic observations of the lunar cavity. *J. Geophys. Res.* **76**, 2940–2957.

Colburn D. S., Currie R. G., Mihalov J. D., and Sonett C. P. (1967) Diamagnetic solar-wind cavity discovered behind moon. *Science* **158**, 1040.

Coleman P. J. Jr., Schubert G., Russell C. T., and Sharp L. R. (1972a) The particles and fields subsatellite magnetometer experiment. In *The Apollo 15 Preliminary Science Report*, NASA SP-289, 22-1.

Coleman P. J. Jr., Lichtenstein B. R., Russell C. T., Schubert G., and Sharp L. R. (1972b) Magnetic fields near the moon. *Proc. Third Lunar Sci. Conf., Geochim. Cosmochim. Acta*, Suppl. 3, Vol. 3, pp. 2271–2286. MIT Press.

Coleman P. J. Jr., Lichtenstein B. R., Russell C. T., Schubert G., and Sharp L. R. (1973) The particles and fields subsatellite magnetometer experiment. In *The Apollo 16 Preliminary Science Report*, NASA SP-315, p. 23-1.

Dyal P., Parkin C. W., and Sonett C. P. (1972a) Lunar surface magnetometer experiment. In *Apollo 15 Preliminary Science Report*, NASA SP-289, p. 9-1.

Dyal P., Parkin C. W., and Cassen P. (1972b) Surface magnetometer experiments: Internal lunar properties and lunar field interactions with the solar plasma. *Proc. Third Lunar Sci. Conf., Geochim. Cosmochim. Acta*, Suppl. 3 Vol. 3, pp. 2287–2308. MIT Press.

Dyal P., Parkin C. W., Colburn D. S., and Schubert G. (1973) Lunar surface magnetometer experiment. In *Apollo 16 Preliminary Science Report*, NASA SP-315, p. 11-1.

Dyal P., Parkin C. W., Sonett C. P., Dubois R. L., and Simmons G. (1971) Lunar portable magnetometer. In *Apollo 14 Preliminary Science Report*, NASA SP-272, p. 13-1.

Gose W. A., Strangway D. W., and Pearce G. W. (1973) A determination of the intensity of the ancient lunar magnetic field. *The Moon* **7**, 196.

Helsley C. E. (1970) Magnetic properties of lunar 10022, 10069, 10084 and 10085 samples. *Proc. Apollo 11 Lunar Sci. Conf., Geochim. Cosmochim. Acta*, Suppl. 1., Vol. 3. pp. 2213–2219. Pergamon.

Helsley C. E. (1971) Evidence for an ancient lunar magnetic field. In *Proc. Second Lunar Sci. Conf., Geochim. Cosmochim. Acta*, Suppl. 2, Vol. 3, pp. 2485–2490. MIT Press.

Lyon E. F., Bridge H. S., and Binsack J. H. (1967) Explorer 35 plasma measurements in the vicinity of the moon. *J. Geophys. Res.* **72**, 6113.

Mihalov J. D., Sonett C. P., Binsack J. H., and Moutsoulas M. D. (1971) Possible fossil lunar magnetism inferred from satellite data. *Science* **171**, 892.

Ness N. F. and Whang Y. C. (1972) Reply, *J. Geophys. Res.*, **77**, 6924.

Ness N. F., Behannon K. W., Scearce C. S., and Cantarano S. C. (1967) Early results from the magnetic field experiment on lunar Explorer 35. *J. Geophys. Res.* **72**, 5769–5778.

Ness N. F., Behannon K. W., Taylor H. E., and Whang Y. C. (1968) Perturbations of the interplanetary magnetic field by the lunar wake. *J. Geophys. Res.* **73**, 3421.

Parkin C. W., Dyal P., and Daily W. D. (1973) Magnetic permeability and iron abundance of the moon from magnetometer measurements (abstract). In *Lunar Science—IV*, p. 584. The Lunar Science Institute, Houston.

Pérez de Tejada H. (1973) On the continuum fluid approach to the solar wind-moon interaction problem. *J. Geophys. Res.* **78**, 1711.

Runcorn S. K. and Urey H. C. (1973) A new theory of lunar magnetism. *Science* **180**, 636.

Russell C. T. and Lichtenstein B. R. (1973) On the source of lunar limb compressions. Submitted to *J. Geophys. Res.*

Schubert G., Lichtenstein B. R., Coleman P. J. Jr., and Russell C. T. (1973) Simultaneous observation of transient events by the Explorer 35 and Apollo 15 subsatellite magnetometers: Implications for lunar electrical conductivity studies. Submitted to *J. Geophys. Res.*

Sharp L. R., Coleman P. J. Jr., Lichtenstein B. R., Russell C. T., and Schubert G. (1973) Orbital mapping of the lunar magnetic field. *The Moon* **7**, 322.

Siscoe G. L. and Mukherjee N. R. (1972) Upper limits on the lunar atmosphere determined from solar-wind measurements. *J. Geophys. Res.* **77**, 6042–6051.

Sonett C. P. and Mihalov J. D. (1972a) Lunar fossil magnetism and perturbation of the solar wind. *J. Geophys. Res.* **77**, 588.

Sonett C. P. and Mihalov J. D. (1972b) Comment on paper by Y. C. Whang and N. F. Ness, 'Magnetic field anomalies in the lunar wake.' *J. Geophys. Res.* **77**, 6922.

Sonett C. P., Smith B. F., Colburn D. S., Schubert G., and Schwartz K. (1972) The induced magnetic field of the moon: Conductivity profiles and inferred temperature. *Proc. Third Lunar Sci. Conf.*, *Geochim. Cosmochim. Acta*, Suppl. 3, Vol. 3, 2309–2336. MIT Press.

Proceedings of the Fourth Lunar Science Conference
(Supplement 4, *Geochimica et Cosmochimica Acta*)
Vol. 3, pp. 2847–2853

Some correlations between measurements by the Apollo gamma-ray spectrometer and other lunar observations

J. I. Trombka

Goddard Spaceflight Center, Greenbelt, MD

J. R. Arnold, R. C. Reedy,* and L. E. Peterson

University of California, San Diego, La Jolla, CA

A. E. Metzger

Jet Propulsion Laboratory, California Institute of Technology, Pasadena, CA

Abstract—Observations by the Apollo 15 and 16 gamma-ray spectrometers are compared with those of a number of other experiments, both compositional and non-compositional. A general correspondence with topography is seen. The Van de Graaff area is a unique farside region with respect to observations by the laser altimeter, the subsatellite magnetometer, and the gamma-ray spectrometer. X-ray and alpha particle orbital measurements show a broad general agreement with gamma-ray data, though results from additional elements in the gamma-ray spectrum are needed to extend the comparison with X-ray data. A comparison of Th concentrations with those found at various landing sites shows generally good agreement, with the orbital values tending to be somewhat higher.

Results of the Apollo orbital gamma-ray spectrometer experiment have begun to show notable correlations with a number of other lunar observations.

Figure 1 shows a profile of lunar topography below the Command and Service Modules (CSM) for single orbits of the Apollo 15 and 16 missions, as observed by the laser altimeter (Sjogren and Wollenhaupt, 1973). Superimposed are points representing the relative concentration of natural radioactivity as derived from the integral gamma-ray count between 0.55 MeV and 2.75 MeV (Metzger *et al.*, 1973). The major contribution to the measured flux arises from the cosmic-ray induced continuum which varies little with chemical composition. Therefore, observed differences in the measured flux actually represent more than an order of magnitude variation in the radioactive content of the lunar surface. The Apollo 16 intensities have been normalized to those of Apollo 15, by using a time-weighted ratio of values within the cross-over areas. The circled gamma-ray intensity values are plotted for every 15° of longitude and represent the local 5°×5° segment of area through which the laser altimeter ground track passes. The gamma-ray data are corrected for changes in spacecraft altitude, as well as for differences in elevation, with the assumption that a 5°×5° area can be characterized by the mean elevation measured by the laser altimeter within that segment. The elevation correction is

---

*Present address: Los Alamos Scientific Laboratory, Los Alamos, NM.

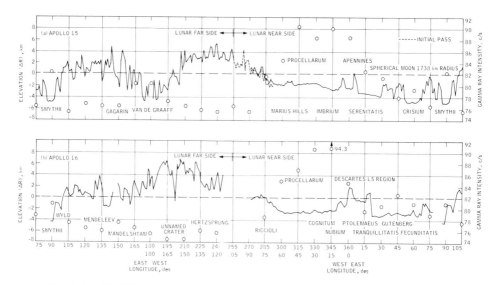

Fig. 1. Profile of lunar topography and natural radioactivity across the Apollo 15 and Apollo 16 ground tracks as measured by the laser altimeter and gamma-ray spectrometer. The elevation is represented by the line, radioactivity plus the underlying continuum by the points.

based on the ratio of the solid angle corresponding to the mean elevation to the solid angle of a spherical moon. Figure 1 displays a strong inverse correlation between elevation and natural radioactivity over the entire 360° track. Regions of high elevation are characterized by low natural radioactivity and vice versa. On the farside, this inverse correlation extends to an observation of greater east-west asymmetry around 180° for the Apollo 16 trajectory than that of Apollo 15, an asymmetry which exists for both elevation and natural radioactivity. The correlation does not apply consistently to an intercomparison among maria. This relationship appears to be reflecting both the nature and extent of major lunar differentiation processes. If the moon is in isostatic equilibrium, the more extensive the early anorthositic differentiation, as characterized by lower densities and lower concentrations of the naturally radioactive nuclides than found in mare regions, the higher the elevation and the lower the radioactivity expected.

The major depression which occurs in the vicinity of the crater Van de Graaff and shows the sharpest contrast in elevation with adjacent highlands along either ground track, i.e., about 8 km, is also the site of the one major farside enhancement in natural radioactivity. The depression extends some 30° in longitude on either side of 180°, and the gamma-ray feature is of comparable extent. Figure 2 shows the Van de Graaff area in detail. The gamma-ray intensities (counts/sec) between 0.55 MeV and 2.75 MeV are displayed in 2° × 2° segments of area north and south of the laser altimeter track. In this case the gamma-ray data have been corrected for spacecraft altitude but not for differences in elevation. Also shown is the lunar surface magnetic field as mapped by the Apollo 15 subsatellite mag-

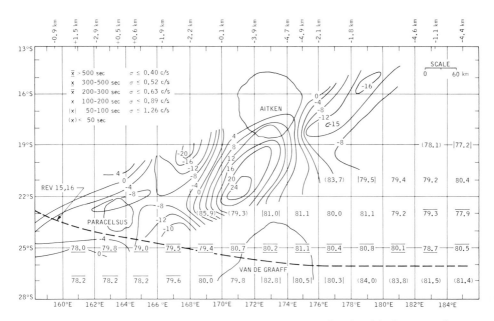

Fig. 2. The region around the crater Van de Graaff, showing the altitude-corrected, gamma-ray intensities between 0.55 MeV and 2.75 MeV in $2° \times 2°$ areas, the location of the laser altimeter ground track with elevation given at the top, and a contour map of the lunar contribution to the solar-directed component of the magnetic field from an elevation of 67 km. The subsatellite magnetometer values are in tenths of gammas. The insert gives the coding which represents the observation time within the $2° \times 2°$ area, and the maximum 1σ standard deviation corresponding to that period.

netometer with contours given in tenths of gammas (Coleman *et al.*, 1973). The elevation track is shown by the dotted line with variations from a mean radius of 1738 km tabulated at the top (Sjogren and Wollenhaupt, 1973). The magnetic feature shown is the strongest seen by either the Apollo 15 or 16 subsatellite magnetometer (Coleman *et al.*, 1973). Ignoring values based on less than 50 sec of counting time, the gamma-ray high is located identically with the minimum in elevation. Thus, over that portion of the lunar surface scanned to date, the largest surface remanent magnetic field, one of the deepest depressions, and the only significant farside enhancement in radioactivity have all been observed within about 5° (150 km) of each other. This area is a notable and to date, quite singular exception to the general conditions prevailing on the farside and, when understood, is likely to contribute significantly to an understanding of lunar evolutionary processes.

Turning to other compositional measurements, a comparison of results from the gamma-ray spectrometer with those of the X-ray spectrometer reflects in both cases an ability to characterize most simply the highland and mare compositions as dominated by anorthositic and a combination of basalt and KREEP (Meyer *et al.*, 1971) components, respectively. This follows from observations by the X-ray spectrometer of distinctly higher Al/Si ratios over highland regions than over the maria (Adler *et al.*, 1972a), while the natural radioactivity is lower over the highlands than over maria regions (Metzger *et al.*, 1972; Metzger *et al.*, 1973). This correspondence does not extend to a comparison of eastern and western maria. The pronounced general enhancement of the KREEP component in the western maria is reflected in the natural radioactivity, the concentration of which is more than a factor of 2 greater in the western maria than in eastern maria regions, while the Al/Si and Mg/Si ratios agree within 30%.

On the other hand, within the western unit of Mare Imbrium and Oceanus Procellarum, two of the three local areas of enhanced natural radioactivity are found south of Fra Mauro and somewhat southeast of Archimedes. Although the concentrations of Th and U are comparable, the Al/Si ratio is significantly higher over the area south of Fra Mauro. These distinctions between the observations of the two experiments are to be expected with a surface composition containing three or more components of characteristic composition, at least two of which can vary independently.

By comparison with returned sample analyses, both X-ray and gamma-ray data demonstrate the existence of more than one significant rock component in the lunar highlands. Ratios of Al/Si above that of anorthositic gabbro suggest an anorthosite or gabbroic anorthosite component such as has been identified in studies of glass fragments from highland sites (Adler *et al.*, 1972b; Ridley *et al.*, 1973). The average Th concentration of the highlands, while lower than that of the maria, is still relatively high and much in excess of the plagioclase rock types identified to date. This suggests the presence of another component in the highlands, but whether it is KREEP, the very high aluminum basalt recently proposed as a distinct lunar type associated with highland regions (Bansal *et al.*, 1973), an as yet unidentified component, or some combination of these remains to be seen. Derivation of K and Fe concentrations from the gamma-ray measurements will be helpful in resolving this point.

The alpha-particle spectrometer has observed the presence of ^{222}Rn, a radioactive daughter of ^{238}U, in concentrations which vary over the lunar surface (Gorenstein and Bjorkholm, 1973). The highest concentrations have been found in the western nearside quadrant in which Mare Imbrium and Oceanus Procellarum make up more than half the ground track, the next highest in the eastern nearside quadrant, followed by the farside (Gorenstein *et al.*, 1973). This agrees with the order of natural radioactivities observed by the gamma-ray spectrometer. Since the source depth of the emanation is limited by the half-life of ^{222}Rn (3.8 d) and the rate of diffusion, the effective source sampling depths for the two measurements should be sufficiently similar to expect this general correlation.

It is notable that the locale of maximum ^{222}Rn emission observed by the alpha-particle spectrometer is in the vicinity of Aristarchus, where one of the three western maria highs in natural radioactivity was found (Trombka *et al.*, 1973; Gorenstein and Bjorkholm, 1973). Present analysis centers the gamma-ray feature less than 5 degrees west of the alpha-particle maximum. Other local gamma-ray features have not been identified as sources of ^{222}Rn.

Absolute concentrations of elements are being obtained both from a computer matrix inversion analysis of the entire spectrum and by analysis of the intensities of the full-energy peaks of individual gamma rays. The periods of spectrum accumulation used for these have been on the order of 30 minutes and greater. Preliminary results of Th for shorter periods of time and therefore over a smaller area [down to the estimated 2.5° (70 km) of resolution when the contrast in concentration is distinctive] have been obtained by applying the ratios of integral count rates from 0.55 to 2.75 MeV for a local area to a large one whose concentration has been obtained via spectral analysis. This permits the comparisons with local sites from returned lunar sample analyses shown in Table 1. An earlier calculation made use of the whole moon as the large area reference (Trombka et al., 1973). Here eastern region (5–60°E) and western region (0–90°W) references have been applied to the local areas as appropriate to diminish the effect of other compositional differences. The areas sampled are either 2° × 2° or 5° × 5° depending on the available counting times.

The Mare Tranquillitatis and Fra Mauro ground track segments of area corresponding to the closest approach to the landing sites are centered some 200 km away. The agreement between the Apollo 11 soil analyses and the Surveyor V alpha scattering measurements, as well as the relative uniformity of response observed in Mare Tranquillitatis by the X-ray spectrometer indicates that the Apollo 11 site represents an excellent region for comparing the results of local and regional measurements. In the case of the Fra Mauro formation, either the abundance of the KREEP component is decreasing towards the south or the very high concentration of Th at the Apollo 14 landing site is a distinctively local characteristic. The Hadley rille site is a transition region between the Apennine highlands and the contiguous mare, so a large variation in composition over a short

Table 1. Thorium concentrations in ppm.

	Mare Tranquil.	Fra Mauro	Hadley	Descartes	Littrow	Luna 16 site
Returned sample analyses of soils	2.2[a]	11.6[b]	3.8[c]	2.0[d]	1.9[e]	1.2[f]
Gamma-ray Spectrometer[g]	2.6[h,i]	8.3[h,j]	4.7[j]	2.3[j]	2.2[i]	1.9[i]

[a]LSPET (1972).
[b]Eldridge et al. (1972).
[c]O'Kelley et al. (1972); Rancitelli et al. (1972).
[d]Eldridge et al. (1973).
[e]Eldridge—personal communication; O'Kelley et al. (1973).
[f]Tera and Wasserburg (1972).
[g]Systematic error estimated at ± 30%.
[h]Centered 6–8° from the landing site.
[i]5° × 5° segment of area.
[j]2° × 2° segment of area.

distance can be expected and was in fact found for Apollo 15 soils and breccias (Schonfeld and Meyer, 1972). Because of this, the value used in Table 1 is the mean of the separate averages obtained by two investigator groups representing a larger number of samples. The spatial resolution of the Apollo gamma-ray spectrometer, 2.5°–5° depending on the degree of contrast, should be kept in mind.

Preliminary thorium analyses of four Apollo 17 soils with an average concentration of 1.2 ppm were reported at the Fourth Lunar Science Conference (O'Kelley *et al.*, 1973). These were all dark soil samples. Subsequent analyses by the ORNL group of seven light soils which are more typical of the general region have yielded an average of 2.3 ppm (Eldridge, J. S., personal communication, 1973). The value used in the table is the average of all eleven light and dark soil samples and is in essential agreement with the orbital result. Littrow, like the Hadley Rille site, is also at the mare-highland interface on the eastern edge of Mare Serenitatis and the Th concentration observed from orbit falls to about 1.5 ppm several degrees both south and east of the Apollo 17 landing site. The degree of difference between the orbital and local sample data at the Luna 16 site is rather surprising as the site is located well into Mare Fecunditatis, and the value reported (Tera and Wasserburg, 1972) is less than that seen for any 5° × 5° unit within the eastern mare. The low Th concentration in Luna 16 soil may therefore be a local characteristic. Agreement at the Descartes site is good.

This paper has discussed the most obvious correlations between the Apollo gamma-ray spectroscopy experiment and other Apollo experiments. These will become better defined and other areas of comparison will emerge as the data are more fully analyzed and understood.

Acknowledgments—The design, construction and testing of the gamma-ray spectrometer was a joint effort on the part of many individuals at the Jet Propulsion Laboratory and the Analog Technology Corporation of Pasadena, California. It is a pleasure to acknowledge our appreciation at their success. The work was constructively monitored by Johnson Space Center (JSC) engineering and science personnel.

Special thanks are extended to the crews of Apollo 15 and 16, particularly Major Al Worden and Commander Ed Mattingly, the Command Module pilots, for their skill in operating the experiment.

Discussions with a number of our colleagues, I. Adler, P. J. Coleman, Jr., P. Gorenstein, C. Meyer, Jr., and W. L. Sjogren, were helpful in the preparation of this paper.

The work described in this paper was carried out in part under NASA Contract No. NAS7-100 at the Jet Propulsion Laboratory, California Institute of Technology, and in part under NASA Contract No. NAS9-10670 at the University of California, San Diego.

REFERENCES

Adler I., Trombka J., Gerard J., Lowman P., Schmadebeck R., Blodgett H., Eller E., Yin L., Lamothe R., Gorenstein P., and Bjorkholm P. (1972a) Apollo 15 geochemical X-ray fluorescence experiment: Preliminary report. *Science* **175**, 436–440.

Adler I., Trombka J., Gerard J., Lowman P., Schmadebeck R., Blodgett H., Eller E., Yin L., Lamothe R., Osswald G., Gorenstein P., Bjorkholm P., Gursky H., and Harris B. (1972b) Apollo 16 geochemical X-ray fluorescence experiment: Preliminary report. *Science* **177**, 256–259.

Bansal B. M., Gast P. W., Hubbard N. J., Nyquist L. E., Rhodes J. M., Shih C. Y., and Wiesmann H. (1973) Lunar Rock Types (abstract). In *Lunar Science—IV*, pp. 48–50. The Lunar Science Institute, Houston.

Coleman P. F., Lichtenstein B. R., Russel C. T., Schubert G., Sharp L. R. (1973) The particles and fields subsatellite magnetometer experiment. *Apollo 16 Preliminary Science Report*, NASA SP-315, 23-1-13.

Eldridge J. S. (1973) Personal communication.

Eldridge J. S., O'Kelley G. D., and Northcutt K. J. (1972) Abundances of primordial and cosmogenic radionuclides in Apollo 14 rocks and fines. *Proc. Third Lunar Sci. Conf., Geochim. Cosmochim. Acta*, Suppl. 3, Vol. 2, pp. 1651–1658. MIT Press.

Eldridge J. S., O'Kelley G. D., and Northcutt K. J. (1973) Radionuclide concentrations in Apollo 16 samples (abstract). In *Lunar Science—IV*, pp. 219–221. The Lunar Science Institute, Houston.

Gorenstein P. and Bjorkholm P. (1973) Detection of radon emanation from the crater Aristarchus by the Apollo 15 alpha particle spectrometer. *Science* **179**, 792–794.

Gorenstein P., Golub L., and Bjorkholm P. (1973) Spatial non-homogeneity and temporal variability in the emanation of radon from the lunar surface: Interpretation (abstract). In *Lunar Science—IV*, pp. 307–308. The Lunar Science Institute, Houston.

LSPET (Lunar Sample Preliminary Examination Team) (1972) The Apollo 15 lunar samples: A preliminary description. *Science* **175**, 363–375.

Metzger A. E., Trombka J. I., Peterson L. E., Reedy R. C., and Arnold, J. R. (1972) A first look at the lunar orbital gamma-ray data. *Proc. Third Lunar Sci. Conf., Geochim. Cosmochim. Acta*, Suppl. 3, Vol. 3, frontispiece. MIT Press.

Metzger A. E., Trombka J. I., Peterson L. E., Reedy R. C., and Arnold J. R. (1973) Lunar surface radioactivity: Preliminary results of the Apollo 15 and Apollo 16 gamma-ray spectrometer experiments. *Science* **179**, 800–803.

Meyer C. Jr., Brett R., Hubbard N. J., Morrison D. A., McKay D. S., Aitken F. K., Takeda H., and Schonfeld E. (1971) Mineralogy, chemistry and origin of the KREEP component in soil samples from the Ocean of Storms. *Proc. Second Lunar Sci. Conf., Geochim. Cosmochim. Acta*, Suppl. 2, Vol. 1, pp. 393–411. MIT Press.

O'Kelley G. D., Eldridge J. S., and Northcutt K. J. (1972) Primordial radioelements and cosmogenic radionuclides in lunar samples from Apollo 15. *Proc. Third Lunar Sci. Conf., Geochim. Cosmochim. Acta*, Suppl. 3, Vol. 2, 1659–1670. MIT Press.

O'Kelley G. D., Eldridge J. S., and Northcutt K. J. (1973) Solar flare induced radionuclides and primordial radioelement concentrations in Apollo 17 rocks and fines (abstract). In *Lunar Science—IV*, pp. 572–574. The Lunar Science Institute, Houston.

Rancitelli L. A., Perkins R. W., Felix W. D., and Wogman N. A. (1972) Lunar surface processes and cosmic ray characterization from Apollo 12–15 lunar sample analyses. *Proc. Third Lunar Sci. Conf., Geochim. Cosmochim. Acta*, Suppl. 3, Vol. 2, 1681–1691. MIT Press.

Ridley W. I., Reid A. M., Warner J., Brown R. W., Gooley R., and Donaldson C. (1973) Major element composition of glasses in two Apollo 16 soils and a comparison with Luna 20 glasses (abstract). In *Lunar Science—IV*, pp. 625–627. The Lunar Science Institute, Houston.

Schonfeld E. and Meyer C. (1972) The abundances of components of the lunar soils by a least-squares mixing model and the formation age of KREEP. *Proc. Third Lunar Sci. Conf., Geochim. Cosmochim. Acta*, Suppl. 3, Vol. 2, pp. 1397–1420. MIT Press.

Sjogren W. L. and Wollenhaupt W. R. (1973) Lunar shape via the Apollo laser altimeter. *Science* **179**, 275–278.

Tera F. and Wasserburg G. J. (1972) U-Th-Pb analyses of soil from the Sea of Fertility. *Earth and Planetary Science Lett.* **13**, 457–466.

Trombka J. I., Arnold J. R., Reedy R. C., Peterson L. E., and Metzger A. E. (1973) Some correlations between measurements by the Apollo gamma-ray spectrometer and other lunar observations (abstract). In *Lunar Science—IV*, pp. 517–519. The Lunar Science Institute, Houston.

Proceedings of the Fourth Lunar Science Conference
(Supplement 4, *Geochimica et Cosmochimica Acta*)
Vol. 3, pp. 2855–2864

Composition and dynamics of lunar atmosphere

R. R. Hodges, Jr., J. H. Hoffman, and F. S. Johnson

The University of Texas at Dallas
Dallas, Texas

D. E. Evans

Johnson Space Center
Houston, Texas

Abstract—The model of lunar atmosphere is updated to take into account new information on the dynamics and amounts of H_2, ^4He, ^{20}Ne, ^{36}Ar, and ^{40}Ar. Helium and neon appear to be in close balance with the solar wind, although ^{36}Ar is depleted in the atmosphere, suggesting that surface materials are not saturated with argon. Atmospheric carbon compounds, which should result from the solar wind influx of carbon, remain undetected, as do nitrogen compounds. However, evidence of a volcanic gas release is presented, which suggests the transient presence of these elements.

Introduction

THE LUNAR atmosphere is so tenuous that it is a collisionless gas, except for molecular encounters with the surface of the moon. In the absence of particle interactions, hydrodynamic processes do not exist. However, the statistical distribution of molecular trajectories over the moon causes pseudo-collective phenomena, similar to tides, waves and winds, to exist.

In the preliminary analysis of data from the Apollo 17 lunar surface mass spectrometer a diurnal tidal oscillation of helium is clearly present (Hoffman *et al.*, 1973). The nighttime concentration of He is about 20 times that in the daytime, in close agreement with the theoretical model of Hodges (1973).

Helium is unique in lunar mass spectrometric data because there is no contaminant source of a substance with mass of 4 amu. Inasmuch as degassing of remnant spaceflight hardware produces artifacts at virtually all other mass numbers in the daytime, recognition of a native species requires correlation of some part of its diurnal variation with a theoretical model. For example, Hodges and Johnson (1968) pointed out that a gas which condenses on the cold surface of the dark side of the moon will form a pocket of gas over the sunrise terminator due to release of adsorbed gases from the rapidly warming surface. This phenomenon includes a presunrise increase in concentration due to particles which travel westward into the nighttime hemisphere from their point of release near the sunrise terminator. In the Apollo 17 mass spectrometer data it is evident that both ^{36}Ar and ^{40}Ar have precisely this presunrise behavior, but the post sunrise data at masses 36 and 40 amu are complicated by the release of contaminants from the spaceflight hardware at the ALSEP site.

In addition to these diurnal effects, it is important to understand the characteristics of volcanic events, and particularly the differences of such events from artifact gas releases.

This paper presents a review of the present state of knowledge of the dynamics of the lunar atmosphere from both theoretical and experimental viewpoints. It includes data on both diurnal variation and transient volcanic events, which are used to update the model of the composition of the lunar atmosphere.

Sources of Lunar Atmosphere

The solar wind is probably the dominant source of lunar atmosphere. Solar wind ions impact the moon with energies the order of 1 keV per amu, which is sufficient to imbed the ions in surface materials. Present abundances of most trapped solar wind gases in returned soil samples are lower than would be expected if a significant fraction of the impinging solar wind were currently being trapped. Thus it is likely that the soil is saturated with trapped gases and that a detailed balance of the solar wind influx and the release of previously trapped gases exists.

An important verification of this hypothesis is the close balance of the solar wind flux and the lunar atmospheric content of helium (Hodges, 1973). A counter example is the surprisingly small amount of atomic hydrogen detected by the Apollo 17 orbital ultraviolet spectrometer (Fastie *et al.*, 1973), which may be explained by postulating that most of the hydrogen released from the soil is molecular (H_2). The solar wind balanced H_2 model of Hodges (1973) is compatible with the lowest upper bound on H_2 that could be inferred from the data reported by Fastie *et al.* (1973). Contaminant H and H_2 in available mass spectrometric data precludes elucidation of this problem, except to set a nighttime upper bound on H_2 at 6.5×10^4 cm^{-3} (Hoffman *et al.*, 1973).

The amounts of nitrogen, carbon, oxygen, neon, and argon in the solar wind are significant, and the soil is apparently saturated with these elements. Mass spectrometric data suggest that neither N nor C exists in atomic form in the lunar atmosphere. Presumedly molecular nitrogen could be formed, but the preponderance of protons in the impinging solar wind would more likely lead to formation of NH_3. A similar process should lead to production of CH_4. Neither ammonia nor methane, which freeze at 196°K and 91°K respectively, appears in the nighttime lunar atmosphere. Their daytime levels are obscured by contaminants, which should eventually dissipate from the Apollo 17 lunar surface mass spectrometer data. The large amounts of oxygen in the soil may lead to formation of CO and possibly NO, although the reactions to produce these gases would probably be reversible. Oxygen ions of the solar wind must react rapidly with the soil, precluding oxygen in the atmosphere. Neon and argon have been detected on the moon (Hodges *et al.*, 1972b, and Hoffman *et al.*, 1973).

It is clear from the lack of a dense atmosphere that the rate of volcanic degassing of the moon is somewhat less than on earth. Hodges *et al.* (1972a) have found an upper bound on lunar venting to be 1.5×10^{-16} g cm^{-2} sec^{-1}, which is several orders of magnitude less than would occur if the release rate were the same per unit mass as for earth (cf. Johnson *et al.*, 1972). Despite this low average level of volcanic activity, rather convincing evidence of currently active, sporadic volcanism in certain regions of the moon is found in the alpha particle data reported by Gorenstein *et al.* (1973). The time scale of this activity is the order of 21 years, the lifetime of the ^{210}Pb in the decay of ^{222}Rn. It is possible that these events occur so infrequently as to have no more than a transient effect on the lunar atmosphere.

The radioactive decay of uranium and thorium in the moon produces alpha particles, and hence helium, at several times the rate of solar wind influx of helium. However the bulk of these atoms must be permanently trapped within the moon. A test of this hypothesis will be obtained with the Apollo 17 lunar surface mass spectrometer when the moon passes through the high latitude part of the geomagnetic tail, during which time the solar wind source of helium will be eliminated. If the atmospheric helium dissipates at the expected rate (about 1 day^{-1}), the solar wind source will be confirmed. Otherwise a lunar source must be considered.

Decay of ^{40}K in surface materials is probably the main source of ^{40}Ar in the lunar atmosphere. Subsequent photoionization and acceleration of the resulting ions by solar wind fields causes some of the ^{40}Ar to be retrapped in the soil along with solar wind ^{36}Ar, so that the isotopic composition of impacted argon in the soil may be indicative of the amount of ^{40}Ar in the atmosphere (Manka and Michel, 1971).

IDENTIFICATION OF NONCONDENSABLE GASES

On the dark side of the moon, where the temperature falls below 100°K, most gases are adsorbed. Analogy with a laboratory cold trap suggests that hydrogen, helium, nitrogen, and neon would not condense at night. Of these only helium is easily identifiable, because of the absence of 4 amu contaminants in mass spectra from the moon. However, neon has also been detected as the excess of the 20 amu measurement when contaminants have been accounted for (Hodges *et al.*, 1972b, and Hoffman *et al.*, 1973).

Figure 1 shows the presently available data on the concentration of ^{20}Ne at the lunar surface, along with a fitted theoretical distribution ($n \propto T^{-5/2}$, as derived by Hodges and Johnson, 1968), plotted as functions of longitude measured from the subsolar meridian. Circles are from the Apollo 16 orbital mass spectrometer, and squares are preliminary data from the Apollo 17 ALSEP mass spectrometer. Water ($H_2^{18}O$) was a large contaminant in the orbital neon data, and the circles represent all of the data where the water contribution was low enough that it could be subtracted with reasonable accuracy. Even so, the lower bound of the statistical uncertainty of each data point includes zero. These data are generally about a factor of 4 lower than the preliminary value of neon concentration given by Hodges *et al.* (1972b), the difference being due to improved laboratory data on the cracking pattern of water in the mass spectrometer ion source. The paucity of data from the Apollo 17 lunar surface instrument reflects the difficulty in

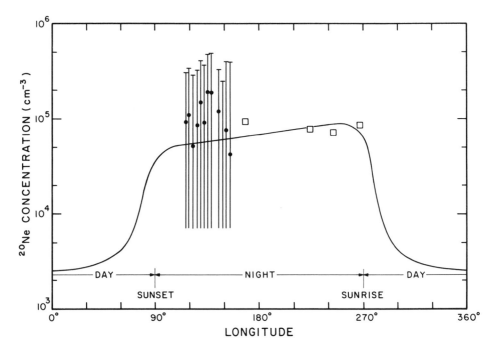

Fig. 1. A hypothetical equatorial distribution of ^{20}Ne at the lunar surface, predicted by the $T^{-5/2}$ law, is shown by the solid curve. Data from the Apollo 16 orbital mass spectrometer are represented by circles, while those from the Apollo 17 lunar surface mass spectrometer are represented by squares.

obtaining a neon measurement. This is accomplished by a complicated command sequence which lowers the ion source temperature sufficiently to condense the contaminant at 20 amu, which is HF (Hoffman *et al.*, 1973).

There is an order of magnitude discrepancy between the predicted neon concentration of Johnson *et al.* (1972); i.e., about $1.3 \times 10^6 \, \text{cm}^{-3}$ at night, and the data given in Fig. 1, necessitating a reexamination of the theory. Subsequent discussion shows that a surprising increase in the calculated rate of loss of neon from the moon arises when the $T^{-5/2}$ concentration distribution is substituted for the previously used stepped model.

The amount of neon in the lunar atmosphere must be in equilibrium with the solar wind influx, i.e.,

$$\pi R^2 \Phi_{sw} = \frac{N_s}{2\tau_i} \tag{1}$$

where R is the lunar radius, Φ_{sw} is the solar wind flux of neon, N_s is the total number of atoms in sunlight, and τ_i is the photoionization lifetime. The factor 2 in the denominator gives an approximate accounting for the fraction of photoions which impact the moon and subsequently return to the atmosphere (Johnson, 1971).

Approximating the vertical distribution of neon as barometric and the surface temperature distribution as symmetrical about the moon–sun axis, N_s is given by

$$N_s = 2\pi R^2 \int_0^{\pi/2} d\psi\, nH \sin\psi + 2\pi \int_{-\infty}^0 dz \int_R^0 d\rho\rho\, ne^{(\sqrt{z^2+\rho^2})/h} \qquad (2)$$

where n is concentration at the surface, H is scale height, ψ is the lunarcentric angle from the subsolar point, while z and ρ are lunarcentric cylindrical coordinates, with z measured toward the sun. The first integral represents the number of atoms in the daytime hemisphere while the second gives those over the night side at altitudes great enough to be in sunlight. Temperature is approximated by radiative equilibrium in daytime ($\propto \cos^{1/4}\psi$) and constant at night. Letting $n \propto T^{-5/2}$, integration of expression 2 results in

$$N_s = 2\pi R^2 n_N H_N \left\{ \frac{8}{5}\left(\frac{T_N}{T_D}\right)^{3/2}\left[1 - \frac{3}{8}\left(\frac{T_N}{T_D}\right)^{5/2}\right] + e^{R/H_N} K_2\left(\frac{R}{H_N}\right) \right\} \qquad (3)$$

where the subscript N denotes nighttime value, T_D is the daytime maximum temperature, and K_2 is the modified Bessel function of the second kind and of order 2. The first of the bracketed terms is due to the daytime integral; and for a day to night temperature ratio of 4 its value is 0.20. The last term represents the nighttime integral; and is 0.21 for $T_N = 95°K$.

The abundance of ^{20}Ne in the solar wind is about 1/570 that of ^{4}He according to Geiss et al. (1972), while the amount of ^{4}He is 0.045 that of hydrogen (Johnson et al., 1972). Assuming an average proton flux of 3×10^8 cm^{-2}sec^{-1}, the solar wind flux of ^{20}Ne should be 2.4×10^4 cm^{-2} sec^{-1}. Using this flux, a photoionization lifetime of 10^7 sec, and the evaluation of expression 3 above in Equation 1, gives 2.3×10^5 cm^{-3} for the nighttime ^{20}Ne concentration. This is about $\frac{1}{5}$ the amount predicted by Johnson et al. (1972), with most of the difference being due to the previous neglect of photoionization near the terminators.

Manka (1972) has suggested that the photoionization lifetime of neon should be 6×10^6 sec, which would further reduce the theoretical nighttime concentration to 1.4×10^5 cm^{-3}. A slight further reduction in the theoretical value can be made on the basis that the moon spends only about 25 of each 29 days in the solar wind (with the other 4 days in the geomagnetic tail), reducing the influx of neon in that proportion, and hence reducing the predicted nighttime concentration of neon to 2.0×10^5 cm^{-3} or 1.2×10^5 cm^{-3}, depending on which photoionization lifetime is used. Table 1 gives the present theoretical estimates and experimental values on the noncondensable gases on the moon. Theoretical values for neon reflect the uncertainty regarding the photoionization lifetime. Atomic hydrogen, nitrogen and noncondensable compounds containing nitrogen and carbon are omitted because they are not present in significant quantities in the nighttime atmosphere.

CONDENSABLE GASES

In available mass spectroscopic data the only obvious condensable gases of the lunar atmosphere are ^{40}Ar and ^{36}Ar, while the presence of ^{222}Rn has been identified in alpha particle data (Gorenstein et al., 1973). Other species, such as ^{38}Ar, NH$_3$, or

Table 1. Noncondensable gases of the lunar atmosphere.

	H$_2$		^{4}He		^{20}Ne
Solar wind flux (cm^{-2}sec^{-1})	3×10^8 (protons)		1.3×10^7		2.4×10^4
Residence time (sec)	6.5×10^3	(a)	7.9×10^4	(a)	$1\text{-}3 \times 10^7$
Surface concentration (cm^{-3})					
Subsolar { theory	2.0×10^3	(a)	1.6×10^3	(a)	$4\text{-}7 \times 10^3$
{ experiment			2×10^3	(b)	
Antisolar { theory	1.2×10^4	(a)	3.8×10^4	(a)	$1\text{-}2 \times 10^5$
{ experiment	$<6.5 \times 10^4$	(b)	4×10^4	(b)	$\sim 10^5$

(a) Hodges (1973)
(b) Hoffman et al. (1973)

CH$_4$, may also exist, but their abundances are not sufficient to overcome the artifact background in the mass spectrum.

Identifiable features of the diurnal variations of a native condensable gas include concentration minima near both the subsolar and antisolar regions, with the former being due to transport effects and the latter due to adsorption on the cold nighttime surface of the moon. Since artifacts dominate the daytime data, only the nighttime behavior has been detected.

To elucidate the nature of a condensable gas it is helpful to use the diffusion approximation of exospheric transport (Hodges, 1972)

$$\Phi_s = \frac{\alpha}{1-\alpha} \frac{n\langle v \rangle}{4} + \Omega \frac{\partial}{\partial \phi} nH - \nabla_h^2 n \langle v \rangle H^2 \qquad (4)$$

where Φ_s is the upward flux near sunrise due to release of adsorbed gas, α is the fraction of the downcoming flux adsorbed by the surface, $\langle v \rangle$ is mean particle speed, Ω is the rate of angular rotation of the moon, ϕ is longitude measured from the subsolar meridian, and ∇_h^2 is the horizontal part of the Laplacian operator in a sun-referenced coordinate system. In the daytime hemisphere α is zero; while at night it must be small as compared to unity for the differential equation to be valid.

To simplify solution of equation 4 it is helpful to restrict the problem to the equator, where meridional flow of the atmosphere must vanish, and then to approximate the flow as though the lunar surface were cylindrical. In addition the release of adsorbed gas is assumed to take place very near the sunrise terminator, so that Φ_s can be approximated as a line source. Then equation 4 becomes 1-dimensional, and its solution can be found easily by integration.

Figure 2 shows a theoretical distribution of ^{40}Ar at the equator, where the longitudinal dependence of the surface adsorption fraction, α, has been chosen to insure that the solution matches the indicated experimental data points from the Apollo 17 lunar surface mass spectrometer. Just prior to sunrise the appropriate value of α is 0.054 while that immediately following sunset is 7×10^{-4}. The difference in these values of α is explainable as an indication of the temperature depen-

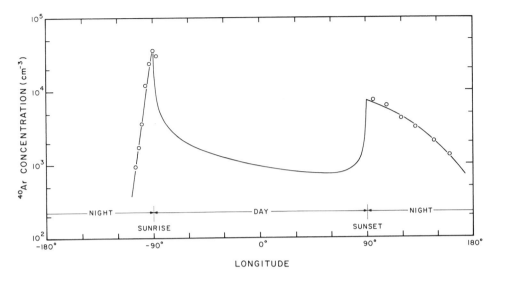

Fig. 2. Equatorial distribution of ^{40}Ar computed from Equation 4, with surface adsorption fraction chosen to fit the Apollo 17 mass spectrometer data.

dence of the adsorption mechanism, wherein adsorption becomes increasingly more likely as the temperature approaches the freezing temperature of argon ($\sim 84°K$).

A similar diurnal variation of ^{36}Ar has been detected, but at a much lower level. In addition, a persistent nighttime contaminant at mass 36 amu precludes accurate determination of the level. However, it is clear that the ratio of ^{36}Ar to ^{40}Ar in the lunar atmosphere is the order of 0.1. This result is perhaps surprising in view of the near equality of ^{36}Ar and excess ^{40}Ar trapped in the soil (cf. Yaniv and Heymann, 1972). It suggests that the soil is not saturated with ^{36}Ar, and hence that most of the impinging solar wind flux of ^{36}Ar is permanently trapped. The amount of ^{20}Ne in returned samples exceeds that of ^{36}Ar by a factor greater than 2 (cf. Eberhardt *et al.*, 1972, or Heymann *et al.*, 1972). Since solar wind argon impacts the moon with almost twice the energy of neon, and hence is implanted deeper, the saturation level of argon might be expected to exceed neon. Thus the hypothesis that the soil is saturated with neon but not with argon is plausible. Owing to the small influx of ^{36}Ar there is no conflict of saturation with the present amount in the soil and a shallow mixing depth (< 10 meters) of the regolith over geologic time.

An estimate of the release rate of ^{40}Ar into the atmosphere can be made by a crude adaptation of the equatorial variation in Fig. 2 to a global distribution, and subsequent determination of the amount of the gas in sunlight, as was done for neon earlier. Using a photoionization lifetime of 1.6×10^6 sec (Manka, 1972), the average flux of ^{40}Ar emanating from the lunar surface is about 2.6×10^3 cm^{-2} sec^{-1}.

EVIDENCE OF LUNAR VOLCANISM

In all of the data from the orbital mass spectrometers on Apollo 15 and Apollo 16, and preliminary data from the Apollo 17 lunar surface instrument, only one probable volcanic event has been discovered. Figure 3 shows measurements at masses 14, 28, and 32 amu from the Apollo 15 orbital mass spectrometer. The sudden excursions of these three masses occurred at 0822 hours GMT on August 6, 1971, as the spacecraft passed over 110.3°W, 4.1°S (i.e., northwest of Mare Orientale and in lunar night). No coincident change occurred at any other mass in the spectrum from 12 to 67 amu. Excursions with amplitudes similar to that at 32 amu would have been detected at all masses except 16, 17, 18, and 44 amu, which were dominated by large contaminant levels (Hodges *et al.*, 1972a). The absence of other substances in this event may be a temporal artifact, caused by a short

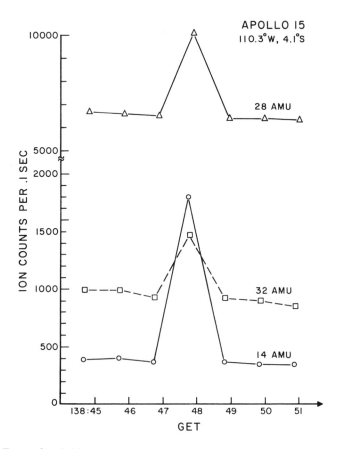

Fig. 3. Event of probable volcanic origin in data from the Apollo 15 orbital mass spectrometer. Ground elapsed time (GET) of 138:48 corresponds to 0822 hours GMT on August 6, 1971. Argon sensitivity corrected for orbital velocity ram effect was 2000 atoms/cc/count/sec, and the ion source used a 70 eV electron beam.

lived disturbance that did not span the entire duration of one sweep of the mass spectrum (62 sec).

It is practical to rule out some conjectured causes of this event. There is no evidence of recurrence of this pattern of gas release that would suggest a spacecraft origin. The lone crew member was asleep when the event occurred, and all monitors of spacecraft operation were nominal. A similar type of perturbation of only 14 and 28 amu shown in Fig. 4 was produced by the release of a large quantity of N_2 from the panoramic camera whenever its control was switched (by the crewman) to "operate." The panoramic camera produced no effect at 32 amu, while the ratio of 28 amu to 14 amu was typical of the cracking pattern of N_2, and different from that of the supposed volcanic event of Fig. 3. Thus accidental release of N_2 from the camera is not a plausible explanation of the event.

While the above comparison seems to indicate that the volcanic gas at 28 amu was not entirely N_2, the absence of a large effect at 12 amu seems to rule out the dominance of CO as well. A mixture of N, N_2, and a small amount of CO would be

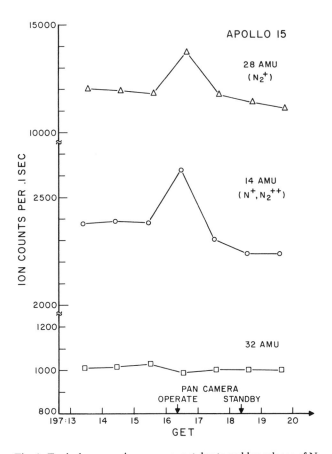

Fig. 4. Typical panoramic camera event due to sudden release of N_2.

plausible, however. Mass 32 amu could have been O_2, or possibly SO_2 if the duration of the event were short enough to have dissipated by the time the instrument measured 64 amu (about 20 sec after the 32 amu measurement).

In a word, the event shown in Fig. 3 asks more questions than it answers. The origin of its component gases is difficult to explain in terms of volcanism. Certainly N and O_2 are unlikely constituents. On the positive side, the rate of gas release necessary to have produced this event can be extrapolated from the work of Hodges et al. (1972a) to be the order of 1 kg/sec, or about 20 kg total, which is small in volcanic terms, albeit a significant contribution to the lunar atmosphere.

Acknowledgments—It is a pleasure to acknowledge the efforts of C. M. Peters and H. D. Hammack in the processing of mass spectrometric data. This research was sponsored by NASA under Contracts NAS9-10410 and NAS9-12074.

REFERENCES

Eberhardt P., Geiss J., Graf H., Grögler N., Mendia M. D., Mörgeli M., Schwaller H., Stettler A., Krähenbühl U., and von Gunten H. R. (1972) Trapped solar wind noble gases in Apollo 12 lunar fines 12001 and Apollo 11 breccia 10046. *Proc. Third Lunar Sci. Conf., Geochim. Cosmochim. Acta,* Suppl. 3, Vol. 2, pp. 1821–1856. MIT Press.

Fastie W. G., Feldman P. D., Henry R. C., Moos V. W., Barth C. A., Lillie C., Thomas G. E., and Donahue T. M. (1973) The Apollo 17 orbital ultraviolet spectrometer experiment (abstract). In *Lunar Science—IV,* p. 233. The Lunar Science Institute, Houston.

Geiss J., Buehler F., Cerutti H., Eberhardt P., and Filleux Ch. (1972) Solar wind composition experiment. *Apollo 16 Preliminary Science Report,* NASA SP-315, pp. 141–149.

Gorenstein P., Golub L., and Bjorkholm P. (1973) Spatial nonhomogeneity and temporal variability in the emanation of radon from the lunar surface: Interpretation (abstract). In *Lunar Science—IV,* p. 307–308. The Lunar Science Institute, Houston.

Heymann D., Yaniv A., and Lakatos S. (1972) Inert gases from Apollo 12, 14, and 15 fines. *Proc. Third Lunar Sci. Conf., Geochim. Cosmochim. Acta,* Suppl. 3, Vol. 2, pp. 1857–1863. MIT Press.

Hodges R. R. Jr. (1972) Applicability of a diffusion model to lateral transport in the terrestrial and lunar exospheres. *Planet. Space Sci.* **20,** 103–115.

Hodges R. R. Jr. (1973) Helium and hydrogen in the lunar atmosphere (submitted to *J. Geophys. Res.*).

Hodges R. R. Jr. and Johnson F. S. (1968) Lateral transport in planetary exospheres. *J. Geophys. Res.* **73,** 7307–7317.

Hodges R. R. Jr., Hoffman J. H., Yeh T. T. J., and Chang G. K. (1972a) Orbital search for lunar volcanism. *J. Geophys. Res.* **77,** 4079–4085.

Hodges R. R. Jr., Hoffman J. H., and Evans D. E. (1972b) Lunar orbital mass spectrometer experiment. *Apollo 16 Preliminary Science Report,* NASA SP-315, pp. 211–216.

Hoffman J. H., Hodges R. R. Jr., and Evans D. E. (1973) Lunar atmospheric composition results from Apollo 17. *Proc. Fourth Lunar Sci. Conf., Geochim. Cosmochim. Acta.* In press.

Johnson F. S. (1971) Lunar atmosphere. *Rev. Geophys. Space Phys.* **9,** 813–823.

Johnson F. S., Carroll J. M., and Evans D. E. (1972) Lunar atmospheric measurements. *Proc. Third Lunar Sci. Conf., Geochim. Cosmochim. Acta,* Suppl. 3, Vol. 3, pp. 2231–2242. MIT Press.

Manka R. H. (1972) Lunar atmosphere and ionosphere. Ph.D. Thesis, Rice University, 1972.

Manka R. H. and Michel F. C. (1972) Lunar atmosphere as a source of lunar surface elements. *Proc. Second Lunar Sci. Conf., Geochim. Cosmochim. Acta,* Suppl. 2, Vol. 2, pp. 1717–1728. MIT Press.

Yaniv A. and Heymann D. (1972) Atmospheric Ar^{40} in lunar fines. *Proc. Third Lunar Sci. Conf., Geochim. Cosmochim. Acta,* Suppl. 3, Vol. 2, pp. 1967–1980. MIT Press.

Proceedings of the Fourth Lunar Science Conference
(Supplement 4, *Geochimica et Cosmochimica Acta*)
Vol. 3, pp. 2865–2875

Lunar atmospheric composition results from Apollo 17

J. H. HOFFMAN, R. R. HODGES, Jr., and F. S. JOHNSON

The University of Texas at Dallas, Dallas, Texas

D. E. EVANS

Johnson Space Center, Houston, Texas

Abstract—The Apollo 17 mass spectrometer has confirmed the existence of helium, neon, argon, and possibly molecular hydrogen in the lunar atmosphere. Helium and neon concentrations are in agreement with model predictions based on the solar wind as a source and their being non-condensible gases. ^{40}Ar and ^{36}Ar both exhibit a pre-dawn enhancement which indicates they are condensible gases on the nightside and are re-released into the atmosphere at the sunrise terminator. Hydrogen probably exists in the lunar atmosphere in the molecular rather than atomic state, having been released from the surface in the molecular form. Total nighttime gas concentration of known species in the lunar atmosphere is 2×10^5 molecules/cm^3.

INTRODUCTION

A STUDY of the composition of the lunar atmosphere is being conducted using the mass spectrometer placed on the moon during the Apollo 17 mission. The instrument, part of the Apollo Lunar Surface Experiments Package (ALSEP), measures the downward flux of gas molecules at the lunar surface but has been calibrated in terms of gas concentrations. Due to the extreme tenuosity of the lunar atmosphere, gas molecules do not collide with each other, but instead, travel in ballistic trajectories between collisions with the lunar surface, the trajectory heights and horizontal travel being functions of the surface temperature and molecular mass.

Previously flown Cold Cathode Gauges from Apollo 14 and 15 have determined an upper bound on the number density of the lunar atmosphere to be about 10^7 cm^{-3} in the daytime and 2×10^5 at night (Johnson *et al.*, 1972). In order to measure such low gas concentrations, the present instrument was designed with an extremely high sensitivity, of the order of 100 molecules per cm^3 for most gases, such as nitrogen and argon. As a result, initial spectra showed a large number of peaks indicative of substantial outgassing of the instrument especially during the daytime when the instrument temperature is high. These high background spectra prevented operation of the instrument during daytime because of the potential degradation of the ion source sensitivity during operation at relatively high pressures (greater than 10^{-9} torr). Consequently, operation has been confined to nighttime except for brief checks during the first two lunar days after deployment. Data have been obtained during the first four lunar nights, the results from which are discussed in this paper.

2865

INSTRUMENTATION

Identification of gas species in the lunar atmosphere and determination of their concentrations is accomplished by a miniature magnetic-deflection mass spectrometer. Gas molecules entering the instrument aperture are ionized by an electron bombardment ion source, collimated into a beam and sent through a magnetic analyzer to the detector system. Three mass ranges are scanned simultaneously by varying the ion-accelerating voltage applied to the ion source. These are 1–4, 12–48, and 27–110 amu, termed low, mid, and high mass ranges, respectively. The mid and high ranges are so related that mass 28 and 64 amu are detected simultaneously. This feature permits the simultaneous measurement of CO and SO_2, which may be candidates for volcanic gases, by inhibiting the ion accelerating voltage sweep and locking onto that voltage corresponding to the focussing conditions for these mass numbers (or for any other set with the mass ratio 28:64).

Standard ion counting techniques employing electron multipliers, pulse amplifiers, discriminators and counters are used, one system for each mass range. The number of counts accumulated per voltage step (0.6 sec) for each channel is stored in 21-bit accumulators until sampled by the telemetry system. Just prior to interrogation, the 21-bit word is converted to a floating point number in base 2, reducing the data to a 10-bit word, consisting of a 6-bit number and a 4-bit multiplier. This scheme maintains 7-bit accuracy (1%) throughout the 21-bit (2×10^6) range of data counts.

Calibration of the instrument was performed at the NASA Langley Research Center (LRC) in a manner similar to the Lunar Orbital Mass Spectrometer flown on Apollo 15 and 16 (Hoffman et al., 1972). A molecular beam apparatus produces a beam of known flux in a liquid helium cryochamber. The instrument entrance aperture intercepts the beam at one end of the chamber. With known beam flux and ion source temperature, instrument calibration coefficients are determined. Variation of gas pressure in the molecular beam source chamber behind a porous silicate glass plug varies the beam flux and provides a test of the linearity of the instrument response. Good linearity was achieved up to 5×10^5 counts/sec where the onset of counter saturation occurs.

Calibrations were done with a number of gases that may be candidates for ambient lunar gases, e.g., Ar, CO_2, CO, Kr, Ne, N_2, and H_2. Helium calibrations are not possible with this system because no helium beam can be formed in the chamber since helium is not cryopumped at the wall temperature. Sensitivity to helium was determined in the UT-Dallas ultrahigh vacuum chamber by using the LRC absolute argon calibration of the instrument as a standard for calibrating an ionization pressure gage. The gage calibration for helium was subsequently infered from the ratio of ionization cross sections for He and Ar (Von Ardenne, 1956). The resulting helium sensitivity is the ratio of the calibrated gage pressure to the helium counting rate.

Further details of the mass spectrometer can be found in Hoffman et al. (1973).

RESULTS

The data presented herein have been mostly taken from "quick-look" strip chart recorder records of mass spectra covering a period from initial turn-on of the instrument, approximately 50 hours after the first sunset after deployment, until the fifth sunset (April 23, 1973). An example of such a record is found in Hoffman et al. (1973). As a result of reading this type of chart, the accuracy of the data presented is considered to be ±30% with the smaller amplitude peaks having a factor of 2 uncertainty.

Figure 1 is a reduced spectrum derived from the raw spectrum by subtraction of background noise from each peak amplitude. Counting rate, accumulated counts per telemetry main frame (0.604 sec), are plotted against atomic mass number. These data were taken at sunrise terminator crossing of the instrument site. Because the instrument is on the floor of a valley, the sun angle must increase 4 degrees before the valley floor becomes illuminated and any local heating occurs. Heating of the site releases many condensed gases and increases outgas-

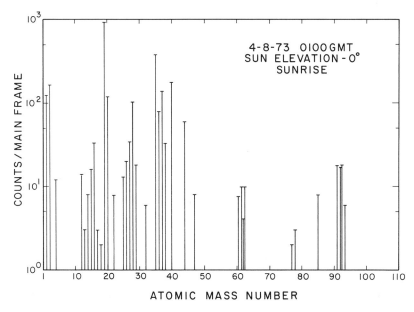

Fig. 1. Reduced mass spectrum taken at sunrise of fourth lunar night after deployment. Counts per telemetry main frame as a function of atomic mass number. Counting rates can be obtained by multiplying peak amplitude by 1.65. Corrections have been applied to each peak for background counting rates.

sing rates by orders of magnitude producing a daytime spectrum containing large amplitude peaks at nearly every mass number. These daytime spectral peaks are all considered artifact at this time, except for helium which will be discussed below.

In order to be assured that a given mass peak is not artifact (produced by outgassing of the ion source) a cool down test of the ion source is performed periodically. Normal nighttime ion source temperature is 270°K as determined by a sensor mounted on the ion source. Removing filament power for nearly an hour causes the ion source temperature to drop to approximately 195°K. Subsequent turn-on of the filament produces a relatively clean spectrum for the first spectrum scan. That is, peaks due to outgassing have decreased markedly, some to zero amplitude, while those gases not originating in the ion source produce essentially unchanged peak amplitudes. Figure 1 is a spectrum taken during one of these cool down tests.

Positive identification of at least three gases native to the lunar atmosphere, helium, neon, and argon, exists at this time. The mass 1 peak is largely produced in the ion source by dissociative ionization of hydrogenated molecules. If there exists any native atomic hydrogen it is certainly masked by that produced in the ion source. The mass 2 peak is a measure of the molecular hydrogen concentration which may exist in the lunar atmosphere. A concentration of 6.5×10^4 molecules/cm^3 has been determined on the nightside at solar zenith angles of $-136°$ during the third night and at $+168°$ and $-89°$ during the fourth lunar night.

All measurements were made during ion source cool down tests. While this may be an upper limit on molecular hydrogen, the existence of hydrogen in molecular form is consistent with the result from the Far UV spectrometer experiment flown on Apollo 17 which reported an upper limit of atomic hydrogen of 10 atoms/cm^3 (Fastie, private communication). Fastie also reported a daytime upper limit of H$_2$ to be less than their detection limit of 6×10^3 molecules/cm^3.

Hodges (1973) performed a Monte Carlo calculation of the H$_2$ distribution around the equatorial region of the moon assuming a symmetric temperature distribution about the subsolar point and a constant nighttime temperature of 95°K. A solar wind source (3×10^8 molecules/cm^2-sec) on the dayside and thermal escape from the entire lunar surface produced an anti-solar to subsolar point ratio of 6. Combining this with Fastie's daytime limit gives a nighttime upper limit of 4×10^4 which is close to the present measurement. Hodges (1973) calculated a daytime concentration of 2.0×10^3, which gives a nighttime value of 1.2×10^4, a factor of 5 lower than the measured values. Earlier predictions of the hydrogen concentration in the lunar atmosphere (Johnson *et al.*, 1972) assumed no recombination at the surface and thus an atmosphere of H. The existence of H$_2$ has not been predicted.

Previous discussions of hydrogen measurements by the Apollo 17 mass spectrometer (Hoffman *et al.*, 1973) indicated H$_2$ was due largely to outgassing from the ion source. However, recent cool down tests of the ion source have confirmed that the H$_2$ source is now largely external to the ion source of the instrument. It is still possible that H$_2$ is being evolved from other parts of the landing site even at the very low nighttime temperature and that the value obtained is an upper limit, or that there is some natural source of hydrogen besides the solar wind. An isotropic source would increase the diurnal ratio by a factor of nearly 2 and the total concentration by an amount depending on the source strength.

Figure 2 shows the helium concentration plotted as a function of solar zenith angle during the first four nights after deployment of the instrument. Included are two daytime (subsolar point) values taken during the first two lunar days. The abscissa scale begins at the subsolar point and progresses through the lunation. Sunset, midnight, and sunrise are identified. The scatter of the nighttime data is indicative of the variations that have been measured (Hundhausen *et al.*, 1970) in the solar wind flux of helium (the source function for helium in the lunar atmosphere). The loss mechanism of helium is thermal escape, mainly from the dayside. Residence time is of the order of 8×10^4 sec (Hodges, 1973) indicating a response time to fluctuations in the solar wind of a few tens of hours. The frequency of variations observed is consistent with this time scale.

The solid curve on Fig. 2 is the equatorial helium distribution from the Monte Carlo calculations similar to those for hydrogen (Hodges, 1973) based on a solar wind flux of 1.3×10^7 He ions cm^{-2} sec^{-1}, a nominal value based on the measurements of Geiss *et al.* (1972). The agreement with the data is seen to be excellent. Maximum calculated diurnal ratio is 22 vs. the measured value of approximately 20. Symmetry of the calculated curve is due to an assumed symmetric temperature distribution about the subsolar point and a constant nighttime temperature,

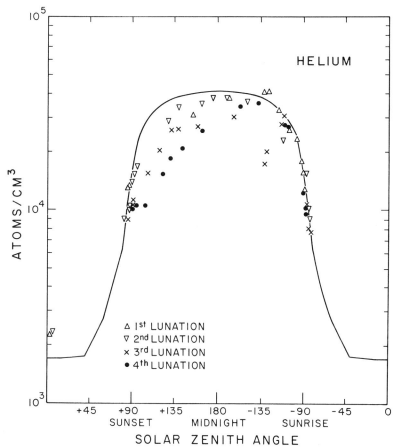

Fig. 2. Diurnal variation of ^{4}He. Concentration as a function solar zenith angle. Sunset, midnight and sunrise are identifield. Data points are from the first four lunar nights. Solid curve is theoretical distribution based on a Monte Carlo calculation. Distribution is typical of a non-condensible gas.

whereas the slight shift of the data maximum to post-midnight results from the continual post-sunset temperature decrease until dawn. In general, a non-condensible gas exhibits a maximum concentration near dawn, the region of coldest lunar surface temperature. However, owing to the large scale size of the helium trajectories significant helium is lost from pre-dawn and post-sunset regions to the dayside from which thermal escape is a rapid loss mechanism.

The mass 20 and 22 amu isotopes of neon have been measured during cool down test at solar zenith angles of $-136°$ and $-122°$ during the third lunar night and at $+168°$ and $-89°$ during the fourth lunar night. Table 1 gives atmospheric concentrations and ^{20}Ne/^{22}Ne ratios obtained. The value of 8×10^4 cm^{-3} is nearly a factor of 20 below that predicted by Johnson *et al.* (1972) and remained somewhat of a puzzle since the orbital mass spectrometer on Apollo 16 gave a similar value (Hodges *et al.*, 1973) near the sunset terminator. However, a reassessment of the

Table 1. Neon concentration.

Solar Zenith Angle	Neon Concentration Molecules/cm^3	^{20}Ne/^{22}Ne
+168°	9.4×10^4	14.4
−136°	7.7×10^4	13.6
−112°	7.1×10^4	14.0
−89°	8.7×10^4	15.0
Ave	8.2×10^4	14.2

neon distribution taking account more recent values of lifetimes against photoionization and the loss rate of neon on the nightside near the terminator, a value of 1.2×10^5 has been obtained (Hodges *et al.*, 1973). This value is reasonably in agreement with the experimental results indicating that the problem of the missing neon has been resolved. The value of 14.2 for the neon isotopic ratio (Table 1) is in good agreement with that in the solar wind, 13.7, as determined by the solar wind composition experiment (Geiss *et al.*, 1972).

A problem associated with the measurement of neon, and also two of the argon isotopes as will be discussed later, is the presence of the large peaks at masses 19, 35, and 37 (Fig. 1). These are believed to be from fluorine and chlorine. The 35 : 37 ratio is that of the chlorine isotopes. Hydrogenated halogen peaks occur at masses 20, 36, and 38, probably formed in the ion source by a reaction with hydrogen as their abundance is a strong function of ion source temperature. Ion source cool down tests reduce the amplitudes of the 36 and 38 amu peaks to essentially zero at certain times of the night, but there is always a residual mass 20 peak which results from the neon isotope. The origin of the halogen peaks is uncertain but some possibilities are the outgassing of warm areas (e.g., ion source, instruments, central station, radioisotope thermal generator) of the landing site, or the natural degassing of lunar materials themselves.

The diurnal distribution of ^{40}Ar is shown in Fig. 3. Coordinates are similar to those of Fig. 2. Data are from the third, fourth, and fifth lunar nights after instrument deployment. Daytime data are masked by hydrocarbon peaks from outgassing of the instrument and site. By the time the sunset terminator has crossed the landing site (6 hours [3°] after the sun has dipped below the mountains to the west), the site has cooled sufficiently that hydrocarbon peaks no longer interfere with the ^{40}Ar measurement. The ^{40}Ar concentration at sunset is 7–8×10^3/cm^3. This steadily decreases through the night as the surface cools reaching a value of 10^2/cm^3 (the instrument detection limit) at $-135°$ solar zenith angle. It appears that the argon freezes out on the cold nighttime surface. At about 20° before sunrise, the ^{40}Ar concentration begins to increase reaching a value of 3.5×10^4 at terminator crossing of the site, a factor of 5 larger than at sunset. Sunrise at the site is delayed 8 hours by shadowing from the mountains to the east which precludes significant local heating before this time. The pre-dawn enhancement of argon results from release of the gas from the warm approaching terminator region and the westward flow of this pocket of gas into the nightside several scale heights before being

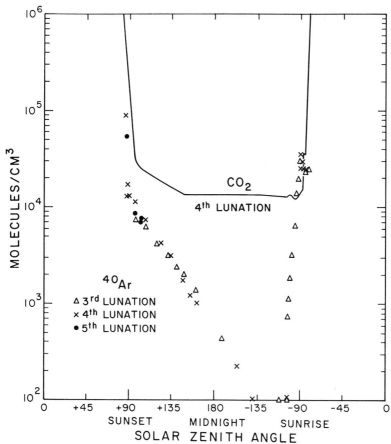

Fig. 3. ^{40}Ar diurnal variation. Coordinates similar to Fig. 2. Pre-dawn enhancement suggests boil-off of adsorbed gases from warm terminator region and transport of these gases into nightside where they are re-adsorbed. CO_2 is shown as an example of a gas not exhibiting a pre-dawn enhancement.

adsorbed on the lunar surface. The concentration of CO_2, which does not exhibit a pre-dawn enhancement, is shown for comparison. After sunrise, rapid heating occurs and hydrocarbon peaks again dominate the spectrum.

A more detailed picture of the pre-dawn and sunrise conditions is given in Fig. 4. Gas concentrations are plotted as a function of solar zenith angle from $-110°$ to $-80°$. T indicates sunrise terminator crossing the landing site; S is the point at which sunlight reaches the valley floor and the onset of heating of the instrument and site occurs. ^{40}Ar exhibits by far the largest pre-dawn enhancement. After terminator crossing there appears to be a slight decrease in ^{40}Ar concentration as though the source were being somewhat depleted of argon.

The general behavior of argon follows that expected of a gas condensible at the nighttime surface temperature. The maximum concentration occurs at sunrise with a decreasing amount towards the subsolar point as the temperature increases.

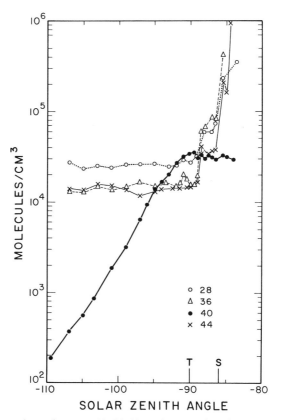

Fig. 4. Concentrations of masses 28, 36, 40, and 44 as a function of solar zenith angle from -110 to $-80°$. "T" is time of terminator crossing landing site longitude. "S" is sunrise at landing site. ^{40}Ar shows clear pre-dawn enhancement, while mass 36 shows only a small pre-dawn enhancement. Masses 28 and 44 have none.

An enhancement would again occur at sunset, but with lower amplitude, with a decrease throughout the night as the adsorption rate on the surface increases with decreasing temperature. The pre-dawn enhancement then comes from release of the adsorbed gas from the suddenly warmed terminator region.

The mass 36 peak appears to exhibit a small pre-dawn enhancement, about 10% that of the ^{40}Ar. This enhanced portion is believed to be ^{36}Ar. The remainder of the mass 36 peak is the HCl as was discussed above. The $^{40}Ar/^{36}Ar$ ratio is then approximately 10, and the ^{36}Ar concentration at sunrise is $3 \times 10^3/cm^3$. The ^{36}Ar source is the solar wind whereas ^{40}Ar is believed to arise from degassing of the regolith.

Other gases shown in Fig. 4, masses 28 and 44, do not appear to exhibit a pre-dawn enhancement. Mass 28, whether it is N_2 or CO, would not be adsorbed at the night lunar temperature and would therefore not show a pre-dawn enhancement. By this same argument, CO_2 (mass 44) would be expected to follow the argon pattern. The negative result places an upper limit on CO_2 at the sunrise terminator of $3 \times 10^3/cm^3$. Mass 32, O_2, also exhibits no pre-dawn enhancement and several ion source cool down tests have shown a zero amplitude peak for O_2. An upper limit is in the low $10^2/cm^3$ range.

At 89° solar zenith angle, masses 28, 36, and 44 all show a marked enhancement. At this time the massifs above the valley floor, where the instrument is deployed, are illuminated by sunlight and are being heated. A small heating of instrument and site could occur by infrared from the massif faces increasing outgassing rates, although the ion source temperature sensor does not so indicate, or release of gases from these warm surfaces could account for these peak amplitude increases. At 86°, the rapid rise in amplitude results from instrument and site outgassing from direct solar heating.

The part of the nighttime spectrum, Fig. 1, above 44 amu is characterized by a paucity of peaks, except for the groups near 92 and 61 amu, which are believed to be multiply charged ions from the isotopes of tungsten and rhenium vaporizing from the ion source filament. All other peaks in the spectrum are considered artifact at this time.

The total gas concentration measured near the anti-solar point on the fourth lunation is 2×10^5 molecules/cm^3 without the mass 19, 35, and 37 peaks included. This is the same value obtained by the Cold Cathode Gauge Experiment on Apollo 14 and 15 (Johnson *et al.*, 1972). However, addition of the 19, 35, and 37 peaks (F and Cl) doubles the nighttime gas concentration. It is unclear whether these gases can be considered lunar or artifact, or whether they were present at the Apollo 14 and 15 site and were being measured as part of the nighttime gas concentration by the Cold Cathode Gauges. Daytime comparisons are not meaningful at this time due to the high outgassing rate still prevailing at the Apollo 17 site. CCGE daytime readings were not made until the ninth lunation.

CONCLUSIONS

Data obtained from the first four lunations have positively confirmed the existence of three gases in the lunar atmosphere, helium, neon, and argon. Additionally, molecular hydrogen measurements are consistent with upper limits placed by the Far UV spectrometer experiment and only a factor of five above predicted levels based on the solar wind as a source. This is not considered a large discrepancy, but of course, an additional source of hydrogen, particularly if isotropic would account for the high empirical value.

Helium concentrations and diurnal ratio are in excellent agreement with predictions based on the solar wind as a source, indicating basic tenets of the theory of a non-condensible gas are correct. The measured neon nighttime concentration of $7-9 \times 10^4/cm^3$ is in reasonable agreement with a reassessment of the predicted value for neon and removes an earlier discrepancy of a factor of nearly 20 between model and measurement.

Argon isotopes at masses 36 and 40 both show a pre-dawn enhancement after being adsorbed on the late night (coldest) lunar surface. The gases are being released from the warm approaching terminator region. The sunrise to sunset ratio

Table 2. Summary of gases in the lunar atmosphere.

Gas	Concentration Molecules/cm³		Model‡ Molecules/cm³	
	Day	Night	Day	Night
H_2	—	6.5×10^4	2.0×10^3	1.2×10^4
^4He	2×10^3	4×10^4	1.6×10^3	3.8×10^4
^{20}Ne	—	8×10^4	$4-7 \times 10^3$	1.2×10^5
^{36}Ar	—	$3 \times 10^{3}*$	—	—
		$3.5 \times 10^{4}*$		
^{40}Ar	—	$7-8 \times 10^{3}†$	—	—
O_2	—	$<2 \times 10^2$	—	—
CO_2	—	$<3 \times 10^3$	—	—

*Sunrise terminator.
†Sunset terminator.
‡Hodges, 1973 and Hodges *et al.*, 1973.

of ^{40}Ar is 5 with concentrations of 3.5×10^4 and $7-8 \times 10^3$ at these locations, respectively. Clearly argon does not behave as a non-condensible gas and, therefore, cannot be compared to models. Furthermore, ^{40}Ar probably comes from ^{40}K decay and subsequent release from lunar surface materials, so its presence is evidence of a truly native lunar gas.

From the lack of a pre-dawn enhancement of CO_2, an upper limit has been set at $3 \times 10^3/cm^3$. The O_2 upper limit is 2×10^2.

The sum of all the known gases in the nighttime lunar atmosphere (H_2, He, Ne, Ar) equals that measured by the Cold Cathode Gauges at the Apollo 14 and 15 sites: 2×10^5 molecules/cm³. Table 2 lists a summary of these gas concentrations. All other peaks in the spectrum are considered artifact at this time. As purging of the site continues other gas species may emerge as being native lunar gases.

Acknowledgments—This work was supported by NASA Contract NAS9-12074 and by Bendix Aerospace Sub-Contract SC-830 of NASA Contract NAS9-5829. The authors express their appreciation to the many people of UT-Dallas, Bendix Aerospace Corporation and NASA who contributed significantly to the design construction and testing of the mass spectrometer experiment. The NASA/Langley Research Center Molecular Beam Calibration Facility was under the direction of Mr. Paul Yeager. Special thanks are due to the crew of Apollo 17 who so successfully deployed the instrument on the moon.

References

Geiss J., Buelher F., Cerutti H., Ebehardt P., and Filleux Ch. (1972) Solar wind composition experiment, *Apollo 16 Preliminary Science Report*, NASA Report SP-315, pp. 141–149.

Hodges R. R. Jr. (1973) Helium and hydrogen in the lunar atmosphere. *J. Geophys. Res.* **78**. In press.

Hodges R. R. Jr., Hoffman J. H., Johnson F. S., and Evans D. E. (1973) Composition and dynamics of lunar atmosphere (abstract). In *Lunar Science—IV*, pp. 374–375. The Lunar Science Institute, Houston.

Hoffman J. H., Hodges R. R. Jr., Johnson F. S., and Evans D. E. (1972) Lunar orbital mass spectrometer experiment. *Apollo 15 Preliminary Science Report*, NASA Report SP-289, pp. 19-1 to 19-7.

Hoffman J. H., Hodges R. R. Jr., Johnson F. S., and Evans D. E. (1973) Lunar atmospheric composition experiment. *Apollo 17 Preliminary Science Report*, NASA Report. In press.

Hundhausen A. J., Bame S. J., Asbridge J. R., and Sydoriak S. J. (1970) Vela 3 observations from July 1965 to June 1967. *J. Geophys. Res.* **75**, 4643.

Johnson F. S., Carroll J. M., and Evans D. E. (1972) Lunar atmosphere measurements. *Proc. Third Lunar Sci. Conf., Geochim. Cosmochim. Acta,* Suppl. 3, Vol. 3, pp. 2231–2242. MIT Press.

Von Ardenne M. (1956) *Tabellen der Elektronenphysik, Ionenphysik and Übermikroskopie,* Vol. 1, p. 488. Deutscher Verlag der Wissenschaften, Berlin.

Proceedings of the Fourth Lunar Science Conference
(Supplement 4, *Geochimica et Cosmochimica Acta*)
Vol. 3, pp. 2877–2887

The electric potential of the lunar surface

M. A. FENNER, J. W. FREEMAN, JR., and H. K. HILLS

Department of Space Science
Rice University
Houston, Texas 77001

Abstract—Acceleration and detection of the lunar thermal ionosphere in the presence of the lunar electric field yields a value of at least +10 volts for the lunar electric potential for solar zenith angles between approximately 20° and 45° and in the magnetosheath or solar wind. An enhanced positive ion flux is observed with the ALSEP Suprathermal Ion Detector when a pre-acceleration voltage attains certain values. This enhancement is greater when the moon is in the solar wind as opposed to the magnetosheath.

INTRODUCTION

THEORETICAL STUDIES of the electric potential of the sunlit lunar surface in the presence of the solar wind have predicted values ranging from +20 volts down to a few volts positive. Originally, Öpik and Singer (1960) proposed a value of about +20 volts. More recently Grobman and Blank (1969) revised this estimate downward by about an order of magnitude and finally Manka (1972) has estimated a value near +10 volts. In the absence of the solar wind, Reasoner and Burke (1972) have reported potentials as high as +200 volts as the moon crosses the magnetospheric tail.

In this paper we discuss evidence for a sunlit potential in the magnetosheath and solar wind of about +10 volts. This evidence comes from the analysis of the energy spectra of lunar ionosphere thermal ions accelerated toward the moon by an artificial electric field. The data are provided by the Apollo Lunar Surface Experiment Package (ALSEP) Suprathermal Ion Detector Experiment (SIDE) deployed at the Apollo 14 and 15 sites.

THE SIDE

The Suprathermal Ion Detector Experiment (SIDE) contains a total ion detector curved plate analyzer and ion mass analyzer designed to measure positive ions down to a few electron volts. To accomplish this in the face of possible lunar surface potentials of the order of several tens of volts, the instrument is equipped with a ground plane electrode (in contact with the lunar surface) whose potential with respect to a wire grid above the ion entrance apertures is stepped through a series of 24 voltages (*see* Fig. 1b). For certain negative (accelerating) voltages, the imposed electric field is able to overcome a positive surface potential allowing thermal ions to be measured by the instrument.

The ground plane electrode is a circular wire grid 65 cm in diameter. Close-up

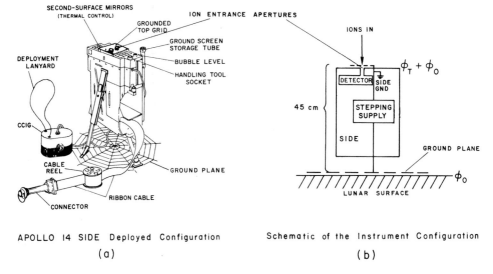

APOLLO 14 SIDE Deployed Configuration   Schematic of the Instrument Configuration

(a)   (b)

Fig. 1. The Suprathermal Ion Detector Experiment (a) as deployed on the lunar surface, and (b) showing the ground plane and top wire grid configuration schematically. Thermal ions are attracted to or repelled from the top of the instrument by the top grid. Φ_T is the voltage produced by the stepping supply. The ground plane is in actual contact with the lunar surface.

photographs show that the grid makes contact with the surface at many points, so good electrical contact will be assumed (*see* Appendix A). The SIDE is D.C. isolated from the ALSEP central station and other experiments.

These features of the SIDE allow the determination of the electric potential of the lunar surface under certain conditions. This is accomplished by examination of the energy spectra of thermal ions accelerated into the SIDE by the ground plane voltage in the presence of the electric field due to the surface charge of the moon.

Figure 1a shows the instrument and the ground plane in the deployed configuration. Further details of the experiment can be found in Freeman *et al.* (1972).

INSTRUMENT FUNCTION

Assuming good electrical coupling between the ground plane and the lunar surface the instrument would be expected to function as follows:

(a) In the case of a near-zero lunar surface potential a negative ground plane stepper voltage accelerates thermal positive ions into the instrument with an energy approximately that of the stepper voltage.

(b) When the lunar surface is substantially positive and the stepper voltage negative, but larger in absolute value, the energy of the detected ions is less than the stepper voltage by an amount equal to the lunar surface potential. When the stepper voltage exactly matches but is of opposite sign from the lunar surface potential, the ions are seen unaccelerated.

(c) When the lunar surface is negative the ions may be repelled by a sufficiently large positive stepper voltage.

We can assume that the ions seen are principally ionized neutral gas in thermal equilibrium with the lunar surface ($T \leqslant 400°K$) and further that most of the ions appear to have come from infinity so far as the energy acquired from the surface electric fields is concerned. Under these conditions, the foregoing can be made more explicit by expressing the ion energy, E, seen at the detector by

$$E = E_i - (\varphi_t + \varphi_0)q \qquad (1)$$

where E_i is the initial ion energy (the neutral thermal energy), φ_t the potential of the top wire grid of the SIDE relative to the ground plane and hence to the lunar surface, φ_0 the lunar surface potential, and q the ion charge (assumed to be $+1$). We will be concerned with cases in which φ_t is negative and for which we assume $q|\varphi_t + \varphi_0| \gg E_i$ so that (1) becomes

$$E \cong -(\varphi_t + \varphi_0)q \qquad (2)$$

Using this equation observable values of $\varphi_0 q$ are displayed in Table 1 for the differential energy channels, E, and stepper voltages, φ_t, of the Apollo 14 instrument. If the initial energy of the ions is not negligible then the tabular values represent $(\varphi_0 q - E_i)$ and hence provide a lower limit on the surface potential; a situation that we consider unlikely (*see* Appendix B).

OBSERVATIONS

A feature of the data when the solar zenith angle is between approximately 20° and 45° is the frequent appearance of narrow low-energy ion flux spectra which show a correlation with the ground plane stepper voltage. That is, ions at certain peak energies recur with specific ground plane stepper voltages. We will refer to this occurrence as a resonance. There is often a complete absence of ions in any energy channel except at the resonant ground plane stepper voltage.

This phenomenon has been seen with the Apollo 14 and 15 SIDEs and exclusively in the solar wind or magnetosheath. Figure 2 shows the location in lunar orbit of these observations. Only one such observation has been made on the dawn side of the tail. It was seen at a time of high local gas pressure following the Apollo 15 LM impact. Each lunar cycle for which Apollo 14 data are available shows the resonance in the afternoon or dusk side magnetosheath and solar wind and some cycles of Apollo 15 data show this resonance in this quadrant as well.

Figure 3 shows an extended set of data from the Apollo 14 SIDE. Here the count rates from the two lowest energy channels have been grouped according to the ground plane stepper voltage. The energies that stand out near the center of the figure are 7 eV at ground plane voltage -16.2 and 17 eV at ground

Table 1. Apollo 14 SIDE—Observable values
of $(\varphi_0\ q)$.

| 0267 | φ_t (volts) | E Total ion detector Energy Channels | |
		7 eV/q	17 eV/q
	Ground Plane Stepper Voltage		
	− 0.6	− 6.4	− 16.4
	− 1.2	− 5.8	− 15.8
	− 1.8	− 5.2	− 15.2
	− 2.4	− 4.6	− 14.6
	− 3.6	− 3.4	− 13.4
	− 5.4	− 1.6	− 11.6
	− 7.8	+ 0.8	− 9.2
	− 10.2	+ 3.2	− 6.8
	− 16.2	+ 9.2	− 0.8
	− 19.8	+ 12.8	+ 2.8
	− 27.6	+ 20.6	+ 10.6
	0.0	− 7.0	− 17.0
	+ 0.6	− 7.6	− 17.6
	+ 1.2	− 8.2	− 18.2
	+ 1.8	− 8.8	− 18.8
	+ 2.4	− 9.4	− 19.4
	+ 3.6	− 10.6	− 20.6
	+ 5.4	− 12.4	− 22.4
	+ 7.8	− 14.8	− 24.8
	+ 10.2	− 17.2	− 27.2
	+ 16.2	− 23.2	− 33.2
	+ 19.8	− 26.8	− 36.8
	+ 27.6	− 34.6	− 44.6

plane voltage − 27.6. From Equation (2) and Table 1 these examples yield a lunar surface potential, φ_0, of approximately + 10 volts.

Ten months of Apollo 14 data have been scanned. Each lunation shows resonances similar to Fig. 3 indicating a surface potential of ∼ + 10 volts for zenith angles between approximately 20° and 45°. A typical segment of data for this region is shown in Fig. 4. Here the counting rates in the 17 eV channel are plotted each time the ground plane stepper reads − 27.6 volts. The counting rates for the same energy channel are plotted when the ground plane stepper is at 0.0 volts as a measure of the background when particles cannot be accelerated into the detector by the stepper potential. This plot clearly shows the increase in the counting rate for the resonant step as the detector moves out into the free-flowing solar wind. The maximum rate, although not sharply defined, usually occurs near a zenith angle of 35°. In all the data that have been scanned, the resonance ends at about 45°. This suggests that at that point the lunar surface potential decreases below the energy level required to produce a resonance in our differential energy passbands.

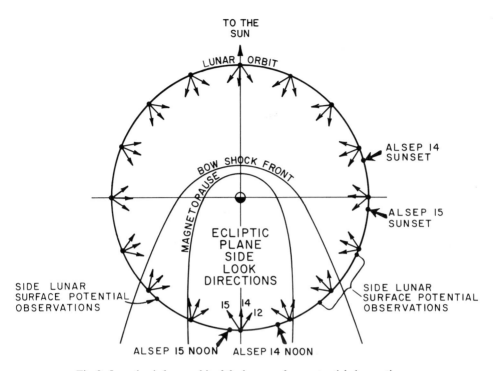

Fig. 2. Location in lunar orbit of the lunar surface potential observations.

These passbands are narrow (±5% FWHM) and not contiguous, so it is possible for the ion peak to disappear between them. Hence the disappearance of the resonance does not imply a discontinuity in the lunar surface potential. In fact it is important to note that the resonance shown here indicates that the surface potential does not change greatly over a large portion of the sunlit surface. At the equatorial site (Apollo 14) the potential remains $+10$ v ± 2 v over the solar zenith angle range 20° to 45°. The same potential has been observed at least once within the same zenith angle range at the Apollo 15 site 26.1° north of the equator.

That the magnitude of the counting rate at resonance is related to the solar wind is further seen in Fig. 5. The solar wind velocity is relatively constant over the seven month interval shown; however, the solar wind number density varies by an order of magnitude. In November an unusually high number density in the solar wind was reflected in a very intense resonance. This observation is consistent with the source of ions considered below.

THE THERMAL LUNAR EXOSPHERE

The foregoing assumes that the thermal lunar ionosphere is being accelerated into the instrument in the presence of a lunar surface electric field and an artificial electric field arising from the SIDE. This hypothesis is not only supported by the

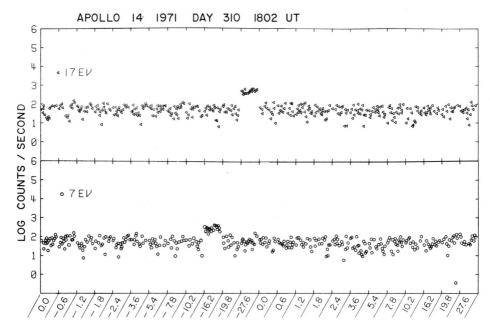

Fig. 3. SIDE total ion detector count rates in two energy channels for a three hour time interval grouped by ground plane stepper voltage. The time given is the start time of the interval.

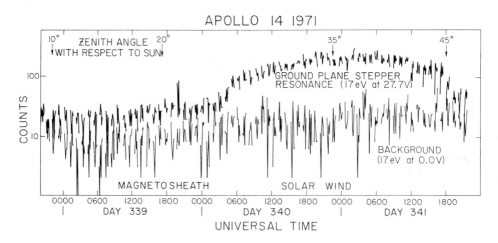

Fig. 4. Typical ground plane stepper resonance seen by Apollo 14 in the lunar afternoon. The counting rates in the resonant channel are compared with those in the 0.0 volt step which were used as a measure of background, i.e., ions not affected by the ground plane voltage. The six data points from $\frac{1}{24}$th of the stepper cycle are plotted on an expanded scale in each period of approximately one hour. Note that on this lunation the resonance begins at a solar zenith angle of 10°.

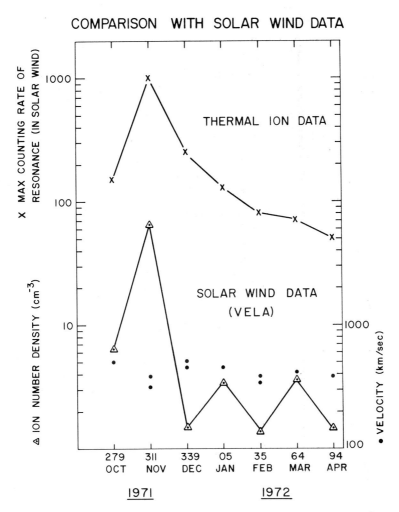

Fig. 5. Comparison of thermal ion data and solar wind data. The maximum counting rate in the resonant step is plotted for each month. Vela data is from *Solar-Geophysical Data*, U.S. Department of Commerce, Boulder, Colorado.

ground plane stepper resonance but also qualitatively by the observed narrow energy spectrum. This spectrum is typically less than the width between detector differential energy channels or 10 eV/q. There is an absence of a low energy tail.

Consider the electric field region to be confined to a screening length, L, analogous to the conventional Debye length. For the solar wind L is of the order of a few meters. At the lunar surface it is probably much less than this due to photoelectrons and electrons from the ionized neutrals. In any event, $L \ll H$, where H is the scale height of the neutral atmosphere to be ionized. As a result the volume above the detector within 1 screening length is very small compared to that within the scale height H. The ions produced in the atmosphere within or above height H of the detector greatly predominate over those produced within height L. A large fraction of these will drift thermally or ballistically into the field region. The ion distribution function in energy space will then peak sharply at the potential of the instrument relative to infinity; as is actually observed. (In reality the drift of the thermal ions will also be somewhat influenced by the interplanetary electric field, however, this must be a second order effect because of the high stability of the resonance fluxes despite major changes in the interplanetary magnetic field.)

Assuming focusing to the detector of about 1 steradian we can use the observed ion flux to obtain an order of magnitude estimate of the requisite neutral number density in the lunar atmosphere. Following a treatment given by Manka (1972), for an exponential atmosphere, the omnidirectional ion flux seen by the detector, F, is related to the neutral surface number density, N_0, by

$$F \cong \frac{PN_0H}{2} \tag{3}$$

where P is the total production rate for ions, and H the scale height for neutrals. Referring to Fig. 3 we take 300 counts/second as typical of the 17 eV ions and calculate a flux of 3×10^6 ions/cm^{2}sec.

We must now make a choice regarding the composition of the atmosphere. Johnson (1971) has shown that the atmosphere is expected to consist largely of neon. Freeman *et al.* (1971) quote an afternoon surface temperature of 360°K yielding an H of 100 km for mass 20. Siscoe and Mukherjee (1972) estimate a total ion production rate of 1.1×10^{-7} ions/sec atom for neon. The requisite surface number density is then $N_0 \sim 5 \times 10^6$ atoms/cm^3. This is in reasonable agreement with the daytime surface number density reported by the ALSEP cold cathode gauges (Johnson, F. S., private communication).

Although the resonance phenomenon has been seen with the SIDE mass analyzer we have not been able to determine directly the ambient ion masses. Statistics are generally poor and the resonance mass analyzer data often fall in a very low energy channel where mass calibration is uncertain. The only occasion to date on which a well defined mass spectrum has been seen due to the ground plane stepper resonance followed shortly after the Apollo 15 mission. At that time the spectrum was typical of an exhaust gas spectrum. Lindeman, however, has found non-resonance SIDE mass analyzer spectra peaking near mass 20, 36, and 40 believed to represent the ambient lunar atmosphere [Lindeman *et al.*, 1973]. The mass 20 N_0 calculation is given here only for illustrative purposes since a correct calculation would require taking into account all of the masses present in their proper abundances.

The enhanced resonance peak in November 1971 (Fig. 5) can be explained by the increased solar wind flux. Bernstein *et al.* (1963) give the charge-exchange cross-section of solar wind protons on neon as 5×10^{-16} cm^2. As seen from Fig. 5 the solar wind flux reached $\sim 2 \times 10^9$ ions/cm^{2}sec for a production rate of

10^{-6} ions/sec atom; easily enough to account for the enhanced resonance during November.

<h2 style="text-align:center">Discussion and Summary</h2>

Acceleration of the lunar thermal ionosphere with known voltages has led to the value of at least $+10$ volts for the dayside lunar surface potential. It should be emphasized this value holds only in the solar wind or magnetosheath plasma. Reasoner and Burke (1972) have reported evidence for potentials as high as $+200$ volts in the greater vacuum of the magnetospheric tail.

This $+10$ volt value appears to hold at least between solar zenith angles of 20° and 45°. The increase in intensity of the resonance during passage from the magnetosheath to the free streaming solar wind may be due to a slight change in the lunar surface potential such that the accelerated ions are more nearly centered on an energy passband. This shift in potential might arise from a changing electron temperature. On the other hand, the solar wind may enhance the ionization process itself since it is clear from Fig. 5 that the solar wind does play a key role in the magnitude of the resonance phenomena.

From SIDE data, Lindeman *et al.* (1973) have found that the potential of the lunar surface near the terminator goes from positive to negative as one approaches the lunar night. Figure 6 shows the composite data on the lunar surface

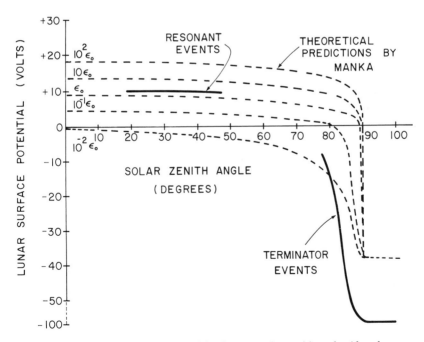

Fig. 6. Composite lunar surface potential values vs. solar zenith angle. Also shown are predicted values computed by Manka (1972) for typical solar wind parameters.

potential together with theoretical calculations by Manka (1972) for typical solar wind conditions. The general shape of the experimental curve is in good agreement with prediction. As pointed out by Manka (private communication), the point at which the potential goes negative and the ultimate night side potential are very sensitive to the assumed solar wind electron temperature.

Acknowledgment—We have profited from discussions with several Rice University scientists; particularly Dr. Richard Vondrak and Dr. Robert Manka. The Rice research was supported by NASA contract NAS 9-5911.

A portion of the research reported in this paper was done while one of us (J. W. F.) was a Visiting Scientist at The Lunar Science Institute which is operated by the Universities Space Research Association under Contract No. NSR 09-051-001 with the National Aeronautics and Space Administration.

REFERENCES

Bernstein W., Fredricks R. W., Vogl J. L., and Fowler W. A. (1963) The lunar atmosphere and the solar wind. *Icarus* **2**, 233.

Freeman J. W. Jr., Hills H. K., and Fenner M. A. (1971) Some results from the Apollo 12 suprathermal ion detector. *Proc. Second Lunar Sci. Conf., Geochim. Cosmochim. Acta*, Suppl. 2, Vol. 3, pp. 2093. MIT Press.

Freeman J. W. Jr., Fenner M. A., Hills H. K., Lindeman R. A., Medrano R., and Meister J., Suprathermal Ions Near the Moon. *Icarus* **16**, 328.

Grobman W. D. and Blank J. L. (1969) Electrostatic potential distribution of the sunlit lunar surface. *J. Geophys. Res.* **74**, 3943.

Johnson F. S. (1971) Lunar atmosphere. *Rev. Geophys. Space Phy.* **9**, 813.

Lindeman R. A., Freeman J. W. Jr., and Vondrak R. (1973) Ions from the lunar atmosphere. *Proc. Fourth Lunar Sci. Conf., Geochim. Cosmochim. Acta.* In press.

Manka, R. H. (1972) Lunar atmosphere and ionosphere. Ph.D. Thesis, Rice University.

Manka R. H. and Michel F. C. (1970) Lunar atmosphere as a source of Argon-40 and other lunar surface elements. *Science* **169**, 278.

Opik E. J. (1962) The lunar atmosphere. *Planet Space Sci.* **9**, 221.

Opik E. J. and Singer S. F. (1960) Escape of gases from the moon. *J. Geophys. Res.* **65**, 3065.

Reasoner D. L. and Burke W. J. (1972) Characteristics of the lunar photoelectron layer while in the geomagnetic tail. Paper presented at the *Sixth ESLAB Symposium*, 26–29 September 1972, Noordwijk, Holland.

Siscoe G. L., and Mukherjee M. R. (1972) Upper limits on the lunar atmosphere determined from solar wind measurements. *J. Geophys. Res.* **77**, 6042.

Strangway D. W., Chapman W. B., Olhoeft G. R., and Carnes J. (1972) Electrical properties of lunar soil dependence on frequency, temperature and moisture. *Earth and Planet. Sci. Lett.* **16**, 275.

APPENDIX A

The question of the D.C. coupling between the ground plane and the lunar surface will be examined.

Close-up photographs show that the grid makes contact with the lunar soil at many points and where there is not direct contact the wires are rarely more than a few centimeters from the surface. Strangway *et al.* (1972) have measured the D.C. conductivity of lunar soil samples and found values below 10^{-13} mhos/m in the temperature range of interest. This is sufficiently low that the soil may be considered a perfect insulator and we can direct our attention to conductivity via the plasma only.

The problem can be recast by considering the relative importance of two impedances R_z and R_{LS}

and thinking of the SIDE as the center terminal in a voltage divider network. R_∞ represents the impedance between the SIDE top wire grid and infinity. R_{LS} represents the impedance between the ground plane and the lunar surface. In either case the impedance, R is approximated by

$$R \cong \frac{L}{\sigma A}$$

where L is the screening length for the plasma, σ its conductivity and A the effective area of the electrode. (Where the ground plane wire is less than L from the surface then some mean distance to the surface might replace L.) We assume that L and σ are the same at both electrodes and find that the relative impedances are determined by the areas of the two electrodes. The condition of small relative potential difference between the ground plane and the lunar surface, which is

$$R_\infty \gg R_{LS}$$

now becomes

$$A_{TG} \ll A_{GP}$$

where A_{TG} and A_{GP} are the effective areas of the top grid and ground plane respectively. The ground plane is approximately 65 cm in diameter and contains over 1300 cm of wire. The top is approximately 10×12 cm and consists of knit wire mesh. We estimate

$$\frac{A_{TG}}{A_{GP}} \sim 0.01$$

and hence conclude that the potential difference between the ground plane and the lunar surface is a small fraction of the lunar surface potential relative to infinity.

Appendix B

In the case of a screening length long compared to the dimension of the SIDE above the lunar surface (~ 50 cm) the lunar surface potential field may overwhelm the SIDE top grid field some distance out and set up a positive potential barrier for positive ions attempting to reach the SIDE. This suggests the possibility of non-negligible initial ion energies, E_i. One must then ask what the initial energy source might be since the resonance is clearly the result of lunar atmospheric ions as is demonstrated by the observed number density and the mass spectra seen on at least one occasion. Ion acceleration by the interplanetary electric field as discussed by Manka and Michel [1970] is frequently seen in the SIDE ion data, however, this leads to a sporadic highly directional and energy varying flux unlike the stable monochromatic flux associated with this resonance phenomenon. No other ion acceleration mechanisms are known that are appropriate to this zenith angle regime, hence we prefer the assumption of a smaller screening length and negligible E_i. In any event, our value for the lunar surface potential stands as a lower limit.

Proceedings of the Fourth Lunar Science Conference
(Supplement 4, *Geochimica et Cosmochimica Acta*)
Vol. 3, pp. 2889–2896

Ions from the lunar atmosphere

ROBERT LINDEMAN, JOHN W. FREEMAN, JR., and RICHARD R. VONDRAK

Department of Space Science
Rice University
Houston, Texas 77001

Abstract—The ionization of neutral atoms in the lunar atmosphere produces an ionosphere around the moon. These ions are accelerated by the interplanetary electric field and local surface fields to energies of 10 to 500 eV. The Suprathermal Ion Detector Experiment (SIDE) has been observing these ions from the lunar atmosphere. The observations have been divided into four categories based on the acceleration mechanism.

INTRODUCTION

THE LUNAR atmosphere is being constantly ionized by the solar UV and the solar wind. These ions are then accelerated by the local and interplanetary electric fields. Thus the observation of these ions formed from the lunar atmosphere should provide information concerning both the neutral atmosphere and the electric and magnetic fields operating near the moon. A significant observation of H_2O^+ ions over a 14 hour period has already been reported by Freeman *et al.* (1972). We here report more extensive observations of ions from the *ambient* lunar atmosphere.

INSTRUMENT

The Suprathermal Ion Detector Experiment (SIDE) consists of two positive ion detectors: the Total Ion Detector (TID) and the Mass Analyzer (MA). A schematic representation of the TID and MA are shown in Fig. 1. The TID measures the energy per unit charge of all ions in the energy range 10 to 3500 eV. The MA measures the mass per unit charge in six energy levels: 0.2, 0.6, 1.8, 5.4, 16.2, and 48.6 eV. The mass range and mass resolution varies between each instrument. Three SIDEs have been deployed during Apollo missions 12, 14, and 15. The 14 MA has a mass resolution ($\Delta m/m$) of about 0.1 in the mass range 4 to 200 amu. The 15 MA has a mass resolution of 0.07 in the mass range 1 to 130 amu.

Both the TID and MA have a narrow field of view, about a 6° square solid angle, which is canted 15° from the local vertical.

For more information concerning the SIDE refer to Freeman *et al.* (1970).

OBSERVATIONS

This paper presents a summary of one phase of the analysis being performed on the SIDE data. Due to various considerations (Manka and Michel, 1970) the

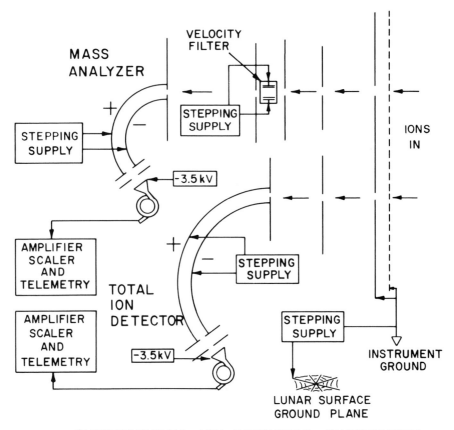

SUPRATHERMAL ION DETECTOR EXPERIMENT

Fig. 1. Schematic diagram of SIDE showing both the Mass Analyzer and the Total Ion Detector.

ions from the lunar atmosphere were expected predominantly to have energies below 100 eV. Thus, a scan of 5 months (August–December, 1971) of TID data from the 14 and 15 SIDEs was conducted to see when low energy (10–100 eV) ions were being observed. Figure 2 shows the periods when these low energy ion were observed. The horizontal lines indicate the periods which the 14 and 15 instruments were on. The blocks show the periods during which the low energy ions were detected. The letters SW, M, and T on the top line indicate solar wind, magnetosheath, and tail respectively.

The low energy ion events can be divided into four types or categories: resonant lunar surface potential event, nonresonant lunar surface potential event, $v \times B$ event, and the geomagnetic tail event. These are discussed below.

The first category of low energy ion events is shown by the open blocks in Fig. 2. This event depends on a resonant effect between the lunar surface potential and the internal voltages of the SIDE to accelerate ions into the instrument. A

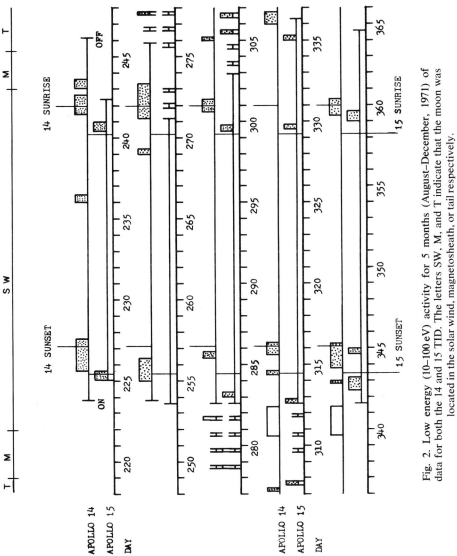

Fig. 2. Low energy (10–100 eV) activity for 5 months (August–December, 1971) of data for both the 14 and 15 TID. The letters SW, M, and T indicate that the moon was located in the solar wind, magnetosheath, or tail respectively.

thorough discussion of these events has been given by Fenner *et al.* (1973). These observations indicate a surface potential of + 10 volts for solar zenith angles of 20° to 45°.

The second type of event relies on a negative surface potential to accelerate positive ions into the instrument. However, no resonance is required between the surface potential and the internal voltages. These events account for most of the activity near the terminators as shown in Fig. 2. This correlation is due to the fact that the surface potential is positive over most of the sunlit hemisphere and becomes negative only near the terminators.

The Apollo 14 data from 5 sunsets crossings are given in Fig. 3. This figure plots the energy channel which contained the maximum flux for each half hour average vs. days before sunset. Note the steady increase in energy of the ions until sunset when the ion energy reaches a steady value of 100 eV. A linear regression analysis of these data is shown in Fig. 4. The sunset data indicate that the lunar surface potential is − 10 V about 1.5 days before sunset and attains a value of − 100 V at sunset. A study of the instrument potential shows that it is near the lunar surface potential in the terminator regions (Lindeman, 1973).

The sunrise data shown in Fig. 4 show much more scatter than the sunset data partially because of the various positions of the bow shock as shown in the figure.

The third type of event utilizes the interplanetary electric field (E_{IP}) to drive ions into the instrument. This electric field is set up by the convection of the interplanetary magnetic field (B_{IP}) past the moon at the solar wind velocity (V_{SW}) and has a value of $E_{IP} = - V_{SW} \times B_{IP}$. When the SIDE is looking perpendicular to V_{SW} (15° from the terminators) and B_{IP} is pointing out of the ecliptic plane, then the SIDE should be able to observe ions accelerated by this electric field. Correlation of SIDE observations with the direction of the interplanetary electric field has been previously reported by Manka (1972).

Since the SIDE must be about 15° from the terminator, these events are often observed simultaneously with the type 2 events. Types 2 and 3 can be differentiated in two ways. Type 2 has a narrow energy spectrum and can last for several days, while type 3 has a wide energy spectrum and lasts only a few hours. Type 3 also has a strong correlation with the component of B_{IP} out of the ecliptic plane. Type 2 shows no correlation with this component of B_{IP}.

Four $V \times B$ events have been compared to Explorer 35 magnetometer data. Only 1 event had flux confined to a narrow cone with a half angle of 10°. The other 3 events had flux in a cone with a half angle of about 60°.

A typical energy spectrum is shown in Fig. 5. The peaks at 10 and 250 eV are common in all four events. Also shown in Fig. 5 is the predicted energy spectrum for single particle dynamics (Manka and Michel, 1970). Since the MA observed a mass of about 20 amu and the surface temperature was about 310°K, a scale height of 80 km is obtained. Note however, that a scale height of 125 km fits the observed data above 250 eV much better than the 80 km. This discrepancy has yet to be adequately explained.

Figure 2 shows that even though the SIDEs were initially cycled off and on for thermal control during the lunar day, both the 14 and 15 instruments regularly

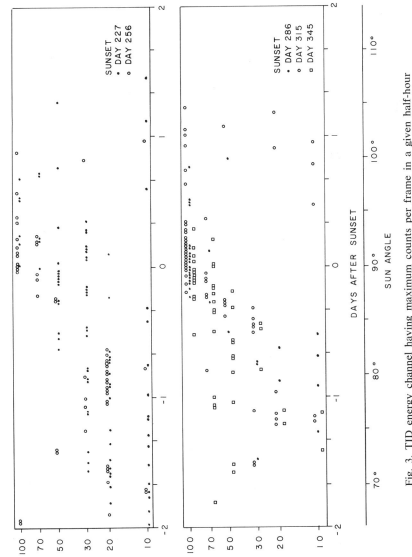

Fig. 3. TID energy channel having maximum counts per frame in a given half-hour period vs. days after sunset for the Apollo 14 SIDE.

R. Lindeman *et al.*

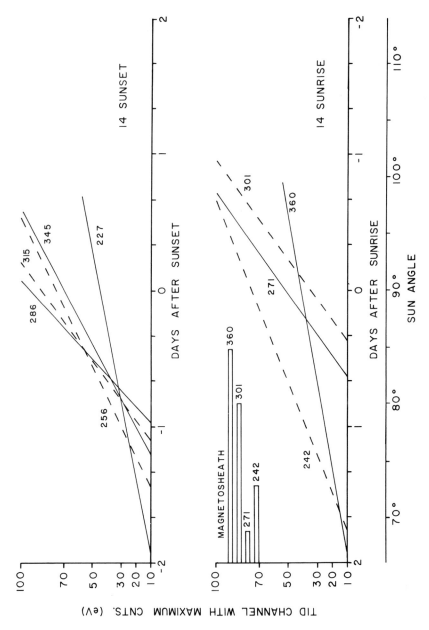

Fig. 4. Linear regression analysis for the Apollo 14 sunrise and sunset events.

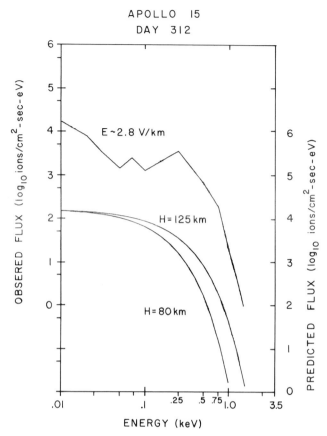

Fig. 5. Predicted and observed differential energy flux for the Apollo 15 event on day
312.

observe low energy ions in the geomagnetic tail. At present the major problem in
interpreting these data is determining the origin of these ions. If they are magnetospheric in origin and are assumed to be isotropic they have a number density
of $20/cm^3$. The energy of the ions varies from 10 to 500 eV. This presents the
problem of how such a dense low energy plasma could accumulate since these
ions are observed both near the plasma sheet and in the high latitude tail.

The more likely answer is that these ions are from the lunar atmosphere. However, we then have the problem of identifying the acceleration mechanism. In the
high latitude tail the surface potential may be greater than +200 volts (Reasoner
and Burke, 1972). Thus the ions require at least 200 eV just to reach the detector.
The cross-tail magnetospheric electric field is pointing almost 90° with respect to
the detectors' field of view. As a result it probably has little to do with observations.

Regardless of the origin these tail events may provide a great deal of
information concerning the plasma dynamics deep in the tail. If the plasma is

magnetospheric in origin then it may provide the clue to the origin of the plasma sheet. If the ions are lunar in origin then we have a good opportunity for determining small scale electric fields deep in the tail.

ION MASS

Data are available from the mass analyzer for a number of the events described herein. These data are all consistent with the noble gases neon or argon serving as the parent atoms. A puzzle that remains is that the SIDE never indicates the presence of ions from both gases simultaneously. We suspect a mass spectrometer effect associated with local electric and magnetic fields and the narrow field of view of the SIDE detectors. An understanding of this effect is essential to a determination of the absolute quantities of the gases present.

CONCLUSIONS

The SIDE observes ions from the lunar atmosphere in a variety of situations. This paper presents a brief summary of the observations. More detailed results will be published in future papers.

Acknowledgments—Discussions with members of the Rice University Space Science Department were useful in the interpretation of these data. This work was supported by NASA contract NAS9-5911.

REFERENCES

Fenner M. A., Freeman J. W., Jr., and Hills H. K. (1973) The electric potential of the lunar surface (abstract), In *Lunar Science—IV*, pp. 234–235. The Lunar Science Institute, Houston.

Freeman J. W. Jr., Balsiger H., and Hills H. K. (1970) Suprathermal Ion Detector Experiment. *Apollo 12 Prelim. Sci. Rep.*, NASA SP-235, pp. 83–92.

Freeman, J. W. Jr., Hills H. K., and Vondrak R. R., (1972) Water vapor whence comest thou? *Proc. Third Lunar Sci. Conf.*, Geochim. Cosmochim. Acta, Suppl. 3, Vol. 3, pp. 2217–2231. MIT Press.

Lindeman R. A. (1973) Observation of ions from the lunar atmosphere. PhD. Thesis, Rice University, Houston, Texas.

Manka R. H. (1972) Lunar atmosphere and ionosphere. Ph.D. Thesis, Rice University, Houston, Texas.

Manka R. H. and Michel F. C. (1970) Energy and flux of ions at the lunar surface (abstract). *Trans. Am. Geophys. Union* **51**, p. 408.

Reasoner D. L. and Burke W. J. (1972) Characteristics of the lunar photoelectron layer in the geomagnetic tail. *J. Geophys. Res.* **77**, 6671–6687.

Proceedings of the Fourth Lunar Science Conference
(Supplement 4, *Geochimica et Cosmochimica Acta*)
Vol. 3, pp. 2897-2908

Lunar ion energy spectra and surface potential

R. H. MANKA

The Lunar Science Institute, Houston, Texas 77058
and
Department of Space Physics and Astronomy
Rice University, Houston, Texas 77001

F. C. MICHEL

Department of Space Physics and Astronomy
Rice University, Houston, Texas 77001

Abstract—The acceleration model for lunar ions and the resulting ionosphere dynamics are reviewed briefly. An application is made to lunar atmosphere trapped in the surface fines, and the enhancement in the Ar^{40}/Ar^{36} ratio in samples from the Apennine Front compared to the adjacent mare is calculated to be ~ 2.0. The predicted lunar ion energy spectra is shown and found to agree well with Suprathermal Ion Detector measurements; from this spectra, the neutral atmosphere scale height can be studied and the neutral atmosphere number density is found to be 1 to 3×10^5 cm^{-3} at the sunrise and sunset terminators. The lunar surface potential is calculated and is found to be several volts positive over much of the sunlit face of the moon but to go tens of volts negative at the terminator.

INTRODUCTION

IN THIS paper we discuss several new aspects of the acceleration of ions formed in the lunar atmosphere. In particular we will look at the predicted and observed energy spectrum of the ions, the lunar surface potential distribution, and a new application of the model for the trapping of lunar atmosphere Ar^{40} in the lunar surface. A model has already been presented for the acceleration of lunar ions by electric and magnetic fields in the solar wind (Manka and Michel, 1970a,b; 1972). It has also been shown that some of the accelerated ions, which impact the lunar surface, can be trapped in the surface; this is likely to be the source of the anomalous Ar^{40} in surface material (Heymann and Yaniv, 1970; Manka and Michel, 1970c). The acceleration model also predicts the flux, energy, and direction of incoming ions measured by ion detectors at the surface; if ion events are selected which have the proper characteristics to indicate that they represent the lunar ionosphere rather than some external source, then the density and pressure of the lunar atmosphere can be calculated (Manka *et al.*, 1972a; Manka and Michel, 1972).

Atmospheric species except hydrogen and helium are gravitationally bound (have lifetimes against gravitational escape much greater than against ionization) and thus form part of the equilibrium lunar atmosphere. The density of each species decreases approximately exponentially with height

$$n(r) \cong n_0 e^{-\left(\frac{r}{h}\right)} \tag{1}$$

2897

where h is the scale height for the species given by $h = kT/mg$ and n_0 is the density per cm^3 at the surface, k is the Boltzmann constant, T and m are species temperature and mass, and g is the lunar gravitational acceleration.

When an ion is formed in the lunar atmosphere, whether it escapes the moon or is accreted to the surface will depend on the interplanetary electric field, and in some cases on electric and magnetic fields at the lunar surface. In a frame of reference at rest with respect to the moon the interplanetary electric field is given by

$$\bar{E}_{sw} = - \bar{V}_{sw} \times \bar{B}_{sw} \qquad (2)$$

where $\bar{V}_{sw}$, $\bar{B}_{sw}$, and $\bar{E}_{sw}$ are, respectively, the solar wind velocity, magnetic field, and electric field. For an ion formed at rest in the lunar frame and accelerated by $\bar{E}_{sw}$ and $\bar{B}_{sw}$, the trajectory is in a plane perpendicular to $\bar{B}_{sw}$ and containing $\bar{E}_{sw}$ and $\bar{V}_D$, where $\bar{V}_D$ is the drift velocity given by (see Manka, 1972a)

$$\bar{V}_D = \frac{\bar{E}_{sw} \times \bar{B}_{sw}}{B_{sw}^2}. \qquad (3)$$

The plane of $\bar{V}_D$ can be thought of as defining a "magnetic" longitude in analogy to the "electric" latitude which will be introduced shortly. In the simplifying case where $\bar{B}_{sw}$ is perpendicular to $\bar{V}_{sw}$ then $\bar{V}_D$ is along $\bar{V}_{sw}$ and the ion drift plane passing through the center of the moon is just the noon-midnight plane.

The equations describing the ion motion have been discussed elsewhere (e.g., Manka and Michel, 1971). The initial motion of an ion is along $\bar{E}_{sw}$ and as the ion gains energy the magnetic force curves the ion in the direction of the solar wind flow with a resulting cycloidal orbit. The height of the cycloid is much greater than lunar dimensions. Thus the ion's trajectory from formation to impact is the initial part of a cycloid, and the motion is nearly parallel to $\bar{E}_{sw}$ and most of the flux of lunar ions to the surface is in a direction perpendicular to the solar wind flow. In general, ions formed in the lower sunlit atmosphere are driven up (with respect to $\bar{E}_{sw}$) into the moon while ions formed at the equator and in the upper hemisphere escape, as illustrated in Fig. 1. Depending upon the direction of $\bar{B}_{sw}$, the interplanetary electric field is generally upward or downward out of the solar ecliptic plane; and, when the direction of $\bar{B}_{sw}$ reverses several times during each solar rotation (as is common due to the sector structure of the interplanetary magnetic field), the direction of $\bar{E}_{sw}$ and the ion flux also reverse.

The energy of the ion at impact is the energy gain along the interplanetary electric field

$$\mathscr{E} = eE_{sw}y_i \qquad (4)$$

where e is the ion charge (we assume single ionization) and y_i is the y-coordinate at impact. The sketch of Fig. 1 illustrates, with some exaggeration, the trajectories of three ions of quite different mass. Actually, hydrogen is not gravitationally bound, scale height does not have a meaning in the usual sense, and most hydrogen ions are formed at much greater heights than for bound species. The solution for the impact coordinates and energy has been discussed elsewhere

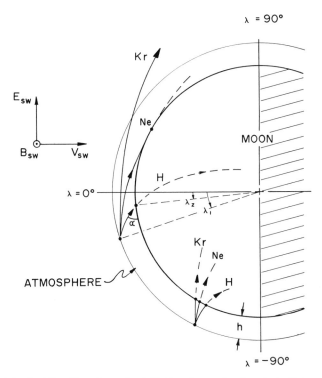

Fig. 1. Sketch of orbits of ions formed outside the plasma sheath and at a distance h from the surface. (Actually krypton and oxygen have different scale heights and hydrogen is not bound.) The sketch illustrates that ions heavier than hydrogen will strike primarily the hemisphere into which they are driven by the interplanetary electric field; when the polarities of $\bar{B}_{sw}$ and $\bar{E}_{sw}$ reverse, the other hemisphere is struck. The trajectories are shown in a plane passing through the center of the moon and containing the drift velocity $\bar{V}_D$ as well as $\bar{E}_{sw}$.

(Manka and Michel, 1971, 1972). In general, impact energies vary from tens of electron volts to a few keV. The impact energies are higher near the equator than at the poles so atmospheric ions should be more easily trapped near the equator. The impact energy is also approximately inversely proportional to the mass of the ion.

APPLICATION TO Ar^{40} TRAPPED ON THE APENNINE FRONT

The analysis of gaseous elements in the lunar samples continues to exhibit unexpected compositions in some gaseous elements. One of the most pronounced anomalies was the excess of surface correlated Ar^{40}, compared to solar wind implanted Ar^{36}, which was discussed by several researchers, including Heymann and Yaniv (1970) who proposed a lunar atmosphere source for the Ar^{40}, originally produced by K^{40} decay in the moon. In a paper showing that the lunar atmosphere is the likely source of the surface Ar^{40}, Manka and Michel (1970c) suggested that the atmosphere would also be the source of other unexpected surface elements, or unexpected concentrations of elements, including solar wind elements which have been cycled through the atmosphere and reimplanted in a manner characteristic of atmospheric ions.

For an ion whose trajectory causes it to impact the moon, whether the ion is firmly implanted in the surface or is quickly released will depend on both the impact energy and the trapping probability $\eta_t(\mathscr{E})$ for the lunar material. Studies at Berne on ion implantation in aluminum foils with their associated oxide layer can be used to estimate the trapping characteristics of lunar grains. The total trapping of ions is calculated by integrating the trapping probability over the incident ion energy spectrum

$$N_t = \int_0^\infty \eta_t(\mathscr{E})n(\mathscr{E})d\mathscr{E} \tag{5}$$

where $n(\mathscr{E})$ is determined from the exponentially decreasing number density in the atmosphere and the resulting trajectory along the interplanetary electric field into the moon.

At present it appears that sufficient Ar^{40} is available for release into the lunar atmosphere, and that the lunar atmosphere supplies the Ar^{40} which is found in lunar samples. This is strongly supported by several types of data. First, lunar ions are observed at the lunar surface with just the direction, energy spectrum, and flux predicted by the model (Manka et al., 1972a), thus the ion flux used in these calculations does exist. Second, the trapping probability appears to be high enough, even for minerals such as olivine, for an adequate amount of Ar^{40} to be deposited. Third, initial results from the Apollo 17 Surface Mass Spectrometer indicate the possibility that even the present lunar atmosphere may have sufficient Ar^{40} to provide the amounts observed in lunar fines: the average density of Ar^{40} in the sunlit lunar atmosphere, which would provide the observed trapped Ar^{40}, is calculated to be 10^2 to 10^3 cm^{-3}; the Surface Mass Spectrometer indicates a possible Ar^{40} peak at sunrise of about 10^4 cm^{-3} (Hoffman et al., 1973). The question of the geologic history of the lunar atmosphere, and its Ar^{40} content, continues to pose many unanswered questions; however, the data mentioned above indicate the possibility that part, or all, of the excess Ar^{40} is from the lunar atmosphere.

Another possible application of the trapping model was brought to our attention by Jordan et al. (1973a) as a result of their study of the Ar^{40}/Ar^{36} ratios in samples from the Apollo 15 site. They report a substantial variation in the Ar^{40}/Ar^{36} ratios, varying from 0.62 to 3.5 with the higher ratios occurring in samples from the Apennine Front and the lower ratios from the Mare. They suggest that one possible explanation may be the relative incoming ion exposure geometry between the flat-lying mare and the north-facing slope of the Apennine Front.

The relative amounts of Ar^{40} and Ar^{36} which are trapped on different slopes can be calculated from Equation (5) above. However, the integral simplifies since the relative energies, of either the lunar atmosphere ions or the solar wind ions, will not change significantly as the slope of the local surface changes, so that $\eta_t(\mathscr{E})$ is equal for the two cases. Likewise, at a given local latitude and longitude, the number of ions as a function of energy which strike a unit area of surface depends only on the relative orientation of the surface. Thus the relative isotopic ratio for a given local orientation of lunar surface can be obtained by integrating the relative fluxes through a lunar day.

The Apollo 15 site is at approximately 26° north selenographic latitude, and on the mare the Ar^{36} flux accrued through the lunar day is proportional to the exposed area

$$\text{Flux (36)} \quad \alpha \int_{-\pi/2}^{\pi/2} \Phi_{36} \cos \lambda \cos \phi \, d\phi = 2 \cos \lambda \Phi_{36} \tag{6}$$

where Φ_{36} is the flux of Ar^{36} in the solar wind, λ is the electric latitude (which on the average can be taken to be the selenographic latitude), and ϕ is the longitude measured from the subsolar point.

Similarly, the flux of Ar^{40} is calculated over the sunlit face since the formation of atmospheric ions requires sunlight or solar wind for ionization. However, lunar ions can only reach the Apollo 15 site about half the time when the interplanetary electric field is pointed southward. Thus, the flux accrued during the lunar day is proportional to

$$\text{Flux (40)} \quad \alpha \int_{0}^{\pi/2} \Phi_{40} \sin \lambda \, d\phi = \frac{\pi}{2} \Phi_{40} \sin \lambda. \tag{7}$$

On the north-facing Apennine Front, assuming a slope γ with respect to the mare, then the accrued solar wind Ar^{36} flux is reduced to

$$\text{Flux (36)} \quad \alpha \int_{-\pi/2}^{\pi/2} \Phi_{36} \cos (\lambda + \gamma) \cos \phi \, d\phi = 2\Phi_{36} \cos (\lambda + \gamma) \tag{8}$$

and the accrued Ar^{40} flux is increased to

$$\text{Flux (40)} \quad \alpha \int_{0}^{\pi/2} \sin (\lambda + \gamma) \, d\phi = \frac{\pi}{2} \Phi_{40} \sin (\lambda + \gamma). \tag{9}$$

Thus the predicted *enhancement* in the Ar^{40}/Ar^{36} ratio on the Apennine Front slope to the ratio on the mare is given by

$$\text{Enhancement in } Ar^{40}/Ar^{36} \text{ Ratio} = \frac{Ar^{40}/Ar^{36}|\text{Slope}}{Ar^{40}/Ar^{36}|\text{Mare}} \cong \frac{\tan (\lambda + \gamma)}{\tan \gamma}. \tag{10}$$

The results are presented in Fig. 2 which shows the enhancement in the Ar^{40}/Ar^{36} ratio as a function of the steepness of the slope. The data for the slope and mare ratios will be presented in a paper by Jordan *et al.* (1973b). From this data an approximate average ratio for samples from the slope was 2.3 while an approximate average ratio for the mare was 0.86; this gives for the Ar^{40}/Ar^{36} ratio a

$$\text{Ratio Enhancement (experimental)} \cong 2.7$$

whereas the calculation predicts for the 20° slope where the sample was collected

$$\text{Ratio Enhancement (predicted)} \cong 2.0$$

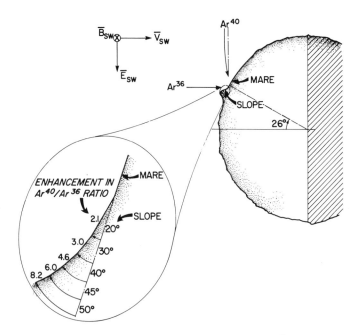

Fig. 2. Sketch of the relative trajectories of lunar atmosphere Ar^{40} and solar wind Ar^{36}. The inset shows the predicted enhancement in the Ar^{40}/Ar^{36} ratio as a function of the slope of the Apennine Front relative to the mare.

Thus the trend is similar and the slightly higher observed enhancement could be because the collected samples were initially further up the slope when the argon was implanted.

LUNAR ION ENERGY SPECTRA

The considerable fluctuations of $\bar{B}_{sw}$ out of the ecliptic plane will rotate the planes of lunar trajectories; also the lunar surface electric and magnetic fields could play important roles in changing an ion's direction and modifying its energy as it approaches the surface. The Suprathermal Ion Detectors (Freeman *et al.*, 1971) are oriented so that they look in the ecliptic plane and see sporadic "clouds" of ions. The fluctuations in the direction of $\bar{E}_{sw}$, may combine with the surface fields to allow the detection of these ions, and give the appearance that only bunches of ions are arriving at the surface.

From Fig. 3, we see that in order for lunar ions to enter one of the SIDE experiments (the Apollo 12 SIDE location is shown), which point in the ecliptic with a narrow look angle, the interplanetary magnetic field must be strongly out of the ecliptic and in a direction such that the interplanetary electric field, $\bar{E}_{sw} = -\bar{V}_{sw} \times \bar{B}_{sw}$, is pointed toward the detector. A further immediate consequence is that since the SIDE's generally look within 15° East/West of the local vertical, and since $\bar{E}_{sw}$ lies in a plane perpendicular to $\bar{V}_{sw}$, then ions from the lunar ionosphere

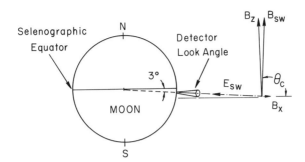

Fig. 3. Sketch of the critical angle θ_c of $\bar{B}_{sw}$ out of the ecliptic for ions to enter one of the Suprathermal Ion Detectors. Here $\bar{B}_{sw}$ is nearly 90° out of the ecliptic plane while $\bar{E}_{sw}$ lies nearly in the ecliptic.

should be observed mostly near local sunrise and sunset at the detector. This apparently is the case, with even the ions having energies higher than the ambient lunar ionosphere tending to peak at sunrise/sunset. (Freeman *et al.*, 1972).

Several ions events have been analyzed for correlation between observed ion flux and direction of $\bar{B}_{sw}$ (Manka *et al.*, 1972a; for a discussion of the Lunar Surface and Explorer 35 Magnetometer data *see* Dyal and Parkin 1971; Colburn *et al.*, 1971). Several events, near local sunrise or sunset, were found when $\bar{B}_{sw}$ was almost 90° out of the ecliptic plane. During this time an intense ion flux was detected by the SIDE, the flux appearing to turn off when $\bar{B}_{sw}$ rotated a few degrees away from this critical angle; these events strongly support the acceleration model. Other events were studied which do not fit the model as well, though $\bar{E}_{sw}$ is generally toward the detector for these events; the energy spectra of these events are not thought to be characteristic of the lunar ionosphere and the ions may be from external sources such as the bow shock.

One energy spectrum for lunar atmosphere ions is illustrated in Fig. 4 (Manka *et al.*, 1972a, 1972b). The solid line is the predicted theoretical spectrum which should be observed by the SIDE, which has an energy band width proportional ($\sim 10\%$) to each observation energy. This spectrum assumes an exponential density distribution in the atmosphere and that all ions accelerated to the surface by $\bar{E}_{sw}$ are collected. The dashed line is the spectrum observed for one of the events where the ion flux correlates with the orientation of B_{sw}. It can be seen that the observed spectrum follows the predicted spectrum fairly closely except for a decreased count rate at low and high energies. This decrease will be discussed in a later paper. It should be noted that the peak at the spectrum occurs approximately at an energy

$$\mathscr{E} = h\,E_{sw} \qquad (11)$$

where h is an effective scale height for the atmospheric species being detected. This provides an additional tool for studying the ionosphere since $\mathscr{E}$ is measured, E_{sw} is often known from magnetometer and solar wind measurements, so that h can be calculated and compared with expected species masses. Furthermore, the

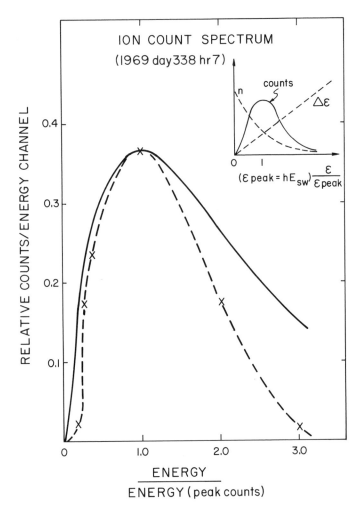

Fig. 4. Plot of a predicted and observed ion count spectrum seen by the Apollo 12 SIDE. The energy is normalized to the energy for which the peak number of counts occurs (250 eV for this event).

integrated ion flux can be related to the neutral density in the lunar atmosphere; assuming the flux to be mostly neon, we calculate an atmospheric number density at the surface of 1 to 3×10^5 cm^{-3} in good agreement with the Cold Cathode Gauge (Johnson et al., 1972).

Lunar Surface Electric Potential

Several authors have studied the lunar surface potential (Opik and Singer, 1960; Bernstein et al., 1963; Grobman and Blank, 1968), particularly for the case of the sunlit face of the moon exposed to the solar wind. Resulting potentials at the subsolar point have usually been calculated to be in the range of a few volts to perhaps 20 volts positive, due to a predominant photoelectric current.

Here we apply the approach of Manka and Anderson (1969) and Manka (1972a,b), and discuss the effects of surface potential on incoming ion energies. These potential calculations differ from preceding calculations in that they extend the potential distribution to the terminator for the case of the moon in the solar wind, and give the potential on both the light and dark sides when the moon is in the various plasma environments of the geomagnetic tail.

The moon can be treated as a body, or probe, immersed in a plasma. In the case of the solar wind, the thermal and flow motions of the plasma must be included while in the geomagnetic tail the thermal motion predominates. The local lunar surface will reach a potential such that the net current to it is zero,

$$I_e + I_i + I_p + I_s = 0 \tag{12}$$

where I_e, I_i, I_p, and I_s are the electron ion, photoelectric and secondary electron currents respectively. Since these currents depend on the surface potential, the equation can be solved for the equilibrium potential. The form of the expressions for currents as a function of potential depends on whether the species described is attracted or repelled.

While all the plasma equations are not presented in this paper, the equations for a few cases are discussed below. For example, in the case of a positive potential, $\phi > 0$, the current density due to repelled thermal plasma ions is

$$I_i = ne \sqrt{\frac{kT_i}{2\pi m_i}} \exp\left(\frac{-e\phi}{kT_i}\right) \tag{13}$$

$$= I_{i0} \exp\left(\frac{-e\phi}{kT_i}\right)$$

where n is the ambient plasma electron density, m is the species mass, k is the Boltzmann constant, and T is the species temperature.

The current density due to attracted thermal plasma electrons is

$$I_e \cong -ne \sqrt{\frac{kT_e}{2\pi m_e}}$$

$$= I_{e0} \tag{14}$$

I_{i0} and I_{e0} are the flux currents when the lunar potential is at the plasma potential (i.e., zero).

On the other hand, for the flowing plasma the thermal contribution to the current is interrelated with the flow contribution, and for example the electron current becomes

$$I_e = \frac{nev_{me}}{2\sqrt{\pi}}[e^{-U^2}{}_e + \sqrt{\pi}\, U_e(1 + \mathrm{erf}\,(U_e))] \tag{15}$$

where $U = V \cos \theta / v_m$, v_m is the species' mean speed, and V is the flow velocity.

The solar wind (flowing) ion current is simply given as

$$I_i \cong neV \cos \theta = I_{i0} \cos \theta \qquad (16)$$

since the incoming proton energy greatly exceeds the expected surface potential.

For the case of a positive potential, the current density of photoelectrons can approximately be written

$$I_p = i_p \cos \theta \exp \left(\frac{-e\phi}{kT_p} \right) \qquad (17)$$

where i_p is the photo current density from an area of the lunar surface at the plasma potential with normally incident sunlight and θ is the polar angle of the local surface with respect to the subsolar point. This expression assumes an equivalent photoelectron temperature, T_p. However, in the case of a negative surface the emitted photoelectron current density will be

$$I_p = i_p \cos \theta. \qquad (18)$$

The results of the calculation of the sunlit surface potential in the case of the moon in the solar wind are shown in Fig. 5. The potential is shown from the subsolar point ($\theta = 0°$) to the terminator ($\theta = 90°$) as a function of different "effective photoemissivities." The effective emissivity ϵ_0 is that emissivity which

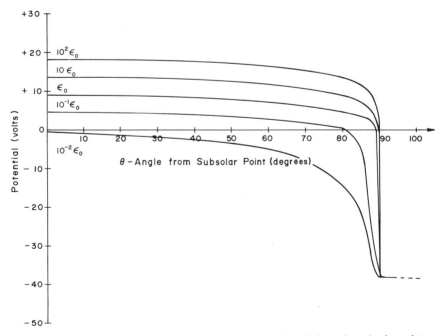

Fig. 5. Plot of the lunar surface potential as a function of angle from the subsolar point. Potentials are calculated for five effective photoelectron emissivities.

would give a photocurrent equal to the current of 5×10^{-9} amp/cm^2 measured from metals exposed to the solar spectrum. The other values, up to two orders of magnitude greater or less, are chosen to cover all likely values of actual lunar photoemissivity. Recent measurements of photocurrents from lunar samples (Feuerbacher *et al.*, 1972) give an integral current of 4.5×10^{-10} amp cm^{-2}. This is almost exactly one-tenth the metal photocurrent; thus the curve corresponding to 10^{-1} ϵ_0 is preferred and subsolar potentials of the order of $+6$ volts are expected. However, near the terminator, the solar wind electron flux begins to dominate over the effectively positive photocurrent and the surface goes negative. In Fig. 5 the potential at the terminator is ~ 40 volts negative if a solar wind electron temperature of 10 electron volts is chosen. However the solar wind temperature is constantly changing and the terminator potential will change with it, a temperature of 20 electron volts giving a potential of about 80 volts negative, etc.

Potential calculations have been completed for the cases of the moon in the plasma sheet and the high latitude regions of the geomagnetic tail and the reader is referred to Manka (1972a,b) for a more complete discussion of the equations and resulting potentials.

SUMMARY

In this paper we have discussed recent data which relate to the acceleration model for lunar ions, and have made further applications of the model. The data showing the correlation between detected ion flux and the magnetic field is important since it shows that at least part of the ambient ion fluxes have the characteristics predicted by the model. A possible new consequence of ion acceleration may be indicated in the slope effect on Ar^{40}/Ar^{36} ratios discussed by Jordan *et al.* (1973a). Another application is to the analysis of ion detector data where the lunar ionosphere energy spectra can be studied and the neutral atmosphere scale height and number density can be calculated; the neutral atmosphere number density at the terminator is found to be 1 to 3×10^5 cm^{-3}. It was shown that under some conditions that the lunar surface potential can have a significant effect on incoming ion energies and for example, that for the moon in the solar wind, the terminator ion energy could be increased by several tens of electron volts.

Acknowledgments—We thank Drs. Freeman, Hills, and Vondrak of Rice for helpful discussions and assistance in interpretation of the SIDE data. We also thank Drs. Colburn, Dyal, Parkin, and Sonett for making available data from the Ames Explorer 35 and Lunar Surface Magnetometers. We thank Drs. D. Heymann, J. L. Jordan, and S. Lakatos (Rice) for bringing to our attention their interpretation that the Apollo 15 argon ratios may indicate a trapping process. This work was partially supported at Rice under Contract NAS 9-5911 and Grant GA-28033X (NSF). One of the authors (RHM) is grateful for support as a Visiting Post-Doctoral Fellow at the Lunar Science Institute which is operated by the Universities Space Research Association under Contract No. NSR-09-051-001 with the National Aeronautics and Space Administration. This paper constitutes the Lunar Science Institute contribution No. 151.

References

Bernstein W., Fredricks R. W., Vogl J. L., and Fowler W. A. (1963) The lunar atmosphere and the solar wind. *Icarus* **2**, 233.

Colburn D. S., Mihalov J. D., and Sonett C. P. (1971) Magnetic observations of the lunar cavity. *J. Geophys. Res.* **76**, 2940.

Dyal P. and Parkin C. W. (1971) Electrical conductivity and temperature of the lunar interior from magnetic transient-response measurements. *J. Geophys. Res.* **76**, 5947.

Feuerbacher B., Anderegg M., Fitton B., Laude L. D., and Willis R. F. (1972) Photoemission from lunar surface fines and the lunar photoelectron sheath. *Proc. Third Lunar Sci. Conf., Geochim. Cosmochim. Acta*, Suppl. 3, Vol. 3, 2655–2663. MIT Press.

Freeman J. W., Hills H. K., and Fenner M. A. (1971) Some results from the Apollo 12 suprathermal ion detector. *Proc. Second Lunar Sci. Conf., Geochim. Cosmochim. Acta*, Suppl. 2, Vol. 3, p. 2093. MIT Press.

Freeman J. W., Fenner M. A., Hills H. K., Lindeman R. A., Medrano R., and Meister J. (1972) Suprathermal ions near the moon. *Icarus* **16**, 328.

Grobman W. D. and Blank J. L. (1969) Electrostatic potential distribution of the sunlit lunar surface. *J. Geophys. Res.* **74**, 3943.

Heymann D. and Yaniv A. (1970) Ar^{40} anomaly in lunar samples from Apollo 11. *Proc. Second Lunar Sci. Conf., Geochim. Cosmochim. Acta*, Suppl. 1, Vol. 2, p. 1261. Pergamon.

Hoffman J. H., Hodges R. R. Jr., and Evans D. E. (1973) Lunar atmospheric composition results from Apollo 17 (abstract). In *Lunar Science—IV*, pp. 376–377. The Lunar Science Institute, Houston.

Johnson F. S., Carroll J. M., and Evans D. E. (1972) Lunar atmosphere measurements. *Proc. Third Lunar Sci. Conf., Geochim. Cosmochim. Acta*, Suppl. 3, Vol. 3, pp. 2231–2242. MIT Press.

Jordan J. L., Lakatos S., and Heymann D. (1973a) Ar-Kr-Xe systematics in Apollo 15 fines (abstract). In *Lunar Science—IV*, pp. 415–417. The Lunar Science Institute, Houston.

Jordan J. L., Heymann D., and Lakatos S. (1973b) Inert gas patterns in the regolith at the Apollo 15 landing site. *Geochim. Cosmochim. Acta.* Pergamon. (In press.).

Manka R. H. (1972a) Lunar atmosphere and ionosphere. Ph.D. Thesis, Rice University, Houston, Texas.

Manka, R. H. (1972b) Plasma and potential at the lunar surface. Presented at the ESLAB Symposium on Photon and Particle Interactions with Surfaces in Space, Noordwijk aan Zee, Holland, Sept. 26–29, 1972. Also to be published in the Conference Proceedings.

Manka R. H. and Anderson H. R. (1969) Lunar Surface Plasma Environment and Electric Potential. *Trans. Am. Geophys. Union* **50**, 217.

Manka R. H. and Michel F. C. (1970a) Energy and Flux of ions at the lunar surface (abstract). *Trans. Am. Geophys. Union* **51**, 408.

Manka R. H. and Michel F. C. (1970b) Lunar atmosphere as a source of Ar^{40} and other lunar surface elements (abstract). *Trans. Am. Geophys. Union* **51**, 344.

Manka R. H. and Michel F. C. (1970c) Lunar atmosphere as a source of argon-40 and other lunar surface elements. *Science* **169**, 278.

Manka R. H. and Michel F. C. (1971) Lunar atmosphere as a source of lunar surface elements. *Proc. Second Lunar Sci. Conf., Geochim. Cosmochim. Acta*, Suppl. 2, Vol. 2, p. 1717. MIT Press.

Manka R. H. and Michel F. C. (1972) Lunar ion flux and energy. Presented at the ESLAB Symposium on Photon and Particle Interactions with Surfaces in Space, Noordwijk ann Zee, Holland, Sept. 26–29, 1972. Also to be published in the Conference Proceedings.

Manka R. H., Michel F. C., Freeman J. W. Jr., Dyal P., Parkin C. W., Colburn D. S., and Sonett C. P. (1972a) Evidence for acceleration of lunar ions (abstract). In *Lunar Science—III*, p. 504. The Lunar Science Institute, Houston.

Manka R. H., Michel F. C., Freeman J. W., Hills H. K. (1972b) Energy spectrum of lunar ions (abstract). *Trans. Am. Geophys. Union* **53**, 439.

Opik E. J. and Singer S. F. (1960) Escape of gases from the Moon, *J. Geophys. Res.* **65**, 3065.

Proceedings of the Fourth Lunar Science Conference
(Supplement 4, *Geochimica et Cosmochimica Acta*)
Vol. 3, pp. 2909–2923

Lunar electromagnetic scattering II.
Magnetic fields and transfer functions for parallel propagation

G. Schubert*

Department Planetary and Space Science
University of California, Los Angeles 90024

K. Schwartz

20858 Collins Street
Woodland Hills, California 91364

C. P. Sonett, D. S. Colburn, and B. F. Smith

NASA Ames Research Center
Moffett Field, California 94035

Abstract—Magnetic field and transfer function amplitudes, resulting from a transverse electromagnetic wave in the interplanetary medium scattering from the moon and its diamagnetic cavity, are presented. Calculations are made using the asymmetric scattering theory of Part I for a spherical two-layer model of the lunar electrical conductivity profile and a nonconducting cylindrical model of the downstream lunar plasma void. Both the field and transfer function magnitudes are calculated as functions of position on the surface of the moon for frequencies relevant to the observations of the lunar surface and orbiting magnetometers. The amplitudes of the magnetic field components on the cavity boundary are also computed as functions of frequency and distance downstream from the lunar limb. Comparisons of the results are made with those of (1) spherically symmetric descriptions of lunar electromagnetic scattering, (2) the quasistatic approximation to asymmetric scattering theory, and (3) observations of the scattering phenomenon by lunar surface and orbiting magnetometers. The quasistatic approximation qualitatively accounts for all aspects of the scattering process except for the propagating cavity surface wave and the interference of different orders of multipole response at high frequency. Asymmetric scattering theory can apparently explain the many features of lunar induction observed by the Apollo and Explorer magnetometers. The numerical finite difference calculation of Riesz *et al.* (1972a,b) for scattering from a cylindrical moon-slab cavity model is either in error or does not describe scattering from a spherical moon-cylindrical cavity model.

Introduction

This paper makes application of an analytic asymmetric theory of lunar electromagnetic induction (Schwartz and Schubert, 1973) for the time dependent electromagnetic fields inside the moon and interior to the diamagnetic cavity in the presence of a fluctuating interplanetary electromagnetic field propagating parallel to the diamagnetic cavity axis.† It was shown in Part I (Schwartz and

---

*John Simon Guggenheim Memorial Foundation Fellow

†Fluctuations of the interplanetary magnetic field are either convected or jointly propagate and convect depending on whether they are MHD structure imbedded in the solar wind or true waves. For the purpose of this paper as well as earlier work no distinction is necessary insofar as the apparent propagation speed is considered. On the other hand for the spatial dispersion of the unit wave normals (not considered in this paper) a distinction is required.

Schubert, 1973) that a surface wave, driven by the interplanetary field, propagates on the cavity boundary, and that the cavity acts as a TE mode waveguide with low frequency cutoff at about 50 Hz. Thus, lunar excited radiation attenuates with distance down the cavity while far downstream from the moon a surface wave propagates on the cavity boundary.

The present application of the theory uses a lunar model with a core of uniform electrical conductivity σ and radius $b < a$ (lunar radius) surrounded by a nonconducting shell. Both magnetic field and transfer function amplitudes are calculated as functions of position on the lunar surface, i.e., local time. The frequencies encompassed are those representative of the lunar surface magnetometers and the orbiting Apollo subsatellite and Explorer 35 magnetometers (see, for example, Coleman et al., 1972; Dyal et al., 1972; Sonett et al., 1972a; Schubert et al., 1973a; Smith et al., 1973a). We also calculate the components of the magnetic field on the cavity boundary as functions of frequency and downstream distance from the lunar limb.

An additional major purpose of this paper is to compare the magnetic field and transfer function amplitudes computed using asymmetric scattering theory with those obtained from spherically symmetric theories (Lahiri and Price, 1939; Schubert and Schwartz, 1969, 1972; Blank and Sill, 1969; Schwartz and Schubert, 1969; Schubert and Colburn, 1971) and the quasistatic approximation to the asymmetric theory (Schubert et al., 1973b). The results presented here and in Part I show that the quasistatic approximation to asymmetric theory describes many of the features of frequency dependent asymmetric scattering from the moon. However, the quasistatic approximation does not account for the cavity surface wave forced by the interplanetary field nor does it account for the complex multipolar interference effects encountered at high frequency.

We also compare the tangential transfer function magnitudes at the subsolar and antisolar positions calculated using asymmetric scattering theory with the empirical transfer functions based on day and night side observations (Sonett et al., 1972a; Schubert et al., 1973c). This comparison can be only qualitative since the lunar electrical conductivity model used here was not obtained through an inversion of that data. Nevertheless the comparison indicates that lunar conductivity models, whose transfer functions simultaneously approximate both day and night side data sets, will be obtainable through inversion of the combined data sets with asymmetric scattering theory. Leavy (1973) has inverted a subset of the available day side data using a subsolar transfer function based on an asymmetric theory similar to that reported in Part I.

We have computed transfer functions for an electrical conductivity model corresponding to that of Riesz et al. (1972a,b). Our results show that their work is incorrect because either their model (a cylindrical moon with a downstream slab cavity) is a poor representation of the spherical moon with its downstream diamagnetic cavity or their numerical finite difference scheme is in error. Thus the criticism by Sonett et al. (1972b) is substantiated.

To facilitate reading of the present paper we present a brief description of the geometry and notation used in Part I. The theory of Part I was derived for a spherical moon model with electrical conductivity σ an arbitrary function of radius; the cavity was modeled by a nonconducting cylinder extending infinitely far downstream. The incident field was given by

$$\begin{Bmatrix} E \\ H \end{Bmatrix}_{\text{incident}} = H_0 e^{i((2\pi z/\lambda)-\omega t)} \begin{Bmatrix} \mu v a_x \\ a_y \end{Bmatrix} \tag{1}$$

where E and H are the electric and magnetic fields, a_x, a_y, and a_z are the unit vectors of a Cartesian coordinate system with origin at the center of the moon, λ is the wavelength in the interplanetary medium, ω is the circular frequency as measured by an observer fixed with the moon, H_0 is the amplitude of the magnetic field oscillation, $v = \lambda\omega/2\pi$ and μ is the magnetic permeability of free space. As in Part I, MKS units are used throughout this paper. The moon-cavity geometry and the geometry of the incident field are shown in Fig. 1. Both spherical (r, θ, ϕ) and cylindrical (ρ, ϕ, z) coordinates are used. The space outside the moon and cavity contains only the incident interplanetary electromagnetic field given by (1). The particular moon model used here is a two-layer one with an insulating outer shell and a core of uniform conductivity σ and radius b. The value, at the lunar surface, of the derivative of the radial part of the transverse electric eigenfunction for the two-layer moon model is given in the Appendix. This formula, together with Part I, is all the information necessary to obtain the results of the present paper.

DISCUSSION OF RESULTS

Magnetic fields

The amplitude of the tangential magnetic field component H_θ at the lunar surface, normalized with respect to that of the incident field, is shown in Fig. 2 as a function of position on the surface with frequency as a parameter. Unless otherwise noted, all the results presented in this paper are for the two-layer moon model with $b = 1600\,\text{km}$ ($b/a = 0.92$), $\sigma = 4 \times 10^{-4}\,\text{mhos/m}$, and $v = 300\,\text{km/sec}$. The lunar model parameters chosen will be seen to provide theoretical transfer functions in qualitative agreement with observations. The value of the propaga-

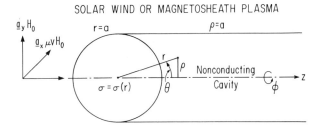

Fig. 1. The geometry of an interplanetary transverse electromagnetic wave scattering from a spherical model of the lunar electrical conductivity profile and a nonconducting cylindrical model of the lunar diamagnetic cavity.

G. Schubert *et al.*

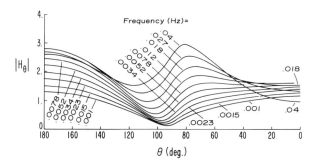

Fig. 2. The amplitude of the tangential magnetic field component H_θ at the lunar surface, normalized with respect to that of the incident field H_0, as a function of position on the surface θ with frequency as a parameter. The model of lunar electrical conductivity is a sphere with an insulating shell and a core of constant conductivity 4×10^{-4} mhos/m and radius 1600 km. The incident interplanetary wave has a phase velocity of 300 km/sec.

tion velocity is a reasonable one considering the direction and wave velocity of solar wind disturbances (Sonett *et al.*, 1972a). The function $|H_\theta|$ is viewed in the plane $\phi = 90°$, i.e., the plane containing the cavity axis and the direction of the incident magnetic field fluctuation which may be arbitrarily oriented with respect to the ecliptic. For fixed frequency $|H_\theta|$ varies strongly with position on the surface of the moon, from the subsolar point, $\theta = 180°$, to the antisolar point, $\theta = 0°$. For all frequencies considered in the figure, $|H_\theta|$ decreases with decreasing θ from its value at the subsolar point to a local minimum which is attained forward of the limb. The position of this local minimum moves toward the subsolar point with increasing frequency.

As θ decreases beyond the position of this minimum and beyond the limb, $|H_\theta|$ increases; for the lower frequencies, e.g., 10^{-3} Hz, the increase is monotonic for all θ to the antisolar point, while for the higher frequencies, e.g., 4×10^{-2} Hz, a local maximum is encountered shortly beyond the limb. At 4×10^{-2} Hz the magnitude of H_θ is an absolute maximum at a position on the surface nearly 10° past the limb and within the cavity! At lower frequencies the absolute maximum amplification occurs at the subsolar point. There are large regions in the neighborhoods of the subsolar and antisolar points where the behavior of $|H_\theta|$ for fixed θ is nonmonotonic with frequency. Figure 2 should be compared with Fig. 5 of Schubert *et al.* (1973b) which is a similar plot for the quasistatic approximation to asymmetric theory. This quasistatic approximation yields variations in $|H_\theta|$ which are in qualitative agreement with the results shown in Fig. 2, even to the extent of predicting the high frequency local maximum in $|H_\theta|$ seen just beyond the limb.

Figure 3 shows $|H_\phi|$ at the lunar surface, normalized with respect to the amplitude of the incident field, as a function of θ for the range of frequencies 0.001 Hz to 0.04 Hz. The variation of $|H_\phi|$ is considered in the plane $\phi = 0°$ i.e., the plane which contains the cavity axis and the normal to the incident magnetic field direction. For a given frequency, $|H_\phi|$ decreases monotonically from a maximum

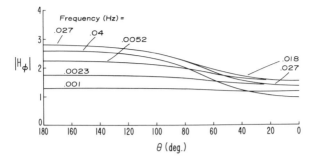

Fig. 3. The magnitude of H_ϕ/H_0 at the lunar surface as a function of θ for frequencies in the range 0.001 Hz to 0.04 Hz. The lunar conductivity model and incident wave speed are specified in Fig. 2. The variation of $|H_\phi/H_0|$ is shown in the plane which contains the cavity axis and the normal to the incident magnetic field direction.

at the subsolar point to a minimum at the antisolar point. For given θ, $|H_\phi|$ at first increases with frequency, but depending on the particular value of θ a frequency will be reached such that $|H_\phi|$ decreases with further increase in frequency. This occurs at about 0.02 Hz at the subsolar point and 0.01 Hz at the antisolar point. Figure 8 of Schubert *et al.* (1973b) shows that the quasistatic approximation to asymmetric scattering theory qualitatively accounts for the variation of $|H_\phi|$ with position on the lunar surface.

Nevertheless there are significant differences between the quasistatic calculations for a moon characterized by a perfectly conducting core and the frequency dependent computations of this paper for a moon with a core of finite conductivity. Figures 2 and 3 show that the fields do not vary monotonically with frequency; the quasistatic results showed monotonic variations with b/a, the parameter which qualitatively represented frequency in the quasistatic approximation. The nonmonotonic behavior with increasing frequency is the result of interference between various orders of multipolar response.

The magnitude of the radial magnetic field component H_r at the moon's surface, normalized with respect to H_0, is shown as a function of θ for frequencies between 0.001 Hz and 0.04 Hz in Fig. 4a and for frequencies between 0.012 Hz and 0.04 Hz in Fig. 4b. We view $|H_r|$ in the plane $\phi = 90°$. Between the subsolar point and the limb, $|H_r|$ is identical to the magnitude of that component of the incident interplanetary field. At surface positions between the limb and antisolar point, $|H_r|$ generally decreases with increasing frequency at low frequencies (Fig. 4a), while it increases with increasing frequency at the higher frequencies (Fig. 4b). From Fig. 4 of Schubert *et al.* (1973b) it can be seen that the quasistatic approximation qualitatively describes the behavior of $|H_r|$ shown in Fig. 4a, but fails to account for the high frequency character of $|H_r|$ shown in Fig. 4b. Once again, the high frequency behavior is attributable to interference between different multipolar orders of lunar induction.

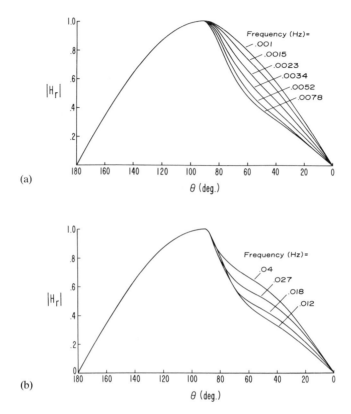

Fig. 4. The amplitude of H_r/H_0 at the moon's surface as a function of θ for (a) frequencies between 0.001 Hz and 0.0078 Hz, (b) frequencies between 0.012 Hz and 0.04 Hz. The lunar conductivity model and incident wave speed are the same as in Fig. 2.

The amplitude of the field components H_z and H_ϕ on the cavity boundary normalized with respect to H_0 are shown in Figs. 5 and 6, respectively, as functions of distance downstream from the limb, in units of lunar radii, with frequency as a parameter. From Fig. 5 we see that at a given z/a, $|H_z|$ increases monotonically with increasing frequency; at a given frequency, $|H_z|$ has a maximum downstream from the limb, the maximum occurring nearer the limb, the higher the frequency. Far downstream, $z/a \rightarrow \infty$, $|H_z|$ decays to the value given by the cylindrical cavity surface field component discussed in Part I. Comparison with Fig. 7 of Schubert *et al.* (1973b) shows that the quasistatic approximation accounts for all the qualitative behavior of H_z except for the far downstream decay of $|H_z|$ to the cylindrical cavity surface field, instead of to zero amplitude. The behavior of $|H_\phi|$ is seen from Fig. 6 to be more complex. In general, at a given frequency, $|H_\phi|$ decays monotonically with downstream distance to the value given by the cylindrical surface wave solution of Part I. However, at the highest frequency investigated, 0.04 Hz, $|H_\phi|$ has a local minimum between 1 and 2 lunar radii downstream. At a

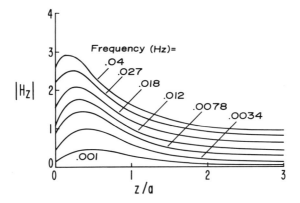

Fig. 5. The magnitude of the field component H_z/H_0 on the cavity boundary as a function of normalized distance z/a downstream from the limb (a is the lunar radius) with frequency as a parameter. The lunar conductivity model and incident wave propagation velocity are as in Fig. 2.

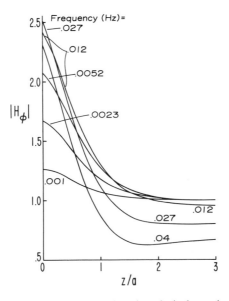

Fig. 6. $|H_\phi/H_0|$ on the cavity boundary as a function of z/a for various frequencies. The conductivity model of the moon and phase speed of the incident wave are given in the caption of Fig. 2.

given downstream location, $|H_\phi|$ depends on frequency in a complex way; sufficiently far downstream, however, $|H_\phi|$ decreases with increasing frequency. Aside from the high frequency behavior of $|H_\phi|$, the quasistatic approximation satisfactorily accounts for the variation of $|H_\phi|$ with distance downstream from the limb (see Fig. 7 of Schubert et al., 1973b).

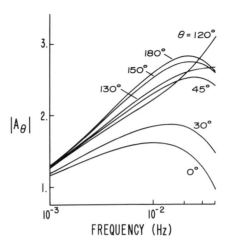

Fig. 7. The amplitude of the tangential transfer function A_θ, i.e., the ratio of H_θ at the lunar surface to H_θ of the incident field, versus frequency with position on the lunar surface θ as a parameter. The lunar conductivity model and incident wave phase speed are as given in Fig. 2.

Transfer functions

The tangential transfer functions A_θ, A_ϕ are, respectively, the ratios of the θ and ϕ components of the magnetic field at the lunar surface to the corresponding components of the interplanetary field. Figure 7 shows the magnitude of A_θ as a function of frequency at a number of positions θ. Tangential magnetic fields at the lunar surface are, in general, amplified above the corresponding components of the interplanetary field by amounts which depend on position and frequency. The amplification generally increases with frequency at a fixed θ; at the higher frequencies, however, a rolloff in amplification usually occurs (Schubert and Schwartz, 1972). At a fixed frequency, amplification ordinarily increases with proximity to the subsolar point $\theta = 180°$. Amplification at the antisolar point $\theta = 0°$ largely results from cavity confinement of induced magnetic fields (Schubert *et al.*, 1973b,c; Smith *et al.*, 1973b; Sill, 1972) whereas at the subsolar point, confinement of induced fields to the near surface lunar interior is the major cause of the larger amplifications. As the curves for $\theta = 180°$, 150°, and 130° show, there is relatively little variation in amplification with position near the subsolar point. On the night side, however, there is considerable change in amplification with distance from the antisolar point. In fact, the transfer function at 45° from the antisolar point is essentially as large as that 50° from the subsolar point and significantly larger than the transfer function predicted by vacuum scattering theory (*see* Fig. 9).

Figure 8 shows the magnitude of A_ϕ as a function of frequency with θ as a parameter. The curves are qualitatively similar to those of $|A_\theta|$ ($|A_\theta| = |A_\phi|$ at $\theta = 0°$ and $\theta = 180°$) exhibiting frequency dependent amplification and high frequency rolloff at all θ. The amplification increases monotonically as position varies between the antisolar and the subsolar points for fixed frequency. As with $|A_\theta|$, $|A_\phi|$

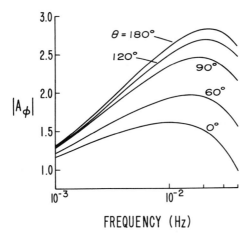

Fig. 8. The magnitude of A_ϕ as a function of frequency with θ as a parameter, computed under the conditions specified in Fig. 2.

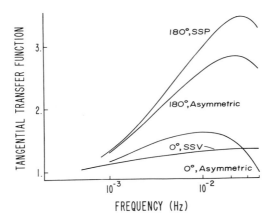

Fig. 9. Comparisons of the tangential transfer function magnitudes at the subsolar ($\theta = 180°$) and antisolar points ($\theta = 0°$) with spherically symmetric plasma confinement (SSP) and spherically symmetric vacuum (SSV) descriptions of lunar electromagnetic scattering. The lunar conductivity model and incident wave phase speed are given in Fig. 2.

shows greater variation with position on the night side than on the day side of the moon. These variations with position are generally smaller, the lower the frequency.

A comparison of $|A_\theta| = |A_\phi|$ at the subsolar point with the spherically symmetric plasma confinement (SSP) calculation of this amplification is made in Fig. 9. The amplification predicted by the asymmetric theory of this paper is smaller than that of the SSP approximation at all frequencies. Such behavior can be intuitively inferred by recalling the earlier argument of Schubert *et al.* (1973b,c) that the effect of asymmetric induction is equivalent to removal of field lines from the sunward hemisphere and substitution of them onto the dark side hemisphere of the moon. Thus the field pressure on the forward hemisphere and therefore the field intensity there is reduced below the value expected for symmetric induction.

Also shown in Fig. 9 is a comparison of $|A_\theta| = |A_\phi|$ at the antisolar point with the spherically symmetric vacuum (SSV) computation of this transfer function. The SSV approximation to the antisolar transfer function cannot exhibit the cavity confinement amplification or the high frequency rolloff seen in the asymmetric calculation of $|A_\theta|$. For the vacuum induction case (SSV) there is a small dependence of induction upon frequency, as expected for a core of finite conductivity, the induction increasing with frequency as the radius at which the inducing currents flow is increased. However, the stronger and nonmonotonic frequency dependence of the observations (see Fig. 11) can be understood only by appeal to a model showing a combination of asymmetric induction, cavity confinement, and interference at high frequency between different orders of multipolar induction.

We next compare the subsolar and antisolar theoretical tangential transfer function amplitudes to empirical values of day and night side tangential transfer functions, keeping in mind that the comparison can be only qualitative at best since the conductivity model used in this paper to compute the theoretical lunar electromagnetic response to interplanetary field fluctuations was rather arbitrarily chosen. For the present, we are mainly interested in the ability of the theoretically determined response using asymmetric theory to account qualitatively for *both* the day and night side magnetic field observations. Later, inversions of the combined day and night side tangential transfer function data sets can yield conductivity models whose theoretical responses will be better fits to the data.

The theoretical subsolar transfer function is compared with day side data (Sonett et al., 1972a) in Fig. 10. The transfer function data labeled as A_{min} in Fig. 10 are, to first approximation, free of contamination of noise introduced by plasma pressure modulation of the local steady field at the Apollo 12 site (Sonett et al., 1972a). We then take A_{min} to represent the best estimate of the lunar day side inductive response. Since the Apollo 12 Lunar Surface Magnetometer is essentially on the lunar equator, the variation of the experimental transfer function can then be studied as a function of θ (Smith et al., 1973b). However, for this qualitative comparison with theory we use the data given in Fig. 10 which represent an average over all the data (67 one- and two-hour swaths and 7 ten-hour intervals) on the sunlit hemisphere. In evaluating the comparison it must be remembered that the theoretical calculation $|A_\theta| = |A_\phi|$ is, however, made at the subsolar point $\theta = 180°$. The theoretical subsolar transfer function can clearly match the frequency dependence of the day side empirical data. Simultaneous inversions to both the day and night side data should better this fit.

The night side data (Schubert et al., 1973c) are compared to the theoretical transfer functions $|A_\theta| = |A_\phi|$ at $\theta = 0°$ and $|A_\theta|$ at $\theta = 30°$ in Fig. 11. The night side tangential transfer functions A_z and A_y are the observed transfer function amplitudes for the local northerly and easterly directions, respectively, at the Apollo 12 site. The empirical data are averaged over 6 cases ranging from 45° from the

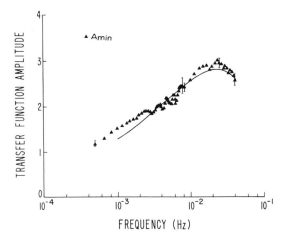

Fig. 10. Comparison of the amplitude of the subsolar transfer function based on the asymmetric scattering theory of Part I with the day side data of Sonett *et al.* (1972a). The day side data, labeled A_{min}, represents the best estimate of the true day side lunar inductive response. The theoretical transfer function is based on the conductivity model of Fig. 2 and a wave phase velocity of 300 km/sec. By inverting the data to yield a conductivity model, a better fit of theory and data could be obtained.

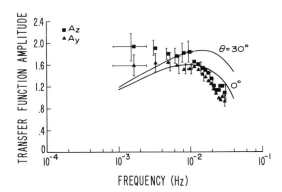

Fig. 11. Comparison of the amplitudes of the antisolar ($\theta = 0°$) and $\theta = 30°$ (A_θ) tangential transfer functions with the night side data of Schubert *et al.* (1973c). The data, A_z and A_y, are the transfer function amplitudes for the local northerly and easterly directions, respectively, at the site of Apollo 12. They represent the average of the observed night side lunar response over positions ranging to 45° from the antisolar point. The conductivity model and incident wave phase speed are as in Fig. 2. The comparison of theory and data is intended to be only qualitative since the data have not yet been inverted.

antisolar point to essentially the antisolar point. Figures 7 and 8 indicate considerable variation in theoretical amplification with position on the lunar night side, a phenomenon which is important to remember when comparing the theory and data shown in Fig. 11. The theoretical antisolar transfer function can match the level of amplification and high frequency rolloff of the night side data. Neither the level of amplification nor the high frequency rolloff could be produced by any vacuum scattering model (*see* Fig. 9 and Schubert *et al.*, 1973c).

Additional points which merit note in connection with the comparisons in Figs. 10 and 11 are that all these data are for both a distribution of wave vector directions and phase velocities, rather than for propagation strictly parallel to the solar wind velocity with speed 300 km/sec, as assumed in the construction of the theoretical functions.

Finally, we compare our results with those of Riesz *et al.* (1972a,b) in Fig. 12. Riesz *et al.* (1972a,b) used a cylindrical two-layer moon model with core conductivity 5×10^{-4} mhos/m, core radius 1400 km and a phase velocity for the interplanetary disturbance of 300 km/sec, together with a nonconducting slab model of the cavity. Their subsolar tangential transfer function, computed using a numerical finite difference scheme, is shown in Fig. 12. We have used the asymmetric theory of Part I together with a spherical two-layer moon model with a core conductivity of 5×10^{-4} mhos/m, a core radius of 1400 km, an insulating shell, a nonconducting cylindrical cavity and a phase velocity of the interplanetary disturbance of 300 km/sec to produce the subsolar and antisolar transfer functions shown in the figure. For convenience, the SSP and SSV transfer functions have also been shown for the same spherical moon model. Figure 12 shows that either the cylindrical moon-slab cavity model used by Riesz *et al.* (1972a,b) is a poor representation of the spherical moon-cylindrical cavity model or their numerical finite difference scheme is in error.

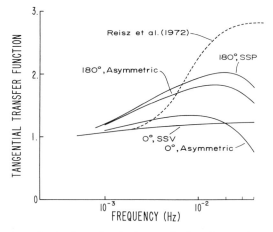

Fig. 12. Comparison of subsolar and antisolar transfer functions to the transfer function of Riesz *et al.* (1972a,b) based on a numerical finite difference calculation using a two-layer cylindrical moon model with a core of conductivity 5×10^{-4} mhos/m and radius of 1400 km and a nonconducting slab cavity. The transfer functions computed here are based on a two-layer spherical moon model with core conductivity 5×10^{-4} mhos/m, radius 1400 km, insulating shell and nonconducting cylindrical cavity. The wave phase velocity was 300 km/sec in both cases. The model of Riesz *et al.* (1972a,b) is either a poor representation of the spherical moon-cylindrical cavity model or their numerical finite difference scheme is in error.

SUMMARY

Using the asymmetric scattering theory of Part I and a two layer model of the lunar electrical conductivity we have calculated the magnitudes of the magnetic field components and the tangential magnetic field transfer functions as functions of frequency and position on the surfaces of the moon and diamagnetic cavity. We have compared the results of the time dependent asymmetric theory with those derived from the quasistatic approximation to this theory (Schubert et al., 1973b) and those derived from the spherically symmetric induction theories. The time dependent asymmetric theory and its quasistatic approximation are in substantial agreement for those values of frequency and wave propagation velocity such that $(2\pi fa/v)$ is less than unity. The time dependent asymmetric theory predicts a nonmonotonic high frequency response, in agreement with the results of Schubert and Schwartz (1972) for the spherically symmetric plasma confinement theory. On the day side of the moon the asymmetric theory yields a tangential amplification which is smaller than that of the SSP theory. The difference, however, is probably not significant in modifying the estimates of the deep lunar electrical conductivity made using SSP theory (Sonett et al., 1972a). The theoretical amplification on the night side is strongly dependent on position within the cavity and is significantly different than that predicted by the SSV theory.

Since the asymmetric theory has not yet been integrated into a systematic inversion using both day and night side data we presented only a qualitative comparison between theoretical and empirical transfer functions for an arbitrarily chosen simple model of the moon's conductivity. Despite the facts that (1) the empirical data represent averages of the response over the entire sunlit hemisphere and over the night side hemisphere as far as 45° from the antisolar point, (2) wave vectors of solar wind disturbances are not simply parallel to the solar wind velocity vector as assumed by the theory, and (3) the propagation speed of solar wind fluctuations is not simply 300 km/sec, the theoretical response of the moon model agrees qualitatively with both day and night side observations. An improved fit of theoretical response to observation should be attainable with more relevant lunar models obtained through inversion, eventually employing a scattering theory which is based on an arbitrary incident wave vector direction with realistic representations of these directions and propagation speeds.

Acknowledgment—This work was supported in part by NASA under contract NAS2-6876.

REFERENCES

Blank J. and Sill W. R. (1969) Response of the moon to the time varying interplanetary magnetic field. *J. Geophys. Res.* **74**, 736–743.

Coleman P. J. Jr., Lichtenstein B. R., Russell C. T., Sharp L. R., and Schubert G. (1972) Magnetic fields near the moon. *Proc. Third Lunar Sci. Conf., Geochim. Cosmochim. Acta*, Suppl. 3, Vol. 3, pp. 2271–2286. MIT Press.

Dyal P., Parkin C. W., and Cassen P. (1972) Surface magnetometer experiments: Internal lunar properties and lunar field interactions with the solar plasma. *Proc. Third Lunar Sci. Conf., Geochim. Cosmochim. Acta*, Suppl. 3, Vol. 3, pp. 2287–2307. MIT Press.

Lahiri B. N. and Price A. T. (1939) Electromagnetic induction in nonuniform conductors, and the determination of the conductivity of the earth from terrestrial magnetic variations. *Phil. Trans. Roy. Soc.* A**237**, 509–540.

Leavy D. (1973) The inverse problem of electrical conductivity of the moon (abstract). Presented at the Symposium on Inversion of Potential Field and Electromagnetic Data, Dept. of Geological and Geophysical Sciences, University of Utah, March 12–14.

Riesz A. C., Paul D. L., and Madden T. R. (1972a) Lunar electrical conductivity. *Nature* **238**, 144–145.

Riesz A. C., Paul D. L., and Madden T. R. (1972b) The effects of boundary condition asymmetries on the interplanetary magnetic field-moon interaction. *The Moon* **4**, 134–140.

Schubert G. and Schwartz K. (1969) A theory for the interpretation of lunar surface magnetometer data. *The Moon* **1**, 106–117.

Schubert G. and Colburn D. S. (1971) Thin highly conducting layer in the moon: Consistency of transient and harmonic response. *J. Geophys. Res.* **76**, 8174–8180.

Schubert G. and Schwartz K. (1972) High frequency electromagnetic response of the moon. *J. Geophys. Res.* **77**, 76–83.

Schubert G., Lichtenstein B. R., Coleman P. J. Jr., and Russell C. T. (1973a) Simultaneous observations of transient events by the Explorer 35 and Apollo 15 subsatellite magnetometers: Implications for lunar electrical conductivity studies. *J. Geophys. Res.* In press.

Schubert G., Sonett C. P., Schwartz K., and Lee H. J. (1973b) The induced magnetosphere of the moon I. Theory. *J. Geophys. Res.* **78**, 2094–2110.

Schubert G., Smith B. F., Sonett C. P., Colburn D. S., and Schwartz K. (1973c) The night side electromagnetic response of the moon. *J. Geophys. Res.* **78**, 3688–3696.

Schwartz K. and Schubert G. (1969) Time dependent lunar electric and magnetic fields induced by a spatially varying interplanetary magnetic field. *J. Geophys. Res.* **74**, 4777–4780.

Schwartz K. and Schubert G. (1973) Lunar electromagnetic scattering I. Propagation parallel to the diamagnetic cavity axis. *J. Geophys. Res.* In press.

Sill W. R. (1972) Lunar conductivity models from the Apollo 12 magnetometer experiment. *The Moon* **4**, 3–17.

Smith B. F., Schubert G., Colburn D. S., Sonett C. P., and Schwartz K. (1973a) Corroborative Apollo 15 and 12 lunar surface magnetometer measurements (abstract). In *Lunar Science—IV*, pp. 683–685. The Lunar Science Institute, Houston.

Smith B. F., Colburn D. S., Schubert G., Schwartz K., and Sonett C. P. (1973b) The induced magnetosphere of the moon II. Experimental results from Apollo 12 and Explorer 35. *J. Geophys. Res.* In press.

Sonett C. P., Smith B. F., Colburn D. S., Schubert G., and Schwartz K. (1972a) The induced magnetic field of the moon: Conductivity profiles and inferred temperature. *Proc. Third Lunar Sci. Conf.*, *Geochim. Cosmochim. Acta*, Suppl. 3, Vol. 3, pp. 2309–2336. MIT Press.

Sonett C. P., Smith B. F., Colburn D. S., Schubert G., and Schwartz K. (1972b) Lunar electrical conductivity—reply. *Nature* **238**, 145–147.

Appendix

In addition to the contents of Part I, the following formulae are required to obtain the results presented herein. For the significance of the quantity $(dG_\ell/dr)_{r=a}$ in asymmetric scattering theory, the reader is referred to Part I.

$$a\left(\frac{dG_\ell}{dr}\right)_{r=a} = \frac{(\ell+1)(1+\ell\eta_\ell) + \ell\xi^{2\ell+1}(1-\eta_\ell(\ell+1))}{1+\ell\eta_\ell - \xi^{2\ell+1}(1-\eta_\ell(\ell+1))} \tag{A.1}$$

where

$$\ell = 1, 2, \ldots,$$

$$\xi = b/a$$

and a is the radius of the moon,

$$\eta_\ell = \frac{j_\ell(kb)}{\frac{d}{db}[bj_\ell(kb)]} \tag{A.2}$$

with

$$k = ((\omega/c)^2 + i\omega\mu\sigma)^{1/2} \tag{A.3}$$

c is the speed of light in vacuum and $j_\ell(x)$ are the spherical Bessel functions. In the form given above $a(dG_\ell/dr)_{r=a}$ is well behaved for any value of η_ℓ and hence any value of kb. Large or small argument approximations are necessary for evaluating η_ℓ. These approximations depend on ℓ.

Proceedings of the Fourth Lunar Science Conference
(Supplement 4, *Geochimica et Cosmochimica Acta*)
Vol. 3, pp. 2925–2945

Surface magnetometer experiments: Internal lunar properties

PALMER DYAL

NASA-Ames Research Center, Moffett Field, California 94035

CURTIS W. PARKIN

University of Santa Clara, Santa Clara, California 95053

WILLIAM D. DAILY

NASA-Ames Research Center, Moffett Field, California 94035

Abstract—Magnetic fields have been measured on the lunar surface at the Apollo 12, 14, 15, and 16 landing sites. The remanent field values at these sites are respectively 38 γ, 103 γ (maximum), 3 γ, and 327 γ (maximum). Simultaneous magnetic field and solar plasma pressure measurements show that the remanent fields at the Apollo 12 and 16 sites are compressed and that the scale size of the Apollo 16 remanent field is $5 \leqslant L < 100$ km. The global eddy current fields, induced by magnetic step transients in the solar wind, have been analyzed to calculate an electrical conductivity profile. From nightside data it has been found that deeper than 170 km into the moon, the conductivity rises from 3×10^{-4} mhos/m to 10^{-2} mhos/m at 1000 km depth. Analysis of dayside transient data using a spherically symmetric two-layer model yields a homogenous conducting core of radius 0.9 R_{moon} and conductivity $\sigma = 10^{-3}$ mhos/m, surrounded by a nonconducting shell of thickness 0.1 R_{moon}. This result is in agreement with the conductivity profile determined from nightside data. The conductivity profile is used to calculate the temperature for an assumed lunar material of periodotite. In an outer layer (~ 170 km thick) the temperature rises to 850–1050°K, after which it gradually increases to 1200–1500°K at a depth of ~ 1000 km. From lunar hysteresis curves it has been determined that the global relative magnetic permeability is $\mu/\mu_0 = 1.029^{+0.024}_{-0.019}$ for the whole moon. This permeability indicates that the moon responds as a paramagnetic or weakly ferromagnetic sphere; lunar iron abundance is calculated for various compositional models in a companion article (Parkin *et al.*, 1973).

INTRODUCTION

APOLLO ASTRONAUTS have deployed five different magnetometers on the lunar surface which measure remanent and induced magnetic fields. Measurements from this network of instruments have been used to calculate the magnetic permeability, electrical conductivity, and temperature of the lunar interior. The fossil remanent magnetic field and its interaction with the solar wind plasma have been measured at three of the landing sites.

The fossil remanent field provides a record of the magnetic field environment that existed at the moon 3.7 to 4.2 billion years ago at the time the crustal material cooled below its Curie temperature. This fossil record indicates the possible existence of ancient large subsurface electrical currents, a solar or terrestrial field much stronger than exists at present, or a lunar dynamo.

The remanent magnetic field is found to be compressed by the solar wind plasma. Properties of the compression are used to study the scale size and topology of the remanent field. The field also causes the solar wind ions and secondary lunar atmospheric ions to be deflected and therefore can be used to study gas accretion in the lunar regolith.

Analysis of the time dependence of eddy-current magnetic fields induced in the moon by solar field transients has yielded a continuous electrical conductivity profile of the lunar interior. The conductivity is related to internal temperature, which is calculated for assumed lunar material compositions.

Magnetic field measurements obtained when the moon is immersed in the steady geomagnetic tail field are used to calculate the magnetic permeability of the lunar interior. The permeability is related to the amount of permeable material and can be used to determine the abundance of iron in the moon.

Experimental Technique

A network of magnetometers has been emplaced on the lunar surface by Apollo astronauts. The vector magnetic field is measured three times per second and transmitted to earth from each of the three sites shown in Fig. 1. Simultaneously magnetometers in lunar orbit, one on board the Explorer 35 (Sonett *et al.*, 1967) and others on board Apollo subsatellites (Coleman *et al.*, 1973) measure the

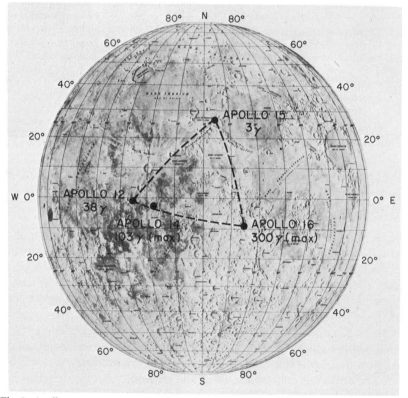

Fig. 1. Apollo magnetometer network on the lunar surface. Maximum remanent magnetic fields measured at each landing site are shown. The maximum field at the Apollo 16 site is 327 γ and not 300 γ as shown in the figure.

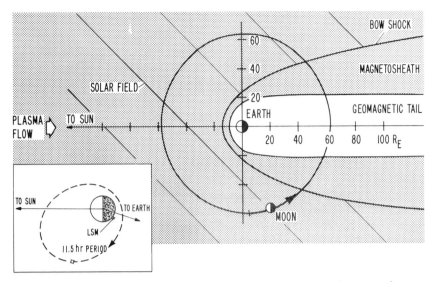

Fig. 2. Orbit of the moon projected onto the solar ecliptic plane. During a complete re-volution around the earth, the moon passes through the earth's bow shock, the mag-netosheath, the geomagnetic tail, and the interplanetary region dominated by solar plasma fields. The insert shows an orbit of the Explorer 35 spacecraft, also projected onto the solar ecliptic plane.

ambient solar or terrestrial field and transmit this information to earth. Portable magnetometers have been used at the Apollo 14 and 16 sites, by the astronauts along their traverse, to measure remanent fields intrinsic to the moon.

The external driving magnetic field in the lunar environment can vary considerably with the lunar orbital position (*see* Fig. 2). Average magnetic field conditions include relatively steady fields of magnitude $\sim 9\gamma$ ($1\gamma = 10^{-5}$ Gauss) in the geomagnetic tail, mildly turbulent fields averaging $\sim 5\gamma$ in the free-streaming solar plasma region, and turbulent fields averaging $\sim 8\gamma$ in the magnetosheath. Average solar wind velocity is ~ 400 km/sec in a direction approximately along the sun-earth line.

Stationary lunar surface magnetometer (LSM)

The stationary magnetometer deployed at the Apollo 16 site in the Descartes region of the Moon is shown in Fig. 3. Characteristics of this and similar instruments deployed at the Apollo 12 and 15 sites (Dyal *et al.*, 1972) are given in Table 1. A more detailed description of the stationary magnetometers is reported by Dyal *et al.* (1970).

Three orthogonal vector components of the magnetic field are measured by three fluxgate sensors located at the ends of 100-cm-long orthogonal booms. The sensors are separated from each other by 150 cm and are 75 cm above the ground. The instrument geometry, shown in Fig. 3, is such that each sensor is directed approximately 35° above the horizontal with the Z-sensor pointed toward the east, the X-sensor approximately toward the northwest, and Y-sensor completing a right hand orthogonal system. Orientation measurements with respect to lunar coordinates are made with two devices. A shadowgraph is used by the astronaut to align and measure the azimuthal orientation with respect to the moon-to-sun line, and gravity level sensors measure tilt angles which are transmitted to earth. Long-term instrument stability is attained by extensive use of digital circuitry and by internally calib-rating the analog portion of the instrument every 18 hours by command from earth. The analog output of the sensor is internally processed by a low-pass digital filter and a telemetry encoder, and the output is transmitted to earth via the central station S-band transmitter. The magnetometer has two data samplers, the analog-to-digital converter (26.5 samples/second) and the central station telemetry

Fig. 3. Apollo 16 magnetometers deployed on the moon at the Descartes landing site. (a) Lunar surface magnetometer (LSM). Sensors are at the top ends of the booms, approximately 75 cm above the lunar surface. (b) Lunar portable magnetometer (LPM), deployed during a magnetic field measurement by astronaut Young.

Table 1. Apollo surface magnetometer characteristics.

Parameter	Apollo 16 stationary magnetometer (LSM)	Apollo 16 portable magnetometer (LPM)
Ranges, gammas (each sensor)	0 to ± 200 0 to ± 100 0 to ± 50	0 to ± 256
Resolution, gammas	0.1	1.0
Frequency response, Hz	dc to 3	dc to 0.05
Angular response	Cosine of angle between field and sensor	Cosine of angle between field and sensor
Sensor geometry	3 orthogonal sensors at ends of 100 cm booms	3 orthogonal sensors in 6 cm cube
Analog zero determination	180° flip of sensor	180° flip of sensor
Power, watts	3.5	1.5 (battery)
Weight, kg	8.9	4.6
Size, cm	$63 \times 28 \times 25$	$56 \times 15 \times 14$
Operating temperature °C	− 50 to + 85	0 to + 50
Commands	10 ground: 1 spacecraft	—

encoder (3.3 samples/second). The prealias filter following the sensor electronics has attenuations of 3 db at 1.7 Hz and 58 db at the Nyquist frequency (13.2 Hz), with an attenuation rate of 22 db/octave. The four-pole Bessel digital filter has an attenuation of 3 db at 0.3 Hz and 48 db at the telemetry sampling Nyquist frequency (1.6 Hz).

The instrument can also be used as a gradiometer by sending commands to operate three motors in the instrument which rotate the sensors such that all simultaneously align parallel first to one of the boom axes, then to each of the other two boom axes in turn. This alignment capability permits the vector gradient in the plane of the sensors to be calculated and also permits an independent measurement of the magnetic field vector at each sensor position.

Lunar portable magnetometer (LPM)

The portable magnetometer developed for the Apollo 16 mission to Descartes is shown in Fig. 3b and the instrument characteristics are given in Table 1. A more detailed description of the instrument is reported by Dyal *et al.* (1973).

The LPM instrument was designed to be a totally self-contained, portable experiment package. Three orthogonally oriented fluxgate sensors are mounted on the top of a tripod, positioned 75 cm above the lunar surface. These sensors are connected by a 15-meter-long cable to an electronics box (mounted on the Lunar Roving Vehicle), which contains a battery, electronics and three digital displays used to read the field output.

Lunar orbiting magnetometers

The ambient steady-state and time-dependent magnetic fields in the lunar environment are measured by the Explorer 35 satellite magnetometer. The satellite, launched in July 1967, has an orbital period of 11.5 hours, aposelene of 9390 km, and periselene of 2570 km (see Fig. 2 insert). The Explorer 35 magnetometer measures three magnetic field vector components every 6.14 sec at 0.4γ resolution; the instrument has an alias filter with 18 db attenuation at the Nyquist frequency (0.08 Hz) of the spacecraft data sampling system. A more detailed description of the instrument is reported by Sonett *et al.* (1967).

The Apollo 15 and 16 subsatellite magnetometers, orbiting approximately 100 km above the lunar surface, have also measured fields intrinsic to the moon. These instruments are described by Coleman *et al.* (1973).

REMANENT MAGNETIC FIELD MEASUREMENTS

Remanent magnetic fields have been measured at the Apollo 12, 14, 15, and 16 landing sites on the lunar surface. At the Apollo 14 Fra Mauro and 16 Descartes sites, portable magnetometers were used by the astronauts to measure the remanent fields at different locations along their traverses. The measured fields and their locations are listed in Table 2. Results show that magnetic fields are lower in the mare regions (3–103 gammas) than in the Descartes highland region (112–327 gammas). The field magnitude gradients vary from 1.2 gammas/kilometer to 370 gammas/kilometer at different locations on the moon.

The source for the measured permanent magnetic field on the lunar surface is generally considered to be the remanence in the near-surface material. Remanent magnetism has been measured in returned lunar samples by many investigators (e.g., Strangway *et al.*, 1973; Nagata, 1972; Runcorn *et al.*, 1970). Also, the Apollo 15 and 16 subsatellite magnetometers have directly measured the lunar remanent fields from orbit and found larger fields associated with the highlands than with the mare regions (Coleman *et al.*, 1973).

Measured remanent field values are much smaller at the subsatellite altitude (1γ at 100 km) than at the surface (327γ at Descartes); since the field magnitudes

Table 2. Summary of lunar surface remanent magnetic field measurements.

Site	Coordinates, deg.	Field Magnitude, Gammas	Magnetic-field components, gammas		
			Up	East	North
Apollo 16:					
ALSEP Site	8.9°S, 15.5°E	234 ± 3	−181 ± 3	−57 ± 3	+136 ± 2
Site 2		189 ± 5	−189 ± 4	+3 ± 6	+10 ± 3
Site 5		112 ± 5	+104 ± 5	−5 ± 4	−40 ± 3
Site 13		327 ± 7	−159 ± 6	−190 ± 8	−214 ± 6
LRV Final Site		113 ± 4	−66 ± 4	−76 ± 4	+52 ± 2
Apollo 15 ALSEP Site	26.1°N, 3.7°E	3.4 ± 2.9	3.3 ± 1.5	0.9 ± 2.0	−0.2 ± 1.5
Apollo 14:	3.7°S, 17.5°W				
Site A		103 ± 5	−93 ± 4	+38 ± 5	−24 ± 5
Site C′		43 ± 6	−15 ± 4	−36 ± 5	−19 ± 8
Apollo 12 ALSEP Site	3.2°S, 23.4°W	38 ± 2	−25.8 ± 1.0	+11.9 ± 0.9	−25.8 ± 0.4

fall off rapidly with distance, the sources must be small in extent and probably are confined to the near-surface crustal materials. The lack of a measurable whole-moon dipolar magnetic field (Russell *et al.*, 1973) is another indication that the crust is the only portion of the moon that is magnetized.

The magnetic fields measured at five different locations at the Descartes landing site are the largest extraterrestrial fields yet measured *in situ.* The largest distance between measurements was 7.1 kilometers. A schematic representation of the measured field vectors is shown in Fig. 4. All vectors point downward except the one at Site 5 near Stone Mountain, which points upward. This suggests that the material underlying Stone Mountain has undergone different geological processes than that underlying the Cayley Plains and North Ray Crater.

Lunar sample measurements and field measurements by orbiting surface instruments all indicate that the moon is covered with remanent magnetic field sources which show a strong regional variation. The lunar crustal material probably was magnetized as it cooled below the Curie temperature in the presence of a strong ambient magnetic field about 4 billion years ago (Gopulan *et al.*, 1971). Sources of this ancient ambient field which no longer exists at the moon, include an extinct lunar dynamo, a stronger solar or terrestrial field, and a field of thermoelectric origin. This latter field could have been generated during the mare basin flooding era.

It has been determined that the thermoremanent magnetization of lunar samples has resulted from their cooling below the Curie point in a field of a few thousand gammas (Gose *et al.*, 1973; Pearce *et al.*, 1973). We have initiated an investigation of thermoelectrically driven currents to account for these fields. Thermoelectric potential is a function of the thermal gradient and electrical properties of the geological material. We model a mare basin as a cylinder which has an axial temperature gradient.

Thermal gradients in the cooling mare lava could produce a Thomson ther-

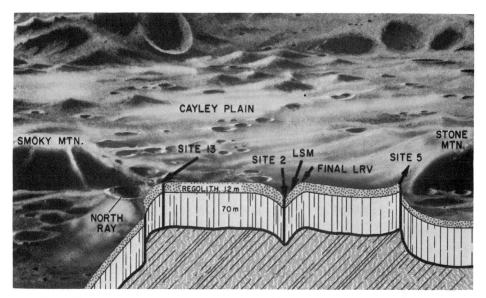

Fig. 4. Pictorial representation of the magnetic field vectors measured at the Apollo 16 Descartes region. The maximum field measured anywhere on the moon to date is that at site 13: 327γ.

moelectromotive force which would drive currents axially through the mare cylinder. The solar wind plasma, highly conducting along magnetic field lines, could provide a return path to complete the electrical circuit from the top surface of the lava to the lunar surface outside the mare and back into the mare through the lunar interior.

The upper limit of the field generated at the mare edge by this process is $B = \mu_0 E \sigma R (\Delta T)/2L$ where R and L are the cylinder radius and length, σ is the electrical conductivity, $\mathbf{E}$ is the thermoelectromotive force and ΔT is the temperature difference across the cylinder. Telkes (1950) reports maximum thermoelectromotive force magnitudes of the order of $10^2 \mu V/\text{deg}$. in geological materials. At the edge of a disk of $R/L = 10^3$, $\Delta T = 10^{3\circ}\text{K}$, $\sigma_{\text{mean}} = 10^{-1}$ mhos/m, and $E = 10^2 \mu V/\text{deg}$., fields of several thousand gammas are possible. Such fields near a mare disk would produce thermoremanent magnetization in the moon of magnitudes measured in lunar samples.

REMANENT MAGNETIC FIELD INTERACTION WITH THE SOLAR WIND

Interaction of the solar wind with the remanent magnetic field has been measured at the Apollo 12 and 16 landing sites. The solar plasma is directly measured at the Apollo 12 and 15 sites (Clay *et al.*, 1972) and simultaneous measurements of the magnetic field at the Apollo 12 and 16 sites show a compression of the steady field as a function of the solar wind pressure. This is schematically shown in the Fig. 5 insert. Time-dependent induction fields are negligible for one-hour averages

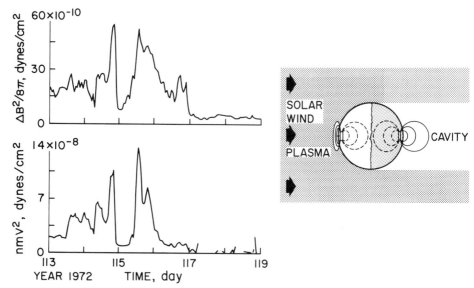

Fig. 5. Compression of the Apollo 16 remanent field by the solar wind plasma. (a) Simultaneous plots of magnetic energy density and plasma bulk energy density. The magnetic energy density shown is computed from the difference between the compressed and uncompressed remanent field at the Apollo 16 site, i.e., $\Delta B^2 = \sum_i (B_{Ai} - (B_{Ei} + B_{Ri}))^2$, $i = x, y, z$. These field components and the ALSEP coordinate system (x, y, z) are defined in the text. Plasma energy density data are from the Apollo 15 solar wind spectrometer, courtesy of C. W. Snyder and D. R. Clay of the Jet Propulsion Laboratory; N is the proton number density, m is the proton mass, and v is the plasma bulk speed. (b) Schematic representation of remanent field compression by a high-density solar wind plasma. The remanent field is unperturbed during nighttime (antisolar side), while on the sunlit side it is compressed.

of LSM field measurements, and the dominant fields measured at the surface (Dyal and Parkin, 1973) are $\mathbf{B}_A = \mathbf{B}_E + \mathbf{B}_R + \Delta\mathbf{B}$. In this equation $\mathbf{B}_A$ is the total magnetic field measured on the lunar surface by an Apollo magnetometer, $\mathbf{B}_E$ is the extralunar field measured by the orbiting Explorer 35 magnetometer, $\mathbf{B}_R$ is the unperturbed remanent field local to the site (measured when the moon is shielded from the solar wind in the earth's magnetotail), and $\Delta\mathbf{B}$ is the change in the remanent field due to its interaction with the solar wind plasma. One-hour average plots in Fig. 5 show that the change in remanent field pressure $\Delta B^2/8\pi$ is related to the plasma pressure nmv^2 (n is proton density, m is proton mass, and v is plasma bulk speed).

The nature of the correlation between magnetic field and plasma pressures is further illustrated in Fig. 6, which shows data from several lunations at the Apollo 12 and 16 LSM sites. The pressures are related throughout the measurement range. The magnitudes of magnetic pressure changes at the Apollo 12 and 16 LSM sites are in proportion to their unperturbed steady field magnitudes of 38γ and 234γ, respectively.

Properties of the remanent field interaction with the solar wind can be used to

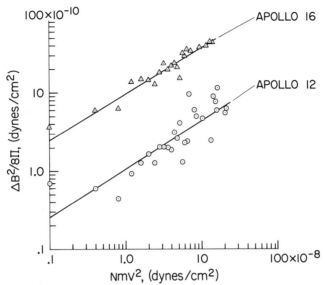

Fig. 6. Magnetic energy density versus plasma energy density at two Apollo sites which have different remanent magnetic fields. Energies are defined in Fig. 5. Uncompressed remanent field magnitudes are 38γ at Apollo 12 and 234γ at Apollo 16. Apollo 12 magnetometer data are plotted versus Apollo 12 solar wind spectrometer (SWS) data, while Apollo 16 magnetometer data are plotted versus Apollo 15 SWS data. SWS data are courtesy of C. W. Snyder and D. R. Clay of the Jet Propulsion Laboratory.

estimate the scale size of the remanent field (Barnes *et al.*, 1971). The Apollo 12 scale size L has previously been calculated to be $2\,\mathrm{km} < L < 200\,\mathrm{km}$ (Dyal *et al.*, 1972). For the Apollo 16 field, portable magnetometer measurements over the lunar roving vehicle traverse showed that $L \geqslant 5\,\mathrm{km}$; the Apollo 16 subsatellite magnetometer showed no field attributable to the Descartes area at orbital altitude, implying that $L < 100\,\mathrm{km}$. Therefore, the Apollo 16 remanent field scale size is $5 \leqslant L < 100\,\mathrm{km}$.

Previous experimental limits (Dyal and Parkin, 1971) of the lunar unipolar induction fields have been supplemented with a more extensive statistical analysis of Explorer 35 and Apollo 12 and 15 LSM magnetometer data to show that the toroidal mode induction fields at the lunar surface are less than a few gammas. The results of this study confirm the idea that the lunar crust is of sufficient electrical resistivity and thickness to quench current flow from unipolar induction.

GLOBAL EDDY CURRENT INDUCTION: ELECTRICAL
CONDUCTIVITY AND TEMPERATURE

Electrical conductivity and temperature of the moon are calculated from global eddy current response to magnetic field transients in the solar wind. When a magnetic discontinuity in the interplanetary magnetic field passes the moon, an eddy current field is induced in the moon which opposes the change in the external field (*see* Fig. 7). The induced field thereafter decays with a time dependence which is a function of the electrical conductivity distribution in the lunar interior. Simultaneous measurements of the transient driving field (by Explorer 35) and the lunar response field (by an Apollo surface magnetometer) allow calculation of the lunar conductivity. Since conductivity is related to temperature, a temperature profile can be calculated for an assumed compositional model of the lunar interior.

Lunar eddy current fields form an induced lunar magnetosphere which is distorted in a complex manner due to flow of solar wind plasma past the moon, as illustrated in Fig. 7. The eddy current field is compressed on the dayside of the moon and is swept downstream and confined to the "cavity" on the lunar nightside. Because of the complexity, earliest models included a theory for transient response of a sphere in a vacuum to model lunar response as measured on the lunar nightside (Dyal and Parkin, 1971) and a harmonic theory of a moon totally confined by the solar wind to model response as measured on the lunar dayside (Sonett *et al.*, 1971). Both harmonic and transient approaches have subsequently been further developed. Transient analysis has evolved to include a continuous conductivity profile (Dyal *et al.*, 1972), and more recently has included effects of

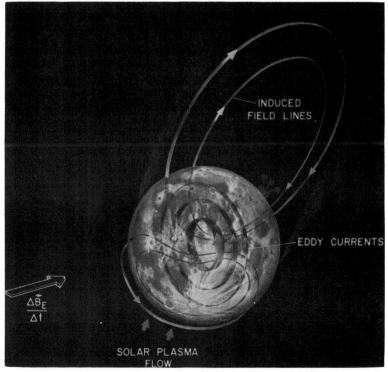

Fig. 7. Eddy currents and corresponding poloidal magnetic field, induced by time-dependent fluctuations in the solar wind magnetic field $\mathbf{B}_E$. This global induced field is confined by the solar wind flowing past the moon; the field is compressed on the lunar sunlit side and confined in the cavity region on the nighttime (antisolar) side of the moon.

cavity confinement on the nightside radial data (Dyal and Parkin, 1973). In this paper we extend the evolutionary process to include effects of cavity confinement on nightside tangential data and to introduce analysis of magnetic step transients measured on the lunar dayside. We find that these additional calculations yield conductivity profiles which are consistent with our earlier results.

Lunar nightside data analysis

Figure 8 shows simultaneous measurements of a step transient by the lunar orbiting Explorer 35 magnetometer and the Apollo 12 LSM located on the lunar nightside. The Apollo data qualitatively show response of a sphere in a vacuum: damping in the radial (B_{Ax}) component and overshoot in tangential components ($B_{Ay,z}$). Averages of normalized decay curves of radial data have been analyzed (Dyal and Parkin, 1973) to yield the conductivity profile illustrated in Fig. 9a; the corresponding temperature profiles for two assumed lunar compositional models are shown in Fig. 9b.

Lunar dayside data analysis

Figure 10 shows an example of a transient measured by the Apollo 12 LSM on the lunar dayside. Comparison with Fig. 8 shows these qualitative differences between dayside and nightside transient data: (1) Apollo dayside radial response is

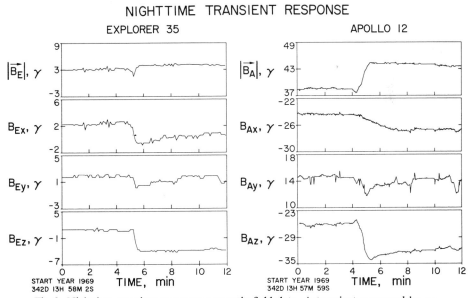

Fig. 8. Nighttime transient response magnetic field data. A transient measured by an Apollo LSM while on the nighttime (antisolar) side of the moon, showing simultaneous external solar wind field data measured by Explorer 35. The ALSEP (x, y, z) coordinate system is defined in the text.

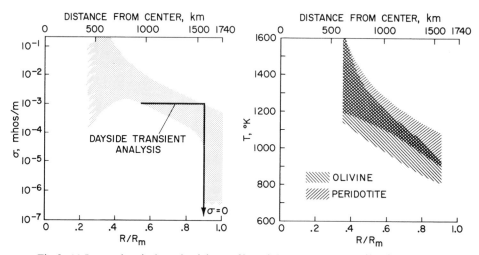

Fig. 9. (a) Lunar electrical conductivity profile and (b) temperature profile of the moon. Shown are previous profiles (Dyal and Parkin, 1973) using vacuum nightside theory and nightside magnetometer data. Superimposed on (a) is a conductivity profile calculated by fitting totally confined 2-layer moon models to the dayside tangential component data illustrated in Fig. 11.

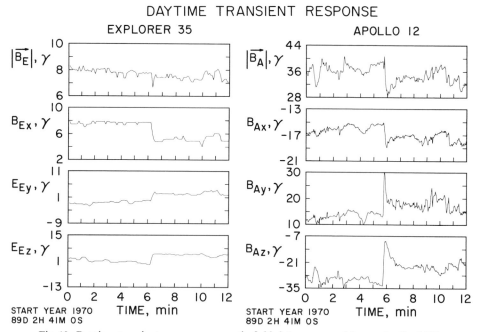

Fig. 10. Daytime transient response magnetic field data, measured by an Apollo LSM while on the dayside (subsolar) side of the moon, showing simultaneous external solar wind field data measured by lunar orbiting Explorer 35. The ALSEP (x, y, z) coordinate system is defined in the text. Note in particular the tangential components: dayside overshoot amplification is much higher relative to the Explorer-measured external step magnitudes than is nighttime amplification (shown in Fig. 8).

not damped, as is the nightside response, but tends to track the Explorer 35 radial-component measurements, and (2) the overshoot in tangential y and z components is much greater in dayside data, illustrating solar wind compression of the eddy current field on the dayside. Theoretical solutions for an eddy current field totally confined to a sphere of homogeneous conductivity σ and permeability μ are derived from Maxwell's equations in the APPENDIX. For the case of an external magnetic step transient of magnitude $|\Delta \mathbf{B}_E|$, applied to the lunar sphere at time $t = 0$, the solutions for the components of the vector field measured on the lunar dayside are expressed as follows:

$$B_{Ax} = B_{Ex} \qquad (1)$$

$$B_{Ay,z} = B_{Ey,z} + \Delta B_{Ey,z} \left[\frac{3\lambda^3}{2(1-\lambda^3)} \right] F(t) \qquad (2)$$

where

$$F(t) = \frac{6}{1-\lambda^3} \sum_s \exp\left[-z_s^2 t / \mu \sigma R_c^2\right] / [z_s^2 + ((h-\tfrac{1}{2})^2 - \tfrac{9}{4})] \qquad (3)$$

In the above equations $\lambda = R_c/R_m$, R_c = radius of conducting core (surrounded by a nonconducting outer shell of thickness $R_m - R_c$); and z_s is a solution of the characteristic equation $\tan z_s = z_s(h-2)/(z_s^2 + (h-2))$. The components of $\mathbf{B}_A$ are expressed in the ALSEP coordinate system which has its origin on the lunar surface at a magnetometer site. The x-axis is directed radially outward from the lunar surface; the y and z axes are tangential to the surface, directed eastward and northward respectively. Equation (1) shows that for this homogeneous-core theory the surface field radial component B_{Ax} tracks the external radial component B_{Ex}; equation (2) illustrates that tangential components will overshoot a step input field with an amplification factor which is a function of the conducting core radius R_c, after which the eddy current field will decay as a function of time, μ, σ, and R_c. Figure 11 shows averages of normalized lunar dayside tangential components in response to seven transient events (error bars are standard deviations). Apollo data have been normalized by dividing by the Explorer 35 step magnitudes to give the response to an effective unit step driving field. The overshoot maximum is amplified by a factor of 5 by solar wind dayside compression. In this homogeneous theory, this amplification factor is used with equation (2) to calculate $\lambda = R_c/R_m = 0.9$. Superimposed on the figure is a family of curves varying with conductivity σ ($\mu = \mu_0$, the permeability of free space). The data show a good fit to a lunar model with a homogeneous core of radius $R_c = 0.9R_m$ and conductivity $\sigma \sim 10^{-3}$ mhos/m. Figure 9 shows this homogeneous conductivity profile superimposed on the continuous conductivity profile calculated from nightside radial magnetic field data; the dayside calculation fits within the error bars of the nightside calculation to depths allowed by the time duration of the events in Fig. 11.

Asymmetric confinement of the lunar induced magnetosphere

Having considered spherically symmetric models to treat nightside and dayside data, we now proceed to the case of asymmetric confinement of the lunar

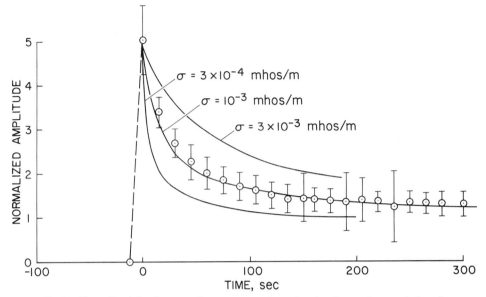

Fig. 11. Normalized daytime transient response data, showing decay characteristics of tangential components ($B_{Ay,z}$) of the total surface field after arrival of a step transient which changes the external magnetic field tangential component by an amount $\Delta B_{Ey,z}$, here normalized to one. The shape of the curve illustrates time-decay characteristics of the induced poloidal eddy current field $\mathbf{B}_P$. Superimposed are conductivity-dependent theoretical response curves for a totally confined moon.

magnetosphere, illustrated in Fig. 7. Previous theoretical treatments of this problem include a two-dimensional time-dependent approach (Reisz *et al.*, 1972) and a three-dimensional quasi-static approach (Schubert *et al.*, 1973). At the time of this writing, a complete time-dependent solution for the lunar magnetosphere has not been developed; therefore our present approach is to show general consistency of daytime and nighttime transient response data with theoretical approximations presently available.

First we model the effects of partial confinement by considering the case of a point dipole confined by a capped cylindrical superconductor (Dyal and Parkin, 1973). Figure 12 shows the theoretical solution for amplification of tangential (transverse) components due to asymmetric confinement of the dipole field by the superconductor. This models the asymmetric solar wind confinement of an induced dipole in the moon. Data points shown in Fig. 12 are laboratory measurements made by placing a small samarium-cobalt magnet inside a helium-cooled superconducting cylinder; data are seen to be in agreement with theory.

This asymmetric confinement yields the value of the overshoot of the tangential eddy current field at the lunar surface antisolar point. At $r/R = 1$ in Fig. 12, analogous to the location of the lunar surface antisolar point, the theoretical confined-to-unconfined ratio is 1.5. In the previous section we found that dayside transient data fit a lunar model with conducting core radius $R_c = 0.9R_m$. In uncon-

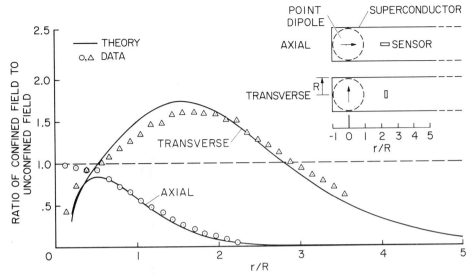

Fig. 12. Confinement of a point dipole magnetic field. Shown are theoretical profiles and experimental data. The inserts schematically show lunar confinement by the solar wind, approximated by a capped-cylinder superconductor enclosing a point-dipole field. The theoretical curves show ratios of confined to unconfined dipolar field versus distance along the cylinder axis. Data are results of a laboratory experiment in which confinement of a small dipole magnet's field by a cylindrical superconductor is measured experimentally.

fined vacuum theory for $R_c/R_m = 0.9$, the tangential component overshoot is $1 + \frac{1}{2}(R_c/R_m)^3 = 1.36$. Multiplying by the confined-to-unconfined ratio of 1.5, we find that for the asymmetric case, the initial overshoot of normalized data measured at the antisolar point should be 2.0. We have averaged tangential events, all obtained within 6° of the antisolar point, and have calculated an overshoot of 2.0 ± 0.3, showing agreement with the confined-dipole theory and laboratory data.

A final check on the overshoot magnitude for the confined case is obtained by comparison with the "quasi-static" theoretical solutions of Schubert *et al.* (1973). This theory considers the response of an infinitely conducting, spherical core confined inside a hemispherically capped cylinder, also of infinite conductivity. The response of the infinitely conducting core is the same as that of our model with a core of finite conductivity, at the instant the external field transient occurs. Figure 5 of Schubert *et al.* (1973) gives the overshoot to be ~ 1.9 for the case $R_c/R_m = 0.9$. This value is well within our nightside data overshoot limits of 2.0 ± 0.3.

In conclusion, we have shown that our earlier electrical conductivity profile (Fig. 9), calculated using radial components of nightside data, is still consistent with more recent calculations which consider effects of solar wind confinement on both dayside and nightside magnetic fields. Further refinement of the conductivity profile of Fig. 9 must await the development of a time-dependent asymmetric theory.

MAGNETIZATION INDUCTION AND GLOBAL MAGNETIC PERMEABILITY

When the moon is in a quiet region of the geomagnetic tail, solar wind interaction fields and the induced eddy current lunar field are negligible (Parkin et al., 1973), and the total field at the lunar surface is $\mathbf{B}_A = \mathbf{B}_E + \mathbf{B}_\mu$. The magnetic moment $\mathbf{m}_\mu$ of the magnetization field $\mathbf{B}_\mu$ is proportional to the external field $\mathbf{B}_E$; that is, $\mathbf{m}_\mu = C\mathbf{B}_E$. (The proportionality constant C in turn depends on the permeability and the dimensions of the permeable region of the moon.) The magnetization of the lunar sphere by the terrestrial magnetic field is shown schematically in Fig. 13.

For the case of a homogeneous permeable lunar sphere of relative permeability μ, the ALSEP components of the total surface magnetic field (Jackson, 1962) are:

$$B_{Ax} = (1 + 2F)B_{Ex} \tag{4}$$
$$B_{Ay,z} = (1 - F)B_{Ey,z} \tag{5}$$

where

$$F = \frac{\mu - 1}{\mu + 2} \tag{6}$$

A plot of B_{Ax} vs. B_{Ex} is in effect a plot of a B–H hysteresis curve. For cases where the ratio B_{Ax}/B_{Ex} is a constant, i.e., for low field B_{Ex}, the hysteresis curve should take the form of a straight line. Figure 14 shows a plot of radial components of Apollo 12 surface field (B_{Ax}) versus the geomagnetic tail field (B_{Ex}) measured by Explorer 35. A least-squares fit and slope calculation determine the factor

EXTERNAL (TERRESTRIAL) FIELD $\vec{B}_E$ INDUCED MAGNETIZATION FIELD $\vec{B}_\mu$

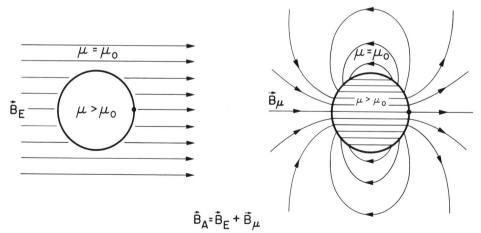

$$\vec{B}_A = \vec{B}_E + \vec{B}_\mu$$

Fig. 13. Induced magnetization field $\mathbf{B}_\mu$. When a homogenous sphere of permeability $\mu > \mu_0$ is immersed in a uniform external magnetic field $\mathbf{B}_E$ (for example, the moon in the steady geomagnetic tail field), a dipolar magnetic field $\mathbf{B}_\mu$ is induced with its dipole moment aligned with $\mathbf{B}_E$. If the magnetizing field $\mathbf{B}_E$ and the magnetic induction ($\mathbf{B}_E + \mathbf{B}_\mu$) are known, a classical B–H hysteresis curve can be constructed from the data.

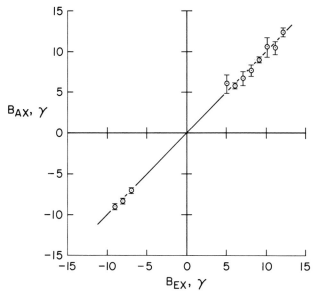

Fig. 14. Hysteresis curve of the moon. The moon is immersed in the steady external magnetizing geomagnetic tail field $\mathbf{B}_E$ (measured by Explorer 35); the total magnetic induction is $\mathbf{B}_A = \mathbf{B}_E + \mathbf{B}_\mu$ measured by the Apollo 12 lunar surface magnetometer, where $\mathbf{B}_\mu$ the induced lunar magnetization field. In this graph only the radial (x) components are plotted. From the slope of the curve the bulk relative permeability is calculated: $\mu/\mu_0 = 1.029^{+0.024}_{-0.019}$.

$F = 0.0095 \pm 0.0060$, which is used in equation (6) to determine the bulk relative magnetic permeability of the moon: $\mu = 1.029^{+0.024}_{-0.019}$.

The relative permeability is greater than 1.0 for both extremes of the error limits, indicating that when immersed in a uniform, steady geomagnetic tail field, the moon is magnetized along the direction of the tail field, causing the moon as a whole to have the properties of a magnetized paramagnetic or weakly ferromagnetic sphere. Iron abundance in the moon is also estimated from lunar hysteresis curves, using various compositional models of the lunar interior. These results are reported in a companion article in this volume (Parkin *et al.*, 1973).

SUMMARY

Remanent magnetic field measurements

The remanent magnetic fields measured on the lunar surface are 38γ at Apollo 12 in Oceanus Procellarum; 103γ and 43γ at two Apollo 14 sites separated by 1.1 km in Fra Mauro; 3γ at the Apollo 15 Hadley Rille site; and 112γ, 113γ, 189γ, 235γ, and 327γ at five Apollo 16 sites in the Descartes region, over a distance of 7.1 km. Simultaneous measurements by a lunar orbiting subsatellite and the Apollo surface magnetometers, and remanence in the returned samples, have shown that the near-surface crustal material is magnetized over most of the lunar globe. The field sources which show a strong regional variation over much of the moon, were probably magnetized as the crustal material cooled below the Curie temperature in the presence of a strong ambient magnetic field about 4 billion years ago. A possible source of this ancient ambient field is thermoelectric currents generated during mare basin flooding.

Remanent magnetic field interaction with the solar wind

Measurements show that the remanent fields at the Apollo 12 and 16 sites are compressed by the solar wind. The 235γ remanent field at the Apollo LSM site is compressed to 265γ by a solar wind pressure increase of 1.5×10^{-7} dynes/cm². The scale size L of the Apollo 16 remanent field has been determined to be in the range $5 \text{ km} \leqslant L < 100 \text{ km}$. The 5 km lower limit is calculated from measurements of the field-plasma interaction, and the 100 km upper limit is deduced from lunar orbiting subsatellite measurements.

Global eddy current induction: Electrical conductivity and temperature

The electrical conductivity and temperature of the moon are calculated from global eddy current response to magnetic field transients in the solar wind. Using a spherically symmetric model, a continuous conductivity profile, with error limits, has been determined from radial step transient response data. Dayside transient response data, compared to the theoretical response of a totally confined sphere, yields a homogeneous conductivity profile ($\sigma \sim 10^{-3}$ for a core of $R_c/R_m = 0.9$) which lies within the error limits of the nightside radial results. Nightside response is adjusted to account for asymmetric confinement of eddy current fields by two methods: using laboratory data obtained by confining a dipole field within a superconducting cylinder, and using a quasi-static theory for confinement of eddy current fields of a spherical source. Initial overshoots of nightside tangential data agree within experimental error with laboratory and theoretical results. An upper limit on the unipolar induction field has been determined which shows that the outer crustal layer, ~ 170 km thick, is a relatively poor electrical conductor compared to the underlying material. Deeper than 170 km, the conductivity rises from 3×10^{-4} mhos/meter to 10^{-2} mhos/m at 1000 km depth. These conductivities are used to calculate temperatures for a lunar model composed of peridotite and show an outer layer, ~ 170 km thick, in which the temperature rises from 250°K to 850–1050°K, then a gradual increase to 1200–1500°K at a depth of ~ 1000 km.

Magnetization induction and global magnetic permeability

Magnetic hysteresis curves have been measured for the moon by simultaneous data obtained from the orbiting Explorer 35 and surface Apollo 12 and 15 magnetometers. From these curves the bulk relative magnetic permeability of the moon is determined to be $\mu = 1.029^{+0.024}_{-0.019}$. This permeability indicates that the moon is a paramagnetic or weakly ferromagnetic sphere and the iron abundance is estimated for various compositional models in a companion article (Parkin *et al.*, 1973).

Acknowledgments—We thank Drs. C. W. Snyder and D. R. Clay of the Jet Propulsion Laboratory for use of solar wind spectrometer data and Drs. C. P. Sonett and D. S. Colburn of NASA-Ames Research Center for use of Explorer 35 data. We have especially appreciated working with Dr. T. J. Mucha of Computer Sciences Corporation in the analysis of dayside magnetic transients. We also acknowledge the efforts of Messrs. J. Keeler and C. Privette for experiment fabrication and testing, and K. Lewis and M. Legg and their team members for help in computer programming and data reduction.

REFERENCES

Barnes A., Cassen P., Mihalov J. D., and Eviatar A. (1971) Permanent lunar surface magnetism and its deflection of the solar wind. *Science* **171**, 716–718.

Clay D. R., Goldstein B. E., Neugebauer M., and Snyder C. W. (1973) Solar-wind spectrometer experiment. In *Apollo 15 Preliminary Science Report*, NASA SP-289, pp. 10-1 to 10-7.

Colburn D. S., Currie R. G., Mihalov J. D., and Sonett C. P. (1967) Diamagnetic solar-wind cavity discovered behind the moon. *Science* **168**, 1040–1042.

Coleman P. J. Jr., Lichtenstein B. R., Russell C. T., Schubert G., and Sharp L. R. (1973) The particles and fields subsatellite magnetometer experiment. In *Apollo 16 Preliminary Science Report*, NASA SP-315, pp. 23-1 to 23-13.

Dyal P. and Parkin C. W. (1971) The Apollo 12 magnetometer experiment: Internal lunar properties from transient and steady magnetic field measurements. *Proc. Second Lunar Sci. Conf., Geochim. Cosmochim. Acta*, Suppl. 2, Vol. 3, pp. 2391–2413. MIT Press.

Dyal P. and Parkin C. W. (1973) Global electromagnetic induction in the moon and planets. *Phys. Earth Planet. Interiors*. In press.

Dyal P., Parkin C. W., and Sonett C. P. (1970) Lunar surface magnetometer. *IEEE Trans. on Geoscience Electronics GE-8* **4**, 203–215.

Dyal P., Parkin C. W., and Cassen P. (1972) Surface magnetometer experiments: Internal lunar properties and lunar surface interactions with the solar plasma. *Proc. Third Lunar Sci. Conf., Geochim. Cosmochim. Acta*, Suppl. 3, Vol. 3, pp. 2287–2307. MIT Press.

Dyal P., Parkin C. W., Sonett C. P., DuBois R. L., and Simmons G. (1973) Lunar portable magnetometer experiment. In *Apollo 16 Preliminary Science Report*, NASA SP-315, pp. 12-1 to 12-8.

Gopulan K., Kaushal S., Lee-Hu C., and Wetherill G. W. (1970) Rb-Sr and U, Th-Pb ages of lunar materials. *Proc. Apollo 11 Lunar Sci. Conf., Geochim. Cosmochim. Acta*, Suppl. 1, Vol. 2, pp. 1195–1205. Pergamon.

Gose W. A., Strangway D. W., and Pearce G. W. (1973) A determination of the intensity of the ancient lunar magnetic field. *The Moon* **7**, 198–201.

Jackson J. D. (1962) *Classical Electrodynamics*. John Wiley and Sons.

Nagata T., Fisher R. M., Schwerer F. C., Fuller M. D., and Dunn J. R. (1972) Rock magnetism of Apollo 14 and 15 materials. *Proc. Third Lunar Sci., Conf. Geochim. Cosmochim. Acta*, Suppl. 3, Vol. 3, pp. 2423–2447. MIT Press.

Parkin C. W., Dyal P., and Daily W. D. (1973) Iron abundance in the moon from magnetometer measurements. *Proc. Fourth Lunar Sci. Conf., Geochim. Cosmochim. Acta*. This volume.

Pearce G. W., Strangway D. W., and Gose W. A. (1972) Remanent magnetization of the lunar surface. *Proc. Third Lunar Sci. Conf., Geochim. Cosmochim. Acta*, Suppl. 3, Vol. 3, pp. 2449–2464. MIT Press.

Reisz A. C., Paul D. L., and Madden T. R. (1972) The effects of boundary condition asymmetries on the interplanetary magnetic field-moon interaction. *The Moon* **4**, 134–140.

Runcorn S. K., Collinson D. W., O'Reilly W. O., Battey M. H., Stephenson A., Jones J. M., Manson A. J., and Readman P. W. (1970) Magnetic properties of Apollo 11 lunar samples. *Proc. Apollo 11 Lunar Sci. Conf., Geochim. Cosmochim. Acta*, Suppl. 1, Vol. 3, pp. 2369–2387. Pergamon.

Russell C. T., Coleman P. J., Lichtenstein B. R., Schubert G., and Sharp L. R. (1973) Apollo 15 and 16 subsatellite measurements of the lunar magnetic field (abstract). In *Lunar Science—IV*, pp. 645–646. The Lunar Science Institute, Houston.

Schubert G., Sonett C. P., Schwartz K., and Lee H. J. (1973) The induced magnetosphere of the moon I. Theory. Submitted to *Rev. Geophys. Space Phys.*

Sharp L. R., Coleman P. J., Lichtenstein B. R., Russell C. T., and Schubert G. (1972) Orbital mapping of the lunar magnetic field. Institute of Geophysics and Planetary Physics, UCLA Publication No. 1092-13.

Sonett C. P., Colburn D. S., Currie R. G., and Mihalov J. D. (1967) The geomagnetic tail; topology, reconnection, and interaction with the moon. In *Physics of the Magnetosphere* (editors R. L. Carovillano, J. F. McClay, and H. R. Radoski). D. Reidel.

Sonett C. P., Schubert G., Smith B. F., Schwartz K., and Colburn D. S. (1971) Lunar electrical conductivity from Apollo 12 magnetometer measurements: Compositional and thermal inferences. *Proc. Second Lunar Sci. Conf., Geochim. Cosmochim. Acta*, Suppl. 2, Vol. 3, pp. 2415–2431. MIT Press.

Strangway D. W., Gose W. A., Pearce G. W., and Carnes J. G. (1973) Magnetism and the history of the moon. *AIP Conf. Proc. No. 10*, pp. 1178–1196. Am. Inst. Phys., New York.

Telkes M. (1950) Thermoelectric power and electrical resistivity of minerals. *Am. Mineralogist* 35, No. 7.

Appendix

We consider a sphere which has a conducting core of radius R_c and electrical conductivity σ, with an insulating crust of thickness $R_m - R_c$, imbedded in a uniform magnetic field $\mathbf{B}_E$. Magnetic permeability everywhere is that of free space, μ_0. At time $t = 0$ the field is terminated such that $b(t) = 1$ for $t \leq 0$; $b(t) = 0$ otherwise. If the induced fields are confined within the radius R_m, azimuthal symmetry and continuity of the radial field across the boundary allow the magnetic vector potential to be written

$$\mathbf{A} = \begin{cases} \boldsymbol{\Phi} \sin \theta U(r, t) & r < R_c \\ \boldsymbol{\Phi} \sin \theta V(r, t) & R_c < r < R_m \end{cases} \tag{1}$$

which must satisfy the Helmholtz equation

$$\nabla^2 \mathbf{A} = \mu_0 \sigma \frac{\partial \mathbf{A}}{\partial t}.$$

The boundary conditions on the radial component of field requires

$$\frac{1}{r \sin \theta} \frac{\partial}{\partial \theta} (\sin \theta \mathbf{A})|_{r=R_m} = \boldsymbol{\Phi} \Delta B_E b(t) \cos \theta. \tag{2}$$

Equations (1) and (2) yield the differential equations

$$r^{-2}[r^2 U_{rr} + 2r U_r - 2U] = \mu_0 \sigma U_t \tag{3}$$

$$U(0, t) \text{ is bounded}$$

where the r subscript denotes $\partial/\partial r$, and

$$r^2 V_{rr} + 2r V_r - 2V = 0$$
$$V(R_m, t) = 0 \tag{4}$$

with the conditions

$$U(R_c, t) + R_c U_r(R_c, t) = V(R_c, t) + R_c V_r(R_c, t)$$
$$U(R_c, t) = V(R_c, t).$$

It can be verified that the bounded solution to equation (3) is

$$U(r, t) = \sum_s a_s j_1(k_s r) \exp\left(-\frac{k_s^2}{\mu_0 \sigma} t\right) \tag{5}$$

and the time-decay solution to equation (4) is

$$V(r, t) = \left(r - \frac{R_m^3}{r^2}\right) \sum_n A_n \exp(-q_n t).$$ (6)

For the boundary conditions to hold at all times we must have

$$q_n = \frac{k_s^2}{\mu_0 \sigma} = \frac{z_s^2}{\pi^2 T}; \quad s = n = 1, 2, 3, \ldots$$

where $z_s = k_s R_c$, $T = R_c^2 \mu_0 \sigma / \pi^2$ with the coefficients A_n and $a_n(s = n)$ defined by

$$a_s[z_s j_1'(z_s) + j_1(z_s)] = A_s \frac{R_m}{\lambda^2}(2\lambda^3 + 1); \quad \lambda = R_c / R_m$$

and

$$a_s j_1(z_s) = A_s \frac{R_m}{\lambda^2}(\lambda^3 - 1)$$

which in turn determine the characteristic equation for z_s. After elimination of the unknown constants a_s, and A_s, the characteristic equation is

$$z_s j_1'(z_s) + h j_1(z_s) = 0$$

Then the eigenvalue z_s is a solution of

$$\tan z_s = \frac{(h-2)z_s}{z_s^2 + (h-2)}$$ (7)

where

$$h = \frac{2 + \lambda^3}{1 - \lambda^3}$$

The initial condition $U(r, 0) = \Delta B_E r/2$ applied to equation (5) requires

$$a_s = \frac{\Delta B_E R_c (1+h)}{[z_s^2 + (h-2)(h+1)]j_1(z_s)}$$

which in turn defines

$$A_s = -\frac{\Delta B_E \lambda^3 (1+h)}{(1-\lambda^3)[z_s^2 + (h-2)(h+1)]}.$$

The vector potential can now be written

$$\mathbf{A} = -\mathbf{\Phi} \frac{\Delta B_E \sin \theta}{2} \left(\frac{\lambda^3}{1-\lambda^3}\right)\left(r - \frac{R_m^3}{r^2}\right) F(t) \quad R_c \leqslant r \leqslant R_m$$ (8)

where z_s are the nontrivial roots of equation (7) and

$$F(t) = \frac{6}{1-\lambda^3} \sum_s [z_s^2 + (h-2)(h+1)]^{-1} \exp(-z_s^2 t/\pi^2 T)$$

The induced eddy current field in the outer shell is given by the curl of equation (8):

$$B_r(r, \theta, t) = -\Delta B_E \cos \theta \left(\frac{\lambda^3}{1-\lambda^3}\right)\left[1 - \left(\frac{R_m}{r}\right)^3\right] F(t)$$

$$B_\theta(r, \theta, t) = \Delta B_E \sin \theta \left(\frac{\lambda^3}{1-\lambda^3}\right)\left[1 + \tfrac{1}{2}\left(\frac{R_m}{r}\right)^3\right] F(t)$$

We rotate the (r, θ, Φ) coordinate system about $\hat{r}$ to coincide with the ALSEP (x, y, z) system (defined in text) and evaluate the fields at $r = R_m$ to get the time-dependent solutions:

$$B_{Ax}(t) = B_{Ex}(t)$$

$$B_{Ay,z}(t) = B_{Ey,z}(t) + \Delta B_{Ey,z} \left[\frac{3}{2}\left(\frac{\lambda^3}{1-\lambda^3}\right)\right] F(t).$$

Proceedings of the Fourth Lunar Science Conference
(Supplement 4, *Geochimica et Cosmochimica Acta*)
Vol. 3, pp. 2947–2961

Iron abundance in the moon from magnetometer measurements

CURTIS W. PARKIN

Department of Physics, University of Santa Clara, Santa Clara, California 95053

PALMER DYAL and WILLIAM D. DAILY

NASA-Ames Research Center, Moffett Field, California 94035

Abstract—Apollo 12 and 15 lunar surface magnetometer data with simultaneous lunar orbiting Explorer 35 data are used to plot hysteresis curves for the whole moon. From these curves a whole-moon permeability $\mu = 1.029^{+0.024}_{-0.019}$ is calculated. This result implies that the moon is not composed entirely of paramagnetic material, but that ferromagnetic material such as free iron exists in sufficient amounts to dominate the bulk lunar susceptibility. From the magnetic data the ferromagnetic free iron abundance is calculated. Then for assumed compositional models of the moon the additional paramagnetic iron is determined, yielding total lunar iron content. The calculated abundances are as follows: ferromagnetic free iron, 5 ± 4 wt.%; total iron in the moon, 9 ± 4 wt.%.

INTRODUCTION

IN THIS paper we calculate lunar iron abundance from simultaneous magnetic field measurements made by instruments on the lunar surface and in orbit near the moon. The total iron abundance is the sum of metallic free iron and chemically combined iron, calculated for suitable compositional models of the lunar interior.

Previous estimates of whole-moon iron abundance have generally been $\sim 10\%$ by weight, often based on meteoritic compositional models, using the lunar density of 3.34 g/cm^3 as a constraint. Urey (1962) reported that the moon, if composed of chondritic material of proper density, would have 11–14% iron by weight, depending upon the high pressure phases present in the moon. Reynolds and Summers (1969), using a mathematical model involving equations of state at high pressure, placed a value of 13% on lunar iron abundance. Later Urey and MacDonald (1971) estimated values of 5% total iron using a model in which all the iron is present as FeO, and 8.65% total iron for another model in which the iron is combined in equal amounts of FeO and FeS. Wänke *et al.* (1973) reported a value of 9%, modeling lunar composition by that of the Allende chondritic meteorite. These previous results are generally consistent with the whole-moon iron abundance calculated in this paper.

We calculate lunar iron content from whole-moon magnetic hysteresis curves. These curves are plots of total surface magnetic field measured by Apollo 12 and 15 surface magnetometers simultaneous with the external field measured by the lunar orbiting Explorer 35 magnetometer. The slopes of the hysteresis curves are used to calculate magnetic permeability of the moon, from which we can calculate total iron content in the lunar interior. A paramagnetic mineral (olivine or orthopyroxene), combined with ferromagnetic free iron, is used with the lunar density constraint to calculate the iron abundance.

Theory

A lunar hysteresis curve is plotted using data obtained during times when the lunar sphere is magnetized by the earth's magnetic field. The total magnetic field **B** measured at a magnetometer site on the lunar surface is, in electromagnetic units,

$$\mathbf{B} = \mathbf{H} + 4\pi\mathbf{M}, \tag{1}$$

where **H** is the steady external (terrestrial) magnetizing field and **M** is the magnetization field induced in permeable lunar material (*see* Fig. 1). Alternately, Equation (1) can be expressed $\mathbf{B} = \mu\mathbf{H}$, where the relative permeability $\mu = 1 + 4\pi k$; k is magnetic susceptibility in emu/cm^3.

The lunar interior is modeled by a homogeneous sphere which has an iron Curie point (T_c) isotherm at some depth R/R_{moon}. The sphere is assumed to be composed of a paramagnetic mineral with free iron distributed throughout. The free iron is ferromagnetic in the outer shell where $T < T_c$, and paramagnetic in the core where $T > T_c$. The permeability μ_2 of the core therefore corresponds to paramagnetic material only, while μ_1 of the shell combines contributions from ferromagnetic iron and a paramagnetic mineral.

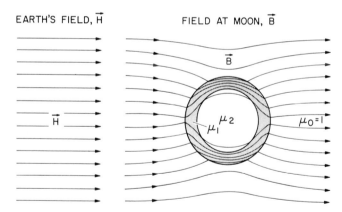

Fig. 1. Induced magnetization of the moon. When a two-layer permeable sphere is immersed in a uniform external magnetic field **H** (in our case the moon in the steady geomagnetic tail field), a dipolar magnetization field **M** is induced with its dipole moment aligned along the direction of **H**; the total field near the sphere is $\mathbf{B} = \mathbf{H} + 4\pi\mathbf{M}$. In the case illustrated here, $\mu_1 > \mu_2 > \mu_0 = 1$ (free space), corresponding qualitatively to the case where the outer shell (of radius R_1) contains both ferromagnetic and paramagnetic material, whereas the core (of radius R_2 and temperature above the Curie point) contains only paramagnetic material. Measurements of **B** and **H** allow construction of a classical B-H hysteresis curve for the sphere, from which permeability and iron abundance can be calculated.

Following the method of Jackson (1962) the total field on the outer surface of the sphere can be expressed in vector component form (*see* Appendix for details):

$$\mathbf{B} = H_x(1 + 2F)\hat{x} + H_y(1 - F)\hat{y} + H_z(1 - F)\hat{z} \qquad (2)$$

where

$$F = \frac{(2\eta + 1)(\mu_1 - 1) - \lambda^3(\eta - 1)(2\mu_1 + 1)}{(2\eta + 1)(\mu_1 + 2) - 2\lambda^3(\eta - 1)(\mu_1 - 1)} \qquad (3)$$

Here $\eta = \mu_1/\mu_2$; μ_1 and μ_2 are relative permeability of shell and core respectively (permeability of free space $\mu_0 = 1$); $\lambda = R_2/R_1$, where R_2 = radius of the core and R_1 = radius of the whole sphere. The components of $\mathbf{B}$ are expressed in the ALSEP coordinate system which has its origin on the lunar surface at a magnetometer site. The x-axis is directed radially outward from the lunar surface; the y and z axes are tangential to the surface, directed eastward and northward, respectively.

Once field measurements of $\mathbf{H}$ and $\mathbf{B}$ have been used to determine a value for F, iron abundances can be calculated for suitable lunar compositional models which must meet the additional constraints imposed on the lunar sphere by the overall lunar bulk density (3.34 g/cm^3) and moment of inertia. The core radius R_2 is determined by the iron Curie temperature where, because of internal temperature and pressure (Bozorth, 1951), iron changes from the ferromagnetic to the paramagnetic state. It is assumed that the core material has only paramagnetic susceptibility k_2 ($\mu_2 = 1 + 4\pi k_2$) and that the temperature variation of the paramagnetic susceptibility varies according to the Langevin theory:

$$k = nm^2/3KT, \qquad (4)$$

where K is the Boltzmann constant, T is absolute temperature, n is the number of ions per gram, and m is the atomic moment. m is of the order of a few Bohr magnetons μ_B; e.g., for the Fe^{2+} ion, $m = 5.25 \mu_B$ to $5.53 \mu_B$ (Nagata, 1961).

In the outer shell where temperature $T < T_c$, the susceptibility $k_1 = k_{a1} + k_{p1}$, where k_{a1} is "apparent" ferromagnetic susceptibility and k_{p1} is paramagnetic susceptibility. The ferromagnetic component is metallic free iron, assumed to be composed of multidomain, noninteracting grains. Furthermore, the expected pressures and temperatures in the outer shell are such that the ferromagnetic susceptibility of iron will not be substantially altered (Bozorth, 1951; Kapitsa, 1955).

The measured ferromagnetic susceptibility of the shell material is an apparent value which differs from the intrinsic ferromagnetic susceptibility of the iron because of self-demagnetization of the iron grains and the volume fraction of iron in the shell. The apparent ferromagnetic susceptibility k_{a1} is related to the intrinsic susceptibility k_{f1} according to

$$k_{a1} = \frac{pk_{f1}}{1 + Nk_{f1}}, \qquad (5)$$

where N is the demagnetization factor and p is the volume fraction of the iron. For spherical iron grains $N = 4\pi/3$; experimentally N is found to range between 3 and 4 (Nagata, 1961). We shall use $N = 3.5$ in our calculations.

EXPERIMENTAL TECHNIQUE

The basic technique uses simultaneous magnetometer measurements of the fields **B** and **H**, made while the moon is in the quiet geomagnetic tail region (*see* Fig. 2). The total surface field **B** is measured by lunar surface magnetometers (LSM) at the Apollo 12, 15, and 16 sites. The LSM instrument properties are described in detail by Dyal *et al.* (1972). Simultaneous measurements of the geomagnetic tail field **H** are made by the lunar orbiting Explorer 35 magnetometer, which orbits the moon with $0.5\,R_{moon}$ periselene and $5\,R_{moon}$ aposelene at a period of 11.5 hours. The Explorer 35 orbit is displayed in a companion paper in this volume (Dyal *et al.*, 1973), and detailed instrument characteristics are outlined by Sonett *et al.* (1967).

Lunar induction in the geomagnetic field

The geomagnetic tail consists of two regions ("lobes") of differing magnetic polarity, associated with the earth's dipole field extended downstream in the solar wind. At about 60 earth radii the lunar orbit intersects the magnetic tail as indicated by the accented portion of the orbit shown in Fig. 2, and the moon is in the geomagnetic tail field about four days of each lunation. The magnetic and plasma morphology of the geotail are especially favorable for observing the steady state induced magnetization field of the moon. The ambient field in the geotail is steadier and more uniform than in any other magnetic region of the lunar orbit. The geomagnetic field shields the moon from the solar plasma, minimizing electromagnetic interaction effects other than the magnetization field.

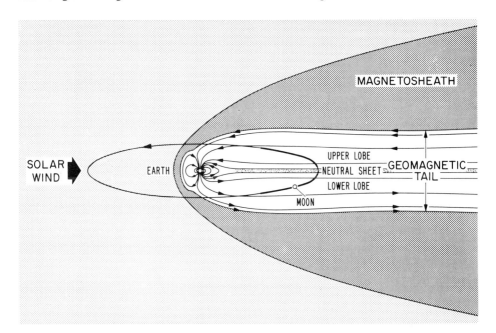

Fig. 2. Magnetic environment of the moon during a complete lunar orbit, with emphasis on the geomagnetic tail region. The plane of the lunar orbit very nearly coincides with the ecliptic plane of the earth's orbit. The earth's permanent dipole field is swept back into a cylindrical region known as the geomagnetic tail; at the lunar distance the field magnitude is ~ 10 gammas or 10^{-4} oersteds. Substructure of the tail consists of two "lobes"; the upper or northward lobe has its magnetic field pointing roughly toward the earth, whereas the lower lobe field points away from the earth. The moon is immersed about four days of each orbit in the tail; the moon can pass through either or both lobes (accented portion of orbit), depending upon the characteristics of the particular orbit.

In general the total magnetic field at the lunar surface is a vector sum of the external magnetic field measured by Explorer 35; the steady remanent field at the LSM site; the poloidal field resulting from transverse electric (TE) induction; the toroidal field resulting from transverse magnetic (TM) induction; the magnetization field induced in the permeable material within the moon; the diamagnetic field associated with the plasma-moon interface currents; and the field associated with the dynamic interaction between plasma flow and the lunar fields. A careful consideration of each of these contributions to the total field at the lunar surface is necessary to justify the data selection and analysis best suited to determine accurately the magnetization fields of the moon in the geomagnetic environment. We will see that some of these fields are important to our analysis while contributions from others can be neglected.

The solution given by Equations (2) and (3) for the magnetization of the moon assumes a uniform external magnetizing field. Under this assumption the induced field is dipolar; higher order multipoles contribute when inhomogeneities are present in the external field. Uniformity of the ambient geotail field cannot be guaranteed with the available data. It is possible to discriminate against data from nonuniform tail fields by requiring magnetic conditions at both LSM and Explorer instruments simultaneously to be steady over long time intervals. If the geomagnetic field and moon are in relative motion with a speed characterized by $V \gtrsim L/T$, where L is the distance between LSM and Explorer 35 magnetometers and T is the time over which the field is steady, then any nonuniformities can be detected as magnetic fluctuations at either instrument. Various criteria on the peak-to-peak fluctuations at both magnetometers have been used to select data. A summary of the data selection criteria will be presented in a later subsection.

Fluctuations of the geotail field ($\partial \mathbf{B}_E/\partial t$) drive time-dependent eddy currents in the lunar interior. These in turn result in a poloidal magnetic field (TE induction) which is probably the largest of the electromagnetic induction fields measured at the surface. In the limit of low-frequency or small-amplitude driving field fluctuations, $\partial \mathbf{B}_E/\partial t \rightarrow 0$ and induction from the poloidal mode vanishes. Therefore the restrictions placed on the peak-to-peak variations of surface and external fields, used to discriminate against nonuniform geotail fields, will simultaneously eliminate data obviously contaminated by eddy current fields. Furthermore, the data averaging procedures used in the analysis tend to filter out the higher frequencies which are the major contributors in this mode.

Relative motion of the moon and the geomagnetic field results in a motional electric field $\mathbf{E}_m = \mathbf{V} \times \mathbf{H}$ in the lunar reference frame, where $\mathbf{V}$ is the relative lunar velocity and $\mathbf{H}$ is the external field. Currents driven by $\mathbf{E}_m$ set up magnetic fields (TM induction) that are toroidal about the electric field. Theoretically this magnetic field is everywhere tangent to the lunar surface so that the ALSEP x-components contain minimal toroidal induced fields.

If the induced magnetization field $\mathbf{B}_0 = 4\pi \mathbf{M}$ is predominately dipolar, the dipole axis will be aligned with the external field, $\mathbf{H}$. At the lunar surface the radial and tangential components of the dipole field are (*see* Appendix) $B_{0r} = 2\alpha \cos \theta$ and $B_{0\theta} = \alpha \sin \theta$, where θ is the angle between the dipole axis and the radius vector to the surface site, and α is the dipole moment. We see that the maximum radial component is along the dipole axis, which for the average geomagnetic tail field is along the earth-sun line. As a result the average surface dipole field at both Apollo 12 and Apollo 15 LSM sites is predominantly in the radial or ALSEP $\hat{x}$ direction.

As plasma particles spiral about magnetic field lines, their motion induces an opposing field resulting in plasma diamagnetism. The solar wind plasma diamagnetism has been detected by Colburn *et al.* (1967) and is seen as a magnetic field increase ($\mathbf{B}_d$) of about 1.5 gammas in magnitude in the plasma void on the antisolar side of the moon. This diamagnetic field can be expressed (Boyd and Sanderson, 1969):

$$B_d \propto f N_i (T_e + T_i)/H.$$

H is magnitude of the external vacuum field, N_i is the plasma ion density, and T_e and T_i are the electron and ion temperatures perpendicular to the magnetic field $\mathbf{H}$. The average tail magnetic field is about 10 gammas while the plasma density is at least two orders of magnitude below average solar wind values. Little is known of the tail plasma temperature, but it is probably $\leqslant 10^{5\circ}$K. An upper limit on B_d in the geotail is therefore ~ 0.03 gamma, which is far below the LSM resolution of 0.2 gamma. Therefore plasma diamagnetism will be neglected in this analysis.

The dynamic interaction between solar wind plasma and the lunar surface fields have been found to be $\leqslant 16$ gammas and $\leqslant 1$ gamma at Apollo 12 and 15 sites respectively (Dyal *et al.*, 1973). Plasma streaming pressure responsible for compression of surface fields is $(\frac{1}{2})N_imv^2$ where m is the proton mass and v is the plasma bulk speed. We assume a typical speed in the tail is characterized by the average magnetopause motions which are about 70 km/sec (Mihalov *et al.*, 1970). Inside the magnetotail lobes the plasma density is below about $0.01/cm^3$. Therefore the plasma-field interaction in the geotail has to be at least 10^3 times weaker than in the free streaming solar wind and can be neglected.

Since remanent surface field compression is negligible, it will be assumed that the steady remanent fields at both LSM sites are constant in the tail, the importance of which will soon be apparent. In addition these fields are assumed to be not measurable by Explorer 35 at periselene. The results of the Apollo subsatellite magnetometer reported by Sharp *et al.* (1972) confirm this assumption. Therefore Explorer 35 measures only the external geomagnetic field.

Data processing criteria

Magnetization effects can be maximized and competing electromagnetic induction minimized by applying data selection criteria. It has been pointed out that at the Apollo 12 and 15 sites the magnetization field is largest in the radial or ALSEP x-axis. Furthermore, toroidal mode fields are not present in the x-component and plasma interaction effects are minimal along the x-direction. Therefore, only the ALSEP x-components of the LSM and Explorer 35 data have been used to calculate magnetization fields. Several averaging intervals (6 minutes, 15 minutes, 30 minutes, 1 hour) were used for analysis of Apollo 12 data, each with different peak-to-peak criteria, to insure external field uniformity and minimize poloidal field contamination. Only 60-minute and 3-hour averages, without peak-to-peak criteria, were useful for the Apollo 15 data since the simultaneous Explorer 35 data were contaminated by a high-frequency satellite spin-modulated tone. A typical 30-minute swath of Apollo 12 data used in the analysis is shown in Fig. 3. The steadiness of these data is characteristic of much of the geomagnetic tail field.

Lunar hysteresis curves

Typical lunar "hysteresis" curves are shown in Fig. 4. Apollo 12 and 15 averages of x-components are plotted on the abscissa with the simultaneous Explorer 35 averages on the ordinate. These data were taken from four lunations of Apollo 12 and two lunations of Apollo 15 data. Bifurcation in the

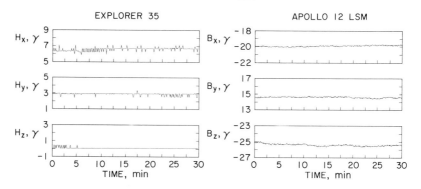

Fig. 3. Simultaneous Apollo 12 and Ames Explorer 35 magnetometer data from a quiet region of the geomagnetic tail, 1970. Vertical scales for the two magnetometers differ due to the existence of a 38 ± 3 gamma remanent magnetic field at the Apollo 12 surface site. Components of the fields are expressed in the ALSEP coordinate system, which has its origin located on the moon's surface at the Apollo 12 site; the x-axis is directed radially outward from the lunar surface, whereas y and z axes are tangential to the surface, directed eastward and northward, respectively.

LUNAR HYSTERESIS CURVES

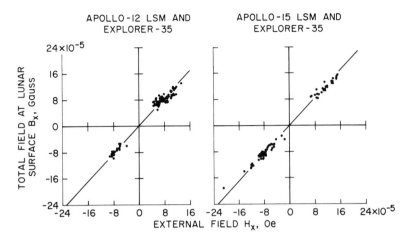

Fig. 4. Hysteresis curves for the moon. For each data point the moon is immersed in the external magnetizing geomagnetic tail field **H** (measured by Explorer 35) which is steady over 30-minute or one-hour time intervals; the total magnetic induction is $\mathbf{B} = \mu\,\mathbf{H}$, measured by an Apollo surface magnetometer. Here only the radial (x) components are plotted. In this low-applied-field regime (~ 10 gamma or 10^{-4} oersteds), the hysteresis curves are linear and are fitted by least-squares straight lines. (a) Simultaneous Apollo 12 and Explorer data; each data point represents a 30-minute average during quiet times in the geomagnetic tail for four lunations, 1969–1970. (b) Simultaneous one-hour averages of Apollo 15 and Explorer data for two lunations, 1971.

distributions is a result of the data selection criteria which eliminate data from near the neutral sheet. The normal "S" shape of the hysteresis curve degenerates at these low field values to a straight line (Ellwood, 1934) intersecting the origin since the x axis intercept (x-component of the remanent field at both sites) has been subtracted from the data.

Statistical analysis

Four lunations of Apollo 12 data and two of Apollo 15 data have been processed for each of the averaging times, employing the data selection criteria described previously. This has resulted in a total of six separate hysteresis curves, one for each averaging interval of both instruments. To each of these data sets a least squares straight line has been fitted. The average and standard deviation values of the slopes are 1.019 ± 0.012. Both extrema are greater than 1.0, implying that the moon, as a whole, acts as a paramagnetic or weakly ferromagnetic sphere.

Scatter in the Apollo 12-Explorer 35 data points of Fig. 4(a) is primarily a result of magnetic inhomogeneities between the moon and Explorer 35, small contributions from eddy current fields, and instrumental noise in the Apollo and Explorer magnetometers. These error sources may introduce small random fluctuations into the data which will not substantially affect the slope or intercept of the least-squares line. An error source which could affect the least-squares slope is a time-dependent heading drift of the Explorer 35 magnetometer. The drift problem, which began after the Apollo 12 and Explorer 35 data used in Fig. 4(a) were acquired, is therefore applicable only to analysis of simultaneous Explorer 35 and Apollo 15 data. The slope of the Apollo 15-Explorer 35 hysteresis curve (1.019 ± 0.004) shown in Fig. 4(b) agrees well with that obtained from the Apollo 12-Explorer 35 curve (1.019 ± 0.013), however, indicating that the Explorer 35 heading error is minimal for the time periods used in Fig. 4(b). Furthermore, the agreement in the analysis of data from these two Apollo surface sites, which are separated by 1000 km, is a good test of our assumption of lunar azimuthal symmetry.

RESULTS

Bulk magnetization properties of the lunar sphere

It is shown in the THEORY section that the slope of the hysteresis curve at very low field values is related to the magnetic susceptibility of the moon. Figure 5 relates F of Equation (3) to the susceptibilities k_1 and k_2 of the lunar shell and core. The curve $\epsilon = k_1/k_2$ corresponds to the case of a magnetically homogeneous sphere of permeability $\mu = 1 + 4\pi k_1$, from which bulk permeability of the moon is calculated to be $\mu = 1.029^{+0.024}_{-0.019}$. Error limits are one standard deviation. Since both upper and lower limits of permeability are above 1.0, the moon as a whole responds magnetically as a paramagnetic or weakly ferromagnetic sphere. The field at the surface is in general a function of both shell and core susceptibilities; as the susceptibility ratio $\epsilon = k_1/k_2$ increases, the core susceptibility k_2 becomes less important magnetically, and the family of curves merge for $\epsilon \geqslant 100$.

A further use of the bulk lunar permeability is to place limits on the magnetic dipole moment induced in the moon by the geomagnetic field. The dipole moment is HR_1^3F (*see* Appendix); it is oriented parallel to the external field, and for $H = 10$ gammas, the induced dipole moment is of magnitude $5.0 \pm 3.2 \times 10^{23}$ gammas-cm^3. Russell *et al.* (1973) using subsatellite magnetic data, have placed an upper limit on the *permanent* lunar dipole moment of 1.7×10^{23} gammas-cm^3. While interesting to compare, these two results are for different phenomena. The averaging techniques performed over many subsatellite orbits to obtain the permanent dipole value may effectively average out the magnetization fields which change magnitude and polarity as a function of the external field **H**.

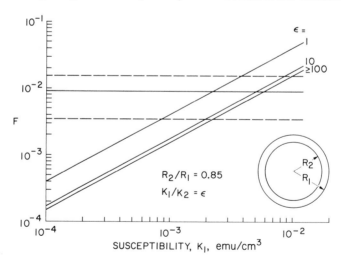

Fig. 5. The function F vs. susceptibility k_1 of the lunar outer shell (refer to Equations (2) and (3)). This figure is for the case where the iron Curie temperature isotherm is at core radius $\lambda = R_2/R_1 = 0.85$, as inferred from temperature models illustrated in Fig. 6. Superimposed on the vertical scale are the best value and limits of F calculated from lunar hysteresis curve slopes; the calculated range of F determines possible ranges for susceptibilities of core and shell for this model.

Lunar composition and iron abundance

In this section we calculate iron abundance in the moon, using various compositional models of the lunar interior. The models are constrained by magnetic permeability, temperature, pressure, density, and moment of inertia determinations.

The lunar moment of inertia is approximately that of a sphere of homogeneous density (Ringwood and Essene, 1970). We will constrain our lunar models to be of uniform density $\rho = 3.34 \, gm/cm^3$ throughout.

We have selected two paramagnetic minerals to model the lunar interior: olivine and orthopyroxene. In the moon's outer shell, free iron is assumed to be interspersed throughout the paramagnetic rock matrix. These minerals, with iron content constrained by the overall density of 3.34, have properties which are consistent with measured seismic velocity profiles at mantle depths (Toksöz et al., 1972) and geochemical considerations (Ringwood and Essene, 1970; Green et al., 1971; Gast, 1972). Furthermore, pyroxenes and olivines have been reported to be major mineral components of lunar surface fines and rock samples (Nagata et al., 1971; Zussman, 1972; Weeks, 1972), with combined iron present as Fe^{2+}. The ferromagnetic component of lunar samples is primarily metallic iron (Nagata et al., 1972; Pearce et al., 1971) most of which is thought to be native to the moon rather than meteoritic in origin (Strangway et al., 1973).

Magnetization of the lunar sphere, as measured by the Apollo-Explorer 35 magnetometer network, imposes a magnetic constraint on the lunar composition. This constraint is determined by the hysteresis-curve slope, represented by the value $F = 0.0095 \pm 0.0060$ (see Equations (2) and (3)); F is in turn related to the internal susceptibility of the moon.

The Curie point of metallic iron will define the boundary between a homogeneous shell of susceptibility k_1 and a homogeneous core of susceptibility k_2. The boundary is located using the temperature profiles proposed for the moon by several authors as shown in Fig. 6. Also plotted is the Curie temperature of iron which is a function of pressure and therefore of depth (Bozorth, 1951).

For the calculations that follow, three temperature models are used in an effort to match "hot," "warm," and "cold" models. For the hot model the Curie point is at $\lambda = R_2/R_1 = 0.9$. We assume shell and core temperatures are 600°C and 1400°C, respectively. For the warm model the shell temperature is 500°C and that of the core is 1300°C, while the boundary is at $\lambda = 0.85$. Temperatures are 300°C and 700°C for shell and core of the cold model, which has $\lambda = 0.7$. The magnetic properties of the core are determined entirely by the paramagnetism of olivine (Nagata et al., 1957) and orthopyroxene (Akimoto et al., 1958) and the Langevin temperature dependence of these minerals (see Equation (4)). Small amounts of free iron will not substantially change the total effective susceptibility of the core since iron is paramagnetic above its Curie temperature.

Crustal susceptibility will be determined by the free and combined iron under the assumption that the free iron is in noninteractive, multidomain grains whose shape demagnetizing factor is on the average 3.5 (Nagata, 1961). Normally the ferromagnetism of free iron is dominated by a dependence on pressure and tem-

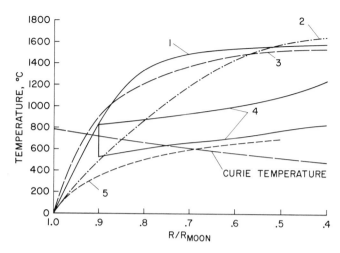

Fig. 6. Temperature profiles for the lunar interior published by various authors: (1) McConnell *et al.* (1967), (2) Fricker *et al.* (1967), (3) Hanks and Anderson (1972), (4) Dyal and Parkin (1972), and (5) Sonett *et al.* (1971). Superimposed is the pressure-dependent Curie temperature for iron (Bozorth, 1951) vs. depth in the moon. The Curie point is reached at a position $R/R_{moon} \sim 0.85$ for the "warm" model and ~ 0.7 for a "cold" model.

perature. However at very low field strengths, such as those of the geomagnetic tail, the susceptibility of iron is not strongly dependent on temperature below the Curie point (Bozorth, 1951). Uniaxial stress on iron changes its susceptibility (Kern, 1961); however, hydrostatic stress should not affect the susceptibility (Kapitsa, 1955) unless, at great pressures, there is a change in volume (Breiner, 1967). Therefore, it will be assumed that the susceptibility of free iron in the shell is independent of pressure and temperature.

Using the constraints outlined above, forward calculations have been made to determine the iron content in the moon. Olivine (10 mole % Fe_2SiO_4) and orthopyroxene (25 mole % $FeSiO_3$) minerals are selected using the bulk lunar density constraint. For a given temperature model Langevin's law yields a range of susceptibilities for each core composition (*see* Fig. 7) from which the paramagnetic iron content of the core is calculated.

Core susceptibility is then used, along with the magnetically determined range of F, to define a range of susceptibility for the shell. Apparent susceptibility of the shell is then a function of the free and combined iron in the shell. In all cases it is found that the shell susceptibility cannot be accounted for by the paramagnetic susceptibility (k_{p1}) alone; rather, the apparent ferromagnetic susceptibility (k_{a1}) is dominant: $k_{a1}/k_{p1} \sim 100$. Therefore, using Equation (5) the shell free iron abundance is determined to account for the range of F values (*see* Fig. 5). Finally, the paramagnetic rock matrix of the shell is chosen so that, containing the free iron, it matches the bulk lunar density. Both the ferromagnetic and paramagnetic iron content of the shell are then determined.

From these calculations it has been found that the paramagnetism of any

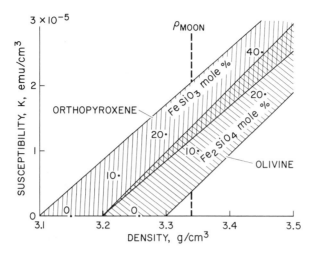

Fig. 7. Magnetic susceptibility vs. density for two minerals at 1600°K, olivine $(x\,Fe_2SiO_4 \cdot (1 - x)Mg_2SiO_4)$ and orthopyroxene $(x\,FeSiO_3 \cdot (1 - x)MgSiO_3)$, used to model the lunar interior. For a homogeneous moon composed of either of these minerals, the iron silicate concentrations with densities greater than $\rho = 3.34\ \text{gm/cm}^3$ are forbidden.

olivine and orthopyroxene cannot account for the measured bulk permeability of the moon; contributions from free iron or other highly permeable material are necessary. When metallic iron is included in a relatively thin outer shell ("hot" model where $\lambda \geq 0.9$), it is difficult to account for the measured bulk permeability of the moon while still maintaining the constraint of homogeneous density throughout the moon.

When the iron Curie point is assumed to be deeper in the moon, i.e., $\lambda = 0.85$ ("warm" model) or $\lambda = 0.7$ ("cold" model), magnetic permeability and density constraints can be easily satisfied by our compositional models. Ferromagnetic iron content in the shell is the same for both olivine and orthopyroxene, but is a function of thermal model. The warm model requires 2 to 9 wt. % free iron in the shell, whereas the cold model requires 1 to 5 wt. %. The total iron abundance, on the other hand, is the same for both thermal models, but different for the two mineral compositional models. For olivine (mole % $Fe_2SiO_4 \leq 10\%$) total iron is 5 to 6 wt. %; for orthopyroxene (mole % $FeSiO_3 \leq 25\%$), total iron is 12 to 13 wt. %. The iron abundance limits reported here all include the one-standard-deviation limits of the measured hysteresis curve slope 1.019 ± 0.012. Results are summarized in Fig. 8.

SUMMARY AND CONCLUSIONS

(1) Apollo 12 and 15 lunar surface magnetometer data, simultaneous with Explorer 35 magnetometer data, are used to plot hysteresis curves for the whole moon; from these curves a whole-moon permeability $\mu = 1.029^{+0.024}_{-0.019}$ is calculated. Both error limits constrain the permeability to be greater than 1.0, implying that the moon as a whole is paramagnetic or weakly ferromagnetic.

C. W. PARKIN *et al.*

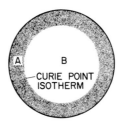

COMPOSITIONAL MODEL	IRON CURIE POINT LOCATION	WT % FREE IRON IN SHELL A	TOTAL WT % IRON IN MOON
OLIVINE AND FREE IRON	$R/R_{moon} = 0.85$ 0.7	2 – 9 1 – 5	5 – 6
ORTHOPYROXENE AND FREE IRON	0.85 0.7	2 – 9 1 – 5	12 – 13

Fig. 8 Iron abundance in the moon. Total iron abundance of the moon is calculated for two compositional models of the lunar interior, each with two different thermal profiles. Total iron includes both free iron and iron combined in the paramagnetic state. The portion of the total iron which contributes most strongly to the measured permeability, ferromagnetic free iron, is listed separately.

(2) The permeability calculation places a constraint on the composition of the lunar interior. When the moon is modeled by homogeneous paramagnetic material (olivine or orthopyroxene) of density 3.34 gm/cm³, it is found that these purely paramagnetic models have magnetic permeabilities far lower than the range determined from magnetometer measurements. We have found no reasonable paramagnetic mineral or combination of minerals, with the correct lunar density, that would have permeability high enough to be consistent with our measured values. From this we infer that the moon must contain some material in the ferromagnetic state, such as free metallic iron, in order to account for the measured global permeability.

(3) The paramagnetic-ferromagnetic combination is investigated using a two-layer lunar model; the boundary between the layers is determined by the location of the iron Curie point. Three thermal models, "hot," "warm," and "cold," are used to establish the Curie isotherm location. "Warm" and "cold" models fit both the density and permeability constraints whereas the "hot" models do not. We conclude that the ferromagnetic free iron abundance of the outer shell is 2 to 9 wt.% for "warm" and 1 to 5 wt.% for "cold" models, yielding an overall range of 5 ± 4 wt.%. The free iron content is dependent upon the volume of lunar material below the Curie temperature and independent of the composition of the non-ferromagnetic portion of the moon.

(4) For two compositional models the total iron abundance is calculated (*see* Fig. 8). Including both the olivine and orthopyroxene models, the total iron is independent of the thermal model used and is within the limits of 9 ± 4 wt.%.

Acknowledgments—We are grateful to Dr. T. E. Bunch and R. T. Reynolds for many helpful discussions and Drs. C. P. Sonett and D. S. Colburn for use of Explorer 35 magnetometer data. M. Legg and her group members of Massey Temporary Services deserve special thanks for their data reduction services, as do Ken Lewis and his group for their programming support. We are pleased to acknowledge the support of research by C.W.P. under NASA grant no. NGR 05 017 027.

REFERENCES

Akimoto S., Horai K., and Boku T. (1958) Magnetic susceptibility of orthopyroxene. *J. Geomag. Geoelect.* **10**, 7–11.

Behannon K. W. (1968) Intrinsic magnetic properties of the lunar body. *J. Geophys. Res.* **73**, 7257–7268.

Boyd T. J. and Sanderson J. J. (1969) *Plasma Dynamics.* Barnes and Noble.

Bozorth R. M. (1951) *Ferromagnetism.* D. Van Nostrand.

Breiner S. (1967) The Piezomagnetic effect on seismically active areas. Final Report No. E 22-76-67(N), Dept. of Geophysics, Stanford University.

Colburn D. S., Currie R. G., Mihalov J. D., and Sonett C. P. (1967) Diamagnetic solar-wind cavity discovered behind the moon. *Science* **168**, 1040–1042.

Dyal P. and Parkin C. W. (1972) Lunar properties from transient and steady magnetic field measurements. *The Moon* **4**, 63–87.

Dyal P., Parkin C. W., and Cassen P. (1972) Surface magnetometer experiments: Internal lunar properties and lunar surface interactions with the solar plasma. *Proc. Third Lunar Sci. Conf., Geochim. Cosmochim. Acta*, Suppl. 3, Vol. 3, pp. 2287–2307. MIT Press..

Dyal P., Parkin C. W., and Daily W. D. (1973) Surface magnetometer experiments: Internal lunar properties. *Proc. Fourth Lunar Sci. Conf., Geochim. Cosmochim. Acta.* In press.

Ellwood W. B. (1934) A new ballistic galvanometer operating in high vacuum. *Rev. Sci. Inst.* **5**, 300–305.

Fricker P. E., Reynolds R. T., and Summers A. L. (1967) On the thermal history of the moon. *J. Geophys. Res.* **72**, 2649–2663.

Gast P. W. (1972) The chemical composition and structure of the moon. In *Lunar Geophysics* (editors Z. Kopal and D. Strangway), pp. 630–657. Lunar Science Institute Contribution No. 86, Houston. D. Reidel.

Green D. H., Ringwood A. E., Ware N. G., Hibberson W. O., Major A., and Kiss E. (1971) Experimental petrology and petrogenesis of Apollo 12 basalts. *Proc. Second Lunar Sci. Conf., Geochim. Cosmochim. Acta*, Suppl. 2, Vol. 1, pp. 601–615. MIT Press.

Hanks T. C. and Anderson D. L. (1972) Origin, evolution and present thermal state of the moon. *Phys. Earth Planet. Interiors* **5**, 409–425.

Jackson J. D. (1962) *Classical Electrodynamics.* John Wiley and Sons.

Kapitsa S. P. (1955) Magnetic properties of igneous rock under mechanical stresses. Bull. (I.V.) Acad. Sci. USSR, Geophys. Ser. No. 6.

Kern J. W. (1961) The effect of stress on the susceptibility and magnetization of a partially magnetized multidomain system. *J. Geophys. Res.* **66**, 3807–3816.

McConnell R. K. Jr., McClaine L. A., Lee D. W., Aronson J. R., and Allen R. V. (1967) A model for planetary igneous differentiation. *Rev. Geophys.* **5**, 121–172.

Mihalov J. D., Colburn D. S., and Sonett C. P. (1970) Observations of magnetopause geometry and waves at the lunar distance. *Planet. Space Sci.* **18**, 239–258.

Nagata T. (1961) *Rock Magnetism.* Maruzen Co. Ltd.

Nagata T., Yukutake T., and Uyeda S. (1957) On magnetic susceptibility of Olivines. *J. Geomag. Geoelect.* **9**, 51–56.

Nagata T., Fisher R. M., Schwerer F. C., Fuller M. D., and Dunn J. R. (1971) Magnetic properties and remanent magnetization of Apollo 12 lunar materials and Apollo 11 lunar microbreccia. *Proc. Second Lunar Sci. Conf., Geochim. Cosmochim. Acta*, Suppl. 2, Vol. 3, pp. 2461–2476. MIT Press.

Nagata T., Fisher R. M., Schwerer F. C., Fuller M. D., and Dunn J. R. (1972) Rock magnetism of Apollo 14 and 15 materials. *Proc. Third Lunar Sci. Conf., Geochim. Cosmochim. Acta*, Suppl. 3, Vol. 3, pp. 2423–2447. MIT Press.

Pearce G. W., Strangway D. W., and Larson E. E. (1971) Magnetism of two Apollo 12 igneous rocks. *Proc. Second Lunar Sci. Conf., Geochim. Cosmochim. Acta*, Suppl. 2, Vol. 3, pp. 2451–2460. MIT Press.

Pearce G. W., Strangway D. W., and Gose W. A. (1972) Remanent magnetization of the lunar surface. *Proc. Third Lunar Sci. Conf., Geochim. Cosmochim. Acta.*, Suppl. 3, Vol. 3, pp. 2449–2464. MIT Press.

Reynolds R. T. and Summers A. L. (1969) Calculations on the composition of the terrestrial planets. *J. Geophys. Res.* **74**, 2494–2511.

Ringwood A. E. and Essene E. (1970) Petrogenesis of lunar basalts and the internal constitution and origin of the moon. *Science* **167**, 607–610.

Russell C. T., Coleman P. J., Lichtenstein B. R., Schubert G., and Sharp L. R. (1973) Apollo 15 and 16 subsatellite measurements of the lunar magnetic field (abstract). In *Lunar Science—IV*, pp. 645–646. The Lunar Science Institute, Houston.

Sharp L. R., Coleman P. J., Lichtenstein B. R., Russell C. T., and Schubert G. (1972) Orbital mapping of the lunar magnetic field. Institute of Geophysics and Planetary Physics, UCLA Publication No. 1092–13.

Sonett C. P., Colburn D. S., Currie R. G., and Mihalov J. D. (1967) The geomagnetic tail; topology, reconnection, and interaction with the moon. In *Physics of the Magnetosphere* (editors R. L. Carovillano, J. F. McClay, and H. R. Radoski). D. Reidel.

Sonett C. P., Schubert G., Smith B. F., Schwartz K., and Colburn D. S. (1971) Lunar electrical conductivity from Apollo 12 magnetometer measurements: Compositional and thermal inferences. *Proc. Second Lunar Sci. Conf., Geochim. Cosmochim. Acta*, Suppl. 2, Vol. 3, pp. 2415–2431. MIT Press.

Strangway D. W., Gose W. A., Pearce G. W., and Carnes J. G. (1973) Magnetism and the history of the moon. Proc. of the 18th Annual Conf. on Magnetism and Magnetic Materials, *J. Applied Phys.* In press.

Toksöz M. N., Press F., Dainty A., Anderson K., Latham G., Ewing M., Dorman J., Lammlein D., Sutton G., and Duennebier F. (1972) Structure, composition, and properties of lunar crust. *Proc. Third Lunar Sci. Conf., Geochim. Cosmochim. Acta*, Suppl. 3, Vol. 3, pp. 2527–2544. MIT Press.

Urey H. C. (1962) The origin of the moon. In *The Moon* (editors Z. Kopal and Z. K. Mikhailov), pp. 133–148. Academic Press.

Urey H. C. and MacDonald G. J. F. (1971) Origin and history of the Moon. In *Physics and Astronomy of the Moon* (editor Z. Kopal), pp. 213–289. Academic Press.

Wänke H., Baddenhausen H., Dreibus G., Quijano-Rico M., Palme H., and Teschki F. (1973) Multi-element analysis of Apollo 16 samples and about the composition of the whole moon. In *Lunar Science—IV*, pp. 761–763. The Lunar Science Institute, Houston.

Weeks R. A. (1972) Magnetic phases in lunar material and their electron magnetic resonance spectra: Apollo 14. *Proc. Third Lunar Sci. Conf., Geochim. Cosmochim. Acta*, Suppl. 3, Vol. 3, pp. 2503–2517. MIT Press.

Zussman J. (1972) The mineralogy, petrology and geochemistry of lunar samples—a review. *The Moon* **5**, 422–435.

Appendix

Consider a radially inhomogeneous two-layer permeable sphere in an initially uniform magnetic field H_0. In the absence of currents $H = -\nabla\Phi$ and since $B = \mu H$, at any point $\nabla \cdot H = 0$. Therefore Φ satisfies Laplace's equation consistent with the continuity of the normal components of B and tangential components of H at the spherical boundaries. In the various regions the potentials must be

$$
\Phi = \begin{cases}
-H_0 r \cos\theta + \displaystyle\sum_{i=0}^{\infty} \frac{\alpha_i}{r^{i+1}} P_i(\cos\theta) & r > R_1 \\[2ex]
\displaystyle\sum_{i=0}^{\infty} \left(\beta_i r^i + \gamma_i \frac{1}{r^{i+1}} \right) P_i(\cos\theta) & R_2 < r < R_1 \\[2ex]
\displaystyle\sum_{i=0} \delta_i r^i P_i(\cos\theta) & r < R_2
\end{cases}
$$

with R_1 and μ_1 the outer radius and permeability, and R_2 and μ_2 the inner radius and permeability. Using Jackson's (1962) notation conventions we match the boundary conditions

$$\frac{\partial \Phi}{\partial \theta}(R_1^+) = \frac{\partial \Phi}{\partial \theta}(R_1^-) \qquad\qquad \frac{\partial \Phi}{\partial \theta}(R_2^+) = \frac{\partial \Phi}{\partial \theta}(R_2^-)$$

$$\frac{\partial \Phi}{\partial r}(R_1^+) = \mu_1 \frac{\partial \Phi}{\partial r}(R_1^-) \qquad \mu_1 \frac{\partial \Phi}{\partial r}(R_2^+) = \mu_2 \frac{\partial \Phi}{\partial r}(R_2^-)$$

All but the $i = 1$ coefficients vanish leaving the four simultaneous equations

$$\alpha - R_1^3 \beta - \gamma = R_1^3 H_0$$
$$2\alpha + \mu_1 R_1^3 \beta - 2\mu_1 \gamma = -R_1^3 H_0$$
$$R_2^3 \beta + \gamma - R_2^3 \delta = 0$$
$$\mu_1 R_2^3 \beta - 2\mu_1 \gamma - \mu_2 R_2^3 \delta = 0$$

We are interested in α, the induced dipole moment, which is given by

$$F \equiv \frac{\alpha}{H_0 R_1^3} = \frac{(2\eta + 1)(\mu_1 - 1) - \lambda^3 (\eta - 1)(2\mu_1 + 1)}{(2\eta + 1)(\mu_1 + 2) - 2\lambda^3 (\eta - 1)(\mu_1 - 1)}$$

where $\eta = \mu_1/\mu_2$ and $\lambda = R_2/R_1$. The resulting dipolar field in an $(\hat{r}, \hat{\theta}, \hat{\Phi})$ spherical coordinate system is

$$\mathbf{B_0} = \frac{\alpha}{R_1^3}(2 \cos \theta \hat{r} + \sin \theta \hat{\theta}).$$

We rotate the coordinate system about the $\hat{r}$ (or ALSEP $\hat{x}$) axis so that $\hat{\theta}$ and $\hat{\Phi}$ correspond to the ALSEP $\hat{y}$ and $\hat{z}$ axes; then

$$\mathbf{B_0} = F(2H_x \hat{x} - H_y \hat{y} - H_z \hat{z})$$

The total field at the surface is the sum of the uniform external field $\mathbf{H}$ and the dipole field $\mathbf{B_0}$:

$$\mathbf{B} = H_x(1 + 2F)\hat{x} + H_y(1 - F)\hat{y} + H_z(1 - F)\hat{z}.$$

Proceedings of the Fourth Lunar Science Conference
(Supplement 4, *Geochimica et Cosmochimica Acta*)
Vol. 3, pp. 2963–2976

Magnetic properties of Apollo 15 and 16 rocks

D. W. Collinson, A. Stephenson, and S. K. Runcorn

Department of Geophysics and Planetary Physics
School of Physics, University of Newcastle upon Tyne
Newcastle upon Tyne NE1 7RU, England

Abstract—The remanent magnetism of samples of Apollo 15 and 16 rocks has been measured and its characteristics studied. The stable remanence, which is found in most of the samples, is carried by iron and in some cases has the properties of a thermoremanence. One of the Apollo 16 samples has provided an exceptionally high estimate of the field in which it acquired its magnetism. A detailed study has been made of the homogeneity of magnetization in a breccia, and the rock magnetism and magnetic mineralogy of some of the samples is described.

1. Introduction

Since the discovery of hard components of natural remanent magnetization (NRM) in rocks returned by the earlier Apollo missions (*see*, for instance, Strangway *et al.*, 1971; Collinson *et al.*, 1972; Pearce *et al.*, 1972), increasing interest is being shown in the magnetism of lunar rocks because of the evidence accumulating that the NRM was acquired in an ancient lunar magnetic field. The possible origin of such a field in a small, molten lunar core by a mechanism similar to that which generates the Earth's field is also receiving increasing attention. An important part of the investigations continues to be the study of the NRM which is described in this paper, together with some work on the magnetic mineralogy of selected samples.

The rock samples which have been investigated are chips from 15085, 15499, 15555, 60015, 62235, and 68416 (crystalline), 15459, 66055, and 67915 ('rock' breccia) and 15086 (soil breccia).

2. Remanent Magnetism

2.1. Apollo 15

Figure 1 shows the characteristics of the NRM of the samples during alternating field (A.F.) demagnetization. Of the crystalline samples, 15085,31 showed approximately constant directions of NRM after removal of the soft components, but 15499,21 showed a steady movement with increasing field; thermal demagnetisation up to 500°C of a chip from 15085,32 produced the same general behavior in the NRM direction as in the A.F. treatment, but at higher temperatures the directions became scattered, and no stable direction of NRM was revealed by thermal demagnetization of 15499,27. The persistence of NRM up to 750°C in both samples indicates that iron is carrying the remanent magnetization. Although

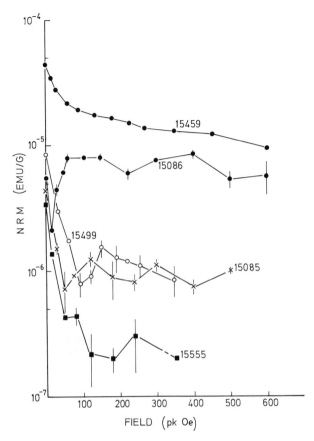

Fig. 1. Alternating field demagnetization of Apollo 15 samples. Vertical bars indicate range of intensities obtained after repeated demagnetization.

there is some evidence of a very weak hard NRM in 15555,75 the directions became very scattered after 75 Oe, due in part to measurement errors because of low signal levels.

Both the breccia samples, 15086,12 and 15459,95 show different behavior to the crystalline rocks on A.F. demagnetization, in the absence of strong soft components of NRM and the high intensity of the hard magnetization. Both samples show a stable direction of NRM above 30 and 100 Oe respectively, with 15459,95 apparently showing further movement above 500 Oe (Fig. 2). Thermal demagnetization of 15459,96 again indicated iron as the carrier of the NRM and an approximately stable direction between 100 and 650°C; at higher temperatures there is a further change in NRM direction which suggests, with the A.F. demagnetization, that there is a second, very hard component of magnetization in this rock.

In samples 15085, 15086, 15499, and 15555 there is some evidence for anomalous variations in intensity during A.F. demagnetization, although not as marked as

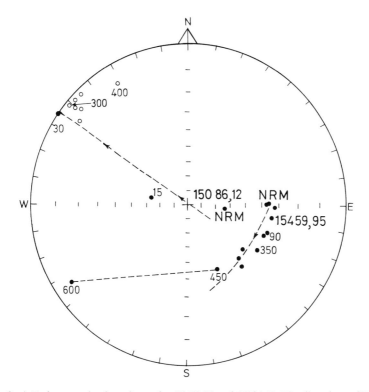

Fig. 2. A.F. demagnetization of samples 15459,95 and 15086,12. The directions of NRM
are referred to arbitrary axes in the rocks.

in some rocks from previous missions (Strangway *et al.*, 1971; Collinson *et al.*, 1972); this feature is discussed in Section 4.

The acquisition of viscous remanent magnetization was investigated in the samples, the results falling into two distinct groups. The applied field was 1.0 or 2.0 Oe, and the growth was proportional to the logarithm of time. A strong VRM was acquired by 15086,12 and 15459,95, intensities of 130 and 214×10^{-6} emu.g^{-1} respectively being acquired in one week; Table 1 shows the viscosity coefficients obtained.

2.2. Apollo 16

The anorthosite sample 60015,49 has a barely detectable initial susceptibility ($<2 \times 10^{-6}$ emu.g^{-1}.Oe^{-1}) and a very weak remanence of 1.1×10^{-6} emu.g^{-1}; on A.F. demagnetization this was reduced to less than 2×10^{-7} emu.g^{-1} without significant change of direction in a field of 90 Oe, and therefore does not appear to possess a measurable hard NRM. As expected, the saturation isothermal remanent magnetization (IRM) is also low, 0.76×10^{-4} emu.g^{-1}.

Sample 68416,23, a gabbroic anorthosite, had an initial NRM of $2.0 \times$

Table 1. Some properties of Apollo 15 samples. Initial susceptibility, χ; saturated IRM, I_0; decay of saturated IRM, $(100/I_0)(\partial I_0/\partial \log t)$; rate of acquisition of VRM in a field of 1 Oe, $\partial I_t/\partial \log t$; induced magnetization, J_i. Dashes indicate samples not measured and sample sub-numbers are shown in brackets.

Rock	15101,75	15459,95,96	15086,12	15085,31,32	15499,21,27	15555,75,92
Description	Fines	Rock Breccia	Soil Breccia	Basalt	Basalt	Basalt
χ (10^{-3} emu.g^{-1}.Oe^{-1})	2.2	1.0(95)	0.82	0.10(32)	0.075(27)	0.075(92)
I_0 (10^{-3} emu.g^{-1})	60	68(95)	21	0.65(32)	0.67(27)	0.70(92)
$\dfrac{100}{I_0}\dfrac{\partial I_0}{\partial \log t}$	2.9%	3.1%(95)	3.6%	—	—	—
J_i(8 KOe)(emu.g^{-1})	—	0.86(96)	—	0.48(32)	0.37(27)	0.42(92)
$\dfrac{\partial I_t}{\partial \log t}$ (10^{-6} emu.g^{-1})	—	47	33	0.30(31)	0.70(21)	0.08(75)

10^{-6} emu.g^{-1}; after removal of a soft component in a field of 60 Oe with a rise of intensity to 3.0×10^{-6} emu.g^{-1} the intensity then decayed steadily to about 6×10^{-7} emu.g^{-1} at 500 Oe, the direction remaining constant. There thus appears to be a hard NRM in this rock nearly opposed in direction to the soft one and of initial intensity of about 2.0×10^{-6} emu.g^{-1}.

The breccia, 67915,47, showed very anomalous intensity variations during A.F. demagnetization, varying between 1.2 and 4.0×10^{-6} emu.g^{-1} up to 400 Oe with an initial value of 3.2×10^{-6} emu.g^{-1}. Some stability of direction was apparent between 15 and 300 Oe but in higher fields considerable movement took place combined with poor repeatability at the same demagnetizing field. A chip from 67915,49 had an initial NRM of 3.2×10^{-6} emu.g^{-1} which was reduced to 0.4×10^{-6} emu.g^{-1} in 30 Oe demagnetizing field and became too weak to measure in higher fields.

More detailed investigations were carried out on samples 62235,53 and 66055,10, in particular regarding homogeneity of the NRM.

66055,10. This is a coarse breccia with a whitish matrix and dark grey clasts typically several millimetres in size. It was divided into 4 chips, 66055,10A, B, F, and R in order to study the homogeneity of the NRM of the original piece and to compare the behavior of the chips during A.F. demagnetization. The intensity of NRM of the original piece (O) was 18.1×10^{-6} emu.g^{-1} and that of the chips was 30.0, 7.9, 7.3, and 15.6×10^{-6} emu.g^{-1} respectively. Fig. 3 shows the direction of NRM of the original and the chips (referred to a common arbitrary direction); they appear to be in reasonable agreement considering the difficulty of preserving a common orientation in each chip after division from the original. There is no obvious correlation between NRM intensity and the proportion of dark clasts in the chips.

Chips A, F, and R all show the presence of a dominant, soft component of magnetization; chip A did not reveal a hard component before becoming too weak to measure, but F and R possessed a hard NRM of intensity about 1.0 and 2.0×10^{-6} emu.g^{-1} respectively, persisting up to 300 Oe. The changes in direction of NRM in the three chips during demagnetization are also shown in Fig. 3, and they are seen to be significantly different.

Although a complete explanation of this behavior is not clear, a possible cause is that the breccia was not heated above the iron Curie point ($\sim 780°C$) during its formation, and the constituent rock fragments were not completely remagnetized in any field then present. The different behavior of the chips can then be explained by the residual hard magnetization of the rock fragments being randomly oriented in the breccia, and by the presence of one or more components of NRM of differing hardness within each of the chips. Sometime between the formation of the breccia and the present it acquired a soft, relatively strong remanence, thus providing the directionally homogeneous NRM observed in the original rock. The variation in intensity is presumably due to variation in the iron content of the chips, and also possibly to the presence of opposed directions of NRM components, e.g., in chip R. The above general explanation may also account for the different behavior of 67915,47 and 67915,49.

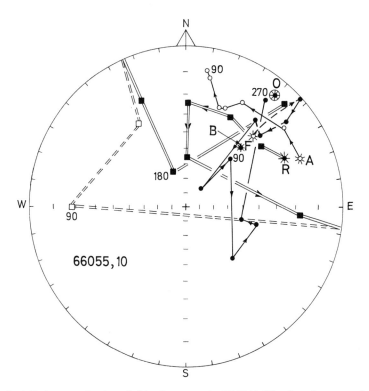

Fig. 3. A.F. demagnetization of chips from sample 66055,10. The directions are referred to a common reference direction in the original rock; full and open circles refer to positive and negative inclinations respectively.

62235,53. This crystalline sample was divided into five chips, A, B, C, D, and R in which the initial NRM intensity was 169, 188, 107, 137, and 133×10^{-6} emu.g^{-1}. The directions of NRM in the chips were close to the direction shown by the original rock piece (OR) (Fig. 4). On A.F. demagnetization the direction in chip D remained nearly constant up to 400 Oe while its intensity decreased to 20×10^{-6} emu.g^{-1}; between 400 and 1200 Oe its direction changed through about 30°, its intensity remaining at about 15×10^{-6} emu.g^{-1} at 1200 Oe. This suggests a component of very high stability in the chip, and this is confirmed by thermal demagnetization of chip A, in which the direction of NRM remained almost constant up to 600°C and then moved through nearly 100° on further heating to 750°C. The relative change in NRM direction was similar to that in chip D. There thus appears to be two stable components of magnetization in this sample, one of which is not only extremely hard, but also of unusually high intensity.

Since the direction of NRM is constant up to about 400°C and the less hard component of NRM is dominant in this region, it seemed worthwhile to test for thermoremanence in this sample by comparison of the loss and gain of TRM in different temperature intervals and possibly to derive a value for the lunar field

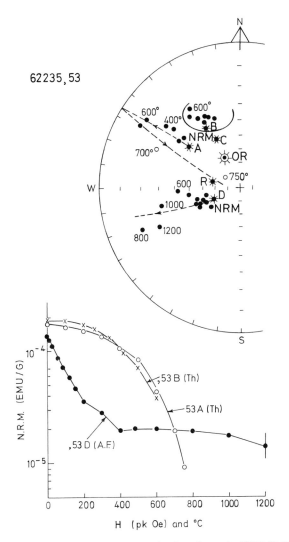

62235,53

Fig. 4. Alternating field and thermal demagnetization of sample 62235,53. The directions
are referred to a common reference direction in the rock.

intensity; this method is now widely used in palaeomagnetism and is based on the
technique originally developed by Thellier and Thellier (1959).

Chip B from the sample was heated to successively higher temperatures in
field free space, as in conventional thermal demagnetization; at each temperature
the PTRM gained by the rock in cooling from that temperature to 20°C in a known
field was measured before heating to the next higher temperature. The result is
shown in Fig. 5. There is strong evidence for a thermoremanent origin of the NRM
of this sample, acquired in a lunar magnetic field of 1.2 Oe. The applied field was

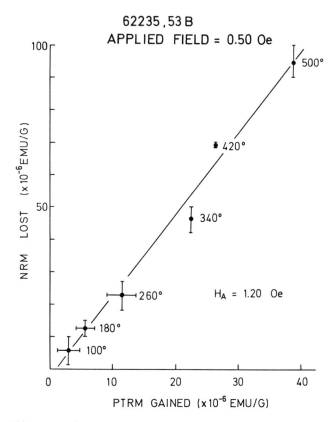

Fig. 5. Field intensity determination on sample 62235,53B. The vertical bars show the range of NRM lost after heating to a given temperature, taken from the two thermal demagnetization curves of Fig. 4. Up to 260°C, a repeat PTRM was determined at each temperature, giving the range shown by the horizontal bars.

0.50 Oe and the heating was carried out in a vacuum of better than 10^{-4} torr. Changes in the rock were monitored by measuring the initial susceptibility and low field (10 Oe) IRM after each heating; the latter changed significantly after heating to 600°C, although the susceptibility remained constant up to 800°C.

3. ROCK MAGNETISM—APOLLO 15

The six samples which were investigated could be divided into two distinct groups. The first group comprised the fines, 15101,75, a rock breccia from Spur Crater 15459,95 and 96, and the soil breccia 15086,12. The second group consisted of crystalline samples 15499,21 and 27,15555,75 and 92, and 15085,31 and 32. The breccia group were characterised by initial susceptibilities about an order of magnitude higher than those of the crystalline group. The saturated IRM of this group were also much higher (Table 1).

3.1. *Breccia group*

A sample of the fines, 15101,75, was investigated to determine whether it contained the single domain grain distribution present in Apollo 11 fines, and also in Apollos 12 and 14. Detailed study of Apollo 11 fines (Stephenson, 1971) has indicated that a distribution law $N(v) = \beta v^{-n}$, where v is the volume of the grain, β is a constant and $N(v)dv$ is the number of iron grains within the volume range v to $v + dv$, provided a good basis for quantitatively explaining a wide range of magnetic measurements. In Apollo 11 samples, n was found to be 2 to the nearest integer, and may be determined from two measurements. The first involves giving the sample a TRM as it is cooled to $-196°C$ in a low field and then measuring the decay curve as the sample is warmed to room temperature in zero field; the second is the measurement of the partial decay of the saturated IRM acquired at $-196°C$ as the sample is warmed to room temperature.

To improve the accuracy of the determination of n for the Apollo 15 fines a larger mass of sample was used (420 mg) and a higher field (0.8 Oe) to produce the TRM. Figure 6 shows the results of the two measurements described above. The curves drawn through the experimental points are both of the form $1n\ (293/T)$ which is the theoretical variation expected for $n = 2.0$ (Stephenson, 1971). The TRM (I_T) varies according to the equation $I_T = 1.67 \times 10^{-3}\ 1n\ (293/T)$ and the IRM (I_R) as $I_R = 0.060 + 0.0654\ 1n\ (293/T)$. From these two results H_c is calculated to be

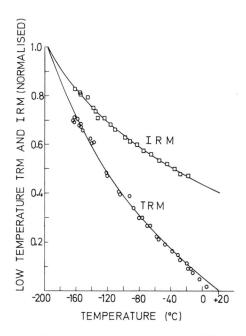

Fig. 6. Decay curves obtained on warming sample in zero field from $-196°C$ of (a) TRM acquired between 20 and $-196°C$ in 0.8 Oe, and (b) saturated IRM acquired at $-196°C$. Intensities at $-196°C$ are 2.24×10^{-3} emu.g^{-1} (TRM) and 0.186 emu.g^{-1} (IRM).

1100 Oe, using the equation

$$H_c = 26\left(\frac{4h_T}{3}\right)\left(\frac{\partial I_R}{\partial T}\Big/\frac{\partial I_T}{\partial T}\right)$$

where h_T is the field producing the low temperature TRM. This value of H_c is of the same order as that found in Apollo 11 fines (1700 Oe). There was also some evidence for this distribution in the other two samples of this group.

The variation of initial susceptibility, χ, with temperature over the range -196 to 800°C was very similar for the fines and 15459 (15086 was not measured). Both gave a curve similar to that obtained for the Apollo 11 fines (Runcorn *et al.*, 1970), with an iron Curie point. After heating to the Curie point in air, 15459,95 developed a secondary Curie point at 333 ± 4°C on cooling. This was confirmed by a second heating of the sample and is possibly due to the formation of pyrrhotite from troilite. The susceptibility of this second phase was about 15% of the susceptibility due to the iron grains, which was 1.0×10^{-3} emu.g^{-1}.Oe^{-1} at 20°C.

All three breccia group samples exhibited rapid acquisition of VRM at room temperature and decay with time of their saturated IRM (Table 1).

3.2. Crystalline group

All three samples exhibited induced magnetization curves which were almost identical and which failed to saturate in 8 kOe, the slope of the graphs at this field being in all cases 31×10^{-6} emu.g^{-1}.Oe^{-1}, still a significant fraction of the initial susceptibility (Table 1). The $J_i - T$ curves for samples 15085,32 and 15555,92 in a field of 4.2 kOe both clearly showed an iron Curie point, with the latter showing the presence of a paramagnetic component above the Curie point.

15085,31 and 15555,72, when given a TRM between room temperature and -196°C, both exhibited a sudden decrease in magnetization at about -150°C when warmed to 20°C (Fig. 7) and a synthetic single crystal whose composition was very near to ulvöspinel gave almost identical results. The basalt sample 15499 did not show this result. On cooling sample 15555,72 to -196°C after giving it a saturation IRM, and then re-warming, an irreversible decrease in magnetization occurred which was not associated with any memory effect, thus eliminating the possibility of magnetite being responsible. This was further checked by warming the demagnetized sample from -196 to 20°C in a low field, when no transition TRM was produced. It is to be noted that the ulvöspinel in the lunar samples can not be detected by measuring the variation of saturation magnetization with temperature. This is because the contribution due to the iron which is present is very much greater than that due to the ulvöspinel.

Ulvöspinel–chromite of a wide range of composition and containing other ions has been found by Haggerty (1973). Although this provides confirmatory evidence for the presence of ulvöspinel it is rather surprising that a single Curie point should be obtained, since the Curie points of ulvöspinel–chromite solid solutions cover the range -153 to -193°C (Schmidbauer, 1971). A value close to -193°C for chromite was also found by Francombe (1957). The Curie point of 150°C reported

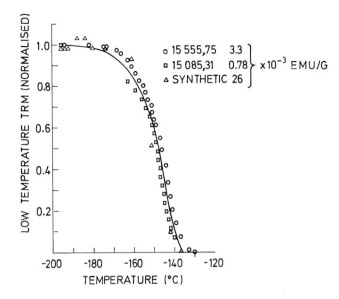

Fig. 7. Decay curves obtained by warming samples in zero field from $-196°C$ for two Apollo 15 samples and a synthetic titanomagnetite crystal very near ulvöspinel in composition. Intensities are given at $-196°C$.

by Banerjee (1972) for a synthetic titanochromite of composition $Fe^{2+}_{1.14}Cr^{3+}_{1.72}Ti^{4+}_{0.14}$ is therefore not in agreement with Schmidbauer's results, which indicate a Curie point of about $-173°C$ for the above composition. Because of the considerable difference in the strengths of the superexchange interactions, on which the Curie point depends, between samples containing only Fe^{2+} ions as carriers of moment (e.g., $Fe^{2+}[Fe^{2+}Ti^{4+}]O_4$, $T_c = -120°K$ $(-153°C)$) and Fe^{3+} ions (e.g., $Fe^{3+}_{0.89}Mg^{2+}_{0.11}[Fe^{3+}_{1.11}Mg^{2+}_{0.89}]O_4$, $T_c = 715°K$ $(442°C)$), it seems possible that the relatively high Curie point ($423°K$ ($150°C$)) of Banerjee's sample is due to a phase containing Fe^{3+}. The 'transition' observed at $100°K$ $(-173°C)$ may possibly be the Curie point of the unoxidized material.

The spinels which Haggerty has analysed, and which do not contain Fe^{3+} are thus incapable of carrying a remanent magnetization at $20°C$ and cannot therefore influence palaeofield determinations or be confused with magnetite in low temperature studies, as suggested by Banerjee.

3.3. ARM experiments

These were carried out in an attempt to obtain estimates of ancient lunar field intensity by an alternative method to that described in Section 2.2. Samples were given an anhysteretic remanent magnetization (ARM) in a low field in a peak alternating field of up to 1360 Oe; this was in most cases sufficient to almost saturate the ARM. The coercivity spectra of the ARM and NRM (previously determined) were then compared. Preliminary experiments done on Apollo 15 samples showed that the ARM produced in a field of 0.6 Oe was, in the three crystalline samples, within a factor of 2 of the NRM. The two types of magnetization showed comparable stability but the method at that time did not permit an ancient field determination. This method is currently being refined and is now giving promising results on Apollo 16 samples. These indicate ancient lunar fields of between 0.1 and 1.5 Oe and, in particular, support the high value found by the Thellier method in sample 62235,53 (Section 2.2). These experiments will be described in detail elsewhere.

4. Discussion

The Apollo 15 samples show a sharp division of properties between the crystalline rocks and the breccias, the latter having many of the properties associated with the Apollo fines. The breccias also show an abnormally high intensity and stability of NRM and 15459 is quite different magnetically from the rock breccias investigated by the authors from Apollo 14 (Collinson *et al.*, 1972), the latter having much weaker rock-magnetic properties and a weaker hard component of NRM; although this probably in part reflects a smaller amount of iron in the Apollo 14 material, there also appears to be a different grain size distribution.

Evidence continues to accumulate from both the Apollo 15 and 16 samples of hard components of NRM and for the presence of a magnetic field on the moon at the time the rocks were formed. The cause of the apparent intensity variations during A.F. demagnetization of some samples is still not clear, although the most likely explanation appears to be one based on the existence of more than one magnetization event in the rocks concerned. The most likely origin of the very soft component of NRM observed in some rocks appears to be magnetic fields in the spacecraft during the missions (Strangway, 1973) which account for the NRM which is eliminated in demagnetizing fields up to about 30 Oe. Higher fields usually reveal a range of coercivities, and where an NRM* direction (e.g., 60015,49), it is likely to have been acquired at or soon after the time of formation of the rock along with any harder component of NRM which it is natural to suppose is a primary thermoremanent magnetization.

It was suggested by Runcorn *et al.* (1970) that the lunar magnetic field was generated in a small iron core, the existence of which was proposed by Runcorn (1967). The subsequent disappearance of the field since 3200 m.y. ago could then either result from the solidification of the core or its magnetic Reynolds number falling below the critical value for dynamo action. An alternative explanation of a lunar dipole field is that it had its origin in the permanent magnetization of the deep interior of the moon, a relic of a process involving an interplanetary magnetic field at the time of origin of the solar system (Runcorn and Urey, 1973). In this case the absence of a present dipole field is attributed to the heating of the moon's deep interior above the iron Curie point sometime during the last 3200 m.y.

---

*Of intermediate stability is identified which on further demagnetization undergoes little or no change of NRM.

Some evidence about the ancient lunar field intensity which is now emerging may be relevant here. Early estimates based on the Thellier method, e.g., that of Helsley (1970) suggested an intensity of between 10^3 and $10^4\gamma$; more recently Gose et al. (1973) obtained a value of $2100\pm80\gamma$ from breccia sample 15498,36. The intensity determination reported by the present authors for sample 62235,53 gives a field some 50 times stronger than those previously found; although further work in this field is desirable this high field is apparently based on good evidence and does suggest that the lunar field may have been highly variable in intensity. This would be easier to explain by a dynamo origin of the field than by the Runcorn–Urey theory, which postulates lunar rocks acquiring a magnetization in a constant surface field arising from the permanent magnetization of the lunar interior.

Variations in the intensity of the lunar field may also explain another feature of some of the Apollo samples, namely the very weak or apparently non-existent hard components of NRM in samples in which it can be shown that grains are present which are capable of holding such an NRM. Examples of this behaviour are the basalt 12020,23 (Runcorn et al., 1971), 15555,75, and 60015,49. In the latter, for instance, although the NRM was no longer measurable after demagnetization in 90 Oe, respectively 28% and 20% of the saturated IRM remained at demagnetizing fields of 300 and 500 Oe. Thus grains of high magnetic stability are present but they may not contribute any of the NRM because of a very weak magnetizing field, although since there is clearly very little iron in this rock there may be a hard component that is too weak to detect.

The small angles that often occur between components of intermediate and high stability (e.g., 14066,24, 14318,31 (Collinson et al., 1972), and 15459,95) can be explained by supposing that some changes in the magnetic grains take place after the initial cooling of the lavas or breccias from above the Curie point, allowing a further magnetization to be picked up perhaps of the order of 10^2–10^4 years later. Alternatively, in the case of the lavas, some re-heating may occur through contact with new flows and a PTRM acquired by some grains in the rock. Thus there is possible evidence here for secular variation of the field which, if substantiated, would more easily be explained on the dynamo theory.

Acknowledgments—One of the authors (A.S.) is in recept of a Senior Research Associateship from the Natural Environment Research Council. We are grateful to NASA for providing the lunar samples under the Lunar Sample Analysis Program, British participation in which is facilitated by the Science Research Council.

REFERENCES

Collinson D. W., Runcorn S. K., Stephenson A., and Manson A. J. (1972) Magnetic properties of Apollo 14 rocks and fines. Proc. Third Lunar Sci. Conf., Geochim. Cosmochim. Acta, Suppl. 3, Vol. 3, pp. 2343–2361. MIT Press.
Francombe M. H. (1957) Lattice changes in spinel-type iron chromites. J. Phys. Chem. Solids 3, 37–43.
Gose W. A., Strangway D. W., and Pearce G. W. (1973) A determination of the intensity of the ancient lunar magnetic field. The Moon. In press.

Haggerty S. E. (1972) Chemical characteristics of spinels in some Apollo 15 basalts (abstract). In *Lunar Science—IV*, pp. 92–97. The Lunar Science Institute, Houston.

Helsley C. E. (1970) Magnetic properties of lunar 10022, 10069, 10084, and 10085 samples. *Proc. Apollo 11 Lunar Sci. Conf., Geochim. Cosmochim. Acta*, Suppl. 1, Vol. 3, pp. 2213–2219. Pergamon.

Pearce G. W., Strangway D. W., and Gose W. A. (1972) Remanent magnetization of the lunar surface. *Proc. Third Lunar Sci. Conf., Geochim. Cosmochim. Acta*, Suppl. 3, Vol. 3, pp. 2449–2464. MIT Press.

Runcorn S. K. (1967) Convection in the moon and the existence of a lunar core. *Proc. Roy. Soc. Lond. A.* **296**, 270–284.

Runcorn S. K., Collinson D. W., O'Reilly W., Battey M. H., Stephenson A., Jones J. M., Manson A. J., and Readman P. W. (1970) Magnetic properties of Apollo 11 lunar samples. *Proc Apollo 11 Lunar Sci. Conf., Geochim. Cosmochim. Acta*, Suppl. 1, Vol. 3, pp. 2369–2387. Pergamon.

Runcorn S. K., Collinson D. W., O'Reilly W., Stephenson A., Battey M. H., Manson A. J., and Readman P. W. (1971) Magnetic properties of Apollo 12 lunar samples. *Proc. Roy. Soc. Lond. A.* **325**, 157–174.

Runcorn S. K. and Urey H. C. (1973) New theory of lunar magnetism. *Science.* In press.

Schmidbauer E. (1971) Magnetization and lattice parameters of Ti substituted Fe–Cr spinels. *J. Phys. Chem. Solids* **32**, 71–76.

Stephenson A. (1971) Single domain grain distributions; I. A method for the determination of single domain grain distributions. II. The distribution of single domain iron grains in Apollo 11 lunar dust. *Phys. Earth Planet. Interiors* **4**, 353–360, 361–369.

Strangway D. W., Pearce D. W., Gose W. A., and Timme R. W. (1971) Remanent magnetization of lunar samples. *Earth Planet. Sci. Lett.* **13**, 43–52.

Strangway D. W., Gose W. A., Pearce D. W., and Carnes J. G. (1973) Magnetism and the history of the moon. *J. Appl. Phys.* In press.

Thellier E. and Thellier O. (1959) Sur l'intensite du champ magnetique terrestre dans le passe historique et geologique. *Ann. Geophys.* **15**, 285–376.

Proceedings of the Fourth Lunar Science Conference
(Supplement 4, *Geochimica et Cosmochimica Acta*)
Vol. 3, pp. 2977–2990

Magnetic properties and granulometry of metallic iron in lunar breccia 14313

D. J. DUNLOP*

Lunar Science Institute, Houston, Texas 77058 and
Department of Physics, University of Toronto, Toronto 5, Canada

W. A. GOSE

Lunar Science Institute, Houston, Texas 77058

G. W. PEARCE

Lunar Science Institute, Houston, Texas 77058 and
Department of Physics, University of Toronto, Toronto 5, Canada

D. W. STRANGWAY

Geophysics Branch, NASA Manned Spacecraft Center
Houston, Texas 77058

Abstract—Based on a detailed study of time-dependent or viscous remanence (VRM), thermoremanence (TRM) and magnetic granulometry of soil breccia 14313, single-domain particles of iron 100–200 Å in size are proposed as the major carriers of natural remanence (NRM) in this rock. The VRM of 14313 is unusually intense and exhibits a logarithmic time decrease of VRM which ceases fairly abruptly after a time about equal to the original exposure to the field, unlike crystalline rocks and high-grade breccias whose logarithmic decay persists to extremely long times. The partial TRM spectrum reveals both a high-blocking-temperature fraction, scarcely affected by AF demagnetization to 1000 Oe, and an unusual concentration of blocking temperatures just above room temperature. The former fraction would contribute a very hard and stable component to any NRM of lunar origin, but the latter fraction, which accounts for the pronounced VRM of 14313 and undoubtedly has imparted a large viscous NRM component in the earth's field, is also surprisingly hard. A substantial portion (20–40%) is not demagnetized by an 800 Oe field. Metallic iron in a single-domain state is the only plausible carrier of such hard remanences and magnetic granulometry using partial TRM data indicates that two distinct populations are involved. The first consists of 120–200 Å roughly equidimensional particles. The number of particles of volume v in this population varies as $v^{-1.8}$, in good agreement with the v^{-2} dependence found by Stephenson for an Apollo 11 soil, and the distribution appears to peak in the superparamagnetic range below 100 Å. The second population contains elongated particles $\leqslant 100$ Å in size, whose microscopic coercive forces are distributed right up to 10,000 Oe, the shape anisotropy limit for single-domain needles of iron.

---

*Visiting Scientist at LSI.

Table 1. Symbols and abbreviations used in text and illustrattions.

SPM	superparamagnetic
SD	single-domain
MD	multidomain
J	magnetization
J_0	initial value of magnetization
J_r	remanent magnetization or remanence
VRM, J_{vr}	viscous remanence
TRM	thermoremanence
partial TRM	partial thermoremanence
NRM	natural remanence
AF	alternating-field (e.g. AF demagnetization)
τ	relaxation time
τ_{min}, τ_{max}	minimum, maximum relaxation time
t	time
t_a	acquision time or time of exposure to a field
t_0	lapse time between field suppression and VRM measurement
t_d	decay time (an extended lapse time)
S	viscosity coefficient, $dJ_{vr}/d \log t$
S_a	acquisition coefficient $dJ_{vr}/d \log t_a$
S_d	decay coefficient, $dJ_{vr}/d \log t_d$
k	Boltzmann's constant
Q	a near constant of order 25 in cgs units
T	temperature
T_B	blocking temperature
J_s	spontaneous magnetization
v	particle volume
H_c	particle critical field or microscopic coercive force
subscript 0 on T, J_s and H_c denotes room temperature	
$f(v, H_{c0})$	distribution of remanence with respect to particle volume and coercive force

INTRODUCTION

LUNAR BRECCIAS contain metallic iron of three general types: igneous, meteoritic, and interstitial. The last type of iron is thought to result from solid-state reduction of glasses and silicate minerals during the heating that follows a major impact (Pearce *et al.*, 1972). Because the size range of the metallic iron is very broad (< 100 Å to $> 100 \mu$m), breccias contain, in varying proportions, particles displaying the three classic types of ferromagnetic behavior, superparamagnetic (SPM), single-domain (SD), and multidomain (MD). Breccias of low metamorphic grade contain a substantial fraction of SD iron just above SPM size and consequently may exhibit a short-term viscous remanence (VRM) comparable in intensity to their stable thermo remanence (TRM). As a result, separation of primary and secondary components of natural remanence (NRM) is particularly difficult in these rocks and a prime incentive of the present study was the hope of establishing optimum demagnetization techniques for extracting a small "signal" from a large viscous "noise" component.

Sample 14313 was chosen for detailed magnetic investigation because its basic properties were well documented (Gose *et al.*, 1972a) and appeared to be representative of those of the abundant low-grade breccias returned by Apollo 14 and 15. 14313 is a fine-grained breccia with a detrital matrix and abundant glass. Opaques are mainly interstitial and of sub-micron size. It belongs to the lowest metamorphic grade in the classification of Warner (1972).

Gose *et al.* (1972a) have reported evidence of a stable component of NRM in chip 14313,25. During AF demagnetization between 0 and 100 Oe, the intensity of a superimposed VRM produced by 51 days storage in the earth's field fell rapidly from its initial value (about twenty times the stable NRM intensity) and the remanent vector swung along a great circle. However, both intensity and direction seemed to stabilize *short* of the inferred direction of the stable NRM after demagnetization between 100 and 200 Oe, suggesting the existence of a component of VRM which is relatively resistant to AF demagnetization.

Viscous decay of shorter-term VRM's (acquisition times t_a of 8 min to 5 days) has been investigated for the same chip (Gose *et al.*, 1972a; Gose and Carnes, 1973). The decay curves are initially linear in $\log t$ and have pronounced tails extending to times as great as $100\ t_a$. A calculation based on the Richter (1937) and Néel (1949) theories of magnetic viscosity suggested that SD iron particles in the range 120 to 160 Å carry these VRM's.

The SD interpretation of VRM decay was confirmed by the rounded SD-type hysteresis curve of 14313,25 (Gose *et al.*, 1972a,b). The relatively high value of reduced saturation remanence, J_{rs}/J_s, (0.066) likewise reflects the importance of SD particles (Gose *et al.*, 1972b).

The VRM and TRM experiments described in the following sections were performed on a companion chip, 14313,29. TRM's were induced in a furnace in which a vacuum of 5×10^{-6} torr or better could be maintained at temperatures up to 800°C. It is therefore unlikely that significant oxidation took place during initial heating to 800°C (this heating preceded the main sequence of experiments and was designed to stabilize the magnetic properties) but some sintering of SD material, or possibly evaporation of fine particles, evidently occurred in the process since the VRM of 14313,29 after heating was several times less than the corresponding VRM of 14313,25. Magnetization measurements were made with a PAR spinner magnetometer and AF demagnetization was carried out with a Schonstedt 400 Hz demagnetizing coil. The furnace, demagnetizing coil and signal pickup assembly for the spinner were located in a shielded room where the ambient field was reduced to the gamma range. All experiments were performed and the sample was stored in the same room.

VISCOUS REMANENT MAGNETIZATION (VRM)

The acquisition and decay of VRM in sample 14313,29 were studied over a time scale of a few seconds to one or two days. As Fig. 1 shows, the intensity of VRM acquired in a field of 2.5 Oe for acquisition times $t_a = 8$ min, 2 hr and 24 hr increases as $\log t_a$. Because of experimental limitations, the VRM cannot be measured at the instant the field is switched off but only beginning a few seconds

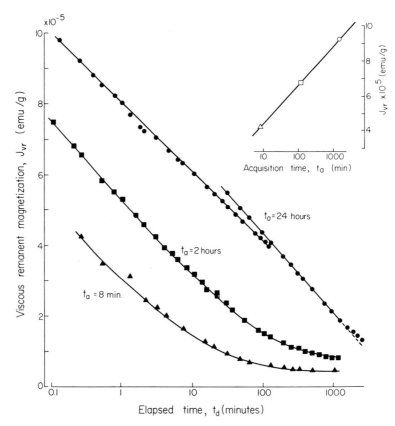

Fig. 1. The acquisition and decay of VRM in breccia 14313,29. VRM was acquired in a field of 2.5 Oe applied for times of 8 min, 2 hr, and 24 hr. The sample was AF demagnetized before each experiment, except in the case of the lower t_d segment of the t_a = 24 hr curve (*see* text). Decay curves are essentially linear in log t_d, up to $t_d \sim t_a$ and become non-linear for longer t_d. The acquisition curve is linear in log t_a, with a slope numerically equal to the initial slopes of the decay curves (*see* Table 1).

later, when some viscous decay has already occurred. The three VRM's shown in Fig. 1 (inset) were measured at a decay time $t_d = 15$ sec, the first instant at which accurate readings of remanence were available in all three experiments. Except for one experiment discussed later in this section, the sample was AF demagnetized to 800 Oe peak field before beginning each new VRM experiment. Any residual moment remaining after demagnetization was subtracted from the total remanence measured at 15 sec and subsequent times.

The viscous decay of the above three VRM's was recorded continuously to a maximum decay time of 24 hr (8 min and 2 hr VRM's) or 48 hr (24 hr VRM). As more detailed experiments of the same type have since been carried out by Gose and Carnes (1973), the results shown in Fig. 1 will be discussed very briefly. The decay curves can be explained satisfactorily by the Richter (1937) theory of magnetic after-effect, which postulates a uniform distribution of $\log \tau$ (τ is relaxation time) between limits τ_{min} and τ_{max}. For t_d's well below τ_{max}, VRM decays linearly with t_d, while for t_d's well above τ_{max}, the decay goes as $t_d^{-1} \exp(-t_d/\tau_{max})$.

Since the $t_a = 24$ hr curve is linear in $\log t_d$ over almost the complete range shown, while the $t_a = 8$ min curve is essentially exponential and the $t_a = 2$ hr curve shows both types of behavior, it is clear that τ_{max} is not limited by the availability of iron particles with long relaxation times. Rather it is the acquisition time t_a which limits the effective τ distribution in each case, since τ's greater than t_a are not activated to any great extent in the original VRM.

It is easy to demonstrate from the data in Fig. 1 that $\tau_{max} \approx t_a$. One can locate τ_{max} approximately by noting that the functions $-\log t_d$ and $t_d^{-1} \exp(-t_d/\tau_{max})$ have incompatible slopes near τ_{max}. Thus although there is a gradual transition between these laws, one can expect a maximum rate of change of slope near τ_{max}. We therefore fit the 8 min and 2 hr decay curves of Fig. 1 by fifth-order polynomials and used the best-fit coefficients to determine the second derivative curves of Fig. 2.

For $t_a = 8$ min, τ_{max} as indicated by the second derivative peak is about 12 min, and for $t_a = 120$ min, $\tau_{max} \cong 160$ min. For $t_a = 24$ hr, the decay curve does not extend to sufficiently long times to determine a maximum rate of change of slope, but the first departure from logarithmic decay occurs, as expected, just below $t_d = 24$ hr. It appears quite certain, therefore, that the maximum relaxation time observed in VRM decay for 14313,29 is determined by t_a and not by a limiting size of iron particles carrying remanence.

Values of the viscosity coefficient, $S = dJ_{vr}/d \log t$, are given in Table 2 for both the acquisition of VRM and the decay of the 2 hr and 24 hr VRM's. The value reported for the 2 hr VRM is of course the initial slope, up to $t_d \cong 3$ min. Figure 1 suggests that the 8 min decay curve initially parallels the 2 hr curve, but the linear region is so transitory and the data are so scattered that no figure is reported for S_d in this case. All VRM experiments in which 14313 was initially AF demagnetized yield the same numerical value, about 0.96×10^{-5} emu/g, for S. (The two segments of the 24 hr curve came about because the first half-hour of the original decay record was lost due to malfunction of the chart recorder and a new VRM was induced *without* first AF demagnetizing the sample. The decay curve of this latter VRM has a distinctly lower slope, $|S| = 0.855 \times 10^{-5}$ emu/g). Numerical equality of acquisition and decay coefficients, as observed here, is predicted by virtually all theories of magnetic viscosity.

THERMOREMANENT MAGNETIZATION (TRM)

Since the stable primary component of NRM of 14313 is believed to be of thermoremanent origin, TRM experiments can be expected to clarify the properties of the primary NRM and of the grains that carry it. Less obvious is the fact that the low blocking temperature fraction of TRM is carried by the same grains that are responsible for secondary NRM of viscous origin. For example, for SD grains in a small field, Néel (1949) has shown that the grains carrying a VRM

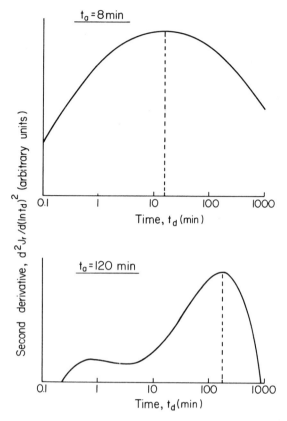

Fig. 2. The rate of change of slope of the two lower decay curves of Fig. 1 as a function of t_d. Zero or near zero values indicate classic log t_d decay. Maximum values, according to the Richter (1937) theory of VRM, are an indication of the upper limit τ_{max} of the relaxation time spectrum activated in the VRM. Experimentally, the maximum falls near

$$t_d = t_a.$$

Table 2. Viscosity coefficients determined from the data of Fig. 1. The acquisition coefficient $S_a = dJ_{vr}/d \log t_a$ and the decay coefficient $S_d = dJ_{vr}/\log t_d$.

Type of curve	S_a or $S_d \times 10^5$ (emu/g)
Acquisition	0.955
Decay, $t_a = 2$ hr	-0.960
Decay, $t_a = 24$ hr[a]	-0.975
Decay, $t_a = 24$ hr[b]	-0.855

[a]VRM acquired following AF demagnetization.
[b]VRM acquired without prior AF demagnetization.

acquired at room temperature T_0 over a time t_a and measured t_0 seconds later have values of the product vH_{co} (v is particle volume; H_{co} is the room-temperature value of particle coercive force, determined by the shape of the particle) lying between $(2kT_0/J_{s0})(Q + \log t_0)$ and $(2kT_0/J_{s0})(Q + \log t_a)$, where k is Boltzmann's constant, $Q \approx 25$ and J_s is spontaneous magnetization. A partial TRM acquired between T_1 and T_0 has vH_{co} lying between $(2kT_0/J_{s0})(Q + \log t_0)$ and $(2kT_1J_{s0}/J_s^2(T_1))(Q + \log t_0)$. The VRM and partial TRM are therefore carried by the same SD particle ensembles (and consequently will exhibit identical coercivity and blocking temperature spectra) if

$$T_1 \left[\frac{J_{s0}}{J_s(T_1)} \right]^2 = \frac{Q + \log t_a}{Q + \log t_0} T_0 \tag{1}$$

In addition, if the applied field was the same in both cases, the VRM and partial TRM will have equal intensities (a prediction that will be shown to be substantiated by experiment later in this section). We conclude that a low-temperature partial TRM is an analog of VRM or by extension, of secondary NRM of viscous origin.

The blocking temperature spectrum of 14313,29 is outlined in Fig. 3. Note that the ordinate in Fig. 3 is magnetization intensity per °C, thereby allowing for the

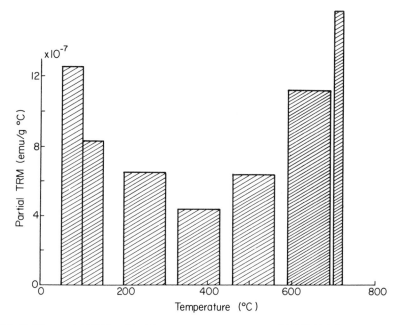

Fig. 3. The partial TRM or blocking temperature spectrum of 14313,29. The remanences have been normalized to take account of the differing temperature intervals over which the TRM's were produced. A notable feature is the abundance of very low blocking temperatures.

differing temperature intervals over which the partial TRM's were acquired. Partial TRM's between 720°C and the Curie point of iron (770°C) were not investigated because it was clear from the 720–710°C partial TRM that these remanences make no contribution to the coercivity spectrum of 14313 in the range of demagnetizing fields (0–1000 Oe) available to us (cf. Fig. 4). The blocking temperature peak above 700°C is a normal feature of ferromagnetic particle dispersions, and has its origin in the rapid decrease of J_s with T within a hundred degrees or so of the Curie point. The abundance of low blocking temperatures, particularly between 150°C and 50°C, is, however, a most unusual feature (in the context of terrestrial rocks) which seems to be characteristic of many lunar materials (Dunn and Fuller, 1972).

If the VRM/partial TRM analogy is real, the low-temperature peak in Fig. 3 explains the high intensity of VRM in 14313 (and other low-grade breccias). We can easily test this idea. From Fig. 1, the intensity of VRM for $t_a = 24\,\text{hr} = 86{,}400\,\text{sec}$, $t_0 = 0.1\,\text{min} = 6\,\text{sec}$ is about $10 \times 10^{-5}\,\text{emu/g}$. From Equation (1), assuming J_s does not vary appreciably between T_0 and T_1 (a reasonable assumption for iron), we have that

$$T_1 \cong \frac{Q + \log 86{,}400}{Q + \log 6}\, T_0 \cong 400°\text{K}$$

The temperature interval between T_1 and T_0 is thus about 100°C. We have no measured value of the partial TRM between 120°C and 20°C, but we do have a reasonable approximation to it, namely the sum of the partial TRM's from 150–100°C and 100–50°C, $(4.16 + 6.31) \times 10^{-5} = 10.47 \times 10^{-5}\,\text{emu/g}$. This figure is within 5% of the intensity of the "corresponding" VRM, providing strong evidence for the equivalence of VRM and low-temperature partial TRM. VRM and TRM (or secondary and primary NRM) are thus not distinct remanences carried by quite different SD grains: VRM is indistinguishable from the low-blocking-temperature fraction of TRM.

Since low-temperature partial TRM's (and by analogy VRM's) are extremely soft to thermal demagnetization, it might be thought that they should be similarly soft to AF demagnetization. Figure 4a shows that this is not the case. In the four lowest temperature partial TRM's AF stability does increase with increasing blocking temperature of the partial TRM in the normal way in the restricted AF range below 150 or 200 Oe, but throughout the much larger region above 200–250 Oe, the trend is exactly reversed. The curves themselves are of a very peculiar form, the 100–50°C and 150–100°C partial TRM's in particular having a pronounced plateau between about 150 and 600 Oe. The unusual behavior of 14313,25 (Gose *et al.*, 1972a) noted in the introduction, with a rapid initial swing of the total remanence vector toward the primary direction as the secondary NRM or a laboratory-induced VRM was AF demagnetized between 0 and 100 or 150 Oe, followed by an apparent stabilization of direction between 150 and 200 Oe, is probably a result of the plateau effect seen in Fig. 4a.

The observed form of the demagnetization curves of Fig. 4a strongly suggests the superposition of a component of low AF stability and one of high stability. If

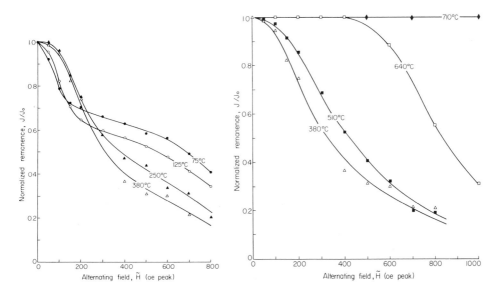

Fig. 4. AF demagnetization curves of the partial TRM's of Fig. 3. The mean blocking temperature is indicated on each curve. The low-temperature partial TRM's suggest the presence of two populations of iron particles, one relatively "soft" to AF demagnetization, the other relatively "hard." The high-temperature partial TRM's follow the normal trend of increasing coercivity with increasing blocking temperature.

the curves of Fig. 4a were replotted on an absolute scale, Fig. 3 suggests that the "soft" component would be of a similar magnitude in all four partial TRM's, while the hard component would be very much larger in the lowest temperature partial TRM's. Indeed, the hard component would be largely responsible for the unusual low-temperature peak in the blocking-temperature spectrum. It could also result in an unexpectedly hard component in any viscous secondary NRM acquired on earth which, if not recognized as such, would partially mask the primary lunar NRM. AF demagnetization is thus far from an ideal method of "cleaning" the NRM of low-grade lunar breccias.

It is interesting that viscous components having extreme resistance to AF demagnetization have been noted in some terrestrial sedimentary rocks (Avchyan and Faustov 1966, Biquand and Prévot 1971), where they are apparently carried by hematite grains with very high H_{co} but very small v. In the following section, we shall demonstrate that metallic iron grains with similar properties probably carry the hard component of partial TRM in 14313,29. Since the thermal process is equivalent to the viscous process, while the AF process is not, viscous components of high AF stability, if their presence is detected or suspected, can readily be erased, as in terrestrial red sediments, by thermal cleaning.

AF curves of the higher temperature partial TRM's shown in Fig. 4b, are simple in form, suggesting the presence of only one SD component, and follow the normal trend of increasing hardness with increasing blocking temperature

(Everitt, 1961, Dunlop and West, 1969). Within 150°C of the Curie point, coercivity increases very markedly, and the 720–700°C partial TRM is unaffected by AF's of 1000 Oe or less. This extreme hardness is not unexpected, since shape anisotropy of elongated iron particles can in principle yield microscopic coercive forces as high as $2\pi J_s \approx 10,500$ Oe.

MAGNETIC GRANULOMETRY

A method of deriving the function $f(v, H_{co}) \, dv \, dH_{co}$ describing the distribution of v and H_{co} in a particular assemblage of SD particles (in other words, the spectrum of particle sizes and shapes) from partial TRM demagnetization data has been described in detail by Dunlop (1965) and Dunlop and West (1969). Only an outline of the method will be given here. Note that f may be either a fraction of the total *number* of particles, or, as assumed here, a fraction of the total *remanence* of the sample. One can of course readily transform one distribution into the other.

The procedure used in our analysis is roughly as follows. A convenient set of 20–30 values of H_{co} was chosen, and for each, the minimum, maximum, and average values of v (corresponding to $(T_B)_{min}$, $(T_B)_{max}$ and $\langle T_B \rangle$ of the particular partial TRM under consideration) were calculated from the formula

$$v = 2kT_B J_{s0}(Q + \log t_0)/J_s^2(T_B)H_{c0} \tag{2}$$

where T_B is blocking temperature. Each $(\langle v \rangle, H_{co})$ is the midpoint of an area $\Delta v \Delta H_{co}$ (the sum of all such incremental areas constitutes the area occupied by the particular partial TRM on a v, H_{co} graph), with $\Delta v = v_{max} - v_{min}$ and ΔH_{co} equal to the increment between successive H_{co} values. The demagnetizing field $\tilde{H}$ corresponding to each $(\langle v \rangle, H_{co})$ is then found from

$$\tilde{H} = H_{c0} - \sqrt{2H_{c0}kT_0(Q' + \log t_0)/\langle v \rangle J_{s0}}$$

the first term being the classical coercive force and the second the fluctuation field. This equation can be simplified by substituting from Equation (2), and ignoring the small difference between Q and Q'. The result is

$$\tilde{H} = H_{c0} - H_{c0} \sqrt{\frac{T_0}{T_B}} \frac{J_s(T_B)}{J_{s0}} \tag{3}$$

From Figs. 3 and 4, the increments of remanence between successive values of $\tilde{H}$, hence associated with a particular $(\langle v \rangle, H_{co})$, were determined and the function $f(\langle v \rangle, H_{co})\Delta v \Delta H_{co}$ constructed. Figure 5 shows the "blocking curve" or locus of $(\langle v \rangle, H_{co})$ for each partial TRM of Fig. 4 together with contours of the remanence distribution function $f(v, H_{co})$ of 14313,29.

The distribution function has a number of interesting features. There is a relatively sharp truncation below $H_{co} = 300$–500 Oe, reflecting the fact that H_{co} has a well-defined minimum value for perfectly equidimensional SD particles set by the crystalline anisotropy of iron. There is a less well-defined cut-off between

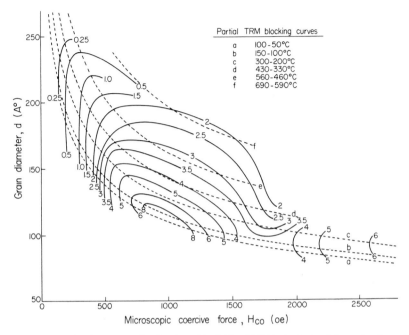

Fig. 5. The grain size-shape distribution, $f(v, H_{co})$ of 14313,29, derived from the data of Figs. 3 and 4 as described in the text. (H_{co} depends directly on particle shape.) Contour values are in arbitrary units. Mean blocking curves of the various partial TRM's are shown as dashed lines. The main distribution peaks just above 100 Å at an intrinsic coercivity of 1000 Oe. The hard component evident in Fig. 4a peaks off the diagram at very high H_{co}.

1500 and 2000 Oe, and between these limits $f(v, H_{co})$ is approximately independent of H_{co}. That is, the contours are roughly parallel to the H_{co} axis.

The distribution appears to have an upper limit near a particle size of 200 Å, which corresponds reasonably well to the critical SD size for iron. $f(v, H_{co})$ increases steadily with decreasing particle size, apparently peaking for a size just above 100 Å and a microscopic coercive force of about 1000 Oe, a reasonable value for only slightly elongated iron particles. The details of the distribution below this size are unknown, but there are certain to be abundant particles in this superparamagnetic region, in agreement with the hysteresis data for 14313.

Stephenson (1971a, b) has studied the size distribution of iron particles in an Apollo 11 dust sample and concludes that a variation as v^{-2} yields the best fit to a variety of experimental data. His result is rather similar to that presented in Fig. 5. $f(v, H_{co})$ for 14313 varies with v approximately as $v^{-0.8}$ according to the data of Fig. 5 and recalling that it is the remanence distribution that is shown, the corresponding number distribution would vary as $v^{-1.8}$.

In addition to the main part of the distribution discussed above, there is an entirely separate portion of the distribution lying above 2000 Oe and extending virtually to the shape anisotropy limit of $\simeq 10,500$ Oe. Because of its extended

form, only a small part is shown in Fig. 5. This distribution describes the population of particles carrying the "hard" component of partial TRM evident in Fig. 4a.
As noted in the previous section, this component is strongest in the lowest temperature partial TRM's, where because the fluctuation field is large, particles with
large values of H_{co} are susceptible to AF's in the range 0–1000 Oe. We are thus
able to state with some confidence that the "hard" component of partial TRM is
even harder in terms of intrinsic coercivities, having values of H_{co} throughout the
range 2000–10,000 Oe. The carriers of this component must thus be elongated,
very small (< 100 Å) iron particles, which would be superparamagnetic except for
their extremely high coercivities.

Grant et al. (1973) have described abundant single-crystal metallic iron spheres
in glasses from samples 10084,85 and 15101,92 whose size distributions bear a
striking resemblance to that of the main population in Fig. 5. As observed in
transmission electron micrographs, the sphere diameters range from 40 Å to 250 Å
with a peak in the mass distribution (which can be compared directly to the
remanence distributions of Fig. 5) near 125 Å. Equally important, Grant et al.
(1973) believe that iron of this type, resulting from subsolidus reduction of glasses,
can account for virtually all the excess metallic iron in lunar soils. Presumably the
same conclusion holds true for soil breccias.

It is also interesting that Tsay et al. (1971), in a computer simulation of the
characteristic ESR resonance of lunar fines, postulated a distribution of fine
($< 1~\mu$m) equidimensional iron particles. Thus similar populations of metallic iron
particles apparently have been detected by three very dissimilar techniques.

The second population shown in Fig. 5 appears to have been observed independently by Carter (1973) in the course of an SEM investigation of the morphology of glass-coated breccia 15015. Elongated to needle-like metallic iron crystals
with cross-section dimensions of 100–150 Å were clearly resolved and still finer
particles beyond the resolution of the SEM may well have been present. An
important point is the high degree of crystal perfection observed (Carter, 1973,
personal communication), for De Blois and Bean (1959) have shown that only iron
whiskers which are practically free of crystal imperfections exhibit coercivities
approaching the theoretical maximum. Thus it seems plausible that particles like
those seen by Carter could have H_{co} values well in excess of 2000 Oe.

DISCUSSION

Although multidomain iron undoubtedly comprises an appreciable fraction of
the magnetic material in 14313, its contribution to remanent properties (VRM and
TRM) seems to be slight. The assumption of single-domain particles made in
deriving the grain distribution of Fig. 5 from the partial TRM data of Figs. 3 and 4
is justified by the resulting distribution, which is concentrated well below the
SD–MD threshold for iron and continues into the superparamagnetic region below
100 Å. The viscous properties of 14313 are likewise consistent with a predominance of SD remanence carriers. As predicted by SD theory, the intensities of
24-hr VRM and 150–50°C partial TRM (both acquired in a field of 2.5 Oe) are very

nearly equal, and the acquisition and decay of VRM depend logarithmically on time, except where decay is observed over a time scale extending well beyond the acquisition time, t_a.

We identify the primary component of NRM of 14313 with TRM acquired between 770°C and perhaps 300°C in a lunar environment, and secondary NRM with a terrestrial VRM, which has erased any surviving lunar TRM with blocking temperatures lower than about 300°C. The viscous component is large compared to the primary NRM for two reasons. First, the lunar field at the time the NRM was acquired was only 1/10th to 1/20th the terrestrial field (Gose *et al.*, 1972c). Second, in common with many other lunar samples (Dunn and Fuller, 1972), 14313 has an unusually high proportion of low blocking temperatures.

Since part of the VRM has an AF stability comparable to that of high-temperature TRM (cf. Fig. 4), and cannot be erased by the maximum available field of 1000 Oe, a clean separation between primary and secondary remanences in lunar soils and breccias can probably only be achieved by thermal demagnetization to 300°C or so. This relatively low temperature produces negligible oxidation or sintering of particles (Pearce, 1973) and scarcely affects the bulk of stable TRM but is sufficient to totally erase any secondary remanence of viscous origin.

Conclusions

(1) The remanent properties of low metamorphic grade breccia 14313 are dominated by single-domain particles of iron. It appears that this result can be generalized for all low-grade breccias and many soils.

(2) The number of iron particles varies with volume v approximately as v^{-2}, as found previously by Stephenson (1971). Few particles larger than 200 Å are indicated magnetically but there are many superparamagnetic particles smaller than 100 Å.

(3) There are two separate populations of particles, having coercive forces H_{c0} above and below 2000 Oe and composed of nearly equidimensional and extremely elongated particles respectively.

(4) The primary NPM of 14313 is probably a lunar TRM, a considerable fraction of which is stable to AF cleaning in 1000 Oe.

(5) Secondary NRM in 14313 is probably a terrestrial VRM. VRM acquired in 24 hours erases any lunar TRM with blocking temperatures of 120°C or less, and since VRM is acquired in proportion to the logarithm of time, blocking temperatures of a few hundred degrees at most are affected by storage in the earth's field for 1 or 2 months. Therefore the most efficient method of selectively erasing VRM is to heat the sample in field-free space to 200–300°C. This is preferable to AF demagnetization because low-temperature partial TRM's and VRM's contain a significant component whose AF stability exceeds 1000 Oe.

Acknowledgments—The authors thank the Lunar Science Institute and the National Research Council of Canada for financial support. The Lunar Science Institute is operated by the Universities Space Research Association under Contract No. NSR 09-051-001 with the National Aeronautics and Space Administration. This paper constitutes the Lunar Science Institute Contribution No. 148.

REFERENCES

Avchyan G. M. and Faustov S. S. (1966) On the stability of viscous magnetization in variable magnetic fields. *Akad. Nauk SSSR, Izv., Earth Physics*, No. 5, pp. 96–104.

Biquand D. and Prévot M. (1971) A.F. demagnetization of viscous remanent magnetization in rocks. *Z. Geophys.* **37**, 471–485.

Carter J. L. (1973) VLS (vapour-liquid-solid) growth on the surface of rock 15015 (abstract). *EOS (Trans. Am. Geophys. Union)* **54**, 506.

De Blois R. W. and Bean C. P. (1959) Nucleation of ferromagnetic domains in iron whiskers. *J. Appl. Phys.* **30**, 225S–226S.

Dunlop D. J. (1965) Grain distributions in rocks containing single domain grains. *J. Geomag. Geoelec.* **17**, 459–471.

Dunlop D. J. and West G. F. (1969) An experimental evaluation of single-domain theories. *Rev. Geophys.* **7**, 709–757.

Dunn J. R. and Fuller M. (1972) On the remanent magnetism of lunar samples with special reference to 10048,55 and 14053,48. *Proc. Third Lunar Sci. Conf., Geochim. Cosmochim. Acta*, Suppl. 3, Vol. **3**, pp. 2263–2386. MIT Press.

Everitt C. W. F. (1961) Thermoremanent magnetization, I. Experiments on single domain grains. *Phil. Mag.* **6**, 713–726.

Gose W. A. and Carnes J. G. (1973) The time dependent magnetization of fine-grained iron in lunar breccias. Submitted to *Earth Planet Sci. Lett.*

Gose W. A., Pearce G. W., Strangway D. W., and Larson E. E. (1972a) On the applicability of lunar breccias for paleomagnetic interpretations. *The Moon* **4**, 106–120.

Gose W. A., Pearce G. W., Strangway D. W., and Larson E. E. (1972b) Magnetic properties of Apollo 14 breccias and their correlation with metamorphism. *Proc. Third Lunar Sci. Conf., Geochim. Cosmochim. Acta*, Suppl. 3, Vol. **3**, pp. 2387–2395. MIT Press.

Gose W. A., Strangway D. W., and Pearce G. W. (1972c) A determination of the intensity of the ancient lunar magnetic field. *The Moon* **7**, 198–201.

Grant R. W., Housley R. M., and Paton N. E. (1973) Origin and characteristics of excess Fe metal in lunar glass welded aggregates (abstract). In *Lunar Science—IV*, pp. 315–316. The Lunar Science Institute, Houston.

Néel L. (1949) Théorie du traînage magnétique des ferromagnétiques en grains fins avec applications aux terres cuites. *Ann. Géophys.* **5**, 99–136.

Pearce G. W. (1973) Magnetism and lunar surface samples. Ph.D. thesis, Univ. of Toronto.

Pearce G. W., Williams R. J., and McKay D. S. (1972) The magnetic properties and morphology of metallic iron produced by subsolidus reduction of synthetic Apollo 11 composition glasses. *Earth Planet. Sci. Lett.* **17**, 95–104.

Richter G. (1937) Über die magnetische Nachwirkung am Carbonyleisen. *Ann. Physik* **29**, 605–635.

Stephenson A. (1971a) Single domain grain distributions, I. A method for the determination of single domain grain distributions. *Phys. Earth Planet. Interiors* **4**, 353–360.

Stephenson A. (1971b) Single domain grain distributions, II. The distribution of single domain iron grains in Apollo 11 lunar dust. *Phys. Earth Plant. Interiors* **4**, 361–369.

Tsay F-D., Chan S. I., and Manatt S. L. (1971) Ferromagnetic resonance of lunar samples. *Geochim. Cosmochim. Acta* **35**, 865–875.

Warner J. L. (1972) Metamorphism of Apollo 14 breccias. *Proc. Third Lunar Sci. Conf., Geochim. Cosmochim. Acta*, Suppl. 3, Vol. 1, pp. 623–644. MIT Press.

Proceedings of the Fourth Lunar Science Conference
(Supplement 4, *Geochimica et Cosmochimica Acta*)
Vol. 3, pp. 2991–3001

Magnetic and Mössbauer studies of Apollo 16 rock chips 60315,51 and 62295,27

Aviva Brecher, David J. Vaughan, and Roger G. Burns

Department of Earth and Planetary Sciences
Massachusetts Institute of Technology
Cambridge, Mass. 02139

K. R. Morash

Department of Metallurgy and Materials Science
Massachusetts Institute of Technology
Cambridge, Mass. 02139

Abstract—Analysis of the Mössbauer spectra of two Apollo 16 rocks showed that 60315,51 is much richer in iron metal and troilite, but poorer in olivine, than 62295,27. The values of magnetic parameters, derived from hysteresis loops at 175 and 300°K, indicate the high metal contents and the predominance of coarse multidomain grains in both rocks. These coexist with a superparamagnetic grain fraction in 60315 and with a small single-domain grain fraction in 62295. The high $Fe°/Fe^{2+}$ ratios, the nonlinear acquisition of laboratory thermoremanence and the drastic changes in magnetic parameters upon heating, support the proposed formation of both rocks from the lunar regolith, with incorporation of shocked meteoritic metal grains during high-temperature impact events and simultaneous acquisition of magnetic remanence. Values estimated for ancient lunar magnetic fields by comparing the natural remanence with laboratory thermoremanence acquired in fields of 0.05 and 0.5 Oe, range from 0.01 to >1 Oe. However, changes in the samples during heating in the laboratory may invalidate such paleointensity estimates.

INTRODUCTION

Magnetic and Mössbauer studies of two Apollo 16 rocks chips (60315,51 and 62295,27) have been undertaken. Rock 60315 has been described as a recrystallized polymict breccia (Bence *et al.*, 1973) of KREEP-like basaltic composition (LSPET, 1973; Bansal *et al.*, 1973). It contains large poikiloblastic grains of orthopyroxene enclosing plagioclase, olivine and some metal; or orthopyroxene rimmed by augite. Augite, olivine and plagioclase are present interstitially. Rock 62295 has been described as a gabbro by Rose *et al.* (1973) or as a spinel-bearing troctolite by Nord *et al.* (1973). This rock contains plagioclase laths, olivine and pyroxene. Both samples have peculiar primordial nuclide chemistry attesting to unusual formation histories (Eldridge *et al.*, 1973).

Optical microscopy revealed a greater content of large metal spherules in 60315,51 than in 62295,27. The compositions of these large metal grains in both rocks have been described as meteoritic (Taylor and McCallister, 1973; Agrell *et al.*, 1973).

Mössbauer Spectroscopy

Rock chips of the Apollo 16 samples were crushed to a fine powder and used to prepare thin-disc Mössbauer absorbers. The Mössbauer source was Co^{57} in Pd and both source and absorber were at room temperature. A computer program employing a least squares fit to Lorentzian lineshapes was used to deconvolute the spectra (Stone, 1967).

The spectra obtained are shown in Fig. 1. In spectrum A (sample 60315,51), two distinct sets of magnetic hyperfine peaks are present and are attributed to metallic iron and troilite, the metallic iron being more abundant. Two intense quadrupole doublets in the central region of the spectrum have isomer shift and quadrupole splitting parameters consistent with olivine and pyroxene, the olivine giving rise to the outer doublet.

The Mössbauer isomer shifts (I.S.) relative to iron metal, quadrupole splittings (Q.S.) and the internal magnetic fields (H_{loc}) for the subspectra are as follows. Iron metal $H_{loc} \sim 337$ KOe; troilite $H_{loc} \sim 317$ KOe; olivine I.S. ~ 1.10 mm/sec, Q.S. ~ 2.09 mm/sec. These parameters are close to those previously reported for lunar

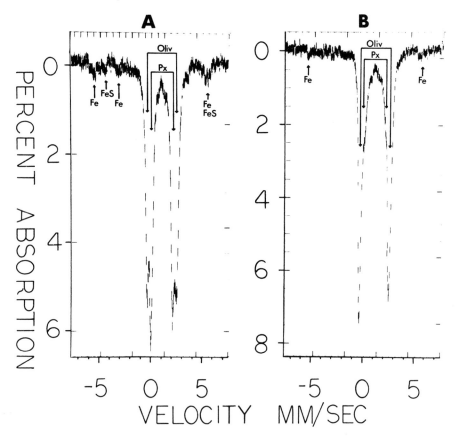

Fig. 1. Mössbauer spectra of whole rock samples of 60315 (A) and 62295 (B) at 300°K.

and terrestrial iron, troilite, olivine and pyroxene (e.g., Gay *et al.*, 1971, Herzenberg *et al.*, 1971, Housley *et al.*, 1970, 1971). Computer analysis of this spectrum suggests the following distribution of iron between phases: pyroxene 52%, olivine 35%, metallic iron 10%, troilite 3%. However, some of the intensity attributed to olivine probably represents Fe^{2+} in the pyroxene M1 site and 35% should be considered an upper limit. Some iron is also present in spinel and ilmenite phases not resolved in the region of closely overlapping innermost hyperfine peaks.

Spectrum B (sample 62295,27) differs from spectrum A both in the reduced intensity of the magnetic peaks and in the considerable reduction in the intensity of the innermost quadrupole doublet arising from iron in pyroxene. Mössbauer parameters are very similar to those for 60315 and as follows. Iron metal $H_{loc} \sim 340$ KOe; olivine I.S. ~ 1.07 mm/sec, Q.S. ~ 1.98 mm/sec. Computer analysis suggested the following iron distribution in 62295,27: olivine 71%, pyroxene 28%, metallic iron 1%. Rock 62295,27 appears to be depleted in iron and troilite and enriched in olivine relative to 60315,51. Traces of troilite, such as a shoulder on the iron peak at positive velocity, are also apparent in this spectrum and the same provisions regarding the enhancement of the olivine doublet and the presence of unresolved spinels and ilmenite apply as for 60315,51. Despite these uncertainties, the distribution of iron between magnetically ordered and paramagnetic phases is considered to have an accuracy better than $\pm 1\%$. No evidence was found in the Mössbauer spectra for the presence of ferric iron in either sample.

MAGNETIC PARAMETERS FROM HYSTERESIS LOOPS

The analysis of magnetization curves (Fig. 2 and Table 1) can provide estimates of absolute and relative amounts of Fe° and Fe^{2+} in lunar samples. In addition, the size spectrum and remanent properties of the metallic iron can also be inferred (cf., Nagata *et al.*, 1972; Gose *et al.*, 1972).

Previously reported results (Brecher *et al.*, 1973) were obtained from the analysis of hysteresis loops measured at 185°K with a PAR Parallel Field Vibrating Sample Magnetometer (VSM), in fields up to 50 KOe provided by a superconducting magnet. The fully saturated magnetization curves obtained at ~ 180°K yielded *saturation magnetization* (J_s/m) values of 3.8 emu/g for 60315 and 0.825 emu/g for 62295, corresponding to iron metal contents of ~ 1.7 wt.% and 0.37 wt.%, respectively. Accurate determinations of the coercivity field (H_c) and the saturation remanence (J_r) were not then possible.

Average results (Fig. 2, Table 1) based on at least four sets of hysteresis curves, each obtained at two temperatures are now reported. These measurements were made at 175°K (the approximate lunar night temperature) and at 300°K with a transverse field VSM, in fields up to 12 KOe provided by an electromagnet. The saturation values and estimated metal contents thus obtained are somewhat lower than before (by $\leqslant 20\%$ for 60315 and by $\sim 15\%$ for 62295), suggesting incomplete saturation. It is possible that the magnetization and remanence are lower and the paramagnetic susceptibility higher, than the true values. However, they are

MAGNETIZATION CURVES

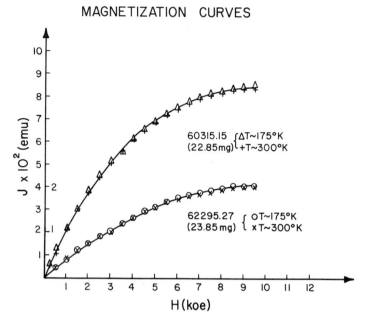

Fig. 2. Average magnetization behavior of 60315 and 62295 rock chips, at 175°K and 300°K. The magnetization of 62295 is plotted on the inner scale.

adequate for the purpose of comparison, since the samples were magnetized in maximum fields (12 KOe) higher than the saturating fields (<7.5 KOe for 62295 and <10 KOe for 60315) previously determined (Brecher *et al.*, 1973). Also, because the hysteresis loops were obtained on small rock chips (22.85 mg for 60315,51 and 23.85 mg for 62295,27), the values reported in Table 1 may not be representative of the bulk material. The differences in saturation values at 300°K and 175°K should indicate the fraction of fine metal grains which are superparamagnetic (*SPM*) at 300°K but are able to block remanence and behave like stable single-domained (*SD*) ferromagnetics at 175°K. To avoid oxidation of the samples, data were not obtained at the lunar day temperature (≥400°K). However, the small temperature-dependence of the magnetization behavior for both

Table 1. Average magnetic parameters.

	T (°K)	J_s/m (emu/g)	f_{ferro} (wt.%)	J_r/m (emu/g)	J_r/J_s	H_c (Oe)	$\chi_p \times 10^6$ (gauss/Oe·g)	f_{para} (wt.%)	$\chi_i \times 10^4$	J_s/χ_i
60315,51	175	2.99	1.36	0.0085	0.0028	10.0	77.6	21.0	9.6	3112
(22.85 mg)	300	2.98	1.35	0.011	0.0037	13.0	77.2	35.9	9.22	3232
62295,27	175	0.695	0.316	0.0065	0.0094	35.0	22.4	6.08	1.5	4633
(23.85 mg)	300	0.613	0.278	0.0066	0.0095	30.5	24.0	11.2	2.05	3088

rocks (Fig. 2), due to the dominance of multidomain (*MD*) grains, supports the applicability of these experimental results over the lunar diurnal temperature cycle.

The values of J_s/m at $\sim 175°K$ in 60315 and 62295 (Table 1) show a much greater abundance of metallic iron in the former. The metal contents (f_{ferro}) may be estimated at 1.36 wt.% for 60315 and ~ 0.32 wt.% for 62295. These are probably lower limits (correct to within 20%) since the values are calculated assuming all the metal is pure Fe (with $J_s/m = 220$ emu/g) rather than Ni–Fe alloys with lower saturation magnetization. While the value for 62295 is typical of lunar breccias and soils, rather than of lunar igneous rocks (cf., Gose *et al.*, 1972), the amount of iron metal in 60315 is large compared to most lunar rocks, but comparable to other Apollo 16 samples (Pearce and Simonds, 1973).

The values of *paramagnetic susceptibilities*, χ_p (Table 1), are within the upper range of values for coarse igneous rocks in the case of 60315 and more like those of fine-grained igneous rocks or fragmental rocks for 62295 (Gose *et al.*, 1972; Nagata *et al.*, 1972). These values correspond to an Fe^{2+} weight fraction (f_{para}) of ~ 21 wt.% for 60315, but only ~ 6.1 wt.% for 62295 at 175°K, derived using the formula

$$\chi_P = f_{para}\frac{6.45 \times 10^{-2}}{T}(emu/g)$$

(Nagata *et al.*, 1972). Values obtained from samples at $\sim 180°K$ and saturated to 50 KOe were $\chi_p \sim 40 \times 10^{-6}$ emu/g and $f_{para} \sim 12$ wt.% for 60315 and $\chi_p \sim 19 \times 10^{-6}$ emu/g and $f_{para} \sim 5.5$ wt.% for 62295.

Thus, the magnetically determined $Fe°/Fe^{2+}$ ratios range from 0.06 at 175° to 0.0375 at 300°K in 60315 and from 0.05 at 175°K to 0.025 at 300°K in 62295 with even higher ratios derived from the fully saturated curves at $\sim 180°K$. The higher values at 175°K include the fraction of *SPM* iron grains with blocking temperatures in the range 175–300°K, and correspond more closely to the true value of $Fe°/Fe^{2+}$. Such high ratios, and the evidence for the predominance of coarser *MD* grains in the size spectrum of both rocks (from values of χ_i and J_s/χ_i or J_r/J_s discussed below), are in very good agreement with the conclusions of Pearce and Simonds (1973). The values of *initial susceptibility*, χ_i (in emu/g) are within the range of fragmental and igneous rock values for 62295, and in the upper range of igneous rock values for 60315 (Table 1). They indicate a larger amount of material with high χ_i values (either *SD* or *SPM* grains) in 60315.

The ratio J_s/χ_i has been correlated directly with the ratio of $MD/(SPM + SD)$ grains (Nagata *et al.*, 1972; Gose *et al.*, 1972). The values of J_s/χ_i (several thousand in both rocks) also imply the dominance of low susceptibility, coarse *MD* grains. This ratio shows an inverse correlation with the ratio of viscous to stable remanence because the latter depends mostly on the amount of *SPM* iron. Indeed, 60315 has a larger χ_i and a lower J_s/χ_i value (by $\sim 30\%$ at 175°K) than 62295, due to the larger fraction of *SPM* grains present. This implies that 60315 can carry a higher viscous remanence component relative to the stable natural remanence (NRM). The generally lower *bulk coercivity* (H_c) values of 60315 compared to

62295, are consistent with the presence of a higher *SPM* grain fraction. Similarly, the higher reduced remanence ratios (J_r/J_s) in 62295 compared to 60315 allow a maximum of 3.5% of all metal grains in 62295 to be *SD* magnetically stable grains, whereas $\leq 1\%$ are allowed in 60315. The *saturation remanence* (J_r/m) values are comparable in both samples in spite of higher overall metal content in 60315. This confirms that some *SPM* grains are mixed mostly with *MD* grains in 60315, whereas a small fraction of *SD* material is mixed with mostly *MD* grains in 62295. A small fraction of *SPM* material has a more drastic effect in reducing both the coercivity (H_c) and the saturation remanence (J_r/m) of a mixture with *SD* material, than an equivalent *MD* fraction (Bean, 1955). The values of J_r/J_s are ≤ 0.01 for both rocks, i.e., in the range of values for high and medium metamorphic grade breccias (Strangway *et al.*, 1973).

REMANENT PROPERTIES

Measurements of magnetic moments were made with a point-contact SQUID magnetometer in a magnetically shielded room, and with P.A.R. and Schonstedt spinner magnetometers (with sensitivities of $\leq 5 \times 10^{-6}$ emu).

The values of natural remanence (NRM) obtained for different fragments of the rock chips 60315,51 and 62295,27 are scattered within an order of magnitude for each rock. For three chips of 60315, NRM values of 2.69×10^{-5} emu/g (for a 0.457 g fragment), 4.06×10^{-5} emu/g (0.279 g fragment) and 4.9×10^{-4} emu/g (~ 0.12 g fragment) were obtained. For 62295, values of NRM ranged from 0.2×10^{-5} emu/g (for a 0.9 g fragment) to 6.44×10^{-5} emu/g (for a 0.25 g fragment). The scatter reflects the inhomogeneous distribution of ferromagnetic material, and a propensity to acquire viscous remanence (VRM). Large VRM was easily acquired by both rocks, as a result of the presence of a large magnetically-soft grain fraction. A rapid decay (by a factor of 2 in 24 hours) of saturation remanence (IRM$_s$) was observed in 60315 after storage in field free space, suggesting Type II viscous behavior (Nagata *et al.*, 1972). This is in agreement with the above conclusion that 60315 contains a sizeable *SPM* metal fraction. Both samples have low Konigsberger ratios $(Q = \mathrm{NRM}/\chi_i)$ in the ranges 0.02–0.5 for 60315 and 0.012–0.32 for 62295, suggesting large unstable remanence components.

Since iron oxide phases have been reported in some Apollo 16 rocks (Taylor *et al.*, 1973), an attempt was made to identify them by their low temperature behavior (Nagata *et al.*, 1967). However, repeated thermal cycling (300–77°K) in field free space showed no evidence for loss of secondary remanence residing in multidomain iron oxide phases (hematite or magnetite) in either of the samples. A large increase in the initial moments was observed on low temperature cycling, by factors of ~ 2 in 60315 and ~ 4 in 62295 after the first cycle, and factors of ~ 4 and ~ 5 after the second cycle. Thermal cycling also appeared to increase the stability of NRM to AF demagnetization in 60315 (Fig. 3), suggesting that soft remanence may be similarly cleaned during the lunar diurnal cycle. A coercivity spectrum analysis for a bulk sample of 60315, was approximated by increments of magnetization in 1 KOe field intervals up to 10 KOe (Dunlop, 1972). This revealed

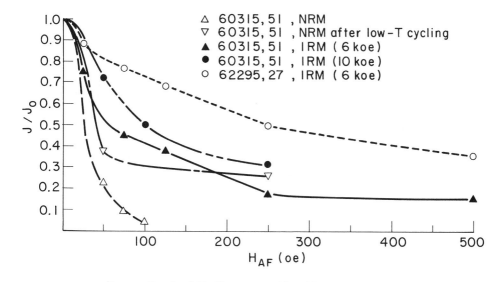

Fig. 3. Relative stability of NRM and IRM to AF demagnetization.

a mean coercivity (H_{cr}) well below 1 KOe, and a high saturating field ($H_s \lesssim$ 10 KOe), consistent with a predominantly *MD* grain population and with the rather linear approach to saturation (Fig. 2). The saturating field in 62295 is somewhat lower, ($\leqslant 7.5$ KOe). The relative intensities of several types of remanence are compared in Table 2 and their relative stability to AF demagnetization is displayed in Fig. 3. Only the AF demagnetization behavior above the spinner noise level is shown in Fig. 3. The saturation remanence (IRM_s) is larger than the NRM by factors of ~ 10 in 60315 and ~ 50 in 62295, and larger than TRM acquired in 0.5 Oe by factors of ~ 35 and ~ 50, respectively.

The stability of saturation remanence (IRM_s) to AF demagnetization may serve as an upper limit to the stability of weak-field thermoremanence (TRM), if the latter resides mostly in *MD* grains (Dunn and Fuller 1972). In 60315, the stability of 10 KOe IRM_s is rather low; 28% of it is lost in 50 Oe fields and 30% remains after 250 Oe cleaning. The NRM is considerably less stable, with 60–80% losses incurred in 50 Oe fields, and with a stable fraction of only $\sim 10\%$. Such stability behavior is typical of breccias and igneous rocks (Gose *et al.*, 1972). In

Table 2. Relative intensities of remanence.

	60315,51 ($m = 0.13$ g)	62295,27 ($m = 0.25$ g)
NRM (emu/g)	4.9×10^{-4}	6.44×10^{-5}
IRM_s (6 KOe) (emu/g)	4.64×10^{-3}	2.94×10^{-3}
TRM (0.05 Oe) (emu/g)	0.63×10^{-4}	4×10^{-5}
TRM (0.5 Oe) (emu/g)	1.34×10^{-4}	6×10^{-5}

62295 only the IRM_s demagnetization curve is available; its high relative stability suggests a greater content of SD grains which are able to carry a small but hard NRM. This is consistent with the conclusions based on hysteresis loop parameters.

Two single-heating experiments were carried out to impart total thermoremanence TRM. They involved heating the samples to $\geqslant 800°C$ in vacuum ($p < 10^{-6}$ torr) and cooling in magnetic fields of ~ 0.05 Oe and ~ 0.5 Oe. The increase of external field strength by a factor of 10, resulted in an increase by only a factor of $\leqslant 2$ in the intensity of TRM. Heating obviously decreased the capability of these samples to carry TRM (Dunn and Fuller, 1972). The intensity of TRM acquired in 0.5 Oe is still too low by a factor of $\leqslant 4$ compared to the NMR in 60315, although it is comparable to the NRM in 62295.

In order to understand such nonlinear behavior of TRM acquisition the effects of laboratory heating must be assessed. It appears that heating, even in a hard-vacuum, produces irreversible changes and that some subsolidus reduction of iron bearing minerals and coalescence of iron grains takes place (Pearce *et al.*, 1972). After heating, the IRM_s values are reduced by 22% in 60315 (from 7.46 to 5.87×10^{-3} emu/g) and by 30% in 62295 (from 2.94 to 2.013×10^{-3} emu/g), indicating the loss of some *SD*, high-remanence grains and an increase in the lower remanence *SPM + MD* grain fraction. Hysteresis loops measured on a chip of 62295 (0.12 g), twice heated to 800°C in vacuum, showed that the saturation magnetization had doubled (from 0.695 to 1.225 emu/g), indicating a major increase in iron metal content. The heat treatment reduced the value of J_r/J_s to ~ 0.003 and the coercivity to $H_c \sim 19$ Oe and increased the initial susceptibility (to $\chi_i \sim 2.76 \times 10^{-4}$) as measured at 300°K. The new metal grain population is therefore of the superparamagnetic type. The lower value of $\chi_p (\sim 16 \times 10^{-6})$ and more rounded shape of the hysteresis loop suggest that newly precipitated *SPM* iron grains formed by subsolidus reduction of Fe^{2+} in various mineral phases (Pearce *et al.*, 1973). In contrast, changes in magnetic parameters produced by similarly heating a chip of 60315 (0.113 g) suggested that mostly coarsening of the metal grains occurred. The loss of some *SPM* material and increase in the *MD* fraction is reflected in the lower χ_p value ($\sim 7 \times 10^{-4}$), the increase in J_r/m (to 0.36 emu), in J_r/J_s (to 0.011), and in the J_s/χ_i ratio (by $\sim 50\%$). Only a small ($<10\%$) increase in the ferromagnetic metal content resulted from heating, leading to an increase of $Fe°/Fe^{2+}$ ratio from 0.036 to 0.057.

If the NRM in these lunar rocks is presumed to be of TRM-type and to obey the Théllier field-proportionality law ($H_0/H_\oplus = NRM/TRM$), then the following can be estimated on the basis of single heating experiments. Namely that 60315 would require initial fields $H_0 \leqslant 0.4$ Oe to $\leqslant 1.8$ Oe from the two TRM values corresponding to 0.05 Oe and 0.5 Oe, respectively, whereas 62295 would require $H_0 \sim 0.082$ to 0.5 Oe.

If, however, we accept only the stable 10% of the NRM as representing the TRM component, (Fig. 3 and Gose *et al.*, 1972; Strangway *et al.*, 1973) then a corresponding fraction of the above field values could imprint an NRM in the observed range of values, (~ 0.04–0.18 Oe for 60315 and ~ 0.01–0.05 Oe for 62295).

These estimates allow a broad range of values for an ancient lunar field, from 1000γ to >1 Oe. The raw paleofield intensities obtained are not uniquely high, Collinson et al. (1973) also arrived at values of $H_0 \sim 0.5$ Oe from single heating experiments on a chip of 62235. These results, however, emphasize the important connection between the thermal history of the samples and its effect on the metal grain population, and the lack of reproducibility of experimental determinations of the presumed lunar magnetic field strength (Strangway et al., 1973). The higher raw fields estimated for the more coarsely crystalline and metal rich 60315, which also carries a more intense but less stable NRM than 62295, support a correlation between the magnetic fields inferred from remanent properties and the degree of thermal metamorphism during impact events (Dunn et al., 1973).

Conclusions

Pearce and Simonds (1973) have suggested that most Apollo 16 rocks formed at high temperatures ($\sim 1000°C$) from the lunar regolith. The high values of the ratio $Fe°/Fe^{2+}$ reported from Mössbauer and magnetic measurements in the present work, and the evidence that coarse (*MD*) metal grains dominate the grain-size spectra in both rocks studied, support their suggestion. These rocks were probably formed by impact melting and compaction of soils. The meteoritic compositions of the metal grains, the evidence of metamorphism and of thermal, mechanical and magnetic properties associated with shock (Dunn et al., 1973; Agrell et al., 1973; Simonds et al., 1973; Taylor and McCallister, 1973), support this theory of their origin. It is unlikely that breccias with such a complex history carry a unique thermoremanence acquired in an indigenous lunar field (Strangway et al., 1973). It is more likely that several generations of TRM were superimposed during various brecciation and compaction events. Alternatively, thermochemical remanence may have been acquired by precipitation of fine *SPM* grains which grew through *SD* or *MD* sizes as they cooled in the presence of magnetic fields (Dunn et al., 1973).

Acknowledgments—Dr. D. Cohen is thanked for providing access to the low-field laboratory at the National Magnet Laboratory, M.I.T. Prof. M. Fuller and J. R. Dunn generously made available facilities in their laboratory at the University of Pittsburgh. Mrs. Virginia Mee Burns provided editorial assistance and Ms. Dorothy Frank typed the manuscript. Support from NASA grants NGR 22-009-551 and NGL 22-009-187 is acknowledged.

References

Agrell S. O., Agrell J. E., Arnold A. R., and Long J. V. P. (1973) Some observations on rock 62295. In *Lunar Science—IV*, pp. 15–17. The Lunar Science Institute, Houston.

Bansal B. M., Gast P. W., Hubbard N. J., Nyquist L. E., Rhodes J. M., Shih C. Y., and Weissmann H. (1973) Lunar rock types. In *Lunar Science—IV*, pp. 48–50. The Lunar Science Institute, Houston.

Bean C. P. (1955) Hysteresis loops of mixtures of ferromagnetic micropowders. *J. Appl. Phys.* **26**, 1381–1383.

Bence A. E., Papike J. J., Sueno S., and Delano J. W. (1973) Pyroxene poikiloblastic rocks from Apollo 16. In *Lunar Science—IV*, pp. 60–62. The Lunar Science Institute, Houston.

Brecher A. and Morash K. R. (1973) Magnetic characteristics of Apollo 17 orange and grey soils. *EOS Trans. Am. Geophys. Union* **54**, 581.

Brecher A., Vaughan D. J., Burns R. G., Cohen D., and Morash K. R. (1973) Magnetic and Mössbauer studies of Apollo 16 rock chips 60315,51 and 62295,27. In *Lunar Science—IV*, pp. 88–90. The Lunar Science Institute, Houston.

Collinson D. W., Runcorn S. K., and Stephenson A. (1973) Magnetic properties of Apollo 16 rocks. In *Lunar Science—IV*, pp. 155–157. The Lunar Science Institute, Houston.

Dunlop D. J. (1972) Magnetic mineralogy of unheated and heated red sediments by coercivity spectrum analysis. *Geophys. J. R. Astr. Soc.* **27**, 37–55.

Dunn J. R. and Fuller M. (1972) Thermoremanence of lunar samples. *The Moon* **4**, 49–62.

Dunn J. R., Fisher R., Treller M., Lally S., Rose F., Schwerer F., and Wasilewski P. (1973) Shock remanent magnetization of lunar soil. In *Lunar Science—IV*, pp. 194–195. The Lunar Science Institute, Houston.

Eldridge J. S., O'Kelley G. D., and Northcutt K. J. (1973) Radionuclide concentrations in Apollo 16 samples. In *Lunar Science—IV*, pp. 219–221. The Lunar Science Institute, Houston.

Gay P., Brown M. G., Muir I. D., Bancroft G. M., and Williams P. G. L. (1971) Mineralogical and petrographic investigation of some Apollo 12 samples. *Proc. Second Lunar Sci. Conf., Geochim. Cosmochim. Acta*, Suppl. 2, Vol. 1, pp. 377–392. MIT Press.

Gose W. A., Pearce G. W., Strangway D. W., and Larson E. E. (1972) On the applicability of lunar breccias for paleomagnetic interpretations. *The Moon* **5**, 106–120.

Herzenberg C. L., Moler R. B., and Riley D. L. (1971) Mössbauer instrumental analysis of Apollo 12 lunar rock and soil samples. *Proc. Second Lunar Sci. Conf., Geochim. Cosmochim. Acta*, Suppl. 2, Vol. 3, pp. 2103–2123. MIT Press.

Housley R. M., Blander M., Abdel-Gawad M., Grant R. W., and Muir A. H. Jr. (1970) Mössbauer spectroscopy of Apollo 11 samples. *Proc. Apollo 11 Lunar Sci. Conf., Geochim. Cosmochim. Acta*, Suppl. 1, Vol. 3, pp. 2251–2268. Pergamon.

Housley R. M., Grant R. W., Muir A. H. Jr., Blander M., and Abdel-Gawad M. (1971) Mössbauer studies of Apollo 12 samples. *Proc. Second Lunar Sci. Conf., Geochim. Cosmochim. Acta*, Suppl. 2, Vol. 3, pp. 2125–2136. MIT Press.

LSPET (1973) Apollo 16. *Science* **179**, 23–34.

Nagata T., Kobayashi K., and Fuller M. (1964) Identification of magnetite and hematite in rocks by magnetic observations at low temperature. *J. Geophys. Res.* **69**, 2111–2120.

Nagata T., Fisher R. M., and Schwerer F. C. (1972) Lunar rock magnetism. *The Moon* **4**, 161–186.

Nord G. L., Lally J. S., Christie J. M., Heuer A. H., Fisher R. M., Griggs D. T., and Radcliffe S. V. (1973) High voltage electron microscopy of igneous rocks from Apollo 15 and 16. In *Lunar Science—IV*, pp. 564–566. The Lunar Science Institute, Houston.

Pearce G. W. and Simonds C. H. (1973) Magnetic properties and mode of formation of Apollo 16 samples. *EOS Trans. Am. Geophys. Union* **54**, 358.

Pearce G. W., Strangway D. W., and Gose W. A. (1973) Magnetic properties and temperature of formation of lunar breccias. In *Lunar Science—IV*, pp. 585–587. The Lunar Science Institute, Houston.

Pearce G. W., Williams R. J., and McKay D. S. (1972) The magnetic properties and morphology of metallic iron produced by reduction of synthetic Apollo 11 composition glasses. *Earth Planet. Sci. Lett.* **17**, 95–104.

Rose H. J. Jr., Carron M. K., Christian R. P., Cuttitta I., Dwornik E. J., and Ligon D. T. Jr. (1973) Elemental analysis of some Apollo 16 samples. In *Lunar Science—IV*, pp. 631–633. The Lunar Science Institute, Houston.

Simonds C. H., Warner J. L., Phinney W. C., and Gooley, R. (1973) Mineralogy and mode of formation of poikilitic rocks from Apollo 16. In *Lunar Science—IV*, pp. 676–678. The Lunar Science Institute, Houston.

Stone A. J. (1967) Appendix to: Bancroft G. M., Maddock A. G., Ong W. K., Prince R. H., and Stone A. J. Mössbauer spectrum of iron (III) Diketone complexes, *J. Chem. Soc.* (A), 1966–1971.

Strangway D. W., Gose W. A., Pearce G. W., and Carnes J. G. (1973) Magnetism and the history of the moon. *AIP Conf. Proc.* **10**, pp. 1178–1196. Am. Inst. Physics, New York.

Taylor L. A., Mao H. K., and Bell P. M. (1973) Apollo 16 "rusty rock" 66095. In *Lunar Science—IV*, pp. 715–716. The Lunar Science Institute, Houston.

Taylor L. A. and McCallister R. H. (1973) Opaque mineral geothermometers as indicators of cooling histories of lunar rocks. In *Lunar Science—IV*, pp. 717–719. The Lunar Science Institute, Houston.

Proceedings of the Fourth Lunar Science Conference
(Supplement 4, *Geochimica et Cosmochimica Acta*)
Vol. 3, pp. 3003–3017

Magnetic effects of experimental shocking of lunar soil

S. Cisowski and M. Fuller

Department of Earth and Planetary Sciences, University of Pittsburgh, Pa. 15213

M. E. Rose

U.S. Naval Weapons Lab., Dahlgren, Virginia 22448

P. J. Wasilewski

Department of Geology, The George Washington University, Washington, D.C.

Abstract—Lunar soil samples from 65901,10 have been experimentally shocked at approximately 50, 75, 100, and 250 kb, using the flying plate technique. On recovery, all samples were found to be lithified. In the 50 kb sample some shock melting was evident, but other regions were undeformed. The range of shock effects are comparable with those found in some lunar breccias. In the 250 kb sample, large amounts of glass were produced by the shock. Some of this glass was clear, and some contained fine particles which appear to be both magnetic α-iron and non-magnetic γ-iron.

The magnetic characteristics of the soil were changed by the shock experiments. In the low shock range, the coercivity (H_c) and saturation isothermal remanent magnetization (IRM$_s$) increased. In the higher shock range, magnetic viscosity increased, but coercivity (H_c) and IRM$_s$ were similar to those of the control material. The 100 kb sample exhibited anisotropy of IRM$_s$.

Shock remanent magnetization (SRM) was acquired as the result of the shock loading. The maximum value, which was that of the 50 kb sample, was 10^{-3} gauss cm^{3}g^{-1}. Assuming a linear dependence of SRM upon field for a given shock pressure, this would give rise to remanence of 10^{-4} and 10^{-5} gauss cm^{3}g^{-1} in fields of $10^3\gamma$ and $10^2\gamma$ respectively. The direction of SRM was not simply related either to the field in which the samples were shocked nor to the fields in which they were recovered. SRM acquired in the low shock range was stable in direction during AF demagnetization. However, SRM acquired in the 100 and 250 kb experiments changed in direction during AF demagnetization. The remanence acquired in 100 kb shock has distributed blocking temperatures.

It is concluded that shock remanent magnetization (SRM) is a relatively efficient mechanism of magnetization and should not be discounted as a source of lunar NRM. SRM may also account for the NRM of certain meteorites. Indeed SRM could be a relatively common phenomena in the early solar system being produced in planetesimal collisions as well as by meteoroid impact phenomena.

Introduction

The origin of the natural remanent magnetization (NRM) of the returned lunar samples is not yet resolved. Neither the mechanisms of magnetization, nor the fields in which the magnetization was acquired has been convincingly demonstrated. Initially, a lunar dynamo was advocated as the source field (e.g., Runcorn *et al.* 1971, Strangway *et al.*, 1971). However, more recently the dynamo model has lost favour (Levy, 1972) and a fossil field model has been suggested by many people (e.g., Urey and Runcorn, 1973, Strangway *et al.*, 1973b, Alfven and Lindberg, 1973). In the latter model, the moon is held to have acquired a remanent mag-

3003

netization early in its history, and it was in this primitive remanent field, which is supposed to be approximately $10^3 \gamma$ at the surface, that the samples were magnetized some 3–4 b.y. ago. Subsequently, the primative remanent field has been lost due to the thermal demagnetization of the moon implied by proposed thermal models (Toksöz *et al.*, 1972).

Both the dynamo and fossil field models face severe difficulty in explaining the NRM of returned samples because of the wide range of inducing fields, which appear to be necessary to account for the observed NRM. Two complete ancient field intensity determinations are available. One, by Strangway *et al.* (1973a), gives $2100 \pm 80 \ \gamma$. The other by Collinson *et al.* (1973), gives 1.2 Oe. In addition to these estimates of the ancient fields, an order of magnitude estimate can be obtained by comparing the ratio of intensity of NRM to saturation isothermal remanent magnetization (IRM_s) for certain rocks (Fuller, 1973). This method is admittedly less precise than one would like, but has the advantage that it can be applied to a large number of samples. It also suggests a wide range of inducing fields from $10^2 \gamma$ to the order of an oersted.

It seems possible that some version of the fossil field model might be viable, if in addition to the background field, local transient perturbations of this field were possible to account for the very strongly magnetized samples. It is for this reason that we have been concerned with lunar surface processes, which might give rise to field amplification and a mechanism of magnetization, which might record the field. One of the most promising of the lunar surface phenomena appears to be meteoroid impact shock, and its related magnetization.

Shock magnetization in the weak shock range of less than a kilobar is a relatively well known phenomena (Shapiro and Ivanov 1967, Shapiro and Alova, 1970, Nagata, 1971). In the stronger shock range little information is available which is directly useful in assessing the possibility of lunar shock remanent magnetization. However, it has been demonstrated that shock loading of powdered fayalite to pressures of 200 kb gives rise to a shock associated remanence (Wasilewski, 1973a). The recovered sample consisted of a dispersion of iron spheres in glass (Sclar, 1969). Thus, new ferromagnetic material was formed and so remanence may be carried by a phase not present in the starting material, whenever porous iron bearing silicates are strongly shocked.

There have been a number of studies of the NRM of materials known to be shocked, either naturally, or in bomb tests. One such study by Hargraves and Perkins (1969) demonstrated certain magnetic effects due to shock, but the lack of control over experimental conditions made the interpretation of the processes involved difficult. Similarly, the studies of the Ries and Rochechouart craters do not give much indication of the detailed relations between shock and magnetization, although they clearly demonstrate that stable NRM was generated by these impacts (e.g., Pohl and Angenheister 1969, Pohl and Soffel 1971). Recently, Wasilewski (1973a) has shown that the NRM of a basalt from the Danny Boy nuclear test was much more stable than NRM of control samples, even though the saturation isothermal remanent magnetization (IRM_s) demagnetization curves were similar for shocked and control samples.

There are numerous indications that strong shock changes the magnetic characteristics of materials (e.g., Rose *et al.*, 1969, Wasilewski 1973b). Although these investigations are not directly indicative of remanence acquired during shock, they are relevant; for example, the latter paper records major and systematic changes in coercive force brought about by shock. Such effects strongly suggest that the remanent magnetization state will be altered by shock.

There is little doubt that impact related shocking has been an important process on the lunar surface, but until the magnetic effects of the shock are calibrated we cannot evaluate its importance as a mechanism of magnetization. The aim of our work is to begin to supply the necessary data to permit this calibration, by studying the magnetic effects of artificially shocked lunar soil.

The artificially shocked samples are of interest as crude calibration standards for the effect of shock (Christie *et al.*, 1973). This aspect of the work has led into considerations of the origin of certain breccias and most particularly of the origin of the excess iron in the soil and breccias.

EXPERIMENTAL TECHNIQUE

An adaptation of a simple technique for shocking metal foils was used in these experiments (Rose and Grace, 1967). The assembly and the explosive train are shown schematically in Fig. 1. This is essentially the flying plate technique described by Duvall and Fowles (1963). A Du Pont sheet explosive line wave generator is placed between two glass sheets and canted at an angle to form a plane

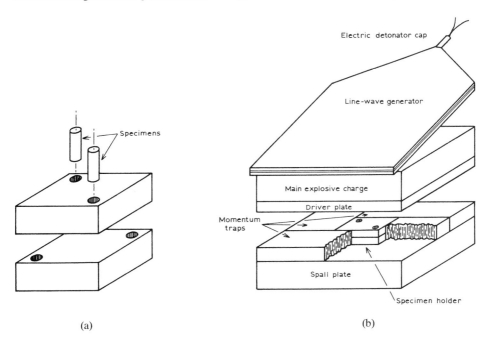

(a)   (b)

Fig. 1. Flying plate technique: (a) Copper blocks and sample retaining pins; (b) Explosive train and sample assembly.

wave generator. When it detonates, a plane of glass particles moves downwards, reaching the top surface of the main explosive charge simultaneously. The main charge then detonates and propels the driver plate, which, upon impacting the sample holder, generates an approximately plane wave. This shock wave propagates through the sample assembly.

The sample holder is shown in more detail in Fig. 1a. The blocks are made of high purity copper and have $\frac{1}{4}''$ holes driven approximately three-quarters of the way through them. An end mill was used to give a plane bottom surface to the hole. The sample is inserted into the hole and copper pins are press fitted into the holes. The sample now forms a small, partially compact disc, between one and five mils thick.

The copper blocks can be made almost any size consistent with good shock wave design. In these experiments, the blocks were $1'' \times 1'' \times 0.25''$ and add mating surfaces were ground to a $\#$ 16 micro-finish. Two blocks were used in each experiment. The physical dimensions of the side momentum traps and the spall plate are determined by the behavior of reflection waves from free surfaces in the system. In general, the lateral dimension of the side confinement is chosen, so that relief waves propagating in from the free surfaces at the sides cannot reach the sample whilst the shock is still in it. Thus, the sample sees a uniform plane wave shock compression followed by relief from the rear portion of the shock wave. Similarly, the dimensions of the spall plate are chosen, so that the entire pressure pulse passes through the sample before the reflected tensile wave from the free surface of the spall plate reaches the interface between the spall plate and sample holder.

In these experiments the samples were stopped in sawdust traps. This allowed the final orientation of the sample with respect to the earth's field to be obtained roughly.

Because of the preliminary nature of the experiments, it was decided to use an earlier calibration to estimate driver plate velocity, rather than to measure it for each experiment. The calibration consisted of the results of approximately fifty planar shock wave experiments, using the same explosives as were used in our experiments. In these experiments, driver plate velocities were accurately determined for several charge mass ratios.

The procedure used for estimating shock pressure was as follows: from accurate measurement of the explosive charge dimensions, the free surface velocity of the driver plate is estimated using the calibration curves. The pressure induced upon impact in the copper assembly is then determined using the cross curve method of impedance matching. From the scatter of the data making up the calibration curve, we think that the errors associated with the estimates of pressure in the copper blocks are $\pm 10\%$. The pressure-time profiles in the sample itself are far more complicated. The Hugoniot for lunar soils is not available and in any case would be composition and porosity dependent. Perhaps, most importantly from the point of view of these experiments, it should be noted that local reflections and focussing of the shock wave are likely to give peak pressures far in excess of the pressure in the copper and to give major inhomogeneites on the scale of micrometers.

In addition to the pressure pulse, there is a thermal pulse associated with the shock compression. We have used the expressions of McQueen and Marsh (1960) to obtain the temperature in the copper in the high pressure state and the residual temperature after the passage of the shock wave. It is evident that the heating of the copper in the 50 kb shock experiment is insignificant. Even at 250 kb we estimate that not more than approximately 200°C is reached in the copper. However, in the soil itself far higher temperatures will be reached locally.

The samples from the first run were studied magnetically in the copper blocks, so as to minimize the rise of modifying the shock remanent magnetization by sample preparation. However, the technique gives rise to difficulties in AF demagnetization due to the skin depth of the copper at 60 cycles. Subsequently, samples have been cut into small cylinders, which avoids the AF demagnetization difficulty. Moreover, it has been found that the remanence is not critically affected. A control experiment also revealed that, although magnetization is acquired due to the static loading of the sample during preparation for the sample assembly, it is small compared with the effects of the shock. After the completion of the studies of remanent magnetization acquired during shocking, basic magnetic observations of saturation magnetization, coercive force and magnetic viscosity were made so that a better understanding of the effect of shock upon magnetic behaviour of the soil could be obtained. Finally, the samples were sectioned and prepared for optical and electron microscopy.

The samples used in the experiments were lunar soil (65901,10), which comes from Station 5 on Stone Mountain at the Apollo 16 Descartes site.

Introduction

The results may be conveniently considered in three categories. First, there are the petrological observations, which indicate the effect of the shock upon the soil. Second, there are magnetic observations such as saturation magnetization and coercivity, which define the changes in magnetic behavior of the soil brought about by the shock. Finally, there are the determinations of remanent magnetization acquired as a result of the shock experiments.

Petrological observations

This aspect of the work is covered in detail in the paper by Christie *et al.* (1973) and so only major points are noted here. The recovered samples were all lithified and hence qualify as "instant rock" produced by the shock. This result is similar to that reported by Fredriksson and De Carli (1964) who artificially shocked material from the Bjurböle chondrite. A difficulty in the assessment of the effect of the artificial shock of the lunar soil is that it initially exhibits a range of natural shock effects. However, optical examination and electron microscopy of the starting material and the series of shocked samples at 50, 100, and 250 kb has revealed clear evidence of effects of the experimental shocking.

In the 50 kb sample moderate shock damage is seen in optical examination in the form of fracturing and displacement of fragments, but some glass spherules are undeformed. Some porosity is evident in the clasts. The 100 kb sample reveals a greater range of shock in the feldspar including veins of green or black isotropic material. Some of the pre-existing glass has been deformed. The plagioclase in the 250 kb samples has been almost completely transformed to an optically clear homogeneous glass.

Transmission electron microscopy demonstrated that two types of glass are present in the 50 kb shock sample (Fig. 2). One contains plentiful small particles, some of which are metallic iron. This glass is similar to that initially in the soil. The other is homogeneous clear glass, which provides the bonding material of the matrix and was produced by the shock experiment. The variability of shock effects in this sample is remarkable; tracks survive in some clasts but total melting has taken place elsewhere. The 100 kb sample contains more of the clear glass. In the larger clasts of plagioclase lamellar structures of probably shock origin are evident. The matrix of the 250 kb sample is glassy. Both clear homogeneous glass and particle bearing glass were produced in the experiment.

The particles were studied by electron diffraction and scanning electron microscopy. Unfortunately, the particles were too small to permit diffraction patterns to be made on individual particles, but regions with high particle concentrations gave *fcc* and *bcc* patterns. The amount of nickel in the region is less than 1% of the amount of iron. Thus the experimentally produced metal, like the metal in the glassy fraction of the soil (Housley 1972), is poor in nickel. Both soil and experimentally produced metal exhibit *fcc* and *bcc* diffraction patterns. It, therefore,

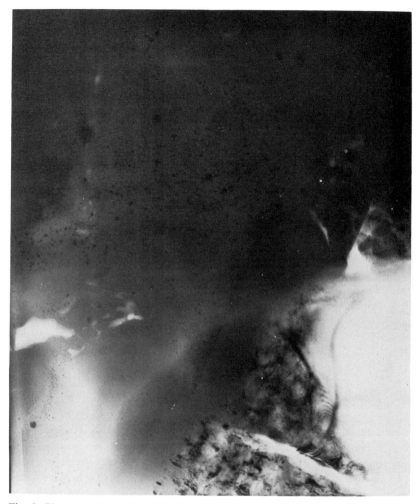

Fig. 2. Electron micrograph of 50 kb shocked lunar soil (×67,000); both clear and particle bearing glass are evident. Some deformation of the particle bearing glass is seen near the centre of the plate. The clear glass forms the bonding material of the artificial rock and was largely formed in the shock experiment. Particle bearing glass was present in soil initially.

appears that both in the soil and in the shocked material α- and γ-iron are present: the γ-phase is austenite and not kamacite.

In summary we note that the 50 kb sample is similar to unmetamorphosed lunar breccias. In contrast, the 250 kb sample is more glassy than lunar breccias. From a magnetic point of view, it is particularly important that fine iron is generated in the 250 kb sample.

Magnetic results

Measurements of saturation magnetization permit an estimate of the amount

of metallic iron present in the sample. Measurements of other magnetic parameters such as saturation remanent magnetization, coercivity and viscosity give
indications of the physical state of this iron, whether it is superparamagnetic,
single-domain, or multi-domain (e.g., Nagata *et al.*, 1972). Thus, by making these
observations, it is possible to define the effect of shock upon the magnetic characteristics of the soil samples. The results are summarized in Table 1.

The comparison between the measurements of the control sample made at
room temperature and at 4.2°K reveals an increase in saturation magnetization,
which is in part due to the temperature dependence of the saturation magnetization of iron, but also suggests the ordering of additional magnetic phases between
room temperature and 4.2°K in the soil. The marked increase of remanence is
likely to be primarily due to the transition of fine α-iron from superparamagnetic
to stable single domain, as is the increase in coercive force. However, the decrease in remanent coercivity is quite unexpected, and indicates the ordering of an
as yet unidentified soft magnetic phase.

The measurements of saturation magnetization at room temperature reveal
that in the 100 and 250 kb experiments iron has been produced. In contrast, the
50 kb result is little different from the control value and suggests the destruction of
a small amount of iron. Measurements of the magnetic viscosity, or time dependent changes of remanence, which in these samples is primarily due to superparamagnetic or near superparamagnetic material, revealed little difference between the control and the 50 kb sample. In contrast, the amount of material with
very short relaxation times increased markedly in the 250 kb sample. This result is
consistent with the production of very fine iron having a grain size of the order of
100 Å.

Neither the saturation isothermal remanence (IRM$_s$), nor the coercivity of the
samples changes in a simple manner with increasing shock pressure. The 50 kb
sample exhibits the highest intensity of isothermal remanence, although it is not
much greater than the 250 kb sample. The ratio of saturation isothermal remanence to saturation magnetization (IRM$_s$/I_s) is, however, considerably larger in the
50 kb sample than it is in any other. Similarly, the coercive force of the 50 kb
sample is the greatest; it is three times that of the control material. Hence, after
50 kb shock pressure the samples are more efficient carriers of isothermal rema-

Table 1. Effect of shock upon the magnetic properties of lunar soil—65901,10.

	Saturation Magnetization (I_s) gauss cm^3 g^{-1}	Saturation Remanent Magnetization IRM$_s$ gauss cm^3 g^{-1}	IRM$_s$/I_s	Coercive Force H_c Oersted	Remanent Coercivity H_{cr} Oersted	H_{cr}/H_c	Wt.% Metallic Iron
Control (RT)	1.39	0.045	0.032	22.5	500	22	0.6
4°K	1.95	0.27	0.14	152	350	2.3	—
50 kb (RT)	1.1	0.09	0.08	61	500	8.2	0.5
100 kb (RT)	1.9	0.066	0.035	46	550	12	0.9
250 kb (RT)	1.85	0.08	0.043	21	450	21	0.8

nent magnetization. However, the similarity of the remanent coercivity and the AF demagnetization curves indicate that the shock is not making the remanence more difficult to demagnetize or harder. Rather it appears to be eliminating some of the superparamagnetic material. This is again consistent with the variation of the ratio of remanence to saturation isothermal remanent magnetization. In the higher shock range of 100 and 250 kb, the magnetic characteristics of the samples tend to resemble those of the control or starting material. Measurements of IRM_s parallel and perpendicular to the plane of the shock wave revealed that 100 kb shock wave induces a magnetic anisotropy with the plane of the wave a preferred plane of magnetization.

In summary we note that there is good evidence for the production of additional fine iron in the higher shock range studied. In the 50 kb shock sample the ratio of IRM_s to saturation magnetization is highest, as is the coercive force. The increase of coercive force and IRM_s coupled with invariance of the remanent coercivity, require a decrease in the amount of superparamagnetic iron and an increase of the capability of the stable single domain iron to carry remanence. The magnetic anisotropy observed after 100 kb shock should be due to shape effect since the saturation magnetization of iron is so large (Uyeda *et al.*, 1963). It may arise due to the deformation of glass bearing iron particles, similar to that seen in the electron micrograph of the 50 kb sample. Such deformation evidently produces preferentially aligned elongate iron particles. The production of more elongate iron particles could also explain the changes in coercivity and IRM_s and the reduction of the amount of superparamagnetic iron, but the suggestion remains to be investigated.

Remanent magnetization acquired due to shock

The magnitude of remanence acquired in these shock experiments in the earth's field was between 2.3×10^{-4} and 1.3×10^{-3} gauss $cm^3 g^{-1}$. The values for the various experiments are given in Table 2. Somewhat surprisingly the 50 kb shock gave rise to the largest remanence. The AF and thermal demagnetization characteristics of the shock remanent magnetization (SRM) have not yet been completely established, but it is already clear that the remanence in this range of shock

Table 2. Shock remanent magnetization—65901,10.

	Magnitude of SRM J gauss $cm^3 g^{-1}$	Median Destructive Field Oersted	Median Destructive Temperature °C
50 kb	1.3×10^{-3}	$\simeq 100$	—
75 kb	3.2×10^{-4}	> 200	—
100 kb (1)	3.8×10^{-4}	—	$\simeq 260$
(2)	2.3×10^{-4}	$\simeq 110$	—
250 kb	5.5×10^{-4}	$\simeq 100$	—

is in part a stable type, like thermo-remanent magnetization (TRM), and not entirely soft, like weak field isothermal remanent magnetization (IRM); the median destructive fields observed were all in the range of 100–200 Oe, (Figs. 3 and 4, Table 2). Figure 3 also shows that the direction of magnetization of 75 kb shock sample did not change on demagnetization to 300 Oe. Similar behavior was previously seen in the 50 kb sample. In contrast, the 100 and 250 kb samples show marked changes in direction in this range of demagnetizing fields (Fig. 4). Stability against thermal demagnetization was determined for one of the two samples shocked at 100 kb, and the SRM was found to have distributed blocking temperatures (Figs. 4 and 5). As can be seen in Fig. 5, the SRM is smaller in value than partial thermo-remanent magnetization (pTRM) acquired between 350°C and room temperature in a 0.5 Oe but exhibits a greater range of blocking temperatures.

The availability of a thermal demagnetization of SRM and of pTRM acquisition in a 0.5 Oe field permits a rough estimate of the field intensity recorded,

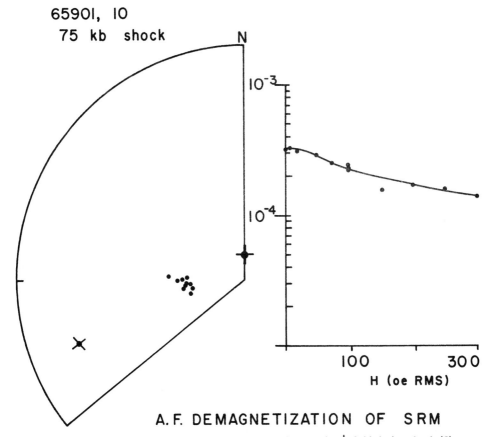

Fig. 3. Alternating field demagnetization of the 75 kb sample, ◆ field during shock, ✖ field during post shock cooling.

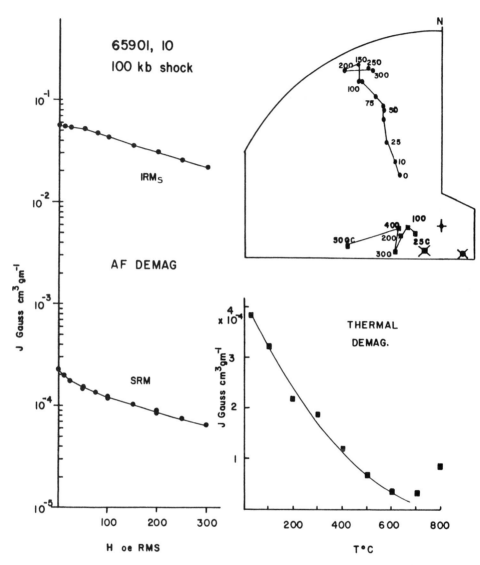

Fig. 4. Alternating field and thermal demagnetization of 100 kb shock sample, ◆ field during shock, ✖ field during post shock cooling of AF demagnetized sample, ♦ field during post shock cooling for thermally demagnetized sample.

making use of the same principles used in the Thellier/Thellier analysis; i.e., a comparison can be made between SRM lost and pTRM gained for particular temperature intervals. The SRM does not give very obviously non-ideal behavior in the low temperature range. The field estimate is low, but within a factor of ten of the local geomagnetic field.

An attempt was made to distinguish between remanence acquired during shock and during post shock cooling, by noting the orientation of the sample dur-

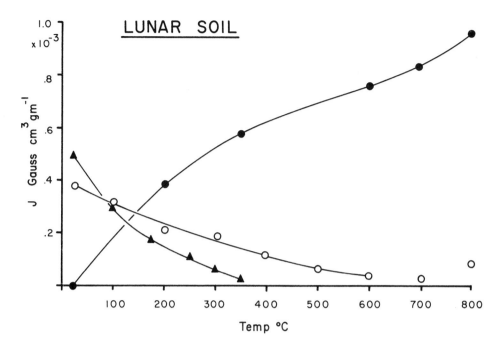

Fig. 5. Thermal demagnetization of shock remanent magnetization (SRM) and of 0.5 Oe partial thermo-remanent magnetization (pTRM) and acquisition of 0.5 Oe pTRM: ○ Demagnetization of SRM, ▲ demagnetization of pTRM, ● acquisition of pTRM.

ing shock and its final orientation in the sawdust trap. Unfortunately, the results did not give an unequivocal answer to the question, and were not simply related to either direction. However, it was evident, as we noted above, that in the low shock range no significant change of direction took place upon demagnetization to 200 Oe. In the case of the 50 kb sample, remanence was acquired in a direction which was somewhat similar to the orientation of the field in which the sample was recovered and dissimilar to the field during shock. The 75 kb sample acquired remanence in a direction intermediate between the field orientations during shock and during cooling (Fig. 3). Both the 100 and 250 kb shock remanence changed in direction during AF demagnetization up to 200 Oe. The initial direction of the 100 kb sample was close to both the field during shock and cooling, which happened to be similar. On demagnetization, the remanence migrates towards the plane parallel to the shock wave front, which is a plane of high anisotropy in the sample.

The most important result from these experiments is that shock remanent magnetization (SRM), in this shock range, is a strong remanence. Moreover, in 50 and 75 kb it is also stable. The 100 kb SRM has somewhat distinctive distributed blocking temperatures. Analysis of the direction in which remanence was acquired has not given any simple relation between the direction of the field during shock or during post-shock. The interpretation of remanence is obscure.

DISCUSSION OF RESULTS

The results of these artificial shocking experiments are of potential significance to the understanding of lunar magnetism. Not only is the calibration of the
shock remanent magnetization itself of importance, but so also is the range of
shock which may give rise to breccias and the production of iron, since they imply
certain conditions under which magnetic phases may have been formed on the
moon.

The results give some indication of the range of pressure in which breccias
could be formed by shock from the lunar soil. The experiment's simulation of the
lunar surface conditions is unfortunately imperfect; the sample is small and en-
cased in copper, so that it cools much more rapidly than it would on the moon.
From the appearence of the 250 kb sample, it seems that cooling was so rapid that
it precluded recovery. Moreover, the artificial shock experiments were carried out
on material packed in air, which clearly will modify the shock effects giving, for
example, extreme local gas temperatures. Nevertheless, the production of a
lithified rock at 50 kb strongly suggests that some lunar breccias may be formed at
pressures of tens of kb. This also implies a low temperature of origin which is of
great paleomagnetic importance, as well as being of obvious interest in other aspects of the lunar breccias, such as gas retention. Such a low shock pressure with
its implied low temperature of origin, would also explain the survival of glass in
these rocks and the wide variability of shock effects. Magnetically, the significance is, of course, that such breccias should carry a shock remanence or some
combination of shock and pTRM and not a true thermo-remanent magnetization.

The production of fine particles of iron in the glass derived from the pyroxene
in the 250 kb sample is consistent with the mechanism proposed by Housely (1972)
to explain the generation of excess iron in the soil. It is curious that both in the
artificially produced glass bearing iron and in the lunar soil glass, α- and γ-iron are
suggested by electron diffraction (Christie *et al.*, 1973). The γ-phase is, of course,
metastable, but may have been stabilized by the fine grain size. The presence of
metastable iron could be important magnetically. Unlike α-iron, it is paramagnetic
at room temperature. Hence, it will, if it inverts to α-iron, give rise to a remanence
recording the field in which the transition took place. If this happens on the moon,
then the weak ambient field will induce some secondary NRM. If it happens during thermal experiments, then it may produce difficulties such as non-ideal behavior in field intensity determinations. The presence of γ-iron may also provide
a means of distinguishing between models of iron generation such as that of
Housley (1972) which depend upon shock heating, from other explanations, such
as that of Pearce *et al.* (1972), who suggest that heating in the low oxygen fugacity
of the moon will give sub-solidus reduction of silicates and hence iron. It will be
interesting to see if this purely thermal mechanism gives rise to both α- and
γ-iron. Thus the suggested occurrence of the metastable γ-iron is of considerable
importance and warrants further investigation (Wasilewski, 1973c).

The magnitude of SRM produced in these experiments in the earth's field
reaches 10^{-3} J gauss $cm^3 g^{-1}$ for the 50 kb sample. Assuming the relationship between field and SRM to be linear for a given shock pressure, this gives SRM of

10^{-4} and 10^{-5} gauss cm^3g^{-1} for 10^3 and $10^2\gamma$ fields. Hence, using these values for ancient lunar fields, the NRM of certain lunar breccias can be explained. It remains to be seen what thermal demagnetization characteristics such breccias may have.

The nature of remanence produced is not yet clear, but we do have some idea of its origin. In the very high stress range, the whole sample exceeds the Curie point of its constituent ferromagnetic materials, so that remanence is of a thermo-remanent type (TRM) and acquired during post-shock cool down. In the shock range of tens of kb the whole sample does not exceed the Curie point of the ferromagnetic material, although locally high temperatures may be achieved for short periods of time. Hence, some partial thermo-remanent type of magnetization should be generated with some true shock remanence. At low shock pressures the process does not involve significant heating, so that only true shock remanence is seen.

50 kb is about the lowest shock pressure for which the flying plate technique can be conveniently used. To get to lower shock will involve modification of present gun techniques so that sample recovery can be achieved. However, it is clearly very important to study shock magnetization in this range. Larger volumes of rock will have been exposed to these lower pressures than to pressures in excess of 100 kb. Magnetization acquired in this range is more likely to be isothermal. Preliminary experiments with the soil in the very low shock range $(10^2$ bars) and fields of the order of an oersted gave rise to viscous magnetization, which is unlikely to be of importance because it relaxes too quickly. Thus in the range of a few tens of kb a transition to the type of remanence which we have investigated with the flying plate technique presumably takes place. Any heating effects in this range should be extremely localized and hence short lived. Thus both the isothermal and partial thermo-remanent type of magnetization may record the ambient field instantaneously.

We have not yet carried out a zero field experiment to see if shock remanent magnetization can be generated in the absence of an ambient field. It seems unlikely that remanence would be generated in such an experiment, but it is important to investigate the point. On the moon impact shocking now takes place either in the solar wind field or a local remanent field. It is possible that field enhancement may take place with either the solar wind field or a local remanent field as an ambient starting field. Such enhancement has been discussed by Hide (1972). More recently it has been pointed out (Inoue, private communication) that the geometry of a plasma motion in ejecta from an impact crater can give rise to a favorable situation for vertical field enhancement. Thus, the combination of a relatively small local field and impact enhancement, recorded by shock magnetization could possibly give rise to some of the larger values of NRM observed.

The possibility of generating some lunar NRM by shock processes is intriguing and could explain some of the puzzling features of lunar NRM such as its inhomogeneity (Dunn and Fuller, 1972). SRM may indeed be of a more general significance. NMR has been reported in many meteorites (Stacey et al., 1961; Pochtarev and Guskova, 1962; Banerjee and Hargraves, 1971, 1972; Brecher, 1971,

1972). Although this NRM is not understood, a shock origin is not implausible, since many meteorites exhibit shock effects and moreover, recent ideas on the origin of chondrules (Urey, 1952, Fredriksson et al., 1973) suggest that impact plays an important role in the origin of chondrites. In addition, collision between planetesismals is likely to be an important aspect of planetary formation. Hence, as already noted by Meadows (1972) and Wasilewski (1973b) shock associated magnetization may be a relatively common phenomena in the solar system of which we see but a very imperfect record, preserved under special circumstances.

Acknowledgments—The authors gratefully acknowledge the assistance of J. R. Dunn, of the Department of Earth and Planetary Sciences, University of Pittsburgh, F. C. Schwerer and J. S. Lally of the U.S. Steel Research Centre, Monroeville, Pennsylvania, and J. M. Christie of the Department of Geology, U.C.L.A..

REFERENCES

Alfven H. and Lindberg L. (1973) Magnetization of primeval earth and moon, *The Moon.* In press.

Banerjee S. K. and Hargraves R. B. (1971) Natural remanent magnetization of carbonaceous chondrites. *Earth Planet. Sci. Lett.* **10**, 392.

Banerjee S. K. and Hargraves R. B. (1972) Natural remanent magnetization of carbonaceous chondrites and the magnetic field in the early solar system. *Earth Planet. Sci. Lett.* **1**, 110.

Brecher A. (1971) Early interplanetary fields and the remanent magnetization of meteorites. *Publ. Astron. Soc. Pacific* **83**, 602.

Brecher A. (1972) The paleomagnetic record in meteorites. *Meteoritics* **8**, 17.

Christie J. M., Heuer A. H., Radcliffe S. V., Lally J. S., Nord G. L., Fisher R. M., and Griggs D. J. (1973) Lunar breccias; An electron petrographic study (abstract). In *Lunar Science—IV*, pp. 133–135. The Lunar Science Institute, Houston.

Collinson D. W., Runcorn S. K., and Stephenson A. (1973) Magnetic properties of Apollo 16 rocks. *Proc. Fourth Lunar Sci. Conf., Geochim. Cosmochim. Acta.* In press.

Dunn J. R. and Fuller M. (1972) Thermo-remanent magnetization (TRM) of lunar samples. *The Moon* **4**, 50–62.

Duvall G. E. and Fowles G. R. (1963) Shock waves. In *High Pressure Physics and Chemistry* (editor R. S. Bradley) Vol. 2, pp. 209–291. Academic Press.

Fredriksson K. and De Carli P. (1964) Shock emplaced argon in a stony meteorite, 1, shock experiment and petrology of sample. *J. Geophys. Res.* **69**, 1403–1406.

Fredriksson K., Noonan A., and Nelen J. (1973) Meteoritic, lunar, and lunar impact chondrules. *The Moon.* In press.

Fuller M. (1973) Lunar magnetism. Submitted to *Rev. Geophys. Space Phys.*

Hargraves R. B. and Perkins W. E. (1969) Investigations of the effect of shock on natural remanent magnetism. *J. Geophys. Res.* **74**, 2576–2589.

Hide R. (1972) Comments on the moon's magnetism. *The Moon* **4**, 213.

Housley R. W., Groni R. W., and Abdel-Gawad M. (1972) Study of excess Fe–Ni in the lunar fines by magnetic separation Mössbauer spectroscopy and microscopic observations. *Proc. Third Lunar Sci. Conf., Geochim. Cosmochim. Acta.*, Suppl. 2, pp. 1, 1065–1070. MIT Press.

Levy E. H. (1972) Magnetic dynamo in the moon: A comparison with earth. *Science* **178**, 52.

McQueen R. G. and Marsh S. P. (1960) Equation of state for nineteen metallic elements from shock wave measurements to two megabars. *J. Appl. Phys.* **31**, 1253–1269.

Meadows A. J. (1972) Remanent magnetization in meteorites. *Nature* **237**, 274.

Nagata T. (1971) Introductory notes on shock remanent magnetization and shock demagnetization of igneous rocks. *Pure Appl. Geophys.* **89**, 159–177.

Nagata T., Fisher R. M., Schwerer F. C., Fuller M., and Dunn, J. R. (1972) Magnetism of Apollo 14 rocks. *Proc. Third Lunar Sci. Conf., Geochim. Cosmochim. Acta.*, Suppl. 1, Vol. 3, p. 3.

Pearce G. W., Williams R. J., and McKay D. S. (1972) The magnetic properties of metallic iron produced by subsolidus reduction of synthetic Apollo 11 composition glasses. *Earth Planet. Sci. Lett.* **17**, 95.

Pochtarev V. I. and Guskova E. G. (1962) The magnetic properties of meteorites. *Geomag. Aeron.* **2**, 626.

Pohl J. and Angenheister G. (1969) Anomalien des Erdmagnetfeldes und Magnetesierung der Gesteine im Nördlinger Ries. *Geol. Bavaria* **61**, 327.

Pohl J. and Soffel H. (1971) Paleomagnetic age determination of the Rochechouart-impact structure. *Zeit. Geophys.* **37**, 857.

Rose M. F. and Grace F. I. (1967) A simple technique for shock deforming metal foils. *Brit. J. Appl. Phys.* **18**, 671–674.

Rose M. F., Villere M. P., and Berger T. L. (1969) Effect of shock waves on the residual magnetic properties of armco iron. *Phil. Mag.* **19**, 39.

Runcorn S. K., Collinson D. W., O'Reilly W., Battey M. H., Stephenson A., Jones J. N., Manson A. J., and Readman P. W. (1970) Magnetic properties of Apollo 11 lunar samples. *Proc. Apollo 11 Lunar Sci. Conf., Geochim. Cosmochim. Acta.*, Suppl. 1, Vol. 3, pp. 2360–2387. Pergamon.

Runcorn S. K., Collinson D. W., O'Reilly W., Stephenson A., Battey M. H., Manson A. J., and Readman P. W. (1971) Magnetic properties of Apollo 12 lunar samples. *Proc. Roy. Soc. London (A)* **325**, 157–174.

Sclar C. B. (1969) Shock wave damage in olivine. *EOS* **50**, 219.

Shapiro V. A. and Ivanov N. A. (1967) Dynamic remanence and the effect of shocks on the remanence of strongly magnetic rock. *Dokl. Akad. Nauk. SSSR* **173**, 1065–1068.

Shapiro V. A. and Alova N. I. (1970) Characteristics of dynamic remanent magnetization of ferromagnetic rocks at temperatures from 0° to 600°C, *Izv. Akad. Nauk SSSR.* (Physics of the Solid Earth) **11**, 743.

Stacey F. D., Lovering J. F., and Parry L. G. (1961) Thermomagnetic properties, natural magnetic moments and magnetic anisotropies of some chondritic meteorites. *J. Geophys. Res.* **66**, 1523.

Strangway D. W., Pearce G. W., Gose W. A., and Timme, R. W. (1971) Remanent magnetization of lunar samples. *Earth Planet. Sci. Lett.* **13**, 43–52.

Strangway D. W., Gose W. A., Pearce G. W., and Carnes J. G. (1973a) Magnetism and the history of the moon. *Proc. 18th Ann. Conf. Magnetism and Magnetic Materials, J.A.P.* In press.

Strangway D. W., Sharpe H. N., Gose W. A., and Pearce G. W. (1973b) Magnetism and the early history of the moon (abstract). In *Lunar Science—IV*, pp. 697–699. The Lunar Science Institute, Houston.

Toksöz N. N., Press F., Dainty A. M., Solomon S. C., and Anderson, K. R. (1973) Lunar structure, compositional inferences and thermal history (abstract). In *Lunar Science—IV*, pp. 734–735. The Lunar Science Institute, Houston.

Urey H. C. (1952) *The Planets*. Yale University Press.

Urey H. C. and Runcorn S. K. (1973) A new theory of lunar magnetism. *Science* **180**, 636–638.

Uyeda S., Fuller M., Belshe J. C., and Girdler R. W. (1963) Anisotropy of susceptibility of rocks and minerals. *J. Geophys. Res.* **68**, 279.

Wasilewski P. J. (1973a) Shock remanent magnetization associated with meteorite impact at planetary surfaces. *The Moon.* In press.

Wasilewski P. J. (1973b) Magnetic remanence mechanisms in iron and iron nickel alloys, metallographic recognition criteria and implications for lunar sample research. *Proc. Fourth Lunar Sci. Conf., Geochim. Cosmochim. Acta.* In press.

Wasilewski, P. J. (1973c) γ-iron in lunar samples. Submitted to *Science.*

Proceedings of the Fourth Lunar Science Conference
(Supplement 4, *Geochimica et Cosmochimica Acta*)
Vol. 3, pp. 3019–3043

Magnetic properties and natural remanent magnetization of Apollo 15 and 16 lunar materials

TAKESI NAGATA

Geophysics Research Laboratory,
University of Tokyo, Tokyo, Japan

R. M. FISHER and F. C. SCHWERER

U.S. Steel Corporation, Research Center,
Monroeville, Pennsylvania

M. D. FULLER and J. R. DUNN

Department of Earth and Planetary Sciences,
University of Pittsburgh, Pittsburgh, Pennsylvania

Abstract—The basic magnetic characteristics and the intensity and stability of the natural remanent magnetization of five Apollo 15 materials (15058, 15418, 15495, 15555, and 15556) and eight Apollo 16 materials (60255, 60315, 65010, 66055, 66095, 67455, 68415, and 68815) have been examined. The anorthositic highland materials containing only 4–9% of FeO have a considerably smaller paramagnetic susceptibility in comparison with the lunar mare materials.

The ferromagnetic constituent in all these highland materials including igneous rocks consists of both essentially pure metallic iron and a considerable amount of kamacite. An appreciable fraction of the ferromagnetic metal grains are in the form of extremely fine particles which behave super-paramagnetically or even pseudo-paramagnetically at room temperature. Hence their saturation remanent magnetization and coercive force increase considerably with decreasing temperature down to 4.2 K, continuously in some samples and discontinuously in others. The smallest estimated size of the fine particles is 40 Å or less in mean diameter.

The natural remanent magnetization (NRM) of Apollo 15 and 16 rocks seems to be composed of a soft component and a hard and stable component. The soft component can be demagnetized easily by AF-fields of 15–30 Oe rms. and sometimes its direction is much different from that of the stable component, so that the soft component can be regarded as a result of magnetic contamination. The intensity and direction of the stable component of NRM is approximately invariant against AF-demagnetization up to 200 Oe rms. or more. It seems that rocks with large remanence coercive force can maintain a larger and more stable remanent magnetization. The intensity of the stable NRM obtained as a result of magnetic cleaning amounts to $(0.9 - 11) \times 10^{-6}$ emu/gm. These stable NRM's can be attributed to the thermoremanent magnetization acquired in a magnetic field of 800–3,000 γ or to the piezoremanent magnetization acquired by a uniaxial compression of 50 Kbars in a field of 200–400 γ.

INTRODUCTION

THE MAGNETIC properties and characteristics of the natural remanent magnetization of Apollo 11, 12, and 14 lunar materials have been systematically examined by the authors (Nagata *et al.*, 1970, 1971, 1972a, 1972b), and those of Apollo 15 lunar materials also have been largely studied (Nagata *et al.*, 1972c, 1973). The results of these studies have clearly shown that the main ferromagnetic constituent in the

3019

lunar materials is an Fe–Ni–Co alloy, in which the amount of Fe present is very much greater than the other two elements, and that the presence of fine super-paramagnetic particles of native iron is common throughout all lunar materials, although the relative amount of fine particles ranges from a few per cent to about half the total native iron abundance. These two remarkable characteristics of the magnetic properties of lunar materials concluded by the authors are in agreement with the results obtained by others studying the magnetization of lunar rocks (e.g., Pearce *et al.*, 1971; Grommé and Doell, 1971; Collinson *et al.*, 1972; Gose *et al.*, 1972).

Apollo 11, 12, and 14 lunar samples are believed to be mostly mare materials while some Apollo 15 samples and most Apollo 16 samples are considered to represent the lunar highlands. Differences in magnetic properties between the mare and highland materials may be expected and, actually, an unusually small value of paramagnetic susceptibility (1.5×10^{-5} emu/gm) is observed for an anorthosite breccia (sample 15418) and a considerable amount of ferromagnetic kamacite is found in a basaltic rock (sample 15556) at the Apollo 15 landing site (Nagata *et al.*, 1972c). In addition, analysis of the natural remanent magnetization (NRM) of lunar samples have indicated that some lunar rocks maintain an intense and highly stable NRM, while many other lunar rocks carry only a small amount of stable NRM (i.e., 10^{-6} emu/gm in the order of magnitude) along with a much larger amount of soft remanent magnetization probably attributable to magnetic contamination acquired after these lunar samples were returned to the earth's surface.

In this connection it is important to recognize that the relaxation time at 450° K for the remanent magnetization of the most probable large single domain particles of metallic iron (i.e., 270 Å in mean diameter) is only about 1.7×10^8 years for lunar rocks with only a small measure of stable NRM, while for rocks with a considerable amount of stable NRM it is estimated to be 4×10^{15} years or more. Since the isotope age of these lunar rocks is $(3-4) \times 10^9$ years, it seems likely that a reliance on NRM as the magnetic fossil should be limited to those rocks whose magnetic relaxation time is considerably larger than their age. Thus, the reliability of NRM of Apollo 16 lunar rocks is examined from this viewpoint in addition to the ordinary AF- and thermal-demagnetization tests in the present report.

DESCRIPTION OF APOLLO 15 AND 16 LUNAR MATERIALS

According to the preliminary descriptions given in Apollo 15 Preliminary Science Report (NASA 1972), the petrographic characteristics of the five samples examined are as follows:

15058: A blocky, angular medium-grain vuggy basalt. (C(FeO) = 19.97%)
15418: A blocky, angular breccia with 5% to 10% leucocratic clasts bigger than 1 mm. (C(FeO) = 5.37%)
15495: A blocky, subangular, vuggy basalt with dark-brown pyroxene prisms to 10 mm long in vugs.

15555: A blocky, subangular, very vuggy, coarse-grained basalt with green and brown mafic silicates and plagioclase with equiangular texture. (C(FeO) = 22.47%)

15556: A blocky, subangular, highly vesicular fine-grained basalt. (C(FeO) = 22.25%)

The relative abundance of Fe (assumed as FeO) from sample analysis also given in the report is shown in the parentheses. The FeO content in sample 15418 is much less than in the others, whereas the content of Al_2O_3 amounts to 26.73% and is less than 10% in the other three samples; namely, sample 15418 is anorthositic.

According to the preliminary descriptions given in Lunar Sample Information Catalog Apollo 16 (NASA 1972) the petrographic characteristics of eight Apollo samples are as follows:

60255: A block glassy breccia with angular white clasts. The overall composition appears to be that of gabbroic anorthosite.

60315: A hornfelsed diabase or basalt porphyry. (C(Fe) = 8.86%, C(Al_2O_3) = 17.24% C(Ni) = 191 ppm)

65010: Soils at station 5: (C(FeO) = 5.87, C(Al_2O_3) = 26.47%, C(Ni) = 356 ppm) for soil sample 65701 sampled at the same station.

66055: A breccia with white matrix: 70% feldspar (fine-grained, white, sugary lustre), 20% pale, yellowish-green mineral (anhedral, fine-grained), 5% black mineral. (C(FeO) = 4.2–7.8%, C(Al_2O_3) = 19.6–24.8%) (McKay *et al.*, 1973)

66095: A medium light gray anorthosite: 57% plagioclase, 40% olivine or pyroxene, 2% opaque mineral and 1% metal and goethite. (C(FeO) = 7.16%, C(Al_2O_3) = 23.55%, C(Ni) = 258 ppm)

67455: A blocky, very friable, white or grayish white breccia. (C(FeO) = 3.41%, C(Al_2O_3) = 30.42%) (Rose *et al.*, 1973)

68415: A greenish-gray anorthositic gabbro: 85% plagioclase, 10% mafic silicate and 2% opaque mineral. (C(FeO) = 4.25%, C(Al_2O_3) = 28.63%, C(Ni) = 49 ppm)

68815: A tough, medium dark gray breccia, most probably a nearly completely melted breccia of anorthositic gabbro composition which has crystallized to a very fine grain crystalline rock with several remnant grains that did not melt: matrix (90% of rock) consists of 30% plagioclase, 35% clinopyroxene and 35% devitrified glass white mineral; clasts (10% of rock, consist of 90% plagioclase and 10% clinopyroxene. (C(FeO) = 4.75%, C(Al_2O_3) = 27.15%, C(Ni) = 206 ppm).

As far as data are available, the relative abundance of FeO, C(FeO), and that of Al_2O_3, C(Al_2O_3), of these Apollo 16 samples are listed in the above descriptions. In general, Apollo 16 lunar materials are anorthositic, being rich in Al_2O_3 and poor in FeO. The relative abundance of Ni, C(Ni), is unusually high in Apollo 16 igneous rocks such as sample 60315 and 68415.

BASIC MAGNETIC PROPERTIES

The basic magnetic properties of the lunar materials examined in the present study are as follows: the initial magnetic susceptibility at room temperature (χ_0)

measured with the aid of an induction susceptibility bridge of 1000 Hz in frequency and 2 Oe in peak intensity. The saturation magnetization (I_s), the saturation remanent magnetization (I_R), the coercive force (H_c), the remanence coercive force (H_{RC}) and the paramagnetic susceptibility (χ_a) at room temperature and at the liquid helium temperature (4.2°K) derived from magnetic hysteresis curves at respective temperatures measured by a vibration magnetometer in a field range from -16 KOe to $+16$ KOe, the magnetic transition temperatures (Θ_c and Θ_c^*) measured by a Curie balance in a constant magnetic field of 5.53 KOe in 10^{-6}–10^{-5} Torr in atmospheric pressure for a temperature range from 20°C to 850°C, where Θ_c^* denotes the lower transition temperature(s) observable only in the cooling process. In addition, magnetization versus temperature curves were observed for a temperature range between 4.2°K and 300°K in a constant magnetic field of 10 KOe with the aid of a vibration magnetometer in order to detect the antiferromagnetic Néel point, if any, in the low temperature range.

Table 1 gives a summary of the observed values of these basic magnetic parameters at room temperature of the five Apollo 15 lunar materials, where two samples (15556,37 and 15556,38) of an igneous rock mass (15556) are separately measured, and further sample 15556,38 has been divided into a dark black colored part (38-1) and a bright gray colored part (38-2) for the magnetic measurements. The observed small differences among the numerical values of the magnetic parameters may represent the range of inhomogeneity of magnetic constituents within the rock mass.

Table 2 shows a summary of the basic magnetic parameters at room temperature of the eight Apollo 16 lunar materials. It may be noted in this table that χ_a-values of all the Apollo 16 materials are considerably smaller than those of Apollo 11 through Apollo 14 materials and the majority of Apollo 15 samples, in which $\chi_a > 2.1 \times 10^{-5}$ emu/gm and that all the samples are associated with the lower magnetic transition temperature (Θ_c^*), which indicates the presence of an

Table 1. Magnetic properties of Apollo 15 lunar materials at room temperature.

				Sample				
Magnetic parameter	15058 -55	15418 -52	15495 -132	15555 -37	15556 -37	15556 -38-1	15556 -38-2	Unit
χ_0	4.54	0.86	—	—	2.58	4.46	2.52	$\times 10^{-4}$ emu/gm
χ_a	3.5	1.5	3.8	3.8	3.8	4.3	4.1	$\times 10^{-5}$ emu/gm
I_s	0.125	0.145	0.165	0.127	0.125	0.124	0.111	emu/gm
I_R	10	13	7.5	10	15	19	7	$\times 10^{-4}$ emu/gm
H_c	10	10	10	7	10	17	10	Oersteds
H_{RC}	—	—	—	—	—	—	415	Oersteds
Θ_c	790	765	—	—	784	786	782	°C
Θ_c^*	—	668	—	—	668	641	668	°C
m(Fe)	0.057	0.067	0.076	0.058	0.057	0.057	0.052	wt.%
Ni/(Fe + Ni)	0	4.0	—	—	4.0	5.5	4.0	wt.%
T^*	110	No sharp change	120	105	110	105	105	°K

Table 2. Magnetic properties of Apollo 16 lunar materials at room temperature.

Magnetic parameter	Sample 60255	60315 -47	65010	66055 -18	66095 -39	67455 -19	68415 -53	68815 -70	Unit
χ_0	12.1	11.5	22.4	15.6	9.7	0.66	2.45	12.9	$\times 10^{-4}$ emu/gm
χ_a	1.7	3.1	1.8	0.4	2.6	0.7	1.0	1.4	$\times 10^{-5}$ emu/gm
I_s	1.02	9.70	1.39	0.82	2.63	0.046	0.66	1.34	emu/gm
I_R	570	7	450	64	63	1.3	30	30	$\times 10^{-4}$ emu/gm
H_c	47	3	22	20	7.5	20	15	12	Oersteds
H_{RC}	680	—	500	330	250	270	410	350	Oersteds
Θ_c	780	775	780	778	768	811	781	772	°C
Θ_c^*	638	663	651	640	643	734	$\left\{ \begin{array}{l} 395 \\ 682 \end{array} \right.$	651	°C
m(Fe)	0.47	4.45	0.64	0.38	1.21	0.021	0.30	0.61	wt.%
Ni/(Fe + Ni)	6.0	5.0	5.5	6.0	5.5	3.0	$\left\{ \begin{array}{l} 12.0 \\ 4.0 \end{array} \right.$	6.0	wt.%
T^*	No sharp change	—	No sharp change	65	—	55	85	No change	°K

appreciable amount of kamacite phase. As shown in Table 2 the H_{RC} of sample 60255 is considerably larger than that of the others, suggesting its remanent magnetization relaxation time is sufficiently long to maintain the remanence throughout the age of the moon (Nagata *et al.*, 1973).

Figures 1 and 2 illustrate the magnetization versus temperature curves in a constant magnetic field of 10 KOe in a low temperature range from 4.2 to 150°K for Apollo 15 and 16 materials, respectively. No conspicuous Néel point peak of ilmenite is observed around 55°K in any of these curves, but a very marked Néel point can be observed at about 25°K in the curve of sample 15058 only. This Néel point could be due to an antiferromagnetic pyrox-ferroite or other pyroxenes (Nagata *et al.*, 1971b), but no direct proof of the presence of such a phase has been made as yet.

MAGNETIC TRANSITION TEMPERATURE

All Apollo 16 lunar materials examined in the present study show a reproducible thermal hysteresis in their magnetization versus temperature relationship in a high temperature range between 20°C and 850°C. Figure 3 illustrates a typical example of reproducible thermal hysteresis for several heating and cooling cycles. In most cases, the thermal hysteresis has a pattern as shown in Fig. 4, where the magnetization curve is composed of a thermally irreversible component (I) and a thermally reversible one (II) as well as the paramagnetic component (III). The thermally irreversible component (I) can be identified to a NiFe alloy where the intensity of magnetization is practically zero at temperature above the $\gamma \rightarrow \alpha$ transition point, while the reversible component (II) may represent an almost pure iron phase. The magnetization as a function of absolute temperature (T) and

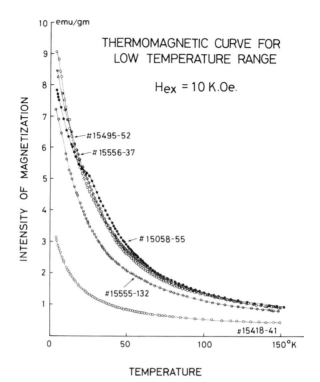

Fig. 1. Thermomagnetic curves of Apollo 15 samples in a temperature range between
4.2°K and 150°K.

strong magnetic field (H) results from a paramagnetic component ($\chi_a(T) \cdot H$),
kamacite (I_k), and almost pure metallic iron (I_F), and is given by

$$I(T, H) = m_p \frac{C}{T} H + m_k I_k(T) + m_F I_F(T), \qquad (1)$$

where χ_a varies inversely with T, $I_k = 0$ above the $\gamma \to \alpha$ transition (Θ_c^*) and m_p, m_k
and m_F represent respectively masses of paramagnetic, kamacite and iron compo-
nents in a unit mass of a specimen. The practical curve fitting analysis of an
observed $I(T) \sim T$ curve with the aid of eq. (1) has been carried out in the
following ways. In the cooling curve,

$$I(T) = m_p \frac{C}{T} H \quad \text{for} \quad T > \Theta_c,$$

$$I(T) = m_p \frac{C}{T} H + m_F I_F(T) \quad \text{for} \quad \Theta_c > T > \Theta_c^*,$$

$$I(T) = m_p \frac{C}{T} H + m_k I_k^c(T) + m_F I_F(T) \quad \text{for} \quad \Theta_c^* > T,$$

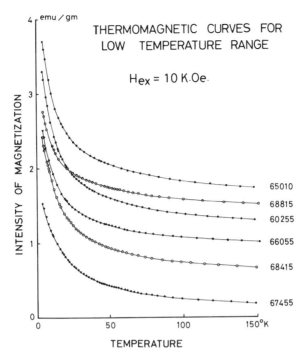

Fig. 2. Thermomagnetic curves of Apollo 16 samples in a temperature range between 4.2°K and 150°K.

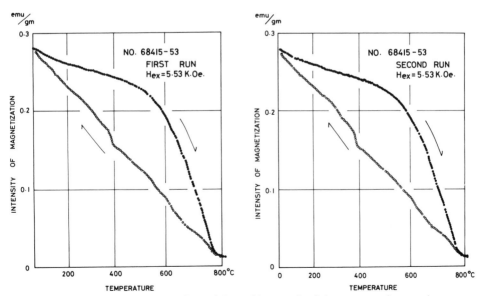

Fig. 3. Example of the comparison of thermal hysteresis of thermomagnetic curve in a temperature range from 20°C to 820°C between the first and second runs of measurement.

while in the heating curve,

$$I(T) = m_p \frac{C}{T} H \quad \text{for} \quad T > \Theta_c,$$

$$I(T) = m_p \frac{C}{T} H + m_k I_k^h + m_F I_F(T) \quad \text{for} \quad \Theta_c > T,$$

where I_k^c and I_k^h represent respectively the magnetization of kamacite component in the cooling and heating processes. Hence the magnitude of $m_p C$ can be estimated from $I(T)$ values for $T > \Theta_c$, and then $m_F I_F(T)$ values are evaluated from $I(T)$ for a temperature range of $\Theta_c > T > \Theta_c^*$ in the cooling curve. Since,

$$I(\text{heating}) - I(\text{cooling}) = m_k I_k^h(T) \quad \text{for} \quad \Theta_c > T > \Theta_c^*,$$

$m_k I_k^h(T)$ also can be estimated for a temperature range of $\Theta_c > T > \Theta_c^*$. Assuming then that the dependence of $I_F(T)$ upon T is subjected to Curie law which is approximately represented by a curve of

$$\left(I(T) - m_p \frac{C}{T} H \right)$$

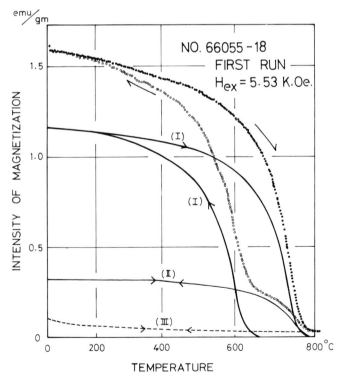

Fig. 4. Example of the resolution or inversion of thermal hysteresis of thermomagnetic curve into three components. (*See* text.)

in the heating process, both $m_F I_F(T)$ and $m_k{}^h I_k(T)$ can be estimated for the whole range of T. Finally, $m_k I_k{}^c(T)$ can be determined from $I(T)$ in the cooling process by subtracting $m_p(C/T)H$ and $m_F I_F(T)$. Curves (I), (II) and (III) in Fig. 4 illustrate respectively $m_k I_k(T)$, $m_F I_F(T)$, and $m_p(C/T)H$ thus determined.

In the case of sample 68415 shown in Fig. 3, however, the cooling curve of $I(T)$ is more complicated than in the other cases, having at least three transition points, i.e., $\Theta_{c1}^* = 395°C$, $\Theta_{c2}^* = 682°C$ and $\Theta_c = 781°C$, and the heating curve of $I(T)$ also seems to have at least two transition points, i.e., $\Theta_{c1} \simeq 700°C$ and $\Theta_{c2} = 781°C$. Results of curve-fitting in this case are shown in Fig. 5, where curves (I) and (II) represent a kamacite component of both high and low Ni content respectively, curve (III) the reversible almost pure iron component while curve (IV) shows the paramagnetic magnetization. Compared with the thermal hysteresis curves of kamacites of various Ni contents, the pattern of (I) is reasonably normal but that of (II) in the cooling process is considerably deviated from the standard cooling curve of kamacite of a single composition. The most reasonable interpretation of this deviation appears to be that Θ_{c2}^* is not a single value but spreads over a temperature range of about 150 degrees below 682°C. As given in Table 2,

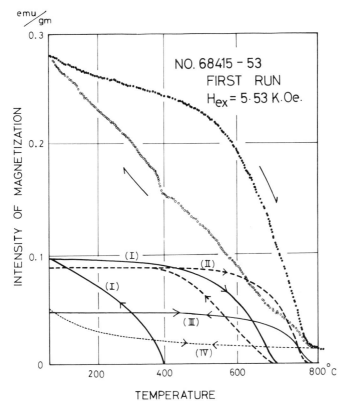

Fig. 5. Example of the inversion of thermal hysteresis of thermomagnetic curve into four components.

$\Theta^*_{c1} = 395°C$ and $\Theta^*_{c2} = 682°C$ correspond to the $\gamma \rightarrow \alpha$ transition temperatures of kamacite phases of 12% and 4.5% respectively in Ni content. Kamacites in sample 68415 therefore appear to have a sharp peak around 12% and a small group with Ni contents from 4% to 9%. Actually, Gancarz et al. (1972) have examined the contents of Ni and Co in seven native iron blebs in sample 68415, showing that those iron blebs' Ni contents are 12–13.5% in three, 9% in one, 7% in one, 5.5% in one and 3% in one. This result seems to be in fairly good agreement with the conclusion derived from the present magnetic analysis.

Ni contents, Ni/(Fe + Ni), in kamacite phases in all Apollo 15 and 16 materials estimated from Θ^*_c values are summarized together with the total metallic iron contents, m(Fe), estimated from I_s values, in Table 1 and 2 respectively. Apollo 11 through Apollo 16 lunar materials which have been confirmed by the authors to contain a kamacite phase are listed in Table 3, where the total abundance of metallic iron (m), the ratio of almost pure iron to kamacite (m_F/m_k) in the metallic iron, and the Ni content in the kamacite phase are all summarized. It may be worthwhile to note that only Apollo 15 and 16 igneous rocks contain kamacite and none has been detected in Apollo 11 and 12 lunar materials listed in Table 3.

A general characteristic of the magnetic transition temperature of Apollo lunar

Table 3. Kamacite component in lunar metallic iron—Total abundance of metallic iron (m); ratio of iron to kamacite (m_F/m_k) and Ni content of kamacite phase (Ni/(Fe + Ni)).

Sample	$m = m_F + m_k$	m_F/m_k	Ni/(Fe + Ni) in kamacite
(Igneous rocks)	(% in weight)		(% in weight)
15556-37	0.057	0.58	4.0 ± 0.5
15556-38	0.057	0.42	5.5 ± 0.5
60315-47	4.45	0.71	5.0 ± 0.5
68415-53	0.11	0.29	$\left\{ \begin{array}{l} 4.5 ± 0.5 \\ 12.0 ± 0.5 \end{array} \right.$
(Breccias)			
14301-65	0.32	0.29	5.5 ± 0.5
14303-35	0.58	0.14	7.0 ± 0.5
14311-45	0.34	0.44	3.0 ± 0.5
15418-41	0.057	0.15	4.0 ± 0.5
60255-	0.47	0.88	6.0 ± 0.5
66055-18	0.44	0.27	6.0 ± 0.5
66095-39	1.21	0.21	6.0 ± 0.5
67455-19	0.024	0.12	2.5 ± 0.5
68815-70	0.61	0.97	5.5 ± 0.5
(Fines)			
65901-10	0.83	1.00	5.5 ± 0.5

Remarks: No kamacite phase is detected in samples 10024-22, 12053-47, 14053-48, 15058-55 (igneous rocks); 10048-55, 14047-47 (breccias); 10084-89, 12070-120, and 14259-69 (fines).

HISTOGRAMS OF MAGNETIC
TRANSITION TEMPERATURES

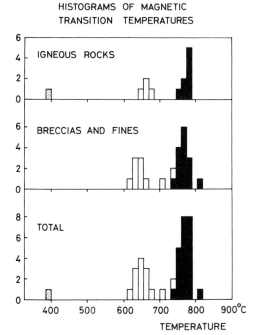

Fig. 6. Histogram of magnetic transition temperature of Apollo 11–16 lunar materials.
Full column: Magnetic transition temperature in the heating process which coincides
with the highest transition temperature in the cooling process (Θ_c).
Blank column: Lower transition temperature which appears only in the cooling process (Θ_c^*).
Gray column; Lowest transition temperature in the cooling process for three component ferromagnetic constituent. (Θ_{c1}^*).

materials may be observed in a histogram of Θ_c and Θ_c^* of all the samples from Apollo 11 through 16 examined by the authors, which is illustrated in Fig. 6, where the magnetic transition temperatures of 23 Apollo lunar materials are summarized. Among 23 lunar samples, 9 samples have only one magnetic transition temperature Θ_c around 770°C, while the other 14 samples show the kamacite $\gamma \rightarrow \alpha$ transition (Θ_c^*) in the cooling process. Table 3 and Fig. 6 seem to statistically indicate that the ferromagnetic constituents in lunar materials are almost pure metallic iron and kamacite of 5–6% Ni content on average, and the former is dominant in the mare materials while the latter is more abundant in the highland materials.

PARAMAGNETIC SUSCEPTIBILITY

The majority of lunar materials returned by the Apollo 16 mission are anorthositic rocks, in which Al_2O_3 is rich and FeO is poor, so that their paramagnetic susceptibility which is mainly due to the paramagnetic Fe^{2+} ion is unusually small as shown in Table 2. Thus, the content of FeO in various lunar materials returned by Apollo 11 through Apollo 16 missions ranges from about 25% to 3%. In Fig. 7, the paramagnetic susceptibility (χ_a) observed at room temperature is plotted against FeO content for all lunar materials whose χ_a-values as well as FeO content are measured. A good positive correlation can be observed between χ_a (300°K) and FeO content in this diagram. However, the observed values of χ_a (300°K) are, in general, considerably larger than the theoretically expected values, where the magnetic moment of Fe^{2+} is assumed to be 5.39 Bohr magnetons, i.e., the average of experimentally observed values. The deviations of observed values of χ_a from the theoretical ones are not random, but take positive values for almost all plots, which suggests that there may be additional effects on observed χ_a values.

The deviations, χ_a(observed) $- \chi_a$(theoretical) $= \Delta\chi_a$, are plotted against the contents of metallic iron, m(Fe), in Fig. 8, where a good positive correlation between $\Delta\chi_a$ and m(Fe) is observed except for two samples. This result may suggest that a fraction of metallic iron is in the form of very fine grains which behave paramagnetically at room temperature. The experimental procedure to determine χ_a-values in the present study is based on a reversible linear relation between the magnetization and the applied magnetic field in a magnetic hysteresis curve for high magnetic fields between 10 and 16 KOe. If the superparamagnetism of extremely fine particles of metallic iron represents the observed $\Delta\chi_a$, then its dependence on the applied magnetic field must be practically linear within the high magnetic field range. This criterion gives the approximate size of the hypothetical fine particles of metallic iron as 20–30 Å.

The linear relationship between $\Delta\chi_a$ and m(Fe) shown in Fig. 8 may suggest

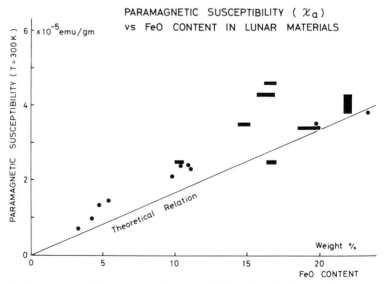

Fig. 7. Dependence of the paramagnetic susceptibility (χ_a) on FeO content in lunar materials. The vertical length and width of rectangular blocks represent respectively a range of variety of χ_a measurements and that of FeO analysis for different specimens taken from a block sample rock.

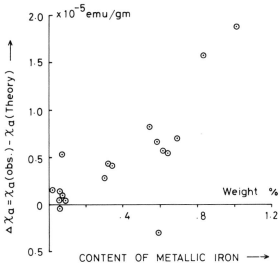

Fig. 8. Dependence on the metallic iron content of the excess value of paramagnetic susceptibility: i.e., the deviation of observed values from the theoretical ones of paramagnetic susceptibility of lunar materials.

that a nearly constant fraction of the total metallic iron is in the form of paramagnetically fine grain size regardless of the variety of total abundance of metallic iron. As no particular reason for such a grain size distribution of metallic iron could be presumed for lunar materials, exceptional cases such as shown in Fig. 8 could naturally take place.

Dependence of Coercive Force and Saturation Remanent Magnetization on Temperature

It has been pointed out (Nagata *et al.*, 1973) that the magnetic coercive force (H_c) and the saturation remanent magnetization (I_R) of Apollo 15 lunar materials generally increase with decreasing temperature, suggesting that the super-paramagnetically fine particles of metallic iron are blocked at low temperatures. The magnetic hysteresis of Apollo 16 materials has been measured at various temperatures between 4.2°K and 300°K to study the dependence of H_c and I_R on temperature in more detail. Figure 9 through Fig. 11 illustrate three types of the observed dependence of H_c and I_R on temperature. The group represented by Fig. 9 shows a sharp increase in H_c and I_R at a certain low temperature, T^*, ($T^* = $ 65°K in sample 66055 and $T^* = 85$°K in sample 68415) with decreasing temperature, but the largest values of H_c and I_R at low temperatures are only two or three times as large as the corresponding values at room temperature. The sharp increase in H_c and I_R at T^* may indicate that the distribution spectrum of fine grains of metallic iron has a sharp peak of grain size whose blocking temperature is represented by T^*. The mean diameter of the representative grain size is

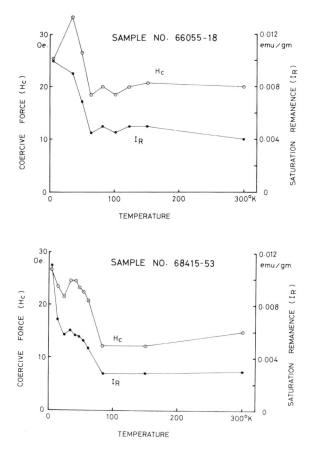

Fig. 9. Examples of Group I of the dependence of H_c and I_R on temperature; there is a sharp increase in H_c and I_R at a critical temperature (T^*), but H_c value in the low temperature side is far smaller than $\frac{1}{2}H_{RC}$.

estimated from observed values of T^* and H_{RC} to be 116 Å and 117 Å, respectively, for samples 66055 and 68415. The observations that the increased values of H_c and I_R at low temperatures are only several times as large as their room-temperature values and, in particular, H_c at the low temperature is far smaller than a half of the remanence coercive force (H_{RC}). This seems to suggest that the superparamagnetism still remains even at 4.2°K and those grains having their blocking temperatures between 4.2°K and T^* (between 40 Å and 115 Å in mean diameter) are rather a small fraction of the total metallic iron.

Figure 10 shows examples of a group whose dependence of H_c and I_R on temperature is represented by a continuous, gradual increase of H_c and I_R with decreasing temperature. In this case, the blocking temperatures of fine grains of metallic iron may be distributed continuously throughout the entire low temperature range, though the distribution density seems likely to become larger with

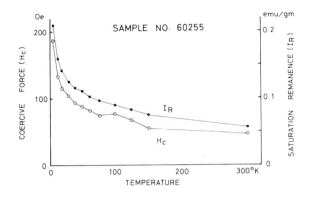

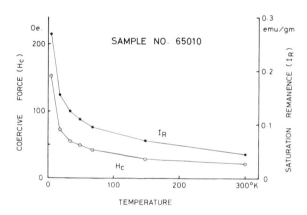

Fig. 10. Examples of Group II of the dependence of H_c and I_R on temperature; H_c and I_R continuously increase with decreasing temperature from 300°K to 4.2°K and $H_c(4.2°K)$ is reasonably close to $\frac{1}{2}H_{RC}$.

decreasing temperature. The largest value of H_c observed at the lowest temperature is still appreciably smaller than $\frac{1}{2}H_{RC}$, but the former is reasonably close to the latter in the order of magnitude. An Apollo 15 anorthosite breccia (sample 15418) also belongs to this group. It may thus be concluded that the majority of the supermagnetically fine particles of metallic iron in this group can be blocked at liquid helium temperature.

Figure 11 shows an example of samples where the dependence of H_c and I_R on temperature is characterized by a sharp increase at $T^*(T^* = 55°K$ for sample 67455) and considerably larger values of H_c and I_R on the low temperature side. Apollo 15 igneous rocks (samples 15058, 15495, 15555, and 15556) also belong to this group (Nagata *et al.*, 1972c, 1973). Table 4 summarizes observed values of H_c, I_R and the saturation magnetization (I_s) of Apollo 15 and 16 samples at 4.2°K and 300°K. It may be noted in the table that I_s values at 4.2°K are several times as large as those at 300°K for this group of lunar materials. This fact may suggest that a

T. Nagata *et al.*

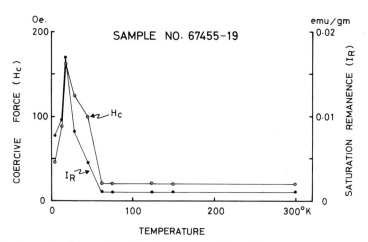

Fig. 11. Example of Group III of the dependence of H_c and I_R on temperature; there is a sharp increase in H_c and I_R at a critical temperature (T^*) and the largest value of H_c at the low temperature is reasonably close to $\frac{1}{2}H_{RC}$.

Table 4. Coercive force (H_c), saturation remanent magnetization (I_R) and saturation magnetization (I_s) of Apollo 15 and 16 samples at 4.2°K and 300°K

Sample	H_c		I_R		I_s	
	4.2°K	300°K	4.2°K	300°K	4.2°K	300°K
	(Oersteds)		(emu/gm)			(emu/gm)
60255	187	46.5	0.21	0.057	1.40	1.02
65010	152	22	0.27	0.045	1.95	1.39
66055	25	20	0.010	0.004	1.05	0.82
67455	46	20	0.0077	0.001	0.22	0.046
68415	27	15	0.011	0.003	0.76	0.66
68815	16	11.5	0.032	0.0030	1.30	1.34
15058	89	10	0.070	0.0010	0.33	0.125
15418	45	10	0.018	0.0013	0.59	0.145
15495	87	10	0.078	0.0008	0.98	0.165
15555	70	7	0.068	0.0010	0.87	0.127
15556-37	90	10	0.080	0.0015	0.57	0.125
-38-1	102	17	0.086	0.0019	0.65	0.124
-38-2	90	10	0.076	0.0007	0.61	0.114

rather large amount of superparamagnetically fine particles of metallic iron are blocked at temperatures below T^* in these lunar materials. In cases of Apollo 15 igneous rocks, magnitudes of H_c and I_R still increase or at least are kept nearly constant with a decrease in temperature below T^*, but in the case of Fig. 11 they decrease with decreasing temperature below 20°K. A possible physical mechanism of this anomalous phenomenon has not yet been experimentally clarified.

In summarizing the dependence of H_c and I_R on temperature of Apollo 15 and

16 lunar materials, listed in Tables 1 and 2, it may be concluded that all the lunar materials more or less contain superparamagnetically fine particles of metallic iron, but with differing size distribution spectra. A continuous spectrum between 40 Å and 160 Å seems to occur in some cases and a sharp peak of spectral density at a certain value less than 120 Å in others. The presence of fine superparamagnetic particles of metallic iron in lunar materials has already been pointed out based on the viscous magnetization (e.g., Nagata and Carleton 1970, Nagata *et al.* 1972a, b), the abnormally small values of H_c and I_R at room temperature (Nagata *et al.*, 1972a) and the anomalously large value of the initial magnetic susceptibility at room temperature (Nagata *et al.*, 1973).

NATURAL REMANENT MAGNETIZATION AND ITS STABILITY AGAINST AF-DEMAGNETIZATION

The natural remanent magnetization (NRM) of Apollo 15 and 16 igneous rocks and breccias, listed in Tables 1 and 2, has been measured and the stability of intensity and direction of their NRM against AF-demagnetization also has been examined. Before the measurements of NRM, all the samples had been stored in a non-magnetic space for at least a week, and the NRM measurements were repeated several times with one day intervals of storing in a non-magnetic space. Thus, any viscous magnetization of short duration (less than 7 days) acquired in the geomagnetic field has been almost completely eliminated. This condition was experimentally confirmed from the constancy of NRM's intensity and direction among the repeated measurements. Although the best possible care has been taken for the received lunar samples to avoid any magnetic contamination, there might be a number of occasions for these samples to be magnetically contaminated during the period from their sampling *in situ* and their receipt by scientific workers.

As shown in Fig. 12, for example, the intensity of NRM sharply decreases during the initial field range of 15–30 Oe rms. in an AF-demagnetization procedure in most cases. The direction of residual remanent magnetization after demagnetizing by an AF-field of 30–10 Oe rms. sometimes differs considerably from the original direction of NRM (e.g., Fig. 14). It seems likely that the soft component of remanent magnetization, which can be easily demagnetized by the comparatively weak AF-field, can be attributed to a magnetic contamination acquired during handling after its mother rock mass was sampled by the astronauts. As seen in Figs. 12, 13 and 14, however, most lunar igneous rocks and breccias sustain a stable component of remanent magnetization of a small magnitude after demagnetizing by AF-fields of 100 Oe rms. or more in intensity. With an increase in the AF-field intensity in an AF-demagnetization process, however, a possibility of an acquisition of the anhysteretic remanent magnetization (ARM) increases, because of inevitable imperfections in any AF-demagnetizer. This effect could be judged from a constancy of residual remanent magnetizations after repeated AF-demagnetization processes by a same field intensity of different demagnetizing times. The criterion adopted in the present study to judge such an artificial effect

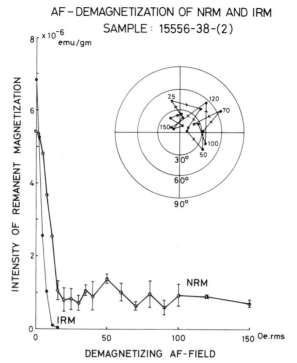

AF – DEMAGNETIZATION OF NRM AND IRM
SAMPLE : 15556-38-(2)

Fig. 12. Example of AF-demagnetization curves of NRM and IRM whose intensity is nearly close to NRM intensity. Error bars represent a range between the largest and the smallest values for each AF-field. Numeral figures in the direction change diagram represent AF-fields in units of Oe rms.

of the ARM acquisition is a comparison of the ARM vector with the vector of residual NRM; it is assumed that the AF-demagnetization becomes meaningless when the former approaches the latter. Another additional criterion for the upper limit of the peak intensity of AF-demagnetization field is not to exceed the observed value of H_{RC} of the sample concerned because of the obvious reason.

In Fig. 12, an error bar represents a range between the largest and the smallest observed values for each value of an applied AF-field; when the largest vector difference between individual measurements of the residual NRM becomes the same order of magnitude as their average magnitude, the artificial noises are regarded as too much in comparison with the signal. This happened at 250 Oe rms. of AF-field in the case of Fig. 12. It seems, however, that the observed residual NRM in a range from about 30 to 150 Oe rms. may represent a stable component of NRM of this sample, which maintains approximately the same intensity and direction during the AF-demagnetization process. An AF-demagnetization curve of the isothermal remanent magnetization (IRM) whose intensity is roughly near the initial NRM intensity also is shown in Fig. 12. It is clear that the stability of the stable component of NRM is much higher than that of the artificially acquired IRM against the AF-demagnetization.

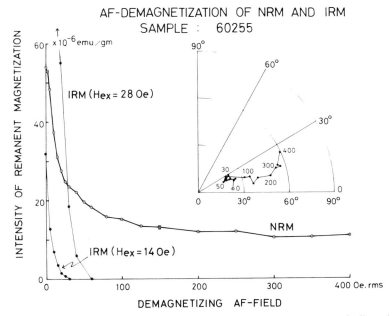

Fig. 13. AF-demagnetization curves of NRM and IRM for the most magnetically stable sample among examined Apollo 16 materials.

Figure 13 shows an AF-demagnetization result of the most stable NRM obtained in the present study. The stable component in this sample is fairly large (about 11×10^{-6} emu/gm) and very stable with respect to its intensity and direction against 100–400 Oe rms. in AF-demagnetization field, so that its stable component of NRM of sample 60255 can be considered a genuine natural remanent magnetization acquired on the lunar surface in some way. It seems that the particularly high stability of remanent magnetization in this sample is generally assured by its large H_{RC}-value (680 Oersteds).

Figure 14 illustrates another example of the AF-demagnetization curve of a comparatively large and stable NRM. In this case, the direction of the soft component of NRM which can be AF-demagnetized by a 100 Oe rms. field intensity is much different from that of the stable one whose direction and intensity are in reasonably good convergency ranges. Hence, the soft component of NRM could be regarded as an effect of some magnetic contamination.

The specific intensity (I_n) of NRM is given in Table 5 for all observed Apollo 15 and 16 rocks. When a residual NRM remains stable with a AF-demagnetization above 100 Oe rms., it is considered to be the "reliable" component of NRM and noted by I_n^0 in Table 5. The reliability criterion is so defined that any error bar of one of the three components of magnetization should not touch the zero line and fluctuations of the direction of stable NRM should be within a circle of 30 degrees in radius. A cross for I_n^0 in the table means that a possible stable component is buried under the noise level for AF-fields smaller than 100 Oe rms. $\tilde{H}_0$ and $\tilde{H}^*$ denote respectively the effective AF-demagnetization field and critical demagnet-

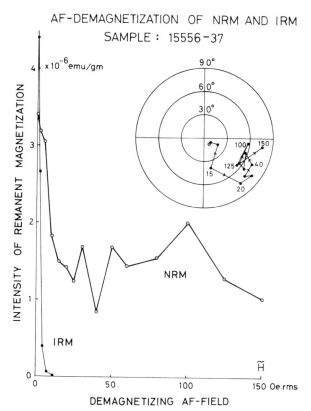

Fig. 14. AF-demagnetization curves of NRM and IRM for an Apollo 16 rock, where the direction of the soft component is much different from that of the stable component of NRM.

Table 5. Natural remanent magnetization and its stability against AF-demagnetization of Apollo 15 and 16 materials.

Sample	I_n (emu/gm)	H_0 (Oe rms.)	$H*$ (Oe rms.)	h (Oersteds)	H_0' (Oe rms.)	I_n^0 (emu/gm)	I_v/I_0
60255	54.9×10^{-6}	45	200	18	7	11×10^{-6}	4.6
60315	7.9×10^{-6}	50	40	5	1.5	4×10^{-6}	0.40
66055	58.6×10^{-6}	13	60	10	1.8	3×10^{-6}	0.063
66095	129.7×10^{-6}	25	60	6	1.5	3×10^{-6}	0.11
67455	2.3×10^{-6}	—	—	22	3	$\times$	~0
68415	3.5×10^{-6}	22	15	12	3	$\times$	0.065
68815	17.9×10^{-6}	13	125	13	3	2×10^{-6}	0.11
15058	18.8×10^{-6}	9	15	13	4	$\times$	~0
15418	6.4×10^{-6}	52	60	19	4	0.8×10^{-6}	0.070
15556-37	3.4×10^{-6}	25	15	9	3	1.2×10^{-6}	0.063
-38-1	88.4×10^{-6}	35	40	38	17	$\times$	0.63
-38-2	5.8×10^{-6}	14	40	17	5	0.9×10^{-6}	0.62

ization field (Nagata *et al.*, 1971) for NRM. Since physical significance is given to only the stable component of NRM in the present study, (the soft component being ignored), $\tilde{H}_0$ and $\tilde{H}^*$ values which are concerned with the initial vector of NRM may now become less significant. Observed values of $\tilde{H}_0$ and $\tilde{H}^*$ are listed in Table 5 only for comparing the apparent stability of NRM of Apollo 15 and 16 samples with that of previously studied samples of Apollo 11 through 14. The value of h in the table represents the intensity of static magnetic field which can produce IRM whose intensity is the same as that of NRM (Nagata *et al.*, 1972a), while $\tilde{H}'_0$ denotes the effective demagnetizing field of IRM thus acquired in h. These two magnetic parameters, h and $\tilde{H}'_0$, permit estimates of the degree of acceptability of any magnetic contamination of individual samples. $\Delta I_v/I_0$ is the ratio of the viscous component to the stable one of IRM whose intensity is approximately the same as that of NRM. It must be noted here that any viscous magnetization of short duration is almost completely eliminated in measuring NRM.

Summarizing various magnetic parameters concerning NRM, IRM and the viscous remanent magnetization (VRM) of Apollo 15 and 16 materials, it may be concluded that most igneous rocks and breccias maintain a stable component of NRM of 10^{-6}–10^{-5} emu/gm in intensity. The strongest and most stable NRM among all the samples in the table is that of sample 60255, in which $H_{RC} = 680$ Oe and $\Delta I_v/I_0 = 4.6$. As discussed in Section 6, sample 60255 contains superparamagnetically fine particles of metallic iron of a continuous grain size spectrum over a mean diameter range from 40 Å to 140 Å. An analysis of the viscous magnetization at room temperature of this sample with the aid of a method described in the previous paper (Nagata *et al.*, 1972a,b) has led to the conclusion that the continuous grain size spectrum extends to at least 220 Å in mean diameter. It may then be considered that the continuous spectrum of grain size is further extended into a range of 250–300 Å in mean diameter, which is the optimum for single-domain metallic iron acquiring TRM. The relaxation time (τ) of the remanent magnetization acquired by an assemblage of such single-domain iron particles of $H_{RC} = 690$ Oe is estimated to be as long as 2×10^{17} years even for 250 Å diameter particles at 450°K, which is approximately the daytime temperature on the lunar surface. It becomes 1.3×10^{26} years for 270 Å diameter particles on the same condition.

On the contrary, the relaxation time for a remanent magnetization of iron particles of $H_{RC} = 400$ Oe is 1.3×10^3, 1.7×10^8 and 2.1×10^{17} years respectively for 250, 270, and 300 Å in diameter at 450°K. This result may suggest that only the largest single-domain particles whose diameter is close to 300 Å in low H_{RC} lunar rocks can maintain their remanent magnetization over the period of $(3$–$4) \times 10^9$ years. Thus, the fact that lunar rocks of larger value of H_{RC} have a larger amount of the stable remanent magnetization is understandable.

ACQUISITION EXPERIMENT OF THERMOREMANENT MAGNETIZATION
AND PIEZOREMANENT MAGNETIZATION

Acquisition experiments of partial TRM were carried out with samples 15556 and 66055 in the temperature range 20°C and 800°C in a magnetic field of 0.5 Oe.

The results are shown in Fig. 15. The partial TRM of sample 15556 increases steadily with temperature to 800°C except for a small bump around 600°C, while the partial TRM of sample 66055 sharply increases between 500°C and 600°C, but it decreases with increasing temperature from 600°C to 800°C. As shown in Tables 1, 2, and 3, both samples contain two ferromagnetic phases, i.e., an almost pure metallic iron component with small amounts of Ni and Co represented by $\Theta_c \simeq$ 780°C and a kamacite component of $\Theta_c^* \simeq 640$°C. If there were only the kamacite composition in these rocks, the partial TRM can be acquired only at temperatures below 640°C. If there is no strong magnetic interaction between the two phases, the partial TRM curve could be represented by a simple addition of two partial TRM curves; one for the Fe-phase reaching the maximum value at 780°C and another for the kamacite phase attaining the maximum at 640°C. This case may correspond to the partial TRM curve of sample 15556.

If, on the other hand, a strong magnetic interaction exists between the two phases, a reverse TRM phenomenon could take place (e.g., Nagata, 1961), that is, the partial TRM increases with temperature to Θ_c^*, but then the partial TRM acquired by cooling from a temperature higher than Θ_c^* consists of the normal partial TRM of Fe and the reverse partial TRM of kamacite so that the total TRM could become smaller than the partial TRM acquired by cooling from Θ_c^*, possibly even becoming negative. It may happen in actual cases that a portion of the neighboring two phases strongly interact but the remainder does not. This case may

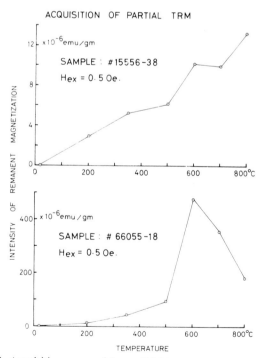

Fig. 15. Acquisition curves of the partial thermoremanent magnetization.

correspond to the partial TRM curve of sample 66055. However, no direct experimental proof of a possible strong magnetic interaction between hypothetical neighboring phases has yet been made.

If we assume that the stable component of NRM (I_n^0) of these two lunar samples, given in Table 5, is attributable to TRM, the ambient magnetic field on the lunar surface is estimated to be 3400γ and 840γ for samples 15556 and 66055 respectively.

Acquisition experiments of the piezoremanent magnetization (PRM) also have been carried out with two breccia samples, 15418 and 68815. As is well known (e.g., Nagata, 1970), PRM (I_p) acquired by a strong uniaxial compression (P) in the presence of a weak magnetic field (H) is approximately expressed as

$$I_p = CP \cdot H$$

The coefficient C is estimated from the experimental results to be 4×10^{-9} and 2.0×10^{-8} emu/gm/Oe bar respectively for samples 15418 and 68815. If we assume that the stable component of NRM of these two samples is due to a uniaxially compressive shock of 50 K bars in strength caused by a meteoritic impact, the ambient magnetic field is estimated to be about 400γ and 200γ respectively for samples 15418 and 68815.

CONCLUDING DISCUSSIONS

In summary, the present experiments have demonstrated that lunar highland materials sampled in the neighbourhood of the Apollo 16 landing site have unusual magnetic characteristics; (1) their paramagnetic susceptibility is much smaller than that of previous Apollo samples because of a smaller abundance of Fe ions in the anorthositic materials, (2) all Apollo 16 samples analyzed contain a distinct kamacite component in addition to metallic iron which sometimes contains a small amount of Ni and/or Co. In previous discussions (Nagata et al., 1972a, b), the origin or kamacite in the lunar materials was considered to be mostly due to contributions from chondritic meteorites because the kamacite phase had been found only in breccias among lunar materials returned by Apollo 11 through 14 missions. Goldstein et al. (1970, 1971, 1972) have extensively studied metallic particles in Apollo 11 through 14 lunar soils and have reached a conclusion that the majority contain Ni and Co in the range of meteoritic metal and therefore are considered to originate from predominantly chondritic projectiles. However, we have now clear evidence that some lunar basalts (samples 15556 and 60315) and an anorthositic gabbro (sample 68415) also contain comparatively large amounts of kamacite. Thus, it is rather natural to consider that the metallic composition of lunar igneous rocks consists of a mixture of almost pure metallic iron (with less than 1% of Co and 0–1% Ni) and kamacite of 3–12% of Ni. If this is true, the metallic component of lunar rocks has a unique characteristic in comparison with chondrites which contain kamacites and taenites of various Ni contents with no, or very little, amounts of almost pure metallic iron.

Careful examination of NRM of lunar igneous rocks and breccias in the present study gave consistent results. All twelve lunar samples, whose NRM and its stability against the AF-demagnetization test were examined, have soft components as well as a hard and stable NRM; the soft component can be AF-demagnetized by 15–30 Oe rms. in field intensity in most cases, while the stable component can maintain its intensity and direction against AF-demagnetization up to 200 Oe rms. or more in field intensity. It seems safe to regard the soft component as due to some artificial magnetic contamination after a rock sample was collected *in situ*, and consequently is disregarded in the present study. Among the twelve lunar samples examined, an anorthositic breccia, sample 60255, has the largest intensity and the highest stability of the stable component of NRM. Examination of the variation of H_c and I_R at low temperatures revealed that it contains a continuous distribution spectrum of grain size of single-domain metallic iron in the range of 37–142 Å in mean diameter; from the viscous magnetization at room temperature it was also found to have an extended continuous spectrum of grain size from 140–220 Å. From these results it is reasonable to believe that the continuous spectrum of grain size extends to about 300 Å which is the maximum possible diameter of a single-domain particle of metallic iron. On the other hand, the remanence coercive force (H_{RC}) of this rock is as large as 680 Oe so that the relaxation time of the remanent magnetization of single-domain iron particles is about 2×10^{17} years for a mean diameter of 250 Å at 450°K which is the highest possible temperature on the lunar surface. The relaxation time of the remanent magnetization is sensitive to H_{RC} and the grain size as well as temperature. It is only 1.3×10^3 years for single-domain iron particles of 250 Å in diameter at 450°K in case of $H_{RC} = 400$ Oe, but it becomes 9×10^9 years in case of $H_{RC} = 500$ Oe. For single-domain iron particles of 300 Å in diameter, however, the relaxation time at 450°K is 2×10^{17} years and 1.7×10^{29} years, respectively, in cases of $H_{RC} = 400$ Oe and $H_{RC} = 500$ Oe.

Since the isotope age of lunar rocks concerned is $(3\text{-}4) \times 10^9$ years, the relaxation time of NRM of lunar rock samples examined for the palaeomagnetic purpose must be appreciably longer than 4×10^9 years. From this viewpoint, the stable component of NRM of sample 60255 is fully reliable for palaeomagnetic purposes, since all fine iron grains of 250–300 Å in diameter have sufficiently long relaxation times. For other lunar rocks summarized in Table 5, only the larger iron grains of diameter close to 300 Å and relaxation time appreciably longer than 4×10^9 years can have maintained the stable NRM. It seems that this is the main reason why the stable component of NRM (I_n^0) of these samples is so small.

It would be worthwhile to note in this regard that among fourteen lunar rocks returned by Apollo 11 through 14 missions, whose NRM characteristics have been previously examined by the authors (Nagata *et al.*, 1972a), two particular breccia samples (10048 and 10085) maintain an extremely stable and comparatively intense NRM. Their H_{RC} values are 520 and 690 Oe respectively for samples 10048 and 10085, while H_{RC} values of the other twelve samples range from 80 Oe to 450 Oe. It may be concluded that NRM's of lunar rocks with H_{RC} significantly greater than 500 Oe are the most reliable for the palaeomagnetic purpose.

REFERENCES

Collinson D. W., Runcorn S. K., Stephenson A., and Manson A. J. (1972) Magnetic properties of Apollo 14 rocks and fines. *Proc. Third Lunar Sci. Conf., Geochim. Cosmochim. Acta*, Suppl. 3, Vol. 3, pp. 2343–2361. MIT Press.

Gancarz A. J., Albee A. L., and Chodos A. A. (1972) Comparative petrology of Apollo 16 sample 68415 and Apollo 14 samples 14276 and 14310 (1972). *Earth. Plan. Sci. Letts.* **16**, 307–330.

Goldstein J. I., Henderson E. P., and Yakowitz H. (1970) Investigation of lunar metal particles. *Proc. Apollo 11 Lunar Sci. Conf., Geochim. Cosmochim. Acta*, Suppl. 1, Vol. 1, pp. 499–512. Pergamon.

Goldstein J. I. and Yakowitz H. (1971) Metallic inclusion and metallic particles in Apollo 12 lunar soils *Proc. Second Lunar Sci. Conf., Geochim. Cosmochim. Acta*, Suppl. 2, Vol. 1, pp. 177–191. MIT Press.

Goldstein J. I., Axon H. J. and Yen C. F. (1972) Metallic particles in the Apollo 14 lunar soils. *Proc. Third Lunar Sci. Conf., Geochim. Cosmochim. Acta*, Suppl. 3, Vol. 1, pp. 1637–1064. MIT Press.

Gose W. A., Pearce G. W., Strangway D. W., and Larson E. E. (1972) Magnetic properties of Apollo 14 breccias and their correlation with metamorphism. *Proc. Third Lunar Sci. Conf., Geochim. Cosmochim. Acta*, Suppl. 3, Vol. 3, pp. 2387–2395. MIT Press.

Grommé C. S. and Doell R. R., (1971) Magnetic properties of Apollo 12 lunar samples 12052 and 12065, *Proc. Second Lunar Sci. Conf., Geochim. Cosmochim. Acta*, Suppl. 2, Vol. 3, pp. 2491–2499. MIT Press.

McKay G., Kridelbaugh S. and Weill D., (1973) A preliminary report on the petrology of microbreccia, 66055 (abstract). In *Lunar Science—IV*, pp. 487–489. The Lunar Science Institute, Houston.

Nagata T. (1961) *Rock Magnetism*, pp. 1–350, Maruzen Co., Tokyo.

Nagata T., (1970) Basic magnetic properties of rocks under the effect of mechanical stress; *Tectonophys.* **9**, 167–195.

Nagata T., Ishikawa Y., Kinoshita H., Kono M., Syono Y., and Fisher R. M. (1970) Magnetic properties and natural remanent magnetization of lunar materials. *Proc. Apollo 11 Lunar Sci. Conf., Geochim. Cosmochim. Acta*, Suppl. 1, Vol. 3, pp. 2325–2340. Pergamon.

Nagata T. and Carleton B. J., (1970) Natural remanent magnetization and viscous magnetization of Apollo 11 lunar materials. *J. Geomag. Geoele.* **22**, 491–506.

Nagata T., Fisher R. M., Schwerer F. C., Fuller M. D., and Dunn J. R. (1971) Magnetic properties and remanent magnetization of Apollo 12 lunar materials and Apollo 11 lunar microbreccia. *Proc. Second Lunar Sci. Conf., Geochim. Cosmochim. Acta*, Suppl. 2, Vol. 3, pp. 2461–2476. MIT Press.

Nagata T., Fisher R. M., and Schwerer, F. C. (1972a) Lunar rock magnetism. *The Moon*, **4**, 160–186.

Nagata T., Fisher R. M., Schwerer F. C., Fuller M. D., and Dunn J. R. (1972b) Rock magnetism of Apollo 14 and 15 materials. *Proc. Third Lunar Sci. Conf., Geochim. Cosmochim. Acta*, Suppl. 3, Vol. 3, pp. 2423–2447. MIT Press.

Nagata T., Fisher R. M., Schwerer F. C., Fuller M. D., and Dunn J. R. (1972c) Summary of rock magnetism of Apollo 15 lunar materials. In *The Apollo 15 Lunar Samples*, pp. 442–445. The Lunar Science Institute, Houston.

Nagata T., Fisher R. M., and Schwerer F. C. (1973) Some characteristic magnetic properties of lunar materials. *The Moon*, **5** In press.

Pearce G. W., Strangway D. W. and Larson E. E. (1971) Magnetism of two Apollo 12 igneous rocks. *Proc. Second Lunar, Sci. Conf., Geochim. Cosmochim. Acta*, Suppl. 2, Vol. 3, pp. 2451–2460. MIT Press.

Rose H. J., Carron M. K., Christian R. P., Cuttitta F., Dwornik E. J., and Ligon D. T. (1973) Elemental analysis of some Apollo 16 samples (abstract). In *Lunar Science—IV*, pp. 631–633. The Lunar Science Institute, Houston.

Proceedings of the Fourth Lunar Science Conference
(Supplement 4, *Geochimica et Cosmochimica Acta*)
Vol. 3, pp. 3045–3076

Magnetic studies on Apollo 15 and 16 lunar samples

G. W. Pearce

Lunar Science Institute
Houston, Texas 77058

and

University of Toronto
Toronto, Ontario, Canada

W. A. Gose

Lunar Science Institute
Houston, Texas 77058

D. W. Strangway

NASA Johnson Space Center
Houston, Texas 77058

Abstract—The magnetic properties of lunar samples are almost exclusively due to rather pure metallic iron. The mare basalt contains about 0.06 wt.% Fe, the soils 0.5–0.6 wt.%, and the breccias 0.3–1.0 wt.%. Most of the additional iron in the soils and breccias is believed to be the result of reduction processes operating on the lunar surface. Whereas the total metallic iron content of the soils from all landing sites is rather constant, the Fe°/Fe^{++} ratio and the average iron grain size increases with the age of the landing site reflecting increasing maturity. The crystalline rocks studied from Apollo 16 have highly variable, but generally, very high metallic Fe content (up to 1.7 wt.% Fe). It is suggested that these rocks are either breccias or igneous samples which have been severely thermally metamorphosed in a highly reducing environment. Many lunar basalts and breccias contain a magnetization which is stable against AF demagnetization, although results on an Apollo 12 sample returned to the moon with the Apollo 16 mission show that the samples can acquire a strong isothermal remanence during handling. A conglomerate test on 16 surface oriented samples shows that their stable remanent magnetization was acquired before they were last put on the lunar surface. The stable magnetization of mare basalts is typically $1–2 \times 10^{-6}$ emu/gm. The stable component in breccias varies over a wide range but there is a class of breccias with a stable remanence of about 1×10^{-4} emu/gm.

Introduction

In this paper we summarize the measurements we have made on samples from the Apollo 15 and 16 missions and show how they relate to concepts of lunar processes along the lines discussed in our previous papers. Briefly these concepts can be summarized as follows:

(1) Mare basalt samples are low in metallic iron content ($\sim 0.06\%$) while soils and breccias are high in metallic iron content (~ 0.3–1.0%). Some of this difference is due to meteoritic components, but much of it is the result of reduction of Fe^{++} from glasses in impact events.

(2) The range of magnetic properties from soil to poorly welded breccias to highly recrystallized breccias and perhaps even to completely remelted rocks follows an important sequence—soils are rich in superparamagnetic iron ($< \sim 150$ Å); moderately recrystallized breccias are rich in single domain iron (~ 150 Å–300 Å) and highly recrystallized breccias are rich in multidomain iron ($> \sim 300$ Å). This is confirmed by the apparent presence of a thermoremanent magnetization even in low metamorphic grade breccias implying cooling from above 770°C in the presence of a magnetic field, and by the distribution of iron particle grain size range.

(3) The presence of natural remanent magnetization (NRM) in almost all returned lunar rock samples implies that there was an ancient lunar magnetic field. This magnetization has many of the characteristics of thermoremanent magnetization (TRM). We have found that those samples which behave dominantly with single domain characteristics are also very strongly magnetized ($\sim 10^{-4}$ emu/gm) and very stable to alternating fields. Tests for stability show that many samples are capable of acquiring a soft remanent magnetization but there usually seems to be a stable component. In cases where there is a large percentage of multidomain iron present, it is often difficult to detect this stable component. The samples in their most recent lunar orientation show random directions of stable magnetization indicating that they have not been remagnetized by any surface process operating uniformly on the samples after they were last put on the lunar surface.

The paper is divided into two parts, the first examining the magnetic properties of lunar samples, the second examining the natural remanent magnetization of the rock samples. Together they present our recent findings concerning the concepts outlined above.

Part A   Magnetic Properties

Curie point determinations

The identification of iron in its various forms and grain size distributions has been one of the important contributions made by magnetic studies. The first step in this process is the analysis of the variation of the saturation magnetization with temperature. There are several basic types of behavior found in lunar samples. These are illustrated in Fig. 1. In Fig. 1a we illustrate the simplest behavior in an Apollo 12 soil sample. This figure shows the dominant effect of essentially pure iron with very low nickel content. In some samples this is complicated by the presence of a reversible effect with a distinct phase lag. This effect, illustrated in Fig. 1b for another soil, shows the transition of iron from the α (magnetic) to the γ (non-magnetic) form and the reverse sluggish transition from the γ to the α form which occurs at lower temperatures. This transition temperature is characteristic of iron with a few percent nickel. This effect was first noted in lunar samples by Strangway *et al.* (1970) and later several examples were given by Nagata *et al.* (1972a). This phase is probably kamacite, often considered to be of meteoritic origin, but also observed as an indigenous lunar phase (Reid *et al.*, 1970).

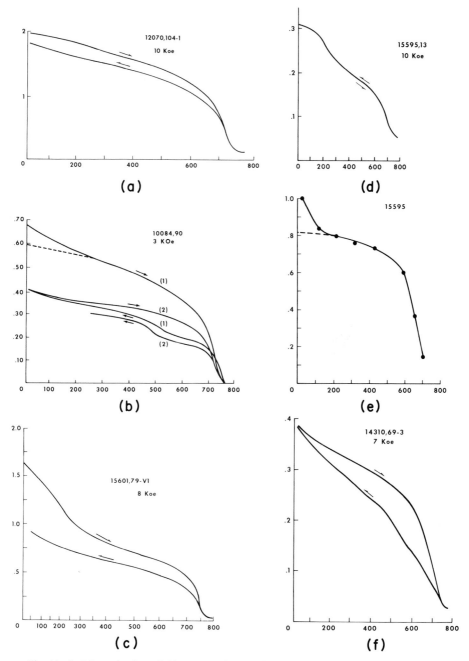

Fig. 1(a–f). Magnetization of 5 lunar samples as a function of temperature. Horizontal axis is in °C, vertical axis in emu/gm. Sample numbers and applied field are shown for each curve except 1e. 1e shows normalized saturation magnetization (determined from the high field portion of a magnetization curve) as a function of the temperature at which the magnetization curve was measured.

Table 1. Curie points of some lunar samples.

Sample	Type	Low T_c	High T_c	Cooling $T_{\gamma-\alpha}$	Ni Content of Metal (wt.%)
10084	S	250	740	570	~7.5
12021	I		780		
12070	S		765		
14049	B		740	550	~8.0
14163	S		750	620	~6.0
14310	crystalline		750	distributed	
14313	B		740		
15498	B	400	760	distributed	
15595	I	200	750		
15601	S	250	760		
60335	crystalline	350	760		
64801	S		745	620	~6.0

S = soil, B = breccia, I = igneous.

A third type of behavior is seen in Fig. 1c where a low Curie temperature phase appears. This behavior in an Apollo 15 soil is probably due to the presence of a small amount of iron with about 30–40% nickel. This effect has been noted in several samples, for example, mare basalt sample 15595 (Fig. 1d, e) appears to have some of this high-nickel iron (note in Fig. 1e that the paramagnetic $1/T$ effect has been subtracted). In a final category of lunar samples we find that there is a distributed thermal hysteresis (Fig. 1f). No specific magnetic transition associated with the α–γ phase change is recognized, but it is likely that there is a distribution of nickel contents in the iron so that one is looking at an assembly of particles with varying iron-nickel compositions.

The dominant magnetic material present in all lunar samples is nearly pure iron. This is demonstrated in Table 1 and in the histogram of Fig. 2 showing that most Curie temperatures are near 750–770°C. (Note: these tabulate data from our studies only and can be added to the set of data reported by Nagata *et al.*, 1972a. The data reported by Nagata *et al.* are quite similar to those in Fig. 2.) In some

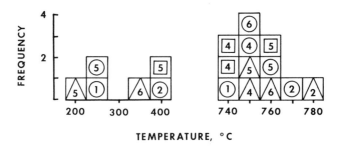

Fig. 2. Histogram of low and high Curie points for lunar samples. Data is from Table 1. Numbers in boxes refer to missions: 1 to Apollo 11, 2 to 12, 4 to 14, 5 to 15, and 6 to 16. △ represent crystalline rocks, □ breccias, and ○ soils.

cases there appears to be a kamacite-phase present with 5–10 wt.% nickel (phase lag at about 600–700°C on cooling). In some cases there is a taenite phase present with about 30–40 wt.% nickel (Curie point of 200–300°C). In still others there appears to be a range of nickel contents. Since kamacite and taenite are found in meteorites, it is tempting to accept the identification of these phases in lunar samples as meteoritic contamination and to use it to estimate the meteoritic fraction of the iron. Unfortunately, this is not possible and one can only estimate the upper limit of the iron–nickel phases present since igneous rocks contain nickel in the iron as well (Reid *et al.*, 1970). If we realize, however, that in most samples the nickel-rich fraction is small relative to the pure iron fraction, and if we realize that much of the excess iron in the soils is in the range of particle sizes < 150 Å, we are forced to conclude that only a small part of the excess iron can be of meteoritic origin.

Hysteresis loops measured at room temperature

A large number of samples have now been studied at room temperature in high fields since from these measurements we can infer, (a) the metallic iron (Fe°) content from the saturation magnetization (J_s), (b) the paramagnetic iron (Fe^{++}) content from the susceptibility in high fields (χ_p), (c) the particle size characteristics of the iron (superparamagnetic, single-domain or multidomain) from a combination of the initial susceptibility, χ_0, the saturation remanence (J_{rs}), the remanence coercivity (H_{rc}) and the coercivity (H_c). These data are tabulated in Tables 2–5 for a variety of the samples we have studied.

Metallic iron content. We have in the past and in the present paper estimated the metallic iron (Fe°) content of lunar samples from the saturation magnetization—J_s. When making this estimate we assume that J_s is due exclusively to metallic iron. This assumption is valid in as much as there are generally no other magnetic phases with Curie points above room temperature. However, the addition of nickel and cobalt to the metallic phase may introduce systematic errors in the computation of the Fe° content.

The effect of cobalt is generally very small due to its low concentration— usually 2 wt.% or less. Nickel, on the other hand, is sometimes present in sufficient quantity to cause a systematic overestimate of the actual iron content. Fig. 3 illustrates this for a simple model which considers a uniform Ni-content distribution between 0 and x wt.% Ni. For example, if the metal contains between 0 and 25 wt.% Ni uniformly distributed in this range, its actual Fe content is 87.5 wt.%; based on the J_s for pure iron, however, a value of 97.2 wt.% Fe would be obtained. As can be seen from the figure the error could be as high as 10%. This error appears to be an unrealistic maximum since as we discussed above, the majority of the metal in lunar samples contains $< 10\%$ Ni. Consider a sample with 80% of its metal in the 0–10% Ni range and 20% with Ni contents around taenite (25–35% Ni). The error in this case would only be 1.4%. It thus may be stated that metallic Fe determinations by magnetic means are systematically too high, but the error is probably less than 3% in general and has no effect on inferences we made in the past related to the metallic iron content.

Table 2. Room temperature magnetic properties of igneous and crystalline rocks.

	J_s (emu/gm)	χ_0 (emu/gm Oe) ×10^6	χ_0 (emu/gm Oe) ×10^4	J_{rs}/J_s	H_{rc} (Oe)	H_c (Oe)	J_s/χ_0 (KOe)	Equiv. % Fe++	Equiv. % Fe°	Chem. % Fe++
14310,69	0.22	13.1	0.57	0.004		10.0	3.9	0.10	6.0	6.5[a]
14321,184-48*	1.62	30.7	4.0	0.001			4.1	0.74	14.1	
-51*	1.97	41.8	4.9	0.001			4.0	0.90	19.2	
15016,29-1	0.14	38.7	0.30	0.005		27.0	4.6	0.065	17.8	17.6[a]
15076,50-1	0.21	33.8	0.53	0.004		11.0	3.9	0.095	15.5	14.5[a]
15555,204	0.14	42.3	0.52	0.0025		25.0	2.7	0.065	19.4	17.6[a]
15595,13-1	0.13	37.4	0.48	0.006		33.0	2.7	0.060	17.2	15.7[b]
15668,2	0.10	38.9						0.045	17.8	
60335,30-1	0.78	8.08	2.26	0.007	200	25.0	3.2	0.36	3.70	3.51[c] 3.58[d]
66095,36	3.13	12.1	8.37	0.0025	210	9.5	3.7	1.44	5.55	5.57[c]
68415,78	1.92	7.14	4.27	0.0017	360	8.0	4.5	0.88	3.27	3.30[c] 3.20[d]

*Igneous clast from breccia 14321.

[a] Apollo 15 Preliminary Examination Team (1972). *Science*, **175**, 363.
[b] Chappel B. W. and Green D. H. (1973). *Earth Planet Sci. Lett.*, **18**, 237–246.
[c] Apollo 16 Preliminary Examination Team (1972). *Science*, **179**, pp. 23–34.
[d] Rose H. J. Jr., Carron M. K., Christian R. P., Cuttitta F., Dwornik E. J., and Ligon D. T. (1973). In *Lunar Science—IV*, p. 631. The Lunar Science Institute, Houston.

Table 3. Room temperature magnetic properties of soils.

	J_s (emu/gm)	χ_p (emu/gm Oe) $\times10^6$	χ_0 (emu/gm Oe) $\times10^4$	J_{rs}/J_s	H_{rc} (Oe)	H_c (Oe)	J_s/χ_0 (KOe)	Equiv. % Fe°	Equiv. % Fe^{++}	Chem. % Fe^{++}
10084,90	0.87		11.4	0.094		36	0.76	0.40		
12070,104	1.49	36.4	22.1	0.096	450	24	0.67	0.69	16.7	12.8[a]
14163,68	1.04	22.9	24.4	0.044	500	22	0.43	0.48	10.5	8.0[b]
14163,131	1.30	24.0	23.5	0.077	480	25	0.51	0.60	11.5	
15301,38	1.07	30.4	27.2	0.047	400	20	0.39	0.48	13.9	10.9[c]
15301,130*	0.60	23.6	18.0	0.055	340	20	0.33	0.28	10.8	
15471,72*	0.50	28.2	9.7	0.040	345	22	0.52	0.23	12.9	12.6[c]
15601,79	1.05	40.2	23.0	0.073	450	38	0.46	0.48	18.5	15.4[c]
15601,117*	0.44	35.4	9.8	0.052	320	24	0.45	0.20	16.2	
60501,29	1.31	9.8	24.8	0.032	500	21	0.53	0.60	4.5	4.24[d]
60600,2	1.1	14.0						0.5	6.0	
61241,25	0.89	7.2	11.9	0.030	550	20	0.75	0.41	3.3	4.15[e], 4.20[f]
64801,29	1.06	10.5	19.1	0.035	490	21	0.55	0.49	4.8	4.44[g]

*<150 mm fraction.

[a]Warner J., unpublished lunar sample curatorial file.

[b]Hubbard N. J., Gast P. W., Rhodes M., and Wiesmann H. (1972). In *Lunar Science—III*, p. 407. The Lunar Science Institute, Houston.

[c]See footnote 1, Table 1.

[d]Morrison G. H., Nadkarni R. A., Jawarski J., Botto R. B., and Roth J. R. (1973). In *Lunar Science—IV*, p. 543. The Lunar Science Institute, Houston.

[e]See footnote 3, Table 1.

[f]See footnote 4, Table 1.

[g]Mason B., Simkin T., Noohan A. F., Switzer G. S., Nelen J. A., Thompson G., and Melson W. G. (1973). In *Lunar Science—IV*, p. 505. The Lunar Science Institute, Houston, average of two separates.

Table 4. Room temperature magnetic properties of breccias.

	J_s (emu/gm)	χ_p (emu/gm Oe) ×10^6	χ_0 (emu/gm Oe) ×10^4	J_{rs}/J_s	H_{rc} (Oe)	H_c (Oe)	J_s/χ_0 (KOe)	Equiv. % Fe°	Equiv. % Fe++	Chem. % Fe++
14049,28	1.16	23.9	14.1	0.039	510	30	0.82	0.53	11.0	7.6[c]
14312,20	0.54	15.7	1.53	0.010			3.5	0.24	7.1	
14313,29	1.02	29.0	9.1	0.066	470	65	1.12	0.47	13.3	
14321,78	0.41	19.7	0.91	0.019		15	4.5	0.19	9.0	8.5[c], 9.5[d]
14321,184-35[a]	0.51	17.2	1.0	0.002			5.0	0.23	7.9	
14321,184-43[a]	0.65	17.9	1.1	0.001			5.9	0.30	8.2	
15426,97[b]	0.05	32.8		0.08				0.02	15.0	
15498,36	0.75	36.9	6.7	0.088	680	75	1.12	0.34	16.9	

[a] Microbreccia-3 fragments from 14321.

[b] Green clod.

[c] Lunar Sample Preliminary Examination Team (1971). *Science*, **173**, 681.

[d] Scoon J. H. (1972). *Proc. Third Lunar Science Conf., Geochim. Cosmochim. Acta*, Suppl. 3, Vol. 2, pp. 1335. MIT Press.

Table 5. Room temperature magnetic properties of small samples (< 10 mg) from soils.

Sample	J_s	χ_0	J_{rs}/J_s	H_c	J_s/χ_0	% Fe Equiv.	Comment
Agglutinates							
14162,40-GA1	5.12	19.7	0.016	48	2.7	2.35	
-GA2	1.33	24.0	0.078	43	0.54	0.61	
-GA3	1.27	18.0	0.019		0.71	0.58	
Glass beads							
67482,14-1	1.30		0.0035				Darkest
-2	0.265		< 0.002				
-3	< 0.020						
-4	< 0.020						
-5	< 0.008						
-6	< 0.010						Lightest

In Figs. 4–7 we have presented our results and values from the literature (see caption to Fig. 4) for metallic iron contents of lunar samples. The Apollo 16 rocks are not as easily catagorized as samples from previous missions and will be treated separately. It is readily seen that the metallic iron content of the mare basalts ranges from less than 0.02 wt.% to about 0.12 wt.% with a distinct peak at about 0.06–0.08 wt.%. We therefore believe that this is a value which is typical for the mare basalts of the lunar surface, and basalt-like samples which deviate greatly from this probably have a complex history. Such may be the case with the two igneous clasts from breccia 14321, which we measured. They are an order of magnitude richer in native iron than the normal igneous rocks (these samples are not shown in the histogram, Fig. 4). Either these fragments represent quite different parent rocks from other mare basalts, or they have been altered after formation probably by the event that formed their parent breccia 14321. The explanation that heating generated the excess iron can also be used to explain why igneous rock 14053 is enriched in metallic iron (Nagata *et al.*, 1972b). Finger *et al.* (1972) have found evidence for the reheating of 14053 to about 840°C at some time after its formation. El Goresy *et al.* (1972) find that much subsolidus reduction of fayalite and spinels has occurred in 14053. This reduction mechanism can produce considerable quantities of coarse-grained iron at temperatures above about 900°C (Pearce *et al.*, 1972b). In Fig. 5 the data from soils measured in the laboratory from all missions through Apollo 16 are shown. The metallic iron content is surprisingly uniform considering the variety of landing sites. A value of about 0.5–0.6% Fe° seems to be characteristic, almost 10 times the value found in the mare basalts. As we have reported earlier (Gose *et al.*, 1972) and as will be discussed shortly in more detail much of this excess iron is in very-fine grains, superparamagnetic (< 150 Å) and single domain (150–300 Å). This is illustrated in Fig. 5 showing the metallic iron content of the greater than 150 μm fraction of three Apollo 15 samples. These fractions, which are predominantly composed of fine igneous fragments, have a significantly smaller total iron content than the typical soils.

The magnetic properties of some individual small silicate particles obtained

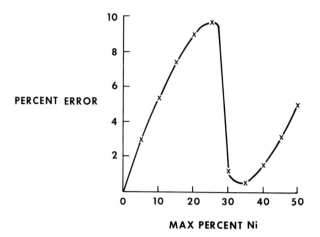

Fig. 3. Systematic error in determining iron content from saturation magnetization due to nickel impurity. See text for explanation.

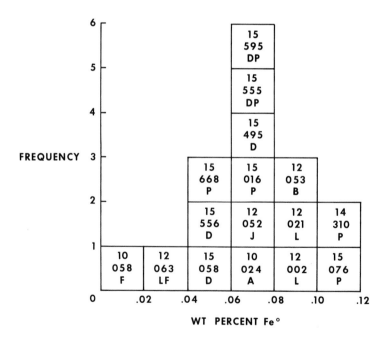

Fig. 4. Histogram of magnetically determined metallic iron contents of lunar crystalline rocks. Numbers inside boxes are sample identification. Letters represent the source of data according to the following scheme: A—Nagata *et al.* (1970), B—Nagata *et al.* (1971), C—Nagata *et al.* (1972a), D—Nagata *et al.* (1972b), E—Nagata *et al.* (1973), F—Hargraves and Dorety (1971), G—Doell *et al.* (1970), H—Runcorn *et al.* (1970), J—Grommé *et al.* (1971), K—Runcorn *et al.* (1971), L—Gose *et al.* (1972), M—Brecher *et al.* (1973), P—present paper. Where several measurements for a single sample are available, an average value was entered in the figure.

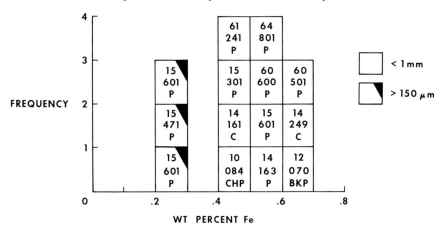

Fig. 5. Histogram of magnetically determined metallic iron contents of lunar soil samples. Numbers and symbols are as in Fig. 4.

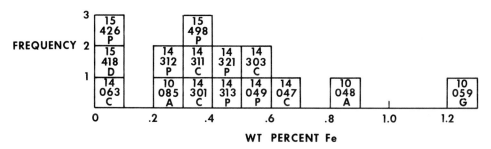

Fig. 6. Histogram of magnetically determined metallic iron contents of lunar breccias. Numbers and symbols are as in Fig. 4.

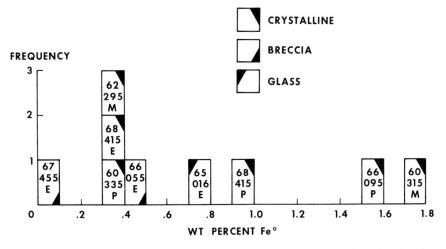

Fig. 7. Histogram of magnetically determined metallic iron contents of Apollo 16 rock samples. Numbers and symbols are as in Fig. 4.

from soil samples are given in Table 5. (χ_p is not given because the samples are too small to provide reliable determinations.) These samples were contributed and described by D. S. McKay of NASA–JSC. These measurements represent the beginning of an attempt to learn where the excess iron, particularly the fine-grained excess, resides in soil samples. The agglutinates are small fragile clusters of rock, glass and mineral fragments held together by glass. All three examples examined contain considerable metallic iron. The glass beads represent a sequence of coloring from a dark grey, cloudy bead to a clear, light bead. The darker beads are the only ones with significant metallic iron and it is probable that the coloration is, at least in part, caused by the presence of submicroscopic iron particles. The presence of high concentrations in these typical soil particles, agglutinates and dark glass beads is in full agreement with Housley *et al.* (1972) who argues that much of the excess metal in the soils is to be found in these particles and that a process of reduction similar to that envisioned for the igneous clasts from 14321 is responsible for it.

Whereas the soils and mare igneous rocks form relatively uniform groups in terms of metallic iron content, the breccias have highly variable iron contents. In general, the values are high and much like those of the lunar soil with many values in the range 0.3–0.6 wt.% Fe°. This variability is undoubtedly the result of: (a) the complex thermal history of the breccias, and (b) the high degree of inhomogeneity within the breccias since the iron content varies considerably within a single breccia. Breccia lithification occurs either as a direct result of shock compaction of regolith material due to the impact of a meteorite or as a product of a hot base surge deposit which may be of volcanic or impact origin (McKay *et al.*, 1970). Many of the Apollo 14 breccias have magnetic properties which suggest that they have been heated to at least 800°C during formation and that they formed in the base surge or ash flow type of deposit (Gose *et al.*, 1972; Pearce *et al.*, 1972b).

Two of the samples in Fig. 6, with the lowest metallic iron contents, are cataclastic anorthosites. These are essentially monomict breccias (matrix and most clasts are composed of one rock type, in this case anorthosite) and are sometimes considered to be a crushed form of an important highland material. Since the brecciation process is likely to increase the iron content of rocks, it appears that unbrecciated anorthosite has a very low metallic iron content.

The other low-Fe° sample is the green "clod" 15426 which is composed almost entirely of green glass. Its low-Fe° content suggests that it was formed in a process more oxidizing than most lunar processes.

Ferrous iron content. Although we have used the paramagnetic susceptibility of lunar samples to estimate the ferrous iron content in the past (Gose *et al.*, 1972), this information has been of secondary interest. However, for many samples the ferrous iron content has not been, cannot be or will not be measured by chemical methods (X-ray fluorescence, neutron activation; wet chemical, etc.) and for these samples this independent magnetic determination is of interest. For this reason an estimate of the accuracy of the method is useful.

We obtain the paramagnetic susceptibility, χ_p, from the high field portion of the magnetization curve (by "high field" we mean sufficiently high to virtually saturate the ferromagnetic material present—for lunar samples this is generally the portion with field greater than about 10 KOe). In the high field region, the magnetization J is given by:

$$J = \chi_p H + \chi_h H + J_s \left(1 - \frac{a}{H} - \frac{b}{H^2} \cdots \right) \tag{1}$$

where J_s is the saturation magnetization of the ferromagnetic component, whether superparamagnetic, single domain or multi-domain. H is the applied field, χ_h is the susceptibility of the sample holder (usually negative) and a, $b \ldots$ are empirically determined constants. Thus, in addition to the linear paramagnetic relation between magnetization and field for high fields, there is also a nonlinear relation caused by the asymptotic approach to saturation of the ferromagnetic constituents of the sample. If J_s itself is not large compared to $(\chi_p + \chi_h) H$, this ferromagnetic variation can be neglected.

For many lunar samples J_s is not large (<0.5 emu/gm) and $\chi_p \gg \chi_h$, so that

$$J(\text{high field}) \cong \chi_p H + J_s. \tag{2}$$

We have used relation 2 in the past to estimate paramagnetic susceptibility for lunar samples. To determine $\%Fe^{++}$ from χ_p, we divide χ_p of the sample by that of Fe^{++} as determined from various ferrous salts. We make the assumption that Fe^{++} is the only paramagnetic ion present, a justifiable assumption for lunar samples.

However, for Apollo 16 samples, J_s is often greater than 0.5 emu/gm and χ_p is relatively low ($\% Fe^{++}$ is low) so that χ_h is not always negligible. Thus for these samples, the accuracy of the estimate of χ_p can be greatly improved by using relation 1. In fact, we have used this more accurate method for all the values of χ_p given in Tables 2–4. The parameter a in the formula was taken as 0.5 KOe, a value representative of various samples of finely disseminated multidomain iron grains measured in our laboratory.

The equivalent Fe^{++} for a variety of samples have been compared with available chemical determinations in Fig. 8. For igneous rocks and the Apollo 16 samples the magnetic determinations are in good agreement with the chemical data. (Note—the values reported here are slightly different in some cases from those reported for the same samples in earlier papers. The present results have been corrected as mentioned here and replace any earlier reports.) However, for the <1 mm soils from Apollo sites 11, 12, 14, and 15 as well as the low grade breccia sample 14049, the magnetic values are too high. These samples contain very small, superparamagnetic iron metal particles which can add to the unsaturated magnetization at higher fields (10–20 KOe) and masquerade as paramagnetic material. Thus the Fe^{++} content determined from χ_p will tend to overestimate that concentration. If a sample with these very small grains is heated to temperatures of the order of 800°C in a good vacuum, these small particles tend to be lost through evaporation or they tend to coalesce into larger grains so that the paramagnetic susceptibility and the resulting estimate of Fe^{++} is expected to

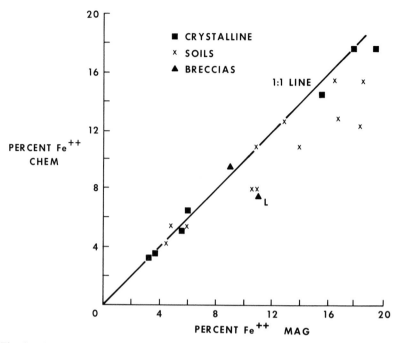

Fig. 8. Plot of Fe^{++} concentration in some lunar samples as determined by chemical methods and by our magnetic studies. (■) are crystalline samples, (X) are soils, (▲) are breccia 14321 and (▲$_L$) is breccia 14049.

approach the chemical determination. Table 6 shows the behavior of three soils we have examined in this manner. The agreement of the magnetic and chemical determinations of Fe^{++} after heating of the sample is striking. Apollo 16 soils do not show a high magnetic value, presumably due to a smaller superparamagnetic fraction.

If the unsieved soils and 14049 are excluded the average ratio Fe^{++} (MAG): Fe^{++} (CHEM) is $1.04 \pm 0.03 : 1$. Thus the magnetic determination of Fe^{++} can be expected to be accurate to 3–4%, provided the sample does not contain appreciable amounts of superparamagnetic grains.

Fe°/Fe^{++} Ratio. Using the magnetic determinations of the metallic and paramagnetic iron content we have plotted their ratio in Fig. 9. This ratio Fe°/Fe^{++}

Table 6.

Soil	% Fe^{++}(MAG)		% Fe^{++}(CHEM)
	Initial	After heating	
12070,104-5	16.8	14.4	12.8
15301,38-2	14.2	10.7	10.9
15601,79-3	18.5	15.7	15.4

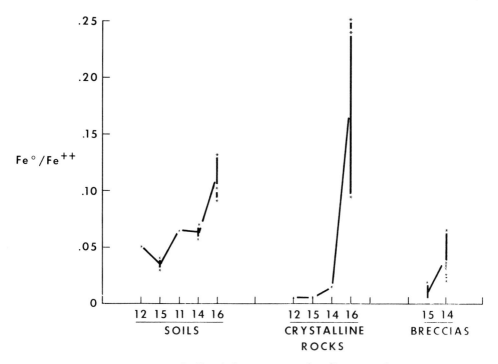

Fig. 9. Plot of Fe°/Fe^{++} ratio for some categories of lunar samples.

can be used as a rough indication of the degree of reduction in the sample. The presence of the Fe° may be due to an actual reduction process resulting from the heating of the sample in an impact event, or, in the case of igneous rocks, due to the state of reduction in the melt. The Fe° may also be due to addition of meteoritic material to soils or breccias with soil components. In the case of soils and breccias, the higher the ratio Fe°/Fe^{++}, the larger the number of impact cycles and/or the higher the temperature reached in their histories. Although the Fe° content of the soils is nearly uniform, Fig. 9 shows that the ratio follows a simple rule. The ratio increases from Apollo 12 and 15, to 11 and 14, to 16 showing that the more mature soils also have the greater relative metallic iron content. A similar pattern can be seen for the mare basalt rocks which have very low ratios (Apollo 12 and 15). The breccias from Apollo 14 and 15 are comparable to the soils while the whole range of rocks from Apollo 16 has a very high ratio, much like the soils from 16.

   Grain-size distributions. Not only can one determine the metallic iron content and ferrous iron content magnetically, but also much information about the grain size distribution of the metallic iron can be derived from the shape of the room-temperature magnetization curve. For example, the ratio of the saturation remanent magnetization (i.e., that remanence left after exposure to very large fields) to the saturation magnetization is an indicator of the amount of single-domain or multi-domain grains present. Large values of the saturation remanent magnetization are associated with particles in the range of 150–300 Å while small values are associated with particles greater than 300 Å or less than 150 Å. The data in Tables 2–5 and in Fig. 10 summarize this. There is a good correlation between the relative grain size of the Fe° particles and the total concentration of iron in the soils. The FeO-poor soils (Apollo 14 and 16) have coarser iron particles on the average than the FeO-rich mare soils. This conclusion is based on the relative values of J_{rs}/J_s (see Table 3 or Fig. 10), the smaller values for the FeO-poor soils showing the presence of fewer of the small (150 Å–300 Å) single domain particles.

In contrast to the soils, which contain metallic iron in a wide range of grain sizes, the mare igneous rocks contain almost exclusively multi-domain iron particles, as is shown by their characteristic magnetization curves and their low values of J_{rs}/J_s (typically 0.005) (Fig. 10), which are not far from the value expected for bulk iron. The breccias which range in character from soil-like to igneous-rock-like show the complete range of J_{rs}/J_s from high values for the low metamorphic grades to igneous-rock-like for the high metamorphic grades.

In addition to the J_{rs}/J_s ratio a separate property has been found useful for characterizing the grain-size distribution. This property is the curvature of the magnetization curve. In Fig. 11 for example, soils from each of several missions are shown. Basically all cases are dominated by fine-grained iron and the curves show the strong curvature characteristic of small grains. (The magnetization curves of multi-domain particles are controlled by demagnetization fields and are generally almost linear to saturation at about $(4\pi/3)J_s$ for equant grains.) In terms of soil maturity it can be seen that the freshest soils (Apollo 12 and 15) are the most curved while the most mature soils (Apollo 14 and 16) are the least curved. This suggests that not only does the $Fe°/Fe^{++}$ ratio increase with maturity, but also the grain size of the iron. Igneous rocks and highly metamorphosed breccias have multidomain-type curves.

As an alternate way of presenting this information, the ratio of J_s/χ_0 is a rough indication of the curvature of the magnetizing curves (Fig. 12). High values of χ_0 are related to the curved loops of superparamagnetic and single domain iron, while small values are characteristic of the straight line curves of multi-domained iron. Thus large values of J_s/χ_0 are characteristic of the multi-domain mare basalt rocks, while soils have very small values. Again the breccias have intermediate characteristics ranging from low-grade metamorphics that are soil-like to high-grade metamorphics that are mare basalt-like.

The Apollo 16 rocks are a varied lot and do not fit in with the simple scheme of basaltic, igneous rocks and breccias that has served to characterize the samples from previous missions. The most obvious difference to be seen in magnetic properties from that expected from previous missions is the very high Fe° content of crystalline rocks (Fig. 7). Four of these "crystalline" samples belong to a group considered by the Preliminary Examination Team (LSPET, 1972) to be highly recrystallized breccias. Thus they may have acquired metal iron from a very iron-rich regolith component and/or through reduction from the silicate and oxide minerals during the reheating process. The other "crystalline" rock is 68415 which is

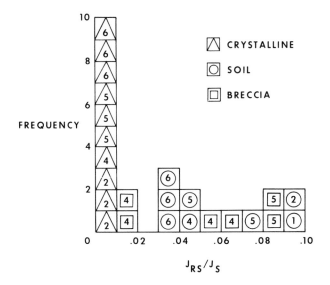

Fig. 10. Histogram of J_{rs}/J_s ratio. Numbers in boxes refer to missions as in Fig. 2. Data from our studies only.

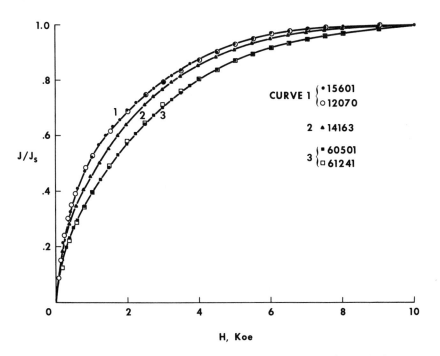

Fig. 11. Magnetization curves at room temperature for some lunar soil samples from our studies. Paramagnetic magnetization has been subtracted from data.

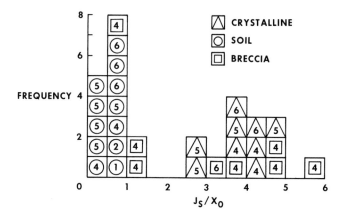

Fig. 12 Histogram of J_s/χ_0 ratio. Numbers in boxes refer to missions as in Fig. 2. Data
from our studies only.

considered by LSPET to be the only fresh igneous rock recovered from Apollo
16. Comparison of the measurements of different portions of 68415 by Nagata
et al. (1973) and by the authors of the present paper show this rock to be very
inhomogeneous, as well as high in Fe° content, compared with other igneous rocks
returned from the moon. While it is risky to generalize from the previous magnetic information, this seems to imply that 68415 is either a highly recrystallized
breccia like the other Apollo 16 "crystalline" samples or, perhaps, an igneous rock
intensely metamorphosed in the manner implied for 14053 and for the two igneous
clasts from 14321.

The Apollo 16 rock samples are quite diverse in their Fe° contents, but are at
least as high on the average as the samples from any previous sites. When the low
FeO content is taken into account, the $Fe°/Fe^{++}$ ratios (Fig. 9) or degree of
reduction is certainly higher than for other missions. The rocks of this site appear
to have been subjected to a much more intense thermal history than is shown by
the Fra Mauro Formation at Apollo 14, which is thought to be ray material ejected
in the impact that produced Mare Imbrium.

PART B NATURAL REMANENT MAGNETIZATION

Portions of our data on Apollo 15 and 16 samples have been published in
various papers and abstracts and we will summarize these data here as well as
present new results. The petrological classification of the Apollo 15 rocks is
relatively straight forward. They are mainly mare basalts from the vicinity of the
landing site and Hadley Rille and breccias from the Apennine Front. The
Apollo 16 rocks, however, are rather difficult to classify. They are almost all
breccias of various degrees of metamorphism. At this point no distinction between "Cayley" formation and "Descartes" formation can be made. In our discussion we will consider the Apollo 16 samples individually rather than in a
specific geological context. The rock descriptions are given by Butler (1972).

In many lunar samples, the natural remanent magnetization (NRM) consists of two components (Strangway *et al.*, 1970). A soft component can usually be removed by AF demagnetization in fields less than 50 Oe and is thought to be an isothermal remanence (IRM) acquired by exposure to magnetic fields in the space craft or laboratory. We will describe results on an Apollo 12 sample which was returned to the moon with the Apollo 16 mission. A second "hard" component is quite stable against AF demagnetization and is considered to be of lunar origin. The high stability can further be demonstrated by doing a conglomerate test using lunar surface oriented samples. Generally, the Apollo 15 and 16 samples confirm previous results.

Apollo 15

We have studied four Apollo 15 basalt samples (15016, 15076, 15555, 15595), one anorthosite (15415) and one breccia (15498).

Breccia 15498 is an unusual sample in that it can acquire a strong viscous rema-nence (Gose and Carnes, 1973), but at the same time it carries an exceedingly sta-ble and strong natural remanence (Fig. 13). The magnetic properties of this rock are dominated by superparamagnetic and stable single domain iron particles as is expected from a breccia of low metamorphic grade (Gose *et al.*, 1972, Pearce *et*

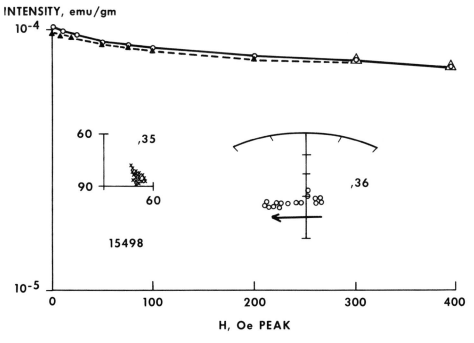

Fig. 13. Change in intensity and direction upon AF demagnetization of samples 15498,35 and 36. Relative orientation is arbitrary. Arrow indicates movement of vector during de-magnetization.

al., 1972). Chip 36 was subjected to a Thellier–Thellier test. The sample proved to be stable against thermal demagnetization and a paleointensity of 2100 gammas was obtained (Gose *et al.*, 1973).

The anorthosite sample 15415,56 is the weakest sample we have measured thus far. On demagnetization (Fig. 14) the intensity decreases to about 2×10^{-7} emu/gm, which seems to be a stable remanence. No systematic variations of the direction were observed and the scatter of points seems to be due to the low intensity which is close to the detection limit of the instrument rather than to instability. This sample is of particular importance in that it represents one of the few anorthositic samples measured to date.

Basalt sample 15595 was broken off a large boulder close to the rim of Hadley Rille. This boulder is believed to be bedrock (Swann *et al.*, 1972). We had the opportunity to measure the NRM of the whole sample 15595,0 and then two chips in detail (15595,10 and 13). In addition, we determined the magnetization of 15596,0, a sample from the same boulder. The directions of their magnetization in lunar coordinates are shown in Fig. 15. The NRM directions of the three oriented

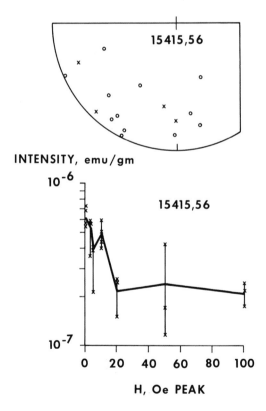

Fig. 14. AF demagnetization of anorthosite sample 15415,56. No systematic changes in direction were observed. Error bars on intensity curve represents absolute error for repeat measurements. (The same definition applies to all similar figures.)

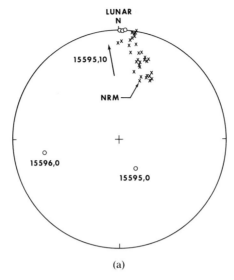

(a)

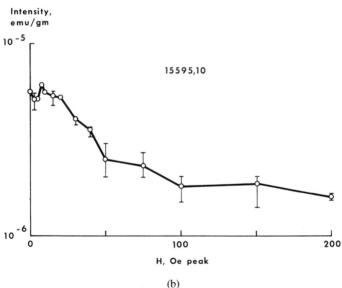

(b)

Fig. 15(a). Direction of magnetization in lunar coordinates of three samples from the same boulder. Only 15595,10 could be demagnetized and the arrow shows the change in direction. (b) Change in intensity during AF demagnetization of 15595,10.

samples, before cleaning in alternating fields, differ considerably and a random, non-lunar, secondary magnetization seems likely. We were only permitted to demagnetize the two chips. The direction of magnetization of chip 10 (Fig. 15a) moves along a great circle but a stable magnetization can clearly be isolated (Fig. 15a, b). Chip 13 unfortunately is not oriented, and since it it is rather small it is hard to measure. Its behavior is similar to that of chip 10. If these samples indeed represent bedrock then the direction of the magnetic field at this site as shown by chip 10 was almost due north and horizontal at the time when the bedrock was formed.

Two chips of each of three additional basalt samples were AF demagnetized. 15555,13 and 204, about 1 gm each, did not yield any meaningful data. The probable explanation is that the magnetic properties of this rock are dominated by multidomain iron grains. Chip 37 and 50 of sample 15076 on the other hand have a stable magnetization (Fig. 16). A soft magnetization is eliminated after cleaning in 25 Oe and no major change in intensity or direction takes place up to 150 Oe. A different type of behavior is observed in sample 15016. Chip 28 demagnetizes systematically and a stable magnetization is indicated (Fig. 17) whereas chip 29, although it contains no soft component, shows only a random scatter of its direction.

Apollo 16

Five samples from the Apollo 16 mission have been studied. Their response to AF demagnetization is quite varied, reflecting their diverse nature. Sample 67016,62, a light gray breccia, contains only one component of magnetization which is fairly stable against AF demagnetization (Fig. 18). On the other hand, sample 66095,36 (anorthosite, "rusty rock") has a very pronounced soft component (Fig. 19). The direction changes along a great circle and a stable direction is obtained at 100 Oe. Sample 60255,19, a black glassy breccia, is currently under investigation. It can acquire a strong VRM. The non-viscous component demagnetizes from 7×10^{-5} emu/gm to 2×10^{-5} emu/gm at 200 Oe and is directionally stable.

Sample 60335 has been described as a completely recrystallized breccia (LSPET, 1972). The opaque minerals make up less than 1% of the rock, but some large (up to 0.25 mm) FeNi and troilite particles are present. Rock 60335 is the "LPM rock." An attempt was made to measure its intensity on the lunar surface by placing it on the Lunar Portable Magnetometer in order to evaluate the "magnetic contamination" by the spacecraft. The intensity proved to be below the resolution of the LPM (Dyal *et al.*, 1972). Upon return to earth, the sample was measured again with a backup model of the LPM in our zero magnetic field chamber. Again, no reading could be obtained. Measurement in our cryogenic magnetometer yielded an intensity of 5.4×10^{-6} emu/gm which is indeed below the resolution of the LPM.

Two individual chips of 60335 have been demagnetized (Fig. 20). The intensity of chip 18 decreases gradually up to 100 Oe, but a stable direction is only obtained at 350 Oe. By contrast, the intensity of chip 30 is stable against AF demagnetization and a stable direction is obtained at 60 Oe. It is probably of importance that chip 30 was stored in a field free chamber for 28 days prior to any measurement. It thus appears that 60335 acquired a viscous remanence prior to its cutting. This VRM is fairly stable against AF demagnetization but not against storage in a

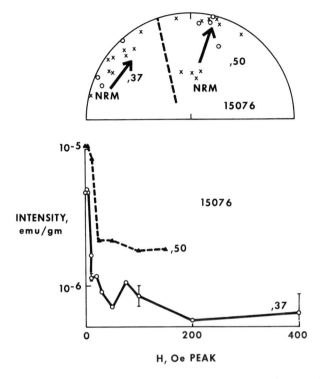

Fig. 16. AF demagnetization of 15076. Arrows show change in direction upon demagnetization.

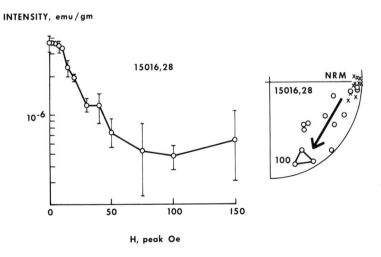

Fig. 17. AF demagnetization of 15016. Arrows show change in direction upon demagnetization.

G. W. PEARCE *et al.*

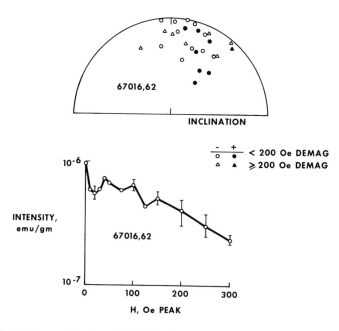

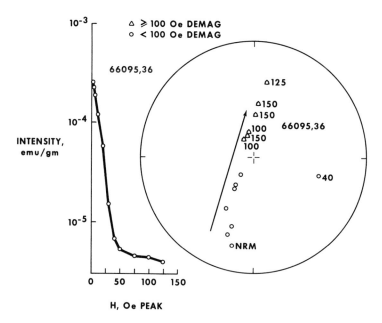

Fig. 18. AF demagnetization of 67016,62. No systematic changes in direction were observed above 200 Oe.

Fig. 19. Change in intensity and direction upon AF demagnetization of sample 60095,36. Arrow indicates directional trend.

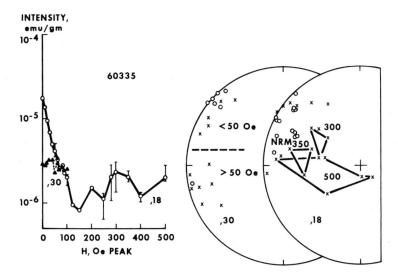

Fig. 20. AF demagnetization of the LPM rock 60335. Relative orientation of the few chips is accurate to within about 30°.

field-free chamber. Since it decayed to a large extent within 28 days, a lunar origin seems unlikely. This behavior is quite similar to results reported by Dunlop *et al.* (1973), on breccia 14313 and further studies are necessary.

Sample 68415 is a crystalline anorthositic gabbro. It is magnetically quite inhomogeneous (Fig. 21). Chip 41 is very stable against AF demagnetization whereas chip 17 has a pronounced soft component, whose direction is different from the stable direction. The direction of the stable component in both chips agrees well. The high metallic iron concentration as well as its inhomogeneous dis-

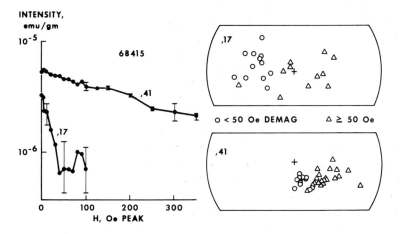

Fig. 21. AF demagnetization of 68415,17 and 41. For fields larger than 50 Oe no systematic changes in direction occurred.

tribution, together with the different response to AF demagnetization, strongly indicates that this sample is a highly recrystallized breccia as proposed by Hörz (Butler, 1972) rather than an igneous rock as suggested by LSPET (1972).

Several rock samples described in this paper contain a "soft" magnetization superposed on a "hard" or stable magnetization. Such a soft component has been observed on many rocks from previous missions and it has been suggested (e.g., Strangway *et al.*, 1971) that it may be an isothermal remanent magnetization (IRM). An IRM could result from exposure to a strong magnetic field in the spacecraft or during sample preparation. In order to check this suggestion a tracer sample was returned to the moon on Apollo 16 to be brought back with the Apollo 16 samples. The tracer chosen was a chip from igneous rock 12002 which possessed a soft component after its original trip, but had no VRM component. The chip was demagnetized up to 400 Oe leaving only the stable magnetization. It was then delivered to the spacecraft in a μ-metal box and unpacked and installed about 30 hrs. before launch. As samples normally do not make the outbound trip, it would have been more appropriate to keep it in the μ-metal box to protect it from magnetic fields, until after landing on the moon. However, as this would have been a difficult procedure, the location of the sample on the spacecraft was chosen to be similar on the outbound and the inbound trip. The outbound trip was in the Lunar Module Ascent Stage and during the return trip the sample was transferred to the Command Module with the lunar samples. The storage place chosen for the sample on the spacecraft was the interim stowage assembly (ISA), a bag of the type in which samples were returned on the Apollo 14, 15, and 16 missions. Its stowage location is quite similar to that of the Surveyor bag on Apollo 12. The sample, inside a small Beta cloth bag, was attached to the flap of the ISA. At recovery the sample was detached from the ISA and returned in one of the padded crates used for lunar samples. On arrival at the Lunar Receiving Laboratory the sample was picked up and brought to the magnetic properties laboratory at the Johnson Space Center where the NRM of the sample was measured several times over a period of four days. During this time it was stored in a field-free room. The NRM showed no significant change during this storage test. Next, the sample was demagnetized by alternating fields in steps to 100 Oe. (Fig. 22). It was possible to eliminate the soft component acquired in the sample in an alternating field of 20 Oe. At this point, the direction and intensity of magnetization was approximately the same as that in the sample before it left the earth (Fig. 22). In Fig. 23 the intensity of the stable component (2×10^{-6} emu/gm) has been subtracted from the NRM so that the magnetization added by the trip can be more easily compared with the tests described below. The component added by the trip is very similar in behavior to the original soft component that this rock had when it first came back from the moon. The intensity of the soft component was somewhat less than that found after the first trip.

In an attempt to simulate the soft component, a series of IRM's were induced in the sample using steady fields of 10, 12, 15, 20, and 40 Oe for a period of 1 min. each. This magnetization was then cleaned by alternating field demagnetization. These demagnetization curves are compared with the original NRM soft compo-

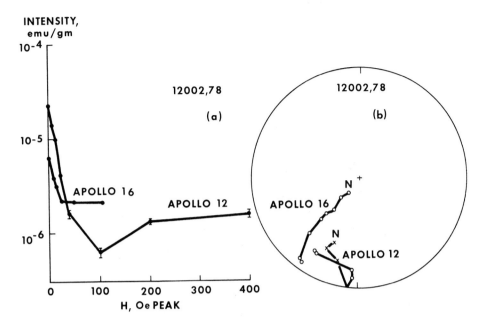

Fig. 22. AF demagnetization of 12002,78 after return from the Apollo 12 and 16 missions. Lines in stereographic projection connect consecutive steps of demagnetization, starting at N. Crosses represent downward inclination, circles upward inclinations.

nent and the Apollo 16 trip-added component in Fig. 23. The shapes of the IRM curves are similar to the trip-added magnetization. Thus the trip-added magnetization is well simulated by an IRM induced by exposure to a steady field of 12–15 Oe, applied for 1 min., while the original NRM would require exposure of the sample to 30–35 Oe fields. This experiment shows that the soft component found in many lunar samples is at least in part an artifact of the trip. AF demagnetization in fields of 50 Oe or less can easily remove this non-lunar magnetization.

Conglomerate test

Several lunar samples have been surface oriented by pitcounts, gamma-ray spectroscopy, track counting, and photography. Most of these techniques can only discriminate between the top and bottom of a sample and thus one can only determine the magnetic inclination with respect to the lunar surface. Based on this we have published previously (Strangway *et al.*, 1971; Pearce *et al.*, 1972) a conglomerate test in order to determine whether the magnetization of these samples is stable on a geological time scale. We can now add three Apollo 15 and three Apollo 16 samples for a total of 16 samples. Figure 24 compares the observed inclinations with the theoretically expected distribution for random orientations. Clearly, the samples have not been remagnetized by any process acting uniformly on the lunar surface.

G. W. Pearce *et al.*

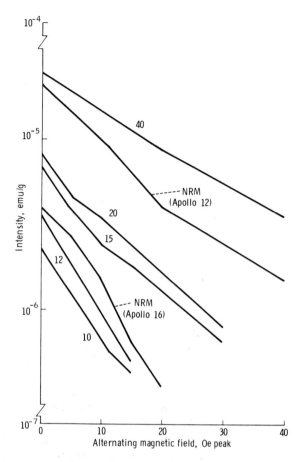

Fig. 23. AF demagnetization of soft component of magnetization of 12022,78 after the Apollo 12 and 16 missions and after exposure to magnetic fields of 10, 12, 15, 20, and 40 Oe.

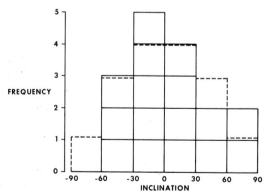

Fig. 24. Histogram of inclinations of lunar surface oriented rocks. Dashed line represents the expected random distribution.

Intensity of magnetization

We now have samples from five different landing sites and it seems interesting to compare the intensity of magnetization of all samples to see whether some general conclusions can be drawn or whether there exist differences related to different rock types or different ages of the samples.

Figure 25 summarizes all our data on the NRM of lunar samples. Included in this histogram are eighteen "zero field" samples from Apollo 15. These are curator samples on which we were only permitted to measure their NRM without spinning them and were measured in a cryogenic magnetometer. These measurements could serve as a reference for future studies and since this constitutes the only measurment before the samples go into "permanent storage" their NRM intensities are listed in Table 7. Samples 15330,0 and 15320,0, both breccias, are very viscous similar to samples 14313 and 15498 (Gose and Carnes, 1973).

The histogram shows a wide range of intensities. Most mare basalts have NRM intensities slightly less than 10^{-5} emu/gm while the breccias span a large range. Most of the samples with intensities above about 5×10^{-5} emu/gm are breccias. After AF demagnetization to 100 Oe, an even clearer separation becomes obvious. All but one Apollo 11 sample in the group around 10^{-4} emu/gm are breccias. The basalts have intensities typically around $1-2 \times 10^{-6}$ emu/gm. Some breccias have low intensities as well, but it seems significant that the high intensities are almost exclusively carried by breccias. No simple correlation of the intensities with age seems to exist.

Table 7.

Sample Number	Intensity ($\times 10^{-6}$ emu/gm)
(a) Basalts	
15115,0	7.2
15119,0	8.6
15595,0	1.7
15596,0	6.4
15606,0	7.0
15614,0	5.4
15630,0	3.3
15664,0	4.1
15636,0	3.2
15639,0	6.7
15648,0	2.5
15658,0	0.7
15660,0	2.7
15665,0	6.5
15675,0	3.8
(b) Breccias	
15148,0	19
15320,0	27
15330,0	48

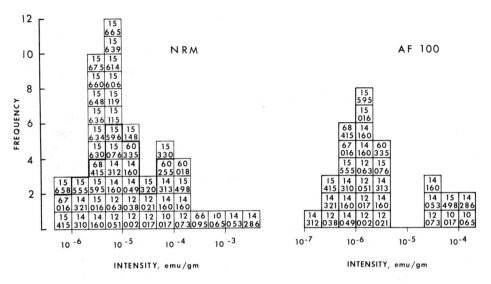

Fig. 25. Histogram of NRM intensities and intensities after AF demagnetization in 100 Oe. Samples above heavy line in the left figure are zero field samples.

Summary

It has been the purpose of this paper to present all our data on Apollo 15 and 16 lunar samples and to put them in context with our previous results. Although many fundamental questions are still unanswered (e.g., the origin of the magnetizing field) some general conclusions can be drawn.

(1) Many lunar rocks contain a stable magnetization of lunar origin. A secondary, "soft" component is most likely an IRM of non-lunar origin.

(2) Mare basalts have stable intensities typically around $1-2 \times 10^{-6}$ emu/gm. The intensity of the breccias varies over a wide range but there is a class of breccias characterized by a stable remanence of about 1×10^{-4} emu/gm.

(3) The metallic iron content of lunar soils and breccias is considerably higher than that of mare basalt. Most of this additional iron is not of meteoritic origin, but rather the result of reduction processes operating on the lunar surface.

(4) Whereas the total metallic iron content of the soils from all missions is rather constant, the Fe°/Fe^{++} ratio increases with the age of the landing site. This increase with soil maturity is paralleled by an increase in the average metallic iron grain size.

(5) The samples from Apollo 16 deviate from the pattern established for mare sites in that the crystalline rocks are generally higher in metallic iron than the soils and several of them are the richest in metallic iron of any bulk sample returned from the moon. It is suggested that they are severely thermally metamorphosed breccias or igneous rocks similarly altered. 68415 appears to be of this category of rock.

Acknowledgments—The authors wish to thank Dr. J. G. Carnes of the Lockheed Electronics Company for measuring the zero field samples. The paper profitted from many discussions with Dr. Carnes and our friends in the Geochemistry and Geology Branches of NASA Johnson Space Center. The technical help of Mr. Ralf Schmidt of the Lockheed Electronics Company is greatly appreciated. Part of this work was supported by the Lunar Science Institute which is operated by the University Space Research Association under Contract no. NSR-09-051-001 with the National Aeronautics and Space Administration. This paper constitutes the Lunar Science Institute contribution number 152.

REFERENCES

Brecher A., Vaughan D. J., Burns R. G., Cohen D., and Morash K. R. (1973) Magnetic and Mössbauer studies of Apollo 16 rock chips 60315,51 and 62295,27 (abstract). In *Lunar Science—IV*, pp. 88–90. The Lunar Science Institute, Houston.

Butler P. (1972) Apollo 16 Sample Catalog, NASA publication MSC 03210.

Doell R. R., Grommé C. S., Thorpe A. N., and Senftle F. E. (1970) Magnetic studies of Apollo 11 lunar samples. *Proc. Apollo 11 Lunar Sci. Conf., Geochim. Cosmochim. Acta,* Suppl. 1, Vol. 3, pp. 2079–2102. Pergamon.

Dunlop D. J., Gose W. A., Pearce G. W., and Strangway D. W. (1973) The magnetic properties and granulometry of metallic iron in lunar breccia 14313. *Proc. Fourth Lunar Sci. Conf.* This volume.

Dyal P., Parkin C. W., Sonett C. P., Dubois R. L., and Simmons G. (1972) The lunar portable magnetometer experiment. *Apollo 16 Preliminary Science Report,* NASA publication SP-315, pp. 11-1 to 11-14.

El Goresy A., Ramdohr P., and Taylor L. A. (1972) Fra Mauro crystalline rocks: Petrology, geochemistry, and subsolidus reduction of the opaque minerals (abstract). In *Lunar Science—III,* pp. 224–226. The Lunar Science Institute, Houston.

Finger L. W., Hafner S. S., Schürmann K., Virgo D., and Warburton D. (1972) Distinct cooling histories and reheating of Apollo 14 rocks (abstract). In *Lunar Science—III,* pp. 259–261. The Lunar Science Institute, Houston.

Gose W. A., Pearce G. W., Strangway D. W., and Larson E. E. (1972) On the applicability of lunar breccias for paleomagnetic interpretation. *The Moon* 5, 106–120.

Gose W. A. and Carnes J. G. (1973) The time dependent magnetization of fine-grained iron in lunar breccias. *Earth Planet. Sci. Lett.* In press.

Gose W. A., Strangway D. W., and Pearce G. W. (1973) A determination of the intensity of the ancient lunar magnetic field. *The Moon* 7, 198–201.

Grommé C. S. and Doell R. R. (1971) Magnetic properties of Apollo 12 lunar samples 12052 and 12065. *Proc. Second Lunar Sci. Conf., Geochim. Cosmochim. Acta,* Suppl. 2, Vol. 3, pp. 2491–2499. MIT Press.

Hargraves R. B. and Dorety N. (1971) Magnetic properties of some lunar crystalline rocks returned by Apollo 11 and 12. *Proc. Second Lunar Sci. Conf., Geochim. Cosmochim. Acta,* Suppl. 2, Vol. 3, pp. 2477–2483. MIT Press.

Housley R. M., Grant R. W., and Abdel-Gawad M. (1972) Study of excess Fe metal in the lunar fines by magnetic separation (abstract). In *Lunar Science—III,* p. 392. The Lunar Science Institute, Houston.

LSPET (1972) Preliminary examination of the Apollo 16 samples. *Science* 179, 23–34.

McKay D. S., Greenwood W. R., and Morrison D. A. (1970) Origin of small lunar particles and breccia from the Apollo 11 site. *Proc. Apollo 11 Lunar Sci. Conf., Geochim. Cosmochim. Acta,* Suppl. 1, Vol. 1, pp. 673–694. Pergamon.

Nagata T., Ishikawa Y., Kinoshita H., Kono M., Syono Y., and Fisher R. M. (1970) Magnetic properties and natural remanent magnetization of lunar materials. *Proc. Apollo 11 Lunar Sci. Conf., Geochim. Cosmochim. Acta,* Suppl. 1, Vol. 3, pp. 2325–2340. Pergamon.

Nagata T., Fisher R. M., Schwerer F. C., Fuller M. D., and Dunn J. R. (1971) Magnetic properties and remanent magnetization of Apollo 12 lunar materials and Apollo 11 lunar microbreccia. *Proc. Second Lunar Sci. Conf., Geochim. Cosmochim. Acta,* Suppl. 2, Vol. 3, pp. 2461–2476. MIT Press.

Nagata T., Fisher R. M., Schwerer F. C., Fuller M. D., and Dunn J. R. (1972a) Rock magnetism of Apollo 14 and 15 materials. *Proc. Third Lunar Sci. Conf., Geochim. Cosmochim. Acta,* Suppl. 3, Vol. 3, pp. 2423–2448. MIT Press.

Nagata T., Fisher R. M., Schwerer F. C., Fuller M. D., and Dunn J. R. (1972b) Summary of rock magnetism of Apollo 15 lunar materials. In *The Apollo 15 Lunar Samples,* pp. 442–445. The Lunar Science Institute, Houston.

Nagata T., Schwerer F. C., Fisher R. M., Fuller M. D., and Dunn J. R. (1973) Magnetic properties and natural remanent magnetization of Apollo 15 and 16 rocks (abstract). In *Lunar Science—IV,* pp. 552–554. The Lunar Science Institute, Houston.

Pearce G. W., Strangway D. W., and Gose W. A. (1972a) Remanent magnetization of the lunar surface. *Proc. Third Lunar Sci. Conf., Geochim. Cosmochim. Acta,* Suppl. 3, Vol. 3, pp. 2449–2464. MIT Press.

Pearce G. W., Williams R. J., and McKay D. S. (1972b) The magnetic properties and morphology of metallic iron produced by subsolidus reduction of synthetic Apollo 11 composition glasses. *Earth Planet. Sci. Lett.* **17,** 95–104.

Reid A. M., Meyer C. Jr., Harmon R. S., and Brett R. (1970) Metal grains in Apollo 12 igneous rocks. *Earth Planet. Sci. Lett.* **8,** 1–5.

Runcorn S. K., Collinson D. W., O'Reilly W., Battey M. H., Stephenson A., Jones J. M., Manson A. J., and Readman P. W. (1970) Magnetic properties of Apollo 11 lunar samples. *Proc. Apollo 11 Lunar Sci. Conf., Geochim. Cosmochim. Acta,* Suppl. 1, Vol. 3, pp. 2369–2387. Pergamon.

Runcorn S. K., Collinson D. M., O'Reilly W., Stephenson A., Battey M. H., Manson A. J., Readman P. W. (1971) Magnetic properties of Apollo 12 lunar samples. *Proc. R. Soc. Lond.* **A325,** 157.

Strangway D. W., Larson E. E., and Pearce G. W. (1970) Magnetic studies of lunar samples—breccia and fines. *Proc. Apollo 11 Lunar Sci. Conf., Geochim. Cosmochim. Acta,* Suppl. 1, Vol. 3, pp. 2435–2451. Pergamon.

Strangway D. W., Pearce G. W., Gose W. A., and Timme R. W. (1971) Remanent magnetization of lunar samples. *Earth Planet. Sci. Lett.* **13,** 43–52.

Swann G. A., Bailey N. G., Batson R. M., Freeman V. L., Hait M. H., Head J. W., Holt H. E., Howard K. A., Irwin J. B., Larson K. B., Muehlberger W. R., Reed V. S., Rennilson J. J., Schaber G. G., Scott D. R., Silver L. T., Sutton R. L., Ulrich G. E., Wilshire H. G., and Wolfe E. W. (1972) Preliminary geologic investigations of the Apollo 15 landing site. *Apollo 15 Preliminary Science Report,* NASA publication SP-289, pp. 5-1 to 5-112.

Proceedings of the Fourth Lunar Science Conference
(Supplement 4, *Geochimica et Cosmochimica Acta*)
Vol. 3, pp. 3077–3091

Dielectric spectra of Apollo 15 and 16 lunar solid samples

DAE H. CHUNG

Weston Geophysical Observatory, Weston, Massachusetts 02193
Massachusetts Institute of Technology, Cambridge, Massachusetts 02139

W. B. WESTPHAL

Massachusetts Institute of Technology, Cambridge, Massachusetts 02139

Abstract—The dielectric constants, losses and AC-conductivities of lunar samples 15065,27, 15459,62, 15555,88, 15415,57, 60015,29, 60315,34, 61016,34, and 62235,17 as they are measured over a range of frequencies from 100 Hz to 10 MHz and temperatures from 77°K to 473°K by two-terminal capacitance substitution methods are presented. The dielectric spectra of lunar gabbroic basalts are typified by the dielectric behavior of samples 15065 and 15555, showing complex relaxation processes which are both frequency- and temperature-dependent. Such feldspar-rich crystalline lunar basalts like samples 15459, 62235, and 60015 showed essentially the same dielectric relaxation mechanisms which are also a function of both frequency and temperature with a very broad relaxation frequency distribution. The nature and a detailed process of these relaxation mechanisms are not yet understood and as more experimental data become available these mechanisms may be separable in terms of a DC-conductivity and dielectric relaxation mechanisms. Highly-brecciated lunar sample, 60315, exhibits rather an unusual dielectric behavior which is dependent upon on field-strength. Incorporating information on the chemical and physical structure of this sample, we have attributed the observed electrical properties to effects arising from numerous particles of metallic free-iron and other opaque alloys dispersed along mineral boundaries.

INTRODUCTION

THIS PAPER, the fourth in our series of reports presenting the electrical properties of the returned lunar samples, presents now the observed dielectric spectra of various lunar samples returned from Apollo 15 and 16 landing sites. Our laboratory characterizations of dielectric behavior of lunar samples received from Apollo 11, 12, and 14 missions have been previously reported (Chung *et al.*, 1970, 1971, 1972). Based on our laboratory data combined with lunar models, electrical properties of the lunar interior down to about 150 km deep were theoretically investigated and the possibility of a low-frequency lunar probing was examined (*see* Chung *et al.*, 1971). The dielectric spectra determined on various lunar samples were analyzed to relate electrical properties to physical and chemical structures of the lunar samples (Chung *et al.*, 1972; Chung, 1972), but no general conclusion as to the observed dielectric relaxation processes was made. Obviously what is needed is more data and realistic theories. In this paper, we present new data on dielectric constants, dielectric losses and AC-conductivities of the following lunar samples: 15065,27, 15459,62, 15555,88, and 15415,57 returned from the Apollo 15 landing site and 60015,29, 60315,34, 61016,34, and 62235,17 from the Apollo 16 landing site. In a subsequent report (Olhoeft and Chung, 1973), our attempt to relate these observed dielectric spectra with chemical and physical processes within lunar samples is made.

Lunar Samples

The chemical and mineralogical description of samples 15065, 15459, 15555, and 15415 may be found in various reports, e.g., LSPET (1972) and Chamberlain and Watkins (1973) and of samples 60015, 60315, 61016, and 62235 in LSPET (1973) and also various reports found in Chamberlain and Watkins (1973); readers are referred to these reports for detailed information about the lunar samples. A brief description of their classification and rock type along with apparent density is made in Table 1.

Methods of transporting, handling, heating, and drying lunar samples, along with other experimental details utilized in the present work were exactly the same as those described earlier by Chung *et al.* (1970, 1971), and, therefore, will not be repeated here. As noted before, our special attention was given to drying and baking lunar samples at about 570°K in a vacuum oven so that the samples would be free from any absorbed moisture.

Dielectric Spectra of Lunar Samples

Values of the dielectric constant κ', dissipation factor $\tan \delta$, and AC-conductivity σ for sample 15065,27 are shown in Figs. 1 through 3. Similar data for sample 15555,88 are shown in Figs. 4 through 6. The electrical properties of these samples 15065 and 15555 are very similar and they, as shown, are typical of the dielectric behavior of gabbroic basalts. Note in Figs. 2 and 5 that some kinds of frequency and temperature-dependent relaxation processes can be clearly seen in these materials. As also evident from Figs. 3 and 6, a temperature- and frequency-dependent dispersion is seen; this phenomenon is particularly apparent at high temperatures and low frequencies. Since, we believe, our test samples are free from absorbed moisture, the observed dispersion must be due to some as yet unidentified relaxation mechanism occurring at low frequencies ($f \leq 1$ MHz) and high temperatures ($T \geq$ about 300°K). These behavioral characteristics yet to be analyzed and explained, are also observed in such lunar samples as 61016,34, as they are shown in Figs. 7 through 9. It seems that there exist at least *two* distinctively different mechanisms operating in this class of lunar samples: one is a loss due to DC-conductivity and the other to dielectric mechanism. As shown

Table 1. Lunar samples and their typical material characteristics.

Sample number	Classification of rock type	ρ_3 (g/cm^3)	High-freq. κ'	High-freq. $\tan \delta$
15065,27	gabbroic basalt	2.86	6–7	0.008–0.01
15459,62	clastic breccia	2.76	6.5–7	0.004–0.008
15555,88	gabbroic basalt	3.10	6	0.008–0.02
15415,57	anorthosite	2.70	4	~0.001
60015,29	igneous; feldspar	2.76	6–8	0.004–0.002
60315,34	metallic breccia†	1.95	~29–56	~0.2
61016,34	igneous; feldspar	2.79	6–8	0.01–0.02
62235,17	clastic breccia	2.78	6–7	0.008–0.01

†This sample contains numerous particles of metallic free-iron and iron-nickel alloys dispersed randomly along mineral boundaries (*see* text for discussion).

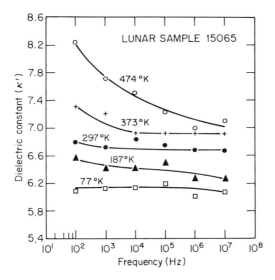

Fig. 1. Dielectric constant of sample 15065,27 as a function of frequency and temperature.

earlier by Chung *et al.* (1972) with reference to the dielectric behavior of lunar sample 14310,75, this dielectric loss must be related to a term given by $\kappa''(\omega, T)/\kappa'(\omega, T)$. The loss due to a DC-conductivity mechanism has the usual expression containing a term which is inversely proportional to $\epsilon_0 \omega \kappa'(\omega, T)$, where $\epsilon_0 = \epsilon/\kappa'$ (*see*, for example, Olhoeft *et al.* (1972)). The loss tangents, as shown in Figs. 2 and 8, are in general frequency-dependent at all temperatures, and this

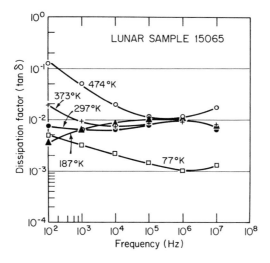

Fig. 2. Dielectric losses in sample 15065,27 as a function of frequency and temperature.

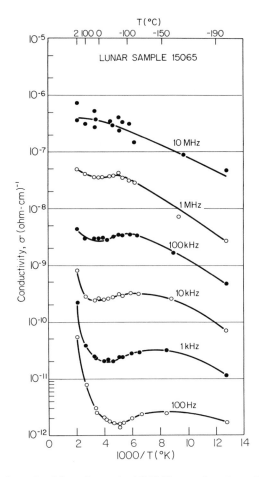

Fig. 3. Electrical conductivity of sample 15065,27 as a function of frequency and temperature.

frequency-dependence arises primarily from a complicated relaxation mechanism not yet understood with certainty. In a separate report, Olhoeft and Chung (1973) have called attention to this subject.

The observed electrical properties of samples 15459,62, 62235,17, and 60015,29 are illustrated in Figs. 10 through 12, Figs. 13 through 15, and Figs. 16 and 17, respectively. The dielectric spectra of these samples are very similar in general, and there seem to be two distinctly separate dielectric relaxation mechanisms operating at the same time. These relaxation mechanisms appear to be both frequency- and temperature-dependent, but it is difficult to discuss these mechanisms with certainty at the present time. The nature and a detailed process of these relaxation mechanisms will be communicated in the near future as more experimental data become available from our laboratory investigations as well as from

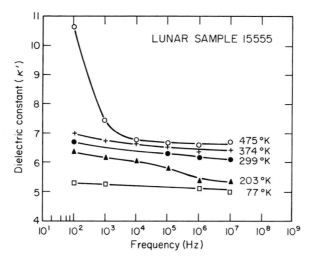

Fig. 4. Dielectric constant of sample 15555,88 as a function of frequency and temperature.

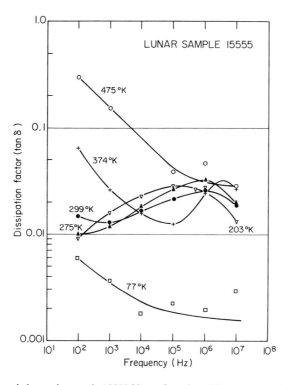

Fig. 5. Dielectric losses in sample 15555,88 as a function of frequency and temperature.

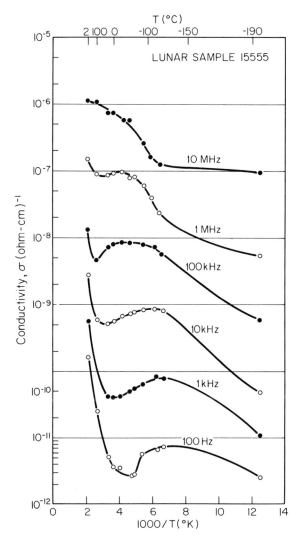

Fig. 6. Electrical conductivity of sample 15555,88 as a function of frequency and temperature.

Olhoeft *et al.* (1973) and Katsube and Collett (1973). Typically, however, the dielectric constants and losses for these samples at 1 MHz frequency, for example, range from 6 to 8 for κ' and 0.008 to 0.01 for tan δ. These electrical properties of samples 15459,62, 62235,17, and 60015,29 are very typical for the dielectric behavior of feldspar-rich crystalline lunar basalts.

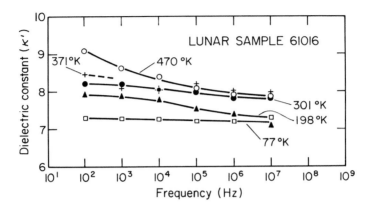

Fig. 7. Dielectric constant of sample 61016,34 as a function of frequency and temperature.

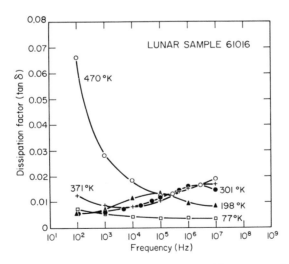

Fig. 8. Dielectric losses in sample 61016,34 as a function of frequency and temperature.

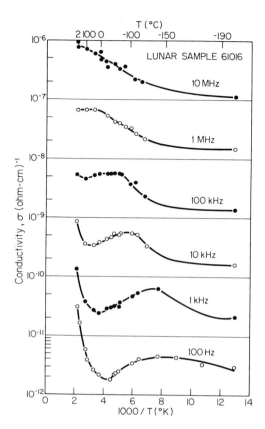

Fig. 9. Electrical conductivity of sample 61016,34 as a function of frequency and temperature.

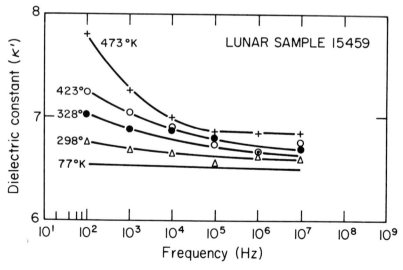

Fig. 10. Dielectric constant of sample 15459,62 as a function of frequency and temperature.

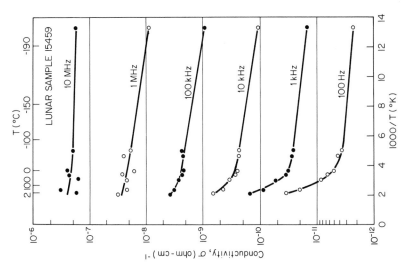

Fig. 12. Electrical conductivity of sample 15459,62 as a function of frequency and temperature.

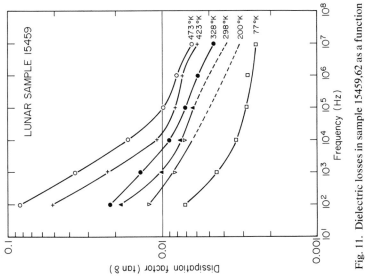

Fig. 11. Dielectric losses in sample 15459,62 as a function of frequency and temperature.

D. H. CHUNG and W. B. WESTPHAL

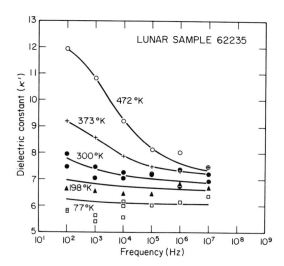

Fig. 13. Dielectric constant of sample 62235,17 as a function of frequency and temperature.

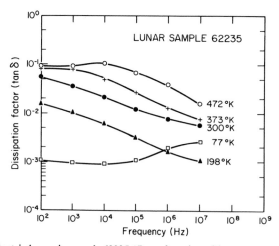

Fig. 14. Dielectric losses in sample 62235,17 as a function of frequency and temperature.

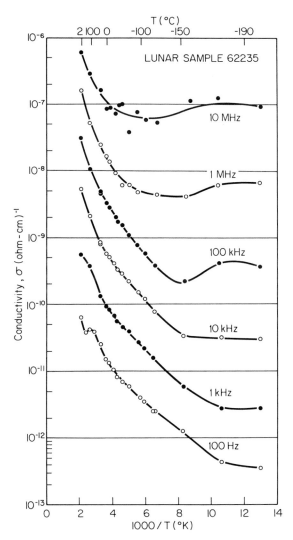

Fig. 15. Electrical conductivity of sample 62235,17 as a function of frequency and temperature.

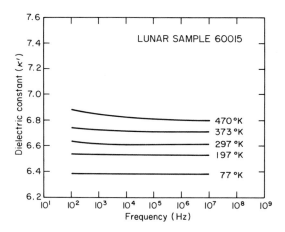

Fig. 16. Dielectric constant of sample 60015,29 as a function of frequency and temperature.

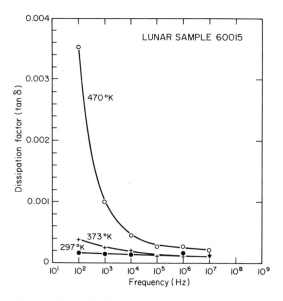

Fig. 17. Dielectric losses in sample 60015,29 as a function of frequency and temperature.

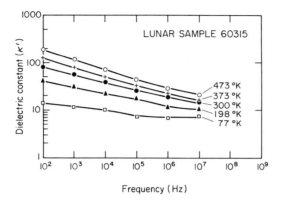

Fig. 18. Dielectric constant of sample 60315,34 as a function of frequency and temperature.

The measured electrical properties for sample 60315,34 are very unusual (Figs. 18 through 20). The high-frequency dielectric constant is in the neighborhood of 29 at 300°K and 1 MHz. The losses are very high, e.g., 0.2 at 1 MHz frequency and 300°K. When these parameters were studied as a function of temperature and also of frequency, this sample exhibited a thermally activated hysteresis on both high and low temperature measurements. Further, at 300°K and at higher temperatures, this sample was seen to be field-strength sensitive. For example, at 300°K and 1 MHz frequency, when the field was increased from 0.1 to 1.0 volt/cm we observed a change of the κ' value from 29 to 56 and σ from 0.87 to 1.09×10^{-8} ohm cm^{-1}. Our examination of a thin section made from this sample revealed that this sample contained numerous particles of metallic free-iron and some unidentified opaque metallic materials dispersed mainly along mineral boundaries.

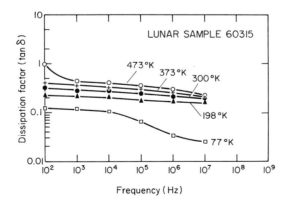

Fig. 19. Dielectric losses in sample 60315,34 as a function of frequency and temperature.

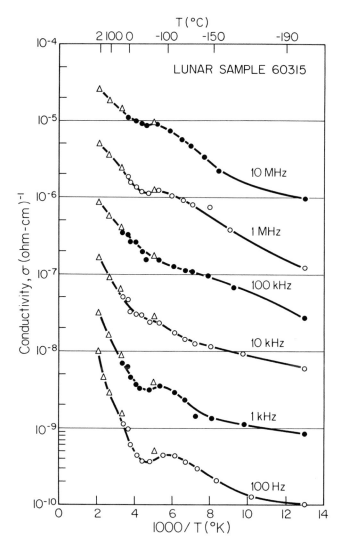

Fig. 20. Electrical conductivity of sample 60315,34 as a function of frequency and temperature.

Acknowledgments—We express our sincere gratitude to Professor A. R. von Hippel for his interest and encouragement during our work on lunar samples. We thank Professor D. W. Strangway for his interest and Gary R. Olhoeft for discussion. This work was supported by NASA under Grant NGR-22-009-597.

REFERENCES

Chamberlain J. W. and Watkins C. (editors) (1973) In *Lunar Science—IV*. The Lunar Science Institute, Houston.

Chung D. H. (1972) Laboratory studies on seismic and electrical properties of the moon. *The Moon* **4**, 356–372.

Chung D. H., Westphal W. B., and Simmons G. (1970) Dielectric properties of Apollo 11 lunar samples and their comparison with earth materials. *J. Geophys. Res.* **75**, 6524–6531.

Chung D. H., Westphal W. B., and Simmons G. (1971) Dielectric behavior of lunar samples: Electromagnetic probing of the lunar interior. In *Proc. Second Lunar Sci. Conf., Geochim. Cosmochim. Acta*, Suppl. 3, Vol. 3, pp. 2381–2390. MIT Press.

Chung D. H., Westphal W. B., and Olhoeft G. R. (1972) Dielectric properties of Apollo 14 lunar samples. In *Proc. Third Lunar Sci. Conf., Geochim. Cosmochim. Acta*, Suppl. 3, Vol. 3, pp. 3161–3172. MIT Press.

Katsube T. J. and Collett L. S. (1973) Electrical and EM propagation characteristics of Apollo 16 samples (abstract). In *Lunar Science—IV*, p. 431. The Lunar Science Institute, Houston.

LSPET (Lunar Sample Preliminary Examination Team) (1972) Preliminary examination of lunar samples from Apollo 15. *Science* **175**, 363–375.

LSPET (Lunar Sample Preliminary Examination Team) (1973) Preliminary examination of lunar samples from Apollo 16. *Science* **179**, 23–34.

Olhoeft G. R. and Chung D. H. (1973) Interpretation of lunar sample electrical properties. To be published in *Earth and Planetary Science Letters*.

Olhoeft G. R., Frisillo A. L., and Strangway D. W. (1972) Lunar soil sample 15301,38: Correlation of electrical parameters with physical properties. *EOS Trans. AGU* **53**, 1034.

Olhoeft G. R., Frisillo A. L., and Strangway D. W. (1973) Electrical properties of lunar solid samples (abstract). In *Lunar Science—IV*, pp. 575–577. The Lunar Science Institute, Houston.

Proceedings of the Fourth Lunar Science Conference
(Supplement 4, *Geochimica et Cosmochimica Acta*)
Vol. 3, pp. 3093-3100

Grain size analysis and high frequency electrical properties of Apollo 15 and 16 samples

T. GOLD, E. BILSON, and M. YERBURY

Center for Radiophysics and Space Research
Space Sciences Building
Cornell University
Ithaca, New York

GRAIN SIZE ANALYSIS

THE METHOD of measuring the sedimentation rate in a column of water was used to determine the particle size distribution in Apollo 15 and 16 surface fines from various locations. Eleven samples, from different depths below the surface, from the Apollo 15 deep-drill core tubes, collected at the ALSEP site have been also analyzed. The sedimentation method has been described earlier (Gold *et al.*, 1971); its accuracy is adequate for comparative measurements in the particle size range 1–100 μm. The simplicity of this method has made the analysis of a great number of samples possible. In Figs. 1 and 2 we have plotted the number of particles per micron particle range vs. particle diameter instead of the more conventionally used cumulative distribution curves. However, this is the most straightforward presentation of our data and our purpose is to compare the distribution curves of samples from different locations and depths rather than the detailed analysis of the individual curves. Figure 1 shows distribution curves for Apollo 15 surface samples from four different stations and for Apollo 16 surface samples from three different stations. The slopes of the Apollo 15 curves are very similar, especially in the range below 10 microns, with the exception of the slope of the 15221 curve, which is significantly steeper than that of the others. Sample 15221 was collected at Station 2, near the St. George Crater, Butler *et al.* (1972) analyzed nine Apollo 15 surface samples and reported the smallest graphic mean size for 15221. Of the three Apollo 16 samples, 67601 seems to be the finest and in general the distribution curves of these samples are very similar to the typical Apollo 15 curves. Figure 2 shows distribution curves for the Apollo 15 deep-drill core samples from the top three tubes. Samples 15006 came from the top tube (0–40 cm below the surface), samples 15005 from the second tube (40–80 cm below the surface) and samples 15004 from the third tube (80–120 cm below the surface). Along with the curves of the 15006 samples, that of 15041 is shown; this latter was collected from the surface at the same location as the core samples. There are some significant differences in the slopes of the distribution curves of these core samples. Especially noticeable is the variation in the distribution curves of such samples as 15005,162; 15005,179; and 15005,186, which come from layers separated only by a few centimeters. (The distance between the layer of 15005,162 and

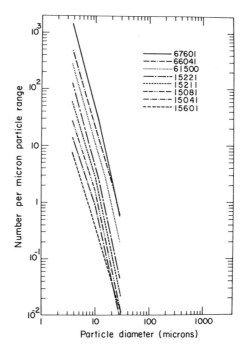

Fig. 1. The differential particle size distribution for several Apollo 15 and 16 fines.

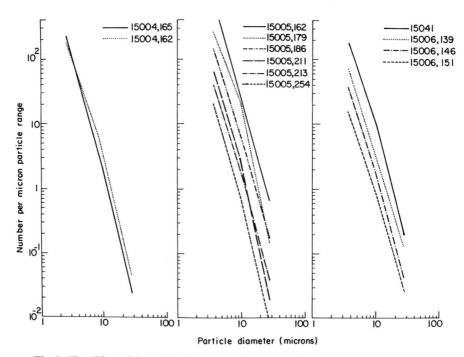

Fig. 2. The differential particle size distribution for Apollo 15 deep-drill core samples
from the top three tubes.

that of 15005,179 is 3 cm; the distance between the layer of 15005,179 and that of 15005,186 is 1.5 cm.) The slopes of the distribution curves of 15004,162 and 15004,165 are also significantly different. These samples originate in layers less than 4 cm apart. The fact that the grain size distribution in the core samples shows significant differences within a few centimeters variation of depth is important for the understanding of the surface transportation processes which are responsible for the deposition of thin layers of different physical and/or chemical origin. The existence of numerous thin textural units in this drill core was established by visual inspection during its dissection by Heiken *et al.* (1972). They propose that the layered structure of the regolith at the Apollo 15 site is formed by ejecta from near and far impact events. It seems difficult, however, to conceive that layers as thin as 1 cm can be formed by ejecta arriving with high speeds dictated by ballistic considerations, without stirring up the soil. A much more gentle mechanism seems to be required. These facts must be taken as evidence for the existence of some slow surface transportation process, as has been discussed in other contexts (Gold, 1955, 1971).

ELECTRICAL PROPERTIES

Eight surface powder samples, 15021.144; 15041.81; 15081.43; 15211.38; 15221.59; 15301.43; 15401.57; and 15601.105; and two rock samples, 15498.39 and 15597.30 from the Apollo 15 mission; six surface powder samples, 61500.7; 62240.5; 63501.25; 66041.13; 67601.22; and 68121.9; and one rock sample, 60017.45, from the Apollo 16 mission have been analyzed by techniques described in detail in a previous publication (Campbell and Ulrichs, 1969). Figure 3 shows

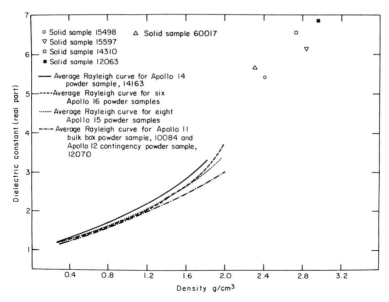

Fig. 3. Dielectric constant vs. density curves for Apollo 16, 15, 14, 12, and 11 powder samples. Dielectric constant vs. density points for solid samples are also shown.

the average dielectric constant vs. density curves for Apollo 15 and 16 powder samples; curves for Apollo 14, 12, and 11 samples are also shown for comparison as well as data points for solid samples. These curves differ only slightly from each other; the Apollo 15 and 16 curves are especially similar. The dielectric constants of the two Apollo 15 rock samples and of the Apollo 16 rock sample lie close to the extrapolated average Rayleigh curves (Campbell and Ulrichs, 1969) of the powder samples. Indeed it is questionable whether the rock samples measured are representative of the soil samples since the soils have glass spheres at various concentrations present and soils from different locations are derived from different rocks (or vice versa). In our use of the Rayleigh formula for the dust samples, we did not make use of the permittivity of the solid samples but rather measured the permittivity of the soil at a certain compaction and derived permittivity values for other densities. However, as the results mentioned above indicate, the dielectric properties of the soil and rock samples seem to follow the same permittivity-density relationship. Figures 4a and b show the Apollo 15 and 16 curves with the

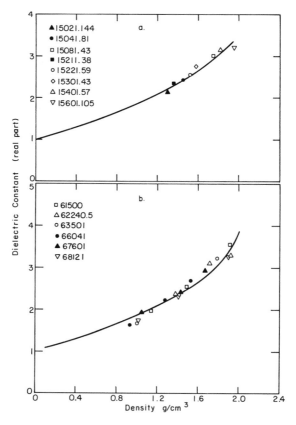

Fig. 4(a). Average Rayleigh curve for Apollo 15 powder samples with the experimental data points. (b) Average Rayleigh curve for Apollo 16 powder samples with the experimental data points.

data points. In the case of the eight Apollo 15 samples the dielectric constant was determined at only one density for each sample, whereas in the case of the six Apollo 16 powder samples, each sample was compacted to three different densities and the dielectric constant was determined for each of these densities. The curve on Fig. 4a is an average of 8 Rayleigh curves (generated from one measurement on each of the 8 samples) and the curve on Fig. 4b is an average of 18 Rayleigh curves (there were 3 curves generated for each of the 6 samples). The error in the dielectric constant measurements is estimated to $\pm 3\%$, whereas the error in the density measurements is less than $\pm 0.5\%$.

The ground-based radar determinations of the dielectric constant (*see* Evans and Hagfors, 1968) are in good agreement with measurements on all of our lunar samples obtained from the Apollo 11 to 16 missions if one assumes a density of about 1.5–1.7 g/cm^3 for the soil at a depth of 20 cm. (Bulk density in the Apollo 12 and 14 core tube samples ranges from 1.60 to 1.98 g/cm^3, Carrier III, 1972.)

The power absorption length at 450 MHz in Apollo 15 and 16 samples is shown in Figs. 5 and 6. (Table 1 lists the loss tangent values for the powder samples.) The error in the loss tangent measurements is estimated to be $\pm 10\%$. We observed again extremely low absorption, namely absorption lengths of 57 wavelengths and

Table 1.

Dust Sample	Density Measured g/cm^{-3}	Loss Tangent
15021.144	1.303	0.00418
15041.81	1.451	0.00486
15081.43	1.746	0.00536
15211.38	1.358	0.00389
15221.59	1.529	0.00274
15301.43	1.576	0.00438
15401.57	1.822	0.00471
15601.105	1.954	0.00251
61500.7	1.143	0.00277
61500.7	1.489	0.00347
61500.7	1.906	0.00503
62240.5	1.383	0.00364
62240.5	1.713	0.00416
62240.5	1.916	0.00482
63501.25	1.014	0.00161
63501.25	1.420	0.00253
63501.25	1.788	0.00341
66041.13	0.932	0.00225
66041.13	1.279	0.00300
66041.13	1.531	0.00388
67601.22	1.151	0.00216
67601.22	1.429	0.00258
67601.22	1.675	0.00290
68121.9	1.014	0.00234
68121.9	1.410	0.00312
68121.9	1.899	0.00445

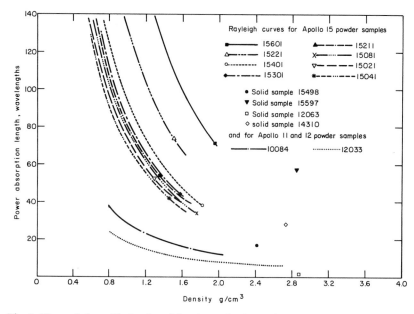

Fig. 5. The variation with density of the absorption length in Apollo 15 powder samples, in the Apollo 11 bulk box powder sample and in the Apollo 12 powder sample, 12033. Absorption length vs. density points for an Apollo 15, 14, and 12 solid samples are also shown.

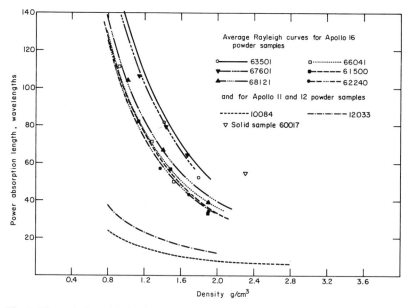

Fig. 6. The variation with density of the absorption length in Apollo 16 powder samples, in the Apollo 11 bulk box powder sample and in the Apollo 12 powder sample, 12033. An absorption length vs. density point for an Apollo 16 solid sample is also shown.

Table 2. Loss tangent.

Rock Sample No.	Katsube *et al.*	Chung *et al.*	Basset and Shackelford	Gold *et al.*
10017	0.021 (at 10^7 Hz)	—	—	—
10022	—	—	—	0.06 (at 450 MHz)
12002	0.015 (at 10^7 Hz)	0.05 (at 10^7 Hz)	—	—
12022	—	0.15 (at 10^7 Hz)	—	—
12063	—	—	—	0.069 (at 450 MHz)
14310	—	0.006 (at 10^7 Hz)	0.0075 (at 9.46 GHz)	0.00454 (at 450 MHz)
15597	—	—	—	0.0022 (at 450 MHz)
15498	—	—	—	0.008 (at 450 MHz)
60017	—	—	—	0.0024 (at 450 MHz)

54 wavelengths in solid rock samples 15597 and 60017 respectively and in general the absorption was less for all Apollo 15 and 16 powder samples than for earlier Apollo samples. Strikingly low absorption, 28 wavelengths, was first observed for the Apollo 14 rock sample 14310. One suspects water contamination as having increased the absorption in early Apollo samples beyond the value on the moon, and we have been informed that indeed the method of handling in the course of distribution of the samples was different. Table 2 shows loss tangent values obtained for rock samples from various lunar sites, in our laboratory as well as by other research groups. There appears to be a definite tendency for lower absorption in rocks collected in more recent missions, tending to confirm the contamination hypothesis. If indeed the results obtained for the Apollo 14–16 samples apply more accurately to lunar materials in general, then our results would indicate that for the case of meter wavelength radar waves, reflections from depths of more than 100 meters generally contribute significantly to the radar echos obtained.

Acknowledgments—We wish to thank Mr. H. J. Eckelmann for technical assistance. Work on lunar samples was carried out under NASA Grant NGR-33-010-137.

REFERENCES

Butler J. C., King E. A., and Carman M. F. (1972) Size frequency distributions and petrographic observations of Apollo 15 samples (abstract). In *The Apollo 15 Lunar Samples*, pp. 45–46. The Lunar Science Institute, Houston.

Campbell M. J. and Ulrichs J. (1969) Electrical properties of rocks and their significance for lunar radar observations. *J. Geophys. Res.* **74**, 5867–5881.

Carrier W. D. III (1972) Core sample depth relationships: Apollo 14 and 15 (abstract). In *Lunar Science—III*, pp. 122–124. The Lunar Science Institute, Houston.

Evans J. V. and Hagfors T. (1968) *Radar Astronomy*. McGraw-Hill.

Gold T. (1955) The lunar surface. *Mon. Not. Roy. Astr. Soc.*, Vol. 115, pp. 585–604.

Gold T. (1971) Erosion, transportation and the nature of the maria. *Proc. I.A.U. Symp.*, No. 47 (editors S. K Runcorn and H. C. Urey), pp. 55–67. D. Reidel.

Gold T., O'Leary B. T., and Campbell M. (1971) Some physical properties of Apollo 12 lunar samples. *Proc. Second Lunar Sci. Conf., Geochim. Cosmochim. Acta,* Suppl. 2, Vol. 3, pp. 2173–2181. MIT Press.

Heiken G., Duke M., Frywell R., Nagle J. S., Scott R., and Sellers G. A. (1972) *Stratigraphy of the Apollo 15 Drill Core.* NASA TM X-58101.

Proceedings of the Fourth Lunar Science Conference
(Supplement 4, *Geochimica et Cosmochimica Acta*)
Vol. 3, pp. 3101–3110

Electrical characteristics of Apollo 16 lunar samples

T. J. KATSUBE and L. S. COLLETT

Geological Survey of Canada, Ottawa, Ontario, Canada

Abstract—Electrical parameters of 1 fine sample (66041) and 4 rock samples (60025, 62295, 66055, and 68815) from Apollo 16 flight have been measured over the frequency range from 10^2 to 1.8×10^8 Hz. The purpose of these measurements is to accumulate data on the electrical characteristics of various rocks, to extend the frequency range of measurement, and to obtain data with sufficient accuracy to characterize the general trend of the electrical parameters. General trends of the 4 samples 66041,8, 60025,55, 62295,17, and 68815,43 are similar to previous measurements by various scientists. The real relative permittivity ranges from 5 to 7 and shows little variation with frequency. The parallel resistivity decreases with frequency and the dissipation factor generally decreases with frequency but shows a minimum and a maximum at about 10^5–10^7 Hz, for certain samples. The breccia sample 66055,7 shows electrical characteristics that are unusual for a lunar sample in many ways. K' is about 10 at 10^2 Hz, and levels off at 3.7 from frequencies above 3×10^5 Hz. The parallel resistivity is about 3×10^7 ohm-m at 10 Hz, and decreases to about 3×10^4 ohm-m at 1.8×10^8 Hz. This parallel resistivity at the lower frequencies is perhaps the lowest ever reported for lunar rocks. D is about 0.6 at 10^2 Hz and decreases to about 0.001 at 1.8×10^8 Hz. These trends for real relative permittivity, parallel resistivity and dissipation factor at the lower frequency resemble a terrestrial pyroxene or serpentinite. The electrical parameters of this rock seem to have a close relation to the apparent matrix of the specimen at the lower frequencies. These parameters at the higher frequencies seem to be related to the true matrix of the sample.

INTRODUCTION

PRODUCTION of data on electrical measurements of lunar samples has increased rapidly during the last two years (Olhoeft *et al.*, 1972, 1973; Sill *et al.*, 1973; Chung *et al.*, 1972; Strangway *et al.*, 1972; Gold *et al.*, 1971, 1972; Golovkin, 1971; Katsube and Collett, 1971). These data cover measurements in dry nitrogen atmosphere at room temperatures, at low and elevated temperatures in vacuum and in moist condition, and of optical reflection. However, there is a limit to the types of rocks that have been measured so far, the applied frequency range is mainly below 10^7 Hz or at isolated frequencies above 10^8 Hz, and the existing data are insufficient to characterize the general trend of the real relative permittivity (K'), parallel resistivity (ρ_p), and dissipation factor (D) of the lunar rocks.

Electrical measurements of 5 lunar samples from Apollo 16 flight have been completed over the frequency range from 10^2 to 1.8×10^8 Hz, at the Geological Survey of Canada (G.S.C.). The measurements were carried out in a dry nitrogen atmosphere and at room temperature. The type of samples ranged from fines (66041), anorthosites (60025), norites (62295) to breccias (66055 and 68815). The purpose of this set of measurements is to (1) contribute in accumulating data on electrical measurements of lunar samples, (2) extend the frequency range of measurements above 10^7 Hz, and (3) to obtain sufficient data in order to characterize the trends of real relative permittivity (K'), parallel resistivity (ρ_p) and dissipation factor (D) of the lunar rocks. It is necessary to characterize the trends of the

electrical parameters of the lunar rocks in order to study the conductive and
dielectric mechanism of the rocks. A report on the results of measurements of
these 5 lunar samples is given in this paper.

Parameters and Equipment

Definitions of K' and D or loss tangent (tan δ) are based on Von Hippel (1954) and ASTM (1968)
and can be found in the paper by Collett and Katsube (1973). Parallel resistivity (ρ_p) appears for the
first time in the paper by Katsube and Collett (1972) but it is the conventional concept of resistivity
which is defined as the reciprocal of the real conductivity (σ'):

$$\rho_p = \frac{1}{\sigma'}$$

and not equal to the real resistivity (ρ'):

$$\rho_p \neq \rho'.$$

In other terms ρ_p can be defined as:

$$\rho_p = \frac{1}{\omega K' \epsilon_0 D}$$

where

$$D = \tan \delta$$

  ω: angular frequency
  ϵ_0:   permittivity of air or vacuum (8.85×10^{-12} F/m)
  K':   real relative permittivity
  D: dissipation factor
tan δ: loss tangent.

A General Radio Capacitance Bridge (GR 716-C) was used for measurements at frequencies from
10^2 to 10^5 Hz, a Marconi Q-meter (TP 1245A) for frequencies between 3×10^7 to 1.8×10^8 Hz. A
Manostat Vacuum glove box (Model 41-907-24) was used for maintaining a dry nitrogen atmosphere
during measurement. Further details of the measuring equipment are given in the paper by Collett and
Katsube (1973). Both Q-meters are placed inside the vacuum chamber for maintaining minimum distance between the sample holder and the measuring apparatus. When using the GR 716-C bridge, only
the sample holder is placed inside the vacuum chamber, and is connected to the bridge by a coaxial
cable (RG-62/U). A nitrogen atmosphere in the vacuum chamber is obtained by first drying the interior
air through a filter, then repeating a procedure of lowering the air pressure (to one third of the
atmospheric pressure) and introducing the dry nitrogen for more than 6 times. In this way a dry
nitrogen atmosphere can be obtained without damaging the electronic equipment.

Type of Samples

It is difficult to construct a table of mineral contents from the currently existing
data (LSIC Apollo 16, 1972) for the 5 lunar samples from Apollo 16. 66041,8 is a
sample of fines. 60025,55 is a white anorthosite which contains 70% plagioclase,
20% olivine, and 10% orthopyroxene (LSIC, 1972, p. 79). 62295,17 is a sample of
norite which contains 57% plagioclase, 26% orthopyroxene, and 19% other minerals (LSIC, 1972, p. 181). 66055,17 is a breccia which consists of a white matrix and
medium gray clasts (LSIC, 1972, p. 264). 68815,43 is also a breccia in which the

matrix (90%) contains 30% plagioclase, 35% clinopyroxene, and 35% devitrified glass and the mineral clasts (10%) contain 90% plagioclase and 10% clinopyroxene (LSIC, 1972, p. 363). These breccia samples have a very complicated texture, so that the specimens for the electrical measurements must be examined very closely in order to determine whether the specimen represents the rock sample.

RESULTS

Results of measurements of the electrical parameters of sample No. 66041,8 over the frequency range from 10^2 to 1.8×10^8 Hz are shown in Fig. 1. This is a

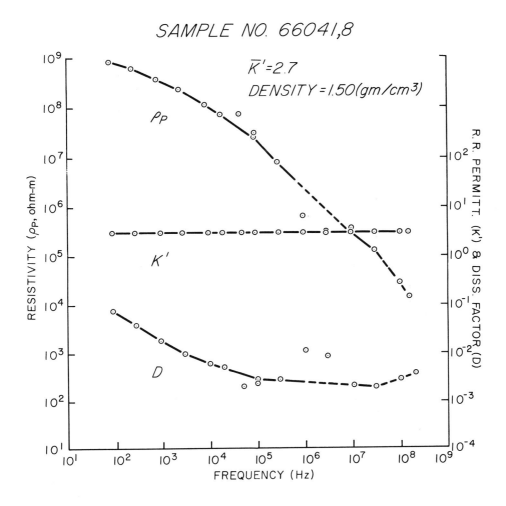

Fig. 1. Real relative permittivity (K'), parallel resistivity (ρ_p), and dissipation factor (D or tan δ) for lunar sample No. 66041,8 (fines).

sample of fines from Station 6. The density at the time of measurement was 1.50 g/cm. The real relative permittivity (K') is 2.7, and is constant over the entire frequency range. This is similar to other measurements on fines at room temperature (Olhoeft *et al.*, 1972; Strangway *et al.*, 1972; Katsube and Collett, 1971). The dissipation factor (D) decreases from about 0.07 at 10^2 Hz to about 0.002 at 10^5 Hz. In Fig. 1, D tends to level off from frequencies above 10^5 Hz. Above 10^5 Hz this trend is perhaps not a true characteristic of the sample, but due to instrumental limitations. If the true values of D continue to decrease above 10^5 Hz at the same rate as they do between 10^2 and 10^5 Hz, D is estimated to be about 10^{-4} at 10^9 Hz.

Results of sample No. 60025,55 are shown in Fig. 2. This sample is a white solid sample of anorthosite from Station 10. K' is 5.0 which is a rather low value for a lunar solid sample, and is more or less constant with frequency. The parallel

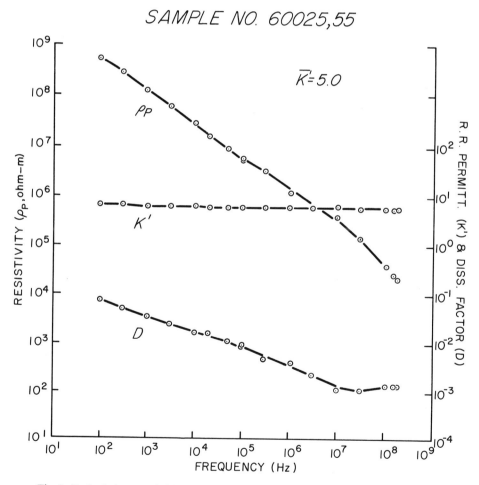

Fig. 2. Real relative permittivity (K'), parallel resistivity (ρ_p), and dissipation factor (D or tan δ) for lunar sample No. 60025,55 (solid, anorthosite).

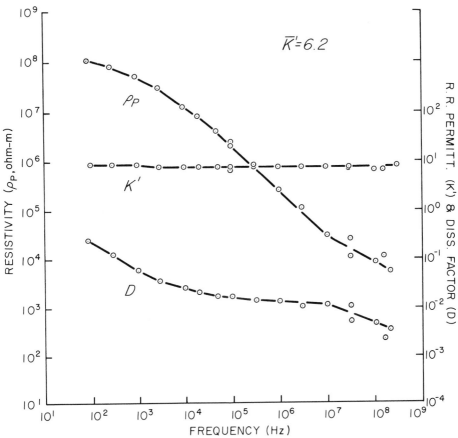

Fig. 3. Real relative permittivity (K'), parallel resistivity (ρ_p), and dissipation factor (D or tan δ) for lunar sample No. 62295,17 (solid, norite).

resistivity (ρ_p) is comparatively high. D decreases from about 0.07 at 10^2 Hz to about 0.0012 at 1.8×10^8 Hz. D tends to level off from frequencies above 10^7 Hz. This trend above 10^7 Hz may also be due to instrumental limitation.

Results for sample No. 62295,17 are shown in Fig. 3. This is a norite sample (crystalline) from Station 2. K' is 6.2 and shows little change with frequency. ρ_p is generally lower than that of sample No. 60025. D is about 0.2 at 10^2 Hz, decreases to about 0.01 from 10^5 Hz to 10^7 Hz, and decreases to 0.003 at 1.8×10^8 Hz. These trends are typical of lunar and terrestrial rocks (Chung *et al.*, 1972; Katsube and Collett, 1971). Measurement accuracy of D decreases from about 0.01 and below, and the actual values of D at frequencies above 3×10^6 Hz may be a small amount larger than what appears in Fig. 3. This would indicate a relative maximum of D at about 10^6 Hz.

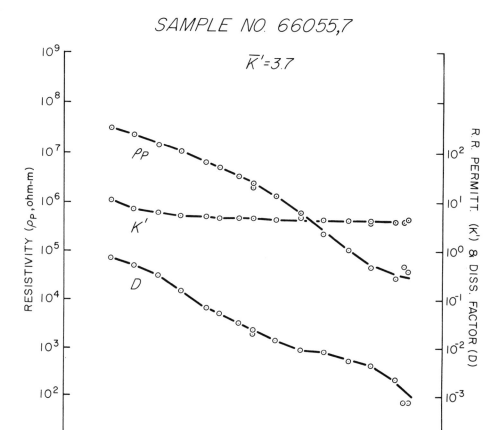

Fig. 4. Real relative permittivity (K'), parallel resistivity (ρ_p), and dissipation factor (D or tan δ) for lunar sample No. 66055,7 (solid, breccia).

Results of sample No. 66055,7 are shown in Fig. 4. This is a breccia sample with white matrix from Station 6. It consists of dark colored minerals and white minerals. The electrical characteristics of this rock are rather unusual in many ways. K' is about 10 at 10^2 Hz, and levels off at 3.7 from frequencies above 3×10^5 Hz. ρ_p is about 3×10^7 ohm-m at 10^2 Hz, and decreases to about 3×10^4 ohm-m at 1.8×10^8 Hz. ρ_p at the lower frequencies is perhaps the lowest ever reported for lunar rocks in dry state. D is about 0.6 at 10^2 Hz, decreases to about 0.001 at 1.8×10^8 Hz.

Results for sample No. 68815,43 are shown in Fig. 5. This is a breccia sample from Station 8. K' is 7.0 and shows little change with frequency. ρ_p is high at the lower frequency end, and comparatively low at the higher frequencies. D generally decreases with frequency but shows a minimum at about 5×10^4 Hz, and a

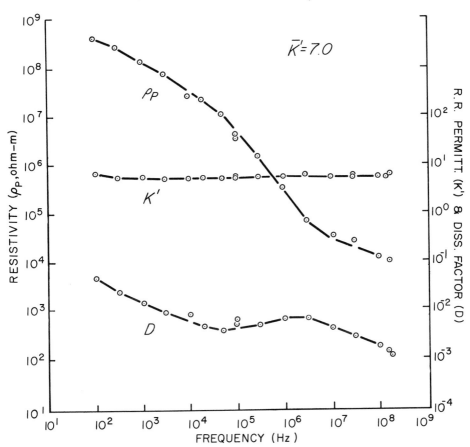

Fig. 5. Real relative permittivity (K'), parallel resistivity (ρ_p), and dissipation factor (D or tan δ) for lunar sample No. 68815,43 (solid, breccia).

relative maximum at about 3×10^6 Hz. This trend is similar to that of sample 14310 (Sill *et al.*, 1973; Chung *et al.*, 1972).

Discussion

The dissipation factor of samples 66041,8 and 60025,55 reach a minimum of 0.001–0.002 at about 10^5 Hz and 3×10^7 Hz respectively, which is perhaps due to instrumental limitations. The maximum sensitivity for measurements of D for the Marconi Q-meter (TF 1245A) is about 10^{-5}. This requires a coil with a quality factor (Q) of about 300, and this type of coil requires a total internal or instrumental capacitance of about 90–100 pico-farads for resonance. The capacitance of the samples that are used in this measurement range from about 0.5 to 1.5 pico-farads. This means that the internal or instrumental capacitance is about 60–200 times that of the sample capacitance, and therefore the maximum sensitivity of the Q-meter for measurements of D is reduced to 6×10^{-4}–2×10^{-3}, depending on the specimen.

The real relative permittivity (K') shows little variation with frequency for all samples except No. 66055,7. This trend is normal for lunar samples and can be seen in the measurements by other scientists (Olhoeft *et al.*, 1972 and 1973; Sill *et al.*, 1973; Strangway *et al.*, 1972; Chung *et al.*, 1971 and 1972; Katsube and Collett, 1971). The large variation of K' for sample No. 66055,7 is unusual for dry lunar rocks and resembles the trends of K' for a terrestrial pyroxenes and serpentinite (Katsube and Collett, 1973b). ρ_p is always seen to decrease with frequency. The values of ρ_p and the rate that ρ_p decreases with frequency varies according to the sample. D generally decreases with frequency, but in some cases (Fig. 5) D shows a minimum and a maximum at the higher frequencies. These trends of D are seen in the work by Katsube and Collett (1971) and Sill *et al.* (1973) and are normal for lunar samples.

Samples 68815 and 62295 contain more basic minerals (mainly pyroxene) than sample 60025. ρ_p of sample 68815,43 and 62295,17 is generally lower than that of sample 60025,55. K' and D are generally higher for samples 68815 and 62295 than those for sample 60025,55. This relation between electrical parameters and mineral content agrees with previous studies by Katsube and Collett (1972 and 1973b). A pyroxene sample is seen to have a considerably large permittivity and dissipation factor, particularly at the lower frequencies. Also K' and D generally seem to increase with increase in the content of basic minerals. The relation seen between K' and ilmenite (Hansen *et al.*, 1973) agrees with this statement since ilmenite is associated with the more basic minerals.

Sample No. 66055,7 shows rather unusual electrical characteristics. ρ_p is comparatively low and D is relatively large at the lower frequencies, and ρ_p is comparatively high and D is relatively low at the higher frequencies. Based on the study by Katsube and Collett (1973a) this trend suggests that the matrix of this sample consists of low resistivity minerals, with the mineral dusts having a relatively high resistivity. According to the photographs and descriptions in the LSIC (1972), the matrix of the rock consists of white minerals with dark mineral clusts contained in it. Since the white minerals can be considered to have a higher resistivity than the dark colored ones, the trends of ρ_p and D do not agree with the petrological description in the Apollo 16 LSIC (1972). Actually, the specimen (66055,7) used for these electrical measurements mainly consists of dark colored minerals: the apparent matrix of the specimen consisting of the dark colored minerals. Therefore, the electrical characteristics agree with the apparent texture of the specimen, but not the true texture of the sample. This is an important fact that should be considered when selecting specimens, or when interpreting the results of the measurements. No explanation is found so far for K' showing a very low value ($K' = 3.7$) at the higher frequencies.

CONCLUSION

The measurements on these 5 lunar samples contribute in generalizing the characteristics of real relative permittivity (K'), parallel resistivity (ρ_p), and dissipation factor (*D*) with frequency extended to 1.8×10^8 Hz. *K'* generally shows little variation with frequency over the frequency range from 10^2 to 1.8×10^8 Hz. *D* tends to decrease to or below 10^{-3} at the higher end of the frequency spectrum. The electrical parameters show a good relation with the apparent texture and mineralogy of the specimen but not with the true texture of the sample, for No. 66055,7. This is an important factor to be considered in future studies.

Acknowledgments—These measurements have been made possible by National Aeronautics and Space Administration supplying the samples. We thank Mr. J. J. Frechette for carrying out the measurements, and for the maintenance of the measuring apparatus.

REFERENCES

ASTM (American Society for Testing Materials) (1968) Standard methods of testing for AC loss characteristics and dielectric constant of solid electrical insulating materials. *ASTM* D 150-68, 29-54.
Collett L. S. and Katsube T. J. (1973) Electrical parameters of rocks in developing geophysical techniques. *Geophysics* **38**, 76–91.
Chung D. H., Westphal W. B., and Simmons, G. (1971) Dielectric behavior of lunar samples: Electromagnetic probing of the lunar interior. *Proc. Second Lunar Sci. Conf., Geochim. Cosmochim. Acta*, Suppl. 2, Vol. 3, pp. 2381–2390. MIT Press.
Chung D. H., Westphal W. B., and Olhoeft G. R. (1972) Dielectric properties of Apollo 14 lunar samples (abstract). In *Lunar Science—III*, pp. 139–140. The Lunar Science Institute, Houston.
Gold T., O'Leary B. T., and Campbell M. (1971) Some physical properties of Apollo 12 lunar samples. *Proc. Second Lunar Sci. Conf., Geochim. Cosmochim. Acta*, Suppl. 2, Vol. 3, pp. 2173–2181. MIT Press.
Gold T., Bilson E., and Yerbury M. (1972) Grain size analysis, optical reflectivity measurements and determination of high frequency electrical properties for Apollo 14 lunar samples (abstract). In *Lunar Science—III*, pp. 318–320. The Lunar Science Institute, Houston.
Golovkin A. R., Dukhovski E. A., Novik G. IA., and Petrochenkov R. G. (1971) Research on the electrical properties of lunar soil and its analogs by the Q-meter. *Reports of the Academy of Science*, USSR, 199, 1271–1273.
Hansen W., Sill W. R., and Ward S. H. (1973) The dielectric properties of selected basalts. *Geophysics* **38**, 135–139.
Katsube T. J. and Collett L. S. (1971) Electrical properties of Apollo 11 and Apollo 12 lunar samples. *Proc. Second Lunar Sci. Conf., Geochim. Cosmochim. Acta*, Suppl 2, Vol. 3, pp. 2367–2379. MIT Press.
Katsube T. J. and Collett L. S. (1972) Electrical and EM propagation characteristics of igneous rocks. Presented at 42nd Annual International Meeting of S.E.G.
Katsube T. J. and Collett L. S. (1973a) Electrical and EM propagation characteristics of Apollo 16 samples and problems in interpreting electrical measurements for planetary EM sounding (abstract). In *Lunar Science—IV*, pp. 429–431. The Lunar Science Institute, Houston.
Katsube T. J. and Collett L. S. (1973b) Measuring techniques for rocks with high permittivity and high loss. *Geophysics* **38**, 92–104.
LSIC (Lunar Sample Information Catalog) (1972) Apollo 16. Manned Spacecraft Center.
Olhoeft G. R., Frissillo A. L., and Strangway D. W. (1972) Frequency and temperature dependence of the electrical properties of a soil sample from Apollo 15. In *The Apollo 15 Lunar Samples*, pp. 477–479. The Lunar Science Institute, Houston.
Olhoeft G. R., Frisillo A. L., Strangway D. W., and Sharpe H. (1973) Analysis of lunar solid sample electrical properties (abstract). In *Lunar Science—IV*, pp. 575–577. The Lunar Science Institute, Houston.

Sill W. R., Hansen W., Ward S. H., Katsube T. J., Collett L. S., and Brown W. E. (1973) The dielectric constant and loss tangent of Apollo 14 and Apollo 16 soil and rock samples. Presented at the Geophys. and Geochem. Exploration of the Moon and Planets Conference, Houston, Texas.

Strangway D. W., Olhoeft G. R., Chapman W. B., and Carnes J. (1972) Electrical properties of lunar soil: dependence on frequency, temperature and moisture. *Earth Planet. Sci. Lett.* **16**, 275–281.

Von Hippel A. R. (1954) *Dielectrics and Waves.* Wiley.

Proceedings of the Fourth Lunar Science Conference
(Supplement 4, *Geochimica et Cosmochimica Acta*)
Vol. 3, pp. 3111–3131

Electrical characteristics of rocks and their application to planetary and terrestrial EM-sounding

T. J. KATSUBE and L. S. COLLETT

Geological Survey of Canada, Ottawa, Ontario, Canada

Abstract—It is essential to have a good understanding of the conductive and dielectric mechanism of the rocks in order to lay a basis for future planetary and terrestrial electrical and EM-sounding. There has been a rapid increase of data on electrical measurements for lunar and terrestrial rocks over the last few years. Based on these data it is possible to characterize the trends for real relative permittivity, parallel resistivity, and dissipation factor for many rocks over the frequency range from about 1.0 to 10^8 Hz. These trends suggest that from a macroscopic view grain boundaries, insulating and low resistivity minerals are important elements in determining the conductive and dielectric mechanism of rocks in general. There are cases where the dielectric relaxation are also thought to be of importance. At frequencies below the critical frequencies, the effect of insulating materials and grain boundaries vertical to the electric current are important for dry rocks, and the effect of the grain boundaries parallel to the current is dominant for rocks which contain water or other liquids. At frequencies above the critical frequency, the effect of low resistive minerals may be dominant for dry rocks and low porosity moist rocks. These studies on the conductive and dielectric mechanism indicate that, (1) there is a possibility for radar techniques to be useful in detecting conductive materials, (2) information on the content and conductivity of moisture or other liquids contained in the rocks might be obtained by a combination of LF and HF electromagnetic sounding methods, and (3) there are indications that grain boundaries and critical frequency concepts apply to rocks at elevated temperatures.

INTRODUCTION

IT IS essential to have a good understanding of the electrical mechanism of rocks not only for planning future electrical and EM exploration on other planets, but also for improvement and development of new and existing terrestrial exploration methods. Though much data on electrical measurements of lunar and terrestrial rocks have appeared during the last few years, work that has been published in relation to studies of the electrical mechanism of rocks is relatively scarce. The first paper related to this subject which appeared in recent literature is perhaps that by Saint-Amant and Strangway (1970). It discusses the Debye-relaxation phenomena and the Maxwell–Wagner effect in rocks. The electrical measurements (Olhoeft *et al.*, 1973) on the orange soil samples returned by the Apollo 17 mission may turn out to be a good example of this phenomena in rocks. Katsube and Collett (1972) emphasize the grain boundary effects. There are also papers by Chung *et al.* (1970, 1972), Katsube and Collett (1971), and Hansen *et al.* (1973) in which relations between electrical characteristics and mineral content are discussed, but do not discuss the electrical mechanism of the rocks.

From observations of the electrical data from recent rock measurements, several questions arise which are thought to be of importance at this point: (1) what is the reason for the (parallel) resistivity to decrease continuously with frequency,

(2) to what extent can the existence of water and other liquids be detected in rocks, (3) what effect does heat have on the electrical characteristics of rocks. In this paper, after reviewing the general trends of the electrical characteristics of rocks, first, an attempt is made to explain certain trends of the electrical characteristics, and second, these explanations are used as a basis for discussion on detection of water and other liquids, and on the effect of high temperature.

PARAMETERS

The parameters that are essential for expressing the general electrical characteristics of the rocks are, dissipation factor (D) or loss tangent (tan δ), complex resistivity (ρ^*), parallel resistivity (ρ_p), real relative permittivity or dielectric constant (K') and critical frequency (ω_{cr} or f_{cr}). For more detailed studies of the electrical characteristics of rocks, parameters related to the dielectric relaxation theories are also important, but not utilized in this paper. Definitions for all these parameters can be found in the papers by Collett and Katsube (1973), ASTM (1968), and Von Hippel (1954). The concept of ω_{cr} or f_{cr} is described well in the book by Jordan (1950). Concepts for ω_{cr} or f_{cr} used here are simplified compared to those of Jordan. ρ_p was perhaps first mentioned in the paper by Katsube and Collett (1972) and is defined by $\rho_p = 1/\sigma'$, where σ' is the real conductivity and σ'' is the imaginary conductivity in the following complex conductivity equation:

$$\sigma^* = \sigma' + j\sigma''$$

This parameter ρ_0 was introduced because the real resistivity (ρ') in the following complex resistivity (ρ^*) equation:

$$\rho^* = \rho' - j\rho''$$

where ρ'' is the imaginary resistivity, is generally not equal to the reciprocal of σ': $\rho' \neq 1/\sigma'$.

GENERAL CHARACTERISTICS OF ROCKS

The real relative permittivity (K') usually decreases with frequency. Its rate of decrease is usually under 10% for the frequency range from 10^2 to 2×10^8 Hz for lunar and terrestrial rocks. Therefore on a bi-logarithmic scale it will appear to be more or less constant with frequency, as can be seen in Fig. 1 through Fig. 7. Examples of this can also be seen in the paper by Strangway *et al.* (1972), and Saint-Amant and Strangway (1970). There are cases where K' rises with decrease in frequency, and which are obvious on a bi-logarithmic scale. When moisture is present, K' shows a dispersion with frequency (Strangway *et al.*, 1972). There are also smaller variations of K' which are obvious on bi-logarithmic scales but are not necessary due to moisture. Examples of these for lunar and terrestrial rocks are given by Katsube and Collett (1973 a, b) and Chung *et al.* (1972).

Parallel resistivity (ρ_p) decreases rapidly with frequency, as can be seen in Fig. 1 through Fig. 7. There are cases where the rate of decrease is not constant, as shown in Fig. 2 and Fig. 4. However ρ_p usually decreases at a constant rate as shown in Fig. 1 and Fig. 5 through Fig. 7.

There are basically three types of trends for the dissipation factor (D), (*see also* Katsube and Collett, 1971). Type 1 is a continuous decrease of D as seen in

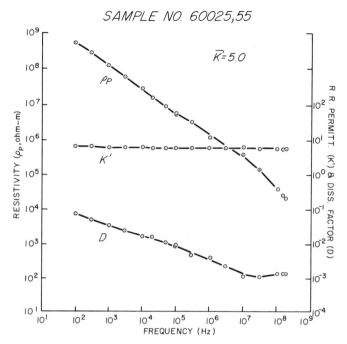

Fig. 1. Real relative permittivity (K'), parallel resistivity (ρ_p), and dissipation factor (D) for lunar sample 60025,55 in dry nitrogen atmosphere and room temperature.

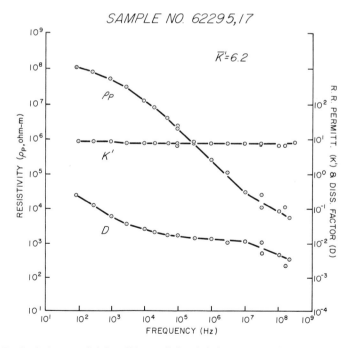

Fig. 2. Real relative permittivity (K'), parallel resistivity (ρ_p), and dissipation factor (D) for lunar sample 62295,17 in dry nitrogen atmosphere and room temperature.

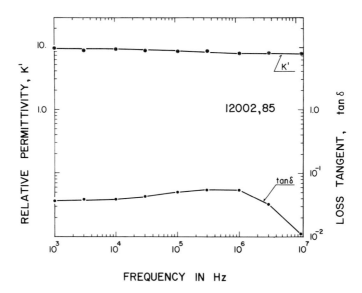

Fig. 3. Real relative permittivity (K') and dissipation factor (D) for lunar sample 12002,84. (Katsube and Collett, 1971) in dry nitrogen atmosphere and room temperature.

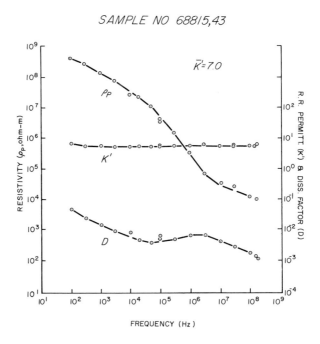

Fig. 4. Real relative permittivity (K'), parallel resistivity (ρ_p), and dissipation factor (D) for lunar sample 68815,43 in dry nitrogen atmosphere and room temperature.

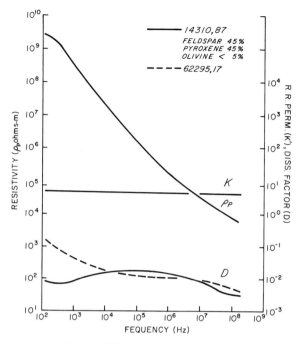

Fig. 5. Real relative permittivity (K'), parallel resistivity (ρ_p), and dissipation factor (D) for lunar sample 14310,87 (Sill *et al.*, 1973) in dry nitrogen atmosphere and room temperature.

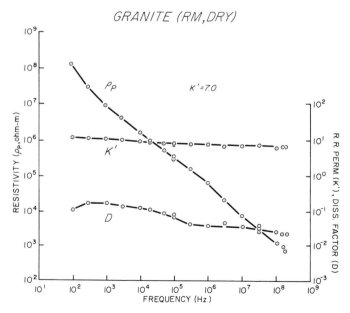

Fig. 6. Real relative permittivity (K'), parallel resistivity (ρ_p), and dissipation factor (D) for a terrestrial granite sample in dry atmosphere and room temperature.

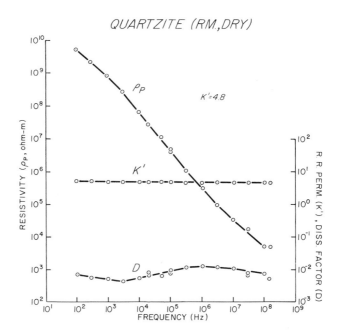

Fig. 7. Real relative permittivity (K'), parallel resistivity (ρ_p), and dissipation factor (D) for a terrestrial quartzite sample in dry atmosphere and room temperature.

Fig. 1, type 2 shows a minimum and a maximum as seen in Figs. 4, 5, and 7, and type 3 is an intermediate of types 1 and 2 as seen in Figs. 2, 3, and 6.

Typical trends of K', ρ_p, and D can be seen in Fig. 5, for lunar sample No. 14310,87, where ρ_p decreases at a more or less constant rate, K' is constant, and D decreases constantly or shows a maximum and a minimum. The problem now is to find an explanation for the cause of these trends.

ELECTRICAL MECHANISM OF DRY ROCKS

When thin sections of rocks are viewed under a microscope a large amount of various types of grain boundaries can be observed (example: Klein *et al.*, 1971; Dence *et al.*, 1971; Brett *et al.*, 1971). These grain boundaries separate mineral grains of similar or different types from one another. Fractures or micro-cracks within a crystal may also be included as grain boundaries. They are filled with air, nitrogen, other gases or are vacuum, depending on the environment of the rock, and have thicknesses which usually range from 1.0 to 20 microns (Deryagin and Zorin, 1968; Roberts, 1973). Actually thicknesses below 1.0 micron are not considered significant here. Grain boundaries hold the water when the rocks are moist and are more or less equivalent to the "joint pores" discussed later in this paper. The thickness of a single grain boundary varies to a certain extent, and partially forms a contact between the grains. These grain boundaries can be generally simulated by a capacitor. However, since they partially consist of contacts which are essentially continuations of material across the grain boundaries and which can be simulated by a parallel RC circuit, the general equivalent circuit of a grain boundary is a parallel RC circuit. The equivalent circuit of a mineral grain and grain boundary would be two parallel RC circuits in series. This resembles the equivalent circuit of a Maxwell–Wagner two-layer condenser, but has a partial difference in the assumptions for the electrical model.

Since the distribution of grain boundaries is very complicated, and therefore will distort the electrical currents flowing through the rock to a great extent, it is not easy to simulate the grain boundary and mineral grains distribution. However, when reducing the complex situation to the single fact that electrical currents traversing a rock flow through mineral grains and grain boundaries which both can be simulated by a RC network, it is possible to simulate a rock by two parallel RC circuits in series, as a first approximation, as shown in Fig. 8. If the average grain size is 1.0 mm in a disc shaped rock sample of 25.4 mm (1″) in diameter and 5.0 mm in thickness, the capacitance due to the grain boundaries would range from 45 to 900 picofarads. If the dielectric constant of the rock is 7.0, the capacitance of the rock sample would be 6.3 picofarads. Therefore, the effect of grain boundaries can be significant. Theoretical curves of ρ_p, K', and D are shown in Fig. 8 for the case where K'_M and ρ_M for the mineral grains is 10, ρ_M is 10^4 ohm meters, while the grain boundary thickness and porosity of the rock is 1.0 micron and 0.1–1.0%, respectively.

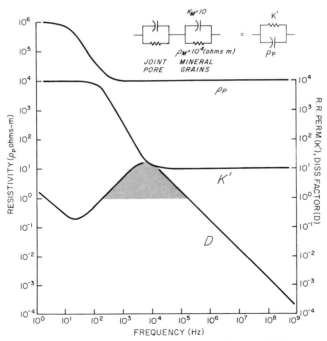

Fig. 8. Theoretical curves of K', ρ_p, and D for a model rock which consists of grain boundaries with single thickness and mineral grains.

In Fig. 8, D shows a minimum and a maximum similar to that of the typical lunar samples in Fig. 4 and Fig. 5. But D rises above 1.0 at 10^4 Hz, K' rises to 10^4 below 10^2 below 10^2 Hz and ρ_p is constant with frequency above 10^3 Hz, which is quite different from the trends in Fig. 4 and Fig. 5. So that the electrical model in Fig. 8 is not sufficient for the lunar samples. Microscopic observations suggest the existence of different thicknesses for different types of grain boundaries. If it is assumed that the grain boundaries have a distributed thickness ranging from 0.1 to 10 micron with the mean being 1.0 micron, the rock with its grain boundaries can be simulated by the equivalent circuit shown in the upper part of Fig. 9, as a second approximation.

In this case, the resistance in the RC parallel circuits are considered constant, and only the capacitance is considered to vary according to the thickness of the boundaries. The capacitance (C) of a single grain boundary is calculated by $C = \epsilon_0 A/d$, where

A is the area of the sample,
d is the thickness of a the grain boundary, and
ϵ_0 is the permittivity of air or vacuum.

Theoretical curves of ρ_p, K', and D are shown in Fig. 9 for the case where K'_M, ρ_M, and porosity are similar to those in Fig. 8, and the total resistance across the grain boundaries of a 1 m^3 of rock specimen is assumed to be about 100 times that of ρ_M.

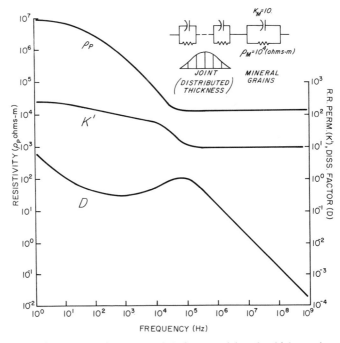

Fig. 9. Theoretical curves of K', ρ_p, and D for a model rock which consists of grain boundaries with distributed thicknesses and mineral grains.

The capacitance of the accumulated grain boundaries was calculated from an equally distributed number of thicknesses from 1 to 10 microns.

In Fig. 9 the gradients of the curve of D between 10^0 and 10^5 Hz is smaller than that of D in Fig. 8, which is closer to the actual trend (Figs. 4 and 5). But the maximum value of D at 10^5 Hz is larger than 1.0 and the trends of ρ_p and K' still differ from those of the actual ones shown in Figs. 4 and 5. Theoretical curves of ρ_p, K, and D are shown in Fig. 10, for a rock which consists mainly of high resistive material (10^7 ohm-meters).The model and equivalent circuit of this rock is shown in Fig. 11. All capacitances (C_{M1}, C_K, C_{M2}) and resistors (R_{M1}, R_K, R_{M2}) of the equivalent circuit are derived from the following equations:

$$C_{M1} = K'_M \epsilon_0 \frac{a^2 - p^2}{a} \qquad R_{M2} = \rho_M \frac{a - p}{p^2}$$

$$R_{ML} = \rho_M \frac{a}{a^2 - p^2} \qquad C_K = K'_k \epsilon_0 p$$

$$C_{M2} = K'_M \epsilon_0 \frac{p^2}{a - p} \qquad R_K = \rho_k / p$$

where a and p are defined in Fig. 11. ρ_M and K'_M are the resistivity and dielectric constant of the insulating mineral. The maximum value of D at 10^7 Hz is much

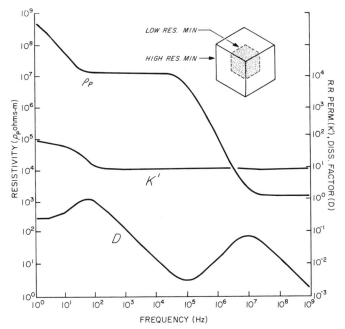

Fig. 10. Theoretical curves of K', ρ_p, and D for a model rock which consists of grain boundaries (distributed thicknesses), 90% of insulating minerals (resistivity: 10^7 Ω-m), and 10% of low resistivity minerals (resistivity: 10^2 Ω-m).

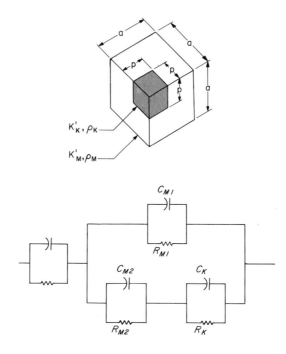

Fig. 11. Equivalent circuit of the model rock described in Fig. 10.

smaller than 1.0 which is similar to the actual case (Figs. 4 and 5). The general trends of ρ_p and K' are also much closer to the actual trends (Figs. 4 and 5) than those were in Figs 8 and 9.

Though there are still many differences between the theoretical and actual curves, some of the basic trends of ρ_p, K', and D are explained by these models. The explanation for ρ_p to decrease far below the resistivity (ρ_M) of the insulating material at the higher frequencies, is perhaps one of the most important of these. It is thought that if a distributed resistivity for the low resistivity minerals is considered in the model, the theoretical curves will be more similar to the actual ones. For the study of more complex models, application of the relaxation theories suggested by Olhoeft *et al.* (1973) may perhaps be more practical.

The fact that ρ_p decreases with frequency and D shows a maximum due to small amounts of conductive minerals inside the rock, indicates that there is a possibility for conductive minerals to be detected by HF or radar techniques.

DETECTION OF MOISTURE AND OTHER LIQUIDS IN ROCKS

Complex resistivity (ρ^*), ρ_p and D of a dry and moist (saturated with water) diorite sample (porosity of about 0.1%) is shown in Fig. 12 (Katsube and Collett, 1972). The broken lines represent the electrical parameters of the rock in dry state, and the solid lines represent the rock in the moist state. It can be seen that there is a large difference between the parameters at frequencies below the critical fre-

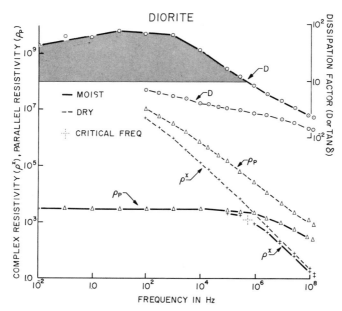

Fig. 12. Complex resistivity (ρ^*), parallel resistivity (ρ_p) and dissipation factor (D) for a dry and moist terrestrial diorite sample (crystalline rock). The dry rock was measured in a dry nitrogen atmosphere. (by Katsube and Collett, 1972).

quency (f_{cr}), but little difference between the parameters at the higher frequencies. There is a particularly small difference between ρ^* for the rock in moist and dry state at frequencies above f_{cr}. It is evident that the existence of water can be detected by measurement of ρ^* at frequencies above and below f_{cr}. Ammonia or other liquids that might exist on other planets would have effects similar to water, as long as the liquids do not behave as insulators.

Models of various rock pores are shown in Fig. 13. These are based on the

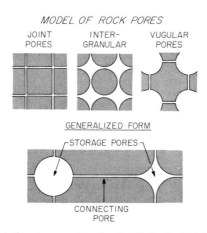

Fig. 13. Model of rock pores, described by Keller (in Parkhomenko, 1967).

descriptions by Keller (Parkhomenko, 1967). Joint pores are seen in crystalline rocks, intergranular pores are in sandstones and other sedimentary rocks, and vugular pores are seen in volcanic rocks. Intergranular and vugular pores partially consist of joint pores. The pores in lunar crystalline rocks are usually joint and vugular pores. When the storage of water or other liquids is considered, the three types of pores can be classified into storage pores and connecting pores, as shown in Fig. 13.

A simplified model of a rock which consists of connecting and storage pores is shown in Fig. 14. The equivalent circuit of the model is shown in the lower part of Fig. 14. The capacitances and resistors in the equivalent circuit are determined by the following equations:

$$C_{W1} = K'_W \epsilon_0 \frac{2ab + b^2}{a + b} \qquad R_{M2} = \rho_M \frac{a}{a^2 - p^2}$$

$$R_{W1} = \rho_W \frac{a + b}{2ab + b^2} \qquad C_{W3} = K'_W \epsilon_0 (1 - a^2)$$

$$C_{W2} = K'_W \epsilon_0 \frac{a^2 - p^2}{b} \qquad R_{W3} = \frac{\rho_W}{1 - a^2}$$

$$R_{W2} = \rho_W \frac{b}{a^2 - p^2} \qquad C_{M3} = K'_M \epsilon_0 \frac{p^2}{1 - p - b}$$

$$C_{M2} = K'_M \epsilon_0 \frac{a^2 - p^2}{a} \qquad R_{M3} = \rho_M \frac{1 - p - b}{p^2}$$

where K'_W and ρ_W are the dielectric constant and resistivity of water, K'_M and ρ_M are the dielectric constant and resistivity of the rock, and a, b, p are the dimensions shown in Fig. 14.

The descriptions by Keller (Parkhomenko, 1967) and the measured results by Archie (1952) are considered when constructing the model of the joint pores. Based on these considerations, the following relations are given for b and p:

$$b = p^m + d \text{ (meters)}$$

where m is related to the state of cementation, and d related to the type of pore. When the rock is well cemented, m is considered large and vice versa when it is poorly cemented. For intergranular pores, d is considered large and small for joint pores. Archie's Law (Keller in Parkhomenko, 1967) was considered in these designations. The equation to determine b was set up by considering the geometrical structure of the joint and storage pores, the shape of the entrance from the joint pores to the storage pores, the contact area of the mineral grains, and the effect of porosity increase on the change of shape of the joint pores. These considerations will be described in a later paper. For this paper only the equation for determining b is shown. For the calculations in Figs. 15 to 20, m was given a value of 6 for poorly cemented rocks and 12 for well cemented rocks. ρ_W is given a value of 10 ohm-meters.

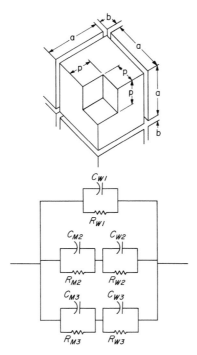

Fig. 14. Model of a rock with storage and joint pores and the equivalent circuit of the model rock.

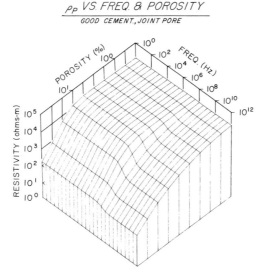

Fig. 15. Theoretical plane for parallel resistivity (ρ_p) vs. frequency (f) and porosity (ϕ) for a rock with well cemented ($m = 12$) joint pores ($d = 0.1$ micron).

The porosity of the rock ϕ is determined from,

$$\phi = 1 - a^3 + \rho^3$$

where

$$a + b = 1 \text{ (meter)}$$

A theoretical plane for parallel resistivity ρ_p vs. frequency and porosity for a rock with well cemented (m: large) joint pores (d: small) is shown in Fig. 15. In Fig. 16, a difference can be seen between ρ_p at the lower frequencies and the higher frequencies, particularly at larger porosities. The same type of ρ_p plane for a poorly cemented rock (m: small) with intergranular pores (d: small) is shown in Fig. 16. There is little difference between ρ_p at the lower and at higher frequencies. The theoretical plane of dielectric constant (K') vs. frequency and porosity is shown in Fig. 17. K' increases with increase in porosity. The rate of increase is comparatively large at the larger porosities. There is generally little difference between K' at the lower and higher frequencies.

A theoretical curve of K' at frequencies above the critical frequency for rocks with various types of pores and state of cementation is shown in Fig. 18. It can be seen that the state of cementation and type of pore has little effect on K' at these frequencies, so that there is a good relation between K' and porosity. Therefore it will be, perhaps, possible to predict the porosity of a water saturated rock by measuring K' at frequencies above f_{cr}. Theoretical curves of ρ_p at frequencies above f_{cr} for rocks with various types of pores and different states of cementation are shown in Fig. 19. Type of pore and state of cementation of the rock affect ρ_p at

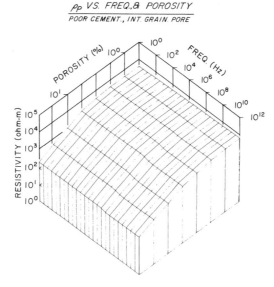

ρ_p VS. FREQ.,& POROSITY

POOR CEMENT., INT. GRAIN. PORE

Fig. 16. Theoretical plane for parallel resistivity (ρ_p) vs. frequency (f) and porosity (ϕ) for a rock with poorly cemented ($m = 6$) intergranular ($d = 0.1$ micron) pores.

K' VS FREQ. POROSITY

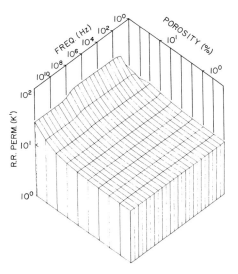

Fig. 17. Theoretical plane for real relative permittivity (K'), frequency (f), and porosity (f).

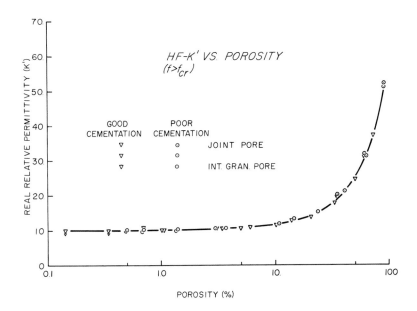

Fig. 18. Theoretical curve for K' at frequencies above the critical frequency for rocks with various types of pores and state of cementation.

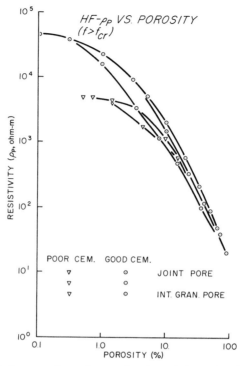

Fig. 19. Theoretical curves of ρ_p at frequencies above f_{cr} for rocks with various types of pores and different states of cementation.

low porosities, but have little effect at the higher porosities. Therefore, a porosity-liquid resistivity (ρ_w) product could be predicted from the measurement of ρ_p when the porosity is large. Since the porosity can be predicted by the value of K' (Fig. 18), ρ_w can be determined from ρ_p. Theoretical curves of ρ_p at frequencies below f_{cr} for rocks with various types of pores and different states of cementation are shown in Fig. 20. Since the porosity and ρ_w can be determined from measurements of K' and ρ_p at frequencies above f_{cr}, a certain amount of information on the cementation and type of pores can be obtained from ρ_p measured at frequencies below f_{cr}, as can be understood from the theoretical curves in Fig. 20.

Based on existing measurements of moist rocks and these theoretical considerations, it is thought that detection of water or other liquids and estimation of liquid content may be possible by an exploration technique using a combination of frequencies above and below f_{cr}.

Effect of High Temperature in Rocks

During the last few years a considerable amount of contributions have been made to the knowledge of electrical properties of lunar and terrestrial rocks at

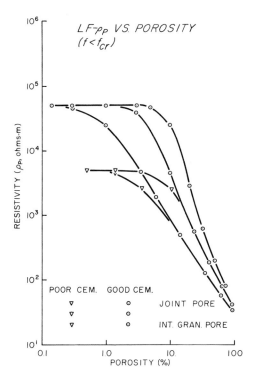

Fig. 20. Theoretical curves of ρ_p at frequencies below f_{cr} for rocks with various types of pores and different states of cementation.

elevated temperatures (Saint-Amant and Strangway, 1970; Olhoeft *et al.*, 1972; Chung *et al.*, 1971, 1972). Von Hippel (1954b, pp. 413, 415, and 424) shows results which agree very well with the Debye-relaxation theory, for several materials. Saint-Amant and Strangway (1970, pp. 631) also show a good example of this type of agreement for powdered augite. However, in data obtained to date the dielectric constant of rocks at the lower frequencies and at elevated temperatures is seldom seen to saturate in the way that Saint-Amant and Strangway (1970) or Von Hippel (1954) show in their examples. Therefore, it has been seldom possible to estimate the static dielectric constant. The dielectric constant is often seen to continue to increase with decrease in frequency at low frequencies as seen in the example for lunar sample 15301,38-A by Olhoeft *et al.* (1972).

Keller (1966) also shows results of electrical measurements of rocks at elevated temperatures. The authors think that the "complex resistivity and critical frequency" concept (Katsube and Collett, 1972) may be applicable to the high temperature effect in rocks. A good example of this is given in Fig. 21 (Keller, 1966). From the shapes of the curves in this figure it is assumed that the term "resistivity" is equivalent to ρ^* and not ρ_p. It can be seen that at elevated temperatures ρ^* is constant until it reaches what can be considered f_{cr}, and then decreases at a rate of 45° on the log–log scale. It is also evident that f_{cr} increases with

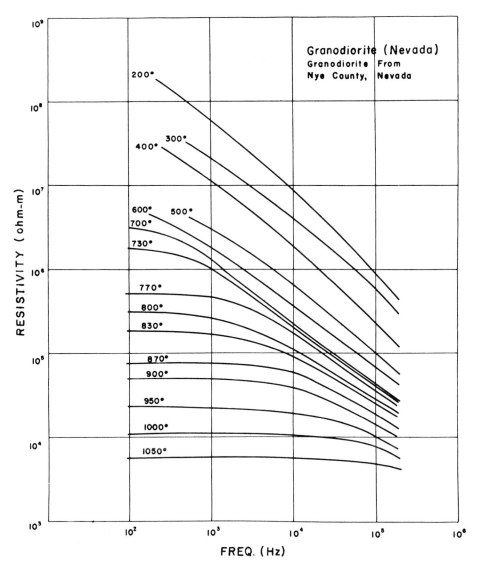

Fig. 21. Resistivity of granodiorite as a function of temperature and frequency (Keller, 1966).

temperature. This trend is similar to ρ^* of the moist diorite sample in Fig. 12. This suggests that when the rock is simulated by RC parallel circuits, the R decreases with increase in temperature. The only difference that is seen between these two diagrams is the shifting of the ρ^* curves above f_{cr}, towards lower values of resistivity as the temperature increases. This indicates an increase of C or K' with temperature. This increase in K' can be seen in many cases such as the work by Chung *et al.* (1972) as shown in Fig. 22, but cannot be seen in that by Olhoeft *et al.* (1972) as shown in Fig. 23.

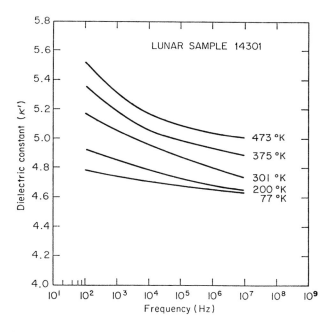

Fig. 22. Real relative permittivity (K') of sample 14301,41 as a function of frequency and temperature (Chung *et al.*, 1972).

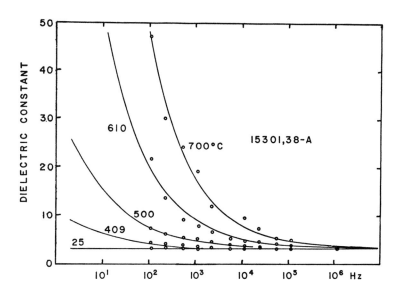

Fig. 23. Real relative permittivity (K') vs. frequency for selected temperatures of sample 15301,38 (Olhoeft *et al.*, 1973).

From Von Hippel (1954) or Collett and Katsube (1973) the relation between ρ^*, D, and K' is

$$|\rho^*| = \frac{1}{\omega K' \epsilon_0 \sqrt{1+D^2}}$$

where K': dielectric constant; D: dissipation factor; ω: angular frequency.

Since $D < 1$ usually for measurements at elevated temperatures such as shown by Olhoeft et al. (1972) or Chung et al. (1972), $|\rho^*|$ seems to decrease with frequency both above and below the critical frequency, for most cases. The results of Keller (1966), which show $|\rho^*|$ to be constant with frequency below the critical frequency, do not seem to agree with these results.

This disagreement may be due to the difference in measuring conditions (room atmosphere, vacuum, etc.), measuring systems, or type of samples used for measurement. However, this disagreement may also be due to measuring techniques including treatment of micro-airgaps (Katsube and Collett, 1973) or grain boundary effects. Therefore, further studies based on the various theories related to the dielectric mechanisms should include considerations of grain boundary effects and measuring techniques.

Conclusions

Usually for dry rocks, the dielectric constant shows little variation with frequency, the parallel resistivity decreases quite rapidly with frequency, and the dissipation factor or loss tangent generally decreases with frequency but often shows a minimum and maximum in the frequency range from 10^4 to 10^7 Hz.

A minor content of low resistive or conductive minerals surrounded by insulating minerals affects the electrical parameters of the rock at the higher frequencies. This suggests that there is a possibility for low frequency radar techniques to be used for detection of conductive minerals.

Theoretically, information on porosity, conductivity of liquid contained in pores, and types of pores may be obtained by measurement and comparison of dielectric constant and parallel resistivity above critical frequency and parallel resistivity below the critical frequency.

There seems to be some interesting problems in relation to measurements at elevated temperatures.

References

Archie G. E. (1952) Classification of carbonate reservoir rocks and petrophysical considerations. *Bull. Am. Assoc. Petrol. Geologists* **36**, 278–298.

ASTM (American Society for Testing Materials) (1968) Standard methods of testing for AC loss characteristics and dielectric constant of solid electrical insulating materials: *ASTM D* 150–68, 29–54.

Brett R., Butler P. Jr., Meyer C. Jr., Reid A. M., Takeda H., and Williams R. (1971) Apollo 12 igneous rocks 12004, 12008, 12009 and 12022: A mineralogical and petrological study. *Proc. Second Lunar Sci. Conf., Geochim. Cosmochim. Acta*, Suppl. 2, Vol. 1, pp. 301–318. MIT Press.

Collett L. S. and Katsube T. J. (1973) Electrical parameters of rocks in developing geophysical techniques. *Geophysics* **38**, 76–91.

Chung D. H., Westphal W. B., and Simmons G. (1970) Dielectric properties of Apollo 11 lunar samples and their comparison with Earth materials. *J. Geophys. Res.* **75**, 6524–6531.

Chung D. H., Westphal W. B., and Olhoeft G. R. (1972) Dielectric properties of Apollo 14 lunar samples (abstract). In *Lunar Science—III*, pp. 139–140. The Lunar Science Institute, Houston.

Dence M. R., Douglas J. A. V., Plant A. G., and Trail R. J. (1971) Mineralogy and petrology of some Apollo 12 samples. *Proc. Second Lunar Science Conf., Geochim. Cosmochim. Acta*, Suppl. 2, Vol. 1, pp. 285–300. MIT Press.

Deryagin B. V. and Zorin Z. M. (1968) Formation of droplets of anomalous water on a flat quartz surface. *Colloid J. U.S.S.R.* **30**, 232–233.

Hansen W., Sill W. R., and Ward S. H. (1973) The dielectric properties of selected basalts. *Geophysics* **38**, 135–139.

Jordan E. C. (1950) *Electromagnetic Waves and Radiating Systems*, pp. 128–131. Prentice-Hall.

Katsube T. J. and Collett L. S. (1971) Electrical properties of Apollo 11 and Apollo 12 lunar samples. *Proc. Second Lunar Sci. Conf., Geochim. Cosmochim. Acta*, Suppl. 2, Vol. 3, pp. 2367–2379. MIT Press.

Katsube T. J. and Collett L. S. (1972) Electrical and EM propagation characteristics of igneous rocks. Presented at 42nd Annual International Meeting of S.E.G.

Katsube T. J. and Collett L. S. (1973a) Measuring techniques for rocks with high permittivity and high loss. *Geophysics* **38**, 92–104.

Katsube T. J. and Collett L. S. (1973b) Electrical and EM propagation characteristics of Apollo 16 samples. Presented at 4th Lunar Science Conf., Houston, Texas.

Keller G. V. (1966) Electrical properties of rocks and minerals. In *Handbook of Physical Constants, Geol. Soc. Amer. Mem.* **97**, (editor S. P. Clark, Jr.) pp. 533–571.

Keller G. V. (1967) "Supplementary guide" in the book *Electrical Properties of Rocks* by Parkhomenko. Pleunum Press.

Klein C. Jr., Drake J. C., and Frondel C. (1971) Mineralogical, petrological and chemical features of four Apollo 12 lunar microgabbros. *Proc. Second Lunar Sci. Conf., Geochim. Cosmochim. Acta*, Suppl. 2, Vol. 1, pp. 265–284. MIT Press.

Lunar Science Institute (LSI) (1972) Post-Apollo Lunar Science. NASA, Houston, Texas.

Olhoeft G. R. and Strangway D. W. (1972) Magnetic relaxation in the electromagnetic response parameter. Presented at SEG Annual International Convention, Anaheim, California.

Olhoeft G. R., Frisillo A. S., and Strangway D. W. (1972) Frequency and temperature dependence of the electrical properties of a soil sample from Apollo 15. In *The Apollo 15 Lunar Samples*, pp. 477–481. The Lunar Science Institute, Houston.

Olhoeft G. R., Frisillo A. L., Strangway D. W., and Sharpe H. (1973) Electrical properties of lunar solid samples (abstract). In *Lunar Science—IV*, pp. 575–577. The Lunar Science Institute, Houston.

Olhoeft G. R. (1973) Personal communication.

Roberts W. N. (1973) Personal communication.

Saint-Amant M. and Strangway D. W. (1970) Dielectric properties of dry geologic materials. *Geophysics* **35**, 624–645.

Schwerer F. C., Huffman G. P., Fisher R. M., and Nagata T. (1973) Electrical conductivity of lunar surface rocks at elevated temperatures (abstract). In *Lunar Science—IV*, pp. 663–665. The Lunar Science Institute, Houston.

Sill W. R., Hansen W., Ward S. H., Katsube T. J., Collett L. S., and Brown W. E. (1973) The dielectric constant and loss tangent of Apollo 14 and Apollo 16 soil and rock samples. Presented at the Geophys. and Geochem. Exploration of the Moon and Planets Conference, Houston, Texas.

Strangway D. W., Olhoeft G. R., Chapman W. B., and Carnes J. (1972) Electrical properties of lunar soil: dependence on frequency, temperature and moisture. *Earth Planet. Sci. Lett.* **16**, 275–281.

Von Hippel A. R. (1954) *Dielectric Material and Applications*. Wiley.

Proceedings of the Fourth Lunar Science Conference
(Supplement 4, *Geochimica et Cosmochimica Acta*)
Vol. 3, pp. 3133–3149

Lunar sample electrical properties

G. R. OLHOEFT

Lockheed Electronics Company
Houston, TX 77058

D. W. STRANGWAY

Physics Branch
NASA Johnson Space Center
Houston, TX 77058

A. L. FRISILLO

N.R.C. Fellow
NASA Johnson Space Center
Houston, TX 77058

Abstract—Electrical conductivity and dielectric constant measurements have been performed in vacuum on solid and soil samples over a wide range of temperatures and frequencies. The temperature dependence and the frequency response of the dielectric properties together with the temperature dependence of the DC conductivity have permitted us to propose a mathematical model describing the mechanisms controlling the electrical properties. In general, each lunar sample has several distributed mechanisms, each mechanism dominant in a particular temperature range.

INTRODUCTION

ELECTRICAL PROPERTIES of lunar materials are important in the interpretation of earth-based radar (Evans and Pettengill, 1963), the Apollo 17 surface electrical properties experiment (SEP) (Simmons *et al.*, 1973), the Apollo Lunar Sounder Experiment (ALSE) (Phillips *et al.*, 1973), and magnetometer derived lunar conductivity profiles (Dyal and Parkin, 1972a, 1972b; Sill, 1972; Sonnett *et al.*, 1972). An introduction to this subject may be found in Strangway (1970).

Chung *et al.* (1972) have reviewed much of the electrical properties data available for lunar materials up to the time of the Third Lunar Science Conference. We intend here to review the measurements we have made since that time and to report on progress in understanding electrical properties mechanisms. We will discuss our basic experimental procedure and present lunar sample data. We will then propose a model of electrical properties and discuss interpretations of the model parameters in terms of possible mechanisms.

EXPERIMENTAL

Our sample handling and measurement techniques are discussed elsewhere (Olhoeft *et al.*, 1973a). Briefly, we use standard three-terminal capacitance bridge techniques for the dielectric measurements

G. R. OLHOEFT *et al.*

and electrometer charging techniques for DC conductivity. We employ radiative heating in an ion-pumped vacuum system which typically is at a pressure in the range 8×10^{-8} to 2×10^{-7} torr. A mass spectrometer monitored residual out gassing of the chamber and sample. The sample holder is made of beryllium oxide and the electrodes are of molybdenum. The typical frequency range measured is 50 Hz to 2×10^6 Hz over a temperature span of 25–840°C. Samples are received triply bagged in dry nitrogen and transferred under a dry nitrogen tent to the vacuum chamber to prevent atmospheric contamination. Table 1 lists the pertinent characteristics of the samples studied to date.

Worst case measurement errors are $\pm 3\%$ for the dielectric constant of soils, $\pm 9\%$ for the dielectric constant of solids (due to geometries), $\pm 2°C$ in temperature (25–900°C), $\pm 0.5\%$ in frequency (50 Hz–1 MHz), $\pm 2\% + 0.0005$ for the loss tangent of both soils and solids, and $\pm 10\%$ for DC conductivity to 10^{-14} mho/meter. Measurement repeatability for the dielectric constant is $\pm 0.5\%$ and $\pm 1\%$ for the loss tangent.

MOISTURE

St. Amant and Strangway (1970) have shown explicitly that small amounts of moisture have a large effect on the electrical properties of dry, terrestrial, geological material. Shown in Figs. 1 and 2 are similar measurements (after Strangway *et al.*, 1972) showing the effect of moisture on lunar soil 14163,131 at room temperature. The distinctive low frequency increase in the dielectric constant and loss tangent is characteristic of moisture effects (usually attributed to interfacial effects such as adsorption on grain surfaces changing the grain–grain interactions). The DC conductivity changed by over four orders of magnitude from 6×10^{-15} mho/m in vacuum, to 1×10^{-10} mho/m when exposed to an atmosphere with 30% relative humidity at 25°C.

EXPERIMENTAL RESULTS

Table 1 lists some chemical properties of the lunar samples we have characterized to date. We have reported elsewhere the results of electrical property measurements of soils 14163,131 (Strangway *et al.*, 1972) and 15301,39 (Olhoeft *et al.*, 1973a). Briefly, in 15301 we found three electrical processes occurring—a repeatable DC conductivity with an apparent quadratic temperature dependent activation energy and two dielectric mechanisms, one an apparently intrinsic electronic mechanism with an activation energy at 2.5 eV and a second mechanism with indeterminate parameters.

Table 1.

Sample	Type	Density	TiO$_2$	FeO
12002,85	B	3.40	—	—
14163,131	D	1.20	1.79[a]	10.35[a]
15301,38	D	1.47	1.17[a]	14.05[a]
65015,6	B	2.70	1.26[b]	8.59[b]

[a]Apollo 15 Preliminary Science Report.
[b]Duncan *et al.* (1973).

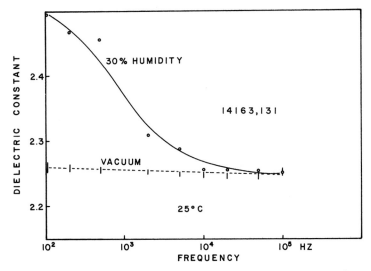

Fig. 1. Lunar soil 14163,131 dielectric constant vs. frequency in vacuum $< 10^{-7}$ torr and 30% humid air at 25°C (error bars are shown for vacuum data; after Strangway *et al.*, 1972).

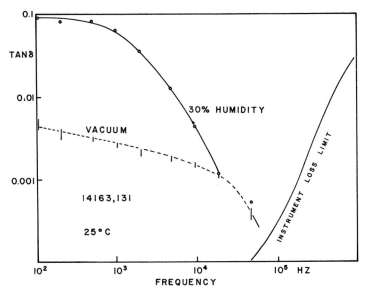

Fig. 2. Lunar soil 14163,131 total loss tangent vs. frequency in vacuum $< 10^{-7}$ torr and 30% humid air at 25°C (after Strangway *et al.*, 1972).

12002,85

Lunar sample 12002,85 is a solid, fine to medium grained holocrystalline basalt previously measured by Katsube and Collett (1971). We received this sample already exposed to air and hence it is the only contaminated lunar sample we have measured. By measuring in vacuum, we found, however, that the sample was only slightly contaminated with water (see discussion in Olhoeft *et al.*, 1973b) when compared with terrestrial materials.

The dielectric data are shown in Figs. 3 and 4 (the room temperature values agree very well with the independent measurements of Katsube and Collett) and the DC conductivity is shown in Fig. 5. The DC conductivity was irreversible on heating and cooling after heating above 400°C and has been discussed elsewhere (Olhoeft *et al.*, 1973b; see also Schwerer *et al.*, 1972, 1973).

In our earlier analysis (Olhoeft *et al.*, 1973b) DC conductivity-temperature data was discussed in terms of an exponential temperature dependence (quadratic temperature dependent activation energy). In Fig. 5 we display the heating only data of 12002,85 showing that it can also be explained by two Boltzmann distributions (see model discussion below) as described in Table 2. The shift in mechanism near 350°C from an activation energy of 0.48–1.09 eV may either be a change in basic conduction mechanism or it may be interpreted in terms of a change in sample chemistry. Below 350°C, the sample reversibly follows the same curve on heating and cooling, but above 350°C the sample has higher conductivities during cooling than during heating (see Olhoeft *et al.*, 1973b).

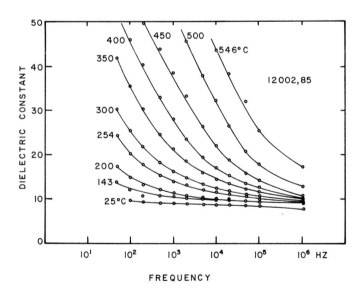

Fig. 3. Lunar basalt 12002,85 dielectric constant vs. frequency (circles are data; solid lines are smoothed data).

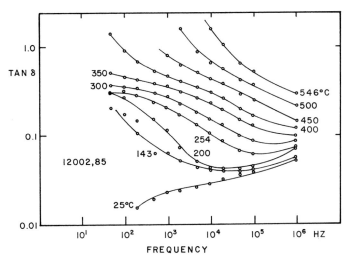

Fig. 4. Lunar basalt 12002,85 total loss tangent vs. frequency (solid lines are smoothed data).

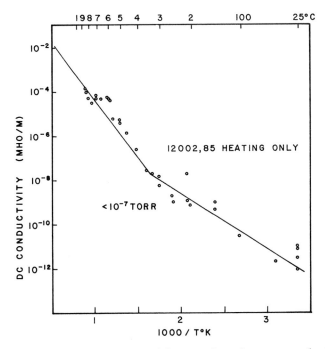

Fig. 5. Lunar basalt 12002,85 DC conductivity vs. reciprocal temperature (heating only, *see* text; circles are data, solid line is model).

65015,6

Lunar sample 65015,6 is a glassy agglutinate of mineral and lithic fragments cemented by "KREEP" glass which was extensively recrystallized with incipient melting of interstitial K-rich material (Albee *et al.*, 1973). The dielectric data are shown in Figs. 6 and 7, and the DC conductivity is shown in Fig. 8.

The DC conductivity exhibited non-linear current-voltage behavior similar to space-charge interfacial phenomena as well as irreversible heating and cooling curves as discussed elsewhere (Olhoeft *et al.*, 1973b). At low frequencies and below 250°C the space-charge like effect caused anomalous dielectric constant dispersion and loss.

MODEL

In an attempt to understand conduction mechanisms in lunar samples, sets of empirical fits have been applied to the data. A distribution of mineral phases, each with its own characteristic electrical response, has been assumed. General relaxation theory (see Gevers, 1945a–e or Shuey and Johnson, 1973) was then used, and a distribution of Debye single relaxations was chosen which best fitted the experimental data over the range of frequency and temperature studied. This distribution proved empirically to be the well known Cole–Cole frequency distribution (Cole and Cole, 1941):

$$\kappa = \kappa' - j\kappa'' = \kappa_\infty + \frac{\kappa_0 - \kappa_\infty}{1 + (j\omega\tau)^{1-\alpha}} \tag{1}$$

in which the distribution parameter $1 - \alpha$ has extreme values of 1.0 (a Debye single

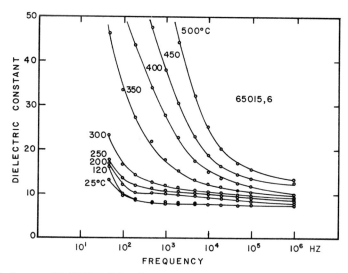

Fig. 6. Lunar solid 65015,6 dielectric constant vs. frequency (solid lines are smoothed data, *see* text).

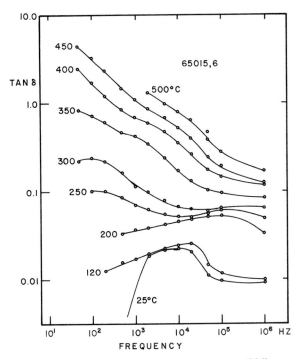

Fig. 7. Lunar solid 65015,6 total loss tangent vs. frequency (solid lines are smoothed data).

relaxation) and 0.0 (a continuous distribution of relaxation times) and where

$j = \sqrt{-1}$
$\kappa' = $ real dielectric constant
$\kappa'' = $ imaginary dielectric constant
$\omega = $ frequency (radians per second)
$\tau = $ the time constant such that at $\omega\tau = 1$, $\kappa' = (\kappa_0 + \kappa_\infty)/2$
$\kappa_0 = \lim\limits_{\omega \to 0} \kappa'$
$k_\infty = \lim\limits_{\omega \to \infty} \kappa'$

To fit the temperature variation, we found it necessary to modify this formula by adding a Boltzmann temperature dependence (Gevers, 1945c) to the relaxation time

$$\tau = \tau_0 e^{E_0/kT} \qquad (2)$$

where $T = $ temperature in °K

$\tau_0 = \lim\limits_{T \to \infty} \tau$
$k = $ Boltzmann's constant $= 8.6176 \times 10^{-5}$ eV/°K
$E_0 = $ activation energy in eV

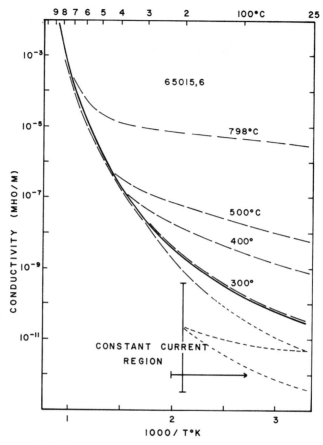

Fig. 8. Lunar solid 65015,6 DC conductivity vs. reciprocal temperature with space charge region indicated, solid line showing exponential (quadratic activation energy) heating dependence; dashed lines show sample behavior on cooling from temperatures indicated (*see* text, after Olhoeft *et al.*, 1973).

In addition, we experimentally observed a temperature dependence in the distribution parameter (similar to that observed by Fuoss and Kirkwood, 1941) which we fit with

$$1 - \alpha = \frac{1}{e^{\beta(T_0 - T)} + 1} \tag{3}$$

where β controls the slope of $1 - \alpha$ versus temperature and

$$\lim_{T \to \infty} (1 - \alpha) = 1.0 \text{ (Debye single relaxation time)}$$

$$\lim_{T \to T_0} (1 - \alpha) = 0.5 \text{ (defines } T_0)$$

$$\lim_{T \to 0} (1 - \alpha) \text{ approaches a constant value.}$$

Further experimental work is required to confirm the general applicability of this expression.

The dielectric loss tangent is defined as

$$\tan \delta_\epsilon = \kappa''/\kappa'$$

and the total loss tangent which includes the DC conductivity contribution as

$$\tan \delta = (\kappa''/\kappa') + (\sigma/\omega\kappa'\epsilon_0)$$

where $\epsilon_0 = 8.854 \times 10^{-12}$ farads/meter and σ is the DC conductivity.

The temperature dependence of the DC conductivity was found to follow the general formula (*see* Adler, 1971)

$$\sigma = \sum_i \sigma_i e^{-E_i/kT}$$

in mho/meter where σ_i is the high temperature limiting value of the ith term, and E_i is the activation energy which may itself be temperature dependent and of the form (Adler, 1971)

$$E = E_1 - E_2(kT)^2$$

(*see* discussion in Olhoeft *et al.*, 1973b).

By studying the temperature variation of the frequency dependence of the dielectric constant and of the DC conductivity it is possible to begin understanding the mechanisms involved. Let us first investigate how we would expect the above model parameters to characterize the frequency and temperature dependence of material electrical properties.

Figure 9 shows the normalized frequency dependence of the Cole–Cole dielectric constant model and the variation of the distribution with the $1 - \alpha$ parameter. The normalization is such that $\kappa_0 = 1.0$ and $\kappa_\infty = 0.0$ with $\omega\tau = 1$ at $\kappa' = (\kappa_0 + \kappa_\infty)/2 = 0.5$. Figure 10 shows the corresponding frequency dependence of the dielectric loss tangent, normalized in such a way that the maximum loss tangent is 1.0 for $1 - \alpha = 1$.

Using the limits of κ' as ω goes to zero and infinity it can be shown for the Cole–Cole equation that κ_0 is the low frequency limit of κ' when $\omega\tau \ll 1.0$ and κ_∞ is the high frequency limit of κ' when $\omega\tau \gg 1.0$ (assuming there are no other higher or lower frequency distributions). From the derivatives of κ' with respect to $\log \omega$ we find the slope of κ' vs. $\log \omega$ to be $-(1 - \alpha)(\kappa_0 - \kappa_\infty)$ at $\omega\tau = 1.0$ or $-(1 - \alpha)$ in the normalized coordinates of Fig. 9.

The derivatives of $\log \tan \delta$ with respect to $\log \omega$ result in slopes of $(1 - \alpha)$ for $\omega\tau > 1.0$ and $-(1 - \alpha)$ for $\omega\tau < 1.0$ as seen in Fig. 10. Similarly by setting the derivative equal to zero we find that the maximum loss tangent occurs at

$$\omega\tau = (1 + \xi)^{1/2(1-\alpha)}$$

with a value of

$$\tan \delta_{\epsilon_{max}} = \frac{\xi \sin (1 - \alpha)\pi/2}{2\sqrt{1+\xi} + (2+\xi) \cos (1 - \alpha)\pi/2}$$

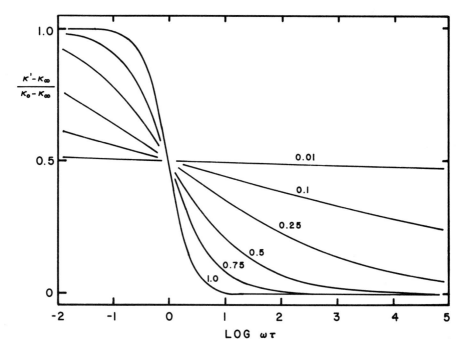

Fig. 9. Normalized real dielectric constant vs. $\omega\tau$ for varying values of $1-\alpha$.

where

$$\xi = (\kappa_0 - \kappa_\infty)/\kappa_\infty$$

Dielectric measurements over a frequency range in which a loss tangent peak is observed give the following model parameters graphically: $1-\alpha$ (from the slope), κ_∞ (at high frequency), the maximum value of loss tangent, and the frequency of the maximum. Thus, even if the dielectric constant does not approach a steady value at low frequencies we can infer an equivalent value for κ_0 and thus determine τ:

$$\xi = 2DS\frac{(C+SD)+\sqrt{1+D^2}}{(S-CD)^2}$$

where $D = \tan \delta_{\epsilon_m} = $ maximum loss tangent at ω_m
$\quad\quad C = \cos(1-\alpha)\pi/2$
$\quad\quad S = \sin(1-\alpha)\pi/2$

and

$$\kappa_0 = \kappa_\infty(1+\xi)$$
$$\tau = \omega_m^{-1}(1+\xi)^{1/2(1-\alpha)}$$

As κ_0 and κ_∞ are independent of temperature, we may vary the temperature and solve for τ_0, E_0, β, and T_0 from the changes τ and $1-\alpha$.

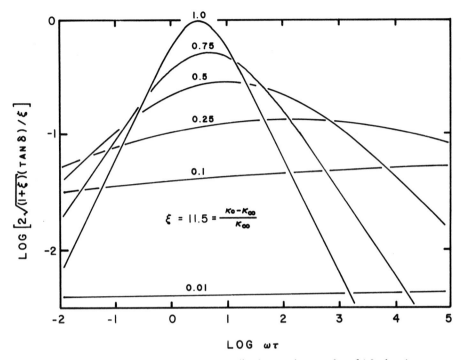

Fig. 10. Dielectric loss tangent vs. $\omega\tau$ (normalized to maximum value of 1.0 when $1 - \alpha =$ 1.0) for varying values of $1 - \alpha$.

In Fig. 9, it should be noted that the distribution parameter has no effect on κ' other than slope, while in Fig. 10, the slope of the loss tangent-frequency changes with $1 - \alpha$ variation as does the peak frequency of maximum loss—the peak shifting to higher frequencies as the distribution broadens ($1 - \alpha \rightarrow 0$). The distribution narrows ($1 - \alpha$ approaches 1.0) as the temperature is increased thus lowering the peak frequency and raising the value of maximum loss tangent. The Boltzmann temperature dependence of the time constant raises the peak frequency of loss and has no effect on the value of the maximum loss tangent.

DISCUSSION

Let us now look at the lunar sample data in light of the preceding model. Sample 12002,85 exhibits a very complex spectrum of electrical properties. The dielectric constant in Fig. 3 varies smoothly with frequency and temperature and by itself does not indicate the presence of different mechanisms. The total loss tangent in Fig. 4 shows distinct transitions in the nature of the frequency response as the temperature is varied. Several distributed processes account for this variation.

One process is indicated at 25°C by the positive slope of log loss tangent vs. frequency ($1 - \alpha = 0.12$). The positive slope indicates that the peak loss occurs at a frequency above our range of measurement. This distributed process appears to cause the positive slopes in the high frequency data up to 350°C.

A second dielectric process is indicated by the negative slope $(1 - \alpha = 0.39)$ at low frequencies in the 143–254°C data. The frequency of the peak loss for this distribution is thus below our range of measurement. As the absolute value of the slope at 254°C decreases (neglecting the low frequency rollover to be discussed below), a third transition is marked (due to the tendency of slope to increase with increasing temperature as noted above in the model discussion). This transition does not appear to mark a third process however, but rather to mark a change in the 143–254°C distribution mechanism. A similar alteration appears to occur above 450°C, (this will be further discussed below). Above 400°C another slope change occurs as the DC conductivity dominates the total loss and causes the low frequency total loss to have a slope higher in absolute value than the higher frequency data. The inflection point at which the slope increases toward lower frequencies occurs at the frequency at which the DC conduction term is equal to the dielectric loss term. Katsube *et al.* (1973) call this frequency, below which the DC term dominates and above which the dielectric term dominates, the critical frequency.

The absence of a loss tangent peak at 25°C at high frequencies and at 143–254°C at low frequencies does not allow us to parameterize these two distributions. However, as the 143–254°C distribution is altered, the peak appears in the 254°C data as the low frequency rollover of the loss tangent. This altered distribution in the region 300–450°C is fitted very well by the model using the parameters listed in Table 2. This fit is illustrated in Figs. 11, 12, and 13. The discrepancies in the fit at and below 254°C indicate that the 143–254°C frequency

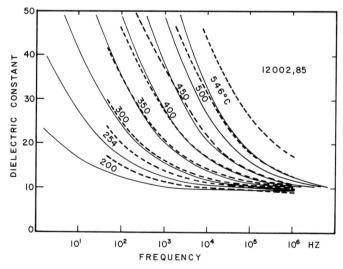

Fig. 11. Lunar basalt 12002,85 dielectric constant vs. frequency showing the model (solid lines) fit to the data (dotted lines) from Fig. 3. Note the large discrepancies above 450°C and below 300°C.

Table 2.

Sample	κ_0	κ_∞	τ_0 (sec)	E_0 (eV)	β	T_0 (°K)	σ_1 (mho/m)	E_1 (eV)	E_2 (eV)$^{-1}$	σ_2	E_3
12002,85	(a) 200.	9.	0.22×10^{-9}	1.1	0.00088	1260.	0.13×10^2	1.09	—	0.18×10^{-3}	0.48
	(b) second mechanism unresolved										
14163,131	—	2.26	further parameters unresolved due to insufficient data								
15301,38	6700.	3.0	0.26×10^{-11}	2.5	0.0018	1000.	0.6×10^{-17}	—	275.	—	—
65015,6 (a)	9.?	8.9?	10^{-8}	0.21	0.01	298.	0.3×10^{-13}	—	267.	—	—
(b)	26.-63.?	9.-11.?	10^{-13}	1.17	0.01	573.					

DC conductivity using the above parameters is described by: $\sigma = \sigma_1 e^{-E_1/kT} e^{+E_2 kT} + \sigma_2 e^{-E_3/kT}$.

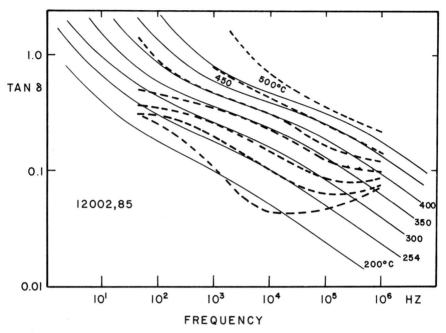

Fig. 12. Lunar basalt 12002,85 total loss tangent vs. frequency showing the model (solid lines) fit to the data (dotted lines) from Fig. 4.

distribution was altered by an increase in the activation energy and a broadening of the distribution. The exact details of the alteration are unknown. Similarly, an increase in the activation energy apparently occurs above 450°C providing a second change in the distribution. These trends are best illustrated in Fig. 13 where an increase in the absolute slope of log conductivity vs. reciprocal temperature corresponds to an increase in the activation energy. The conductivity plotted in Fig. 13 is the total conductivity given by

$$\sigma_T = \sigma_{DC} + \omega \kappa' \epsilon_0 \tan \delta_\epsilon.$$

Further resolution and parameterization are not possible due to the inadequacies of the data (inadequate low and high frequency coverage to observe the distribution peaks) and due to the inadequate understanding of the way multiple distribution mechanisms mix and interact. This is particularly true with regard to the temperature variation and alteration of distributions.

Sample 65015,6 has an exceedingly complex, multiply distributed, electrical properties spectrum. The abrupt change in the temperature activation of the dielectric constant frequency response between 300 and 350°C in Fig. 6 indicates a change from one mechanism to another. However, as in 12002,85, the loss tangent data is much more informative.

As seen in Fig. 7, the obvious loss tangent peak near 9 kHz at 25°C is one distribution. As the temperature is increased, this distribution moves toward

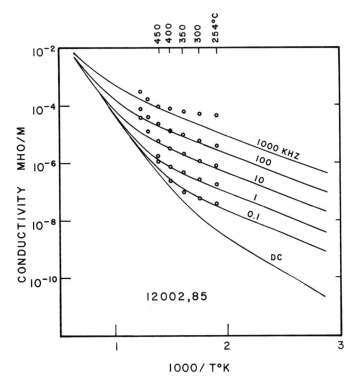

Fig. 13. Lunar basalt 12002,85 total AC and DC conductivity vs. reciprocal temperature
showing the model (solid lines) fit to the data (circles) replotted from Figs. 3 and 4.

higher frequencies with an activation energy of 0.21 eV and can be parameterized
as listed in Table 1. This relaxation mechanism can be recognized by peaks near
9 kHz, 20 kHz, 80 kHz, and 200 kHz at 25°, 120°, 200°, and 300°C respectively. As
the magnitude of the peak is also increasing with temperature, this indicates a
narrowing of the distribution with increasing temperature.

A second loss tangent peak appears at 300°C near 150 Hz, moving to higher
frequencies as the temperature is increased with an activation energy of 1.19 eV
and parameters as listed in Table 2. This peak is seen near 150 Hz, 1 kHz, 3 kHz,
8 kHz, and 30 kHz at 300°, 350°, 400°, 450°, and 500°C respectively. The increase in
slope below these frequencies at their respective temperatures is caused by the
appearance of the DC conductivity as the dominant mechanism.

The difficulty in modeling this sample, even though both dielectric processes
and the DC conductivity are well separated and distinct, lies in the nature of the as
yet unknown mixing relationship of multiple distributions. A completely unique
solution is thus not possible until the mixing process is better understood. As
such, the parameterization of 65015,6 as listed in Table 2 is extremely tentative
with only the activation energies for the two dielectric processes known with any
degree of confidence (± 0.1 eV).

CONCLUSIONS

We are just beginning to model some of the dielectric and DC conduction data for lunar samples. As we do this, we have eventually to go a long way toward identifying mechanisms or classes of mechanisms. We are now capable of parameterizing data with regard to limiting cases of frequency dependence and with regard to temperature variations of the frequency dependence.

We have shown the major effect of moisture on the electrical properties of lunar samples. We have proposed models which fit frequency and temperature dependencies and outlined the minimum data requirements needed to parameterize the electrical properties. Lunar samples appear to have multiple mechanisms, with very wide $(1 - \alpha < 0.4$ at 25°C) distributions of relaxation times. These distributions change shape as well as shift in frequency as the temperature is changed. Activation energies have been found in the range from less than 0.05 to 2.5 eV in dielectric mechanisms and from 0.48 to 1.09 eV in DC conduction mechanisms. There also appears to be a quadratic temperature dependent activation energy in some of the DC conductivity data.

Even in the absence of water-related complications, lunar sample electrical properties are indeed complex. A simple model, as outlined above, requires six parameters per dielectric mechanism and 2 (3 if a temperature dependent activation energy) per DC conduction mechanism. Thus a sample like 12002,85 required 15 to 16 parameters to characterize its electrical properties. Also, due to the unknown nature of multiple distribution interactions, the characterization is not necessarily a unique model.

REFERENCES

Apollo 15 Preliminary Science Report (1972) NASA SP-289, U.S. Government Printing Office, Washington, D.C.

Adler D. (1971) *Amorphous Semiconductors.* CRC Press, Cleveland.

Albee A. L., Gancarz A. J., and Chodos A. A. (1973) Sanidinite facies metamorphism of Apollo 16 sample 65015 (abstract). In *Lunar Science—IV*, pp. 24–26. The Lunar Science Institute, Houston.

Chung D. H., Westphal W. B., and Olhoeft G. R. (1972) Dielectric properties of Apollo 14 lunar samples. *Proc. Third Lunar Sci. Conf., Geochim. Cosmochim. Acta*, Suppl. 3, Vol. 3, pp. 3161–3172. MIT Press.

Cole K. S. and Cole R. H. (1941) Dispersion and absorption in dielectrics. *J. Chem. Phys.* 9, p. 341.

Duncan A. R., Ahrens L. H., Elbank A. J., Willis J. P., and Gurney J. J. (1973) Composition and inter-relationships of some Apollo 16 samples (abstract). In *Lunar Science—IV*, pp. 190–192. The Lunar Science Institute, Houston.

Dyal P. and Parkin C. W. (1972a) Lunar properties from transient and steady magnetic field measurements. *The Moon* 4, 63–87.

Dyal P., Parkin C. W., and Cassen P. (1972b) Surface magnetometer experiments: Internal lunar properties and lunar field interactions with the solar plasma. *Proc. Third Lunar Sci. Conf., Geochim. Cosmochim. Acta*, Suppl. 3, Vol. 3, pp. 2287–2307. MIT Press.

Evans J. V. and Pettengill G. H. (1963) The scattering behavior of the moon at wavelengths of 3, 6, 68, and 785 centimeters. *J. Geophys. Res.* 68, 423–447.

Fuoss R. M. and Kirkwood J. G. (1941) Electrical properties of solids, VIII. Dipole moments in polyvinyl chloride-diphenyl systems. *Am. Chem. Soc. J.* 63, pp. 385–394.

Gevers M. (1945a–e) The relation between the power factors and the temperature coefficient of the dielectric constant of solid dielectrics. *Philips Res. Repts.*, Vol. 1 (in 5 parts), p. 197(a), 279(b), 298(c), 362(d), 447(e).

Katsube T. J. and Collett L. S. (1971) Electrical properties of Apollo 11 and Apollo 12 lunar samples. *Proc. Second Lunar Sci. Conf.*, *Geochim. Cosmochim. Acta*, Suppl. 2, Vol. 3, pp. 2367–2379. MIT Press.

Katsube T. J. and Collett L. S. (1973) Electrical and EM propagation characteristics of igneous rocks. *Geophysics*. In press.

Olhoeft G. R., Frisillo A. L., and Strangway D. W. (1973) Electrical properties of lunar soil sample 15301,38. *J. Geophys. Res.* In press.

Olhoeft G. R., Frisillo A. L., Strangway D. W., and Sharpe H. N. (1973b) Temperature dependence of electrical conductivity and lunar temperatures. *The Moon.* In press.

Phillips R. J., Adams G. F., Brown W. E. Jr., Eggleton R. E., Jackson P., Jordan R., Peeples W. J., Procello L. J., Ryn J., Schaber T., Sill W. R., Thomson T. W., Watd S. H., and Zelenka J. S. (1973) Preliminary results of the Apollo lunar sounder experiment (abstract). In *Lunar Science—IV*, pp. 590–591. The Lunar Science Institute, Houston.

Saint-Amant M. and Strangway D. W. (1970) Dielectric properties of dry, geologic materials. *Geophysics* **35**, 624–645.

Schwerer F. C., Huffman G. P., Fisher R. M., and Nagata T. (1972) Electrical conductivity and Mössbauer study of Apollo lunar samples. *Proc. Third Lunar Sci. Conf.*, *Geochim. Cosmochim. Acta*, Suppl. 3, Vol. 3, pp. 3173–3185. MIT Press.

Schwerer F. C., Huffman G. P., Fisher R. M., and Nagata T. (1973) Electrical conductivity of lunar surface rocks at elevated temperatures (abstract). In *Lunar Science—IV*, pp. 663–665. The Lunar Science Institute, Houston.

Shuey R. T. and Johnson M. (1973) On the phenomenology of electrical relaxation in rocks. *Geophysics* **38**, 37–48.

Sill W. R. (1972) Lunar conductivity models from the Apollo 12 magnetometer experiment. *The Moon* **4**, 3–17.

Simmons G., Baker R. H., Bannister L. A., Brown R., Kong J. A., LaTorraca G. A., Tsang L., Strangway D. W., Cubley D. W., Annan A. P., Redman J. D., Rossiter J. R., Watts R. D., and England A. W. (1973) Electrical structure of Taurus Littrow (abstract). In *Lunar Science—IV*, pp. 675. The Lunar Science Institute, Houston.

Sonnett C. P., Smith B. F., Colburn D. S., Schubert G., and Schwartz K. (1972) The induced magnetic field of the moon: Conductivity profiles and inferred temperature. *Proc. Third Lunar Sci. Conf.*, *Geochim. Cosmochim. Acta*, Suppl. 3, Vol. 3, pp. 2309–2336. MIT Press.

Strangway D. W. (1970) Possible electrical and magnetic properties of near-surface lunar materials. In *Electromagnetic Exploration of the Moon* (editor Linlor). Mono Press, Baltimore.

Strangway D. W., Olhoeft G. R., Chapman W. B., and Carnes J. (1972) Electrical properties of lunar soil—dependence of frequency, temperature, and moisture. *Earth and Plan. Sci. Lett.* **16**, 275–281.

Proceedings of the Fourth Lunar Science Conference
(Supplement 4, *Geochimica et Cosmochimica Acta*)
Vol. 3, pp. 3151–3166

Electrical conductivity of lunar surface rocks at elevated temperatures

F. C. Schwerer, G. P. Huffman, and R. M. Fisher

U.S. Steel Research Laboratory, Monroeville, Penna. 15146

Takesi Nagata

Geophysical Institute, University of Tokyo, Japan

Abstract—The electrical conductivity of several lunar surface rocks has been measured in the temperature range 20–850°C in atmospheres adjusted to be either reducing or oxidizing with respect to the original material. Specific samples studied were from an Apollo 11 breccia (10048) and from a basalt and a recrystallized breccia from the Apollo 15 suite (rocks 15555 and 15418, respectively). Similar measurements were made for several terrestrial pyroxenes. For all samples studied, the electrical conductivity was observed to depend in a complex fashion on the furnace atmosphere and prior thermal history; however, the data obtained for specified sets of conditions were reproducible. Data for lunar samples measured under potentially reducing conditions were similar to those measured during the initial heating of the samples and are presented as representative of the pristine Apollo material. The conductivity values measured under these conditions indicate that some of the returned lunar rocks are among the least electrically conducting natural bulk materials reported. Mössbauer spectroscopy (Fe^{57} indicator) was used to characterize the dependence of the electrical conductivity on furnace atmosphere, and the results support the expectation that the conductivity of the anhydrous silicate phases is influenced principally by the concentration, oxidation state, and distribution of multivalent cations, most notably iron. The distribution of iron among various mineral phases and oxidation states for several Apollo 16 samples was determined by Mössbauer spectroscopy.

INTRODUCTION

SEVERAL RECENT efforts to interpret lunar surface and interior phenomena assume or generate values for the electrical conductivity of lunar materials (or their supposed analogs) (Anderson and Hanks, 1972; Criswell, 1972; Dyal *et al.*, 1973; Smith *et al.*, 1973, and references therein). Experimental data on silicates vary widely as the electrical properties of low-conductivity materials are sensitive to the moisture content and the degree of hydration as well as the oxidation state. The electrical conductivity of anhydrous silicates appears to be affected more by the concentration, oxidation state, and distribution of multivalent cations, most notably iron, than by the crystal structure. Consequently, the lunar samples by virtue of the almost complete absence of hydrated phases and ferric iron represent a unique suite of materials. Previously reported laboratory measurements of the electrical conductivity of lunar surface rocks at elevated temperatures (Schwerer *et al.*, 1971, 1972; Olhoeft *et al.*, 1973) were distinguished by large, irrecoverable increases in conductivity after heating above moderately elevated temperatures. Furthermore, data from the former set of studies indicated conductivity values which were considerably larger than were expected either from measurements at lunar surface temperatures of a-c electrical properties of returned lunar rocks (Chung *et al.*, 1972, and references therein), or from d-c electrical-conductivity measurements of terrestrial mafic silicates at elevated temperatures (Duba, 1972; Housley and Morin, 1972). Recent laboratory measurements of the d-c and low-frequency (5 Hz) a-c electrical conductivity of several lunar surface rocks and terrestrial pyroxenes at elevated temperatures are presented here. The salient features of the present results are that the data were found to be completely reproducible provided the measurements were made with the samples in specific environments, and that the conductivity values for the lunar samples were considerably lower than some of those reported previously (Schwerer *et al.*, 1971, 1972).

Electrical-Conductivity Measurements

Previous determinations of the electrical conductivity of lunar rocks at elevated temperatures by Schwerer *et al.* (1971) were marked by large irreversible increases in conductivity after heating to above approximately 400°C. A subsequent report (Schwerer *et al.*, 1972) described partial recovery of this thermal hysteresis by heating at low temperatures in oxidizing atmospheres. In these earlier studies, heating of the samples involved various vacuum systems using oil diffusion pumps, liquid-nitrogen cold traps, and neoprene O-rings and valve packings. It now appears that the hysteresis in electrical conductivity observed in these studies resulted from sample contamination—most probably from decomposed hydrocarbons from O-rings and valve packings—and that the beneficial effect of the oxidation treatment was removal of a carbon surface film. Carbon contamination is often encountered in demountable, dynamic vacuum systems and is apparently associated with the evolution of hydrocarbons from vacuum-system components. Deposition of a graphite surface film about 1000 Å in thickness could account for the observed conductivity increases. Identification of contamination as the source of changes in conductivity as large as a factor of 10^6 was complicated by several apparently contraindicating observations including (i) observation of similar phenomena in measurements conducted independently in the authors' two laboratories and (ii) the absence of contamination on adjacent insulating parts of the sample holder (boron nitride) and on certain test samples. Apparently, lunar samples are more sensitive to this form of contamination; perhaps their low oxygen activities or the presence of very fine metallic particles facilitates decomposition of hydrocarbons.

The major modification in experimental procedure for the present studies was the use of a high-purity-gas buffering system in a glass and metal furnace chamber in which the only breakable joints were sealed with teflon gaskets. Electrical contact was made by attaching platinum electrodes to the sample with silver conductive point (Silver Print, GC No. 21-1, GC Electronics Corp., Rockford, Ill.). The contacts were cured in flowing dry He gas at room temperature and at 120°C for several days. The electrical conductivity was measured by a standard three-electrode technique that in principle enables the separation of high-conductivity surface contributions from bulk effects. It is difficult to evaluate the effectiveness of the guard electrode—in a few cases it was obvious that the separation was effective to a few parts in 10^5; however, in the general case, the currents collected at the "surface" and "bulk" electrodes were approximately in the ratio of the effective electrode areas, an indication that an anomalous surface condition was not dominating the measurements. Both a-c (nominally at 5 Hz) and d-c conductivities were determined by simple voltage-current measurements; for the a-c conductivity, phase-sensitive detectors were used to separate resistive and capacitive components. Equivalent residual or leakage conductivities for this experimental arrangement are shown in several of the figures; for example, the solid and dashed lines in Fig. 1 represent the d-c and a-c leakage conductivities, respectively.

Preliminary information on the effects of furnace atmosphere on the electrical conductivity was obtained by introducing either O_2 or dry H_2 into a purified He carrier to volume fractions of 0.2 and 0.95, respectively. The gas mixture was flowed through the furnace at a pressure of about 1.1 atm. Obviously, neither mixture will provide the appropriate oxygen activity to buffer the lunar samples

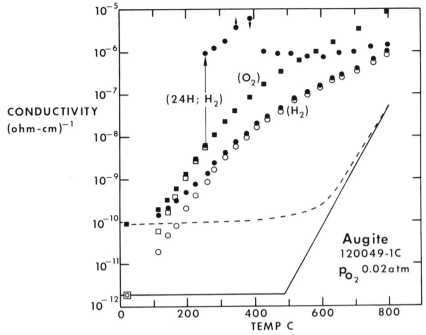

Fig. 1. Electrical conductivity of terrestrial augite (National Museum Number 120049) in reducing (H_2) and oxidizing atmospheres (O_2). Full and open symbols represent d-c and a-c (5 Hz) conductivities, respectively. The effect of reduction at 260°C for 24 hours in hydrogen following oxidation is indicated by the labeled vertical arrow. Time-dependent characteristics along the upper branch are indicated. Dashed and solid lines represent equivalent leakage conductivities for the a-c and d-c measurements, respectively.

(Sato and Hickling, 1973); however, on the basis of evidence presented subsequently, it will be argued that the changes which occur in the lunar samples in the reducing atmosphere (also detailed below) do not significantly alter the values of electrical conductivity, and that such data are representative of pristine Apollo samples.

With a proviso concerning changes during the initial heating of the samples and the extent to which each was studied, the effects of the various furnace atmospheres on the conductivity were qualitatively the same for terrestrial pyroxenes as for lunar rocks. For this reason and because of the relative availability of the two types of materials, the most exhaustive studies to date have been studies of a terrestrial augite. These results are described here in detail to characterize the sort of phenomena observed for the lunar rocks.

ELECTRICAL CONDUCTIVITY OF A TERRESTRIAL AUGITE

The electrical conductivity was measured at elevated temperatures for several samples from a terrestrial augite (National Museum sample No. 120049). This augite is basically a homogeneous single crystal with sparse sub-boundaries and cracks. Second phase iron oxides and iron sulfides are present at estimated concentrations of the order of 0.05%. The chemical composition is shown in Table 1 and typical data are presented in Fig. 1. (In general, the full symbols represent a-c

Table 1. Chemical analyses of conductivity samples.*

	Terrestrial†		Lunar Samples‡	
	Augite 120049	Breccia 10048	Breccia 15418	Basalt 15555
SiO_2	48.72	39.88	44.97	44.24
Al_2O_3	8.20	12.40	26.73	8.48
CaO	16.91	11.03	16.10	9.45
MgO	13.70	7.17	5.38	11.19
MnO		0.21	0.08	0.29
Cr_2O_3		0.30	0.11	0.70
FeO	4.98	16.34	5.37	22.47
Fe_2O_3	3.60	0	0	0
TiO_2	2.28	8.77	0.27	2.26
Na_2O	1.80	0.48	0.31	0.24
K_2O	0.09	0.17	0.03	0.03
Total	100.28	96.75	99.35	99.35

*Reported as weight percent oxides.
†Analyses provided by National Museum of Natural History, Smithsonian Institution, Washington D.C.
‡Analyses provided by Lunar Receiving Laboratory, NASA, Houston.

(5 Hz) conductivities and the open symbols represent d-c conductivities; where only one symbol is shown, both a-c and d-c conductivity values plot within the area of the symbol.) The measured values of the conductivity depend on the prior thermal and chemical treatment of the sample. For example, after a sample had been heated at 810°C in an atmosphere of He–H$_2$ until the conductivity showed no change with time (several hours), subsequent measurements of the conductivity at lower temperatures yielded the values shown by the lower branch of circles in Fig. 1. For these measurements the furnace atmosphere was a reducing He–H$_2$ mixture and the data were reproducible for the several cycles investigated. After a sample had been heated at 810°C in an oxidizing He–O$_2$ atmosphere ($P_{O_2} \sim 0.02$–0.15 atm), during subsequent runs in the same atmosphere the conductivity values represented by squares in Fig. 1 were measured. When samples were reduced at low temperatures subsequent to having been oxidized, the value of the conductivity again increased. For example, the effect of heating the sample in a He–H$_2$ mixture at 260°C is indicated by the vertical arrow in Fig. 1. After sufficient time at these low temperatures the conductivity attained an apparent steady-state value; whereupon, subsequent thermal cycling below a certain maximum temperature in a He–H$_2$ atmosphere yielded conductivity data lying along a single curve, which is partly defined in Fig. 1 by the upper branch of circles. Data for two thermal cycles in this region are presented in greater detail in Fig. 2. When the temperature exceeded about 400°C, time-dependent decreases in conductivity were observed, and eventually the conductivity values described previously for a sample reduced at high temperatures were recovered. The same qualitative features were observed for all samples studied and for subsequent thermal and chemical treat-

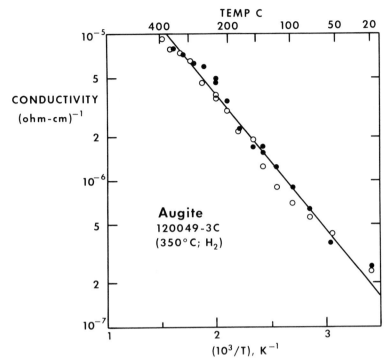

Fig. 2. Electrical conductivity of terrestrial augite (National Museum Number 120049) for sample reduced at low temperatures after oxidation. Full and open circles are for separate temperature cycles.

ments of the same sample. Quantitatively, data for a sample were reproducible with estimated differences of the order of 10 percent.

Qualitatively similar behavior has been observed in other systems, for example in lead silicate glasses (Holland, 1964). In these cases, the oxidized sample is the glass with the cationic network modifiers oxidized and in solution. Low-temperature heating in hydrogen causes a reduction of the cations through the introduction of oxygen vacancies. However, at these low temperatures the reduced metal atoms remain dispersed and provide low-activation-energy electron sources for extrinsic semiconduction. At higher temperatures the metal atoms become mobile and can diffuse to form more stable metallic clusters. Presumably, the dominant low-temperature conductivity mechanism in the oxidized glass is associated with the metal oxides, and their removal during high-temperature reduction causes a net decrease in conductivity.

Mössbauer spectroscopy of samples of the augite following each of the three major types of thermochemical treatment suggests that the observed changes in conductivity are related to the distribution of iron. The percentages of total Fe contained in the various oxidation states are given in Table 2, and typical spectra shown in Fig. 3. Computer analysis of each of these spectra was based on the contributions from the silicate phase being described by four independent

Table 2. Distribution and oxidation of Fe in terrestrial
augite 120049.*

Sample treatment	Silicate†		Metallic Fe	Fe_2O_3
	Fe^{2+}	Fe^{3+}		
As received	83.0	17.0	0	0
↓ He–H_2 at 800°C	83.3	14.9	1.8	0
↓ He–O_2 at 800°C	77.0	21.3	0	2.1
↓ He–H_2 at 350°C	78.5	21.5	n.d.‡	n.d.‡

*Determined by Fe^{57} Mössbauer spectroscopy and
reported as percentages of total Fe.
†Standard deviations for the changes in Fe^{3+}, Fe^{2+}
content are ∼ 1%.
‡n.d.—none detected; minimum detectable amount
for magnetic phases is ∼ 0.5%.

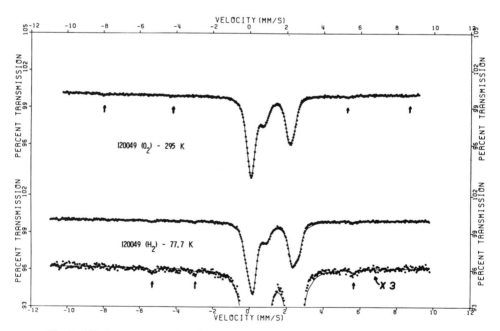

Fig. 3. Mössbauer spectra of augite 120049 after annealing at 800°C in He–O_2 and He–H_2
atmospheres. The arrows denote the positions of the weak metallic Fe and Fe_2O_3 peaks.
(To display these peaks, the vertical scale for the latter curve has been increased by a
factor of three, curve labeled ×3.) In all Mössbauer spectra, velocities are measured with
respect to metallic Fe; the absorber temperature is indicated in the figure, and the solid
curves are the results of least-squares computer analyses.

Lorentzian peaks for the Fe^{2+} components and a single quadrupole doublet for the Fe^{3+} components with equal heights and widths for the peaks of the doublet. This procedure yielded values for quadrupole splittings and isomer shifts of about 0.6 and 0.5 mm/sec, respectively, for the Fe^{3+} doublet. Although this fitting technique cannot be expected to produce accurate absolute values for the percentage of Fe^{3+} and Fe^{2+}, it should produce relative values of sufficient accuracy for monitoring changes in a sample following thermochemical treatments.* The principle features of Table 2 are as follows: (i) high-temperature reduction produces metallic Fe in clusters large enough to be ferromagnetic and decreases slightly the amount of Fe^{3+} in the silicate phase; (ii) high-temperature oxidation converts the metallic Fe to Fe_2O_3 and increases the Fe^{3+} in the silicate; and (iii) low-temperature reduction of the oxidized sample effectively wipes out any magnetic peaks and produces only very minor changes in the Fe^{2+}, Fe^{3+} distribution of the silicate phase. The absence of any magnetic peaks may result from reduction of Fe_2O_3 into several states (e.g., Fe_3O_4, FeO, Fe), none of which contain enough total Fe to be detectable. Absorption peaks of any very finely dispersed $Fe°$ produced during the low-temperature reduction would appear near zero velocity and be obscured by the large Fe^{2+} and Fe^{3+} silicate peaks. It must be emphasized that for the atmospheres, temperatures, and time used here it is not expected that the samples are in equilibrium states; the Mössbauer analyses clearly indicate this. However, the measurements reported here were made after the conductivity of the samples had reached values that were nearly time-independent.

Similar effects were observed for a terrestrial enstatite with a low ferric iron content; however, the conductivity values and characteristic temperatures were considerably lower than those for the augite. These results will be presented in greater detail elsewhere.

ELECTRICAL CONDUCTIVITY OF LUNAR SURFACE ROCKS

Electrical conductivity has been measured for three types of Apollo rocks: 10048—a coherent breccia; 15555—a porous, friable basalt; and 15418—an almost completely recrystallized breccia of anorthositic composition. Chemical analyses of these rocks as reported by the Lunar Receiving Laboratory are given in Table 1. During initial heating the conductivity was independent of furnace atmosphere up to temperatures of about 250°C for the porous basalt in purified He, and to temperatures of about 400°C for the more coherent breccias, 10048 in He and 15418 in He–H_2. To date the most complete conductivity study has been for the breccia 15418; typical d-c conductivity data are shown in Fig. 4. The curves labeled (Initial), (H_2), (O_2), and (H_2; 470°C) correspond to data obtained during initial heating in a He–H_2 atmosphere, after reduction at 850°C in He–H_2, after oxida-

---

*It is seen that the Fe^{3+}/Fe^{2+} ratio of Table 2 is different from that of Table 1. Similar differences between Fe^{3+} content as determined chemically and by Mössbauer spectroscopy have been observed previously (Bancroft et al., 1967) and it has been suggested that the discrepancy may arise because oxidation of Fe^{2+} by OH^- takes place during wet chemical analysis (Schreyer and Chinner, 1966).

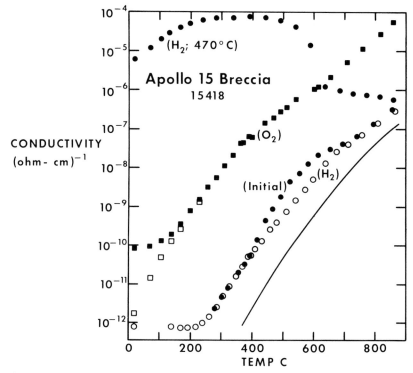

Fig. 4. Electrical conductivity (dc) of lunar breccia (Apollo 15418) measured during initial heating, in reducing (H_2) and oxidizing (O_2) atmospheres, and in a reducing atmosphere after oxidation followed by low-temperature reduction (H_2; 470°C). Solid curve represents equivalent d-c leakage conductivity.

tion at 850°C in He–O_2, and after reduction at 470°C in He–H_2 subsequent to the high-temperature oxidation, respectively. The same qualitative features described above for the terrestrial augite can readily be identified in this figure; however there are obvious quantitative differences with respect to characteristic temperatures and conductivity parameters.

As stated above, it is recognized that none of the furnace atmospheres used in these conductivity studies provide the proper activities to buffer the lunar samples. Some of the changes which occur in these samples under the conditions of the conductivity study have been characterized by Mössbauer spectroscopy and are presented in a subsequent section. However, with specific reference to the conductivity measurements, several observations suggest that the values of electrical conductivity of the samples measured in He–H_2 atmospheres are probably not very different from the values for pristine Apollo material. Chief among these observations is the coincidence, over the temperature range indicated above, of the conductivity values measured during the initial heating of each of the samples with the values measured after heating at high temperatures in the He–H_2 atmosphere. Further support for favoring the data obtained for samples in a potentially reducing, as opposed to oxidizing, atmosphere is the general observation of the dramatic increase in conductivity which accompanies the introduction of ferric iron into intrinsically low-conductivity materials and the almost total absence of detectable ferric iron in pristine Apollo samples.

For convenience, the electrical-conductivity data for the three lunar samples as measured in the reducing environment were analyzed in terms of the familiar expression

$$\sigma(T) = \sigma_0^{(1)} \exp(-E^{(1)}/kT) + \sigma_o^{(2)} \exp(-E^{(2)}/kT) \qquad (1)$$

where the symbols have the customary meanings. Values for $\sigma^{(i)}$ and $E^{(i)}$ obtained by a least-squares fit of the data with a logarithmic metric are given in Table 3 along with the temperature ranges, number of points fitted, and root-mean-square values for $\Delta\sigma/\sigma$. The data are compared with the analytical expression in Figs. 5–7. (In these figures the lower branches represent data for samples measured in reducing atmosphere; the upper branches represent data for samples measured in an oxidizing atmosphere.) It is to be emphasized that at this stage, Equation (1) is used merely for convenience. No formal identification has been made regarding conductivity mechanisms, and consequently the parameters $E^{(i)}$ and $\sigma_0^{(i)}$ have only mathematical significance.

Furthermore, it also should be emphasized that the long times required for solid-state diffusion of the various atomic species at the temperatures used in this study means that the chemical reactions are occurring mainly in the more accessible regions of the sample, for example, near internal boundaries such as microcracks. This comment is to be compared with the previously mentioned observation of the similarity of bulk and surface conductivities, which indicates that there are no apparent anomalous conductance paths on the external surface of the sample, and that data were recorded only after the conductivities had reached nearly steady-state values. Consequently, the changes in electrical conductivity associated with the various thermochemical treatments should not be considered to be representative of a homogeneous equilibrated sample but to be qualitative indicators of the effects of these treatments.

Table 3. Electrical-conductivity parameters for lunar rocks.*

$$\sigma(T) = \sum_{i=1}^{2} \sigma_0^{(i)} \exp(-E^{(i)}/kT)$$

Sample	$\sigma_0^{(1)}$	$E^{(1)}$	$\sigma_o^{(2)}$	$E^{(2)}$	RMS	Number of points	Temperature range (°C)
10048 AC	5.18×10^{-5}	0.533	5.09×10^{-2}	0.867	0.037	14	211–850
DC	2.66×10^{-5}	0.559	3.50×10^{-2}	0.896	0.054	15	163–850
15555 AC	3.18×10^{-6}	0.420	2.16×10^{-1}	0.993	0.050	21	169–797
DC	1.27×10^{-4}	0.604	3.68×10^{-1}	1.04	0.059	21	169–797
15418 AC	4.39×10^{-7}	0.514	1.35×10^{-1}	1.260	0.055	20	284–861
DC	9.84×10^{-4}	0.971	1.37×10^{-0}	1.509	0.106	24	194–861

*Measured in a He–H$_2$ environment.
$\sigma_0^{(i)}$ are in (ohm-cm)$^{-1}$, $E^{(i)}$ in eV, and the RMS value is for $\Delta\sigma/\sigma$.
Determined by least-squares analyses.

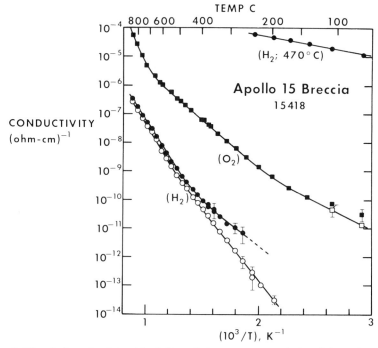

Fig. 5. Electrical conductivity (dc, full symbols; ac, open symbols) of lunar breccia (Apollo 15418) in various atmospheres (*see* Fig. 4). Solid lines are results of least-squares fit to Equation 1 (*see* text).

One interesting feature of these data which has not been investigated in sufficient detail, but does warrant some comment, is the apparent low-frequency dispersion observed in the conductivity measured for samples in reducing environments, *see* Figs. 5–7. This dispersion was most notable at low temperatures for lunar rock 15418 and for a terrestrial enstatite; these samples were also the lowest-conductivity samples studied. This dispersion is accompanied by a long-time constant decay of transient currents following the application of a constant voltage to the sample. The long times required to reach steady-state values result in an apparent dependence of the conductivity on prior voltage polarities and magnitudes and consequently produce an apparent non-ohmic behavior. In one case this dispersion appeared to be associated with a $\sigma \propto \omega^{\eta}$, $0.5 < \eta < 1$ type behavior. The source of these effects has not yet been identified.

MÖSSBAUER ANALYSES OF LUNAR SAMPLES

Some of the changes in lunar samples due to annealing in reducing and oxidizing atmospheres were monitored by Mössbauer spectroscopy. The results are tabulated in Table 4 and several typical spectra are shown in Figs. 8–10. In all spectra, the silicate peaks were fitted using four independent Lorentzians; the

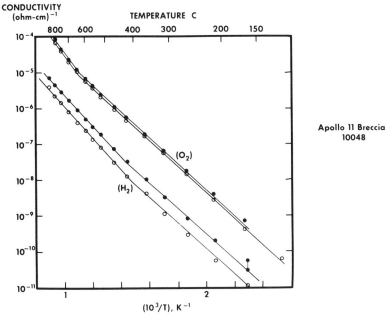

Fig. 6. Electrical conductivity (dc, full symbols; ac, open symbols) for lunar breccia (Apollo 10048) in reducing and oxidizing atmospheres (lower and upper sets of curves, respectively). Solid lines are results of least-squares fit to Equation 1.

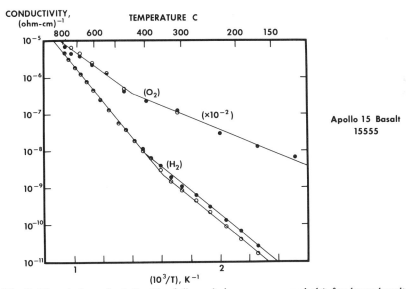

Fig. 7. Electrical conductivity (dc, full symbols; ac, open symbols) for lunar basalt (Apollo 15555) in reducing and oxidizing atmospheres (lower and upper sets of curves, respectively). Solid lines are results of least-squares fit to Equation 1. (For purposes of presentation, data for the oxidizing environment have been reduced by a factor of 100.)

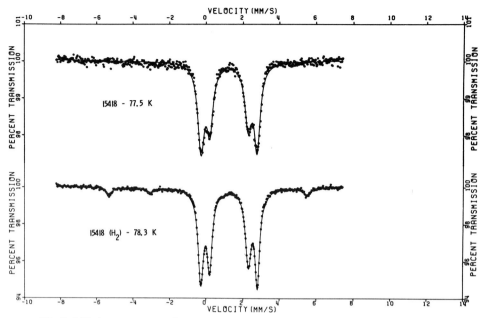

Fig. 8. Mössbauer spectra of Apollo sample 15418 as received and after annealing at 800°C in He–H₂.

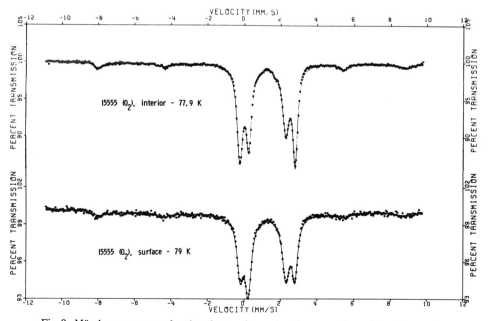

Fig. 9. Mössbauer spectra of an interior portion and surface scrapings of Apollo sample 15555 after annealing at 800°C first in He–H₂, then in He–O₂.

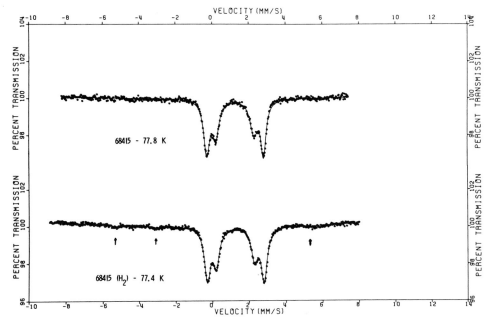

Fig. 10. Mössbauer spectra of Apollo sample 68415 in the as-received state and after annealing in He–H₂.

Table 4. Distribution of Iron Compounds in Selected Lunar Samples. (Results reported as percentages of total Fe, minimum detectable amount $\sim 0.5\%$. (H₂) indicates $\geqslant 800°C$ anneal in He-H₂; (H₂), (O₂) indicates $\geqslant 800°C$ anneal first in He–H₂, then in He–O₂.)

Sample	Total Silicate	Pyroxene	Olivine	Ilmenite	Metallic Fe	Fe_2O_3, Fe_3O_4
15418	100.0	59.5	40.4	0	0	0
15418(H₂)	90.5	59.4	31.1	0	9.5	0
15555	98.5	77.6	20.9	1.3	<0.5	0
15555(H₂), (O₂); interior	85.8	66.3	19.5	1.0	0	13.2
15555(H₂), (O₂); surface	78.0	76.0	~2.0	1.0	0	21.1
						Other
68415	100.0	54.3	45.7	0	0	
68415(H₂)	96.7	51.4	45.2	0	3.3	
67455	100.0	56.5	43.5	0	0	
66055	95.2	37.8	57.4	~0.5	4.3†	
60255	95.1	64.9*	30.2	2.9	2.1†	
68815	91.0	31.3*	59.7	0	7.0†	2.0(FeS ?)
66095	96.5	51.8	44.7	1.5	2.0†	

*Petrographic analyses (Butler, 1972) and Mössbauer linewidths indicate appreciable glass content; the percentages given in the pyroxene column for these samples therefore include any Fe^{2+} in glass.

†The hyperfine fields are ~ 345–350 kG at 77°K indicating up to 5% Ni in the metallic Fe phase.

method by which the olivine percentages were determined is outlined in Appendix A. In general, the results are qualitatively similar to those observed for the terrestrial augite sample and corroborate the conductivity mechanisms discussed previously. Several points rate more specific discussion. First, it is clear from Table 4 and the figures that significantly more metallic Fe is produced by reduction in lunar samples than in the augite; this is undoubtedly because of the absence of any Fe^{3+} in the original lunar material. However, from the conductivity results, it is evident that there is not enough metallic Fe present to produce any contiguous low-resistance paths even in the most severe case (15418). Second, there are some surface effects present. A sample scraped from the surface of 15555 after high-temperature reduction ($\geq 800°C$) followed by high-temperature oxidation showed the surface material to be depleted in olivine and somewhat enriched in Fe_2O_3 and Fe_3O_4 relative to the interior (*see* Fig. 9 and Table 4). The initial high-temperature anneal in He–H_2 apparently reduces olivine preferentially with respect to pyroxene in the surface layers. Preferential reduction of olivine is also indicated for 15418 (Fig. 8), whereas the interior sample of 15555 (Fig. 9) and the Apollo 16 sample, 68415 (Fig. 10) both show a slight trend in the opposite direction, with the pyroxene percentage being slightly decreased. The changes in the silicate phase of these three samples, however, may not be too significant, since the error in our method of determining the relative amounts of olivine and pyroxene is at least 5% (*see* Appendix A).

Several other Apollo 16 samples have also been studied, and are notable for their high olivine content and low total Fe percentage. The Fe phase distributions for a number of these are also listed in Table 4. Conductivity measurements are in progress for several of these samples.

SUMMARY

Electrical conductivities of lunar surface rocks were measured at temperatures from 20 to 850°C. Although conductivity was observed to be sensitive to the furnace atmosphere at temperatures as low as 250°C, reproducible data were obtained under specific sets of conditions. Data obtained for samples heated in He–H_2 gas atmospheres are presented as being most typical of *in situ* lunar surface rocks. The conductivity values measured under these conditions for the several lunar samples reported here bracket values obtained by Olhoeft *et al.* (1973) during the initial heating of lunar rocks in high vacua. The conductivity values for rock 15418 are just slightly less than those obtained by Duba (1972) for a terrestrial olivine ($Fa_{9.4}$) with negligible ferric iron, and as such, this rock probably qualifies as the least-conducting natural bulk material ever measured. A comparison of the present conductivity results for lunar rocks and terrestrial pyroxenes with recent data for terrestrial olivines (Duba, 1972; Housley and Morin, 1972) indicates that electrical conductivity of these anhydrous silicates is determined principally by the concentration, oxidation state, and dispersion of the multivalent cations—and consequently the chemical activities of that environment—rather than by the silicate crystal structure. In this regard it should be noted that Wright

(1971) has already suggested that the dependence of the electrical conductivity on oxidation state may be a significant factor in determining the conductivity profile of the lunar interior.

Future studies will be concerned with extending the suite of lunar rocks and terrestrial pyroxenes for which conductivity data are available, defining the conductivity mechanisms, and examining in a more controlled manner the dependence of conductivity on the oxygen activity of the samples.

Acknowledgments—The authors gratefully acknowledge the transfer of samples of terrestrial pyroxenes by J. S. White, Jr., of the National Museum of Natural History. This work was supported in part by the National Aeronautics and Space Administration under Contract NAS 9-12271.

REFERENCES

Anderson D. L. and Hanks T. C. (1972) Is the Moon hot or cold? *Science* **178**, 1245–1249.
Bancroft G. M., Maddock A. G., and Burns R. G. (1967) Applications of the Mössbauer effect to silicate mineralogy—I. Iron silicates of known crystal structure. *Geochim. Cosmochim. Acta* **31**, 2219–2246.
Butler P. (1972) *Lunar Sample Information Catalog, Apollo 16.* NASA MSC 03210.
Chung D. H., Westphal W. B., and Olhoeft G. R. (1972) Dielectric properties of Apollo 14 lunar samples. *Proc. Third Lunar Sci. Conf., Geochim. Cosmochim. Acta*, Suppl. 3, Vol. 3, pp. 3161–3172. MIT Press.
Criswell D. R. (1972) Lunar Dust Motion. *Proc. Third Lunar Sci. Conf., Geochim. Cosmochim. Acta*, Suppl. 3, Vol. 3, pp. 2671–2680. MIT Press.
Duba A. (1972) Electrical conductivity of olivine. *J. Geophys. Res.* **77**, 2483–2495.
Dyal P., Parkin C. W., and Daily W. D. (1973) Surface magnetometer experiments: Internal lunar conductivity profile and steady field measurements (abstract). In *Lunar Science—IV*, pp. 205. The Lunar Science Institute, Houston.
Hien P. Z. and Shpinel V. L. (1963) Dependence of the γ-quantum resonance absorption spectrum on crystal temperature. *Soviet Physics JETP* **17**, 268–270.
Holland L. (1964) *The Properties of Glass Surfaces*, pp. 481–491. Chapman and Hall.
Housley R. M. and Morin F. J. (1972) Electrical conductivity of olivine and the lunar temperature profile. *The Moon* **4**, 35–38.
Huffman G. P., Schwerer F. C., and Fisher R. M. (1972) Mössbauer analyses of Apollo 15 samples. In *The Apollo 15 Lunar Samples*, pp. 440–441. The Lunar Science Institute, Houston.
Olhoeft G. R., Frisillo A. L., Strangway D. W., and Sharpe H. (1973) Temperature dependence of the electrical conductivity and lunar temperatures. In *Proc. Conf. on Geophysical and Geochemical Exploration of the Moon and Planets*, January 1973, Houston, *The Moon* (to be published).
Sato M. and Hickling N. (1973) Oxygen fugacity values of some lunar rocks (abstract). In *Lunar Science—IV*, pp. 650–652. The Lunar Science Institute, Houston.
Schreyer W. and Chinner G. A. (1966) Staurolite-quartzite bands in kyanite quartzite at Big Rock, Rio Arriba County, New Mexico. *Contr. Mineral. Petrol* **12**, 223–244.
Schwerer F. C., Huffman G. P., Fisher R. M., and Nagata T. (1972) Electrical conductivity and Mössbauer study of Apollo lunar samples. *Proc. Third Lunar Sci. Conf., Geochim. Cosmochim. Acta*, Suppl. 3, Vol. 3, pp. 3173–3185. MIT Press.
Schwerer F. C., Nagata T., and Fisher R. M. (1971) Electrical conductivity of lunar surface rocks and chondritic meteorites. *The Moon* **2**, 408–422.
Smith B. F., Schubert G., Colburn D. S., Sonett C. P., and Schwartz K. (1973) Corroborative Apollo 15 and 12 lunar surface magnetometer measurements (abstract). In *Lunar Science—IV*, pp. 683–685. The Lunar Science Institute, Houston.
Wright D. A. (1971) Electrical conductivity of lunar rock. *Nature Phys. Sci.* **231**, 169–170.

APPENDIX A

Method for determining percentage of total Fe *contained in olivine*

At room temperature (300°K), the M1 (outer) and M2 (inner) quadrupole doublets in pyroxene Mössbauer spectra are split by about 2.5–2.6 mm/sec and 2.0–2.05 mm/sec, respectively. On cooling to liquid nitrogen temperatures (78°K), splitting of the M1 doublet increases to approximately 3.0 mm/sec, while the M1 separation, which increases to not more than 2.1 mm/sec, shows little temperature dependence. The M1 and M2 doublets are thus well resolved at 78°K but overlap each other at 300°K. Consequently, computer analysis of pyroxene spectra obtained at 300°K will nearly always attribute part of the area and effective thickness associated with the M1 doublet to the larger M2 doublet.* If the M1 and M2 effective thicknesses at 78°K are denoted by X_{M1} and X_{M2}, then for a pure pyroxene sample,

$$X_1 = X_{M1} \qquad X_2 = X_{M2} \tag{1}$$

$$X_1'/X_2' = f X_{M1}/\{X_{M2} + (1-f)X_{M1}\} \tag{2}$$

where X_1 and X_2 are the computer-calculated effective thicknesses for the outer and inner doublet at 78°K, X_1' and X_2' are the corresponding quantities at 300°K, and f is the fraction of the true M1 site effective thickness ascribed to the outer doublet at 300°K. The ratio of Equation 2 eliminates the temperature dependence of the recoilless fraction, which is assumed to be equal for M1 and M2 sites. Values of f ranging from about 0.4 to 0.6, with an average of 0.49, were obtained using Equations 1 and 2 and previous analyses of a number of lunar samples in which pyroxene is the dominant silicate phase (Huffman *et al.*, 1972 and references therein).

Olivine exhibits a single doublet with a splitting of 2.9–3.0 mm/sec at 300°K (Bancroft *et al.*, 1967) which increases to not more than about 3.1 mm/sec at 78°K. Consequently, a mixture of olivine and pyroxene will exhibit a two doublet spectrum with the outer doublet widely split at both 300 and 78°K; furthermore, the outer doublet will show an anomalous increase in intensity at 78°K due to the outward shift of the pyroxene M1 doublet (*see* Fig. 1, Huffman *et al.*, 1972). If the olivine effective thickness is denoted X_{0l} then for a mixture of olivine and pyroxene, Equations 1 and 2 become

$$X_1 = X_{0l} + X_{M1} \qquad X_2 = X_{M2} \tag{3}$$

$$X_1'/X_2' = (X_{0l} + f X_{M1})/\{X_{M2} + (1-f)X_{M1}\} \tag{4}$$

(In obtaining these equations it was assumed that the recoilless fractions for olivine and pyroxene are equal.) For $f = 0.49$ and with experimental values of X_1, X_2 and X_1'/X_2', Equations 3 and 4 can be solved to yield values for X_{0l} and X_{M1}. The results are not very sensitive to the value of f, which clearly is the least well-determined parameter in this treatment. For example, a value for f of 0.4 (0.6) gives olivine percentages that are about 5% higher (lower) than the values shown in Table 4.

Fe^{2+} in lunar glass produces a two-doublet spectra somewhat similar to pyroxene, with the splitting of the outer doublet more temperature dependent than that of the inner doublet. It is therefore not unreasonable to apply the method outlined here to obtain the percentage of Fe contained in olivine for samples containing significant amounts of glass in addition to pyroxene and olivine, for example, lunar sample 60255. However, the broadness of the peaks arising from glassy phases probably makes the results somewhat less reliable.

---

*It is well known that peak area and intensity do not depend linearly on effective thickness. Therefore, in all determinations of Fe phase percentages, the effective thickness was calculated from the experimental peak area using the appropriate theoretical expressions (Hien and Shpinel, 1963).

Proceedings of the Fourth Lunar Science Conference
(Supplement 4, *Geochimica et Cosmochimica Acta*)
Vol. 3, pp. 3167–3174

Polarimetric properties of the lunar surface and its interpretation Part 6: Albedo determinations from polarimetric characteristics

E. Bowell and A. Dollfus

Observatoire de Paris-Meudon, 92190—Meudon, France

B. Zellner

University of Arizona, Tucson, Arizona 85721, USA, and Observatoire de Paris-Meudon

J. E. Geake

UMIST, Manchester M60 1QD, England

Abstract—Optical polarization results for 35 lunar samples from Apollos 11 to 15 and from Lunas 16 and 20 are analyzed in terms of 2 relationships: (1) Lunar fines (with the exception of 15401,62), and also the lunar surface itself, show an inverse linear relationship between geometric albedo and the maximum value of polarization, thereby providing a method of determining albedos. Lunar rocks and breccias do not usually fit this relationship. (2) All lunar samples, and all observations of atmosphereless objects in the solar system, show an inverse linear relationship between geometric albedo and the slope of the polarization curve for small positive values of polarization. This enables albedos (and hence diameters) to be determined for objects where the polarization curve is only observable for small phase angles, even when these objects cannot be resolved telescopically.

Introduction

This paper continues the lunar sample work described previously (Geake *et al.*, 1970; Dollfus *et al.*, 1971b; Bowell *et al.*, 1972). We have now investigated the polarization characteristics at 5 wavelengths in the visible and UV regions for 35 lunar samples from Apollos 11 to 15 and from Lunas 16 and 20; these consisted of 25 fines samples (including 8 from core tubes), 5 crystalline rocks and 12 different zones on 5 breccias.

Figure 1 defines the polarization parameters used. We also define the geometric albedo of a reflecting object as the ratio of its brightness to that of a flat, normally illuminated Lambert screen at the same position and having the same projected area. The quantity A is thus uniquely defined for a body or surface element of any shape, orientation, scattering properties or conditions of illumination. Laboratory albedos have been determined with respect to a MgO screen at a phase angle of 5°; the surge in brightness close to zero phase (the opposition effect) is thereby practically avoided, as is the effect of shadowing which dominates the albedos of intricate surfaces at larger phase angles (Dollfus *et al.*, 1971b).

We have shown elsewhere (Dollfus *et al.*, 1971a; Dollfus and Titulaer, 1971) that it is useful to summarize the results on a diagram in which the albedo of the sample is plotted against the maximum value of the polarization curve, for each wavelength used. When this is done, lunar telescopic observations and most lunar

E. Bowell *et al.*

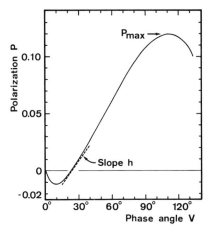

Fig. 1. Definition of polarization parameters used.

fines samples lie near to a straight line, but at least one fines sample stands out as different from the main group, and we shall attempt to explain why. Bowell *et al.* (1972) showed that all lunar rocks and breccias also depart from this linear relationship, implying that the lunar surface cannot have appreciable areas of rock or breccia that are not covered by at least a thin coating of dust. We have also attempted to explain the polarization characteristics of lunar rocks and breccias in terms of their surface structure as observed visually, and by means of a scanning electron microscope (*see* e.g., Bowell *et al.*, 1972). Comparison may then be made with telescopic polarization measurements of the lunar surface itself, and also of atmosphereless planets, satellites, and asteroids for which samples are not available, in order to deduce their surface properties.

In addition, it has recently been shown (Veverka and Noland, 1973; Bowell and Zellner, 1973), that there is a linear relationship between the geometric albedo and the slope of the curve of polarization vs. phase angle (for small positive values of polarization). The present paper gives further evidence that the slope-albedo law applies both to telescopic observations and to laboratory measurements of all samples, including fines, rocks and breccias, and for all wavelengths used. This slope-albedo law is now used to determine the albedo and hence the diameter of objects such as asteroids and planetary satellites, as described below.

POLARIZATION CHARACTERISTICS

In Fig. 2 the albedo A is plotted against the maximum polarization P_{max} (on a log–log scale) for lunar fines samples (for orange light only). The full line encloses the approximate domain occupied by telescopic observations of the moon; the dotted line indicates the linear relationship that best fits the main group of lunar sample measurements. The equation of this line is:

$$\log A = -0.733 \log P_{max} - 1.726$$

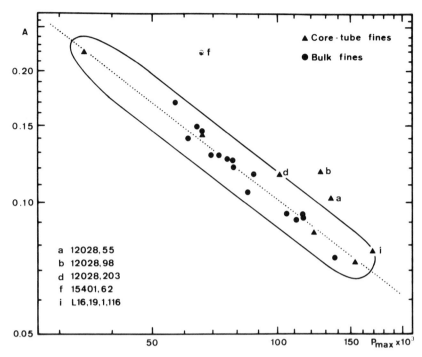

Fig. 2. Albedo vs. polarization maximum for lunar fines at $\lambda = 0.585 \ \mu$m.

The position of this "calibration line" initially involved an empirical re-calibration of the then-accepted lunar geometric albedos, on polarimetric grounds. However, the addition of new lunar sample data has improved the laboratory calibration of albedos, and recent telescopic measurements (Pohn and Wildey, 1970) have provided more accurate lunar regional albedos. The two calibrations are in reasonable agreement, but we still consider the calibration deduced from laboratory measurements of lunar samples to provide more reliable albedos for the lunar surface. It can be seen that at least 5 of the lunar samples show significant departures from the main group that were used to produce the calibration line. Of these, 4 (a, b, d, and i) are core-tube samples. Unfortunately, our polarization measurements of the Apollo core-tube samples were less reliable than usual owing to the small amount of sample available. Sample f (15401,62), on the other hand, is a lunar surface sample which is well known to have other unusual properties (e.g., Nelen *et al.*, 1972; Fleischer and Hart, 1973); through a microscope it is seen to contain a high proportion of green glass spherules. The polarimetric effect of these spherules is seen more clearly in Fig. 3 which extends the same plot to 5 wavelengths; the line for each sample is drawn through the 5 points, although for clarity the actual points are only shown for sample f. For most of the samples these lines are parallel to the direction of the linear relationship in Fig. 2;* this is even true for

---

*Previously published albedos for core-tube sample 12028,98 (Bowell *et al.*, 1972; Fig. 5) have recently be revised.

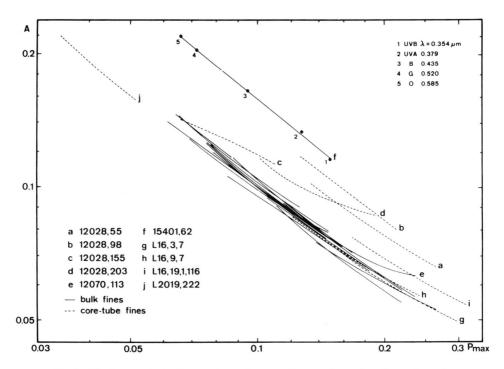

Fig. 3. Albedo vs. polarization maximum for lunar fines, at 5 wavelengths as shown by
lines.

sample f, suggesting that its displacement is caused by the addition of high polarization due to specular reflection by the glassy spherules (which is not wavelength
dependent). This has a large effect on the polarization, which would otherwise be
small for a sample of such high albedo, and it also increases the albedo itself; both
of these increases have the effect of shifting the line away from the calibration line
but parallel to it (Bowell, 1973). The low opacity of the spherules could also cause
a displacement from the calibration line; however, since the spherules are highly
colored, we would expect the displacement to vary with wavelength, resulting in a
line inclined to the calibration line. We infer that the effect of low opacity is in this
case swamped by specular reflection.

The Slope-Albedo Law

Figure 4 shows (on a log–log scale) the slope h of the polarization curve at
small positive values of polarization (*see* Fig. 1) plotted against the geometric albedo A for lunar fines, rocks and breccias, for 5 wavelengths. Figure 5 shows the
same plot for lunar fines only, with the 5 points for each sample at different
wavelengths joined by lines. There is an overall linear relationship between slope
and albedo, within experimental error. Figure 6 shows the plot extended to all

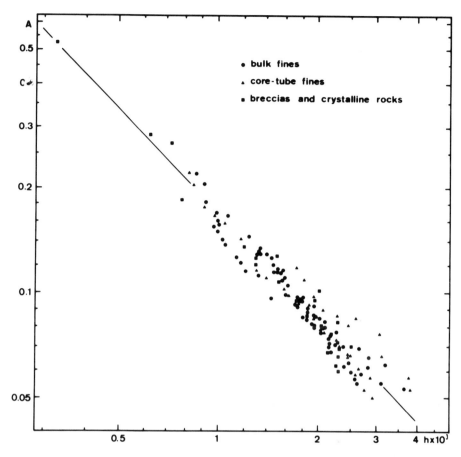

Fig. 4. Albedo vs. polarization slope for lunar materials, at 5 wavelengths.

available data for atmosphereless objects in the solar system, including telescopic observations of the moon. These results may be stated in the form of a slope-albedo law: $\log A = -c_1 \log h + c_2$ where c_1 and c_2 are dimensionless constants to be determined empirically, as summarized in Table 1; c_1 is evidently about 1; if we take c_1 as exactly 1.000, this leads to very consistent values of c_2, averaging -3.767. These values result in the broken line in Fig. 6. Albedos of solar-system objects result from a linear extrapolation of their photometric functions to zero phase, as recommended by Gehrels (1970). That these "astronomical" and the "laboratory" albedos measured at 5° phase are in good agreement is borne out by the consistency of the coefficient c_2 (Table 1) for all the scattering surfaces considered. For high albedo objects the implied equivalence between geometric albedo and normal reflectance need not hold, and this might lead to systematic errors in diameters obtained from "polarimetric" albedos; however, the fact that data points for Europa and Ganymede lie close to the slope-albedo calibration line (Fig. 6) seems to indicate that the error will in general be small.

E. BOWELL *et al.*

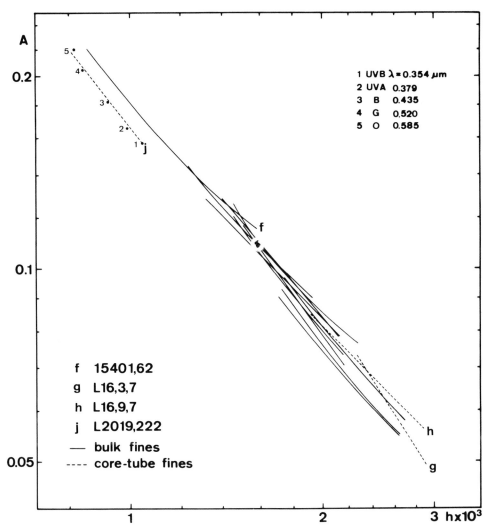

Fig. 5. Albedo vs. polarization slope for lunar fines at 5 wavelengths, as shown by lines.

Table 1. Calibration of the slope-albedo law.

Material	Number of slopes (all wavelengths)	Weight allocated	Albedo limits	c_1	c_2	c_2 if $c_1 = 1$
35 lunar fines and rocks	135	110	0.05–0.50	1.05	−3.89	−3.780
72 terrestrial samples	81	67	0.03–1.0	0.99	−3.64	−3.756
37 telescopically-observed lunar regions	139	96	0.04–0.25	1.03	−3.86	−3.760
5 solar-system objects	9	9	0.06–0.75	1.10	−4.07	−3.768
Weighted mean ($\pm 1\,\sigma$)				1.03 ±0.03	−3.83 ±0.11	−3.767 ±0.011

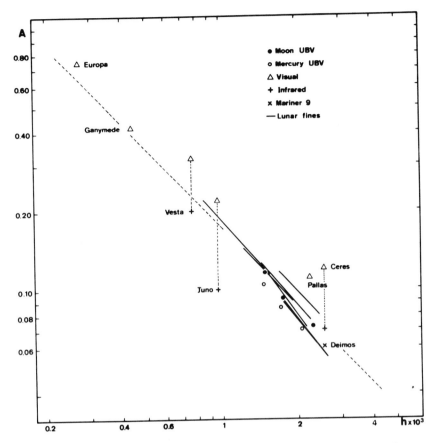

Fig. 6. Albedo vs. polarization slope for solar system objects and for some lunar fines.

The slope-albedo law, thus calibrated, may now be used to determine the geometric albedo of any surface whose polarization curve has a negative branch and for which enough polarization data are available to measure the slope. It is now possible to use this relationship to determine the albedo for solar-system objects, particularly asteroids and planetary satellites, whose polarizations are only measurable at small phase angles. From this polarimetric albedo and the visual flux received from the object, it is then possible to determine its diameter; this can be done for bodies too small to be resolved telescopically. Results already obtained by using this method are given in Table 2 (cf. Bowell and Zellner, 1970, for data sources and discussion).

Acknowledgments—We thank CNRS and CNES who supported the work at Meudon Observatory through the "Recherche Coopérative Programmée" for the study of lunar samples. We are also grateful for financial support from ESRO (EB), NSF and NASA (BZ), and SRC (JEG).

Table 2. Asteroid diameters in kilometers.

Asteroid	V(1,0)[a]	Polarimetric A_V[a]	Polarimetric Diameter	Visual Diameter	Infrared Diameter
1 Ceres	3.40	0.068	1060 ± 130	770	1025 ± 75
2 Pallas	4.53	0.073	600 ± 75	490:	
3 Juno	5.62	0.203	220 ± 30	195	290
4 Vesta	3.54	0.225	550 ± 70	460	580 ± 60
8 Flora	6.62	0.14	165		
9 Metis	6.42	0.17	165		
15 Eunomia	5.49	0.13	295		
89 Julia	7.40	0.11	130		
433 Eros	11.54	0.16:	16:		
1566 Icarus	16.82	0.19	1.3		
1620 Geographos	15.15	0.22:	2.6		
1685 Toro	14.7:	0.15	4:		

[a]V(1,0) is the apparent visual magnitude at unit heliocentric and geocentric distances and at zero phase angle. A_V is the geometric albedo at the V wavelength. If d is the diameter in km and h the polarization slope in $\deg^{-1}$, then

$$\log d = 3.12 - 0.2\, V(1,0) - 0.5 \log A_V,$$
$$\text{or } \log d = 5.01 - 0.2\, V(1,0) + 0.5 \log h.$$

REFERENCES

Bowell E. (1973) Analyse polarimétrique de la Lune, des roches terrestres et des échantillons lunaires avec application aux astéroïdes et satellites. D. ès Sc. Thesis, University of Paris.

Bowell E., Dollfus A., and Geake J. E. (1972) Polarimetric properties of the lunar surface and its interpretation, Part 5. Apollo 14 and Luna 16 lunar samples. *Proc. Third Lunar Sci. Conf., Geochim. Cosmochim. Acta*, Suppl. 3, Vol. 3, pp. 3103–3126. MIT Press.

Bowell E. and Zellner B. (1973) Polarizations of asteroids and satellites. In *Planets, Stars, and Nebulae Studied with Photopolarimetry* (editor T. Gehrels). University of Arizona Press.

Dollfus A., and Titulaer C. (1971) Polarimetric properties of the lunar surface and its Part 3. Volcanic samples in several wavelengths. *Astron. Astrophys* **12**, 199–209.

Dollfus A., Bowell E., and Titulaer C. (1971a) Polarimetric properties of the lunar surface and its interpretation, Part 2. Terrestrial samples in orange light. *Astron. Astrophys* **10**, 450–466.

Dollfus A., Geake J. E., and Titulaer C. (1971b) Polarimetric properties of the lunar surface and its interpretation, Part 4. Apollo 11 and Apollo 12 lunar samples. *Proc. Second Lunar Sci. Conf., Geochim. Cosmochim. Acta*, Suppl. 2, Vol 3, pp. 2285–2300. MIT Press.

Fleisher R. L. and Hart H. R. (1973) Surface history of lunar soil and soil columns (abstract). In *Lunar Science—IV*, pp. 251–253. The Lunar Science Institute, Houston.

Geake J. E., Dollfus A., Garlick G. F. J., Lamb W., Walker G., Steigmann G. A., and Titulaer C. (1970) Luminescence, electron paramagnetic resonance and optical properties of lunar material from Apollo 11. *Proc. Apollo 11 Lunar Sci. Conf., Geochim. Cosmochim. Acta*, Suppl. 1, Vol. 3, pp. 2127–2147. Pergamon.

Gehrels T. (1970) Photometry of Asteroids. In *Surfaces and Interiors of Planets and Satellites* (editor A. Dollfus), pp. 319–375. Academic Press.

Nelen J., Noonan A., and Fredriksson K. (1972) Lunar glasses, breccias, and chondrules. *Proc. Third Lunar Sci. Conf., Geochim. Cosmochim. Acta*, Suppl. 3, Vol. 1, pp. 723–737. MIT Press.

Pohn H. A. and Wildey R. L. (1970) A photoelectric-photographic study of the normal albedo of the Moon. *Geological Survey Professional Paper* **599-E**, US Govt. Printing Office.

Veverka J. and Noland M. (1973) Asteroid reflectivities from polarization curves: calibration of the "slope–albedo" relationship. *Icarus* **19**, 230–239.

Proceedings of the Fourth Lunar Science Conference
(Supplement 4, *Geochimica et Cosmochimica Acta*)
Vol. 3, pp. 3175–3180

Fluidization of lunar dust layers and effect on optical polarization of the diffuse reflectance of light

G. F. J. GARLICK, G. A. STEIGMANN, and W. E. LAMB

Department of Physics, University of Hull, England

J. E. GEAKE

Department of Physics, University of Manchester Institute of Science and Technology, Manchester,
England

Abstract—Previous studies indicated a large change in the reflectance of lunar dust layers when fluidized so that intergrain cohesion disappeared. Changes have now been measured in the degree of polarization of the diffusely reflected light on fluidization of the dust layer. The typical curves of polarization vs. phase angle are enhanced, except at small phase angles; the fractional changes often being greater than the corresponding increase in reflectance. The characteristics vary from sample to sample and show differences in dependence on the wavelength of the reflected light. In the Dollfus type plot of albedo vs. polarization, fluidized dust layers give a distribution of points which are displaced from the distribution for static dust layers and lunar surface measurements. Changes in polarization due to surface disturbance may provide a sensitive method of detecting and observing incipient dust storms on the Martian surface.

INTRODUCTION

IN A PREVIOUS paper Garlick *et al.* (1972) examined the diffuse reflection of light from lunar dust layers and showed that the reflectance of such a layer is enhanced, for phase angles greater than 20°, when an initially roughened surface is fluidized. This work has now been extended to cover the polarization characteristics of light reflected from roughened static layers of lunar dust and also fluidized layers of lunar dust. The degree of polarization (P) of light reflected from layers of lunar dust varies in a characteristic way with the phase angle V (the angle between the incident and reflected light) and is related to the "fairy castle" structure of the dust layer in its normal lunar regolith state. Previous work by Geake *et al.* (1970) has shown that excellent agreement exists between laboratory measurements, on roughened layers of lunar dust, and corresponding astronomical polarization measurements on various regions of the lunar surface.

POLARIZATION OF THE DUST LAYERS

The vibrator system used in our previous experiments (Garlick *et al.*, 1972) was mounted in a Lyot visual fringe polarimeter in order to measure the degree of polarization exhibited by the samples. In these measurements a fixed viewing direction making an angle of 50° to the normal to the sample surface was adopted, the phase angle being altered by varying the angle of incidence. Typical polariza-

tion curves, obtained using white light incident on roughened static layers of dust
and fluidized dust surfaces, for a range of lunar samples are given in Fig. 1.
Without fluidization the curves for the roughened layers are in agreement with
those previously obtained (Geake *et al.*, 1970; Dollfus *et al.*, 1971). However,
when the dust layers are subjected to vertical vibrations of frequency 208 Hz and
approximate amplitude 0.07 mm, complete fluidization occurs. This is accompanied by a marked increase in the degree of polarization (for white light) at phase
angles greater than around 25°. For phase angles less than about 22° (the *inversion*
angle on the polarization curve) little change is observed, although for some samples the depth of the negative branch in the polarization curve (occurring at
$V \sim 10°$) is slightly decreased.

The effect of fluidization of the dust layer is to cause a relative smoothing of
the surface. This has the following effects on the nature of the light reflected from
the dust layer: At large phase angles ($V \sim 100°$) the surface has characteristics
similar to those of a smooth static dust layer. The fluidized layer gives rise then to
increased reflectance and polarization (Garlick *et al.*, 1972). The increase in the
degree of polarization comes from the increased reflections from grains at the top
surface of the layer, whilst the depolarizing effect of multiple scattering between
grains deep in the layer is relatively insignificant. This condition gives a closer

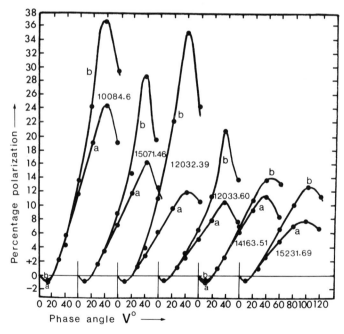

Fig. 1. Effect of fluidization of lunar dust samples on the degree of polarization of
reflected light from dust layers as a function of the phase angle *V* (viewing direction at
40° to the plane of the sample surface).
  (a) Curves for static, roughened layers.
  (b) Curves for fluidized layers.

approach to the polarization effects associated with the Brewster angle conditions of specular surfaces than is the case for a roughened layer of dust. At small phase angles ($V < 20°$) the negative branch in the polarization curve arises from multiple scattering between grains in contact (Bowell *et al.*, 1972). Fluidization of the dust separates the grains which are in contact producing, at these small phase angles, increased intergrain reflection deep in the layer. The effect of this multiple scattering is to slightly depolarize the light thus reducing the depth of the negative branch. This effect is, however, only minor compared to the increase in the positive region of the polarization curve at large phase angles when fluidization occurs.

VARIATION OF P_{max} WITH WAVELENGTH

The variation of the maximum polarization P_{max}, occurring at $V \sim 100°$, has been examined for two Apollo 15 samples as a function of wavelength. This has been carried out for the samples both in the rough and the fluidized states, the measurements being presented in Fig. 2. These show the expected decrease in P_{max} with increasing wavelength (Geake *et al.*, 1970); however, the variation for the static roughened layer yields a curve, whereas a linear relationship exists for

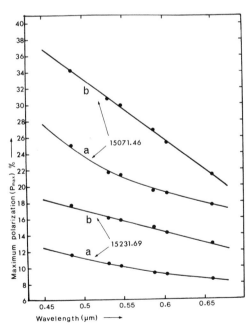

Fig. 2. Variation in peak value of polarization (P_{max}) with wavelength for two Apollo 15 samples.
(a) Static, roughened layer.
(b) Fluidized layer.
(Accuracy in P_{max} ±0.2%).

the fluidized dust layer. This means that the fractional increase in P_{max}, in going from the static roughened state to the fluidized state, is greatest at shorter wavelengths for sample 15071, 46 whereas for sample 15231, 69 the increase is almost independent of wavelength. Both these samples were collected in the vicinity of Hadley Rille. Sample 15071 is a relatively coarse-grained, dark gray material, collected from a small area 25 m east of the rim crest of Elbow Crater. This material was easily fluidized in the vibrator. Sample 15231 on the other hand is very much finer grained and more cohesive, requiring a much greater (~ 0.15 mm) amplitude of vibration to achieve satisfactory fluidization. This sample was collected from beneath a large boulder on the rim of a secondary impact crater and is much lighter in color than sample 15071.

THE RELATION BETWEEN POLARIZATION AND ALBEDO

The laboratory measurements of albedo may be directly compared with the astronomical albedo of solar system objects, at zero phase angle, by correcting for the increase in brightness, which occurs at small phase angles ($V < 5°$) in the laboratory measurements and which is due to the so-called *opposition effect* (Dollfus *et al.*, 1971).

Dollfus and Titulaer (1971) and Bowell *et al.* (1972) have shown that when the albedo A, and the maximum value of polarization P_{max}, are plotted logarithmically against each other, the points for lunar samples fall into a well defined domain. Figure 3 shows the polarization measurements and the albedo data presented in

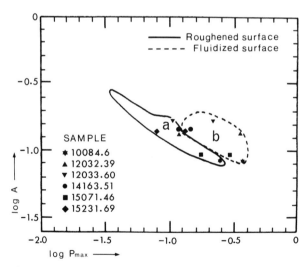

Fig. 3. Plot of log A vs. log P_{max} for lunar materials showing domains characteristic of layers of
(a) static, roughened dust.
(b) fluidized dust.
In the figure the domain (a) contains data for the lunar samples examined here together with data given by Dollfus and Titulaer (1971).

this form. When the lunar dust samples are fluidized the albedo (defined as the reflectance with respect to a MgO screen measured at a phase angle of 5° and corrected for the opposition effect) remains virtually unchanged whereas the value of P_{max} shows a large increase. This means that upon fluidization the sample points move parallel to the horizontal axis to form a new domain. This is shown in Fig. 3 as region (b), which is displaced from the domain (a) characteristic of the roughened dust.

Further recent work by Bowell *et al.* (1973) has revealed a relationship between the slope h of the polarization vs. phase angle curve, at the inversion angle, and the albedo A for various celestial bodies and lunar samples. This relationship is of the form

$$\log A = -c_1 \log h + c_2$$

where c_1 and c_2 are constants. Hence the polarization curve for a given celestial body can be used to determine both the albedo A and the value of P_{max}. For the fluidized samples of lunar material examined here the value of h is the same as for the rough static layer. This is in agreement with the fact that the albedo, measured at $V = 5°$, is unchanged when the sample layer is fluidized. In the case of celestial bodies if A and P_{max} are obtained from the polarization curve and referred to a log A vs. log P_{max} diagram, then the physical state of the dusty covering of the planetary surface might be determined.

CONCLUSION

It is clear that the observation of changes in the polarization of light reflected from lunar and planetary surfaces can provide a sensitive method of detecting the initial stages of dust disturbance or flow. In the special case of Mars, measurements of the polarization of reflected sunlight using an orbiting probe could be employed to detect the onset of dust storms or to observe minor dust flows and disturbances.

Arising from an interest in polarization changes in other cases of terrestrial surfaces and encouraged by the fluidization effects found in lunar dust samples, the authors have designed a polarization detector (Garlick *et al.*, 1973) which provides an image form display of the degree of polarization in a scene independent of light intensity variations across the scene. Any changes in the polarization of reflected light greater than 0.5% can be detected. For lunar or Martian surface observations the amount of disturbed dust which is capable of detection is of the order of 10 mg/cm^2. The minimum area detectable of such a disturbance will depend mainly on the resolution of the optics associated with the detector. Light levels for sunlit planetary surfaces present no problems. Such a device is also capable of detecting polarization variations for light scattered from planetary atmospheres and other structures such as the rings of Saturn.

Acknowledgment—We thank Dr. A. Dollfus of the Observatoire de Paris, Meudon, France, for useful discussion in the course of this work, and are grateful to NASA for the supply of lunar samples.

REFERENCES

Bowell E., Dollfus A., and Geake J. E. (1972) Polarimetric properties of the lunar surface and its interpretation. Part V: Apollo 14 and Luna 16 lunar samples. *Proc. Third Lunar Sci. Conf.*, *Geochim. Cosmochim. Acta*, Suppl. 3, Vol. 3, pp. 3103–3126. MIT Press.

Bowell E., Dollfus A., Geake J. E., and Zellner B. (1973) Optical polarization of lunar samples, with applications to the study of asteroids and planetary satellites (abstract). In *Lunar Science—IV*, pp. 85–87. The Lunar Science Institute, Houston.

Dollfus A. and Titulaer C. (1971) Polarimetric properties of the lunar surface and its interpretation. Part III: Volcanic samples in several wavelengths. *Astron. and Astrophys.* **12**, 199–209.

Dollfus A., Geake J. E., and Titulaer C. (1971) Polarimetric properties of the lunar surface and its interpretation. Part IV: Apollo 11 and Apollo 12 lunar samples. *Proc. Second Lunar Sci. Conf.*, *Geochim. Cosmochim. Acta*, Suppl. 2, Vol. 3, pp. 2285–2300. MIT Press.

Garlick G. F. J., Steigmann G. A., Lamb W. E., and Geake J. E. (1972) An explanation of transient lunar phenomena from studies of static and fluidized lunar dust layers. *Proc. Third Lunar Sci. Conf.*, *Geochim. Cosmochim. Acta*, Suppl. 3, Vol. 3, pp. 2681–2687. MIT Press.

Garlick G. F. J., Steigmann G. A., and Lamb W. E., (1973) Brit. Prov. Pat. 22586/73 May 1973.

Geake J. E., Dollfus A., Garlick G. F. J., Lamb W. E., Walker G., Steigmann G. A., and Titulaer C. (1970) Luminescence, electron paramagnetic resonance and optical properties of lunar material from Apollo 11. *Proc. Apollo 11 Lunar Sci. Conf.*, *Geochim. Cosmochim. Acta*, Suppl. 1, Vol. 3, pp. 2127–2147. Pergamon.

Proceedings of the Fourth Lunar Science Conference
(Supplement 4, *Geochimica et Cosmochimica Acta*)
Vol. 3, pp. 3181–3189

Luminescence of lunar, terrestrial, and synthesized plagioclase, caused by Mn^{2+} and Fe^{3+}

J. E. Geake, G. Walker, and D. J. Telfer

U.M.I.S.T., Manchester, England

A. A. Mills

University of Leicester, England

G. F. J. Garlick

University of Hull, England

Abstract—Luminescence emission spectra have been obtained for 24 different samples of lunar fines, rocks, and breccias, using proton or electron excitation. A representative selection is shown. Plagioclase is the main luminescent component, and its dominant emission is a Mn^{2+}-activated peak at about 5600 Å. Lunar samples differ from similar terrestrial materials mainly in the strength of their near-infrared emission; this is dominant for most terrestrial plagioclases, but is weak or absent for lunar materials. By doping pure synthesized plagioclase we now have experimental confirmation that this emission is probably due to activation by Fe^{3+}. The weakness of this emission from lunar samples may be due to oxygen scarcity at formation giving dominant Fe^{2+}, which quenches luminescence.

Introduction

We have now obtained luminescence emission spectra for 24 different lunar samples from Apollos 11, 12, 14, and 15, including 12 fines samples, 10 rocks and 2 breccias, both from Apollo 11 (Geake *et al.*, 1970, 1971, 1972a, 1972b). Initially we used 60 KeV protons for excitation, and investigated the effects of proton damage as well; more recently we have used 10 KeV electrons, which produce comparable emission with negligible damage, and give the same spectral profiles. Our proton excitation apparatus was described by Geake *et al.* (1972a); the apparatus we now use to record emission spectra and to measure luminescence decay times was described in our previous paper (Geake *et al.*, 1972b).

The aim of the present paper is to collect together a range of typical lunar sample emission spectra, and to give the results of our investigation into the causes of the characteristic emission peaks. This has involved doping terrestrial and synthesized samples with activators suggested by theoretical reasoning. The main advance since our previous paper has been our experimental confirmation that Fe^{3+} is the cause of the near-infrared emission peak, an explanation for which we were only able to give circumstantial evidence last year. This is of interest because the strength of this infrared emission is the main difference that we have found between lunar materials and similar terrestrial ones. The explanation of this emission therefore gives information about the different conditions under which these materials were formed.

LUMINESCENCE EMISSION SPECTRA

Figure 1 shows a representative selection of luminescence emission spectra for lunar samples. Figure 1(a) shows spectra for 4 samples of fines, from 4 different missions; they are very similar in profile, and differ only in efficiency, covering a range of about 10 (roughly from 10^{-4} to 10^{-5}). Their efficiencies are in the same order as their albedos, probably because the luminescence comes mostly from a white component, so both the overall efficiency and the albedo depend on the proportion of it present. Figures 1(b), (c), and (d) show spectra for some lunar rocks, and for one breccia (10023); those in (b) and (c) are typical, and none of the ones that we have not shown differ significantly from these. All the spectra shown in Fig. 1 contain the same features, at different relative intensities; the component mainly responsible for the luminescence is plagioclase [(Na, Ca)(Al, Si)$_4$O$_8$] as shown earlier (Geake *et al.*, 1970; Nash and Greer, 1970; Sippel and Spencer, 1970). This is illustrated by the spectrum for separated lunar plagioclase shown in Fig. 2(a), and is supported by the evidence shown in Fig. 3, where the luminescence emission photograph most resembles the electron microprobe image for aluminum. Most lunar plagioclase shows lumines-

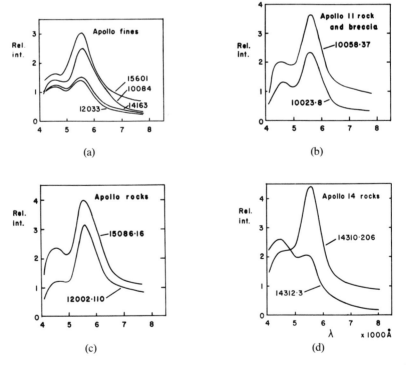

Fig. 1. Luminescence emission spectra for lunar materials, corrected for the spectral response of the instrument. The relative intensity scales for different samples are unrelated.

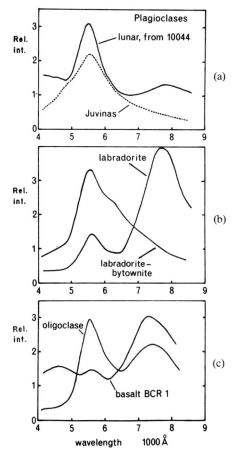

Fig. 2. Corrected luminescence emission spectra for (a) a plagioclase-rich meteorite (the eucrite Juvinas), and for separated plagioclase from a lunar rock (10044), separated by density from the dust produced when sawing a section; (b) and (c) terrestrial materials.

cence emission that appears bluish-white, which is what the eye makes of the spectra shown in Fig. 1. However, some grains which showed orange luminescence were found to be vitrified plagioclase; a microprobe scan for phosphorus was also carried out, on the supposition that orange or red luminescence might be due to apatite, but none was found. We find the luminescence efficiency of lunar plagioclase to be about 10^{-3}, and suppose that the actual efficiencies found for lunar rock and breccia chips depend mainly on their plagioclase content. For lunar fines samples it is probable that both the efficiency and the albedo are reduced by radiation damage and sputtering effects incurred while each grain was exposed on the lunar surface (Borg et al., 1971; Hapke et al., 1971). Lunar plagioclase is usually calcium-rich, i.e., toward the anorthite end of the series. Figure 2 also shows spectra for a plagioclase-rich meteorite (the eucrite Juvinas), and for some terrestrial plagioclases.

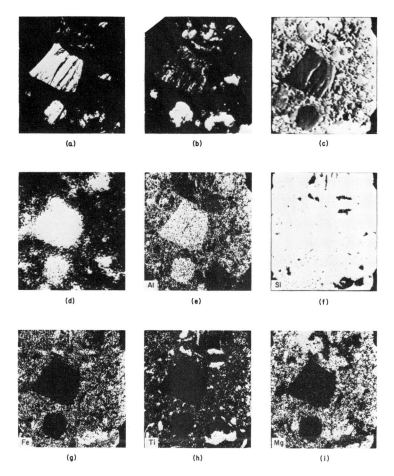

Fig. 3. 9 views of the same area, about 250 μm across, in a thin section of Apollo 11 breccia 10059.36, (a) with a visual microscope, in transmitted light, (b) the same, with crossed polarizers, (c) by back-scattered electrons, (d) luminescence emission with 6 KeV electron excitation, (e)–(i) electron microprobe images, for the elements stated. The luminescence image resembles the Al image and also comes in bright regions in the Si image, indicating plagioclase. The 'holes' in the Si image show white on the Fe and Ti images, indicating ilmenite. Some pyroxene can be seen from the Fe, Mg, and Si images.
A phosphorus image gave negative results, indicating the absence of apatite.

The characteristic luminescence emission spectrum of plagioclase contains three main peaks—a blue one at about 4500 Å, a green one at about 5600 Å, and sometimes one in the near infrared at between 7000 and 7800 Å. The blue peak is the one found for most silicates, and it is probably due to lattice defects rather than to any particular activator (Sippel and Spencer, 1970). The green peak, which is usually dominant for lunar samples, has been shown to be caused by Mn^{2+} as an activator (Geake *et al.*, 1971); we demonstrated this by doping some terrestrial plagioclase (labradorite), and in the present paper we give some further evidence obtained by doping pure synthesized anorthite. We have only found one lunar sample for which this green peak is not dominant, i.e., rock 14312 (Fig. 1(d)) for which the blue peak is the strongest, possibly because of shock-enhancement (Sippel and Spencer, 1970). The infrared peak is usually strong and often dominant for terrestrial plagioclases, as shown in Fig. 2, but it is weak or absent for lunar materials, and this constitutes a major difference between them. Some of the lunar samples in Fig. 1 show a weak flat "tail" of emission in the near infrared, but the only lunar sample to show a significant peak there is the separated plagioclase in Fig. 2(a). In contrast, nearly all terrestrial plagioclases show a strong infrared peak; the only exception we have found has been the labradorite-bytownite whose spectrum is shown in Fig. 2(b). We now find that this emission in the near infrared is probably caused by Fe^{3+} as an activator, and our experimental evidence for this is discussed in the next section.

We may therefore summarize the luminescence emission properties of lunar materials by saying that their spectra are dominated by the Mn-activated green peak of plagioclase; individual samples differ somewhat in their spectral profiles, depending mainly on the Mn^{2+} content of their plagioclase and on shock effects, and they show different efficiencies, depending on their plagioclase content and (in the case of fines) on their radiation history. It is interesting that Mn should be the dominant activator for lunar materials, as we earlier found it to be responsible for the most efficient luminescence from meteorite material, where it caused bright red luminescence from enstatite (Geake and Walker, 1966). Mn thus produces the most efficient luminescence from all the extra-terrestrial samples at present available to us. This may be partly because of the ease with which it substitutes for Ca and Mg, and also because it is widely available.

THE NATURE OF LUMINESCENCE CENTERS IN PLAGIOCLASE

In order to elucidate the causes of the observed luminescence emission-bands in plagioclases, a series of synthetic anorthites, labradorites, and albites have been produced from gels, using hydrothermal techniques (Hamilton and Henderson, 1968). These synthetic samples have been selectively doped during gelling with Mn^{2+} or Fe^{3+}, or both.

We have previously shown that Mn^{2+}, probably in Ca^{2+} sites, is responsible for the green emission peak which occurs in both lunar and natural terrestrial samples of plagioclase (Geake *et al.*, 1971). The nature of this luminescence center was predicted on theoretical grounds by a simple semi-quantitative application of crystal field theory, and vindicated by doping a natural terrestrial sample of labradorite with Mn^{2+}, which resulted in a twenty-fold increase in the intensity of the green emission peak. Synthetic anorthites have now been doped with Mn^{2+} and again a large green emission peak is obtained, as shown in Fig. 4.

Our recent work has been mainly concerned with the cause of the emission peak in the near I.R., in view of the interest in it as a difference between lunar and terrestrial materials. We have previously given some circumstantial evidence that this peak might be caused by Fe^{3+} present as an impurity in the plagioclase (Geake

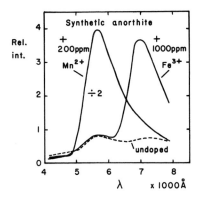

Fig. 4. Corrected luminescence emission spectra for synthetic anorthite, before doping,
and after adding either Mn^{2+} or Fe^{3+}.

et al., 1972b); this suggestion was supported by luminescence decay-time measurements and by crystal field theory considerations. A marginal enhancement of
the I.R. peak was indeed produced by doping terrestrial labradorite with Fe^{3+}, but
this was difficult to interpret as the material already showed a large I.R. peak
before doping. However, a sample of synthetic anorthite has now been produced
of sufficient purity to show a negligible I.R. peak; when this material is doped with
1000 ppm of Fe^{3+} a large peak at about 7000 Å is produced, as shown in Fig. 4.

In the synthesis of these materials a major difficulty has been contamination by
Cr^{3+}; this occurs by the diffusion of chromium into the sample from the metal of
the hydrothermal bomb during the firing process, and it gives rise to an emission
band in the near I.R. rather near to the Fe^{3+} band. Even so, no actual confusion
arises as the chromium band is easily recognized by its fine structure; also, the
presence of Fe^{3+} tends to quench it. However, it is better not to have it there, and
this contamination has now been largely prevented by using double-layer encap-
sulation with a silver foil barrier between two layers of gold, and by firing at a
slightly lower temperature ($\sim 800°C$ for anorthites). We have recently synthesized
both labradorites and albites as well as anorthites, with up to 1% Fe^{3+} added, and
in each case strong I.R. emission is induced.

In both natural and synthetic samples there is some variation in the I.R.
emission peak position from sample to sample; this variation is found to be at least
500 Å ($\sim 1000 \text{ cm}^{-1}$) for terrestrial samples, and there is a tendency for the emission from albites to be at longer wavelengths than that from anorthites. Figure 5
shows a plot of I.R. peak position (uncorrected for spectral response of the
instrument) against anorthite composition. The considerable variation in the position of the I.R. emission peak is in marked contrast to that of the Mn^{2+} emission
band which shows only slight variation in position (about 100 Å or 200 cm^{-1}). The
position of the Fe^{3+} emission peak should depend on its environment in the
crystal; if Fe^{3+} were in a similar lattice site to Mn^{2+} (i.e., the Ca^{2+} site) then a
positive correlation of spectral shifts of the emission bands of the isoelectronic
species Mn^{2+} and Fe^{3+} would be expected. (The converse of this statement is of

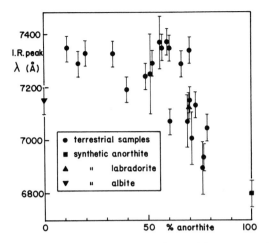

Fig. 5. The wavelength of the near-infrared emission peak in uncorrected spectra plotted against percentage anorthite, for a suite of terrestrial plagioclases of known composition.

course not necessarily true.) However, Fig. 6 shows that, for a large proportion of natural plagioclases examined, the I.R. emission band position shows no correlation with the Mn^{2+} emission peak position, the former varying whilst the latter remains constant. Whilst some positive correlation is present for a proportion of the samples investigated, such a correlation does not of course mean the Fe^{3+} and Mn^{2+} are necessarily in similar sites. There is much evidence in the literature to suggest that Fe^{3+} would tend to be in Al^{3+} sites. In agreement with Pott and McNicol (1970) we have found that Fe^{3+} produces an I.R. emission band in γ-Al_2O_3 at about 8000 Å. In a further paper Pott and McNicol (1971) have shown that Fe^{3+} produces an emission band in $LiAl_5O_8$ at about 6800 Å.

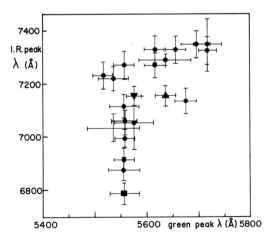

Fig. 6. The wavelength of the near-infrared emission peak plotted against that of the green peak, using uncorrected spectra, for the same samples as in Fig. 5. The same key applies.

Although the *average* size of a tetrahedral site in plagioclases is smaller for albite than for anorthite, the aluminum-rich tetrahedral sites are not appreciably different in size. Thus there does not appear to be a simple correlation between the size of the largest available tetrahedral site (Fe^{3+} is larger than Al^{3+}) and the I.R. peak position. Investigation of this problem continues, and further evidence as to the environment of the Fe^{3+} ion is to be sought by the measurement of excitation spectra.

It is still not certain why it is that lunar plagioclase does not usually show an Fe^{3+} emission band—the only lunar sample that we have found to give a significant peak in the near I.R. is the separated lunar plagioclase as shown in Fig. 2(a). Fe^{3+} has, however, been detected in lunar plagioclase by using E.S.R. techniques (Weeks, 1972), although the precise concentration does not appear to be known. We have found that the Fe^{3+} emission band for synthetic plagioclases is just detectable when only 100 ppm of Fe^{3+} have been added; preliminary results show that the intensity of this band increases with the amount of Fe^{3+} added, up to about 1 percent, beyond which there is no further intensity increase, probably because self-quenching becomes important. Furthermore, the proportion of the added Fe^{3+} that actually enters the appropriate lattice sites to produce luminescence centers will depend on the conditions during preparation, in a way that is not yet understood. Probably the most important factor is the $Fe^{2+}:Fe^{3+}$ ratio, since Fe^{2+} is likely to quench Fe^{3+} emission; we have no means of knowing what this ratio is for lunar plagioclase, although one might expect less Fe^{3+} and more Fe^{2+} than in terrestrial samples, in view of the scarcity of free oxygen when the lunar surface materials were formed (Geake *et al.*, 1972b). It is certainly difficult to prepare synthetic samples that are doped with Fe^{3+} without adding some Fe^{2+} ions as well. The strength of the Fe^{3+} emission peak may therefore give an indication of the amount of free oxygen available at formation.

Acknowledgments—We are grateful to Prof. J. Zussman and Prof. W. S. MacKenzie of the Geology Department, University of Manchester, for lending us the sample of separated lunar plagioclase and the samples of terrestrial plagioclases, respectively, and for permission to use their facilities for hydrothermal synthesis; to Dr. D. Hamilton and Mr. G. Norris, of the same department for their advice and assistance; to Mr. P. Suddaby of the Geology Department, Imperial College, University of London, for the electron microprobe analysis; to Mr. M. L. Gould and Mr. F. Kirkman for technical and photographic assistance; to the Science Research Council for financial support, and to NASA, and especially to all the Apollo astronauts, for providing the lunar samples.

References

Borg J., Maurette M., Durrieu L., and Jouret C. (1971) Ultramicroscopic features in micron-sized lunar dust grains and cosmophysics. *Proc. Second Lunar Sci. Conf., Geochim. Cosmochim. Acta,* Suppl. 2, Vol. 3, pp. 2027–2040. MIT Press.

Geake J. E. and Walker G. (1966) The luminescence spectra of meteorites. *Geochim. Cosmochim. Acta* **30**, 929–937.

Geake J. E., Dollfus A., Garlick G. F. J., Lamb W., Walker G., Steigmann G. A., and Titulaer C. (1970) Luminescence, electron paramagnetic resonance and optical properties of lunar material from Apollo 11. *Proc. Apollo 11 Lunar Sci. Conf., Geochim. Cosmochim. Acta,* Suppl. 1, Vol. 3, pp. 2127–2147. Pergamon.

Geake J. E., Walker G., Mills A. A., and Garlick G. F. J. (1971) Luminescence of Apollo lunar samples. *Proc. Second Lunar Sci. Conf., Geochim. Cosmochim. Acta*, Suppl. 2, Vol. 3, pp. 2265–2275. MIT Press.

Geake J. E., Walker G., and Mills A. A. (1972a) Luminescence excitation by protons and electrons, applied to Apollo lunar samples. *Proc. IAU Symp.* **47**, *The Moon* (editors Urey and Runcorn). Reidel.

Geake J. E., Walker G., Mills A. A., and Garlick, G. F. J. (1972b) Luminescence of lunar material excited by electrons. *Proc. Third Lunar Sci. Conf., Geochim. Cosmochim. Acta*, Suppl. 3, Vol. 3, pp. 2972–2979. MIT Press.

Hamilton D. and Henderson C. M. B. (1968) Preparation of silicate compounds by a gelling method. *Min. Mag.* **36**, (382), 832–838.

Hapke B. W., Cassidy W. A., and Wells E. N. (1971) Albedo of the Moon: evidence for vapor-phase deposition processes on the lunar surface. Presented at the Second Lunar Science Conference, Houston.

Nash D. B. and Greer R. T. (1970) Luminescence properties of Apollo 11 lunar samples and implications for solar-excited lunar luminescence. *Proc. Apollo 11 Lunar Sci. Conf., Geochim. Cosmochim. Acta*, Suppl. 1, Vol. 3, pp. 2341–2350. Pergamon.

Pott G. T. and McNicol B. D. (1970). The phosphorescence of Fe^{3+} ions in γ-alumina. *Chem. Phys. Lett.* **6**, 623–625.

Pott G. T. and McNicol B. D. (1971) The phosphorescence of Fe^{3+} ions in oxide host lattices. Zero-phonon transitions in $Fe^{3+}/LiAl_5O_8$. *Chem. Phys. Letts.* **12**, 63–64.

Sippel R. F. and Spencer A. B. (1970) Luminescence petrography and properties of lunar crystalline rocks and breccias. *Proc. Apollo 11 Lunar Sci. Conf., Geochim. Cosmochim. Acta*, Suppl. 1, Vol. 3, pp. 2413–2426. Pergamon.

Weeks R. A. (1972) Magnetic phases in lunar material. *Proc. Third Lunar Sci. Conf., Geochim. Cosmochim. Acta*, Suppl. 3, Vol. 3, pp. 2503–2517. MIT Press.

Proceedings of the Fourth Lunar Science Conference
(Supplement 4, *Geochimica et Cosmochimica Acta*)
Vol. 3, pp. 3191–3196

Infrared spectra of Apollo 16 fines

John W. Salisbury, Graham R. Hunt, and Lloyd M. Logan

Terrestrial Sciences Laboratory, Air Force Cambridge Research Laboratories, L. G. Hanscom Field, Bedford, Mass. 01730

Abstract—Mid-infrared spectra of Apollo soil samples are interpreted in terms of their mineralogy, and hence the average mineralogy of local parent material. Soil samples apparently derived from Cayley material vary in mineralogical composition from anorthositic to ultramafic, in part because of sampling bias. Two soil samples from Stone Mountain that may be derived from Descartes material are not strikingly different from typical Cayley material. A comparison of Apollo 11 and Apollo 16 soil spectra indicates that cross-contamination effects have not disguised compositional differences.

Introduction

Early balloon-borne observations of lunar mid-infrared emission spectra suggested that this spectral range could be used to map compositional differences on the lunar surface (Salisbury *et al.*, 1970). Subsequent laboratory studies of returned lunar soil samples confirmed the validity of the remote observations (Logan *et al.*, 1972).

Unlike visible and near-infrared spectral features which are caused by electronic transitions in constituent or impurity transition elements, the mid-infrared features near 10μm are caused by molecular vibration phenomena involving the fundamental structure of silicate materials. Thus, rather than identifying the presence of a specific element such as iron or titanium, from which can be inferred the presence of particular mafic minerals or titanium-rich glasses in the soil, mid-infrared spectral features provide information concerning average soil mineralogy. In addition, although spectral information in the visible and near-infrared may be dominated by the presence of dark glasses in fines and breccias, the vitrification process has relatively little effect on the information content of mid-infrared spectra. This is illustrated in Fig. 3, in which a spectrum of Apollo 11 soil (severely darkened by titanium-rich glass), can be compared with spectra of the much lighter Apollo 16 soils. The darkness of the Apollo 11 soil has resulted in a reduction of spectral contrast. Vitrification has, however, no effect on the location of the emissivity peak.

Information is available from lunar surface spectra in the form of two parameters: (1) the position, and (2) the spectral contrast of the mid-infrared emissivity maxima (Logan *et al.*, 1972). The effect of varying the environmental conditions on these two parameters has been discussed elsewhere (Logan *et al.*, 1973). For a given environment, such as the lunar surface, it may be assumed on the basis of other abundant surface morphological evidence, that differences in these parameters between two sites (or surface samples) are caused only by differences in composition. Soil composition, in turn, may be affected to some extent by regional cross-contamination (Adams and McCord, 1972), but is primarily controlled by the composition of nearby parent material.

This paper explores the degree of compositional heterogeneity in the Apollo 16 landing site area represented by spectra of different soil samples. We evaluate, from mid-I.R. spectral evidence, the significance of cross-contamination effects on these spectra.

EXPERIMENTAL BACKGROUND

Infrared emission spectra of lunar soil samples and terrestrial standards were obtained in a simulated lunar environment in a manner previously described (Logan *et al.*, 1972). The two parameters chosen to characterize composition, the position, and the contrast of the mid-I.R. emissivity maximum, are plotted against each other in Fig. 1. Each symbol represents the spectral behavior of an individual soil or rock sample, and those for rocks belonging to any one of the four major igneous rock divisions are connected to the group centric. The names of the rock types assigned to each division under our classification scheme are listed in Fig. 1.

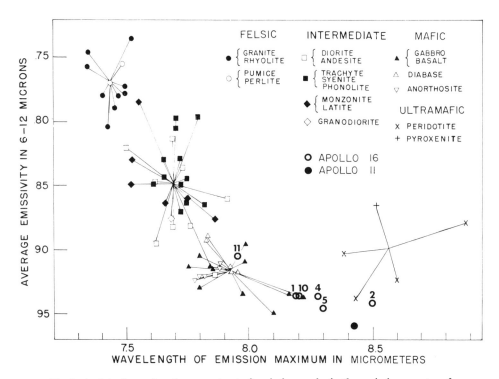

Fig. 1. A plot of wavelength vs. contrast of emission peaks in the emission spectra of different terrestrial rocks and the lunar soil samples discussed in this paper. The terrestrial standards have been ground to a particle size less than 74 microns. (Modified from Logan *et al.*, 1972.)

The accuracy with which the wavelength of the emissivity maximum can be measured is determined by the contrast in the spectrum and the noise level on the data. For a spectrum recorded with the experimental parameters used here, an accuracy of ~ 0.02 microns is easily achievable. Thus, the variability within each of the four rock divisions is a function more of mineralogical variability of different samples of the same rock type than of measurement error. In fact, useful mineralogical generalizations can be made about members of each of the four groups, based upon the distribution of points within the group. In the intermediate rock division, for example, those rocks which display both a lower average emissivity and a shorter wavelength peak generally tend to have a higher quartz and potash feldspar content than do those with a higher emissivity and a longer wavelength peak. Likewise, in the mafic rock division, those towards the left in Fig. 1 tend to have more sodic plagioclase and significant quartz, while those to the right tend to contain calcic plagioclase and more olivine and/or pyroxene.

APOLLO 16 RESULTS

Figure 2 shows the source locales of our soil samples with respect to major landmarks and astronaut traverse lines at the Apollo 16 landing site.

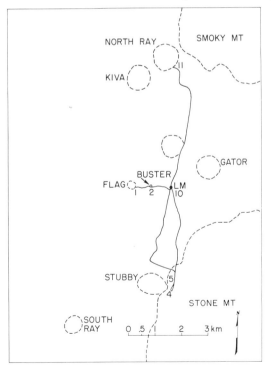

Fig. 2. Sketch map of the Apollo 16 landing site showing landmarks (dashed lines), astronaut traverses (solid lines) and station numbers where samples studied in this paper were obtained.

The sample collected at Station 11 on the rim of North Ray Crater is probably not typical of surrounding soil, being derived at least in part from rake disaggregation of an underlying white friable boulder (Muehlberger *et al.*, 1972a). On inspection it appears to be composed almost entirely of calcic plagioclase, which is consistent with its spectrum in Fig. 3. No doubt a good deal of anorthositic breccia is exposed at the rim of North Ray Crater, but it is doubtful that the mineralogy of this sample reflects the average rock composition in the vicinity of its collection. Thus, little can be concluded from this soil spectrum beyond the well-known fact that the Cayley Formation contains anorthositic material.

From the astronauts' descriptions, photographs, and geomorphological evidence it appears that the soil samples taken from Stations 1, 2, and 10 are mature,

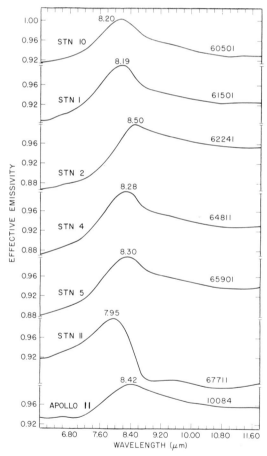

Fig. 3. Emission spectra of Apollo soil samples. The wavelength of each emission peak, sample numbers and sample sources are indicated for each spectrum. Effective emissivity records the ratio between energy emitted by the sample and a black body, the temperature of which is chosen to match the maximum (monochromatic) brightness temperature of the sample in the spectral range covered.

typical of surrounding soils and are derived primarily from the underlying Cayley Formation (Muehlberger *et al.*, 1972a). The spectra of soils collected from Stations 1 and 10 indicate similar mineralogy; a mineralogy that is consistent with that indicated by chemical analyses of other mature soils derived from the Cayley Formation (LSPET, 1973). In contrast, the soil from Station 2 on the rim of Buster Crater, which lies less than a kilometer away from Stations 1 and 10, displays a spectrum which is decidedly different. It has, in fact, an emission maximum characteristic of terrestrial ultramafic rocks. Thus, if this sample is representative of soil around Buster, this crater may be one source of the rare ultramafic rocks found at the Apollo 16 site (LSPET, 1973). Interestingly, Buster is the only crater the Apollo crew visited that appears to be flat-bottomed, with coarse blocks covering its floor and part of one wall. It is apparently a relatively recent crater that penetrated the blocky substrate which appears to underlie the regolith at a depth on the order of 10–15 m (Muehlberger *et al.*, 1972b). The source area of the ultramafic parent material of Buster soil is uncertain, but it may be in this blocky layer, or at greater depth.

Our only clues to the composition of the Descartes Formation lie in soil samples from Stations 4 and 5, the spectra of which are identical within measurement error. These samples were selected so as to minimize contamination with South Ray Crater ejecta, the sample from Station 5 being taken at the bottom of a shallow rake trench. The spectral peaks in emission from these samples lie within approximately 0.1μm of those found for the soils collected at Stations 1 and 10. Thus, with the exception of Station 2, there seems to be no striking difference between the Cayley and Descartes soils, which is in agreement with available chemical analyses of these soils by the LSPET (1973). Unfortunately, the Station 2 soil has not yet been analyzed. We suggest that such analysis together with a closer inspection of the associated Station 2 rocks should prove very interesting.

Included in Fig. 3 is a spectrum of Apollo 11 soil from Mare Tranquillitatis, one of the maria closest to the Apollo 16 landing site. It is clear from a comparison of the Apollo 16 and Apollo 11 spectra that regional cross-contamination has not disguised compositional differences. On a shorter time and distance scale, local cross-contamination in the Apollo 16 landing area has also failed to disguise differences between individual Apollo 16 soil samples.

References

Adams J. B. and McCord T. B. (1972) Optical evidence for regional cross-contamination of highlands and mare soils (abstract). In *Lunar Science—III*, pp. 1–3. The Lunar Science Institute, Houston.

Logan L. M., Hunt G. R., Balsamo S. R., and Salisbury J. W. (1972) Mid-infrared emission spectra of Apollo 14 and 15 soils and remote compositional mapping of the moon. *Proc. Third Lunar Sci. Conf.*, *Geochim. Cosmochim. Acta*, Suppl 3, Vol. 3, pp. 3069–3076. MIT Press.

Logan L. M., Hunt G. R., Salisbury J. W., and Balsamo S. R. (1973) Compositional implications of Christiansen frequency maxima for infrared remote sensing applications. *J. Geophys. Res.* In press.

LSPET (Lunar Sample Preliminary Examination Team) (1973) The Apollo 16 lunar Petrographic Petrographic and chemical description. *Science* **179**, 23–24.

Muehlberger W. R., Bailey N. G., Batson R. M., Freeman V. L., Hait M. H., Hodges C. A., Jackson E. D., Larson K. B., Reed V. S., Schaber G. G., Stuart-Alexander D. E., Sutton R. L., Swann G. A., Tyner R. L., Ulrich G. E., Wilshire H. G., and Wolfe E. W. (1972a) Documentation and environment of the Apollo 16 samples: A preliminary report. Interagency Report: *Astrogeology* **51**, prepared under NASA Contract No. T-5874A.

Muehlberger W. R., Bailey N. G., Batson R. M., Boudette E. L., Duke C. M., Eggleton R. E., Elston D. P., England A. W., Freeman V. L., Hait M. H., Gall T. A., Head J. W., Hodges C. A., Holt H. E., Jackson E. D., Jordan J. A., Larson K. B., Milton D. J., Reed V. S., Rennilson J. J., Schaber G. G., Schafer J. P., Silver L. T., Stuart-Alexander D., Sutton R. L., Swann G. A., Tyner R. L., Ulrich G. E., Wilshire H. G., Wolfe E. W., and Young J. W. (1972b) Preliminary geological investigation of the Apollo 16 landing site. Interagency Report: *Astrogeology* **54**, to be published as Chapter Six in NASA Special Paper 315.

Salisbury J. W., Vincent R. K., Logan L. M., and Hunt G. R. (1972) Infrared emissivity of lunar surface features. *J. of Geophys. Res.* **75**, 2671–2682.

Proceedings of the Fourth Lunar Science Conference
(Supplement 4, *Geochimica et Cosmochimica Acta*)
Vol. 3, pp. 3197–3212

Some physical parameters of micrometeoroids

D. E. Brownlee

Department of Astronomy, University of Washington, Seattle, Washington 98195

Friedrich Hörz

Johnson Space Center, Houston, Texas 77058

J. F. Vedder and D. E. Gault

NASA-Ames Research Center, Moffett Field, Ca. 94035

J. B. Hartung

Max Planck Institut für Kernphysik, Heidelberg, Germany

Abstract—Detailed morphological parameters (depth/diameter ratio, circularity index) of microcraters in the 0.2–100 μm diameter range were obtained via SEM techniques for three lunar glass surfaces. The depth/diameter ratios are typically 0.5–0.8 with a range of 0.3–1.3. The circularity index varies from 0.4–1.0 with a pronounced maximum at 0.7–0.9. These parameters are compared with microcraters produced in the laboratory via electrostatic particle accelerators. The following conclusions are drawn:

The great majority of observed crater depths are compatible with micrometeoroid densities of 2–4 g/cm³; crater depths are incompatible for projectile densities < 1 g/cm³ and > 7 g/cm³. The circularity index of microcrater pits indicates rather equidimensional if not spherical projectiles. Needles, platelets and other highly irregular shapes can be excluded. Less than 5% of all craters observed may offer different conclusions. These results differ in part significantly from popular views about physical parameters of micrometeoroids.

Introduction

A variety of arguments have been presented in the past which substantiate the origin of microcraters on lunar surface materials as due to the impact of primary, interplanetary matter (Hörz *et al.*, 1971; Hartung *et al.*, 1972). Such primary craters can be defined as having a melted, glass-lined depression—termed "pit"—with or without a surrounding spall zone, depending on the absolute pit diameter. Statistical studies of microcrater populations have already yielded valuable information concerning the mass-frequency and the absolute flux of interplanetary particles (Hartung *et al.*, 1972; Morrison *et al.*, 1972; Schneider *et al.*, 1973; and others). In this study, however, we will attempt to demonstrate that lunar microcraters may also contribute to our understanding of some physical properties of interplanetary dust.

Lunar glass surfaces 15286, 67115 and a Luna 16 glass-spherule (G12-1c-4) were investigated via scanning electron microscope (SEM) techniques to obtain detailed morphological data concerning the central pit. Observational conditions are optimized on glass-surfaces as compared to crystalline and breccia materials. Furthermore experimental cratering data were predominantly produced on glass surfaces.

Such laboratory cratering experiments using electrostatic particle accelerators demonstrated that crater morphology may be indicative of—if not truly sensitive to—certain projectile parameters (Vedder, 1971; Mandeville and Vedder, 1971; Kerridge and Vedder, 1972; Bloch *et al.*, 1971; Neukum, 1971; Schneider, 1972). Though far from being complete, these experiments were performed over a wide range of projectile velocities (1–60 km/sec), projectile densities (1–7 g/cm^3), projectile masses (10^{-9}–10^{-15} g), and a variety of impact angles as well as target materials. It is important to note that these conditions resulted in essentially identical crater-sizes as those which will be described here from lunar materials, thus excluding crater scaling effects to the best of our present knowledge. Figure 1 illustrates the definition of the morphological parameters used in this study. For photomicrographs of typical microcraters see McKay *et al.* (1971); Bloch *et al.* (1972); Hartung *et al.* (1972); and others.

CRATER DEPTH

The depth of the pit bottom below the surrounding uncratered surface was measured from crater profiles produced via SEM techniques. For small craters ($<20\ \mu$m) profiles were produced by the contamination line technique (Vedder and Lem, 1972). Profiles of larger craters were derived from SEM stereo pairs using a modified version of the graphical technique outlined by Malhotra and

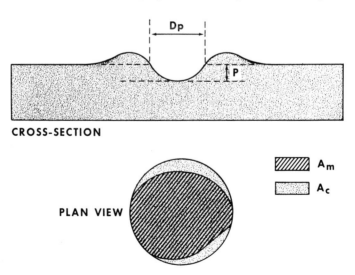

Fig. 1. Schematic cross section and plan view of a microcrater illustrating the definitions of the morphological parameters measured. A_m is the area bounded by the inside of the pit rim and A_c is the area of the smallest circle which just encloses A_m.

Brandon (1971). Differential tilt angles were nominally 25° resulting in measurement errors of 10% or less for the pit depth. Craters with pit rims altered by spallation were not analyzed because of the uncertainty in determing the mean pit diameter. Exclusion of these craters introduces no significant bias as such craters are very rare on the glasses studied.

The resulting depth/diameter ratios (P/D_p) for 68 craters are plotted in Fig. 2, There is no strong dependence of P/D_p on pit diameter although there appears to be a larger dispersion among the smaller craters. Because some of this increased scatter may be caused by increased measurement difficulties for small features, a relative frequency plot of P/D_p for craters larger than 4 μm is separately shown in Fig. 3. It is seen that for the micron sized craters there is a strong peaking of P/D_p around the value 0.7.

Preliminary results of laboratory experiments using spherical particles ranging in density from 1–8 g/cm³ and impacting soda-lime-glass targets at normal incidence is illustrated in Fig. 4 (Vedder and Mandeville, 1973). Impact velocities ranged up to 13 km/sec for aluminum and polystrene projectiles, however, only up to 7 km/sec for the iron particles. A clear relationship of projectile density and impact velocity is apparent. Accordingly the depth/diameter ratios illustrated in Fig. 2 and especially Fig. 3, seem to define limits on the density of micrometeoroids, provided their impact velocities are known.

According to a summary by Hartung *et al.* (1972), the mean impact velocity of micrometeoroids seems to be in the 20±5 km/sec range. Recent *in situ* velocity measurements of individual micrometeoroids by Pioneer 8/9 (Berg and Gerloff, 1971; Berg and Grün, 1973; Gerloff and Berg, 1973) indicate on the one hand that this mean velocity is somewhat higher, however, they also—in accordance with previous measurements— substantiated that a significant fraction of cosmic dust has orbits which produce impact velocities at the lunar surface on the order of

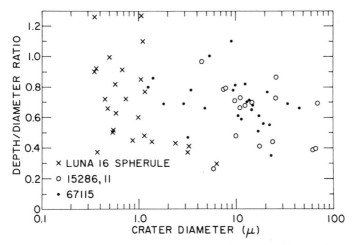

Fig. 2. Relations of depth and diameter of microcrater pits. Note an increased spread for the smaller structures as well as the similarity in scatter for two different glass-surfaces.

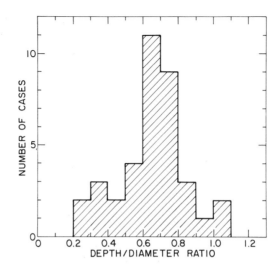

Fig. 3. Frequency of depth/diameter values for craters larger than 4 μm pit diameter.

5–20 km/sec. This low velocity region has been adequately simulated to date. It can be seen from Fig. 4 that if micrometeoroids had densities less than unity, the low velocity component of the micrometeoroid complex should produce many craters with $P/D_p < 0.25$. Contrary to this, all observed ratios of P/D_p are greater than 0.25. As demonstrated in addition (below), approximately half of the shallow

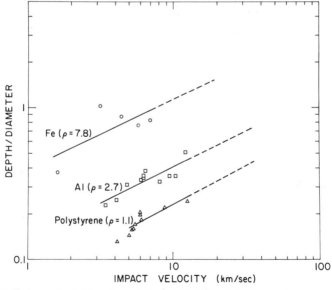

Fig. 4. Preliminary depth/diameter values for experimental microcraters on soda-lime glass using a variety of projectile materials and impact velocities. Significant, density dependent, differences for velocities up to 13 km/sec are observed, though at present the extrapolation (dashed line) to very high impact velocities is tentative.

craters ($P/D_p \approx 0.3$–0.5) are due to oblique impact, rather than low projectile density. As a consequence the lack of shallow craters presents strong evidence that projectiles of densities below $1 \, g/cm^3$ were nonexistent or extremely rare during the times the three lunar surfaces investigated were exposed. This is a significant finding and stands in strong contrast to most meteoroid deceleration data (Verniani, 1969 and others) which suggest mean densities of 0.2–$0.5 \, g/cm^3$ as well as theoretical considerations concerning the micro-meteoroid complex, e.g., Donn (1964), Baldwin and Shaeffer (1971).

On the other hand the rarity of truly deep craters (P/D_p above 1.0) presents evidence against particle densities as high as iron. Clearly the simulation data for iron projectiles go only up to $7 \, km/sec$ (Fig. 4) and thus an extrapolation to 20–$30 \, km/sec$ may not be readily justified. Nevertheless the measured P/D_p values certainly suggest that very high density objects (e.g., iron) cannot be typical and are probably rare in the interplanetary medium. This interpretation is not only supported by the lack of metallic projectile residue upon visual inspection of the pit glass liner (Hörz et al., 1971) but also by our recent attempts to search for projectile remnants via energy dispersive X-ray analysis, thus confirming the findings of Bloch et al. (1971). Because laboratory craters by Vedder (1971) and Dietzel et al. (1972) yielded substantial projectile residue in the crater interiors, using iron projectiles up to 13 km/sec, we conclude that an insignificant portion of the micrometeoroid complex consists of iron particles.

According to Figs. 3 and 4 our measurements are consistent with mean densities of 2–$4 \, g/cm^3$ and mean impact velocities of $20 \pm 5 \, km/sec$.

CRATER CIRCULARITY

In the laboratory non-circular craters are produced by either irregular projectiles or by oblique impact. In the first case the resulting crater shape resembles the irregular outlines of the projectile (Kerridge and Vedder, 1972) basically because for the crater sizes under consideration a linear relation of pit diameter (D_p) and micrometeoroid diameter (D_m) exists. Typical values for 20 km/sec impact velocities are $D_p/D_m = 2$ (Mandeville and Vedder, 1971; Bloch et al., 1971). Thus any gross-irregularity of the projectile will be reflected by crater shape. For oblique impact and spherical projectile, an elongation and ovoid form of the crater is observed which possesses bilateral symmetry along a plane defined by the impact trajectory (Vedder, 1971; Mandeville and Vedder, 1971; Schneider, 1972; Gault, 1973). Both effects combined, i.e., an irregular projectile shape and oblique impact, would most likely result in even more asymmetric craters.

The circularity index (CI) measured on lunar craters was defined as A_m/A_c (see Fig. 1). A_m is defined as the area actually measured along the intersection of a hypothetical plane with the inside of the pit rim. A_c is the area of the smallest circle which just encloses the boundary of A_m.

The circularity indices for 131 pits ranging in size from $0.2 \, \mu m$ to $100 \, \mu m$ are plotted in Fig. 5. Though not illustrated, there is no dependence of circularity with

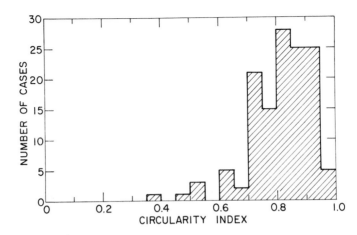

Fig. 5. Histogram of the circularity index of 131 microcraters ranging in size from 0.2 to 80 μm in diameter.

size. However, it is obvious from Fig. 5 that more than 90% of micrometeoroid craters have *CI*'s >0.7. Extreme anisometries were not observed at all.

Both small scale and large scale cratering experiments (Mandeville and Vedder, 1971; Gault, 1973) provide a basis to qualitatively discriminate between the effects of projectile irregularity and inclined impact angle. Craters produced with spheres at oblique projectile trajectories are noncircular and systematically associated with shallower crater depth, because the energy transfer into the target is less efficient as compared to vertical incidence.

These systematics were qualitatively observed for many irregular, shallow craters. They are also supported by quantitative measurements of the random selection of craters shown in Fig. 6, where *CI* is plotted vs. P/D_p. The statistics of the noncircular craters ($CI < 0.7$) are not good, however it is seen that the two most noncircular craters are shallow and that as a group noncircular craters are shallower than the more circular ones. In addition the deepest craters are very circular. This is taken as evidence that many of the noncircular craters are due to oblique impact rather than irregular projectile shape. This relation also strengthens the above arguments concerning projectile density, because many, although not all, shallow craters seem to be the result of oblique impact rather than irregular projectile shape or low density. The deviation of most of the craters from perfect circularity is in good agreement with Gault (1973), who considered the statistical probability of oblique impact assuming an isotropic particle flux.

The absence of highly noncircular craters and the possibility that noncircular craters can be produced by oblique impact leads us to the conclusion that the majority of micrometeoroids have equidimensional shapes. Highly nonspherical shapes such as rods and platelets (Donn, 1964) can conclusively be discarded for the crater population investigated. If dust grains were modeled as ellipsoids then the average ratio of major to minor axis would have to be less than 2.0.

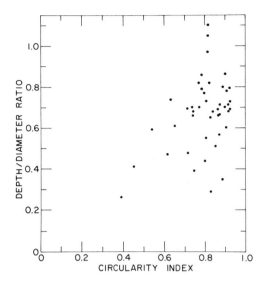

Fig. 6. Comparison of the depth/diameter values and the circularity index. Note that the more asymmetric craters are systematically associated with shallower craters and conversely the deep structures are more circular in outline.

FROTHY RIM

Approximately half of the pits between 1 and 10 μm had vesicular rim structures, indicative of the degassing or surface tensional features of a melt (Fig. 7). This "frothyness" is variable and ranges from just a few bubbles near the outer rim boundary to occasionally remarkable porous structures encompassing the entire rim. Equivalent structures have not been produced in the laboratory to date.

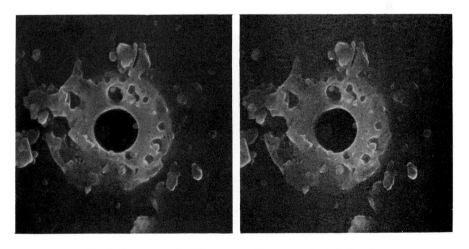

Fig. 7. Stereo pair of a 1.5 μ crater with an excessively frothy rim.

The frothyness may be caused by a variety of independent processes or any combination thereof:

(a) Strictly surface tensional features, i.e., the shock produced melt "film" possessing the low viscosity typical of lunar materials was simply thinned out sufficiently to break up. No degassing is required at all for this case.

(b) Strong degassing of melt because of sufficient volatiles trapped in the glass target (Gibson and Moore, 1972; Bogard and Nyquist, 1972).

(c) Strong degassing with the necessary volatiles exclusively derived from the projectile.

(d) Impact velocity regimes presently inaccessible in laboratory studies.

At present, no definite conclusions can be reached. However, possibilities (a) and (b) may perhaps be ruled out because of the existence of craters only a few microns apart with the same pit diameter that are otherwise morphologically similar except for the existence of frothy rims. Accordingly, "frothyness" seems not to be caused by target properties, but must be an effect of projectile characteristics. The present impact-simulations range up to velocities of 60 km/sec, though admittedly not on lunar glasses. In no case, however, was "frothyness" observed. This leads us to favor the intriguing possibility that a substantial portion of micrometeoroids is rich in volatiles. The greatest source of volatiles would be represented by hydrated minerals, a suggestion which is consistent with the formation of lunar goethite (Fe O(OH)) (Agrell *et al.*, 1972; El Goresy, 1973) due to extra lunar H_2O, though the above authors have indicated that H_2O may also be of endegeneous origin.

UNUSUAL CRATERS

The preceding results, discussions and conclusions including those below are pertinent to 95% of all micrometeoroid craters observed. However, there are structures which offer and indicate different interpretations. They are briefly described in this section. No statistical frequency data are given, because of the small number of cases observed, but collectively the "unusual" craters make up no more than 5% of a given microcrater population.

Multiple depression craters

Figure 8a, b and c illustrate examples of craters which do not have a single central, glass-lined pit, but instead the pit-area consists of a variety of parasitary depressions. The absence of overlapping flow structures and the continuity of an incipient spall zone common to the entire structure argues against individual, overlapping impact events.

In our laboratory experiments similar craters have been produced by aggregates of bonded spheres. Accordingly we tentatively ascribe these features to projectiles of aggregate structure, i.e., of heterogeneous mass distribution. However, in some cases one can also envision that they are caused by grains adhering to the target surface and which get pushed "into" the developing pit. (see Vedder, 1972).

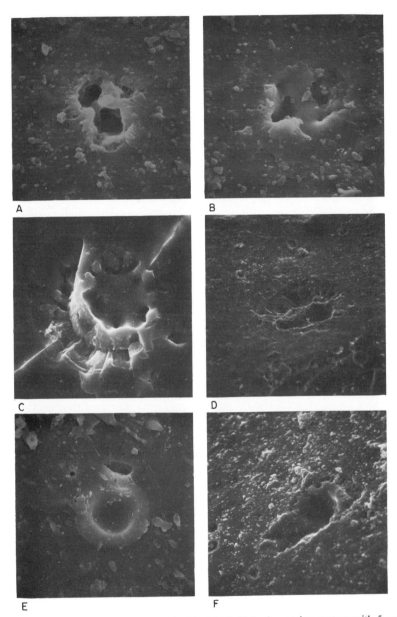

Fig. 8. Unusual craters (see text): (a), (b), (c) Multiple depression craters with 5 μm, 7 μm and 25 μm mean dimensions respectively. (d) Irregular crater, 32 μm mean pit dimension. (e), (f) craters with unusually small spall zones, 7 and 25 μm mean pit diameter respectively.

Highly irregular craters

Figure 8d illustrates a highly irregular crater. Because it is identical in all major aspects to those produced in the laboratory by highly irregular projectiles, it is concluded that the micrometeoroid responsible for this structure was indeed of very irregular shape.

Undersized spall zone

Existence, type and degree of fracturing and spallation around a central pit is primarily determined by pit size for lunar craters (Hartung *et al.*, 1972b). A threshold impact velocity typically above 4 km/sec (Vedder, 1971; Mandeville and Vedder, 1971) reproduces, in the laboratory, the spall features observed on lunar craters. Typical spall diameter to pit diameter (D_s/D_p) ratios of 2 are observed on lunar glasses for pits 5–50 μm in diameter.

Figure 8e and f illustrates craters which have unusually small spall zones. At present the interpretation is subject to discussion because such effects are not simulated adequately and could be caused by low velocity impacts, low density projectiles, target irregularities or any combination thereof.

SIZE-FREQUENCY OF CRATERS

Detailed crater statistics were obtained by optical technique on the entire glass-coating of 15286 and random portions of a 7 mm square chip (15286,11) cut out of 15286 and studied in the SEM. The surface quality of this glass was excellent and considerable caution was exercised not to overlook shallow or irregular craters.

The crater frequency on the larger (optical) area matched that on the smaller area in slope but not in absolute crater density. This is an artifact of slightly different exposure conditions, probably solid angle. In Fig. 9 this effect was compensated for by multiplying the optical points by 0.5.

The results are plotted in Fig. 9 together with data generated by Schneider *et al.* (1973) for glass coating 15205. These are the best documented surfaces to date because of good statistical data and a relatively large range of pit diameters covered.

In general the two curves agree very well over the 10 orders of magnitude of meteoroid mass covered. However, there are subtle differences, most notably an apparent depletion of pits having 1–20μm diameter on rock 15205 which is not as readily seen in the 15286 data. This depletion is of considerable interest because it implies bimodal size distribution and potentially two different source areas for the corresponding micrometeoroid mass regimes. Because curve 15205 is based on approximately 1800 craters, the depletion observed is beyond statistical error and a reality. At present we cannot conclusively demonstrate that the subtle difference of 15286 is either an artifact of the counting process or evidence for a time variation of the micrometeoroid complex. However, the following conclusions may be drawn:

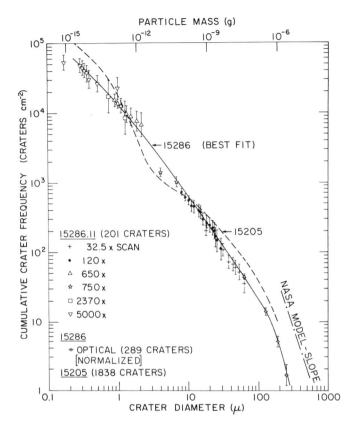

Fig. 9. Size frequency distribution of microcraters measured on 15286, compared with 15205. Error bars indicate uncertainty due to counting statistics only. The mass scale was computed assuming a ratio of $D_p/D_m = 2$ and spherical particles of density 2.5 g cm^{-3}.

(a) If the celebrated radiation pressure cut off for 10^{-12} g micrometeoroid masses were real, we would observe a flattening of the crater size distribution approximately at 1 μm pit diameter. The established lack of such a cut off can be interpreted that radiation pressure never exceeds gravity for interplanetary particles as predicted for silicates of densities > 1 g/cm^3 by Gindilis *et al.* (1969). The concept of a minimum projectile mass (= cut off at 10^{-12} g) is not consistent with our results (see Dohnanyi, 1972). This does not, however, necessarily imply that all craters are produced by particles in bound orbits. It may well be that many craters are produced by particles on escape orbits generated at cometary perihelion (Harwit, 1963).

(b) The lack of slope increase in the 0.5–1 μm region seems to rule out the possibility of 0.1 μm diameter particles being blown out of the solar photosphere in large numbers during the exposure time of rock 15286 as well as 15205. If the radiation pressure force exceeded gravity by only 10% then these particles would impact the moon at velocities in excess of 100 km/sec. Impact of a 0.1 μm parti-

D. E. Brownlee *et al.*

cles at such velocities would produce craters approximately 0.8 μm in diameter (Schneider, 1972). Such craters were not observed to be overabundant and we therefore conclude that the role of radiation pressure forces may have been significantly overestimated in the past or that refractory particulates are not formed in the solar photosphere.

(c) The meteoroid population index undergoes a nearly discontinuous change at 5×10^{-7} g. This observed change in the cumulative crater frequency allows micrometeorite data to be accurately matched to that for larger bodies (Anon., 1969).

DISCUSSION

There is evidence based on the analysis of orbital parameters that like the somewhat larger meteoroids which produce meteors, the micrometeoroids discussed here also have a cometary source (Dohnanyi, 1972 and many others). Recent additional arguments are derived from in flight orbit determinations by Pioneer 8/9 (Berg and Grün, 1973; Gerloff and Berg, 1973) as well as the apparent flux decrease with solar distance measured by Pioneer 10. Furthermore, based on chemical analysis of lunar soil samples, Anders *et al.* (1973) have shown that the bulk of micrometeorites that impact the moon have unfractionated solar abundances similar to Type 1 carbonaceous (Cl) chondrites. Thus a cometary origin of such particles appears realistic, though not entirely unambiguous.

The mass densities indicated by our crater analysis are entirely consistent with Cl chondrites which if fragmented would produce particles with effective densities from 1.5 g/cm³ for aggregates of phyllosilicate grains to 5 g/cm³ for the magnetite spheres which are very abundant in Cl's (Jedwab, 1971). The abundance of frothy rimmed craters is consistent with impacts of hydrated phyllosilicates. The apparent lack of particles as dense as iron is evidence that the source material of interplanetary dust is not the same as that of ordinary chondritic meteorites.

If a cometary origin of micrometeoroids as well as the reported densities for the somewhat larger cometary meteors of 0.2–0.5 g/cm³ are correct, then our results imply that the microstructure of cometary meteoroids is such that the porosity cannot exist in particles smaller than 50 μm. A model of loose, open aggregates of fairly equidimensional particles larger than 50 μm and densities larger than 1 g/cm³ results. A similar hypothesis, based on the duration of meteor flares, was proposed by Simonenko (1968), who estimated the size of the individual solid structural elements to be between 60 and 220 μm. Although perhaps coincidental, this size also coincides with the only major change in the meteoroid population index over more than 25 orders of magnitude of particle mass (Dohnanyi, 1972).

Let us now turn to the equidimensional character of micrometeoroids. If comets are accretional mixtures of silicates and "ice" grains or simply ice coated silicate grains, then it seems reasonable that the silicate grains released during comet sublimation should be similar in shape to those found in primitive meteorites. This suggested similarity is based on the traditional proposition that the grains in both objects represent original condensates from similar environments in the solar nebula. Another possibility is that the grains in comets are not solar system products at all but are interstellar grains which were incorporated into the outer regions of the solar nebula but never vaporized (Cameron, 1972).

The shapes of cometary grains implied by crater circularities are, however, not consistent in a straightforward way with those found in primitive meteorites. Carbonaceous chondrites contain highly non-spherical silicate grains resembling platelets. Figure 10a is an optical micrograph of a 0.5 μ thin section of the Allende (C3) meteorite illustrating a typical assemblege of ground mass particles. The grains in Allende are typically 1 μ or 2 μ thick and 5–15 μ long. In Cl chondrites the matrix grains are only 100 Å to 1000 Å in size and are hydrated but the shapes are still highly platelet like and qualitatively similar in shape to Allende (*see* Kerridge, 1964). If interplanetary grains had such shapes then noncircular craters (Fig. 10b) would be very common and Fig. 6 would show a few very deep elongated craters. Highly asymmetric shapes such as needles, platelets and dendritic whiskers have also been envisioned as original condensates on the basis of grain growth considerations (Donn, 1964).

If interplanetary dust consists of condensates similar to those accumulated into primitive meteorites then it must be in the form of dense aggregates of grains

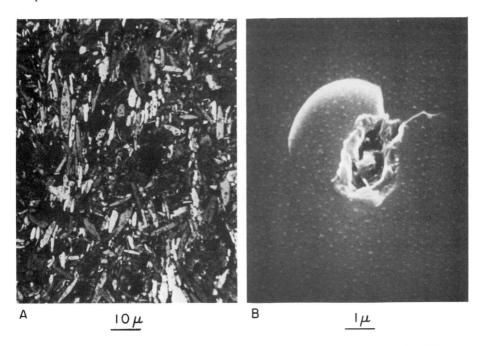

A   10 μ B   1 μ

Fig. 10(a). Photomicrograph of an ultra thin section (0.5 μm thick) of the matrix of the Allende C3 meteorite. Note the oblong shapes of component grains, i.e., olivine, which are interpreted to be solar nebula condensates. The thin section was prepared by R. Beauchamp at Battelle Laboratories, Richland, Washington: (b) Irregular crater in soda-lime glass produced by the hypervelocity impact of a kaolinite platelet similar in shape to the Allende grains.

smaller than 1000 Å, the aggregation process possibly occurring in the crust region enveloping cometary nuclei. To accommodate reported meteor densities one then has to make larger bodies (mm to cm in diameter) by loosely aggregating the dense micrometeoroid particles.

Alternately, isometry of interplanetary dust is suggestive of a history for cometary silicates in accordance with recent developments about the early history of meteorites, largely stimulated by the investigation of lunar breccias, (Anders *et al.*, 1973; Fredriksson *et al.*, 1972; King *et al.*, 1972; Kurat *et al.*, 1972; Kurat, 1973). While there is general agreement that the geochemical features of meteorites truly reflect various phases of a condensation process, the textural aspects and in particular the brecciated nature of many meteorites seems to indicate that considerable mechanical reworking took place either during the accretion phase or shortly thereafter. The dominant process which is responsible for this mechanical reworking appears to be the collision of individual bodies, i.e., features produced by the propagation of shock waves in silicates and rocks. The conspicuous absence of needles and other irregular forms for even the smallest present-day micrometeoroids, which supposedly are released from the matrix of comets due to comet sublimation, is an indication that possibly even the silicates included in cometary matrixes are not composed of crystallites of the primordial condensation process but materials which seem to be already mechanically processed by meteoroid impact in an environment roughly analogous to the lunar regolith.

CONCLUSIONS

(1) Micrometeoroids have densities greater than 1.0 g cm^{-3} and probably less than 7.0 g cm^{-3}.

(2) Micrometeoroids are roughly equidimensional.

(3) The simplistic model of the micrometeoroid as a sphere of density 2.5 g cm^{-3} is valid.

(4) The slope of the micrometeoroid cumulative flux curve increases gradually from 10^{-15} g to $5 \times 10^{-7} \text{ g}$ where it undergoes an abrupt change and matches the 1969 NASA meteoroid model.

(5) Micron and submicron interplanetary particles do not have shapes similar to grains found in primitive meteorites. This result is interpreted as indicating that grains released from comets have one or more of the following past histories.

  (a) primordial condensates < 1000 Å aggregated to form dense micron sized particles

  (b) they were at some time reworked by collisional processes

  (c) they are equidimensional interstellar grains never strongly heated in the solar nebula.

Acknowledgments—We are grateful to H. Lem and R. Canterna for assistance in data gathering and analysis. We also thank R. Beauchamp of Battelle Northwest for the Allende thin section photography and preparation. This work was supported by various NASA grants.

REFERENCES

Agrell S. O., Scoon J. H., Long J. V. P., and Coles J. N. (1972) The occurrence of goethite in a microbreccia from the Fra Mauro formation (abstract). In *Lunar Science—III*, pp. 7–9. The Lunar Science Institute, Houston.

Anders E., Ganapathy R., Krähenbühl U., and Morgan J. W. (1973) Meteoritic material on the Moon. *The Moon*. In press.

Anonymous (1969) Meteoroid environment Model-1969 (near Earth of lunar surface), NASA SP-8013.

Arrhenius G. and Alven H. (1971) Asteroidal theories and observation. In *Physical Studies of Minor Planets* (editor T. Gehrels). *Proc. 12th Coll. International Astronomical Union*, Tucson, Arizona, March 8–10.

Baldwin B. S. and Sheaffer Y. (1971) Ablation and breakup of large meteoroids during atmospheric entry. *J. Geophys. Res.* **76**, 4653–4668.

Berg O. and Gerloff U. (1971) More than 2 years of micrometeorite data from two pioneer satellites. *Space Research* **11**, 225–235. Akademie Verlag.

Berg O. and Grun E. (1973) Evidence of hyperbolic cosmic dust particles. In *Space Research*. **13**. Akademie Verlag. In press.

Bloch M. R., Fechtig H., Gentner W., Neukum G., and Schneider E. (1971) Meteorite impact craters, crater simulations, and the meteoroid flux in the early solar system. *Proc. Second Lunar Sci. Conf., Geochim. Cosmochim. Acta*, Suppl. 2, Vol. 3, pp. 2639–2652. MIT Press.

Bogard D. D. and Nyquist L. E. (1972) Noble gas studies on regolith materials from Apollo 15 and 16. *Proc. Third Lunar Sci. Conf., Geochim. Cosmochim. Acta*, Suppl. 3, Vol. 2, pp. 1797–1819. MIT Press.

Cameron A. G. W. (1972) Models of the primitive solar nebula. *Symposium on the Origin of the Solar System, Proceedings* (editor Hubert Teeves). National Center for Scientific Research, Paris.

Dohnanyi J. S. (1972) Interplanetary objects in review: Statistics of their masses and dynamics. *Icarus* **17**, 1–48.

Donn B. (1964) The origin and nature of solid particles in space. *Ann. New York Academy of Sciences* **119**, Art. 1, pp. 5–16.

El Goresy A. (1973) Personal communication.

Fredriksson K., Nelen J., and Nooan A. (1973) Lunar, terrestrial and meteoritic impact breccias (abstract). In *Lunar Science—IV*, pp. 263–265. The Lunar Science Institute, Houston.

Gault D. E. (1973) Displaced mass, diameter and effects of oblique trajectories for impact craters formed in dense crystalline rocks. *The Moon* **6**, 32–44.

Gerloff U. and Berg O. E. (1973) The orbits of 14 elliptical and 6 hyperbolic micrometeoroids derived from Pioneer 8 and 9 measurements. *Space Research* **13**. Akademie Verlag. In press.

Gibson E. K. and Moore G. W. (1972) Inorganic gas release and thermal analysis study of Apollo 14 and 15 soils. *Proc. Third Lunar Sci. Conf., Geochim. Cosmochim. Acta*, Suppl. 3, Vol. 2, pp. 2029–2040. MIT Press.

Gindilis L. M., Divari N. B., and Reznova L. V. (1969) Solar radiation pressure on particles of interplanetary dust. *Soviet Astronomy, A. J.* **13**, 114–119.

Hartung J. B., Hörz F., and Gault D. E. (1972) Lunar microcraters and interplanetary dust. *Proc. Third Lunar Sci. Conf., Geochim. Cosmochim. Acta*, Suppl. 3, Vol. 3, pp. 2735–2753. MIT Press.

Hartwit M. (1965) Origins of the zodiacal dust cloud. *J. Geophys. Res.* **68**, 2171–2180.

Hörz F., Hartung J. B., and Gault D. E. (1971) Micrometeorite craters on lunar rock surfaces. *J. Geophys. Res.* **76**, 5770–5798.

Jedwab J. (1971) La magnetite de la Meteorite D'Orgueil, vue au microscope electronique a Balayage. *Icarus* **15**, 319–340.

Kerridge J. F. (1964) Low-temperature minerals from the fine-grained matrix of some carbonaceous meteorites. *Ann. New York Academy of Sciences*, **119**, Art. 1., pp. 41–53.

Kerridge J. F. and Vedder J. F. (1972) Accretionary process in the early solar system: An experimental approach. *Science* **177**, 161–162.

King E., A., Butler J. C., and Carman M. E. (1972) Chondrules in Apollo 14 samples and size analyses of Apollo 14 and 15 fines. *Proc. Third Lunar Sci. Conf., Geochim. Cosmochim. Acta*, Suppl. 3, Vol. 1, pp. 673–686. MIT Press.

Kurat G., Keil K., Prinz M., and Nehru C. G. (1972) Chondrules of lunar origin. *Proc. Third Lunar Sci. Conf., Geochim. Cosmochim. Acta*, Suppl. 3, Vol. 1, pp. 707–721. MIT Press.

Kurat G. (1973) The Lance chondrite: Further evidence for the complex development of chondrites. *Meteoritics* **8**, 51–52.

Malhotra R. C. and Brandow V. D. (1971) Compilation of profiles and contours of micro-craters using stereo micrographs obtained from a scanning electron microscope. *Technical Report LEC/HASD* No. 640-TR-047.

Mandeville J-C. and Vedder J. F. (1971) Microcraters formed in glass by low density projectiles. *Earth Planet. Sci. Lett.* **11**, 297–306.

Marti K. (1973) Ages of Allende chondrules and inclusions: Do they set a time scale for the formation of the Solar System. *Meteoritics* **8**, 55–56.

McKay D. S., Greenwood W. R., and Morrison D. A. (1970) Origin of small lunar particles and breccia from the Apollo 11 site. *Proc. Apollo 11 Lunar Sci. Conf., Geochim. Cosmochim. Acta*, Suppl. 1, Vol. 1, pp. 673–694. Pergamon.

Morrison D. A., McKay D. S., and Heiken G. H. (1972) Microcraters on lunar rocks. *Proc. Third Lunar Sci. Conf., Geochim. Cosmochim. Acta*, Suppl. 3, Vol. 3, pp. 2767–2791. MIT Press.

Neukum G. (1971) Untersuchungen über Einschlagskrater auf dem Mond. Ph.D. Thesis, Heidelberg.

Schneider E., Storzer D., Mehl A., Hartung J. B.., Fechtig H., and Gentner W. (1973) Microcraters on Apollo 15 and 16 samples and corresponding cosmic dust fluxes (abstract). In *Lunar Science—IV*, pp. 656–658. The Lunar Science Institute, Houston.

Schneider E. (1972) Mikrokrater auf Mondgestein und deren Labor-Simulation. Ph.D. Thesis, Heidelberg.

Simonenko A. N. (1968) The separation of small particles from meteor bodies and its influence on some parameters of meteors. In *Physics and Dynamics of Meteors* (editors L. Kresak and P. M. Millman), pp. 207–216. D. Reidel.

Vedder J. F. (1971) Microcraters in glass and minerals. *Earth Planet. Sci. Lett.* **11**, 291–296.

Vedder J. F. (1972) Craters formed in mineral dust by hypervelocity micro-particles. *J. Geophys. Res.* **77**, 4304–4309.

Vedder J. F. and Lem H. (1972) Profiling with the electron microscope. *Photogrammetric Eng.* **38**, 243–244.

Vedder J. F. and Mandeville J-C. (1973) Microcraters formed in glasses by projectiles of various densities. In preparation.

Verniani F. (1969) Structure and fragmentation of meteoroids. *Space Science Reviews* **10**, 230–261.

Proceedings of the Fourth Lunar Science Conference
(Supplement 4, *Geochimica et Cosmochimica Acta*)
Vol. 3, pp. 3213–3234

The development of microcrater populations on lunar rocks

Jack B. Hartung

Max-Planck-Institut für Kernphysik Heidelberg, West Germany

Friedrich Hörz

NASA Johnson Space Center, Houston, Texas 77058

F. Kenneth Aitken

Department of Geology, Rice University, Houston, Texas 77001

D. E. Gault

NASA Ames Research Center, Moffett Field, California 94035

D. E. Brownlee

Department of Astronomy, University of Washington, Seattle, Wash. 98195

Abstract—A Monte Carlo model was developed to simulate microcrater population development on lunar rocks. The model produces craters according to a selected flux and size distribution of impacting particles and destroys them by the superposition of larger, subsequent impacts. Using a model cratering rate corresponding to 2.5 craters with pit diameters larger than 0.05 cm per cm^2 per m.y., which is based on a conservative or low estimate of the meteoroid flux, a production crater size distribution can exist not more than 1 m.y. The half-life of a population of craters with pit diameters of 0.01 cm is about 1 m.y. Longer exposure times permit lunar rock surfaces to approach an equilibrium condition with respect to the cratering process.

The measured areal crater densities for 26 lunar rocks with surfaces judged to be in equilibrium are all less than that predicted by the Monte Carlo model. This difference is attributed to increased spallation associated with larger craters and the presence of unrecognized pitless craters. These effects vary with rock type, thus indicating that different areal crater densities on lunar rock surfaces not clearly in production are due to different mechanical properties of the target rock and are not necessarily related to its exposure time. Determinations of exposure ages exceeding 1 m.y. based on microcrater statistics are subject to serious question.

Introduction

The moon's surface may be viewed as being subjected to and shaped by a more or less continuous "rain" of solid, interplanetary material. The most numerous particles are those less than 10^{-4} cm in diameter. However, those particles contributing relatively the most kinetic energy, and consequently most mass and thus having important moon-wide effects, are 0.01 to 0.1 cm in diameter (Hartung *et al.*, 1972a). The impact of these mm-sized grains on the moon causes extensive ionization, vaporization, and melting of lunar surface and meteoritic material (Gault *et* al., 1972). Considerable erosion and lateral transport or diffusion of lunar surface material may be expected from this process but not a great deal of vertical mixing. Vertical mixing is caused by larger, but much less frequent, impact events.

A long range goal in this field is to describe the above-mentioned processes quantitatively. We undertake in this paper to describe the development of micro-crater populations on lunar rocks.

The approach is first to present observational information and data developed during the last three years that contribute to our understanding of microcrater population development. Then, a statistical model for the production and destruction of craters with time, using Monte Carlo techniques, is developed. Finally, results of such a Monte Carlo simulation are compared with other data for real microcrater populations to improve our understanding of the process.

OBSERVATIONS

Single microcraters

Single microcraters have been described in some detail previously (Bloch et al., 1971; Hörz et al., 1971, and others). Glass-lined "pits" larger than about 5 μm in diameter are surrounded by spallation zones, which can be easily delineated on smooth glass surfaces. However, on crystalline rock and breccia surfaces, the extent of the spall zone can be observed for only the most recent events, and then only part of the spalled area can be clearly established.

It is important, from the standpoint of describing quantitatively the microcrater production and destruction processes discussed in this report, to establish the amount of spallation associated with each observed central pit. For about 15% of all craters at least one radius to the boundary of the spall zone can be measured and an average spall-to-pit diameter ratio (D_s/D_p) calculated. A value for D_s/D_p of 4.5 was obtained from measurements using Apollo 12 crystalline samples (Hörz et al., 1971). Data from Apollo 14 rocks (Morrison et al., 1972) and Apollo 16 rocks (Neukum et al., 1973) yielded values ranging from 3.0 to 4.5 for D_s/D_p. Morrison et al. (1973) delineated the following relationship: $D_s = 2.37 \times D_p^{1.07}$.

The spread of D_s/D_p values for a given rock may be considerable. Values from 2.5 to 7 were observed. We estimate for the averages given an uncertainty of a factor of 1.4 which corresponds to an uncertainty in the area renewed or eroded of about a factor of 2.

The D_s/D_p values obtained are clearly a function of rock type with breccias (in particular those of medium metamorphic grades, LSPET, 1973), generally having lower values than hard recrystallized breccias or igneous rocks. The particular rock properties controlling this correlation are not well established, but they are probably related to a trade-off involving the tensile strength of the sample and the efficiency of shock wave propagation at or near the sample surface.

Microcraters without central glass-lined pits also occur on lunar rocks (McKay, 1970; Hörz et al., 1971). Evidence has been presented supporting the view that these craters may also be formed by the impact of extralunar particles

(Hartung and Hörz, 1972). The relative numbers of pitless craters is an important factor influencing the statistical data for heavily cratered surfaces. Their total abundance within a given crater population may vary from 5% on glasses (12054) up to 50% in extremely unusual cases (14310). These phenomena are presently subject to more detailed studies.

We conclude that both the spall-to-pit diameter ratio and the tendency to form pitless craters depend to a considerable extent on the properties of the target material. We further suggest that these effects increase in significance as the size of the cratering event increases, because neither the astronauts' first hand observations nor the analyses of stereoscopic lunar surface photographs yielded any glass lined pits larger than about 2 cm in diameter.

Size distribution

The size distribution of microcrater pits from 0.005 to over 0.05 cm in diameter was measured for lunar rock 12054 (Hartung et al., 1972b). The shape of this distribution has been recently confirmed by observations of rock 60015 (Neukum et al., 1973). Both surfaces are large glass coatings, relatively smooth, and free from vesicles or other distracting features. These data probably represent the best in this size range that can be obtained using lunar microcrater measurements. An uncertainty of less than 20% over the entire size range is estimated for these data. These measurements together with those for smaller craters (Schneider et al., 1973; Brownlee et al., 1973) and a summary of a variety of interplanetary dust measurements corresponding to larger crater sizes (Whipple, 1967; NASA SP-8013, 1969) yield size distribution data throughout the range of sizes affecting lunar rocks.

From the standpoint of microcrater population development, the important characteristic of the size distribution obtained is flattening. The slope of the *log*-cumulative-number vs. *log*-pit-diameter curve changes from about -3 to -1 as the pit diameter changes from 0.05 to 0.005 cm. Assuming a constant D_s/D_p for all crater sizes, this slope change means that craters with pits in this size range, or slightly larger, are most effective in exposing fresh surfaces on lunar rocks. Smaller craters, though more numerous, do not cover a significantly large surface area. Somewhat larger craters may be effective in removing a greater volume of material, but there are too few of these to contribute importantly to exposing new surfaces. Because of measurement and counting uncertainties at larger crater sizes and the increased possibility of irregular spallation and pitless craters, the uncertainties related to this result for larger craters is greater than that for smaller craters.

Crater production rate

Determining the crater production rates or, alternatively, the impacting particle flux, using data from lunar rocks, has been a prime objective (Hartung et al., 1972a; Morrison et al., 1972; Neukum et al., 1972; Schneider et al., 1973; Morrison *et al.*, 1973; Neukum, 1973). The method used by all workers is to measure the areal crater density on lunar rock surfaces not in equilibrium with respect to the cratering process and to correlate these results with independently measured surface exposure times for the same rocks.

Shown in Fig. 1, an updated version of a figure presented earlier, is the areal crater density of pits with diameters >0.05 cm (500 μm) vs. exposure age for all rocks having both types of data available. A pit diameter of 0.05 cm was chosen because this was the largest size, and therefore most likely in the production state, for which reasonably good statistical data could be obtained for all rocks of interest. A cumulative minimum of 10 craters was judged acceptable for this purpose.

In Table 1 are given the actual counts for pits >0.05 cm in diameter and the type of exposure time measurement used for each rock. Uncertainties for surface exposure times are estimated to be a factor of 2 for the solar flare track, galactic cosmic ray track, and cosmogenic noble gas measurements. Gamma-ray counting of cosmogenic aluminum-26 yields only a minimum exposure time of about 1.5 m.y. when the samples are found to be in equilibrium with respect to radioactive decay.

The slope in Fig. 1 is the average number of craters with pit diameters >0.05 cm formed per cm^2 per m.y. A minimum crater production rate of 2.5 craters/cm$^2 \times$ m.y. was previously obtained based essentially on crater counts and galactic cosmic ray track data for rock 12017 (Hartung *et al.*, 1972a).

Assuming an uncertainty of a factor of two for the age measurement, a line with a slope indicating a production rate of craters with pit diameters >0.05 cm of 5.0 craters/cm$^2 \times$ m.y. may be drawn within the limits of uncertainty for rocks 12017, 12038, 12054, and 14301. Because of the lack of a clearly established pro-

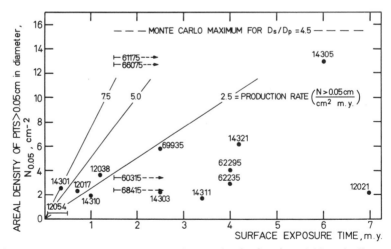

Fig. 1. Correlation of exposure ages and crater density for pits >0.05 cm in diameter. The slope of a straight line through the origin is the production rate of craters or the flux of impacting particles. Data points lying below these lines represent no longer surfaces in production for pits >0.05 cm.

Table 1. Surface exposure ages and crater counts for pits >0.05 cm in diameter.

Rock	No. of Craters Counted	Surface Exposure Age, m.y.	Type of Technique	Exposure Age References	Crater Count References
12017	12	0.7	tracks	Fleischer et al., 1971	Hörz et al., 1971
12021	101	7	tracks	Price et al., 1971	Hörz et al., 1971
12038	30	1.3	tracks	Lal, 1973	Hörz et al., 1971
12054	14	0.05 to 0.5	cosmogenic Al^{26}	Schönfeld, 1971	Hartung et al., 1972
14301	54	0.34	tracks	Hart et al., 1972	Morrison et al., 1972
14303	10	2.5	tracks	Bhandari et al., 1972	This work
14305	125	6	cosmogenic Mn^{53}	Herr et al., 1972	Morrison et al., 1972
14310	30	1.0	tracks	Lal, 1973	Hartung et al., 1972
14311	10	3.4	tracks	Hart et al., 1972	Morrison et al., 1972
14321	28	4.2	tracks	Lal, 1973	Morrison et al., 1972
60315	14	>1.5	cosmogenic Al^{26}	Keith, 1973*	Neukum et al., 1973
61175	196	>1.5	cosmogenic Al^{26}	O'Kelly et al., 1973*	Neukum et al., 1973
62235	75	4	tracks	Goswami et al., 1973*	Neukum et al., 1973
62295	34	4	tracks	Goswami et al., 1973*	Neukum et al., 1973
66075	147	>1.5	cosmogenic Al^{26}	O'Kelly et al., 1973*	Neukum et al., 1973
68415	57	>1.5	cosmogenic Al^{26}	Rancitelli et al., 1973*	Neukum et al., 1973
69935	85	2.5	tracks	Goswami et al., 1973*	Morrison et al., 1973

*Presented at the Fourth Lunar Science Conference or personal communications.

duction (non-equilibrium) surface for two of the four samples, this result should still be considered a minimum average value for the crater production rate. Data points for other rocks lie below the production rate curves indicating equilibrium has been reached for these cases. Based on an analysis of satellite-borne meteoroid detection experiments, a minimum value for the corresponding particle flux or crater production rate was found to be about 8 to 10 craters/cm^2 × m.y. (Gault et al., 1972). Thus agreement within an order of magnitude is obtained for the average crater production rate using lunar rock data and in situ meteoroid detection data.

Equilibrium populations

It was recognized early that most rocks exposed on the lunar surface possess a population of microcraters in equilibrium with respect to the cratering process and that crater destruction resulted from the superposition of subsequent craters in the mm size range (Hörz et al., 1971). Later it was shown that the areal density of craters on surfaces not obviously in production, i.e., those possibly in equilibrium with respect to the cratering process, could vary as much as a factor of four (Morrison et al., 1972; Neukum et al., 1973). At present a controversy exists regarding the explanation of this variation.

One side argues that all rocks, regardless of petrologic type, should possess the same equilibrium areal density of craters and that differences in crater densities directly reflect different durations of surface exposure (Morrison et al., 1972;

1973). Furthermore, though an equilibrium population may exist for small crater sizes, somewhat larger craters are not necessarily in equilibrium, and the densities of these craters reflect different durations of surface exposure (Neukum, 1973). In opposition to these interpretations is the view presented throughout this paper that such differences in crater population for clearly non-production surfaces are predominantly caused by different mechanical properties of various rocks. It will be argued that the crater populations observed are less a function of time than of mechanical rock properties, i.e., for identical projectiles different amounts of material may be excavated, and craters may be destroyed with different efficiencies on various rock materials.

Large microcraters and catastrophic rupture

There is always a maximum crater size that can possibly be observed on any given rock because a slightly more energetic event would have resulted in catastrophic rupture of the rock (Fig. 2). Based on cratering experiments for kinetic energies and crater diameters of the same order of magnitude as those involved in the destruction of lunar rocks, Gault *et al.* (1972) arrive at the following expression:

$$E_{RS} = 2.5 \times 10^6 \, S_c r^{-0.225} \tag{1}$$

Fig. 2. Rock 73155 containing a large crater and major related fractures. The energy related to the 2-mm-diameter glass-lined pit ($\approx$ 1-cm-diameter spall) was just short of that required to catastrophically rupture the 4-cm-diameter rock.

where E_{RS} is the energy in ergs required to rupture 1 g of a target rock with radius, r, in cm. (The original reference gives an erroneous value of 2.5×10^5 for the coefficient, which is corrected here.) Using a conservative (high) value of 3 for the compressive strength S_c, in kilobars of crystalline rocks, the kinetic energy E_{kin} of a projectile capable of destroying a rock with a given mass may be calculated. A crater spall diameter, D_c (destructive), corresponding to this E_{kin} is obtained by Gault (1973):

$$D_c = 10^{-2.823} \times \rho_p^{1/6} \times \rho_t^{-1/2} \times E_{kin}^{0.370} \qquad (2)$$

Cgs units are used throughout. Taking $3\,g/cm^3$ for both projectile and target densities, ρ_p and ρ_t, respectively, these diameters were calculated, and are shown in Fig. 3 as a function of E_{kin} or rock mass able to be destroyed.

Also contained in Fig. 3 are the observed maximum pit diameters on lunar rocks measured using a binocular microscope or close-up rock photographs. Of particular interest are those craters formed by impacts which cause extensive fracturing but not complete rupturing of the rock. A line D_p (max. observed) may be constructed through points corresponding to these craters. Increasing the pit diameter by a factor of 1.5 which corresponds to an approximately 4-fold increase in E_{kin}, the line D_p (destructive) was derived. This line is considered a conservative estimate of the maximum pit diameter observable on lunar rocks of a given mass. Applying a factor of 4.5, the D_s/D_p ratio, to this line, results in a remarkable agreement between experimental and observational results.

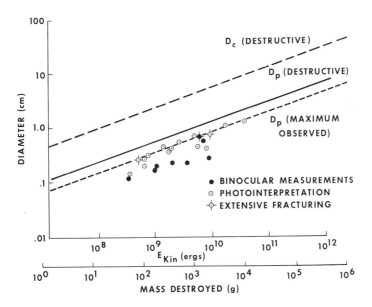

Fig. 3. Spall and pit diameters required for catastrophic rupture of a given rock mass based on experimental and observational results. D_p (destructive) is considered an upper limit for pit diameters observable on lunar rocks.

MONTE CARLO MODEL

Several workers have developed analytical models describing the development of crater populations where existing craters were destroyed by simple superposition of subsequent craters (Marcus, 1964, 1966; Walker, 1967; Neukum and Dietzel, 1971). These studies considered only crater size distributions of the form,

$$N_D = CD^\alpha \tag{3}$$

where N is the areal density of craters with a diameter equal to or greater than D, and C and α are assumed constant. However, in the size range of interest the *log-log* cumulative size distribution is not a straight line, i.e., α is not constant (Hartung et al., 1972b; Neukum et al., 1973). This non-linear behavior may be approximated by two, or possibly more, straight-line segments for some purposes, but the observational data available justify a more detailed model to describe the process. We have developed a model based on the Monte Carlo method which permits the use of any type of size distribution and thus enables study of the transition range between steep ($\alpha = -3$) and flat ($\alpha = -1$) distributions. This work was suggested by studies of Gault (1970), who studied crater population development in loose regolith-type materials.

Model description

In general terms, we assume a mass distribution of impacting particles and allow these different-sized particles to "impact" in a random sequence at random locations on a reference surface. Should the spall zone of a new crater overlap the center (pit) of an equal or smaller sized old crater, the old crater is considered destroyed and is deleted from the list of "observable" craters. Information available for analysis is the size, number, position, and formation interval of all "observable" and "destroyed" craters.

To compare our results with analytical models we made several runs using straight-line production size distributions. Two examples are shown in Fig. 4 for slopes of -3 and -1.

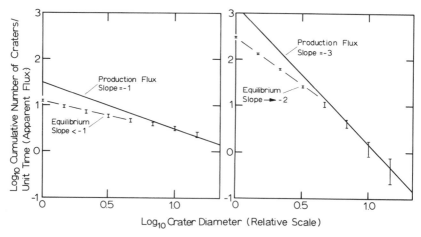

Fig. 4. Production and resulting equilibrium crater size distributions for two different production size distributions based on Monte Carlo model.

A -3 production slope corresponds to a distribution dominated by smaller craters. In this case, craters are destroyed by other craters similar in size. After some time the smallest craters have been destroyed many times over, and thus, in this size range, equilibrium has been reached. The number of larger craters still corresponds approximately to the number produced. The slope of the equilibrium portion of the size distribution is -2, which was predicted by the analytical models mentioned above and observed for regolith crater populations (Shoemaker *et al.*, 1970; Soderblom, 1970).

For crater populations with production slopes much less than -2, destruction of craters tends to be dominated by the largest craters permitted to impact the surface. The slope of the size distribution which results after some time is shown to be less than the production slope of -1, but this result may be somewhat artificial because the effects of still larger craters are arbitrarily excluded. In this case the development of a crater population is very sensitive to the upper size limit considered. A state of equilibrium apparently cannot be reached for this type of size distribution because the surface is being continuously renewed by large-sized craters. Analytical models which permit the upper limit of the crater size to approach infinity, predict "equilibrium" distributions with the same slope as the production distribution. Our results are not inconsistent with this finding.

To model the real development of crater populations on lunar rocks we have adopted the following input data and simplifying conditions: the size distribution of craters corresponding to impacting particle masses from 10^{-7} to 10^{-5} g, the flux of these particles, and the relationship between impacting particle mass and resulting crater size are taken from Hartung *et al.* (1972a). The size distribution of larger particles was based on an analysis of satellite-borne meteoroid detection experiments and radar meteor data (NASA SP-3018). Seven size classes with intervals corresponding to a factor of the square root of ten were selected to cover the particle mass range from 10^{-7} to $10^{-3.5}$ g. The reference area, 3.23 cm², resulted from the imposed requirements that one crater of the largest size class should be expected to occur on the surface in each one-million-year interval, and that the side of the square reference area be 5 times the diameter of this largest crater (*see also* Oberbeck *et al.*, 1973). A summary of the actual input data used is given in Table 2.

For all the examples illustrated in this paper a D_s/D_p ratio of 4.5 was used. Use of another ratio would change the relationships between particle mass and crater size significantly, but the resulting *equilibrium* areal density of craters will not be affected substantially. In effect, the cumulative equilibrium size distribution curve will be shifted parallel to a line of -2 slope by changes in D_s/D_p.

A size-class zero was established to include all particles or cratering "events" larger than those in size class 1. These large events are rare for size distributions with slopes much steeper than -2, but there is a finite chance of their occurrence on the reference surface. On lunar rocks the impact of such a large particle ruptures the sample or otherwise destroys the observable population of smaller craters. Therefore, our model records the occurrence of large, size-class-zero, events, but does not permit these craters to actually form on the model surface and consequently they do not participate in the

Table 2. Size distribution and flux data used in Monte Carlo model.

Size class	Impacting particle mass, micrograms	Spall diameter of resulting crater, cm	No. of impacts per cm² per 10^6 yr*
7	0.100	0.0360	95.5
6	0.316	0.0528	58.1
5	1.000	0.0774	34.2
4	3.162	0.0137	16.3
3	10.00	0.1668	5.98
2	31.62	0.2449	1.66
1	100.0	0.3594	0.310
0	316.2 and greater	—	0.104

*Based on a crater production rate corresponding to 2.5 craters with pit diameters larger than 0.05 cm per cm² per 10^6 yr. (*see* Fig. 1).

destruction of smaller craters. The number of large events recorded during a run may be interpreted as a rough estimate of the number of times an *average* rock would have been broken up by impacts during the simulated time interval.

Results

The principal result of a single Monte Carlo run is a time history of the number of observable craters in each size class. Data from one such run using the above described input conditions are shown in Fig. 5. For the average impacting particle flux given by Hartung *et al.* (1972a), 12 printout intervals are equivalent to one million years. Should this absolute micrometeoroid flux be revised, the conversion from intervals to years may be easily made.

According to Fig. 5, smaller craters size (classes 6 and 7) are in a production state, and no large craters are formed during the first few time intervals. After about 10 intervals, however, about one-half of the smaller craters produced have already been destroyed. This condition is considered a "transition" state between production and equilibrium distributions. Equilibrium, i.e., no change in number

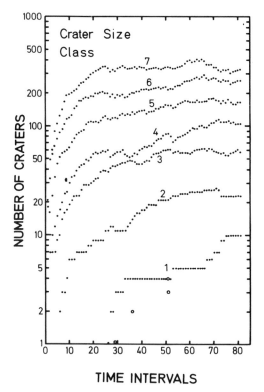

TIME INTERVALS

Fig. 5. Time history of the numbers of different-sized craters observable on a reference surface of 3.23 cm². Crater size classes are defined in Table 2. Twelve time intervals correspond to a period of 1 m.y. using the flux of Hartung *et al.* (1972a). Open circles correspond to "large" events which may be interpreted to have broken up the rock.

of craters with time, is approached in the smaller size classes at about the same time, after 20 to 30 intervals, because the size distribution is flat for these sizes and the total number of craters is controlled by the effects of larger craters. Larger sized craters approach equilibrium at successively later times because in this size range the size distribution steepens and craters are destroyed by other craters similar in size.

To obtain a visual impression of the various crater densities, the microcrater populations after intervals 1, 8, and 38 are illustrated in Fig. 6. These intervals represent respectively production, transition, and equilibrium distributions. Even in the very early stages some crater destruction may be expected.

A comparison between the number of craters produced and the number of craters that survive as observable after 72 intervals, i.e., a model elapsed time of about 6 million years, is shown in Fig. 7. About 4 to 5 times more small craters were formed than remain as observable. Craters in this range are clearly in equilibrium. Somewhat fewer larger craters survived destruction than were originally produced. Craters in this size range are considered in a transition state. Only the largest craters can possibly be considered in production, and this results in part from the arbitrary selection of an upper limit for the crater size (*see also* Shoemaker *et al.*, 1970). The curvature of the equilibrium distribution reflects the curvature of the production distribution, but the local slope of the equilibrium distribution is everywhere flatter than the corresponding slope for the production distribution.

Absolute formation age measurements of single microcraters can, in principle, be accomplished on lunar rocks, but to date statistically significant results for a population of microcraters from a single lunar rock surface have not been obtained. Because the computer maintains a record of the times of production and destruction of individual craters, an expected age distribution of individual lunar microcraters is another result of these Monte Carlo studies. Shown in Fig. 8 are the expected differential age distributions for craters of different sizes assuming a constant crater production and destruction rate. Because the destruction of craters of a given age is a random process dependent only on the number of craters

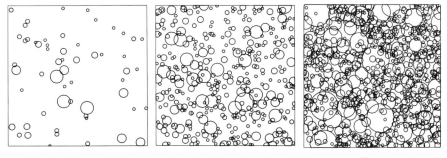

Fig. 6. "Observable" craters after time intervals 1, 8, and 38, corresponding to production, transition, and equilibrium populations and to model exposure times of 8×10^4, 7×10^5, and 3×10^6 yr, respectively. The model length along one side of the illustrations is 1.8 cm. The circles shown to scale, correspond to microcrater spall zones.

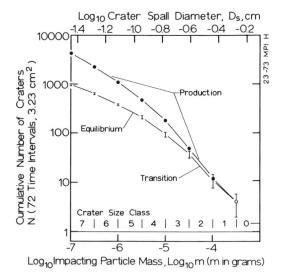

Fig. 7. Comparison of the number of craters produced and the number that remains as observable after a Monte Carlo model elapsed time of about 6 million years.

remaining, the age distribution function has the same form as that for radioactive decay,

$$N = ke^{\lambda t} \tag{4}$$

where N is the number of craters of a given age, t is the time or age, and k and λ are constants. The values for these constants are undoubtedly different for different rocks and crater sizes considered. Nevertheless, a curve corresponding to the expected half-lifes of different-sized craters may be indicated on Fig. 8. The half-life of smaller sized craters is fairly constant and near one million years. Larger craters may be expected to survive destruction by superposition a much longer, but less well-defined, time. Destruction of crater populations due to rock break-up has been treated earlier (Gault *et al.*, 1972) and should also be considered in this connection.

Statistical variation and the selection effect

Referring to the time history of crater numbers shown in Fig. 5, considerable variation exists in the observable number of craters in a given size class, even after equilibrium is reached. For example, the number within size classes 6 and 7 varies by about 30% after equilibrium is reached. Compared to this the 5 to 8% "counting uncertainty," based on the square root of the number in each single size class, is small. These variations may be due to the occurrence of isolated, but still random, large impacts or simply to the controlling statistical process. The occurrence of several size-class-1 impacts does have a noticeable effect on the corresponding number of smaller sized craters. In any case, the same processes operate

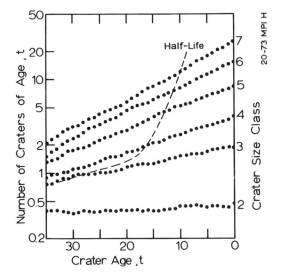

Fig. 8. Computed age distribution of different-sized microcraters assuming a constant impacting particle flux. The half-life of craters against destruction by superposition is also indicated.

on lunar rocks and suggest the statistical quality of measurements on lunar rocks may not be as great as previously expected. Selection of "representative" lunar rock surface areas is therefore a problem, not only for microcratering studies, but also for other investigations of lunar surface processes using mm to cm-sized rock chips.

The survival time of rocks on the lunar surface is controlled by the catastrophic rupture. Predictions for mean survival times have been presented earlier (Gault *et al.*, 1972). Some variation around the statistical mean value is expected. Because rocks destroyed early in their exposure lifetime are not collected by the astronauts, we can expect to recover a disproportionately large number of rocks which have survived longer than the expected *mean* lifetime. Thus, most rocks collected may not have received their fair share of larger craters and have enjoyed a statistical selection effect. As a consequence, those large craters not accounted for would cause an artificial steepening of the cumulative size distribution curve. Therefore, lunar rock surfaces which have already reached an equilibrium state for small craters, do not necessarily yield valid production distributions for larger craters. Shown in Fig. 9 is a hypothetical *log-log* cumulative size distribution diagram with a reference equilibrium line with a slope of −2. Additional lines are shown that indicate how the *equilibrium* distribution would appear for different numbers of unaccounted large craters due to the selection effect.

To illustrate the simplest example of the selection effect, when plotting data for a cumulative size distribution, there must always exist one largest crater size which corresponds to a count of one. No crater exists for the next larger size intervals, so the curve on the *log-log* diagram must artificially steepen toward zero on the vertical axis.

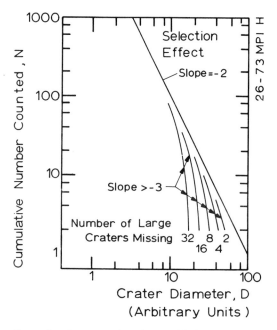

Fig. 9. Diagram illustrating the steepening of an equilibrium size distribution curve due to selection effects.

At present there is no definite procedure to quantitatively correct for selection effects and the artificial steepening of the *log-log* cumulative size distribution curve. Therefore, genuine production crater populations for large events (>0.03 cm pit diameter) are not readily obtained.

DISCUSSION

Model and actual microcrater populations

Shown in Fig. 10 are cumulative size distribution data for a variety of lunar rock surfaces considered in equilibrium together with that resulting from the Monte Carlo run described above, using a $D_s/D_p = 4.5$. A line corresponding to 100% of saturation (Gault, 1970) is shown for a reference. The error bars for the Monte Carlo data represent the limits of variation of the data after equilibrium was reached. We suggest similar uncertainty values should be applied to the data for equilibrium microcrater populations without regard for the number of craters actually counted. The rocks, for which data are shown, were selected because they represented different rock types and possessed the highest quality statistical data (Hörz *et al.*, 1971; Neukum *et al.*, 1973).

The shapes of the Monte Carlo and real size distributions are very similar throughout the size range, but the Monte Carlo crater densities are significantly higher than those for nearly all lunar rock surfaces. The maximum Monte Carlo

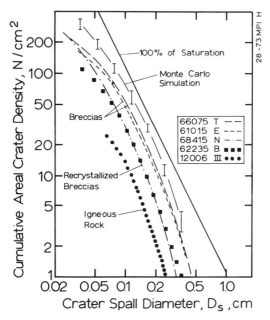

Fig. 10. Comparison of Monte Carlo and natural lunar microcrater size distributions for probable equilibrium surfaces. Differences in vertical position of the curves are judged due to variations in target rock properties. The error bars correspond to the range of variation simulated by the Monte Carlo model after equilibrium had been reached.

crater densities agree very well with the "limiting frequency value," i.e., equilibrium density, given by Morrison *et al.* (1972) based on data for 0.8 cm² of rock 14306, Surface 1. We suggest that because of the small area studied which was furthermore selected because of high crater density this sample may be somewhat nonrepresentative for an "average" density in the sense discussed earlier.

The present Monte Carlo model produces a crater population on an idealized rock surface of 3.23 cm². None of the 26 rocks for which surface areas larger than 3 cm² were studied exhibits crater densities as high as the model predicts. Differences in D_s/D_p ratios cannot account for the observed deviations from the model and the differences related to different rock types (Neukum *et al.*, 1973). We attribute the different areal crater densities for rock or model surfaces in equilibrium to increased effectiveness of spallation associated with larger craters and to greater numbers of unrecognized pitless craters on certain rock types. We suggest that spallation and pitless crater formation are controlled by target rock physical properties, which may vary considerably for different rocks.

The following arguments lend support to this hypothesis.

(a) Extremely soft soil breccias from Apollos 14 and 16 are well rounded. No lunar surface process other than micrometeoroid impact has been suggested to produce such a degree of rounding. Still, very few craters are observed on these surfaces. The amount of mass wasting required to round these rocks is such that no original rock surface is preserved. We are led to the conclusion that these surfaces must have reached a state which is equivalent to equilibrium but the observed areal crater density is, paradoxically, extremely low. Accordingly, observed equilibrium crater densities on lunar materials may vary from almost no craters to the limiting frequency range of Morrison *et al.* (1972).

(b) Even among "hard" rocks, systematic differences between igneous rocks and breccias are observed. Igneous rocks are typically a factor of 2 less densely cratered than breccias (Neukum *et al.*, 1973).

(c) The most densely cratered rocks observed are moderately coherent to coherent breccias. It may be that impacts into these materials generate proportionally more heat because of the increased porosity of the target and thus produce proportionally more glass which in turn causes a more secure bonding of the pit to the underlying material. This bonding may not be accomplished as effectively in more friable materials and in less friable crystalline rocks.

(d) A large range of physical properties for lunar rocks is established e.g., sound velocities (Toksöz *et al.*, 1972) or porosities (Chao *et al.*, 1972). However, correlation of these properties with measured crater densities is not possible at present because such measurements are not available for the samples under consideration.

(e) Different erosion rates measured on igneous rocks and friable breccias (Rancitelli, 1973) indicate that similar erosion mechanisms may have considerably different efficiencies on different rocks.

These arguments together with the evidence based on cratering experiments that physical properties govern the cratering process (Gault *et al.*, 1972), lead us to the conclusion that the areal crater density for equilibrium lunar rock surfaces may vary considerably. The highest crater densities observed to date fall within the "limiting frequency" range of Morrison *et al.* (1972), but in contrast to these authors we do not require that *all* equilibrium microcrater populations fall within this range.

Lunar rock exposure times from microcrater statistics

Other workers have taken the areal crater density variations from rock to rock discussed above to be due to different times of exposure to the meteoroid bombardment. Gault (1970) and Shoemaker *et al.* (1970) introduced techniques to establish the formation age of large scale lunar surfaces using regolith crater populations. Variations of these techniques accounting for specific differences associated with microcrater populations were applied to lunar rock surfaces by Morrison *et al.* (1972, 1973) and Neukum (1973). The two approaches used are schematically illustrated in Fig. 11.

Approach 1 essentially uses the areal density of microcraters of a constant diameter, D_x, as a measure of the rock exposure age. A line with the production slope, α, of -3 is established through the measured data points in the cumulative size distribution plots in order to arrive at a consistent presentation of the areal crater density. The density of craters, $N(D_x)$, where $D_x = 0.1$ cm spall diameter, is then taken as a measure of exposure age. The essential point of this technique is

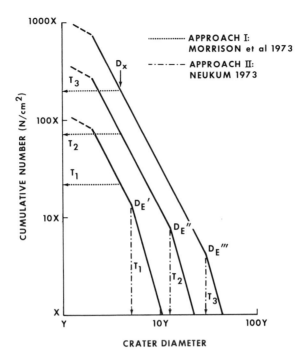

Fig. 11. Basic approaches used in attempts to determine exposure ages using micro-crater populations (*see* text).

that crater densities and exposure times are linearly related as long as the crater density is below the "limiting frequency value" (Morrison *et al.*, 1972). We accept the validity of this approach as long as the rock surface of concern is *demonstrably* in a production state. However, this approach does not account for the destruction of any crater with a spall diameter >0.1 cm before the "limiting frequency value" is reached. As a consequence, this approach implies the development of a crater population which is contrary not only to our Monte Carlo calculations, but also to other analytical models, e.g., Marcus (1966), Walker (1967), Neukum and Dietzel (1971), Neukum (1973), Oberbeck *et al.* (1973), as well as experimental results (Gault, 1970). Furthermore, "transition" populations *have been* identified on lunar rocks (Neukum *et al.*, 1973). Crater destruction by superposition is readily seen on lunar rocks at crater densities significantly lower than the "limiting frequency value." Indeed, evidence was presented above which suggests that some rock surfaces may never reach the "limiting frequency value" no matter how long they are exposed.

Approach II utilizes a concept similar to that applied by Shoemaker *et al.* (1970) and others to derive the formation ages of large scale lunar surfaces. A steepening of the slope of the cumulative crater size distribution at larger crater sizes is interpreted to represent a production slope. This production slope

($\alpha = -3.6$) intercepts the equilibrium slope ($\alpha = -2$) at a crater size D_E. This size is then taken as the basic measure for exposure age.

Such an approach strongly depends on the positive identification of equilibrium and production slopes. As demonstrated previously (Neukum *et al.*, 1973), there is considerable ambiguity in positively associating a specific D^{-2} crater distribution with a genuine equilibrium state, basically because the production distribution in the crater size range of concern is equal or close to D^{-2}. Thus, a slope of -2 is not a certain indication of an equilibrium condition, though extended D^{-2} plateaus certainly are suggestive of it. A production slope of -3.6 for pits > 0.03 cm is used in this approach based on the average of observed slopes, which ranged from -2.5 to about -6. Such a large spread in actual slopes is in our opinion more compatible with an artificial steepening as discussed in the above section on "statistical variations and the selection effect."

A steepening of the slope for the largest craters observed, regardless of crater size, is common to all crater-counting studies. As stated, cumulative size distribution must approach zero at a finite crater size. Steepening caused by such effects is readily observed at crater diameters significantly less than 0.03 cm, e.g., rocks 60325, 64455, 61156 T (Neukum *et al.*, 1973), 12017 G, and 12073 (Hörz *et al.*, 1971). These effects are also present in statistical data of significantly larger regolith craters (Gault, 1970; Shoemaker *et al.*, 1970). Such steepening effects are especially pronounced if the total crater size range under consideration covers only a decade. The crater sizes used to establish the -3.6 "production" slope range over an even smaller span, typically between D_E and $5 D_E$.

Accordingly, we maintain that approach II does not yield absolute exposure age information because an equilibrium slope cannot be recognized with confidence and the postulated production slope of -3.6 is not established. The establishment of both conditions is required before the dating technique of Shoemaker *et al.* (1970) can be applied properly.

Additional arguments which indicate that absolute exposure age determinations based on microcrater densities are premature at present and which apply to both approaches I and II are as follows:

(a) Neither approach accounts for the differences in physical rock properties and the resulting consequences on the observed crater populations (Fig. 12).

(b) By definition, neither approach allows any crater larger than D_x or D_E to be destroyed. Estimates of the total erosion rate of lunar rocks—purposely excluding those which are based on microcrater studies—yield values of 0.01 to 0.2 cm/m.y. with an average of about 0.05 cm/m.y. (Barber *et al.*, 1971; Fleischer *et al.*, 1971; Wahlen *et al.*, 1972; Crozaz *et al.*, 1972; Rancitelli, 1973.) Taking conservative crater geometries (pit diameter/depth = 3) one must conclude that significant numbers of craters with diameters, D_x and D_E (depth ≈ 0.01 cm), have been eroded in less than 1 m.y. and large numbers of craters must have been eroded for the claimed exposure times of 10 m.y. (Neukum, 1973).

(c) Referring to Fig. 1, assuming the most conservative crater-production rate of 2.5 craters with pit diameters > 0.05 cm per cm^2 per m.y., it is readily seen that rocks 60315, 68415, 14311, 62295, 62235, and 12021 have lost significant numbers

Fig. 12. Rock 72235: Differences in microcrater populations on different target materials. Note significantly more abundant large craters in "soft" breccia matrix (right) as opposed to a "hard" inclusion (left).

of these craters. Such craters are larger than those actually used in approaches I and II. Notice that rocks 12017, 14311, and 12021 have very similar crater densities and yet exposure ages ranging from 0.7 to 7 m.y.

(d) The absolute exposure age is dependent on the knowledge of the absolute flux of micrometeoroids which is presently uncertain by at least a factor of 2.

CONCLUSIONS

1. Our estimate of the minimum production rate of microcraters with pit diameter > 0.05 cm is 5 craters/cm$^2 \cdot$m.y. averaged over the last million years. This represents an increase by a factor of 2 over the value presented by us previously and is in reasonable agreement with estimates based on satellite-borne meteoroid detection experiments.

2. The size of the largest craters observed on lunar rocks agrees very well with the size expected from laboratory impact experiments. Events somewhat more energetic than those observed would have catastrophically ruptured the rock.

3. The development of microcrater populations on lunar rocks is quantitatively described using a Monte Carlo model. The statistical variation of the equilibrium number of craters of a given size was found to be about 30% and independent of the total number of craters present.

4. Many lunar rocks returned may have enjoyed a natural selection effect because these rocks survive catastrophic rupture by impact. Thus, crater size distributions measured on lunar rocks are depleted in large craters, and an artificial steepening occurs in the cumulative size distribution curve.

5. Monte Carlo results and independent erosion rate studies indicate appreciable crater destruction must have occurred for exposure ages of 1 m.y.

6. The wide range of areal crater densities observed on rock surfaces not obviously in a production state is probably due to irregular spallation and the formation of unrecognized pitless craters. Both of these processes depend on the physical properties of the target rock. Such different densities do not reflect differences in the exposure time of the rock on the lunar surface. Accordingly, we consider the determination of exposure times for periods in excess of 1 m.y. to be premature.

Acknowledgments—We gratefully acknowledge the support of the NASA Johnson Space Center Computation and Analysis Division and especially the programming support of J. W. Blackmon. Though we do not share some of their views, fruitful discussions with D. E. Morrison and G. Neukum had a significant input to our thinking. Part of this work was performed while one of us (F.K.A.) held a NRC Post-doctoral Fellowship at the NASA Johnson Space Center, Houston.

References

Barber D. J., Cowsik R., Hutcheon I. D., Price P. B., and Rajan P. S. (1971) Solar flares, the lunar surface and gas-rich meteorites. *Proc. Second Lunar Sci. Conf., Geochim. Cosmochim. Acta*, Suppl. 2, Vol. 3, 2705–2714. MIT Press.

Bhandari N., Goswami J. N., Gupta S. K., Lal D., Tamhane A. S., and Venkatavaradan V. S. (1972) Collision controlled radiation history of the lunar regolith. *Proc. Third Lunar Sci. Conf., Geochim. Cosmochim. Acta*, Suppl. 3, Vol. 3, 2811–2829. MIT Press.

Bloch M. R., Fechtig H., Gentner W., Neukum G., Schneider E., and Wirth H. (1971) Natural and simulated impact phenomena: A photo-documentation. Max-Planck-Institut für Kernphysik. Heidelberg, Germany (private printing).

Brownlee D. E., Hörz F., Vedder J. F., Gault D. E., and Hartung J. B. (1973) Some physical properties of micrometeoroids. *Proc. Fourth Lunar Sci. Conf., Geochim. Cosmochim. Acta*. This volume.

Chao E. C. T., Minkin J. A., and Best J. B. (1972) Apollo 14 breccias: general characteristics and classification. *Proc. Third Lunar Sci. Conf., Geochim. Cosmochim. Acta*, Suppl. 3, Vol. 1, 645–659. MIT Press.

Crozaz G., Drozd R., Hohenberg C. M., Hoyt H. P., Ragan D., Walker R. M., and Yuhas D. (1972) Solar flare and galactic cosmic ray studies of Apollo 14 and 15 samples. *Proc. Third Lunar Sci. Conf., Geochim. Cosmochim. Acta*, Suppl. 3, Vol. 3, 2917–2931. MIT Press.

Fleischer R. L., Hart H. R. Jr., Comstock G. M., and Evwaraye A. O. (1971) The particle track record of the Ocean of Storms. *Proc. Second Lunar Sci. Conf., Geochim. Cosmochim. Acta*, Suppl. 2, Vol. 3, 2559–2568. MIT Press.

Gault D. E. (1970) Saturation and equilibrium conditions for impact cratering on the lunar surface: Criteria and implications. *Radio Sci.* 5, 273–291.

Gault D. E. (1973) Displaced mass, depth, diameter and effects of oblique trajectories for impact craters formed in dense crystalline rocks. *The Moon* 6, 32–44.

Gault D. E., Hörz F., and Hartung J. B. (1972) Effects of microcratering on the lunar surface. *Proc. Third Lunar Sci. Conf., Geochim. Cosmochim. Acta*, Suppl. 3, Vol. 3, 2713–2734. MIT Press.

Hart H. R. Jr., Comstock G. M., and Fleischer R. L. (1972) The particle track record of Fra Mauro. *Proc. Third Lunar Sci. Conf., Geochim. Cosmochim. Acta*, Suppl. 3, Vol. 3, 2831–2844. MIT Press.

Hartung J. B. and Hörz F. (1972) Microcraters on lunar rocks. *Proc. 24th Int. Geol. Cong., Montreal, Section 15, Planetology,* 48–56.

Hartung J. B., Hörz F., and Gault D. E. (1972a) Lunar microcraters and interplanetary dust. *Proc. Third Lunar Sci. Conf., Geochim. Cosmochim. Acta,* Suppl. 3, Vol. 3, 2735–2753. MIT Press.

Hartung J. B., Hörz F., and Gault D. E. (1972b) Lunar rocks as meteoroid detectors. In *Proc. Internat. Astron. Union Colloq.,* No. 13, *The Evolutionary and Physical Problems of Meteoroids.* In press.

Herr W., Herpers U., and Woelfle R. (1972) Study on the cosmic ray produced long-lived Mn-53 in Apollo 14 samples. *Proc. Third Lunar Sci. Conf., Geochim. Cosmochim. Acta,* Suppl. 3, Vol. 2, 1763–1769. MIT Press.

Hörz F., Hartung J. B., and Gault D. E. (1971) Micrometeorite craters on lunar rock surfaces. *J. Geophys. Res.* **76,** 5770–5798.

Lal D. (1973) Hard rock cosmic ray archeology. *Space Sci. Rev.* **14,** 3–102.

LSPET (Lunar Sample Preliminary Examination Team) (1973) The Apollo 16 lunar samples: Petrographic and chemical description. *Science* **179,** 23–33.

McKay D. S. (1970) Microcraters in lunar samples. In *Proc. Twenty-Eighth Annual Meeting Electron Microscopy Soc. of America* (editor C. J. Arceneaux), pp. 22–23. Claitors.

Marcus A. H. (1964) A stochastic model of the formation and survival of lunar craters, 1. *Icarus* **3,** 460–472.

Marcus A. H. (1966) A stochastic model of the formation and survival of lunar craters, 2. *Icarus* **5,** 165–177.

Morrison D. A., McKay D. S., Moore H. J., and Heiken G. H. (1972) Microcraters on lunar rocks. *Proc. Third Lunar Sci. Conf., Geochim. Cosmochim. Acta,* Suppl. 3, Vol. 3, 2767–2791. MIT Press.

Morrison D. A., McKay D. S., and Moore H. J. (1973) Microcraters on Apollo 15 and 16 rocks. *Proc. Fourth Lunar Sci. Conf., Geochim. Cosmochim. Acta.* This volume.

NASA SP-8013 (1969) Meteoroid Environment Model—1969 (Near Earth to Lunar Surface).

Neukum G. (1973) Micrometeoroid flux, microcrater population development and erosion rates on lunar rocks, and exposure ages of Apollo 16 rocks derived from crater statistics (abstract). In *Lunar Science—IV,* pp. 558–560. The Lunar Science Institute, Houston.

Neukum G. and Dietzel H. (1971) On the development of the crater population on the moon with time under meteoroid and solar wind bombardment. *Earth Planet. Sci. Lett.* **12,** 59–66.

Neukum G., Schneider E., Mehl A., Storzer D., Wagner G. A., Fechtig H., and Bloch M. R. (1972) Lunar craters and exposure ages derived from crater statistics and solar flare tracks. *Proc. Third Lunar Sci. Conf., Geochim. Cosmochim. Acta,* Suppl. 3, Vol. 3, 2793–2810. MIT Press.

Neukum G., Hörz F., Morrison D. A., and Hartung J. B. (1973) Crater populations on lunar rocks. *Proc. Fourth Lunar Sci. Conf., Geochim. Cosmochim. Acta.* This volume.

Oberbeck V. R., Quaide W. L., Mahan M., and Paulson J. (1973) Monte Carlo calculations of lunar regolith thickness distributions. *Icarus* In press.

Price P. B., Rajan R. S., and Shirk E. K. (1971) Ultra-heavy cosmic rays in the moon. *Proc. Second Lunar Sci. Conf., Geochim. Cosmochim. Acta,* Suppl. 2, Vol. 3, 2621–2627. MIT Press.

Rancitelli L. A. (1973) Personal communication.

Schneider E., Storzer D., Mehl A., Hartung J. B., Fechtig H., and Gentner W. (1973) Microcraters on Apollo 15 and 16 samples and corresponding cosmic dust fluxes. *Proc. Fourth Lunar Sci. Conf., Geochim. Cosmochim. Acta.* This volume.

Schönfeld E. (1971) Personal communication (*see* LSPET (Lunar Sample Preliminary Examination Team) (1970)) Preliminary examination of the lunar samples from Apollo 12, *Science* **167,** 1325–1339.

Shoemaker E. M., Hait M. H., Swann G. A., Schleicher D. L., Schaber R. L., Sutton R. L., Dahlem D. H. Goddard E. N., and Waters A. C. (1970) Origin of the lunar regolith at Tranquillity Base. *Proc. Apollo 11 Lunar Sci. Conf., Geochim. Cosmochim. Acta,* Suppl. 1, Vol. 3, 2399–2412. Pergamon.

Soderblom L. A. (1970) A model for small impact erosion applied to the lunar surface. *J. Geophys. Res.* **75,** 2655–2661.

Toksöz M. N., Press F., Dainty A., Anderson K., Latham G., Ewing M., Dorman J., Lammlein D., Sutton G., and Duennebier F. (1972) Structure, composition and properties of lunar crust. *Proc. Third Lunar Sci. Conf., Geochim. Cosmochim. Acta,* Suppl. 3, Vol. 3, 2527–2544. MIT Press.

Wahlen M., Honda M., Imamura M., Fruchter S. S., Finkel R. C., Hohl C. P., Arnold J. R., and Reedy

R. C. (1972) Cosmogenic nuclides in football-sized rocks. *Proc. Third Lunar Sci. Conf., Geochim. Cosmochim. Acta*, Suppl. 3, Vol. 2, 1719–1732. MIT Press.

Walker E. H. (1967) Statistics of impact crater accumulation on the lunar surface exposed to a distribution of impacting bodies. *Icarus* **7**, 233–243.

Whipple F. L. (1967) On maintaining the meteoritic complex. *The Zodiacal Light and the Interplanetary Medium* (editor J. L. Weinberg), 409–426, NASA SP-150.

Proceedings of the Fourth Lunar Science Conference
(Supplement 4, *Geochimica et Cosmochimica Acta*)
Vol. 3, pp. 3235–3253

Microcraters on Apollo 15 and 16 rocks

D. A. Morrison, D. S. McKay, and R. M. Fruland

Johnson Space Center Houston, Texas 77058

H. J. Moore

U.S. Geological Survey, Menlo Park, Calif. 94025

Abstract—Microcrater frequency distributions, determined for 11 Apollo 16 rocks and three Apollo 15 rocks, fall into four categories. Category 1 rocks (68415, 68416, 69935, 62235) are angular, cratered on one side only and have moderate crater densities. Category 2 rocks (60016, 66075, 61175) are subrounded, cratered on all sides and have distributions suggestive of the steady state. Category 3 rocks (61015, 62295) are subangular and cratered on only one side but the crater frequency distributions have some of the characteristics of category 2 rocks. Category 4 rocks (15015, 15017, 15076, 60335) are angular, cratered on only one side and have moderated to very low crater densities.

The crater frequency distributions of categories 1 and 4 have properties indicating the possibility of estimating the time they were exposed to micrometeor bombardment. Category 1 rocks appear to have been exposed for 2–3 m.y. These rocks particularly 68415, 68416, and 69935 may be ejecta from South Ray Crater, indicating an age of 2–3 m.y. for South Ray Crater. Category 4 rocks have been exposed for much shorter periods. We estimate exposure times of $3-7 \times 10^2$ yrs. for 15017 (glass surface), 2×10^4 yrs. for 15015 (glass surface), and approximately 8×10^5 yrs. for 60335 and 15076. Total exposure time for category 2 rocks is estimated to be in excess of 20 m.y. If any category 2 rocks are North Ray ejecta then North Ray Crater formed more than 20 m.y. ago.

The crater frequency distributions for glass surfaces 15015 and 15017 show a reduction in the frequency of craters less than 30 microns in diameter. Our data show a relatively flat slope of -1.3 in the range 10–30 microns (μm) central pit diameter, approximately -2.5 below 10 μm central pit diameter, and approximately -2.0 below 1.0 μm central pit diameter.

Small particles of varying morphology appear to have deposited continuously on the surfaces of 15015 and 15017 with deposition rates significantly larger than microcratering rates. The abundance of accretionary particles on the surface of 15015 is sufficient to produce a steady-state crater distribution by superposition of particles on craters. This steady-state distribution would be substantially different than the steady-state distribution of craters that result from destruction of craters by other craters.

Spalling is a function of pit diameter. Craters less than 3 microns show no spall. For larger craters the relation between pit (P_d) and spall (S_d) diameter is: $S_d = 2.37 \, P_d^{1.07}$.

Data from rock 14301 suggest a cumulative flux equation of the form $\phi = 10^{-12.26} M^{-1.21}$ per cm^2 per year for a micrometeor mass (M) of 10^{-6} g and larger. For masses down to 10^{-10} g the exponent on M is -0.5, (i.e. the crater diameter range 10–30 μm). Below this exponents increase to -1.0 and then decrease to -0.79.

Introduction

The purpose of this paper is to present frequency distributions of microcraters on Apollo 15 and 16 rocks, interpret the distributions, estimate the time surfaces have been exposed to meteoroid impacts, draw attention to accretionary processes on some lunar glass surfaces, and discuss spalling around central pits of microcraters.

Location and surface orientations of all the rocks examined have been determined (Sutton et al., 1972; Swann et al., 1972; Muehlberger et al., 1972). This is
important because interpretations of the crater distributions, correlations with
cosmic particle track exposure ages, and interpretations of history and
provenance of the samples require detailed knowledge of the orientation and
location of the samples on the lunar surface.

Two glass surfaces on Apollo 15 rocks were examined with a scanning electron
microscope (SEM) to magnifications of 60,000 X. Apollo 16 rocks were examined
with binocular microscopes partly in cooperation with G. Neukum and F. Hörz.
Experimental data and procedures for the Apollo 16 rocks appear in Neukum et
al. (this volume).

Rocks and their surfaces described below can be placed in four categories: (1)
angular rocks with moderately cratered surfaces, (2) rounded to subrounded rocks
with densely cratered surfaces, (3) angular rocks with moderately cratered surfaces but crater frequency distributions we do not understand, and (4) rocks with
exceptionally low crater frequencies.

Before proceeding to a description of the crater frequency distributions, it is
useful to make some brief comments about the limiting or "steady-state" crater
frequency distribution first demonstrated empirically using rock 14306 (Morrison
et al., 1972b) and later confirmed by theoretical considerations (Hartung et al.,
1973). The steady-state distribution describes a cratered surface that has reached
a condition where craters are destroyed as rapidly as they are formed (see, for
example, Moore, 1964; Soderblom, 1970; Marcus, 1970). Theoretical considerations (Marcus, 1970) indicate this steady-state distribution should be of the form:

$$N = CD^{-2} \tag{1}$$

where N is the cumulative number of craters per unit area with spall diameters
equal to D and larger and C is a constant. Additionally, the exponent on D for the
crater production distribution must be less than -2 (i.e., -2 to -4 but not -1.9 to
-1.0) when craters are destroyed by super-position by one upon another. Evaluation of cumulative frequencies of craters with spall diameters larger than about
0.08 cm for rock 14306 confirm the form of Equation 1 and other data satisfy the
condition for the crater production frequency distribution for which the exponent
on D is near -3. From 14306, the value of C is estimated to be between $10^{-0.096}$
and $10^{-0.398}$ (Morrison et al., 1972b). Equation 1 and $C = 10^{-0.398}$ imply the cumulative frequency of craters with pit diameter 0.1 cm and larger will not exceed 2.5
and 4.5 when the spall-pit diameter ratios are 3.9 and 3.0 respectively. When C is
$10^{-0.096}$, these frequencies are twice as large. Thus, when the cumulative frequency
of pit diameters is less than about 2.5 craters/cm^2, for craters with pit diameters
0.1 cm (1000 μm) and larger, the steady-state has not been reached for craters this
size and larger. When the cumulative frequency exceeds 2.5 craters/cm^2 at 0.1 cm
diameters, the steady-state may have been attained for craters this size and smaller. There may be evidence for a change in slope of the frequency distributions
to confirm the steady-state if statistical fluctuations in the distributions are not
pronounced. Finally, it should be pointed out that Equation 1 does not apply to

craters with pits less than about 0.003 m (30 μm) in diameter because of a flexure in the crater production distribution.

CRATER FREQUENCY DISTRIBUTIONS

As noted in the introduction, the frequency distributions fell into four categories (Table 1). For the Apollo 16 rocks, we find these categories are entirely consistent with field data and rock descriptions which delineate two groups of rocks—angular and rounded. Categories 1, 3, and 4 belong to the angular group while 2 belongs to the rounded group. Category 3 is somewhat artificial since we decline to fully interpret the distributions.

Category 1 distributions (60315, 62235, 68415, 68416, 69935)

Rocks and chips with category 1 distributions are angular and microcraters are found only on half the surfaces of each rock. Thus, there is no evidence for extensive erosion by micrometeors and tumbling of the small rocks (60315, 62235) on the surface. A cumulative frequency of 10 per cm² corresponds to craters with pit diameters between 0.03 and 0.04 cm (Fig. 1). Slopes of the crater frequency distributions for craters larger than about 0.05 cm are steep and near -3. The distributions either intersect the 0.1 cm pit diameter ordinate at 0.2 to 0.6 craters/cm² or require a modest extrapolation to intersect between these values (Table 1). Thus, the steady-state has not been attained at this size. Slopes of the distributions flatten below 0.02 cm pit diameters. This flattening is partly due to incomplete counts of the smallest craters and is characteristic of distributions obtained with optical microscopes. The frequency of craters is also sufficiently large that the steady-state may have attained for these smaller craters and the flattening may result from these two factors. When spall diameters are plotted using measured pit to spall diameter ratios for each rock, the cumulative frequency distributions fall well below the limiting or steady-state frequency distributions for craters with spall diameters larger than about 0.08 cm. For smaller craters, the curves become tangent to the limiting distribution. Visual graphic fits of lines with -3 slopes (see Fig. 1) to the cumulative frequency distributions for the larger craters intersect the 0.1 cm spall diameter ordinate at cumulative frequencies of 19 to 32 craters/cm². These values are always near the intersection of the actual plots.

Some variation in the intercept at the 0.1 cm spall diameter ordinate may result from the orientation of the surface of the sample. In particular, 68416 was collected from the upper surface of a 0.5 m feldspathic boulder on the rim of a 5 m crater at Station 8 whereas 68415 was chipped from a north-facing surface of the same boulder below the position of 68416 (Muehlberger et al., 1972). Such a difference in orientation on the lunar surface could partly account for the difference in intercepts for 68416 (28 craters/cm²) and 68415 (19 craters/cm²). The remaining surfaces in this category (60315, 62235, 69935) have essentially the same distributions. They were small rocks perched on the lunar surface without fillets

Table 1. Exposure age estimates.

Rock	Crater Freq. D_p 0.1 cm	Crater Freq. D_s 0.1 cm	Exposure Ages				Best Guess	D_s/D_p
			ϕ_1	ϕ_3	ϕ_k	ϕ_o		
Category 1 populations								
68416	0.5	28	1.2×10^5	5.4×10^5	2.6×10^6	0.7 to 3.0×10^7		3.9
68415	0.2	19	0.8×10^5	3.7×10^5	1.8×10^6	0.32 to 1.0×10^7		3.9
	0.2	20	0.85×10^5	4.0×10^5	1.9×10^6	0.34 to 1.1×10^7		3.0
60315	0.5	28	1.2×10^5	5.4×10^5	2.6×10^6	0.7 to 3.0×10^7	2–3×10^6	3.7
62235	0.3	32	14.0×10^5	6.4×10^5	3.0×10^6	0.85 to 3.4		4.2
	0.6	32	14.0×10^5	6.4×10^5	3.0×10^6	0.85 to 3.4		
69935	0.6	30	1.3×10^5	6.0×10^5	2.8×10^6	0.8 to 3.2×10^7		3.3
Category 2 populations								
60016	2.0	79	3.8×10^5	16.0×10^6	7.2×10^6	—		3.4
61175	2.5	75	3.5×10^5	1.5×10^6	7.0×10^6	—		3.0
66075	2.9	100	4.6×10^5	1.9×10^6	9.5×10^6	—		3.4
	1.9	75	3.5×10^5	1.5×10^6	7.0×10^6	—	> 7–10×10^6	
	2.5	—	—	—	—	—		
	2.3	100	4.6×10^5	1.9×10^6	9.5×10^6	—		
	2.2	—	—	—	—	—		
Category 3 populations								
61015	0.7	40	—	—	—	—		
	1.2	35	—	—	—	—		
62295	0.6	40–70	—	—	—	—		
Category 4 populations								
15015	0.007	0.1–0.2	4–8×10^2	2–4×10^3	1–2×10^4	0.4 to 1.0×10^3	1–2×10^4	3.0
15076	0.4	8–18	3–7×10^4	1.6–36×10^5	0.8–1.7×10^6	1.7 to 3.5×10^6	8–17×10^5	4.0
60335	0.3	6–8	2.4–3.2×10^4	1.2–1.6×10^5	6–8×10^5	0.8 to 1.7×10^6	6–8×10^5	3.0
15017	—	—	—	—	3–7×10^2	—	3–7×10^2	3.0

Estimated 0.1 cm diameter intercepts and times of exposure to micrometeors for Apollo 15 and 16 rocks using various models. Models are: ϕ_1—Soberman's (1971) micrometeor flux, ϕ_3—Whipple's curve B, ϕ_k empirically calibrated variable flux, ϕ_o empirically calibrated constant flux. Column labeled "best guess" based upon ϕ_k. Total exposure time for category 2 rocks is at least twice the values shown.

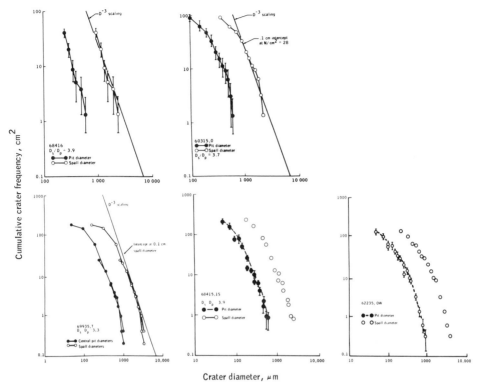

Fig. 1. Cumulative frequency distributions for category 1 rocks: (a) 68415, (b) 68416, (c) 60315, (d) 62235, (e) 69935. Lines show D^{-3} scaling and determination of the 0.1 cm intercept is shown by 69935.

and photographic data do not indicate that their crater frequency distribution would be affected by their orientations.

Category 2 distributions (60016, 61175, 66075)

Rocks with category 2 distributions are typically rounded to subrounded and microcraters are found on all surfaces of each rock. Thus, the rocks have tumbled and rounding may have resulted from extensive erosion by micrometeors. Cumulative frequencies of 10 per cm^2 corresponds to craters with pit diameters between 0.05 and 0.06 cm (Fig. 2). Slopes of the crater frequency distributions for craters larger than about 0.05 cm are not as steep as those of category 1 rocks and are near -2. In contrast with category 1 distributions, category 2 distributions intersect the 0.1 cm pit diameter ordinate at about 2.0 to 2.9 craters/cm^2 (Table 1). This suggests the rocks are in or close to the steady-state for craters with pit diameters at least as large as 0.1 cm and spall diameters near 0.3 to 0.4 cm. Flattening of the distributions for the smallest craters shown is chiefly the result of resolution difficulties.

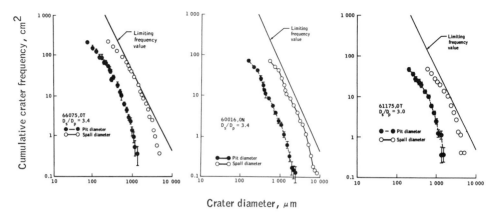

Fig. 2. Cumulative frequency distributions for category 2 rocks: (a) 60016, (b) 61175, (c) 60075. The limiting frequency (or steady-state) line is a plot of Equation 1 with $C = 10^{-0.398}$.

The form of these distributions indicate that a steady-state has attained for craters with spall diameters of 0.25 to 0.35 cm and smaller. The slope of the distribution for larger craters on some of the surfaces changes from -2 to -3 or so. Assuming the frequency of these larger craters represents a production distribution, graphical curves with a slope of -3 tangent to the distributions intersect the 0.1 cm spall diameter ordinate at 75 to 100 craters/cm^2.

Rocks 60016 and 66075 had poorly developed fillets before they were collected. Apparently 61175 did not have a fillet.

Category 3 distributions (61015, 62295)

Rocks with category 3 distributions have characteristics of both categories 1 and 2. Both rocks were angular and perched on the lunar surface at the time of collection. Microcraters are found only on half the surfaces. Thus, the rocks have not tumbled or been extensively eroded by micrometeors. Cumulative frequencies of 10 per cm^2 correspond to pit diameters between 0.025 and 0.05 cm. Some of the frequency distributions are markedly undulatory suggesting there may be some problem associated with the counts and areas used to obtain the distributions. Alternatively, there may be significant statistical fluctuations. We do not believe these rock surfaces represent the steady-state because the intercepts at pit diameters of 0.1 cm are about 1 and substantially less than the lower bound steady-state intercept of 2.5. Frequencies elsewhere are also too low. Cumulative frequencies are shown in Neukum *et al.* (this volume).

Category 4 distributions (60335, 15015, 15017, 15076)

These distributions have exceptionally low frequencies of craters. Crystalline rocks vary from angular (60335) to subrounded (15076). 15015 is a breccia coated with smooth, vesicular glass; the smooth glass surfaces are contorted, complex, and irregular and have produced local shielding from micrometeors. 15017 was an uneroded glass sphere. Thus, with the exception of 15076, the macroscopic character of the rocks indicates they are young and uneroded. For 60335, a cumulative frequency of 10 per cm^2 corresponds to pit diameters of 0.025 cm; the frequency for a pit diameter of 0.1 cm is near 0.3 craters/cm^2; and, the cumulative frequency of about 8 per cm^2 corresponds with spall diameters of 0.1 cm and larger. These are the lowest frequencies of craters on Apollo 16 rocks reported here.

Our microscopic data for 15076 suggest the crater frequency distribution for 15076 is essentially the same as that of 60335. Also included in Fig. 3 are the results of Schneider *et al.* (1972) for 1.0 μm craters on a chip of 15076 which indicate that the cumulative frequencies of craters are inversely proportional to

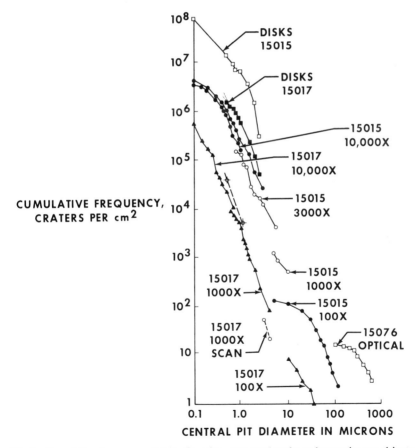

Fig. 3. Cumulative frequency distributions for category 4 rocks and accretionary objects on 15017 and 15015. Numbers refer to magnifications of SEM mosaics. 15076 SEM points from Schneider *et al.* (1972) shown as crossed circles. Open circles for diameters between 1 and 10 microns represent visual scans at 1000×.

the square of the pit diameter between 0.1 and 0.5 μm. These counts when combined with our whole rock microscopic data indicate that a large flexure in the distribution occurs between pit diameters of 1000 μm and about 10 μm if both surfaces are of the same age or else sample related problems.

Cumulative frequency distributions for 15017 and 15015 were obtained from photomosaics with magnifications of 100, 600, 1000, and 10,000× (Fig. 3). Additional frequencies were obtained by visual scanning at 1000× of sequential frames in order to cover areas up to 1 mm². Surfaces were gold coated for SEM examination. Visible microcraters and areas studied under larger magnifications were mapped and referenced to a grid system on 100× photo maps of the entire surface of each sample. The total areas examined at magnifications larger than 100× were dictated by requiring approximately 30 craters for each magnification to yield acceptable statistics (Brownlee *et al.*, 1972). Areas were calculated from direct measurement of the mosaics and corrected to allow for surface curvatures, vesicles, and accretionary particles superposed on the surface. SEM magnifications were calibrated using a diffraction grid with 54,865 lines/inch. We consider the results for 15017 the most reliable because the surfaces are smooth, vesicles are absent, and the surface has few superposed glass splashes. This was not the case for 15015.

The cumulative frequency distribution for 15017 has a relatively flat slope (approximately -1.3) from pit diameters of 10 μm to 30 μm. Below 10 μm the slope steepens (approximately -2.5) and then flattens (approximately -2.0) again below 1.0 μm. Throughout the length of the distribution, the frequencies are everywhere more than three orders of magnitude below the steady-state distribution (*see* Fig. 3).

Counts for 15015 were found to be erratic from area to area (Fig. 3). For example, the frequencies obtained from 1000× mosaics are about 1/6 of those obtained at 3000×. The frequencies from 10,000× mosaics were found to differ from each other (*see* Fig. 3).

ACCRETIONARY OBJECTS

The glass surfaces of 15017 and 15015 were found to have small positive objects 0.1 to 1000 μm across adhering to and superposed on their surfaces. These positive objects (Fig. 4), which are the antithesis of craters, have several morphologies: (1) spheroids, (2) disks, (3) rayed and stellate forms, (4) angular to subrounded shapes, primarily glasses, and (5) crystals. Spheroidal objects are abundant in the 0.1 to 0.2 μm size range. They are spherical except where they wet the glass surface. Discoid objects are the most abundant form on the surfaces. They vary in diameter from less than 0.1 μm to millimeter sizes but most are less than 10 μm across. Diameter-thickness ratios vary considerably from slightly less than 1 to very large values. Disk edges are typically round and indicate they were once molten. Thin disks conform to the underlying topography as though they deformed plastically upon impact with the surface, and disks superposed on rims of pre-existing disks conform to the rounded rims of the pre-existing disk. Most

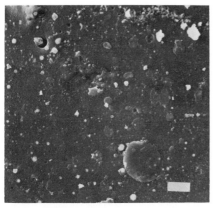 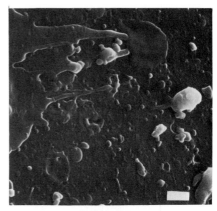

Fig. 4a   Fig. 4b

Fig. 4(a). 1000× SEM photograph of accretionary objects, primarily disks on 15015. Scale bar is ≈ 10 μm. (b) 10,000× SEM photograph of accretionary objects on 15015. A rayed glass splash is shown in upper left. Small disks are superposed on the splash. A large disk of somewhat irregular shape is superposed on a ray of the glass splash. Several sub-micron microcraters are also visible. Scale bar is ≈ 1 μm.

disks on 15017 are elliptical, in contrast to those on 15015 which are more circular. This suggests more disks impacted obliquely on 15017 than on 15015, a result consistent with the difference in symmetry of microcraters on the two rocks. Disks are gradational with rayed and stellate objects which are characterized by thin finger-like extensions radiating from a disk-like center. These objects are the result of splashes of molten materials. Angular to subrounded objects have a variety of shapes but there is no evidence they were entirely molten. Some fragments and crystals occur in the centers of impact pits. The deformation and morphologies of accretionary objects indicates they were molten to solid ejecta from lunar impact craters.

Glass objects and crystals are superposed on the glass surface, on the microcraters, and upon each other. Discoid objects are cratered to varying degrees. Neither disks nor spheroids occur on surfaces shielded from micrometeors. Therefore, microcratering and particle accretion occurred continuously over a like interval of time on exposed surfaces.

The frequencies of superposed accretionary objects exceed those of microcraters of comparable size (Fig. 3). Cumulative frequencies of disks on 15017 are two orders of magnitude larger than the cumulative frequency of microcraters in the 0.3 to 3 μm size range. Although data on the frequency distribution of the accretionary objects for 15017 are limited, the slope of the curve is steep and comparable to the slope of the microcrater distribution. Similar results are found for disks on 15015. Both the frequency of the microcraters and disks varies from place to place on this sample; and, an increase in frequency of one is matched by an increase of the other. Comparison of the disk distribution with the highest microcrater frequencies on 15015 suggests disks are about 6 times more frequent than craters in the interval 0.1 to 2.5 μm. Although very small accretionary objects are more frequent than very small microcraters, we find no evidence to indicate particles larger than 100 μm are as frequent as craters of the same size; thus, accretionary particles do not affect steady-state distributions of craters larger than 100 to 200 μm.

Spalling

Microcraters on the glass surface of rock 15015 may be grouped into five types. The first type are circular to oval shaped glass-lined pits (<3 μm across) rimmed by upward protruding bulbous shapes (Fig. 5a) with no evidence for spall

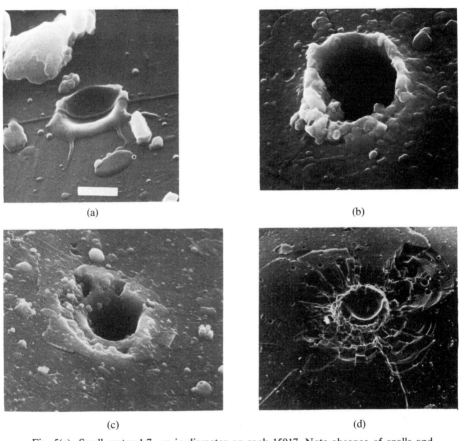

(a)   (b)

(c)   (d)

Fig. 5(a). Small crater 1.7 μm in diameter on rock 15017. Note absence of spalls and raised rim. Contamination line emphasizes morphology of the crater. Accretionary objects including disks, spherules and angular glass and crystal fragments are also visible. Bar corresponds to one micron. 20,000$\times$ SEM photo. (b) Microcrater with spalls that have not separated. Central pit is 3.6 μm in diameter. (c) Microcrater with partial spall separation. Central pit is 11.6 μm in diameter. Arc of separation is 120°. (d) Microcrater with multiple spall zones and complete separation of spalls from within spall zone. Central pit is 110 μm in diameter.

or radial fractures. The second type are pits 4 μm to 12 μm across with surrounding surfaces that slope outward to distances near 10 μm where discontinuous fractures occur (Fig. 5b). Radial fractures are found near the pit. Thus, the spalls did not separate. One crater had two rings of concentric arcs and radial fractures. Craters with both outward sloping fracture surfaces and inward sloping spall surfaces represent the third type and are common for craters larger than a few micrometers. Only part of the spalls have ejected and a second spall zone is present (Fig. 5c). The fourth type of crater, which is prevalent for the larger sizes, has complete separation of spalls from an innermost zone and complete to partial separation of spalls from outer zones (Fig. 5d). In one region, concentric arcs and radial fractures are found beyond the spall surfaces. The final type of crater, where there is evidence for two or more spall zones, has been previously illustrated (Figs. 5c and 5d).

Grouping of the craters into four size classes and estimating the degrees of arc where spall separation has occurred shows the extent of spall separation as a function of crater size. For craters of types 1 and 2 the arc of spall separation is zero since there is no evidence for separation (Figs. 5a and 5b). Spalls have separated from about 120° of arc for the crater in Fig. 5c and spalls have separated through 360° of arc in the first zone for the crater in Fig. 5d. Estimates of arcs of spall separation from the first or innermost zone (Fig. 6) for 121 craters shows: (1)

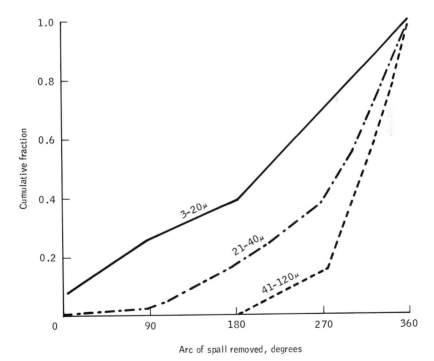

Fig. 6. Cumulative frequency of craters with arcs of spall separation equal to or less than angle indicated for three size classes of craters.

thirty-eight percent of the craters between 3 and 20 μm had spalls ejected through 180° and less, (2) seventeen percent of the craters 21 to 40 μm across had spalls ejected through 180° of arc and less, and (3) all the craters 41 to 110 μm across had spalls ejected through 180° of arc or more. Thus, the separation and ejection of spalls is a function of the size of the central-pit and other variables.

Pit and spall diameters of the microcraters in 15015 were measured from SEM photographs. For the measurements, the limit of spall represented by the outer concentric fractures around the central pit was measured even though separation of the spall did not occur. A least squares fit for 117 microcraters with pit diameters (P_d) between 3 and 110 μm and spall diameters (S_d) between 8.6 and 390 μm yields:

$$S_d = 2.37 \, P_d^{1.07} \tag{2}$$

Here, S_d and P_d are in microns and the standard error of the estimate of S_d on P_d is $\pm\log_{10} 1.38$. The results here have been changed from those of our previous abstract (Morrison *et al.*, 1973) because several previous measurements were found to be in error.

The types of craters on the glass surfaces of rock 15015 are consistent with previous results for microcraters on glass (Hartung *et al.*, 1972; Neukum *et al.*, 1972, p. 2794; Vedder, 1971) in that the amount of spalling and spall separation decreases with decreasing size; and in some respects, the small microcraters behave like large craters produced by hyper-velocity impacts with aluminum targets (Denardo, 1962).

The relationship between pit and spall diameters are in reasonable agreement with experimental data for low density microparticle impacts with glass (Mandeville and Vedder, 1971), although there are differences. Predictions of spall diameters for craters with central pits 3 μm across are essentially the same as those of Mandeville and Vedder (Table 2). For pit diameters of 100 μm, disagreement is large—particularly for the highest velocity experimental data (12–14 km/sec) but reasonable agreement is obtained for intermediate velocities (7–9 km/sec, *see* Table 2). For the experimental data at velocities of 5.3–6.4 km/sec, predicted spall diameters are larger than for 15015.

Table 2. Measured and predicted spall diameters.

Data Source	Equation S_D	S_D for $P_D = 3 \, \mu$m	S_D for $P_D = 100 \, \mu$m
Lunar rock 15015	2.37 $P_D^{1.07}$	7.8	330
Experiment ($V_P = 12$–14 km/sec)	1.45 $P_D^{1.44}$	7.3	1100
Experiment ($V_P = 7$–9 km/sec)	1.82 $P_D^{1.15}$	6.9	380
Experiment ($V_P = 5.3$–6.4 km/sec)	1.59 $P_D^{1.27}$	6.6	550

MICROMETEOR EXPOSURE AGES

Previous work has suggested it may be possible to estimate the time rock surfaces have been exposed to micrometeor impacts (Morrison *et al.*, 1972a,b; Gault *et al.*, 1972). Such estimates are model dependent and we have estimated these times or micrometeor exposure ages using four models (Table 1). The first two are theoretical and combine a hyper-velocity impact equation for rock targets with (a) Soberman's (1971) best estimate of the micrometeor flux (ϕ_1, Table 1), and (b) Whipple's (1963) curve B (ϕ_3, Table 1). The second two, first proposed by Morrison *et al.* (1972a) are empirical and combine data on crater frequencies for rock surfaces with cosmic ray particle track exposure ages for the same rocks. Because available data are for rocks with cosmic ray exposure ages near 0.3 to 1 m.y., two models are considered: (1) micrometeor flux has been constant for 10 or so m.y. (ϕ_k, Table 1), and (2) the micrometeor flux has increased during 10 or so m.y. to the present value (ϕ_v, Table 1). These second two models rely heavily on data for rock 14301. The models are portrayed graphically in Fig. 7 and more detail on them appears in Morrison *et al.* (1972b). We use a graphic technique and estimate micrometeor ages by obtaining the cumulative frequency of craters with spall diameter 0.1 cm (1000 μm) and larger which in turn corresponds to a micrometeor exposure age as shown in Fig. 7. It should be noted that we correctly predicted the cosmic particle track exposure age for 14311 using unsophisticated graphical methods and the empirical ϕ_k model. At this time, we feel such methods are consistent with the data which will yield only approximate ages. Our crater frequency distributions for category 1 rocks indicate they were exposed to micrometeors for essentially the same length of time and that cumulative frequency

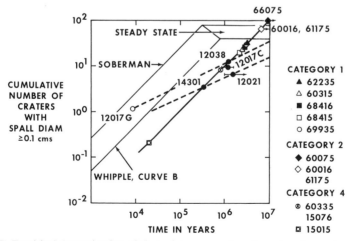

Fig. 7. Empirical determination of time of exposure of surfaces to micrometeors. D^{-3} scaling of cumulative frequency curve determines Y-axis point. Intercept with flux models determines exposure age. Upper solid line represents Soberman's (1971) best flux estimate (ϕ_1) and lower solid line represents Whipple's (1963) curve B model (ϕ_2). Line through points 14301 and 14308 represents ϕ_k model. Dotted lines represent ϕ_v. Shaded area is steady-state condition defined by rock 14306.

of craters with spall diameters 0.1 cm and larger is near 19 to 32 per cm^2. Orientation of 68415 on the lunar surface suggests a value of 19 per cm^2 may be low because of shielding and thus it represents a lower bound. The exposure times of the surfaces vary from 0.08 (ϕ_1) to 34 (ϕ_v) m.y. depending on the model used (*see* Table 1). The empirical correlation, ϕ_k, appears to be the best estimate of exposure age because it is empirically correlated to 14301, correctly predicts the exposure age of 14311, and the Passive Seismic Experiment (Latham *et al.*, 1973) indicates meteor fluxes are lower than previous estimates—a result consistent with the relationship between ϕ_1, ϕ_3, and ϕ_k. Thus, we place the duration of microcratering for rock 68415 near 2 to 3 m.y. Since only the uppermost surfaces were cratered, this would represent the total exposure time.

Category 2 surfaces appear to have steady-state distributions for most crater sizes and they are cratered on all sides. Estimates of frequencies of craters with spall diameters larger than 0.1 cm can be only made for some of the distributions by somewhat tenuous extrapolations. Such extrapolations yield numbers of 7.0 to 9.5 m.y. for ϕ_k. Clearly, these represent lower bounds for a single surface. Since all sides of the rocks are cratered, the rock exposure times are more than twice the time for a single surface. Thus, the ϕ_k model indicates they have been exposed to micrometeors for more than 14.0 to 19.0 m.y. As for category 3 surfaces, we feel they are akin to category 1; but, because of the undulatory nature of the distributions we do not estimate their micrometeor ages.

Category 4 distributions, particularly 15015 and 15017, present special problems because of the limited number of craters larger than 100 μm and the probable existence of a flexture in the frequency distribution below but near this size. The distribution for 15076 crosses the 0.1 cm spall diameter ordinate near 8 craters/cm^2 and a curve with a slope of -3 drawn tangent to the distribution intersects at 18 craters per cm^2. These intercepts yield ages near 0.8 to 1.7 m.y. The 60335 distribution crosses the 0.1 cm spall diameter near 6 per cm^2 and a tangent curve with a slope of -3 intersects at 8 per cm^2. Thus, 60335 is near 0.6 to 0.8 m.y. Intercepts for 15015 are between 0.1 to 0.2 using -3 slope curves that bracket the distribution of the largest craters. This yields best guess micrometeor ages of 0.01 and 0.02 m.y. Such an extrapolation is rather large but it is clear that 15015 has been exposed a relatively short period of time in comparison with the other rocks.

The distribution of 15017 stops at crater sizes larger than 35 μm and extrapolations are even more tenuous than that for 15015. We believe the exposure age of 15017 is 1/30 that of 15015 because ratios of crater frequencies for the two rocks corresponding to crater diameters of 20 to 30 μm are about 1/30. This would indicate that 15017 is near 300 to 700 years old.

DISCUSSION

Our results for Apollo 16 suggest there is a population of angular rocks with similar moderate crater densities (category 1); and, our best estimate for the micrometeor exposure age is near 2–3 m.y. The samples were cratered on one side only, and two come from a large boulder at Station 8. It seems likely that this group of rocks are part of the ejecta from South Ray Crater because it is the youngest feature of its size in the immediate vicinity of the 16 site and rays from it can be traced across Station 8, to the LM-Station 1 region, and on northward to North Ray Crater (Muehlberger *et al.*, 1972). Rocks of category 2, which were exposed more than about 20 m.y., cannot be assigned to any particular crater, although North Ray is a likely candidate. If they are from North Ray, North Ray is older than 20 m.y.

It is also noteworthy, that cosmic particle track ages for one of the rocks, 68415, is about 4 m.y. and because of the uncertainties Behrmann *et al.* (1973) estimate 2 to 4 m.y. We consider this to be in reasonable agreement with our independent estimate of 2 or 3 m.y. for 68415 and other rocks in category 1.

We know of no reliable particle track ages for category 4 rocks. Schneider *et al.* (1972) report that an uncratered surface of rock 15015 (15015,24) has a solar flare track age between 4 and 40 yr. This result does not appear to represent 15015 because we find it has craters. Additionally, we find a wide spatial-variation in crater frequencies below 10 μm diameter. It seems likely that the solar flare track age corresponds to part of 15015 that was not exposed to micrometeors through-out the total time the rock spent on the surface. Alternatively, the small craters below 10 μm may be the result of a non-steady flux of secondary particles.

Consideration of solar flare track ages for 15076 yield similar results. Such track ages are 4×10^4 to 1.6×10^5 yrs. (Schneider *et al.*, 1972) which are smaller than our "best guess" estimates from larger craters: 8×10^5 to 17×10^5 yrs. Discrepancies are further illustrated when Schneider's (1972) counts for craters less than 1.0 μm are compared with our counts for 15017. Frequencies for 15076 are about twice those of 15017 and imply the solar flare track age of 15017 should be 2×10^4 to 8×10^4 yrs. Yet, consideration of frequency of craters 20 to 30 μm yield micrometeor ages less than 10^3 yrs. for 15017.

It is noteworthy that some agreement is found between our results and Schneider *et al.* (1972). In particular, slopes for both frequency distributions are near -2.5. This slope is substantially different from that reported by Hartung *et al.* (1972) which is about -1.

At this time, our best estimate of the micrometeor flux is represented by the data on rock 14301 for which a cumulative frequency of 3.5 craters/cm^2 with spall diameters 0.1 cm and larger corresponds to 3.4×10^5 yrs. If the projectile density is taken as 0.44 g/cm^3 (Morrison *et al.*, 1972b), a crater with a spall diameter of 0.1 cm (1000 μm) corresponds to a 10^{-6} g particle. Slopes on most carefully determined crater frequency distributions are near -3.0 for craters larger than 100 μm or so and an equation for the flux with a form like Soberman's (ϕ_1) is reasonable:

$$\phi = KM^{-1.21} \tag{3}$$

14301 implies $\phi = 1 \times 10^{-5}$ craters/cm^2/yr when $M = 10^{-6}$ g. Thus, K is about $10^{-12.26}$. Latham *et al.* (1973) obtain:

$$N = 10^{-11.62} M^{-1.16} \text{ craters/cm}^2/\text{yr}. \tag{4}$$

from their seismic data for the meteor mass range of 10^2 to 10^6 g. The values of the cumulative frequencies for 1.0 g meteoroids from Equations 3 and 4 agree within a factor of 4.5. Latham's equation predicts the frequency of 10^{-6} g meteors to be 2.2×10^{-5} which is remarkably close to the results from 14301. Thus, Latham's curve may be as reasonable an estimate of the micrometeor flux from 10^6 to 10^{-6} g as any estimate. It predicts an exponent on the crater production curve of -2.93 which is close to -3. Although we assume low density micrometeors, we note Latham's calibration data were produced by low density projectiles, i.e., LM ascent stages and SIVB impacts. Increase of micrometeor densities to 3.5 and 8 would reduce fluxes by approximately 1/3.5 and 1/5.5 in Equation (3).

Although we feel that there are unresolved problems with crater frequency distributions below crater diameters of 30 μm, we have interpreted the distributions for 15015 and 15017 in terms of a mass-flux distribution (Fig. 8) using our "best estimate" of exposure ages and the particle mass-spall relationship of Morrison *et al.* (1972b, Equation (6), p. 2778). Exponents for M in flux distributions (*see* Equations (3) and (4)) can be calculated using equations in Morrison *et al.* (1972b) and our results for 15017. For the diameter ranges 10–30 μm, and 1–10 μm and 0.1–1 μm exponents on M should be near -0.5, -1.0, and -0.78 respectively. Our results on the form of the mass distribution deviate significantly from the form of the spacecraft distributions (Gault *et al.*, 1972) below a mass of

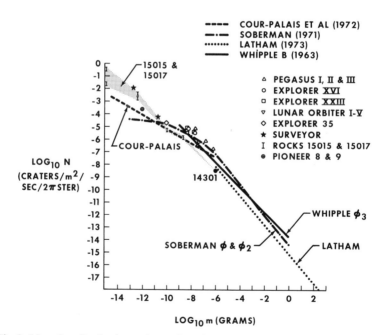

Fig. 8. Mass-flux distribution estimates based upon rocks 15015, 15017, and 14301. Data for orbital spacecraft after Gault *et al.* (1972). Predictions using the equation of Latham *et al.* (1973) for meteors is also shown. Surveyor data from Brownlee *et al.* (1971).

10^{-10} g. Between 10^{-10} and $10^{-8.6}$ g the distributions are parallel. The distribution below 10^{-10} g could represent a population composed chiefly of secondary particles, which, unlike the accretionary objects, have sufficient velocities to produce craters. Some models predict that small secondary impacts should be about 0.1 as frequent as primary craters in the same size (*see*, for example, Moore, 1964). Such a relationship would produce the flexure shown in Fig. 10 if the frequency of primary particles stopped at $10^{-8.7}$ g (Whipple, 1963) or decreased rapidly (Soberman, 1971). In terms of flux, our results are similar to those of the Apollo Window Meteoroid Experiment (Cour-Palais *et al.*, 1972) for the interval 10^{-8} to 10^{-11} g (Fig. 8). Frequencies exceed those of Pioneer 8 and 9 but are everywhere less than the limits established by studies of Surveyor III (Brownlee *et al.*, 1971).

CONCLUSIONS

1. Crater frequency distributions for Apollo 16 rocks fall into four categories. One of these (category 1) may represent ejecta from South Ray which formed about 2–3 m.y. ago.

2. Crater counts on rocks 15015, 15017, 15076, and 60335 indicate they have been exposed to micrometeors for a relatively short time. We have made "best guess" estimates of these times, but there are no reliable particle track ages to substantiate or refute these estimates. Thus, we cannot distinguish between constant and variable flux models for micrometeors.

3. Some Apollo 16 rocks (category 2) have been exposed to micrometeors for periods in excess of 20 m.y.

4. The origin, flux, and frequency distributions of craters 0.1 to 30 μm pit diameters are poorly understood at this time. Whether or not these small craters are produced by primary or secondary particles is unresolved.

5. Accretionary particles, chiefly disks, are deposited continuously on exposed surfaces and may produce the darkening typical of exposed lunar rocks. Accretion rates of small objects on lunar rock surfaces exceed the rate of microcratering in the micron and submicron size range. This may result in a steady-state for which small craters are destroyed by superposition of disks, other objects, and craters. This steady-state, like that of the regolith, could be substantially different than the steady-state for larger craters (> 100 μm) on rock surfaces. Accretionary particles are too small to affect craters larger than 100 μm.

6. Spalling around microcrater pits is a function of pit diameter. Both the ratio of spall to pit diameter and completeness of spalling increase with increasing size. Spalling does not occur around crater pits less than about 3 μm.

7. Our estimate of the micrometeor flux for 10^{-6} g masses and larger agrees well with extrapolations of the equation of Latham *et al.* (1973) for much larger micrometeors. There may be a population of secondary particles of masses 10^{-10} g and smaller although primary particles cannot be excluded.

REFERENCES

Behrmann C., Crozaz G., Drozd R., Hohenberg C., Ralston C., Walker R., and Yuhas D. (1973) Radiation history of the Apollo 16 site (abstract). In *Lunar Science—IV*, pp. 54–56. The Lunar Science Institute, Houston.

Brownlee D., Bucher W., and Hodge P. (1971) Micrometeoroid flux from Surveyor glass surfaces. *Proc. Second Lunar Sci. Conf., Geochim. Cosmochim. Acta*, Suppl. 2, Vol. 3, pp. 2781–2789. MIT Press.

Brownlee D. E., Hörz F., Hartung J. B., and Gault D. E. (1972) Micrometeoroid craters smaller than 100 microns. In *The Apollo 15 Lunar Samples*, pp. 407–409. The Lunar Science Institute, Houston.

Cour-Palais B. G., Flaherty R. E., Brown M. L., and McKay D. S. (1972) Apollo window meteoroid experiment, S-176 (abstract). In *Lunar Science—IV*, pp. 157–159. The Lunar Science Institute, Houston.

Denardo B. P. (1962) Measurements of momentum transfer from plastic projectile to massive aluminum targets at speeds up to 25,600 feet per second. *NASA Tech Note Tn D-1210*, 30 pp.

Gault D. E., Hörz F., and Hartung J. B. (1972) Effects of microcratering on the lunar surface. *Proc. Third Lunar Sci. Conf., Geochim. Cosmochim. Acta*, Suppl. 3, Vol. 3, 2713–2734. MIT Press.

Hartung J. B., Hörz F., and Gault D. E. (1972) Lunar microcraters and interplanetary dust. *Proc. Third Lunar Sci. Conf., Geochim. Cosmochim. Acta*, Suppl. 3, Vol. 3, pp. 2735–2753. MIT Press.

Hartung J. B., Aitken F. K., Blackman J. W., and Hörz F. (1973) Microcrater population development on lunar rocks—a Monte Carlo approach (abstract). In *Lunar Science—IV*, p. 3395. The Lunar Science Institute, Houston.

Latham G., Dorman J., Duennebier F., Ewing M., Lammlein D., and Nakamura Y. (1973) Moon quakes, meteoroids, and the state of the lunar interior (abstract). In *Lunar Science—IV*, pp. 457–459. The Lunar Science Institute, Houston.

Mandeville J. C. and Vedder J. F. (1971) Microcraters formed in glass by low density projectiles. *Earth Planet. Sci. Lett.* **11**, pp. 297–307.

Marcus A. H. (1970) Comparison of equilibrium size distribution for lunar craters. *J. Geophys. Res.* **75**, 2977–2984.

Moore H. J. (1964) Density of small craters on the lunar surface. *U.S. Geol. Survey, Astrogeologic studies annual progress report*, August 25, 1962 to July 1, 1963, pt. D, pp. 34–51.

Morrison D. A., McKay D. S., Moore H. J., and Heiken G. H. (1972a) Microcraters on lunar rocks (abstract). In *Lunar Science—III*, pp. 558–560. The Lunar Science Institute, Houston.

Morrison D. A., McKay D. S., Moore H. J., and Heiken G. H. (1972b) Microcraters on Lunar Rocks. *Proc. Third Lunar Sci. Conf., Geochim. Cosmochim. Acta*, Suppl. 3, Vol. 3, pp. 2767–2791. MIT Press.

Morrison D. A., Moore H. J., and McKay D. S. (1973) Microcraters on Apollo 15 and 16 rocks (abstract). In *Lunar Science—IV*, pp. 540–542. The Lunar Science Institute, Houston.

Muehlberger W. R., Batson R. M., Boudette E. L., Duke C. M., Eggleton R. E., Elston D. P., England A. W., Freeman V. L., Hait M. H., Hall T. A., Head J. W., Hodges C. A., Holt H. E., Jackson E. D., Jordan J. A., Larson K. B., Milton D. J., Reed V. W., Rennilson J. J., Schaber G. G., Schafer J. P., Silver L. T., Stuart-Alexander D. E., Sutton R. L., Swann G. A., Tyner R. L., Ulrich G. E., Wilshire H. G., Wilfe E. W., and Young J. W. (1972) Preliminary geologic investigation of the Apollo 16 landing site. *NASA SP-315*, pp. 6-1 to 6-81.

Neukum G., Schneider E., Mehl A., Storzen D., Wagner G. A., Fechtig H., and Block M. B. (1972) Lunar craters and exposure ages derived from crater statistics and solar flare tracks. *Proc. Third Lunar Sci. Conf., Geochim. Cosmochim. Acta*, Suppl. 3, Vol. 3, pp. 2793–2810. MIT Press.

Neukum G., Hörz F., Morrison D. A., and Hartung J. B. (1973) Crater populations on lunar rocks. *Proc. Fourth Lunar Sci. Conf., Geochim. Cosmochim. Acta*. This volume.

Schneider E., Storzen D., and Fechtig H. (1972) Exposure ages of Apollo 15 samples by means of microcrater statistics and solar flare particle tracks. In *The Apollo 15 Lunar Samples*, pp. 415–419. The Lunar Science Institute, Houston.

Soberman R. K. (1971) The terrestrial influx of small meteoric particles. *Rev. Geophys. and Space Phys.* **9**, 239–258.

Soderblom L. A. (1970) A model for small impact erosion applied to the lunar surface. *J. Geophys. Res.* **75**, 2655–2661.

Sutton R. L., Hait M. H., Larson K. B., Swann G. A., Reed V. S., and Schaber G. G. (1972) Documentation of Apollo 15 samples. *U.S. Geol. Survey Interagency Report: Astrogeology* **47**, 257 pp.

Swann G. A., Bailey N. G., Batson R. M., Freeman V. L., Hait M. H., Head J. W., Holt H. E., Howard K. A., Irwin J. B., Larson K. B., Muehlberger W. R., Reed V. S., Rennelson J. J., Schaber G. G., Scott D. R., Silver L. T., Sutton R. L., Ulrich G. E., Wilshire H. G., and Wolfe E. W. (1972) Preliminary geologic investigation of the Apollo 15 landing site. *NASA* SP-289, pp. 5-1 to 5-112.

Vedder J. F. (1971) Microcraters in glass and minerals. *Earth Planet. Sci. Lett.* **11**, pp. 291–296.

Whipple F. L. (1963) On meteorite and penetration. *J. Geophys. Res.* **68**, 4929–4939.

Proceedings of the Fourth Lunar Science Conference
(Supplement 4, *Geochimica et Cosmochimica Acta*)
Vol. 3, pp. 3255–3276

Crater populations on lunar rocks

GERHARD NEUKUM*

Lunar Science Institute, Houston, Texas 77058

FRIEDRICH HÖRZ and DONALD A. MORRISON

NASA Johnson Space Center, Houston, Texas 77058

JACK B. HARTUNG

Max-Planck-Institut für Kernphysik, Heidelberg, Germany

Abstract—Approximately 10,000 microcraters were investigated using binocular microscope techniques on fifteen Apollo 16 rocks: "crystalline" rocks 60315, 60335, 61156, 62235, 62295, and 68415; "breccias" 60016, 61015, 61175, 66075, and 69935; and glass surfaces 60015, 60095, 60135, and 64455. Diameter measurements of the central glass-lined pits (D_P) and surrounding spall zones (D_S) were made. Ratios of spall to pit diameters may range from 3.0 to 4.5 for different rock surfaces.

Crater size distributions obtained for production surfaces confirm and extend to larger crater sizes data published previously. The crater size distribution on lunar rocks in the pit diameter range, 10 to 1000 microns, is shown to depend on the average angle of impact which is a function of the exposure geometry.

In contrast to results of earlier studies, a wide range of crater densities was observed on relatively heavily cratered surfaces. The highest crater densities observed for lunar breccias are about a factor of 2 higher than that for crystalline rocks, which, in turn, appear up to 4 times more densely cratered than loose regolith in equilibrium.

Analytical models yield the expression for the cumulative equilibrium crater density, $N_E = AD_S^{-2}$, which has been adapted to microcratering on lunar rocks. A minimum value for the coefficient, A, is 0.15 assuming the largest measured spall-to-pit-diameter ratio of 4.5. This minimum is consistent with measurements.

Four independent criteria for recognizing equilibrium populations, (1) absolute crater densities, (2) constant crater densities for different exposure angles, (3) extent of D^{-2} slope, and (4) erosional state of surface, were applied to nine non-production Apollo 16 rocks, but only populations from two rocks (62235, 66075) satisfied all four criteria and were unambiguously shown to be in an equilibrium state.

INTRODUCTION

INVESTIGATION of impact craters caused by primary micrometeoroids on returned lunar materials is of interest for a variety of lunar studies. Such investigations may yield the flux of particulate interplanetary matter in the 10^{-5} to 10^{-15} g mass range in near lunar space (Neukum *et al.*, 1972; Morrison *et al.*, 1972; Hartung *et al.*, 1972; Schneider *et al.*, 1972) and also contribute to an improved understanding of small scale lunar surface processes, such as ionization, vaporization, melting, erosion, and transport—laterally and vertically—of regolith materials (Gault *et al.*,

---

*Permanent address: Max-Planck-Institut für Kernphysik, Heidelberg, Germany.

1972). At present these processes are understood only in a rather qualitative manner because the mass distribution and flux of meteoritic material over geologic times are not known accurately. The purpose of this paper is to describe and discuss the significance of microcrater populations on a variety of Apollo 16 rocks. Detailed models of meteoroid mass distributions and fluxes will be presented in subsequent papers together with implications for small scale lunar surface processes and exposure age information.

Two major objectives were pursued in these studies. The first objective was to determine the size distribution of crater populations in the "production" state (Shoemaker *et al.*, 1969; Gault, 1970) and thereby improve our knowledge about the mass distribution of micrometeoroids. This first objective requires rock surfaces of low absolute crater density.

In contrast, the second objective was to characterize crater size distributions typical of the highest observable crater densities on individual rock surfaces, i.e., those in or approaching equilibrium with respect to the cratering process (Shoemaker *et al.*, 1970; Gault, 1970). These investigations were stimulated by a systematic difference in absolute crater density between Apollo 12 basaltic rocks (Hörz *et al.*, 1971) and Apollo 14 breccias (Morrison *et al.*, 1972). The highest crater densities measured on Apollo 12 rocks are approximately a factor of 3 lower than the highest values observed on Apollo 14 breccias. Consequently, a variety of Apollo 16 crystalline rocks and breccias with high crater densities were investigated. An understanding of these highest crater densities is paramount to the interpretation of exposure histories as well as erosion mechanisms and rates on whole lunar rocks.

Observational Procedures

A total of fifteen Apollo 16 rocks were investigated, the selection criteria being to obtain a variety of rock types and a variety of crater densities as judged with the naked eye, according to the two main objectives outlined above. The following rocks were selected: glass surfaces: 60015, 60095, 60135, 64455; "crystalline" rocks: 60315, 60335, 61156, 62235, 62295, 68415; "breccias": 60016, 61015, 61175, 66075, 69935. For this discussion "crystalline" is defined to include truly igneous and high grade metamorphic rocks; "breccias" include both polymict clastic as well as partially molten breccias. This division may be greatly over-simplified and is subject to revision as detailed petrographic and, most importantly, data on physical properties governing the process of crater formation become available. Pertinent data on these rocks are summarized in Table 1.

The rocks selected were studied with a binocular microscope in a similar fashion as described by Hörz *et al.* (1971), and Morrison *et al.* (1972). The lowest magnification used was 4×, the highest magnification was 40×. Under the microscope a crater was identified by the presence of a glass-lined central pit. The diameter, D_P, of each pit observed was recorded.

Because the spall zone surrounding a central pit is eroded and degraded in more than 80% of all cases, the measurement of the spall zone diameter, D_S, was confined in this study to very fresh craters. Morrison *et al.* (1972) found that the ratio D_S/D_P varies significantly with rock types. Because the effective crater diameter is that of the spall zone rather than the pit, a precise characterization of D_S/D_P is necessary in order to convert the measured pit diameters of older craters, without recognizable spall zones, into spall diameters. Consequently, emphasis was placed on obtaining accurate D_S/D_P values on a given rock.

A variety of impact features were of the "pitless" crater type (McKay, 1970). Hartung *et al.* (1972) presented observational evidence suggestive of their origin by primary micrometeoroids. The total

Table 1. Summary of PET descriptions of rocks used in this study.

Rock	Rock Type[a]	Mass[b] (g)	Coherence[a]	Mode[c] (%)						Size[d] (mm)	Surface[e] History
				Opx	Cpx	Plag	Ol	Glass	Aphanitic		
61156	III	58	2	18	—	65	12	—	—	—	Complex
60315	III	788	2	55	20	15	5	—	—	0.1	Simple†
68415	III	371	2	—	15.2	80	5	—	—	0.1	Simple‡
60335	III	318	2	—	—	70	25	3	—	0.1	Simple†
62235	III	320	2	45	—	45	—	—	—	1.0	Simple†
62295	III	251	2	24	—	57	—	—	16	0.4	Simple†
69935	IV	128	2	—	40	15	—	××	××	—	?
61015	IV	1803	1	—	15	55	—	—	—	0.2	?
61175	I	543	2	—	—	80	—	—	—	—	Simple†
66075	I	347	1	—	50	45	—	—	××	0.1	Complex†
60016	I	4307	2	—	—	45	—	—	—	0.5	Complex
60095	V	46	2	—	—	—	—	97	—	—	Complex
60015	II	5574	2	—	—	—	—	100*	—	—	Complex
60135	II	138	2	—	—	—	—	100*	—	—	Simple
64455	II	57	2	—	—	—	—	100*	—	—	Simple†

*Large glass coatings.
†Good lunar surface documentation.
‡Excellent lunar surface documentation.
[a] Classification according to LSPET (1972). Type I: Polymict breccia with clastic matrix. Type II: Cataclastic anorthosite. Type III: Igneous and high grade metamorphic rocks. Type IV: Partially molten breccias. Type V: Genuine glass.
[b] According to Lunar Sample Information Catalog, Apollo 16 (1972). (1) Friable. (2) Tough.
[c] According to LSPET (1972) Thin Section Descriptions. Mode of breccias refers to matrix only. (Opx = orthopyroxene, Cpx = clinopyroxene, Plag = plagioclase, Ol = olivine, × × present).
[d] Average grain size according to PET descriptions. There are significant ranges of grain sizes in individual samples.
[e] According to microcrater distributions (Hörz et al., 1972). Complex: Tumbled repeatedly. Simple: Did not tumble; has uncratered surfaces.

abundance of these features generally is less than 5% of all impact craters. Some few rocks (e.g., 60015), however, displayed up to 20% pitless craters for the largest sizes. If such abundances occurred, the pitless crater diameters, i.e., the effective spall zones were measured and original diameters reconstructed via the measured D_S/D_P relationship.

Another phenomenon primarily observed on crystalline surfaces was the occurrence of large "halo" zones. Material was spalled off and the lighter microfractured underlying rock was easily visible. Because these zones could not be related to single impacts, we excluded these areas in the data reductions if more than 5% of the area of a field of view was so occupied.

The data reduction was identical to our previous studies. Only those fields of view which were representative of individual faces of a given rock, i.e., of similar geometry, were combined. Each individual surface investigated approximately represents the surface visible in the orthogonal photos obtained in the Lunar Receiving Laboratory (LRL) during the Preliminary Examination period. The "location" of a particular surface is described by the letters, T, B, N, S, E, and W, which correspond to similar indicators in these photos.

RESULTS

The detailed pit counts obtained at various magnifications are listed in Table 2 together with the total surface area observed at a given magnification. The specific location of this area with reference to a rock face shown in the LRL orthogonal photography is also given. The resulting cumulative size distributions are illustrated in Figs. 1 through 5. Both actual pit diameter distributions and the corresponding spall diameter distributions are shown. Collectively, a total of about 10,000 microcraters was counted.

Uncertainties in these data arise from a variety of sources. A comparison was made of the counting of different observers. An example of the agreement between two individuals is illustrated in Fig. 6. In general, the differences between observers was less than 20%, not including counts of the smaller craters. The statistical or counting uncertainties are indicated for each individual surface with error bars based on the square root of the sum of all craters larger than a given diameter. Other uncertainties are related to the selection and measurement of the areas studied, to the pit diameter measurement itself, and to the D_S/D_P determinations. These uncertainties refer to only the observed number of craters per unit area. Additional factors, such as some irregular spallation or a significant number of unobserved pitless craters, are difficult to evaluate. A conservative or high estimate, however, of the over-all uncertainty in the data presented is 50%.

INTERPRETATION

Production populations

A production surface is defined by a population of craters where destruction of pre-existing craters by subsequent impacts can be excluded. Therefore, they are of significant interest because size distribution of such a crater population directly reflects the energy distribution of the impacting micrometeoroids. Assuming certain standard velocities and densities for these micrometeoroids, these crater diameters eventually may be converted via empirical impact experiments into projectile masses and thereby into mass distributions of micrometeoroids.

Table 2. Cumulative frequencies of pit diameters at various magnifications on Apollo 16 rocks (μm).

4× Magnification

Rock	Location	D_S/D_P	500	750	1000	1250	1500	1750	2000	2250	2500	2750	3000	3250	3500	3750[b]	(cm^2)[c]
60016	N	3.4	—	155	80	45	25	13	7	6	5	4	4	3	3	2	39.27
60315	T_1[d]	3.7	—	46	22	14	8	4	2	2	1	1	1				39.27
600315	T_1[d]	3.7	46	22	13	8	5	2	2	2	1	1	1				19.64

5× Magnification

Rock	Location	D_S/D_P	600	800	1000	1200	1400	1600	(cm^2)[c]
69935	T	3.3	86	40	16	3	1	1	25.13

8× Magnification

Rock	Location	D_S/D_P	250	375	500	625	750	875	1000	1125	1250	1375	1500	1625	1750	1875	2000[b]	(cm^2)[c]
60015	N	3.4	75	32	15	5												12.76
60016	N	3.4		211	129	87	50	36	23	17								14.73
60315	T_2	3.7	56	28	14	5	4											7.85
61015	T	3.3		235	124	66	28	22	10									14.73
61015	E	3.3		294	163	93	58	39	23	19	16	13	12	8				19.63
61175	T	3.0		319	196	148	89	59	38	19	18	6	6	2	2	2	2	14.73
62235	B	4.2		94	45	24	15	12	9	8	7	2	2	2	2	2	2	15.71
62235	W	4.2	114	76	30	13	6	3	1									9.82
62295	B	4.5		73	34	19	9	7	7	4	3							8.34
62295	S	4.5		20	20	6	3	3										4.91
68415	S	3.9	59	29	8	8												4.91
68415	N	3.9	299	162	59	24	11		6									24.54
69935	T	3.3	352	196	85	42	24	14										14.73

10× Magnification

Rock	Location	D_S/D_P	200	300	400	500	600	700	800	900	1000	1100	1200	1360	1400	1500	1600[b]	(cm^2)[c]
60016	N	3.4	47	44	40	36	17	13	11	9	5	4						3.14
66075	T	3.4		293	203	147	95	71	51	33	27	15	11	6	4	4	2	11.94
66075	N	3.4		240	183	135	92	71	53	40	33	21	17	11	7	7	2	11.31
66075	S	3.4		159	108	74	47	37	30	22	16	11	10	8	7	5	2	6.28
66075	E	3.4		134	93	66	45	34	24	17	12	7	5	4	3	3	3	6.28
68415	S	3.9	65	30	18	10	4											4.58

Table 2. (continued)

16× Magnification

Rock	Location	D_S/D_P	62.5	125	187.5	250	312.5	375	437.5	500[b]	(cm²)[c]
61156	E	4.4	69	38	16	7	4				3.18
61156	W	4.4	65	42	18	8	4				0.98
61156	T	4.4	—	96	90	56	32	16	9	1	3.82
61156	S	4.4	234	97	42	19	12	6	5	3	5.00

20× Magnification

Rock	Location	D_S/D_P	50	100	150	200	250	300	350	400	450	500	550	600	650	700	750	(cm²)[c]
60015	A[a]	3.4		318	176	103	65	36	23	12	10	7						7.62
60016	N	3.4			114	81	69	43	29	18								1.57
60095	A[a]	3.4	37	22	14	6	4											0.39
60315	T_2	3.7	78	73	47	25	17	8	6	4	4	3	2	2	2	1		1.57
60315	T_1	3.7		71	48	38	26	16	12	9	7	5	2	1				0.78
61015	T	3.3			282	174	129	82	62	46	39							3.14
61015	E	3.3			263	174	122	84	65	46								3.14
61175	T	3.0				155	85	64	61	46	37		15					2.36
62235	B	4.2					111	78	21	44	8	34						6.36
62235	W	4.2		138	99	67	47	32	21	10	14	7	2					2.36
62295	B	4.5			126	78	54	32	23	16	3	14	5					2.36
62295	S	4.5				14	10	7	4	4	8	3	1					0.78
64455	B	3.7	173	130	85	57	42	27	20	12	27	4	3	2	1	1		2.83
66075	T	3.4			143	104	85	59	42	30	39	23	15					1.57
66075	W	3.4					74	13		43		32	23	21	16	13	8	
68415	S	3.9		105	67	35	20											1.33
68415	N	3.9		246	172	106	77	48	30									3.93
69935	T_F	3.3	165	63	35	20	11	8	5	4	3	2	2					1.42
69935	T	3.3		631	407	257	179	109	83	60	47			1				4.71

40× Magnification

Rock	Location	D_S/D_P	25	50	75	100	125	150	175	200	225	250	275	300	325	350	375	(cm²)[c]
60015	A[a]	3.4		457	305	210	161	117	85	59	45	33						3.93
60095	A[a]	3.4	53	40	21	14	9	7	4									0.70
60135	A[a]	3.4		351	231	138	92	62	40	23	15	8	4	1	1			1.37
60315	T_2	3.7		70	48	35	31	24	18	16	12	11				1	1	0.59
61015	T	3.3			250	180	153	125	104	81	64	58	38	37	31			1.37
61015	E	3.3			65	52	42	36	29									0.39

66075	N	3.4	—	—	—	213	205	189	174	152	139	131	99	93	81	73	68	52	51	48	47	45	2.75
66075	E	3.4	—	—	—	116	103	94	87	78	66	48	44	36	34	32	20	20	19	18			1.18
66075	W	3.4	—	—	—	53	50	45	41	38	34	34	30	30	27	25							0.78
68415	S	3.9	—	41	31	19	15	9															0.20
68415	N	3.9	—	89	63	37	26	21	12														0.39
69935	T	3.3	—	299	194	122	83	64	49														0.79
69935	T_F	3.3	127	80	37	24	15	11	6	3	—	1											0.52
			Counts at various Magnifications																				
			50	100	150	200	300	350	400	450		500	550	600	700	800	900	100[b]					
60335	T[d]	3.0	67	64	50	45	29	24	17	15	8	1	1	1	1	1	1	1					3.14
61175	T[d]	3.0	—	97		84		70		50		40		34	23	20	17	9					3.14
69935	T[d]	3.3	852	702		279		98		45		30		18	13	6	2	1					4.89

[a]Average for entire rock (= A).
[b]Pit diameter (μm).
[c]Surface area per magnification and individual rock surface.
[d]Repeated counts by different observer (*see* Fig. 6).

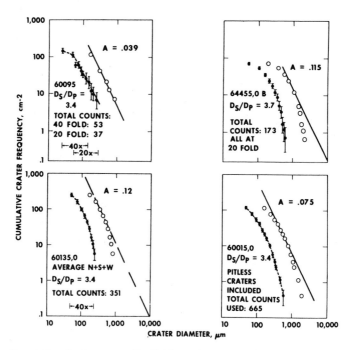

Fig. 1. Cumulative size frequency distributions of crater populations typical of Apollo 16 glass-surfaces in the production state. Dots represent the actual pit diameters measured; open circles represent the size-frequency distributions of the corresponding spall diameters as calculated from the measured D_S/S_P values. The best fit of a -2 slope through the spall diameters was graphically constructed to obtain the "areal density coefficient" (A), which is a numerical measure of the absolute number of craters per unit area. The following figures are constructed in identical fashion.

Smooth glass surfaces are by far the most suitable materials on which to observe such production conditions. We complemented our previous data with additional counts on rocks 60015, 60095, 60135, and 64455. These counts are illustrated in Fig. 1. Because of its large size, specimen 60015 was well suited for obtaining statistically significant data for pit diameters up to 625 μm. Production populations were also found on crystalline rocks, particularly on various surfaces of rock 61156 (N, E, T, S), as illustrated in Fig. 2. An additional production surface (Fig. 3) on breccia 69935 was investigated (T_F, indicating a freshly fractured area within the large surface, T).

The production distributions on Apollo 16 samples show a behavior similar to that seen earlier on Apollo 12 and 14 rocks (Bloch *et al.*, 1971; Hörz *et al.*, 1971; Hartung *et al.*, 1972; Neukum *et al.*, 1972; Morrison *et al.*, 1972). Figure 7 summarizes in a normalized way the statistically most significant production size distributions obtained to date on lunar glass surfaces. The slope of the log–log size distribution curve is about -2 at a pit diameter of 100 μm to 200 μm and steepens to -3 at a pit diameter of about 300 μm. The slope flattens to values of -1.0 to -1.5 at a pit diameter less than 100 μm.

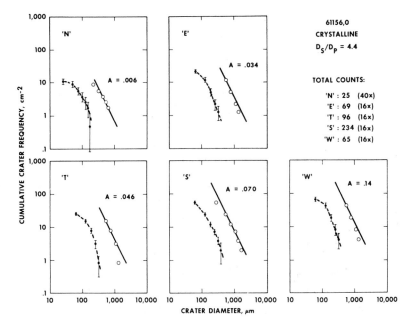

Fig. 2. Cumulative size frequency distribution of microcraters obtained on various sides of rock 61156,0. Note the various areal density coefficients for different surfaces. Surfaces N, E, T, and S are typical of production populations on crystalline rocks. The quality of the statistical data suffers from the small size of specimen 61156,0.

Despite the general similarity of the curves shown for different rocks, there are some subtle differences, which may be significant for the interpretation of the micrometeoroid complex. For example, sample 60135 shows a significantly steeper distribution than does rock 60015. Though this difference may be due to observational effects, it may also be attributed to different average impact angles which correspond to various solid angles of exposure. This effect may be understood analytically by considering a size distribution having a variable log–log slope. A straight-line size distribution at a given time may be represented as follows (Gault, 1970; Neukum and Dietzel, 1971), assuming a vertical impact angle,

$$N(D) = \frac{a\beta}{\alpha} k^{\alpha} D^{-\alpha}$$

where $N(D)$ is the cumulative number of craters greater than diameter D; a, β, and k are distribution and scaling constants; and α is the slope of the log–log crater size distribution. Based on laboratory experiments which take into account variable impact angles (Gault, 1973), the above equation may be modified to

$$N(D) = \frac{\alpha\beta}{\alpha} k^{\alpha} D^{-\alpha} (\sin i)^{\epsilon\alpha}$$

where i is the average impact angle, and ϵ is an empirical constant. A curved

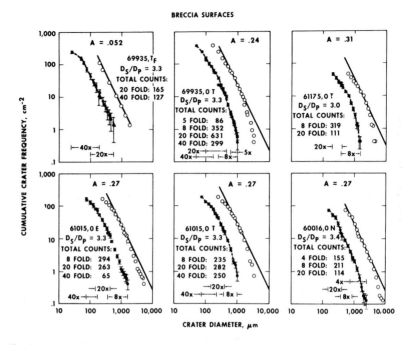

BRECCIA SURFACES

Fig. 3. Cumulative size frequency distributions of microcraters obtained on various Apollo 16 breccias. Note the difference in areal density for surfaces 69935,0 T, and 69935,0 T_F, the latter one being a typical production surface. Also note the identical areal density coefficients for surfaces 61015,0 E and 61015,0 T.

distribution may be represented by allowing α to be a variable, namely $\alpha = \alpha(D)$. For two diameters, D_1 and D_2, and an impact angle, i_1, we get

$$N_1(i_1) = \frac{a\beta}{\alpha_1} k^{\alpha_1} D_1^{-\alpha_1} (\sin i_1)^{\epsilon\alpha_1}$$

and

$$N_2(i_1) = \frac{a\beta}{\alpha_2} k^{\alpha_2} D_2^{-\alpha_2} (\sin i_1)^{\epsilon\alpha_2}$$

For an impact angle i_2, we get the analogous expressions, i_1 being replaced by i_2. For a *constant* shape of the distribution the following relationships must be valid.

$$\frac{N_1(i_1)}{N_1(i_2)} = \frac{N_2(i_1)}{N_2(i_2)}$$

and from above

$$\left(\frac{\sin i_1}{\sin i_2}\right)^{\epsilon\alpha_1} = \left(\frac{\sin i_1}{\sin i_2}\right)^{\epsilon\alpha_2}$$

which is true only when $i_1 = i_2$ or $\alpha_1 = \alpha_2$.

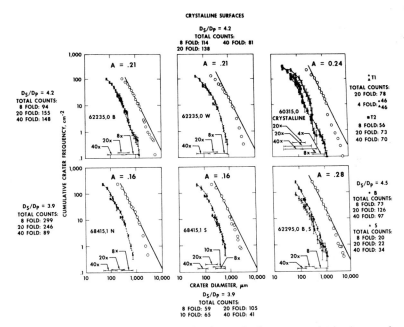

Fig. 4. Cumulative size frequency distributions of microcraters obtained on various Apollo 16 crystalline rocks. Note the identical areal density coefficients for various surfaces of rock 68415 (N, S) and 62235 (B, W) as well as the differences between 62235 B and S.

However, the slope, α, has been shown not to be constant, and we have postulated different average angles of impact, i. Therefore, the above relationships are not valid under these conditions. As a consequence, different impact angles in a size range where the production distribution slope is changing causes different shapes of the observed crater size distribution. For an assumed difference in $\sin i_1/\sin i_2$ of 1.4 (e.g., $i_1 = 45°$, $i_2 = 30°$) over the range of pit diameters from 50 to 500 μm, corresponding to slopes from about -1 to -3, a difference of a factor of 2 craters may be expected at one extreme of the distribution normalized at the other extreme.

Another somewhat more speculative prospect is that if interplanetary dust motion is confined to some degree to the ecliptic plane, then the shape of the size distribution of pits in this size range may be dependent upon the selenographic direction which the surface under study was facing. Unfortunately, however, such data are not available at present for the rocks investigated.

Equilibrium populations

After a sufficiently long exposure time, pre-existing craters on a surface will be destroyed by subsequent impacts. It is suggested that the cumulative crater density, $N(D)$, will change with time as shown qualitatively in Fig. 8. After some time the number of craters observed will be less than the extrapolated production

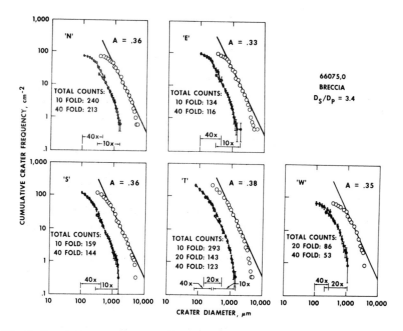

Fig. 5. Cumulative size frequency distributions of microcraters obtained on various surfaces of breccia 66075,0. Note the essentially identical areal density coefficients on all five surfaces.

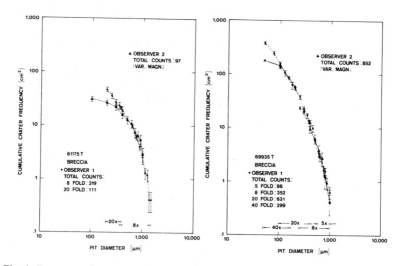

Fig. 6. Representative examples of cumulative size frequency distributions obtained by different observers on identical surfaces. The agreement is normally within 20%, except for the smallest sizes.

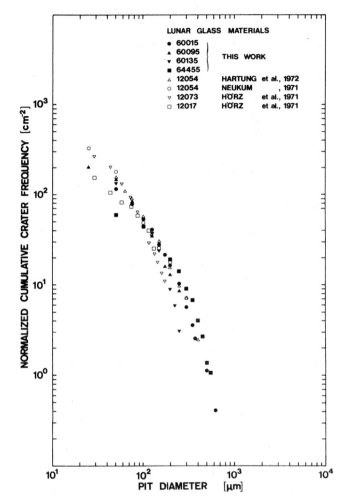

Fig. 7. Cumulative frequencies of microcraters obtained on a variety of glass-surfaces representing production populations for microcraters with pit diameters $>25\ \mu$m. The consistent data for 12054 and 60015 are considered to be the most reliable because of the large number of craters counted in both cases. Note the relative steepening of the production slope at about 200 μm pit diameter resulting in a production slope steeper than -2. The curves are normalized at a pit diameter of 100 μm.

number. The population density gradually approaches an upper limit, $N_E(D)$, after passing through some intermediate transition stage. The boundaries between production, transition, and equilibrium are somewhat arbitrary, which is indicated by the overlapping ranges in the figure. In practice, however, a production state can usually be recognized visually because one can observe the failure of craters to overlap one another.

It is important at this point to establish criteria to be used as a basis for recognizing equilibrium crater populations. In the following sections we will

G. NEUKUM *et al.*

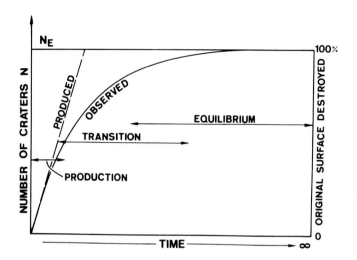

Fig. 8. Schematic development of a crater population where the only destruction mechanism operating is that of crater superposition. Note the gradual transition between production, transition, and equilibrium populations.

discuss four independent approaches toward establishing a particular crater size distribution as an equilibrium distribution.

1. *Absolute crater densities.* An analytical model has been developed which may be used to predict the crater density expected for equilibrium distributions. Hörz *et al.* (1971) pointed out that microcraters on hard lunar rocks are exclusively destroyed by direct superposition of subsequent impacts. Ballistic sedimentation can be neglected, though it is the dominant process of crater obliteration on the loose, lunar regolith surface according to Gault (1970), Soderblom (1970), and others. Consequently, models of the development of crater populations neglecting ballistic sedimentation (Marcus, 1964, 1966, 1970; Walker, 1967; Chapman *et al.*, 1968; Neukum and Dietzel, 1971) are better suited to describe the development of microcrater populations on hard lunar rocks over geologic periods of time. Based on these models the cumulative equilibrium density, $N_E(D)$, is a well-defined quantity and can be derived from the general expression for destruction by superposition as time approaches infinity. The expression from Neukum and Dietzel (1971) reads $N_E(2(\alpha - 2))/(\pi \cdot b^{(2-\alpha)/\beta}) \times D^{-2}$, where D can be either D_P or D_S. The quantity b can be defined by the relation $m_{min} = bm$ which in terms of crater diameter gives $D_{min} = b^{1/\beta} \times D$. D represents the diameter of the target crater and D_{min} is the diameter of the crater formed by m_{min} able to destroy the target crater. The above formula gives a general expression for the crater size distribution in equilibrium for a constant $\alpha > 2$. The case $\alpha < 2$ is not considered here. Therefore, the concept, as defined above, may be applied only to craters with pit diameter $> 250 \mu$m. In addition, the expression derived for the equilibrium crater density is valid in this form only for an infinite distribution ($m_{max} \to \infty$). Because we have finite distributions on lunar rocks, i.e., the largest crater has a diameter,

$D_{\max} < \infty$, the relation will be approximately correct for a limited size range. This limited range we call the "D^{-2} range of equilibrium". For an idealized representation of the relationships discussed, see Fig. 9. The end of the equilibrium distribution corresponding to large craters may be close to the production state because impact events are not frequent enough to destroy the large craters in this range. The opposite end of the equilibrium distribution reflects the flattening of the production distribution at smaller sizes.

To establish the criteria for destruction of an existing crater by a later impact, we set

$$b^{1/\beta} = \frac{D_S}{D_P}$$

This means an impact event with a spall diameter the same size as the pit diameter of the target crater will destroy the target crater. This is a conservative assumption in the sense that it would probably require a larger impact event to destroy the target crater. Thus, the minimum equilibrium crater size distribution for a lunar rock surface where only pit diameters can be quantitatively recorded may be written,

$$N_E = \frac{2(\alpha - 2)}{\pi} \left(\frac{D_P}{D_S}\right)^{\alpha - 2} D_S^{-2}$$

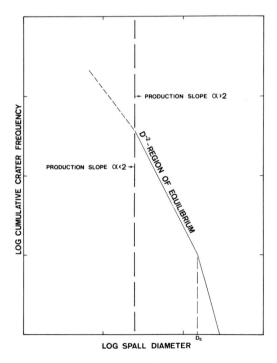

Fig. 9. Schematic of a lunar rock crater size frequency distribution in the equilibrium state.

In order to compare the crater densities obtained using this analytical expression with measured densities from various rocks, we define a standard quantity as follows by simplifying the above relationship:

$$N_E = A\,D_S^{-2}$$

where A is the equilibrium areal crater density coefficient, or simply areal density coefficient, and is given by

$$A = \frac{2(\alpha - 2)}{\pi} \left(\frac{D_P}{D_S}\right)^{\alpha - 2}$$

On rocks with different D_S/D_P ratios we must expect different values of A. The D_S/D_P ratio on one rock may furthermore vary with exposure time because of increasing degree of microfracturing of surface materials. Thus the areal density coefficient varies not only from rock to rock but it may also vary for one given specimen even though an equilibrium state, i.e., an extended D^{-2}-region of equilibrium, was reached earlier.

The absolute value of A for a given D_P/D_S is dependent on the production slope, α. Thus, for numerical calculations it must be known. In reality the slope, α, is not quite constant in the range, $D_P > 250$ μm. However, a variation from $\alpha = 2.7$ to $\alpha = 3.5$ will not significantly influence the areal density coefficient, A. For D_S/D_P ratios between 2.7 and 4.5 we have a variation in A by $\lesssim 20\%$. Thus, within the limits of the accuracy of our measurements, we can treat the problem as if we deal with a constant slope, $\alpha = 3$.

The analytically derived values of $A(A_{\text{anal.}})$ based on an α of 3.0 and an average of measured values of D_P/D_S for a given rock, may be compared with measured values of $A(A_{\text{meas.}})$. This quantity was determined by drawing a line of -2 slope through the distributions on the log–log plot and reading $A_{\text{meas.}}$ on the cumulative number scale where the line intersected the crater-diameter coordinate at $D_S = 1$ cm. These lines and the corresponding equilibrium areal crater density coefficients, or areal density coefficients, are also shown on Figs. 1 through 5. Although A is defined in terms of an equilibrium distribution having a -2 slope, it may also be applied as a measure of production or transition distributions which have a -2 slope. For the greatest measured value $D_S/D_P = 4.5$ (crystalline rocks) the analytical model gives a value of $A_{\text{anal.}} = 0.15$. This is the minimum equilibrium areal density coefficient to be expected for crater populations in equilibrium on lunar rocks, if no larger D_S/D_P ratio occurs.

The comparison between analytically derived and measured densities for Apollo 12, 14, and 16 rocks is shown in Table 3. Agreement is fair, however, in a few cases, especially for breccias, there is considerable deviation. With the quality of the observational data in mind, we may classify surfaces with $A_{\text{meas.}}/A_{\text{anal.}}$ in the range 0.7–1 as probably not being in equilibrium and surfaces with values between 1 and 1.5 as being in equilibrium. In some cases $A_{\text{meas.}}/A_{\text{anal.}}$ exceeds 1.5. For these, and perhaps other cases, we suggest that the surfaces are in equilibrium and the effective destruction diameter for a given cratering event may be less than D_S.

Such a case is indicated by the close proximity of two glass-lined pits shown in

Table 3. Comparison of measured and calculated maximum areal density coefficients on Apollo 12, 14, and 16 rocks.

Rock	D_S/D_P	$A_{meas.}$	$A_{anal.}$	$\dfrac{A_{meas.}}{A_{anal.}}$
"Crystalline"				
12006 (III)	4.5*	0.12	0.15	0.8
12017 (I)	4.5*	0.12	0.15	0.8
12021 (II)	4.5*	0.11	0.15	0.7
12038 (II)	4.5*	0.19	0.15	1.3
12047 (I)	4.5*	0.12	0.15	0.8
12051 (I)	4.5*	0.13	0.15	0.9
12053,37	4.5*	0.17	0.15	1.1
12063,106	4.5*	0.18	0.15	1.2
14053	3.6	0.13	0.18	0.7
60315 T	3.7	0.24	0.18	1.3
61156 W	4.4	0.14	0.15	0.9
62235 B	4.2	0.21	0.16	1.3
62235 W	4.2	0.21	0.16	1.3
62295 B	4.5	0.28	0.15	1.9
68415,1N	3.9	0.16	0.17	0.9
68415,1S	3.9	0.16	0.17	0.9
"Breccia"				
12073	4.5	0.18	0.15	1.2
14303,13	3.4†	0.21	0.20	1.1
14305, (D₃)	3.5	0.44	0.19	2.3
14306 (1)	3.4	0.64	0.20	3.2
14318 (B₂)	3.5	0.39	0.19	2.1
14321 (I)	4.5	0.40	0.15	2.7
60016 N	3.4	0.27	0.20	1.4
61015 T	3.3	0.27	0.20	1.4
61015 E	3.3	0.27	0.20	1.4
61175 T	3.0	0.31	0.22	1.4
66075 T	3.4	0.38	0.20	1.9
66075 S	3.4	0.36	0.20	1.8
66075 N	3.4	0.36	0.20	1.8
66075 E	3.4	0.33	0.20	1.7
66075 W	3.4	0.35	0.20	1.8
69935 T	3.3	0.24	0.20	1.2

*Average for all Apollo 12 crystalline rocks.
†Not determined; typical value for breccias.

the photograph and the idealized sketch in Fig. 10. One might expect that the spall zone of the more recent event would have destroyed a nearby older pit. Thus, there appears to be another unknown parameter which governs the ease with which a rock retains central glass pits. The parameters causing such an apparent decrease in the effective destruction radius are unknown at present. Therefore, our model, which assumes the measured D_S is the diameter of destruction, cannot be expected to explain all effects and give exact values for equilibrium crater densities on different types of rocks.

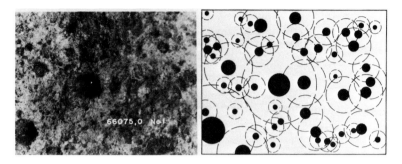

Fig. 10. Close-up photograph of rock 66075,0 considered to represent an equilibrium population. (Width of frame approximately 1 cm.) For comparison a sketch depicting actual central pit locations and size (black dots) is added and the actual, as well as eroded, spall zones (dashed circles) are indicated using the measured D_S/D_P value. Note the close proximity of many events demonstrating that the effective destruction diameter operating in this surface is D_S. Note also the relative paucity of small events within the (somewhat younger) spall zone of the big crater in the center.

2. *Constant crater density for different exposure angles.* If the absolute crater densities on different surfaces of the same rock are different from one another, one must postulate—based on exposure geometry arguments—that only the crater population with the highest density observed can potentially correspond to equilibrium. It may, of course, still be in the transition state while crater populations with the lower densities must either be in the production or transition state. Examples of varying crater densities for individual faces are rocks 61156 (surfaces N,E,T,S,W), 62295 (surfaces B,S) and 60315 (surfaces T_1 and T_2).

If, however, a well documented rock displays identical crater densities on surfaces of drastically different exposure geometries, the following conclusions may be drawn: the surface with the optimum solid angle of exposure must always display more craters than all the other surfaces; only after it reaches "equilibrium" will the number of craters remain constant. However, during this period, surfaces of less favorable exposure geometries will still accumulate more craters until they also finally reach equilibrium. This stage is indicated when all surfaces have the same crater densities. We suggest that the following rocks belong to this category: 62235 (surfaces T and W), and 66075 (surfaces T,S,N,E,W). These criteria are, of course, valid only if one can demonstrate that such rocks did not tumble repeatedly, i.e., that they possess at least one surface which is not cratered at all. As indicated earlier, the uncertainty associated with the crater densities may be as high as 50%, though as a general rule they can be judged to be better than 20%. Nevertheless, the lack of better accuracy may result in ambiguous interpretations.

3. *Extent of D^{-2} slope.* As an additional criterion, the extent of the D^{-2} slope may be considered to establish "equilibrium" conditions. The more the D^{-2} slope deviates from production distributions, by extending to spall zone diameters significantly greater than 1000 μm, the more likely it is that such surfaces are in

equilibrium. Surfaces of this category are: 60016 N, 61175 T, 61015 T and E, and 66075 T, N, E, S, W. Here again, however, unambiguous interpretations are limited by the quality of statistical data available, especially at larger crater sizes.

4. *Erosional state of surface.* Another criterion is based on the optical inspection of a given rock surface for either evidence of abundant overlap of cratering events or the presence of original uncratered rock surface. If the cratered surfaces of a given rock differ drastically in their degree of roundness with the uncratered surfaces, it may be concluded that mass-wasting has proceeded so far that no original rock-surface has been preserved. Thus, well rounded surfaces may possess "equilibrium" crater populations. Surfaces of this category are: 60315 T_1, 62235 B and W, 62295 B, 60016 N, 61175 T, 61015 E and T, 68415 N and S, and 66075 T, N, E, S, W.

In Table 4 the observational data are compared with these four criteria for equilibrium. The only surfaces which qualify in all aspects are 62235 B, W and 66075 T, N, E, S, W. All other surfaces remain ambiguous in their interpretation.

COMPARISON OF CRATER DENSITIES

Maximum measured areal density coefficients for a number of different rocks with different physical properties are plotted in Fig. 11 together with the limiting frequency value of Morrison *et al.* (1972). Gault's (1970) saturation percent values are given for comparison with the areal density coefficient values. We find, especially for the Apollo 16 rocks, a wide range of crater densities. Low maximum values for "glass" surfaces occur because we measured essentially populations in the

Table 4. Evaluation of various criteria to establish equilibrium conditions on densely cratered Apollo 16 rocks.

Criterion		1	2			3	4	
Rock	Locations	$\dfrac{A_{meas.}}{A_{anal.}}$ (>1)	Exposure History Simple (no tumbling)	Geometry Different	Equal A	Extended D^{-2} Region	Rounded Surface	Equilibrium
60016	N	×		—	—	×	×	likely
60315	T_1, T_2	×	×	×	—	×(T_1)	×	T_1: doubtful, T_2: no
61015	T, E	×	?	?	×	×	×	likely
61175	T	×	—	—	—	×	×	likely
62235	B, W	×	×	×	×	×	×	yes
62295	B, S	×	×	×	—	×(B)	×	B: doubtful S: no
66075	T,N,E,S,W	×	×	×	×	×	×	yes
68415	N, S	×	×	?	×	×	×	doubtful
69935	T	×	×	—	—	×	?	doubtful

× indicates criterion for equilibrium satisfied.

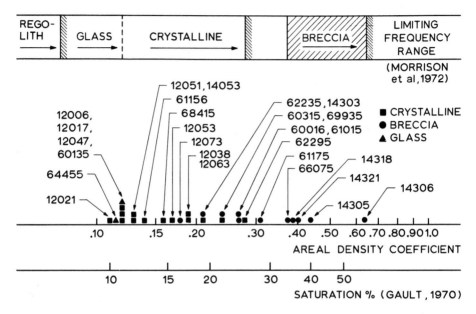

Fig. 11. Comparison of highest areal density coefficients observed to date on various lunar surface materials. Only the highest densities are plotted. Note the systematic difference of the areal density coefficients for various lunar rocks. These differences may be attributed to exposure age, rock properties, or a combination of both. The areal density coefficients for glass do not necessarily mean the highest approachable densities because only production populations have been investigated. For comparison, coefficients for loose regolith are indicated. Also shown are saturation percent values defined by Gault (1970).

production state. Other populations with similar values may be considered also in production provided D_S/D_P is the same. Regarding the higher areal density values we have concluded that these populations are in or approaching the equilibrium state. We find no simple relationship between areal density values for populations close to equilibrium and physical properties of the rocks. For example, rocks 60016 and 61175 are polymict breccias with only moderate intergranular coherency (PET), whereas 62295 is a crystalline rock (metaclastic) with tough intergranular coherency (PET), but the crater density coefficients are nearly the same. Similarly 68415 and 61156 are "crystalline" rocks very similar to 62295 in terms of intergranular coherence but their crater density coefficients are much lower than 62295. In general, however, crystalline rocks seem to have lower maximum densities than breccias.

None of the areal densities exceeds the limiting frequency range of Morrison *et al.* (1972); but all are greater than densities observed for the lunar regolith (Shoemaker *et al.*, 1970; Soderblom, 1970; Oberbeck and Quaide, 1968) and for cratering experiments in loose sand (Gault, 1970) (denoted "Regolith" in Fig. 11).

These low densities in loose materials are explained by the fact that erosion and ballistic sedimentation are the most effective mechanisms for crater obliteration in such target materials. The higher density of craters on lunar rocks demonstrates that microcraters are destroyed by the superposition of individual events.

CONCLUSIONS

Crater populations evolve in terms of cumulative size frequency distributions starting with the production state and reach equilibrium crater densities at larger crater size after passing through a transition stage.

Counts on four glass surfaces have yielded reliable statistics for production size distributions. Some differences in the shape of the distributions observed may be due to different exposure geometry of the surfaces studied.

Four criteria to distinguish between production and equilibrium cumulative crater size frequencies are: (1) absolute crater densities, (2) a constant crater density for surfaces of varying exposure geometry on the same rock suggest that equilibrium has been achieved, (3) an extended -2 slope for the cumulative crater frequency distribution particularly at the largest crater diameters measured, suggests equilibrium, (4) the erosional state of a surface, i.e., evidence for extensive rounding and mass removal may indicate equilibrium. Only populations from two rocks satisfied all four criteria and were unambiguously shown to be in an equilibrium state.

Our results confirm that destruction by superposition is the dominant process for crater obliteration on lunar rocks. The application of the analytical model yields for the cumulative equilibrium crater density, $N_E = A \cdot D_S^{-2}$. For spall-to-pit diameter ratios no greater than 4.5 there exists a minimum areal density coefficient $A = 0.15$ (equivalent to 14% saturation) for populations in equilibrium. This is consistent with our measurements.

The wide range of areal density coefficients in excess of the theoretical lower limit is at present not fully understood. These differences may be attributed to exposure age, rock properties, or a combination of both. These different views require separate discussions.

Acknowledgment—The research reported here was done while the first author was a Visiting Scientist at the Lunar Science Institute, which is operated by USRA under NASA contract NSR 09-051-001. This paper constitutes Lunar Science Institute Contribution No. 150.

REFERENCES

Bloch M. R., Fechtig H., Gentner W., Neukum G., and Schneider E. (1971a) Meteorite impact craters, crater simulations, and the meteoroid flux in the early solar system. *Proc. Second Lunar Sci. Conf., Geochim. Cosmochim. Acta*, Suppl. 2, Vol. 3, pp. 2639–2652. MIT Press.

Chapman C. R., Pollack J. B., and Sagan C. (1968) An analysis of the Mariner 4 photography of Mars. *Smithsonian Astrophys. Observ. Spec. Rep.* **268**.

Gault D. E. (1970) Saturation and equilibrium conditions for impact cratering on the lunar surface: Criteria and implications. *Radio Sci.* **5**, 273–291.

Gault D. E. (1973) Displaced mass, depth, diameter, and effects of oblique trajectories for impact craters formed in dense crystalline rocks. *The Moon* **6**, 32–44.

Gault D. E., Hörz F., and Hartung J. B. (1972) Effects of microcratering on the lunar surface. *Proc. Third Lunar Sci. Conf., Geochim. Cosmochim. Acta*, Suppl. 3, Vol. 3, pp. 2713–2734. MIT Press.

Hartung J. B., Hörz F., and Gault D. E. (1972) Lunar microcraters and interplanetary dust. *Proc. Third Lunar Sci. Conf., Geochim. Cosmochim. Acta*, Suppl. 3, Vol. 3, pp. 2735–2753. MIT Press.

Hörz F., Hartung J. B., and Gault D. E. (1971) Micrometeorite craters on lunar rock surfaces. *J. Geophys. Res.* **76**, 5770–5798.

Hörz F., Carrier W. D., Young J. W., Duke C. M., Nagle J. G., and Fryxell R. (1972) Apollo 16 special samples. NASA-SP. In press.

Lunar Sample Preliminary Examination Team (1972) The Apollo 16 lunar samples: Petrographic and chemical description. *Science* **179**, 23–34.

Marcus A. H. (1964) A stochastic model of the formation and survival of lunar craters, 1. *Icarus* **3**, 460–472.

Marcus A. H. (1966) A stochastic model of the formation and survival of lunar craters, 2. *Icarus* **5**, 165–177.

Marcus A. H. (1970) Comparison of equilibrium size distributions for lunar craters. *J. Geophys. Res.* **75**, 4977–4984.

McKay D. S. (1970) Microcraters in lunar samples. *Proc. 28th Annual Mtg. Electron Microsc. Soc. Amer.* (editor C. J. Arceneaux), pp. 22–23. Claitors.

Morrison D. A., McKay D. S., Moore H. J., and Heiken G. H. (1972) Microcraters on lunar rocks. *Proc. Third Lunar Sci. Conf., Geochim. Cosmochim. Acta*, Suppl. 3, Vol. 3, pp. 2767–2791. MIT Press.

Neukum G. (1971) Untersuchungen über Einschlagskrater auf dem Mond. Doctor's thesis, University of Heidelberg.

Neukum G. and Dietzel H. (1971) On the development of the crater population on the moon with time under meteoroid and solar wind bombardment. *Earth Planet. Sci. Lett.* **12**, 59–66.

Neukum G., Schneider E., Mehl A., Storzer D., Wagner G. A., Fechtig H., and Bloch M. R. (1972) Lunar craters and exposure ages derived from crater statistics and solar flare tracks. *Proc. Third Lunar Sci. Conf., Geochim. Cosmochim. Acta*, Suppl. 3, Vol. 3, pp. 2793–2810. MIT Press.

Oberbeck V. R. and Quaide W. L. (1968) Genetic implications of lunar regolith thickness variations. *Icarus* **9**, 446–465.

Schneider E., Storzer D., and Fechtig H. (1972) Exposure ages of Apollo 15 samples by means of microcrater statistics and solar flare particle tracks. In *The Apollo 15 Lunar Samples*, pp. 415–419, The Lunar Science Institute, Houston.

Shoemaker E. M., Batson R. M., Holt H. E., Morris E. C., Rennilson J. J., and Whitaker E. A. (1969) Observations of the lunar regolith and the earth from the television camera on Surveyor 7. *J. Geophys. Res.* **74**, 6081–6119.

Shoemaker E. M., Hait M. H., Swann G. A., Schleicher D. L., Schaber R. L., Sutton R. L., Dahlem D. H., Goddard E. N., and Waters A. C. (1970) Origin of the lunar regolith of Tranquillity Base. *Proc. Apollo 11 Lunar Sci. Conf., Geochim. Cosmochim. Acta*, Suppl. 1, Vol. 3, pp. 2399–2412. Pergamon.

Soderblom L. A. (1970) A model for lunar impact erosion applied to the lunar surface. *J. Geophys. Res.* **75**, pp. 2655–2661.

Walker E. H. (1967) Statistics of impact crater accumulation on the lunar surface exposed to a distribution of impacting bodies. *Icarus* **7**, 233–243.

Proceedings of the Fourth Lunar Science Conference
(Supplement 4, *Geochimica et Cosmochimica Acta*)
Vol. 3, pp. 3277–3290

Microcraters on Apollo 15 and 16 samples and corresponding cosmic dust fluxes

E. Schneider, D. Storzer, J. B. Hartung, H. Fechtig, and W. Gentner

Max-Planck-Institut für Kernphysik, Heidelberg, Germany

Abstract—Crystalline, glass, and metal samples from Apollo 15, 16, and Luna 16 missions have been investigated for micrometeorite impact craters using scanning electron microscope and electron microprobe techniques. At magnifications of 10,000–20,000 the range of crater detection extends down to 0.1 μm diameter. All samples show production state crater distributions. Surface exposure ages ranging roughly between 10 and 10^5 years have been determined using the solar flare track method. Crater number densities together with exposure ages allowed a calculation of the flux of micron and submicron sized particles onto the moon. Within an uncertainty of about a factor of two all samples yield a constant value for the flux of particles. Compared to the trend of the size distribution of larger particles there exists exceptionally large numbers of micron and submicron sized particles, thus suggesting a bimodal size distribution for the interplanetary dust. Neumann bands were found near a 30 μm diameter microcrater on a meteoritic Fe–Ni particle which was separated from coarse fines, 60502,17. Laboratory cratering experiments indicated the Neumann bands were produced during the cratering event.

Introduction

Microcrater studies on lunar samples (Hörz *et al.*, 1971; Morrison *et al.*, 1972; Bloch *et al.*, 1971; Brownlee *et al.*, 1972; and others) together with laboratory impact experiments (Gault and Heitowit, 1963; Fechtig *et al.*, 1972; Vedder, 1971; Schneider, 1972; and others) have contributed to our understanding of lunar surface processes, as well as the flux of interplanetary particles over long periods of time. Quantitative descriptions of lunar surface processes, such as cratering, regolith production, abrasion and catastrophic rupture of rocks (Gault and Wedekind, 1969), depend on the size distribution and flux of interplanetary particles. In this paper we obtain flux values based on crater number densities on lunar rocks and exposure ages derived from solar flare track densities using the same rock surfaces.

Crater Size Distribution

Selected surface areas of samples 15076,31; 15927,3; 15015,24; 15301,79; 15205,51; 60502,17; and one sample of Luna 16 have been investigated for micrometeorite impact craters down to 0.1 μm crater diameter using a scanning electron microscope (SEM). The surfaces of the samples selected were generally extremely smooth and clean and were in a production state with respect to cratering.

To obtain quantitative crater counts in the submicron size range of diameters SEM scanning magnifications up to 20,000 were used. With scan areas of the order

0.01 to 1 mm² the absolute number of craters counted per sample range from about 100 to over 1000. Figure 1 shows a plot of cumulative numbers of craters per unit area vs. pit diameter. Each point represents the number of craters larger than a given diameter. All samples show a relatively greater increase of number densities going to smaller diameters which starts between about 5 and 1 μm. This increase of slopes has already been pointed out by Schneider *et al.* (1972).

As an example we will discuss in detail the crater counting procedures used and results found for sample 15205,51. The principle advantage provided by this sample is that it was sufficiently large and exposed a long enough time that a

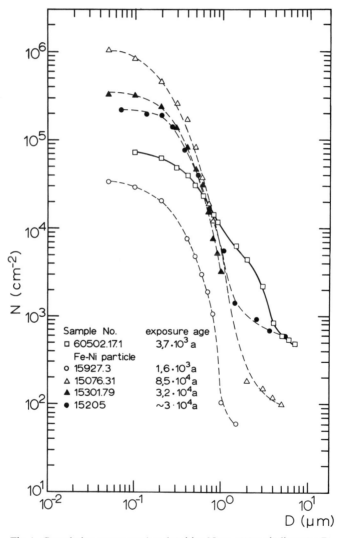

Fig. 1. Cumulative crater number densities *N* vs. crater pit diameter *D*.

statistically significant population of craters occurred over the entire pit diameter range from 0.2 to 200 μm.

The rock specimen was taken from the top corner of a meter-sized boulder, thus minimizing the effects of secondary glass splashes and dust coatings that may occur along the soil line of lunar rocks.

The approach to counting craters on this sample was to observe successively smaller craters on smaller sample areas at higher magnifications. Studies at a single magnification may yield distorted size distribution data because of selection effects and poor statistics at the larger crater sizes as well as recognition problems or incomplete counting at the smallest crater sizes observed. The envelope of the data taken over a range of magnifications is judged to be the best representation of the size distribution actually existing.

A cm-sized chip of 15205 was selected for detailed study. During chipping the selected sample broke into seven smaller fragments, designated 15205,51 and 15205,52. Three of these fragments (area $= 0.76 \, \text{cm}^2$) were covered with glass splashes and welded dust and, therefore, were not used in this study. Pits were counted and diameters measured on the four remaining fragments (area $= 0.39 \, \text{cm}^2$) at a magnification of 50 times. Areas were estimated using the calibrated scale in the microscope field of view.

One fragment, designated 15205,51,1, was selected for SEM examination. About one-half of this surface was rejected for SEM study because of the micro-vesicular character of the surface and the presence of large spall-zone-surfaces exposed some time after the original exposure of the smooth glass coating. The remaining area, about $0.08 \, \text{cm}^2$, was scanned at a magnification of 200 times. This same area was scanned again at a magnification of 1000 times. The next scan, at a magnification of 5000 times, was done over only one-fifth of the sample area, in order to limit the time required to a reasonable level. The scan was done in such a way that every fifth line of "data" was recorded using the entire surface, rather than recording all of the "data" from only one part of surface. Although this procedure eliminated possible area-selection effects, at this scale the quantity of glass splashes and welded dust seriously limited the quality of the data at the smallest crater sizes observed, especially towards one end of the sample. So, for the final scan, at a magnification of 10,000 times, the cleanest smooth glass surface on the sample was selected and entirely scanned (area $= 0.0020 \, \text{cm}^2$).

In order to reduce, but probably not eliminate, the recognition problem at the small crater sizes a lower limit for detectable pit diameters was established at each magnification used. For the optical measurements at a magnification of 50 times, a lower limit of 20 μm was set. This corresponds to about 1% of the field of view diameter. For the SEM data a lower limit corresponding to 2% of the field width was set for all magnifications. It should be noted that the actual feature observed was from two to five times greater in diameter than the measured pit diameter due to spall zones in the optical case and raised lips in the SEM cases.

A sufficient number of craters were counted in each size interval so that the data may be effectively displayed differentially, rather than in the usual cumulative form. The size of the interval chosen corresponds to a factor of the square root of 2. In Fig. 2 is the raw data obtained during this study plotted as a *log-log* histogram for each magnification used. Data for a magnification of 200 times is not included, because it covers the same range as the optical data, is of inferior statistical quality, and is essentially duplicated by the 1000 times magnification data. In this presentation there is no normalization with respect to area. The optical counts are most numerous in the pit diameter range, 200 down to 20 μm. The SEM data at a magnification of 1000 times are best down to 2 μm, and below that the data taken at a magnification of 5000 times are superior. Data taken at a magnification of 10,000 times are included to show the increased range of uncertainty at pit sizes less than 0.5 μm. Because the area scanned at a magnification of 10,000 was not as clean from dust particles as surfaces of other samples used for crater counting in the submicron size range, these crater density values below one micron crater diameter should be considered minimum values.

The slopes of the size distributions for all magnifications in the region, greater than 10 μm pit diameter, are essentially the same and equal to about $-2/3$ on the *log-log* plot. Between pit diameters of 1 and 5 μm the slopes, were measurable, are much steeper and consistent with a value between -2 and -3. Thus, a break, or steepening, of the distribution curve, indicating a bimodal size distribution, occurs in the pit diameter range around about 5 μm.

To permit direct comparison of the relative number of different-sized craters two approaches are possible. One can simply divide by the area over which measurements were taken for each magnification to obtain an areal density for craters of all different sizes. However, because considerable uncertainty is attributable to area selection and measurement, and because the important parameter to be

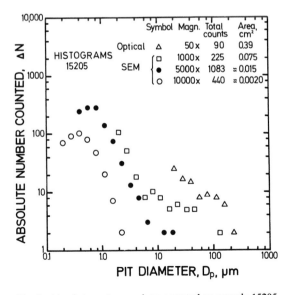

Fig. 2. Absolute crater numbers counted on sample 15205.

emphasized is the relative numbers of different-sized craters, we have chosen to proceed differently. We simply assume the data for a magnification of 1000 times to be correct in the range 2 to 20 μm, and adjust the data corresponding to magnifications of 50, 5000, and 10,000 times so that no discontinuities occur at the points of connection. The resulting crater size distribution is shown in Fig. 3. The ranges, where the different sets of data are used, are indicated. The vertical range of the data bars correspond to the square root of the number of counts within a *single* size interval and thus represent a conservative estimate of the statistical uncertainties involved in this study. It is easily seen that the bimodal character of the distribution cannot be due to "poor statistics."

For submicron-sized craters there exists no spallation zone as an indication of a genuine impact crater (Hartung *et al.*, 1972; Neukum *et al.*, 1972), thus the unambiguous detection of these craters is problematic. For this study circular to slightly oval depressions with lips or rims raised above the level of the sample surface were judged genuine impact craters. Figure 4 shows natural lunar and laboratory-produced submicron-sized impact craters (Schneider, 1972). There may be other processes that produce similar crater forms. For example, molten ejecta droplets containing bubbles may impact the target, thus causing the bubbles to burst and leave a characteristic pit. Such processes have not been simulated in the laboratory, and their possible occurrence produces some additional uncertainty of the data in this size range.

Another question is the rôle of secondary impacts. It cannot be ruled out that secondaries may contribute to crater counts, but their percentage should be low according to Hartung *et al.* (1972). Light gas gun experiments have been started to

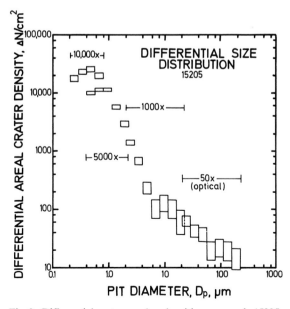

Fig. 3. Differential crater number densities on sample 15205.

E. Schneider *et al.*

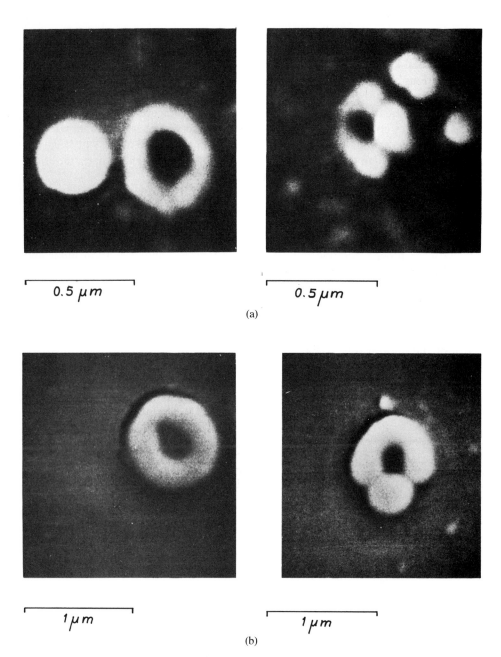

(a)

(b)

Fig. 4. (a) Submicron-sized craters on sample 15076,31. (SEM-photograph at a primary magnification of 100,000). (b) Laboratory produced submicron-sized craters on Duran-glass. Projectile: Iron sphere (mass: $3.1 \cdot 10^{-15}$ g, velocity: 60 km/s) SEM-photograph at a primary magnification of 50,000).

study those effects. Glass and metal targets have been mounted around primary targets to collect crater ejecta. Preliminary data show that areal densities of secondary craters on metal targets are higher by a factor of 10^4 than those on glass targets. Thus, most of the secondary particles are not energetic enough to produce craters on silicate material. These experiments will be continued in order to obtain more quantitative results about secondary effects.

Solar Flare Exposure Ages

In order to convert crater populations on a given sample into the interplanetary dust flux the sample's residence time at the lunar surface must be known. Therefore, solar flare exposure ages were determined for samples 15076,31, 15015,24, 15301,79, 15205,51,1, 60502,17, and a glass lined crater pit found in the Luna 16 core (C 19.118; 20–22 cm). For track observation, both optical and scanning electron microscopy were used. Detailed procedures are described elsewhere (Storzer *et al.*, 1973).

In all but sample 15076 the depth dependence of solar flare tracks was studied in the glass phase. As already noted (Neukum *et al.*, 1972; Storzer and Wagner, 1972), in lunar glasses the solar flare tracks are annealed to considerable extent. Accordingly, the track sizes are diminished as well as the track densities reduced (Storzer and Wagner, 1969). However, this very fact allows accurate track counting in a sample's near surface high track density region and to evaluate the depth dependence of solar flare tracks. Thereafter, the track densities were corrected according to the respective size shrinkage of the track etch pits (Storzer and Wagner, 1969) and/or normalized to the track densities of pyroxenes which were trapped within the glass phases.

In all samples studied (except for sample 60502, probably due to a 100 μm Ni–Fe-cover) the track densities P decrease with depth R from the surface according to $P = const \times R^{-\alpha}$. The α-values range between 2.5 and 2.8 (Storzer *et al.*, 1973) which agrees well with the track record of the uneroded Surveyor-III glass lens (Crozaz and Walker, 1971; Fleischer *et al.*, 1971; Barber *et al.*, 1971). We therefore conclude that erosion affected our samples only to a minor degree. Furthermore the surfaces were irradiated and cratered having a solid angle of exposure of essentially 2π. The solar flare exposure ages are complied in Table 1. For their calculation conventional track production rates were used (Barber *et al.*,

Table 1. Exposure ages determined by means of solar flare tracks.

Sample No.	15015,24*	Luna 16*	15927,3	15205,51,1	15301,79	15076,31	60502,17†
Solar flare exp. age (a)	13	$1.9.10^2$	$1.6.10^3$	$\sim 3.10^4$	$3.2.10^4$	$8.5.10^4$	$3.8.10^3$

*No microcraters have been found on these samples.
†Considering the higher stopping power of iron we corrected our data to an equivalent shielding of stony material.

1971) which are based on the Surveyor-III track record. The Surveyor-III space-craft was on the moon during the period of maximum solar activity and for a fraction of the present solar cycle only. On the other hand, the flux data must not be representative for the long-time averaged solar activity. This uncertainty in applying the Surveyor-III record to our lunar samples results in an age uncertainty of about a factor of 2. An additional uncertainty arises as the solar flare flux derived from the Surveyor-III track record probably has to be lowered (Storzer *et al.*, 1973). This would imply that our solar flare exposure ages are systematically too low by a factor up to 3.

INTERPLANETARY DUST SIZE DISTRIBUTION AND FLUX

Taking the crater size distributions together with corresponding surface exposure ages, the interplanetary dust flux in the micron and submicron size range can be calculated.

The particle mass distribution may be obtained from the crater size distribution on silicate targets by applying the empirically derived value of 2 for the ratio of the diameters of the microcrater pit and the impacting projectile (Bloch *et al.*, 1971) and assuming a density of 3 g/cm^3 for the projectile. This pit to projectile diameter ratio of 2 corresponds to an assumed mean impact velocity of 20 km/sec (Cour-Palais, 1969; Berg and Gerloff, 1971; Miller, 1970). Based on recent laboratory experiments a pit to projectile diameter value of 5 was used for the metallic sample No. 60502,17. The resulting interplanetary dust flux as a function of particle mass based on data for sample 15076,31 is shown in Fig. 5 along with other

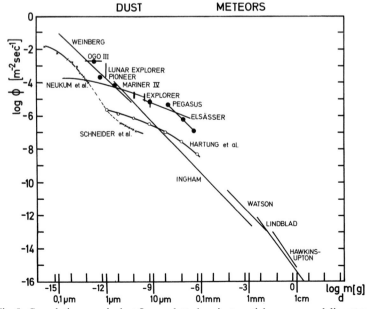

Fig. 5. Cumulative cosmic dust fluxes plotted against particle masses and diameters.

curves based on lunar sample studies and satellite-borne particle detection experiments. Sample 15076,31 has been chosen for flux calculations because its surface showed an absolute minimum of dust covering thus yielding best data in the submicron and lower micron size range of crater diameters. Up to now no flux values could be derived from sample 15205 for its solar flare track age is still preliminary.

The data of sample 15076,31 given in Fig. 1, which were obtained at high SEM-magnifications (10,000–20,000 fold) show poor statistics in the range $>1 \mu$m. Therefore data obtained at lower scanning magnifications (2000–5000 fold), yielding better statistics in the size range $>1 \mu$m, have been used for calculations of the particle flux in the upper size range.

There is evidence that the interplanetary dust flux in the micron and submicron size range is not isotropic in space (Berg and Gerloff, 1971), a result which recently has been confirmed by preliminary HEOS data (Hoffmann, 1972). Accounting for this anisotropy and the rotation of the moon, we increased our flux data by a factor of π.

Flux values based on lunar microcrater studies are generally less than those based on direct measurements made by satellite-borne detectors. This discrepancy is not readily resolved but may be due to one or more of the following factors.

(a) A systematic error may exist in the solar flare track method, possibly related to our present day knowledge of the solar flare particle flux.

(b) Systematic differences may exist related to the calibration of lunar microcraters and satellite-borne detectors.

(c) A real difference may exist such that satellite detectors record an enhanced flux due to particles ejected from the lunar surface.

(d) The present flux, measured by satellite detectors, may be higher than the long-term average, which is derived from the lunar microcrater data.

The number density of pits 0.8 μm in diameter or greater is plotted against the solar flare track exposure age for each sample in Fig. 6. A line having a slope of 1 on this *log-log* presentation corresponds to a flux that is constant in time. Such a line may be drawn through the data within the uncertainty limits, thus supporting the existence of a constant flux over the last 10^5 years, however, a best fit line through the data points has a flatter slope suggesting a lower flux may have existed in the past than in more recent times. Alternatively, a higher solar flare flux may have existed in the past, thus producing the same results.

Near one micron particle diameter the slope of the flux curve steepens. This corresponds directly with the steepening slope of the crater size distribution in this size range and indicates a bimodal distribution of the interplanetary dust. Such a bimodal distribution may be explained if larger particles spiralling towards the sun due to Poynting–Robertson effect are, at some critical distance, reduced in size and driven away from the sun (Belton, 1967). Another possibility is that the population of meteoroids is enhanced in the submicron size range by cometary dust particles which would be released as more volatile material is evaporated while the comet is near the sun.

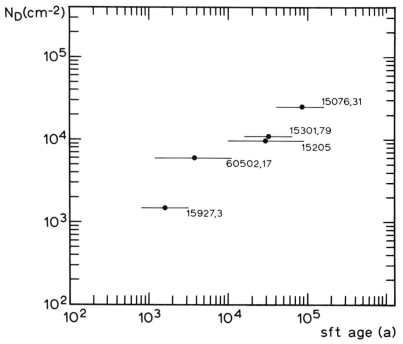

Fig. 6. Cumulative crater number densities N_D at a crater diameter $D = 0.8 \mu$m vs. solar flare track ages.

Fe–Ni Sample

On a small metallic particle ($\sim 1 \times 2 \times 1$ mm) separated from coarse fines, 60502,17, microcraters were observed using an optical microscope. Therefore it was scanned for micron- and submicron-sized craters using the SEM. Figure 7 shows SEM-photographs of microcraters on this sample and laboratory-produced microcraters on metal targets (Rudolph, 1968). Measurements of the bulk chemical composition of the metal particle using electron microprobe methods (95% Fe, 4.7% Ni, 0.38% Co, and 0.12% P) clearly indicated it was of meteoritic origin. Several larger microcraters were investigated qualitatively for Si, Mg, and Ca, in an effort to detect residual material from the impacting particles. No positive results were obtained. We offer one or more of the following explanations for this.

Projectile material did not contribute to the wall or floor of pits; the dilution of residual projectile material in impact-melted target material is so strong that the percentage of projectile material is below the detection limit; or the craters were produced by iron meteoroids. These results differ from those reported by Fredriksson *et al.* (1970) who concluded that similar craters in a metallic particle were formed by the impact of lunar ejecta particles based on the presence of lunar composition material inside the pits.

The size distribution of craters on this sample (*see* Fig. 1) is different from those found for silicate samples because of a different pit to projectile diameter

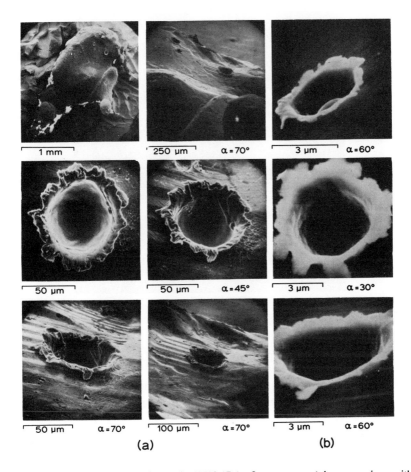

Fig. 7. Microcraters on Fe–Ni sample 60502,17 (a, first two rows) in comparison with laboratory produced craters on metal targets (b, third row) (SEM-photographs, α = observation angle).

ratio for metal targets. Impacting a polished section of an iron meteorite with 20 km/s-projectiles from an electrostatic dust accelerator produced a pit to projectile diameter ratio which is about a factor of 2.5 higher than that for silicate target material. Therefore, the characteristic steepening of the slope of the crater size distribution occurs already at pit diameters larger than 5 μm.

In Fig. 7 several sets of parallel striations believed to be Neumann bands may be seen in the vicinity of the relatively large craters. To test whether these structures are impact-related or not, a polished iron meteorite target was impacted with microparticles in an electrostatic dust accelerator. The impacts were small and did not produce any comparable striated structures. Subsequently, using a light gas gun, the section was impacted by an iron projectile 1.5 mm in diameter with a velocity of about 4 km/s. In this case similar parallel bands shown in Fig. 8

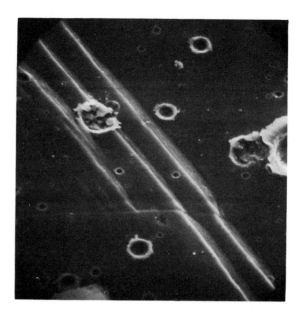

10 μm

Fig. 8. Laboratory produced Neumann bands. (SEM-photograph)

occurred in the vicinity of the impact site. The impact crater produced by the 1.5 mm-projectile is not to be seen in the photograph.

CONCLUDING REMARKS

Production state distributions of microcraters on lunar samples in connection with solar flare track exposure ages provide a means to calculate the interplanetary particle flux onto the moon within a limited size range. Concordant results have been obtained for several lunar samples within the error limits of the crater counting method and the uncertainty of the solar flare flux. A steepening of the slopes of crater size-frequency distributions, which occurs at some microns going to crater diameters has been found on all samples investigated. This result is not in agreement with data obtained by Brownlee *et al.* (1973) who find a smooth distribution curve down to the submicron size range, slightly bending over at smaller and smaller crater diameters.

Our exposure age determinations for breccia 15015,24 (4–40 years) and crystalline sample 15076,31 ($4 \times 10^4 - 1.6 \times 10^5$ years) seem to be in contradiction with age values given by Morrison *et al.* (1973) who find 1.9×10^4 years for 15015 and 9.2×10^5 years for 15076. They apply micrometeoroid model fluxes to production state crater distributions on whole rock surfaces for calculating respective exposure ages. Our ages however refer to very small surface areas of small sample chips, which might be part of younger spallation zones of larger microcraters covering the whole chip, or, on 15015,24 for instance, our data were obtained from a glass-splash covering part of the sample. In both cases our exposure ages would be younger therefore than those given by Morrison *et al.* (1973) and the disagreement should be expected.

The long-term-average solar flare flux might be not represented sufficiently by the 31-months exposure time of the Surveyor-III glass lens which serves as a calibration standard. Long time changes of this flux would cause the derived interplanetary dust flux to shift to either larger or smaller values, but would not influence the characteristic shape of the flux curve.

Acknowledgments—We thank the National Aeronautics and Space Administration for providing lunar samples. We are indebted to Dr. H. Reichenbach, Dr. A. Stilp, and K. Götting for help and advice on impact experiments performed at the light gas gun facility of the Ernst-Mach-Institut, Freiburg (Germany). Mrs. S. Papp and W. Weiss provided help with scanning work and technical assistance, which we gratefully acknowledge. Also we would like to thank H. Weber and his team for the photographic work.

REFERENCES

Barber J., Cowsik R., Hutcheon I. R., Price P. B., and Rajan R. S. (1971) Solar flares, the lunar surface and gas-rich meteorites. *Proc. Second Lunar Sci. Conf., Geochim. Cosmochim. Acta*, Suppl. 2, Vol. 3, pp. 2705–2714. MIT Press.

Belton M. J. S. (1967) Dynamics of interplanetary dust particles near the sun. NASA Sp-150, pp. 301–306.

Berg O. E. and Gerloff U. (1971) More than two years of micrometeorite data from two Pioneer satellites. *Space Res.* **XI**, 225–235. Akademie-Verlag, Berlin.

Bloch M. R., Fechtig H., Gentner W., Neukum G., and Schneider E. (1971) Meteorite impact craters, crater simulations, and the meteoroid flux in the early solar system. *Proc. Second Lunar Sci. Conf., Geochim. Cosmochim. Acta*, Suppl. 2, Vol. 3, pp. 2639–2652. MIT Press.

Brownlee D. E., Hörz F., Hartung J. B., Gault D. E. (1972) Micrometeoroid craters smaller than 100 microns. In *The Apollo 15 Lunar Samples*, pp. 407–409. The Lunar Science Institute, Houston.

Brownlee D. E., Hartung J. B., Hörz F., and Gault D. E. (1973) Morphology of micron sized impact craters on lunar glasses (abstract). In *Lunar Science—IV*, pp. 97–99. The Lunar Science Institute, Houston.

Cour-Palais B. G. (1969) Meteoroid environment model—1969. NASA SP-8013, pp. 1–31.

Crozaz G. and Walker R. M. (1971) Solar particle tracks in glass from the Surveyor 3 spacecraft. *Science* **171**, 1237–1239.

Fechtig H., Gault D. E., Neukum G., and Schneider E. (1972) The simulation of lunar craters in the laboratory. *Die Naturwissenschaften* **4**, 151–157. Springer.

Fleischer R. L., Hart H. R., and Comstock G. M. (1971) Very heavy solar cosmic rays: Energy spectrum and implications for lunar erosion. *Science* **171**, 1240–1242.

Fredriksson K., Nelen J., Melson W. G., Henderson E. P., and Andersen C. A. (1970) Lunar glasses and microbreccias: Properties and origin. *Science* **167**, 664–666.

Gault D. E. and Heitowit E. D. (1963) The partition of energy for hypervelocity impact craters formed in rock. *Proc. 6th Symp. Hypervelocity Impact*, Vol. 2, pp. 419–456.

Gault D. E. and Wedekind J. A. (1969) The destruction of tektites by micrometeoroid impact. *J. Geophys. Res.* **74**, 6780–6794.

Hartung J. B., Hörz F., and Gault D. E. (1972) Lunar microcraters and interplanetary dust. *Proc. Third Lunar Sci. Conf., Geochim. Cosmochim. Acta*, Suppl. 3, Vol. 3, pp. 2735–2753. MIT Press.

Hoffmann H. J. (1972) Private communication.

Hörz F., Hartung J. B., and Gault D. E. (1971) Micrometeorite craters on lunar rock surfaces. *J. Geophys. Res.* **76**, 5770–5798.

Miller C. D. (1970) Empirical analysis of unaccelerated velocity and mass distributions of photographic meteors. NASA TN D-5710.

Morrison D. A., McKay D. S., Heiken G. A., and Moore H. J. (1972) Microcraters on lunar rocks. *Proc. Third Lunar Sci. Conf., Geochim. Cosmochim. Acta*, Suppl. 3, Vol. 3, pp. 2767–2791. MIT Press.

Morrison D. A., McKay D. S., Moore H. J. (1973) Microcraters on Apollo 15 and 16 rocks (abstract). In *Lunar Science—IV*, pp. 540–542. The Lunar Science Institute, Houston.

Neukum G., Schneider E., Mehl A., Storzer D., Wagner G. A., Fechtig H., and Bloch M. R. (1972) Lunar craters and exposure ages derived from crater statistics and solar flare tracks. *Proc. Third Lunar Sci. Conf., Geochim. Cosmochim. Acta*, Suppl. 3, Vol. 3, pp. 2793–2810. MIT Press.

Rudolph V. (1968) Untersuchungen an Kratern von Mikroprojektilen im Geschwindigkeitsbereich von 0.5–10 km/s. *Z. F. Naturforschung* **24a**, 326–331.

Schneider E. (1972) Mikrokrater auf Mondgestein und deren Laborsimulation. Doctor's thesis, University of Heidelberg.

Schneider E., Storzer D., and Fechtig H. (1972) Exposure ages of Apollo 15 samples by means of microcrater statistics and solar flare particle tracks. In *The Apollo 15 Lunar Samples*, pp. 415–419. The Lunar Science Institute, Houston.

Storzer D. and Wagner G. A. (1969) Correction of thermally lowered fission track ages of tektites. *Earth Planet. Sci. Lett.* **5**, 463–468.

Storzer D. and Wagner G. A. (1972) Track analyses and uranium content of Apollo 14 glasses. *Trans. Amer. Nucl. Soc.* **15**, 119–120.

Storzer D., Poupeau G., and Krätschmer W. (1973) Track exposure and formation ages of some lunar samples. *Proc. Fourth Lunar Sci. Conf., Geochim. Cosmochim. Acta*. This volume.

Vedder J. F. (1971) Microcraters in glass and minerals. *Earth Planet. Sci. Lett.* **11**, 291–296.

Lunar Sample Inventory for
Apollo 11, 16, and 17

Following is a complete inventory of lunar samples as returned from Apollo 11, 16, and 17. The weight of each sample is the best known value; previous lists, as in sample catalogues, are not as accurate. The indicated "Sample Type" serves only as a *general* guide. Location and comments are *generally* stated. More details on the specific location of each sample may be obtained from various publications, especially the Sample Catalogues.

The first digit (6,7) is the designation of the Apollo 16 or 17 mission. The numbers are grouped by sampling site, which is designated by the second digit (e.g., 73xxx is Apollo 17 between station 3 and 2). For a more detailed explanation of the numbering system, refer to the sample catalogue for each mission. A glossary of terms for this inventory follows:

GLOSSARY

ALSEP: Apollo Lunar Surface Experiment Package: a location on the lunar surface specific for each mission.

ALSRC: Aluminum, vacuum sealable (indium seal) container used to transport rocks from the Lunar surface to the LRL.

Basalt: Used here to signify a crystalline rock with an igneous texture.

Bio Prep: A laboratory in the LRL used for processing rocks in a nitrogen atmosphere.

Bio Prime: Approx. 100 gm of lunar material used for biological protocol in the LRL.

Bio Pool: Approx. 500 gm of lunar material used for biological protocol in the LRL.

BSLSS: Buddy Secondary Life Support System (Bag)

Bulk sample: Lunar material in the first ALSRC collected during Apollo 11.

Chip: A lunar sample less than 50 gm and greater than 1 cm.

Contingency sample: Sample collected during Apollo 11, 12, and 14 near LM.

Comprehensive sample: Approx. 50 chips and fines collected from a small area during EVA-1 on Apollo 14.

Core tube: Old name for a drive tube.

CSVC: Core Sample Vacuum Container.

DB: Documented Bag: A small numbered teflon bag that lunar rocks are put into on the lunar surface. DBs are then put into SCBs or ALSRCs for transport to the LRC.

Documented sample: Lunar material in the second ALSRC collected during Apollo 11.

Drill stem: Tube of returned lunar regolith collected using a rotary drill: *see* Drive stem.

Drive stem: Tube of returned lunar regolith collected by hammering a hollow tube into the ground: *see* Drill stem.

DSB: Drill Stem Bag.

EVA: Extravehicular activity: Lunar traverse.

F-201: Vacuum chamber in the LRL used for processing samples during Apollo 11 and 12.

Fines: Less than 1 cm lunar material.

GASC: Gas Analysis Sample Container: one of several special sample containers that were filled and sealed on the lunar surface using an indium seal.

LESC: Lunar Environment Sample Container: *see* GASC.

LM: Lunar Module: a location on the Lunar surface specific for each mission.

LRL: Lunar Receiving Laboratory.

Residue: Chips and fines of lunar material collected from the bottom of ALSRCs, SCBs, DBs, and LRL processing chambers.

Rock: A lunar sample greater than 50 gm. In the Apollo 14 list, all rocks are breccias unless otherwise stated.

SCB: Sample Collection Bag: cloth bag used to transport rocks from the lunar surface to the LRL.

SESC: Surface Environment Sample Container, *see* GASC.

SRC: Sample Return Container ("Rock Box").

Sta: Station.

Sweepings: Lunar material recovered from LRL sample processing chambers.

Weigh bag: Old name for a SCB.

York mesh: Stainless steel meshing used as packing material in ALSRCs.

APOLLO 11 INVENTORY

Sample Number	Weight (grams)	Sample (type)	Comment
10001	————		Documented Sample Container (ALSRC) studied in F-201 (vacuum laboratory)
10002	~ 350.		Bulk (selected) ALSRC; Includes chips and fines; fines distributed as 10084-10086
10003	213.	Basalt	Documented sample
10004	52.5	Fines	Core tube-second taken on lunar surface
10005	64.	Fines	Core tube-first taken on lunar surface
10006	No sample		Solar wind experiment foil
10007	————	Fines	Gas reaction cell; from 10001; renumbered 10015
10008	————	Fines	Bio prime sample; from 10001
10009	112.	Breccia	Documented sample
10010	504.	Fines	Contingency sample
10011	~ 200.	Fines	Documented sample; vacuum fines; distributed as 10087
10012	————	————	Organic moniter of York mesh; not a lunar sample
10013	————	————	Part of Bio prime sample; from 10002
10014	50.	Chips	Documented sample
10015	————	Fines	Gas reaction cell; from 10001
10016	Number not used		
10017	973.	Basalt	Documented sample
10018	213.	Breccia	Documented sample
10019	297.	Breccia	Documented sample
10020	425.	Basalt	Documented sample
10021	255.0	Breccia	Contingency sample
10022	95.6	Basalt	Contingency sample; located
10023	66.0	Breccia	Contingency sample; located
10024	68.1	Basalt	Contingency sample
10025	8.6	Breccia	Contingency sample
10026	9.3	Breccia	Contingency sample
10027	9.9	Breccia	Contingency sample
10028	3.5	Breccia	Contingency sample
10029	5.5	Basalt	Contingency sample
10030	1.8	Breccia	Contingency sample
10031	2.7	Breccia	Contingency sample
10032	3.1	Basalt	Contingency sample
10033	1.1		Contingency sample
10034	‹1.	Residue	Contingency sample
10035	1.	Residue	Contingency sample
10036 10037 10038	————		Material used for Bio prime sample. It is a mixture of several samples.
10039 10040 10041 10042 10043			Residue from splits of many samples studied by PET.

Apollo 11 Inventory *(continued)*

Sample Number	Weight (grams)	Sample (type)	Comment
10044	247.5	Basalt	Bulk sample
10045	185.	Basalt	Bulk sample
10046	663.	Breccia	Bulk sample; located
10047	138.	Basalt	Bulk sample
10048	579.	Breccia	Bulk sample
10049	193.	Basalt	Bulk sample
10050	114.5	Basalt	Bulk sample
10051			
10052			Material used for Bio pool sample.
10053			It is a mixture of several samples.
10054			
10055	———	Residue	From F-201
10056	186.0	Breccia	Bulk sample
10057	919.0	Basalt	Bulk sample
10058	282.0	Basalt	Bulk sample
10059	188.0	Breccia	Bulk sample
10060	722.	Breccia	Documented sample
10061	346.	Breccia	Documented sample
10062	78.5	Basalt	Documented sample
10063	146.	Breccia	Documented sample
10064	65.	Breccia	Documented sample
10065	347.	Breccia	Documented sample
10066	43.	Basalt	Documented sample
10067	69.3	Breccia	Documented sample
10068	218.	Breccia	Documented sample
10069	119.5	Basalt	Documented sample
10070	64.	Breccia	Documented sample
10071	189.5	Basalt	Documented sample
10072	447.	Basalt	Documented sample
10073	124.5	Breccia	Documented sample
10074	55.5	Breccia	Documented sample
10075	48.2	Breccia	Documented sample
10076	———	Residue	From F-201
10077	———	Residue	From Bio-Prep
10078	———	Residue	From Bio-Prep
10079	———	Residue	From Bio-Prep
10080	Number not used		
10081	Number not used		
10082	50.5	Breccia	Documented sample
10083	Number not used		
10084	10391.	1mm fines	Bulk sample—formally part of 10002
10085	10391.	1mm-1cm fines	Bulk sample—formally part of 10002
10086	10391.	Fines	Part of 10084 distributed for organic analysis
10087	———	Fines	Distributed part of 10011
10088	Number not used		
10089	~50.	Breccia	Chip from 10002
10090	6.	Breccia	Chip from 10002
10091	~50.	Breccia	Chip from 10002

Apollo 11 Inventory *(continued)*

Sample Number	Weight (grams)	Sample (type)	Comment
10092	Number not used		
10093	Number not used		
10094	Number not used		
10095	Number not used		
10096	Number not used		
10097	Number not used		
10098	Number not used		
10099	Number not used		
10100	———	Residue	Fines, chips, sweepings from curator processing
10101	———	Residue	Fines, chips, sweepings from curator processing
10102	———	Residue	Fines, chips, sweepings from curator processing
10103	———	Residue	Fines, chips, sweepings from curator processing

APOLLO 16 INVENTORY

Sample Number	Weight (grams)	Description	Location	SRC/DB or SCB/DB
60001	30.1	Core bit	Sta 10-ALSEP	
60002	211.9	Core stem	Sta 10-ALSEP	
60003	215.5	Core stem	Sta 10-ALSEP	
60004	202.7	Core stem	Sta 10-ALSEP	
60005	76.1	Core stem	Sta 10-ALSEP	
60006	165.6	Core stem	Sta 10-ALSEP	
60007	105.7	Core stem	Sta 10-ALSEP	
60009	759.8	Core 54 lower	Sta 10-ALSEP	SRC2/
60010	635.3	Core 45 upper	Sta 10-ALSEP	SRC2/
60013	757.3	Core 32 lower	Sta 10-ALSEP	SCB7/
60014	570.3	Core 27 upper	Sta 10-ALSEP	SCB7/
60015	5574.0	Crystalline	Sta 10-ALSEP	SCB5/
60016	4307.0	Breccia	Sta 10-ALSEP	SCB7/
60017	2102.0	Breccia	Sta 13	SCB7/
60018	1501.0	Breccia	Sta 10-ALSEP	SCB7/
60019	1887.0	Breccia	Sta 10-ALSEP	SCB4/
60020	51.91	SCB residue	Sta 10-ALSEP	SCB3/
60025	1836.0	Anorthosite	Sta 10-ALSEP	SCB3/
60030	79.26	DB residue	Sta 10-ALSEP	SRC1/351
60035	1052.0	Anorthosite gabbro	Sta 10-ALSEP	SRC1/351
60040	12.12	SCB residue	Sta 10-ALSEP	SCB5/
60050	3.27	Unsieved fines	Sta 10-ALSEP	SRC1/355
60051	195.3	<1mm fines	Sta 10-ALSEP	SRC1/355
60052	11.43	1-2mm fines	Sta 10-ALSEP	SRC1/355
60053	7.41	2-4mm fines	Sta 10-ALSEP	SRC1/355
60054	8.40	4-10mm fines	Sta 10-ALSEP	SRC1/355
60055	35.48	Anorthosite	Sta 10-ALSEP	SRC1/355
60056	16.07	Breccia	Sta 10-ALSEP	SRC1/355
60057	3.10	Anorthosite	Sta 10-ALSEP	SRC1/355
60058	2.12	Breccia	Sta 10-ALSEP	SRC1/355
60059	1.05	Anorthosite	Sta 10-ALSEP	SRC1/355
60070	79.83	DB residue	Sta 10-ALSEP	SRC1/373
60075	183.8	Breccia	Sta 10-ALSEP	SRC1/373
60090	0.10	DB residue	Sta 10-ALSEP	SCB1/004
60095	46.60	Glass	Sta 10-ALSEP	SCB1/004
60110	2.08	DB residue	Sta 10-ALSEP	SCB1/381
60115	132.5	Breccia	Sta 10-ALSEP	SCB1/381
60130	1.07	DB residue	Sta 10-ALSEP	SCB6/430
60135	137.7	Crystalline, glass coated	Sta 10-ALSEP	SCB6/430

Apollo 16 Inventory *(continued)*

Sample Number	Weight (grams)	Description	Location	SRC/DB or SCB/DB
60210	12.78	Residue	Sta 10-ALSEP	SCB6/13
60215	385.8	Anorthositic cataclasite	Sta 10-ALSEP	SCB6/13
60230	39.61	DB residue	Sta 10-ALSEP	SCB6/15
60235	70.13	Crystalline	Sta 10-ALSEP	SCB6/15
60250	18.31	DB residue	Sta 10-ALSEP	SCB6/17
60255	871.0	Breccia	Sta 10-ALSEP	SCB6/17
60270	37.26	DB residue	Sta 10-ALSEP	SCB7/18
60275	255.2	Breccia	Sta 10-ALSEP	SCB7/18
60310	2.02	DB residue	Sta 10-ALSEP	SCB7/20
60315	787.7	Crystalline	Sta 10-ALSEP	SCB7/20
60330	2.51	Residue	Sta 10-ALSEP	SCB6/331
60335	317.8	Breccia	Sta 10-ALSEP	SCB6/331
60500	234.4	Unsieved fines (raked)	Sta 10-ALSEP	SCB4/350
60501	433.8	<1mm fines (raked)	Sta 10-ALSEP	SCB4/350
60502	17.69	1-2mm fines (raked)	Sta 10-ALSEP	SCB4/350
60503	9.94	2-4mm fines (raked)	Sta 10-ALSEP	SCB4/350
60504	6.63	4-10mm fines (raked)	Sta 10-ALSEP	SCB4/350
60510	7.67	DB residue	Sta 10-ALSEP	SCB4/349
60515	16.74	Anorthositic (rake)	Sta 10-ALSEP	SCB4/349
60516	7.91	Anorthositic (rake)	Sta 10-ALSEP	SCB4/349
60517	1.23	Anorthositic (rake)	Sta 10-ALSEP	SCB4/349
60518	1.12	Anorthositic (rake)	Sta 10-ALSEP	SCB4/349
60519	.50	Anorthositic (rake)	Sta 10-ALSEP	SCB4/349
60525	12.84	Crystalline (rake)	Sta 10-ALSEP	SCB4/349
60526	8.42	Crystalline (rake)	Sta 10-ALSEP	SCB4/349
60527	7.36	Crystalline (rake)	Sta 10-ALSEP	SCB4/349
60528	2.94	Glass (rake)	Sta 10-ALSEP	SCB4/349
60529	1.24	Glass (rake)	Sta 10-ALSEP	SCB4/349
60535	7.23	Breccia (rake)	Sta 10-ALSEP	
60600	182.6	Unsieved fines (rake)	Sta 10-ALSEP	SCB4/348
60601	330.2	<1mm fines (rake)	Sta 10-ALSEP	SCB4/348
60602	14.93	1-2mm fines (rake)	Sta 10-ALSEP	SCB4/348
60603	8.57	2-4mm fines (rake)	Sta 10-ALSEP	SCB4/348
60604	3.94	4-10mm fines (rake)	Sta 10-ALSEP	SCB4/348
60610	34.74	DB residue	Sta 10-ALSEP	SCB4/347
60615	32.97	Crystalline (rake)	Sta 10-ALSEP	SCB4/347
60616	3.40	Crystalline (rake)	Sta 10-ALSEP	SCB4/347

Apollo 16 Inventory *(continued)*

Sample Number	Weight (grams)	Description	Location	SRC/DB or SCB/DB
60617	2.77	Crystalline (rake)	Sta 10-ALSEP	SCB4/347
60618	21.67	Crystalline (rake)	Sta 10-ALSEP	SCB4/347
60619	28.00	Anorthosite (rake)	Sta 10-ALSEP	SCB4/347
60625	117.00	Crystalline (rake)	Sta 10-ALSEP	SCB4/347
60626	15.87	Crystalline (rake)	Sta 10-ALSEP	SCB4/347
60627	12.09	Crystalline (rake)	Sta 10-ALSEP	SCB4/347
60628	6.86	Anorthosite (rake)	Sta 10-ALSEP	SCB4/347
60629	4.92	Anorthosite (rake)	Sta 10-ALSEP	SCB4/347
60635	15.05	Crystalline (rake)	Sta 10-ALSEP	SCB4/347
60636	35.65	Crystalline (rake)	Sta 10-ALSEP	SCB4/347
60637	7.98	Breccia, clastic-matrix	Sta 10-ALSEP	SCB4/347
60638	.72	Breccia, clastic-matrix	Sta 10-ALSEP	SCB4/347
60639	175.1	Breccia, clastic-matrix	Sta 10-ALSEP	SCB4/347
60645	33.5	Breccia, heterogeneous	Sta 10-ALSEP	SCB4/347
60646	3.39	Breccia, heterogeneous	Sta 10-ALSEP	SCB4/347
60647	1.76	Breccia, heterogeneous	Sta 10-ALSEP	SCB4/347
60648	2.84	Breccia, heterogeneous	Sta 10-ALSEP	SCB4/347
60649	1.03	Breccia, heterogeneous	Sta 10-ALSEP	SCB4/347
60655	8.63	Breccia, heterogeneous	Sta 10-ALSEP	SCB4/347
60656	11.23	Breccia, heterogeneous	Sta 10-ALSEP	SCB4/347
60657	6.05	Breccia, heterogeneous	Sta 10-ALSEP	SCB4/347
60658	5.47	Breccia, heterogeneous	Sta 10-ALSEP	SCB4/347
60659	22.20	Breccia, heterogeneous	Sta 10-ALSEP	SCB4/347
60665	90.1	Glass, vesicular (rake)	Sta 10-ALSEP	SCB4/347
60666	15.95	Glass, vesicular (rake)	Sta 10-ALSEP	SCB4/347
60667	7.66	Glass, vesicular (rake)	Sta 10-ALSEP	SCB4/347
60668	2.91	Glass, vesicular (rake)	Sta 10-ALSEP	SCB4/347
60669	2.54	Glass, vesicular (rake)	Sta 10-ALSEP	SCB4/347
60675	1.30	Glass, vesicular (rake)	Sta 10-ALSEP	SCB4/347
60676	8.92	Glass, vesicular (rake)	Sta 10-ALSEP	SCB4/347
60677	5.23	Glass, vesicular (rake)	Sta 10-ALSEP	SCB4/347
60678	1.25	Glass, vesicular (rake)	Sta 10-ALSEP	SCB4/347
60679	2.96	Glass, vesicular (rake)	Sta 10-ALSEP	SCB4/347
61010	64.19	Residue	Sta 1	SRC1
61015	1804.	Breccia veined	Sta 1	SRC1
61016	11729.	Anorthosite	Sta 1	BSLSS
61017	2.62	Breccia, friable	Sta 1	SRC1/
61130	12.51	DB residue	Sta 1	SRC1/362
61135	245.1	Breccia	Sta 1	SRC1/362
61140	74.13	Unsieved fines	Sta 1-Traverse plum	SRC1/363
61141	134.7	<1mm fines	Sta 1-Traverse plum	SRC1/363

Apollo 16 Inventory *(continued)*

Sample Number	Weight (grams)	Description	Location	SRC/DB or SCB/DB
61142	9.43	1-2mm fines	Sta 1-Traverse plum	SRC1/363
61143	5.38	2-4mm fines	Sta 1-Traverse plum	SRC1/363
61144	5.71	4-10mm fines	Sta 1-Traverse plum	SRC1/363
61150	16.13	DB residue	Sta 1-Traverse plum	SRC1/371
61155	47.59	Breccia	Sta 1-Traverse plum	SRC1/371
61156	58.46	Breccian, annealed	Sta 1-Traverse plum	SRC1/371
61157	11.26		Sta 1-Traverse plum	SRC1/371
61158	14.79		Sta 1-Traverse plum	SRC1/371
61160	52.79	Unsieved fines	Sta 1-Surface	SRC1/356
61161	90.0	<1mm fines	Sta 1-Surface	SRC1/356
61162	5.12	1-2mm fines	Sta 1-Surface	SRC1/356
61163	3.60	2-4mm fines	Sta 1-Surface	SRC1/356
61164	2.16	4-10mm fines	Sta 1-Surface	SRC1/356
61170	16.32	DB residue	Sta 1-Surface	SRC1/364
61175	542.7	Breccia	Sta 1-Surface	SRC1/364
61180	93.40	Unsieved fines	Sta 1-N of LRV	SRC1/369
61181	156.2	<1mm fines	Sta 1-N of LRV	SRC1/369
61182	9.43	1-2mm fines	Sta 1-N of LRV	SRC1/369
61183	6.23	2-4mm fines	Sta 1-N of LRV	SRC1/369
61184	6.09	4-10mm fines	Sta 1-N of LRV	SRC1/369
61190	16.61	DB residue	Sta 1-N of LRV	SRC1/002
61195	587.9	Microbreccia, glassy	Sta 1-N of LRV	SRC1/002
61220	191.6	Unsieved fines (white)	Sta 1-Below surface	SRC1/357
61221	61.0	<1mm fines (white)	Sta 1-Below surface	SRC1/357
61222	6.36	1-2mm fines (white)	Sta 1-Below surface	SRC1/357
61223	9.61	2-4mm fines	Sta 1-Below surface	SRC1/357
61224	10.58	4-10mm fines	Sta 1-Below surface	SRC1/357
61225	3.52	Micro-crystalline	Sta 1-Below surface	SRC1/357
61226	1.53	Plagioclase	Sta 1-Below surface	SRC1/357
61240	160.8	Unsieved fines, upper gray soil	Sta 1-Below surface	SRC1/352
61241	247.1	<1mm fines, upper gray soil	Sta 1-Below surface	SRC1/352
61242	17.26	1-2mm fines, upper gray soil	Sta 1-Below surface	SRC1/352
61243	13.80	2-4mm fines, upper gray soil	Sta 1-Below surface	SRC1/352
61244	13.25	4-10mm fines, upper gray soil	Sta 1-Below surface	SRC1/352
61245	8.25		Sta 1-Below surface	SRC1/352
61246	6.05		Sta 1-Below surface	SRC1/352
61247	2.48		Sta 1-Below surface	SRC1/352
61248	1.71		Sta 1-Below surface	SRC1/352
61249	1.17		Sta 1-Below surface	SRC1/352
61255	1.13		Sta 1-Below surface	SRC1/352
61280	68.49	Unsieved fines	Sta 1-Plum	SRC1/368

Apollo 16 Inventory *(continued)*

Sample Number	Weight (grams)	Description	Location	SRC/DB or SCB/DB
61281	169.6	<1mm fines	Sta 1-Plum	SRC1/368
61282	10.20	1-2mm fines	Sta 1-Plum	SRC1/368
61283	6.74	2-4mm fines	Sta 1-Plum	SRC1/368
61284	3.48	4-10mm fines	Sta 1-Plum	SRC1/368
61290	7.70		Sta 1-Plum	SRC1/353
61295	187.00	Breccia	Sta 1-Plum	SRC1/353
61500	267.8	Unsieved fines	Sta 1-Crater rim	SRC1/354
61501	466.9	<1mm fines	Sta 1-Crater rim	SRC1/354
61502	27.43	1-2mm fines	Sta 1-Crater rim	SRC1/354
61503	20.08	2-4mm fines	Sta 1-Crater rim	SRC1/354
61504	12.70	4-10mm fines	Sta 1-Crater rim	SRC1/354
61505	1.651		Sta 1-Crater rim	SRC1/354
61510	38.88	DB residue	Sta 1-Crater rim	SRC1/372
61515	2.00	Breccia, clastic (rake)	Sta 1-Crater rim	SRC1/372
61516	2.38	Breccia, clastic (rake)	Sta 1-Crater rim	SRC1/372
61517	.47	Breccia, clastic (rake)	Sta 1-Crater rim	SRC1/372
61518	.16	Breccia, clastic (rake)	Sta 1-Crater rim	SRC1/372
61519	.33	Breccia, clastic (rake)	Sta 1-Crater rim	SRC1/372
61525	10.35	Breccia, clastic (rake)	Sta 1-Crater rim	SRC1/372
61526	4.08	Breccia, clastic (rake)	Sta 1-Crater rim	SRC1/372
61527	.52	Breccia, clastic (rake)	Sta 1-Crater rim	SRC1/372
61528	.24	Breccia, clastic (rake)	Sta 1-Crater rim	SRC1/372
61529	.28	Breccia, clastic (rake)	Sta 1-Crater rim	SRC1/372
61535	.23	Breccia, clastic (rake)	Sta 1-Crater rim	
61536	85.99	Breccia, clastic (rake)	Sta 1-Crater rim	
61537	6.62	Breccia, clastic (rake)	Sta 1-Crater rim	
61538	4.76	Breccia, clastic (rake)	Sta 1-Crater rim	
61539	5.78	Breccia, clastic (rake)	Sta 1-Crater rim	
61545	3.61	Breccia, clastic (rake)	Sta 1-Crater rim	
61546	110.7	Glass, vesicular (rake)	Sta 1-Crater rim	
61547	17.93	Glass, vesicular (rake)	Sta 1-Crater rim	
61548	24.18	Glass, vesicular (rake)	Sta 1-Crater rim	
61549	3.76	Glass, vesicular (rake)	Sta 1-Crater rim	
61555	3.46	Glass, vesicular (rake)	Sta 1	SRC1/372
61556	2.23	Glass, vesicular (rake)	Sta 1	SRC1/372
61557	.93	Glass, vesicular (rake)	Sta 1	SRC1/372
61558	3.00	Glass, vesicular (rake)	Sta 1	SRC1/372
61559	.62	Glass, vesicular (rake)	Sta 1	SRC1/372
61565	.88	Glass, vesicular (rake)	Sta 1	SRC1/372
61566	.66	Glass, vesicular (rake)	Sta 1	SRC1/372
61567	.19	Glass, vesicular (rake)	Sta 1	SRC1/372
61568	19.32	Crystalline (rake)	Sta 1	SRC1/372
61569	12.02	Crystalline (rake)	Sta 1	SRC1/372

Apollo 16 Inventory *(continued)*

Sample Number	Weight (grams)	Description	Location	SRC/DB or SCB/DB
61575	5.26	Crystalline (rake)	Sta 1	SRC1/372
61576	5.87	Plagioclase (rake)	Sta 1	SRC1/372
61577	.21	Plagioclase (rake)	Sta 1	SRC1/372
62230	4.64	Unsieved fines	Sta 2	SRC1/005
62231	86.74	‹1mm fines	Sta 2	SRC1/005
62232	6.96	1-2mm fines	Sta 2	SRC1/005
62233	5.32	2-4mm fines	Sta 2	SRC1/005
62234	8.46	4-10mm fines	Sta 2	SRC1/005
62235	319.6	Basalt	Sta 2	SRC1/005
62236	57.27	Breccia, anorthositic	Sta 2	SRC1/005
62237	62.35	Breccia, anorthositic	Sta 2	SRC1/005
62238	1.565		Sta 2	SRC1/005
62240	162.4	Unsieved fines	Sta 2-Edge of Buster	SRC1/006
62241	243.4	‹1mm fines	Sta 2-Edge of Buster	SRC1/006
62242	21.74	1-2mm fines	Sta 2-Edge of Buster	SRC1/006
62243	19.60	2-4mm fines	Sta 2-Edge of Buster	SRC1/006
62244	16.37	4-10mm fines	Sta 2-Edge of Buster	SRC1/006
62245	6.03	Hornfels	Sta 2-Edge of Buster	SRC1/006
62246	4.59	Anorthosite	Sta 2-Edge of Buster	SRC1/006
62247	2.11	Breccia	Sta 2-Edge of Buster	SRC1/006
62248	1.61	Breccia	Sta 2-Edge of Buster	SRC1/006
62249	1.41	Breccia	Sta 2-Edge of Buster	SRC1/006
62250	18.43	DB residue	Sta 2-Edge of Buster	SRC1/007
62255	1192.	Breccia	Sta 2-Edge of Buster	SRC1/007
62270	22.17	DB residue	Sta 2-Edge of Buster	SRC1/009
62275	443.0	Breccia	Sta 2-Edge of Buster	SRC1/009
62280	143.0	Unsieved fines	Sta 2-Edge of Buster	SRC1/011
62281	218.5	‹1mm fines	Sta 2-Edge of Buster	SRC1/011
62282	21.71	1-2mm fines	Sta 2-Edge of Buster	SRC1/011
62283	13.11	2-4mm fines	Sta 2-Edge of Buster	SRC1/011
62284	14.30	4-10mm fines	Sta 2	SRC1/011
62285	3.524		Sta 2	SRC1/011
62286	2.917		Sta 2	SRC1/011
62287	2.474		Sta 2	SRC1/011
62288	1.939		Sta 2	SRC1/011
62289	1.135		Sta 2	SRC1/011
62290	27.70	DB residue	Sta 2	SRC1/010
62295	250.8	Crystalline	Sta 2	SRC1/010
62305	.810		Sta 2	SRC1/011
62315	.77	Breccia	Sta 2	SRC1/006
63010	51.3	SCB residue	Sta 13	SCB6/

Apollo 16 Inventory *(continued)*

Sample Number	Weight (grams)	Description	Location	SRC/DB or SCB/DB
63320	320.0	Unsieved fines	Sta 13	SCB6/426
63321	25.67	‹1mm fines	Sta 13	SCB6/426
63322	2.65	1-2mm fines	Sta 13	SCB6/426
63323	2.02	2-4mm fines	Sta 13	SCB6/426
63324	1.14	4-10mm fines	Sta 13	SCB6/426
63335	65.4	Breccia	Sta 13	SCB6/428
63340	149.7	Unsieved fines	Sta 13	SCB6/427
63341	25.88	‹1mm fines	Sta 13	SCB6/427
63342	2.52	1-2mm fines	Sta 13	SCB6/427
63343	2.13	2-4mm fines	Sta 13	SCB6/427
63344	.96	4-10mm fines	Sta 13	SCB6/427
63350	23.49		Sta 13	SCB6/429
63355	68.24	Breccia	Sta 13	SCB6/429
63500	201.8	Unsieved fines (rake)	Sta 13	SCB4/346
63501	342.5	‹1mm fines (rake)	Sta 13	SCB4/346
63502	25.29	1-2mm fines (rake)	Sta 13	SCB4/346
63503	14.53	2-4mm fines (rake)	Sta 13	SCB4/346
63504	17.34	4-10mm fines (rake)	Sta 13	SCB4/346
63505	5.41	Breccia, anorthositic	Sta 13	SCB4/346
63506	4.9	Crystalline	Sta 13	SCB4/346
63507	2.78	Breccia, soil	Sta 13	SCB4/346
63508	2.61	Anorthosite	Sta 13	SCB4/346
63509	2.05	Crystalline	Sta 13	SCB4/346
63515	1.32	Crystalline	Sta 13	SCB4/346
63520	22.08	DB residue (rake)	Sta 13	SCB4/345
63525	6.68	Crystalline (rake)	Sta 13	SCB4/345
63526	2.91	Crystalline (rake)	Sta 13	SCB4/345
63527	6.10	Crystalline (rake)	Sta 13	SCB4/345
63528	4.12	Crystalline (rake)	Sta 13	SCB4/345
63529	23.48	Crystalline (rake)	Sta 13	SCB4/345
63535	6.85	Crystalline (rake)	Sta 13	SCB4/345
63536	1.02	Crystalline (rake)	Sta 13	SCB4/345
63537	4.78	Crystalline (rake)	Sta 13	SCB4/345
63538	35.06	Crystalline (rake)	Sta 13	SCB4/345
63539	.39	Crystalline (rake)	Sta 13	SCB4/345
63545	15.95	Crystalline (rake)	Sta 13	SCB4/345
63546	9.23	Crystalline (rake)	Sta 13	SCB4/345
63547	4.90	Crystalline (rake)	Sta 13	SCB4/345
63548	1.13	Crystalline (rake)	Sta 13	SCB4/345
63549	26.57	Crystalline (rake)	Sta 13	SCB4/345
63555	3.38	Crystalline (rake)	Sta 13	SCB4/345
63556	18.10	Crystalline (rake)	Sta 13	SCB4/345

Apollo 16 Inventory *(continued)*

Sample Number	Weight (grams)	Description	Location	SRC/DB or SCB/DB
63557	7.53	Crystalline (rake)	Sta 13	SCB4/345
63558	7.09	Crystalline (rake)	Sta 13	SCB4/345
63559	6.04	Glass, vesicular (rake)	Sta 13	SCB4/345
63565	.94	Glass, vesicular (rake)	Sta 13	SCB4/345
63566	19.61	Glass, vesicular (rake)	Sta 13	SCB4/345
63567	3.21	Glass, vesicular (rake)	Sta 13	SCB4/345
63568	4.06	Glass, vesicular (rake)	Sta 13	SCB4/345
63569	.43	Glass, vesicular (rake)	Sta 13	SCB4/345
63575	4.72	Glass, vesicular (rake)	Sta 13	SCB4/345
63576	1.23	Glass, vesicular (rake)	Sta 13	SCB4/345
63577	12.41	Breccia	Sta 13	SCB4/345
63578	19.60	Breccia	Sta 13	SCB4/345
63579	11.35	Breccia	Sta 13	SCB4/345
63585	32.62	Breccia	Sta 13	SCB4/345
63586	1.98	Breccia	Sta 13	SCB4/345
63587	20.51	Breccia	Sta 13	SCB4/345
63588	2.40	Breccia	Sta 13	SCB4/345
63589	13.51	Breccia	Sta 13	SCB4/345
63595	2.10	Breccia	Sta 13	SCB4/345
63596	6.40	Glass, vesicular (rake)	Sta 13	SCB4/345
63597	5.67	Glass, vesicular (rake)	Sta 13	SCB4/345
63598	12.66	Glass, vesicular (rake)	Sta 13	SCB4/345
64001	752.3	Core 38	Sta 4	SCB3/
64002	584.1	Core 43	Sta 4	SCB2/
64420	112.2	Unsieved fines	Sta 4-Trench bottom	SCB3/399
64421	206.9	‹1mm fines	Sta 4-Trench bottom	SCB3/399
64422	6.17	1-2mm fines	Sta 4-Trench bottom	SCB3/399
64423	3.76	2-4mm fines	Sta 4-Trench bottom	SCB3/399
64424	2.06	4-10mm fines	Sta 4-Trench bottom	SCB3/399
64425	14.62	Breccia (black & white rock)	Sta 4-Trench bottom	SCB3/399
64430	28.22	DB residue	Sta 4-Trench bottom	SCB1/394
64435	1079.0	Breccia	Sta 4-Trench bottom	SCB1/394
64450	1.57		Sta 4-Trench bottom	SCB3/397
64455	56.68	Anorthosite, glass coated	Sta 4-Trench bottom	SCB3/397
64470	27.09	DB residue	Sta 4-Trench bottom	SCB3/398
64475	1032.0	Breccia	Sta 4-Trench bottom	SCB3/398
64476	125.1	Breccia	Sta 4-Trench bottom	SCB3/398
64477	19.32	Breccia	Sta 4-Trench bottom	SCB3/398
64478	12.34	Breccia	Sta 4-Trench bottom	SCB3/398
64500	320.6	Unsieved fines (rake)	Sta 4-Trench bottom	SCB1/396

Apollo 16 Inventory *(continued)*

Sample Number	Weight (grams)	Description	Location	SRC/DB or SCB/DB
64501	495.7	‹1mm fines	Sta 4-Trench bottom	SCB1/396
64502	28.38	1-2mm fines (rake)	Sta 4-Trench bottom	SCB1/396
64503	24.11	2-4mm fines	Sta 4-Trench bottom	SCB1/396
64504	24.15	4-10mm fines	Sta 4-Trench bottom	SCB1/396
64505	5.392		Sta 4-Trench bottom	SCB1/396
64506	5.079		Sta 4-Trench bottom	SCB1/396
64507	4.474		Sta 4-Trench bottom	SCB1/396
64508	4.168		Sta 4-Trench bottom	SCB1/396
64509	3.150		Sta 4-Trench bottom	SCB1/396
64515	3.761		Sta 4-Trench bottom	SCB1/396
64516	2.929		Sta 4-Trench bottom	SCB1/396
64517	1.546		Sta 4-Trench bottom	SCB1/396
64518	1.490		Sta 4-Trench bottom	SCB1/396
64519	1.124		Sta 4-Trench bottom	SCB1/396
64525	1.107		Sta 4-Trench bottom	SCB1/396
64530	102.8	DB residue (rake)	Sta 4-Trench bottom	SCB1/395
64535	256.6	Breccia (rake)	Sta 4-Trench bottom	SCB1/395
64536	177.5	Breccia (rake)	Sta 4-Trench bottom	SCB1/395
64537	124.3	Breccia (rake)	Sta 4-Trench bottom	SCB1/395
64538	30.03	Breccia (rake)	Sta 4-Trench bottom	SCB1/395
64539	17.76	Breccia (rake)	Sta 4-Trench bottom	SCB1/395
64545	14.09	Breccia (rake)	Sta 4	SCB1/395
64546	12.80	Breccia (rake)	Sta 4	SCB1/395
64547	10.90	Breccia (rake)	Sta 4	SCB1/395
64548	8.49	Breccia (rake)	Sta 4	SCB1/395
64549	6.47	Breccia (rake)	Sta 4	SCB1/395
64555	5.29	Breccia (rake)	Sta 4	SCB1/395
64556	5.15	Breccia (rake)	Sta 4	SCB1/395
64557	4.790	Breccia (rake)	Sta 4	SCB1/395
64558	3.130	Breccia (rake)	Sta 4	SCB1/395
64559	21.82	Crystalline (rake)	Sta 4	
64565	14.73	Crystalline (rake)	Sta 4	SCB1/395
64566	14.13	Crystalline (rake)	Sta 4	SCB1/395
64567	13.86	Crystalline (rake)	Sta 4	SCB1/395
64568	9.379	Crystalline (rake)	Sta 4	SCB1/395
64569	14.32	Crystalline (rake)	Sta 4	SCB1/395
64575	6.837	Crystalline (rake)	Sta 4	SCB1/395
64576	6.916	Crystalline (rake)	Sta 4	SCB1/395
64577	5.692	Crystalline (rake)	Sta 4	SCB1/395
64578	5.596	Crystalline (rake)	Sta 4	SCB1/395
64579	4.802	Crystalline (rake)	Sta 4	SCB1/395
64585	4.696	Crystalline (rake)	Sta 4	SCB1/395

Apollo 16 Inventory *(continued)*

Sample Number	Weight (grams)	Description	Location	SRC/DB or SCB/DB
64586	3.337	Crystalline (rake)	Sta 4	SCB1/395
64587	7.180	Breccia (rake)	Sta 4	SCB1/395
64588	2.546	Breccia (rake)	Sta 4	SCB1/395
64589	4.039	Anorthosite (rake)	Sta 4	SCB1/395
64800	166.3	Unsieved fines (rake)	Sta 4-Crater rim	SCB3/400
64801	286.8	‹1mm fines	Sta 4-Crater rim	SCB3/400
64802	10.96	1-2mm fines	Sta 4-Crater rim	SCB3/400
64803	8.09	2-4mm fines	Sta 4-Crater rim	SCB3/400
64804	7.89	4-10mm fines	Sta 4-Crater rim	SCB3/400
64810	102.14	Unsieved fines	Sta 4-Crater rim	SCB3/401
64811	174.7	‹1mm fines	Sta 4-Crater rim	SCB3/401
64812	9.53	1-2mm fines	Sta 4-Crater rim	SCB3/401
64813	9.10	2-4mm fines	Sta 4-Crater rim	SCB3/401
64814	5.34	4-10mm fines	Sta 4-Crater rim	SCB3/401
64815	20.90	Ultramafic, crushed (rake)	Sta 4-Crater rim	SCB3/401
64816	3.83	Crystalline, fine-grained	Sta 4-Crater rim	SCB3/401
64817	8.98	Crystalline, fine-grained	Sta 4-Crater rim	SCB3–401
64818	15.98	Crystalline, fine-grained	Sta 4-Crater rim	SCB3/401
64819	11.76	Anorthosite, fine-grained (rake)	Sta 4-Crater rim	SCB3/401
64825	21.50	Breccia (rake)	Sta 4	SCB3/401
64826	11.33	Breccia (rake)	Sta 4	SCB3/401
64827	8.11	Breccia (rake)	Sta 4	SCB3/401
64828	.97	Breccia (rake)	Sta 4	SCB3/401
64829	2.20	Breccia (rake)	Sta 4	SCB3/401
64835	2.32	Breccia (rake)	Sta 4	SCB3/401
64836	1.76	Breccia (rake)	Sta 4	SCB3/401
64837	2.18	Breccia (rake)	Sta 4	SCB3/401
65010	42.37	SCB residue	Sta 5	SCB1/
65015	1802.0	Crystalline	Sta 5	SCB3/
65016	21.02	Glass sphere	Sta 5	SCB1/
65030	45.0	DB residue	Sta 5	SCB1/404
65035	446.1	Breccia	Sta 5	SCB1/404
65050	32.29	DB residue	Sta 5	SCB3/337
65055	500.8	Crystalline	Sta 5	SCB3/337
65056	64.78	Glass agglutinate	Sta 5	SCB3/337
65070	5.05	DB residue	Sta 5	SCB1/403
65075	107.9	Breccia	Sta 5	SCB1/403
65090	28.26	DB residue	Sta 5	SCB3/336
65095	560.1	Breccia, anorthosite	Sta 5	SCB3/336
65310	45.08	DB residue	Sta 5	SCB1/405

Apollo 16 Inventory *(continued)*

Sample Number	Weight (grams)	Description	Location	SRC/DB or SCB/DB
65315	300.4	Breccia (put into rake DB)	Sta 5	SCB1/405
65325	67.87	Anorthositic (rake)	Sta 5	SCB1/405
65326	36.40	Anorthositic (rake)	Sta 5	SCB1/405
65327	6.97	Anorthositic (rake)	Sta 5	SCB1/405
65328	1.28	Anorthositic (rake)	Sta 5	SCB1/405
65329	1.92	Anorthositic (rake)	Sta 5	SCB1/405
65335	1.63	Anorthositic (rake)	Sta 5	SCB1/405
65336	.63	Anorthositic (rake)	Sta 5	SCB1/405
65337	11.57	Breccia (rake)	Sta 5	SCB1/405
65338	2.65	Breccia (rake)	Sta 5	SCB1/405
65339	1.62	Breccia (rake)	Sta 5	SCB1/405
65345	.86	Breccia (rake)	Sta 5	SCB1/405
65346	.80	Breccia (rake)	Sta 5	SCB1/405
65347	.43	Breccia (rake)	Sta 5	SCB1/405
65348	11.66	Glass, vesicular (rake)	Sta 5	SCB1/405
65349	7.58	Glass, vesicular (rake)	Sta 5	SCB1/405
65355	4.94	Glass, vesicular (rake)	Sta 5	SCB1/405
65356	2.53	Glass, vesicular (rake)	Sta 5	SCB1/405
65357	18.76	Crystalline (rake)	Sta 5	SCB1/405
65358	7.02	Crystalline (rake)	Sta 5	SCB1/405
65359	2.53	Crystalline (rake)	Sta 5	SCB1/405
65365	2.16	Crystalline (rake)	Sta 5	SCB1/405
65366	8.48	Glass, fragment (rake)	Sta 5	SCB1/405
65500	413.0	Unsieved fines (rake)	Sta 5	SRC2/333
65501	150.0	‹1mm fines (rake)	Sta 5	SRC2/333
65502	9.50	1-2mm fines (rake)	Sta 5	SRC2/333
65503	23.24	2-4mm fines (rake)	Sta 5	SRC2/333
65504	22.48	4-10mm fines (rake)	Sta 5	SRC2/333
65510	171.3	Unsieved fines (rake)	Sta 5	SRC2/332
65511	190.2	‹1mm fines (rake)	Sta 5	SRC2/332
65512	14.68	1-2mm fines (rake)	Sta 5	SRC2/332
65513	20.21	2-4mm fines (rake)	Sta 5	SRC2/332
65514	13.98	4-10mm fines (rake)	Sta 5	SRC2/332
65515	50.25	Soil clods (rake)	Sta 5	SRC2/332
65516	10.49	Soil clods (rake)	Sta 5	SRC2/332
65517	11.85	Soil clods (rake)	Sta 5	SRC2/332
65518	9.477	Soil clods (rake)	Sta 5	SRC2/332
65519	10.58	Soil clods (rake)	Sta 5	SRC2/332
65525	7.483	Soil clods (rake)	Sta 5	SRC2/332
65526	3.545	Soil clods (rake)	Sta 5	SRC2/332
65527	2.890	Soil clods (rake)	Sta 5	SRC2/332
65528	3.082	Soil clods (rake)	Sta 5	SRC2/332

Apollo 16 Inventory *(continued)*

Sample Number	Weight (grams)	Description	Location	SRC/DB or SCB/DB
65529	2.55	Soil clods (rake)	Sta 5	SRC2/332
65535	2.658	Soil clods (rake)	Sta 5	SRC2/332
65536	1.575	Soil clods (rake)	Sta 5	SRC2/332
65537	2.426	Soil clods (rake)	Sta 5	SRC2/332
65538	2.342	Soil clods (rake)	Sta 5	SRC2/332
65539	2.180	Soil clods (rake)	Sta 5	SRC2/332
65545	1.797	Soil clods (rake)	Sta 5	SRC2/332
65546	1.346	Soil clods (rake)	Sta 5	SRC2/332
65547	1.587	Soil clods (rake)	Sta 5	SRC2/332
65548	3.023	Soil clods (rake)	Sta 5	SRC2/332
65549	2.094	Soil clods (rake)	Sta 5	SRC2/332
65555	2.202	Soil clods (rake)	Sta 5	SRC2/332
65556	1.170	Soil clods (rake)	Sta 5	SRC2/332
65557	1.114	Soil clods (rake)	Sta 5	SRC2/332
65558	1.695	Soil clods (rake)	Sta 5	SRC2/332
65559	1.533	Soil clods (rake)	Sta 5	SRC2/332
65565	.852	Soil clods (rake)	Sta 5	SRC2/332
65566	1.998	Soil clods (rake)	Sta 5	SRC2/332
65567	1.289	Soil clods (rake)	Sta 5	SRC2/332
65568	.808	Soil clods (rake)	Sta 5	SRC2/332
65569	.873	Soil clods (rake)	Sta 5	SRC2/332
65575	.907	Soil clods (rake)	Sta 5	SRC2/332
65576	.906	Soil clods (rake)	Sta 5	SRC2/332
65577	.706	Soil clods (rake)	Sta 5	SRC2/332
65578	.320	Soil clods (rake)	Sta 5	SRC2/332
65579	.612	Soil clods (rake)	Sta 5	SRC2/332
65585	9.294	Glassy agglutinates (rake)	Sta 5	SRC2/332
65586	6.763	Glassy agglutinates (rake)	Sta 5	SRC2/332
65587	2.141	Glassy agglutinates (rake)	Sta 5	SRC2/332
65588	9.629	Plagioclase (rake)	Sta 5	SRC2/332
65700	92.30	Unsieved fines (rake)	Sta 5	SCB1/402
65701	171.3	<1mm fines (rake)	Sta 5	SCB1/402
65702	4.89	1-2mm fines (rake)	Sta 5	SCB1/402
65703	1.58	2-4mm fines (rake)	Sta 5	SCB1/402
65704	1.39	4-10mm fines (rake)	Sta 5	SCB1/402
65710	91.23	Rake sample	Sta 5	SCB1/334
65715	31.36	Breccia (rake)	Sta 5	SCB1/334
65716	14.28	Breccia (rake)	Sta 5	SCB1/334
65717	7.415	Breccia (rake)	Sta 5	SCB1/334
65718	10.61	Breccia (rake)	Sta 5	SCB1/334
65719	7.04	Breccia (rake)	Sta 5	SCB1/334

Apollo 16 Inventory *(continued)*

Sample Number	Weight (grams)	Description	Location	SRC/DB or SCB/DB
65725	6.67	Breccia (rake)	Sta 5	SCB1/334
65726	5.19	Breccia (rake)	Sta 5	SCB1/334
65727	4.30	Breccia (rake)	Sta 5	SCB1/334
65728	4.22	Breccia (rake)	Sta 5	SCB1/334
65729	3.81	Breccia (rake)	Sta 5	SCB1/334
65735	4.26	Breccia (rake)	Sta 5	SCB1/334
65736	2.74	Breccia (rake)	Sta 5	SCB1/334
65737	.85	Breccia (rake)	Sta 5	SCB1/334
65738	1.17	Breccia (rake)	Sta 5	SCB1/334
65739	.95	Breccia (rake)	Sta 5	SCB1/334
65745	7.76	Breccia (rake)	Sta 5	SCB1/334
65746	4.19	Breccia (rake)	Sta 5	SCB1/334
65747	.82	Breccia (rake)	Sta 5	SCB1/334
65748	.97	Breccia (rake)	Sta 5	SCB1/334
65749	.95	Breccia (rake)	Sta 5	SCB1/334
65755	1.42	Breccia (rake)	Sta 5	SCB1/334
65756	.77	Breccia (rake)	Sta 5	SCB1/334
65757	26.20	Breccia (rake)	Sta 5	SCB1/334
65758	5.95	Breccia (rake)	Sta 5	SCB1/334
65759	3.11	Breccia (rake)	Sta 5	SCB1/334
65765	1.12	Breccia (rake)	Sta 5	SCB1/334
65766	1.01	Breccia (rake)	Sta 5	SCB1/334
65767	17.51	Agglutinates, glassy (rake)	Sta 5	SCB1/334
65768	3.25	Agglutinates, glassy (rake)	Sta 5	SCB1/334
65769	2.74	Agglutinates, glassy (rake)	Sta 5	SCB1/334
65775	3.50	Agglutinates, glassy (rake)	Sta 5	SCB1/334
65776	2.33	Agglutinates, glassy (rake)	Sta 5	SCB1/334
65777	16.53	Crystalline (rake)	Sta 5	SCB1/334
65778	12.22	Crystalline (rake)	Sta 5	SCB1/334
65779	12.71	Crystalline (rake)	Sta 5	SCB1/334
65785	5.16	Crystalline (rake)	Sta 5	SCB1/334
65786	83.02	Breccia (rake)	Sta 5	SCB1/334
65787	8.28	Breccia (rake)	Sta 5	SCB1/334
65788	9.32	Breccia (rake)	Sta 5	SCB1/334
65789	12.24	Anorthosite, granular (rake)	Sta 5	SCB1/334
65795	6.84	Anorthosite, gabbroic (rake)	Sta 5	SCB1/334
65900	233.2	Unsieved fines (rake)	Sta 5-15 cm below surface	SCB1/406
65901	393.2	<1mm fines (rake)	Sta 5-15 cm below surface	SCB1/406
65902	14.84	1-2mm fines (rake)	Sta 5-15 cm below surface	SCB1/406

Apollo 16 Inventory *(continued)*

Sample Number	Weight (grams)	Description	Location	SRC/DB or SCB/DB
65903	11.40	2-4mm fines (rake)	Sta 5-15 cm below surface	SCB1/406
65904	9.51	4-10mm fines (rake)	Sta 5-15 cm below surface	SCB1/406
65905	12.08		Sta 5	SCB1/406
65906	6.584		Sta 5	SCB1/406
65907	4.658		Sta 5	SCB1/406
65908	2.162		Sta 5	SCB1/406
65909	2.024		Sta 5	SCB1/406
65915	2.060		Sta 5	SCB1/406
65916	0.994		Sta 5	SCB1/406
65920	12.06	Rake sample	Sta 5	SCB1/335
65925	3.82	Breccia (rake)	Sta 5	SCB1/335
65926	3.03	Breccia (rake)	Sta 5	SCB1/335
65927	.72	Breccia (rake)	Sta 5	SCB1/335
66030	50.49	Unsieved fines	Sta 6	SCB1/407
66031	75.6	<1mm fines	Sta 6	SCB1/407
66032	2.99	1-2mm fines	Sta 6	SCB1/407
66033	2.16	2-3mm fines	Sta 6	SCB1/407
66034	3.36	4-10mm fines	Sta 6	SCB1/407
66035	211.4	Breccia	Sta 6	SCB1/407
66036	4.384	Breccia	Sta 6	SCB1/407
66037	3.718	Breccia	Sta 6	SCB1/407
66040	166.5	Unsieved fines (gray soil)	Sta 6	SRC2/338
66041	357.4	<1mm fines (gray soil)	Sta 6	SRC2/338
66042	19.5	1-2mm fines (gray soil)	Sta 6	SRC2/338
66043	15.5	2-4mm fines (gray soil)	Sta 6	SRC2/338
66044	11.3	4-10mm fines (gray soil)	Sta 6	SRC2/338
66050	30.36	DB residue	Sta 6	SCB1/408
66055	1306.0	Breccia	Sta 6	SCB1/408
66070	8.06	DB residue	Sta 6	SRC2/409
66075	347.1	Breccia	Sta 6	SRC2/409
66080	106.1	Unsieved fines (white patch on regolith)	Sta 6	SRC2/339
66081	177.3	<1mm fines (white patch on regolith)	Sta 6	SRC2/339
66082	9.85	1-2mm fines (white patch on regolith)	Sta 6	SRC2/339
66083	4.53	2-4mm fines (white patch on regolith)	Sta 6	SRC2/339
66084	3.13	4-10mm fines (white patch on regolith)	Sta 6	SRC2/339

Apollo 16 Inventory *(continued)*

Sample Number	Weight (grams)	Description	Location	SRC/DB or SCB/DB
66085	3.66		Sta 6	SRC2/339
66086	2.027		Sta 6	SRC2/339
66090	9.47	DB residue	Sta 6	SCB1/410
66095	1185.0	Anorthosite	Sta 6	SCB1/410
67010	459.5	Residue	Sta 11	SCB7/
67015	1194.0	Breccia	Sta 11	SCB7/
67016	4262.0	Breccia	Sta 11	BSLSS
67020	357.6	Residue	Sta 11	BSLSS
67025	16.06	Anorthosite	Sta 11	BSLSS
67030	.77	Fragments	Sta 11	SCB7/382
67031	52.73	‹1mm fragments	Sta 11	SCB7/382
67032	13.30	1-2mm fragments	Sta 11	SCB7/382
67033	14.88	2-4mm fragments	Sta 11	SCB7/382
67034	14.55	4-10mm fragments	Sta 11	SCB7/382
67035	245.2	Breccia	Sta 11	SCB7/382
67050	18.54	DB residue	Sta 11	SCB7/383
67055	221.88	Breccia	Sta 11	SCB7/383
67070	.70	DB residue	Sta 11	SCB7/384
67075	219.2	Anorthosite	Sta 11	SCB7/384
67090	3.86	DB residue	Sta 11	SCB7/385
67095	339.8	Breccia	Sta 11	SCB7/385
67110	17.28	DB residue	Sta 11	SCB7/386
67115	240.0	Breccia	Sta 11	SCB7/386
67210	276.9	Breccia	Sta 11	SCB6/PDB1
67230	938.34	Breccia	Sta 11	SCB6/PDB2
67410	58.72	DB residue	Sta 11	SCB6/387
67415	174.9	Breccia, anorthosite	Sta 11	SCB6/387
67430	23.26	DB residue	Sta 11	SCB6/415
67435	353.5	Breccia	Sta 11	SCB6/415
67450	217.2	DB residue	Sta 11	SCB6/416
67455	942.2	Breccia	Sta 11	SCB6/416
67460	123.7	Unsieved fines (fillet soil)	Sta 11	SCB6/417
67461	222.2	‹1mm fines	Sta 11	SCB6/417
67462	17.4	1-2mm fines	Sta 11	SCB6/417
67463	6.24	2-4mm fines	Sta 11	SCB6/417
67464	.70	4-10mm fines (fillet soil)	Sta 11	SCB6/417

Apollo 16 Inventory *(continued)*

Sample Number	Weight (grams)	Description	Location	SRC/DB or SCB/DB
67475	175.1	Breccia	Sta 11	SCB6/418
67480	87.05	Unsieved fines (ref soil)	Sta 11	SCB6/419
67481	132.7	<1mm fines	Sta 11	SCB6/419
67482	14.65	1-2mm fines	Sta 11	SCB6/419
67483	8.37	2-4mm fines	Sta 11	SCB6/419
67484	6.02	4-10mm fines	Sta 11	SCB6/419
67485	6.55	Crystalline, aphanitic	Sta 11	SCB6/419
67486	5.80	Glass	Sta 11	SCB6/419
67487	2.65	Crystalline, aphanitic	Sta 11	SCB6/419
67488	2.25	Crystalline, aphanitic	Sta 11	SCB6/419
67489	2.06	Crystalline, aphanitic	Sta 11	SCB6/419
67495	1.34	Breccia	Sta 11	SCB6/419
67510	9.22	Unsieved fines (rake soil)	Sta 11	SCB6/420
67511	59.5	<1mm fines	Sta 11	SCB6/420
67512	14.46	1-2mm fines	Sta 11	SCB6/420
67513	19.39	2-4mm fines	Sta 11	SCB6/420
67514	31.03	4-10mm fines (rake soil)	Sta 11	SCB6/420
67515	60.8	Anorthositic, granulated (rake)	Sta 11	SCB6/420
67516	14.38	Anorthositic, granulated (rake)	Sta 11	SCB6/420
67517	9.65	Anorthositic, granulated (rake)	Sta 11	SCB6/420
67518	3.74	Anorthositic, granulated (rake)	Sta 11	SCB6/420
67519	2.04	Anorthositic, granulated (rake)	Sta 11	SCB6/420
67525	2.52	Anorthositic, granulated (rake)	Sta 11	SCB6/420
67526	2.44	Anorthositic, granulated (rake)	Sta 11	SCB6/420
67527	2.40	Anorthositic, granulated (rake)	Sta 11	SCB6/420
67528	1.24	Anorthositic, granulated (rake)	Sta 11	SCB6/420
67529	1.13	Anorthositic, granulated (rake)	Sta 11	SCB6/420
67535	.99	Anorthositic, granulated (rake)	Sta 11	SCB6/420
67536	1.20	Anorthositic, granulated (rake)	Sta 11	SCB6/420
67537	1.29	Anorthositic, granulated (rake)	Sta 11	SCB6/420
67538	1.77	Anorthositic, granulated (rake)	Sta 11	SCB6/420
67539	2.12	Anorthositic, granulated (rake)	Sta 11	SCB6/420
67545	1.88	Anorthositic, granulated (rake)	Sta 11	SCB6/420
67546	1.50	Anorthositic, granulated (rake)	Sta 11	SCB6/420
67547	.83	Anorthositic, granulated (rake)	Sta 11	SCB6/420
67548	1.36	Anorthositic, granulated (rake)	Sta 11	SCB6/420
67549	43.1	Breccia, heterogeneous (rake)	Sta 11	SCB6/420
67555	3.54	Breccia, heterogeneous (rake)	Sta 11	SCB6/420
67556	8.21	Breccia, heterogeneous (rake)	Sta 11	SCB6/420
67557	3.30	Breccia, friable	Sta 11	SCB6/420
67558	2.56	Breccia, friable	Sta 11	SCB6/420
67559	32.9	Crystalline	Sta 11	SCB6/420

Apollo 16 Inventory *(continued)*

Sample Number	Weight (grams)	Description	Location	SRC/DB or SCB/DB
67565	10.43	Crystalline	Sta 11	SCB6/420
67566	4.31	Crystalline	Sta 11	SCB6/420
67567	11.51	Glass, vesicular	Sta 11	SCB6/420
67568	11.05	Glass, vesicular	Sta 11	SCB6/420
67569	7.27	Glass, vesicular	Sta 11	SCB6/420
67575	4.47	Glass, vesicular	Sta 11	SCB6/420
67576	3.98	Glass, vesicular	Sta 11	SCB6/420
67600	2.17	Unsieved fines (rake soil)	Sta 11-Crater rim	SCB6/422
67601	161.8	<1mm fines (rake soil)	Sta 11-Crater rim	SCB6/422
67602	13.45	1-2mm fines	Sta 11-Crater rim	SCB6/422
67603	6.16	2-4mm fines	Sta 11-Crater rim	SCB6/422
67604	2.62	4-10mm fines	Sta 11-Crater rim	SCB6/422
67605	44.52	Breccia	Sta 11-Crater rim	SCB6/422
67610	66.83	DB residue (rake)	Sta 11-Crater rim	SCB6/421
67615	8.77	Crystalline (rake)	Sta 11-Crater rim	SCB6/421
67616	21.29	Crystalline (rake)	Sta 11-Crater rim	SCB6/421
67617	14.32	Crystalline (rake)	Sta 11-Crater rim	SCB6/421
67618	11.17	Crystalline (rake)	Sta 11-Crater rim	SCB6/421
67619	6.15	Crystalline (rake)	Sta 11-Crater rim	SCB6/421
67625	6.72	Crystalline (rake)	Sta 11-Crater rim	SCB6/421
67626	19.19	Glass, vesicular (rake)	Sta 11-Crater rim	SCB6/421
67627	79.64	Glass, vesicular (rake)	Sta 11-Crater rim	SCB6/421
67628	49.71	Glass, vesicular (rake)	Sta 11-Crater rim	SCB6/421
67629	32.84	Glass, vesicular (rake)	Sta 11-Crater rim	SCB6/421
67635	9.12	Anorthositic, granulated (rake)	Sta 11-Crater rim	SCB6/421
67636	3.23	Anorthositic, granulated (rake)	Sta 11-Crater rim	SCB6/421
67637	2.34	Anorthositic, granulated (rake)	Sta 11-Crater rim	SCB6/421
67638	7.23	Breccia (rake)	Sta 11-Crater rim	SCB6/421
67639	7.34	Breccia (rake)	Sta 11-Crater rim	SCB6/421
67645	.84	Breccia (rake)	Sta 11-Crater rim	SCB6/421
67646	3.94	Breccia (rake)	Sta 11-Crater rim	SCB6/421
67647	47.72	Breccia (rake)	Sta 11-Crater rim	SCB6/421
67648	7.88	Breccia (rake)	Sta 11-Crater rim	SCB6/421
67649	1.60	Breccia (rake)	Sta 11-Crater rim	SCB6/421
67655	4.11	Breccia (rake)	Sta 11-Crater rim	SCB6/421
67656	1.93	Breccia (rake)	Sta 11-Crater rim	SCB6/421
67657	1.70	Breccia (rake)	Sta 11-Crater rim	SCB6/421
67658	1.35	Breccia (rake)	Sta 11-Crater rim	SCB6/421
67659	1.62	Breccia (rake)	Sta 11-Crater rim	SCB6/421
67665	5.88	Breccia (rake)	Sta 11-Crater rim	SCB6/421
67666	5.47	Breccia (rake)	Sta 11-Crater rim	SCB6/421
67667	7.89	Ultramafic, crushed (rake)	Sta 11-Crater rim	SCB6/421

Apollo 16 Inventory *(continued)*

Sample Number	Weight (grams)	Description	Location	SRC/DB or SCB/DB
67668	3.58	Crystalline (rake)	Sta 11-Crater rim	SCB6/421
67669	12.54	Breccia (rake)	Sta 11-Crater rim	SCB6/421
67675	1.07	Glass, vesicular (rake)	Sta 11-Crater rim	SCB6/421
67676	2.33	Crystalline (rake)	Sta 11-Crater rim	SCB6/421
67700	142.6	Unsieved fines (rake soil, white)	Sta 11	SCB4/388
67701	235.0	<1mm fines	Sta 11	SCB4/388
67702	21.69	1-2mm fines	Sta 11	SCB4/388
67703	13.71	2-4mm fines	Sta 11	SCB4/388
67704	7.47	4-10mm fines	Sta 11	SCB4/388
67705	6.57	Breccia	Sta 11	SCB4/388
67706	2.08	Breccia	Sta 11	SCB4/388
67707	1.84	Breccia	Sta 11	SCB4/388
67708	1.43	Breccia	Sta 11	SCB4/388
67710	133.39	Unsieved fines (rake)	Sta 11	SCB4/423
67711	205.3	<1mm fines	Sta 11	SCB4/423
67712	34.84	1-2mm fines	Sta 11	SCB4/423
67713	22.45	2-4mm fines	Sta 11	SCB4/423
67714	12.66	4-10mm fines (rake)	Sta 11	SCB4/423
67715	9.44	Breccia (rake)	Sta 11	SCB4/423
67716	17.02	Breccia (rake)	Sta 11	SCB4/423
67717	5.56	Breccia (rake)	Sta 11	SCB4/423
67718	41.05	Breccia (rake)	Sta 11	SCB4/423
67719	2.13	Breccia (rake)	Sta 11	SCB4/423
67725	5.85	Breccia (rake)	Sta 11	SCB4/423
67726	4.53	Breccia (rake)	Sta 11	SCB4/423
67727	1.80	Glass, vesicular (rake)	Sta 11	SCB4/423
67728	9.25	Glass, vesicular (rake)	Sta 11	SCB4/423
67729	73.2	Glass, vesicular (rake)	Sta 11	SCB4/423
67735	13.30	Crystalline, fine-grained (rake)	Sta 11	SCB4/423
67736	14.92	Crystalline, fine-grained (rake)	Sta 11	SCB4/423
67737	4.56	Crystalline, fine-grained (rake)	Sta 11	SCB4/423
67738	5.84	Crystalline, fine-grained (rake)	Sta 11	SCB4/423
67739	2.03	Crystalline, fine-grained (rake)	Sta 11	SCB4/423
67745	3.53	Crystalline, fine-grained (rake)	Sta 11	SCB4/423
67746	3.47	Crystalline, fine-grained (rake)	Sta 11	SCB4/423
67747	6.30	Crystalline, fine-grained (rake)	Sta 11	SCB4/423
67748	4.74	Crystalline, fine-grained (rake)	Sta 11	SCB4/423
67749	11.47	Breccia (rake)	Sta 11	SCB4/423
67755	3.53	Breccia (rake)	Sta 11	SCB4/423
67756	4.82	Breccia (rake)	Sta 11	SCB4/423
67757	4.83	Breccia (rake)	Sta 11	SCB4/423
67758	4.06	Breccia (rake)	Sta 11	SCB4/423
67759	4.56	Breccia (rake)	Sta 11	SCB4/423

Apollo 16 Inventory *(continued)*

Sample Number	Weight (grams)	Description	Location	SRC/DB or SCB/DB
67765	1.73	Breccia (rake)	Sta 11	SCB4/423
67766	5.47	Breccia (rake)	Sta 11	SCB4/423
67767	1.67	Breccia (rake)	Sta 11	SCB4/423
67768	.99	Breccia (rake)	Sta 11	SCB4/423
67769	3.05	Breccia (rake)	Sta 11	SCB4/423
67775	6.58	Breccia (rake)	Sta 11	SCB4/423
67776	3.10	Breccia (rake)	Sta 11	SCB4/423
67910	180.3	Residue	Sta 11	SCB4/
67915	2559.0	Breccia	Sta 11	SCB4/
67930	8.51	Unsieved fines	Sta 11	SCB4/389
67935	108.9	Breccia	Sta 11	SCB4/389
67936	61.82	Breccia	Sta 11	SCB4/389
67937	59.67	Breccia	Sta 11	SCB4/389
67940	27.22	Unsieved fines	Sta 11-E-W split in boulder	SCB4/390
67941	105.9	‹1mm fines	Sta 11-E-W split in boulder	SCB4/390
67942	12.23	1-2mm fines	Sta 11-E-W split in boulder	SCB4/390
67943	9.36	2-4mm fines	Sta 11-E-W split in boulder	SCB4/390
67944	8.59	4-10mm fines	Sta 11-E-W split in boulder	SCB4/390
67945	4.37	Metaclastic	Sta 11-E-W split in boulder	SCB4/390
67946	3.20	Breccia	Sta 11-E-W split in boulder	SCB4/390
67947	2.43	Breccia	Sta 11-E-W split in boulder	SCB4/390
67948	1.59	Crystalline	Sta 11-E-W split in boulder	SCB4/390
67950	8.21		Sta 11-E-W split in boulder	SCB4/390
67955	162.6	Breccia	Sta 11-E-W split in boulder	SCB4/390
67956	3.70	Crystalline	Sta 11-E-W split in boulder	SCB4/390
67957	1.73	Breccia	Sta 11-E-W split in boulder	SCB4/390
67960	12.11	Unsieved fines	Sta 11-E-W split in boulder	SCB4/391
67970	3.15	DB residue	Sta 11-E-W split in boulder	SCB4/392
67975	446.6	Breccia	Sta 11-E-W split in boulder	SCB4/392

Apollo 16 Inventory *(continued)*

Sample Number	Weight (grams)	Description	Location	SRC/DB or SCB/DB
68001	840.7	Core 36	Sta 8	
68002	583.5	Core 29	Sta 8	
68030	2.85	DB residue	Sta 8	SCB3/413
68035	20.96	Breccia	Sta 8	SCB3/413
68110	35.76	DB residue	Sta 8	SRC2/340
68115	1191.0	Breccia	Sta 8	SRC2/340
68120	90.49	Unsieved fines	Sta 8	SRC2/374
68121	141.9	‹1mm fines	Sta 8	SRC2/374
68122	10.92	1-2mm fines	Sta 8	SRC2/374
68123	7.36	2-4mm fines	Sta 8	SRC2/374
68124	8.65	4-10mm fines	Sta 8	SRC2/374
68410	1.46		Sta 8	SRC2/341
68415	371.2	Crystalline	Sta 8	SRC2/341-2
68416	178.4	Allivatite	Sta 8	SRC2/341
68500	304.5	Unsieved fines (rake soil)	Sta 8	SCB3/412
68501	521.1	‹1mm fines	Sta 8	SCB3/412
68502	37.80	1-2mm fines	Sta 8	SCB3/412
68503	25.10	2-4mm fines	Sta 8	SCB3/412
68504	17.27	4-10mm fines	Sta 8	SCB3/412
68505	1.30	Breccia	Sta 8	SCB3/412
68510	17.48	Crystalline	Sta 8	SCB3/411
68515	236.1	Breccia (rake)	Sta 8	SCB3/411
68516	34.04	Breccia (rake)	Sta 8	SCB3/411
68517	13.13	Breccia (rake)	Sta 8	SCB3/411
68518	29.82	Breccia (rake)	Sta 8	SCB3/411
68519	10.56	Breccia (rake)	Sta 8	SCB3/411
68525	38.96	Crystalline, fine-grained (rake)	Sta 8	SCB3/411
68526	7.21	Crystalline, fine-grained (rake)	Sta 8	SCB3/411
68527	3.03	Crystalline, fine-grained (rake)	Sta 8	SCB3/411
68528	1.08	Breccia (rake)	Sta 8	SCB3/411
68529	7.03	Glass, fragment (rake)	Sta 8	SCB3/411
68535	8.04	Crystalline, fine-grained (rake)	Sta 8	SCB3/411
68536	1.85	Crystalline, fine-grained (rake)	Sta 8	SCB3/411
68537	1.41	Crystalline, fine-grained (rake)	Sta 8	SCB3/411
68810	72.3	Residue	Sta 8	SRC2/
68815	1826.0	Breccia	Sta 8	SRC2/343
68820	83.73	Unsieved fines (fillet)	Sta 8	SCB1/375
68821	123.9	‹1mm fines (fillet)	Sta 8	SCB1/375
68822	7.35	1-2mm fines (fillet)	Sta 8	SCB1/375
68823	3.52	2-4mm fines (fillet)	Sta 8	SCB1/375

Apollo 16 Inventory *(continued)*

Sample Number	Weight (grams)	Description	Location	SRC/DB or SCB/DB
68824	1.50	4-10mm fines (fillet)	Sta 8	SCB1/375
68825	8.658		Sta 8	SCB1/375
68840	154.46	Unsieved fines (reference soil)	Sta 8	SCB1/344
68841	266.6	‹1mm fines (reference soil)	Sta 8	SCB1/344
68842	14.36	1-2mm fines (reference soil)	Sta 8	SCB1/344
68843	8.89	2-4mm fines (reference soil)	Sta 8	SCB1/344
68844	5.01	4-10mm fines (reference soil)	Sta 8	SCB1/344
68845	4.556		Sta 8	SCB1/344
68846	2.284		Sta 8	SCB1/344
68847	2.854		Sta 8	SCB1/344
68848	1.770		Sta 8	SCB1/344
69001	558.3	Core 34	Sta 9	SRC2/CSVC
69003	0.0	Surface sampler 1	Sta 9	SCB1/
69004	0.0	Surface sampler 2	Sta 9	SCB1/
69920	.71	Unsieved fines (skim soil)	Sta 9	SCB3/376
69921	12.9	‹1mm fines	Sta 9	SCB3/376
69922	2.8	1-2mm fines	Sta 9	SCB3/376
69923	1.7	2-4mm fines	Sta 9	SCB3/376
69924	1.3	4-10mm fines	Sta 9	SCB3/376
69930	4.13	DB residue	Sta 9	SCB3/378
69935	127.57	Breccia	Sta 9	SCB3/378
69940	149.4	Unsieved fines (scoop soil)	Sta 9	SCB3/377
69941	254.7	‹1mm fines	Sta 9	SCB3/377
69942	11.85	1-2mm fines	Sta 9	SCB3/377
69943	8.07	2-4mm fines	Sta 9	SCB3/377
69944	4.47	4-10mm fines	Sta 9	SCB3/377
69945	6.88	Crystalline	Sta 9	SCB3/377
69950	3.77	DB residue	Sta 9	SCB3/380
69955	75.94	Anorthosite	Sta 9	SCB3/380
69960	171.0	Unsieved fines	Sta 9-under boulder	SCB3/379
69961	307.9	‹1mm fines	Sta 9-under boulder	SCB3/379
69962	13.93	1-2mm fines	Sta 9-under boulder	SCB3/379
69963	9.93	2-4mm fines	Sta 9-under boulder	SCB3/379
69964	4.80	4-10mm fines	Sta 9-under boulder	SCB3/379
69965	1.11	Breccia, glass veined	Sta 9-under boulder	SCB3/379

APOLLO 17 INVENTORY

Sample Number	Weight (grams)	Description	Location	Containers Outer/Inner
70001	29.78	Drill core bit	ALSEP	DSB
70002	207.8	Drill core stem	ALSEP	DSB
70003	237.8	Drill core stem	ALSEP	DSB
70004	238.8	Drill core stem	ALSEP	DSB
70005	240.7	Drill core stem	ALSEP	DSB
70006	234.2	Drill core stem	ALSEP	DSB
70007	179.4	Drill core stem	ALSEP	DSB
70008	261.0	Drill core stem	ALSEP	DSB
70009	143.3	Drill core stem (top)	ALSEP	DSB
70010	3.92	Fines outside stem	ALSEP	DSB
70011	440.7	SESC	LM	SCB5/
70012	485.0	Drive tube (52)	LM	BSLSS/
70017	2957.	Coarse basalt	LM	BSLSS/
70018	51.58	Dark matrix breccia	LM	SRC1/SCB1
70019	159.9	Agglutinate	Between station and LM	SRC2/469
70030	33.92	SCB residue	EVA 1	SCB2/
70035	5765.	Coarse basalt	SEP	SCB2/
70040	2.494	Fragments	All EVA's	Suit pocket
70050	573.4	BSLSS residue (unsieved)	EVA 3	BSLSS/
70051	1438.	BSLSS residue, <1mm	EVA 3	BSLSS/
70052	67.76	BSLSS residue, 1-2mm	EVA 3	BSLSS/
70053	86.93	BSLSS residue, 2-4mm	EVA 3	BSLSS/
70054	94.41	BSLSS residue, 4-10mm	EVA 3	BSLSS/
70060	0.24	Dust and sweepings	All EVA's	SCB7/15E
70061	60.80	<1mm fines	All EVA's	SCB7/15E
70062	2.12	1-2mm fines	All EVA's	SCB7/15E
70063	1.24	2-4mm fines	All EVA's	SCB7/15E
70064	0.86	4-10mm fines	All EVA's	SCB7/15E
70070	0.11	Dust and sweepings	?	BSLSS/108
70075	5.64	Fine basalt	?	BSLSS/108
70130	12.18	DB residue	ALSEP	SCB2/10E
70135	446.3	Coarse basalt	ALSEP	SCB2/10E
70136	10.65	Coarse basalt	ALSEP	SCB2/10E
70137	6.16	Coarse basalt	ALSEP	SCB2/10E
70138	3.66	Coarse basalt	ALSEP	SCB2/10E
70139	3.16	Coarse basalt	ALSEP	SCB2/10E
70145	3.07	Coarse basalt	ALSEP	SCB2/10E
70146	1.71	Coarse basalt	ALSEP	SCB2/10E
70147	1.35	Coarse basalt	ALSEP	SCB2/10E
70148	0.92	Coarse basalt	ALSEP	SCB2/10E
70149	0.95	Coarse basalt	ALSEP	SCB2/10E
70155	0.77	Coarse basalt	ALSEP	SCB2/10E
70156	0.63	Coarse basalt	ALSEP	SCB2/10E
70157	0.57	Coarse basalt	ALSEP	SCB2/10E
70160	106.1	Unsieved fines	ALSEP	SRC1/SCB1/474
70161	197.7	<1mm fines	ALSEP	SRC1/SCB1/474
70162	5.14	1-2mm fines	ALSEP	SRC1/SCB1/474

Apollo 17 Inventory *(continued)*

Sample Number	Weight (grams)	Description	Location	Containers Outer/Inner
70163	3.43	2-4mm fines	ALSEP	SRC1/SCB1/474
70164	1.66	4-10mm fines	ALSEP	SRC1/SCB1/474
70165	2.143	Coarse basalt	ALSEP	SRC1/SCB1/474
70170	42.31	DB residue	ALSEP	SCB5/55Y
70175	339.6	Dark matrix breccia	ALSEP	SCB5/55Y
70180	93.25	Unsieved fines	ALSEP	SRC1/SCB1/475
70181	157.1	‹1mm fines	ALSEP	SRC1/SCB1/475
70182	4.63	1-2mm fines	ALSEP	SRC1/SCB1/475
70183	3.12	2-4mm fines	ALSEP	SRC1/SCB1/475
70184	1.68	4-10mm fines	ALSEP	SRC1/SCB1/475
70185	466.6	Coarse basalt	ALSEP	SRC1/SCB1/475
70215	8110.	Fine basalt	SEP-LM	BSLSS
70250	62.04	DB residue	SEP	SCB8/22E
70255	277.2	Fine basalt	SEP	SCB8/22E
70270	70.46	Unsieved fines	SEP	SCB8/23E
70271	116.1	‹1mm fines	SEP	SCB8/23E
70272	2.97	1-2mm fines	SEP	SCB8/23E
70273	1.46	2-4mm fines	SEP	SCB8/23E
70274	2.33	4-10mm fines	SEP	SCB8/23E
70275	171.40	Medium basalt	SEP	SCB8/23E
70290	56.36	DB residue	SEP	SCB7/45Y
70295	361.2	Dark matrix breccia	SEP	SCB7/45Y
70310	6.82	DB residue	LRV 12	SCB5/54Y
70311	106.5	‹1mm fines	LRV 12	SCB5/54Y
70312	4.20	1-2mm fines	LRV 12	SCB5/54Y
70313	3.21	2-4mm fines	LRV 12	SCB5/54Y
70314	5.25	4-10mm fines	LRV 12	SCB5/54Y
70315	148.6	Coarse basalt	LRV 12	SCB5/54Y
70320	78.24	Unsieved fines	LRV 12	SCB5/53Y
70321	141.6	‹1mm fines	LRV 12	SCB5/53Y
70322	5.420	1-2mm fines	LRV 12	SCB5/53Y
70323	4.100	2-4mm fines	LRV 12	SCB5/53Y
70324	4.00	4-10mm fines	LRV 12	SCB5/53Y
71010	32.84	Residue in SRC1 and SCB1	EVA 1	SRC1/
71030	36.28	DB residue	Sta 1A	SRC1/SCB1/476
71035	144.8	Medium basalt	Sta 1A	SRC1/SCB1/476
71036	118.4	Medium basalt (refrigerated)	Sta 1A	SRC1/SCB1/476
71037	14.39	Medium basalt	Sta 1A	SRC1/SCB1/476
71040	94.89	Unsieved fines	Sta 1A	SRC1/SCB1/455
71041	137.8	‹1mm fines	Sta 1A	SRC1/SCB1/455
71042	7.21	1-2mm fines	Sta 1A	SRC1/SCB1/455
71043	6.19	2-4mm fines	Sta 1A	SRC1/SCB1/455
71044	12.84	4-10mm fines	Sta 1A	SRC1/SCB1/455
71045	11.92	Medium basalt	Sta 1A	SRC1/SCB1/455
71046	3.037	Medium basalt	Sta 1A	SRC1/SCB1/455
71047	2.780	Coarse basalt	Sta 1A	SRC1/SCB1/455
71048	2.457	Fine basalt	Sta 1A	SRC1/SCB1/455

Apollo 17 Inventory *(continued)*

Sample Number	Weight (grams)	Description	Location	Containers Outer/Inner
71049	1.860	Fine basalt	Sta 1A	SRC1/SCB1/455
71050	4.000	DB residue	Sta 1A	SRC1/SCB1/454
71055	669.6	Medium basalt	Sta 1A	SRC1/SCB1/454
71060	199.4	Unsieved fines	Sta 1A	SRC1/SCB1/456
71061	229.2	<1mm fines	Sta 1A	SRC1/SCB1/456
71062	20.74	1-2mm fines	Sta 1A	SRC1/SCB1/456
71063	22.79	2-4mm fines	Sta 1A	SRC1/SCB1/456
71064	34.35	4-10mm fines	Sta 1A	SRC1/SCB1/456
71065	28.83	Fine basalt	Sta 1A	SRC1/SCB1/456
71066	19.96	Fine basalt	Sta 1A	SRC1/SCB1/456
71067	4.245	Medium basalt	Sta 1A	SRC1/SCB1/456
71068	4.208	Medium basalt	Sta 1A	SRC1/SCB1/456
71069	4.058	Fine basalt	Sta 1A	SRC1/SCB1/456
71075	1.563	Medium basalt	Sta 1A	SRC1/SCB1/455
71085	3.402	Medium basalt	Sta 1A	SRC1/SCB1/456
71086	3.329	Fine basalt	Sta 1A	SRC1/SCB1/456
71087	2.200	Fine basalt	Sta 1A	SRC1/SCB1/456
71088	2.064	Fine basalt	Sta 1A	SRC1/SCB1/456
71089	1.733	Medium basalt	Sta 1A	SRC1/SCB1/456
71095	1.483	Medium basalt	Sta 1A	SRC1/SCB1/456
71096	1.368	Medium basalt	Sta 1A	SRC1/SCB1/456
71097	1.355	Medium basalt	Sta 1A	SRC1/SCB1/456
71130	49.51	Unsieved fines	Sta 1A	SRC1/SCB1/477
71131	86.4	<1mm fines	Sta 1A	SRC1/SCB1/477
71132	3.99	1-2mm fines	Sta 1A	SRC1/SCB1/477
71133	3.22	2-4mm fines	Sta 1A	SRC1/SCB1/477
71134	0.91	4-10mm fines	Sta 1A	SRC1/SCB1/477
71135	36.85	Fine basalt	Sta 1A	SRC1/SCB1/477
71136	25.39	Fine basalt	Sta 1A	SRC1/SCB1/477
71150	1.565	DB residue	Sta 1A	SRC1/SCB1/478
71151	57.6	<1mm fines	Sta 1A	SRC1/SCB1/478
71152	2.60	1-2mm fines	Sta 1A	SRC1/SCB1/478
71153	2.36	2-4mm fines	Sta 1A	SRC1/SCB1/478
71154	1.37	4-10mm fines	Sta 1A	SRC1/SCB1/478
71155	26.15	Fine basalt	Sta 1A	SRC1/SCB1/478
71156	5.420	Fine basalt	Sta 1A	SRC1/SCB1/478
71157	1.466	Fine basalt	Sta 1A	SRC1/SCB1/478
71170	16.38	DB residue	Sta 1A	SRC1/SCB1/479
71175	207.8	Medium basalt	Sta 1A	SRC1/SCB1/479
71500	359.5	Unsieved fines	Sta 1A	SRC1/SCB1/459 (rake soil)
71501	600.9	<1mm fines	Sta 1A	SRC1/SCB1/459 (rake soil)
71502	22.68	1-2mm fines	Sta 1A	SRC1/SCB1/459 (rake soil)
71503	17.58	2-4mm fines	Sta 1A	SRC1/SCB1/459 (rake soil)
71504	13.13	4-10mm fines	Sta 1A	SRC1/SCB1/459 (rake soil)

Apollo 17 Inventory *(continued)*

Sample Number	Weight (grams)	Description	Location	Containers Outer/Inner
71505	29.45	Fine basalt	Sta 1A	SRC1/SCB1/459 (rake soil)
71506	12.11	Fine basalt	Sta 1A	SRC1/SCB1/459 (rake soil)
71507	3.962	Medium basalt	Sta 1A	SRC1/SCB1/459 (rake soil)
71508	3.423	Coarse basalt	Sta 1A	SRC1/SCB1/459 (rake soil)
71509	1.690	Coarse basalt	Sta 1A	SRC1/SCB1/459 (rake soil)
71515	1.635	Agglutinate	Sta 1A	SRC1/SCB1/459 (rake soil)
71520	48.16	DB residue	Sta 1A	SRC1/SCB1/457 & 458 (rake)
71525	3.900	Fine basalt	Sta 1A	SRC1/SCB1/457 & 458 (rake)
71526	12.91	Fine basalt	Sta 1A	SRC1/SCB1/457 & 458 (rake)
71527	2.186	Fine basalt	Sta 1A	SRC1/SCB1/457 & 458 (rake)
71528	11.25	Fine basalt	Sta 1A	SRC1/SCB1/457 & 458 (rake)
71529	6.025	Medium basalt	Sta 1A	SRC1/SCB1/457 & 458 (rake)
71535	17.71	Coarse basalt	Sta 1A	SRC1/SCB1/457 & 458 (rake)
71536	5.322	Coarse basalt	Sta 1A	SRC1/SCB1/457 & 458 (rake)
71537	12.25	Fine basalt	Sta 1A	SRC1/SCB1/457 & 458 (rake)
71538	8.038	Fine basalt	Sta 1A	SRC1/SCB1/457 & 458 (rake)
71539	10.90	Fine basalt	Sta 1A	SRC1/SCB1/457 & 458 (rake)
71545	17.26	Fine basalt	Sta 1A	SRC1/SCB1/457 & 458 (rake)
71546	150.7	Fine basalt	Sta 1A	SRC1/SCB1/457 & 458 (rake)
71547	12.54	Medium basalt	Sta 1A	SRC1/SCB1/457 & 458 (rake)
71548	25.46	Medium basalt	Sta 1A	SRC1/SCB1/457 & 458 (rake)
71549	7.903	Medium basalt	Sta 1A	SRC1/SCB1/457 & 458 (rake)
71555	4.547	Medium basalt	Sta 1A	SRC1/SCB1/457 & 458 (rake)
71556	29.14	Coarse basalt	Sta 1A	SRC1/SCB1/457 & 458 (rake)
71557	40.35	Coarse basalt	Sta 1A	SRC1/SCB1/457 & 458 (rake)

Apollo 17 Inventory *(continued)*

Sample Number	Weight (grams)	Description	Location	Containers Outer/Inner
71558	15.81	Coarse basalt	Sta 1A	SRC1/SCB1/457 & 458 (rake)
71559	82.16	Coarse basalt	Sta 1A	SRC1/SCB1/457 & 458 (rake)
71565	24.09	Coarse basalt	Sta 1A	SRC1/SCB1/457 & 458 (rake)
71566	415.4	Coarse basalt	Sta 1A	SRC1/SCB1/457 & 458 (rake)
71567	146.0	Coarse basalt	Sta 1A	SRC1/SCB1/457 & 458 (rake)
71568	10.02	Coarse basalt	Sta 1A	SRC1/SCB1/457 & 458 (rake)
71569	289.6	Fine basalt	Sta 1A	SRC1/SCB1/457 & 458 (rake)
71575	2.113	Fine basalt	Sta 1A	SRC1/SCB1/457 & 458 (rake)
71576	23.54	Fine basalt	Sta 1A	SRC1/SCB1/457 & 458 (rake)
71577	234.7	Fine basalt	Sta 1A	SRC1/SCB1/457 & 458 (rake)
71578	353.9	Medium basalt	Sta 1A	SRC1/SCB1/457 & 458 (rake)
71579	7.937	Medium basalt	Sta 1A	SRC1/SCB1/457 & 458 (rake)
71585	13.86	Medium basalt	Sta 1A	SRC1/SCB1/457 & 458 (rake)
71586	26.92	Medium basalt	Sta 1A	SRC1/SCB1/457 & 458 (rake)
71587	41.27	Medium basalt	Sta 1A	SRC1/SCB1/457 & 458 (rake)
71588	48.98	Medium basalt	Sta 1A	SRC1/SCB1/457 & 458 (rake)
71589	6.860	Medium basalt	Sta 1A	SRC1/SCB1/457 & 458 (rake)
71595	25.21	Medium basalt	Sta 1A	SRC1/SCB1/457 & 458 (rake)
71596	61.05	Medium basalt	Sta 1A	SRC1/SCB1/457 & 458 (rake)
71597	12.35	Coarse basalt	Sta 1A	SRC1/SCB1/457 & 458 (rake)
72010	76.92	SCB residue	EVA 2	SCB8/
72130	79.91	Unsieved fines	LRV 1	SCB8/26E
72131	107.9	<1mm fines	LRV 1	SCB8/26E
72132	8.53	1-2mm fines	LRV 1	SCB8/26E
72133	10.95	2-4mm fines	LRV 1	SCB8/26E
72134	13.18	4-10mm fines	LRV 1	SCB8/26E
72135	336.9	Dark breccia of basalt fragments	LRV 1	SCB8/26E
72140	115.0	Unsieved fines	LRV 2	SCB6/27E

Apollo 17 Inventory *(continued)*

Sample Number	Weight (grams)	Description	Location	Containers Outer/Inner
72141	225.9	‹1mm fines	LRV 2	SCB6/27E
72142	5.32	1-2mm fines	LRV 2	SCB6/27E
72143	1.88	2-4mm fines	LRV 2	SCB6/27E
72144	2.73	4-10mm fines	LRV 2	SCB6/27E
72145	1.25	Dark matrix breccia	LRV 2	SCB6/27E
72150	53.29	DB residue	LRV 3	SCB6/28E
72155	238.5	Medium basalt	LRV 3	SCB6/28E
72160	80.0	Unsieved fines	LRV 3	SCB8/29E
72161	162.5	‹1mm fines	LRV 3	SCB8/29E
72162	4.018	1-2mm fines	LRV 3	SCB8/29E
72163	2.538	2-4mm fines	LRV 3	SCB8/29E
72164	0.946	4-10mm fines	LRV 3	SCB8/29E
72210	1.83	DB residue	Sta 2	SCB6/514
72215	379.2	Layered light-gray breccia	Sta 2	SCB6/514
72220	136.2	Unsieved fines	Sta 2	SCB8/496
72221	225.8	‹1mm fines	Sta 2	SCB8/496
72222	11.13	1-2mm fines	Sta 2	SCB8/496
72223	7.92	2-4mm fines	Sta 2	SCB8/496
72224	7.51	4-10mm fines	Sta 2	SCB8/496
72230	1.66	DB residue	Sta 2	SCB6/515
72235	61.91	Layered light-gray breccia	Sta 2	SCB6/515
72240	113.3	Unsieved fines	Sta 2	SCB8/497
72241	186.0	‹1mm fines	Sta 2	SCB8/497
72242	11.20	1-2mm fines	Sta 2	SCB8/497
72243	7.93	2-4mm fines	Sta 2	SCB8/497
72244	3.99	4-10mm fines	Sta 2	SCB8/497
72250	7.74	DB residue	Sta 2	SCB8/494
72255	461.2	Layered light-gray breccia	Sta 2	SCB8/494
72260	100.60	Unsieved fines	Sta 2	SCB8/498
72261	161.9	‹1mm fines	Sta 2	SCB8/498
72262	7.70	1-2mm fines	Sta 2	SCB8/498
72263	4.40	2-4mm fines	Sta 2	SCB8/498
72264	4.40	4-10mm fines	Sta 2	SCB8/498
72270	26.11	DB residue	Sta 2	SCB8/495
72275	3640.	Layered light-gray breccia	Sta 2	SCB8/495
72310	1.09	DB residue	Sta 2	SCB6/516
72315	131.4	Vesicular, poikilitic clast	Sta 2	SCB6/516
72320	26.17	Unsieved fines	Sta 2	SCB8/500
72321	77.3	‹1mm fines	Sta 2	SCB8/500
72322	1.38	1-2mm fines	Sta 2	SCB8/500
72323	0.50	2-4mm fines	Sta 2	SCB8/500
72324	0.96	4-10mm fines	Sta 2	SCB8/500
72330	0.36	DB residue	Sta 2	SCB6/517
72335	108.9	Vesicular poikilitic clast	Sta 2	SCB6/517
72350	1.70	Dust and sweepings	Sta 2	SRC2/518
72355	367.4	Green-gray breccia	Sta 2	SRC2/518
72370	0.02	Dust and sweepings	Sta 2	SRC2/519

Apollo 17 Inventory *(continued)*

Sample Number	Weight ·(grams)	Description	Location	Containers Outer/Inner
72375	18.16	Green-gray breccia	Sta 2	SRC2/519
72390	20.63	DB residue	Sta 2	SCB8/499
72395	536.4	Green-gray breccia	Sta 2	SCB8/499
72410	52.00	DB residue	Sta 2	SCB8/503
72415	32.34	Brecciated dunite clast	Sta 2	SCB8/503
72416	11.53	Brecciated dunite clast	Sta 2	SCB8/503
72417	11.32	Brecciated dunite clast	Sta 2	SCB8/503
72418	3.55	Brecciated dunite clast	Sta 2	SCB8/503
72430	1.45	Dust and sweepings	Sta 2	SCB8/504
72431	72.0	<1mm fines	Sta 2	SCB8/504
72432	3.62	1-2mm fines	Sta 2	SCB8/504
72433	2.33	2-4mm fines	Sta 2	SCB8/504
72434	1.47	4-10mm fines	Sta 2	SCB8/504
72435	160.6	Blue-gray breccia	Sta 2	SCB8/504
72440	161.6	Unsieved fines	Sta 2	SCB8/505
72441	267.3	<1mm fines	Sta 2	SCB8/505
72442	10.60	1-2mm fines	Sta 2	SCB8/505
72443	7.98	2-4mm fines	Sta 2	SCB8/505
72444	2.91	4-10mm fines	Sta 2	SCB8/505
72460	0.51	Dust and sweepings	Sta 2	SCB8/506
72461	113.7	<1mm fines	Sta 2	SCB8/506
72462	5.14	1-2mm fines	Sta 2	SCB8/506
72463	3.90	2-4mm fines	Sta 2	SCB8/506
72464	1.76	4-10mm fines	Sta 2	SCB8/506
72500	325.5	Unsieved fines	Sta 2	SCB8/502(rake soil)
72501	687.2	<1mm fines	Sta 2	SCB8/502(rake soil)
72502	24.13	1-2mm fines	Sta 2	SCB8/502(rake soil)
72503	12.94	2-4mm fines	Sta 2	SCB8/502(rake soil)
72504	7.96	4-10mm fines	Sta 2	SCB8/502(rake soil)
72505	3.09	Green-gray breccia	Sta 2	SCB8/502(rake soil)
72530	18.14	DB residue	Sta 2	SCB8/501 (rake)
72535	221.4	Blue-gray breccia	Sta 2	SCB8/501 (rake)
72536	52.30	Blue-gray breccia	Sta 2	SCB8/501 (rake)
72537	5.192	Blue-gray breccia	Sta 2	SCB8/501 (rake)
72538	11.09	Blue-gray breccia	Sta 2	SCB8/501 (rake)
72539	11.22	Blue-gray breccia	Sta 2	SCB8/501 (rake)
72545	4.055	Blue-gray breccia	Sta 2	SCB8/501 (rake)
72546	4.856	Blue-gray breccia	Sta 2	SCB8/501 (rake)
72547	5.045	Blue-gray breccia	Sta 2	SCB8/501 (rake)
72548	29.29	Blue-gray breccia	Sta 2	SCB8/501 (rake)
72549	21.00	Green-gray breccia	Sta 2	SCB8/501 (rake)
72555	10.48	Green-gray breccia	Sta 2	SCB8/501 (rake)
72556	3.861	Green-gray breccia	Sta 2	SCB8/501 (rake)
72557	4.559	Green-gray breccia	Sta 2	SCB8/501 (rake)
72558	5.713	Green-gray breccia	Sta 2	SCB8/501 (rake)
72559	27.84	Feldspathic breccia	Sta 2	SCB8/501 (rake)
72700	295.2	Unsieved fines	Sta 2	SCB8/508(rake soil)
72701	557.3	<1mm fines	Sta 2	SCB8/508(rake soil)
72702	17.70	1-2mm fines	Sta 2	SCB8/508(rake soil)

Apollo 17 Inventory *(continued)*

Sample Number	Weight (grams)	Description	Location	Containers Outer/Inner
72703	8.05	2-4mm fines	Sta 2	SCB8/508 (rake soil)
72704	4.76	4-10mm fines	Sta 2	SCB8/508 (rake soil)
72705	2.39	Anorthosite breccia and glass	Sta 2	SCB8/508 (rake soil)
72730	1.98	DB residue	Sta 2	SCB8/507 (rake)
72735	51.11	Green-gray breccia	Sta 2	SCB8/507 (rake)
72736	28.73	Tan breccia	Sta 2	SCB8/507 (rake)
72737	3.33	Tan breccia	Sta 2	SCB8/507 (rake)
72738	23.75	Blue-gray breccia	Sta 2	SCB8/507 (rake)
73001	809.0	Drive tube (L46, lower)	Sta 3	SRC2/CSVC
73002	429.7	Drive tube (U31, upper)	Sta 3	SRC2/
73010	34.56	SCB residue	EVA 2	SCB6/
73120	100.2	Unsieved fines	Sta 2A	SCB6/30E
73121	179.7	<1mm fines	Sta 2A	SCB6/30E
73122	5.25	1-2mm fines	Sta 2A	SCB6/30E
73123	2.03	2-4mm fines	Sta 2A	SCB6/30E
73124	0.50	4-10mm fines	Sta 2A	SCB6/30E
73130	77.20	Unsieved fines	Sta 2A	SCB8/31E
73131	132.3	<1mm fines	Sta 2A	SCB8/31E
73132	10.38	1-2mm fines	Sta 2A	SCB8/31E
73133	8.58	2-4mm fines	Sta 2A	SCB8/31E
73134	9.61	4-10mm fines	Sta 2A	SCB8/31E
73140	121.6	Unsieved fines	Sta 2	SCB6/40Y
73141	191.4	<1mm fines	Sta 2	SCB6/40Y
73142	11.69	1-2mm fines	Sta 2	SCB6/40Y
73143	7.84	2-4mm fines	Sta 2	SCB6/40Y
73144	4.47	4-10mm fines	Sta 2	SCB6/40Y
73145	5.60	Dark matrix breccia	Sta 2	SCB6/40Y
73146	3.01	Brecciated anorthosite	Sta 2	SCB6/40Y
73150	52.56	Unsieved fines	Sta 2	SCB6/32E
73151	101.2	<1mm fines	Sta 2	SCB6/32E
73152	3.57	1-2mm fines	Sta 2	SCB6/32E
73153	1.31	2-4mm fines	Sta 2	SCB6/32E
73154	0.31	4-10mm fines	Sta 2	SCB6/32E
73155	79.3	Blue-gray breccia	Sta 2	SCB6/32E
73156	3.15	Fine crystalline	Sta 2	SCB6/32E
73210	37.89	Unsieved fines	Sta 3	SCB6/527
73211	51.95	<1mm fines	Sta 3	SCB6/527
73212	3.47	1-2mm fines	Sta 3	SCB6/527
73213	2.80	2-4mm fines	Sta 3	SCB6/527
73214	2.47	4-10mm fines	Sta 3	SCB6/527
73215	1062.	Light-gray breccia	Sta 3	SCB6/527
73216	162.2	Green-gray breccia	Sta 3	SCB6/527
73217	138.8	Blue-gray breccia	Sta 3	SCB6/527
73218	39.67	Blue-gray breccia	Sta 3	SCB6/527
73219	2.88	Fine basalt	Sta 3	SCB6/527
73220	20.8	Unsieved fines	Sta 3	SCB6/520
73221	48.11	<1mm fines	Sta 3	SCB6/520
73222	2.71	1-2mm fines	Sta 3	SCB6/520
73223	2.61	2-4mm fines	Sta 3	SCB6/520
73224	1.65	4-10mm fines	Sta 3	SCB6/520

Apollo 17 Inventory *(continued)*

Sample Number	Weight (grams)	Description	Location	Containers Outer/Inner
73225	3.66	Crystalline (green-gray breccia?)	Sta 3	SCB6/520
73230	21.34	DB residue	Sta 3	SCB6/524
73235	878.3	Blue gray breccia	Sta 3	SCB6/524
73240	114.7	Unsieved fines	Sta 3	SCB6/521
73241	192.7	‹1mm fines	Sta 3	SCB6/521
73242	14.94	1-2mm fines	Sta 3	SCB6/521
73243	14.38	2-4mm fines	Sta 3	SCB6/521
73244	22.25	4-10mm fines	Sta 3	SCB6/521
73245	1.60	Brecciated anorthosite clast	Sta 3	SCB6/521
73250	15.25	DB residue	Sta 3	SCB6/525
73255	394.1	Light gray or blue-gray breccia	Sta 3	SCB6/525
73260	103.5	Unsieved fines	Sta 3	SCB6/522
73261	194.8	‹1mm fines	Sta 3	SCB6/522
73262	12.01	1-2mm fines	Sta 3	SCB6/522
73263	9.47	2-4mm fines	Sta 3	SCB6/522
73264	6.45	4-10mm fines	Sta 3	SCB6/522
73270	22.43	DB residue	Sta 3	SCB6/526
73275	429.6	Green-gray breccia	Sta 3	SCB6/526
73280	53.54	Unsieved fines	Sta 3	SCB6/523
73281	95.75	‹1mm fines	Sta 3	SCB6/523
73282	5.38	1-2mm fines	Sta 3	SCB6/523
73283	4.74	2-4mm fines	Sta 3	SCB6/523
73284	7.14	4-10mm fines	Sta 3	SCB6/523
73285	2.58	Glass coated gray friable breccia	Sta 3	SCB6/523
74001	1072.	Drive tube (L44, lower)	Sta 4	SRC2/
74002	909.6	Drive tube (U35, upper)	Sta 4	SRC2/
74010	22.52	SRC residue	EVA 2	SRC2/
74110	92.12	Unsieved fines	LRV 5	SCB8/41Y
74111	116.8	‹1mm fines	LRV 5	SCB8/41Y
74112	11.12	1-2mm fines	LRV 5	SCB8/41Y
74113	12.11	2-4mm fines	LRV 5	SCB8/41Y
74114	13.26	4-10mm fines	LRV 5	SCB8/41Y
74115	15.36	Extremely friable light-gray breccia	LRV 5	SCB8/41Y
74116	12.68	Extremely friable light-gray breccia	LRV 5	SCB8/41Y
74117	3.69	Extremely friable light-gray breccia	LRV 5	SCB8/41Y
74118	3.59	Extremely friable light-gray breccia	LRV 5	SCB8/41Y
74119	1.79	Extremely friable light-gray breccia	LRV 5	SCB8/41Y
74120	124.1	Unsieved fines	LRV 6	SCB8/42Y
74121	252.0	‹1mm fines	LRV 6	SCB8/42Y
74122	6.65	1-2mm fines	LRV 6	SCB8/42Y
74123	2.73	2-4mm fines	LRV 6	SCB8/42Y
74124	0.39	4-10mm fines	LRV 6	SCB8/42Y
74220	1180.	Unsieved fines	Sta 4	SRC2/509
74230	0.70	DB residue	Sta 4	SCB8/12E
74235	59.04	Basalt vitrophyre	Sta 4	SCB8/12E

Apollo 17 Inventory *(continued)*

Sample Number	Weight (grams)	Description	Location	Containers Outer/Inner
74240	544.9	Unsieved fines	Sta 4	SRC2/510
74241	307.3	<1mm fines	Sta 4	SRC2/510
74242	22.50	1-2mm fines	Sta 4	SRC2/510
74243	27.67	2-4mm fines	Sta 4	SRC2/510
74244	21.95	4-10mm fines	Sta 4	SRC2/510
74245	64.34	Fine or devit. basalt	Sta 4	SRC2/510
74246	28.81	Dark matrix breccia	Sta 4	SRC2/510
74247	7.761	Fine or devit. basalt	Sta 4	SRC2/510
74248	5.682	Fine or devit. basalt	Sta 4	SRC2/510
74249	4.183	Fine basalt	Sta 4	SRC2/510
74250	22.56	DB residue	Sta 4	SCB6/512
74255	737.3	Coarse basalt	Sta 4	SCB6/512
74260	526.7	Unsieved fines	Sta 4	SRC2/511
74270	9.61	DB residue	Sta 4	SCB4/461
74275	1493.	Fine basalt	Sta 4	SCB4/461
74285	2.212	Medium basalt	Sta 4	SRC2/510
74286	2.102	Medium basalt	Sta 4	SRC2/510
74287	1.568	Fine basalt	Sta 4	SRC2/510
75010	9.25	DB residue	Sta 5	SCB6/462
75015	1006.	Coarse basalt	Sta 5	SCB6/462
75030	2.63	DB residue	Sta 5	SCB6/463
75035	1235.	Medium basalt	Sta 5	SCB6/463
75050	2.5	Dust and sweepings	Sta 5	SRC6/464
75055	949.4	Coarse basalt	Sta 5	SRC6/464
75060	0.527	DB residue	Sta 5	SRC2/465
75061	157.9	<1mm fines	Sta 5	SRC2/465
75062	8.520	1-2mm fines	Sta 5	SRC2/465
75063	6.280	2-4mm fines	Sta 5	SRC2/465
75064	11.63	4-10mm fines	Sta 5	SRC2/465
75065	1.263	Medium basalt	Sta 5	SRC2/465
75066	0.980	Dark gray breccia	Sta 5	SRC2/465
75070	7.260	DB residue	Sta 5	SRC2/466
75075	1008.	Medium basalt	Sta 5	SRC2/466
75080	524.2	Unsieved fines	Sta 5	SRC2/467
75081	932.4	<1mm fines	Sta 5	SRC2/467
75082	38.92	1-2mm fines	Sta 5	SRC2/467
75083	30.88	2-4mm fines	Sta 5	SRC2/467
75084	23.31	4-10mm fines	Sta 5	SRC2/467
75085	4.298	Medium basalt	Sta 5	SRC2/467
75086	2.323	Medium basalt	Sta 5	SRC2/467
75087	2.321	Medium basalt	Sta 5	SRC2/467
75088	1.992	Fine basalt	Sta 5	SRC2/467
75089	1.718	Fine basalt	Sta 5	SRC2/467
75110	122.5	Unsieved fines	LRV 7	SCB8/43Y
75111	235.0	<1mm fines	LRV 7	SCB8/43Y
75112	10.20	1-2mm fines	LRV 7	SCB8/43Y
75113	6.76	2-4mm fines	LRV 7	SCB8/43Y
75114	6.87	4-10mm fines	LRV 7	SCB8/43Y
75115	2.60	Fine basalt	LRV 7	SCB8/43Y

Apollo 17 Inventory *(continued)*

Sample Number	Weight (grams)	Description	Location	Containers Outer/Inner
75120	126.6	Unsieved fines	LRV 7	SCB8/44Y
75121	240.3	<1mm fines	LRV 7	SCB8/44Y
75122	5.208	1-2mm fines	LRV 7	SCB8/44Y
75123	2.147	2-4mm fines	LRV 7	SCB8/44Y
75124	0.956	4-10mm fines	LRV 7	SCB8/44Y
76001	711.6	Drive tube (L48)	Sta 6	SCB7/
76010	20.31	SCB residue	EVA 3	SCB4/
76015	2819.	Green-gray breccia	Sta 6	SCB4/
76030	16.06	DB residue	Sta 6	SCB5/49Y, 48Y
76031	152.6	<1mm fines	Sta 6	SCB5/49Y, 48Y
76032	5.71	1-2mm fines	Sta 6	SCB5/49Y, 48Y
76033	4.58	2-4mm fines	Sta 6	SCB5/49Y, 48Y
76034	2.01	4-10mm fines	Sta 6	SCB5/49Y, 48Y
76035	376.2	Blue-gray breccia	Sta 6	SCB5/49Y, 48Y
76036	3.95	Blue-gray breccia	Sta 6	SCB5/49Y, 48Y
76037	2.52	Medium basalt	Sta 6	SCB5/49Y, 48Y
76055	6412.	Green-gray breccia	Sta 6	SCB5/49Y, 48Y
76120	107.0	Unsieved fines	LRV 9	SCB5/46Y
76121	188.1	<1mm fines	LRV 9	SCB5/46Y
76122	4.72	1-2mm fines	LRV 9	SCB5/46Y
76123	2.49	2-4mm fines	LRV 9	SCB5/46Y
76124	1.61	4-10mm fines	LRV 9	SCB5/46Y
76130	19.57	DB residue	LRV 10	SCB5/47Y
76131	146.1	<1mm fines	LRV 10	SCB5/47Y
76132	6.79	1-2mm fines	LRV 10	SCB5/47Y
76133	5.21	2-4mm fines	LRV 10	SCB5/47Y
76134	3.10	4-10mm fines	LRV 10	SCB5/47Y
76135	133.5	Green-gray breccia	LRV 10	SCB5/47Y
76136	86.6	Medium basalt	LRV 10	SCB5/47Y
76137	2.46	Fine grained crystalline	LRV 10	SCB5/47Y
76210	2.74	DB residue	Sta 6	SCB4/535
76215	643.9	Green-gray breccia	Sta 6	SCB4/535
76220	196.7	Unsieved fines	Sta 6	SCB7/534
76221	390.4	<1mm fines	Sta 6	SCB7/534
76222	13.65	1-2mm fines	Sta 6	SCB7/534
76223	8.26	2-4mm fines	Sta 6	SCB7/534
76224	3.83	4-10mm fines	Sta 6	SCB7/534
76230	6.63	DB residue	Sta 6	SCB4/556
76235	26.56	Brecciated olivine norite	Sta 6	SCB4/556
76236	19.18	Brecciated olivine norite	Sta 6	SCB4/556
76237	10.31	Brecciated olivine norite	Sta 6	SCB4/556
76238	8.21	Brecciated olivine norite	Sta 6	SCB4/556
76239	6.23	Brecciated olivine norite	Sta 6	SCB4/556
76240	450.7	Unsieved fines	Sta 6	SCB4/312
76241	21.14	<1mm fines	Sta 6	SCB4/312
76242	1.20	1-2mm fines	Sta 6	SCB4/312
76243	1.23	2-4mm fines	Sta 6	SCB4/312
76244	1.53	4-10mm fines	Sta 6	SCB4/312

Apollo 17 Inventory *(continued)*

Sample Number	Weight (grams)	Description	Location	Containers Outer/Inner
76245	8.24	Green-gray breccia	Sta 6	SCB4/312
76246	6.50	Green-gray breccia	Sta 6	SCB4/312
76250	4.63	DB residue	Sta 6	SCB4/536
76255	406.6	Banded tan and blue-gray breccia	Sta 6	SCB4/536
76260	96.6	Unsieved fines	Sta 6	SCB4/313
76261	170.7	‹1mm fines	Sta 6	SCB4/313
76262	8.55	1-2mm fines	Sta 6	SCB4/313
76263	6.57	2-4mm fines	Sta 6	SCB4/313
76264	8.76	4-10mm fines	Sta 6	SCB4/313
76265	1.75	Green-gray breccia	Sta 6	SCB4/313
76270	0.46	DB residue	Sta 6	SCB4/537
76275	55.93	Blue-gray fragment breccia	Sta 6	SCB4/537
76280	153.0	Unsieved fines	Sta 6	SCB4/472
76281	251.8	‹1mm fines	Sta 6	SCB4/472
76282	14.27	1-2mm fines	Sta 6	SCB4/472
76283	12.71	2-4mm fines	Sta 6	SCB4/472
76284	10.69	4-10mm fines	Sta 6	SCB4/472
76285	2.208	Agglutinate	Sta 6	SCB4/472
76286	1.704	Brecciated troctolite	Sta 6	SCB4/472
76290	9.65	DB residue	Sta 6	SCB4/538
76295	260.7	Banded tan and blue-gray breccia	Sta 6	SCB4/538
76305	4.01	Brecciated olivine norite	Sta 6	SCB4/556
76306	4.25	Brecciated olivine norite	Sta 6	SCB4/556
76307	2.49	Brecciated olivine norite	Sta 6	SCB4/556
76310	25.39	DB residue	Sta 6	SCB7/539
76315	671.1	Blue gray breccia	Sta 6	SCB7/539
76320	260.3	Unsieved fines	Sta 6	SCB7/557
76321	502.7	‹1mm fines	Sta 6	SCB7/557
76322	23.10	1-2mm fines	Sta 6	SCB7/557
76323	15.84	2-4mm fines	Sta 6	SCB7/557
76324	11.80	4-10mm fines	Sta 6	SCB7/557
76330	418.6	DB residue	Sta 6	BSLSS/560
76335	352.9	Friable anorthositic breccia	Sta 6	BSLSS/560
76500	345.2	Unsieved fines	Sta 6	SCB4/559 (rake soil)
76501	630.7	‹1mm fines	Sta 6	SCB4/559 (rake soil)
76502	22.76	1-2mm fines	Sta 6	SCB4/559 (rake soil)
76503	10.09	2-4mm fines	Sta 6	SCB4/559 (rake soil)
76504	10.72	4-10mm fines	Sta 6	SCB4/559 (rake soil)
76505	4.69	Greenish-gray breccia	Sta 6	SCB4/559 (rake soil)
76506	2.81	Friable dark matrix breccia	Sta 6	SCB4/559 (rake soil)
76530	70.27	DB residue	Sta 6	SCB4/558 (rake)
76535	155.5	Coarse norite	Sta 6	SCB4/558 (rake)
76536	10.26	Brecciated norite	Sta 6	SCB4/558 (rake)
76537	26.48	Fine basalt	Sta 6	SCB4/558 (rake)
76538	5.870	Medium basalt	Sta 6	SCB4/558 (rake)
76539	14.80	Vitrophyric basalt	Sta 6	SCB4/558 (rake)
76545	7.676	Dark vitreous matrix breccia	Sta 6	SCB4/558 (rake)
76546	24.31	Dark vitreous matrix breccia	Sta 6	SCB4/558 (rake)
76547	10.05	Dark vitreous matrix breccia	Sta 6	SCB4/558 (rake)
76548	2.527	Dark vitreous matrix breccia	Sta 6	SCB4/558 (rake)

Apollo 17 Inventory *(continued)*

Sample Number	Weight (grams)	Description	Location	Containers Outer/Inner
76549	9.175	Dark vitreous matrix breccia	Sta 6	SCB4/558 (rake)
76555	8.435	Crystalline matrix-rich basalt	Sta 6	SCB4/558 (rake)
76556	7.396	Crystalline matrix-rich basalt	Sta 6	SCB4/558 (rake)
76557	5.592	Crystalline matrix-rich basalt	Sta 6	SCB4/558 (rake)
76558	0.683	Crystalline matrix-rich basalt	Sta 6	SCB4/558 (rake)
76559	0.747	Crystalline matrix-rich basalt	Sta 6	SCB4/558 (rake)
76565	11.60	Friable dark matrix breccia	Sta 6	SCB4/558 (rake)
76566	2.639	Friable dark matrix breccia	Sta 6	SCB4/558 (rake)
76567	5.490	Friable dark matrix breccia	Sta 6	SCB4/558 (rake)
76568	9.477	Basalt-rich breccia	Sta 6	SCB4/558 (rake)
76569	4.207	Crystalline breccia (blue-gray?)	Sta 6	SCB4/558 (rake)
76575	16.25	Crystalline breccia, clast rich	Sta 6	SCB4/558 (rake)
76576	5.327	Crystalline light-gray breccia	Sta 6	SCB4/558 (rake)
76577	13.54	Crystalline light-gray breccia	Sta 6	SCB4/558 (rake)
77010	93.65	SCB residue	EVA 3	SCB7/
77017	1730.	Brecciated olivine gabbro	Sta 7	SCB7/541
77035	5727.	Green-gray breccia	Sta 7	BSLSS/
77070	9.28	DB residue	Sta 7	SCB7/544
77075	172.4	Dark gray dike	Sta 7	SCB7/544
77076	13.97	Dark gray dike	Sta 7	SCB7/544
77077	5.45	Dark gray dike	Sta 7	SCB7/544
77110	0.15	Dust and sweepings	Sta 7	SCB4/561
77115	115.9	Blue-gray breccia	Sta 7	SCB4/561
77130	1.42	DB residue	Sta 7	SCB4/562
77135	337.4	Green-gray breccia	Sta 7	SCB4/562
77210	111.7	DB residue	Sta 7	SCB7/543
77215	846.4	Brecciated norite	Sta 7	SCB7/543
77510	77.57	Unsieved fines	Sta 7	SCB7/540
77511	118.1	<1mm fines	Sta 7	SCB7/540
77512	2.45	1-2mm fines	Sta 7	SCB7/540
77513	1.19	2-4mm fines	Sta 7	SCB7/540
77514	1.24	4-10mm fines	Sta 7	SCB7/540
77515	337.6	Green-gray breccia	Sta 7	SCB7/540
77516	103.7	Medium basalt	Sta 7	SCB7/540
77517	45.6	Feldspathic breccia	Sta 7	SCB7/540
77518	42.5	Green-gray breccia	Sta 7	SCB7/540
77519	27.4	Green-gray breccia	Sta 7	SCB7/540
77525	1.19	Feldspathic breccia	Sta 7	SCB7/540
77526	1.07	Feldspathic breccia	Sta 7	SCB7/540
77530	82.76	Unsieved fines	Sta 7	SCB7/542
77531	126.6	<1mm fines	Sta 7	SCB7/542
77532	3.13	1-2mm fines	Sta 7	SCB7/542
77533	2.51	2-4mm fines	Sta 7	SCB7/542
77534	4.46	4-10mm fines	Sta 7	SCB7/542
77535	577.8	Coarse basalt	Sta 7	SCB7/542
77536	355.3	Coarse basalt	Sta 7	SCB7/542
77537	71.7	Green-gray breccia	Sta 7	SCB7/542
77538	47.2	Light gray breccia	Sta 7	SCB7/542
77539	39.6	Tan-gray breccia	Sta 7	SCB7/542
77545	29.5	Green-gray breccia	Sta 7	SCB7/542

Apollo 17 Inventory *(continued)*

Sample Number	Weight (grams)	Description	Location	Containers Outer/Inner
78120	75.78	Unsieved fines	LRV 11	SCB5/50Y
78121	121.6	<1mm fines	LRV 11	SCB5/50Y
78122	4.43	1-2mm fines	LRV 11	SCB5/50Y
78123	2.49	2-4mm fines	LRV 11	SCB5/50Y
78124	5.64	4-10mm fines	LRV 11	SCB5/50Y
78130	3.62	DB residue	Sta 8	SCB4/563
78135	133.9	Medium basalt	Sta 8	SCB4/563
78150	0.65	Dust and sweepings	Sta 8	SCB4/567
78155	401.1	Gabbroic breccia	Sta 8	SCB4/567
78220	108.3	Unsieved fines	Sta 8	SCB7/545
78221	227.1	<1mm fines	Sta 8	SCB7/545
78222	5.21	1-2mm fines	Sta 8	SCB7/545
78223	2.69	2-4mm fines	Sta 8	SCB7/545
78224	1.48	4-10mm fines	Sta 8	SCB7/545
78230	82.98	Unsieved fines	Sta 8	SCB4/564
78231	122.7	<1mm fines	Sta 8	SCB4/564
78232	2.68	1-2mm fines	Sta 8	SCB4/564
78233	1.42	2-4mm fines	Sta 8	SCB4/564
78234	0.72	4-10mm fines	Sta 8	SCB4/564
78235	199.0	Coarse norite	Sta 8	SCB4/564
78236	93.06	Coarse norite	Sta 8	SCB4/564
78238	57.58	Coarse norite	Sta 8	SCB4/564
78250	50.57	Unsieved fines	Sta 8	SCB4/546
78255	48.31	Coarse norite	Sta 8	SCB4/546
78420	97.94	Unsieved fines	Sta 8	SCB4/548
78421	186.2	<1mm fines	Sta 8	SCB4/548
78422	4.16	1-2mm fines	Sta 8	SCB4/548
78423	2.41	2-4mm fines	Sta 8	SCB4/548
78424	1.91	4-10mm fines	Sta 8	SCB4/548
78440	81.38	Unsieved fines	Sta 8	SCB4/551
78441	162.8	<1mm fines	Sta 8	SCB4/551
78442	3.78	1-2mm fines	Sta 8	SCB4/551
78443	2.44	2-4mm fines	Sta 8	SCB4/551
78444	1.19	4-10mm fines	Sta 8	SCB4/551
78460	138.1	Unsieved fines	Sta 8	SCB7/550
78461	264.5	<1mm fines	Sta 8	SCB7/550
78462	5.328	1-2mm fines	Sta 8	SCB7/550
78463	2.787	2-4mm fines	Sta 8	SCB7/550
78464	1.303	4-10mm fines	Sta 8	SCB7/550
78465	1.039	Dark matrix breccia	Sta 8	SCB7/550
78480	89.33	Unsieved fines	Sta 8	SCB4/549
78481	173.9	<1mm fines	Sta 8	SCB4/549
78482	2.69	1-2mm fines	Sta 8	SCB4/549
78483	1.21	2-4mm fines	Sta 8	SCB4/549
78484	0.32	4-10mm fines	Sta 8	SCB4/549
78500	391.1	Unsieved fines	Sta 8	SCB4/566(rake soil)
78501	718.7	<1mm fines	Sta 8	SCB4/566(rake soil)
78502	21.38	1-2mm fines	Sta 8	SCB4/566(rake soil)
78503	16.41	2-4mm fines	Sta 8	SCB4/566(rake soil)
78504	19.16	4-10mm fines	Sta 8	SCB4/566(rake soil)
78505	506.3	Coarse basalt	Sta 8	SCB4/566(rake soil)

Apollo 17 Inventory *(continued)*

Sample Number	Weight (grams)	Description	Location	Containers Outer/Inner
78506	55.97	Coarse basalt	Sta 8	SCB4/566(rake soil)
78507	23.35	Coarse basalt	Sta 8	SCB4/566(rake soil)
78508	10.67	Friable dark matrix breccia	Sta 8	SCB4/566(rake soil)
78509	8.68	Basalt	Sta 8	SCB4/566(rake soil)
78515	4.76	Coherent dark matrix breccia	Sta 8	SCB4/566(rake soil)
78516	3.18	Friable dark matrix breccia	Sta 8	SCB4/566(rake soil)
78517	1.82	Friable white breccia	Sta 8	SCB4/566(rake soil)
78518	0.88	Friable dark matrix breccia	Sta 8	SCB4/566(rake soil)
78525	5.11	Agglutinate	Sta 8	SCB4/565 (rake)
78526	8.77	Breccia with green glass vein	Sta 8	SCB4/565 (rake)
78527	5.16	Brecciated gabbroic rock	Sta 8	SCB4/565 (rake)
78528	7.00	Fine basalt	Sta 8	SCB4/565 (rake)
78530	88.92	DB residue	Sta 8	SCB4/565 (rake)
78535	103.4	Coherent dark matrix breccia	Sta 8	SCB4/565 (rake)
78536	8.67	Coherent dark matrix breccia	Sta 8	SCB4/565 (rake)
78537	11.76	Coherent dark matrix breccia	Sta 8	SCB4/565 (rake)
78538	5.82	Coherent dark matrix breccia	Sta 8	SCB4/565 (rake)
78539	3.73	Coherent dark matrix breccia	Sta 8	SCB4/565 (rake)
78545	8.60	Coherent dark matrix breccia	Sta 8	SCB4/565 (rake)
78546	42.66	Coherent dark matrix breccia	Sta 8	SCB4/565 (rake)
78547	29.91	Friable dark matrix breccia	Sta 8	SCB4/565 (rake)
78548	15.95	Friable dark matrix breccia	Sta 8	SCB4/565 (rake)
78549	16.09	Friable dark matrix breccia	Sta 8	SCB4/565 (rake)
78555	6.64	Friable dark matrix breccia	Sta 8	SCB4/565 (rake)
78556	9.50	Friable darc matrix breccia	Sta 8	SCB4/565 (rake)
78557	7.19	Friable dark matrix breccia	Sta 8	SCB4/565 (rake)
78558	3.78	Friable dark matrix breccia	Sta 8	SCB4/565 (rake)
78559	3.05	Friable dark matrix breccia	Sta 8	SCB4/565 (rake)
78565	3.50	Friable dark matrix breccia	Sta 8	SCB4/565 (rake)
78566	0.77	Friable dark matrix breccia	Sta 8	SCB4/565 (rake)
78567	18.88	Friable dark matrix breccia	Sta 8	SCB4/565 (rake)
78568	3.57	Friable dark matrix breccia	Sta 8	SCB4/565 (rake)
78569	14.53	Friable dark matrix breccia	Sta 8	SCB4/565 (rake)
78575	140.0	Coarse basalt	Sta 8	SCB4/565 (rake)
78576	11.64	Coarse basalt	Sta 8	SCB4/565 (rake)
78577	8.84	Coarse basalt	Sta 8	SCB4/565 (rake)
78578	17.13	Coarse basalt	Sta 8	SCB4/565 (rake)
78579	6.07	Medium basalt	Sta 8	SCB4/565 (rake)
78585	44.60	Fine basalt	Sta 8	SCB4/565 (rake)
78586	10.73	Fine basalt	Sta 8	SCB4/565 (rake)
78587	11.48	Fine basalt	Sta 8	SCB4/565 (rake)
78588	3.77	Fine basalt	Sta 8	SCB4/565 (rake)
78589	4.10	Fine basalt	Sta 8	SCB4/565 (rake)
78595	4.19	Fine basalt	Sta 8	SCB4/565 (rake)
78596	7.55	Fine basalt	Sta 8	SCB4/565 (rake)
78597	319.1	Fine basalt	Sta 8	SCB4/565 (rake)
78598	224.1	Fine basalt	Sta 8	SCB4/565 (rake)
78599	198.6	Fine basalt	Sta 8	SCB4/565 (rake)
79001	743.4	Drive tube (50, lower)	Sta 9	SCB7/
79002	409.4	Drive tube (37, upper)	Sta 9	SCB7/
79010	87.05	SCB residue	EVA 3	SCB5/
79035	2806.	Dark matrix breccia	Sta 9	BSLSS/

Apollo 17 Inventory *(continued)*

Sample Number	Weight (grams)	Description	Location	Containers Outer/Inner
79110	66.30	DB residue	Sta 9	SCB5/568
79115	346.3	Dark matrix breccia	Sta 9	SCB5/568
79120	116.4	Unsieved fines	Sta 9	SCB5/569
79121	214.4	<1mm fines	Sta 9	SCB5/569
79122	13.97	1-2mm fines	Sta 9	SCB5/569
79123	13.14	2-4mm fines	Sta 9	SCB5/569
79124	14.48	4-10mm fines	Sta 9	SCB5/569
79125	1.91	Dark matrix breccia	Sta 9	SCB5/569
79130	3.99	Dust and sweepings	Sta 9	SCB5/480
79135	2283.	Dark matrix breccia	Sta 9	SCB5/480
79150	5.63	DB residue	Sta 9	SCB5/571
79155	318.8	Coarse basalt	Sta 9	SCB5/571
79170	43.42	DB residue	Sta 9	SCB7/481
79175	677.7	Agglutinate	Sta 9	SCB7/481
79190	13.38	DB residue	Sta 9	SCB7/482
79195	368.5	Dark matrix breccia	Sta 9	SCB7/482
79210	5.55	DB residue	Sta 9	SCB7/486
79215	553.8	Brecciated troctolite	Sta 9	SCB7/486
79220	93.49	Unsieved fines	Sta 9	SCB5/483
79221	152.6	<1mm fines	Sta 9	SCB5/483
79222	7.22	1-2mm fines	Sta 9	SCB5/483
79223	6.24	2-4mm fines	Sta 9	SCB5/483
79224	9.75	4-10mm fines	Sta 9	SCB5/483
79225	7.42	Friable dark matrix breccia	Sta 9	SCB5/483
79226	6.73	Friable dark matrix breccia	Sta 9	SCB5/483
79227	5.57	Disaggregated clod	Sta 9	SCB5/483
79228	2.50	Disaggregated clod	Sta 9	SCB5/483
79240	113.3	Unsieved fines	Sta 9	SCB5/484
79241	174.3	<1mm fines	Sta 9	SCB5/484
79242	11.32	1-2mm fines	Sta 9	SCB5/484
79243	10.46	2-3mm fines	Sta 9	SCB5/484
79244	10.85	4-10mm fines	Sta 9	SCB5/484
79245	10.11	Crystalline	Sta 9	SCB5/484
79260	118.9	Unsieved fines	Sta 9	SCB5/485
79261	187.8	<1mm fines	Sta 9	SCB5/485
79262	11.74	1-2mm fines	Sta 9	SCB5/485
79263	11.46	2-4mm fines	Sta 9	SCB5/485
79264	15.85	4-10mm fines	Sta 9	SCB5/485
79265	2.6	Basalt	Sta 9	SCB5/485
79510	107.6	Unsieved fines	Sta 9	SCB5/570
79511	179.2	<1mm fines	Sta 9	SCB5/570
79512	11.32	1-2mm fines	Sta 9	SCB5/570
79513	9.94	2-4mm fines	Sta 9	SCB5/570
79514	12.24	4-10mm fines	Sta 9	SCB5/570
79515	33.00	Medium basalt	Sta 9	SCB5/570
79516	23.92	Medium basalt	Sta 9	SCB5/570
79517	10.23	Dark matrix breccia	Sta 9	SCB5/570
79518	5.20	Dark matrix breccia	Sta 9	SCB5/570
79519	3.65	Dark matrix breccia	Sta 9	SCB5/570

Apollo 17 Inventory *(continued)*

Sample Number	Weight (grams)	Description	Location	Containers Outer/Inner
79525	3.03	Dark matrix breccia	Sta 9	SCB5/570
79526	2.93	Dark matrix breccia	Sta 9	SCB5/570
79527	2.65	Dark matrix breccia	Sta 9	SCB5/570
79528	2.38	Dark matrix breccia	Sta 9	SCB5/570
79529	1.84	Dark matrix breccia	Sta 9	SCB5/570
79535	1.69	Dark matrix breccia	Sta 9	SCB5/570
79536	1.66	Dark matrix breccia	Sta 9	SCB5/570
79537	1.05	Dark matrix breccia	Sta 9	SCB5/570

Author Index

Author Index